d — p

REPRESENTATIVE ELEMENTS

NOBLE GASES

B	VIII	VIII	VIII	I B	II B	III A	IV A	V A	VI A	VII A	0
										1.0080 **H**[1] 1	4.00260 **He**[2] 2
						10.81 **B**[5] 3	12.011 **C**[6] 4	14.0067 **N**[7] 5	15.9994 **O**[8] 6	18.9984 **F**[9] 7	20.179 **Ne**[10] 8
						26.9815 **Al**[13] 3	28.086 **Si**[14] 4	30.9738 **P**[15] 5	32.06 **S**[16] 6	35.453 **Cl**[17] 7	39.948 **Ar**[18] 8
380 [25] ,2	55.847 **Fe**[26] 14,2	58.9332 **Co**[27] 15,2	58.71 **Ni**[28] 16,2	63.545 **Cu**[29] 18,1	65.37 **Zn**[30] 18,2	69.72 **Ga**[31] 18,3	72.59 **Ge**[32] 18,4	74.9216 **As**[33] 18,5	78.96 **Se**[34] 18,6	79.90 **Br**[35] 18,7	83.80 **Kr**[36] 18,8
) 43] ,1	101.07 **Ru**[44] 15,1	102.906 **Rh**[45] 16,1	106.4 **Pd**[46] 18	107.868 **Ag**[47] 18,1	112.40 **Cd**[48] 18,2	114.82 **In**[49] 18,3	118.69 **Sn**[50] 18,4	121.75 **Sb**[51] 18,5	127.60 **Te**[52] 18,6	126.905 **I**[53] 18,7	131.30 **Xe**[54] 18,8
.2 75] 3,2	190.2 **Os**[76] 32,14,2	192.22 **Ir**[77] 32,15,2	195.09 **Pt**[78] 32,17,1	196.967 **Au**[79] 32,18,1	200.59 **Hg**[80] 32,18,2	204.37 **Tl**[81] 32,18,3	207.12 **Pb**[82] 32,18,4	208.981 **Bi**[83] 32,18,5	(209) **Po**[84] 32,18,6	(210) **At**[85] 32,18,7	(222) **Rn**[86] 32,18,8
1) 7]	[108]										

f

5) [61] ,2	150.4 **Sm**[62] 24,8,2	151.96 **Eu**[63] 25,8,2	157.25 **Gd**[64] 25,9,2	158.925 **Tb**[65] 27,8,2	162.50 **Dy**[66] 28,8,2	164.930 **Ho**[67] 29,8,2	167.26 **Er**[68] 30,8,2	168.934 **Tm**[69] 31,8,2	173.04 **Yb**[70] 32,8,2	174.97 **Lu**[71] 32,9,2
7) 93] ,2	(244) **Pu**[94] 24,8,2	(243) **Am**[95] 25,8,2	(247) **Cm**[96] 25,9,2	(247) **Bk**[97] 26,9,2	(251) **Cf**[98] 28,8,2	(254) **Es**[99] 29,8,2	(253) **Fm**[100] 30,8,2	(256) **Md**[101] 31,8,2	(253) **No**[102] 32,8,2	(257) **Lw**[103] 32,9,2

A value (of a relative weight) given in parentheses denotes the mass number of the isotope of longest known half-life. Elements required for health are shown in color.

An Introduction to Physical Science

An Introduction to Physical Science

FIFTH EDITION

James T. Shipman
Ohio University

Jerry L. Adams

Jerry D. Wilson
Lander College

D.C. HEATH AND COMPANY
Lexington, Massachusetts Toronto

Published simultaneously in Canada.

Printed in the United States of America.

International Standard Book Number: 0-669-12022-7

Library of Congress Catalog Card Number: 86-81269

Acquisitions Editor: Mary Le Quesne
Production Editor: Antoinette Tingley Schleyer
Design Coordinator: Victor A. Curran
Production Coordinator: Mike O'Dea
Photo Researcher: Toni Jurras
Text Permissions Editor: Margaret Roll

Cover photo: Marc Bernheim/Woodfin Camp and Associates. Desert mountains at sunset—or are they? This photomicrograph of an organic crystal reveals that there can be remarkable similarities in structure and appearance among objects in the physical world, in spite of vast differences in scale.

Chapter Opening Photographs
1 J. T. Shipman **2** Eastman Kodak Company **3** NASA **4** Bloom/Monkmeyer Press Photo Service **5** Monkmeyer Press Photo Service **6** Richard Megna/Fundamental Photographs **7** Jeff Thiebauth/Lightwave **8** ©Tom Ives, 1982 **9** Photo by Johan Hagemeyer, Bancroft Library. Courtesy AIP Niels Bohr Library **10** Stanford University/Photo Researchers, Inc. **11** Richard Wood/Taurus Photos **12** Rita Rivera/Fundamental Photographs **13** Fundamental Photographs **14** E. R. Degginger **15** A. Avis/Visuals Unlimited **16** Fundamental Photographs **17** K. G. Murti/Visuals Unlimited **18** NASA **19** Mark Antman/The Image Works **20** NASA **21** Mount Wilson & Palomar Observatories **22** NOAA **23** NOAA **24** The Boston Globe **25** E. R. Degginger **26** Los Angeles County Air Pollution Control District **27** Werner H. Muller/Peter Arnold, Inc. **28** Donald Burrows/Taurus Photos **29** Grant Heilman **30** Tom Stack and Associates **31** USDA

Preface

WE HAVE DESIGNED and written *An Introduction to Physical Science, Fifth Edition,* to provide today's students with sufficient knowledge of the fundamental concepts of physical science to understand the nature of their physical world and meet the challenge of advancing technology.

This textbook, written for the first-year college non-science major, presents basic concepts in the five major areas of physical science—astronomy, chemistry, geology, meteorology, and physics. We have presented the concepts in a logical rather than historical fashion. Chapter 1 logically begins with the fundamental concepts of measurement—length, mass, time, and electric charge. From these fundamentals, the concepts of motion, force, energy, wave motion, heat, electricity, magnetism, and modern physics are explained. These concepts are then used to develop the principles of chemistry, astronomy, meteorology, and geology.

The treatment of the subject matter is both descriptive and quantitative, with the relative emphasis of these two approaches left to the instructor. For the instructor who wishes to take only the descriptive approach in teaching physical science, we suggest that all problems at the end of the chapters be omitted and that only the questions be used. In this edition, most derivations of equations have been placed in the Appendix, to allow greater flexibility in teaching the course. The text is readily adaptable for either a one- or two-semester course.

Changes in the Fifth Edition

We have thoroughly revised the text, incorporating many suggestions offered by instructors familiar with previous editions. Chapters 1 through 12, which present the necessary principles of classical and modern physics for a one-semester course, have been rearranged; some additions have been made to give the instructor a broad choice of topics to meet his or her requirements. Chapter 5 (in the Fourth Edition) on wave motion has been rewritten and divided into two new chapters for greater clarity and flexibility. Chapter 5 now covers waves, and Chapter 6, wave effects.

Many new photographs have been added, including four additional pages of color plates. We have also added new line drawings to clarify specific concepts.

We present the most up-to-date information received from the Voyager II spacecraft on the planet Uranus and its moons, with photographs (both color and black and white). We also have the latest physical and chemical data on Comet Halley, including a close-up color photograph from the Giotto spacecraft.

In Chapter 18, a new section on solar systems other than ours has been added, and new ideas concerning Planet X, which may be a member of our solar system, are discussed. We have also presented new thoughts on the origin of the universe.

The material on winds and clouds has been rewritten and combined into one chapter.

The chapters on meteorology reflect the current emphasis on topics such as climatic changes, the greenhouse effect, acid rain, and tornado effects. Information is included on the new Doppler radar used to detect tornadoes. Special boxed features are introduced on lightning and tornado safety. The climatic effects of El Nino and El Chichón are discussed.

The geology chapters now cover the modern theory of sea floor spreading and plate tectonics.

Supplements

A very comprehensive *Instructor's Guide* has been prepared to supplement the instructor's background in the different physical science areas. The guide includes the following on a chapter-by-chapter basis:

- An introduction to the chapter and suggestions for teaching it.
- The answer to every given question.
- The solution to every given problem.
- Many questions of various types that can be used for quizzes and examinations.
- A list of available films and possible demonstrations.
- Selected references.

We have updated the *Study Guide* with additional questions and problems. The programmed summary instructions have been rewritten. In addition to the chapter reviews, the guide also includes a review of basic mathematics, lists of fundamental constants and conversion factors, and a discussion of significant figures.

The *Laboratory Guide* has been expanded by six new experiments, bringing the total number to 45 experiments in the five major areas of physical science. The new larger page size of the guide allows for bigger, clearer data tables and report pages.

New to this edition of the laboratory guide is an *Instructor's Resource Manual* that includes the following for each experiment: (1) an introductory discussion, (2) answers to all questions asked at the end of each experiment, and (3) additional or alternate questions.

Archive, a computerized test generator for Apple® IIe and IBM PC® microcomputers,* contains over 1800 multiple-choice and completion-type questions arranged by chapter. A printed version of the test item file is also available.

A set of 50 *transparencies* is available to the instructor upon adoption of the text.

Acknowledgments

We wish to thank our colleagues and students for the many contributions made to this edition of *An Introduction to Physical Science*. We would also like to thank the following reviewers for their suggestions and comments:

Aaron W. Todd, Middle Tennessee State University;
Alvin D. Aurand, Broward Community College;
J. C. Ledbetter, Middle Georgia College;
Jolly Rahman, Heidelberg College;
Ted Morishige, Central State University (Oklahoma);
Bernard Paradis, University of Maine at Fort Kent.

We also appreciate the help of Thomas G. Read, Martin College; James Gruber, Harrisburg Area Community College; H. H. Daws, Northeast Mississippi Junior College; Linda Moore, Georgia Military College; Murray Goldman, Community College of Philadelphia; Bob Nerbun, University of South Carolina; B. S. Hansra, Daley College; Jonathan Cole, Mira Costa College; Ross Sears, Lake City Community College; Arthur Bycer, Valley Forge Military Junior College; Rayford Keys, Coahoma Junior College; Joseph H. Purser, Polk Community College; Eunice R. Knouse, Spartanburg Methodist College; Dorothy R. Carpenter, DeKalb Community College; Donald E. Johnson, Copiah-Lincoln Junior College; J. E. Tyler, Holyoke Community College; Lucius deYampert, Broward Community College; Nicholas Petraglia, Sullivan County Community College; D. Schexnayder, Copiah-Lincoln Junior College; Kenneth J. Smith, State University of New York at Syracuse; Bill Gene Smith, South Florida Junior College; C. W. O'Neill, Edison Community College; James Griner, Abraham Baldwin Community College; Larry Rabideau, Kankakee Community College; Caroline P. Phillips, Georgia Military College; George Laux, Wallace State Community College; C. R. Adkins, Hinds Junior College; George Mulfinger, Bob Jones University; John E. Rives, University of Georgia; C. M. Holcomb, Mississippi University for Women; B. Jane Greever, Delta State University; John Reitti, Alcorn State University; Frank Robinson, South Carolina State College; Ronnie C. Barnes, Lambuth College; Eugene A. Kirst, Northwestern College (Wisconsin); Kenneth Wireman, Southeastern College (Florida); Duane L. Sea, Bemidji State University; Virginia L. Rawlins, North Texas State University; Do Ren Chang, Averett College; Henry Kuhlman, Southern Missionary College; T. Marshall, Suffolk University; R. H. Dameron, Covenant College; L. H. Adams, Polk Community College; Floyd O'Neal, Augusta College; Jerry Skelton, Oakland City College; Joseph T. Bohanon, Evangel College; John Newoll, Shaw University; Whit Marks, Central State University; George Canty, Jr., The Fort Valley State College; Kyle Forinash, Indiana University Southeast; and S. D. Lewis, University of Maine at Fort Kent.

We also express our gratitude to those individuals and organizations who contributed photographs, illustrations, and other information; in particular, we wish to acknowledge the assistance of the American Institute of Physics, the U.S. Department of Agriculture, the U.S. Geological Survey, the Department of Energy, the U.S. Air Force, the National Aeronautics and Space Administration, the National Oceanic and Atmospheric Administration, the U.S. Park Service, and Clyde Baker, Ohio University. To Mary Le Quesne and Toni Schleyer, a special thanks for their support and help in many ways in preparing the manuscript for publication.

J.T.S.
J.L.A.
J.D.W.

---

*Apple is a registered trademark of Apple Computer, Inc. IBM PC is a registered trademark of International Business Machines Corp.

Contents

Introduction xiii

1 Measurement 2

1.1 The Senses 3
1.2 Fundamental Properties 4
1.3 Standard Units 6
1.4 Derived Units and Conversion Factors 9
1.5 Measurement of Circles 11
1.6 Experimental Error 12
1.7 Powers of 10 Notation 14
Learning Objectives Important Words and Terms Questions Problems

2 Motion 18

2.1 Straight-Line Motion 19
2.2 Speed and Velocity 20
2.3 Acceleration 22
2.4 Acceleration in Circular Motion 25
2.5 Projectile Motion 27
Learning Objectives Important Words and Terms Questions Problems

3 Force and Motion 32

3.1 Newton's First Law of Motion 33
3.2 Newton's Second Law of Motion 35
3.3 Newton's Law of Gravitation 39
3.4 Newton's Third Law of Motion 42
3.5 Conservation Laws 44
Learning Objectives Important Words and Terms Questions Problems

4 Work and Energy 48

4.1 Work 49
4.2 Power 52
4.3 Kinetic and Potential Energy 53
4.4 Law of Conservation of Energy 56
4.5 Forms of Energy 58
4.6 Energy Consumption 61
Learning Objectives Important Words and Terms Questions Problems

5 Heat 66

5.1 Temperature 67
5.2 Heat 69
5.3 Specific Heat 70
5.4 Thermodynamics 72
5.5 Phases of Matter 75
5.6 The Gas Laws 76
5.7 Kinetic Theory 77
5.8 Heat Transfer 79
Learning Objectives Important Words and Terms Questions Problems

6 Waves 84

6.1 Wave Properties 86
6.2 Electromagnetic Waves 87
6.3 Sound Waves 89
6.4 Standing Waves 93
6.5 The Doppler Effect 97
Learning Objectives Important Words and Terms Questions Problems

7 Wave Effects 100

7.1 Reflection 101
7.2 Refraction and Dispersion 103
7.3 Diffraction, Interference, and Polarization 107
7.4 Spherical Mirrors 110
7.5 Lenses 114
Learning Objectives Important Words and Terms Questions Problems

8 Electricity and Magnetism 120

8.1 Electric Charge 122
8.2 Electricity 125
8.3 Magnetism 129
8.4 Electromagnetism 133
8.5 Electronics 137
Learning Objectives Important Words and Terms Questions Problems

9 Relativity 142

9.1 Moving Systems of Reference 143
9.2 Einstein's Postulates 146
9.3 The Lorentz Transformation 146
9.4 Mass-Energy Relationship 149
9.5 General Theory of Relativity 150
9.6 Gravitational Collapse 152
Learning Objectives Important Words and Terms Questions Problems

10 Nuclear Physics 156

10.1 The Atomic Nucleus 158
10.2 Nuclear Stability 160
10.3 Radioactive Decay 164
10.4 Fission 168
10.5 Fusion 171
Learning Objectives Important Words and Terms Questions Problems

11 The Atom 176

11.1 The Dual Nature of Light 177
11.2 Bohr's Equations for the Hydrogen Atom 179
11.3 Energy Levels for the Hydrogen Atom 181
11.4 Photon Emission and Absorption 183
11.5 Energy Levels for Other Systems 186
Learning Objectives Important Words and Terms Questions Problems

12 Quantum Mechanics 190

12.1 Matter Waves 191
12.2 The Schrödinger Equation 193
12.3 The Heisenberg Uncertainty Principle 195
12.4 Atomic Quantum Numbers 196
12.5 Many-Electron Atoms and the Periodic Table 201
Learning Objectives Important Words and Terms Questions Problems

13 The Periodic Table 208

13.1 Elements 209
13.2 The Periodic Table 212
13.3 Classification of Elements 214
13.4 Families of Elements 215
13.5 Periodic Characteristics 219
Learning Objectives Important Words and Terms Questions Problems

14 Compounds, Molecules, and Ions 222

14.1 Principles of Compound Formation 223

14.2 Ionic Compounds 224
14.3 Properties of Ionic Compounds 226
14.4 Covalent Compounds 228
14.5 Properties of Covalent Compounds 231
14.6 Oxidation Number 232
14.7 Naming Compounds 233
14.8 Some Simple Organic Compounds 233
Learning Objectives Important Words and Terms Questions Problems

15 Some Chemical Principles 240

15.1 Types of Matter 241
15.2 Early Chemical Laws 242
15.3 Molecular Weights and Volumes 245
15.4 Avogadro's Number 248
15.5 Solutions 249
Learning Objectives Important Words and Terms Questions Problems

16 Chemical Reactions 254

16.1 Basic Concepts 255
16.2 Energy and Rate of Reaction 257
16.3 Oxidation-Reduction Reactions 260
16.4 Electrochemical Reactions 262
16.5 Acids and Bases 264
16.6 Acids and Bases in Solution 268
Learning Objectives Important Words and Terms Questions Problems

17 Complex Molecules 272

17.1 Common Organic Compounds 273
17.2 The Ingredients of Life and the Genetic Code 280
17.3 Artificial Molecules—Plastics 283
17.4 Drugs 285
Learning Objectives Important Words and Terms Questions Problems

18 The Solar System 290

18.1 The Planet Earth 291
18.2 The Solar System 294
18.3 The Inner Planets 302
18.4 The Outer Planets 308
18.5 The Origin of the Solar System 319
18.6 Other Solar Systems 321
Learning Objectives Important Words and Terms Questions Problems

19 Place and Time 324

19.1 Cartesian Coordinates 325
19.2 Latitude and Longitude 326
19.3 Time 327
19.4 The Seasons 331
19.5 Precession of the Earth's Axis 334
19.6 The Calendar 335
Learning Objectives Important Words and Terms Questions Problems

20 The Moon 340

20.1 General Features 341
20.2 History of the Moon 344
20.3 Lunar Motion 346
20.4 Phases 347
20.5 Eclipses 350
20.6 Tides 351
Learning Objectives Important Words and Terms Questions Problems

21 The Universe 356

21.1 The Sun 357
21.2 The Celestial Sphere 362
21.3 Stars 364

21.4 Galaxies 370
21.5 Quasars 377
21.6 Cosmology 378
Learning Objectives Important Words and Terms Questions Problems

22 The Atmosphere 384

22.1 Composition 385
22.2 Origin 386
22.3 Vertical Structure 387
22.4 Energy Content 391
22.5 Atmospheric Measurements 398
Learning Objectives Important Words and Terms Questions Problems

23 Winds and Clouds 408

23.1 Causes of Air Motion 409
23.2 Local Winds and World Circulation 412
23.3 Jet Streams 416
23.4 Cloud Classification 416
23.5 Cloud Formation 423
23.6 Condensation and Precipitation 426
Learning Objectives Important Words and Terms Questions Problems

24 Air Masses and Storms 434

24.1 Air Masses 435
24.2 Fronts and Cyclonic Disturbances 437
24.3 Local Storms 440
24.4 Tropical Storms 449
Learning Objectives Important Words and Terms Questions Problems

25 Weather Forecasting 456

25.1 The National Weather Service 457
25.2 Data Collection and Weather Observation 461
25.3 Weather Maps 469
25.4 Folklore and the Weather 475
Learning Objectives Important Words and Terms Questions Problems

26 Air Pollution and Climate 483

26.1 Air Pollutants 483
26.2 Sources of Air Pollution 489
26.3 The Costs of Air Pollution 492
26.4 Weather and Climatic Effects of Air Pollution 496
Learning Objectives Important Words and Terms Questions

27 Geology and Time 502

27.1 The Earth's Structure 503
27.2 Continental Drift and Sea Floor Spreading 505
27.3 Plate Tectonics 510
27.4 Geologic Time 515
Learning Objectives Important Words and Terms Questions

28 Rocks and Minerals 518

28.1 Igneous Rock 519
28.2 Sedimentary Rock 522
28.3 Metamorphic Rock 528
28.4 Minerals 529
Learning Objectives Important Words and Terms Questions

29 Internal Processes 536

29.1 Volcanoes 537
29.2 Earthquakes 545
29.3 Mountain Building 552
Learning Objectives Important Words and Terms Questions

30 Surface Processes 558

30.1 Weathering and Mass Wasting *559*
30.2 Agents of Erosion *563*
30.3 The Earth's Water Supply *571*
30.4 The Oceans and Sea Floor Topography *576*
Learning Objectives Important Words and Terms Questions

31 Land and Water Pollution 584

31.1 Land Pollution *585*
31.2 Water Pollution *590*
31.3 The Cost of Clean Water *600*
Learning Objectives Important Words and Terms Questions

Appendixes

I *The Seven Base Units of the International System of Units (SI)* 607

II *Measurement and Significant Figures* 607

III *Rules and Examples of the Four Basic Mathematical Operations Using Powers of Ten Notation* 609

IV *Problem Solving: Five Major Steps* 609

V *Derivation of Bohr's Equations* 613

VI *Psychrometric Tables* 614

***Glossary* 617**

***Index* 631**

Introduction

The object of all science is to coordinate our experiences and to bring them to a logical system.
–Albert Einstein

THE WORLD IN WHICH WE LIVE is changing continually, and people are continually seeking to discover and understand these changes. Work in this area is resulting in advances in technology related to all aspects of our daily life, making it most important that each individual knows and understands the physical concepts behind this advancing technology. It is only through an understanding of the basic physical concepts that the individual today can hope to remain abreast of this technology concerning the natural environment.

The English word *science* is derived from Latin *scientia,* meaning knowledge. Physical science is the organized knowledge of our physical environment and the methods used to obtain it. Physical science is classified into five major divisions: Astronomy, the science of the universe beyond our planet; chemistry, the science of matter and its changes; geology, the science of the Earth and its history; meteorology, the science of climate and weather; and physics, the science of energy and matter. Physical science studies the nonliving matter of the universe, while biological science studies the living matter.

The knowledge acquired and applied by individuals over the past 7000 years through discovery and experimentation, fused with reflective thinking in the formation of theories and laws, has produced our modern civilization. Even after 7000 years of learning, there is still much we do not know. For example, the origin of human beings and how we came to be where we are today may never be known with certainty.

Much of our present knowledge about the environment is actually an approximation. The age of matter in the solar system is based on the analysis of rocks found on the surface of our planet, lunar rocks returned by the Apollo astronauts, and meteorites. Through the use of radioactive decay techniques scientists believe that the solar system, including the Earth, formed over a hundred-million-year period beginning about 4.6 billion years ago.

We do not know how life began, but the fact that we exist proves that the right ingredients, in the right amounts, and arranged in the correct way, did come together at some time in the past. The best approximation is that it happened at least 2.5 to 3 billion years ago.

The appearance of human beings is fairly recent with respect to the geological time scale. The earliest evidence indicates they have been in existence at least one million years. It has been suggested that human beings' ability to reason was first put to use during an ice age, when they were forced from the trees into a cold environment and reasoned that to survive they must hunt in groups. Possessing intelligence, they invented stone tools and developed a crude language in order to live together and to fight their enemies better. Early people fought not only the wild beasts and the forces of nature, but made war against their neighbors as well. They discovered fire and learned to control it. The need for refuge from the fear of certain natural phenomena gave rise to religious rites and the worship of gods. As the climate became warmer, the hunters became shepherds who, with the passing of time, became farmers. Through necessity the plow was invented, animals were domesticated, the use of clay was developed, and weaving wool and flax was learned. Copper was discovered and then tin, which when it was combined with copper made bronze. Iron and other metals were discovered and used in numerous ways. Industries were formed, roads were built, laws were made, medicine was practiced. These and many other progressions, with the passing of time, formed the complex groups of many thousands of people who built the cities that were the beginning of our early civilizations.

Democritus

Our life on Earth may be just beginning (it could be ending, if we are to believe the views of some ecologists). A mere three hundred generations have lived since the early civilizations of the Egyptians and the Babylonians around 4000 B.C. Located within a few degrees of the 30° N parallel along the rich valley lands of the Nile, Indus, Yangtze, Euphrates, and Tigris Rivers, these early civilizations were the first to develop a knowledge of mathematics, astronomy, medicine, engineering, and agriculture and so were the originators of science. They invented writing and writing materials, thus providing a way of recording the knowledge acquired by one generation and a means of passing it on to the next.

By the middle of the fourth millennium B.C., the Egyptians in the Nile Valley and the Babylonians in the Tigris and Euphrates valleys had acquired a language, were able to write, and possessed technical skills in engineering and mathematics. They studied the stars and knew of the motion of the planets, Sun, and moon, from which they fixed the solar year of 12 months.

These civilizations flourished until the fifteenth century B.C., at which time they were invaded by barbarian tribesmen from the north. Thereafter, they experienced a gradual decline.

The pursuit of scientific knowledge is an activity that serves both as a history of those who made important discoveries concerning nature and as a means for the development of new ideas. The following discussion includes some of the individuals whose discoveries or ideas have benefited the human race.

The development of physical science as we know it today began around 600 B.C. with the Greeks, who attempted to understand their immediate environment and the universe as a whole from a purely naturalistic point of view.

One of the earliest known scientists was a Greek named Thales of Miletus, who lived about 600 B.C. He was the first Greek mathematician and astronomer, and a number of geometric propositions are credited to him.

Pythagoras (580–500 B.C.), who is considered to be the founder of science, was also a Greek. He made many important discoveries in arithmetic, geometry, and astronomy. The idea of abstract numbers was established by Pythagoras, and he was the first to point out that science provided a means for a true understanding of our environment.

Anaxagoras (500–428 B.C.) is credited with the discovery of the true cause of eclipses and with the view that the moon shines by reflected light from the Sun. Many historians also credit him with originating the atomic hypothesis, which was formulated by Democritus a generation later. According to Anaxagoras, matter is composed of an infinite number of so-called seeds, which also exist in an infinite variety.

Empedocles (484–424 B.C.) chose four ''primary substances'' (fire, air, water, and earth), from which he contended all other substances were made. His famous experiment with the clepsydra (water clock) proved that air was something that occupied space and exerted force. Through his experiments, he proved that the true nature of the environment could be gained by analyzing nature at first hand.

Democritus (460–370 B.C.) supplied ancient scientists with the fundamental building block of matter. He expanded the atomic hypothesis of Anaxagoras by postulating the theory that the universe consists of a void

Aristotle

(empty space) and an infinite number of invisible and indivisible particles, which are alike in substance but differ in form, size, shape, position, and arrangement. Democritus is also credited with the first thoughts concerning the conservation of matter. His ideas that matter cannot be created from nothing and that nothing that is can cease to exist were revolutionary.

Aristotle (384–322 B.C.), considered to be the greatest of the Greek philosophers, made contributions to many branches of science, including biology, logic, ethics, psychology, metaphysics, and natural science. Because his high social position prevented him from doing experiments, which were considered work for slaves, his scientific views originated only from reason and logic. As one would expect, many were incorrect. In the field of physical science, Aristotle contended that a heavy body dropped from a given height would reach the ground before a lighter one dropped from the same height at the same time. This was considered true until the seventeenth century, when Galileo proved it to be false by actual experiment. Aristotle was held in such high esteem and his authority was so great that his views concerning the structure of the universe, which placed the Earth at rest as the center of the system, overruled the views of other scientists for nearly 2000 years.

Archimedes (287–212 B.C.) represents the present-day notion of the scientist. He experimented to establish the truth of a physical concept, and he combined the results with mathematics to formulate theories. A few of the concepts with which Archimedes is credited are density, specific gravity, center of gravity, and the lever.

Aristarchus of Samos (310–230 B.C.), a Greek astronomer, was the first known person of authority to hold the theory that the Earth revolves around the Sun. He was unable, however, to prove his theory, nor could he gain the support of his colleagues; thus it was that his idea was forgotten until the sixteenth century.

Brahe

Eratosthenes (276–192 B.C.) of Alexandria, librarian and scientific writer, calculated the circumference of the Earth and came close to the value we use today.

Hipparchus of Nicaea (second century A.D.), a Greek astronomer and mathematician, compiled the first star catalog and discovered the precession of the equinoxes.

Ptolemy of Alexandria (about 70–147 A.D.), another Greek astronomer, developed Aristotle's (incorrect) ideas that the Sun, moon, and planets revolve in perfect circles about the Earth, which he believed was the center of the universe.

Generally speaking, during the next 1200 years (200–1400 A.D.) after Ptolemy, very little scientific advancement was made. Roger Bacon (about 1214–1291 A.D.), the British philosopher, scientist, and educator, believed and taught that the true facts of nature could be revealed in experimentation and stressed the importance of mathematics in formulating scientific theories. Leonardo da Vinci (1452–1519 A.D.), the Italian painter whose work in the scientific field has only recently come to be known, also believed strongly in the experimental approach. These two scientists were largely responsible for what advancement in physical science there was during the years 1230–1520.

The sixteenth century, however, was the beginning of an intellectual awakening. Known as the Renaissance, the era brought about advancements in the physical sciences and in other areas.

The new thinking in astronomy began with Nicholas Copernicus (1473–1543), a Polish astronomer who made the necessary mathematical calculations to prove the

Copernicus

Kepler

Sun-centered, or heliocentric, theory put forth by Aristarchus and to disprove the theory of Aristotle.

Tycho Brahe (1546–1601), a Dane who was considered the greatest practical astronomer since the Greeks, spent his entire life observing and recording data on the planets and stars. His work might have gone unnoticed had not Johannes Kepler fallen heir to Brahe's large collection of data. Kepler (1571–1630), a German astronomer and mathematician, believed in Copernicus's heliocentric theory and, with Brahe's data, developed three laws of planetary motion which eliminated the geocentric, or Earth-centered, theory and opened the way for our present views on the solar system. He discovered that the planets move in elliptical orbits around the Sun.

Galileo Galilei (1564–1642), Italian astronomer, mathematician, and physicist, is regarded as the first true experimentalist and the father of modern physics. He was the first to view the moon, planets, Sun, and stars through a telescope. He set forth the basic ideas of motion and laid the groundwork for the laws of motion to be formulated later by Newton; laws which are the foundations of the branch of physics known as mechanics.

Galileo designed and built his own telescope about 1609, after hearing about a Dutch lensmaker's success in magnifying distant objects with two lenses. His first telescope had a magnification of three, which he later improved to a power of more than 30. During the year 1610 he used his telescope to discover the craters and mountains of the moon, the spots on the Sun, which rotated every 27 days, the disk appearance of the planets, the four largest moons of Jupiter, the phase changes (the moon changes phase monthly) of Mercury and Venus, the rings of Saturn, and the multitude of stars in the Milky Way. His discoveries contributed more new information about the universe than those of any other single individual.

Sir Isaac Newton (1642–1727), the English physicist and mathematician regarded by many as the greatest scientist of all times, was born the year Galileo died. He lived at a time when a theory was needed to unify the discoveries made by Copernicus, Brahe, Kepler, and Galileo. Newton formulated such a theory and greatly advanced the branch of mechanics known as dynamics.

Newton was born on December 25, 1642, in a farmhouse near Grantham in Lincolnshire, England. He obtained his formal education at Trinity College, Cambridge University, receiving the M.A. degree in January 1665. The Great Plague closed the University in the fall of 1665, and Newton spent the next two years at home in Lincolnshire, at which time he developed the calculus, formulated his three laws of motion and the law of gravitation, and discovered the nature of white light. In 1667 Newton returned to teach at Cambridge University, where he spent most of his life.

Newton's theories were published many years after they were developed, and they might have gone unpublished had it not been for associates like Edmund Halley, who urged Newton to publish his work. Newton's *Mathematical Principals of Natural Philosophy,* or *The Principia* as it was called, appeared in 1687, over 20 years after he had formulated his laws of motion and gravitation. Following his recovery from a nervous breakdown in 1692, he did very little scientific work. He died in 1727.

Galileo

The seventeenth century also produced Robert Boyle (1627–1691), an Englishman who enunciated the gas law that bears his name and who is considered to be one of the founders of chemistry; Christian Huygens (1629–1695), a Dutch physicist who developed the wave theory of light and helped formulate the law of conservation of momentum; and Robert Hooke (1635–1703), an English physicist who made many scientific contributions but is best known for his law of elasticity.

During the eighteenth century, theories on heat, light, electricity, and magnetism appeared. Advances were made in all fields of physical science by an increasing number of scientists. Fahrenheit developed the temperature scale which bears his name, Celsius developed the centigrade scale (called in recent years the Celsius scale), Black made the distinction between heat and temperature, Bernoulli and Euler discovered the law of angular momentum, Bradley elucidated the aberration of light, Priestly discovered oxygen, Lavoisier, known as the father of modern chemistry, developed the analytical balance, proved the law of conservation of mass, and formulated the oxygen theory of combustion, J. L. Prout stated the law of definite proportions, Benjamin Franklin proved that lightning is electrical in nature, Cavendish proved the inverse square law of electrostatics, and Coulomb began quantitative research in electrostatics.

Included in the numerous achievements in the physical sciences during the nineteenth century are Dalton's atomic theory, Gay-Lussac's law of combining volumes, Avogadro's law for gases, Mendeleev and Meyer's periodic law of elements, and Graham's law of diffusion. Also, the wave theory of light was developed by Young; heat and thermodynamic theories were advanced by Kelvin, Joule, Rumford, Gibbs, and Carnot; additions to the theories of electricity and magnetism were made by Maxwell, Laplace, Green, and Poisson; and Ampere, Coulomb, Faraday, Henry, Oersted, Ohm, and Volta made fundamental electrical discoveries. Hamilton presented a new form of equation of motion to add to the mechanics field, radioactivity was discovered by Becquerel, X rays by Roentgen, and electrons by Thomson.

Newton

Avogadro

The year 1900 proved to be the beginning of a completely new era in physics. It was in this year that Max Planck originated his quantum hypothesis. The word quantum means a discrete amount, and Planck's hypothesis stated that certain forms of energy could exist only in certain discrete amounts, or quanta, rather than in any amount chosen or desired.

In 1905 Albert Einstein anounced his theory of relativity, which completely altered our ideas of space and time. Although Einstein is best known for his theory of relativity, he also related the mass of a particle to its total energy with his famous equation $E = mc^2$. He received the Nobel Prize in physics in 1921 for his work on the quantum theory of radiation.

In 1913 Niels Bohr, using a variation of Planck's hypothesis, developed a quantitatively correct model of the simplest atom, hydrogen. It is evident that by this time, the science revolution had begun in earnest.

The next big step was taken in 1925, when Louis de Broglie introduced the concept of matter waves. The notion of duality of light—that light acts sometimes as a wave and sometimes as a particle—was extended by de Broglie to the basic model of Bohr's atom. The electrons and protons in the atom were shown to act like waves as well as particles.

Erwin Schrödinger in the year 1925 developed a wave

Einstein

equation which described the behavior of particles having mass. In the next seven years some of the greatest achievements in physics and chemistry were made. Schrödinger's equation was applied to atoms, and with the help of Wolfgang Pauli's exclusion principle, the periodic table of Mendeleev became understood from a submicroscopic point of view. Schrödinger's equation was applied to the structure of molecules and solids, and the modern theories of these structures began to take shape. The neutron was first found experimentally by James Chadwick in 1932, and the basic ideas of nuclear structure were then complete.

The impact of the discovery in 1939 by Otto Hahn and Fritz Strassman, German chemists, of the phenomenon of nuclear fission has not yet been measured. With this discovery the possibility of converting mass into great quantities of energy became very real. The explosion of the first atomic bomb on a lonely desert in New Mexico in 1945 ushered in the atomic age, the age of science and technology in which we now live.

Many astronomers have made contributions to science during the twentieth century. The following are a few who have made major contributions to our knowledge in the composition, size, shape, and distribution of stars and galaxies. Edwin P. Hubble (1889–1953), an American astronomer, was the first to give evidence that our galaxy (the Milky Way) was but one of the many galaxies that exist in the universe. He showed that the large-scale structure of the universe is homogeneous, and he discovered the first evidence that the universe is expanding.

Harlow Shapley (1885–1972), American astronomer, made extensive studies of the distribution of globular clusters in the Milky Way, which revealed the true extent of the Milky Way and the position of our solar system in it. He also was one of the first astronomers to use Cepheid variables (stars that vary in brightness) to determine distances within our galaxy.

Henrietta Swan Leavitt (1868–1921), American astronomer, discovered the period-luminosity relationship for Cepheid variables. This relationship provided Hubble the knowledge to show that observed white patches of light (nebulae) were actually galaxies beyond the Milky Way.

Ejnar Hertzsprung (1873–1967), Danish astronomer, was the first to recognize the dissimilarity between dwarf and giant stars. Hertzsprung specialized in properties of stars and star clusters. He is noted for his discovery of a regular pattern that appears when the absolute magnitude of stars are plotted against their color index or their temperature. Henry Norris Russell (1877–1957), an American astronomer, is also recognized for the independent discovery of the same pattern of stars that Hertzsprung discovered. This plotted graph is known as the Hertzsprung-Russell diagram or simply the H-R diagram.

As the principles of physics and chemistry developed, they were applied to the phenomena occuring in the environment. Efforts to understand weather phenomena and their causes gave rise to the science of meteorology. Probably no other branch of physical science is more readily observable and has a more direct effect on our daily activities than meteorology. From empirical weather observations and scientific principles has come a better understnding of the workings of our atmosphere. As a result, reasonably accurate predictions can be made of future weather conditions. The dynamics of weather is

Chadwick

governed by physical laws. Although the physical principles are known, the vastness and the many variables of the atmospheric mass do not permit the age-old desire to completely undersand and control the weather.

On an even larger scale, especially that of time, is the study of our Earth—geology. Most of the fundamental geological principles that we use today were established in the seventeenth and eighteenth centuries. Very little knowledge of scientific value was known concerning the Earth and its origin before this time. In 1669 Nicolaus Steno (1638–1687), Danish physician, published a small pamphlet concerning his observations and investigations of rocks and the fossils found in them. From his studies he concluded it should be possible to infer deductively the order in which sedimentary rocks are formed. His reasoning led to the Law of Sequences.

Hutton

In 1788 James Hutton (1726–1797), Scottish physician, lawyer, and geologist, published a paper entitled "Theory of the Earth." Hutton established the groundwork for the modern science of geology in his conclusion concerning the principle of uniformity of process or the principle of uniformitarianism. The principle states that "the present is the key to the past."

The work of Hutton was advanced by Sir Charles Lyell (1797–1875), English geologist, who used Hutton's principle of uniformitarianism as the unifying theme in his classic publication *Principles of Geology*. Lyell is credited by many as having advanced the knowledge of geology more than any other person.

The work of William Smith (1769–1839), English civil engineer and geological hobbyist, concerning the relationship between rock strata in different localities led to the principle of faunal and floral succession. Smith noted that the rock layers could be identified by the fossil shells and skeletons they contain. This initiated the idea that major rock divisions as well as units of time might be based on the fossil animals and plants found in the rock.

Hubble

Perhaps the most important advance in our understanding of the geology of the Earth during the twentieth century has been the development of the theory of continental drift and plate tectonics. Alfred Wegener, German meteorologist, in a publication in 1915, presented the concept that about 200 million years ago the present continents were one huge land mass or supercontinent, which he called Pangaea. This land mass broke up and drifted apart, leading to present day positions of the continents, which still are slowly moving. Earth scientists have long regarded the Earth as a rigid body, but the discoveries that the Earth's interior may be slowly deformable led to the conclusion that the surface is mobile. The results of geological research in the last 20 years support Wegener's theory. But how could the large separation of the plates take place?

In 1960 Harvey Hess, of Princeton University, proposed that the sea floor cracks open along the crest of the mid-ocean ridges and that new sea floor forms there and spreads apart. The process is known as sea floor spreading. Today most geologists support the theory of plate tectonics, which includes continental drift, sea floor spreading, and subduction.

It is evident that the development and advancement of scientific knowledge has been rapid since the sixteenth century. This accelerated growth was the result of many factors, not the least of which is the procedure scientists

Lyell

use in acquiring knowledge. Although scientists do not follow a given set of rules, they do follow a general plan fo attacking a problem. Such a plan is referred to as the **scientific method** and includes the following, in general terms:

1. *The observation of phenomena and the recording of facts*. The "phenomena" are defined as whatever happens in the environment, and "facts" as accurate descriptions of what is observed.
2. *The formulation of a theory from the generalization of the phenomena*. A "theory" is a description of a certain behavior of nature that extends beyond what has been observed—usually stated in general terms.
3. *The development of the theory that is used to predict new data and new phenomena*. A theory comprises a general scheme of thought that explains the nature or behavior of the phenomena and correlates the known facts in such a manner that new thoughts and relations initiate the prediction of new phenomena. Einstein's theory of relativity and the kinetic theory of gases are examples.
4. *Experimentation is conducted to confirm the new data or phenomena predicted by the theory*.
5. *Confirmation, modification or disposal of the theory*. Further predictions are then made. Steps four and five are then repeated.

A major factor in the development of physical science has been the formulation of concepts used to describe phenomena. A concept is a meaningful idea that can be used to describe phenomena. Mass, length, time, speed, chemical element, and temperature are examples. Concepts develop as a result of experimentation and are subject to change as new experimentation dictates. They are fundamental in the sense that they form the rules for understanding nature, but they are also interconnected. For example, the distance (length) you travel when running is proportional to your speed.

The meaning of a concept is established by a working definition. The definition may be stated in words, symbolic notation, or by means of a mathematical formula. In whatever way the definitions are stated, we must learn them, explain them, and interpret their meaning in respect to our environment.

An understanding of our physical environment begins with a study of fundamental concepts.

Smith

An Introduction to Physical Science

1

Measurement

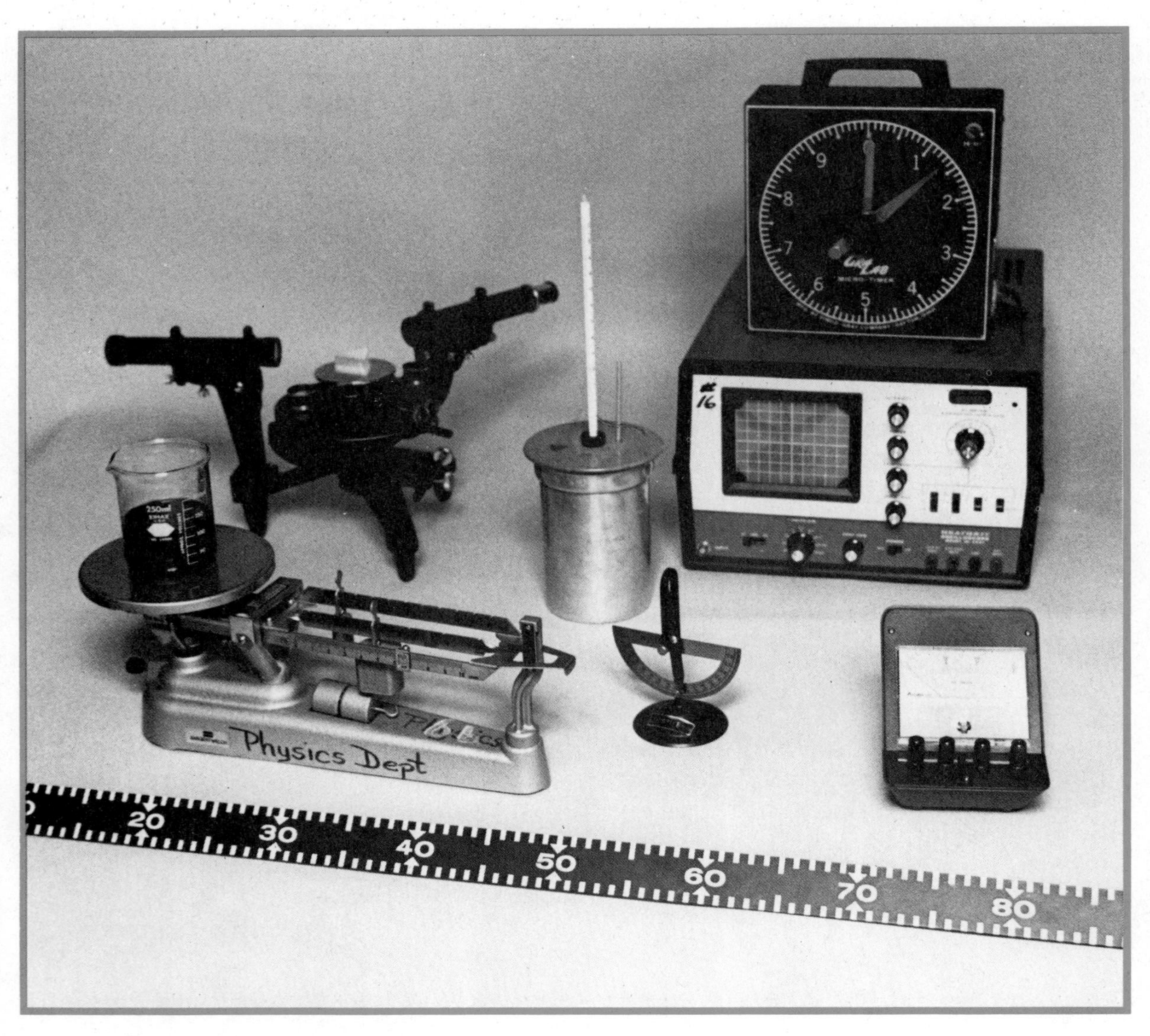

When you can measure what you are speaking about and express it in numbers, you know something about it; but when you cannot measure it, when you cannot express it in numbers, your knowledge is of a meager and unsatisfactory kind.

–Lord Kelvin

A FIRST STEP in understanding our environment is to find out about the physical world through measurements. Over the centuries, human beings have developed increasingly sophisticated methods of measurements, and scientists make use of the most advanced of these. The student should gain both a working knowledge of these methods and an awareness of their limitations.

We are continually making measurements in our daily life. Each day we plan our work, play, and rest schedules as a function of time. With watches and clocks we measure the time for events to take place.

Every 10 years we take the census and determine (measure) the population. We count our money, our homes, our food, the minutes, hours, days, and years of our life.

Some of us keep accurate measurements of food and drugs taken into the body because of illness. Many lives depend on accurate measurements being made by the medical doctor, laboratory technician, and pharmacist in the diagnosis and treatment of disease.

Meteorologists measure the many elements (temperature, pressure, humidity, precipitation, wind) that make up the weather. This information is relayed to millions by the communications media, which must at all times measure all phases of their operation to stay within standards designed to protect the rights of others.

Human beings, in their efforts to understand their total environment, must measure the very small and the very large. Scientists probe farther inward to examine smaller and smaller particles, and explore outward to discover a larger and larger universe.

◄ Instruments used for making measurements of physical properties.

At one time it was thought that all things could be measured with exact certainty. However, as we measured smaller and smaller objects, it became evident that the very act of measuring distorted the measurement. This uncertainty in making measurements of the very small will be discussed in detail in Chapter 12.

The ability of the scientist to know and predict is a function of accurate measurements. From accurate measurements taken on the moon's surface the geologist obtains new knowledge for understanding continental drift.

Our conquest of the moon and planets is largely due to our ability to make accurate measurements on the hardware that makes up the space vehicle and to program computers that make accurate and continuous measurements of position, velocity, and the numerous other factors involved in space travel.

From these examples, we see relevance in all measurements and recognize the need to know the concepts of measurement. Understanding measurement is the first step in the understanding of our physical environment.

1.1 The Senses

Our environment stimulates our senses, either directly or indirectly. The five senses (sight, hearing, smell, touch, taste) make possible our knowledge of the environment. Therefore, they are a good starting point in the study and understanding of the physical world.

Most information about our environment comes through sight. This information is not always a true representation of the facts because the eyes, and therefore the mind, can be fooled. There are many well-known optical illusions, such as those in Fig. 1.1. Many people are quite

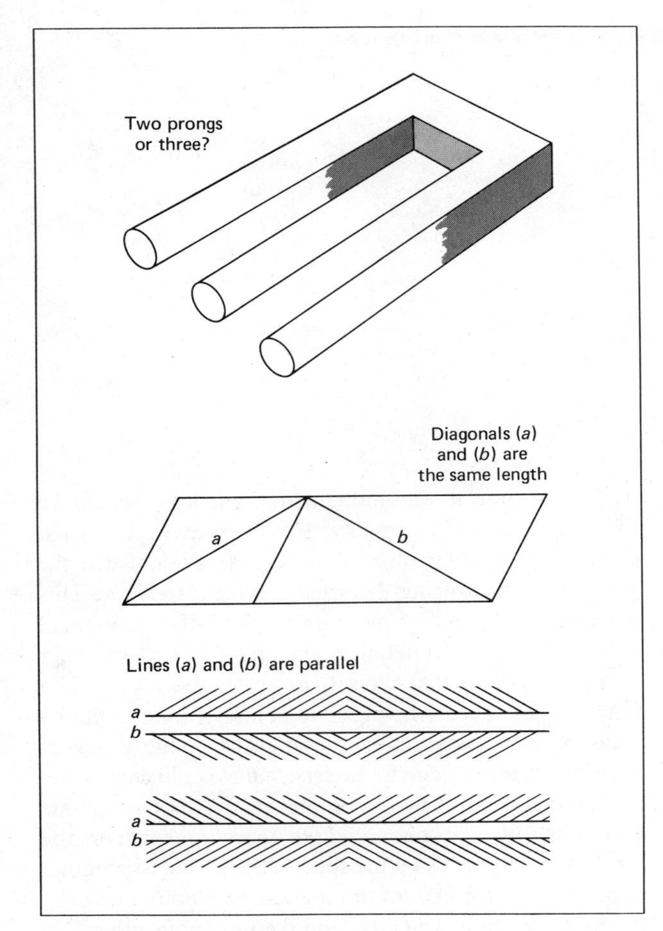
Two prongs
or three?
Diagonals (a)
and (b) are
the same length
a
b
Lines (a) and (b) are parallel
a
b
a
b

Figure 1.1 Some optical illusions. We can be deceived by what we see.

convinced that what they see in such drawings actually exists as they perceive it.

Hearing ranks second to sight in supplying the brain with information about the external world. The senses of touch, taste, and smell, although very important for good health and happiness, rank well below sight and hearing in providing environmental information.

All the senses can be deceived and thus provide false information about our environment. Anyone who has gone to the beach to swim during the early morning hours when the air is cold knows how warm the water feels to the body. Later in the afternoon when the air becomes hot, the water, which has remained at practically the same temperature, feels cold. Thus, if we were asked to judge the temperature of the water, our answers would vary according to the temperature of the air.

Not only can the senses be deceived, but they also have their limitations. The naked eye is unable to distinguish the stars of our galaxy from the planets of our solar system. In fact, the word *planet* meant "wandering star."

The other senses have similar limitations. For instance, if you were asked to distinguish between two identical sounds produced $\frac{1}{100}$ second apart, you would fail.

The handicaps of the senses can be conquered by close scrutiny of phenomena with measuring instruments. For example, in the diagram of optical illusions (Fig. 1.1), diagonals (a) and (b) can be measured with a ruler, and their length accurately determined.

We extend our ability to measure our environment with many specialized instruments, or tools. These, too, have their limitations. The wristwatch is a precision instrument, but it cannot be used to measure time intervals of less than $\frac{1}{10}$ second.

Even the most precise instruments have limitations. More accurate information about our physical environment can be obtained by comparing the measurement of one instrument against another. Later in our study of the microscopic world we shall learn of other limitations concerning the measurement of physical quantities.

1.2 Fundamental Properties

Physical characteristics of observed phenomena can be expressed in terms of fundamental properties, which are basic to our comprehension of the physical world. Scientists have identified four physical properties, which they specify as fundamental—length, mass, time, and electric charge. These properties are fundamental because they form the foundation for other properties needed to obtain order and meaning for the physical sciences.

How would you describe your environment? As you begin to observe, you would ask questions like "Where is the book store?" "When does it open?" "How much do you want to buy?" and so forth. These questions of "Where?" "When?" and "How much?" refer to the basic concepts of space, time, and matter.

The description of space might refer to a location or to the size of an object. To measure locations and sizes we use the fundamental property of **length**, which is defined as the measurement of space in any direction.

Space has three dimensions, each of which can be measured by a length (see Fig. 1.2). This can easily be seen by considering a rectangular object. It has a length, width, and height, but each of these dimensions is a length. A sphere, such as the Earth, has a radius, a

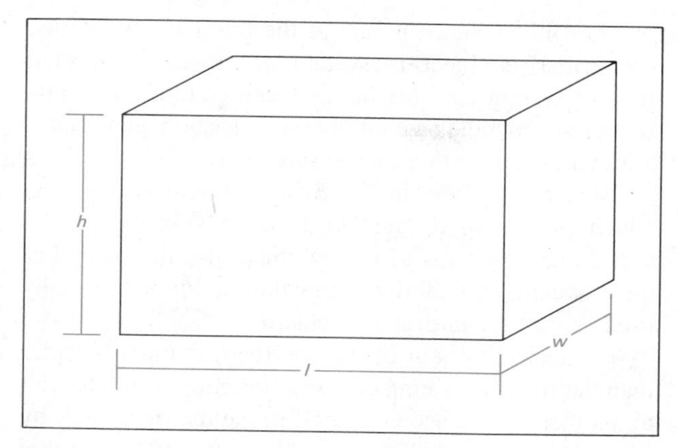

Figure 1.2 The dimensions of a box are commonly given in terms of its length (l), width (w), and height (h), but all are measurements of *length*.

diameter, and a circumference. Again, all of these are easily described by a length measurement.

Once we know where something is, we are frequently interested in what is happening to it. "Is the car moving?" "When will the next plane leave?" "What day will you be going home?"—all of these questions can be answered using the fundamental property of time.

Each of us has an idea of what time is, but we would probably find it difficult to define or to explain it. Some terms that are often used in referring to time are duration, period, or interval. We will define **time** as the continuous, forward flowing of events.

Without events or happenings of some sort, there would be no perceived time (Fig. 1.3). The mind has no innate awareness of time, merely the awareness of events taking place in time. That is, we do not perceive time as such, only the events that take place in time.

Figure 1.3 Time can be described as the continuous forward flowing of events. Here the events are the dropping of sand particles in an hourglass.

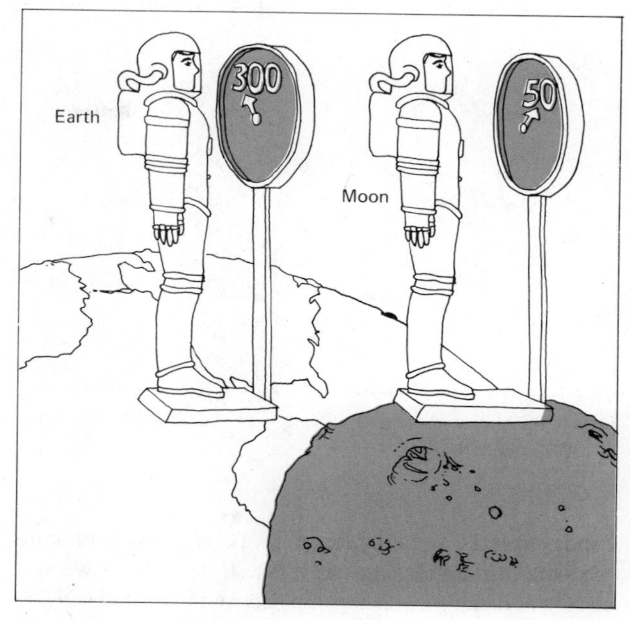

Figure 1.4 The weight of an astronaut on the moon is $\frac{1}{6}$ what it is on the Earth. The mass, however, is the same.

Einstein, in his theory of relativity, has shown that space and time are inextricably linked together into what is called space-time. In his theory, time joins the three dimensions of space as a fourth dimension. Einstein's theory of relativity is the topic of Chapter 9. For the most part, however, we can use our intuitive ideas, which tend to regard space and time as separate fundamental properties.

When we ask questions concerning the quantity of matter, we need a third fundamental property known as mass. To define mass precisely, we need to understand the concepts of force and acceleration. These are discussed in Chapters 2 and 3. For now, let us simply say that **mass** refers to the quantity of matter an object contains.

Since we live on Earth, many of us tend to measure matter in terms of weight. However, in the metric system of measurement (see Section 1.3), weight is not a fundamental property. An astronaut who weighs 300 pounds on Earth will weigh $\frac{1}{6}$ of that amount, or 50 pounds, on the moon (Fig. 1.4). However, the astronaut's mass will be the same on the Earth and the moon.

Weight is related to the force of gravity, which changes depending on where we are in the universe. On the other

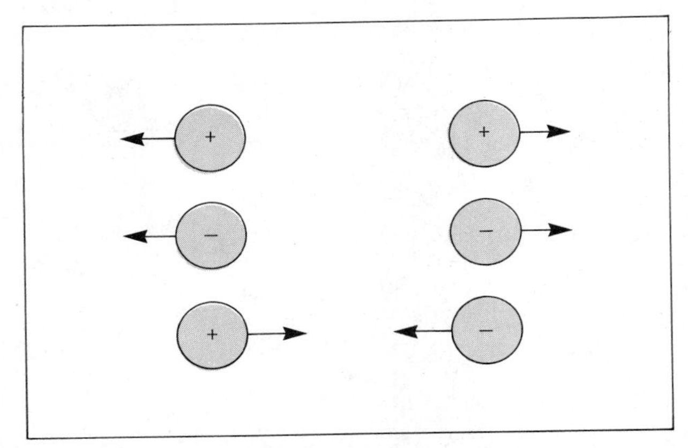

Figure 1.5 Like electrical charges repel. Unlike electrical charges attract.

hand, mass is a fundamental property, which remains the same throughout the universe. In Chapter 3 we will discuss in detail the relationship between mass and weight.

A fourth fundamental property is electric charge. There are two kinds of **electric charge**, called positive and negative. Two positive charges or two negative charges will repel each other, while opposite charges will attract (Fig. 1.5).

Electric charge is an important property of matter since all atoms are composed of electrically charged particles and most matter is composed of atoms. Electric current is simply the flow of electrical charge. The concepts of electric charge and electric forces will be discussed in Chapter 8.

The four **fundamental properties** of nature are length, time, mass, and electric charge. These four properties and their various combinations are all that we need when we try to measure many of the complicated aspects of nature. Of course, there may be unknown fundamental properties as yet undiscovered, but for now, let us discuss how we measure these four.

1.3 Standard Units

In order to measure the fundamental properties and their various combinations, we need to refer them to reference or standard measurements. Each reference is called a standard unit. A **standard unit** has a fixed and reproducible value for the purpose of taking accurate measurements. There are several systems of measurement used around the world. Each system uses different standard units.

The United States is one of the few nations that use the British engineering system of measurement. The **British system** uses the familiar unit of foot for length. It uses the pound as the unit of weight, and is thus called a gravitational system of measurement.

The unit of mass in the British system is the slug, which weighs approximately 32 pounds on the Earth's surface. All systems of measurement use the second as the standard unit of time, and the coulomb is usually used as the standard unit of charge.

The metric system of measurement is much simpler than the British system because converting from one unit to another can be accomplished in the metric system by using factors of 10. For example, in the metric system 1 kilometer is 1000 meters, whereas in the British system 1 mile is 5280 feet. Memorizing the various conversions, such as 12 inches in one foot, 3 feet in one yard, 5280 feet in one mile, etc., makes the British system unwieldy compared to the simplicity of the metric system.

There are actually two metric systems in common usage. One is called the mks system, the other is the cgs system. The letters **mks** stand for *m*eter, *k*ilogram, and *s*econd, while **cgs** stand for *c*entimeter, *g*ram, and *s*econd. Table 1.1 lists the standard units for the mks, cgs, and British systems of measurement.

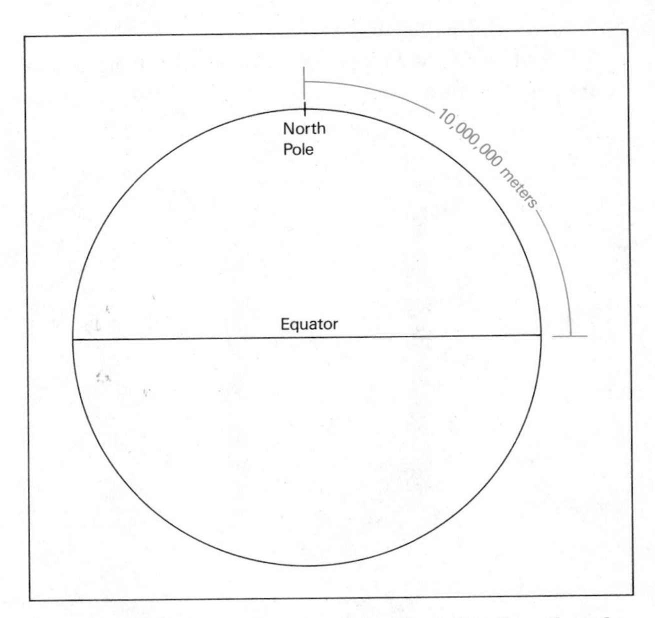

Figure 1.6 The meter was originally defined so that the distance from the equator to the North Pole would be 10,000,000 meters.

Table 1.1 Standard Units for the Metric and British Systems of Measurement

	Absolute Systems		*Gravitational System*
Fundamental Property	*Metric (mks)*	*Metric (cgs)*	*British*
Length	Meter (m)	Centimeter (cm)	Foot (ft)
Mass	Kilogram (kg)	Gram (g)	Slug
Time	Second (s)	Second (s)	Second (s)
Electric charge	Coulomb (C)	Coulomb (C)	Coulomb (C)

The standard unit of length in the meter-kilogram-second (mks) system is the **meter** (from the Greek, *metron*, to measure), which was originally intended to be one ten-millionth of the distance from the Earth's equator to the geographic north pole. This is illustrated in Fig. 1.6. The unit was first adopted by the French in 1793, and it is now used in scientific measurements of length throughout the world.

From 1889 to 1960 the standard meter was a platinum-iridium bar kept in the vaults at the International Bureau of Weights and Measures near Paris, France. However, the stability of this bar was questioned, and so new standards were fixed in 1960 and, most recently, in 1983.

The current definition of the meter links it to the speed of light in a vacuum. One meter is defined to be the distance light travels in 1/299,792,458 of a second. That is, the speed of light in a vacuum is defined to be 299,792,458 meters per second.

From this basic standard, other units of length are defined. For example, the millimeter is defined as $\frac{1}{1000}$ meter, the centimeter as $\frac{1}{100}$ meter, and the kilometer as 1000 meters.

The standard unit of time in both the metric system and the British system is the second. For many years the **second** was defined as a fractional part (1/86,400) of the average solar day. The average day was used because the length of the day (as measured by when the Sun is directly overhead) varies slightly, since the Earth's path around the Sun is not a perfect circle. See Fig. 1.7.

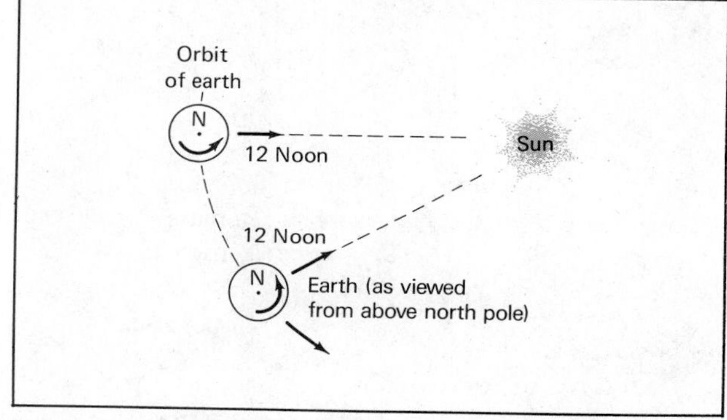

Figure 1.7 Diagram illustrating one solar day. One solar day is the elapsed time between successive crossings of a meridian by the Sun.

Today's scientists use an atomic definition of the second, based upon the vibrations of a cesium 133 atom as it radiates a certain wavelength of light. In this definition, the cesium atom vibrates 9,192,631,770 times each second. (For the complete definition, see section 19.3.)

Although the meter and second are now accurately defined in terms of the speed of light and atoms, the definition of the kilogram is not so precise. It is associated with the meter.

Originally, one **gram** was defined as the mass of one cubic centimeter of pure water at its maximum density. The **kilogram** was taken to be 1000 grams, or the mass of 1000 cm^3 of water. Since one liter is 1000 cm^3, a kilogram was taken to be the mass of one liter of water at its maximum density.

In everyday terms, a kilogram is the mass of one liter of water. Since most drinkable liquids (soft drinks, milk, juices, etc.) have about the same density as water, a liter of any soft drink will have a mass of approximately one kilogram. This is illustrated in Figure 1.8. Also, since a liter is approximately a quart (1 L = 1.056 qt), a quart of milk has a mass of approximately one kilogram.

A definition of the kilogram in terms of the properties of water is not exact enough for precision measurements. Currently, the kilogram is defined to be the mass of a cylinder of platinum-iridium kept at the International Bureau of Weights and Measures near Paris. The United

Figure 1.8 A one-liter bottle of soft drink will have a mass of approximately one kilogram. (Photo courtesy Peter Bogen.)

Figure 1.9 Prototype kilogram number 20, shown in photograph, is the United States standard unit of mass. The prototype is a platinum-iridium cylinder, 39 mm in diameter and 39 mm high. Prototype Number 20 was furnished to the United States by the International Bureau of Weights and Measures in accordance with the treaty of 1875. (Courtesy National Bureau of Standards.)

Figure 1.10 The metric system is being used more in some states than in others. (Courtesy Ohio State Department of Transportation.)

States prototype is kept at the U.S. Bureau of Standards in Washington, D.C. This is shown in Fig. 1.9.

Most countries now use the metric system. In fact, the United States is the only major country not to use the metric system officially. However, the usage of metric units such as grams, kilograms, liters, meters, and kilometers is becoming more common in the United States. Figure 1.10 shows a highway sign given in metric units.

To make comprehension and exchange of ideas among people of different nations as simple as possible, an International System of Units (SI) has been established. The **SI** is a modernized version of the metric system and contains seven base units: the meter (m), the kilogram (kg), the second (s); the ampere (A), to measure rate of flow of electric charge; the kelvin (K), to measure temperature; the mole (mol), for measuring the amount of substance; and the candela (cd), to measure luminous intensity. A definition of each of these seven units is given in Appendix I. All other SI units are derived from the base units and two supplementary units (radian and steradian: a measure of plane and solid angle, respectively) by means of the established scientific laws relating the respective physical quantities.

1.4 Derived Units and Conversion Factors

The appropriate units for most phenomena in nature involve more than one fundamental property. Thus, they are measured by combinations of the standard units. These combinations of standard units are called **derived units**. For example, the speed of an automobile may be measured in miles per hour (mi/h or mph), kilometers per hour (km/h), meters per second (m/s), centimeters per second (cm/s), feet per second (ft/s), or any other combination of length per unit of time. All of these combinations of units are derived units. Some other simple examples are as follows:

Property	*Derived Units*
Area (length^2)	m^2, cm^2, ft^2, etc.
Volume (length^3)	m^3, cm^3, ft^3, etc.
Density (mass per volume)	kg/m^3, g/cm^3

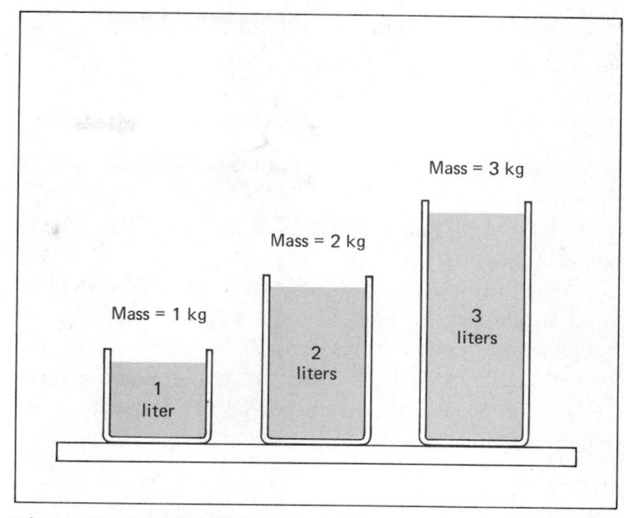

Figure 1.11 Equal densities. Whether you have one, two, or three liters of water, having masses of one, two, or three kilograms, the density is the same. In this example, the density of water in each of the three containers is 1 kg/liter.

The density of an object refers to how compacted a substance is with matter. More precisely, **density** is the mass per unit volume of an object, and its formula can be written

$$\text{density} = \frac{\text{mass}}{\text{volume}}$$

Thus, something with a mass of 20 kg that occupies a volume of 5 m^3 would have a density of 20 kg/5 m^3 = 4 kg/m^3. In this example the mass was measured in kilograms and the volume in cubic meters.

If mass is uniformly distributed throughout the volume, then the density of the matter will remain constant. Figure 1.11 shows that if you have a uniform substance, such as water, the density remains the same no matter how much of the substance is in question.

The density of water is 1 g/cm^3, 1 kg/liter, or 1000 kg/m^3. If density is expressed in units of grams per cubic centimeter, it is easy to compare any density to that of water. For example, a rock might have a density of 3.3 g/cm^3, pure iron has a density of 7.9 g/cm^3, and the Earth as a whole has an average density of 5.5 g/cm^3.

Densities of liquids such as blood or alcohol can be measured by means of a hydrometer. A **hydrometer** consists of a weighted glass bulb that floats in the liquid. The higher the glass bulb floats, the greater the density of the liquid.

When a medical technologist checks a sample of urine, one test he or she runs is for density. Urine has a density in a healthy person of from 1.015 to 1.030 g/cm^3. That is, it consists mostly of water and dissolved salts. If the density is greater or less than this normal range, it may be due to an excess or deficiency of dissolved salts caused by an illness.

A hydrometer is used to test for antifreeze in a car radiator, as shown in Fig. 1.12. The hydrometer is calibrated directly in degrees rather than actual density. The closer the density is to 1.00 g/cm^3, the closer the antifreeze and water solution is to being pure water. When the density corresponds to 1.00 g/cm^3, the hydrometer will read a temperature of 0°C or 32°F, the freezing point of water. The further the mixture is from being pure water, the lower the temperature reading will be.

The units on both sides of an equation must always be similar. For example, we cannot equate two quantities with fundamentally different units. We cannot equate something with units of velocity (e.g., meters per second) to something with units of area (e.g., square meters).

When a combination of units gets very complicated, we frequently give it a name of its own. Consider the following examples which will be discussed in later chapters:

$$\text{joule} = \text{kg} \times \text{m}^2/\text{s}^2$$

$$\text{newton} = \text{kg} \times \text{m}/\text{s}^2$$

$$\text{watt} = \text{kg} \times \text{m}^2/\text{s}^3$$

There are many more, but the point is that it is easier to talk about watts than kg $\times$ m^2/s^3. When a particular combination of units gets to be important, it is usually given its own name.

Frequently we want to convert from one system of units to another in order to make comparisons. For instance, we frequently want to make comparisons between the metric and the British systems. Many of these conversion factors are listed on the inside back cover. For instance,

$$1 \text{ in} = 2.54 \text{ cm}$$

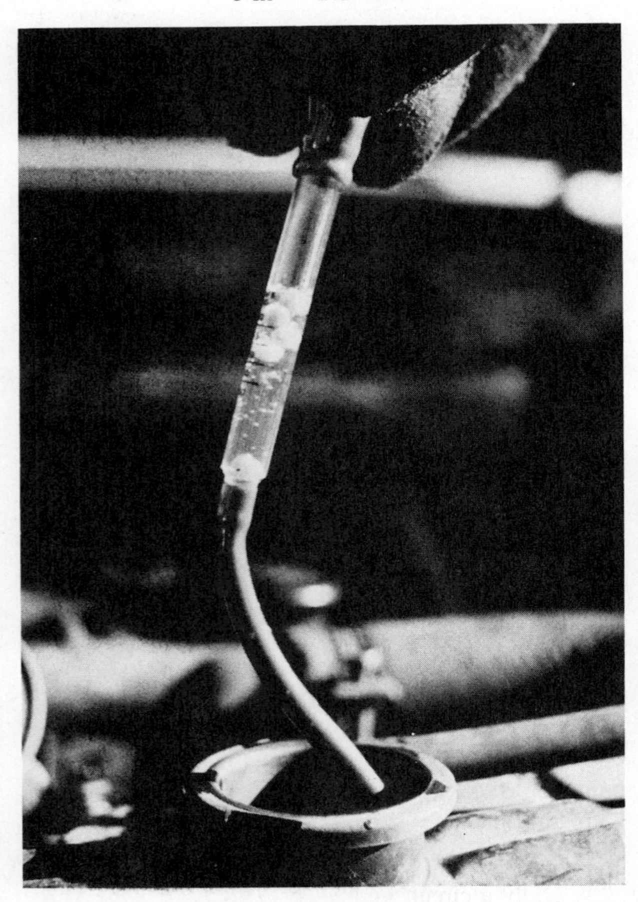

Figure 1.12 A hydrometer is used to measure the density of antifreeze. The freezing temperature of the antifreeze-water solution can then be determined. In the hydrometer shown, the number of balls floating at the top indicates the freezing point of the mixture. Note the four balls floating and the one not floating at the bottom of the glass tube. (Photo courtesy Peter Bogen.)

This conversion factor is frequently used. For example, if you want to know your height in centimeters and you are 5 ft 6 in tall, or 66 in, your height in centimeters is given by (see Fig. 1.13)

$$66 \text{ in} = 66 \text{ in} \times 2.54 \frac{\text{cm}}{\text{in}} = 167.6 \text{ cm}$$

In the metric system your height would be 167.6 cm. If you ask someone what his or her height is and the reply is a number such as 165 or 180 or 190, this would imply so many centimeters. To convert these numbers to inches, you would have to divide by 2.54.

A similar exercise must be done when converting from mass to weight or vice versa on the Earth's surface. Strictly speaking, mass and weight refer to two different quantities. They are not equal, but a given mass does have an equivalent weight here on Earth. The appropriate conversion factor on the Earth's surface is

$$1 \text{ kg mass} = 2.2 \text{ lb weight}$$

To find your mass in kilograms, simply divide your weight by 2.2 (actually 2.2 lb/kg). For instance, if you weigh 132 lb, then your mass in kilograms is

$$132 \text{ lb} \times \frac{1 \text{ kg}}{2.2 \text{ lb}} = 60 \text{ kg}$$

If a person's mass is known in kilograms, his or her equivalent weight on Earth can be found by multiplying

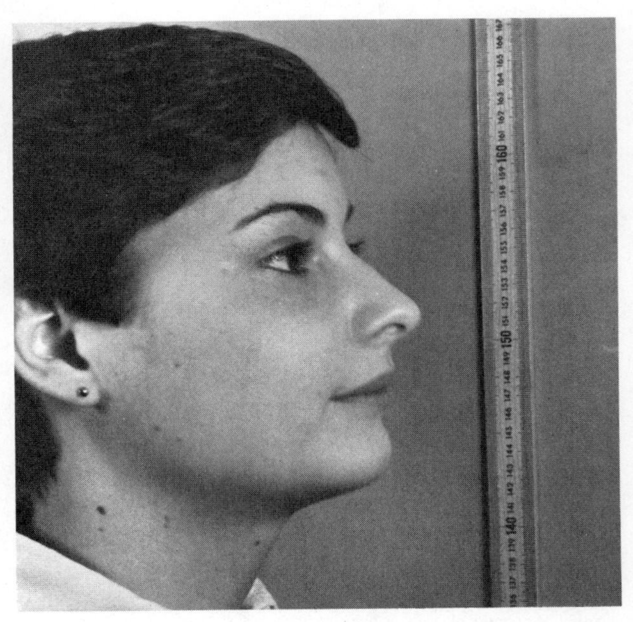

Figure 1.13 A person who is 5 ft 6 in. tall is about 167.6 cm tall. (Photo courtesy Jimmy Adams.)

Table 1.2 Conversion of Some Common Units

To convert from inches to centimeters multiply by 2.54.
To convert from centimeters to inches divide by 2.54.
To convert from kilograms to pounds multiply by 2.2.
To convert from pounds to kilograms divide by 2.2.
To convert from meters per second to miles per hour multiply by 2.24.
To convert from miles per hour to meters per second divide by 2.24.

by 2.2 (actually 2.2 lb/kg). For example, a mass of 70 kg would be an equivalent weight of

$$70 \text{ kg} \times 2.2 \frac{\text{lb}}{\text{kg}} = 154 \text{ lb}$$

Often we are concerned with converting from the metric unit of speed, which is meters per second, to the more familiar miles per hour. The conversion factor necessary to do this is

$$1 \text{ m/s} = 2.24 \text{ mi/h}$$

To convert from meters per second to miles per hour we simply multiply by 2.24 (actually 2.24 mi/h ÷ m/s). As an example, 20 m/s can be converted as follows:

$$20 \text{ m/s} \times 2.24 \frac{\text{mi/h}}{\text{m/s}} = 44.8 \text{ mi/h}$$

In Fig. 1.14, a speedometer is calibrated in both mi/h and m/s. In Table 1.2 we summarize how to convert the quantities we have discussed. A more extensive list is given on the inside back cover.

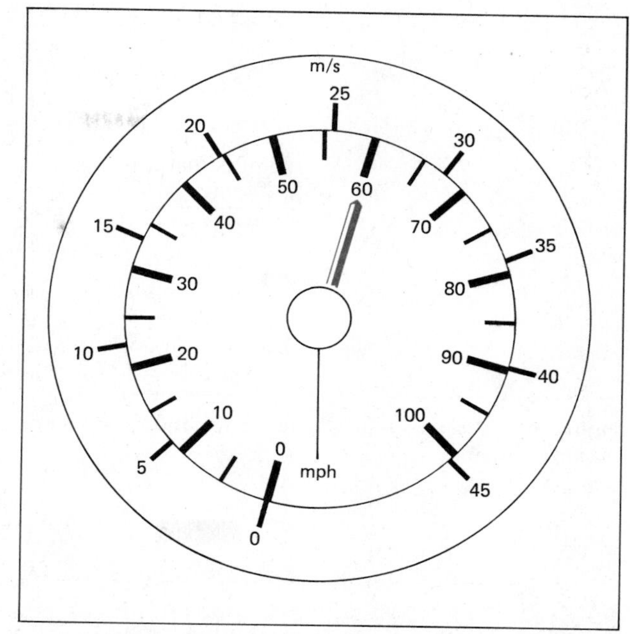

Figure 1.14 A speedometer showing both miles per hour and meters per second. 25 m/s = 56 mi/h.

1.5 Measurement of Circles

The units of time probably originated with the Babylonians, who reckoned the year as 360 days. They were aware that, in any circle, a chord that is equal to the radius subtends an arc equal to 60 degrees (usually designated 60°), as shown in Fig. 1.15 (p. 12).

This interesting property may have been the basis of their sexagesimal number system and perhaps accounts for the division of the **degree** into 60 minutes, and minutes into 60 seconds (see Fig. 1.16, p. 12). Since the Babylonians used the apparent motion of the Sun to tell the passing of daylight hours, their sexagesimal system may also be the basis for our present method for reckoning time, that is, 60 seconds equal 1 minute and 60 minutes equal 1 hour.

The early Greeks were extremely interested in geometry and the properties of circles. They were particularly interested in the ratio of the circumference of a circle to its diameter. This ratio they designated as π (pi). We can write π as

$$\pi = \frac{\text{Circumference of a circle}}{\text{Diameter of a circle}} \tag{1.1}$$

This ratio is always the same for every circle. The value of π is given by

$$\pi = 3.14159\ldots \tag{1.2}$$

and is usually rounded off to 3.14 or $\frac{22}{7}$. From the definition of π, we find that the circumference of a circle is just $\pi \times$ diameter. But, since the diameter is twice the radius, we get

$$\text{Circumference} = 2 \times \pi \times \text{radius} \tag{1.3}$$

EXAMPLE The Earth goes around the Sun in approximately a circular orbit with a radius of 93 million miles. What is the circumference of the Earth's orbit?

The circumference is given by

$$\begin{aligned}\text{Circumference} &= 2 \times 3.14 \times 93{,}000{,}000 \text{ mi}\\ &= 584{,}040{,}000 \text{ mi}\end{aligned}$$

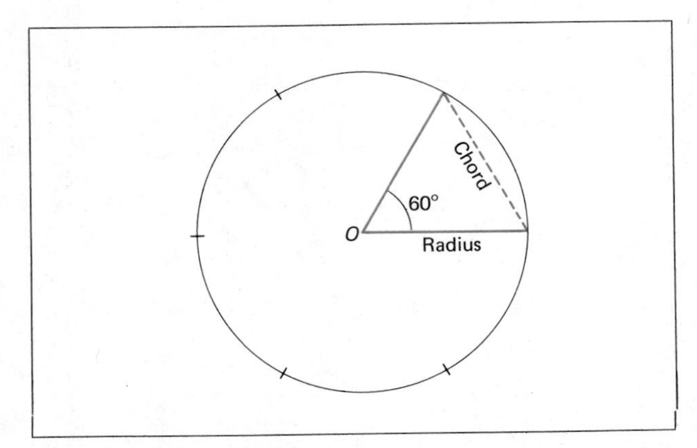

Figure 1.15 A chord that is equal to the radius subtends an angle of 60°.

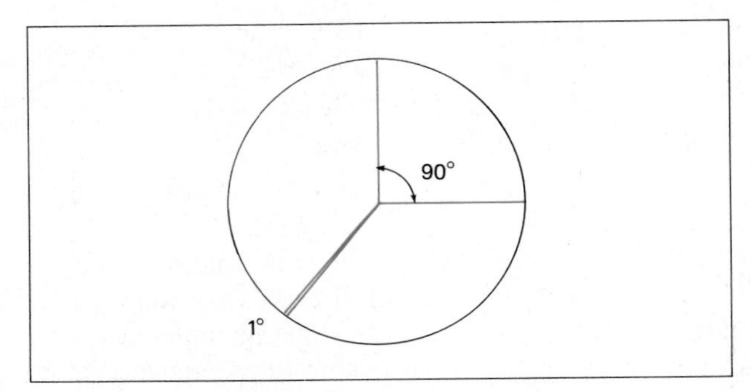

Figure 1.16 A circle is divided into 360°. A right angle has 90°. Each degree is divided into 60 minutes of angle (designated 60′), and each minute of angle is divided into 60 seconds of angle (designated 60″). Thus, there are 3600 seconds of angle in 1 degree of angle.

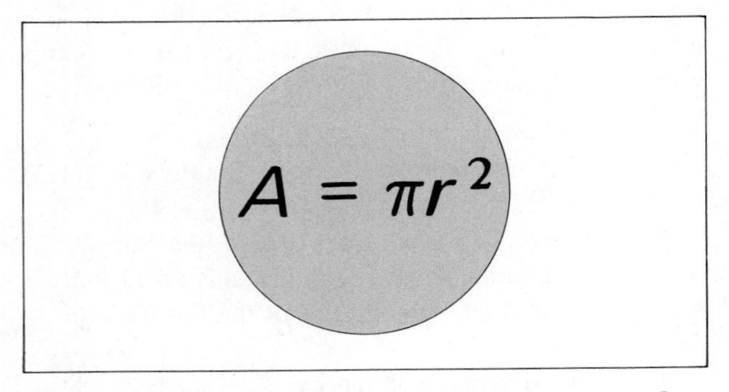

Figure 1.17 The area of a circle is given by $A = \pi r^2$.

We see, then, that each year the Earth travels over 500 million miles in its journey around the Sun.

The quantity π is also important when finding the area of a circle. The area of a circle is given by (see Fig. 1.17)

$$\text{Area} = \pi \times (\text{radius})^2$$

or

$$A = \pi r^2$$

1.6 Experimental Error

Making measurements of physical quantities requires the use of some device that will give a number denoting the ratio of the observed quantity to one of the standard units. A **measurement** is a comparison of the physical quantity with the standard unit.

The process of taking any measurement by any known means always involves some uncertainty. This uncertainty is usually called **experimental error**. In taking measurements, an effort is made to keep the experimental error to a minimum.

Experimental errors are classified as either systematic or random. Systematic errors are always in the same direction; that is, the magnitude, or size, of the number obtained is always too small or too large, as with a watch that runs too fast or too slow, or a speedometer that shows too many or too few miles per hour.

The observer may also be the cause of systematic errors, because of bad vision or other difficulties with his or her senses. An example of systematic experimental error is shown in Fig. 1.18. Not all such errors are so apparent, however.

Random errors, on the other hand, are in either direction and result from accidental variations of the observed physical quantity or of the measuring instruments. Random errors are caused by small variations in any direct or indirect physical quantity associated with the measurement. They can result from such causes as temperature or pressure variations.

When random errors are small, the measurement is said to have high precision. **Precision** refers to the degree of reproducibility of a measurement, that is, to the maximum possible error of the measurement, and may be expressed as a plus or minus correction. For example, if the length of the sample is expressed as 44.4 cm $\pm$ 0.1 cm (read plus or minus $\frac{1}{10}$ of 1 centimeter), we know the sample is somewhere between 44.3 and 44.5 cm long.

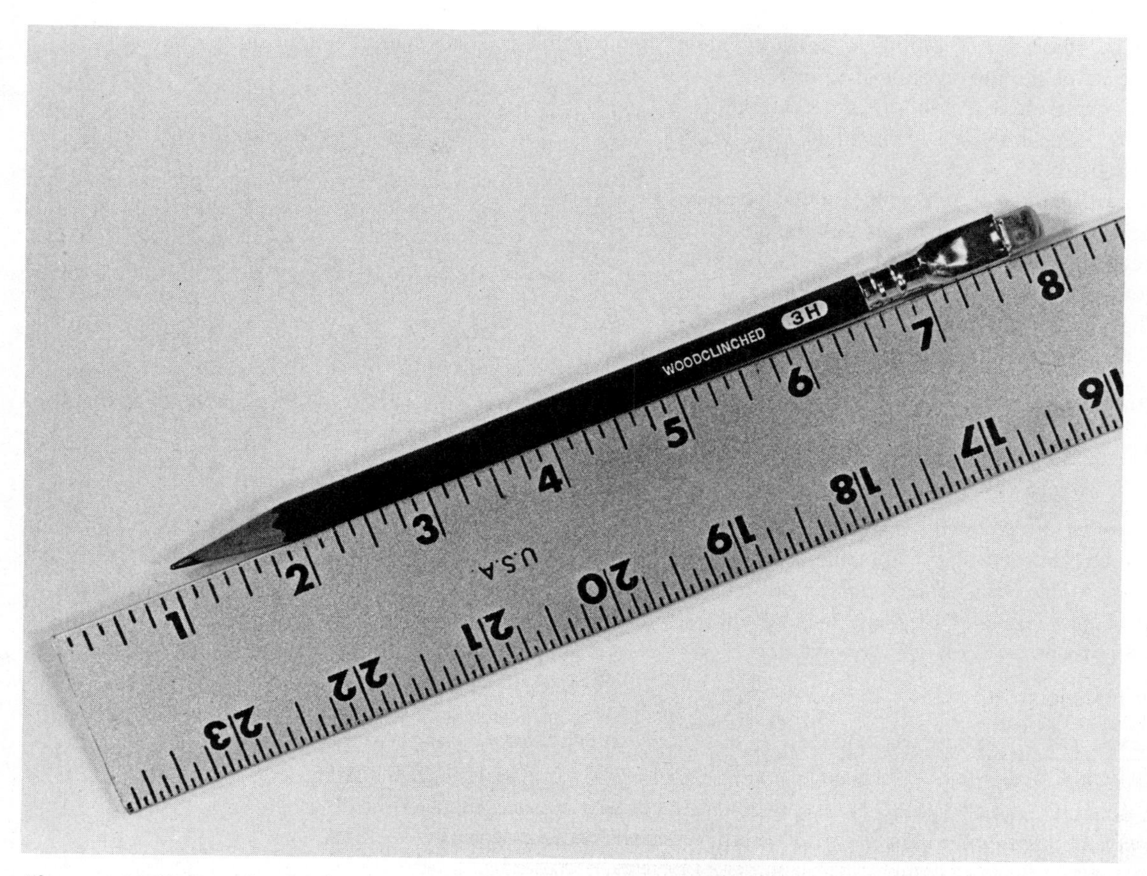

Figure 1.18 If the scientist does not realize that the zero point of the measuring instrument is incorrect, he or she will be making a systematic error. (Photo courtesy James Crouse.)

A measurement having high precision does not necessarily have high accuracy. For example, one could determine the boiling point of a liquid with a precision of $\pm 0.01°C$ (read plus or minus $\frac{1}{100}$ of 1 degree Celsius), but any impurity in the liquid would prevent the accurate determination of the true boiling point of the liquid.

With experimental error always appearing in our measurements, how close an approximation can we obtain in making a measurement of any physical quantity; that is, how close can we come to the true value? The term **accuracy** refers to how close the measurement comes to the true value.

How is the true value determined? The true value is the value currently accepted by the best scientists in the field. This value is subject to change as measurements are made with better methods and better instruments.

It is obvious from the preceding discussion that the terms precision and accuracy should never be used interchangeably. They have different meanings and should be used only where they apply.

For example, suppose a manufacturer advertises a thermometer that will measure to the nearest hundredth of a degree Celsius. Thus, using this thermometer, one can measure with a precision of $\pm 0.01°C$. Now, suppose an experimenter wants to determine the boiling point of pure water to the nearest 0.01°C with this thermometer. If the experimenter has pure water, he or she can perform the experiment and obtain an accurate answer. If, however, the water is not pure, an "accurate" measurement of the boiling point of pure water cannot be made, even though a high "precision" thermometer is used.

Since error is impossible to eliminate, methods have been devised to calculate the amount of error that exists in a given measurement. Calculation of the amount of error is done by either one of two methods, depending on the circumstances.

If there is an accepted or true value of the physical quantity, then a calculation known as percentage error is made. **Percentage error** is defined as the ratio of the absolute difference* between the experimental and the accepted values to the accepted value, expressed as a percentage. This can be written as

$$\text{Percentage error} = \frac{|\text{Exp. value} - \text{Accepted value}|}{\text{Accepted value}} \times 100$$

If there is no accepted value, then a percentage difference is obtained. **Percentage difference** is defined as the ratio of the absolute difference between the experimental values to an average of the experimental values, expressed as a percentage. This can be written as

$$\text{Percentage difference} = \frac{|\text{1st exp. value} - \text{2nd exp. value}|}{\text{Average of the two values}} \times 100$$

A more useful value of measurement in experimental work is often the average value. For example, in finding the length of a table it would be better to take three separate measurements, find the sum of the three measurements, and then divide by three to obtain the average. When three or more measurements are made, the percentage difference is computed, using the average value and the measured values farthest from the average.

A better understanding of errors will be obtained when measurements are taken on real objects and the accuracy of the measurements determined. Generally, this kind of activity occurs in the laboratory.

1.7 Powers of 10 Notation

In physical science many numbers are very big or very small. In order to express very big or very small numbers, the **powers of 10** notation is frequently used. When the number 10 is squared or cubed we get

$$10^2 = 10 \times 10 = 100$$

$$10^3 = 10 \times 10 \times 10 = 1000$$

It is easy to see that the number of zeroes is just equal to the power of 10. As an example, 10^{23} is a one followed by 23 zeroes.

Negative powers of 10 can also be used. For example,

$$10^{-2} = \frac{1}{10^2} = \frac{1}{100} = 0.01$$

From this we see that if a number has a negative exponent we shift the decimal place to the left once for each power of 10. Thus, one micrometer, which is 10^{-6} m, equals 0.000001 m.

We can also multiply numbers by powers of 10. Table 1.3 shows a wealth of examples of various large and small numbers expressed in powers of 10 notation.

There are many standard prefixes that are used to represent powers of 10. These are listed in Table 1.4. These prefixes are frequently used in expressing various units. The most important are **mega, kilo, milli,** and **micro**. Some examples are

1 *mega*ton = 10^6 tons
1 *kilo*meter = 10^3 meters
1 *milli*gram = 10^{-3} grams
1 *micro*second = 10^{-6} seconds

It is possible to represent a number in powers of 10 notation many different ways—all correct. For example,

Table 1.3 Numbers Expressed in Powers of 10 Notation

Number	*Powers of 10 Notation*
0.025	2.5×10^{-2}
0.0000408	4.08×10^{-5}
0.0000001	1×10^{-7}
0.00000000000000000016	1.6×10^{-19}
247	2.47×10^{2}
186,000	1.86×10^{5}
4,705,000	4.705×10^{6}
9,000,000,000	9×10^{9}
30,000,000,000	3×10^{10}
602,300,000,000,000,000,000,000	6.023×10^{23}

*"Absolute difference" is the result obtained when the smaller value is subtracted from the larger. It is designated by two vertical lines enclosing the subtraction.

the distance from the Earth to the Sun is 93 million miles. This can be represented as 93,000,000 miles, or 93×10^6 miles, or 9.3×10^7 miles, or 0.93×10^8 miles, etc.

Scientists generally pick the power of 10 so that it is multiplied by a number between 1 and 10. Thus 9.3×10^7 miles would probably be chosen. However, any of the given representations of 93 million miles is correct.

It can be seen from the above example that changing the power of 10 causes a change in the number it multiplies. When the power of 10 is increased by one, the decimal must be moved one space to the left; when the power of 10 is decreased by one, the decimal must be moved one space to the right. Examples are

$$16 \times 10^{-6} = 1.6 \times 10^{-5} = 0.16 \times 10^{-4}$$

$$24 \times 10^3 = 2.4 \times 10^4 = 0.24 \times 10^5$$

$$13 \text{ milligrams} = 13 \times 10^{-3} \text{ grams} = 1.3 \times 10^{-2} \text{ g}$$

Knowing how to express a measured quantity by using powers of 10 notation is not sufficient for solving problems in the physical sciences. One must know how to use the notation in the simple operations of addition, subtraction, multiplication, and division. See Appendix III for rules and examples.

Table 1.4 Prefixes Representing Powers of 10

Multiple	*Name*	*Abbreviation*
10^{18}	exa	E
10^{15}	peta	P
10^{12}	tera	T
10^9	giga	G
10^6	mega	M
10^3	kilo	k
10^2	hecto	h
10	deka	da
10^{-1}	deci	d
10^{-2}	centi	c
10^{-3}	milli	m
10^{-6}	micro	μ
10^{-9}	nano	n
10^{-12}	pico	p
10^{-15}	femto	f
10^{-18}	atto	a

Learning Objectives

After reading and studying this chapter, you should be able to do the following without referring to the text:

1. State four fundamental properties of nature.
2. State the units in which these fundamental properties are measured in three different systems of measurements.
3. Explain the concept of density and tell how it is different from mass.
4. State how the gram, meter, and second were originally defined.
5. Be able to convert from centimeters to inches, from pounds to kilograms, and from meters per second to miles per hour, or vice versa, using Table 1.2.
6. State the origin of π.
7. State the formulas for the circumference and area of a circle if the radius is known.
8. Distinguish between percentage error and percentage difference.
9. Express any number in powers of 10 notation.
10. State the definitions of the prefixes mega, kilo, milli, and micro.

Important Words and Terms

length
time
mass
electric charge
fundamental properties
standard unit
British system
mks system
cgs system
meter
second
gram
kilogram
SI
derived units
density
hydrometer
degree
π
measurement
experimental error
precision
accuracy
percentage error
percentage difference
powers of 10
mega
kilo
milli
micro

Questions

1. Do all measurements ultimately depend on our senses?
2. What are the four fundamental properties we seek to measure?
3. Would time exist if there were no motion?
4. Are there four dimensions or three?
5. Does a 60-kg astronaut have the same mass on the moon? the same weight?
6. What are the advantages of the metric system of measurement over the British system?
7. What is the origin of the (a) meter? (b) kilogram? (c) second?
8. Why have new standards of length and time been recently adopted?
9. What is the mass of 3 liters of water?
10. Which is more dense, a kilogram of iron or a kilogram of feathers? Which has more mass?
11. How is antifreeze in a car radiator tested?
12. What is the origin of π?
13. Why is a very precise stopwatch sometimes very inaccurate in timing a 100-m dash?
14. Give an example of when you would use percentage difference instead of percentage error.
15. Why would you write Avogadro's number (6.02×10^{23}) in powers of ten notation?

Problems

1. An astronaut has a mass of 60 kg.
   (a) What is this mass on the Earth and on the moon?
   (b) What is this weight on the Earth and on the moon in pounds?
2. Compute the density in grams per cubic centimeter of a rock that has a mass of 1 kg and a volume of 280 cm^3. *Answer:* 3.57 g/cm^3
3. What is the volume of a piece of iron (density 7.86 g/cm^3) that has a mass of 2.30 kg? *Answer:* 292.6 cm^3
4. Compute the height in feet and inches of a person who is 200 cm tall. *Answer:* 6 ft 7 in.
5. Compute the height in centimeters of a woman who is 5 ft 6 in tall.
6. Compute the mass in kilograms of a man who weighs 150 lb. *Answer:* 68.2 kg
7. Compute the weight in pounds of a 2-kg package.
8. Compute the speed in meters per second of an auto traveling at 40 mi/h. *Answer:* 17.9 m/s

9. Compute the speed in miles per hour of an auto traveling at 30 m/s.
10. The moon is approximately 240,000 mi from the Earth. Assuming the moon travels in a circle around the Earth (this assumption is not quite correct):
    (a) What is the circumference of its path around the Earth?
    (b) What is the area of this circle in square miles?
    (c) Express your answer to (b) in powers of 10 notation.
    Answer: (c) 1.81×10^{11} mi^2
11. Compute the area in square inches of a pizza with a diameter of
    (a) 7 in (b) 10 in (c) 14 in (Note: be sure to realize the diameter is given, not the radius.)
    (d) The 14-in pizza is how many times as big as the 7-in pizza? *Answer:* (d) 4 times as big
12. When Eratosthenes measured the circumference of the Earth over 2000 years ago, he got an answer of 24,000 mi. What was his percentage error if the correct answer is 24,900 mi? *Answer:* 3.6%
13. A student measures π to be 3.16. What is the percentage error of this measurement?
14. Fill in the blank with the correct power of 10.
    (a) 0.15 megaton = 1.5 × ______ tons
    (b) 10.5 kilovolts = 1.05 × ______ volts
    (c) 72 milligrams = 7.2 × ______ gram
    (d) 0.65 microwatt = 6.5 × ______ watt
    Answer: (c) 7.2×10^{-2} g
15. State the following quantities in powers of 10 notation.
    (a) 24 megawatts
    (b) 16 kilotons
    (c) 31 micrograms
    (d) 0.1 millimeter
    Answer: (c) 31×10^{-6} g or 3.1×10^{-5} g

2

Motion

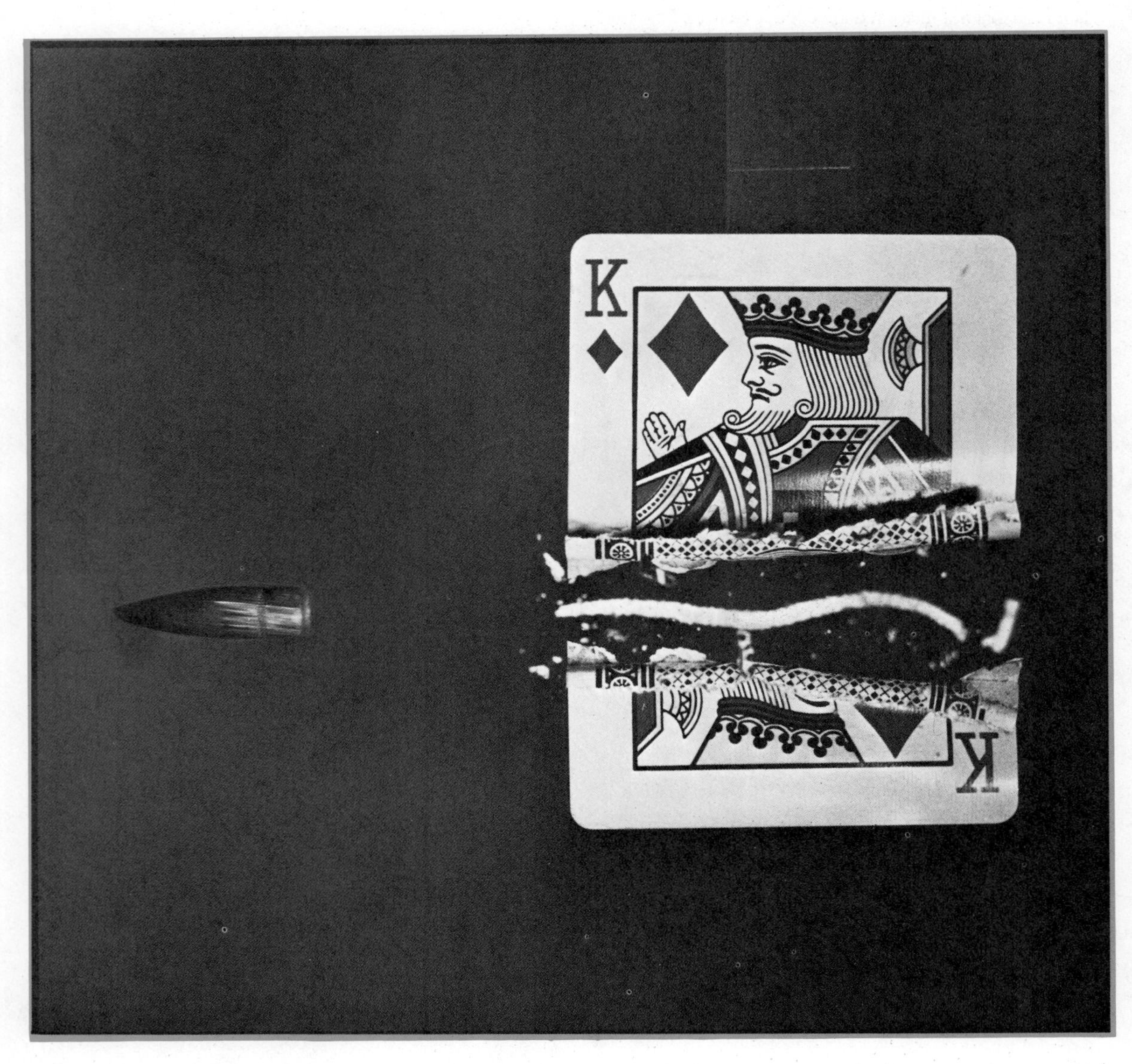

. . . Scientific truth should be presented in different forms, and should be regarded as equally scientific whether it appears in the robust form and the vivid colouring of a physical illustration, or in the tenuity and paleness of a symbolical expression. . . .

–James Clerk Maxwell

MOTION IS EVERYWHERE. We walk to class. We drive to the store. Birds fly. The wind blows the trees. The rivers flow. Even the continents drift.

In the larger environment the Earth rotates on its axis. The moon revolves around the Earth. The Earth revolves around the Sun. The Sun moves in the galaxy. The galaxies move with respect to one another.

This chapter focuses on the description of motion in our environment, with definitions and discussion of terms such as *speed, velocity*, and *acceleration*. We will study these concepts without considering the forces involved, reserving that discussion for Chapter 3.

Two basic kinds of motion are straight-line motion and circular motion. We experience examples of these each day. For instance, we know that driving around a curve gives a different sensation from driving in a straight line. Understanding acceleration is the key to understanding these basic kinds of motion.

2.1 Straight-Line Motion

The term **position** refers to the location of an object. To give the position of an object, another object or reference system must also be stated or implied.

The book is on the table. Atlanta is in Georgia. The shop is at the corner of Broadway and Fifth Avenue. In all these cases a reference system is stated or implied. If an object changes its position, we say that motion has occurred. When an object is undergoing a continuous change in position, we say the object is moving or is in **motion**.

◀ High-speed photography shows a card the instant it is torn apart by a speeding bullet.

Since position is relative, motion must also be relative. For instance, the statement "The student is walking 2 meters per second" indicates that the student is changing position at the rate of 2 meters for each second of time, relative to the sidewalk, ground, or floor. Similarly, if we say, "The car is traveling at the rate of 40 miles per hour," we are using the road as the frame of reference for the car's motion.

An example of straight-line motion is an automobile traveling north on a level highway. The motion of the automobile may or may not be at a constant rate. In either case, the motion is described using the fundamental units of length and time.

Suppose the car travels 20 meters for each second of time. This is illustrated by the diagram shown in Fig. 2.1.

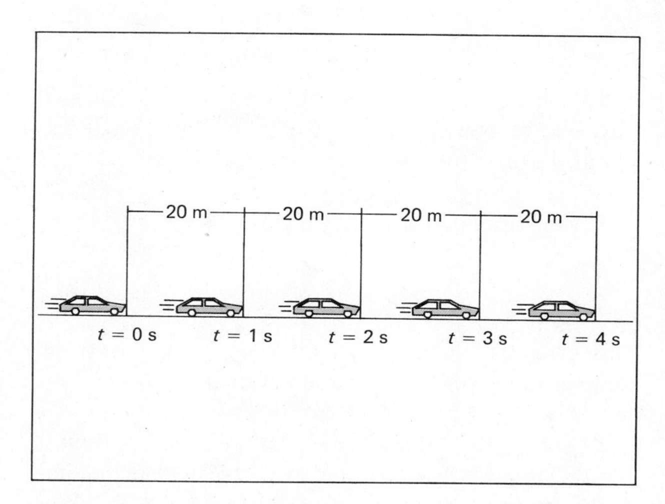

Figure 2.1 Straight-line motion. Equal distances are traveled in equal periods of time by an auto moving straight down a highway at constant speed.

2.2 Speed and Velocity

The terms speed and velocity are often used interchangeably by nonscientists. In physical science, however, the words speed and velocity have distinct meanings. The **average speed** of an object is the total distance traveled divided by the time it takes to travel the total distance.

Speed is a scalar quantity. A **scalar** quantity is one that has magnitude (including units) only, but no direction. A **vector** quantity is one that has both magnitude (including units) and direction. Velocity is a vector quantity. **Average velocity** is displacement divided by time where **displacement** is the straight-line distance between the initial and final positions.

A good example of the difference between speed and velocity is found in the measurement of wind. A wind speed would be 10 mi/h. The wind velocity also includes the direction of the wind; a wind velocity would be 10 mi/h from the west.

In straight-line motion, speed and velocity are very similar. Their magnitudes are the same, and the only distinction between them is that a direction must be specified when discussing velocity.

In physical science, the letter v is frequently used for speed or the magnitude of the velocity. Since the speed of an object is given by the distance traveled divided by the time it takes to travel that distance, we can write the formula

$$\text{speed} = \frac{\text{distance traveled}}{\text{time taken}}$$

or

$$v = \frac{d}{t} \tag{2.1}$$

EXAMPLE 1 Consider the car in Fig. 2.1. It traveled 80 meters in 4 seconds, so its speed was (80 m/4 s) or 20 m/s.

If we rearrange Eq. 2.1, we get

$$t = \frac{d}{v} \tag{2.2}$$

This gives us a formula for the time it takes to travel some distance when going at a constant speed.

EXAMPLE 2 If we know that the speed of light is 186,000 mi/s or 3×10^8 m/s, we can calculate the time it takes for the Sun's rays to reach us using Eq. 2.2 (See Fig. 2.2.)

Compute the time it takes for the Sun's rays to reach the Earth.

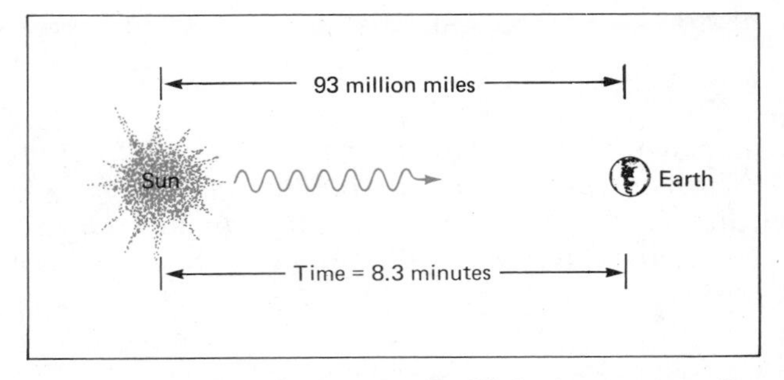

Figure 2.2 Although the speed of light is 186,000 miles per second, it still takes over 8 minutes for light from the Sun to reach us. See Example 2.

Since the Sun is 93 million miles away, we get

$$t = \frac{d}{v}$$

$$t = \frac{93{,}000{,}000 \text{ mi}}{186{,}000 \text{ mi/s}}$$

$$t = 500 \text{ s}$$

From this we realize that although light travels very fast, it still takes 500 s or about 8.3 min for it to arrive at the Earth after it leaves the Sun. This is illustrated in Fig. 2.2.

EXAMPLE 3 Find the average speed of the Earth as it orbits the Sun.

We can solve this problem using Eq. 2.1. We refer to Fig. 2.3. The speed of the Earth on its journey around the Sun can be computed by calculating the total number of miles traveled in one year. This is just the circumference of the circle in which it moves. (We assume it moves in a perfect circle. This is a good approximation.) We recall that the circumference of a circle is $2\pi r$, where r is the radius which is 93 million miles in our example.

The time taken to travel once around the Sun is one year or 365.25 days. Since there are 24 hours in a day, the total number of hours in one year is 365.25 × 24 hours. We can summarize as follows:

$$v = \frac{d}{t} = \frac{2\pi r}{t}$$

$$v = \frac{2\pi \times 93{,}000{,}000 \text{ mi}}{365.25 \times 24 \text{ hours}}$$

$$v = 66{,}600 \text{ mi/h}$$

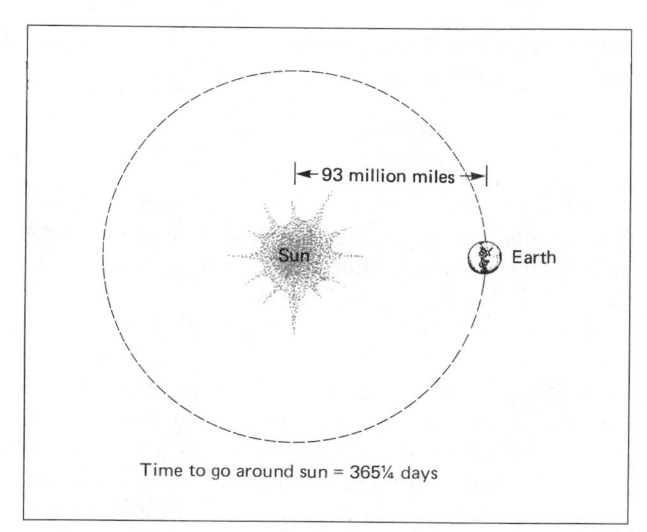

Figure 2.3 The Earth is 93 million miles from the Sun. The total distance traveled around the Sun is $2\pi \times 93$ million miles. It takes one year or $365\frac{1}{4}$ days to complete one revolution.

This is a fantastic result! It says that we are hurtling through space at a speed of 66,600 miles per hour. Can you believe that? All of us are moving at an exceedingly high speed, and yet it doesn't seem to bother us. (We return to this question in Section 2.4.)

One of the powerful aspects of physical science is that it is *quantitative*. Almost everyone is aware that the Earth is going around the Sun. This is a qualitative fact. The fact that we can calculate the speed of this motion as being 66,600 miles per hour quantifies our knowledge and gives a spectacular result. Thus, we begin to see the power and importance of going beyond qualitative statements regarding our physical environment.

A practical application of these concepts of speed, distance, and time would be running around an oval track when the wind is blowing. Although you run faster when the wind is blowing from behind, you will run more slowly when it is blowing toward you. Do these two effects cancel out when running in a circle or oval? No, the wind slows you down when you run in a loop of some kind.

Why does the wind slow you down? If you were to run at 8 m/s for a while and then at 4 m/s, what would be your average speed? If you ran equal amounts of time at each speed, you would average 6 m/s. But if you ran most of the time at the lower speed (4 m/s), then your average speed would be less than 6 m/s.

When running around a track, you run the same distance with and against the wind; but you run for a longer time against the wind. Since more of your time is spent running against the wind at the lower speed, your average speed is reduced.

Great track athletes such as distance runner Mary Decker (Fig. 2.4) want a calm day when they attempt a world record. A wind would slow them down.

Several concepts are needed to explain and understand motion. Thus far we have presented two—average speed and velocity. The speed at any instance of time may be different from the average speed. The *instantaneous speed* of an object is the speed of the object at any instant in time. A familiar example of instantaneous speed is the speed registered on an automobile speedometer such as is shown in Fig. 2.5. It is the speed at which the automobile is traveling right then, at that instant of time.

Figure 2.4 Athletes, such as Mary Decker Slaney, who are attempting to set world records around an oval track want a calm day. A wind slows them down. (Focus on Sports.)

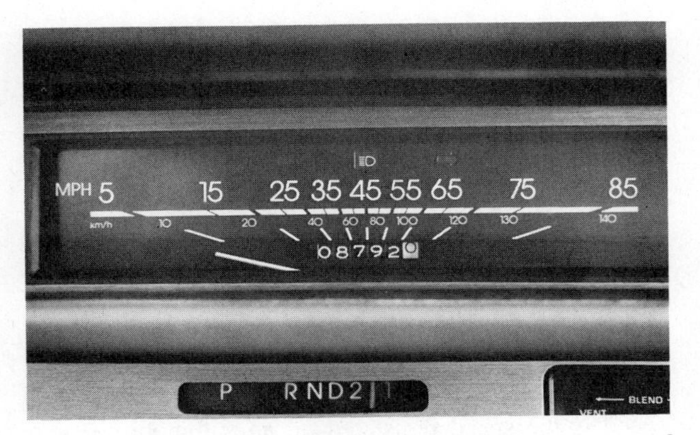

Figure 2.5 The speed indicated by the speedometer of an automobile is an example of instantaneous speed: the speed at that instant. (Jeff Thiebauth/Lightwave.)

The concept of instantaneous velocity is very similar to that of instantaneous speed. The **instantaneous velocity** is the velocity at any instant of time. In uniform motion, the instantaneous velocity and the average velocity are the same. In accelerated motion, discussed in the following sections, we will focus on how the instantaneous velocity changes.

2.3 Acceleration

When you drive down an interstate highway and suddenly increase your speed from 20 m/s (45 mi/h) to 28 m/s (63 mi/h), you feel yourself being forced back against the seat. When you drive in a circle on a cloverleaf you get forced to the outside of the circle. This force to the outside is felt because the direction of your velocity is changing.

There are three ways that you can change the velocity of an object. You can (1) increase or (2) decrease its magnitude when traveling in a straight line, and (3) you can change the direction of the velocity vector. When any of these changes occur, we say that the object is accelerating. The faster the change in the velocity occurs, the greater the acceleration.

Acceleration is defined as the time rate of change of velocity. If we take the symbol Δ to mean "change in," this can be written as

$$\text{acceleration} = \frac{\text{change in velocity}}{\text{time for change to occur}} = \frac{\Delta v}{t}$$

Of course, the change in the velocity is just the final velocity v_f minus the original velocity v_o. Thus, in symbols, we can define the acceleration as

$$a = \frac{\Delta v}{t} = \frac{v_f - v_o}{t} \qquad (2.3)$$

The units of acceleration in the SI system are m/s^2. These units may be confusing at first. It should be kept in mind that an acceleration is a measure of a change in velocity during a given time period.

Consider an acceleration of 9.8 m/s^2. This means that the velocity changes by 9.8 m/s each second. Thus, for straight-line motion with forward acceleration, as the number of seconds increases, the velocity goes from 0 to 9.8 m/s during the 1st second, to 19.6 m/s during the 2nd second, to 29.4 m/s during the 3rd second, and so forth, adding 9.8 m/s each second. This is illustrated in Fig. 2.6 for a falling object that falls at an acceleration of 9.8 m/s^2.

We can also rewrite Eq. 2.3 to give a formula for the final velocity of an object if its original velocity and acceleration are known.

$$v_f - v_o = at$$

or

$$v_f = v_o + at \qquad (2.4)$$

This formula is useful for working problems in which the quantities a, v_o, and t are all known and we wish to

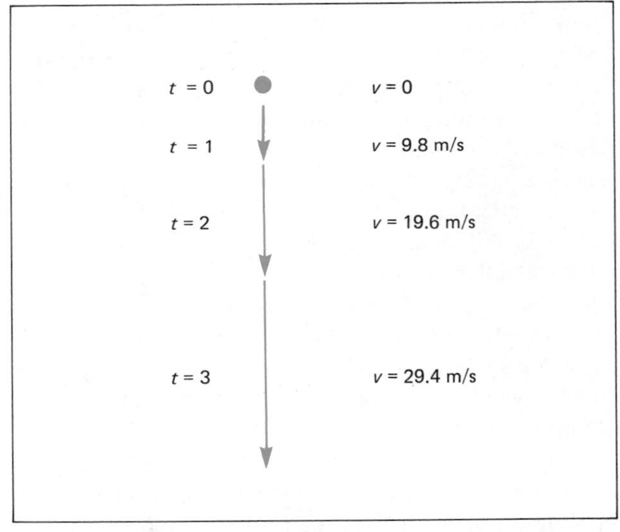

Figure 2.6 For a constant downward acceleration of 9.8 m/s^2 in straight-line motion, the velocity increases by 9.8 m/s each second. The direction of each arrow indicates the direction of the velocity vector. The increasing lengths of the arrows indicate an increasing velocity.

find v_f. If the original velocity $v_o = 0$, then

$$v_f = at$$

Since velocity is a vector quantity, acceleration is also a vector quantity. In fact, we can represent each term in Eq. 2.4 as an arrow or vector. The length of the arrow represents the magnitude, and the direction of the arrow represents the direction of the vector quantity.

Since t is a scalar quantity, the quantity at is a vector. Various illustrations of Eq. 2.4 are shown in Fig. 2.7 with positive, negative, and other types of acceleration.

In Fig. 2.7, part (a), we have an illustration similar to Fig. 2.6. Here both v_o and at are in the same direction, and because of this, v_f is longer than v_o. Part (a) might represent a ball falling to the Earth.

In part (b) of Fig. 2.7 the acceleration vector points in the opposite direction of the velocity and serves to decelerate the object. Part (b) might represent a ball on its way up. Its velocity decreases because gravity tends to slow it down.

Part (c) shows that even though the speed (or magnitude of the velocity) is a constant, an acceleration can change the direction of the velocity. To put it another way, an object moving at constant speed will be accelerating if its direction is changing. An example of this situation would be an object going in a circle at constant speed.

The **acceleration of gravity** at the Earth's surface is directed downward and is denoted by the letter g. Its magnitude in the SI system is

$$g = 9.80 \text{ m/s}^2$$

This corresponds to 980 cm/s^2 or about 32 ft/s^2.

The acceleration of gravity varies slightly depending on such factors as how far you are from the equator and how high up you are. However, the variation is very small, and for our purposes, we will take g to be the same everywhere on the Earth's surface.

The great Italian physicist Galileo Galilei (1564–1642) was the first scientist to assert that all objects fall downward with the same acceleration. Of course, this assumes that frictional effects are negligible. We can state this assertion of Galileo as follows:

If frictional effects can be disregarded, every freely falling object near the Earth's surface accelerates downward at the same rate, regardless of the mass of the object.

The correctness of Galileo's assertion can be seen by dropping a small mass, such as a coin, and a large mass, such as a book, at the same time. The two objects will hit at the same time. It is believed that Galileo himself performed such experiments (see Fig. 2.8).

The effect of air friction can be demonstrated by dropping a piece of tissue paper and a coin. The air friction will prevent the tissue paper from falling as fast as the coin. If the tissue paper is wadded up into a very small ball to minimize air friction, it will fall at the same acceleration as the coin.

On the moon there is no atmosphere, so there is no air friction. One of the astronauts dropped a feather and a hammer at the same time (see Fig. 2.9). They both hit the surface of the moon at the same time because neither the feather nor the hammer was slowed by air friction. This shows that Galileo's assertion applies on the moon as well as on the Earth. Of course, on the moon all objects fall at a slower rate than do objects on the Earth's

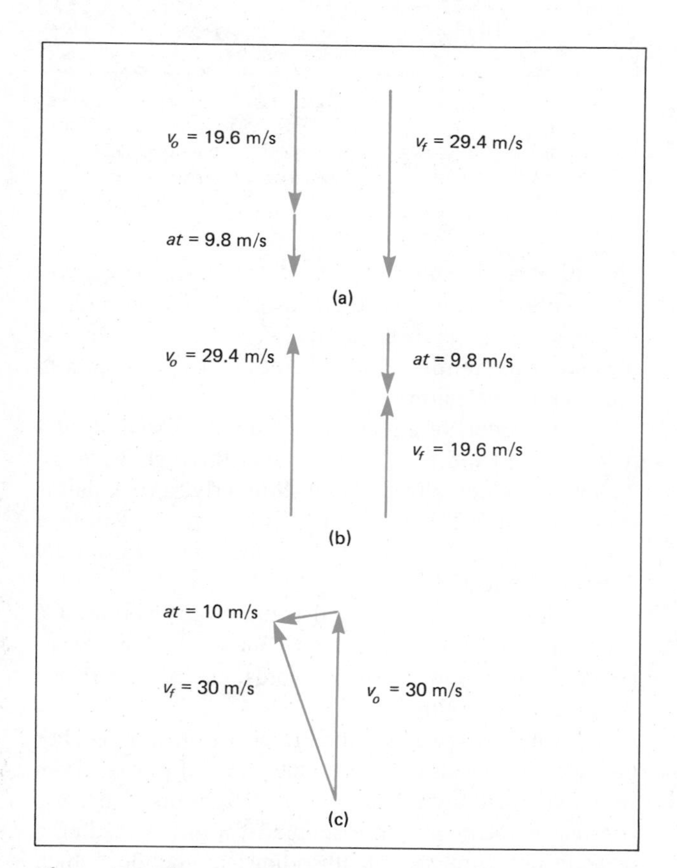

Figure 2.7 The equation $\mathbf{v}_f = \mathbf{v}_0 + a\mathbf{t}$ is illustrated. The arrows $\mathbf{v}_0$ and at combine together to give the resultant arrow $\mathbf{v}_f$. Note that $\mathbf{v}_f$ can be longer, shorter, or the same length as $\mathbf{v}_0$.

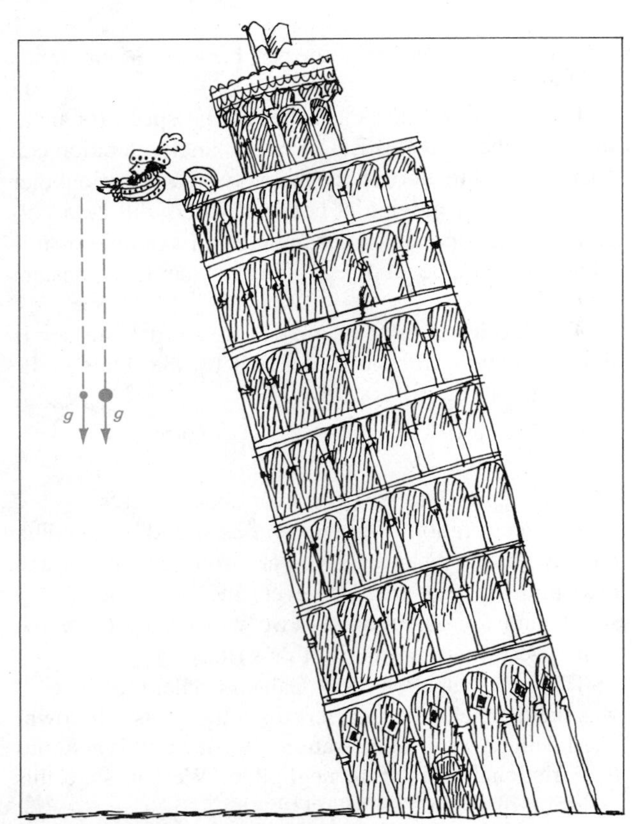

Figure 2.8 All freely falling objects near the Earth's surface have an acceleration $g = 9.8\ m/s^2$ if friction can be disregarded. Galileo is alleged to have shown this by dropping cannon balls of different masses from the Leaning Tower of Pisa. The two cannon balls would have struck the ground at almost exactly the same time.

Figure 2.9 Astronaut Scott demonstrated that a feather and hammer fall at the same rate on the moon. There is no air on the moon to slow down the feather, so it accelerates at the same rate as the hammer.

surface. This is because the acceleration of gravity is less on the moon than on the Earth.

When an object is dropped from a tall building, its velocity increases by 9.8 m/s each second as it falls. The distance it travels downward can be computed using the formula $d = \frac{1}{2}gt^2$.* In Fig. 2.10 the motion of a freely falling object near the Earth's surface is illustrated.

From Fig. 2.10 we can see that after 1 s, the object has fallen 4.9 m. After 2 s the distance fallen has increased to 19.6 m. In general, the distance fallen can be calculated using the formula $d = \frac{1}{2}gt^2$.

If we throw an object up, the velocity decreases by 9.8 m/s each second. For instance, if a rock is thrown upward at an initial speed of 29.4 m/s, its speed will decrease to 19.6 m/s at the end of the first second, to 9.8 m/s at the end of the second second, and to zero at the end of the third second. Then the rock will begin to fall back to the Earth.

The velocities of a rock thrown up at a speed of 29.4 m/s are shown in Fig. 2.11. It takes the rock 3 s to go up and 3 s to come back down. Similarly, a rock that is gently tossed 1.225 m (about 4 ft) into the air will take $\frac{1}{2}$ s to go up and another $\frac{1}{2}$ s to come down, for a total time in the air of 1 s.

So far, we have discussed straight-line motion at a constant speed and acceleration. But what do we experience during motion? In other words, can we experience the effects of motion?

We cannot experience the effects of all motion. For instance, we cannot experience the effect of motion when we move at a constant speed in a straight line, i.e., at a constant velocity. As an example, if a girl is reading a book in her family's smooth-riding automobile, which is going straight down a highway at 55 mph with the windows closed, she will be unaware of the motion of the automobile.

*In general, the distance an object travels when it starts at an original velocity v_o and is accelerated at an acceleration a is $d = v_o t + \frac{1}{2}at^2$.

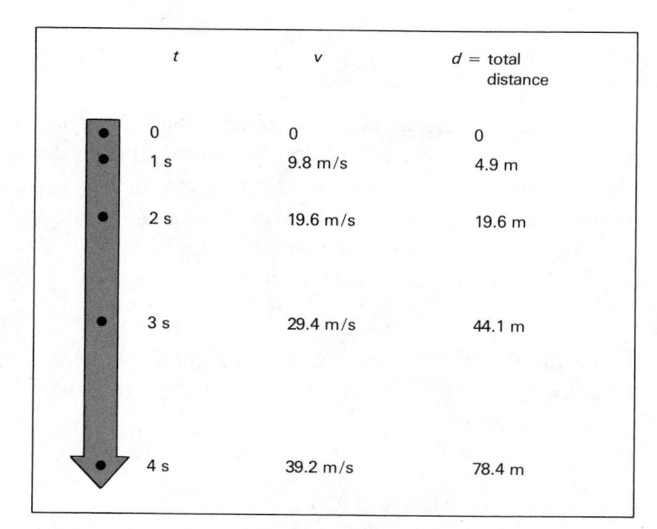

t
v
d = total distance
0
0
0
1 s
9.8 m/s
4.9 m
2 s
19.6 m/s
19.6 m
3 s
29.4 m/s
44.1 m
4 s
39.2 m/s
78.4 m

Figure 2.10 The velocity and distance traveled by a freely falling object at the end of each second. The velocity increases by 9.8 m/s downward each second. The distance traveled is computed by using the formula $d = \frac{1}{2}gt^2$.

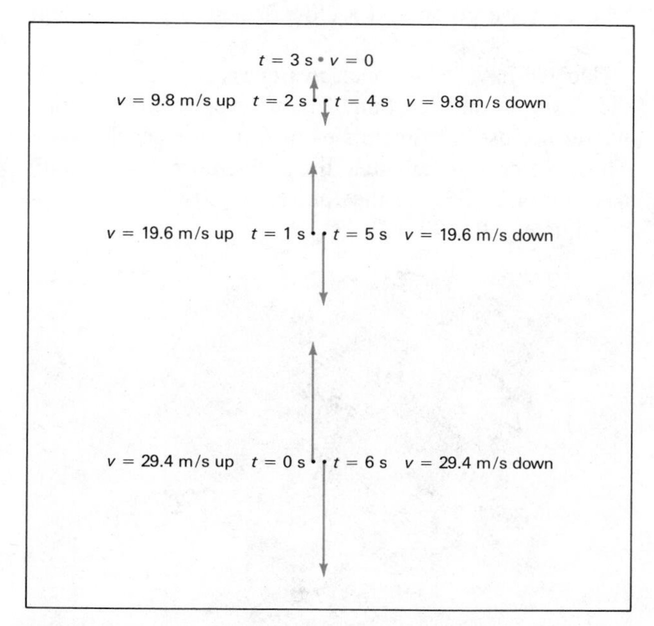

Figure 2.11 A rock is thrown up at an initial velocity of 29.4 m/s. The rock's position and velocity at each second are shown. Going up, the velocity in the upward direction decreases by 9.8 m/s each second. Coming down, the velocity down increases by 9.8 m/s each second. It takes 3 seconds for the rock to go up. It stops for an instant; then it takes 3 more seconds for it to come down. The acceleration at all times is 9.8 m/s² downward.

What we can experience is the effect of an acceleration. Of course, the effects of larger accelerations are easier to experience. If an acceleration is very small, the effect may not be experienced. We are not aware of all motions, but we can experience the effect of motion with sufficient acceleration.

2.4 Acceleration in Circular Motion

Whenever an object travels at a constant speed in a circle, its velocity vector is constantly changing direction. Because its velocity vector is changing, there is an acceleration.

Consider a particle held by a string that is moving in a circle at constant speed. In Fig. 2.12, the particle is at position P_1 and is directed to the right as shown. The velocity vector in Fig. 2.12 is in a direction tangent to the circle at that point. If the string should break at this instant, the particle would fly off, tangent to the circle at that instant.

Assuming the string holds, at some later instant the particle is at point P_2, moving with the same speed but in a direction tangent to the circle at point P_2. The two velocity vectors are drawn the same length to indicate that the speed has remained the same. However, the velocity vectors have different directions and this indicates the *velocity* of the particle has changed in going from position P_1 to position P_2.

Since the velocity of the particle has changed, it has accelerated. The direction of the change in velocity is

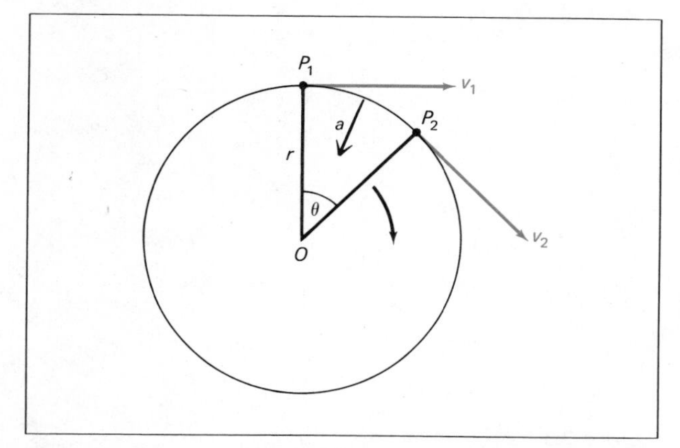

Figure 2.12 The change in direction of a particle moving with constant speed in a circle. The direction of the velocity vector is continually changing.

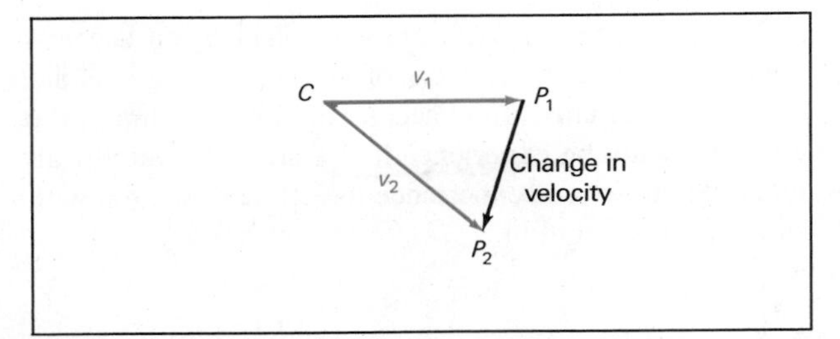

Figure 2.13 Vector diagram for Fig. 2.12. The change in velocity due to a change in direction of **v** is shown. Note that this is similar to Fig. 2.7 part (c).

illustrated in Fig. 2.13. This is also the direction of the acceleration vector. In circular motion at constant speed the direction of the acceleration is toward the center of the circle. Acceleration directed toward a central point is called **centripetal**. The word centripetal was introduced by Isaac Newton and has the meaning "towards the center." It may seem contradictory for a particle moving at a constant speed in a circle to be undergoing an acceleration, but speed and acceleration are different physical quantities.

Some examples of circular motion are: the Earth revolving around the Sun, a wheel turning about an axis, and a car turning through a circular portion of the highway. When a car turns through a curve on a level road, a centripetal force is needed to keep the car in motion in the circular path.

In general, whenever an object moves in a circle with radius r at a constant speed v, the magnitude of the acceleration vector is given by the formula

$$a_c = \frac{v^2}{r} \tag{2.5}$$

Note here that the acceleration increases with the square of the speed v. Also, the smaller the radius, the greater the acceleration. Since the speed v is constant, it is important to realize that for uniform circular motion the acceleration of the object results from the changing direction of the velocity vector of the moving object.

EXAMPLE 4 Determine the magnitude of the acceleration of a car going at 12 m/s on a cloverleaf with a radius of 50 m.

To find this acceleration, we use Eq. 2.5:

$$a = \frac{v^2}{r} = \frac{(12\ \text{m/s})^2}{50\ \text{m}} = 2.9\ \text{m/s}^2$$

This value of 2.9 m/s^2 is about 30% of the acceleration g = 9.8 m/s^2. This is a fairly large acceleration, and we definitely experience the effects of this acceleration whenever we go around a cloverleaf in a car. See Fig. 2.14.

Do we have a personal awareness of the motion of the Earth around the Sun? The answer is no providing we do not use instruments to measure the acceleration. Thus, we need to calculate the acceleration of the Earth as it revolves around the Sun at 66,600 mi/h, or approximately 30,000 m/s.

Figure 2.14 An acceleration is necessary in order to go around a cloverleaf at a constant speed. (Photo courtesy James Crouse.)

To do this calculation we use Eq. 2.5 and the fact that the radius r involved is just the distance between the Earth and the Sun—93 million miles. We also make use of the conversion list on the inside back cover to find that there are 1609 m in one mile.

$$a = \frac{v^2}{r}$$

$$a = \frac{(3 \times 10^4 \text{ m/s})^2}{(93 \times 10^6 \text{ mi})(1609 \text{ m/mi})}$$

$$a = 0.006 \text{ m/s}^2$$

Once again the power of quantitative calculations becomes evident. We see from the calculation that even though the Earth's speed around the Sun is 66,600 mi/h, the resultant acceleration is 0.006 m/s^2, which is not noticeable. For instance, compare this acceleration to g, the acceleration of gravity, which is 9.8 m/s^2.

2.5 Projectile Motion

If a body is thrown out horizontally, it will follow the path shown in Fig. 2.15. It can be seen from this photograph that an object thrown horizontally will fall down at the same rate as an object that is dropped.

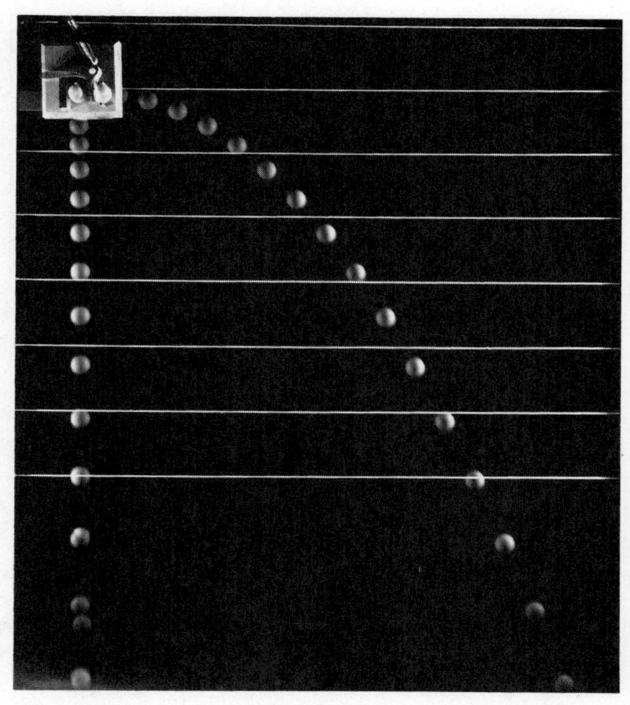

Figure 2.15 A flash photograph of two golf balls, one projected horizontally at the same instant the other was dropped. The strings are 15 cm apart, and the interval between flashes was $\frac{1}{30}$ s. (From PSSC *Physics*, 3rd ed., Heath, Lexington, Mass., 1971.)

We see from this that the horizontal and vertical directions are independent of one another. That is, the velocity in the horizontal direction does not affect the velocity and acceleration in the vertical direction.

As an example, consider a rifle that is fired horizontally (see Fig. 2.16). If a bullet is dropped simultaneously, the bullet fired from the rifle will hit the ground at the same time as the dropped bullet. Of course, this assumes that frictional effects are negligible and the ground is level. The bullets will hit simultaneously because the horizontal velocity of the fired bullet has no effect on its vertical motion. The vertical motion of the fired bullet is identical to that of the dropped bullet.

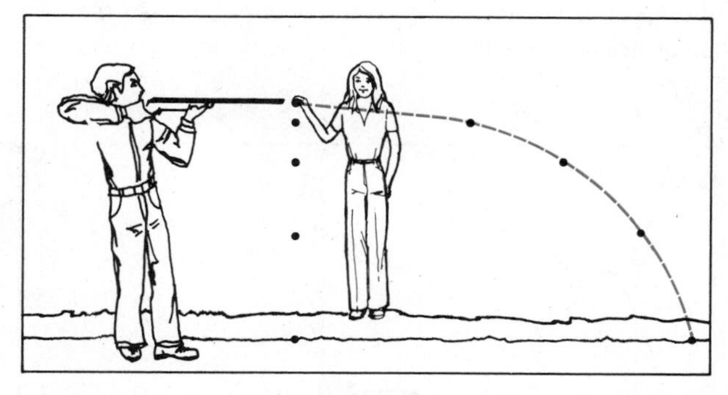

Figure 2.16 If a rifle is fired horizontally and a bullet is simultaneously dropped, both bullets will hit the ground at the same time.

Occasionally a sports announcer claims that a hard-throwing quarterback can throw the football so many yards "on a line," meaning a straight line. This, of course, must be false. All objects thrown horizontally begin falling as soon as they are thrown.

Figure 2.17 illustrates the trajectories of thrown footballs, baseballs, or other objects. Paths A and B represent fast and slow horizontal throws. Path C represents an actual trajectory of a typical football pass. The ball is thrown slightly upward, not horizontally.

Of course, as can be seen in Fig. 2.17, Path A more closely approximates a line than does Path B. Because of gravity, however, straight-line projectile motion cannot be achieved on the Earth's surface.

If a ball or other object is thrown at various angles, the path that it takes will depend on the angle at which it is thrown. If the ball is tossed gently, air friction will

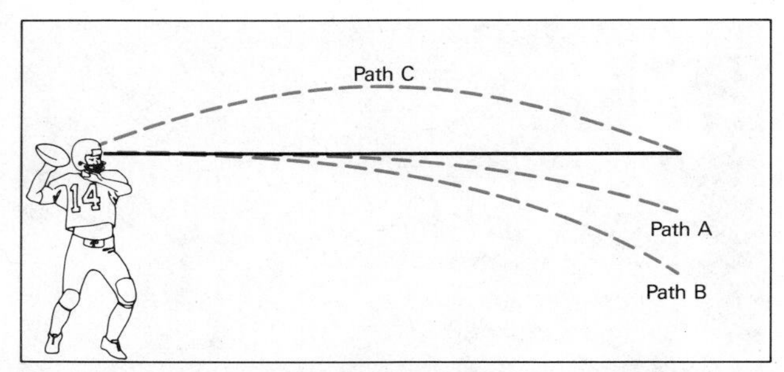

Figure 2.17 The trajectories of thrown footballs are shown. Paths A and B show footballs thrown horizontally. Path A represents a hard throw and Path B a soft throw. Path C represents the trajectory of a typical football pass. The ball is thrown slightly upward, not horizontally.

be negligible, and the path will resemble one of those shown in Fig. 2.18.

This figure demonstrates that the distance the ball travels is maximum when the ball is projected at an angle of 45° relative to level ground. Of course, these considerations also hold for track events such as discus and javelin throwing. In Fig. 2.19 a javelin thrower is shown hurling a javelin at a 45° angle to get the maximum distance.

The trajectories of the paths shown in Fig. 2.18 are all symmetrical about a vertical line that passes through the apex of the trajectory. This occurs because there is no air friction, so the horizontal velocity does not change.

Figure 2.19 An athlete throwing the javelin at an angle of 45° to try to get the maximum distance for his throw.

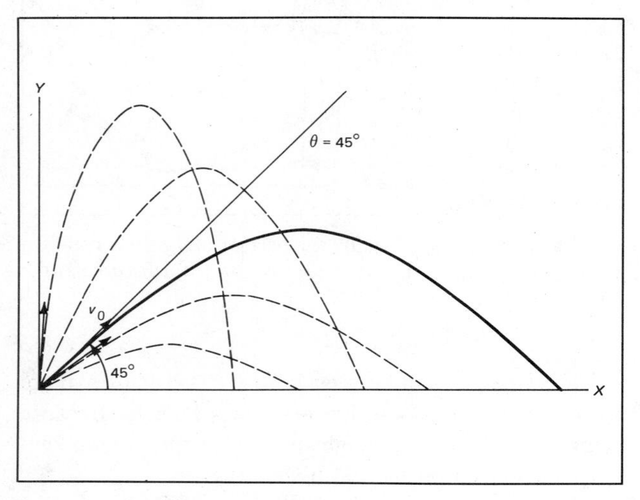

Figure 2.18 The maximum range of a projectile with a given initial velocity is 45°. This assumes that there is no air friction.

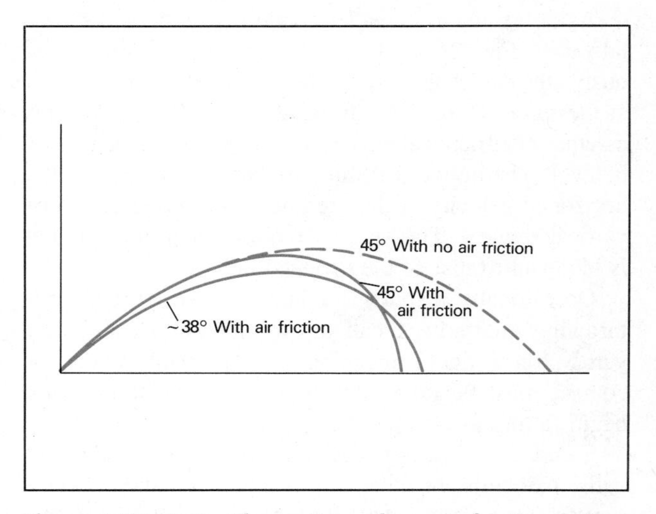

Figure 2.20 Projectile motion when air friction is important. Hard-hit baseballs follow trajectories similar to those shown in this figure.

When a ball is thrown or hit hard, air friction becomes important. The path that the ball takes now resembles one of those shown in Fig. 2.20. The trajectories of these paths are now no longer symmetrical. The horizontal velocity slows down due to air friction. The maximum distance traveled now occurs at an angle less than 45°.

Athletes such as football quarterbacks, baseball players, and track and field competitors are aware of the best angle at which to throw in order to get the maximum distance.

Learning Objectives

After reading and studying this chapter, you should be able to do the following without referring to the text:

1. Explain how you would find the speed of an object in terms of the distance traveled and the time.
2. Differentiate between scalars and vectors and give an example of each.
3. Differentiate between speed and velocity.
4. Explain why a wind slows you down when running in a circle.
5. Differentiate between average velocity and instantaneous velocity and give an example of each.
6. Define the term "acceleration," and give an example referring to throwing a ball upward.
7. Give the units of acceleration in the SI, and explain their meaning.
8. State Galileo's observation concerning freely falling objects.
9. Compute the speed of an object that has fallen for 1, 2, and 3 s ($g = 9.8\ m/s^2$).
10. Explain the meaning of each term in the equation $a = v^2/r$.
11. Give the mks units of each term in the equation $a = v^2/r$.
12. Explain how something can be traveling at a constant speed and still be accelerating.
13. Explain why we can experience the effect of being accelerated around a cloverleaf, but we cannot experience the effect of the motion of the Earth around the Sun.
14. Explain why two bullets, the first dropped and the second simultaneously fired horizontally, will hit the ground at the same time (if friction can be disregarded).

Important Words and Terms

motion	vector	instantaneous velocity
position	velocity	acceleration
speed	displacement	centripetal acceleration
scalar	instantaneous speed	acceleration of gravity

Questions

1. Explain the difference between scalar and vector quantities.
2. Why is it easier to set a world record in the 1500-km run on a calm day?
3. How is the Earth moving in relation to the Sun?
4. Is a speedometer reading a measurement of instantaneous speed, average speed, instantaneous velocity, or average velocity?
5. Can a car be moving at 30 mi/hr and still be accelerating? Explain.

6. Can an object have an instantaneous velocity of zero and still be accelerating? Explain.
7. Can an object have an instantaneous velocity of 9.8 m/s and simultaneously have an acceleration of 9.8 m/s^2? Explain.
8. Can you feel the motion of the Earth as it spins on its axis?
9. When Galileo dropped two objects from a high point, they hit the ground at almost exactly the same time. Why was there a slight difference in when they hit?
10. A ball is dropped from a building whose height is 100 m (32 floors).
    (a) What is the ball's velocity after 1 s?
    (b) After 2 s?
11. Are we accelerating due to the Earth's spinning on its axis?
12. What is the direction of the acceleration vector due to the Earth's spinning on its axis?
13. Can a baseball pitcher throw a fastball on a straight horizontal line? Why or why not?
14. Does a football quarterback throw a long pass at a greater or smaller angle than a short pass? Explain.

Problems

1. The moon is about 240,000 mi from the Earth, and an astronaut on the moon communicates using radio waves that travel at the speed of light, which is 186,000 mi/s. How long does it take for the astronaut's communication to reach the Earth? *Answer:* 1.3 s
2. When Jupiter is 6.0×10^{11} m away from the Earth, how long does it take a radio signal traveling at the speed of light (3×10^8 m/s) to reach us from a satellite passing near Jupiter?
3. A light year is the distance light travels in one year. How many miles is a light year if light travels at 186,000 mi/s?
4. A car is driven for 2 h at 60 mi/h.
    (a) How far did the car go?
    The car is then driven for 2 h at 40 mi/h.
    (b) How far did the car go?
    (c) Find the average speed for the entire trip.
    Answer: (c) 50 mi/h
5. (a) A car is driven 120 mi in 2 h. What is the average speed?
    (b) The return trip takes 3 h. What is the average speed?
    (c) The total distance traveled for the entire trip is 240 mi, and the total time is 5 h. What is the average speed for the entire trip? (Note: This problem is different from Problem 4 because more time is spent traveling at the slower speed.)
    Answer: (c) 48 mi/h
6. A ball is thrown upward at a speed of 19.6 m/s. (a) What is its speed at the end of the first second? (b) What is its speed after 2 seconds? (c) What happens next?
7. A ball is thrown downward with an initial speed of 6 m/s. (a) What is its speed after the first second? (b) What is its speed after 2 seconds have elapsed?
    Answer: (b) 25.6 m/s.
8. A rock is dropped. (a) What is its initial speed? (b) What is its initial acceleration?
9. A rock is dropped. One second elapses. (a) What is its speed? (b) What is its acceleration now?
10. A ball is thrown upward at a speed of 39.2 m/s. After one second, (a) What is the magnitude and direction of its velocity vector? (b) What is the magnitude and direction of its acceleration vector?
    Answer: (a) 29.4 m/s up
11. The Earth's circumference is approximately 25,000 mi. A person on the equator spins with the Earth and makes a circle whose circumference is 25,000 mi in one day.
    (a) Compute the person's speed in mi/h.
    (b) Convert this to m/s. *Answer:* (b) 465 m/s
12. Refer to problem 11, part (b). As the Earth spins, the velocity of a person on the equator is 465 m/s.
    (a) Calculate the acceleration of the person in m/s^2. Use 6.4×10^6 m (approximately 4000 mi) as the radius of the Earth.

(b) What is the direction of this acceleration?
(c) Compare this to 9.8 m/s^2, the acceleration of gravity.

Answer: (a) 0.034 m/s^2

13. A person drives around a cloverleaf at 10 m/s (22.4 mi/h). The cloverleaf's radius is 70 m (230 ft).
(a) What is the acceleration in m/s^2?
(b) What is the direction of this acceleration?
(c) Compare this to 9.8 m/s^2, the acceleration of gravity. Could you feel the acceleration calculated in part (a)?

14. If you drop an object from a height of 1.225 m (4 ft), it will hit the ground in $\frac{1}{2}$ second. If you throw a baseball horizontally at 30 m/s (67.2 mi/h) from this same height, how long will it take to hit the ground?

15. An airplane flies horizontally west at a speed of 200 m/s (448 mi/h). It drops a package of food. (a) What is the package's initial velocity? Give both speed and direction. (b) What is the magnitude and direction of the package's acceleration?

3

Force and Motion

The whole burden of philosophy seems to consist in this—from the phenomena of motions to investigate the forces of nature, and from these forces to explain the other phenomena.

–Sir Isaac Newton

WHAT IS FORCE? What is it that sets a moving object in motion? What stops it? These are fundamental questions to be dealt with in this chapter on force and motion. We will center our efforts on Newton's three laws of motion, his law of universal gravitation, and the laws of conservation of linear and angular momentum.

Galileo (1564–1642) was one of the first scientists to make a formal statement concerning objects at rest and in motion. But it remained for Sir Isaac Newton, who was born the year Galileo died, to actually formulate the laws of motion mathematically and to explain the phenomena of moving objects on the surface of the Earth plus the motion of the planets and other celestial bodies.

Newton was only 25 years old when he formulated his discoveries in mathematics and mechanics. His book *Mathematical Principles of Natural Philosophy* (commonly referred to as the *Principia*), published in 1687 when he was 45, is considered by many to be the most important publication in the history of physics. Certainly it established Newton as one of the greatest scientists of all time.

3.1 Newton's First Law of Motion

Before the time of Galileo and Newton, scientists asked themselves, What is the natural state of motion? The Aristotelian view of motion had held sway for centuries before Galileo and Newton showed it to be incorrect. Newton (Fig. 3.1) answered the question by stating his **first law of motion**. This law can be stated in several different ways. One of these is as follows:

> **A body will remain at rest or in uniform motion in a straight line unless acted upon by an external unbalanced force.**

◄ A tremendous force is applied to place the Space Shuttle in motion.

Figure 3.1 Sir Isaac Newton (1642–1727). Along with Einstein, Newton is generally regarded as one of the two greatest scientists who ever lived. A modest man, Newton wrote shortly before his death, "I seem to have been only a boy playing on the seashore and diverting myself in now and then finding a smoother pebble or a prettier shell than ordinary, whilst the great ocean of truth lay all undiscovered before me." A more accurate description of Newton's contribution to our knowledge was made by Alexander Pope who wrote, "Nature and nature's laws lay hid in the night. God said 'let Newton be,' and all was light." (Courtesy AIP Niels Bohr Library, W. F. Meggers Collection.)

Uniform motion in a straight line means that the velocity is constant. Thus, another way to state Newton's first law is to say that the natural state of motion is at a constant velocity. If the constant velocity is equal to zero, we say that a body is at rest.

From Newton's law we see that if objects are to speed up, slow down, or change direction (e.g., circular motion), they must be acted upon by an external unbalanced force. For instance, a rocket in outer space will continue in uniform motion in a straight line. If it is heading in the correct direction, it doesn't have to burn fuel to continue on to its destination. Only if it needs to alter its course is it necessary to apply a force by burning fuel.

What is a force? We all have an intuitive feeling for the concept of force. A force is simply a push or pull. In terms of Newton's first law we see that a **force** is any quantity that is capable of producing motion or a change in motion. This does not necessarily mean that a change in motion is produced, only that the capability is there.

In a tug-of-war (Fig. 3.2, top), a lot of force is being applied but there is no motion. The forces in this example are large, but they are balanced. Movement only occurs when the forces are unbalanced (see Fig. 3.2, bottom). For a change in motion or acceleration to occur, there must be an unbalanced or *net force* acting.

Forces act in particular directions as illustrated in Fig. 3.2. Therefore, force is a *vector quantity*.

You have experienced the effects of unbalanced forces and inertial masses when riding in an automobile. When the car is started quickly, you are thrown back against

Figure 3.2 (Top) When forces are balanced, a change in motion is not produced. (Bottom) When F_2 is greater than F_1 the forces are unbalanced, and a change in motion occurs. (Courtesy J. D. Wilson, *Physics Concepts and Applications*, Heath, Lexington, Mass., 1981.)

the seat; when it is stopped quickly, you are thrown forward. Your body's inertia resists the change in motion that is taking place.

The property of matter that resists any change in motion is known as **inertia**. Inertia is a property that provides a means for measuring the quantity of matter present in an object. This quantity of matter is called mass. **Mass** is a measure of inertia. The greater the mass the greater the inertia, and vice versa.

Another example of inertia is shown in Fig. 3.3. The stack of coins has an inertia that causes it to resist being moved when it is at rest. If the paper is quickly jerked, the inertia of the coins will prevent them from toppling.

Newton's first law provides an answer for the observed effects we discussed in Chapter 2 concerning circular motion and centripetal acceleration.

The natural state of motion is at a constant velocity. Thus, when you are riding in a car, you tend to go in a straight line. If the car goes around a curve, as in Fig. 3.4, you tend to go straight. Relative to the car you go to the outside.

In this case an unbalanced force is acting on the car from the road to turn the car from its straight path. When you tend to go straight, you are pressed against the side of the car. The side of the car exerts a force on you to change the direction of your motion.

Figure 3.3 A demonstration of inertia. If the strip of paper is given a quick jerk, the inertia of the stack of quarters prevents it from toppling. (Courtesy J. D. Wilson, *Physics Concepts and Applications*, Heath, Lexington, Mass., 1981.)

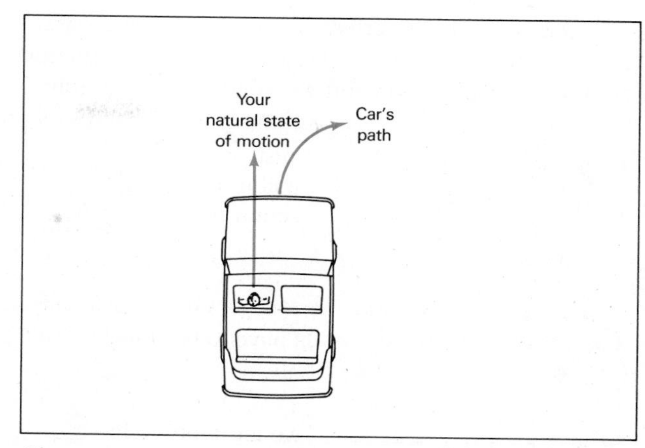

Figure 3.4 According to Newton's first law of motion, when you ride in a car, you tend to travel in a straight line. If the car turns, you tend to keep going straight and thus appear to be thrown to the outside of the car.

3.2 Newton's Second Law of Motion

We defined acceleration in Chapter 2 as being the change in velocity divided by the change in time. This defines acceleration, but it does not tell us what causes acceleration. In Newton's first law of motion, we see that no acceleration will occur unless there is an external unbalanced force acting. From this, we see that force causes acceleration.

Newton realized that force causes acceleration, but he went much further. He quantified the relationship and found that acceleration is directly proportional to the unbalanced force. In Fig. 3.5 and Fig. 3.6 we see an example of this. In Fig. 3.6 the force is three times what it is in Fig. 3.5. The acceleration is then three times as great.

Acceleration of a mass does not depend on the force alone. It also depends on the mass. Mass is a means of measuring inertia. Inertia was defined as resistance to a change in motion. Thus, it seems reasonable to assume that the greater the mass, the smaller will be the change in motion.

Another way to state this is to say that acceleration is inversely proportional to the mass: the smaller the mass the greater the acceleration, and vice versa. Comparing Fig. 3.5 and Fig. 3.7 we see that if an equal force is applied the larger mass is accelerated less.

If we combine the effects of force and mass on acceleration and realize that both force and acceleration are vectors, we get **Newton's second law of motion**:

> **The acceleration produced by an unbalanced force acting on a mass is directly proportional to the unbalanced force, in the direction of the unbalanced force, and inversely proportional to the total mass being accelerated by the unbalanced force.**

If we express Newton's second law mathematically, it is greatly simplified. Using the proportionality sign ($\propto$), we get

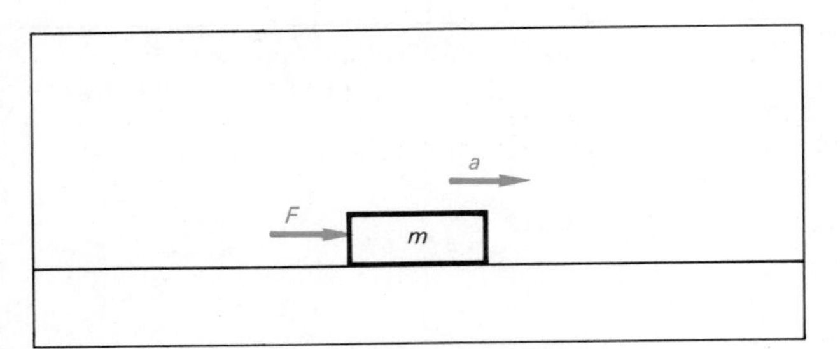

Figure 3.5 An unbalanced force causes an acceleration.

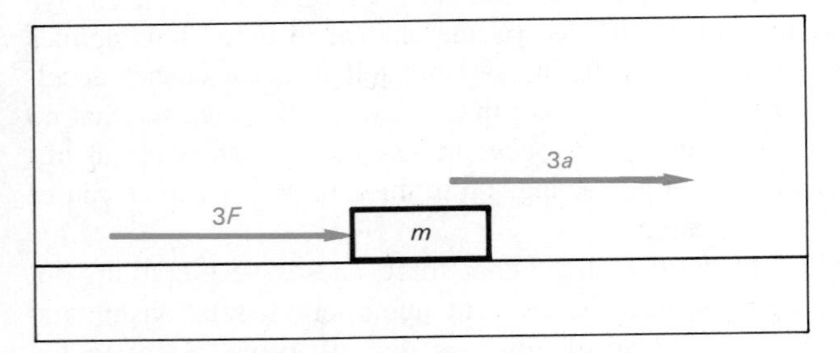

Figure 3.6 Three times the force causes three times the acceleration.

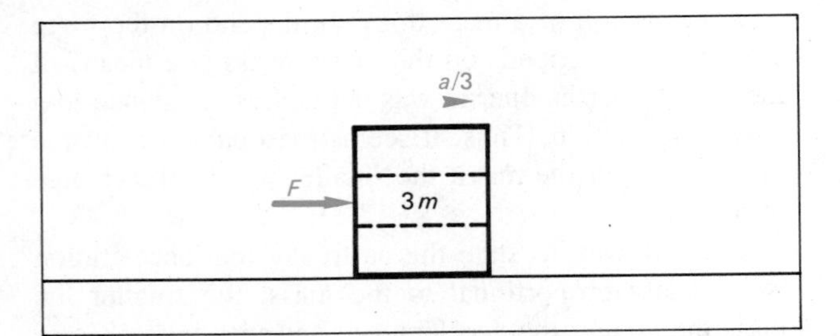

Figure 3.7 The same force as in Fig. 3.5 causes one-third the acceleration since the mass here is three times larger.

$$\text{acceleration} \propto \frac{\text{unbalanced force}}{\text{mass}}$$

or

$$a \propto \frac{F}{m}$$

When appropriate units are used, we can replace the proportionality sign with an equals sign and write

$$a = \frac{F}{m}$$

Newton's second law is usually written in symbol notation as

$$F = ma \qquad (3.1)$$

Equation 3.1 is a mathematical statement of Newton's second law.

In examining Newton's first and second laws of motion we find that the first law is a special case of the second. This can be seen from the answer to a very easy question: What is the acceleration of a mass on which the unbalanced force is zero? The answer is zero; and if the acceleration is zero, then the mass must be moving with a constant velocity. But this is just Newton's first law: a mass that has no unbalanced force acting on it moves at a constant velocity.

The equation $F = ma$ is extremely important in physics. It relates acceleration to force and mass. Recall that acceleration is defined as the change in velocity divided by the time taken. This definition tells us how to measure acceleration. Newton's second law, $F = ma$, tells us what causes the acceleration. An unbalanced force causes acceleration.

One gains an insight into $F = ma$ by considering gas mileage. Cars get better gas mileage on the highway than in the city. Why is this? Newton's second law provides an explanation.

Since $F = ma$, every time a car is accelerated it must burn fuel to supply the unbalanced force causing the acceleration. When a car is driven in city traffic, it must be accelerated back up to speed after it slows down or stops. All this acceleration burns a lot of gasoline.

It does not take as much fuel to keep a car going at a constant speed down a level highway. The less often you accelerate your car, the better the mileage you will get.

Since units on both sides of an equation must be equivalent, the units of force must be the same as the units of mass times acceleration. In the British system the unit of force is the pound. The British units of acceleration, (ft/s^2) and mass (slug), are rarely used, and we will not employ them in this text.

The newton is the unit of force in the metric (SI) system. One **newton**, abbreviated N, is equal to the force necessary to accelerate one kilogram by one meter per second each second. That is,

$$F = ma$$

$$1 \text{ N} = 1 \text{ kg} \times 1 \text{ m/s}^2$$

Since we will see shortly that weight is a force, a newton is also a unit of weight equal to 0.225 lb or 3.6 oz. In Fig. 3.8, the apple shown weighs 3.6 oz. Since the (probably true) story is told that Isaac Newton was struck by an apple while meditating on the concept of gravity, it is easy to remember that an apple weighs approximately one newton.

A dyne is a very small unit of force that is rarely used. It is equal to 1 gm $\times$ cm/s^2. The dyne is only 1/100,000 as large as a newton.

The mass of an object is independent of the Earth or any other object in the universe. Hence, a measure of inertial mass of a body is a quantity that that body alone possesses. This is not true of the weight of a body. The **weight** of a body is the gravitational force acting upon it. On Earth the weight is the force of attraction between a body and the Earth. This force of attraction varies, becoming smaller as the distance between the object and the Earth increases.

Figure 3.8 An average-sized apple weighs about one newton, 3.6 oz, or 0.225 lb. (Photo courtesy James Crouse.)

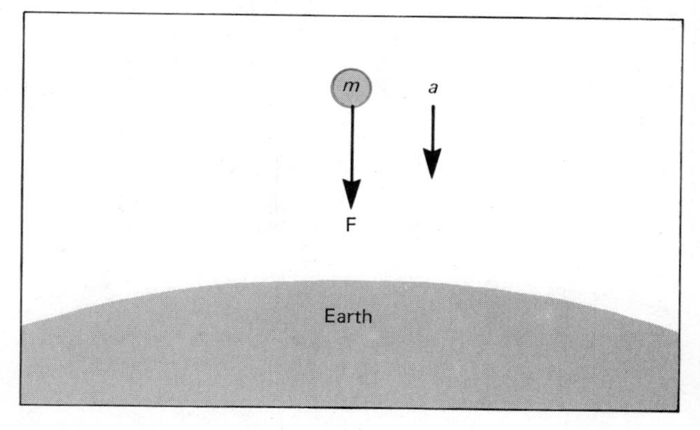

Figure 3.9 The Earth's force of gravity on an object causes the object to have weight. The force F on the object of mass m causes an acceleration a downward.

As an example, a bowling ball with a mass of 5 kg has a weight of 11 lb when placed on a spring scale located at the surface of the Earth. If the ball is taken a trillion miles from the Earth, the gravitational attraction between the ball and the Earth will, for all practical purposes, be zero; hence, the weight will be zero. However, since the quantity of matter has remained the same, the mass will still be 5 kg.

The weight of an object on Earth has been defined as the force of attraction between the Earth and the object. If we have an object, or mass, located near the surface of the Earth, the force of attraction between the object and the Earth will accelerate the object, providing the object is free to move. See Fig. 3.9.

The relation $F = ma$ states that an unbalanced force acting on the mass m produces the acceleration a. But the force acting in the case of a falling object is the weight w of the object. Therefore, we can write

$$w = F$$

$$w = ma$$

EXAMPLE 1 An orange weighing 2 N falls to the ground. What is the unbalanced force acting on it as it falls?

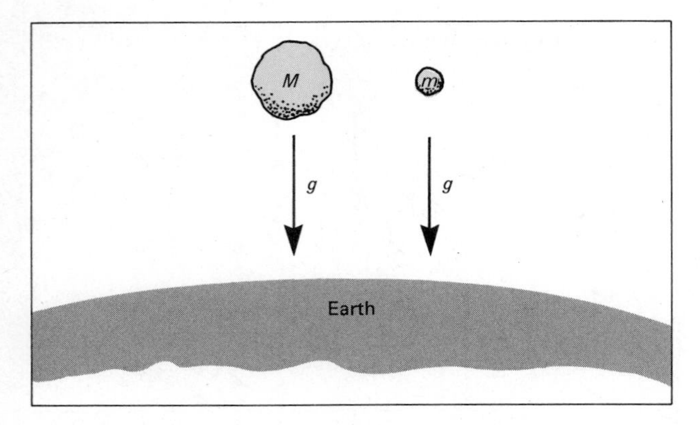

Figure 3.10 Two rocks fall at the same acceleration. The rock that weighs more has more force on it. The small rock has less inertia.

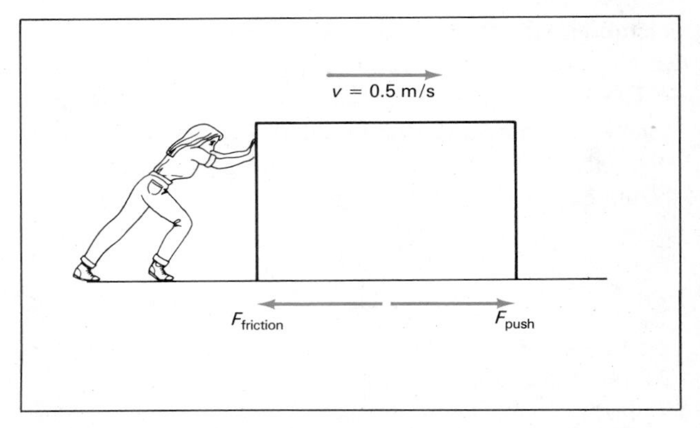

Figure 3.11 When a heavy box is pushed at a constant velocity of 0.5 m/s, the net force on it is zero since it is not accelerated. The pushing force is balanced by the opposing force of friction.

Since the weight is defined to be the gravitational force on an object due to the Earth, the unbalanced force acting on the orange is 2 N.

Any mass falling freely over a short distance, near the surface of the Earth, will accelerate downward due to gravity at a constant value of 9.8 m/s². The acceleration due to gravity is a constant and is given the symbol **g**. Using this symbol for acceleration in the preceding relation, we obtain

$$w = ma = mg \qquad (3.2)$$

or

$$w = mg$$

which states the relationship between mass and weight.

EXAMPLE 2 Compute the weight of a 1-kg mass.

We use Eq. 3.2 to get

$$w = mg$$

$$w = 1 \text{ kg} \times 9.8 \text{ m/s}^2$$

$$w = 9.8 \text{ N}$$

As our country converts from the British system to the metric or SI system, some confusion is bound to occur. One problem concerns the terminology for mass and weight.

Physicists normally distinguish between mass and weight. A 1-kg mass weighs 9.8 N, or 2.2 lb, on Earth. However, you will see the weight expressed in grams or kilograms on many packages because on the Earth's surface mass and weight are proportional. Technically speaking, a package with a net "weight" of 1 kg has a mass of 1 kg and a weight of 9.8 N.

A weight is a force, and things that weigh more are more strongly attracted to the Earth than are things that weigh less. If this is true, why do all things fall to the Earth at the same rate? The answer can be seen by considering the mass of objects as well as the force on them.

If you have two rocks as shown in Fig. 3.10 the larger one has more inertia, and thus it is more difficult to move from rest. The smaller one is much easier to move from rest. Thus, we have two factors involved that exactly cancel. The big rock weighs more, but it also has more inertia. The little rock weighs less, but it has less inertia. Since the two effects cancel, both rocks fall with the same acceleration. Mathematically, we can write

$$a = \frac{F}{\mathrm{m}}$$

or

$$a = \frac{w}{m}$$

or

$$a = \frac{mg}{m}$$

or

$$a = g$$

We see that the mass due to weight in the numerator exactly cancels the inertial mass in the denominator, and the acceleration of all falling objects is the same and equal to g as Galileo stated (Section 2.3). Remember that we are ignoring frictional effects.

Force, like acceleration, is a vector quantity, and forces are added in the way vectors are. For instance, in Fig.

3.2 (top) the forces F_1 and F_2 are equal and opposite, so the total or net force is zero. In Fig. 3.2 (bottom) F_1 and F_2 are both large and in opposite directions. This time, however, F_2 is slightly larger than F_1, so there is a small acceleration in the direction of F_2.

When an object is at rest or moving at a constant velocity, the unbalanced force acting on it is zero. For instance, if you push a heavy box along a floor at a constant velocity, the unbalanced force on the box is zero. The pushing force that you exert is balanced by the opposing force of friction between the box and the floor. (See Fig. 3.11.)

When a mass is held on a string, the force the string exerts is called the tension. Tension is frequently designated by the letter T.

EXAMPLE 3 Compute the tension in a string supporting a 3-kg mass.

Fig. 3.12 is a diagram of this situation. The weight of the mass is given by

$$w = mg$$

$$w = 3 \text{ kg} \times 9.8 \text{ m/s}^2$$

$$w = 29.4 \text{ N}$$

The mass hanging from the string is not accelerating. Thus, the total force on it must be zero.

There are two forces acting on the mass. The weight, mg, is a force acting downward, and the tension in the string acts in the upward direction. Since the total force is zero, these two forces must be equal and opposite. The tension, T, acts upward with a magnitude equal to the weight of the mass. Thus, the tension in the string is also 29.4 N.

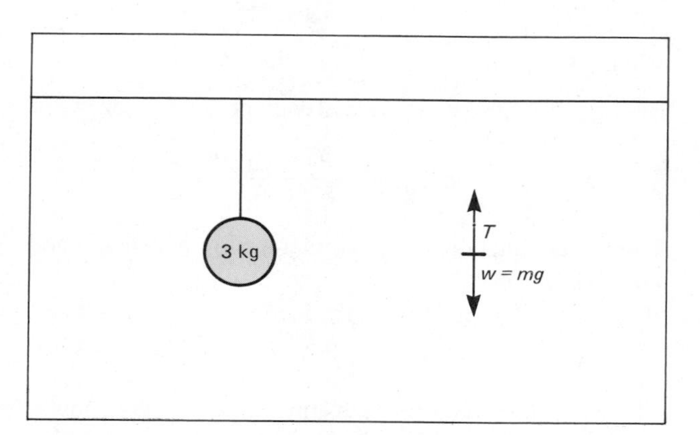

Figure 3.12 A 3-kg mass hung by a string experiences no acceleration. The force down (weight w) is balanced by the force up (tension T).

3.3 Newton's Law of Gravitation

The law governing the force of attraction between two bodies was formulated by Newton from his studies on planetary motion. Known as the **law of universal gravitation**, it is

> **Every mass in the universe attracts every other mass with a force that is directly proportional to the product of their masses and inversely proportional to the square of the distance between their centers.***

Using the proportionality sign ($\propto$), this can be written as

$$\text{Force} \propto \frac{\text{First mass} \times \text{Second mass}}{\text{Sq. of dist. between centers}}$$

or as

$$F = G\frac{m_1 m_2}{r^2} \tag{3.3}$$

where F = the gravitational force of attraction,
m_1 = the mass of one body,
m_2 = the mass of the second body,
r = the distance between the centers of the two masses,
G = a proportionality constant known as the universal gravitational constant, the numerical value of which depends upon the units of F, m, and r.

Newton's law of universal gravitation (Eq. 3.3) applies to point particles and spheres. If a mass is a large sphere, such as the Earth, moon, or Sun, the distance r is referred to the center of the sphere. The other mass must then be external to the sphere.

The force of attraction between any object and the Earth is what causes the object to fall to the Earth when the object is released above the Earth's surface. Newton said the same kind of force (gravitation) causes the planets to revolve around the Sun, and the moon to revolve around the Earth.

*This assumes spherical masses.

Using his law of universal gravitation and his second law, Newton correctly calculated the time it would take for the moon to go around the Earth. His calculation was done without knowing the value of G. In fact, the correct value of G was unknown during Newton's lifetime.

The value for G was found experimentally in 1797, 70 years after Newton's death. Henry Cavendish (1731–1810), an English scientist, performed an experiment in which he succeeded in measuring the value of G. The value for G is very small and is given in Table 3.1. This table tells us what G is and is not.

In Newton's law of universal gravitation the units on both sides must be equivalent. Thus, the units of G must be force units times length units squared divided by mass units squared. With these units for G, both sides of Eq. 3.3 have force units.

Note that with SI units, G is very small and small masses have small gravitational forces, as illustrated by the following example.

EXAMPLE 4 If we had two books, each with a mass of 1 kg (corresponding to a weight of 9.8 N or 2.2 lb), placed 1 m apart, what would be the force between them?

The solution is easy if we use Newton's law of universal gravitation (Eq. 3.3):

$$F = G\frac{m_1 m_2}{r^2}$$

$$F = 6.67 \times 10^{-11}\frac{\text{N} \times \text{m}^2}{\text{kg}^2}\frac{1\text{ kg} \times 1\text{ kg}}{(1\text{ m})^2}$$

$$F = 6.67 \times 10^{-11}\text{ N}$$

which is a very, very small number. According to Newton's law of universal gravitation, everything that has mass is attracted to everything else that has mass. However, since G is such a small number, at least one of the masses involved must be large before the gravitational force between two masses is appreciable. This is illustrated in Fig. 3.13.

Table 3.1 What is G?

$G = 6.67 \times 10^{-11}\text{ N} \times \text{m}^2/\text{kg}^2$
G is a very small quantity in SI units.
G is a universal constant. It is the same throughout the universe.
G is not g.
G is not a force.
G is not an acceleration.

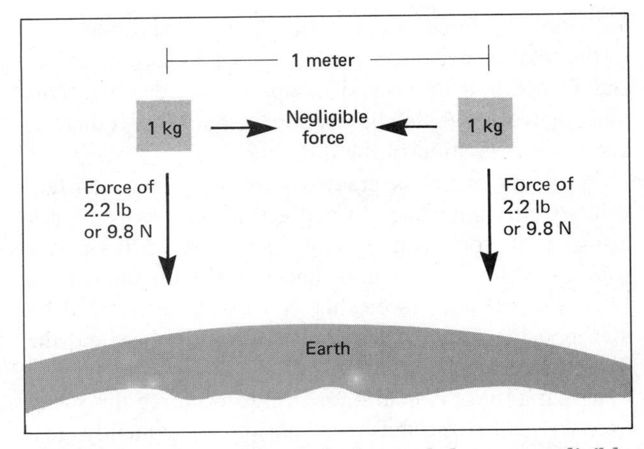

Figure 3.13 Two masses of 1 kg each have a negligible force of gravitational attraction when they are 1 m apart. However, because the Earth's mass is quite large, they are both attracted to the Earth with a force of 2.2 lb or 9.8 N. We call this force the weight of the masses.

When a mass is attracted to the Earth, we call the force involved the *weight of the mass*. Let us now work out the relationship between weight and the properties of the Earth.

If a mass at the Earth's surface is attracted to the Earth, then the distance between the centers of the masses is approximately the radius of the Earth. We can find the force of attraction using Eq. 3.3,

$$F = \frac{G m_{\text{object}}\, m_{\text{Earth}}}{R^2_{\text{Earth}}}$$

where R_{Earth} = radius of the Earth

This force is just what we call weight, so we can write

$$w = m_{\text{object}}\, g = \frac{G m_{\text{object}}\, m_{\text{Earth}}}{R^2_{\text{Earth}}}$$

Using this equation we get a relationship between g and G.

$$g = \frac{G m_{\text{Earth}}}{R^2_{\text{Earth}}} \tag{3.4}$$

This is an important relationship. First, it shows how g and G are related. Second, it shows how the acceleration of gravity on Earth, g, is related to the Earth's mass and radius.

One exciting thing that Newton's law shows is that if we know g, G, and the Earth's radius, the mass of the Earth can be calculated using Eq. 3.4. Once G was found in 1797, scientists were able to calculate the Earth's mass—about 6×10^{24} kg.

The acceleration of gravity on the moon can be computed using a formula similar to Eq. 3.4:

$$g_{\text{moon}} = \frac{Gm_{\text{moon}}}{R^2_{\text{moon}}}$$

From our knowledge of the moon's mass and radius, we get

$$g_{\text{moon}} = \frac{1}{6} g$$

Basically, g_{moon} is small because the moon's mass is so much smaller than the Earth's mass.

The weight of objects on the moon is just equal to mg_{moon}, so we get

$$\text{weight on moon} = \frac{1}{6} \times \text{weight on Earth}$$

Thus, things on the moon fall at one-sixth the acceleration of the Earth's gravity. Objects on the moon also weigh one-sixth as much although they will have the same mass (Fig. 3.14). If you want to lose weight fast, go to the moon.

Similarly, the acceleration of gravity on the surface of any planet can be calculated from

$$g_{\text{planet}} = \frac{Gm_{\text{planet}}}{R^2_{\text{planet}}}$$

In general, objects weigh more on bigger planets such as Jupiter than on smaller planets such as Mercury.

Why doesn't the moon fall down? Anything dropped seems to fall down, so why doesn't the moon? A good way to understand this problem is through examining Fig. 3.15.

Let us suppose we are on top of a tall building throwing tennis balls out the window. At first they take path A in Fig 3.15, but when we throw harder they take path B. If we could throw them hard enough and there were no atmosphere, they would accelerate toward Earth with acceleration equal to g. However, they would take path C and never hit the Earth. They would simply accelerate all the way around the Earth, forming a circular path.

If we remember that anything going in a circle accelerates with $a = v^2/r$, toward the center of the circle,

Figure 3.14 Objects on the moon have the same mass as on the Earth, but they weigh only one-sixth as much. (Courtesy NASA.)

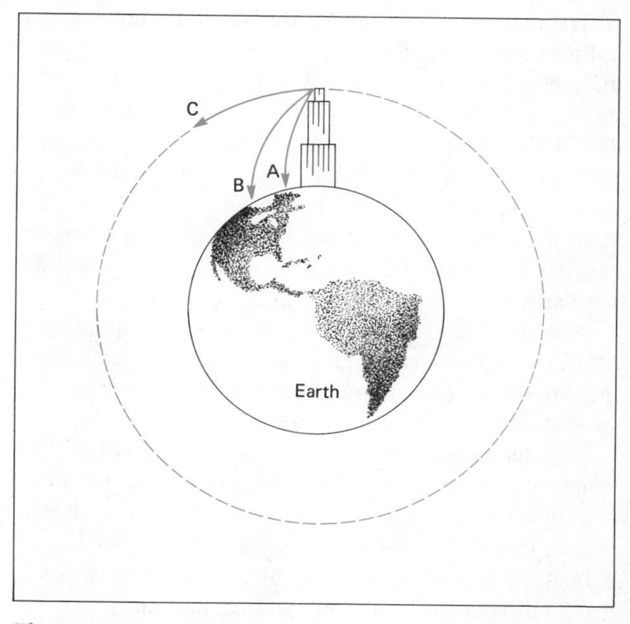

Figure 3.15 An exaggerated drawing illustrating what happens when you throw something out of a building. If there were no atmosphere and you "threw" an object hard enough it would "fall around" the Earth as shown in path C. The moon "falls around" the Earth in a similar manner.

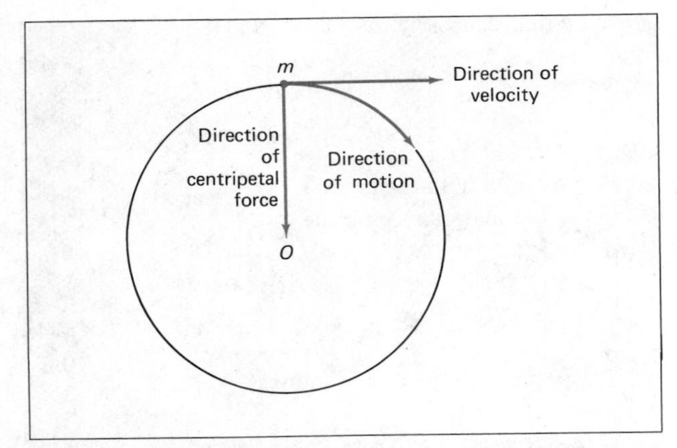

Figure 3.16 The direction of motion, the velocity, and the centripetal force for a mass undergoing uniform circular motion.

we can understand that the moon is being constantly accelerated toward the Earth. Newton's second law becomes

$$F_{\text{grav}} = ma = m\frac{v^2}{r}$$

We understand this as follows: The Earth's gravity pulls on the moon with a strong force. This force causes the moon to travel in a nearly circular orbit. This is the major force on the moon—the force of the Earth's gravity. This unbalanced force causes an acceleration toward the Earth. This acceleration causes a circular motion and the moon does not escape from orbit. Newton was able to analyze this motion quantitatively, and he correctly calculated the time it takes for the moon to "fall around" the Earth.

Whenever an object is going in circular motion, such as the moon going around the Earth, the force that causes the circular motion is called the **centripetal force**. The word centripetal means "center-seeking."

The centripetal force is always directed toward the center of the circle. The centripetal force and the acceleration it causes are both vector quantities and the direction for each is toward the center of the circle. Figure 3.16 is a diagram showing the direction of the velocity, the centripetal force, and the path of motion.

In our example of the moon going around the Earth, the centripetal force was the force of gravity between the Earth and the moon. When an automobile goes around a curve, the centripetal force is the frictional force exerted on the tires by the pavement. If the pavement is wet or icy, the friction may not be enough to provide the required centripetal force, and the automobile may go off the road.

3.4 Newton's Third Law of Motion

Newton's third law of motion is sometimes called the law of action and reaction. **Newton's third law** states:

> **For every action there is an equal and opposite reaction.**

In this statement the words "action" and "reaction" refer to forces. A more precise statement would be the following:

> **Whenever one mass exerts a force upon a second mass, the second mass exerts an equal and opposite force upon the first mass.**

Rocket motion (Fig. 3.17) is an example of Newton's third law. The exhaust gases are accelerated out of the base of the rocket, and the rocket accelerates in the op-

Figure 3.17 A rocket blast-off illustrates Newton's third law. (Photo courtesy NASA.)

Figure 3.18 Isometrics are a good example of Newton's third law. During an isometric exercise, different muscles of the body push against each other. One muscle provides the action; another muscle provides the reaction. (Photo courtesy James Crouse.)

posite direction. A common misconception is that the exhaust gases push against the launch pad to accelerate the rocket. This is nonsense. If this were true, there would be no space travel since there is nothing to push against in space. The correct explanation is one of action (gases going out the back) and reaction (rocket propelled forward).

Another illustration was given by Newton: "If you press on a stone with your finger, the finger is also pressed upon by the stone." Figure 3.18 shows a person doing an isometric exercise as an example of Newton's third law.

Newton's third law can be expressed mathematically as

$$F_1 = -F_2 \qquad (3.5)$$

where F_1 = force exerted on object 1 by object 2

F_2 = force exerted on object 2 by object 1

The minus sign indicates that F_2 is in the opposite direction from F_1. Using Newton's second law, we find

$$m_1a_1 = -m_2a_2 \qquad (3.6)$$

From Eq. 3.5, we see that if m_2 is much bigger than m_1, then a_1 is much bigger than a_2.

As an example, consider dropping a book to the floor. As the book falls, it has a force acting on it (Earth's gravity) that causes it to accelerate. What is the equal and opposite force? The equal and opposite force is the force of the book's gravitational pull on the Earth. As illustrated in Fig. 3.19, the Earth actually rises up to meet the book! However, since the Earth's mass is so huge compared with the book's, the Earth's acceleration is minuscule and cannot be measured.

There are many other examples of action and reaction occurring in everyday life. In fact, from Newton's third law it is clear that every time there is a force, there must be a reaction force also.

It is important to understand the difference between Newton's third law and the concept of balanced forces. If a book weighing 6 N lies on a table, the book exerts a force of 6 N down on the table. The table, in turn, exerts an upward force of 6 N on the book. This satisfies Newton's third law.

Now focus on the 6-N book alone. What are the two forces acting on the book? First, there is the 6-N force of the Earth's gravity acting down. Second, there is the 6-N upward force that the table exerts on the book. From Newton's second law, both the total force and the acceleration of the book are zero.

If the book is dropped, it falls to the floor. In terms of Newton's second law, there is only one force acting on the book. This force is the Earth's 6-N force of gravity. This single force causes the book to accelerate downward.

How is Newton's third law satisfied in the case of the falling book? The Earth accelerates upward as previously discussed and shown in Fig. 3.19.

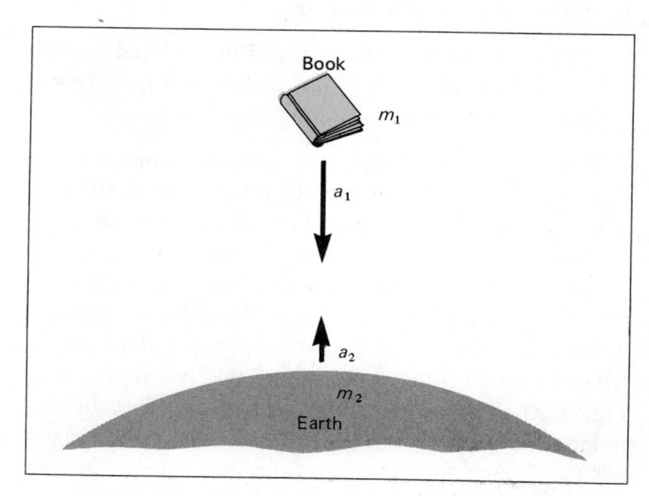

Figure 3.19 As the book is accelerated downward, the Earth is accelerated upward. Because the Earth's mass is so large, the Earth's acceleration is immeasurably small.

Newton's third law relates two equal and opposite forces that act on two separate objects. Newton's second law concerns how forces acting on a single object can cause an acceleration. If these forces acting on a single object are equal and opposite, then there will be no acceleration.

3.5 Conservation Laws

It is difficult to stop a speeding bullet because it has a high velocity. It is also difficult to stop a slowly moving oil tanker because it has a large mass. In general, the combination of mass and velocity is called the linear momentum. That is,

$$\textbf{linear momentum} = \text{mass} \times \text{velocity}$$

or in symbols,

$$\text{linear momentum} = mv$$

Both a speeding bullet and a slowly moving oil tanker have a large linear momentum. We can compare them by actually calculating the linear momentum of each. If we have a system of masses, the linear momentum of the system is found by adding up the linear momentum vectors of all the individual masses.

The linear momentum of a system is important because, if there are no external unbalanced forces, it is conserved; it does not change with time. This property makes it extremely important in analyzing the motion of various systems. The **law of conservation of linear momentum** can be stated as follows:

> **The total linear momentum of an isolated system of masses remains the same if there are no external unbalanced forces acting on the system.**

Both mass and velocity are involved in momentum. A small car and a large truck both traveling at 50 mi/h have the same velocity, but the truck has more momentum because it has a much larger mass.

Another important quantity that Newton found to be conserved is angular momentum. Angular momentum arises when objects go in paths around a fixed point. Consider a comet that goes around the Sun as shown in Fig. 3.20. The **angular momentum** is given by the product mvr, where m is the mass, v is the velocity, and r is the distance of the object from the center of motion which in this case is the Sun.

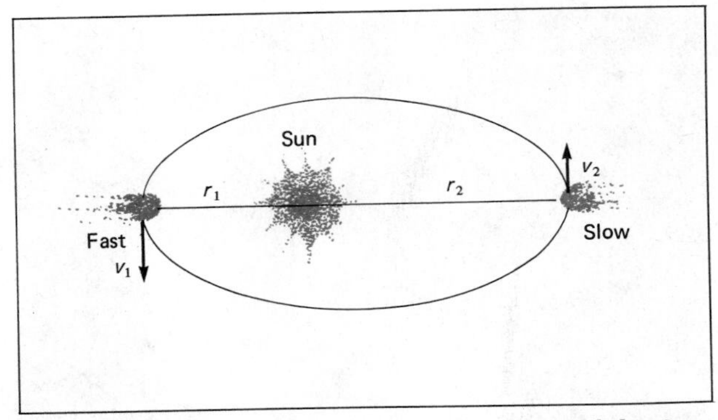

Figure 3.20 The path of a comet going around the Sun. The angular momentum of the comet is given by mv_1r_1 or mv_2r_2. Since the angular momentum is conserved, we get $mv_1r_1 = mv_2r_2$. As the comet gets closer to the Sun, the radial distance r decreases so v must increase. The comet moves fastest when closest to the Sun and slowest when farthest from the Sun.

The linear momentum of a system can be changed by the introduction of an external unbalanced force; the angular momentum of a system can be changed by an external unbalanced torque. A **torque** is a twisting effect caused by one or more forces. For example, in Fig. 3.21 a torque on a steering wheel is caused by two equal and opposite forces acting on different parts of the steering wheel. This causes a twist, or torque, and causes the steering wheel to turn. In general, a torque tends to produce a rotational motion.

The **law of conservation of angular momentum** states that **the angular momentum of an object remains constant if there are no external unbalanced torques.** That is,

$$mv_1r_1 = mv_2r_2 \tag{3.7}$$

In our example of a comet, the angular momentum, mvr, remains the same. As the comet gets closer to the Sun, r decreases so the speed v increases. For this reason, comets spend much less time close to the Sun (where they can be seen more easily from the Earth) and much longer periods of time far from the Sun.

EXAMPLE 5 A comet at its farthest point from the Sun is 900 million miles away and traveling at 6,000 mi/h. What is its speed at its closest point to the Sun, which is 30 million miles away?

We know v_2, r_2, and r_1 so we can calculate v_1 as follows:

$$mv_1r_1 = mv_2r_2$$

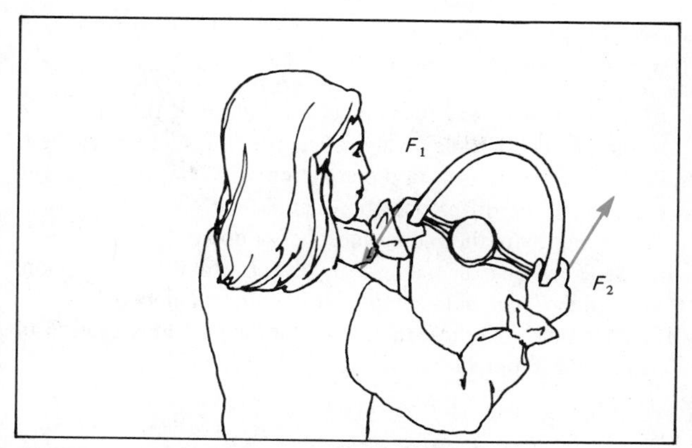

Figure 3.21 A person driving a car twists the steering wheel if two equal and opposite forces are applied. Note that the net force is zero but that there is still a torque, or twisting effect.

Figure 3.22 Conservation of angular momentum. (a) When the student holds the weights away from himself, the speed of rotation is slow. (b) When the weights are brought inward toward the body, the speed of rotation increases. The result is a fast, dizzy spin as indicated by the facial expression of the student. (Courtesy J. D. Wilson, *Physics Concepts and Applications*, Heath, Lexington, Mass., 1981.)

or $$v_1 r_1 = v_2 r_2$$

or $$v_1 = \frac{v_2 r_2}{r_1}$$

$$v_1 = \frac{(6 \times 10^3 \text{ mi/h})(900 \times 10^6 \text{ mi})}{30 \times 10^6 \text{ mi}}$$

$$v_1 = 180{,}000 \text{ mi/h}$$

Thus we see that the comet moves much faster when it is close to the Sun than it does when it is far away from the Sun.

Another example of the conservation of angular momentum is demonstrated in Fig. 3.22. Here a student holds weights on a spinning stool. With the weights held out, the spinning rate is slow because the radial distance r for the weight is large. When the weights are brought inward, r decreases, so the spinning rate must increase.

Ice skaters use the same principle to spin faster on the ice. The skater extends both arms and perhaps one leg and obtains a slow rotation, then the arms and leg are drawn inward and rapid angular velocity is gained because of the decrease in the radial distance of the mass.

The student can illustrate the law by performing the following experiment: Take a small object like a piece of chalk and tie a length of string about a foot long to the chalk. Hold the other end of the string by your forefinger and thumb and whirl the chalk around in a vertical circle. Extend your forefinger so that the string will wind itself around this finger and observe the increase in speed of the chalk as the distance between the chalk and finger becomes smaller.

The law of conservation of angular momentum would cause a helicopter with a single propeller to rotate. In order to conserve angular momentum, the body of the helicopter would have to rotate in the direction opposite that of the rotor. To prevent this, some helicopters have two oppositely rotating propellers. It is more common, however, for the helicopter to have a small ''anti-torque'' propeller (Fig. 3.23) on the tail which produces a rotation opposite to the rotation of the body of the helicopter. With this configuration of propellers, the helicopter is stable.

Figure 3.23 The body of a helicopter with a single rotor would rotate in a direction opposite to that of the propeller. To prevent this, most helicopters have an auxiliary propeller on the tail. (Courtesy Bell Helicopter Textron.)

Learning Objectives

After reading and studying this chapter, you should be able to do the following without referring to the text:

1. State Newton's first law of motion and explain what the natural state of motion is.
2. Define the terms *force, inertia,* and *mass*.
3. State Newton's second law of motion in words and with an equation.
4. Give the units of force in the SI and British system and give an everyday example of the SI unit.
5. Define the term *weight* and give its units in the SI and British system.
6. Explain why any two objects will drop at the same rate no matter what they weigh if air friction can be neglected.
7. Explain each term in the equation $F = Gm_1m_2/r^2$.
8. Explain the differences between G and g.
9. Explain why the moon doesn't fall down.
10. State and give an example of Newton's third law of motion.
11. Define *linear momentum* and *angular momentum*.
12. State and give an example of the law of conservation of angular momentum.

Important Words and Terms

Newton's first law of motion
force
inertia
mass
Newton's second law of motion
newton
weight
g
Newton's law of universal gravitation
G
centripetal force
Newton's third law of motion
linear momentum
law of conservation of linear momentum
angular momentum
torque
law of conservation of angular momentum

Questions

1. What is the natural state of motion?
2. Why does it appear that you are thrown to the outside when a car goes around a curve?
3. What driving habit will improve gas mileage?
4. Which of the following would change for a person who went to the moon? (a) weight (b) inertia (c) mass
5. Explain the relationship between (a) force and acceleration (b) mass and acceleration.
6. Is it possible for a body to be at rest if forces are being applied to it? Explain.
7. If no forces are acting on an object, can the object be in motion?
8. What is a newton?
9. What is the weight of a 3-N hammer?
10. If you dropped a 10-lb rock and a 1-lb rock simultaneously, (a) why would you expect the 10-lb rock to hit first? (b) Why would you expect the 1-lb rock to hit first? (c) Explain what actually happens.
11. Discuss the quantities G and g and tell how they are different.
12. When applying $F = Gm_1m_2/r^2$ to a weight problem here on Earth, what is r?
13. Explain why apples fall off trees, but the moon doesn't fall down.
14. How many forces are acting on the moon?
15. Use Newton's third law to explain how a rocket blasts off.
16. Why are linear momentum and angular momentum important concepts?
17. Several keys on a string are whirled around your forefinger. Explain what happens as the string winds around your forefinger and why.
18. Why are comets, such as Halley's comet, so seldom seen?
19. Why do helicopters have a small propeller on the back?

Problems

1. Determine the force necessary to give a 4-kg mass an acceleration of 6 m/s^2. *Answer:* 24 N
2. A force of 2800 N is exerted on a rifle bullet with a mass of 0.014 kg. What will be the bullet's acceleration?
3. What is the unbalanced force on a car moving at a constant velocity of 20 m/s (45 mi/h)?
4. What is the weight of a 6-kg package of nails? *Answer:* 58.8 N
5. What is the force in newtons acting on a 6-kg package of nails that is falling off a roof?
6. Calculate your weight in newtons if you weigh 110 lb.
7. Calculate the weight in newtons of a 500 gram package of cereal.
8. A vertical fish line supports 5 kg of fish when they are held for a picture. Compute the tension in the fish line. *Answer:* 49 N
9. Use $g_{moon} = \frac{1}{6}g$ to find g_{moon} in m/s^2.
10. Calculate the acceleration of gravity in mks units. Use the formula $g = Gm_{Earth}/R^2_{Earth}$. Use $m_{Earth} = 5.96 \times 10^{24}$ kg and $R_{Earth} = 6.37 \times 10^6$ m.
11. Calculate the acceleration of gravity on the moon in mks units. Use the formula $g_{moon} = Gm_{moon}/R^2_{moon}$. Use $m_{moon} = 7.35 \times 10^{22}$ kg and $R_{moon} = 1.74 \times 10^6$ m. *Answer:* 1.6 m/s^2
12. How much would a person weigh on the moon whose weight is 180 lb on Earth?
13. Calculate the momentum of the following:
    (a) A truck of mass 15,000 kg that is traveling at 20 m/s (45 mi/h), east
    (b) A small car of mass 1,000 kg that is going at 30 m/s (67 mi/h), north *Answer:* (a) 300,000 kg·m/s, east
14. A comet goes around the Sun in a noncircular path. At its farthest point, 600 million miles from the Sun, it is traveling at a speed of 15,000 mi/h. How fast is it traveling when it is at its closest point, 100 million miles from the Sun?
15. An asteroid goes around the Sun in a noncircular path. At its farthest point, 8×10^{11} m from the Sun, it is traveling at a speed of 3×10^4 m/s. How fast is it traveling when it is at its closest point, 2×10^{11} m from the Sun?

4

Work and Energy

I like work; it fascinates me.
I can sit and look at it for hours.
—Jerome K. Jerome

THE COMMON MEANING of the word *work* refers to the accomplishment of some task or job. When work is done, energy has been expended. Hence, work and energy are related.

A student performs a certain amount of work during the day and becomes tired. He or she must obtain rest and food in order to continue the work. We know that rest alone is not sufficient to keep the student going; thus, the food must serve as fuel to supply the necessary energy.

The technical meaning of the word *work* is quite different from the common meaning. A student standing at rest and holding several books is doing no work, although he or she will feel tired after a time. Technically speaking, work is accomplished only when a force acts through a distance.

Energy has been defined as stored work, and our mastery of energy has produced today's modern civilization. From the control of fire to the mastery and control of nuclear energy, we have advanced our standard of living through our ability to use and control the flow of energy.

This aspect of the physical environment is probably the single most important concept we have developed in our search for physical knowledge. Energy takes many forms, such as mechanical, heat, chemical, electrical, radiant, nuclear, and gravitational energy. These, as we shall learn, are classified in the more general categories of kinetic and potential energy.

Our main source of energy is the Sun, which radiates into space each day an enormous amount of radiant energy; only a small portion of this energy is received by the planet Earth. Radiant energy supports all life on our planet and provides us with the means for making nature work for the Earth's inhabitants.

◀ Work is done by falling water at Grand Coulee Dam to produce electric energy.

4.1 Work

Work, as a physical concept, is defined as the product of the average applied force and the parallel distance through which the force acts. This can be written as

$$\text{Work} = \text{Av. applied force} \times \text{Parallel distance}$$

or as

$$W = Fd \tag{4.1}$$

Figures 4.1, 4.2, and 4.3 (p. 50) illustrate the concept of work. In Fig. 4.1, a force is being applied to the wall, but no work is being done since the wall is not moving through any distance. Figure 4.2 shows a mass being moved through a distance d by a force F, and the work done will be the product of the force and distance. The illustration also shows the force F and the distance d parallel to one another. When F and d are not parallel to one another, the component of the force F that is parallel to the distance d must be used to calculate work done. For example, when mowing the lawn with a push-type lawnmower only the force component parallel to the lawn (the horizontal force) is used to accomplish work. The vertical component of the force is doing no useful work, since this force is tending to push the lawnmower into the ground (Fig. 4.3).

When the applied force is perpendicular to the distance through which the object is moved, no work is done (Fig. 4.4). The work done on the mass M is zero, because the force F is vertical and the distance d is horizontal. Thus, there is no distance component parallel

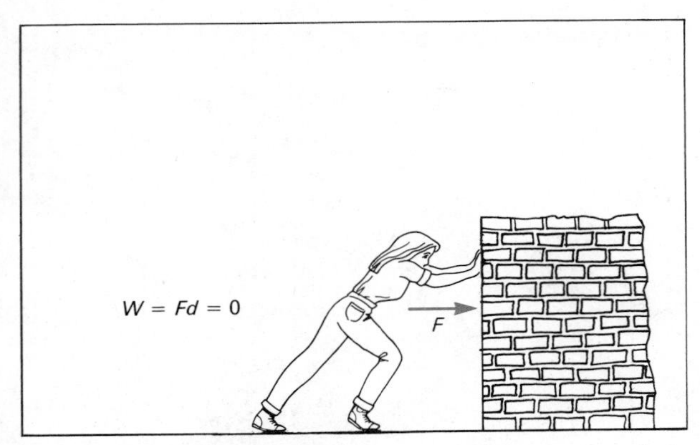

Figure 4.1 A force is applied to a wall but no work is done since $d = 0$.

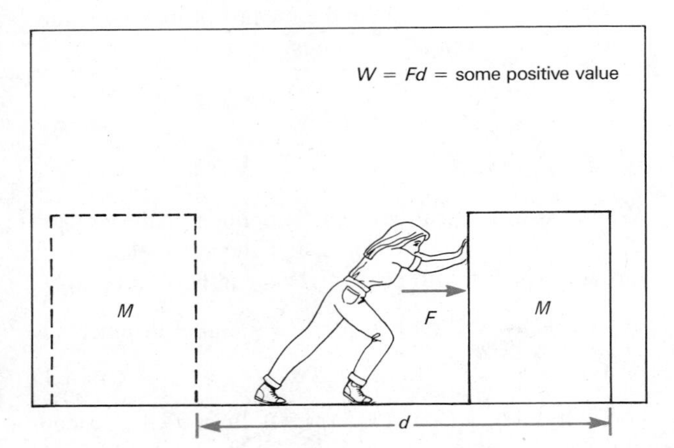

Figure 4.2 A force F is applied to mass M moving the mass through the distance d. Work equals the force times the distance the mass moves in the direction of the applied force.

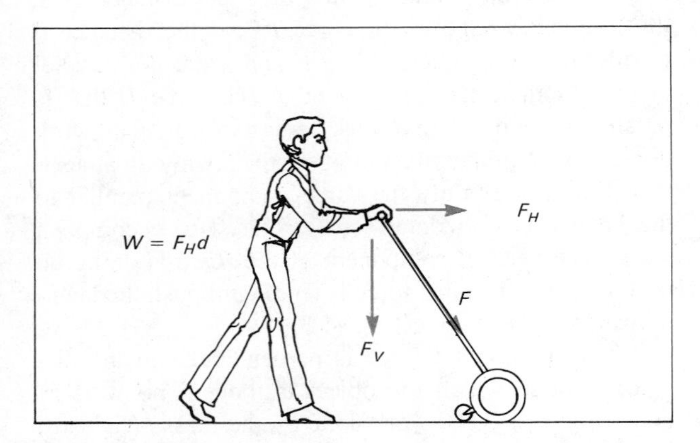

Figure 4.3 Only the horizontal component F_H of the applied force F is used to do work in this example.

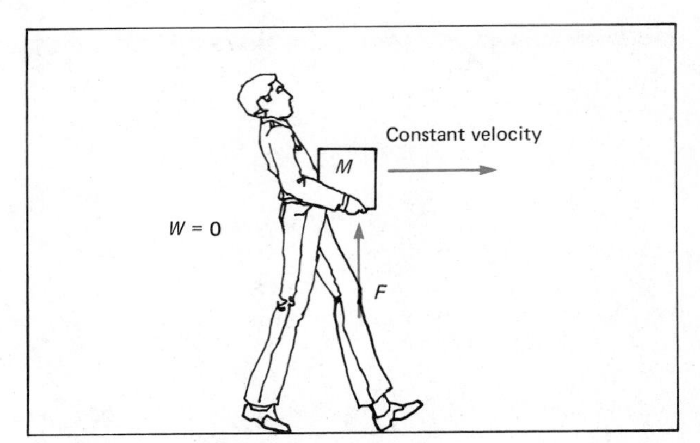

Figure 4.4 Person walking with constant velocity carrying mass M through some distance d. No work is done on the mass M since F is perpendicular to the horizontal distance traveled.

to the upward applied force F. (Work was done, however, when the mass was lifted.)

One of the most important properties of work is that it is a scalar quantity. Both the force and the parallel distance have directions associated with them, but work does not. Work is expressed only as a magnitude with the proper units. It has no direction associated with it.

The units of work are listed in Table 4.1. Since work is the product of a force and a distance, the units of work are the units of force times length.

The joule (pronounced jool) is the unit of work in SI units. The foot-pound is the work unit in the British system. As we will see later in this chapter, these work units are also energy units.

As seen in Table 4.1, one **joule** (J) is the amount of work done in lifting an apple weighing 1 N to a height of 1 m. Similarly, the **foot-pound** is the work done by a force of 1 lb acting through a distance of 1 ft. The name given to one dyne-cm is the **erg**. One erg is 10^{-7} J.

Work can be done on objects in many different ways, among which are

(a) *Work against inertia*. We have learned that any object at rest remains at rest unless acted upon by some

Table 4.1 Work Units (Energy Units)

	Fundamental Units	*Force × Distance Units*	*Name*
SI	kg × m^2/s^2	newton × m	joule (J)
British	lb × ft	lb × ft	foot-pound

outside force, or, if the object is in motion, it will remain in motion at constant velocity in a straight line unless acted upon by some outside force.

Newton referred to the properties of an object to stay at rest or in constant straight-line motion as *inertia*. When a force is applied to change the velocity or direction of an object, work will be done against inertia. Figure 4.5 illustrates this. A mass is shown at rest on a frictionless air table. In order to increase the velocity of the mass from 0 to 1 m/s, a force must be applied through the distance d, thus performing work. The work done has been against inertia.

(b) *Work against gravity*. If we push something up at a (slow) constant velocity, there is no net force on it since it is not accelerating. The weight of the object mg presses down, and we push up with an equal and opposite force equal to mg. The distance parallel to our upward force is the height h which we lift the object.

Thus, the work done against gravity is

$$W = Fd$$

$$W = mgh$$

If we lift a 10-kg mass a distance of 2 m (see Fig. 4.6), the work done is

$$W = mgh$$

$$W = 10 \text{ kg} \times 9.8 \text{ (m/s}^2) \times 2 \text{ m}$$

$$W = 196 \text{ J}$$

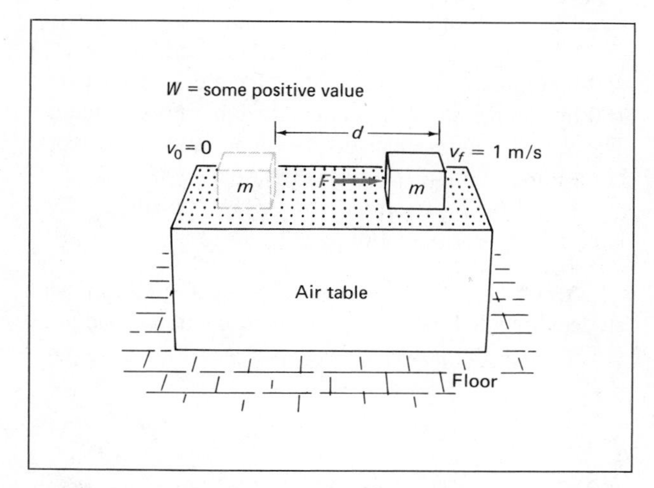

Figure 4.5 Work done against inertia. A force F is applied to move mass m from rest to a velocity of 1 m/s. The air table provides an almost frictionless surface for the mass.

Figure 4.6 Work against gravity. The upward force F is equal to the downward force mg. The distance parallel to F is the height h. The work is mgh.

The concept of work against gravity helps to explain why it is much more tiring to walk up stairs than it is to walk on a level surface. When you walk on level ground your center of gravity remains approximately constant, and you do very little lifting of your body (Fig. 4.7). Of course, there is some work being done by your muscles as you walk, but it is not enough to tire you very much. In fact, most healthy people can easily walk several miles on a level surface.

When you walk up stairs, you have to lift your whole body up. You do an additional amount of work equal to mgh where mg is your weight and h is the height you go up.

(c) *Work against friction*. Examine Fig. 4.8. A horizontal force of 10 N is needed to slide a 10-kg mass a distance of 2 m along the top of a table. What is the work done?

$$W = Fd$$

$$W = 10 \text{ N} \times 2 \text{ m}$$

$$W = 20 \text{ N} \cdot \text{m}$$

$$W = 20 \text{ J}$$

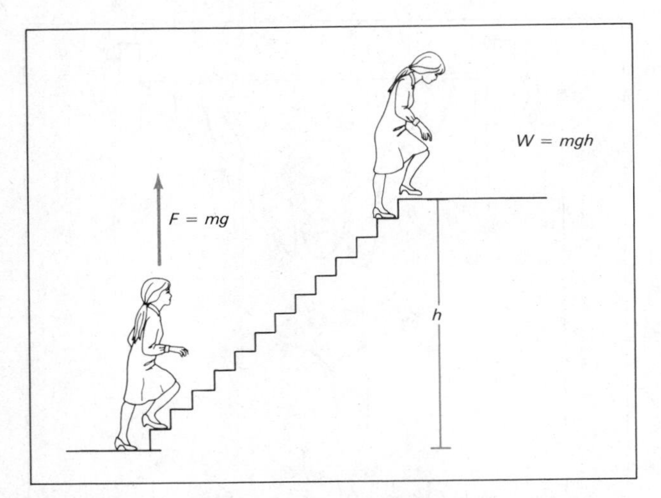

Figure 4.7 Walking up stairs requires much more work than walking on a level surface. When you go up, the work you do is increased by an amount *mgh* over what you would do if you walked on level ground.

You will notice it takes much less work to slide the 10-kg mass along the table 2 m than to lift the same 10-kg mass to a height of 2 m.

In Fig. 4.8 the force of friction opposes the applied force F. If the object is being moved at a constant velocity, the net force must be zero (since there is no acceleration). Thus, the applied force F is equal and opposite to the force of friction.

4.2 Power

When a family moves into a second-floor apartment, a lot of work must be done to carry their belongings up the stairs. In fact, each time the steps are climbed, the movers must carry not only the furniture, books, and so on, but also their own weight up the stairs.

If the movers do all the work in three hours, they have not worked as rapidly as if the work had been done in one hour. The same amount of work is done in each case, but the rate at which work is done is different.

The rate of doing work is called **power**. It is calculated by dividing the work done by the time required to do it. This can be written as

$$\text{Power} = \frac{\text{Work}}{\text{Time}}$$

or as

$$P = \frac{W}{t}$$

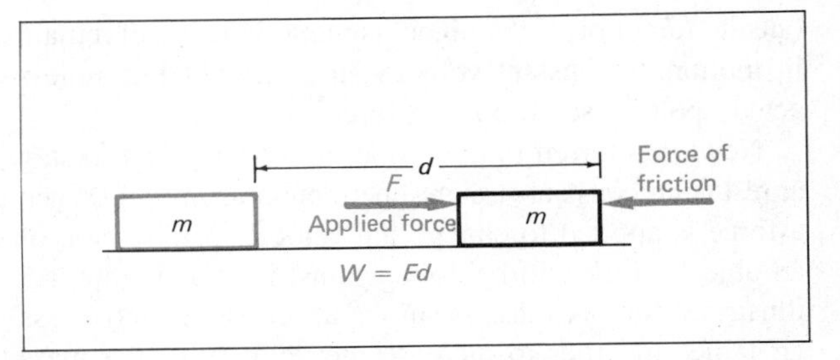

Figure 4.8 Work done against friction. The mass m is being moved at a constant velocity by the force F against the force of friction.

Since work is the product of force and distance, the common formula used to find power is

$$P = \frac{Fd}{t}$$

In SI units, in which work is measured in joules and time in seconds, power is measured in watts, one **watt** (W) equaling one joule per second.

In the British system, power is expressed in *foot-pounds per second* or *horsepower*. One **horsepower** (hp) is equal to 550 foot-pounds per second or 746 watts.

$$1 \text{ horsepower} = \frac{550 \text{ ft} \cdot \text{lb}}{\text{s}} = 746 \text{ watts}$$

The term "horsepower" was originated by the British during the days when coal was mined by hand and brought to the surface with the use of horses. A strong horse was capable of doing, on the average, an amount of work equal to 550 foot-pounds per second.

The greater the power, the faster work is done. A 450-hp engine can accomplish work much faster than a 160-hp engine in a theoretical or laboratory situation. The smaller engine can do the same amount of work, but a longer period of time is required.

The following example shows how power is calculated in a typical situation.

EXAMPLE 1 A force of 50 N is required to push a student's stalled motorcycle 10 m along a flat road in 20 s. Calculate the power in watts.

$$P = \frac{W}{t}$$

$$P = \frac{50 \text{ N} \times 10 \text{ m}}{20 \text{ s}}$$

$$P = 25 \text{ W}$$

The student should be careful not to confuse the meaning of the letter W. In the formula $P = W/t$, W stands for "work." In the statement "$P = 25$ W," the W stands for "watts." Also, recall that w stands for "weight."

In the next section we will see that work produces a change in energy. Thus, power can be thought of as the energy produced or consumed divided by the time taken, and we can write

$$\text{Power} = \frac{\text{energy produced or consumed}}{\text{time taken}}$$

or

$$P = \frac{E}{t}$$

From this we can see that

$$E = P \times t$$

This formula is useful in computing the amount of energy consumed in a home. In particular, since energy units must equal power units multiplied by time units, a watt times hour is a unit of energy. One thousand watt times hours is a **kilowatt-hour,** abbreviated kWh. It is important to remember that a kWh is an energy unit, not a power unit.

Electric energy consumed is frequently measured in kilowatt-hours. The following example will show how this can be calculated.

EXAMPLE 2 Let us compute the energy consumed if we burn a 100-W light bulb for 80 h.

$$E = P \times t$$

$$E = 100 \text{ W} \times 80 \text{ h}$$

$$E = 8000 \text{ W} \times \text{h} = 8 \text{ kWh}$$

In this example, the light bulb consumed 8 kWh of electric energy.

Almost all residential electric bills are based on the total amount of energy used. An electric power company is more of an energy company. The typical charge for electric energy is 5¢ to 10¢ for 1 kWh. Most commercial and industrial users pay a bill based partly on total energy consumption (kilowatt-hours) and partly on maximum power (kilowatts) used.

This 5¢ to 10¢ rate for 1 kWh of electric energy is quite low. In fact, electric energy is one of the best bargains in our technological society.

For instance, if you burn a 100-W light bulb overnight for 12 h, the energy consumed is 100 W × 12 h = 1200 Wh = 1.2 kWh. At 8¢ per kWh this would cost you 8¢ × 1.2 or 9.6¢. Thus, it costs only about 10¢ to keep a light on overnight. If you left it on every night for a 30-day month, it would cost 9.6¢ × 30 or $2.88.

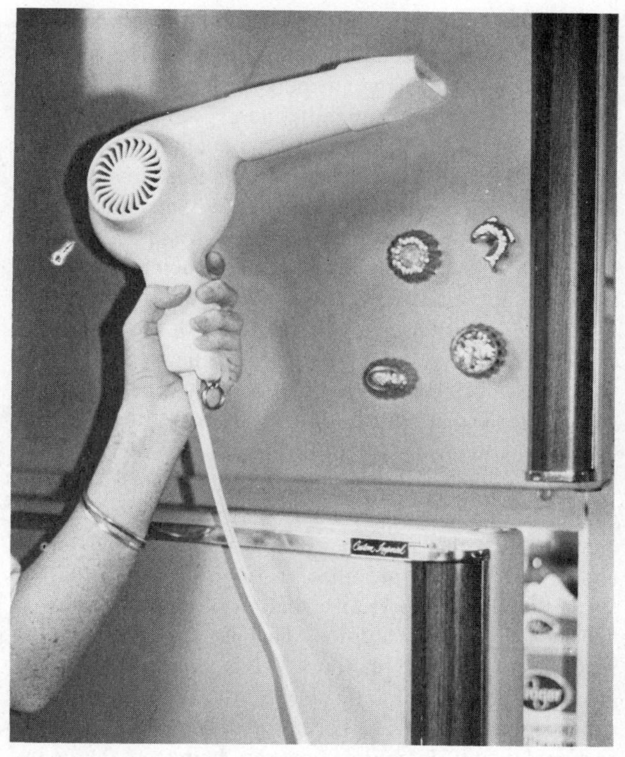

Figure 4.9 Appliances such as refrigerators that run most of the time consume far more energy than appliances such as hair dryers that are not in frequent use. (Photo courtesy Peter Bogen.)

In general, the appliances that run much of the time cost the most to operate. These are refrigerators (see Fig. 4.9), hot water heaters, and air conditioning or heating units. A hair dryer or light bulb are relatively inexpensive because they are used less frequently.

4.3 Kinetic and Potential Energy

When work is done against inertia, an object's speed is changed. When work is done against gravity, an object's height is changed. When work is done against friction, heat is produced, and there is usually a change in the temperature. In all these examples, something is changed when work is done.

The concept of energy helps us to unify all the possible changes that occur when work is done. When work is done, there is a change in energy. The amount of work done is equal to the change in energy.

Energy is one of the most fundamental concepts in science. It is a quantity that is possessed by objects: Work is *done;* objects *have* energy. When work is done on a system, the amount of energy possessed by the system changes. In general, we can define **energy** as the ability to do work or simply as stored work.

Energy occurs in many forms. Two of the most fundamental forms of energy are kinetic energy and potential energy. **Kinetic energy** is energy that a body possesses due to its motion. **Potential energy** is energy that a body possesses due to its position.

Since work produces kinetic and potential energy changes, kinetic and potential energy, and all forms of energy, have the same units as work. Using SI units, energy is measured in joules.

It is important to realize that work and all forms of energy are scalar quantities. That is, they have no direction associated with them. Thus, various forms of energy can be added and subtracted as numbers. Forces can become complicated because of their vector or directional nature. In contrast, energy, being a scalar quantity, is easy to use in computational problems.

The kinetic energy of an object can be written as

$$\text{kinetic energy} = \frac{1}{2} \times \text{mass} \times (\text{velocity})^2$$

or

$$E_k = \frac{1}{2} mv^2 \tag{4.2}$$

Written in this way, the kinetic energy is equal to how much work would have to be done to bring the object to rest. This can be seen from the equations in Chapter 2 where we had $d = \frac{1}{2}at^2$ and $v = at$. The work done against inertia is $W = Fd$, and since $F = ma$ we get

$$W = Fd = mad$$

$$W = ma\left(\frac{1}{2} at^2\right) = \frac{1}{2} m(at)^2$$

$$W = \frac{1}{2} mv^2$$

As an example of doing work to create kinetic energy, consider a pitcher throwing a baseball (Fig. 4.10). The amount of work needed to accelerate a baseball from rest to a speed v is just equal to the baseball's kinetic energy, $\frac{1}{2}mv^2$.

Figure 4.10 The work necessary to increase the velocity of a mass is equal to the increase in kinetic energy. (Assume no energy loss.)

When work is done on a moving body, the work changes the kinetic energy of the body. We get the formula

$$\text{work} = \text{change in kinetic energy}$$

This is an important formula. It shows the relationship between work and kinetic energy.

Because kinetic energy is proportional to the velocity squared, doubling the velocity will cause a fourfold increase in kinetic energy since $2^2 = 4$.

EXAMPLE 3 By what factor is the kinetic energy increased when the speed of a car is increased from 20 mi/h to 30 mi/h?

To solve this we note that the kinetic energy is proportional to the speed squared. Thus, the kinetic energy increases by a factor of $(30)^2/(20)^2$ or 2.25.

To stop a moving automobile takes an amount of work equal to the kinetic energy of the vehicle. In other words, the work to stop an object is equal to the work originally exerted to get it moving. The work to stop an automobile is supplied by friction between the road and the tires.

When cars travel on highways, the braking distance can be defined as the distance a car travels once its brakes are applied. The work done to stop a moving car is equal to the braking force times the braking distance. If the braking force is assumed to be constant, then the braking distance will be proportional to the initial kinetic energy of the car. Thus, from Example 3 we see that the braking distance for a car going at 30 mi/h is 2.25 times that for a car going at 20 mi/h.

This braking distance concept explains why 55 mi/h is so much safer than 65 mi/h. Since the kinetic energy depends on the square of the speed, a 65 mi/h speed is $(65/55)^2 = 1.40$ times as dangerous as 55 mi/h. That is, the braking distance for 65 mi/h is 40% greater than for 55 mi/h.

For instance, if the braking distance at 55 mi/h is 45 meters, the braking distance at 65 mi/h would be 1.40×45 or 63 meters (Fig. 4.11). This braking distance of 63 m is 40% longer than 45 m. Of course, the original reason for having a 55 mi/h speed limit was to conserve gasoline. A welcome by-product has been the saving of lives due to slower speeds.

An object may have energy due to its position. Such energy is called potential energy because the object has the potential to do work by changing its position.

For example, if a book (mass = 1 kg) at rest on the floor is lifted a height of 1 m to the top of the table, work is done. To find the amount of work done, the mass of 1 kg is first changed to force units by multiplying the mass by g, the acceleration due to gravity. ($1 \text{ kg} \times 9.8 \text{ m/s}^2 = 9.8$ newtons.) Then,

$$W = Fd$$

$$W = 9.8 \text{ N} \times 1 \text{ m}$$

or

$$W = 9.8 \text{ J}$$

In this case, work has been done against the force of gravity. The book on the table now possesses energy and the ability to do work, due to its height. The energy is called **gravitational potential energy.** If the book were allowed to fall back to the floor, it would do work on the floor. From the equations

$$W = Fd$$

and

$$F = mg$$

the work it would do can be determined by

$$W = mgh$$

since the distance d is the height h through which the book is lifted.

From this example, we see that it makes sense to define gravitational potential energy, E_p, as

gravitational potential energy = weight × height

or

$$E_p = mgh \qquad (4.3)$$

When work is done with or against gravity or some other force, the potential energy changes. We can write this as

work = change in potential energy

This is true for all kinds of potential energy. For instance, if a bowstring is drawn back, work must be done to pull it back. The bow and bowstring bend and acquire a potential energy. This potential energy is then capable of doing work on an arrow.

For gravitational potential energy, the work done is equal to the weight lifted times the change in height. Thus, we can write

$$E_p = mgh$$

as long as we remember that the h in this formula stands for the *change in* height or the height as measured from an appropriate reference level.

Fig. 4.12 illustrates what happens when a book is lifted from the floor to a table top. The work done on

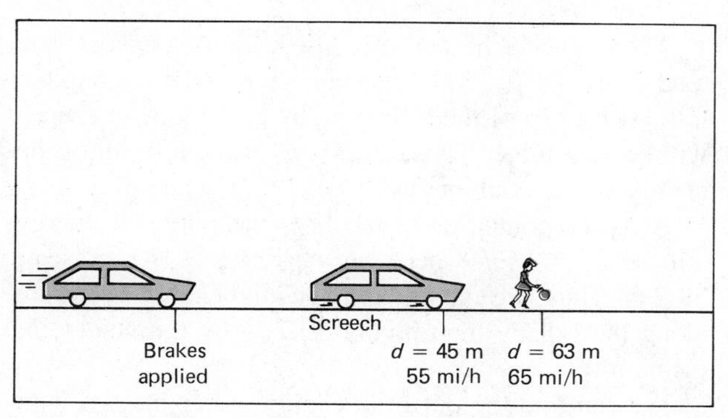

Figure 4.11 An auto moving at 55 mi/h goes 45 m before stopping while one going 65 mi/h goes 63 m before stopping. Of course, the actual distances depend on a variety of factors such as road conditions, tire quality, and so on.

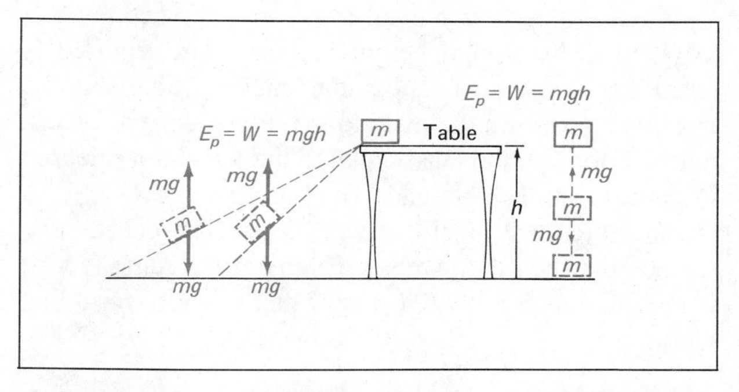

Figure 4.12 Drawing illustrating the potential energy E_p gained by a mass transferred via three different paths from the floor to the top of the table. The E_p of m at the table top is independent of the path taken from the floor. The force acting upward is mg, and it acts vertically upward at all positions.

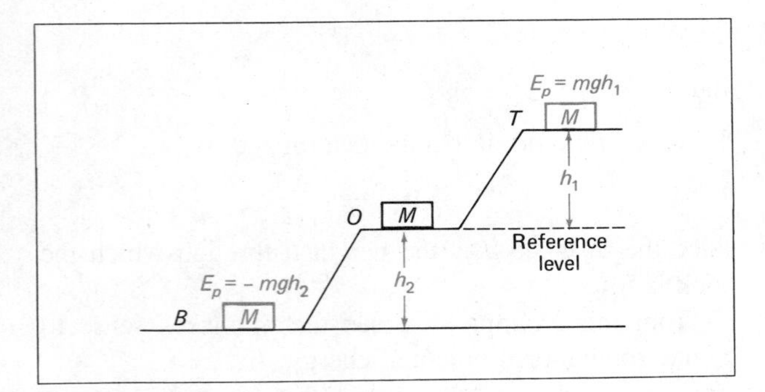

Figure 4.13 Diagram illustrating the idea that potential energy may be positive or negative. Whether the energy is positive or negative depends on the position of the reference level.

the book is independent of the path taken in lifting the book from the floor to the table top. The force necessary to lift the book is the force needed to overcome gravity, which is acting downward; therefore, the component of the applied force used to lift the book must be upward, no matter what path the book takes in arriving at the top of the table.

The potential energy a body possesses depends upon the reference level from which the height is measured, and the value of the energy may take on either a positive or negative value. For example, in Fig. 4.13 where surface O is taken as the reference level, the potential energy the mass m possesses at position T is equal to the work done in moving the mass from the reference level through the vertical height h_1. The magnitude equals mgh_1 and its value would be positive. When the mass is returned from position T to the reference level O, the mass releases energy equal to the amount received when raised through the height h_1.

When the mass is moved from the reference level O to position B through height h_2, the work required is equal to $-mgh_2$. The potential energy value will be negative. To return the mass to the reference level, work must be done on the mass equal to the amount it released in going from the reference level to position B.

Any reference level can be chosen. The energy difference between the levels will remain the same, and it is this *change in potential energy* that is useful for doing work.

4.4 Law of Conservation of Energy

Although the meaning is always the same, the law of conservation of energy is stated in many different ways. Examples of this are as follows: *the total energy of an isolated system remains constant; energy can be neither created nor destroyed; in changing from one form to another, energy is always conserved.*

The **law of conservation of energy,** as mentioned above, can be stated: The total energy of an isolated system remains constant. This means that although energy may be changed from one form to another, energy is not lost.

In equation form we would state the law of conservation of energy as

$$(\text{Total energy})_{\text{time 1}} = (\text{Total energy})_{\text{time 2}} \qquad (4.4)$$

That is, the total energy does not change with time. If the only forms of energy involved are kinetic and potential energy, we would write

$$\left(\frac{1}{2}mv^2 + mgh\right)_{t_1} = \left(\frac{1}{2}mv^2 + mgh\right)_{t_2}$$

In writing this equation we are assuming that no energy is lost in the form of heat energy due to frictional effects. These heat energy effects are frequently important, and are discussed in Chapter 5. However, in Example 4 we will assume that the frictional effects are negligible.

EXAMPLE 4 The law of conservation of energy can be illustrated by Susannah, who is sledding down the snowy north side of Longview Heights (Fig. 4.14). Susannah starts from rest at a height of 25 m above the lower level of the hill. What will be the kinetic and potential energies of Susannah plus the sled at the heights indicated on the drawing if she and the sled have a total mass of 40 kg? (Assume no losses due to friction.)

According to the law of conservation of energy, the total energy, E_T, will be the same everywhere along the hill. At the top of the hill, $v = 0$, so the kinetic energy will be zero there. Thus, at the top of the hill, all of the energy is potential energy.

At other points along the hill, the potential energy will be $E_p = mgh$ where $mg = 40 \text{ kg} \times 9.8 \text{ m/s}^2$ or 39.2 N. Thus, we can find the potential energy anywhere along the hill by multiplying 39.2 N by the height, h, of the hill.

We can find the potential energy at heights of 25 m, 15 m, 10 m, and 0 m as follows:

$$E_p = mgh$$

at $h = 25$ m $\quad E_p = 39.2 \text{ N} \times 25 \text{ m} = 9880 \text{ J}$

at $h = 15$ m $\quad E_p = 39.2 \text{ N} \times 15 \text{ m} = 5880 \text{ J}$

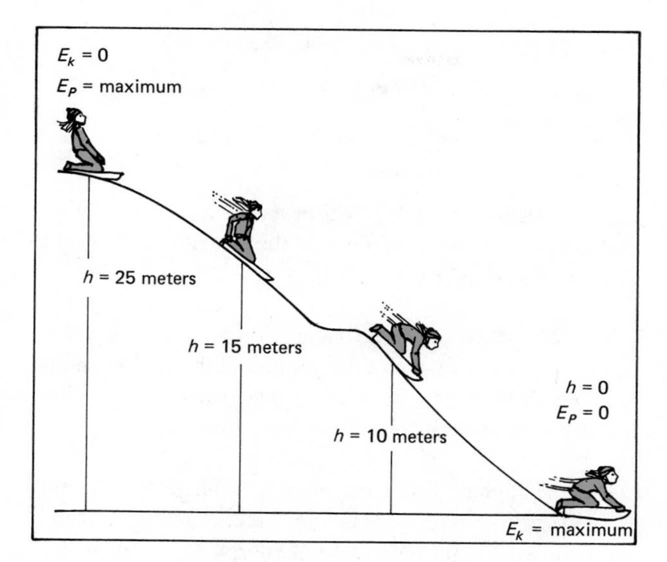

Figure 4.14 The law of conservation of energy. All the available energy at the top of the hill is potential because the velocity of the sled and rider is zero. As the sled and rider move down the hill, the potential energy decreases and the kinetic energy increases. At the bottom of the hill all the available energy is kinetic, the velocity is maximum, and the potential energy is zero.

$$\text{at } h = 10 \text{ m} \quad E_p = 39.2 \text{ N} \times 10 \text{ m} = 3920 \text{ J}$$

$$\text{at } h = 0 \text{ m} \quad E_p = 39.2 \text{ N} \times 0 \text{ m} = 0 \text{ J}$$

The kinetic energy, E_k, can be found by realizing that at every point we have $E_T = E_p + E_k$, where E_T is a constant which in this problem is 9800 J. Thus, the values for E_k can be found by subtracting E_p from 9800 J.

A summary of these results is shown in Table 4.2. At any point along the slope of the hill, the sum of the potential and kinetic energies is equal to the total energy. Observe at the top, where the velocity is zero, the kinetic energy is zero, but h is a maximum, and the potential energy is maximum and equal to the total energy of the system. As the sled (and Susannah) travels down the hill, the potential energy becomes less because the height is decreasing. As the sled descends and loses potential energy, it gains in velocity and increases in kinetic energy. At the bottom, where h is zero, all the energy of the system is in the form of kinetic energy, and the velocity of the sled is at its maximum value.

Table 4.2 Energy of Susannah's Sled

Height	E_p	E_k	E_T
25 m	9800 J	0 J	9800 J
15 m	5880 J	3920 J	9800 J
10 m	3920 J	5880 J	9800 J
0 m	0 J	9800 J	9800 J

Note that the shape of the hill makes no difference in these calculations (if we continue to assume that friction can be neglected). This fact makes the law of conservation of energy a very useful tool. For any system that starts out at rest and then changes its height by an amount h, we can write

$$(E_T)_{\text{top}} = (E_T)_{\text{bottom}}$$

or

$$\left(\frac{1}{2}mv^2 + mgh\right)_{\text{top}} = \left(\frac{1}{2}mv^2 + mgh\right)_{\text{bottom}}$$

but the velocity at the top is zero and the height at the bottom is zero, so our formula becomes

$$0 + mgh = \frac{1}{2}mv^2 + 0$$

or

$$\frac{1}{2}v^2 = gh$$

Solving for v gives

$$v = \sqrt{2gh} \qquad (4.5)$$

Remember in this formula that the symbol h stands for the *change* in the height.

The simple pendulum (Fig. 4.15) also illustrates many of the same features of the law of conservation of energy. A **simple pendulum** consists of a heavy mass (called a bob) attached to a light string and supported so that the bob can swing back and forth, or oscillate, freely. The pendulum is called "simple" because most of the swinging mass is in the bob. The string, for all practical purposes, can be neglected. An object such as a freely swinging meter stick attached at one end is called a **compound pendulum.** The mass in this system is not concentrated at a point but distributed throughout the oscillating system.

The simple pendulum, as shown in Fig. 4.15, can be set in motion by displacing the bob to one side (angle θ) and releasing it. The bob will swing back and forth with periodic motion, undergoing energy transformations.

When the bob is at position a or c, the velocity will be momentarily zero, and the kinetic energy will be zero. The potential energy, however, will be maximum, because the mass is at its maximum height above the lowest part of the swing, or reference level.

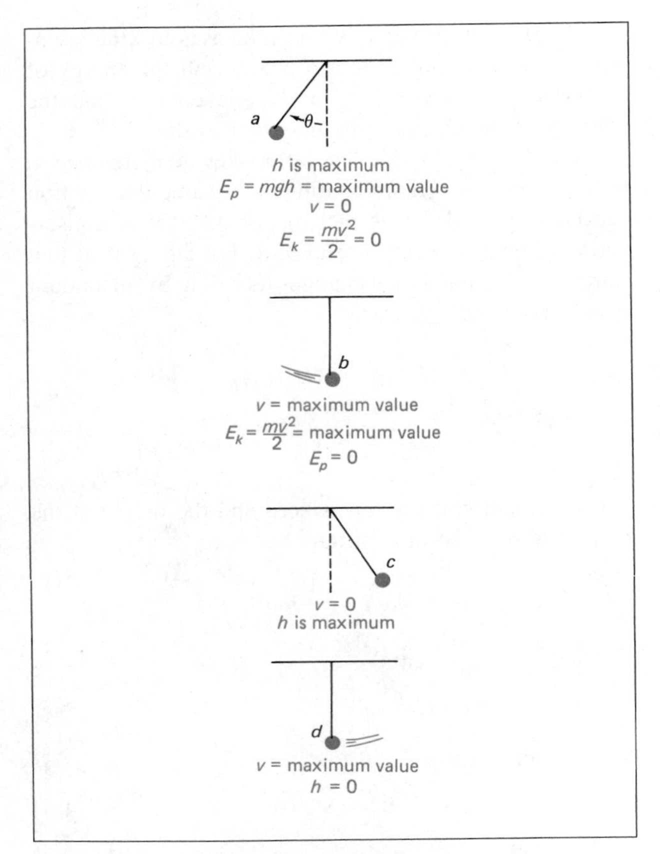

Figure 4.15 A simple pendulum. The pendulum swinging back and forth illustrates energy transformation.

When the bob swings down from position a, the potential energy will decrease and the kinetic energy will increase correspondingly as the mass gains velocity. When the bob is at point b, it will have zero potential energy, because h will have become zero. The kinetic energy at this point will be maximum and the mass will be traveling at its maximum velocity.

As the bob swings upward toward point c, the velocity will start to decrease, and the potential energy will begin to increase. As the bob swings higher, the velocity will continue to decrease until it stops momentarily at point c, where the kinetic energy will have decreased to zero and the potential energy will have reached the maximum. Thus, as the pendulum swings back and forth, the energy is converted from potential to kinetic, back to potential.

The velocity of the bob at point b can be calculated by applying the law of conservation of energy as follows:

At points a or c, $E_T = E_p$ because $E_k = 0$

At points b or d, $E_T = E_k$ because $E_p = 0$

Therefore, E_k at the bottom of the swing equals E_p at the top of the swing:

$$(E_k)_{\text{bottom}} = (E_p)_{\text{top}}$$

and the derivation of Eq. 4.5 proceeds as before. Thus, the velocity of the pendulum at the bottom of its swing is $v = \sqrt{2gh}$ as given by Eq. 4.5.

EXAMPLE 5 A child swings on a rope that is 4 m long. She is 1.5 m above the ground at the highest point and 0.5 m above the ground at the lowest point. What are her velocities at the highest and lowest points?

At the highest point the energy is all potential, and the velocity is zero because the kinetic energy is zero. At the lowest point the potential energy, $E_p = mgh$, has been converted into kinetic energy, and the velocity is a maximum and is given by Eq. 4.5. In this equation we must remember that h stands for the *change* in height which in this problem is 1.5 m − 0.5 m, or 1.0 m.

The child's maximum velocity is then

$$v = \sqrt{2gh}$$

$$v = \sqrt{2 \times (9.8 \text{ m/s}^2) \times 1.0 \text{ m}}$$

$$v = 4.43 \text{ m/s}$$

This speed corresponds to about 10 mi/h.

4.5 Forms of Energy

If we take the law of conservation of energy seriously, we must consider the examples of Section 4.4 in more detail. For instance, we know that both the sled and the pendulum will finally stop, and we must ask the question, Where did the energy go?

Once again, friction is involved. In most practical situations the kinetic or potential energy of large objects eventually ends up as heat energy. Heat energy will be studied at some length in Chapter 5, but for now, let us say that heat energy is kinetic and potential energy on a microscopic level. As things get hot, the energies of the atoms and molecules that make up different substances increase.

There are many forms of energy. Besides the macroscopic kinetic and potential energy of objects, there are the following forms of energy: heat, radiant, sound, chemical, nuclear, and others. All of these are related to microscopic kinetic and potential energies that act together to give macroscopic effects. In addition we will see in Chapter 9 that mass itself is a form of energy.

The main unifying concept when considering energy is the law of conservation of energy. We cannot create or destroy the energy of a single system. However, we can transform it from one form to another.

On a global scale, the source of practically all of our energy is the Sun. Wood, coal, petroleum, and natural gas all come about because of the Sun's radiant energy, which strikes the Earth. This radiant energy provides us with an external source of energy that should be with us for billions of years. The Earth's energy balance is discussed in more detail in Chapter 22.

As we go about our daily lives, each of us is constantly using energy and giving off energy in the form of body heat. The source of this energy is food (Fig. 4.16). An average adult radiates heat energy at about the same rate as a 100-watt light bulb. This explains why a crowded room soon gets hot. In winter, our clothing helps to keep this heat energy within our bodies. In summer, sweating helps us to give off the heat energy that our body produces.

Figure 4.16 We get our energy fuel from food. The body uses this energy to perform our daily tasks and to maintain our body temperature. Each of us radiates heat energy at about the same rate as a 100-watt light bulb. (Photo courtesy Jimmy Adams.)

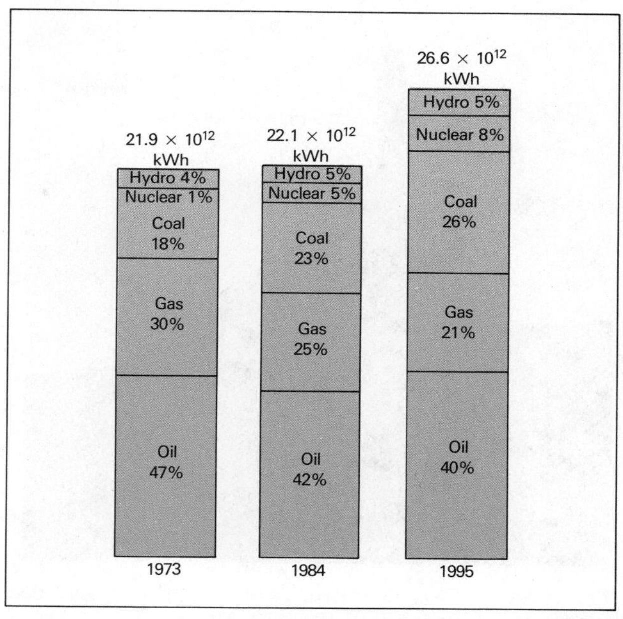

Figure 4.17 Sources of energy in the United States in 1973 and 1984 and projected sources in 1995. (Data from Energy Information Administration's annual energy review and outlook, 1984.)

The commercial sources of our energy on a national scale are mainly coal, oil (petroleum), and natural gas.* Figure 4.17 shows the amounts for each of these sources in 1973 and 1984. In addition, government projections for the year 1995 are given. Nuclear and hydroelectric energy are the only other significant commercial sources of energy in the United States.

The total energy supply has been projected by the federal government to increase slowly from 1984 to 1995, with a lower dependence on oil and natural gas and an increase in the supply of energy from coal and nuclear power plants.

Coal is used primarily for the generation of electricity. Figure 4.18 shows a stockpile of coal for use at a power plant. For over 50 years coal has been an important energy source in the United States, and in the 1970s additional large supplies were found in the western half of the country. Energy companies are depleting these abundant supplies, and coal is expected to increase in importance for at least the next 10 years.

Unfortunately, coal usage brings with it a variety of problems. Coal miners who work underground

*We are ignoring food.

Figure 4.18 Our chief source of electrical energy is coal. (Photo courtesy James Crouse.)

Figure 4.20 Natural gas is widely used in home furnaces. Space heating needs comprise about one-fifth of our national energy needs. (Photo courtesy James Crouse.)

Figure 4.19 Petroleum products such as gasoline make up 42% of our nation's energy supply. Transportation needs comprise about one-fourth of our national energy needs. (Photo courtesy James Crouse.)

frequently develop "black lung" disease; overloaded coal trucks can heavily damage secondary roads; and acid mine drainage, discussed in Chapter 31, pollutes lakes and streams. The biggest problem, however, comes from the effluents discharged from coal-fired power plants. Many scientists believe that acid rain is caused primarily by the sulfur dioxide that is emitted from coal-fired power plants. The increasing acidity of rain is causing major environmental problems (See Chapter 26).

Petroleum (oil) is our most widely used source of energy. It is used mainly in transportation (Fig. 4.19), but it is also widely employed in the residential, commercial, and industrial sectors of the economy. Petroleum is used not only as a fuel but also as the starting material for the petrochemical industry. Plastics and many other synthetic materials, which are discussed in Chapter 17, are made from petroleum.

A large fraction (30% in 1984) of our oil supply is imported each year. This is a cause of great concern to many, and it has resulted in governmental policies that seek to conserve gasoline, such as that concerning the increased fuel efficiency of domestic cars.

Natural gas is a cleanly burning fuel used in home furnaces (see Fig. 4.20) as well as in business and industry. The burning of natural gas results in a minimum of pollution problems, and it is vitally important to our nation's energy picture.

Recently, large deposits of natural gas have been found deep underground. Advances in drilling techniques have made it feasible to recover this gas. Experts are currently divided on the economics of recovery, but it is likely that in the near future such gas will be produced at a competitive cost. Thus, these reserves may prove to be the most important fuel source for the next century.

Natural gas could partially replace both coal and petroleum. It could be burned in place of coal to power our electric-generating stations. Natural gas in a liquid form can also be used in vehicles as a replacement for gasoline. The drawback here is that the range of a full tank is only about 150 mi compared to 300 mi for a car powered by gasoline. However, the technology for such vehicles is already in place and such cars are used in various parts of the world (e.g., British Columbia in Canada and Italy). The increased usage of natural gas may be the most significant energy resource change that occurs in the next century.

Hydroelectric energy also presents a minimum of pollution problems. Unfortunately, most of the feasible locations for hydroelectric power plants have already been utilized, so further exploitation of this resource is not practical.

Nuclear energy is used in nuclear power plants to produce electricity. Its proponents point to its lack of pollution problems, the abundance of uranium fuel, and the safety record of nuclear power plants. Its opponents point to problems of operation and of decommissioning old plants, the enormous capital costs, and radioactive waste disposal problems that many consider to be serious. Nuclear energy promises to remain controversial.

4.6 Energy Consumption

An energy flow diagram for U.S. energy supply and consumption in 1984 is shown in Fig. 4.21. The numbers shown are in units of quadrillion Btu, each of which is the same as 293 $\times$ 10^9 kWh or 1.05 $\times$ 10^{18} J of energy.

Figure 4.21 illustrates several points concerning energy supply and demand in the United States. First, as has previously been pointed out, a large amount of crude

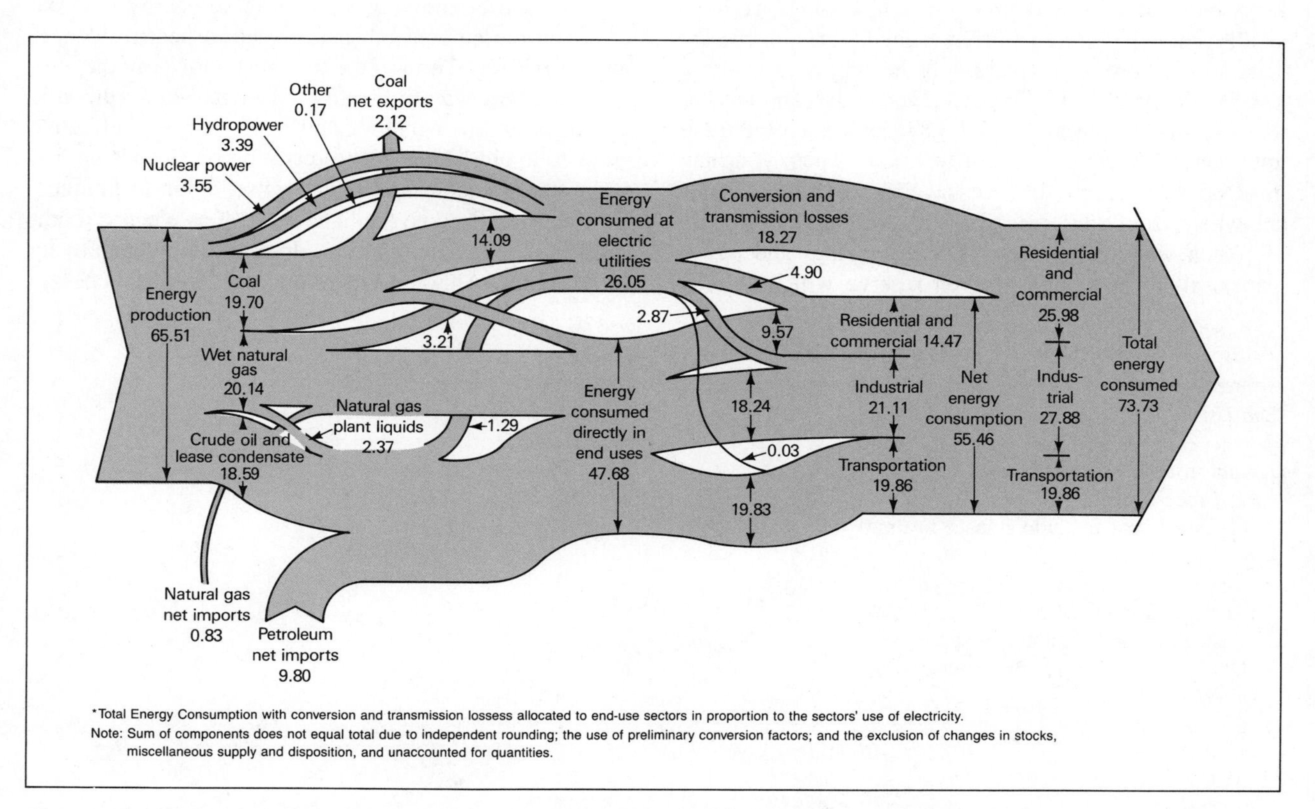

Figure 4.21 An energy flow diagram for the United States in 1984. All numbers shown are in quadrillion Btu, a unit equivalent to 293 $\times$ 10^9 kWh of energy. (From Energy Information Administration's annual energy review, 1985.)

oil is imported (shown in the lower left of Fig. 4.21). Second, note the large amount of energy consumed by electrical utilities; 70% (18.27 out of 26.05 quadrillion Btu) of this energy is lost as waste heat. That is, when a fuel such as coal is burned at an electricity-generating power station, 70% of the energy is lost and only 30% of the energy is available for use by appliances in the home or by business. Most of this loss is unavoidable and is due to a basic law of physics known as the second law of thermodynamics, which is discussed in Chapter 5.

How long will our energy reserves last? The answer to this question depends heavily on how quickly we use up a given resource. That is, it depends on the growth rate of our consumption of that resource.

As a graphic example, consider the following illustration. If we have a million years' supply of a given resource (e.g., aluminum), assuming the same usage of that resource every year, how many years will the supply last if instead we use 3% more each year? The answer is an astonishing 349 years!

A more practical example concerns coal. If we have a 400-year supply of coal in the United States, at current usage rates how many years will the supply last with a (a) 1% (b) 3% (c) 5% increase each year? The answer is 162 years at a 1% growth rate, 87 years at a 3% growth rate, and 63 years at a 5% growth rate. Thus, you may read or hear reports that we have in the United States anywhere from a 60-year to a 400-year supply of coal.

When you read statements in magazines and newspapers about how long a given reserve will last, read carefully to see what assumptions have been made. For instance, a 5% growth rate leads to a drastically different result from a 1% growth rate.

Energy is used in the United States for a variety of purposes. Table 4.3 gives a breakdown of the various sectors of our society that consume energy. Industry uses 37% of our energy. Fuel for transportation consumes 27%. (This is the greatest amount for any individual category.) Heating of our homes and businesses uses 18%, and other residential and commercial categories account for 18% of our energy consumption.

From Table 4.3 we can get some general ideas of how we can best cut down on energy consumption. In terms of individual habits, a decrease in the use of gasoline for transportation is one of the most effective ways of conserving energy. This can be done in many ways. Among them are using cars that are more fuel-efficient; keeping cars in tune; avoiding driving by a variety of methods such as car pooling, using mass transit, and bicycling; and driving at or less than 55 mi/h on the highway.

A second effective way to conserve energy is to cut down on the amount of energy used to heat residences and businesses. Turning the thermostat lower in the winter is the best way to do this. Also, insulating the attic and using storm windows and doors are very effective ways to avoid losing heat already produced. Using less hot water is another important conservation technique.

Many small home appliances, such as electric toothbrushes, telephones, and hair dryers, use comparatively little energy. Although you should never waste energy,

Table 4.3 Energy Requirements in the United States According to End Use

End Use	*% of National Energy Requirements*	
1. Industrial		37
a. Process steam	16	
b. Direct heat (including space heating)	11	
c. Electric motors	7	
d. Other	3	
2. Transportation (fuel only)		27
3. Residential and commercial space heating		18
4. Other residential and commercial uses		18
a. Water heating	4	
b. Air conditioning	2.5	
c. Refrigeration	2	
d. Lighting	1.5	
e. Other	8	
	Total	100

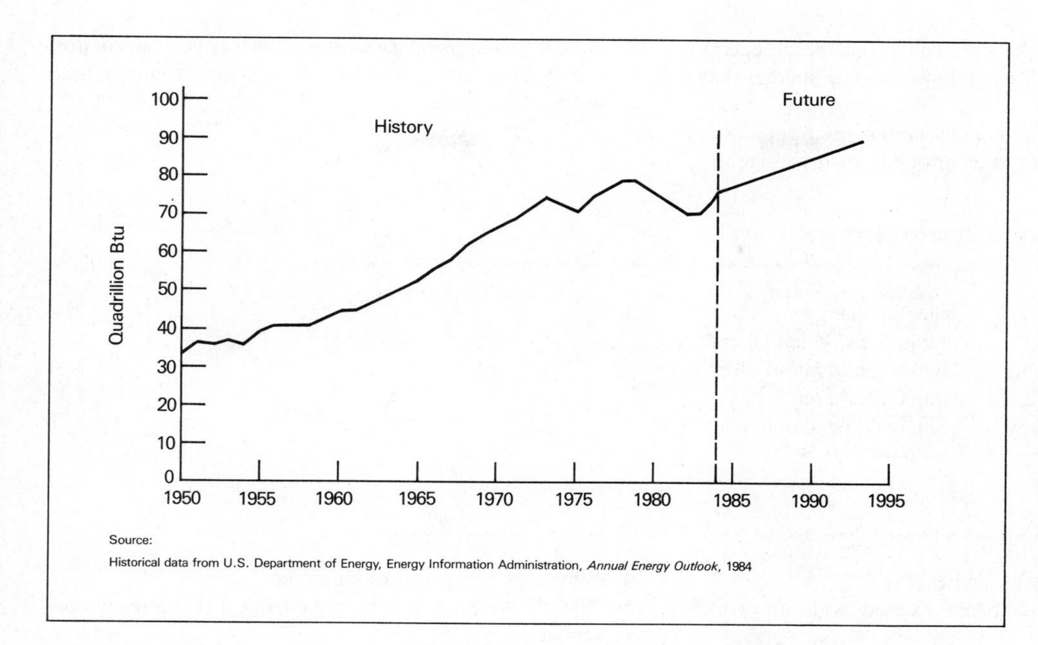

Figure 4.22 Trends in U.S. energy consumption. Historical curve is given for 1950–1984 and the forecast is given for the years 1985–1995. (From U.S. Energy Information Administration's annual energy outlook, 1984.)

you should be aware that using or not using these small appliances makes little impact, while gasoline and heating conservation measures can be important.

Past and predicted trends in U.S. energy consumption are shown in Fig. 4.22. In 1973 an oil embargo imposed by the major oil exporting countries served to do two things. First, it drastically raised the price of oil and consequently of all energy sources. Second, it caused people in the United States and all over the world to become suddenly conscious of the importance of energy conservation. In fact, an immediate drop in energy consumption can be seen after 1973.

After 1973 total energy consumption fluctuated for 11 years around the 1973 level. The decreases in the amount of energy consumed during this period (1973–1984) were due to an 11% drop in per capita energy consumption and a 24% drop in the percentage of gross national product due to energy consumption.

This increased efficiency of our energy usage is good news to all of us. We all need energy to live comfortably, but it is important that we learn to control and reduce our energy usage.

The government's projected increase in our future energy consumption is based on a projected growth in our economy during the years shown in Fig. 4.22. It is difficult to make accurate predictions since so many assumptions, such as the price of world oil, need to be made.

Over the next 50 years, our present energy consumption patterns will slowly change. New technologies such as deep-drilling techniques and solar energy will develop. Assuming world peace and a sound economy, these new technological developments will enable us to have a ready supply of energy available to meet our global needs for many generations to come.

Learning Objectives

After reading and studying this chapter, you should be able to do the following without referring to the text:

1. Define the terms work, energy, and power and give their units in SI.
2. Explain why it is more tiring to walk up stairs than to walk on a level surface.
3. Explain why potential energy can be positive or negative.
4. Explain why 55 mi/h is so much safer than 65 mi/h.
5. State the law of conservation of energy, and explain why it is important.
6. List at least five different forms of energy.
7. Explain why a crowded room gets hot.
8. List the main sources of energy in the United States.

9. State the problems, limitations, and advantages of energy obtained from coal, oil, natural gas, nuclear sources, and hydroelectric sources.
10. State why it is important to conserve energy resources.
11. State why it is important to control our energy usage.
12. State the two principal end uses of energy you can control, and describe ways to reduce the amount of energy used in these categories.

Important Words and Terms

work	watt	potential energy
joule	horsepower	gravitational potential energy
erg	kilowatt-hour	law of conservation of energy
foot-pound	energy	simple pendulum
power	kinetic energy	compound pendulum

Questions

1. Distinguish between work and energy.
2. Can you have a force being exerted with no work? Explain.
3. Which of the following are scalar quantities: force, distance, work, energy, power?
4. Why do doctors sometimes tell recent heart attack victims not to climb stairs?
5. Do we pay for electric energy or power?
6. Is a kWh a unit of energy or power?
7. Approximately how much does it cost to leave a light on overnight?
8. Distinguish between kinetic energy and potential energy.
9. Can you have energy without motion? Explain.
10. Can you have negative amounts of energy? Explain.
11. Why is 55 mi/h so much safer than 65 mi/h?
12. When is the law of conservation of energy invalid?
13. Why do pole vaulters run so fast before jumping?
14. Why does a crowded room get hot?
15. Why do we need to conserve energy if it is already conserved?
16. List five different forms of energy other than kinetic or potential energy.
17. What are the three principal sources of energy in the United States?
18. Discuss the problems, limitations, and advantages of coal, oil, natural gas, nuclear, and hydroelectric as energy sources.
19. What are some of the principal uses of energy in the United States?
20. When and why did the rapid growth of energy consumption stop?
21. How long will our energy reserves last?
22. How can we best conserve our energy resources?

Problems

1. A 6-kg mass is at rest on a table top.
   (a) Calculate the work required to lift the mass 2 m.
   (b) Calculate the work that would be required to lift the mass 2 m if it were on the moon.
   Answer: (b) 19.6 J
2. A 6-kg mass is at rest on a table top. Calculate the work required to move the mass 3 m along the table if the force of friction is 24 N.
3. A 6-kg mass is at rest on a table top.
   (a) Assuming there is no friction, calculate the work required to give the mass a velocity of 5 m/s.
   (b) Assuming that the mass is on the moon and there is no friction, calculate the work required to give the mass a velocity of 5 m/s. *Answer:* (b) 75 J
4. A student weighing 600 N climbs a stairway (vertical height of 4 m) in 20 s.
   (a) How much work was done?
   (b) What was the power output during the 20 s?
5. What is the horsepower necessary to carry 200 lb of books to a height of 22 ft in 40 s? *Answer:* 0.2 hp
6. How many kilowatt-hours of energy are consumed by a 900-W hair dryer that is run for 20 min?
   Answer: 0.3 kWh
7. Calculate the kinetic energy of an automobile with a mass of 1200 kg that is traveling at 31.5 m/s (70.5 mi/h).
   Answer: 595,350 J
8. By what factor is the kinetic energy increased when the speed of a car is increased from 30 mi/h to 40 mi/h?

9. What is the potential energy of a child's swing with a mass of 3 kg that is stationary and 0.5 m above the ground? Explain.
10. What is the potential energy of a 3-kg mass that is in a well 3 m deep? Explain.
11. How much work is required to lift a 3-kg mass out of a hole that is 3 m deep? *Answer:* 88.2 J
12. For a comet orbiting the Sun, answer the following questions.
    (a) Where is its kinetic energy the greatest?
    (b) Where is its potential energy the greatest?
    (c) Where is its total energy the greatest?
13. A child is swinging on a rope that is 5 m long. At the lowest point the child is 0.5 m above the ground. At the highest point the child is 1.0 m above the ground. What is the child's speed:
    (a) At the highest point?
    (b) At the lowest point? *Answer:* (b) 3.1 m/s
14. A sled and rider with a combined weight of 700 N are at the top of a hill that is 5 m high.
    (a) What is their total energy at the top of the hill (use the bottom of the hill as the reference level)? *Answer:* (a) 3500 J
    (b) Assuming there is no friction, what is the total energy halfway down the hill?
    (c) They slide down the hill and eventually slow down and stop due to friction. What happens to the energy?
15. How many watts are given off in a crowded room with three 100-W light bulbs and 30 people?

5

Heat

Then cold and hot and moist and dry,
In order to their stations leap.
—John Dryden

THE HEATING EFFECTS produced by fire, the sensation received when a piece of ice is held, and the warmth produced when hands are rubbed together are well known. Explaining what takes place in each of these cases, however, is not easy. Both heat and temperature are commonly used when referring to hotness or coldness. However, heat and temperature are not the same thing. They have different and distinct meanings as we shall see.

Count Rumford (1753–1814), an American (born Benjamin Thompson), was the first to recognize the relation between mechanical work and heat. In 1789 he published the results of an experiment in which he used a blunt boring tool to drill a cannon barrel immersed in water. The temperature of the cannon and water rose to the boiling point of water in two and one-half hours. This convinced Rumford that large quantities of heat were produced by friction. He concluded from this that heat was a form of energy and appeared to be due to the motion of the drill.

Later, James Prescott Joule (1818–1878), an Englishman, determined the quantitative relationship between mechanical energy and heat. He also established the important law of conservation of energy and originated many of the basic ideas for the kinetic theory of gases.

The concepts of heat and temperature play an important part in our daily lives. We like our coffee hot and our ice cream cold. The temperature of our living and working quarters must be carefully adjusted to our bodies' heat demands. The daily temperature reading is probably the most important datum given in a weather report. How cold or how hot it will be affects the clothes we wear and the plans we make.

◄ Heat is vital in the production of steel. Here molten steel is poured into ingot molds.

We will learn in detail in Chapter 22 how the Sun provides heat to our Earth. The heat balance between various parts of the Earth and its atmosphere gives rise to wind, rain, and other weather phenomena. The thermal pollution of rivers caused by hot water from nuclear power plants is a recent cause for concern because of its effect on the ecology of these rivers.

On a more cosmic scale, the temperature of various stars gives clues to their ages and the origin of the universe. What is temperature? What is heat? What causes heat? How is heat transferred? The answers to these questions will be examined in this chapter. Knowing these answers, we can explain many phenomena occurring around us.

5.1 Temperature

You may already be aware that a **molecule** is an uncharged particle of an element or compound. The molecules of all substances are in constant motion. **Temperature** is a measure of the average kinetic energy of these molecules.

Sometimes temperature is thought of as measuring hotness or coldness. However, temperature is not necessarily a measure of heat. A simple experiment will illustrate this: Place ice cubes and water in a pan and heat. As heat is applied the ice melts. However, when the temperature of the ice and water mixture is checked with a thermometer, we find the temperature remains the same. Even with continued heating (and stirring) there is no change in temperature *until all the ice is melted.* Obviously, heat was added, but it didn't change the temperature. We see, then, that temperature does not necessarily measure heat.

The quantitative measurement of temperature is accomplished through the use of a thermometer. A **thermometer** is a precise and reliable instrument that utilizes the physical properties of materials for the purpose of accurately determining temperature.

The most common thermometer is the mercury-in-glass type. This instrument consists of a thin-walled glass bulb attached to a slender capillary glass tube, both containing mercury.* The upper end of the glass tube is sealed off; and, in most cases, the air above the mercury is removed. A scale is attached or etched on the glass for measuring the expansion of the mercury. When heat is applied, the volume of the mercury increases and more of the slender tube is filled. The glass also expands, but the mercury expands much more; thus, the instrument utilizes the difference between the two expansions to measure the temperature.

After a thermometer is constructed, it is calibrated with a standard to ensure proper temperature readings. Figure 5.1 is a drawing of a clinical thermometer used for measuring body temperature.

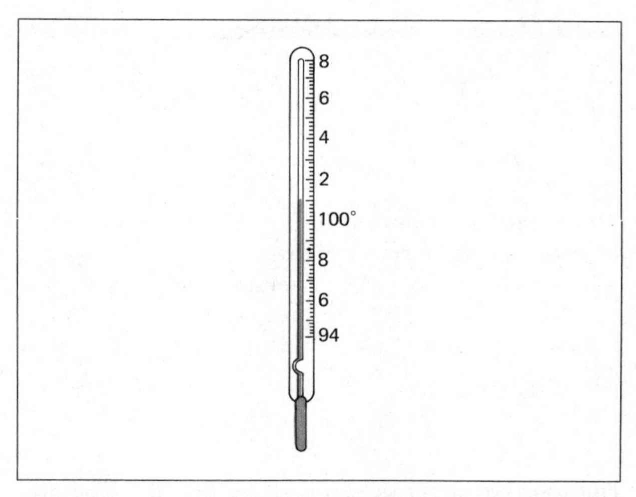

Figure 5.1 Drawing of a clinical thermometer. The reading on the thermometer is 101°F, indicating the patient has a fever.

In this type of thermometer a constriction is made in the capillary tube just above the glass bulb. When the thermometer is removed from the body, the mercury starts to contract but breaks at the constriction, leaving the mercury in the capillary tube. This serves the purpose of holding the reading of the thermometer at its highest value and allowing time for reading the thermometer. The mercury must be forced back down by shaking the thermometer.

The construction of a temperature scale requires two reference points and the choice of a unit. The reference points chosen are the ice point and the steam point. The **ice point** is the temperature of a mixture of ice and air-saturated water at normal atmospheric pressure. The **steam point** is the temperature at which pure water, at normal atmospheric pressure, boils.

Two common temperature scales are the Fahrenheit and Celsius scales. The **Fahrenheit scale** is now based on an ice point of 32° (read 32 degrees) and a steam point of 212°. The difference between the ice point and steam point is evenly divided into 180 units. Each unit is called a degree. Thus, a **degree Fahrenheit,** abbreviated °F, is $\frac{1}{180}$ of the temperature change between the ice point and the steam point.

The **Celsius scale** is based on an ice point of 0° and a steam point of 100°. There are 100 equal units or divisions between these points. So, a **degree Celsius,** abbreviated °C, is $\frac{1}{100}$ of the temperature change between the ice point and the steam point.

Have you ever wondered what doubling the temperature means? On both the Fahrenheit and Celsius scales, this is a confusing question. For instance, what is double zero degrees, or what is double $-10°$? These questions are meaningless unless there is an absolute limit on how cold an object can become. There is such a limit. It occurs at about $-273°C$, or $-460°F$.

The **Kelvin scale** is based on this limit, using zero as the lowest temperature possible, which is $-273°C$. This scale is sometimes called the absolute scale. A **kelvin,** abbreviated K, is the same as a degree Celsius, but doubling the temperature has meaning only on the Kelvin scale. The three temperature scales are shown in Fig. 5.2.

It is easy to convert from the Celsius scale to the Kelvin scale. Simply add 273 to the Celsius temperature and you have the Kelvin temperature. Mathematically,

$$K = C + 273$$

where K = temperature on the Kelvin scale
C = temperature on the Celsius scale

As examples, zero degrees Celsius equals 273 K, and a Celsius temperature of 34°C equals 307 K. It is easy to see that the ice point and the steam point are at 273 K and 373 K. Converting from Fahrenheit to Celsius or vice versa is not so easy. The equations used are

*Some home thermometers contain an alcohol, which often is colored red.

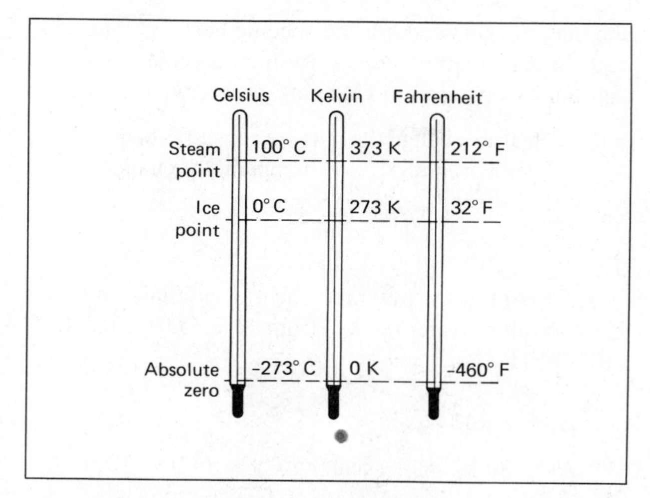

Figure 5.2 Three temperature scales. The Celsius, Kelvin, and Fahrenheit scales are shown.

$$C = \frac{5}{9}(F - 32)$$

and

$$F = \frac{9}{5}C + 32$$

where C = temperature on the Celsius scale
F = temperature on the Fahrenheit scale

As examples, 59°F equals 15°C and 20°C equals 68°F. Of course, F or C or both can be negative numbers.

5.2 Heat

Heat is a form of energy. So far we have studied two kinds of energy—kinetic energy and potential energy. In a way, heat is a third kind of energy. However, if we were to examine the microscopic sources of heat energy, we would discover that the kinetic and potential energy of molecules are the ultimate sources of heat. Thus, heat can be viewed as either another form of energy, or as the manifestation of molecular kinetic and potential energy on a macroscopic scale. In any event, heat is energy.

When we do work, there is frequently a substantial amount of heating caused by friction. The conservation of energy tells us that the work done plus the heat energy produced must equal the original amount of energy available. That is,

Total energy = work done + heat produced (usually by friction)

or

$$E_T = W + H$$

EXAMPLE 1 A sled and the child on it, having a total mass of 20 kg, slide down a hill that is 4 m high, starting from rest. Their final velocity is 8 m/s. How much heat was produced by the friction between the sled runners and the snow?

The total energy initially is potential energy, mgh. The work done against inertia is $\frac{1}{2}mv^2$, where v is the final velocity. The amount of heat produced is then given by

$$H = E_T - W = mgh - \tfrac{1}{2}mv^2$$

$$H = (20 \text{ kg} \times 9.8 \text{ m/s}^2 \times 4 \text{ m}) - \left(\frac{1}{2} \times 20 \text{ kg} \times (8 \text{ m/s})^2\right)$$

$$H = 144 \text{ J}$$

A traditional unit for measuring heat energy is the calorie. A **calorie** (cal) is defined as the amount of heat necessary to raise one gram of pure liquid water by one degree Celsius at normal atmospheric pressure.

In SI units

$$1 \text{ cal} = 4.186 \text{ J}$$

For historical reasons, heat measurements have been made in calories rather than joules. We are now in a transition period going from calories to joules, so both will be used in this book.

The calorie that we have defined is not the same as the one used when discussing diets and nutrition. A kilocalorie is 1000 calories as we have defined it. That is, a **kilocalorie** is the amount of heat necessary to raise the temperature of one kilogram of water by 1°C.

A diet Calorie is equal to one kilocalorie and should be written with a capital C to avoid confusion. That is,

1 food Calorie = 1000 calories

1 food Calorie = 4186 joules

A food Calorie indicates the energy released when the food is consumed in the body.

The unit of heat in the British system of units is the British thermal unit, or Btu. One **Btu** is the amount of heat required to raise one pound of water one degree Fahrenheit at normal atmospheric pressure.

Table 5.1 Relationships among some Common Energy Units

1 cal = 4.186 J
1 kcal = 4186 J = 3.97 Btu = 0.00116 kWh
1 Btu = 1055 J = 0.25 kcal = 0.00029 kWh
1 kWh = 3.6×10^6 J = 862 kcal = 3422 Btu

Some relations between Btu's, Calories, joules, and kWh are given in Table 5.1.

One very well known effect of heating a material is that of expansion. Almost all matter—solids, liquids and gases—expands when heated. As a general rule, a substance expands when heated and contracts when cooled. The most important exception to this rule is water. If water is frozen, it expands. That is, ice at 0°C occupies a larger volume than the same mass of water at 0°C.

The change in length or volume of a substance due to a change in temperature is a major factor in the design and construction of items ranging from steel bridges and automobiles to watches and dental cements. The cracks in a highway are designed so that in summer the concrete will not buckle as it expands due to the heat. Large cracks are designed into bridges for the same reason. Heat-expansion characteristics are used to control such things as the flow of water in car radiators and the flow of heat in homes through the operation of thermostats.

5.3 Specific Heat

One kilocalorie is the amount of heat necessary to raise the temperature of one kilogram of water one degree Celsius. Other substances require different amounts of heat to raise the temperature of one kilogram of the substance one degree. The **specific heat** of a substance is the amount of heat in kilocalories necessary to raise the temperature of one kilogram of the substance one degree Celsius.

The units of specific heat are kcal/kg°C or, in SI units, J/kg°C. The specific heat of pure water is 1.000 kcal/kg°C. This is 4186 J/kg°C. The specific heats of a few common substances are given in Table 5.2.

When you heat substances without changing phase, the temperature rises. The greater the specific heat of the material, the more heat that is required. Also, the greater the mass of the substance or the greater the temperature change desired, the more the heat that is required. Experiments show that the heat necessary to change the temperature of a mass is proportional to three factors: the mass (designated m), the specific heat (designated c) and the temperature change (written as ΔT). Thus, for substances that are not changing phase, we can write

$$\text{Heat to change temperature} = \text{mass} \times \text{specific heat} \times \text{temperature change}$$

or

$$H = mc\Delta T$$

EXAMPLE 2 How much heat does it take to heat a bathtub full of water (80 kg) from 12°C (53.6°F) to 42°C (107.6°F)?

$$H = mc\Delta T$$

$$H = 80 \text{ kg} \times 1 \text{ (kcal/kg°C)} \times (42 - 12)\text{°C}$$

$$H = 2400 \text{ kcal}$$

Each kilocalorie corresponds to 0.00116 kWh so this amount of heat is

$$H = 2400 \text{ kcal} \times .00116 \frac{\text{kWh}}{\text{kcal}} = 2.78 \text{ kWh}$$

At 8¢ per kWh the cost is 2.78 × 8¢ = 22¢. Four people each taking a bath a day for one month would cost 4 × 30 × 22¢ = $26.40.

The **heat capacity** of a substance is the amount of heat (expressed in kilocalories) required to raise the temperature of a substance one degree Celsius. Heat capacity is calculated by finding the product of the mass and the specific heat. In the preceding example, the mass was 80 kg and c was 1 kcal/kg°C, so the heat capacity of the water was 80 kcal/°C.

Table 5.2 Specific Heat

	Specific Heat (20°C)	
Substance	*kcal/kg°C*	*J/kg°C*
Aluminum	0.22	920
Copper	0.092	385
Glass	0.16	670
Human body (average)	0.83	3470
Ice	0.50	2100
Iron	0.105	440
Silver	0.056	230
Soil (average)	0.25	1050
Steam (at constant volume)	0.50	2100
Water	1.000	4186
Wood	0.4	1700

Substances found in our environment are usually classified as being in either the solid, liquid, or gaseous phase. Refer to Fig. 5.3. We observe how the temperature of the different phases changes as heat is applied.

In the lower left-hand corner the substance is represented in the solid phase. As heat is applied the temperature rises. When point A is reached, adding more heat does not change the temperature. Instead, the heat energy is used to change the solid into a liquid. The amount of heat necessary to change one kilogram of a solid into a liquid at the same temperature is called the **latent heat of fusion** of the substance. In Fig. 5.3 this is simply the amount of heat necessary to go from A to B.

When point B has been reached, the substance is all liquid. The temperature of the substance at which this change from solid to liquid takes place is known as the **melting point.** After point B, further heating again causes a rise in temperature. The temperature continues to rise as heat is added until point C is reached.

From B to C the substance is in the liquid phase. When point C is reached, adding more heat does not change the temperature. The added heat is now changing the liquid into a gas. The amount of heat necessary to change one kilogram of a liquid into a gas is called the **latent heat of vaporization** of the substance. In Fig. 5.3 this is simply the amount of heat necessary to go from C to D.

When point D has been reached, the substance is all in the gas phase. The temperature of the substance at which this change from liquid to gas phase occurs is known as the **boiling point.** After point D is reached, further heating again causes a rise in the temperature.

From Fig. 5.3 we get an idea of the differences between temperature and heat. We see that the temperature of a body rises as heat is added only if the substance is not undergoing a change in phase.

When heat is added to ice, the ice melts without changing its temperature. The more ice there is, the more the heat needed to melt it. In general, the heat needed to change a solid into a liquid is just the mass of the substance times its heat of fusion. So we can write

Heat needed to melt a substance = mass × latent heat of fusion

or

$$H = mH_f$$

Similarly, if we want to change a liquid into a gas at the boiling point, the amount of heat needed can be written as

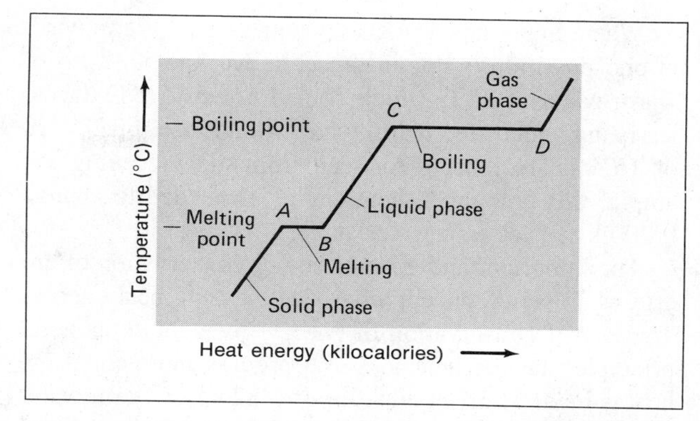

Figure 5.3 Graph of temperature versus heat energy for a typical substance.

Heat needed to boil a substance = mass × latent heat of vaporization

or

$$H = mH_v$$

For water, the latent heats are

$$H_f = 80^* \text{ kcal/kg} = 333{,}000 \text{ J/kg}$$

$$H_v = 540 \text{ kcal/kg} = 2.26 \times 10^6 \text{ J/kg}$$

These are very large numbers compared to the specific heat. It takes 80 times more heat energy to melt 1 kg of ice than to raise its temperature by 1°C. Similarly, boiling 1 kg of water takes 540 times as much energy as raising its temperature by 1°C.

An everyday example of the use of these formulas occurs in the melting of ice.

EXAMPLE 3 Calculate the amount of heat necessary to change 0.200 kg of ice at 0°C into water at 2°C.

The total heat necessary is found in two steps. The ice melts at 0°C, and then warms up from 0°C to 2°C.

$$H = H_{\text{melt ice}} + H_{\text{change } T}$$

$$H = mH_f + mc\Delta T$$

$$H = 0.200 \text{ kg} \times 80 \frac{\text{kcal}}{\text{kg}}$$

$$+ 0.200 \text{ kg} \times 1 \frac{\text{kcal}}{\text{kg°C}} \times (2 - 0)\text{°C}$$

$$H = 16.4 \text{ kcal}$$

---

*H_f = 79.6 kcal/kg to three significant figures, but we will round off to 80 kcal/kg.

Where might this heat energy come from? Warm water is one possibility. For instance, if we had 0.82 kg of warm water at 22°C which cooled down to 2°C due to ice being added, the warm water would lose 0.82 × 20 or 16.4 kcal of heat. As seen from the preceding example, this amount of heat energy is needed to change 0.200 kg of ice into water at 2°C.

The important and general law of conservation of energy is a useful concept when dealing with heat energy. The law of conservation of energy gives us the general principle that the heat lost in a process must equal the heat gained. Thus, in our case of 0.82 kg of warm water at 22°C and 0.200 kg of ice at 0°C, we get

$$\text{Heat lost by warm water} = \text{Heat gained by ice}$$

$$16.4 \text{ kcal} = 16.4 \text{ kcal}$$

Another example of the use of this law is the determination of the specific heat of an unknown substance. If a hot metal pellet is dropped into cool water in a glass we can say, assuming that no heat is lost to the surrounding air,

$$\text{Heat lost by metal} = \text{Heat gained by water} + \text{Heat gained by glass}$$

If the masses and the initial and final temperatures are all known, the specific heat of the metal can be found.

EXAMPLE 4 A 0.045-kg aluminum pellet is heated to 98.5°C and dropped into an 0.080-kg glass beaker ($c = 0.20$ kcal/kg°C) containing 0.200 kg of water at 20°C. The final maximum equilibrium temperature is 23.5°C. Determine the specific heat of aluminum.

$$H_{\text{aluminum}} = H_{\text{water}} + H_{\text{glass}}$$

or $$(mc\,\Delta T)_{\text{aluminum}} = (mc\,\Delta T)_{\text{water}} + (mc\,\Delta T)_{\text{glass}}$$

$$0.045 \text{ kg} \times c \times (98.5°\text{C} - 23.5°\text{C})$$
$$= 0.200 \text{ kg} \times 1 \text{ kcal/kg} \times (23.5°\text{C} - 20.0°\text{C})$$
$$+ 0.080 \text{ kg} \times 0.2 \text{ kcal/kg} \times (23.5°\text{C} - 20.0°\text{C})$$

Since everything is in the proper units, we can now solve for c, the specific heat of aluminum. Its units will be kcal/kg°C. Solving yields

$$c = 0.22 \text{ kcal/kg°C}$$

5.4 Thermodynamics

Thermodynamics means the dynamics of heat, that is, the flow, production, and conversion of heat for use by people. We use heat energy for most of the work done in our environment. The operation of refrigerators and other heat engines is based upon the laws of thermodynamics. These laws are important because they give relations between heat energy and energy associated with mechanical work.

When a balloon is heated, the air temperature inside the balloon increases, and the balloon expands. The increase in air temperature is due to an increase in the internal energy of the air inside the balloon. The air also does work when it expands the balloon.

The preceding example illustrates the first law of thermodynamics, which is based on the law of conservation of energy. The **first law of thermodynamics** states that heat added to a closed system can change the internal energy of the system and/or do work. It can be stated as

$$\text{Heat added to a system} = \text{internal energy change of the system} + \text{work done by system}$$

or

$$H = \Delta E_i + W \tag{5.1}$$

A process in which H remains constant, that is, no heat is added or removed from the system, is known as an **adiabatic process.** In the above equation, if $H = 0$, then $\Delta E_i = -W$. This means that a system that is doing work will have a decrease in its internal energy and the temperature of the system decreases.

The **second law of thermodynamics** can be stated in several ways. One of these is the following:

> **It is impossible for heat to flow spontaneously from an object having a lower temperature to one having a higher temperature.**

This law is well known to all of us. If a cold object and a hot object are placed in contact, the hot object cools down and the cold object heats up (Fig. 5.4). The reverse never happens, even though there is nothing in the first law of thermodynamics to indicate that it could not happen. For example, a cold drink always warms up to room temperature. It doesn't get colder while the room gets hotter! In all of our experience the second law seems to hold.

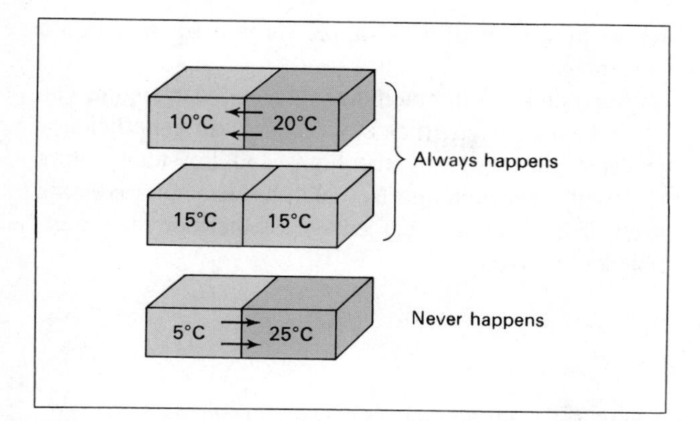

Figure 5.4 Heat always flows from a substance with a higher temperature to one with a lower temperature. A cold object never gets colder when placed in contact with a warmer object.

A stone on a hillside will of its own accord roll down the hill to a level of lower potential energy. If we want the stone to roll uphill, we must apply a force and do work on the stone to raise it to a higher potential level. The stone will not, of its own accord, roll up hill.

Heat will flow of its own accord from a higher temperature to a lower temperature, but work must be done to transfer heat energy from a low-temperature reservoir to a high-temperature reservoir. A device used to perform this task is called a **heat pump.** (See Figure 5.5)

Common heat pumps are the household refrigerator and air conditioner. Heat is transferred from inside the volume of the refrigerator to the outside by the use of electrical energy. The heat transferred, however, is much greater than the electrical energy needed to perform the work.

A heat pump for the home is used to extract heat from the ground or from the air outside in order to heat the air inside the home. The heat pump is used more extensively in the South where the climate is milder. When installed in a home where the winter months are very cold, an auxiliary heating unit (usually an electric heater) must be used in conjunction with the heat pump to supply extra heat energy when needed. Heat pump installation in a home is usually more expensive than a regular furnace.

A **heat engine** is a device that uses heat to do useful work. Heat engines operate between a high temperature, such as the combustion temperature, and a low temperature, such as the temperature of the air. A common example is a car engine.

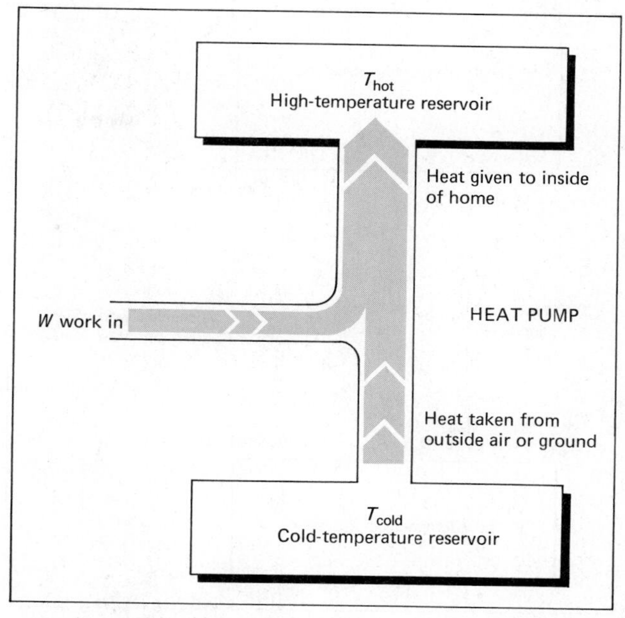

Figure 5.5 Schematic diagram of a heat pump. The substance used for the heat exchange is usually Freon which is pumped through the cold and hot reservoirs.

Normally, a heat engine operates in a cycle. During the cycle, the heat engine takes heat energy from the high-temperature reservoir, does useful work, and then gives off heat energy to the lower-temperature sink. See Figure 5.6. The laws of thermodynamics tell us how much useful work can be done in an ideal heat engine cycle.

In a cyclical heat engine, the system comes back to its original state. Thus, the temperature and internal energy of the system are unchanged. Since $\Delta E_i = 0$, the first law of thermodynamics (Eq. 5.1) then becomes

$$\Delta H = \Delta E_i + W$$

$$\Delta H = W$$

Since the net heat added to the system is the heat coming in from the high-temperature reservoir (H_{hot}) minus the heat rejected to the low-temperature reservoir (H_{cold}), we can write this as

$$H_{hot} - H_{cold} = W$$

or

$$H_{hot} = H_{cold} + W \qquad (5.2)$$

A cyclic heat engine is schematically diagrammed in Fig. 5.6.

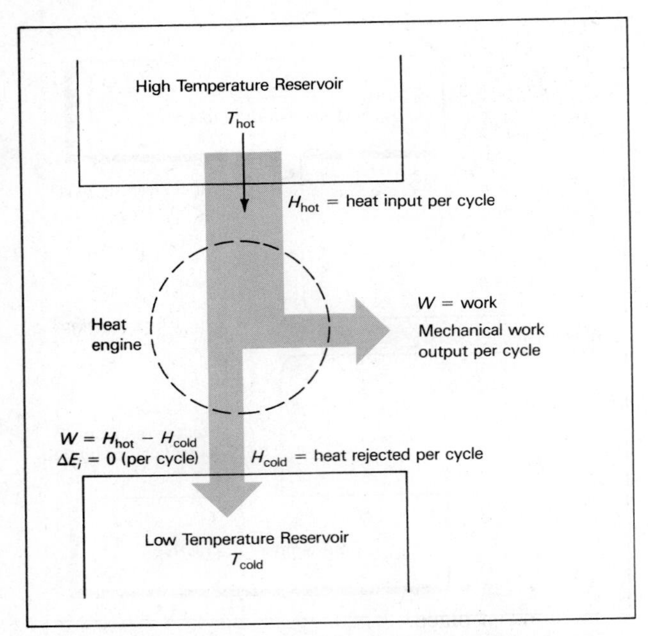

Figure 5.6 A cyclic heat engine. A heat engine takes heat from a high-temperature reservoir, converts some of it to useful work, and rejects the remainder to a low-temperature reservoir.

The second law tells us something about this last equation. Another equivalent statement of the second law of thermodynamics is

No heat engine operating in a cycle can convert its heat input entirely into work.

That is, in Eq. 5.2, H_{cold} is always greater than zero.

At electrical energy stations powered by coal or nuclear energy, giant steam engines are producing useful work in the form of electrical energy. The heat from the burning of coal or nuclear reactions produces a high-temperature reservoir. The low-temperature reservoir is the air, or more often, water in cooling towers that then give off heat into the air. The cooling towers are necessary because of the large amounts of heat given off by the steam engine.

The maximum efficiency that a heat engine can possess is also governed by the laws of thermodynamics. The **efficiency** just tells what percentage of the heat added from the high-temperature reservoir goes into useful work. That is,

$$\%\ \text{efficiency} = \frac{\text{work}}{\text{heat added}} \times 100$$

or

$$\%\ \text{eff.} = \frac{W}{H_{hot}} \times 100$$

where the factor of 100 simply turns a fraction into a percentage.

According to thermodynamics, no heat engine can exceed a maximum efficiency. This maximum efficiency is determined only by the high- and low-temperature reservoirs and nothing else, such as what fuel is being used. The maximum percentage of thermodynamic efficiency is given as

$$\text{max. } \%\ \text{eff.} = \frac{T_h - T_c}{T_h} \times 100 \qquad (5.3)$$

where T_h = temperature of hot reservoir in K
T_c = temperature of cold reservoir in K

Since T_c can never be zero (third law), the efficiency must always be lower than 100%.

The actual efficiency of a heat engine will always be less than that given in Eq. 5.3 because of frictional effects and design considerations.

EXAMPLE 5 What is the maximum thermodynamic efficiency of a coal-fired steam power plant operating between 300° and the boiling point of 100°C?

We use Eq. 5.3 and convert the °C to K.

$$300°\text{C} = (273 + 300)\ \text{K} = 573\ \text{K}$$

$$100°\text{C} = (273 + 100)\ \text{K} = 373\ \text{K}$$

$$\text{max. } \%\ \text{eff.} = \frac{573 - 373}{573} \times 100$$

$$\text{max. } \%\ \text{eff.} = 35\%$$

When frictional effects are taken into account, an actual efficiency of 32% or so is typical at a power plant. Transmission heat losses then account for another percentage point or so loss in efficiency.

The second law can also be expressed in terms of entropy. **Entropy** is a measure of the disorder of a system. When heat is added to an object, its entropy increases because the added energy increases the disordered motion of the molecules. As a natural process takes place, the disorder increases. For example, when a solid melts, the molecules are freer to move in a random motion in the liquid phase than in the solid phase. Likewise, when evaporation takes place, there is greater disorder and an increase in entropy.

In terms of entropy, the second law of thermodynamics can be stated:

The entropy of an isolated system never decreases.

Processes that are left to themselves tend to become more and more disordered, never the reverse. A heat engine is able to change some disorder (heat) into order but the engine cannot transform all of it. A student's dormitory room left to itself becomes disordered . . . never the reverse. True, the room can be cleaned and items put in order and the entropy seems to decrease. But to put things in order, energy from the food we eat must be changed into other forms of motion, thus increasing the total entropy of the universe.

The second law of thermodynamics has important implications. From the first formulation we see that if the law is universally valid for all time, then the universe—the stars and galaxies—will eventually cool down to a final common temperature, at which it will remain forever. This possible fate for the universe is sometimes referred to as its "heat death."

Another less cosmic implication of the second law is that perpetual motion machines are impossible. The fault with many perpetual motion machines is that their "inventors" assume that heat can be transformed completely into work with no increase in the machine's internal energy. This is in violation of the second law.

There is also a **third law of thermodynamics.** It states that a temperature of absolute zero can never be attained. However, temperatures as low as 0.000001 K (one-millionth kelvin unit) have been reached.

5.5 Phases of Matter

The three common phases of matter are the solid phase, the liquid phase, and the gaseous phase. All substances exist in all three phases of matter at some temperature and pressure. At normal room temperatures a substance will be in one of the three phases. For instance, at room temperature oxygen is a gas, water a liquid, and copper a solid.

The principal distinguishing features of solids, liquids, and gases can be understood if we look at the various phases from a molecular point of view. Most substances are made up of very small particles called molecules. As stated earlier, a molecule is the smallest division of a substance which still retains the properties of the substance. In a pure **solid,** the molecules are usually arranged in a specified three dimensional manner. This orderly arrangement of molecules is called a **lattice.** Figure 5.7 shows an example of a lattice in only two dimensions. The molecules (represented in the figure by small circles) are bound to each other by electric forces.

Upon heating, the molecules gain kinetic energy and vibrate about their positions in the lattice. The more heat that is added, the stronger the vibrations become, and the molecules move farther apart, and as shown diagrammatically by Fig. 5.7, the solid expands.

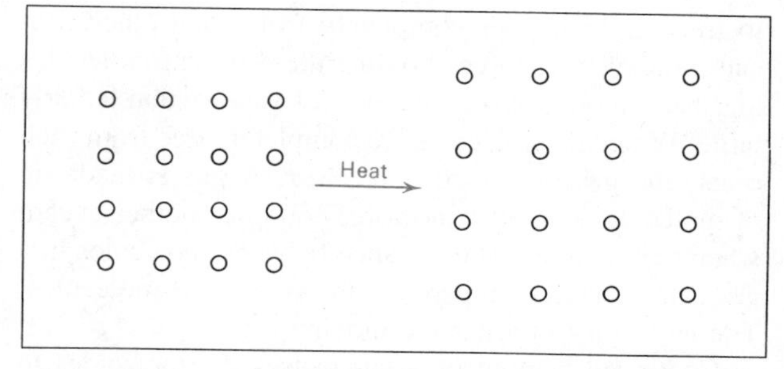

Figure 5.7 Schematic diagram of a typical crystal lattice in two dimensions. Heating causes the molecules to vibrate with greater amplitude about their position in the lattice, thus increasing the volume of the solid.

When the melting point of a solid is reached, the heat energy is great enough to break apart the bonds that hold a molecule in place. When this happens, a hole is produced in the lattice, and a nearby molecule can move to that hole. As more and more holes are produced, the lattice becomes significantly distorted.

Figure 5.8 shows an arrangement of the molecules in a liquid. There are many holes in the liquid, which has only a slight lattice structure. A molecule can easily move to a new spot in the lattice, since there are so many holes in the lattice. A **liquid** is an arrangement of molecules with many holes, so that the molecules are free to move and assume the shape of the container. Upon heating a liquid, the individual molecules gain kinetic energy, and even more holes are produced. This results in the expansion of the liquid.

When the boiling point is reached, the heat energy is sufficient to break the molecules completely apart from each other. The heat of vaporization is the heat necessary

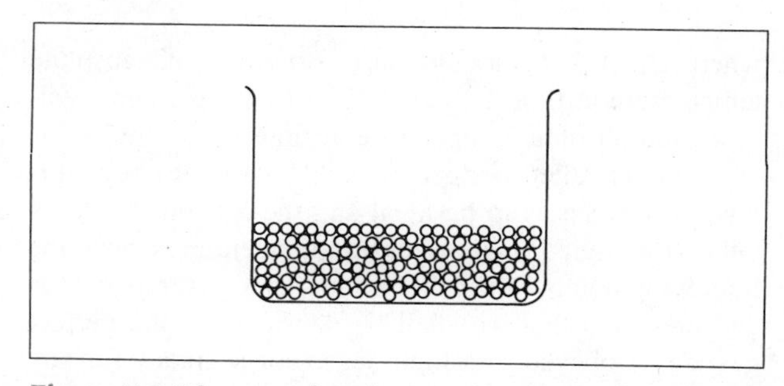

Figure 5.8 Diagram showing arrangement of molecules in a liquid. The small circles indicate molecules. They are packed closely together and form only a slight lattice structure.

to free the molecules completely from each other. Because the electric forces holding the different molecules together are quite strong, the heat of vaporization is fairly large. When the molecules are completely free from each other, the gaseous stage is reached. A **gas** is made up of molecules that exert no forces on one another except when they collide. The distance between molecules in a gas is quite large compared to the size of the molecules. The molecules in a gas are moving rapidly.

Continued heating of a gas causes the molecules to move faster and faster. Eventually, the molecules and atoms are ripped apart by collisions with one another. Inside hot stars, such as our Sun, atoms and molecules do not exist, and other phases of matter occur. Some of these will be discussed in Chapter 21.

We have seen that when a substance is in a single phase, heating increases the kinetic energy of the molecules of which it is composed. When a substance is changing phases, heating supplies the energy to overcome the attractive forces holding the different molecules together. Since the temperature rises only when the substance is not changing phase, we conclude that **temperature** can be defined as a measure of the average kinetic energy of the molecules.

5.6 The Gas Laws

The **pressure** of a gas is the force that the moving gas molecules exert upon a unit area. A common unit of pressure is newton per square meter, or N/m^2, called a pascal (Pa). The normal atmospheric pressure is 1.013×10^5 N/m^2. An important relationship exists among the pressure, volume, and temperature of a gas at any two times, which we shall call 1 and 2. This relation is known as the **ideal gas law** and can be stated as

$$\frac{P_1V_1}{T_1} = \frac{P_2V_2}{T_2}$$

where P_1, V_1, T_1 are pressure, volume, and absolute temperature at time 1, and P_2, V_2, T_2 are pressure, volume, and absolute temperature at time 2.

A gas is called perfect, or ideal, when it obeys this law. A gas ceases to be ideal and the law breaks down either for high pressures or for temperatures near the liquefying point of the gas. For the range of temperatures and pressures that we shall be considering, the perfect gas law is obeyed to a high degree of accuracy for any gas. It doesn't matter if the gas is hydrogen, oxygen, nitrogen, or a combination of gases, such as air.

If the volume of a gas is held constant and we plot the pressure against the temperature of a typical gas such as helium, we get a plot similar to that of Fig. 5.9. From the perfect gas law we see that if $V_1 = V_2$, then

$$\frac{P_1}{T_1} = \frac{P_2}{T_2}$$

or rearranged,

$$\frac{P_1}{P_2} = \frac{T_1}{T_2}$$

This equation means that an increase in temperature results in a proportional increase in pressure. Thus, a plot of pressure and temperature data produces the straight-line graph shown in Fig. 5.9. If we extrapolate the straight line to a pressure of zero (broken line on the graph), we see that the temperature is 0 K, or $-273°C$. This can be seen from the last equation. If $P_1 = 0$, then T_1 must also equal zero.

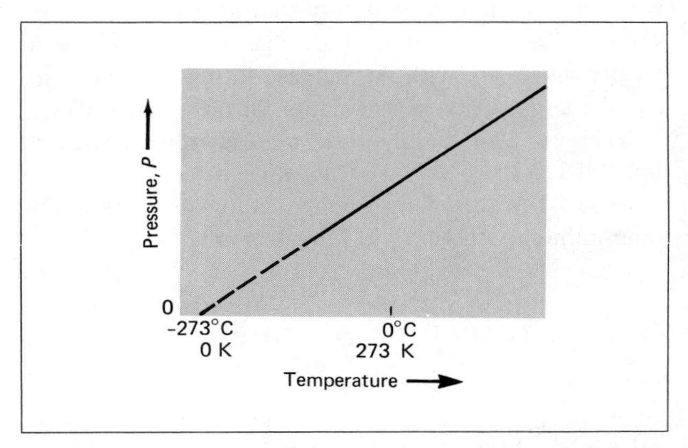

Figure 5.9 Graph of pressure versus temperature for a typical gas. The volume is constant.

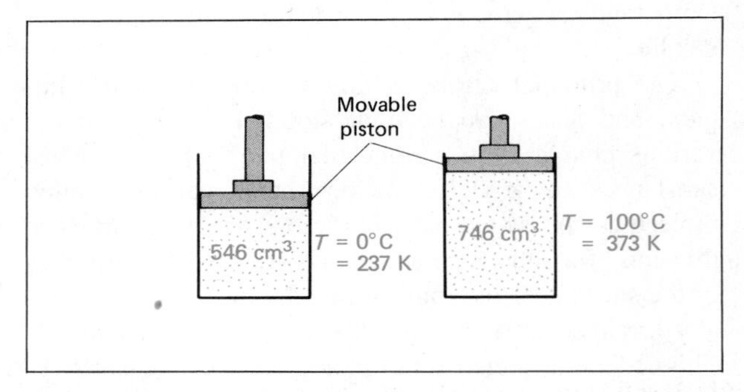

Figure 5.10 The relationship between volume and temperature at constant pressure. As the temperature increases, the volume increases in a direct proportion. This is known as Charles' law.

A similar relation exists between the volume and temperature of a gas. Consider a gas at 0°C (273 Kelvin, or absolute) confined in a tube with a movable piston as in Fig. 5.10. The piston serves to keep the gas at a constant pressure. As the gas is heated to 100°C (373 K), the piston is forced upward. If the gas occupied 546 cm^3 initially, it will expand to 746 cm^3, forcing the piston upward (Fig. 5.10).

Thus, we see that the volume of a gas at a constant pressure is in direct proportion to the absolute temperature. This proportionality is known as **Charles' law,** after the French physicist who discovered it in the eighteenth century. From the perfect gas law when $P_1 = P_2$, we get Charles' law, which can be written as

$$\frac{V_1}{T_1} = \frac{V_2}{T_2}$$

The relationship between pressure and volume, when the temperature is kept constant, was discovered by the British physicist Robert Boyle over three centuries ago. A process in which the temperature remains constant is called an **isothermal process.**

Suppose a column of a gas occupies 400 cm^3 when a 50-g mass is placed on top of it, as in Fig. 5.11 (left). If we now keep the temperature constant and double the pressure by placing another 50 g on top of the column, the volume of the gas is cut in half, as shown in Fig. 5.11 (right). This relationship is known as **Boyle's law** and can be written from the perfect gas law with $T_1 = T_2$ as

$$P_1V_1 = P_2V_2$$

The perfect gas law is a powerful tool for computing unknown pressures, volumes, or temperatures. Some examples of its use follow.

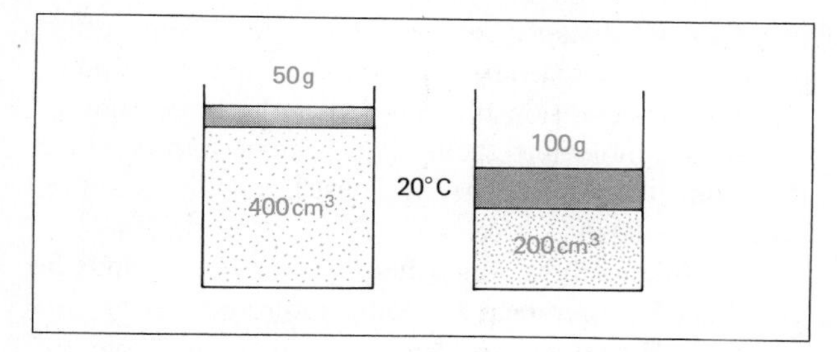

Figure 5.11 The relationship between pressure and volume at constant temperature. The volume is inversely proportional to the pressure. This is known as Boyle's law.

EXAMPLE 6 A steel oxygen tank on a space ship has a volume of 0.50 m^3. It is filled with oxygen at a pressure of 4×10^6 N/m^2. How big a volume would the oxygen occupy at a normal atmospheric pressure of 1.013×10^5 N/m^2 at the same temperature?

We use the perfect gas law, with $T_1 = T_2$

$$\frac{P_1V_1}{T_1} = \frac{P_2V_2}{T_1}$$

and solve for V_2.

$$V_2 = \frac{P_1V_1}{\cancel{T_1}}\frac{\cancel{T_1}}{P_2}$$

$$V_2 = \frac{4 \times 10^6(\text{N/m}^2) \times 0.50 \text{ m}^3}{1 \times 10^5(\text{N/m}^2)}$$

$$V_2 = 20 \text{ m}^3$$

Thus we see that by storing the oxygen at great pressure, we can significantly increase the supply of oxygen available to the astronauts.

EXAMPLE 7 A stratosphere balloon is launched with a volume of 4000 m^3 and a pressure and temperature of 1×10^5 N/m^2 and 27°C. What was the balloon's volume when it reached a much higher altitude where the pressure and temperature were 2×10^4 N/m^2 and −33°C?

We use the perfect gas law with $T_1 = 300$ K and $T_2 = 240$ K, and then we solve for V_2.

$$V_2 = \frac{P_1V_1}{T_1}\frac{T_2}{P_2}$$

$$V_2 = \frac{1 \times 10^5\text{N/m}^2 \times 4000 \text{ m}^3 \times 240}{300 \times 2 \times 10^4\text{N/m}^2}$$

$$V_2 = 16{,}000 \text{ m}^3$$

Thus we see that the balloon will expand to four times its original volume.

5.7 Kinetic Theory

So far we have concerned ourselves mainly with the macroscopic properties of substances such as heat, temperature, and pressure. These things can be perceived by our senses quite easily. When substances are considered from a microscopic point of view as a collection of rapidly moving molecules, we reach the domain of what is known as *kinetic theory*.

According to kinetic theory, a gas is made up of a fantastically large number of molecules, moving at very high speeds (typically 1000 mi/h) and continually colliding with and bouncing off each other (about one billion collisions each second!), as well as bombarding other nearby objects. For instance, the molecules in the air are continually colliding with themselves and everything around them—tables, chairs, walls, people, and so on. When they bounce off any surface, they exert a force per unit area. This is what we perceive as air pressure. (See Fig. 5.12.)

The absolute temperature of a substance is directly related to the average kinetic energy of the molecules in the substance. The faster the molecular movement, the higher the absolute temperature. Conversely, as the motion of the molecules slows down, the absolute temperature decreases. In a simple model of solids, the temperature of absolute zero would then be explained as the point at which molecules cease to move. In the modern theory of matter, however, it is known that there is a slight amount of energy and motion left, even at absolute zero. This is called the *zero point energy*.

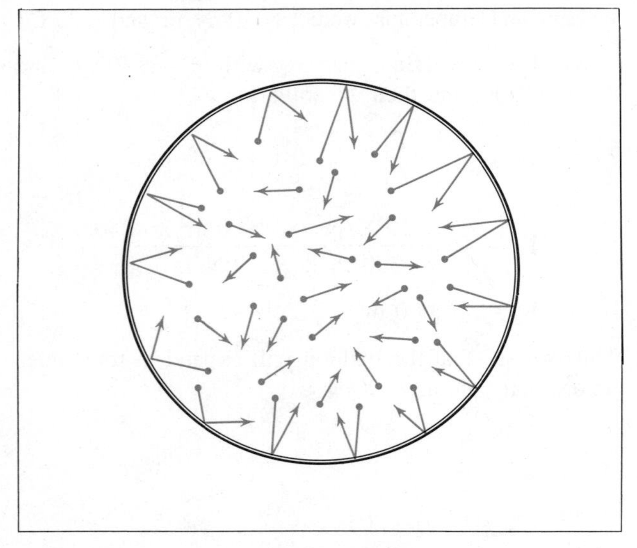

Figure 5.12 The air molecules inside a basketball are continually hitting and bouncing off the walls, exerting pressure and forcing the walls outward. As long as the pressure is greater on the inside than on the outside, the ball remains inflated.

The molecules in a gas move in a random fashion, and, because of their high speeds and very small mass, the effect of gravity is negligible on a small scale. For example, inside any building there are practically as many molecules near the ceiling as there are near the floor. On the scale of the Earth's atmosphere, however, gravity has a noticeable effect, and this results in more molecules being held nearer the Earth's surface. Thus, as we go up in altitude, the number of molecules decreases, and the air pressure undergoes a corresponding decrease.

With the kinetic theory we can now better explain the gas laws of Section 5.6. First, let us consider the relationship between the pressure and temperature of a gas. If we heat a rigid container full of gas, the molecules gain kinetic energy; that is, their velocities increase. This increase in velocity enables them to hit the walls of the container with a greater force, and the evidence of this is an increase in pressure. So, the higher the temperature the higher the pressure. This corresponds to Fig. 5.9.

Charles' law was illustrated by Fig. 5.10. As the temperature was increased, the gas molecules gained kinetic energy, and this increase in kinetic energy caused work to be done on the piston. The piston was raised, and this changed the volume of the gas, although the pressure remained the same. So at constant pressure an increase in temperature increased the volume of the gas.

Boyle's law can also be understood by using the kinetic theory. If we compress a gas, the molecules are confined to a smaller volume. They then bounce off the wall more often, and this is detected as an increase in the pressure. Making the volume smaller results in a pressure increase.

We noted in Section 5.6 that the perfect gas law was no longer valid at either very high pressures or temperatures near the liquefying point of a gas. This can now be explained from a kinetic theory viewpoint. At very high pressures, the gas molecules are squeezed together so that they no longer occupy a negligible volume compared to the distance between them. The same effect occurs for gases at temperatures near the liquefying point. Under these conditions the intermolecular forces and the sizes of the molecules themselves become important. At this point, the perfect gas law must be modified to take account of these factors.

The kinetic theory also helps us to understand the difference between heat and temperature. Heat is a form of energy. It is related to both the kinetic and potential energy of molecules. If a substance is being heated, and the velocity and kinetic energy of the molecules are in-

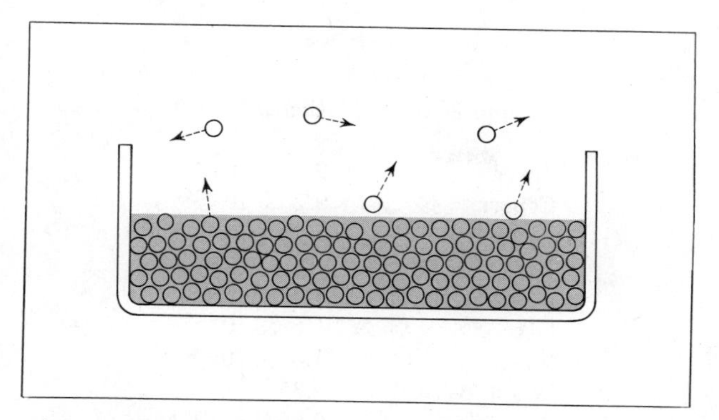

Figure 5.13 Evaporation of water. Water molecules are held together by intermolecular forces that are electrical in nature. When a molecule on the surface has enough kinetic energy to break free, it becomes water in the vapor stage. The latent heat of vaporization represents the energy necessary for evaporation to take place.

creasing, then the temperature will be rising. If, however, the heat energy goes into changing the potential energy of the substance, then the substance is changing phase and no temperature rise will be detected. Thus we see that absolute temperature is a measure of the average kinetic energy. Temperature is not related to the potential energy of the molecules.

Not all of the molecules in a substance are moving at the same speed. That is why we defined absolute temperature as a measure of the *average* kinetic energy of a molecule. Some of the molecules are moving faster and some are moving slower than average. When water (or any liquid) evaporates, it is due to the escape of the more rapidly moving molecules. The slower moving molecules cannot escape from the liquid's surface, but the faster ones can.

We can now easily see why stepping out of a shower produces a cooling sensation. The cooling occurs because of the evaporation of the water droplets on the skin. The heat from your body is used to supply the water molecules with enough energy to break away from the skin's surface and go into the air. Because heat is being drawn out of your body, a cooling sensation is experienced. (See Fig. 5.13.)

Our bodies are subjected to the bombardment of air molecules all the time. The faster the motion of the molecules, the higher the temperature. When the temperature changes and the molecules impinging upon us move faster or slower, our body perceives a change. We call that change hotter or colder.

5.8 Heat Transfer

The transfer of heat can be accomplished by one or more of the following three methods: convection, conduction, and radiation.

The transfer of heat by **convection** requires the movement of a substance, or mass, from one position to another. The movement of air or water is an example of heat transfer by convection.

Most homes are heated by convection (movement of hot air). The air is heated at the furnace, then circulated throughout the house by way of metal ducts. When the air has "lost its heat," it passes through a cold air return on its way back to the furnace to be reheated and recirculated. See Fig. 5.14.

Circulation of air in a room occurs naturally because cold air is heavier than warm air: therefore, it will sink to the floor and displace the warm air, which is forced to rise. The cold air duct is located at floor level to

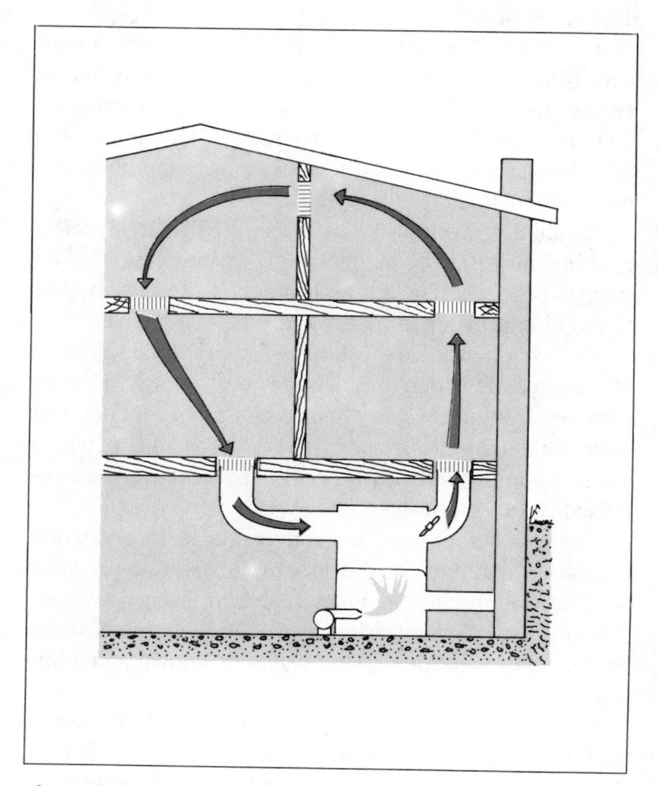

Figure 5.14 Heat transfer by convection. The air heated by the furnace rises and circulates throughout the house to return to the furnace when it has "lost its heat."

provide a path for the air to return to the furnace to be reheated. The transfer of heat energy by means of convection currents, including both air and water currents, is essential in distributing heat energy over the surface of the Earth.

Conduction is the transfer of heat energy by molecular activity. The kinetic energy of the molecules is transferred from one molecule to another through collisions. The heat flows from the higher temperature to the lower temperature, with the rate of flow being directly proportional to the temperature difference, cross-sectional area of the conductor, and the type of material.

Some substances are very good conductors of heat, others are not. The **thermal conductivity** of a substance is a measure of its ability to conduct heat. Metals such as silver, copper, and aluminum are good conductors of heat; thus they have high thermal conductivities. Other materials such as wood, paper, and air are poor conductors of heat; their thermal conductivities are low. Substances that are poor conductors of heat are known as **thermal insulators.** Table 5.3 gives the thermal conductivities of several substances.

Table 5.3 Thermal Conductivities

Substance	*kcal/m · s · °C*
Silver	9.9×10^{-2}
Copper	9.2×10^{-2}
Aluminum	5.0×10^{-2}
Iron	1.6×10^{-2}
Lead	8.4×10^{-3}
Glass	1.6×10^{-4}
Water	1.4×10^{-4}
Wood (oak)	0.35×10^{-4}
Air	0.057×10^{-4}

The thermal conductivities of substances are important in many day-to-day living situations. For instance, a silver spoon will feel colder than a piece of wood when they are both at the same temperature. This is because the silver spoon conducts heat away from the body at a much higher rate.

As another example, it is warmer to wear layers of clothing in winter than one very heavy coat. The air trapped between the clothing layers helps to make layers of clothing a better insulator than one heavy coat.

The conduction of heat plays an important role in making ice. In order to make ice, heat energy must be removed from the water. Since metal trays conduct heat quite well, they are frequently used. The rate at which ice is made depends on how fast the heat can be conducted out of the water.

The transfer of heat by convection and conduction requires a medium for the process to take place. Heat from the Sun reaches the Earth via electromagnetic waves. The process of transferring heat energy through space by means of electromagnetic waves is known as **radiation.**

Electromagnetic waves carry energy and can travel through a vacuum. The heat we get from the Sun is transmitted through the vacuum of space by radiation. All objects continuously emit and absorb energy by radiation. The rate at which an object radiates energy is directly proportional to the surface area and proportional to the fourth power of the Kelvin temperature.*

Black objects are good absorbers of heat energy, while objects that are white are good reflectors and very poor absorbers. For this reason, light-colored clothing is cooler in summer than dark-colored clothing. Similarly, dark clothing is warmer in winter.

A good example of the methods used to impede heat transfer is shown in Fig. 5.15. The vacuum bottle is designed to prevent heat from escaping or entering the

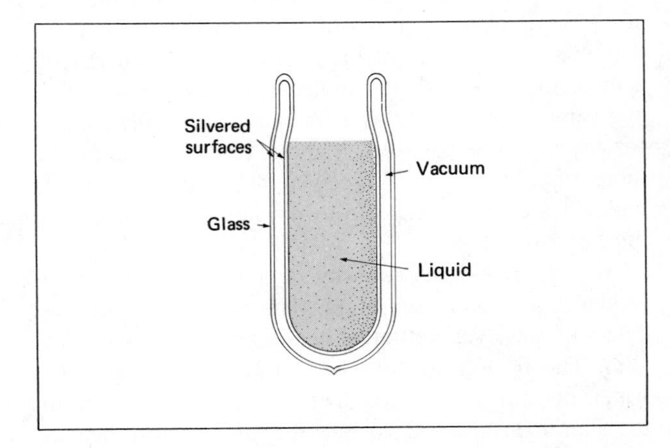

Figure 5.15 A vacuum bottle.

*The rate of radiation is given by the Stefan-Boltzmann law:

$$H/t = e\sigma A[T_o^4 - T_s^4]$$

T_o is the temperature of the object; T_s is the temperature of the surroundings.

glass bottle holding the hot or cold liquid. Heat transfer by convection and conduction is prevented by the evacuated space. By removing most of the mass (air molecules) in this volume of space, the means for convection and conduction are removed.

To prevent transfer by radiation, the inner surfaces of the glass bottle are silvered. These silvered surfaces reverse the direction of the heat radiation. Thus, wanted heat energy is kept in for hot coffee or kept out for cold tea.

Learning Objectives

After reading and studying this chapter, you should be able to do the following without referring to the text:

1. Explain how the Fahrenheit, Celsius, and Kelvin temperature scales are constructed.
2. Convert from one temperature scale to another using the appropriate formulas.
3. Explain the concept of absolute zero temperature.
4. Explain what a food Calorie is.
5. Compute the amount of heat energy necessary to melt ice, boil water, or change the temperature of water, knowing how much mass is involved.
6. State the second law of thermodynamics in three different ways.
7. Compute the maximum thermodynamic efficiency of a power plant knowing its operating temperatures.
8. Discuss the molecular structure of the solid, liquid, and gaseous forms of water.
9. Work problems using the perfect gas law.
10. Describe a gas in terms of kinetic theory.
11. Explain the concept of pressure in terms of the kinetic theory of gases.
12. Explain the difference between heat and temperature in terms of kinetic and potential energy.
13. Describe three methods of heat transfer and give an example of each.

Important Words and Terms

molecule
temperature
thermometer
ice point
steam point
Fahrenheit scale
degree Fahrenheit
Celsius scale
degree Celsius
Kelvin scale
kelvin
heat
calorie
kilocalorie

Btu
specific heat
heat capacity
latent heat of fusion
melting point
latent heat of vaporization
boiling point
first law of thermodynamics
adiabatic process
second law of thermodynamics
heat pump
heat engine
entropy
third law of thermodynamics

solid
lattice
liquid
gas
pressure
ideal gas law
Charles' law
isothermal process
Boyle's law
convection
conduction
thermal conductivity
thermal insulators
radiation

Questions

1. What does a temperature of absolute zero mean?
2. What are the ice point and steam point on the three temperature scales: Fahrenheit, Celsius, and Kelvin?
3. What is a calorie and a kilocalorie? What is a food Calorie?
4. Why does it take a long time for lakes and rivers to freeze when the air temperature is below freezing?
5. One hundred years ago people would place a large tub of water in their root cellars to prevent canned foods from freezing in the winter months. Explain.
6. When does the addition of heat not result in a temperature increase?
7. Why do bridges and sidewalks have cracks built into them?
8. What causes water pipes to break sometimes during cold weather?
9. Which common substance has the highest specific heat?
10. Why do continents have extreme weather (hot in summer, cold in winter), whereas most islands have mild weather?
11. By expending one joule of work, a home heat pump can provide two joules of heat to a room. Explain why this does not violate the law of conservation of energy.
12. Why is approximately two-thirds of the heat generated by a coal or nuclear power plant wasted? What happens to this waste heat?
13. In terms of entropy, why do occupied rooms naturally tend to get messy?
14. How fast are air molecules moving? How many collisions do they make each second?
15. In terms of kinetic and potential energy of molecules, distinguish between heat and temperature.
16. Why does the pressure inside an auto tire increase during long continuous driving?
17. Explain in terms of kinetic theory why a basketball stays inflated.
18. Why does sweating keep you cool?
19. Why does a plastic tile floor feel colder than a rug even though both are at the same temperature?
20. Why are furnaces put in the basement or lowest floor of a building?

Problems

1. Normal body temperature is 98.6°F. What is the equivalent temperature on the Celsius scale? *Answer:* 37°C
2. Normal room temperature is about 68°F. What is the equivalent temperature on the Celsius scale? On the Kelvin scale?
3. The temperature of outer space is about 3 K. What is the equivalent temperature on the Celsius scale?
4. A skier comes down a 10-m-high slope, reaching the bottom at a speed of 10 m/s. If the skier's mass is 60 kg, how much heat did she produce during her run?
5. A college student produces about 100 kcal of heat per hour during an average day. How many watts is this? *Answer:* 116 W
6. Twenty joules of work are expended in stirring 400 g of water. If the initial temperature of the water was 25°C, what will be the final temperature of the water? Assume all the work went to heat the water.
7. (a) How much heat does it take to heat 1 kg of water in a pan (ignore the pan) from room temperature (20°C) to the boiling point? (b) At 8¢ per kWh, how much does it cost to heat up 1 kg of hot water for tea? *Answer:* 0.7¢ (less than 1¢)
8. To make iced tea, 1.6 kg of hot tea at 98°C needs to be cooled to 0°C. What is the minimum amount of ice necessary to do this?
9. It has been proposed to use the temperature difference between the top and lower portions of the ocean to run a power plant. If the surface of the ocean is at 15°C and the lower depths are at 4°C, what is the maximum percentage efficiency possible?
10. A proposed nuclear-fired power plant will have T_h = 350°C and T_c = 100°C. What is its maximum percent efficiency? *Answer:* 40%
11. A coal-fired power plant has T_h = 320°C and T_c = 100°C. What is its maximum percent efficiency?

12. An automobile tire contains a volume of air at a temperature of 22°C and a gauge pressure of 2.0×10^5 N/m^2. If the volume remains constant and the pressure increases to 2.5×10^5 N/m^2, determine the new temperature in degrees Celsius. (Note: gauge pressure is pressure above atmospheric pressure of 1×10^5 N/m^2.)

    Answer: 96°C

13. A balloon is blown up to a pressure of 29 lb/in^2 so that it occupies a volume of 900 cm^3 at room temperature of 17°C. The next day the temperature of the room is up to 27°C, and the balloon has expanded to 1000 cm^3. What is the new pressure in the balloon?
14. A sample of air occupies a volume of 10 m^3 at a temperature of 0°C and an absolute pressure of 1×10^5 N/m^2. Determine the volume of the air sample at 40°C and 4×10^5 N/m^2.
15. A container is filled with helium gas at 22°C. What is this on the Kelvin scale? If heat is applied so that the average kinetic energy of the gas molecules doubles, what is the new temperature of the gas in degrees Celsius and Kelvin units?

6

Waves

When it came night, the white waves paced to and fro in the moonlight, and the wind brought the sound of the great sea's voice to those on shore, and they felt that they could then be interpreters.

–Stephen Crane

SINCE BEGINNING OUR story of energy, much has been said about the forms of energy, the relation of energy to work, and the law of energy conservation—with many questions raised and answered. Other interesting questions remain, however. For example, how is energy changed from one form to another? How is energy transferred to the ball in a baseball game? How do our bodies obtain energy? How is energy transferred to a train, a car, a space ship, a particle of any size? How do we get energy from the Sun?

A partial answer to these questions has been found in terms of particle collisions and wave motion, two important ways for transferring energy (see Fig. 6.1). If a particle applies a force to another particle through a distance, then a transfer of energy has taken place through particle collision, one particle will gain energy, and the other will lose energy. If we are to believe the law of conservation of energy, and we know of no instance where the law has failed, then the increase of energy of one particle will be equal to the decrease of energy of the other, assuming we neglect heat losses. In all cases, we find a transfer of energy takes place whenever two or more particles collide with one another.

When matter is disturbed, energy emanates from the disturbance; this emanation of energy is known as wave motion. For example, a stone dropped on the surface of a pond of water will disturb the water, and energy will be transferred outward from the disturbance as wave motion. Notice that only energy is transferred, not matter (the water).

A similar situation can occur in a solid. For example, during an earthquake a disturbance takes place because of a slippage or other cause, and this disturbance is transmitted to all parts of the Earth as wave motion. Again, this is a transfer of energy, not matter. The transfer of energy takes place with or without a medium.

Sound waves in the air and waves upon stretched strings and steel wires are examples of energy transmissions that require a medium (air, strings, and wires in the above cases). The neighboring particles of the media react upon one another to transfer the disturbance. Electromagnetic waves, including radio, infrared, light, and X rays, are transferred without a medium. We say these disturbances are radiated through space.

The data from which we learn about the planets, the Sun, and other stars come to us by means of light. Later, when quantum mechanics is discussed, we shall study the dual nature of light. Strangely enough, light, which

◀ When a drop of water hits, energy is transferred outward in a circular pattern.

Figure 6.1 Examples of transferring energy.

we consider to be a wave motion, has some of the properties of matter. This will be discussed in Chapter 11.

Our eyes and ears are two wave-detecting devices that serve to link us to our environment. Since a study of wave motion seems relevant to an understanding of our physical environment, a knowledge of wave motion is essential to the understanding of many scientific principles.

6.1 Wave Properties

The disturbance generating a wave motion is usually periodic; that is, the disturbance is repeated again and again at regular intervals. The plucking of a guitar string or the blowing of a whistle sets up periodic waves (Fig. 6.2). However, the disturbance, and the resulting wave motion, does not necessarily have to be periodic; it may be a simple pulse, or shock wave, such as the one originating from a book hitting the floor, or from a jet plane passing through the sound barrier.

A disturbance in an elastic body will set up wave motion. For example, in Fig. 6.3 a stretched spring is

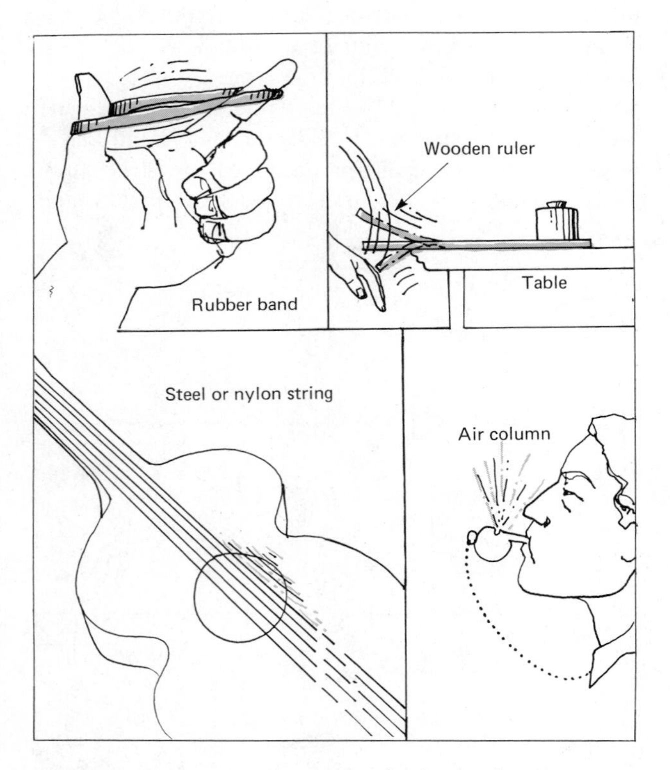

Figure 6.2 Some common methods used to generate wave motion.

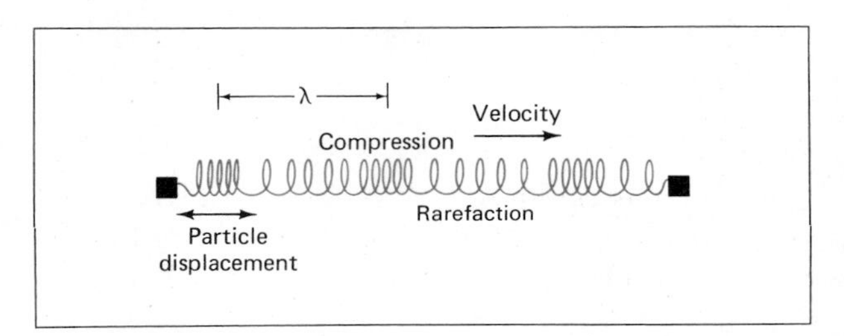

Figure 6.3 A longitudinal wave in a stretched spring.

attached between two fixed points. When four or five coils at one end of the spring are compressed and released, the disturbance is propagated along the length of the coiled spring; that is, the disturbance moves with the velocity vector parallel to the direction in which the particles were displaced. The velocity is known as **wave velocity;** when the particle displacement and the wave motion are in the same direction the wave is called a **longitudinal wave.** Sound waves are of this type.

The term **transverse wave** is used to denote wave motion in which the individual particles are displaced perpendicular to the direction of the wave velocity vector. Figure 6.4 is an illustration of a transverse wave in a stretched rubber cord. The cord is disturbed from its equilibrium position by moving the end of the cord up and down. Moving the cord end from side to side—or in any direction—will also produce a transverse wave. The wave motion in this example is to the right, as indicated by the velocity vector.

Transverse waves, unlike longitudinal waves, can be polarized, a physical property that is used to distinguish between longitudinal and transverse waves. A *polarized wave* is one in which the particle disturbance occurs in only one direction, perpendicular to the direction the wave is traveling. Polarization will be discussed in more detail in the next chapter.

The transfer of energy by transverse waves can take place in the absence of a medium. All electromagnetic radiation is of this type of wave motion. The study of wave motion has resulted in certain terms that are used to explain the action of all waves (Fig. 6.5). Velocity describes the direction and magnitude of the wave motion. The velocity of a longitudinal wave depends upon the properties of the medium. The **frequency** is the number of oscillations, or complete waves, that occur during a given period of time, usually one second. The units of frequency are cycles per second. This unit is given the name hertz. One **hertz** (Hz) is one cycle per second. For example, if five complete wave crests pass a given

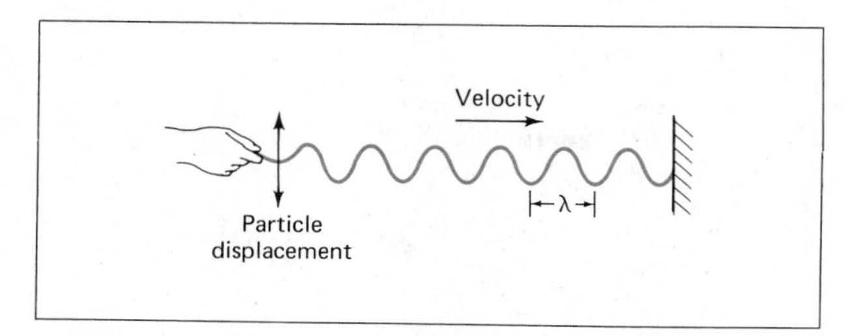

Figure 6.4 A transverse wave in a stretched cord.

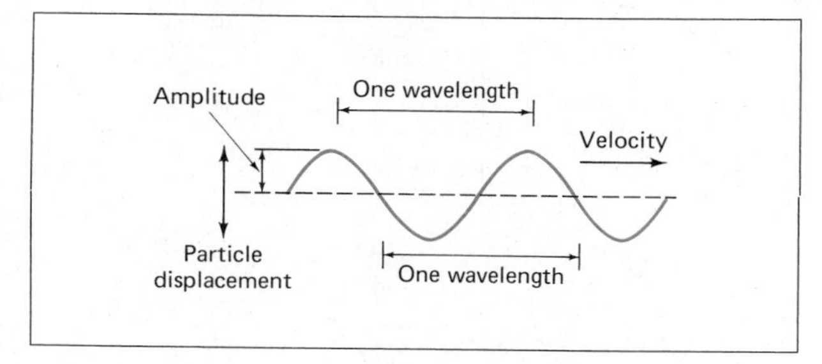

Figure 6.5 The terms used to describe transverse wave motion.

spot in one second, the frequency would be five cycles per second or five hertz.

The **period** of a wave is the time it takes for one complete wave oscillation. If five crests pass by a given point in one second, one crest or complete cycle would pass in one-fifth of a second, and the period would be $\frac{1}{5}$ second. The frequency would be 5 cycles per second. From this example it is easy to see that

$$\text{frequency} = \frac{1}{\text{period}}$$

or

$$f = \frac{1}{T}$$

The wavelength of a wave is measured in units of length. One **wavelength** is the distance from any point on a wave to an identical point on the adjacent wave as shown in Fig. 6.5.

A simple relation between wave velocity, wavelength, and period (or frequency) exists. This can be written as

$$v = \frac{\lambda}{T}$$

or

$$v = \lambda f \tag{6.1}$$

where v = wave velocity measured in meters per second or any other velocity units,
λ = wavelength, measured in length units,
T = period of the wave, usually measured in seconds,
f = frequency of the wave, measured in cycles per second or hertz.

EXAMPLE 1 Consider sound waves with a velocity of 344 m/s and a frequency of (a) 20 Hz (b) 20,000 Hz. Find the wavelength of each of these sound waves.

We rearrange Eq. 6.1 and solve

$$v = \lambda f$$

or

$$\lambda = \frac{v}{f}$$

(a)

$$\lambda = \frac{344 \text{ m/s}}{20 \text{ Hz}}$$

$$\lambda = 17.2 \text{ m}$$

(b)

$$\lambda = \frac{344 \text{ m/s}}{20{,}000 \text{ Hz}}$$

$$\lambda = 0.017 \text{ m}$$

The frequencies of 20 to 20,000 Hz given in Example 1 cover the range of audible sound wave frequencies. Thus, the wavelengths of sound cover the range from 1.7 cm for the highest-frequency sound we can hear up to 17.2 m for the lowest-frequency sound we can hear. In British units sound waves go from wavelengths of approximately $\frac{1}{2}$ in. up to about 50 ft.

The **amplitude** of a wave refers to the maximum displacement of any part of the wave from its equilibrium position. The amplitude of the wave does not affect the wave velocity. The energy transmitted by a wave is related to the square of its amplitude.

6.2 Electromagnetic Waves

When charged particles such as electrons are vibrating, energy is radiated away from them in the form of **electromagnetic waves.** Electromagnetic waves consist of vibrating electric and magnetic fields, which will be more thoroughly studied in Chapter 8. These are vector fields. In electromagnetic waves they radiate outward at the speed of light, which is 3×10^8 m/s.

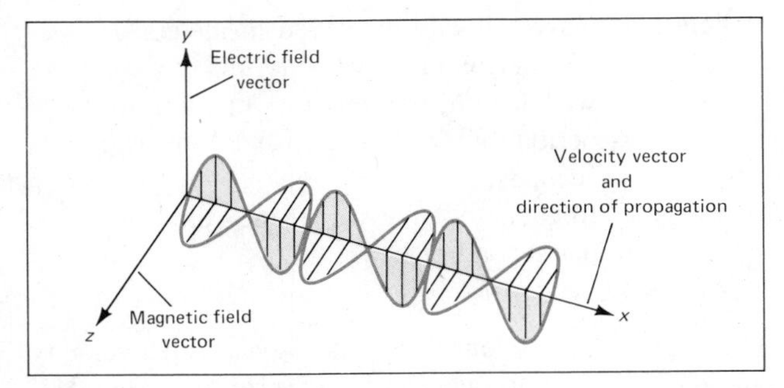

Figure 6.6 Diagram of the vector components of electromagnetic waves. Electromagnetic waves consist of two force fields (electric and magnetic) perpendicular to one another and moving with a velocity vector that is perpendicular to both force fields.

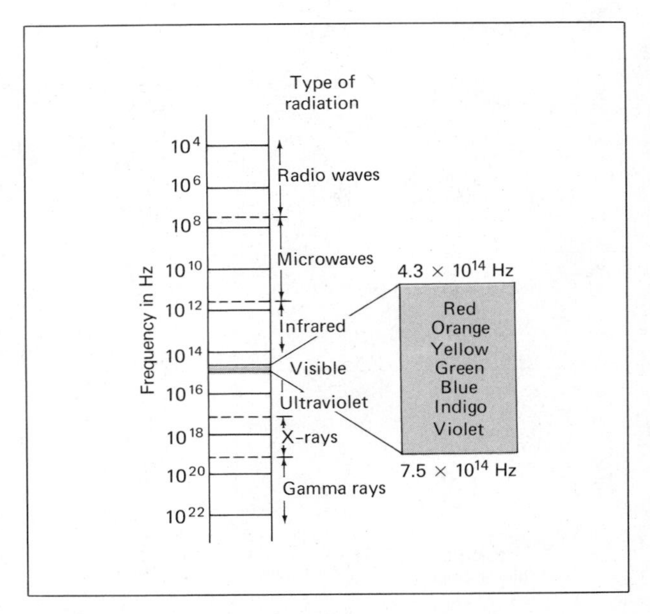

Figure 6.7 The electromagnetic spectrum.

A drawing of an electromagnetic wave is shown in Fig. 6.6. The drawing shows the wave traveling in the x direction. The electric and magnetic field vectors are at angles of 90° to one another, and the velocity vector of the wave is at an angle of 90° to both of the field vectors.

Charged particles are accelerated in many different ways to produce electromagnetic waves of various frequencies. Waves with low frequencies, or long wavelengths, are known as *radio waves* and are produced primarily by causing electrons to oscillate, or vibrate, in a resonant circuit. The frequency of oscillation is controlled by the physical dimensions and other properties of the tuned circuit.

The production of electromagnetic waves with frequencies greater than radio waves is accomplished by molecular excitation. In such cases, radiation occurs from the collision of molecules in hot gases and solids. Since the molecules carry charged particles that are greatly accelerated as the molecules vibrate, the particles will radiate electromagnetic waves ranging from 10^{12} Hz to 4.3×10^{14} Hz. This portion of the electromagnetic spectrum is called the *infrared* region.

As the temperature of gases and solids is increased to higher and higher values, the atoms composing the molecules become more excited and electromagnetic radiation in the *visible* and *ultraviolet* regions of the spectrum is emitted. Still more energy applied to the atom will generate waves of higher frequencies, called *X rays*, which range from 3×10^{17} to 3×10^{19} Hz.

If sufficient energy is applied to the atom to disturb the nucleus, radiation known as *gamma rays* is emitted. The major portion of the electromagnetic spectrum is shown in Fig. 6.7. The portion of the spectrum visible to the human eye falls between the infrared and ultraviolet wavelengths.

The word *light* is commonly given to visible electromagnetic radiation. However, only the frequency (or wavelength) distinguishes visible electromagnetic radiation from the other portions of the spectrum. Our human eyes are only sensitive to certain frequencies or wavelengths, but other instruments can detect other portions of the spectrum. For example, a radio receiver can detect radio waves.

Radio waves are not sound waves. They are electromagnetic waves that are detected and then amplified by the radio frequency circuits of the radio receiver. The radio frequency signal is then demodulated. That is, the audio signal is separated from the radio frequency carrier. The audio signal is amplified, then applied to the speaker system that produces sound waves.

Electromagnetic radiation consists of transverse waves. These waves can travel through a vacuum. For instance, radiation from the Sun travels through the vacuum of space before arriving at the Earth. All electromagnetic waves travel at the same speed in a vacuum. This speed is called the speed of light. The **speed of light** in a vacuum is designated by the letter c and is equal to the following:

$$c = 186{,}000 \text{ mi/s}$$

$$c = 3 \times 10^8 \text{ m/s}$$

This is a very high speed; and, according to Einstein's theory of relativity, nothing can travel faster than c.

We can use the formula $v = \lambda f$ to find the wavelength of electromagnetic radiation if we know the frequency. Visible wavelengths of light are very short (about 10^{-6} m) and are expressed in units called angstroms, named for a Swedish scientist. One **angstrom** (Å) is equal to 10^{-10} m. To find the wavelength if the frequency is known, we use the relation

$$\lambda = \frac{c}{f}$$

EXAMPLE 2 What is the wavelength of red light that has a frequency of 4.4×10^{14} Hz? Express your answer in both meters and angstroms (Å).

We use

$$\lambda = \frac{c}{f}$$

$$\lambda = \frac{3 \times 10^8 \text{ m/s}}{4.4 \times 10^{14} \text{ cycles/s}}$$

$$\lambda = 6.82 \times 10^{-7} \text{ m}$$

Since 1 Å = 10^{-10} m,

$$6.82 \times 10^{-7} \text{ m} \times \frac{1 \text{ Å}}{10^{-10} \text{ m}} = 6820 \text{ Å}$$

Now let us calculate the wavelength of a typical radio wave and compare it to the wavelength of visible light just calculated.

EXAMPLE 3 What is the wavelength in meters of radio waves produced by a station that has a frequency of 1200 kilohertz?

Since *kilo* means one thousand and *hertz* means cycles per second, we have

$$f = 1200 \times 10^3 \text{ cycles/s}$$

$$f = 1.2 \times 10^6 \text{ cycles/s}$$

$$\lambda = \frac{3 \times 10^8 \text{ m/s}}{1.2 \times 10^6 \text{ cycles/s}}$$

$$\lambda = 250 \text{ m}$$

We see from these two examples that radio waves have very much longer wavelengths than those of visible light.

The color of a visible light wave depends upon the frequency of the wave. The brightness depends upon the amount of energy being transferred. Thus, the brightness of a light wave depends upon the square of the amplitude.

6.3 Sound Waves

Sound is a physical phenomenon originated by varying the degree of particle displacement and propagated by varying pressures and stresses in an elastic medium. Sound waves may originate from particle displacement in any kind of matter: solid, liquid, or gas. Some common sources include vibrating strings, tuning forks, ocean waves, tornadoes, and the human vocal cords. Transmission of sound through any medium is by longitudinal waves. Transmission of sound waves through air is illustrated in Fig. 6.8.

The wave motion of sound in any medium consists of a series of wave compressions and rarefactions, displaced along the direction of the wave velocity. A **compression** is that portion of a longitudinal wave in which the vibrating particles are relatively close. A **rarefaction** is that portion of a longitudinal wave in which the vibrating particles are relatively far apart.

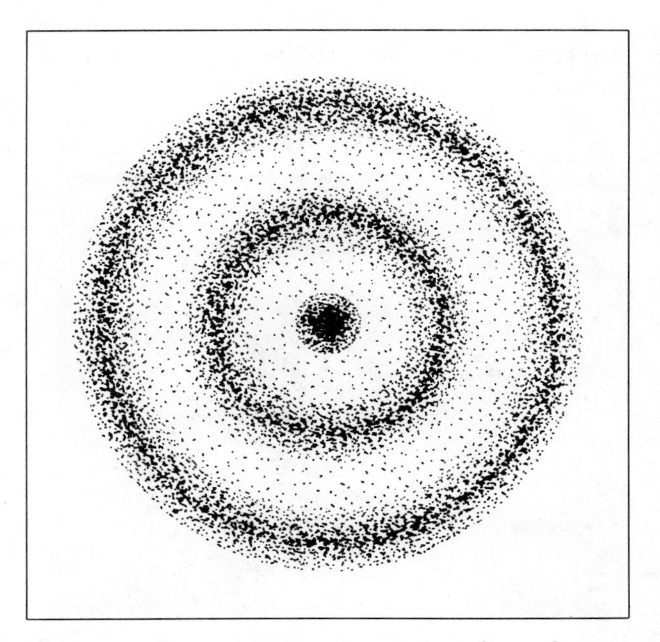

Figure 6.8 Illustrating the transmission of sound waves through air. Note the varying particle density along the radii, illustrating compressions and rarefactions.

Sound waves detected by the ear are caused by variations in air pressure. The eardrum vibrates because of these pressure variations, and the subsequent interpretation by the brain gives us the sensation of sound.

Note that there is no sound on the moon. Because there is no atmosphere on the moon, there is nothing to cause pressure waves to be picked up by the ear. The astronauts on the moon communicated using radio waves—similar to a CB radio conversation.

Sound waves are generated with frequencies ranging from a few hertz up to frequencies in the millions of hertz. Frequencies audible to the human ear range from a minimum of about 20 Hz to a maximum of about 20,000 Hz.

The audible range is a function of the intensity of the sound as well as the frequency. The human ear is most sensitive to a frequency of about 3000 Hz. **Pitch** is the highness or lowness of a sound and is the consequence of the frequency, or the number of hertz, received by the ear.

Ultrasound is the term for sound waves whose frequencies are greater than 20,000 Hz. These waves oscillate too quickly to be detected by the human ear. These exceedingly rapid vibrations have many uses.

One of the most important uses of ultrasound is to look at parts of the body in a manner analogous to X rays. The ultrasonic waves allow different materials such as tissue and bone to be "seen" by bouncing waves off the object to be examined. These *echograms* can be used to examine unborn babies as shown in Fig. 6.9. X rays might harm the fetus and cause birth defects, but ultrasound simply sets up vibrations and has given no evidence of harming the fetus.

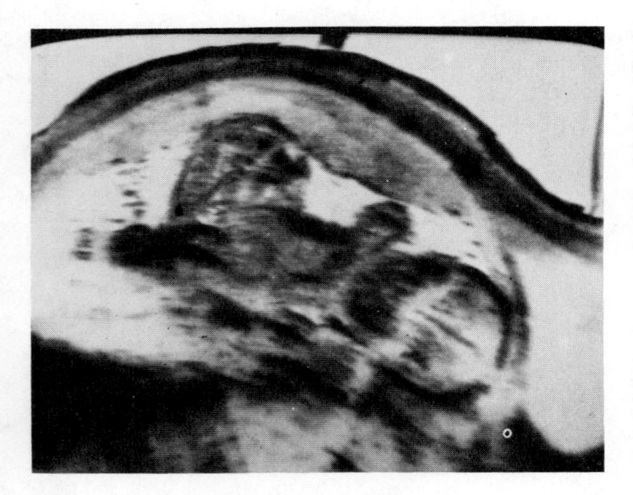

Figure 6.9 A medical use of ultrasound. An echogram of a fetus in the third trimester of pregnancy. (Yale University Medical School scan, courtesy Picker Corp.)

Ultrasound can also be used as a cleaning technique. Minute foreign particles can be removed from objects placed in a liquid bath through which ultrasound is passed. The wavelength of the ultrasound is of the same order of magnitude as the particle size; and the wave energy vibrates, or "scrubs," the particles free. Ultrasound can be especially useful in cleaning objects with hard-to-reach recesses such as teeth. Ultrasonic cleaning baths for false teeth are commercially available.

The speed of sound waves in air at 20°C is

$$v_{\text{sound}} = 344 \text{ m/s}$$

$$v_{\text{sound}} = 770 \text{ mi/h}$$

At higher temperatures this speed increases slightly and at lower temperatures it decreases. This speed is much less than the speed of light.

The relatively slow speed of sound produces an effect that makes marching bands sometimes sound dissonant. Since the speed of sound is 344 m/s, it takes about $\frac{1}{6}$ s for sound to travel about 60 m. If a band is spread out over a football field, parts of the band can easily be as far as 60 m (65 yd) from other parts. Therefore, some instruments will sound $\frac{1}{6}$ s or so before or after they would normally be heard. Thus, the music can sound dissonant rather than harmonious.

Another effect noticeable to someone sitting in an upper grandstand is that the musicians don't seem to begin when directed by the conductor. The band members really do begin playing when so directed, but it takes a fraction of a second for the sound to travel the 100 m or so to the fans in the distant parts of the stadium.

Using the velocity and frequency of sound waves, we can compute their wavelengths.

EXAMPLE 4 What is the wavelength of a sound wave whose frequency is 2200 hertz?

We use the relation

$$\lambda = \frac{v_{\text{sound}}}{f}$$

$$\lambda = \frac{344 \text{ m/s}}{2200 \text{ cycles/s}}$$

$$\lambda = 0.156 \text{ m}$$

This wavelength corresponds to about $\frac{1}{2}$ ft.

From Examples 1, 2, and 4, we see that wavelengths of visible light (10^{-6} m) are much shorter than wavelengths of audible sound.

The *loudness* of a sound wave relates to the amount of energy being transferred. The energy transferred is the amount of energy passing through unit area in unit time, and the amount of energy is directly proportional to the square of the wave amplitude. When we speak of loudness, we are also speaking of how our ear perceives a sound. The measurable quantity that corresponds to this is called intensity.

The **intensity** (I) of a sound wave is defined as the time rate of transfer of energy per unit area of the wave front (I = energy/area/time = joules/meter^2/second). When the mks system of units is used, the energy transfer per unit time is expressed in watts (joules per second) and the wave front area in square meters (I = power/area = joules/second/meter^2 = watts/meter^2).

The intensity of the sound wave decreases as the distance from the source increases. At a distance (R) from a point source of power, the intensity (I) in watts per square meter is directly proportional to the power of the source and inversely proportional to the square of the distance from the source. This is the inverse square law illustrated in Fig. 6.10. The minimum intensity that can be detected by the average human ear is 10^{-12} W/m^2. The maximum intensity (sound becomes painful) is about 1 W/m^2. This is a very wide range of intensities for the human ear to respond to. Because of this wide range of intensities and since the human ear responds, roughly, proportional to the logarithm* of the intensity, a better and more convenient method of measuring the intensity level of a sound uses the **decibel** scale, which is a logarithmic scale.

The intensity level of a sound of intensity I is defined as follows:

Intensity level in decibels (dB) = 10 logarithm (I/I_0)

where I = intensity of the sound in question,
I_0 = reference intensity of 10^{-12} W/m^2 (minimum intensity detected by the human ear)

The decibel is equal to 0.1 bel. The bel is named after Alexander Graham Bell, inventor of the telephone. Figure 6.11 (p. 92) shows the intensity levels in decibels of sounds from various sources.

*The logarithm of a number to a given base (the base 10 is used here) is the power to which the base must be raised to equal the number.

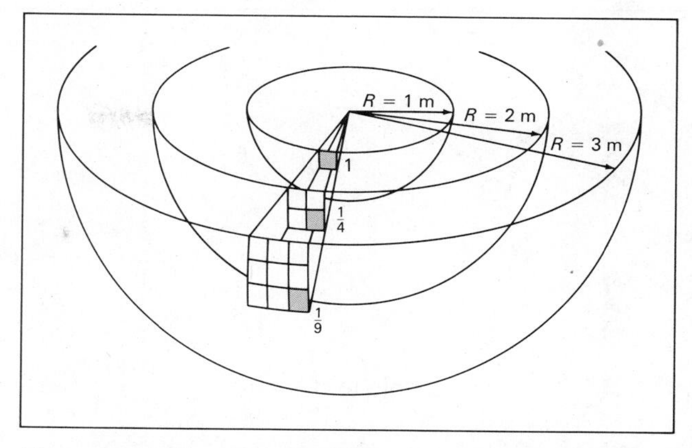

Figure 6.10 Illustration of the inverse square law. The intensity decreases inversely proportional to the square of the distance. When R = 1 meter, I = 1 unit; when R = 2 meters, I = 1/4 unit; when R = 3 meters, I = 1/9 unit.

Each increase in decibel (dB) level of 10 is equal to an increase of 10 in intensity. Thus, a sound of 93 dB is 10 times louder than a sound of 83 dB and 100 times louder than a sound of 73 dB. Similarly, a sound of 65 dB is 1/10,000 times as loud as a sound of 105 dB.

Sound waves that come in contact with objects in the environment tend to produce vibrations in these objects. For example, the sonic boom of a jet plane flying at speeds greater than sound, the vibration of human vocal cords, a radio speaker, or other source of sound waves all cause our eardrums to vibrate. The reaction of a particular object to sound waves depends upon its own frequency of vibration.

Objects similar in physical properties to the vibrating source tend to vibrate more noticeably than objects that are not similar (Fig. 6.12, p. 93). The two tuning forks in this figure are constructed to vibrate at the same frequency and are mounted a few meters apart, as shown. The tuning fork on the left is set in motion by striking it with a rubber hammer. This tuning fork will send out sound waves, which will be received by the tuning fork on the right. Since both forks have been constructed to vibrate at the same frequency, the fork on the right will start oscillating and continue to oscillate on its own. This condition of sympathetic vibration is known as **resonance,** which can be defined as a tuned condition existing whenever an external frequency corresponds to an object's natural frequency of vibration.

There are many instruments designed to use the principle of acoustic (sound) resonance. The structure of the throat and cavities of the head gives the human voice a

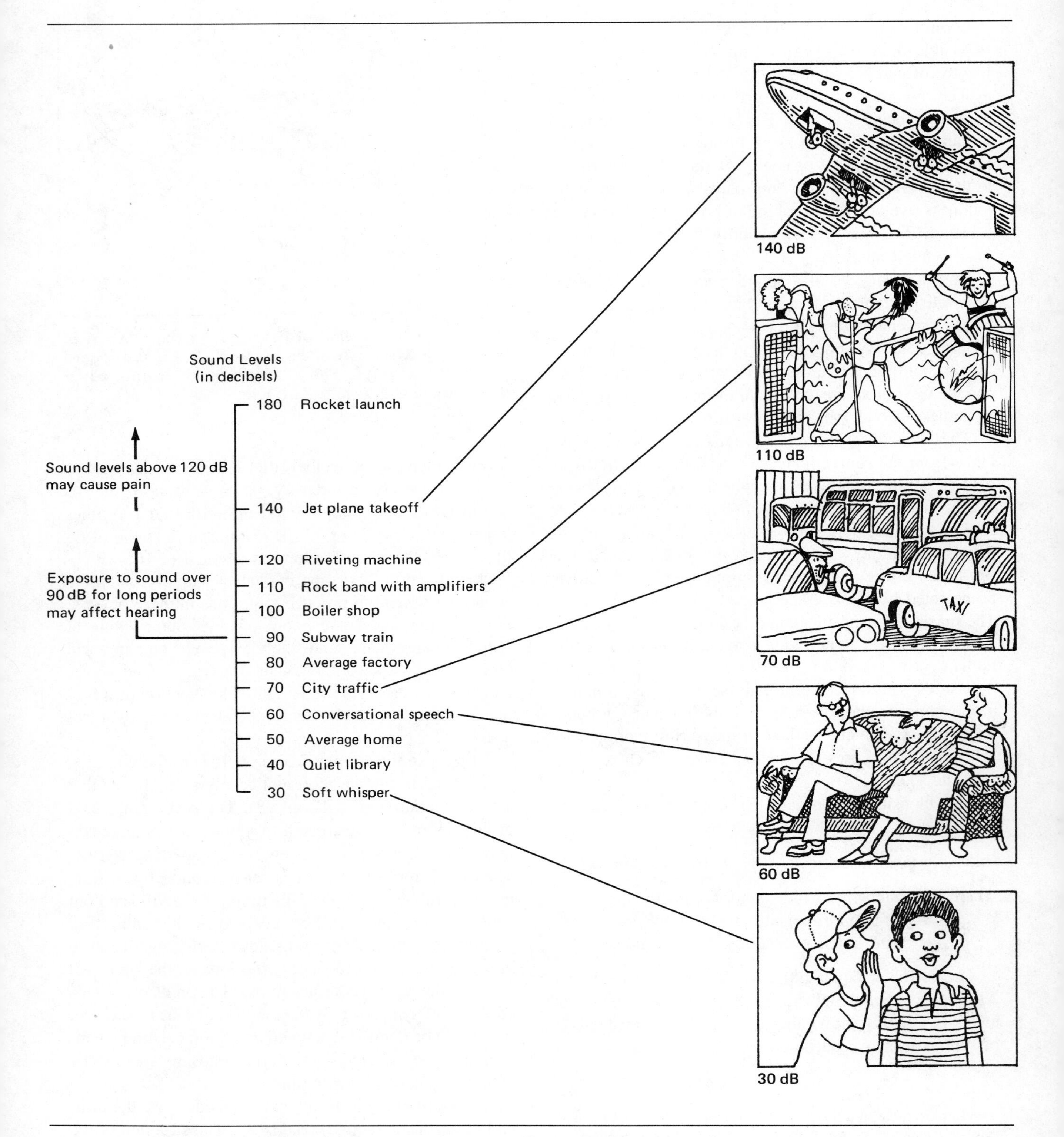

Figure 6.11 The intensity levels, in decibels, of sounds from various sources.

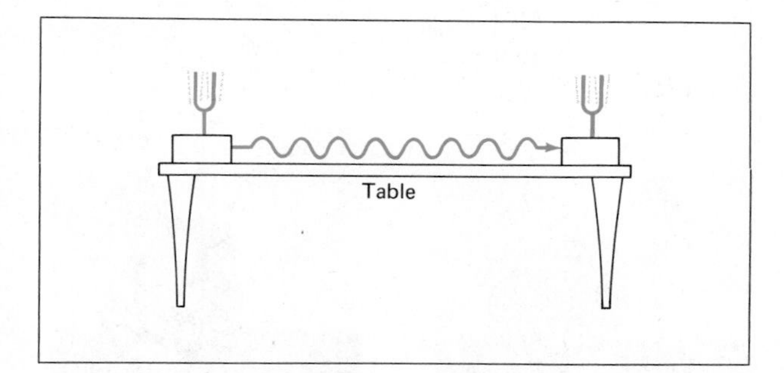

Figure 6.12 A demonstration of resonance. The tuning fork on the right will vibrate when the tuning fork on the left is made to vibrate by striking it with a rubber hammer.

better tone due to resonance. Resonance can produce a harmful effect at times. Examples of mechanical resonance occur when the wind causes large vibrations to build up in a steel bridge to the point where the bridge will buckle and collapse, or when parts of an automobile vibrate and sometimes break as a result of resonance, especially at particular velocities. Loudspeaker enclosures on some radios tend to vibrate at the lower frequencies because of the resonance effect, producing unwanted sounds.

6.4 Standing Waves

If a rubber string is held at one end and shaken, a standing wave pattern can be set up as shown in Fig. 6.13 (p. 94). The points of the string that remain stationary are called the **nodes,** and the points of maximum displacement are called the **antinodes.** The distance between two nodes or two antinodes is one-half the wavelength of the standing wave.

In order to set up the standing waves shown, the string must be moved back and forth at a certain frequency. The standing waves are set up only at certain frequencies. These characteristic frequencies result in an integral number of half-wavelengths in the string. In Fig. 6.13 there are 1/2, 1, 3/2 and 2 wavelengths in the standing waves from left to right.

Since the length of the string is known, the wavelengths are given by

$$\lambda = \frac{2L}{n} \qquad n = 1, 2, 3, 4, \ldots \qquad (6.2)$$

That is, in Fig. 6.13 the wavelengths are $\lambda = 2L, 2L/2, 2L/3$, and $2L/4$.

The frequencies of the standing waves are given by rearranging Eq. 6.1.

$$f = \frac{v}{\lambda}$$

or, substituting for λ in Eq. 6.2 we get

$$f_n = \frac{nv}{2L} \qquad n = 1, 2, 3, 4, \ldots \qquad (6.3)$$

The lowest possible frequency is when $n = 1$

$$f_1 = \frac{v}{2L} \qquad (6.4)$$

This frequency, f_1, is called the **fundamental frequency** or **first harmonic.** The higher frequencies $f_2, f_3, \ldots$ (when $n = 2, 3, 4, \ldots$) are multiples of the fundamental frequency and are called the **second harmonic, third harmonic,** and so on. Taken together, the harmonics higher than the first are called **overtones.**

From Eq. 6.4 we see that the frequency depends on the wave velocity and the length of the string. The wave velocity in turn depends on the tension in the string and the mass per unit length of the string. Note that this wave velocity is the velocity of sound in the stretched string, not the velocity of sound in air.

EXAMPLE 5 If the frequency of the first harmonic is 440 Hz, what are the frequencies of the second and third harmonic?

To solve this we use Eq. 6.3 with $n = 1, 2$, and 3. We realize that since the first harmonic occurs when $n = 1$, then $\frac{v}{2L} = 440$ Hz. Thus,

$$f_1 = 1 \times 440 \text{ Hz} = 440 \text{ Hz (first harmonic)}$$

$$f_2 = 2 \times 440 \text{ Hz} = 880 \text{ Hz (second harmonic)}$$

$$f_3 = 3 \times 440 \text{ Hz} = 1320 \text{ Hz (third harmonic)}$$

The pitch of a stringed instrument, such as a piano, is determined by the frequency. A piano tone (note) is made by striking a key that controls a hammer. The hammer strikes a stretched string (actually wire) that vibrates at the different harmonics determined by Eq. 6.3. The frequencies depend on (1) the length of the piano string; (2) the tension in the string; and (3) the mass per unit length or thickness of the piano string.

Each of these three conditions can be changed. To get a higher note you can (1) shorten the length; (2)

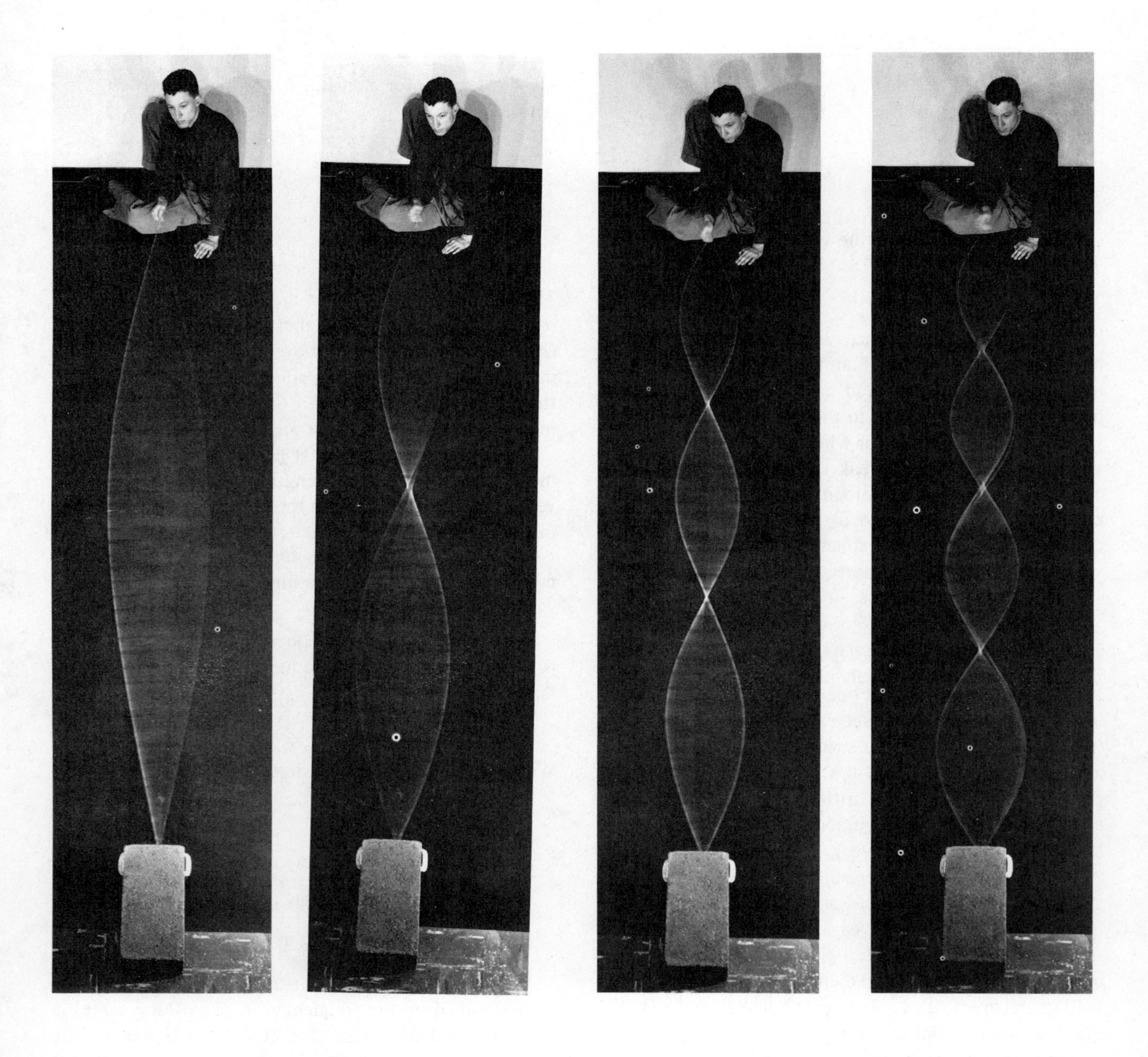

Figure 6.13 **Standing waves.** As one end of the rubber string is moved from side to side with increasing frequency, patterns of more and more loops are formed. However, the standing wave patterns shown are produced only for certain definite frequencies. (Reprinted, by permission, from PSSC *College Physics,* Education Development Center, Newton, Mass.)

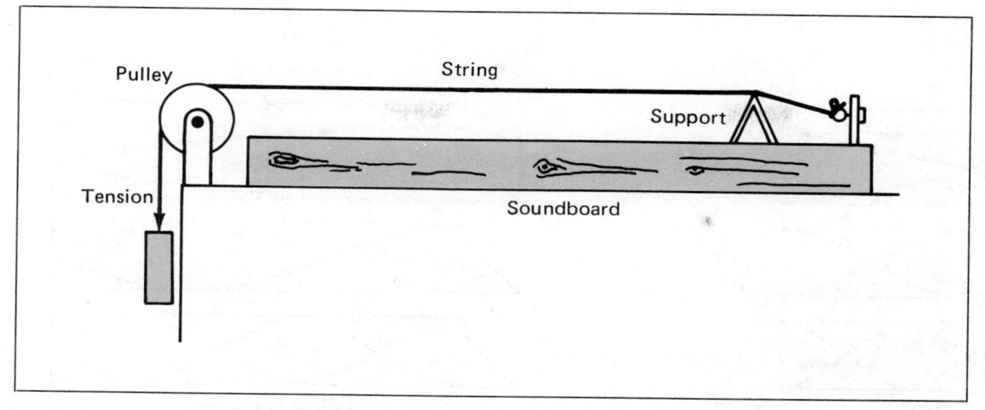

Figure 6.14 To make the sound produced by vibrating strings more audible, the strings are coupled to soundboards. Examples of this application are stringed musical instruments such as the violin, guitar, and piano.

increase the tension; (3) decrease the mass per unit length. If you look inside a piano you will see that the lengths of the strings are different and the lower notes are thicker. To tune a piano you adjust the tension in the strings.

All stringed instruments follow these principles. In the violin family of instruments the strings are tuned by adjusting the tension, and different notes are played by holding the strings down at different spots to change the length.

A single vibrating string does not produce much sound since it is so thin. A harp uses mostly string power. Violins, pianos, and most other stringed instruments are coupled to a soundboard.

Fig. 6.14 is a schematic diagram of a vibrating string connected to its soundboard. The vibration of the string sets the board vibrating, and the vibrations of the soundboard cause the sound to be much louder. In a violin the support is called the bridge and both the top and bottom of the violin act as soundboards (coupled by internal sound posts). Of course, the soundboards must be able to vibrate at the same frequencies as the strings.

The modern musical scale is constructed by arranging the various notes to play at particular frequencies. A portion of a labeled piano keyboard is shown in Fig. 6.15. The white keys are labeled A through G and repeat themselves. Each new repetition is called an **octave.** The frequencies of the octave starting at middle C are listed in Table 6.1. The same notes in two adjacent octaves have a frequency ratio of 2 to 1. Thus, in Table 6.1 the note labeled C′ has a frequency of 523.2 Hz, twice that of middle C, which has a frequency of 261.6 Hz. Similarly, A′ would be 2 × 440 Hz or 880 Hz, A″ below A would be $\frac{1}{2}$ × 440 Hz, or 220 Hz, and so on.

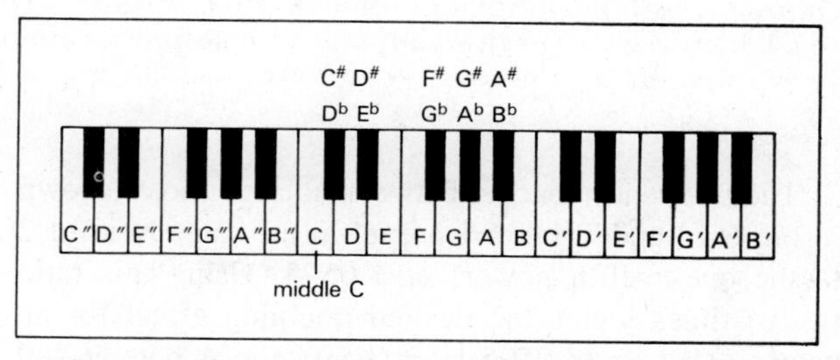

Figure 6.15 A portion of the piano keyboard. The white keys are labeled A through G. Each higher A (or B, C, . . .) is double the frequency of the one before. The twelve divisions between two As (or two Bs, etc.) have a frequency ratio of $2^{1/12}$ or 1.0595. That is, each key plays a frequency 1.0595 times the frequency of the preceding key.

Table 6.1 Frequencies of the Musical Scale from Middle C up One Octave

Note	*Frequency (Hz)*
middle C	261.6
C#, D♭	277.2
D	293.7
D#, E♭	311.1
E	329.6
F	349.2
F#, G♭	370.0
G	392.0
G#, A♭	415.3
A	440.0
A#, B♭	466.2
B	493.9
C′	523.2

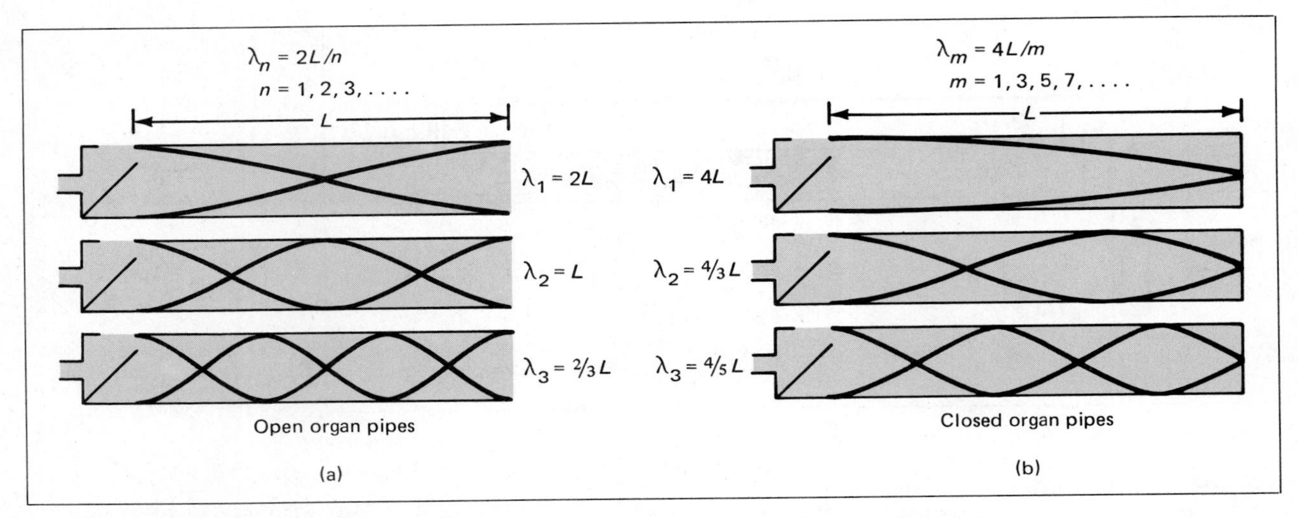

Figure 6.16 An illustration of the modes of vibration in open and closed organ pipes. (a) In an open organ pipe, the vibrating air column has antinodes at each end. (b) In a closed organ pipe, the vibrating air column has a node at the closed end of the pipe.

The frequency spacing between adjacent notes shown in Fig 6.15 and Table 6.1 are all in the same ratio of 2 to the one-twelfth power, or 1.0595. Using this ratio twelve times yields the desired doubling effect for an octave. That is, $(1.0595)^{12} = 2.00$. The A note at 440 Hz is the standard note, and all the other frequencies are found using this standard.

Another way to produce standing waves is by using a vibrating column of air. The musical instruments known as wind instruments, including the organ, use vibrating columns of air to produce sound.

The air column can be open at both ends, called an **open pipe,** or closed at one end, called a **closed pipe.** Organ pipes, illustrated in Fig. 6.16, are examples of open and closed pipes.

The pipe acts to make the sound resonate. The pipe has natural frequencies of vibration determined by its length. If the pipe is open at both ends, as in Fig. 6.16 (left side), the air molecules are free to move at the ends of the pipe. This implies that open pipes have antinodes at each end. The allowable wavelengths are $\lambda_n = 2L/n$, so the allowed frequencies for an open pipe are

$$\text{(open pipe)}\ f_n = \frac{v}{\lambda_n} = \frac{nv}{2L} \quad n = 1, 2, 3, \ldots \tag{6.5}$$

Note that these are the same frequencies as a vibrating string that has a node at each end.

A closed pipe does not allow for air molecules to vibrate at the closed end. Thus, there are antinodes at the open end and nodes at the closed end. From Fig. 6.16 (right side) it can be seen that closed pipes have allowed wavelengths of $\lambda_m = 4L/m$ where m = odd integers. The allowed frequencies (harmonics) are then

$$\text{(closed pipes)}\ f_m = \frac{v}{\lambda_m} = \frac{mv}{4L} \qquad m = 1, 3, 5, 7, \ldots \tag{6.6}$$

Thus, only the odd harmonics are present in the notes heard.

EXAMPLE 6 A closed organ pipe has a fundamental frequency of 392 Hz. What is its length?

We can rearrange Eq. 6.6 to get

$$L = \frac{mv}{4f_m}$$

for the fundamental frequency, $m = 1$, so

$$L = \frac{344 \text{ m/s}}{4 \times 392 \text{ Hz}}$$

or

$$L = 0.22 \text{ m}$$

When a stream of air enters the organ pipe, the air passes over an edge that disturbs the air stream. The air is broken up into many different frequencies, but only certain frequencies resonate with the natural frequencies of the pipe. The sound from these pipe frequencies is amplified, and these are the frequencies heard coming from the pipe.

We have already discussed how the pitch of a piano note (or any stringed instrument) is determined by the length and mass per unit length (thickness) of the string as well as its tension. In wind instruments it is the length and shape that determine the possible frequencies that are heard.

The effective length of wind instruments can be changed by using valves or by opening and closing holes along their lengths. In the trombone the actual length of the instrument is changed by sliding a portion of the tube in and out. The different sizes and shapes of the various wind instruments produce different sounds.

Musical notes are determined by the pitch or fundamental frequency that is present. For example, a fundamental frequency of 440 Hz produces an A. We know, however, that not all As sound alike.

The **quality** of a tone is due to the relative amounts of the various overtones present in the note. If we plot the waves given off by various A notes, we might get shapes similar to those shown in Fig. 6.17.

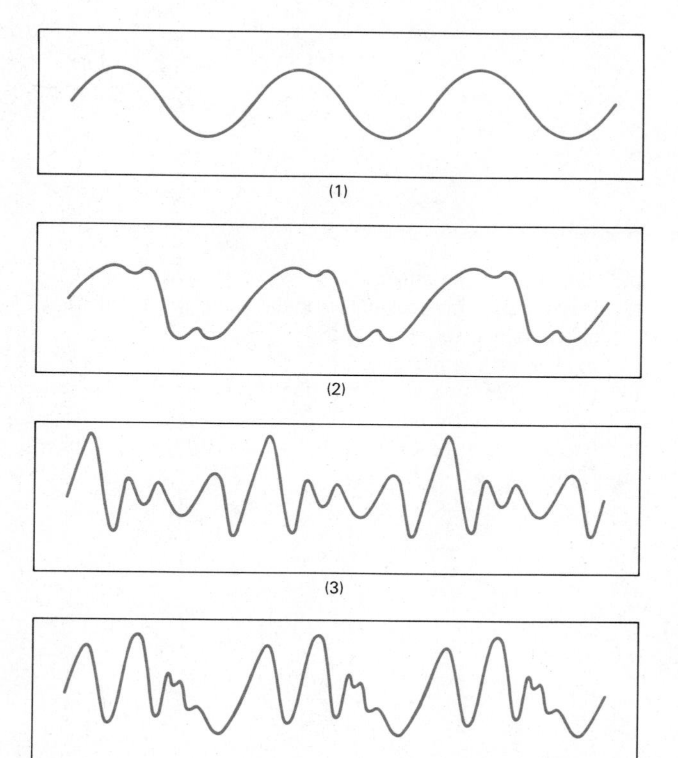

Figure 6.17 The musical note A sounded by (1) a tuning fork, (2) a flute, (3) a saxophone, and (4) a human voice.

A nearly pure tone would be given off by a tuning fork. This tone consists almost totally of the fundamental frequency with no overtones. A flute might give a sound wave similar to Fig 6.17(2). This tone has some of the harmonics mixed in. By comparison, more complicated A tones are given by the saxophone and the human voice, also shown in Fig 6.17.

6.5 The Doppler Effect

Whenever the source of a wave is moving relative to the observer, the frequency (and wavelength) that is observed changes. This is easily seen by referring to Fig. 6.18. As the source moves along to the right, the waves at Observer A are closer together than normal while the waves perceived by Observer B are farther apart. Observer A would perceive a higher frequency than the source is actually emitting. Observer B would hear a lower frequency.

If the source were stationary and the observer moved toward or away from it, a shift in frequency would also be observed. The **Doppler effect** is the apparent change in frequency resulting from the relative motion of the source or the observer.

The Doppler effect is a general effect that occurs for all kinds of waves, such as water waves, sound waves, and light waves. When a racing car with a loud engine noise approaches, a higher than usual frequency is heard. As it passes by, the frequency suddenly shifts lower and

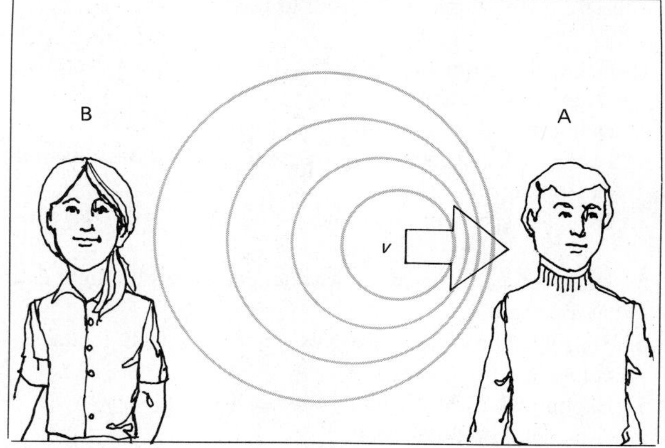

Figure 6.18 If a wave source moves to the right at velocity v, the wave crests will be closer together on the right side and farther apart on the left side. Observer A will hear a higher than usual frequency while Observer B will hear a lower than usual frequency.

a "whoom" sound is heard. You may have heard a similar noise as a siren passes by.

When the Doppler effect occurs with visible light, the frequencies are shifted higher or toward the blue when the light source (such as a star) is approaching. Since blue is a high-frequency color, we say a *blue shift* has occurred. When a light source moves away from us, the frequency is shifted toward the red, and we say a **red shift** has occurred. The amount of blue shift or red shift will tell us how fast the light source, such as a distant star or galaxy, is moving toward or away from us. Red shifts are extremely important in astronomy because they tell us how fast distant galaxies are moving away from us. As we will see in Chapter 21, this speed is related to how far away the galaxy is from us.

Whenever the speed of the source exceeds the wave speed, a **shock wave** is formed. This is illustrated in Fig. 6.19. When this happens with an airplane exceeding the speed of sound, the resulting effect is a "sonic boom." A sonic boom occurs because the compressed air from the shock wave travels out and downward behind the plane. The "boom" travels across the ground, and an observer hears it only once. This effect is not observed for light waves in a vacuum, since nothing can travel faster than the speed of light in a vacuum. However, there are other common examples of shock waves, such as a speed boat that travels faster than the water waves.

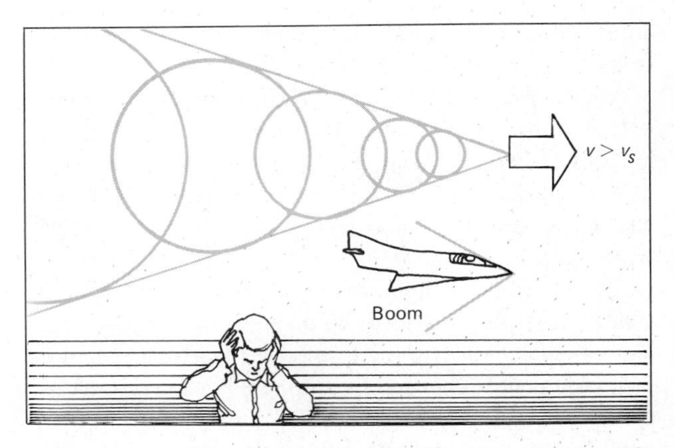

Figure 6.19 When the velocity of a jet plane, v, exceeds the velocity of sound, v_s, a shock wave is formed. A sonic boom is heard as the shock wave passes.

Learning Objectives

After reading and studying this chapter, you should be able to do the following without referring to the text:

1. Define the two main types of waves, and give an everyday example of each.
2. Define five properties of every wave.
3. List in order of increasing frequency the various types of electromagnetic waves.
4. Give the speed of light and the speed of sound in mks units.
5. Differentiate between the wavelengths of visible light and audible sound.
6. State what properties of waves give rise to pitch, loudness, color, and brightness.
7. Explain what causes sound waves, and explain why there is no sound on the moon.
8. Explain why marching bands sometimes sound dissonant.
9. State three factors which determine the pitch of a piano string.
10. Sketch the fundamental and first two overtones for closed and open organ pipes.
11. Explain how the musical scale is constructed.
12. Discuss why musical notes of the same pitch can have different quality.
13. Explain what a red shift is.
14. Discuss how we know that the distant galaxies are all moving away from us.
15. Explain what causes (a) a shock wave (b) the crack of a whip.

Important Words and Terms

wave motion
wave velocity
longitudinal wave
transverse wave
frequency
hertz
period
wavelength
amplitude
electromagnetic waves
speed of light
angstrom
sound
compression
rarefaction
pitch
ultrasound
intensity
decibel
resonance
nodes
antinodes
fundamental frequency
first harmonic
second harmonic
overtones
octave
open pipe
closed pipe
tone quality
Doppler effect
red shift
shock wave

Questions

1. What are the mks units of (a) wavelength? (b) frequency? (c) period?
2. How long is (a) a sound wave? (b) a visible light wave?
3. List in order of increasing frequency the various types of electromagnetic waves.
4. Are radio waves sound waves? Explain.
5. What causes audible sound waves?
6. What wave property distinguishes between (a) different colors? (b) different sound pitches? (c) different radio stations?
7. What wave property causes (a) varying loudness? (b) varying brightness?
8. Is there sound on the moon?
9. What is ultrasound used for?
10. Why do marching bands sometimes sound dissonant?
11. What three factors determine the pitch of a violin string?
12. What factor determines the pitch of a trombone note?
13. How is the musical scale constructed?
14. What gives a tone its quality?
15. How do we know the distant galaxies are all moving away from us?
16. What causes (a) a sonic boom? (b) the crack of a whip?

Problems

1. Waves moving on Lake Michigan have a speed of 2 m/s. The distance between adjacent crests is 5 m.
   (a) Determine the frequency of the waves.
   (b) Find the period of the wave motion.
2. The velocity of sound in water is 1530 m/s. What is the wavelength of a 2000 Hz sound wave?
3. Compute the wavelength in air of ultrasonic sound waves vibrating at 60,000 Hz. *Answer:* 5.7×10^{-3} m or 0.57 cm
4. Compute the wavelength of radio waves emitted by an AM radio station operating at 1340 kHz.
5. Compute the frequency of blue light with a wavelength of 4200 Å. *Answer:* 7.1×10^{14} Hz
6. Compute the wavelength in Å of an X ray with a frequency of 10^{18} Hz. *Answer:* 3×10^{-10} m or 3 Å
7. During a thunderstorm, 4 s elapses between a flash of lightning and the corresponding thunder. Approximately how far away was the lightning?
8. A subway train has a sound intensity of about 90 dB, and a rock band has a sound intensity of about 110 dB. The intensity of the sound of the band is how many times the intensity of the subway? *Answer:* 100 times
9. A rock band with an intensity of 123 dB turns its speakers down to 1% (1/100) of their previous value. What is the new intensity level in dB?
10. A speaker is playing at 80 dB. If it is turned up to 10,000 times the previous power, how many dB will it sound? *Answer:* 120 dB
11. A new jackhammer has only 1/10 the sound intensity of its predecessor, which was rated at 118 dB. How many dB is the new jackhammer?
12. A closed organ pipe has a fundamental frequency of 440 Hz at 20°C. What is the length of the pipe? *Answer:* 0.2 m
13. A closed organ pipe has a length of 2 m. At 20°C, what is the frequency of its (a) first harmonic? (b) third harmonic? (c) fifth harmonic? *Answer:* (b) 129 Hz
14. What length of open organ pipe is needed for a fundamental frequency of 440 Hz?
15. Refer to Fig. 6.15 and Table 6.1 and compute the frequencies of C″ and A″ (the C and A below middle C).

7

Wave Effects

The effects of waves reveal the beauty and mysteries of nature.

THE EFFECTS OF waves—particularly sound and light waves—are all around us. We are aware of many of these effects, but they are such a part of our experience that we take them for granted and rarely try to analyze them. For instance, if you speak loudly to someone in another room, you know the person hears you. Sound waves travel around corners; however, visible light waves do not. You can be heard in the next room, but not seen.

Similarly, when we look up at the sky, we see a blue sky, white clouds, and sometimes even a rainbow. These are natural phenomena that owe their description and understanding to the effects of electromagnetic waves interacting with air and water droplets in the atmosphere. We are well aware of these phenomena, but we rarely stop to consider the wave effects that cause them.

Mirrors and lenses also are based on the wave effects of reflection and refraction. These two effects will be discussed and then the basic principles of lenses and mirrors will be described. This will lead to an understanding of many common optical devices such as the human eye, slide projectors, and eyeglasses.

There are many wave effects. In fact, two of them—resonance and the Doppler effect—have already been discussed. In this chapter we will discuss most of the other important wave phenomena that affect us all the time.

7.1 Reflection

Waves travel through space in a straight line and will continue to do so unless forced to deviate from their original direction. A change in direction takes place when light strikes a surface or the boundary between two media. A change in direction by this method is called **reflection.** We see most of the objects in our environment because of reflected light. The reflection taking place from the surface of the paper of this text is called **diffuse reflection.** The rays (in this discussion of light we can represent its motion by the use of a straight line called a ray) of light falling on the surface of the paper, when reflected, are not leaving the surface parallel to one another because the surface is not perfectly smooth. When the reflection takes place from a smooth surface, the reflected rays are parallel to one another, and this is called **regular reflection.** This occurs from mirrors and highly polished surfaces. See Fig. 7.1 for an illustration of the two types of reflection.

◀ Wave effects such as refraction are quite common in our daily lives.

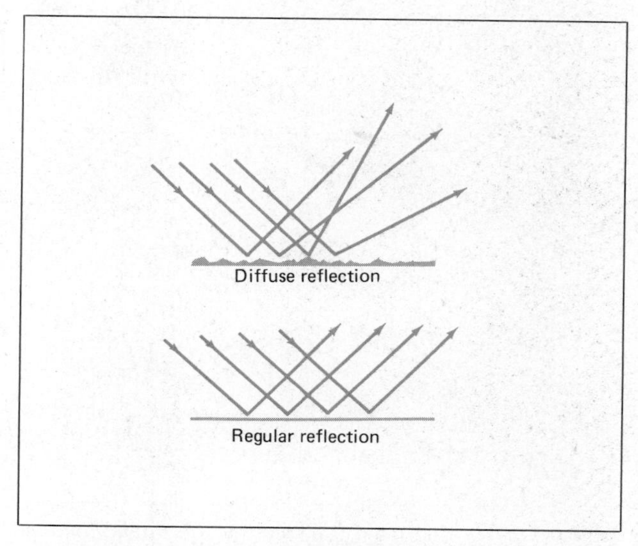

Figure 7.1 Two types of reflected light.

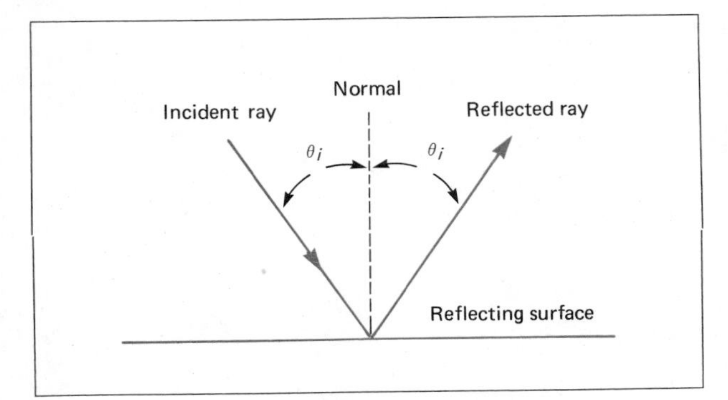

Figure 7.2 When light is reflected from a smooth surface, the angle of reflection equals the angle of incidence θ_i.

Figure 7.2 is an illustration of the two laws governing reflection which state: (1) the angle of reflection is equal to the angle of incidence; and (2) the reflected ray and the incident (or incoming) ray are in the same plane. The *angle of incidence* is defined as the angle between the incident ray of light and the normal, or perpendicular. The *reflected angle* is the angle between the reflected ray and the normal.

A diagram illustrating a method for determining the apparent location of an image formed by a mirror is shown in Fig. 7.3. The image can be located by drawing any two rays of light emitted by the object and observing the laws of reflection. For a plane mirror, the image will be located behind the mirror plane the same distance as the object is in front of the plane of the mirror. Figure 7.4 is a ray diagram showing the light rays involved when the woman sees the extremes of her image. Application of the laws of reflection reveals that for the woman to see her total image in a plane mirror (distance

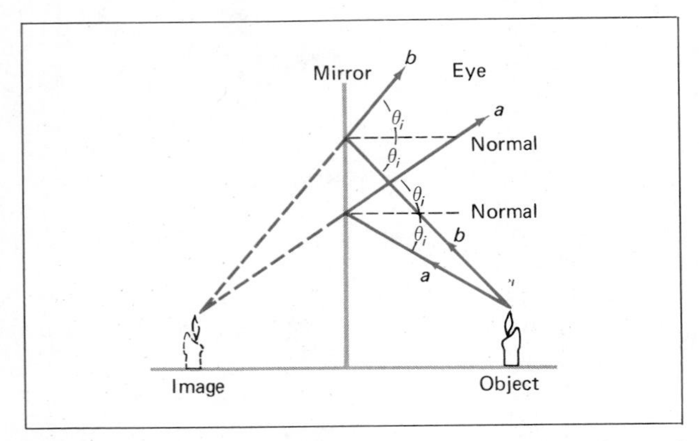

Figure 7.3 Method used for locating the image formed by a plane mirror.

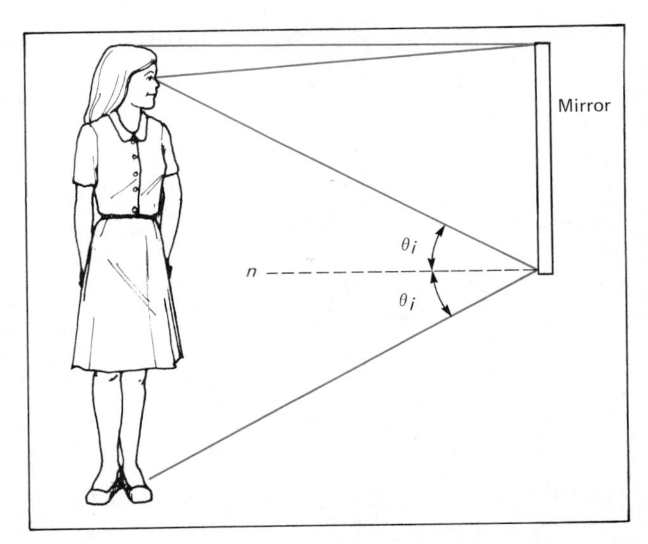

Figure 7.4 Image formation in a plane mirror. Note that the vertical height of the mirror equals one-half the height of the woman.

from the mirror is not a factor) the mirror has to be only half her height.

It is the reflection of light that allows us to see things. Look around you. What you see is light reflected from walls, ceilings, floors, and the objects around you. Of course, there must be one or more sources of light present. These are generally in the form of light bulbs or the Sun. If you are in a dark room with no light present, then there is no light to reflect from anything, and the room is black.

The sky is blue because of sunlight scattering from air molecules. See Fig. 7.5. (On the moon, where there is no atmosphere, the sky is black.) The air molecules are much smaller than the wavelength of visible light. Because of this, the shorter wavelengths, such as violet and blue, are scattered much more than are the long wavelengths such as red and orange.

Violet light is scattered more than blue light. We see blue instead of violet for two reasons. First, more blue than violet light is emitted by the Sun. Second, our eyes are more sensitive to blue than to violet.

Clouds are composed of water droplets, which are quite small compared to many things, but are much larger than a wavelength of visible light. Thus, they scatter all visible wavelengths of sunlight equally. The result is a white color since white is the sum of all the colors of the visible spectrum.

When small dust particles are present due to air pollution, the dust particles also reflect all visible wave-

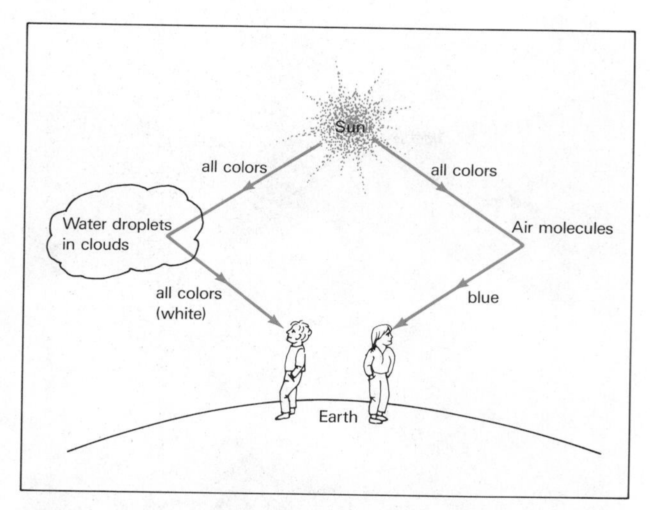

Figure 7.5 The sky is blue because our eyes detect more reflected blue light scattered from air molecules than any other color. The clouds are white because the water droplets reflect all colors equally.

lengths, and the sky appears whitish or hazy. The bluer the sky, the less dust, pollen, and so on, that are present. For further discussion on why the sky is blue, see Section 22.4.

7.2 Refraction and Dispersion

Another method of changing the direction of a wave is to produce a change in its speed. This occurs whenever the wave penetrates from one medium, such as air, into another medium, such as water.

When light goes from one medium to another, its speed changes. The change in speed depends upon the two media involved and on the frequency of the light. Because of the change in speed, the light wave is deviated from its original path. The bending of light waves because of a speed change is called **refraction.** A pencil placed in a glass of water looks magnified and bent, because the light rays are refracted (that is, deflected from their normal directions) since their speed is different in water than in air. The ratio of the speed change, known as the **index of refraction,** can be written as

$$\text{Index of refraction} = \frac{\text{speed of light in vacuum}}{\text{speed of light in medium}}$$

or as

$$n = \frac{c}{c_m}$$

Table 7.1 Indices of Refraction of Some Common Substances

Substance	*n*
Water	1.33
Crown glass	1.52
Diamond	2.42
Air (0°C, 1 atm)	1.00029

The index of refraction is a pure number, since c and c_m are measured in the same units. Indices of refraction for some common substances are shown in Table 7.1.

The way in which bending of light is caused by a change in speed can be seen by examining Fig. 7.6. In Fig. 7.6 the rays of light are designated as a and b. Any line perpendicular to one of these rays, or any line perpendicular to both (as shown in this case with line OO') designates the wave front. As the wave front in this illustration approaches the glass, ray b arrives at point O' when ray a is at point O, the surface of the glass.

If we start recording time at this instant, and continue until ray b arrives at the surface of the glass, the elapsed time will be Δt, and the distance traveled by ray b will be $c\ \Delta t$, which will equal $O'P'$. Similarly, the distance OP will be $c_m \Delta t$. Since c_m is less than c, the distance $O'P'$ will be greater than the distance OP by a factor of c/c_m. This slowing down of ray a in the new medium, while ray b is still traveling in the air, and the subsequent slowing of the other rays between a and b, deviate the wave front toward the normal, or perpendicular.

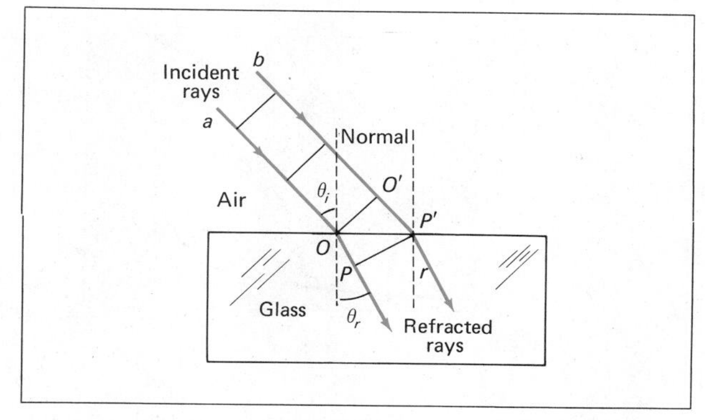

Figure 7.6 The refraction of a wave front. The distance OP is shorter than the distance $O'P'$ because light travels more slowly in glass than in air.

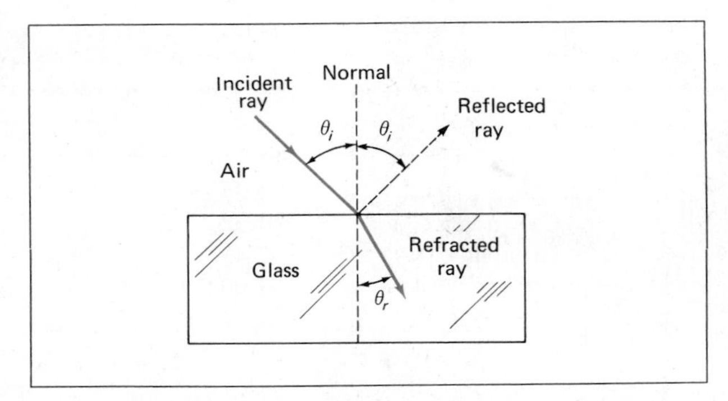

Figure 7.7 Diagram illustrating the deviation of a light ray from its original path as the ray passes from air into glass. A deviation of this type is called refraction. Some of the light is also reflected.

When light enters a new medium, it is bent as shown in Fig. 7.7. In this figure the refracted ray is bent toward the normal because the speed of light in glass is less than the speed of light in air. The amount of bending is due to the ratio of the indices of refraction of the two different media involved. Whenever refraction occurs, some reflection also occurs. This is designated by the dashed line in Fig. 7.7.

The amount of reflection versus the amount of refraction that occurs depends on the angle of incidence. If the light is coming straight through so that $\theta_i = 0$, very little light is reflected and almost all is refracted. The amount of light reflected is generally quite small until close to $\theta_i = 90°$ when all of it is reflected.

A sketch of the amount of light reflected from water versus the angle of incidence is shown in Fig. 7.8. When you look straight down into water, only 2% of the light

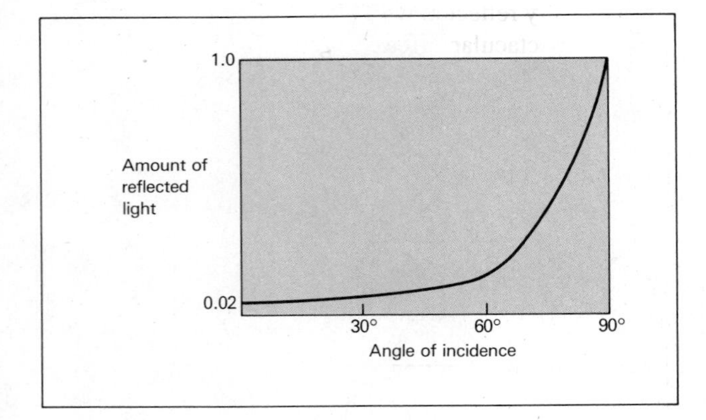

Figure 7.8 Amount of light reflected from water versus angle. Only 2% is reflected when the light ray is perpendicular to the surface of the water.

Figure 7.9 When the angle of incidence is large, much of the light is reflected from a lake.

is reflected. When you look across a calm body of water, the angle of incidence is large, and around 50% to 100% of the light is reflected as shown in Fig. 7.9.

You have probably noticed that at night you can see your reflection in a window pane, but only rarely does this occur in daylight. The explanation for this is found in the reflection and refraction of light through a window pane. When light is incident on a window pane so that $\theta_i = 0$, approximately 4% is reflected at each surface for a total of 8% reflection. The other 92% goes through the window pane. (Absorption of light can be ignored here.)

During the day, much more light is usually coming in than is reflected, so the reflected light is usually masked and cannot be seen. During the night, there is little or no light coming in from the outside, so the small amount (8%) of reflected light can now be seen. Thus your reflection, which is always present, is usually masked during the day, but not at night. This is illustrated in Fig. 7.10.

Sound waves also show the effects of reflection and refraction. An echo is caused by reflection of sound waves. The refraction of sound waves in air can occur if there are air layers at different temperatures, but this is not a common effect.

When light goes from a medium with a high index of refraction to one with a low index of refraction (e.g., from glass to air), an interesting phenomenon occurs. This is illustrated in Fig. 7.11. If the angle is θ_1, then

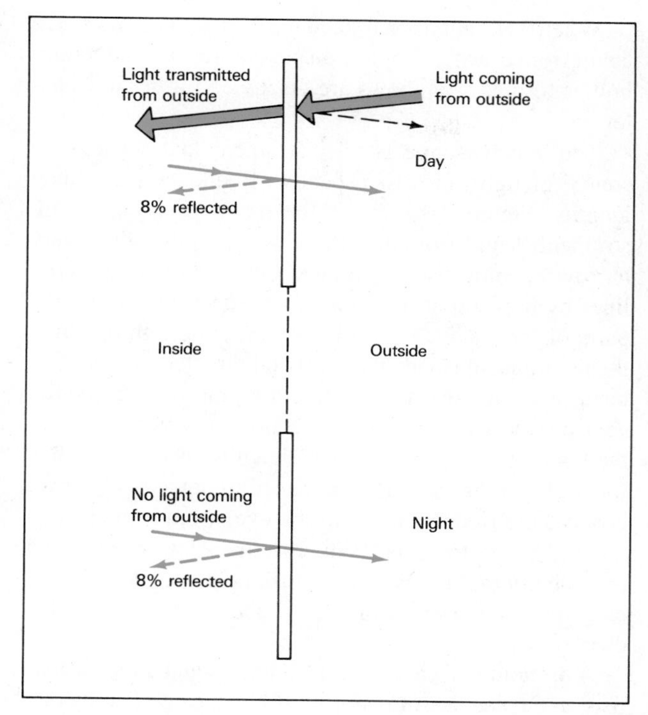

Figure 7.10 Light transmitted and reflected from a window pane. During the day you see light being transmitted from the outside. At night with no light coming through from outside, you can see the reflected light which is always present.

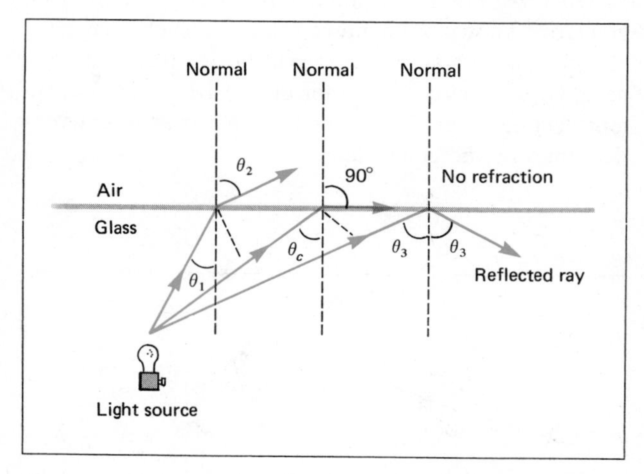

Figure 7.11 Internal reflection. When light goes from glass to air, it is refracted away from the normal. At a certain critical angle θ_c, the angle of refraction is 90°. When the angle of incidence exceeds θ_c, all the light is reflected internally.

Figure 7.12 Refraction, reflection, and total internal reflection. The light enters from the left and is reflected and refracted by the glass. The two beams at the bottom of the glass are totally reflected internally. (Reprinted by permission from PSSC College Physics, 4th ed. Education Development Center, Newton, Mass., 1976.)

the light is refracted as shown. At the angle θ_c, called the *critical angle,* the refracted ray makes an angle of 90° with the normal. At angles greater than θ_c, the light is totally reflected, and none is refracted. This phenomenon is called **total internal reflection.** An illustration of reflection, refraction, and total internal reflection is shown in Fig. 7.12.

Internal reflection enhances the brilliance of cut diamonds. A diamond is cut so that most of the light entering the diamond is internally reflected. The light then emerges from only certain portions with a brilliance far exceeding that of the incoming light. This effect creates the diamond's beautiful sparkle.

Another example of total internal reflection occurs when a fountain of water is illuminated from below. The light is totally reflected within the streams of water, providing a spectacular effect. Figure 7.13 illustrates how this is done with a single stream.

Light can be made to travel along transparent plastic tubes, sometimes called ''light pipes,'' by using the phenomenon of total internal reflection. The tubes are flexible, and the technique has been given the name ''fiber optics.'' One important use has been to pipe the light from hard-to-reach places such as the inside of a person's stomach or heart. Fiber optics also have applications in communications and industry.

The angle of refraction of light depends somewhat on the light's frequency or wavelength. The fact that different frequencies are refracted at slightly different angles is called **dispersion.** When white light (electromagnetic radiation containing all wavelengths visible to the

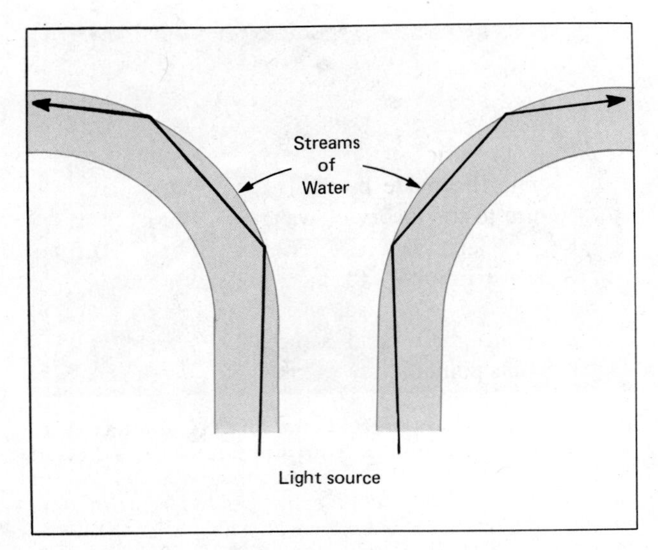

Figure 7.13 When a light source illuminates a water fountain from below, the light is trapped inside the streams of water due to total internal reflection as shown.

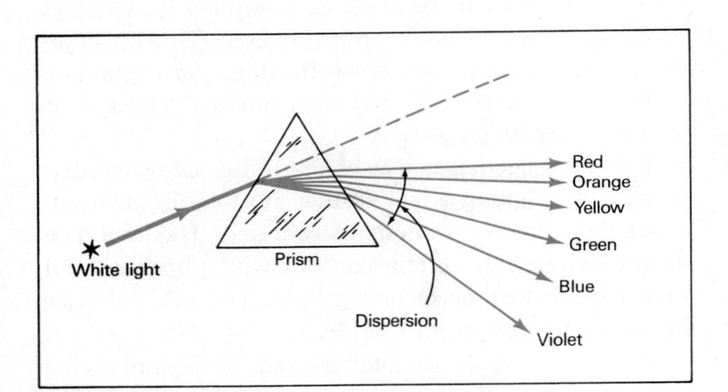

Figure 7.14 Dispersion of light. The white light is dispersed into a spectrum of colors by the glass prism. The short wavelengths such as blue are refracted more than the long wavelengths such as red.

human eye) passes through a glass prism as illustrated in Fig. 7.14, the light rays are refracted on entering the glass and refracted again on leaving the glass. Since the speed changes on entering (decreases) and on leaving (increases), the directions will change accordingly.

The amount of refraction is a function of the wavelength, with the shortest wavelengths being deviated from their path by the greatest amount. Violet light has a shorter wavelength than red light. A prism used in this fashion produces the visible color spectrum, and we say the white light has been dispersed.

When sunlight is refracted, reflected, and refracted again from a water droplet, dispersion results and a rainbow is formed. Rainbows are discussed further in Chapter 23.

The **spectroscope** is an instrument that separates a source of light into its respective frequencies or wavelengths. Figure 7.15 shows the major parts of the instrument. Light from a source falls upon the slit, a very narrow opening used to reduce blurring of the spectral lines by preventing overlaps of the spectrum from other parts of the incoming beam of light. The light beam is then collimated (made parallel) and directed as a parallel beam upon the prism (a diffraction grating [see Section 7.3] can be used in place of a prism) which separates the light into its respective wavelengths. The light is then focused and directed upon a detecting device. When we observe the frequencies with the eye, we use the instrument as a spectroscope. When frequencies are recorded on a photographic plate, the instrument is called a spectrograph, and a spectrometer if the data are recorded electronically.

Astronomers, chemists, physicists, and other scientists use these instruments to study the total universe. When heated, every substance gives off characteristic frequencies. These frequencies can be detected, their spectra produced and studied.

The study of spectra is called spectroscopy, and the study of the universe with these instruments has given us more basic information about ourselves and the universe than any other instrument.

When we observe a photograph taken by a spectrograph of a known or unknown source of electromagnetic radiation, we obtain an image of the slit that is illuminated by the source (see color plate VIII). The images, representing definite wavelengths, appear as bright vertical lines on the color plate.

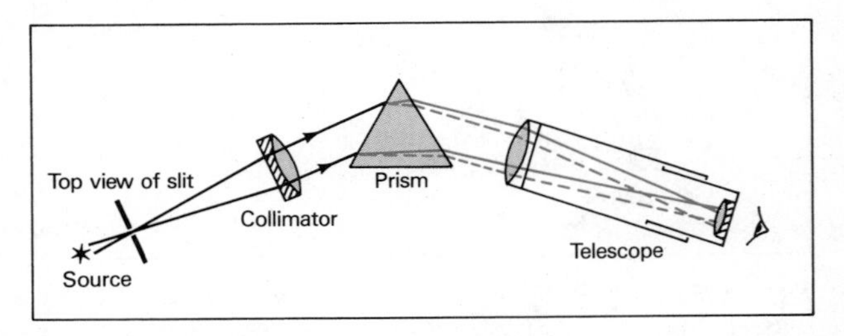

Figure 7.15 Prism spectroscope. The solid color lines represent red wavelengths, the dotted color lines violet wavelengths. See Section 7.2 for an explanation.

7.3 Diffraction, Interference, and Polarization

Another method by which waves can be deviated from a straight-line path is to let them pass through a narrow slit, where a small-scale interaction takes place with the edges of the slit to produce bright and dark fringes. The effect also occurs when light from a fairly bright source passes the edge of an opaque object. Close observation will show a shadow at the edges surrounded by narrow bands, or fringes. This phenomenon can easily be seen by holding two fingers close together to form a narrow slit and observing a bright light from a distance. If a fluorescent lamp is available, position the fingers parallel to the length of the lamp and try it. In all these cases, the deviation of light waves is referred to as **diffraction.**

The diffraction of a water wave as it passes through a small slit is shown in Fig. 7.16. Note how the waves bend around the slit as they pass through. All waves—sound, light, and so on—show this type of bending as they go through small slits. Sound is heard around corners due to diffraction as well as to reflection and refraction.

When a cheerleader yells through a megaphone, her or his voice is focused in the direction the megaphone points. The way in which the sound waves go out through the megaphone is similar to Fig. 7.16. This is a good example of diffraction.

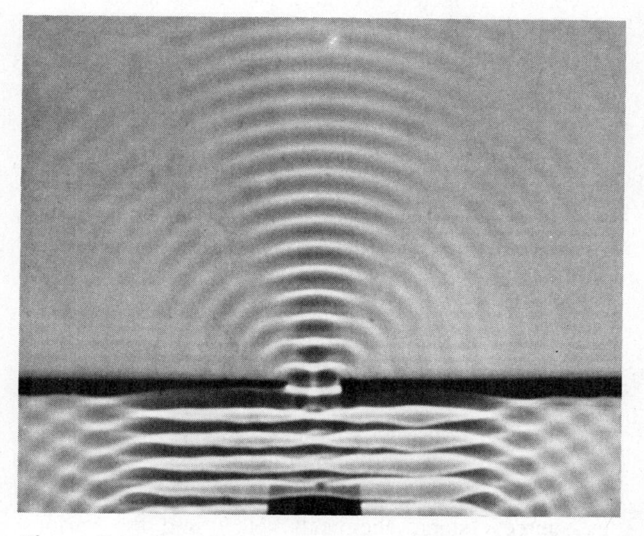

Figure 7.16 Straight waves passing through an opening. Since the width of the opening is near the size of the wavelength, diffraction is easily seen by noting how the waves curve around the ends of the barriers. (From PSSC Physics, 3rd ed., D. C. Heath, 1971.)

In our study of reflection and refraction, the location of a mirror image was done by drawing straight lines; that is, we solved the problem by using geometric methods. The study of optics using these methods is known as **geometric optics.** An explanation of diffraction, however, cannot be made by using geometric optics. We must return to the theory of wave motion to give a satisfactory explanation of the effects observed. Diffraction—and also interference—are the effects produced when a number of waves interact with one another because of difference in "phase" and "amplitude." A study of optics from this point of view is known as **physical optics.**

Effects of physical optics occur when effects such as diffraction become important. The key ratio to a quantitative understanding of diffraction is

$$\frac{\text{wavelength of wave}}{\text{size of opening or obstacle causing diffraction}}$$

or

$$\frac{\lambda}{d}$$

If λ/d is much less than one, very little diffraction occurs. If λ/d is approximately one or greater, diffraction effects are easily observed. Figure 7.16 is a good example of this effect using water waves.

We are well aware that sound waves bend around everyday objects while light waves do not. For instance, if we hold a newspaper in front of our face and speak, the sound of our voice can be heard, but our face cannot be seen. The phenomenon of diffraction and the ratio of λ/d help to explain why.

We know that audible sound waves have wavelengths of centimeters to meters while visible light waves have wavelengths of around 10^{-6} m. Ordinary objects have dimensions d of centimeters to meters. Thus, we get

$$\left(\frac{\lambda}{d}\right)_{\substack{\text{visible}\\ \text{light}}} \cong 10^{-6}$$

$$\left(\frac{\lambda}{d}\right)_{\substack{\text{audible}\\ \text{sound}}} \geq 1$$

From this we see that diffraction readily occurs for audible sound, but not for visible light.

When you sit in a lecture room, a movie theater or a concert hall, the sound will easily diffract around the people in front of you, but the light will not. This is because sound wavelengths are so long and light wavelengths are so short compared to the size of people.

Radio waves are electromagnetic waves of very long wavelength, some being hundreds of meters. Therefore

(λ/d) for radio waves is much greater than one and radio waves are easily diffracted around houses, trees, and so on.

Interference effects are defined as changes in wave motion produced by phase and amplitude relations of two or more waves. The amplitude of a wave has been defined (Section 6.1) as the maximum displacement of the disturbance from its midposition. Two or more waves may reinforce one another when they are changing in the same direction, or tend to cancel one another when varying in opposite directions. When two or more waves have their displacement at all times in the same direction, they are said to be in phase with one another. If their displacements at all times are in opposite directions, they are completely out of phase with one another. Figure 7.17 illustrates the concept of phase.

When interference occurs between waves, the two waves must remain exactly in phase for constructive interference (when waves reinforce each other) and 180° out of phase for destructive interference (when they completely cancel each other). For complete destructive interference, the two waves must have the same amplitude. They must also have the same wavelength and be traveling in the same direction.

Consider light waves hitting a thin film of oil on top of water. By thin we mean that the oil's thickness is the same order of magnitude as a wavelength of visible light. A diagram of two rays of light is shown in Fig. 7.18. The top ray bounces off (reflects from) the air-oil surface, and the bottom ray is refracted at the air-oil surface and bounces off (reflects from) the oil-water surface. The two waves will then be observed by someone's eye as shown.

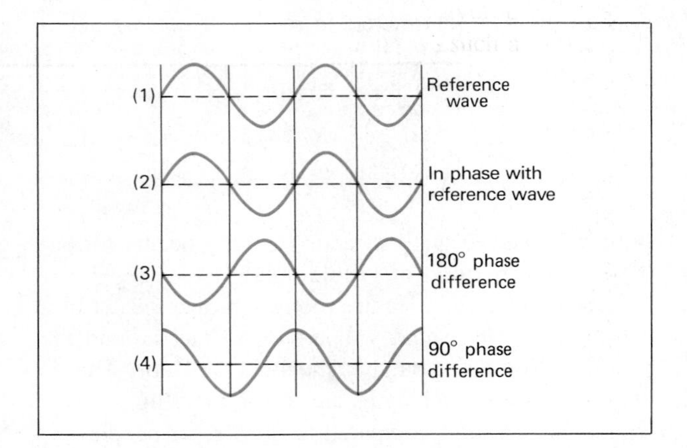

Figure 7.17 Diagram showing when waves are in and out of phase with one another.

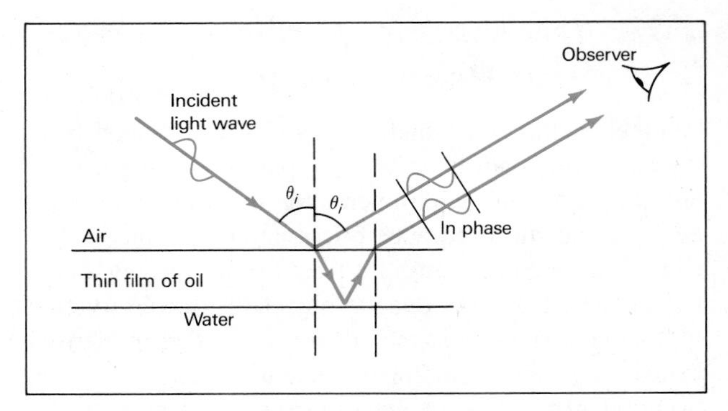

Figure 7.18 Interference occurs between two rays of incident light reflecting from a thin film of oil on water. The bottom ray is in phase with the top ray in this drawing. Usually they will be out of phase. The intensity of reflected light reaching the observer will be quite weak unless constructive interference occurs. Constructive interference occurs for only one visible color at each angle, and that is the color that is seen.

The two outgoing waves can be in phase, totally out of phase, or somewhere in between. In the diagram, the waves are shown in phase, but this will occur only for certain combinations of angles of observation, wavelength of light (color), and thickness of the oil film.

At certain angles and oil thicknesses, only one wavelength of light shows constructive interference. The other visible wavelengths interfere destructively. Thus, only one color is reflected strongly. The different colors we see by looking at different angles are the result of constructive interference.

A diffraction grating works on the principle of interference as well as diffraction. A **diffraction grating** consists of many narrow parallel slits spaced very close together. When light of a particular wavelength goes through the slits, some of the light is bent due to diffraction and will constructively interfere at an angle θ as shown in Fig. 7.19.

The observer sees light of a particular wavelength apparently coming from a position off to the side from the incident beam. If the incident light is a beam of white light, such as a light bulb, different visible wavelengths will interfere constructively at different angles and a spectrum will be produced and observed.

Diffraction gratings are used to split light from various sources (stars, chemicals, etc.) into the various wavelengths that are present. When this is done, the composition of the source can be analyzed.

As you now know, the difference between transverse waves and longitudinal waves is the direction of the par-

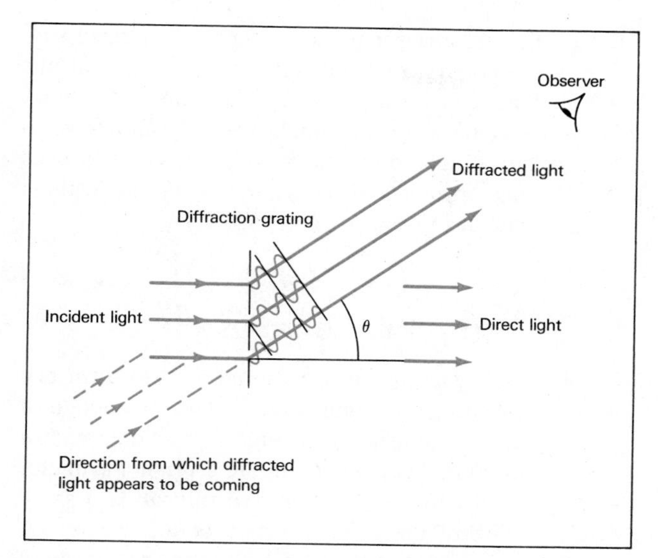

Figure 7.19 A diffraction grating. As light passes through the narrow slits, some is bent due to diffraction. At some angle θ all the rays show constructive interference for the wavelength λ. Each visible wavelength λ constructively interferes at a different angle, and a spectrum is produced.

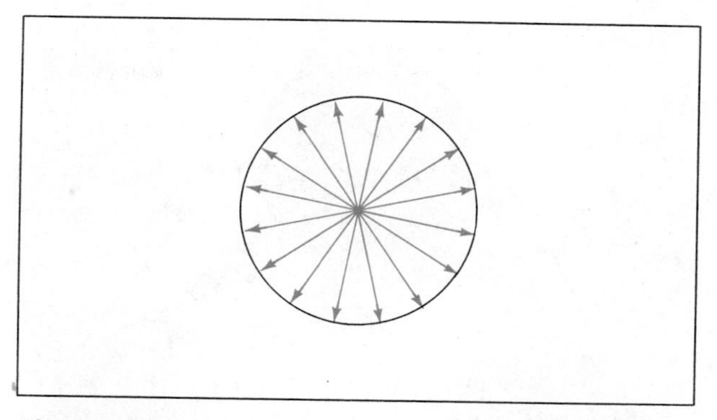

Figure 7.20 A "head-on-view" of a beam of unpolarized light.

ticle displacement relative to the velocity vector. In a transverse wave the particle displacement is perpendicular to the wave motion. Viewing a transverse light wave from the front, we observe a pattern similar to Fig. 7.20, which shows the electric field vectors in a plane perpendicular to the direction of travel. There are an infinite number of orientations of these electric field vectors in the plane, all pointing in different directions. We call such a wave **unpolarized.** If we restrict the vibrations of the electric field vector to one direction, then the wave is said to be **linearly polarized.**

A wave may be polarized by allowing the wave motion to take place in one direction only, using a Polaroid film. Figure 7.21 is an illustration of how an unpolarized light wave is polarized in the vertical plane. The polarizer allows only the vertical component to pass. The vibrations of the other components are restricted and not allowed to pass through the polarizer.

If a second polarizer is placed in front of the light wave at 90° with the first, then all radiation will theoretically stop (Fig. 7.22). The polarization of light is strong experimental proof that light is a transverse wave motion. Longitudinal waves cannot be polarized.

Polaroid sunglasses (Fig. 7.23) consist of a single sheet of Polaroid material that cuts down the total amount of light passing through it. Only vertically polarized light

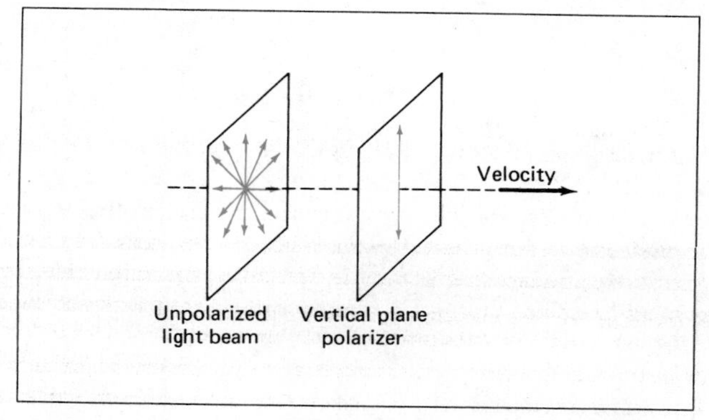

Figure 7.21 Diagram illustrating how a beam of unpolarized light is polarized.

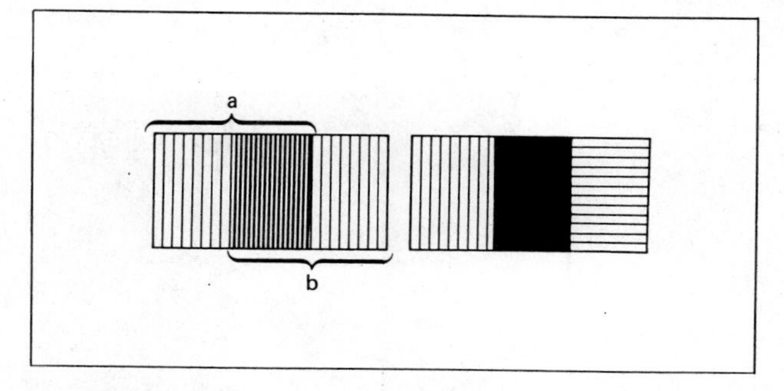

Figure 7.22 The polarization of light using Polaroid material. The left drawing illustrates the planes of polarization overlapping but parallel to one another. The right drawing illustrates the planes of polarization overlapping but perpendicular to one another. The black in the right drawing is achieved through complete absorption of the waves on the overlapping portions of the Polaroid material.

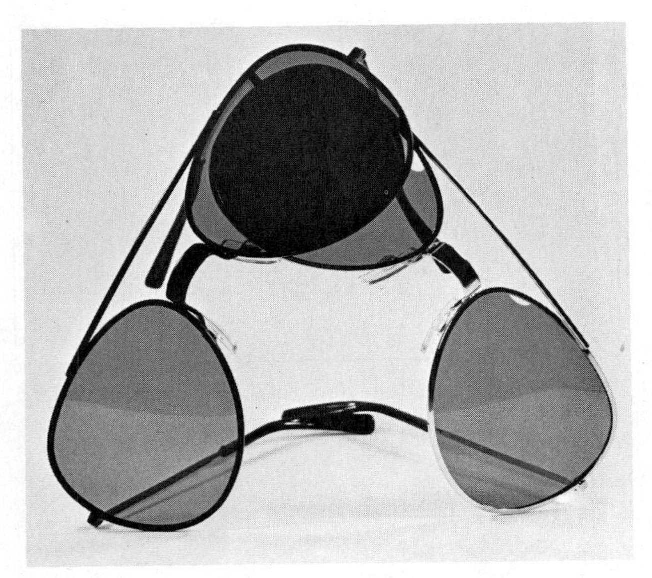

Figure 7.23 Polarized sunglasses consist of a single piece of Polaroid material. When two pieces of Polaroid are placed at right angles, no light can pass through.

Figure 7.24 Polaroid sunglasses reduce the glare of reflected light. A scene is viewed on the left without Polaroid sunglasses. The scene on the right shows how using Polaroid sunglasses blocks out the reflected glare.

can go through the sunglasses. When unpolarized sunlight falls upon a horizontal surface such as a road, the reflected light is polarized in the horizontal direction. Since the Polaroid in the sunglasses only allows vertically polarized light to pass, the reflected glare is blocked out. An example of a scene viewed with and without Polaroid sunglasses is shown in Fig. 7.24.

7.4 Spherical Mirrors

A spherical reflecting surface can be used to form real images from divergent light waves. The construction of such a mirror is illustrated in Fig. 7.25. Some appropriate and important optical terms are given in the figure. The spherical surface of a **concave mirror** is a small inside section of the total curvature of the sphere. (A **convex mirror** has a reflecting surface on the outside of the spherical surface.) The center point of a concave mirror is called the *vertex* v, and a line drawn through v and the center of curvature of the mirror C is called the *principal axis* of the mirror. For a concave mirror, rays of light coming in parallel to the principal axis will be reflected and brought to a focus on the principal axis at a point F known as the *focal point*. The distance from the focal point to the vertex is called the **focal length.** The geometric construction of the concave mirror makes the focal length one-half the value of the radius of curvature, where the radius of curvature is the radius of the spherical surface. Expressed in symbols

$$f = \frac{R}{2}$$

where f = the focal length,
R = the radius of curvature.

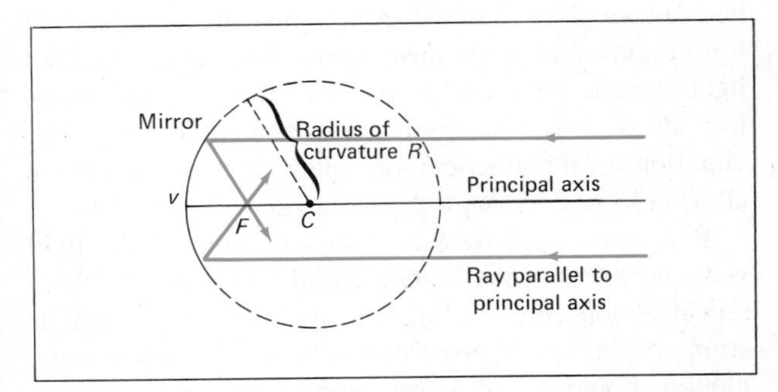

Figure 7.25 Light rays from a concave spherical mirror when incoming light rays are parallel to the principal axis and the image is at focal point F. The vertex is labeled v, and the center of curvature is C.

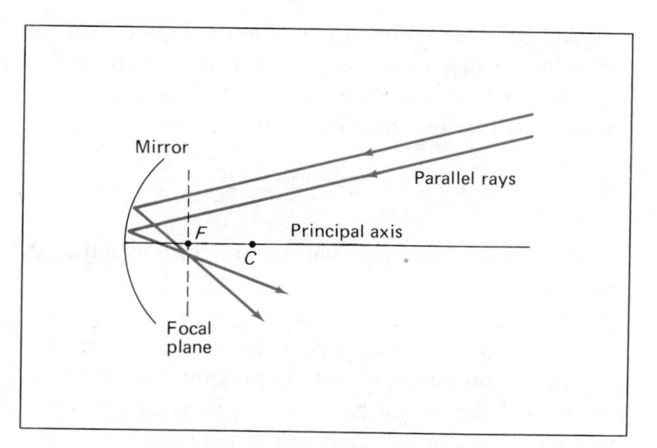

Figure 7.26 Light rays from a concave spherical mirror when incoming light rays are not parallel to the principal axis. The image is formed on the focal plane under these conditions.

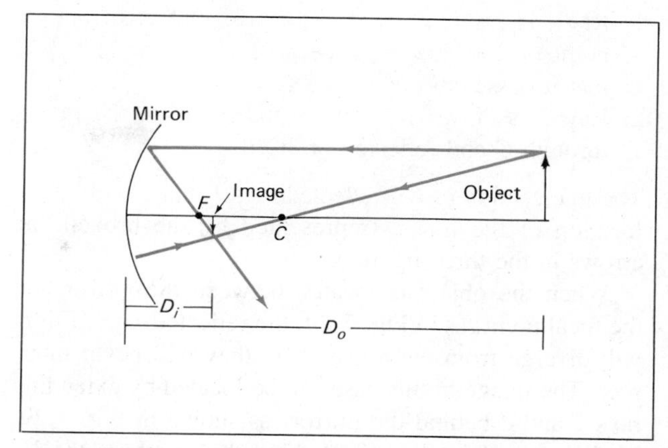

Figure 7.27 The graphic solution for locating the image formed by a concave spherical mirror when the object is beyond the center of curvature C.

Rays of light that come in parallel to one another but not parallel to the principal axis are reflected and brought to a clear image in the *focal plane,* as shown in Fig. 7.26.

The characteristics of the images formed by concave mirrors may be real or virtual; erect or inverted; larger, smaller, or the same size as the object; in front of or behind the mirror. A **real image** is defined as one that can be brought to a focus on a screen, and a **virtual image** is one that cannot. A virtual image is always behind the mirror, and one must look into the mirror to see the image.

The relationship between the object distance, the image distance, and the focal length is given by the equation

$$\frac{1}{D_0} + \frac{1}{D_i} = \frac{1}{f} \tag{7.1}$$

where D_0 = object distance from vertex of mirror,
D_i = image distance from vertex of mirror,
f = focal length.

The values for the object distance D_0 and the focal length f are considered to be positive for a concave mirror. (For a convex mirror, f is given a negative value.) If D_i is positive, then the image is real, inverted, and in front of the mirror. If D_i has a negative value, the image is virtual, erect, and behind the mirror.

Information concerning the location of the image can be obtained through Equation 7.1 or by means of the graphic solutions shown in Figs. 7.27, 7.28, and 7.29. In all three of these illustrations, the location of an image (represented by an arrow) can be determined by drawing two light rays so that the angle of incidence equals the angle of reflection.

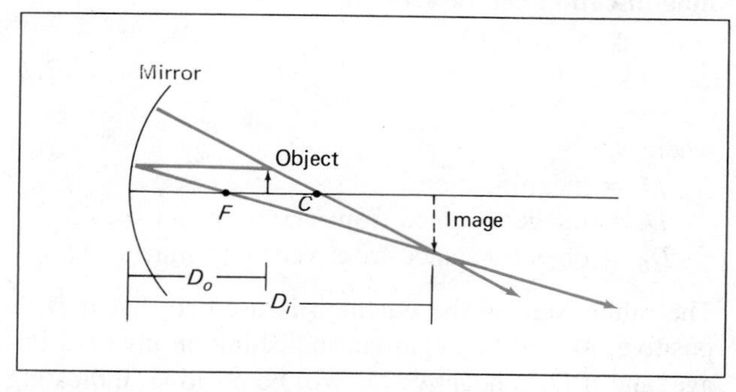

Figure 7.28 The graphic solution for locating the image formed by a concave spherical mirror when the object is between F and C.

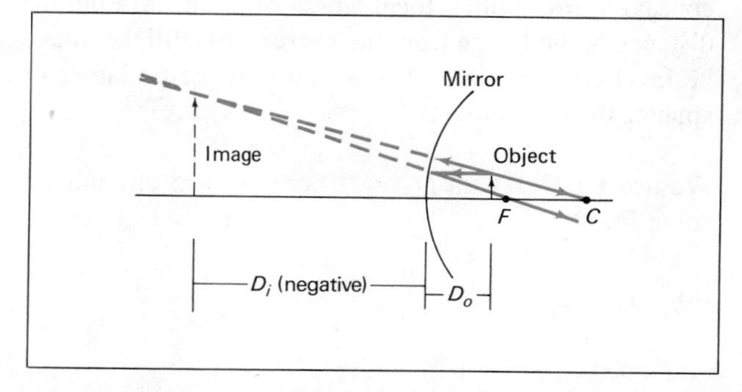

Figure 7.29 The graphic solution for locating the image formed by a concave spherical mirror when the object is between the mirror and F.

1. Ray 1 is drawn parallel to the principal axis from the point on the object to the mirror, then reflected so that it passes through F.
2. Ray 2 is drawn from the same point on the object through C and reflected on itself.

The intersection of the reflected rays 1 and 2 will be the location of the image (represented by the broken-line arrows in the three figures).

When the object is located between the mirror and the focal point, as in Fig. 7.29, the reflected rays of light will diverge from the mirror, and they will never intersect. The image in this case can be located by extending rays 1 and 2 behind the mirror, as shown in Fig. 7.29. Their point of intersection behind the mirror is the location of the image. In Eq. 7.1, D_i will be negative.

When the image is formed behind the concave mirror, it is larger than the object. The equation expressing the **magnification** can be written

$$M = -\frac{D_i}{D_0} \qquad (7.2)$$

where

M = magnification,
D_i = image distance from vertex of mirror,
D_0 = object distance from vertex of mirror.

The minus sign in the equation is used so that if D_i is positive, M will be negative, indicating an inverted image; and if D_i is negative, M will be positive, indicating an upright image.

EXAMPLE 1 An object is placed 20 cm from a concave mirror with a focal length of 5 cm. (a) Find the distance of the image from the mirror. (b) Will the image be inverted or upright? (c) Will the image be larger or smaller than the object?

We use Eq. 7.1 with D_o = 20 cm, f = 5 cm and we solve for D_i.

(a) $$\frac{1}{D_i} = \frac{1}{f} - \frac{1}{D_o}$$

$$\frac{1}{D_i} = \frac{1}{5\text{ cm}} - \frac{1}{20\text{ cm}}$$

$$= \frac{4}{20\text{ cm}} - \frac{1}{20\text{ cm}} = \frac{3}{20\text{ cm}}$$

$$D_i = \frac{20}{3}\text{ cm} = 6.67\text{ cm}$$

(b) and (c) This problem is similar to Figure 7.27. By referring to this figure we see that the image will be inverted and smaller than the object. Thus, the magnification will be less than one and negative.

$$M = -\frac{D_i}{D_o} = -\frac{20/3\text{ cm}}{20\text{ cm}} = -\frac{1}{3}$$

This shows the image will be one-third as tall as the object.

EXAMPLE 2 Consider a concave mirror with f = 50 cm. If you put an object 20 cm from the mirror, (a) where will the image be? (b) will it be upright or inverted? (c) will it be magnified or not?

Again, we use Eq. 7.1 and solve for D_i.

(a) $$\frac{1}{D_i} = \frac{1}{f} - \frac{1}{D_o}$$

$$\frac{1}{D_i} = \frac{1}{50\text{ cm}} - \frac{1}{20\text{ cm}}$$

$$= \frac{2}{100\text{ cm}} - \frac{5}{100\text{ cm}} = -\frac{3}{100\text{ cm}}$$

$$D_i = -\frac{100}{3}\text{ cm} = -33.3\text{ cm}$$

The negative sign for D_i means that the image is behind the mirror instead of in front. Thus, the image will appear to be 33.3 cm behind the mirror.
(b) This situation is similar to Fig. 7.29. From this figure we see that the image is erect and magnified.
(c) From Eq. 7.2 we can compute the magnification M

$$M = \frac{-D_i}{D_o} = -\frac{100/3\text{ cm}}{20\text{ cm}} = +\frac{5}{3} = 1.67$$

so the image will be magnified by a factor of 1.67. The positive sign indicates that the image will be upright.

A makeup mirror works similarly to Fig. 7.29 and Example 2. It is a concave mirror with a focal length of about 50 cm. When you get approximately 20 cm away as in Example 2, each dimension of your face is magnified by a factor of 5/3.

Make-up mirrors such as this are sold commercially and are quite common (see Fig. 7.30).

If the spherical surface constructed in Fig. 7.25 had been silvered on the outside, then a convex mirror would have been formed. The mirror equation can be used to determine the image position by making f a negative

Figure 7.30 A makeup mirror is a concave mirror that magnifies your face when you get closer to the mirror than its focal length. (Photo courtesy Peter Bogen)

value. Geometrical constructions also can be used for image location (see Fig. 7.31).

As before, the rays are drawn using the law of reflection. The first ray bounces off the mirror and up to the left. It is extended through F as shown. The second ray is extended through C and bounces straight back. The image is formed where the two extended rays meet behind the mirror.

We see from Fig. 7.31 that images in a convex mirror are always virtual. They are always upright and smaller than the object. Of course, left and right are also reversed. An example of what is seen in a convex mirror is shown in Fig. 7.32. Convex mirrors provide an expanded field of view. Consequently, they are often used as side mirrors for large trucks and for store monitoring as in Fig. 7.32.

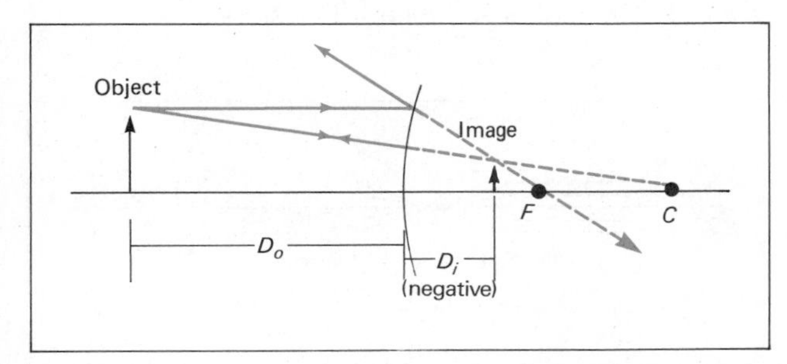

Figure 7.31 The graphic solution for locating the image formed by a convex spherical mirror.

EXAMPLE 3 What is the magnification of an image in a convex mirror for an object 6 m away if the focal length of the mirror is 2 m?

We use Eq. 7.1 with $f = -2$ m and $D_o = 6$ m.

$$\frac{1}{D_i} = \frac{1}{f} - \frac{1}{D_o}$$

$$\frac{1}{D_i} = \frac{1}{-2\text{ m}} - \frac{1}{6\text{ m}} = -\frac{2}{3\text{ m}}$$

$$D_i = -1.5\text{ m}$$

and

$$M = -\frac{D_i}{D_o} = -\frac{-1.5\text{ m}}{6\text{ m}} = +\frac{1}{4}$$

Figure 7.32 A spherical convex mirror gives an expanded field of view. The images are smaller than the original objects and have right and left reversed. (Courtesy J. D. Wilson, *Physics Concepts and Applications*, Heath, Lexington, Mass., 1981.)

This shows that the image is not magnified, but is one-fourth as tall as the object. This can be seen in Fig. 7.32. The positive value for M indicates that the image is upright.

7.5 Lenses

A lens consists of material such as a transparent piece of glass or plastic that refracts light waves to give an image of an object. Lenses are extremely useful and are found in eyeglasses, telescopes, magnifying glasses, cameras, and many other optical devices.

In general there are two main classes of lenses. A **converging lens** or **convex lens** is thicker at the center than at the edge. A **diverging lens** or **concave lens** is thicker at the edges. These two classes and some of the possible shapes for each are illustrated in Fig. 7.33.

Light passing through a lens is refracted twice—once at each surface. The lenses most commonly used are known as thin lenses. This means that when constructing ray diagrams, we can neglect the thickness of the lens, and we can assume that the two surfaces that cause refraction are in essentially the same plane.

The principal axis for a lens goes through the center of the lens and is labeled in Fig. 7.34. Rays coming in parallel to the principal axis are refracted toward the principal axis by a converging lens (see Fig. 7.34). For a converging lens the rays are focused at a point F known as the focal point. For a diverging lens the rays are refracted away from the principal axis and appear to emanate from the focal point F as shown in Fig. 7.34.

The way in which lenses refract light to form images can be obtained by a graphic procedure similar to what

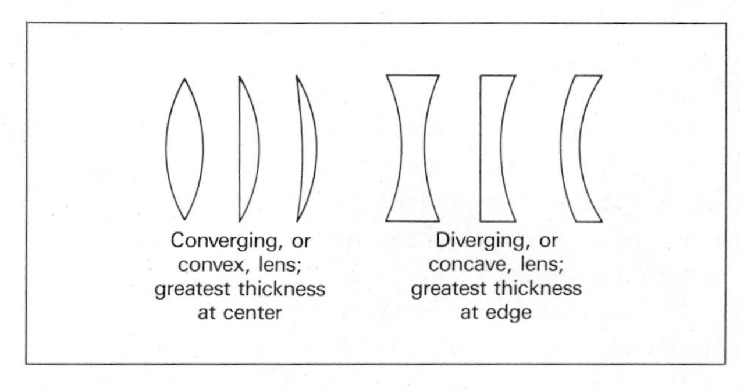

Figure 7.33 Classes and types of lenses.

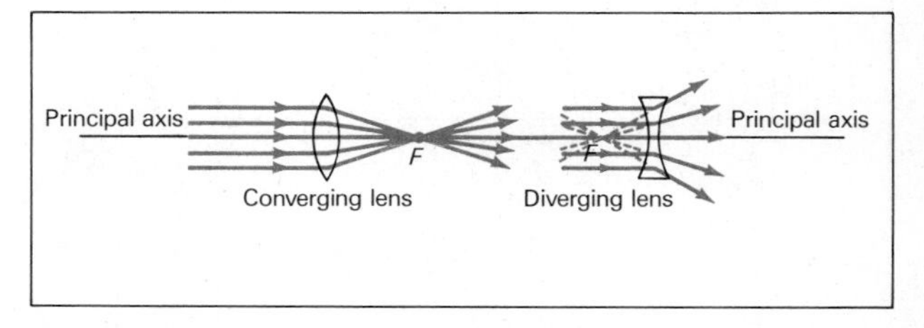

Figure 7.34 The focal point for converging and diverging lenses.

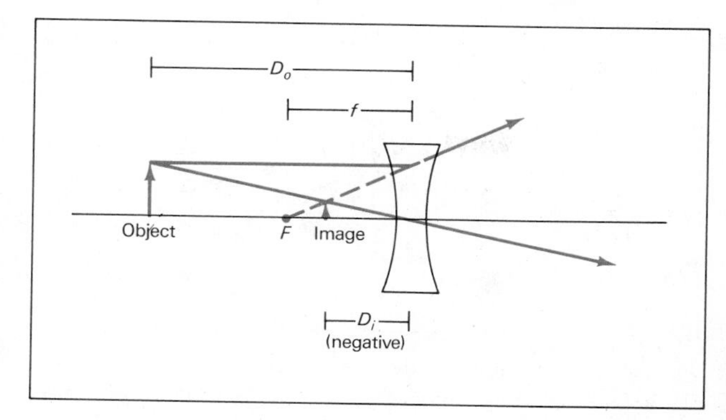

Figure 7.35 Diagram illustrating the image formed by a concave lens. The image is always upright and smaller than the object.

was done with mirrors. In this procedure two rays are drawn from the tip of the object.

1. The first ray goes straight through the center of the lens and is not bent.
2. The second ray goes parallel to the principal axis and then is bent by the lens along a line drawn through a focal point of the lens.

The image of the tip of the arrow occurs where these two rays meet.

An example of this procedure for a concave lens is shown in Fig. 7.35. Note that the image is upright and smaller than the object. This is always the case for a concave lens. When we look through a concave lens we see images as shown in Fig. 7.36.

Possibly you have seen a lens such as the flat plastic lens of Fig. 7.36. They are sold commercially and sometimes used on the back of vans to get a better rear view of traffic.

The lens shown in Fig. 7.36 is a piece of plastic that works like a concave lens. It is flat on one side and has circular grooves on the other. The grooves are made as shown in Fig. 7.37. The grooves cause the light to refract at the surface just like an ordinary concave lens.

When examining the flat grooved lens closely, you can feel and see the circular grooves in it. To make these out of plastic is quite simple. Once a suitable mold has been made, they can be stamped out and mass produced. A similar grooving process can be used to make a flat grooved convex lens.

In working out lens problems we usually want to know the object and image distances. These are the distances between the lens and the object or the image. They are shown in Fig. 7.35. The object distance is labeled

Figure 7.36 Looking through a concave lens we see images that are upright and smaller than the objects. (Photo courtesy Peter Bogen)

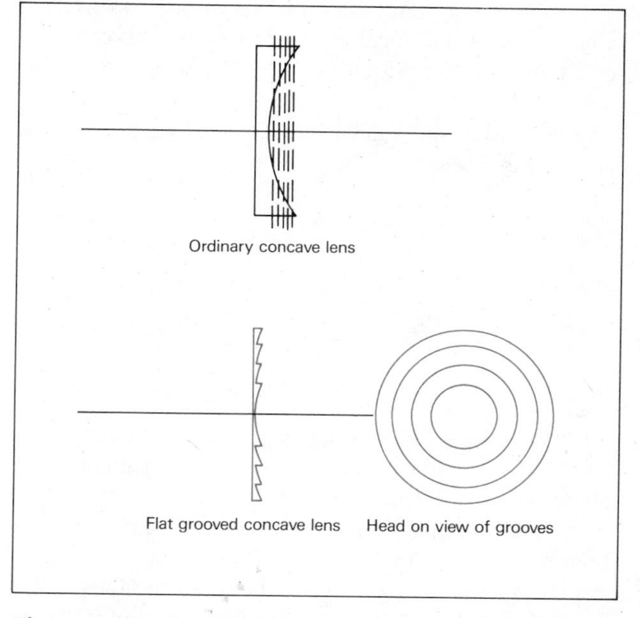

Figure 7.37 A flat grooved concave lens is made with the same curvature as an ordinary concave lens.

D_o, and the image distance is labeled D_i. In the one-lens situations which we will meet, D_o will always be positive.

The image distance D_i is positive if the image is on the opposite side of the lens from the object. Conversely, D_i is negative if the image is on the same side as the object. We have the latter case in Fig. 7.35, so D_i is labeled "negative."

The distance between F and the lens is known as the focal length f. The focal length f is taken to be positive for a convex lens and negative for a concave lens.

Thin-lens problems can be solved analytically with a single equation as long as the proper signs are used. The lens equation is

$$\frac{1}{D_o} + \frac{1}{D_i} = \frac{1}{f} \tag{7.3}$$

When D_i is positive, the image will be a real image, that is, it can be brought to a focus on a screen; when D_i is negative, the image will be virtual and can be viewed only by looking through the lens.

The magnification of a lens is given by the same formula we had for mirrors

$$M = -D_i/D_o \tag{7.4}$$

The following examples will illustrate the use of the lens equation.

EXAMPLE 4 An object is placed 6 cm in front of a convex lens, the focal length of which is 10 cm. Calculate the image position.

Substituting the values into the lens equation given above (Eq. 7.3),

$$\frac{1}{D_i} = \frac{1}{f} - \frac{1}{D_o}$$

$$\frac{1}{D_i} = \frac{1}{10\text{ cm}} - \frac{1}{6\text{ cm}}$$

so $$D_i = -15\text{ cm}$$

Since D_i has a negative value, the image is virtual, erect, larger than the object, and located on the same side of the lens as the object.

A simple magnifying glass is a convex lens that functions as illustrated in Example 4. The object to be viewed and magnified must be located between the focal point and the lens. See Fig. 7.38. The magnification is given as $M = -D_i/D_o$. In Example 4 the magnification is $M = -(-15\text{ cm})/6\text{ cm} = +2.5$.

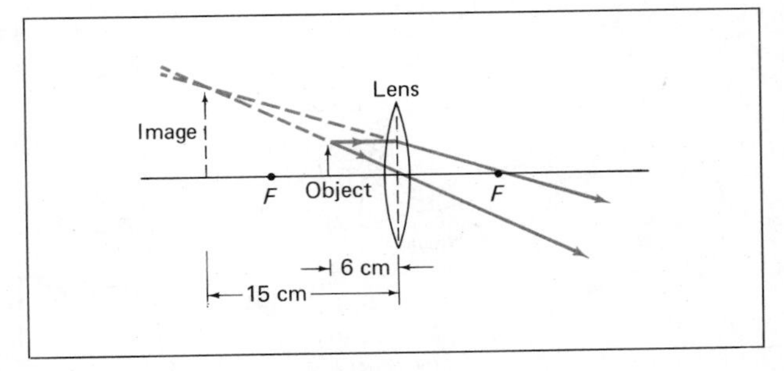

Figure 7.38 Diagram illustrating the formation of the image by a convex lens acting as a simple magnifying glass.

EXAMPLE 5 A concave lens with a focal length of −80 cm is used to view an object 120 cm from the lens. (a) Where will the image be and (b) what will be its magnification?

The situation is similar to that shown in Fig. 7.35. To solve it analytically, we use Eqs. 7.3 and 7.4 with $D_o = 120$ cm and $f = -80$ cm.

(a) $$\frac{1}{D_i} = \frac{1}{f} - \frac{1}{D_o} = \frac{1}{-80\text{ cm}} - \frac{1}{120\text{ cm}}$$

or $$D_i = -48\text{ cm}$$

Since D_i is negative, we know it is on the same side as the object as shown in Fig. 7.35.

(b) $$M = \frac{-D_i}{D_o} = \frac{-(-48\text{ cm})}{120\text{ cm}}$$

$$M = +0.4$$

Thus, we see that the image will be 0.4 times as tall as the object. In this example, the image is upright and smaller than the object as shown in Fig. 7.35.

A slide projector consists of a convex lens and a light source to make the slide bright. The slide is placed in a position that is slightly farther from the lens than the focal length. Since the image will be inverted, the slide is placed upside down as indicated in Fig. 7.39. The image is brought to focus on a screen placed a distance D_i from the lens. As can be seen in Fig. 7.39, the image will be much larger than the slide itself.

EXAMPLE 6 A slide projector has a convex lens with a focal length of 20 cm. The slide is placed upside down 21 cm from the lens. How far away should the screen be from the slide projector's lens so that the slide is in focus?

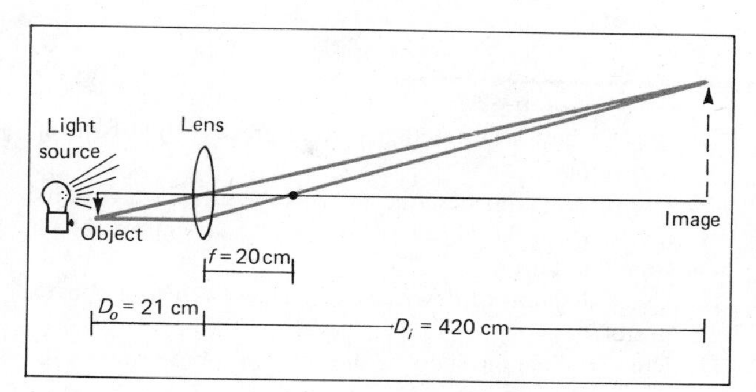

Figure 7.39 The graphical solution for locating the image formed by a convex lens in a slide projector.

To solve this we recognize that the slide will be in focus when the screen is placed at the distance D_i from the lens. We use the lens equation to compute D_i with $D_o = 21$ cm and $f = 20$ cm.

$$\frac{1}{D_i} = \frac{1}{f} - \frac{1}{D_o}$$

$$\frac{1}{D_i} = \frac{1}{20 \text{ cm}} - \frac{1}{21 \text{ cm}}$$

$$D_i = 420 \text{ cm}$$

The image is upside down and real, and it can be placed on a screen. In our example, the screen should be placed 4.2 m from the slide projector's lens. The magnification will be

$$M = -\frac{420}{21} = -20$$

The image will thus be 20 times bigger than the slide in each dimension. That is, a 2 × 2 slide will appear on a screen 4.2 m away as a 40 × 40 image. The minus sign on the magnification indicates that the image is upside down in relation to the slide.

The human eye can be thought of as a convex lens. In Fig. 7.40 the normal path of light is shown. The lens and other parts of the eye have different indices of refraction to give the effect of a convex lens. The light is focused on the **retina,** to which is attached the optic nerve that relays the visual signals to the brain.

Because the distance between the lens and the retina does not vary, we have a situation in which D_i is a constant. Since D_o varies, the focal length f of the eye lens must be varied. This is done by the ciliary muscles. By varying the focal length by only a slight amount, the distances at which objects can be seen can range from infinity to a few centimeters.

Defects of vision occur when the image does not fall on the retina. If the image falls in front of the retina, as in Fig. 7.41, the eye is said to be **nearsighted.** As the object moves closer, the image distance moves back to the retina. Hence, near objects can be seen clearly by a nearsighted person. To correct this defect, glasses consisting of diverging lenses are worn.

The opposite effect occurs when the image falls behind the retina (see Fig. 7.42). This effect causes the eye to be **farsighted,** and distant objects are seen clearly. To correct this defect, glasses consisting of converging lenses are worn.

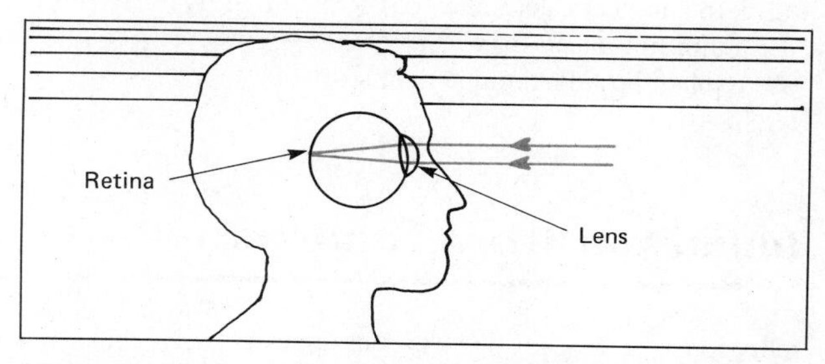

Figure 7.40 The human eye has a lens with an index of refraction which is greater than the surrounding parts of the eye. This convex lens focuses light onto the retina.

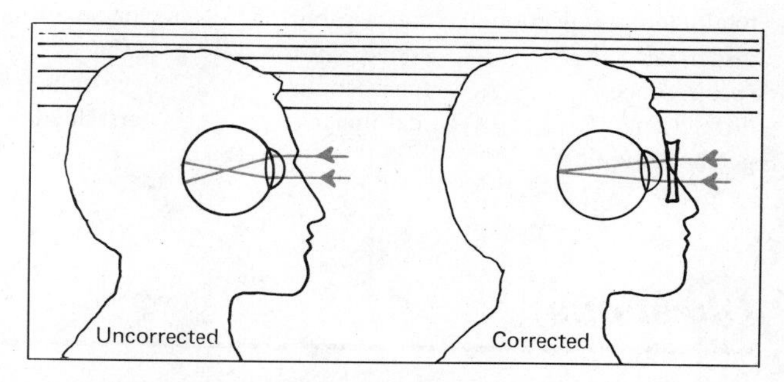

Figure 7.41 Nearsightedness is corrected by placing glasses with a diverging lens in front of the eye.

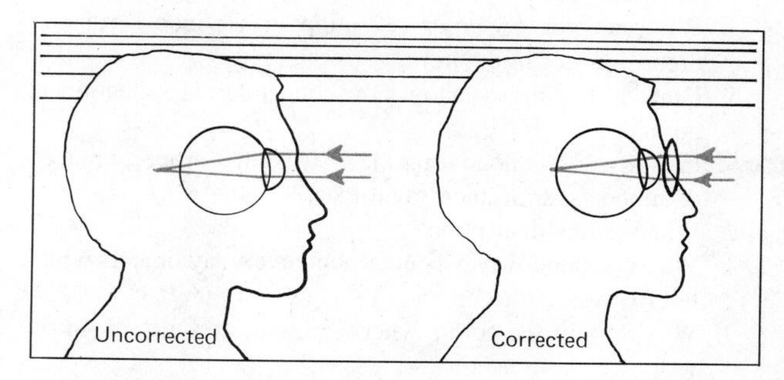

Figure 7.42 Farsightedness is corrected by placing glasses with a converging lens in front of the eye.

Learning Objectives

After reading and studying this chapter, you should be able to do the following without referring to the text:

1. State the two basic laws of reflection.
2. Explain why the sky is blue while clouds are white.
3. Draw diagrams illustrating the principles of refraction and total internal reflection.
4. Explain why you can easily see your reflection in a window pane at night, but only rarely during the day.
5. Discuss the difference between sound waves and visible light waves with regard to diffraction and polarization.
6. Explain what causes the colors when light is reflected off a thin film of oil on water.
7. Explain how Polaroid sunglasses work.
8. Explain how a diffraction grating breaks white light up into colors.
9. Use diagrams to describe the images seen in a concave mirror when the object is (a) close to the mirror (b) far from the mirror.
10. Use a diagram to describe the images seen in a convex mirror.
11. Draw a diagram showing how a slide projector works using a convex lens.
12. Use a diagram to show how a concave lens works, and explain how a flat plastic concave lens is made.
13. Explain how the human eye works and how farsightedness and nearsightedness can be corrected.

Important Words and Terms

reflection
diffuse reflection
regular reflection
refraction
index of refraction
total internal reflection
dispersion
spectroscope
diffraction
geometric optics
physical optics
interference effects
diffraction grating
unpolarized
linearly polarized
concave mirror
convex mirror
focal length
real image
virtual image
magnification
converging lens
convex lens
diverging lens
concave lens
retina
nearsighted
farsighted

Questions

1. What are the two basic laws of reflection?
2. Why is the sky blue while clouds are white?
3. What causes refraction?
4. Why can you see your reflection in a window pane at night, but not in daylight?
5. Draw a diagram to show how total internal reflection is possible.
6. (a) Why do diamonds sparkle? (b) What property makes diamonds sparkle more than glass?
7. What causes dispersion?
8. Why do sound waves bend around everyday objects while light waves do not?
9. Why do you see colors when you look at a film of oil on water?
10. (a) What is a diffraction grating? (b) What does it do?
11. (a) What is unpolarized light? (b) What is polarized light? (c) Can you have polarized sound waves?
12. How do Polaroid sunglasses work?
13. Describe the images seen in a concave mirror at various object distances.
14. Describe the images seen in a convex mirror at various object distances.
15. Describe the images seen through a convex lens at various object distances.
16. How does a flat plastic concave lens work?
17. How does a magnifying glass work?
18. How does a slide projector work?
19. (a) How does the human eye work? (b) What causes farsightedness and how can it be corrected?

Problems

1. Let the dimension of an obstacle be $d = 1$ m. Refer to Chapter 6 and find λ/d for
   (a) visible light at 6000 Å,
   (b) 2000 Hz sound waves,
   (c) 1200 kilohertz radio waves.
   (d) Which of these are easily diffracted around the 1-m object? *Answer:* (c) 250
   (d) sound and radio waves
2. From the equation for index of refraction, determine the speed of light in water.
3. An object 3 cm high is placed 50 cm from a concave spherical mirror with a focal length of 20 cm. Determine graphically the position and height of the image.
   Answer: $D_i = 33.3$ cm
   height = 2 cm
4. Work problem 3 analytically (using the correct equation).
5. An object 4 cm high is placed 15 cm from a concave makeup mirror with a focal length of 20 cm. Determine graphically the position and height of the image.
6. Work problem 5 analytically.
7. An object 6 cm high is placed 10 cm from a convex spherical mirror with a focal length of -3 m. Determine graphically the position and height of the image.
8. Work problem 7 analytically.
9. A slide projector has a convex lens with a focal length of 10 cm. A slide is placed 12 cm from the lens. How far from the lens should a screen be placed so that the slide is in focus? Determine your answer graphically.
   Answer: $D_i = 60$ cm
10. Work problem 9 analytically.
11. A simple magnifying glass has a focal length of 8 cm. An object is placed 6 cm in front of it. Graphically determine (a) the image distance, (b) the magnification.
    Answer: (b) $M = 4$
12. Work problem 11 analytically.
13. A concave lens has a focal length of -2 cm. A 3-cm-high object is placed 5 cm from the lens. Graphically determine
    (a) the image distance,
    (b) the height of the image. *Answer:* (b) 0.86 cm
14. Work problem 13 analytically.
15. A human eye has a focal length of 2.4 cm to 2.2 cm and an image distance of 2.4 cm. What is the closest an object can be and still be in focus? *Answer:* 26.4 cm

8

Electricity and Magnetism

Like charges repel, and unlike charges attract each other, with a force that varies inversely with the square of the distance between them.... Frictional forces, wind forces, chemical bonds, viscosity, magnetism, the forces that make the wheels of industry go round—all these are nothing but Coulomb's law....

–J. R. Zacharias

THE MAGNETIC PROPERTIES of the mineral lodestone (magnetite, Fe_3O_4) were known to the Greeks as early as 600 B.C. Thales of Miletus (640–546 B.C.), an early Greek mathematician and astronomer, was aware of the properties of attraction and repulsion of lodestone with similar pieces of lodestone; he also knew of an electrostatic effect called the amber effect, that is, the attraction of bits of straw to an amber rod that had been rubbed with wool.

The word "magnet" seems to have been derived from Magnesia, a province in Asia Minor, where the Greeks first discovered lodestone. The Chinese were probably the first to use the lodestone as a compass, both on land and sea. Early records indicate that ships sailing between Canton, China, and Sumatra as early as 1000 A.D. were navigated by the use of the magnetic compass.

In the thirteenth century, a Frenchman, Petrus P. de Maricount, described the magnetic compass in some detail and applied the term "pole" to the regions on the compass where the fields of influence were the strongest. The north-seeking pole he called N, and the south-seeking pole he called S. The attraction of unlike poles, the repulsion of like poles, and the formation of new unlike poles when a magnet was broken into two pieces were also described by de Maricount.

In 1600 Dr. William Gilbert (1540–1603), court physician to Queen Elizabeth, published his book on magnetism, *De Magnete*. The book contained all information then known about electricity and magnetism, plus experiments carried out by Gilbert. These experiments included information on the dip (the angle the Earth's magnetic field makes with the Earth's surface) and declination (the angle the compass needle deviates from the geographical north) of the compass, the loss of magnetism by a magnet when heated, and experiments with a sphere-shaped magnet, which led him to the conclusion that the Earth acts like a huge magnet.

Gilbert was also aware of the amber effect. We now know that this effect is due to repulsion and attraction of electric charges. Gilbert carried out many experiments on the amber effect with an instrument he called a versorium (Latin, *verso,* to turn around). The versorium was nothing more than a slender arrow-type nonconducting material balanced on a pivot point so as to give a high degree of sensitivity to the force of attraction when an amber rod or other substances were placed in its vicinity. With his versorium he discovered that many substances possess the amber effect. Gilbert is responsible for the word "electron," which is very familiar today. He classified those substances possessing the amber effect as "electrics" (Greek, *electron,* amber).

A similar major study of electricity and magnetism took place in the eighteenth century, when Charles Coulomb (1736–1806) established the inverse square law of attraction and repulsion between electrostatic charges, and verified the same law to hold for magnetic poles.

Electric and magnetic phenomena were studied in detail during the 1800s. In 1820 Hans Christian Oersted (1777–1851) found that a compass is deflected by a current-carrying wire. In 1831 Michael Faraday (1791–1865) and Joseph Henry (1797–1878) independently found that a magnet plunged into a coil of wire would induce an electric current.

The laws of electricity and magnetism discovered by Coulomb, Oersted, Faraday, Henry, and others were studied in detail by James Clerk Maxwell (1831–1879),

◀ One of the most spectacular effects of electricity is lightning.

a Scottish physicist. Maxwell wondered why the physical laws were not symmetric when expressed in mathematical form. By applying the concept of symmetry he discovered an additional law which completed the equations of electromagnetism in 1865.

Maxwell studied these equations and found that, according to the equations, light is made up of electromagnetic waves. Radio waves were also predicted at a much lower frequency than visible light waves. The theoretical fact that electromagnetic waves of radio frequency are possible led to experiments in which radio waves were generated and detected. This opened up a new era of wireless communication and eventually brought us commercial radio and television and a host of other devices. Many of these practical developments will be described in this chapter.

8.1 Electric Charge

All matter, according to modern theory, consists of very small particles called atoms that are composed in part of negatively charged particles called **electrons,** positively charged particles called **protons,** and particles called **neutrons** that carry no electric charge. Table 8.1 summarizes the fundamental properties of these atomic particles.

As can be seen from the table, all three of these particles have mass, but only the electrons and protons possess electric charge. The magnitudes of electric charge on an electron and a proton are equal, but their signs are different. When we have the same number of electrons and protons, the total charge is zero, and we have a *neutral* situation. Hence, the name *neutron* for the small particle with zero electric charge. Neutrons have virtually no effect in atomic physics. They become important, however, when nuclear physics is discussed in Chapter 10.

The unit of electric charge is called the **coulomb,** named after Charles Coulomb, the French scientist. Electric charge is usually designated by the letter q. When electric charges move, they give rise to an electric current. **Current** is defined as the rate of flow of electric charge. Current is usually designated by the letter I. Current is measured in terms of amperes. One **ampere** is equal to one coulomb per second.

In symbol notation we can relate electric current and electric charge by the formula

$$I = \frac{q}{t} \tag{8.1}$$

where I = electric current, measured in amperes,
q = electric charge flowing past a given point, measured in coulombs,
t = time for the charge to move past the point, measured in seconds.

From Eq. 8.1, we can write

$$q = I \times t \tag{8.2}$$

or 1 coulomb = 1 ampere × 1 second.

This equation is important because the experimental quantity used as the fundamental unit is the ampere. To be precise, the *coulomb* is defined as the amount of charge that flows past a given point in one second when the current is one ampere.

Electric charge cannot be created or destroyed. All protons have the same charge, 1.6×10^{-19} coulomb, and all electrons have the same charge of -1.6×10^{-19} coulomb. So far, all particles found in nature have multiples of this amount of charge. For example, no one has ever found a charge of $\frac{3}{2}$ or 0.43 times this fundamental amount.

Many physicists now believe in the existence of particles called quarks. **Quarks** have charges of $\pm\frac{2}{3}$, and $\pm\frac{1}{3}$. In the currently accepted theory, quarks combine (three at a time) to form protons and neutrons. So far,

Table 8.1 Some Properties of Atomic Particles

Particle	*Symbol*	*Mass*	*Charge*
Electron	e^-	9.11×10^{-31} kg	-1.6×10^{-19} coulomb
Proton	p^+	1.67×10^{-27} kg	$+1.6 \times 10^{-19}$ coulomb
Neutron	n	1.67×10^{-27} kg	0

however, not a single free quark has been found, which has led many scientists to believe that free quarks do not exist.

It has long been known that like charges repel and unlike charges attract. It was Charles Coulomb who first quantified this attraction and repulsion in what is now known as Coulomb's law for electric charges.

Coulomb's law states:

The force of attraction or repulsion between two charged bodies is directly proportional to the product of the two charges and inversely proportional to the square of the distance between them.

This can be written as

$$F = k\frac{q_1 q_2}{r^2} \tag{8.3}$$

where F = the force of attraction or repulsion,
q_1 = magnitude of first charge,
q_2 = magnitude of second charge,
r = distance between charges,

k is a proportionality constant given by

$$k = 9 \times 10^9 \frac{\text{newton} \times \text{meter}^2}{\text{coulomb}^2}$$

Coulomb's law is very similar to Newton's law of universal gravitation. The similarity comes from the fact that both are dependent on the square of the distance between the centers of the objects. One obvious difference between the two laws is that Coulomb's law depends on charge while Newton's law depends on mass.

There are two other important differences. One is that Coulomb's law can give rise to either an attractive or a repulsive force, depending on whether the two charges are different or the same. The force of gravitation, on the other hand, is *always* attractive.

The other important difference is that the electric forces are generally much stronger than the gravitational forces for ordinary objects. When dealing with atoms and the properties of matter, the gravitational forces can be ignored compared to the electric forces of attraction and repulsion.

An object with an excess of electrons is said to be negatively charged, and an object with a deficiency of electrons is said to be positively charged. A negative charge can be placed on a rubber rod by stroking the rod with fur. In Fig. 8.1 a rubber rod that has been rubbed with fur is shown suspended by a thin thread that allows the rod to swing freely. When a similar rubber rod that has been stroked with fur is brought close to the suspended rod, it will swing away from the second rubber rod; that is, the charged rods repel one another.

The same procedure using two glass rods that have been stroked by silk will show similar results (Fig. 8.2). Although the experiment shows repulsion in both cases, the charge on the glass rods is different from the charge on the rubber rod. This is shown in Fig. 8.3, which indicates that the charges on the stroked rubber and glass rods attract one another. The charge placed on the glass rod is positive; on the rubber rod the charge is negative.

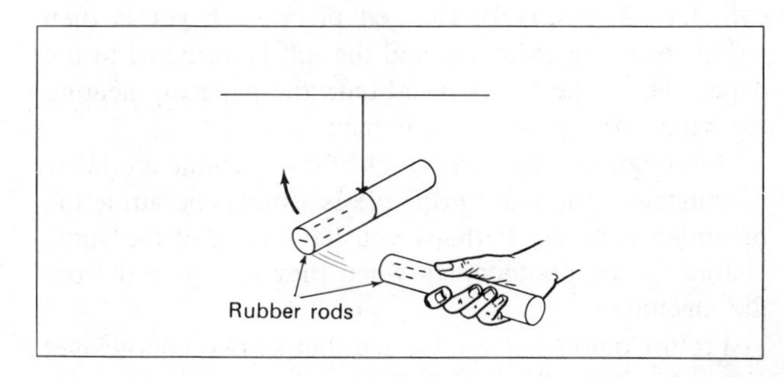

Figure 8.1 Two negatively charged bodies repel one another.

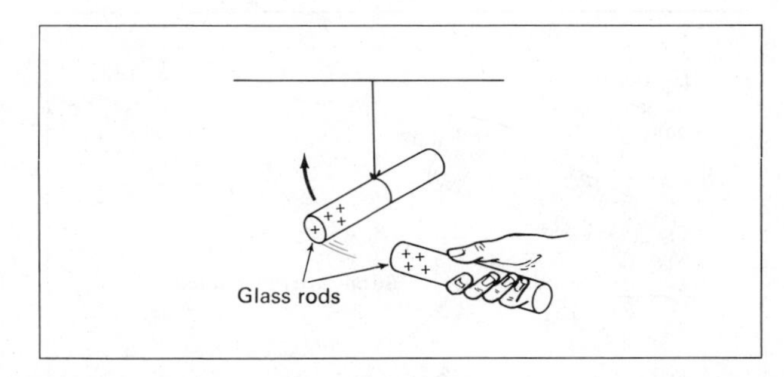

Figure 8.2 Two positively charged bodies repel one another.

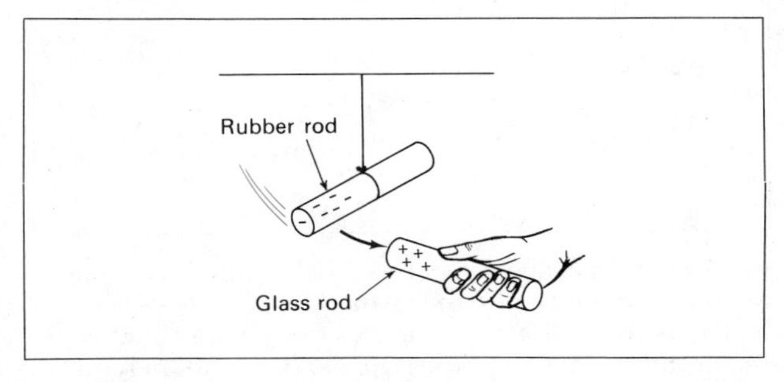

Figure 8.3 A negatively charged body and a positively charged body attract one another.

The important principle of electrostatics—that unlike charges attract—is the basis of photocopying machines such as a Xerox copier. In these machines the material to be copied is placed face down on a transparent glass plate, and a copy is produced.

To copy a page, light is reflected off the page onto a positively charged cylinder or plate made of selenium. Selenium is a photoconductor and charge leaks away from the areas struck by the light, leaving an electrostatic "image" of the dark regions or print of the copy page. Next, negatively charged black ink powder is dusted on the cylinder, sticking to the charged part of the selenium cylinder. A positively charged piece of paper is then passed over the cylinder, and the ink is attracted to the paper. Then, the ink is fused onto the paper by heating the paper. The paper copy is then ejected.

Although the inner workings of the machine are fairly complicated, the main principle is simply the attraction of unlike charges. Perhaps you have noticed the static charge on the photocopies when they are ejected from the machine.

From Coulomb's law we see that as two charges get closer together, the force of attraction or repulsion increases. This important effect can give rise to nonzero electric forces as shown in Fig. 8.4. When a negatively charged rubber comb approaches a small piece of paper, the charges within the paper arrange themselves so that the positive charges move toward the comb and the negative charges are repelled away. Since the positive charges are closer to the comb, the attractive forces are stronger than the repulsive forces. Thus, there is a net attraction between the comb and the piece of paper. Remember, however, that the paper is uncharged.

A similar effect occurs when a negatively charged rubber rod is brought close to a stream of water. The molecules of water arrange themselves so that the positive ends are closer than the negative ends. Then the water is attracted to the rod and the stream is bent as shown in Fig. 8.5.

Since atoms are fundamental parts of matter, it follows that electrons and protons are always present in any material. The electrons, which are located in the outer part of the atom, move within the material with varying degrees of freedom. The protons, located in the inner part of the atom, are much heavier and are not free to move. In metals, such as copper, silver, and aluminum, some of the electrons can move easily. Such metals are called **conductors.** Materials such as wood, paper, cork, glass, and plastics are made such that the electrons cannot move freely. These materials are called **insulators.**

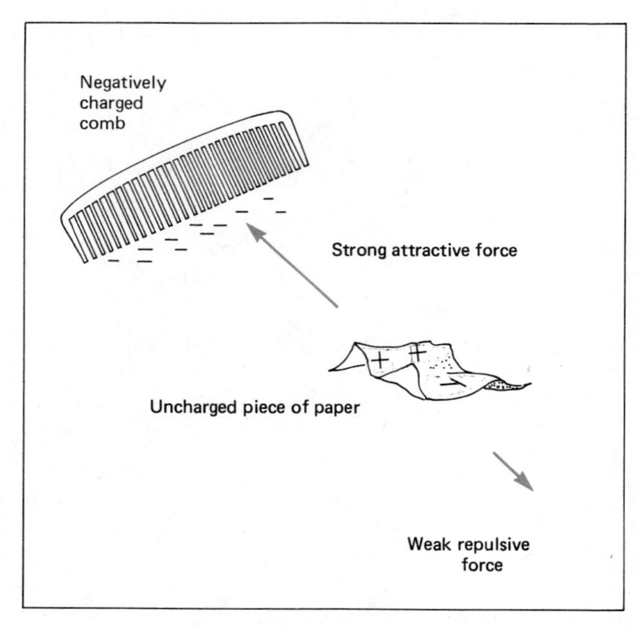

Figure 8.4 A negatively charged rubber comb is brought near a small piece of paper which has no net charge on it. The comb will attract the positive charges and repel the negative charges in the paper. The comb then exerts a net attractive force on the paper because the positive charges are closer to it.

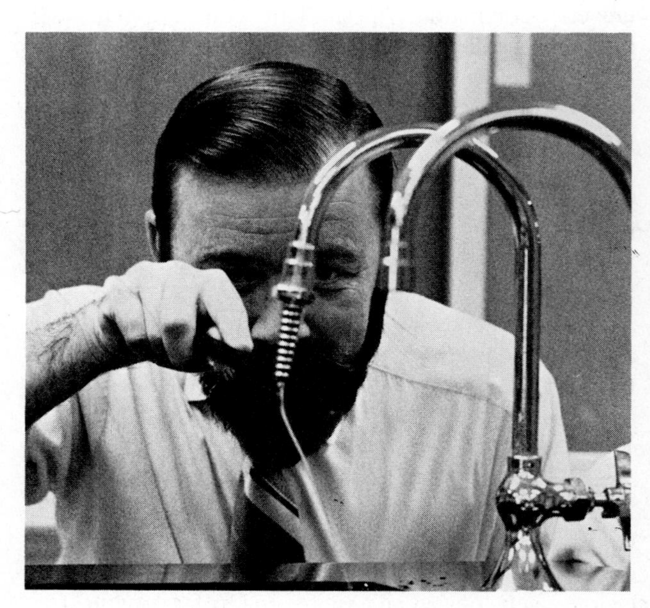

Figure 8.5 A charged rubber rod brought close to a stream of water will bend the stream. (Courtesy J. D. Wilson, *Physics Concepts and Applications,* Heath, Lexington, Mass., 1981.)

Materials that are neither good conductors nor good insulators are called *semiconductors*.

8.2 Electricity

The effects produced by moving charges give rise to what we call **electricity.** In order to get charges to move, they must be acted upon by positive or negative charges.

Consider the situation shown in Fig. 8.6. We start out with some unseparated charges and then begin to separate them. It takes very little work to pull the first negative charge to the left and the first positive charge to the right. When the next negative charge is moved to the left, it is repelled by the negative charge already there, so more work is needed. Similarly, it takes more work to move the second positive charge to the right. As we separate more and more charges, it takes more and more work.

Since work is done in separating the charges, we are creating **electric potential energy.** If we insert a negative charge between the separated charges, it will move to the right as shown in Fig. 8.6; that is, it will acquire a velocity and kinetic energy.

Instead of speaking of electric potential energy, we usually speak of a potential difference or voltage. The voltage is defined as the amount of work it would take to move a charge between two points, divided by the value of the charge. That is, **voltage** is the work per unit charge or the electrical potential energy per unit charge. That is,

$$\text{voltage} = \frac{\text{work}}{\text{charge}}$$

or

$$V = \frac{W}{q} \tag{8.4}$$

The **volt** is the unit of voltage and is equal to one joule per coulomb. A voltage is caused by a separation of charge. Once the charges are separated, a current can be set up, and we get electricity.

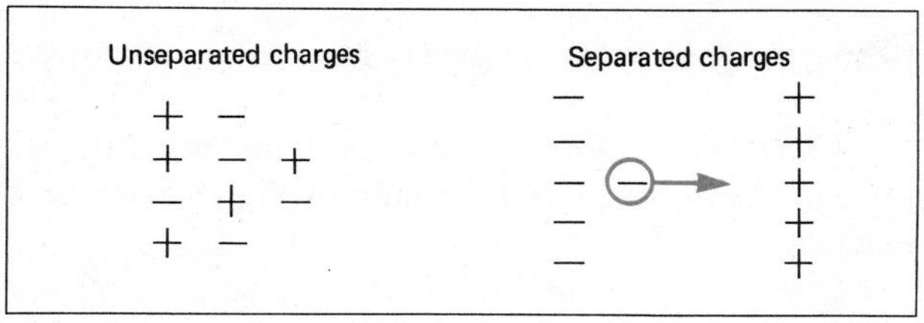

Unseparated charges

Separated charges

Figure 8.6 It takes work to separate positive and negative charges. When they are separated, a free negative charge such as an electron will move from the negative to the positive side.

When you have a current, it meets with some opposition. This opposition to the flow of charge is called **resistance.** It is measured in units called ohms (Ω). There is a simple relationship between voltage, current, and resistance that was formulated by George Ohm (1787–1854), a German physicist. **Ohm's law** can best be stated as the equation

$$\text{voltage} = \text{current} \times \text{resistance}$$

or

$$V = IR \tag{8.5}$$

From this equation we see that one **ohm** is a volt per ampere.

An example of a very simple electric circuit is shown in Figure 8.7. The battery provides the voltage. A con-

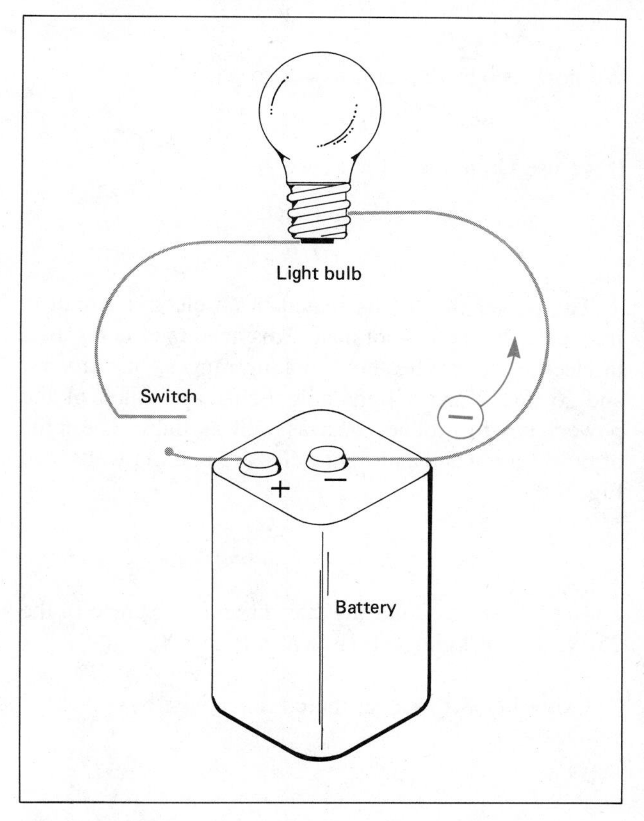

Figure 8.7 A simple electric circuit. A battery provides the voltage. The electrons can be thought of as flowing from the negative to the positive terminals. The light bulb provides the resistance. When the switch is open, as shown, nothing happens. When the switch is closed, the electrons move and the bulb lights up.

ducting wire is placed between the positive and negative terminals, and electrons in the wire can move away from the negative end toward the positive end. If we put a light bulb along the path of the electrons, it provides resistance and will light up.

The switch in the circuit allows the path of the electrons to be closed or open. When the switch is open, as shown in Fig. 8.7, the electrons will not move and nothing will happen. When the switch is closed, the circuit is completed and the bulb will light up.

When current exists in a circuit, work is done to overcome the resistance of the circuit and power is expended. Recall that one definition of power was

$$P = \frac{W}{t}$$

If we substitute qV for W, we get

$$P = \frac{q}{t}V$$

We now recognize that $q/t = I$ to get

$$P = IV \tag{8.6}$$

If we use Ohm's law for V, we get

$$P = I(IR)$$

$$P = I^2R \tag{8.7}$$

The power that is dissipated in an electric circuit is frequently in the form of heat. This heating effect is used in electric stoves, heaters, cooking ranges, hair dryers, and so on. When a light bulb lights up, much of the power goes to produce heat as well as light. The units of power are watts and light bulbs are rated in watts (see Fig. 8.8).

Figure 8.8 Light bulbs are rated in units of power or watts. This tells us how much electric energy is being used per second to produce heat and light. (Courtesy J. D. Wilson, *Physics Concepts and Applications*, Heath, Lexington, Mass., 1981.)

EXAMPLE 1 Find the current and resistance in the 25-W, 120-V light bulb shown in Fig. 8.8.

Using Eq. 8.6 we get the current given by

$$I = \frac{P}{V}$$

or

$$I = \frac{25 \text{ W}}{120 \text{ V}}$$

or

$$I = 0.21 \text{ A}$$

We can then use Eq. 8.5 to get the resistance.

$$R = \frac{V}{I}$$

or

$$R = \frac{120 \text{ V}}{0.21 \text{ A}}$$

or

$$R = 571 \ \Omega$$

An electric company delivers electric energy. Our electric utility bill is computed on the number of kilowatt-hours (kWh) used. This is a unit of energy, since work produces energy and

$$W = P \times t \tag{8.8}$$

Hence, we pay a power company for energy, not power.

EXAMPLE 2 How much does it cost at $0.08 per kWh to leave the 25-W light bulb of Fig. 8.8 on for a month?

We use Eq. 8.8 to figure the total energy in kilowatt-hours used in one month. The number of hours in a 30-day month is $24 \times 30 = 720$ h. So, the total energy consumed using Eq. 8.8 is

$$E = P \times t$$

$$E = 25 \text{ W} \times 720 \text{ h}$$

$$E = 18{,}000 \text{ Wh}$$

$$E = 18 \text{ kWh}$$

So the cost is

$$18 \text{ kWh} \times \$0.08/\text{kWh} = \$1.44$$

There are two principal forms of electric current. Figure 8.7 shows a simple circuit where a battery is used. This type of current is called **dc** or direct current. We use dc in battery-powered devices such as flashlights, radios, and automobiles.

Much more common is **ac** or alternating current, which is produced by constantly changing the voltage from positive to negative to positive, and so on. The frequency of changing from positive to negative is usually at the rate of 60 cycles per second or 60 Hz. Alternating current is produced by electric companies and is used in homes. The formulas for Ohm's law (Eq. 8.5) and power (Eqs. 8.6 and 8.7) apply to both dc and ac circuits containing only resistances.

One of the principal reasons we use alternating current in our homes is that the voltages of alternating current can be easily increased or decreased by means of **transformers.** This is important because in transmitting electrical power, some of the electrical energy is lost as heat when the transmission wires heat up.

Note that Eq. 8.6 tells us that the same amount of power can be transmitted using either a high voltage and a low current or a low voltage and a high current. Equation 8.7 tells us that for a given resistance we want to minimize the current (and maximize the voltage) to minimize the heat loss from the transmission wires. Thus, a power company usually transmits its energy at a very high voltage (as much as 765,000 volts) to minimize heat losses from the wires. The high voltages are attained by using transformers to increase the voltage. This high voltage is then stepped down, again using transformers (Fig. 8.9), until it reaches the home at approximately 220 to 240 volts. In Section 8.4 we will discuss the principles of transformers.

As the electricity is transmitted at these very high voltages, insulators are used to keep the wires away from the supporting towers and apart from each other. These insulators are shown in Fig. 8.10.

Once electricity enters the home or business, it is used in electric circuits. There are two simple kinds of electric circuits that contain only resistance. In Fig. 8.11, a portable electric heater illustrates a series circuit. In a **series circuit,** all of the current passes through all of the resistors, and the various resistances add up. The total voltage is given by

$$V = IR_1 + IR_2 + IR_3 + \cdots$$

Figure 8.9 The transformer shown decreases the voltage from the high-voltage power line to a lower voltage for residential use. (Courtesy J. D. Wilson, *Physics Concepts and Applications,* Heath, Lexington, Mass., 1981.)

Figure 8.10 High-voltage electric transmission lines are used to minimize heat losses as power is transmitted. Long insulators keep the lines away from metal support towers and insulating separators keep the wires from touching. (Courtesy J. D. Wilson, *Physics Concepts and Applications,* Heath, Lexington, Mass., 1981.)

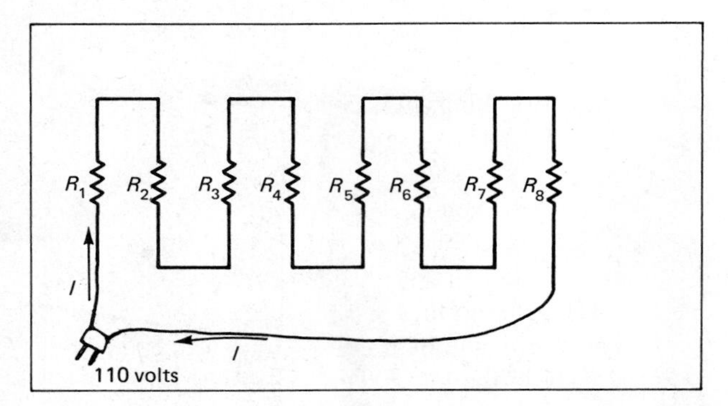

Figure 8.11 In a series circuit all of the current passes through each resistance. The total resistance is just the sum of the individual resistances.

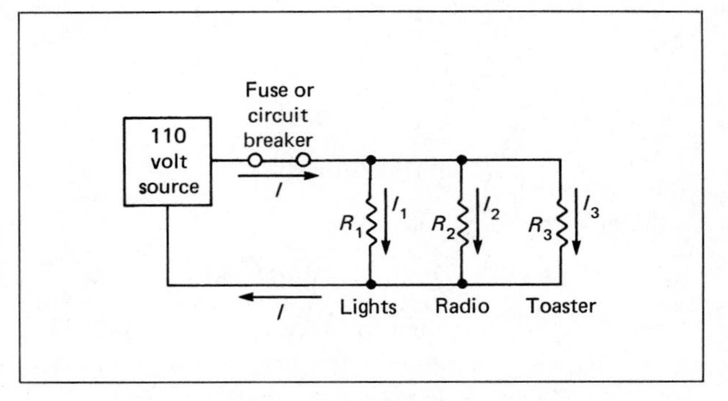

Figure 8.12 A home electric circuit. All electric components are connected in parallel. Each appliance has 110 volts available to it.

or $$V = IR_s$$

$$R_s = R_1 + R_2 + R_3 + \cdots \tag{8.9}$$

where V = voltage of the circuit,
I = current passing through the various resistors,
R_1 = first resistance,
R_2 = second resistance, etc.,
R_s = total resistance of the series circuit.

When using formulas such as $P = I^2R$ for the whole circuit, R_s is the value used for the resistance.

EXAMPLE 3 Compute the current used and power consumed by a portable electric heater with 8 heating elements in series, each of which has a resistance of 2 Ω. The heater is plugged into a wall socket of 110 V.

The total resistance is the sum of all the resistances (Eq. 8.9), which is 8 × 2 Ω = 16 Ω. The current is given by Ohm's Law (Eq. 8.5) as

$$I = \frac{V}{R}$$

$$I = \frac{110\text{ V}}{16\ \Omega}$$

$$I = 6.875\text{ A}$$

The power is then given by Eq. 8.6.

$$P = IV$$

$$P = 6.875\text{ A} \times 110\text{ V}$$

$$P = 756\text{ W}$$

Another common type of simple circuit is called a parallel circuit. An illustration is shown in Fig. 8.12. In a **parallel circuit,** the voltage across each resistance is the same but the current through each resistance varies. The total current is equal to the sum of all the individual currents through the resistances. Thus we get

$$I = I_1 + I_2 + I_3 + \cdots$$

or $$I = \frac{V}{R_1} + \frac{V}{R_2} + \frac{V}{R_3} + \cdots$$

or $$I = \frac{V}{R_p}$$

where $$\frac{1}{R_p} = \frac{1}{R_1} + \frac{1}{R_2} + \frac{1}{R_3} + \cdots \tag{8.10}$$

Note that for two resistors in parallel this becomes

$$R_p = \frac{R_1R_2}{R_1 + R_2}$$

EXAMPLE 4 What is the total resistance of the circuit shown in Fig. 8.12 if R_1 = 100 Ω, R_2 = 20 Ω, and R_3 = 10 Ω?

We use Eq. 8.10.

$$\frac{1}{R_p} = \frac{1}{R_1} + \frac{1}{R_2} + \frac{1}{R_3}$$

$$\frac{1}{R_p} = \frac{1}{100\ \Omega} + \frac{1}{20\ \Omega} + \frac{1}{10\ \Omega}$$

or $$R_p = 6.25\ \Omega$$

Home appliances are wired in parallel. There are two major advantages to the parallel circuit.

1. The same voltage (110–120 V) is available throughout the house. This makes it much easier to design the appliances.
2. If one appliance fails to operate, the others are not affected. In a series circuit, if one fails, none of the others will operate.

As more and more appliances are turned on, more and more current flows and the wires get hotter and hotter. The fuse shown in Fig. 8.12 is a safety device which prevents the wires from getting too hot. When too much current flows, the fuse gets so hot that it melts through and acts as an open switch. This prevents any current from flowing until the fuse is replaced. A circuit breaker has the same function as a fuse. When the current gets too high, the circuit breaker triggers a switch, which opens, halting any current from flowing. When the trouble is corrected, the circuit breaker can be pushed in. This closes the switch so current can again flow in the circuit.

An electric shock can be very dangerous and touching exposed electric wires should always be avoided. Many people are killed every year through electric shocks. The danger is proportional to the amount of electric current that goes through the body. The amount of current going through the body is given by Ohm's law as

$$I = \frac{V}{R_{\text{body}}} \tag{8.11}$$

where R_{body} = resistance of the body.

A current of 0.001 A can be felt as a shock, and a current as low as 0.05 A can be fatal. A current of 0.1 A is nearly always fatal.

The amount of current can be seen from Eq. 8.11 to be very dependent on the body's resistance. The body's resistance varies considerably, mainly due to the contact at the skin. Since our bodies are mostly water, the skin resistance makes up most of the body's resistance.

A dry body can have a resistance as high as 500,000 ohms, and the current from a 110-volt source will be 0.00022 ampere. The danger occurs when the skin is moist or wet. Then the resistance of the body can go as low as 100 ohms, and the current will rise to 1.1 amperes. Injuries and death from shocks usually occur when the skin is wet. For instance, touching an exposed household wire while sitting in a bathtub will frequently be fatal. You should avoid touching even a wall switch when your hands or feet are wet.

8.3 Magnetism

One of the first things we notice in examining a bar magnet is that it has two regions of magnetic concentration or poles at the ends of the magnet. We call these two poles north-seeking, N, and south-seeking, S, because a compass needle was one of the earliest applications of magnetism. The N pole of a magnet, such as a compass, points north, and the S pole points south.

The basic principle of magnetic force has been mentioned already in the introduction to this chapter.

Like poles repel and unlike poles attract.

We say that for N and S poles, N-S, attract and N-N or S-S poles repel each other. The strength of the attraction or repulsion depends on the strength of the magnetic poles. Also, in a manner similar to Coulomb's law, the strength of the magnetic force is inversely proportional to the distance between the magnetic poles. Figure 8.13 shows some toy magnets that seem to defy gravity due to their magnetic repulsion.

Every magnet produces a force on every other magnet. In order to discuss these force effects, it is common to introduce the concept of a magnetic field. A **magnetic field** is a set of imaginary lines that indicates the direction a small compass needle would point if it were placed

Figure 8.13 Small toy magnets show a repulsive effect that causes them to be suspended in air. (Photo courtesy James Crouse.)

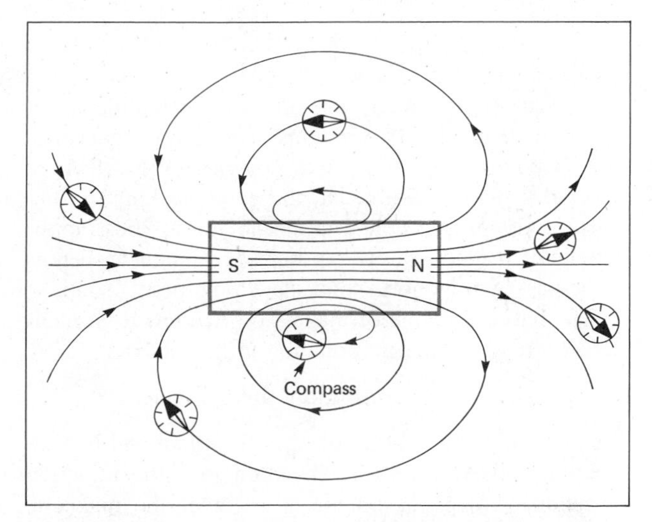

Figure 8.14 Magnetic field lines may be found by using a very small compass as shown. The N pole of the compass points in the direction of the field.

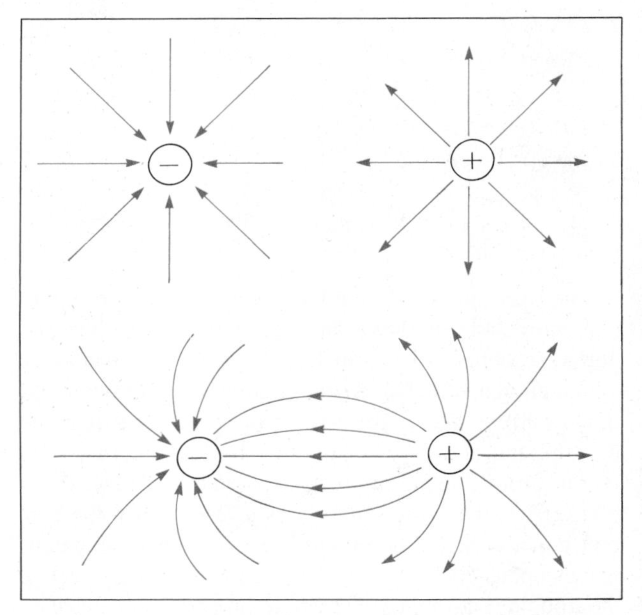

Figure 8.15 Electric field lines for a single positive charge, a single negative charge, and an electric dipole which is a positive and negative charge together.

at a particular spot. Figure 8.14 shows the magnetic field lines around a simple bar magnet. The arrows in the field lines indicate the direction in which a compass N pole would point. The closer together the field lines, the stronger the magnetic force on the imaginary compass.

The field concept also applies to the electric force. An **electric field** is a set of imaginary lines that indicates the direction a small positive charge would move, if it were placed at a particular spot. The electric fields are indicated in Fig. 8.15 for a single positive charge, a single negative charge, and a positive and negative charge together. This last configuration is called an *electric dipole* because there are two poles involved. A common magnet such as that shown in Fig. 8.14 would be called a *magnetic dipole* because it also has two poles. Note the similarity and difference between the fields for an electric dipole and a magnetic dipole.

Two facts about magnetic and electric fields are evident from Figs. 8.14 and 8.15. First, field lines never cross. If they did, it would indicate that at the point of crossing the compass pointed in two directions or the positive charge would move in two directions. Second, electric fields begin and end on electric charges, whereas magnetic field lines are continuous. Magnetic field lines have no beginning or end. All magnets occur with two poles, i.e., as *di*poles. Electric charge can be thought of as isolated single poles or *mono*poles. A **magnetic monopole** would consist of a single N or S pole without the other. There is nothing to prevent a magnetic monopole from existing, but so far none has been found. The discovery of a magnetic monopole would be an important fundamental development.

Electric and magnetic fields have directions associated with them, so they are *vector quantities*. Electromagnetic waves, discussed in Chapter 6, are caused by electric and magnetic fields that vary with time.

Magnetic field patterns are frequently generated by using small iron filings. The iron filings are magnetized and act like small compasses. The magnetic field produced in this way using two pieces of naturally magnetic lodestone (Fe_3O_4) is shown in Fig. 8.16.

Electricity and magnetism are discussed together in this chapter because they are linked. In fact, the source of magnetism can be understood as moving and "spinning" electric charges. Oersted first discovered in 1820 that a compass needle was deflected by an electric current-carrying wire. To see this effect, it is necessary to have a fairly strong electric current.

Current-carrying wires cause different magnetic field configurations. Some of these are shown in Fig. 8.17. A straight wire produces a field in a circular pattern around the wire. A single loop of wire acts like a small bar magnet, and a coil of wire creates a field similar to a bar magnet.

Individual electrons in an atom have a magnetic field due to what is best understood as a spinning motion. This causes all electrons to act like magnetic dipoles or

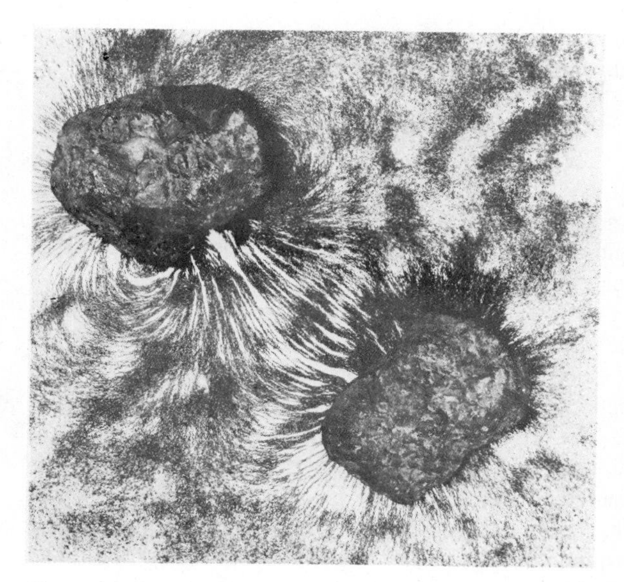

Figure 8.16 An iron filing pattern near two pieces of naturally magnetic lodestone.

very small bar magnets. The electrons also go in orbit around the atom's nucleus. This effect is similar to a single loop of wire and again creates a magnetic field similar to a small bar magnet. There are thus magnetic effects due to an electron's movement and due to its spinning motion. Most atoms have many electrons and the magnetic effects of all these electrons usually cancel out. So most materials are not magnetic or only slightly magnetic. In some atoms, however, the magnetic effect can be strong.

Materials that have large magnetic fields associated with them are called **ferromagnetic.** These materials include the elements iron, nickel, cobalt, several rare Earth elements, and certain alloys of these and other elements. In ferromagnetic materials many atoms combine their magnetic fields in the same direction. This gives rise to magnetic **domains** or local regions of alignment. A single magnetic domain is very small and acts like a tiny bar magnet.

In iron, the domains can be aligned or nonaligned. A piece of iron with the domains randomly oriented is not a magnet and is sometimes referred to as a piece of soft iron. This effect is shown in Fig. 8.18. When the soft iron is placed in a magnetic field such as produced by a current-carrying coil of wire, the domains line up, and the iron is magnetized as shown in Fig. 8.18.

When the magnetized iron is taken out of the magnetic field, the domains will tend to return to a mostly

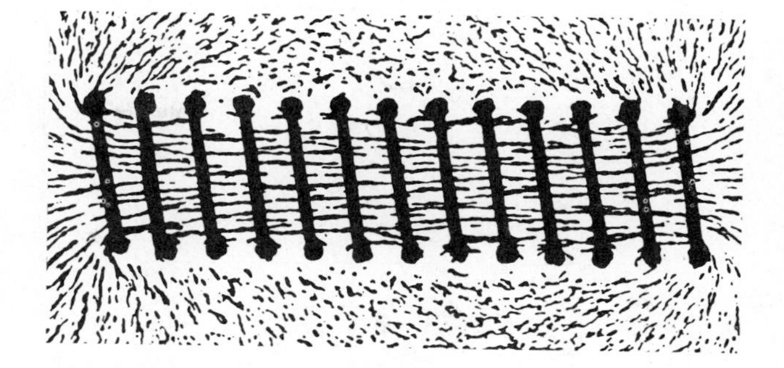

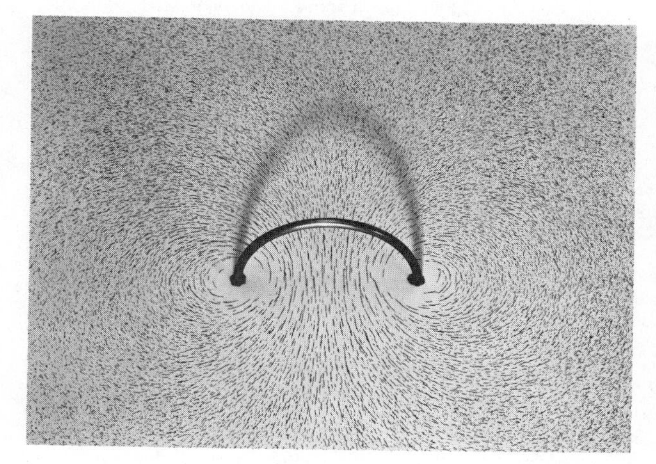

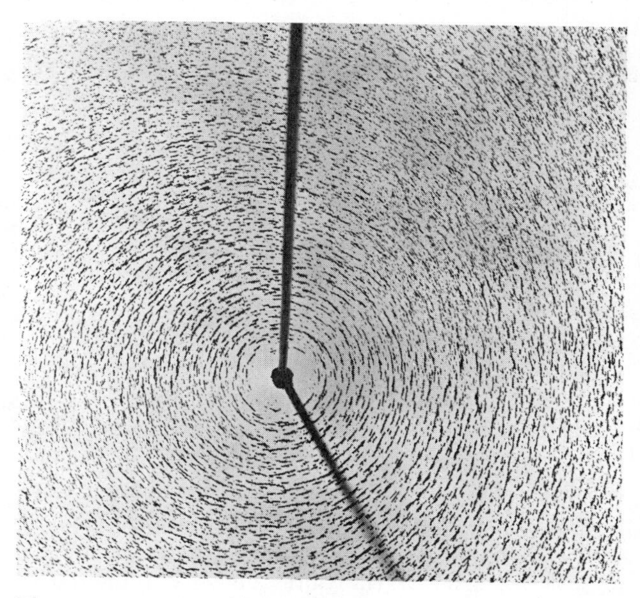

Figure 8.17 Iron filings near current-carrying wires create magnetic field patterns for a straight wire (bottom), a single loop (middle), and a coiled wire (top). (*Top photo:* From Kronig, *Textbook on Physics.* Courtesy of Pergamon Press, Oxford. *Lower photos:* From PSSC Physics, 3rd ed., Heath, Lexington, Mass., 1971.)

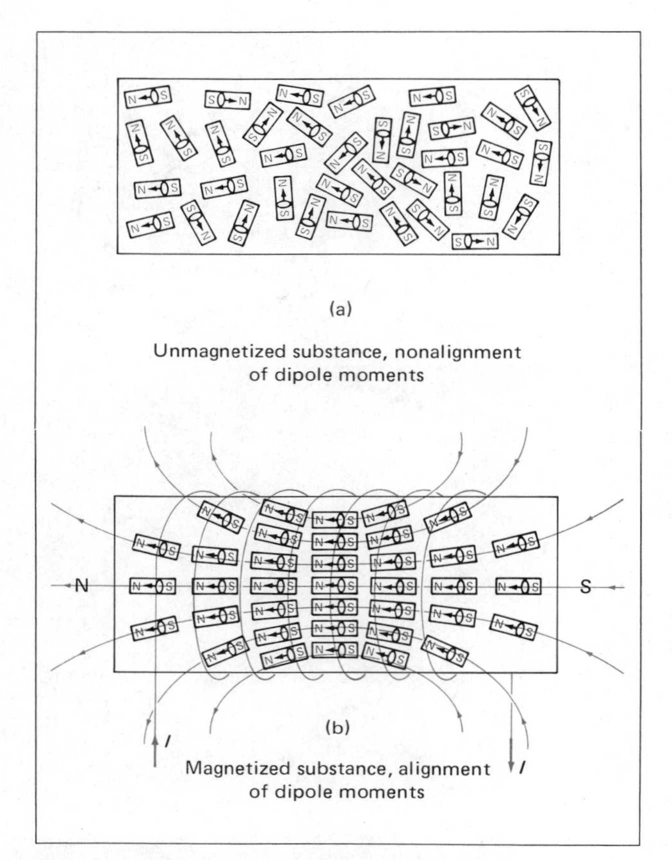

Figure 8.18 In a piece of soft iron the domains are aligned randomly (top). A current-carrying coil of wire wrapped around the iron creates a magnetic field as shown. The domains then line up (bottom).

random arrangement due to heat effects that cause disordering. The amount of alignment remaining after the iron is taken out depends on the strength of the magnetic field in which the iron was placed.

Certain alloys of iron, nickel, cobalt, and other elements are known as "hard" magnetic materials. These alloys, once magnetized, retain their magnetic field for a long time and are known as permanent magnets. When permanent magnets are heated or dropped, the domains are shaken from their aligned positions, and the magnet becomes weaker. In fact, above a certain high temperature, called the **Curie temperature,** the material ceases to be ferromagnetic. The Curie temperature of iron is 770°C.

A permanent magnet is made by aligning the domains inside the material. One way to do this is by heating a piece of hard ferromagnetic material above the Curie temperature and then placing it in a strong magnetic field. As the material cools, the domains line up with the applied magnetic field, and a permanent magnet results.

At the beginning of the seventeenth century, William Gilbert believed the Earth acted like a huge magnet. Today, through knowledge gained from navigation, we know that the magnetic effect does indeed exist over the entire surface of the Earth. Experiments have shown a magnetic field to exist within the Earth and many hundreds of miles out into space surrounding the Earth. The aurora borealis, or northern lights, a common sight in the higher northern latitudes, is associated with the Earth's magnetic field. This effect is discussed in Section 22.3.

The origin of the Earth's magnetic field is not known, but the most acceptable theory indicates that the field is caused by internal currents of charged particles deep within the Earth, rather than by a huge mass of magnetized magnetite. The Earth's interior is believed to be much too hot to be ferromagnetic. Also, the magnetic poles continue to slowly change their positions, which suggests currents.

Presently, the geomagnetic poles are about 16° from the corresponding geographic poles. The north magnetic pole is located at approximately 94°W and 74°N. Figure 8.19 is an illustration of the magnetic field of the Earth.

In general, the magnetic meridian (an imaginary line running from the magnetic north pole to the magnetic

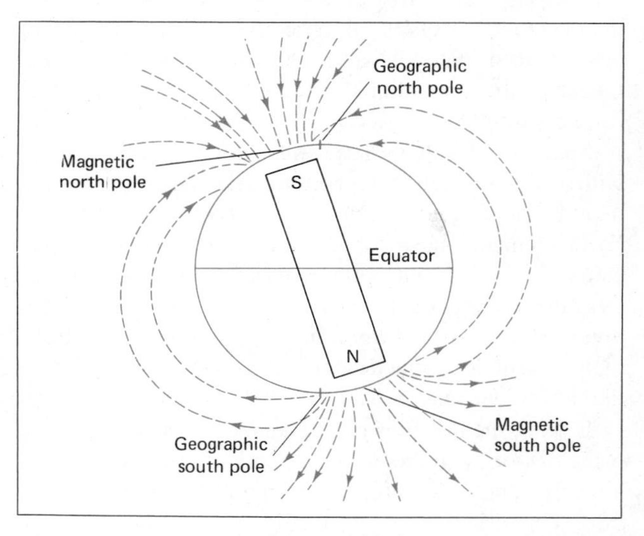

Figure 8.19 The Earth's magnetic field is probably caused by internal currents. It can be thought of as a huge bar magnet as shown. The pole of the bar magnet near geographic north is labeled S since a "north-seeking" compass needle is labeled N and opposite poles attract. Note the variation between geographic and magnetic poles.

south pole) does not coincide with the geographic meridian. The variation between the two is called the **magnetic declination,** or the variation of the compass from true, or geographic, north. The declination may vary east or west of the geographic north-south meridian. See Figs. 8.20 and 8.21.

In addition to the variation of the compass from true north, the compass needle points down from the horizontal plane at all positions on the Earth's surface except at the magnetic equator, providing the compass needle is free to move in a vertical plane. The angle the needle varies downward from the horizontal plane is called the **magnetic dip.** At the magnetic poles, the dip is 90°, and at the magnetic equator, 0°. A dip needle is illustrated in Fig. 8.22.

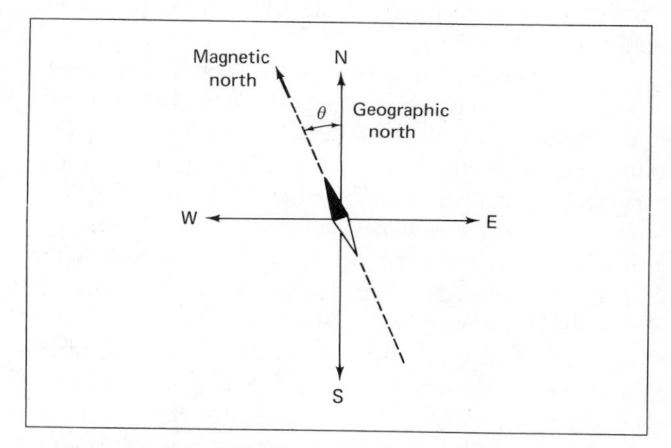

Figure 8.20 Magnetic declination. The angle θ is measured in degrees east or west of geographic north.

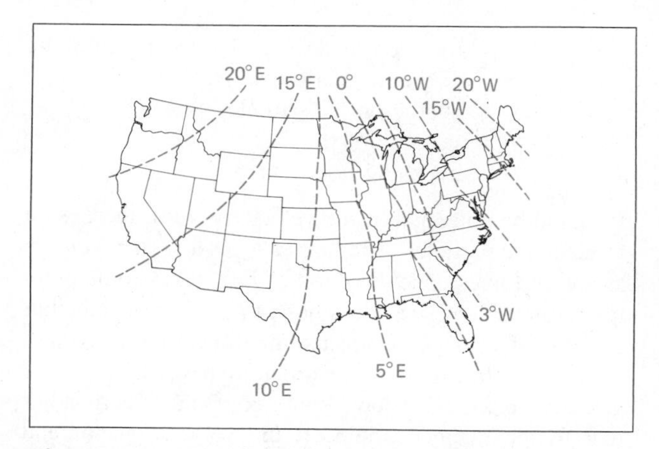

Figure 8.21 Map showing some isogonic (same magnetic declination) lines for the United States.

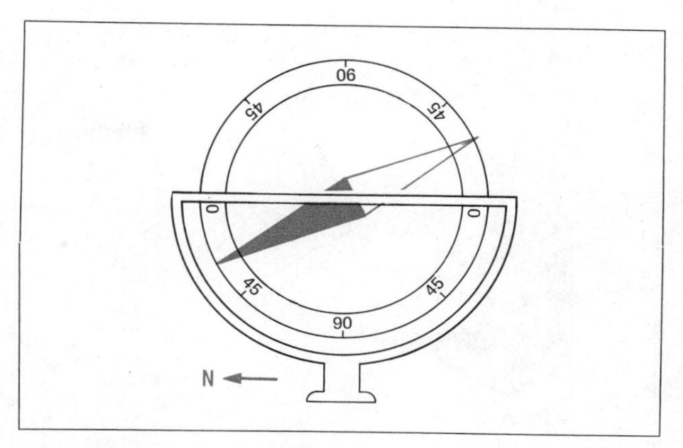

Figure 8.22 A dip needle measures the direction of the Earth's magnetic field in the vertical direction.

The magnetic field of the Earth is fairly weak compared to that of magnets used in the laboratory. However, even the Earth's weak magnetic field is used by certain animals to orient themselves. For instance, migratory birds and homing pigeons are believed to use the Earth's magnetic field to aid them in their homeward flights. Iron compounds have been found in their brains. Honeybees also are believed to use the direction of the Earth's magnetic field as well as the Sun's direction to communicate location of a good food source.

8.4 Electromagnetism

The interaction of electrical and magnetic effects is known as **electromagnetism.** Electromagnetism is one of the most important concepts in physical science, and much of our current technology is directly related to this crucial interaction. We can summarize the two basic principles of electromagnetism as follows:

1. Moving and spinning electric charges create magnetic fields.
2. A magnetic field will deflect a moving charge.

The first idea was mentioned in Section 8.3 when we discussed the source of magnetism. One of the most important applications of this principle is called an electromagnet. An **electromagnet** consists of a current-carrying coil of insulated wire wrapped around a piece of soft iron. This is illustrated in Fig. 8.23. When the current is turned on, a magnetic field is created inside the coil (see Fig. 8.17, top picture). The soft iron is magnetized by this field and makes the magnetic field about 2000 times stronger. When the current is turned off, the

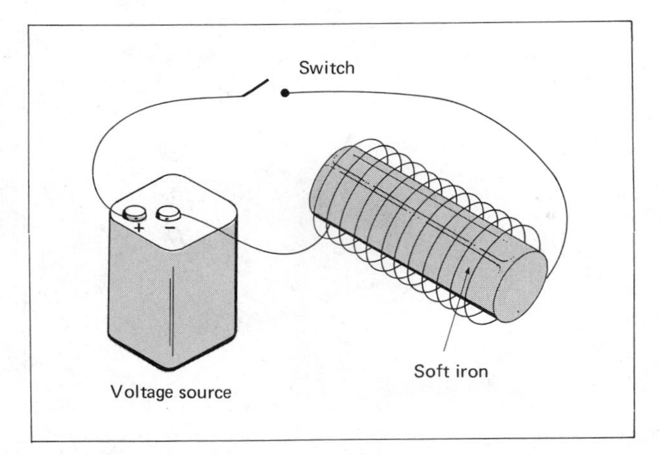

Figure 8.23 An electromagnet consists of a coil of wire wrapped around a piece of soft iron. When the switch is closed, charge flows in the wire and magnetizes the iron, creating a magnet. When the switch is open, no charge flows and the iron is not magnetized.

iron loses most of its magnetism. The advantages of the electromagnet are many. First, it can be switched on and off. Second, the strength of the electromagnet is controlled by the amount of current flowing in the wire. Third, the poles of the magnetic field can be reversed by reversing the current.

Electromagnets are found in a variety of devices and appliances such as doorbells and telephones, and in devices used for moving scrap metal. The mechanism in a telephone receiver will serve as an example of one use. A simplified diagram of a telephone circuit is shown in Fig. 8.24.

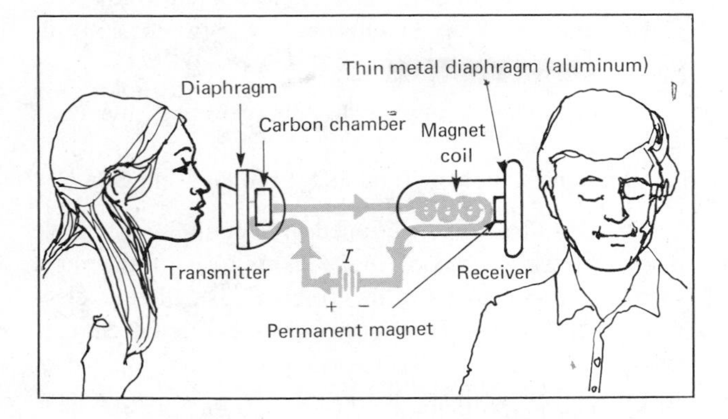

Figure 8.24 A simplified diagram of a (one-directional) telephone circuit. Sound waves are converted to a varying electric current in the transmitter. The varying current travels along the phone lines to the receiver where it is converted back to sound waves.

When a telephone number is dialed, a circuit is completed with another telephone's bell. When the ringing phone is lifted, the circuit between the speakers and receivers of the two telephones is completed. A telephone conversation consists of converting the sound waves into a varying electric current. This varying current travels along the wire to the other telephone's receiver where it is converted back to sound waves.

The transmitter of a phone contains a diaphragm that vibrates due to the sound waves that are spoken. The diaphragm vibrates against a chamber that contains carbon granules. As the diaphragm vibrates, the pressure on the carbon granules varies, and this causes more or less electric resistance in the circuit. The resistance is low when the granules are pressed together and increases as they spread apart. Because of Ohm's law, this varying resistance causes a varying electric current to be transmitted.

At the receiver end there is an electromagnet and a permanent magnet which is attached to a disk. The varying electric current gives the electromagnet a varying magnetic strength. The electromagnet attracts the permanent magnet with a variable force. The changing force on the permanent magnet and disk causes the receiver disk to vibrate. This vibration sets up sound waves that closely resemble the original sound waves, and the voice is heard at the other telephone.

The second basic principle of electromagnetism has been stated in a qualitative way: A magnetic field will deflect moving charges. Quantitatively, we want to understand how the moving charges are deflected. The formula for the force on a moving charge due to a magnetic field is given by

$$F_{mag} = qv \times B \tag{8.12}$$

where F_{mag} = force on a moving charged particle due to the magnetic field B,
q = charge on the particle,
v = velocity of the particle,
B = magnetic field.

It should be noted that if the velocity is zero, there is no force on the particles. The force F_{mag} is a *vector quantity* as are all forces. The direction of F_{mag} takes some practice to understand. The direction of F_{mag} is perpendicular to both the velocity vector and the direction of the magnetic field. In addition, in order to have a force, the velocity vector must have some component perpendicular to the magnetic field. If the velocity vector and magnetic field vector are parallel or antiparallel, there will be no force.

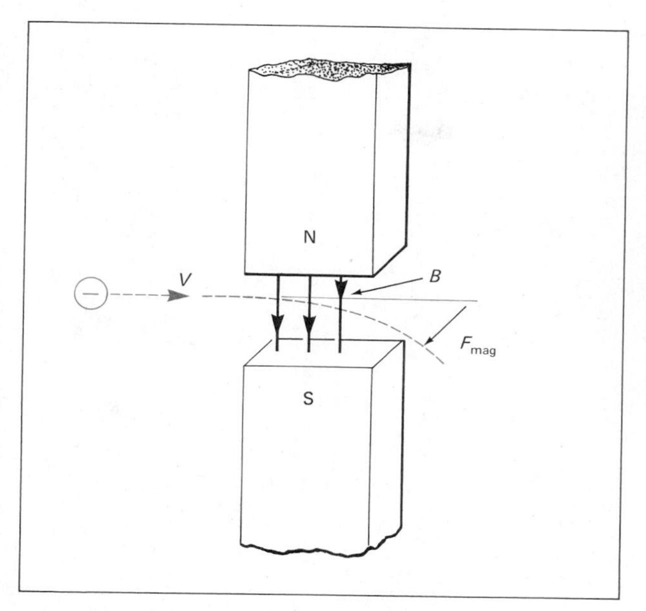

Figure 8.25 Electrons approaching a vertical magnetic field B, as shown, experience a force F_{mag} that deflects them out of the page. See text for explanation.

The best way to understand this deflecting force is by looking at some examples. Figure 8.25 shows a beam of electrons traveling from left to right. The magnetic-field direction is vertical. The force on the electrons is then out from the paper or toward the viewer. Two photographs of an electron beam are shown in Fig. 8.26. In the upper photo, the beam is undeflected. In the lower photo, a vertical magnetic field is set up by the bar magnet and the beam is deflected out toward the viewer similar to Fig. 8.25.

The electrons in conducting wires also experience effects caused by magnetic fields. In Fig. 8.27, a wire is shown in a magnetic field. Since the electrons are not moving, there is no force on the wire. In Fig. 8.28, the electrons in the wire are moving to the right. This is a similar situation to Fig. 8.25 except now the force causes the whole wire to be forced out of the paper toward the viewer. This demonstrates the general principle that a current-carrying wire placed in a magnetic field will experience a force.

In Fig. 8.29, a wire is pushed into a vertical field as shown. The velocity caused by the mechanical push causes the electrons to be deflected to the right, and we say that a current has been induced; that is, an electric current can be set up by an entirely mechanical method. In Fig. 8.29, we have created a current without any batteries, plugs, or other external voltage sources. This is the basic principle behind a generator.

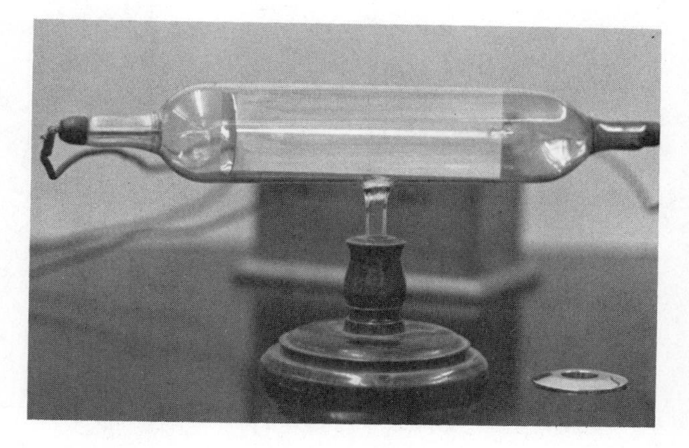

Figure 8.26 An undeflected beam of electrons is shown at the top. A magnet produces a magnetic field which deflects the electrons out of the page similar to Fig. 8.25.

A **generator** converts mechanical work or energy into electric energy. Generators are used at power plants to create electricity. Falling water and pressurized steam from burning coal or fuel oil are two of the sources of mechanical energy. The electricity is carried to homes where it is converted back into mechanical energy or heat energy.

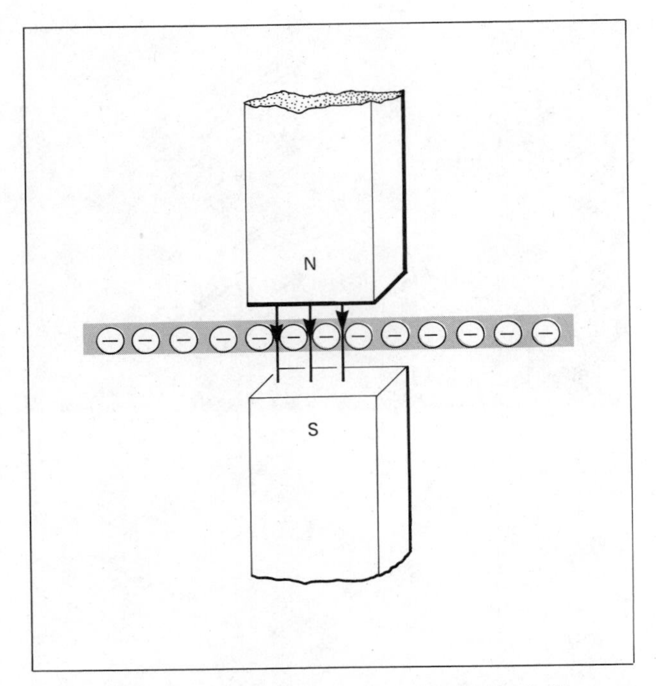

Figure 8.27 A stationary wire in a magnetic field experiences no force on it because the electrons are not moving.

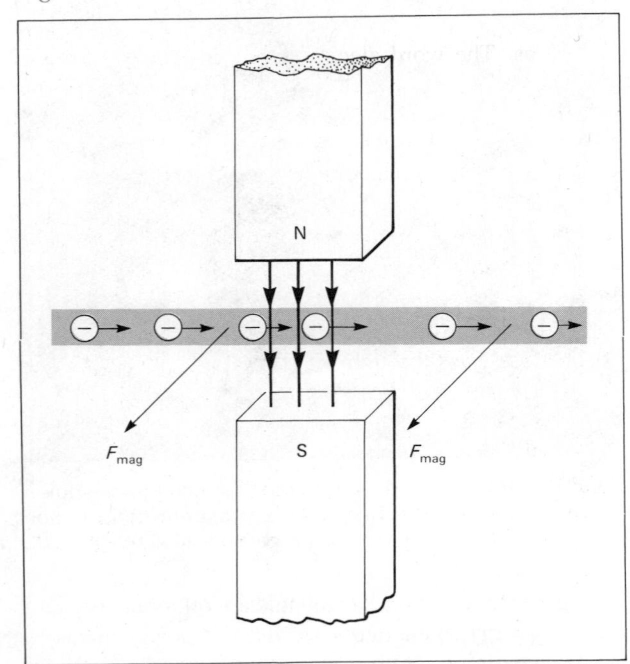

Figure 8.28 A current-carrying wire in which the electrons are moving to the right as shown will be deflected out of the page by the vertical magnetic field. This is the basic principle of an electric motor.

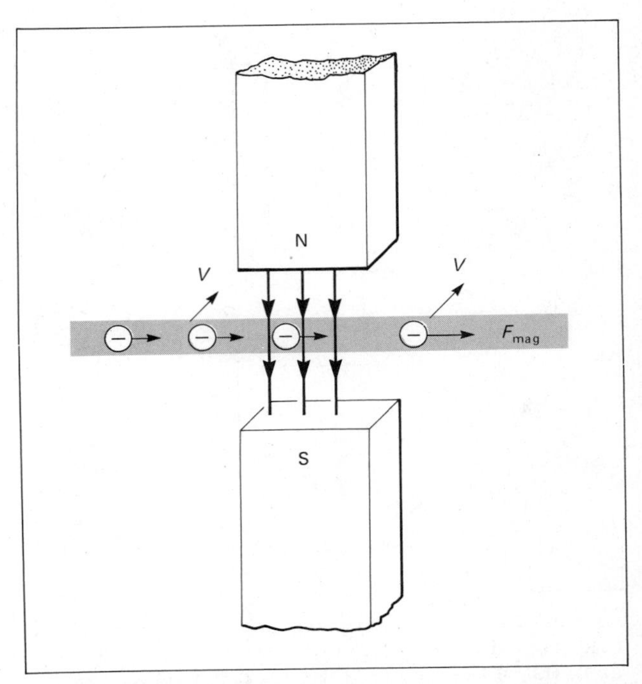

Figure 8.29 A wire is pushed into a vertical magnetic field as shown. This sets up a current in the wire. This is the basic principle of a generator.

A **motor** is a device that converts electric energy into mechanical energy. The basic principle involved is that a current-carrying wire will move in a magnetic field. This principle was illustrated in Fig. 8.28. Motors generally have many windings of wire around a piece of iron in order to enhance the basic effect shown in Fig. 8.28. The conversion of electric energy into mechanical energy is enhanced by more loops of wire and stronger magnets. Motors are heavy because of the iron that is used to concentrate and strengthen the magnetic field.

In the formula for the force due to a magnetic field, the velocity v is the relative velocity between the charged particles and the magnetic field. So far, all of our examples have been with stationary magnetic fields. Michael Faraday, in 1831, did a famous experiment in which he moved the magnet rather than the wire in order to create or induce an electric current. A diagram of this experiment is shown in Fig. 8.30.

We discussed in Section 8.2 how transformers were used to increase or decrease voltages associated with alternating current. Transformers are a good example of the Faraday effect shown in Fig. 8.30.

The principle behind a transformer is illustrated in Fig. 8.31. Alternating current creates a varying magnetic field in the iron core. The changing magnetic field also

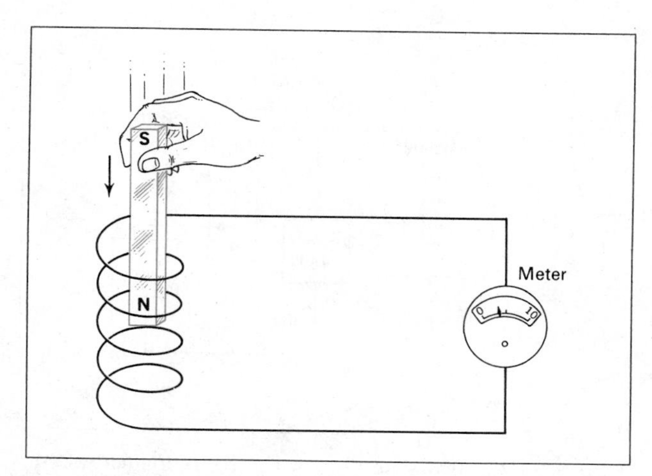

Figure 8.30 A diagram of Faraday's experiment showing electromagnetic induction. The reading on the meter indicates a current in the circuit as the magnet is plunged downward into the coil.

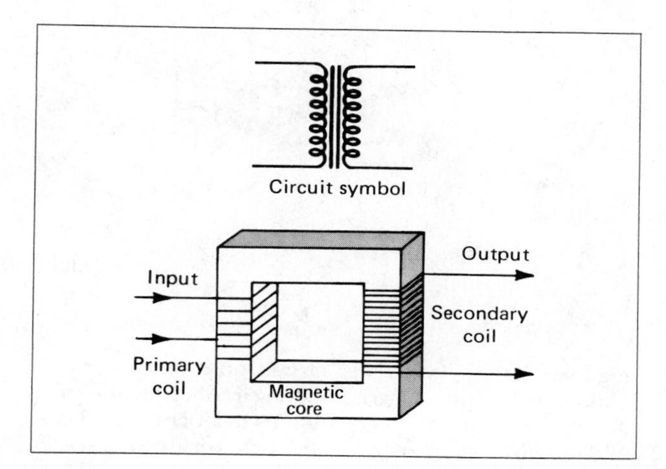

Figure 8.31 The basic features of a transformer consist of a piece of iron, a primary coil, and a secondary coil. The alternating current in the primary coil creates a varying magnetic field in the iron which in turn induces an alternating current in the secondary coil.

passes through the secondary coil where it induces a current. The voltage and current in the secondary coil are different from the primary coil. The amount of voltage is directly proportional to the number of turns in the coil. The current is inversely proportional to the number of turns in the coil. By putting more turns in the secondary coil than in the primary, we can step up the voltage. Similarly, we can decrease the voltage by putting fewer turns in the secondary than in the primary. The formula that we use is

$$V_2 = \frac{N_2}{N_1} V_1 \tag{8.13}$$

where V_2 = secondary voltage,
V_1 = primary voltage,
N_1 = number of turns in primary coil,
N_2 = number of turns in secondary coil.

EXAMPLE 5 A transformer has 500 windings in its primary coil and 25 in its secondary coil. If the primary voltage is 4400 V, find the secondary voltage.

We use Eq. 8.13 to get

$$V_2 = \frac{25}{500}(4400 \text{ V})$$

$$V_2 = 220 \text{ V}$$

This is typical of a step-down transformer on a utility pole near residences.

8.5 Electronics

We live in an age of high technology. Television, computers, and calculators are just three examples of recent developments of the electronics era in which we find ourselves. The word electronics refers to the branch of physics and engineering that deals with the emission and control of electrons. It can take place in an evacuated tube, such as a television picture tube, or in solid semiconductor materials that are the basis of solid-state electronics.

A television picture tube is an example of a cathode-ray tube, or **CRT.** In Fig. 8.32 the basic features of a CRT are shown. The electron gun is a heated filament that ejects electrons. The electrons are accelerated by a voltage and pass through a narrow opening, forming an electron beam. The beam of electrons is then deflected by either electric or magnetic fields. In our example (Fig. 8.32) there are electromagnets. The input current to the electromagnets determines the strength of the magnetic fields and thereby regulates the deflection of the beam.

The deflected electron beam strikes a phosphorescent screen. The intensity of the electron beam controls the brightness of the spot where the electron beam strikes the screen. The beam sweeps across 700 dots to form a line and then goes back to form a new line. It produces every other line and then goes back and fills in the alternate lines to prevent a noticeable flicker. Thirty pictures are made every second to create the illusion of

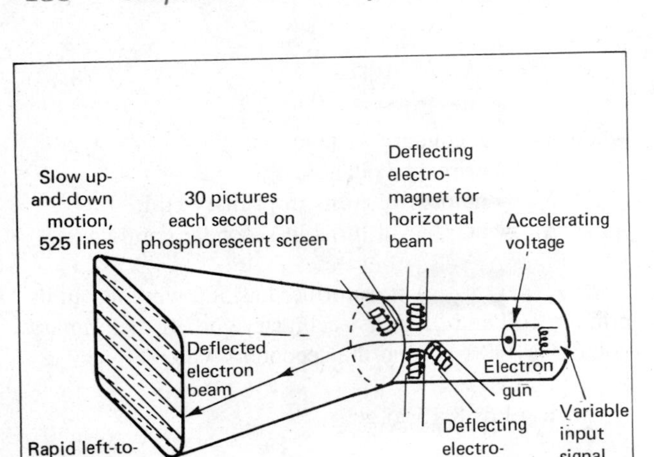

Figure 8.32 A simplified diagram of a cathode ray tube (CRT). The electron gun ejects electrons, which are accelerated by a voltage. The electrons form a beam that is deflected by electromagnets in both the horizontal and vertical directions. The beam sweeps across and up and down the screen. The variable-input signal controls the intensity of the electron beam and hence the brightness of each dot on the screen. Thirty pictures are made each second to give a moving image.

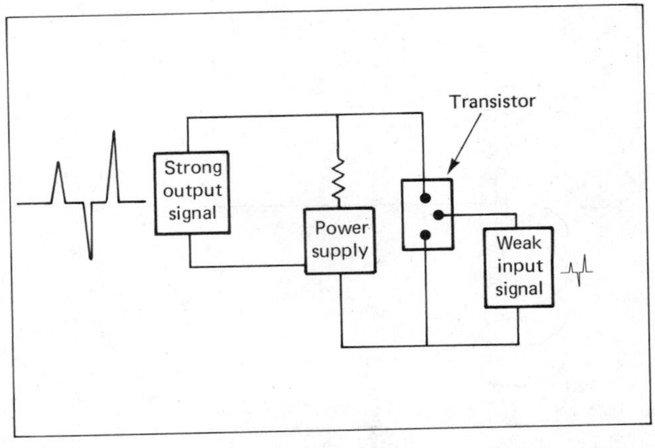

Figure 8.33 A transistor has three parts instead of the usual two. It can be used to amplify a weak input signal, shown on the right, into a strong signal, shown on the left. A power supply such as a battery is needed to provide the energy for amplification.

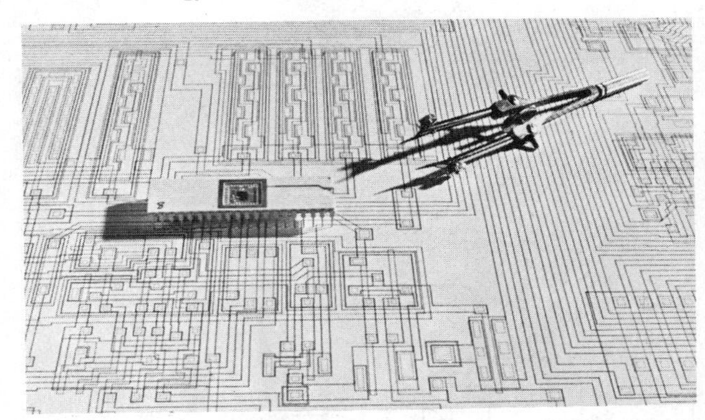

Figure 8.34 An integrated circuit on a silicon chip is shown resting on a diagram of the circuit. Note the small size of this chip, which contains 1800 devices (e.g., diodes and transistors). (Photo courtesy Bell-Northern Research Ltd., Ottawa, Canada.)

movement on the screen. The input signal to the beam is a variable current that controls the intensity of the electron beam and hence the brightness of each dot on the screen. The input signal is received from the antenna, which gets it from a TV transmitter.

A color TV has three electron guns, and each dot on the screen consists of a triad of red, green, and blue dots. By varying the intensities of the three beams, each dot can be given a different color so that a color picture is produced.

Two basic electronic components used in TV sets and other electronics devices need to be mentioned. One is the diode; the other is the transistor. A **diode** is a device that allows current to flow in only one direction. It has a negligible resistance in one direction and a practically infinite resistance in the reverse direction. Diodes have many uses. For instance, they can be used to change alternating current into direct current.

A **transistor** is a device whose primary purpose is to amplify an input signal. It needs an external voltage source such as a battery to supply the power to allow it to operate. In Fig. 8.33 we have a simplified diagram of an amplifying circuit. It consists of a weak input signal, a power supply, and a transistor that amplifies the weak input signal into a much stronger output signal. A typical transistor can amplify a signal by a factor of 15 to 20.

Using diodes, transistors, and resistors, many types of electronics circuits, called integrated circuits, can be made. These are now made in miniature by techniques such as depositing conducting metal onto layers of material made primarily from silicon, a good semiconductor. The silicon layers can be used to make diodes, transistors, and resistors. There is plenty of silicon available since sand is mainly silicon dioxide.

Techniques have been developed to make the integrated circuits smaller and smaller. Integrated circuits

Figure 8.35 A hand-held calculator consists of silicon chips, a power source, and a display. The reduction in size is limited by our need to punch only one key at a time with a finger. (Photo courtesy Jimmy Adams.)

Figure 8.36 A typical computer console. (Photo courtesy Jimmy Adams.)

made in this way are commonly called "chips," and they can be quite complex even if quite small. A picture of a chip is shown in Fig. 8.34.

Chips are used to make small hand-held calculators such as those shown in Fig. 8.35. Calculators such as these can be made extremely small, but most are still made about the size of a person's hand for one simple reason. To be used easily, the spacing of the keys must be far enough apart so that a finger will press only one key at a time.

A typical computer console (Fig. 8.36) uses one or more silicon chips to form the logic circuits, and a CRT to allow information to be displayed to the operator. As our knowledge of electronics grows, computers are being made smaller and more powerful.

Learning Objectives

After reading and studying this chapter, you should be able to do the following without referring to the text:

1. Give the mks system units of electric charge, electric current, voltage, resistance, energy, and power.
2. State the basic force law for (a) electric charges (b) magnetic poles.
3. State one way in which Coulomb's law and Newton's law of universal gravitation are similar. State three important ways in which they are different.
4. Explain briefly why some materials are conductors and some are insulators.
5. Explain why electrons, rather than protons, move along wires.
6. State Ohm's law in equation form.
7. Explain the difference between ac and dc and tell why ac is used in homes.
8. Explain what transformers do and how they work.
9. Draw diagrams of a series circuit and a parallel circuit, and explain why home appliances are wired in parallel rather than series.
10. Explain why a shock from a 110-volt ac source may or may not be fatal.
11. Sketch the magnetic field around the Earth.
12. Briefly give an atomic description of what causes iron to be magnetic.
13. Explain how one magnetizes a piece of iron.
14. Briefly describe how an electromagnet works and give three advantages they have over permanent magnets.
15. Draw a simple diagram to show the basic idea behind an electric generator.
16. Draw a simple diagram to show the basic idea behind an electric motor.
17. Explain how images are produced on TV screens.
18. State one use for a (a) diode (b) transistor (c) chip.

Important Words and Terms

electrons
protons
neutrons
coulomb
current
ampere
quarks
Coulomb's law
conductors
insulators
electricity
electric potential energy
voltage
volt
resistance
Ohm's law
ohm
dc
ac
transformers
series circuit
parallel circuit
magnetic field
electric field
magnetic monopole
ferromagnetic
domains
Curie temperature
magnetic declination
magnetic dip
electromagnetism
electromagnet
generator
motor
CRT
diode
transistor

Questions

1. State the three basic constituents of an atom. Give their fundamental properties.
2. What is a quark?
3. Give the constants k and G in the mks system. What does this indicate about the "strength" of the electric and gravitational forces?
4. Explain how a charged rubber comb attracts bits of paper.
5. Why do conductors carry a current but insulators do not?
6. Draw a diagram of a simple electric circuit, including a switch, and tell how it works.
7. Why is ac used in homes?
8. What would happen if power were transmitted from the generating station to your home at 110 volts. What is done to prevent this from happening?
9. Why are home appliances wired in parallel rather than in series?
10. What is a fuse and why is it used?
11. What causes electric shock and what precautions should be taken to avoid it?
12. What is the difference between an electric dipole and a magnetic dipole?
13. What happens if a bar magnet is cut in half?
14. Why are some materials ferromagnetic?
15. How does a permanent magnet attract a piece of soft iron?
16. What happens to a piece of iron above the Curie temperature?
17. Distinguish between magnetic dip and declination.
18. How does an electromagnet work?
19. How do telephone transmitters and receivers work?
20. What basic principle makes a motor work?
21. What basic principle makes a generator work?
22. Why are motors heavy?
23. How are images produced on a TV screen?
24. What does a diode do?
25. Why are transistors so important?
26. What limits the size of a calculator?

Problems

1. How many electrons make a single coulomb of charge?
2. What happens to the force of attraction between an electron and a proton if the distance between them is tripled?
3. Use Eq. 8.3 to find the force of attraction between a proton and an electron that are 5.3×10^{-11} m apart. (This is the situation in the hydrogen atom.)

   Answer: 8.2×10^{-8} N
4. (a) How much current is drawn by a flashlight using batteries that add up to 3 V and produce a power of 0.5 W?

   (b) What is the resistance of the light bulb?

   Answer: (b) 18 Ω
5. How much does it cost to run a 1500-W hair dryer 30 minutes each day for one month at a cost of $0.08 per kWh?

   Answer: $1.80

6. A refrigerator motor using 1000 W runs $\frac{1}{3}$ of the time. How much does this cost each month at $0.08 per kWh?
7. In Fig. 8.12, the voltage is 110 V.
   (a) What will be the current in a light bulb that has a resistance of 100 Ω?
   (b) What will be the current in a toaster with a resistance of 10 Ω?
   (c) What will be the total current if only the light bulb and toaster are turned on? *Answer:* (c) 12.1 A
8. A portable electric heater has 6 resistances of 1.5 Ω each that are wired in series. The voltage is 120 V.
   (a) Compute the current flowing in the heater.
   (b) Compute the power output of the heater.
9. The heating element of an electric iron operates on 110 V at a current of 10 A.
   (a) What is the resistance of the iron?
   (b) What power is dissipated by the iron? *Answer:* (b) 1100 W
10. (a) What is the total resistance in Fig. 8.12 if the three resistors in parallel have $R_1 = 60\ \Omega$, $R_2 = 30\ \Omega$, and $R_3 = 20\ \Omega$? *Answer:* (a) 10 Ω
    (b) Use Ohm's law and the resistance found in (a) to compute the total current if $V = 110$ V. *Answer:* (b) 11 A
11. A person with dry skin has a body resistance of 50,000 Ω.
    (a) If he or she touches the two terminals of a 2-V dry cell, what will be the current through the body?
    (b) The person's fingers are wet, and the body resistance drops to 2000 Ω. What will be the current now?
12. Sketch the magnetic field (a) between a north and south pole, and (b) between two north poles.
13. Use Fig. 8.21 to find the magnetic declination where you live.
14. A power company runs current along 765,000-V transmission lines. This voltage is changed using a transformer that has 900 turns in the primary coil and 150 turns in the secondary coil. What is the secondary voltage produced?
15. A transformer with 1000 turns in the primary coil has to decrease the voltage from 4400 V to 220 V for home use. How many turns should be in the secondary coil? *Answer:* 50
16. From Fig. 8.32 compute the total number of individual phosphorescent spots created on a TV screen each second. *Answer:* 11 million

9

Relativity

A precocious student, quite bright,
Could travel much faster than light.
She departed one day
In an Einsteinian way,
And arrived on the previous night.
–Anonymous

MANY STATEMENTS IN our conversations involve the "relativeness" of things. For example, we say: Beauty is in the eye of the beholder. A boring class seems to run on and on, while time spent with someone we love flies by. The Earth goes around the Sun or the Sun goes around the Earth, depending on how you look at it.

Of course, we know that it is much simpler to describe the solar system by saying that the Earth and other planets revolve around the Sun and spin as they go. The spinning of the Earth causes day and night. However, to us who live on the Earth, it certainly does appear that the Sun goes around the Earth. In fact, we still speak of sunrise and sunset. We see, then, that when we talk about things that occur in our daily life, it is important to make clear what our frame of reference is. The things that we observe and the values we hold depend on a frame of reference.

When we discussed Newton's laws of motion in Chapter 3, we based our thinking on our common experience of space and time. For example, an astronaut can set a watch with a clock in Houston and fly to the moon and back at high speeds. Upon returning, the watch and the clock in Houston are still synchronized although scientists can detect a very slight change. This illustrates that time doesn't noticeably change even when we travel at speeds as high as 25,000 miles per hour. Our ideas of space are based on similar experiences at low speeds.

In 1905 Albert Einstein (opposite) published a landmark paper in which he put forth a theory that revolutionized our concept of space and time. For instance, he showed that if the astronaut were to travel at speeds approaching the speed of light, the watch would no longer keep almost the same time as the stationary clock. In other words, time is relative. It is affected by the velocity of the timepiece. (Of course, contrary to the limerick at the start of the chapter, it is impossible for time to go backward under any circumstances.)

◀ Albert Einstein's theory of relativity drastically altered our way of thinking about nature.

Einstein's theory deals with space and distances in a similar fashion. They too are relative. They depend on the speed of the observer and are not absolute. Einstein's theory is quite mathematical, but the results that arise are so fascinating as to be well worth the reader's effort to understand them.

9.1 Moving Systems of Reference

Whenever a scientist makes a measurement in a laboratory, he or she employs a system of reference. A **system of reference** is something to which a measurement is referred. Another scientist, making the same measurement but doing so when moving with a constant velocity, would get different results because he or she uses a different system of reference. The Doppler effect is one example of this phenomenon. The frequency of a sound that a person hears from a moving source is different from the actual frequency of the sound at the source. Thus, when a speeding ambulance approaches a person waiting on a street corner, that person hears a sound from the siren with a higher frequency than does the driver of the ambulance.

As another example, let us assume that Dr. Albert is standing in a train station as a train whizzes past at 60 miles per hour. Dr. Baker is on the train and walking at a velocity of 5 miles per hour toward the front of the train. Dr. Baker would comment that she was traveling at a velocity of 5 miles per hour. However, Dr. Albert would contend that Dr. Baker was traveling at a velocity of 65 miles per hour relative to him, since the train was already running at 60 miles per hour and Dr. Baker's walking speed* should be added to that. Who is right? Of course, both are correct. We see, then, that before we make comments about Dr. Baker's velocity, we must specify whether we are talking about velocity relative to the train or relative to the train station.

In our study of motion, thus far, we have been concerned with objects going at velocities that are relatively very low compared to the speed of light. Just about everything that we have experienced is moving slowly relative to the speed of light (186,000 miles per second!). At these relatively low velocities, it is not hard to change from one system of reference to another. Suppose we are in a system of reference that is moving with velocity v_R with respect to a fixed system, say the Earth. Now suppose that during the time change, Δt, we change our position in our moving system by $\Delta d'$. What will the total change in our position be? Let us refer to Fig. 9.1. Because we were on a moving system, we have changed our position by at least $v_R \Delta t$. If we had remained motionless at O', our displacement relative to O would have been exactly $v_R \Delta t$. But since we changed our position in the moving system from O' to A', there is an additional change of $\Delta d'$. The total change in our position in the fixed system is given by

$$\Delta d = v_R \Delta t + \Delta d'$$

We can also write this as

$$\Delta d' = \Delta d - v_R \Delta t \tag{9.1}$$

where $\Delta d'$ = change in position in the moving system,
Δd = change in position in the fixed system,
Δt = change in time,
v_R = velocity of moving system relative to fixed system.

Equation 9.1 is a good example of relativity. **Relativity** relates the measurements made in two different systems of reference. In Eq. 9.1 we have related distances and times in two different reference systems.

In most of our experience, time seems to be absolute; that is, it does not depend on the coordinate system. (Shortly, we will see that this is not always true.) If for

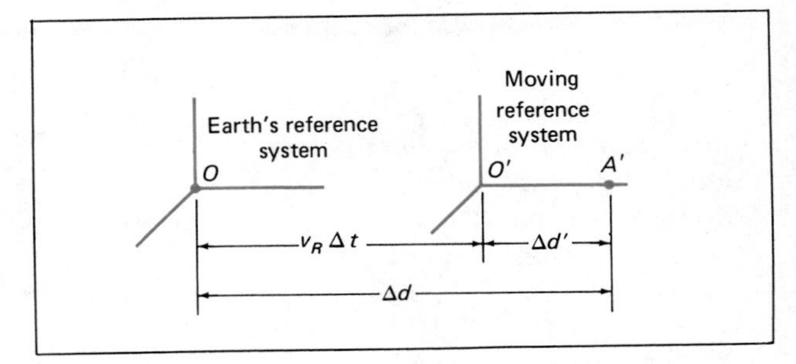

Figure 9.1 Distance relationships in moving reference systems, O and O'. $\Delta d = v_R \Delta t + \Delta d'$. This equation is true only if v_R is much lower than the velocity of light.

the moment, we assume that time is absolute, clocks in both the moving and fixed systems will run at the same speed. So we have

$$\Delta t' = \Delta t \tag{9.2}$$

where $\Delta t'$ = change in time in the moving system,
Δt = change in time in the fixed system.

For low speeds we know that Eq. 9.2 is adequate, since a commercial jet pilot's watch keeps the same time as a clock on the ground.

The velocities in the two reference systems of Fig. 9.1 can now be related by dividing Eq. 9.1 by Eq. 9.2

$$\frac{\Delta d'}{\Delta t'} = \frac{\Delta d}{\Delta t} - v_R \frac{\Delta t}{\Delta t}$$

or

$$v' = v - v_R \tag{9.3}$$

where v' = velocity in the moving system,
v = velocity in the fixed system,
v_R = velocity of the moving system with respect to the fixed system.

This concept of addition of velocities is illustrated in Fig. 9.2.

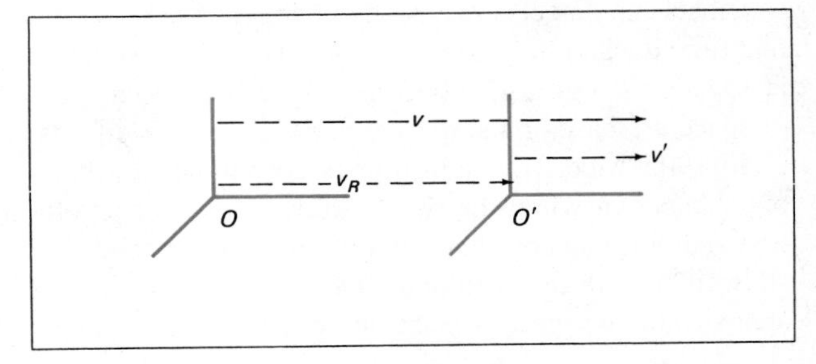

Figure 9.2 Velocity relationships in moving reference systems $v = v_R + v'$. This equation is true only if v_R is much smaller than the velocity of light.

*We will be using "speed" and "velocity" interchangeably throughout this chapter.

In our train example, Dr. Baker's velocity (v') in the moving system (the train) was 5 mi/h. The train was running at a velocity of $v_R = 60$ mi/h. So the velocity according to Dr. Albert in the fixed system (the train station) was $v = 65$ mi/h. Thus, in Eq. 9.3

$$5 \text{ mi/h} = 65 \text{ mi/h} - 60 \text{ mi/h}$$

Equations 9.1, 9.2, and 9.3 are the foundations of relativity at low velocities. We will next examine what happens as we consider velocities that are comparable in magnitude to the speed of light.

A good example of Eq. 9.3 is the shooting of the space shuttle shot toward the east rather than the west. In order for the space shuttle to go around the Earth (at a height of 150 to 400 km), it must be accelerated to a speed of approximately 18,000 mi/h (see Fig. 9.3). The Earth, however, is already spinning toward the east at a speed of about 900 mi/h at Cape Canaveral.

Figure 9.3 The space shuttle at lift-off. The space shuttle and satellites are usually shot out over the Atlantic Ocean to take advantage of the Earth's speed of rotation. Less fuel is required to shoot them toward the east. (Courtesy NASA.)

By shooting the space shuttle toward the east, the Earth's spinning speed is added to the satellite's speed, so not as much fuel is needed. It would be expensive and wasteful of fuel to shoot the space shuttle toward the west. A westward shot would mean that additional fuel would be needed to overcome the Earth's spinning motion. NASA scientists take relative motion into account in order to save the taxpayer's dollar.

9.2 Einstein's Postulates

Between 1850 and 1900 several experiments were performed* which gave very strange results if interpreted in terms of Eqs. 9.1, 9.2, and 9.3. But in 1905, Albert Einstein explained these experiments by postulating two things:

Einstein's Postulates

1. **The physical laws of nature (such as Coulomb's law) are the same in all reference systems that are moving with constant velocity with respect to each other.**
2. **The velocity of light in a vacuum is always the same, independent of the motion of the observer and the source.**

These postulates form the basis of special relativity. Einstein's **special theory of relativity** holds only for systems moving at a constant velocity with respect to one another. In Section 9.5 we will discuss what happens when the systems are accelerating with respect to each other.

As an example of the first postulate, an experiment on a speeding spaceship concerning Coulomb's law of electric charges will yield the same results as an identical experiment carried out in a laboratory on the Earth.

The second postulate seems strange at first. It is in direct contradiction to Eq. 9.3. According to Eq. 9.3, if a star is moving away from the Earth at three-fourths the velocity of light (as in Fig. 9.4),† the speed of light from the star when it reaches the Earth should be only one-fourth the velocity of light. If worked out, using Eq. 9.3, we have

*Some of the experiments that baffled description were Fizeau's experiments on the aberration of starlight, and the Michelson-Morley experiment that measured the speed of light in different directions.

†We let c equal the speed of light in empty space.

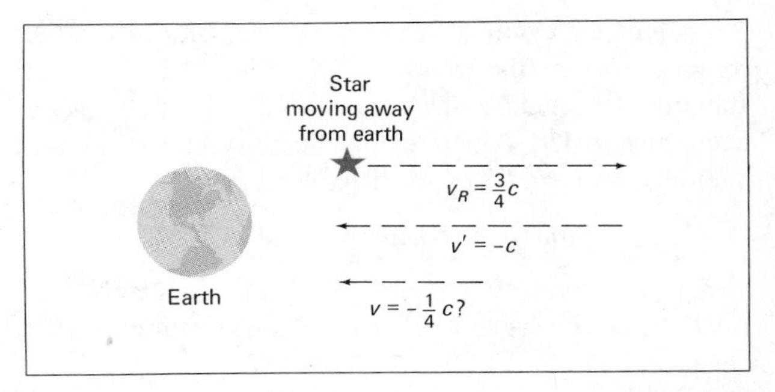

Figure 9.4 The non-Einsteinian way of adding high velocities. This method works for low velocities but not for high velocities.

$$v' = v - v_R$$

or

$$v = v_R + v'$$

Given

$$v_R = \tfrac{3}{4}c$$

and

$$v' = -c$$

Then,

$$v = \tfrac{3}{4}c - c$$

$$v = -\tfrac{1}{4}c \qquad (WRONG)$$

(The minus signs mean that the velocities in Fig. 9.4 are in the opposite direction to those shown in Fig. 9.2.)

According to Eq. 9.3, the speed of light reaching the Earth from distant stars will be less than c. Einstein's second postulate, on the other hand, says that the speed of light reaching the Earth will always be equal to c, no matter how fast the stars move away from us. Something is wrong, and it is Eq. 9.3. This means that Eqs. 9.1 and 9.2 are also inadequate at very high speeds, since we derived Eq. 9.3 from them. If Einstein's postulates are correct, we must therefore replace Eqs. 9.1, 9.2, and 9.3 with more general ones.

9.3 The Lorentz Transformation

Einstein, using his second postulate, showed that Eqs. 9.1 and 9.2 should be rewritten as

$$\Delta d' = \frac{\Delta d - v_R \Delta t}{\sqrt{1 - (v_R^2/c^2)}} \tag{9.4}$$

$$\Delta t' = \frac{\Delta t - \Delta d v_R/c^2}{\sqrt{1 - (v_R^2/c^2)}} \tag{9.5}$$

Equations 9.4 and 9.5 are known as the **Lorentz transformation.** They tell how distances and times in systems moving relative to each other are related. Note that distances depend on times and vice versa. Also, note that if the ratio of v_R to c is very small (i.e., $v_R/c \approx 0$), we obtain Eqs. 9.1 and 9.2.

An important consequence of Eqs. 9.4 and 9.5 is that distances and times in the two coordinate systems are no longer the same. The relationship between distances in the two systems, measured at the same time ($\Delta t = 0$), is

$$\Delta d' = \frac{\Delta d}{\sqrt{1 - (v_R^2/c^2)}}$$

or

$$\Delta d = \Delta d'\sqrt{1 - (v_R^2/c^2)} \qquad (9.6)$$

Equation 9.6 means that a moving meter stick appears to be shorter to an observer who is standing still than to an observer moving with the meter stick. This phenomenon is known as **length contraction.**

EXAMPLE 1 Dr. Baker has a meter stick in her rocket. The rocket is traveling at a speed of 9/10 the speed of light. As it whizzes past Dr. Albert's space station, how long does the meter stick look to Dr. Albert? Refer to Fig. 9.5.

This is solved by recognizing that in the moving system, $\Delta d' = 1$ m. We want to solve for Δd. The speed $v_R = 0.9c$. We use Eq. 9.6.

$$\Delta d = \sqrt{1 - (v_R^2/c^2)} \times \Delta d'$$
$$= \sqrt{1 - ((.9c)^2/c^2)} \times 1 \text{ m}$$
$$= 0.4359 \text{ m}$$

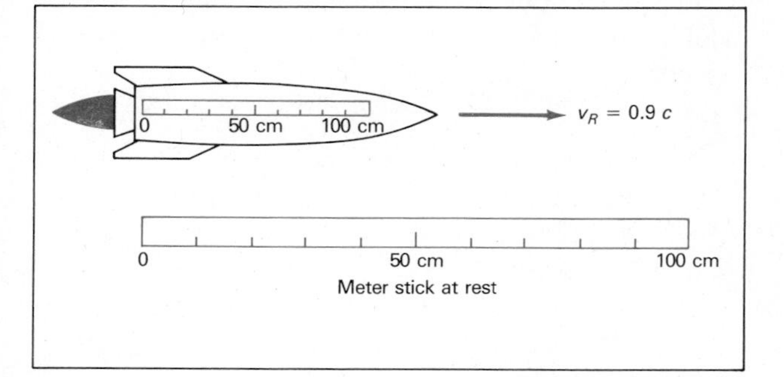

Figure 9.5 A meter stick moving at $v_R = 0.9c$ appears to be only 0.4359 m to an observer at rest.

So, the meter stick appears to be only 0.4359 m to Dr. Albert. Dr. Baker would say that it was 1 m. Of course, when Dr. Baker's rocket stops ($v_R = 0$), Dr. Albert will then say that it is 1 m long.

Now let us consider how the times in the two systems are related. We will assume that a single moving clock keeps the time $\Delta t'$. This time is then compared with two or more synchronized clocks that are at rest. Since there is only one moving clock, we set $\Delta d' = 0$ in Eq. 9.4. The quantity Δd is the distance between the two synchronized clocks, and it is equal to $v_R\ \Delta t$ from Eq. 9.4.

The relationship between Δt and $\Delta t'$ is then found from Eq. 9.5 to be

$$\Delta t' = \Delta t\sqrt{1 - (v_R^2/c^2)}$$

or

$$\Delta t = \frac{\Delta t'}{\sqrt{1 - (v_R^2/c^2)}} \qquad (9.7)$$

According to Eq. 9.7 the single moving clock does not keep the same time as the synchronized clocks! The single moving clock runs slow as measured by the two synchronized clocks. This phenomenon is called **time dilation.**

A good example of the reality of time dilation is provided by the lifetime of small particles called **muons,** or mu mesons. When muons are created in laboratory experiments, they have an average lifetime of $T_o = 2.2 \times 10^{-6}$ s. Within this lifetime they would travel an average distance of $\Delta d = vT_o$ before disappearing by decaying into other particles. Even at the speed of light ($v = c$), the distance traveled (Δd) during T_o would be only about $\frac{4}{10}$ mi. However, muons created by cosmic ray photons at altitudes of about 15 mi reach the Earth's surface. The reason that this is possible is that the muon's "clock" goes slower than an Earth clock. In the muon's reference frame, the lifetime is still $T_o = 2.2 \times 10^{-6}$ s. But in the Earth's frame of reference, the lifetime T_e is given by Eq. 9.7 as

$$T_e = \frac{T_o}{\sqrt{1 - (v_R^2/c^2)}}$$

where T_e = muon's lifetime according to an Earth clock,
T_o = muon's lifetime,
v_R = velocity of muons relative to the Earth.

If $v_R = 0.9998\ c$ (almost the speed of light), we find that $T_e = 50\ T_o$, and the muons will not decay until they have traveled 20 mi. Thus, they will reach the Earth before decaying. Figure 9.6 shows the difference between the two "clocks" in a schematic fashion.

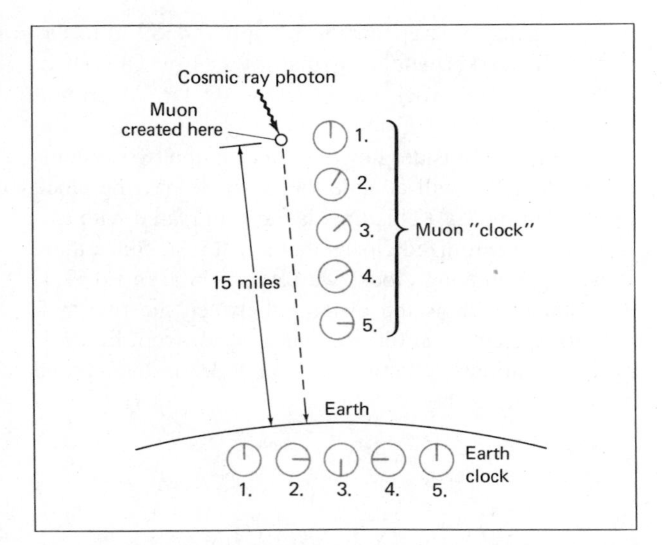

Figure 9.6 Diagram of time differences at high velocities. The muon "clock" runs slower than an Earth clock because the muons have very high velocities.

Equation 9.7 needs to be understood in terms of relative motion. Consider a rocket ship moving toward the Earth. According to an observer on a rocket ship, it is the Earth that is moving. Which is really moving? Of course, that is relative. To find a time difference, comparisons of clocks must be made, so there must be at least two synchronized clocks in either the rocket system or the Earth system. These two (or more) synchronized clocks will be compared to a single clock in the other system. It is the single moving clock that will always run slow according to measurements made by an observer stationary with the synchronized clocks.

Equation 9.7 can be very confusing if we are not extremely careful in interpreting Δt and $\Delta t'$. Consider the following "mind-bending" situation. Assume that Ruth and Walter are twins. Ruth gets in a spaceship and leaves the Earth, traveling at a speed 0.866 times the velocity of light. When 1 h has elapsed on Ruth's ship, the amount of time elapsed on the Earth is

$$\Delta t = \frac{1 \text{ h}}{\sqrt{1 - \frac{(0.866c)^2}{c^2}}}$$

$$= \frac{1 \text{ h}}{\sqrt{\frac{1}{4}}} = 2 \text{ h} \tag{9.7a}$$

Thus, Ruth's clock runs twice as slow as Walter's clock on Earth. But now if we consider Ruth's spaceship as fixed and consider that the Earth is moving away from it, the roles of Δt and $\Delta t'$ are reversed. The Δt would now refer to Ruth's clock on the spaceship and $\Delta t'$ to the Earth clock of Walter. Using Eq. 9.7, we get

$$1 \text{ h} = \frac{\Delta t'}{\sqrt{1 - \frac{(0.866c)^2}{c^2}}} = \frac{\Delta t'}{\sqrt{\frac{1}{4}}} = 2\,\Delta t'$$

or $$\Delta t' = \tfrac{1}{2}\text{h} \tag{9.7b}$$

According to this result, when Ruth's clock has run for 1 h, the Earth clock will have run for $\frac{1}{2}$h, and so Ruth's clock now runs twice as fast as Walter's. But we just concluded that Ruth's clock ran twice as slow as Walter's. This dilemma is sometimes called the **twins paradox.** If Ruth's spaceship traveled for 20 Earth years, Walter, using the reasoning of Eq. 9.7a, would say that Ruth had aged only 10 years. But if Ruth used the reasoning of Eq. 9.7b, she would calculate that 40 rather than 10 of her spaceship years corresponded to 20 Earth years. If they were twins, which twin would really be older? This question has plagued many a student of relativity.

The explanation of the twins paradox is quite subtle—and quite profound, for it calls into question the very basis of science. This is the question of how a measurement is made. When Ruth leaves the Earth, she can set her clock to coincide with Walter's clock, but how do the two ever get back together to compare their clocks again or to see who is older? Eqs. 9.4, 9.5, and 9.7 hold only for systems that move at constant velocities relative to one another. They do not hold for systems that accelerate. Thus, when Ruth's spaceship accelerates, new equations must be used. The fact that Ruth and Walter can never compare their clocks or ages more than once without acceleration is the solution to the twins paradox. In Section 9.5 this problem will be considered again.

The Lorentz transformation Equations 9.4 and 9.5 yield an interesting result when applied to the relativistic addition of velocities. If we divide Eq. 9.4 by Eq. 9.5 we obtain

$$v' = \frac{v - v_R}{1 - \frac{v v_R}{c^2}}$$

If we let $v = v_1$ and $-v_R = v_2$, this can be written as

$$v_3 = \frac{v_1 + v_2}{1 + \frac{v_1 v_2}{c^2}} \tag{9.8}$$

where v_3 = relative velocity,
v_1, v_2 = two velocities to be added.

Equation 9.8 can now be used to show that no velocity can ever exceed c, the speed of light in empty space. As an example, we consider the problem of two spaceships approaching each other, each ship traveling with a velocity of three-fourths the speed of light (Fig. 9.7).

According to our classical equation (Eq. 9.3), occupants in each spaceship would claim that the other ship is approaching it at a velocity of $1.5c$. However, Eq. 9.8 yields a different result. Using this equation, we obtain

$$v_3 = \frac{\frac{3}{4}c + \frac{3}{4}c}{1 + \frac{\frac{3}{4}c\,\frac{3}{4}c}{c^2}} = \frac{1.5c}{1 + \frac{9}{16}}$$

$$v_3 = 0.96c$$

So, occupants in each ship now claim that the other is approaching it at a speed of 96% of the speed of light in empty space. No matter how fast two objects approach each other, Eq. 9.8 will yield a relative velocity of c or less. An interesting result of Einstein's postulates is that no particle or wave can exceed the speed of light in empty space.

The speed of light in empty space has been measured to be the following:

$$\text{mks system, } c = 3.0 \times 10^8 \text{ m/s}$$

$$\text{British system, } c = 186{,}000 \text{ mi/s}$$

This, of course, is much faster than any speed with which we are acquainted. However, it is not particularly fast when we consider interstellar distances. Light from the Sun takes about eight minutes to reach the Earth, and light from the other stars takes years to reach us.

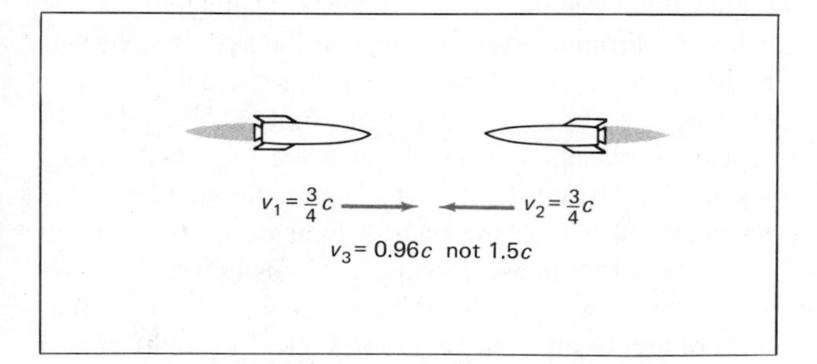

Figure 9.7 **Illustrating relative velocity. The relative velocity of two spaceships, each moving at 0.75*c*, is not 1.5*c*. It is 0.96*c*.**

To say that nothing can go faster than the speed of light is an imprecise statement. We should say that no particle or wave can go faster than c, the speed of light in empty space. When light traverses matter, its velocity is c/n where n is the index of refraction, which is always greater than one. In substances such as water, where the speed of light is $\frac{3}{4}c$, it is possible for subatomic particles such as electrons to go faster than light through water. That is, the electrons go slower than c, but faster than $\frac{3}{4}c$. This effect—of particles going faster than light in a medium—has been experimentally verified.

9.4 Mass-Energy Relationship

Another important relationship that follows from Einstein's postulates is the famous mass-energy relationship. It is given by

$$E = mc^2 \tag{9.9}$$

where E = total energy residing in a particle at rest of mass m,
m = mass of particle,
c = speed of light.

What this equation says is that a very small mass can produce a fantastically large quantity of energy. For instance, the energy in 1 g of matter at rest is given by

$$\begin{aligned} E &= mc^2 \\ &= 0.001 \text{ kg} \times (3 \times 10^8 \text{ m/s})^2 \\ &= 9 \times 10^{13} \text{ J} \end{aligned}$$

This is equivalent to 25 million kilowatt hours of energy, or enough energy to drive an automobile around the Earth 400 times. It is now easy to see how a lump of uranium can supply so much nuclear energy. Incidentally, only a small fraction of the uranium's mass is converted into energy.

The mass-energy relation tells us that mass is a form of energy. The energy that a particle has because it has a mass is called its **rest energy.** If a particle is moving with a velocity v, its total energy is given by

$$E = \frac{mc^2}{\sqrt{1 - (v^2/c^2)}} \tag{9.10}$$

For low velocities, this is the same as

$$E = mc^2 + \tfrac{1}{2}mv^2$$

(for v small compared to c)

which is just the rest energy plus kinetic energy. Often the equation for the total energy of a particle is written

$$E = m'c^2$$

where
$$m' = \frac{m}{\sqrt{1 - (v^2/c^2)}} \tag{9.11}$$

Writing the equation like this puts it on a similar footing to Einstein's equation $E = mc^2$, where m is the mass of the particle at rest, or the **rest mass.** When the particle is in motion, we can still write $E = m'c^2$, but now if we say m' is the mass of the particle, then m' is bigger than the rest mass. As v increases, m' also increases.

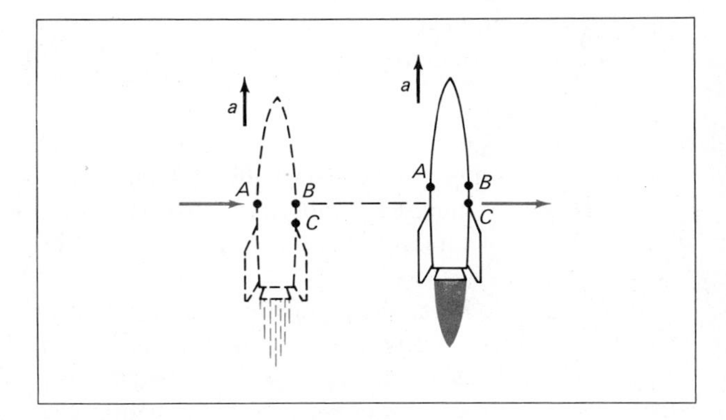

Figure 9.8 Diagram showing the path of light through an accelerating rocketship. For an observer, a light beam entering the rocket at point A should hit point B. However, because of the acceleration of the rocket, the beam hits point C.

9.5 General Theory of Relativity

So far, we have been considering what is sometimes called Einstein's special theory of relativity. The word "special" refers to the fact that it holds only for systems that are moving at a constant velocity with respect to one another.

The more general case would involve systems that are accelerating with respect to one another. The extremely complicated theory that describes such systems is called the **general theory of relativity.** An important aspect of this theory is that the force of gravity produces an acceleration equivalent to any other acceleration. So, any effect that occurs on Earth with its acceleration due to gravity of 9.8 m/s^2 will also occur in a spaceship traveling with an acceleration of 9.8 m/s^2, and vice versa.

An important corollary to this result is that light is bent by gravitational forces. This can be seen if we consider the light traversing a spaceship, which is accelerating as shown in Fig. 9.8. If a beam of starlight passes through point A, it would normally pass through point B, which is almost directly across from point A. However, because the rocket ship is accelerating, by the time the light has reached the other side of the ship, the ship has accelerated upward as shown, and the light passes through point C. An occupant of the spaceship would observe that the light was bent as shown in Fig. 9.9. Now, since a gravitational acceleration should have the same effect as a rocket ship acceleration, we conclude that light is bent in a gravitational field.

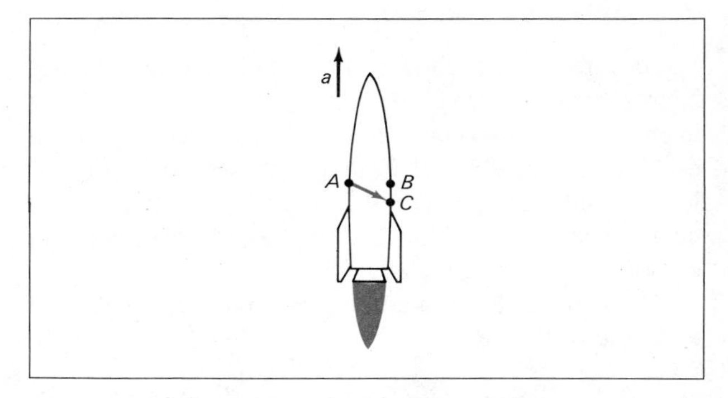

Figure 9.9 Illustrating the bending of light. To an observer on the rocket the light appears to bend from A to C. At the same time the observer feels a downward pull caused by the acceleration of the rocket. This downward pull being similar to gravitational acceleration leads the observer to conclude that gravity caused the light to bend.

Einstein's prediction that light is bent in a gravitational field was verified in 1919 during an eclipse of the Sun. A star was known from astronomical predictions to be at a spot in the sky so as to appear to be at the edge of the Sun during the eclipse. (If the Sun had not been eclipsed, the star, of course, could not have been seen.) Figure 9.10 shows a schematic diagram of the situation. From its known course in the sky, the star was predicted to be at B. However, during the eclipse it was observed as apparently being at point A. The only acceptable explanation assumed that as the star's light passed by the Sun, it was bent by the gravitational pull of the Sun, so that it took the path designated by the colored line. The difference in the apparent position and the actual position was predicted by Einstein to be 1.75 seconds of arc, or an angle of about 0.0005°. The difference was later found to be 1.61 ± 0.30 seconds of arc, thus verifying Einstein's theory.

Because light is bent by a gravitational field, space is sometimes spoken of as being curved. The curvature

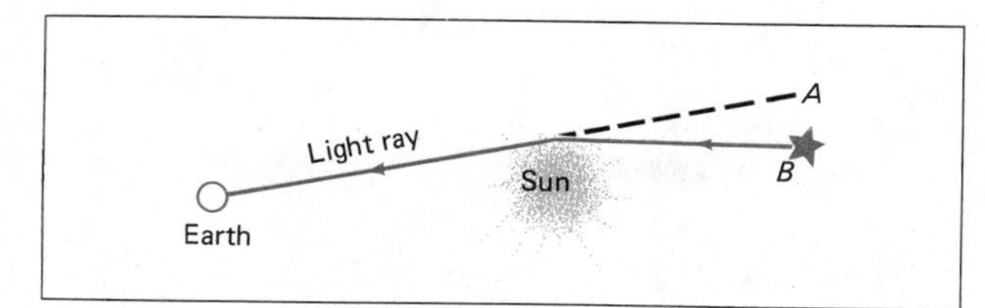

Figure 9.10 Starlight passing near the Sun is bent by its gravitational field. The star is actually at *B*, but its light is bent by the Sun's gravity so that it appears to be at *A*. The drawing is not to scale, by any means.

is caused by masses such as stars. In Einstein's theory, space and time become interconnected in a four-dimensional world of **space-time.** The mathematics of this theory is quite complicated, but some of the results are quite fascinating.

One of the questions this theory considers is the shape and future of the universe. When considering the universe, the space aspects must be considered as just three dimensions of a four-dimensional universe. We can get a taste of what this means by considering two dimensions of our usual three-dimensional space.

The surface of the Earth is a two-dimensional area embedded on a three-dimensional object—a sphere. The surface of the Earth is not flat, and strange phenomena appear on it. For instance, triangles do not have 180° as shown in Fig. 9.11, and two people walking north will eventually meet at the north pole. The geometry of surfaces such as the Earth's surface is different from the plane geometry we learn in school.

Is the surface of the Earth infinite? It is in the sense that you can travel forever on it and never discover an edge. It is finite in the fact that it has a definite measurable area. A spherical surface is a two-dimensional space embedded in a three-dimensional space—the sphere. We say the surface of the Earth is **finite but unbounded.** You can travel on it forever and never come to the end, but you will return to places where you have previously been.

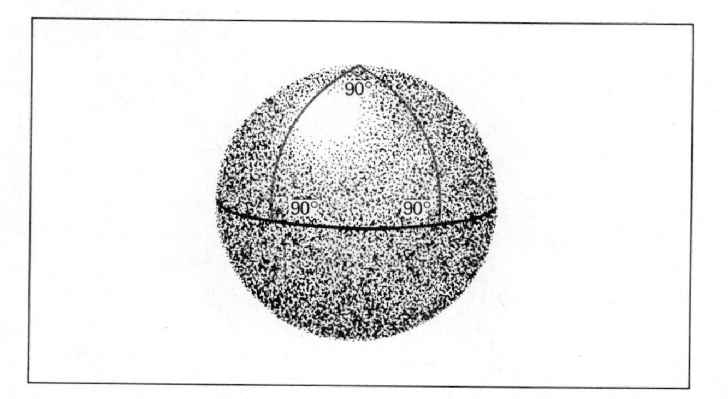

Figure 9.11 Triangles on a sphere do not have 180°. Here is a triangle with 270°.

The space of the universe can be thought of in a similar way. The space of the universe makes up only three dimensions of Einstein's four-dimensional space-time which is curved—just as the Earth's surface is curved.

Another interesting result of Einstein's general theory of relativity is the slowing of time for clocks subjected to a strong gravitational field or other acceleration. This effect (illustrated in Fig. 9.12) is very small for gravitational fields near the Earth's surface, but it has been measured.

The phenomena of time dilation in special relativity and the slowing of time for clocks with large accelerations lead to some interesting thought experiments. In one such experiment, Ruth and Walter are twins. Ruth leaves the Earth in a rocket ship, travels at a speed near the speed of light, accelerates to turn around, and travels back to Earth near the speed of light before decelerating and landing. Let us say Ruth is gone for 10 years according to her biological time and quartz crystal wristwatch. Back on Earth, Walter's biological time and wristwatch will record a much longer time, according to the theory of relativity. Thus, Ruth will find that her twin brother is much older than she.

We see from the above example that times in different systems are truly different. Space travelers will find older people back home! The philosophical, scientific, and sociological implications of this fact are quite thought provoking. Of course, our example was labeled as a thought experiment because such space travel is not yet possible.

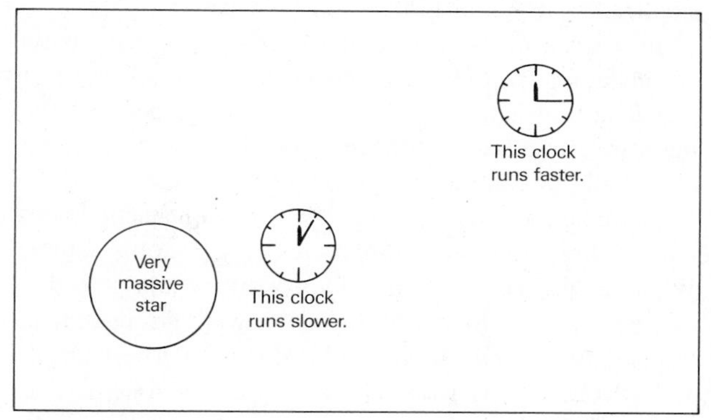

Figure 9.12 Clocks in stronger gravitational fields run slower than clocks in weaker gravitational fields. Thus, clocks close to massive stars run slower than those far from these same stars.

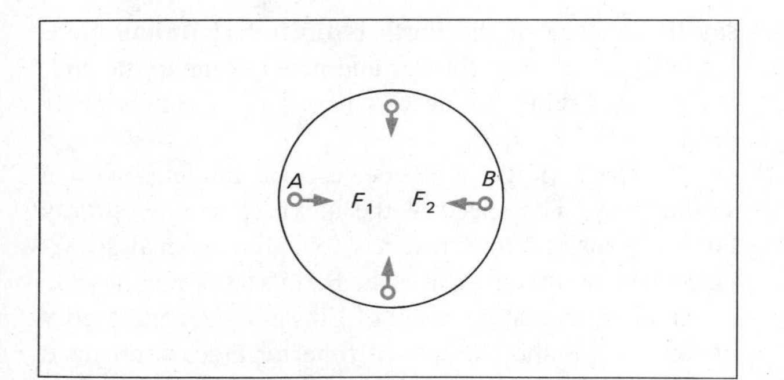

Figure 9.13 The beginning of gravitational collapse. The force of gravitation attracts all particles toward the center of the object.

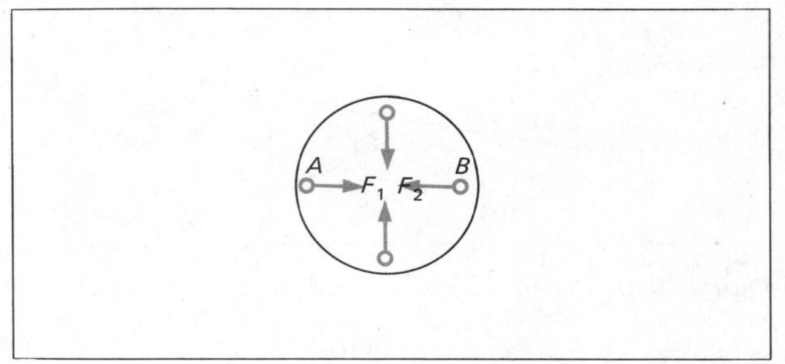

Figure 9.14 Gravitational collapse. The smaller the distances between the particles, the greater the force of gravity becomes.

9.6 Gravitational Collapse

The possibility of a rather remarkable phenomenon has been proposed to explain some strange astronomical findings. This phenomenon may or may not explain these discoveries, but it is interesting in its own right. The phenomenon is called **gravitational collapse** and is the collapse of a very massive body because of its attraction for itself. Recall that Newton's law of universal gravitation was given as

$$F = G\frac{m_1 m_2}{r^2} \tag{9.12}$$

where F = force of gravity between m_1 and m_2,
m_1 = a mass,
m_2 = a second mass,
G = a universal constant,
r = distance between the centers of m_1 and m_2.

From Eq. 9.12 it is evident that if the distance r is made smaller, the force of gravitation increases.

In a large mass such as the Earth or the Sun, parts are always attracting each other. In Fig. 9.13, parts A and B are attracting each other. All other parts of the mass are also attracting one another, so that the net result is that all parts are attracted toward the center.

Normally, in matter such as this, electromagnetic forces tend to keep the various particles (e.g., atoms) apart. What would happen if the body were so massive that the gravitational forces of attraction were stronger than the electromagnetic forces of repulsion? Parts A and B and all other parts would be drawn closer together, as in Fig. 9.14. But according to Eq. 9.12, the forces F_1 and F_2 increase as the distance r decreases, and so A and B would move closer and closer together. But the closer they are, the greater the gravitational force becomes. The whole mass would continue to contract until nothing was left but a fantastically massive point called a **singularity.**

The singularity is surrounded by a surface known as the **event horizon.** Any matter or radiation within the event horizon cannot escape the influence of the singularity. See Fig. 9.15 for a sketch diagram of a singularity and its event horizon. The value R, the radial distance the event horizon is located from the singularity, can be determined by equating the escape velocity formula to the velocity of light. When this is done, we obtain

$$R = 2GM/c^2$$

where G = the universal gravitational constant,
M = the mass,
c = the velocity of light.

Anything located within the event horizon cannot escape. The event horizon is a one-way surface in that

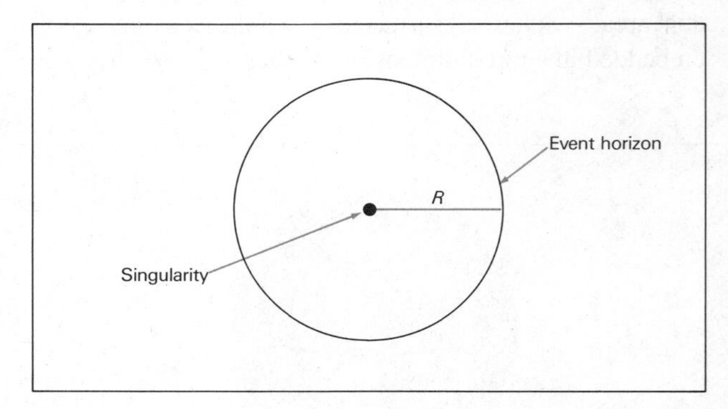

Figure 9.15 This sketch characterizes a black hole. The black dot represents a singularity surrounded by the event horizon at a distance *R*. Anything located within the event horizon cannot escape. Thus, space inward from the event horizon is known as a black hole.

matter and radiation can enter but cannot leave. Thus, space inward from the event horizon is a **black hole.**

Since most stars are rotating, a black hole will probably also be rotating. Since angular momentum is conserved, the rotating rate will increase as the star gets smaller, so the rotation of the black hole will be extremely rapid. Thus, rather than being spherical, a black hole is likely to be disk shaped.

The bending of light or any electromagnetic radiation near a black hole is shown in Fig. 9.16. Radiation within a black hole would be so strongly attracted that it could not escape and light coming within the event horizon would be attracted inward and captured by the black hole. However, the black hole could be detected by observing the bending of light that does not fall within the event horizon. Also, in addition to mass and angular momentum, a black hole might possess a detectable electric charge.

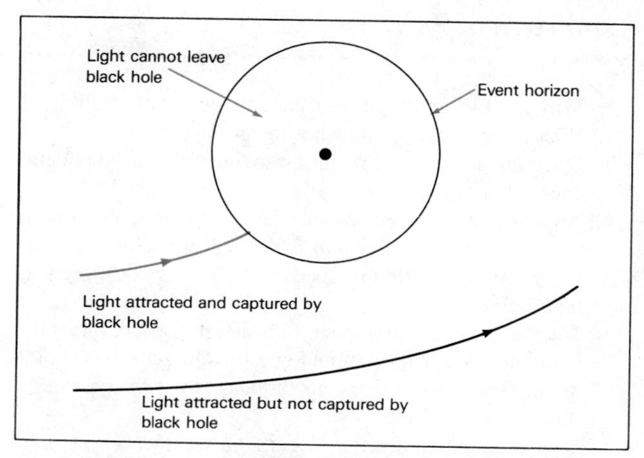

Figure 9.16 The effect of a black hole on electromagnetic radiation such as visible light. Although the black hole does not allow light to escape, its presence can be detected by how it bends nearby light rays.

Learning Objectives

After reading and studying this chapter you should be able to do the following without referring to the text:

1. Explain what is meant by a system of reference.
2. State the two postulates upon which Einstein's special theory of relativity is based.
3. State the basic difference between Einstein's special theory and his general theory of relativity.
4. Describe the concept of length contraction.
5. Describe the concept of time dilation.
6. Explain how electrons can sometimes go faster than light.
7. Explain why $E = mc^2$ is so important.
8. State how light interacts with gravity according to Einstein's general theory of relativity.
9. Discuss the term *finite but unbounded* in relation to the Earth's surface.
10. State how gravitational fields affect clocks.
11. Explain how a black hole is formed.
12. Explain how a black hole is detected.

Important Words and Terms

system of reference
relativity
Einstein's postulates
special theory of relativity
Lorentz transformation
length contraction
time dilation
muons
twins paradox
rest energy
rest mass
general theory of relativity
space-time
finite but unbounded
gravitational collapse
singularity
event horizon
black hole

Questions

1. Why are Einstein's theories called theories of relativity?
2. Why is the space shuttle shot toward the east?
3. Does an object at rest on a moving train have kinetic energy?
4. Approximately, what was the fastest speed attained by a person at the time Einstein published his theory of relativity (1905)? Could this speed cause relativistic effects to be observed?
5. Suppose there are two spaceships in outer space, one traveling at 5,000,000 mi/h and one traveling at 10,000,000 mi/h. How will experiments on board the two spaceships differ in their results?
6. Does a length contraction depend on whether a system is moving away from or toward an observer at rest?
7. Is there any evidence that clocks keep different times due to their relative motion?
8. A newspaper article states that electrons can go faster than light in water. How do you explain this to a 13-year-old who insists that nothing can go faster than light?
9. Give two very important concepts contained in Einstein's formula $E = mc^2$.
10. Is every light beam bent by every mass? Explain.
11. Explain this statement: The surface of the Earth is finite but unbounded.
12. It is the year 2100, and you have lived to be over 130 years old. You see an article in the paper about a 28-year-old astronaut who claims to be the mother of a 46-year-old man. How would you explain this to your great-great-grandchildren?
13. How do clocks behave in gravitational fields?
14. How can we detect a black hole?

Problems

1. If Dr. Baker throws a baseball forward at 40 mi/h while in a train car that is moving at 60 mi/h, how fast will the baseball appear to be moving to Dr. Albert who is on the station platform?
2. If Dr. Baker throws a baseball backward at 20 mi/h while in a train car that is moving forward at 60 mi/h, how fast and in what direction will the baseball appear to be moving according to Dr. Albert who is on the station platform?
   Answer: 40 mi/h forward
3. When Dr. Baker turns on a flashlight pointed forward, the light beam travels at the speed of light. If Dr. Baker is on a train traveling at 60 mi/h, how fast is the light beam traveling, as seen by Dr. Albert, who is on the station platform? *Answer:* Also the speed of light
4. A star moves away from the Earth at a speed of 1/2 the speed of light ($v = 0.5c$). It emits light toward the Earth at the speed c. How fast does the light appear to be approaching Earth according to an observer on the Earth?
5. An observer moves with a muon going at $v = 0.9998c$. An Earth distance of 15 miles is how long according to the observer on the muon? *Answer:* 0.3 mi
6. If a meter stick passes Sarah, who is at rest, with a velocity of $0.6c$ in a direction parallel to its length, what would be its length as observed by Sarah?
7. As a spectator at the Universal Relativistic Olympics, you are watching a pole vaulter sprint down the runway at 3/5 times the speed of light. When he started from rest, his pole was 20 ft long. How long does his pole appear to you while he is running past? *Answer:* 16 ft
8. A dispute arises at the Universal Relativistic Olympics. A runner is timed by a judge standing at the finish line as using 10 s for the 1,000,000-mi dash. The runner insists that her watch reads 8.43 s and that she has broken the galactic record. Who is correct?
9. How long does it take for light to reach us from (a) the moon, 240,000 mi away, (b) Jupiter, when it is 558,000,000 mi away, (c) a star, which is 12 light years away?
   Answer: (b) 3000 s, or 50 min
10. Each second the Sun gives off 4×10^{26} J of energy. How much mass is being lost by the Sun each second?
    Answer: 4.44×10^9 kg
11. Assume that the Sun is losing 4.44×10^9 kg of mass each second as found from problem 10. The Sun's mass is now about 2.000×10^{30} kg. At the stated rate of mass loss, how long will it be before the Sun's mass is reduced to 1.999×10^{30} kg?
    Answer: about 7×10^9 years

12. One reason that no particle can exceed c is that as the velocity of a particle increases, its mass also increases. Thus, since $a = F/m'$, it takes more and more force to maintain an acceleration. What is the mass of a particle moving at $4/5$ the speed of light if its rest mass is 1 g?
13. How fast is an atomic particle moving when its relativistic mass is double its rest mass? *Answer:* $0.866c$
14. An electron has a rest mass of 9.11×10^{-31} kg. What is the electron's relativistic mass when moving with a velocity of $0.8c$?
15. When 8 g of oxygen are united chemically with 1 g of hydrogen, 9 g of water are formed and approximately 120,000 J of heat given off. For this reaction we say "the mass of the product is the same as the mass of the reactants." Knowing that mass and energy are equivalent, determine how much mass is lost in the reaction.

10

Nuclear Physics

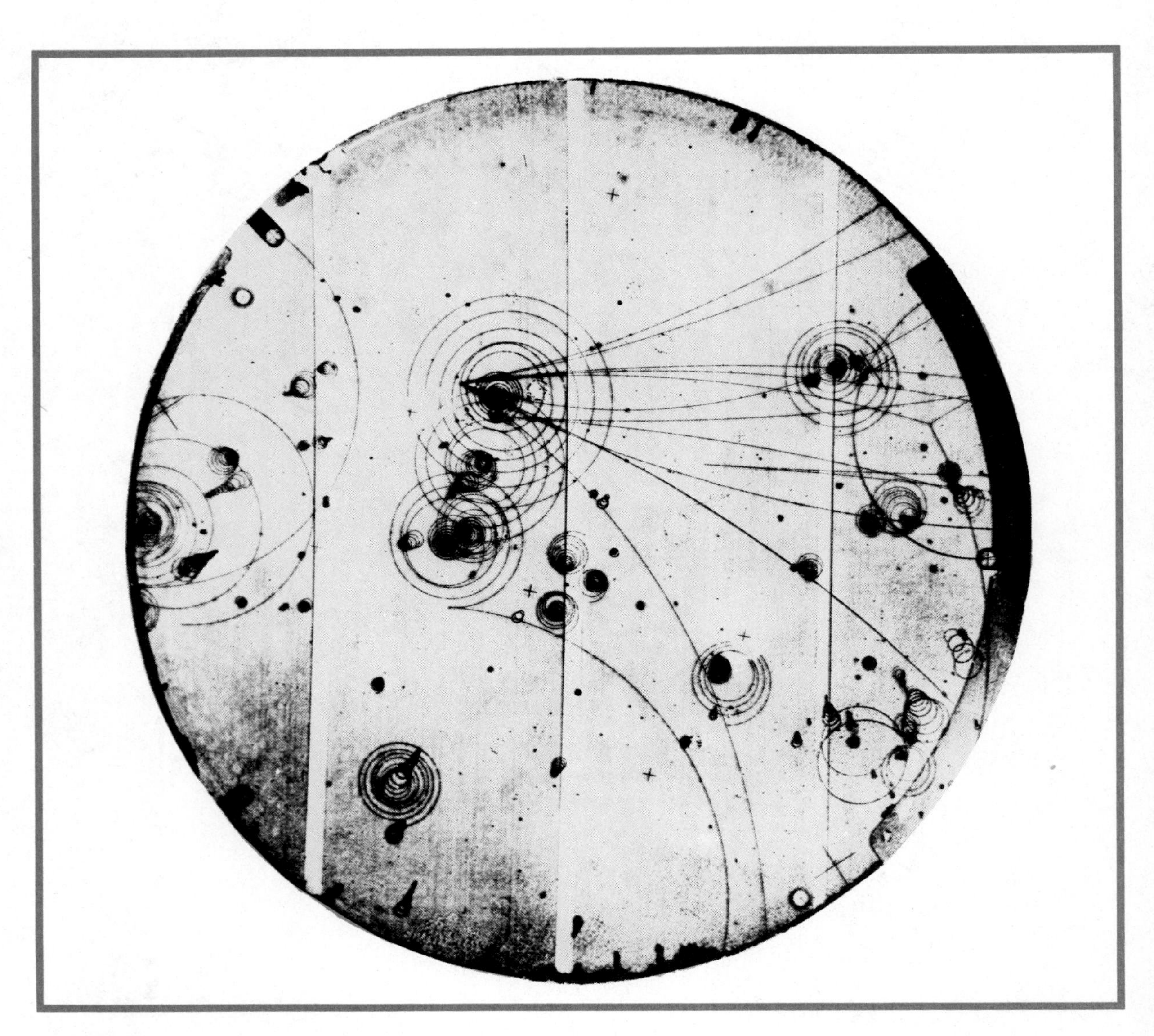

Children of yesterday, heirs of tomorrow
What are you weaving? Labor and Sorrow?
Look to your looms again.
Faster and faster
Fly the great shuttles prepared by the Master.

–Mary A. Lathbury

EINSTEIN'S EQUATION ($E = mc^2$), which relates mass to energy, has been known since 1905. But its importance to the human race was not realized until 1939. In December of 1938 the German scientists Otto Hahn and Fritz Strassman showed that, under appropriate conditions, uranium atoms would split into two approximately equal parts with the accompanying release of large amounts of energy.

Hahn conveyed the startling discovery to Lise Meitner, his co-worker of twenty-five years. Miss Meitner, forced to flee Germany in 1938, and her physicist nephew, Otto Frisch, then worked out the theory of the splitting of the atom during Christmas vacation in Sweden. Frisch communicated their results to the great Danish physicist Neils Bohr, whose atomic theory will be covered in the next chapter. Bohr was just setting out for a scientific meeting in Washington, D.C., and so he conveyed to America the startling news Frisch had told him.

In 1939 the Italian physicist Enrico Fermi (Fig. 10.1) first proposed that through the splitting of the atom large amounts of mass energy could be released for useful work. Soon after, because of the world political climate, the first censorship of scientific information took place. In 1945 the awesome results of atom splitting became known when the first atomic bomb was exploded above a New Mexico desert. The world has never been the same since.

Figure 10.1 Enrico Fermi (1901–1954). Fermi was one of the few scientists capable of both great theoretical and experimental work. After winning the 1938 Nobel Prize in physics, he persuaded the Italian dictator Mussolini to let him take his family to Stockholm for the presentation ceremony. He never returned to Italy but sailed instead to America. Thus it was that an "enemy alien" was in charge of the research that brought about the first controlled nuclear chain reaction in 1942, and later the atomic bomb. (Courtesy American Institute of Physics, Meggers Gallery of Nobel Laureates.)

◀ The V-shaped tracks at left center of this photograph show the resulting products when a high-energy particle of electromagnetic radiation strikes a proton. Nuclear physicists use such reactions to probe the fundamental properties of forces and matter.

Table 10.1 Constituents of an Atom

Particle	*Mass*	*Charge*	*Comments*
Electron	0.91×10^{-30} kg	-1.6×10^{-19} coulomb	Goes in orbit around nucleus
Proton	1673×10^{-30} kg	$+1.6 \times 10^{-19}$ coulomb	1840 times more massive than electron
Neutron	1675×10^{-30} kg	None	Slightly more massive than proton

10.1 The Atomic Nucleus

All matter encountered in day-to-day living is made up of atoms. An **atom** is the basic building block of matter. It is composed of negatively charged particles, called **electrons,** which surround a positively charged nucleus. The **nucleus** is the central core of the atom. It consists of **protons,** which are positively charged, and **neutrons,** which have no electrical charge. The protons and neutrons residing in a nucleus are called **nucleons.** The basic properties of protons, neutrons, and electrons are given in Table 10.1.

The nucleus forms a very small core of the atom, having a diameter of about 10^{-14} m. In contrast to this, the electrons may be thought of as traveling in orbits* that are about 10^{-10} m from the nucleus. These numbers are too small for us to comprehend, but note that 10^{-10} m is 10,000 times larger than 10^{-14} m. This tells us that the diameter of an atom is approximately 10,000 times the diameter of a nucleus.

A schematic diagram of a helium atom is shown in Fig. 10.2. It is not drawn to scale because, if the nucleus were drawn an inch in diameter, the electrons would have to be drawn about one-fifth of a mile away. Nuclei are extremely small, even compared to atoms.

Between the electrons and the nucleus there is nothing. Most of the atom's volume consists of empty space. It is a sobering thought to realize that most of matter consists of a void. If nuclei could be packed together, the resulting matter would be about 10^{12} times as dense as lead. This can occur in dense stars, as mentioned in Section 9.6, but not in atoms on the Earth.

In terms of size, electrons determine the size of atoms. In terms of mass, it is the nucleons that are important, as a study of Table 10.1 shows. The nucleons contribute over 99.9% of the mass of an atom. Of course, this is why the electrons orbit the nucleus instead of vice versa.

In previous chapters we have studied two forces of nature—the gravitational force and the electromagnetic force. The electromagnetic force is the only important force on the electrons in an atom. The electromagnetic force between a proton and an electron in an atom ($r = 10^{-10}$ m) is 10^{39} times greater than the corresponding gravitational force. It is the electromagnetic force that is responsible for the structure of atoms, molecules, and hence, matter in general. The fact that paper tears, rocks are hard, grass is green, and pencils write is due to the action of the electromagnetic force. In fact, about 95% of life's experiences are governed by this one force.

In order for an atom to be neutral (have a total charge of zero), the number of electrons and protons must be the same. Since it is the electromagnetic force that is important in the structure of the atom, atoms are distin-

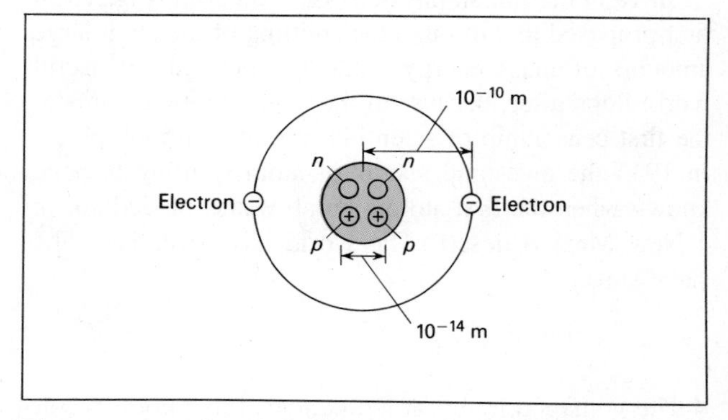

Figure 10.2 Schematic diagram of a helium atom (not to scale). The nucleus is made up of two protons and two neutrons (designated p and n, respectively). The electrons orbit the nucleus.

*We will see in Chapters 11 and 12 that the idea of electrons traveling in orbits is not precisely correct. However, it is a convenient model to use.

guished primarily by the number of charged particles (protons) in the nucleus. An **element** is a substance that has the same number of protons in all of its atoms. As an example, each atom of the element oxygen has eight protons in its nucleus, and each neutral oxygen atom has eight electrons in motion about its nucleus. The atomic properties are thus distinguished primarily by the number of protons in the nucleus.

In a nucleus the protons are packed closely together. According to Coulomb's law, like charges should repel each other, and the protons should not stick together. In fact, according to Coulomb's law, the closer the protons, the greater the repulsive force. Since the protons are closer than 10^{-14} m from each other, the repulsive electric forces are huge.

For the protons to stick together, we need another force—a strong nuclear force. This strong nuclear force is not so easy to understand as the electric force. The **strong nuclear force** is very strongly attractive at short range and is reduced to zero at longer range. This nuclear force is strong enough to hold many nuclei together. However, not all combinations of nuclei will stick together. The question of which combinations stick together and which ones fly apart is discussed in Section 10.2.

The exact formula for the strong nuclear force is not known. It is too complicated to have a simple formula. However, two things about the strong nuclear force are well known. For distances between nuclear particles much greater than the diameter of a nucleus ($r \cong 10^{-14}$ m), the force is zero. For the very short nuclear distances, it is very strongly attractive. So we can write its *approximate* formula

$$\begin{aligned} F_{\text{nucl}} &= \text{Very strongly attractive,} \\ &\quad \text{if } r \text{ is less than about } 10^{-14}\text{ m} \qquad (10.1) \\ F_{\text{nucl}} &= 0 \text{ if } r \text{ is greater than about } 10^{-14}\text{ m} \end{aligned}$$

where r = distance between two particles inside the nucleus.

This simplified formula for the strong nuclear force does not tell the whole story. For a few nucleons, such as only two protons or only two neutrons, this force is not attractive. However, a combination of two protons and two neutrons will strongly attract each other. Many other combinations of protons and neutrons are also attractive.

A **stable** nucleus is one that will not separate of its own accord into two or more groups of particles. An **unstable** nucleus will spontaneously fly apart because the repulsive forces inside the nucleus are greater than the attractive forces. Nuclear stability will be discussed in Section 10.2.

Nuclei with many protons experience both strong electric repulsive forces from all the other protons and strongly attractive nuclear forces. A typical large nucleus is shown in Fig. 10.3. A proton on the edge of the nucleus is attracted only by the six or seven nearest nucleons. Since the strong nuclear force is a short-range force, only the nearby nucleons contribute to the attractive force. There is no attraction from all the other nucleons.

The electric repulsive force is long range. As more and more protons are added to the nucleus, the electric repulsive forces increase while the nuclear attractive forces remain the same. When there are more than 83 protons in the nucleus, the electric forces of repulsion overcome the nuclear attraction, and the nucleus splits apart (usually into two unequal parts) of its own accord.

The **atomic number** of an atom is the number of protons in the nucleus of the atom. The **neutron number** is, as the term indicates, the number of neutrons in the nucleus. The **mass number** is the number of protons plus neutrons in the nucleus. The properties of the nucleus itself depend not only on the number of protons it has but also on the number of neutrons in it.

Isotopes are atoms whose nuclei have the same number of protons but different numbers of neutrons. As an example, an oxygen atom may have seven, eight, nine, or ten neutrons. Each of these possibilities, in combination with eight protons and eight electrons, results in an isotope of oxygen.

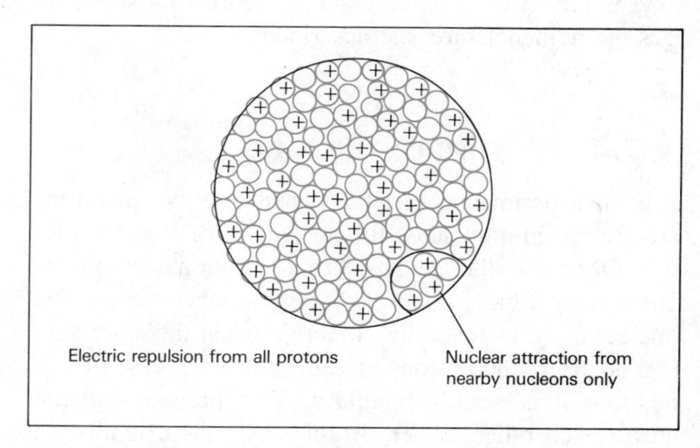

Figure 10.3 A nucleus with many nucleons. The protons on the edge of the nucleus are attracted by only the six or seven nearby nucleons, but they are repelled by all the other protons. When the number of protons exceeds 83, the electric repulsion of the other protons is stronger than the nuclear attraction, and the nucleus splits apart.

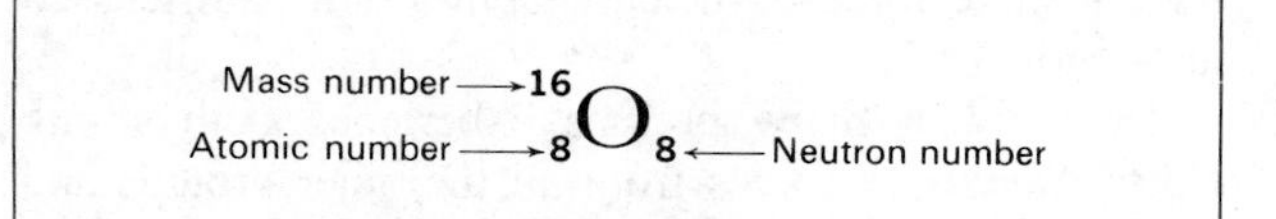

Figure 10.4 Atomic nomenclature. The symbol for the chemical element is shown in the center.

The nomenclature used to distinguish isotopes in nuclear and atomic physics is best described by an example. Let us consider the oxygen isotope $^{16}_{8}O_{8}$. The letter O is the chemical symbol for oxygen.* The left superscript designates the mass number (many textbooks use a right superscript instead). The left subscript designates the atomic number and the right subscript the neutron number. (The right superscript position is left vacant because often there is a charge on the atom and the charge is indicated at this position. For instance, an oxygen atom with only seven electrons would have a total positive charge of 1 because it would have one more proton than electrons.)

The atomic number plus the neutron number must equal the mass number by definition. Usually, the neutron number is omitted from an isotopic symbol, as it can be easily determined by subtracting the atomic number from the mass number. The atomic number and the chemical symbol specify the same thing since all atoms of the same element have the same number of protons. However, only a few tenacious souls have memorized the number of protons in every element, so it is usually specifically designated. Some other possible isotopes of oxygen are $^{15}_{8}O_{7}$, $^{17}_{8}O_{9}$, and $^{18}_{8}O_{10}$. Figure 10.4 summarizes the nomenclature just described.

10.2 Nuclear Stability

It is an experimental fact that when only two neutrons are present in a nucleus, the nuclear force is not attractive. Of course, there is always a gravitational attraction, but it is negligible compared to the nuclear force. The nuclear force is generally attractive when there are several neutrons and protons present, but in the case of two neutrons it is weakly repulsive. Two protons will not attract each other, either. In this case, there is also an electric repulsion present. However, for a combination

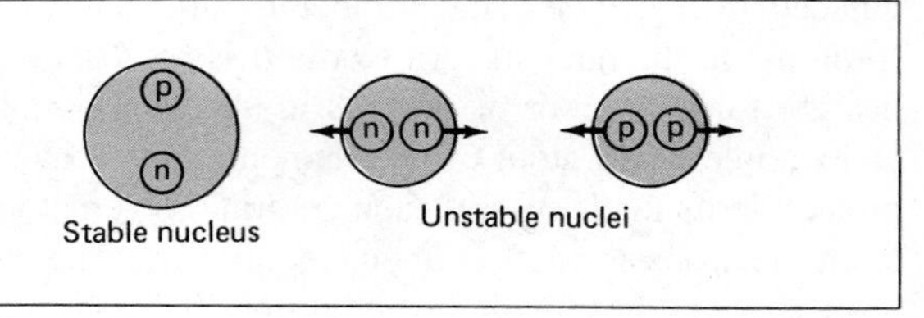

Figure 10.5 Drawing illustrating the force between two nuclear particles. Two neutrons or two protons will not attract each other, but a neutron and a proton will attract. In heavier nuclei the converse effect appears. Two protons or two neutrons inside a large nucleus seem to be a more stable combination than one neutron and one proton.

of a neutron and a proton, commonly called a **deuteron,*** the force is attractive and the nucleus is stable. Figure 10.5 summarizes these facts.

When more neutrons and protons are grouped together, certain combinations will be stable while others will be unstable. For low mass numbers of about 40 or less, most combinations are unstable unless the number of neutrons is about the same as the number of protons. Examples of stable nuclei are $^{4}_{2}He_{2}$, $^{12}_{6}C_{6}$, $^{23}_{11}Na_{12}$, $^{27}_{13}Al_{14}$, $^{40}_{18}Ar_{22}$.

For higher mass numbers, the number of neutrons must be greater than the number of protons to have a stable nucleus. This is because when many protons are present, the repulsive electric forces are very large, and so many neutrons are necessary to make the attractive nuclear force greater than the repulsive electric force. Examples of elements with stable nuclei in this category are $^{44}_{20}Ca_{24}$, $^{62}_{28}Ni_{34}$, $^{89}_{39}Y_{50}$, $^{114}_{50}Sn_{64}$, $^{165}_{67}Ho_{98}$, $^{208}_{82}Pb_{126}$, and $^{209}_{83}Bi_{126}$. This last isotope, $^{209}_{83}Bi_{126}$, has the highest proton number and mass number of all stable nuclei. All nuclei with more than 83 protons are unstable, because of the large repelling electric forces. A chart of all the combinations of protons and neutrons forming stable nuclei is given in Fig. 10.6.

An important fact that is not too evident in Fig. 10.6 is that there are many stable isotopes where both the neutron number and the proton number are even numbers, and very few stable nuclei where they are both odd numbers. The fact that a nucleus with an even number of protons and an even number of neutrons is more stable than one with uneven numbers is a result of what is called the **pairing effect.** In Table 10.2, the number of stable isotopes is divided into three categories, depend-

*The chemical symbols for all the elements are given in the periodic table in Chapter 13.

*Hydrogen with deuterons as the nuclei is called *deuterium*.

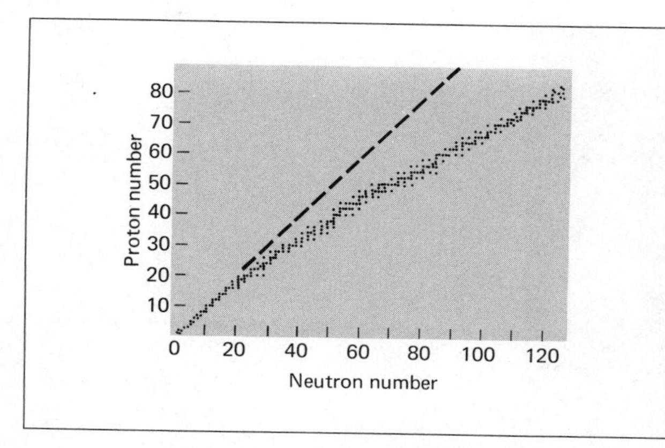

Figure 10.6 A plot of the combinations of proton and neutron numbers resulting in stable nuclei. Each dot represents a stable nucleus. The dashed line indicates where the proton number equals the neutron number. The plot stops at the proton number 83. There are no stable nuclei with more than 83 protons.

ing on whether the number of protons and neutrons is even or odd. There are 165 stable isotopes in which both of these numbers are even, 105 in which one is even and one odd, and only four in which both are odd. The four odd-odd stable isotopes are all low in mass number. They are ${}^{2}_{1}H_1$, ${}^{6}_{3}Li_3$, ${}^{10}_{5}B_5$, and ${}^{14}_{7}N_7$. What the pairing effect says, then, is that in a nucleus consisting of many protons and neutrons, the protons and neutrons tend to pair up among themselves. That is, two protons pair up and two neutrons pair up, but one proton will not pair up with one neutron. An example of the pairing effect is that ${}^{23}_{11}Na_{12}$ is stable, but ${}^{22}_{11}Na_{11}$ is not stable.

The general criteria for nuclear stability can be summarized as follows:

1. Isotopes with proton numbers greater than 83 are unstable.
2. Many even-even nuclei are stable.
3. Many even-odd and odd-even nuclei are stable.
4. Only four odd-odd nuclei are stable.
5. Stable nuclei with a mass number less than 40 have approximately the same number of protons and neutrons.
6. Stable nuclei with a mass number greater than 40 have more neutrons than protons.

So far, we have considered only the qualitative aspects of nuclear stability. In order to describe quantitatively this phenomenon, it is necessary to introduce the concept of energy into the discussion.

Table 10.2 The Pairing Effect in Stabilizing Nuclei

Proton Number	Neutron Number	Number of Stable Nuclei
1. Even	Even	165
2. Even / Odd	Odd / Even	105
3. Odd	Odd	4

Albert Einstein introduced his special theory of relativity in 1905. This was discussed in the previous chapter. One of the results of Einstein's assumptions was that mass is a form of energy. If it were possible to convert a very small amount of mass completely into energy of other forms, the total energy of other forms produced would be tremendous. Einstein's famous equation relating mass and energy is

$$E = mc^2 \tag{10.2}$$

where E = energy,
m = mass,
c = speed of light in a vacuum.

Since $c = 3 \times 10^8$ m/s, c^2 is 9×10^{16} m^2/s^2 and 1 kilogram of mass is equivalent to 9×10^{16} J of energy. This is a fantastically large amount of energy. It can be readily perceived that if we could convert even a minute amount of mass energy into heat energy, we would get a lot of heat out of the process.

EXAMPLE 1 A typical nuclear power plant (see Fig. 10.7) generates 3600 MW of power (in the form of heat energy per second) of which about 1000 MW becomes usable electric power. How much (uranium) mass energy is converted into heat energy each day?

To work this we must use Eq. 10.2 and realize that 3600 MW is 36×10^8 J/s. So every second, 36×10^8 J of mass energy are consumed. The amount of mass consumed each second is then

$$m = \frac{E}{c^2}$$

$$m = \frac{36 \times 10^8 \text{ J}}{9 \times 10^{16} \text{ J/kg}}$$

$$m = 4 \times 10^{-8} \text{ kg}$$

Figure 10.7 A nuclear reactor and electric generating power plant located at Peach Bottom, Pennsylvania. (Courtesy Philadelphia Electric Company.)

We see that the mass energy converted into heat energy is 4×10^{-8} kg each second or, since there are 3600 × 24 seconds in a day, we multiply 4×10^{-8} (kg/s) × 3600 × 24 (s/day) to get 3.5×10^{-3} kg/day. About 3.5 grams of mass energy is converted into heat energy each day in a typical nuclear power plant.

The units of mass and energy commonly used in nuclear physics are quite different from those used in preceding chapters. Because protons and neutrons are so small they are usually measured in very small units of mass called atomic mass units (amu). An **amu** is defined as $\frac{1}{12}$ the mass of the ^{12}C isotope. That is

$$\text{Mass of } ^{12}\text{C} = 12.0000 \text{ amu} \qquad (10.3)$$

One amu is equal to 1.66×10^{-27} kg. The masses of nuclear particles in terms of these units are

$$\text{Mass of proton} = 1.00783 \text{ amu}$$

$$\text{Mass of neutron} = 1.00867 \text{ amu}$$

The unit of energy commonly used in nuclear physics is the **MeV**. This stands for one million electron volts. An **electron volt** is the amount of kinetic energy an electron (or proton) acquires when it is accelerated through an electric potential of 1 volt. One electron volt (eV) is equal to 1.6×10^{-19} joule. If one amu of mass could be converted into energy, 931 MeV would be obtained. So we have the relation

$$(1 \text{ amu})c^2 = 931 \text{ MeV} \qquad (10.4)$$

The concept of *binding energy* plays an important role in the discussion of nuclear stability. The **total binding energy** of a nucleus is the energy necessary to separate a nucleus into protons and neutrons that are free to move independently of one another. The **binding energy per nucleon** is the total binding energy divided by the number of nucleons in the nucleus. Figure 10.8 illustrates how the binding energy per nucleon varies with the total number of nucleons in the nucleus. Note the steep rise at the beginning of the curve then a general leveling off at about 8 MeV per nucleon for mass number 20 and above.

The **neutron separation energy** and the **proton separation energy** are defined as the amounts of energy necessary to separate one neutron or one proton from the rest of the nucleus. As an example of these concepts, the amount of energy necessary to separate a deuteron is 2.22 MeV. See Fig. 10.9. Thus, the total binding energy is 2.22 MeV; the neutron separation energy and

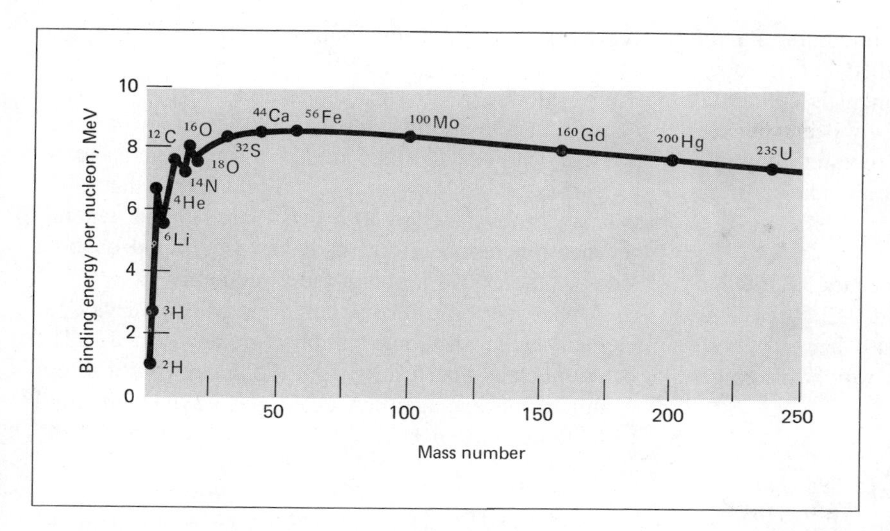

Figure 10.8 Binding energy per nucleon versus mass number.

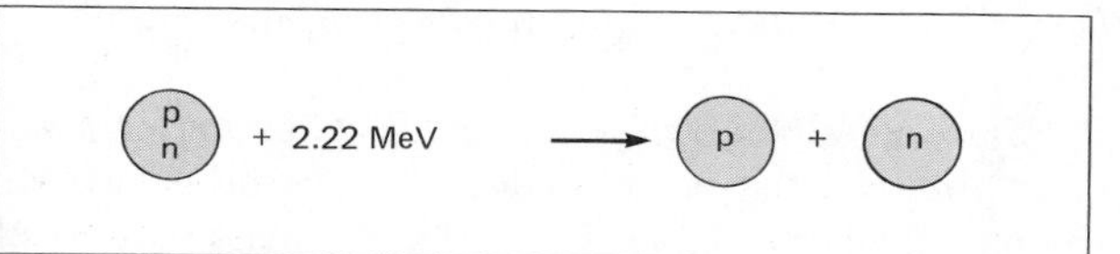

Figure 10.9 The binding energy of the deuteron. It takes 2.22 MeV to split the deuteron into a proton and a neutron.

proton separation energy are the same, in this very simple case, and the binding energy per particle is 2.22 MeV ÷ 2, or 1.11 MeV.

A more instructive example of binding energy is provided by the alpha particle, or ${}_2^4He_2$. This nucleus consists of two protons and two neutrons and is very stable. In Fig. 10.10, the various binding and separation energies are shown. The total binding energy of the alpha particle is 28.3 MeV. Its neutron and proton separation energies are 20.6 MeV and 19.8 MeV, respectively, and its binding energy per particle is 28.3 MeV ÷ 4, or 7.08 MeV.

Thus, we see that it would take 28.3 MeV of energy to split an alpha particle up into two protons and two neutrons, which are sufficiently far apart that they are no longer attracted to each other by nuclear forces. One way in which we can convert mass into other forms of energy is, in effect, to turn the arrow around in Fig. 10.10. For instance, if we could put two protons and two neutrons together to form ${}_2^4He_2$, we could get 28.3 MeV of energy out of the process. Processes of this sort, in which light nuclei come together to form more stable heavier nuclei, thus releasing energy in the process, are called **fusion** reactions.

From the definition of total binding energy in Fig. 10.10, we can find a relation for the mass of any nucleus. It is

$$\text{Nuclear mass} = \text{Mass of all protons} + \text{Mass of all neutrons} - \text{Total binding energy in equivalent mass units} \quad (10.5)$$

Thus, we see that any stable nucleus weighs less than the total weight of the protons and neutrons contained in it. When a nucleus possesses negative binding energy, it is unstable and will fly apart spontaneously.

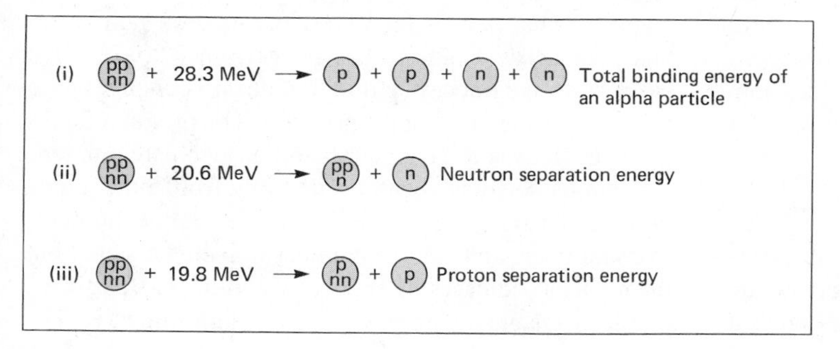

Figure 10.10 The binding energy of an alpha particle.

Each isotope has a specified mass. In nature, a given element usually occurs as several different isotopes. Hence, the **atomic weight** of an element is defined as the *average* mass (weight) of a naturally occurring element. The atomic weight is computed by multiplying the mass of each isotope by its fractional abundance in nature and adding the products.

EXAMPLE 2 Oxygen naturally occurs 99.759% of the time as ^{16}O, 0.037% as ^{17}O, and 0.204% as ^{18}O. The masses of these isotopes are 15.99491 amu, 16.99914 amu, and 17.99916 amu. The atomic weight of oxygen is then

$$\begin{aligned}\text{Atomic wt of}&\\ \text{oxygen} &= 0.99759 \times (15.99491)\\ &\quad + 0.00037 \times (16.99914)\\ &\quad + 0.00204 \times (17.99916)\\ &= 15.9994\end{aligned}$$

The atomic weight of each element is given in the periodic table of Chapter 13.

10.3 Radioactive Decay

So far we have limited our discussion to stable nuclei. What about nuclei that are unstable, that is, those that tend to fly apart of their own accord? These nuclei that disintegrate of their own accord are called **radioactive.**

Radioactive decay was discovered in 1896 by Henri Becquerel, who found that uranium compounds spontaneously emit some very penetrating radiation. In 1898 the husband and wife team of Pierre and Marie Curie (Fig. 10.11) announced the discovery of two new radioactive elements, radium and polonium. Becquerel and the Curies were awarded the 1903 Nobel Prize in physics for their important work on radioactive nuclei.

We now know that there are four common ways by which a radioactive nucleus can disintegrate. These are referred to as alpha decay, beta decay, gamma decay, and fission. In all of these radioactive decay processes, energy is given off. This energy is usually in the form of kinetic energy of the products of the decay process.

Alpha decay is the disintegration of a nucleus into an alpha particle ($^4_2He_2$ nucleus) and the nucleus of another element. An example of this is

$$^{232}_{90}Th_{142} \longrightarrow {}^{228}_{88}Ra_{140} + {}^4_2He_2$$

Note that in this decay, the sums of the atomic numbers and the mass numbers on both sides of the arrow are equal. This is a general law which holds for all nuclear processes. For the atomic numbers, $90 = 88 + 2$, and for the mass numbers, $232 = 228 + 4$.

The atomic numbers must add up to be the same because of a law that states that we cannot gain or lose electric charge. The mass numbers must add up to be the same because of a similar conservation law that states that we cannot gain or lose mass number. In general, the neutron numbers do not have to be equal on both sides of the arrow although they frequently are.

Alpha decay is a very common type of decay for elements with an atomic number greater than 82. The heavy nucleus that results from the decay (^{228}Ra in our example), decays again. A chain of decays results until a stable nucleus such as $^{208}_{82}Pb_{126}$ is formed.

EXAMPLE 3 The nucleus $^{238}_{92}U$ decays by alpha decay. What are the resulting products of this decay?

Since an alpha decay implies the emission of a nucleus of ^4_2He, the other product nucleus must have an atomic number of $92 - 2 = 90$ and a mass number of $238 - 4 = 234$. Thus, the decay is

$$^{238}_{92}U \longrightarrow {}^{234}_{90}Th + {}^4_2He$$

Beta decay is the disintegration of a nucleus into a beta particle (electron) and the nucleus of another element. An example is

$$^{14}_6C_8 \longrightarrow {}^{14}_7N_7 + {}^{0}_{-1}e_0 + \nu$$

In this decay we have represented an electron as $^{0}_{-1}e_0$. Since its electric charge is opposite to that of a proton, it has a charge number of -1. Also, since its mass is negligible compared to that of a proton (see Table 10.1) its mass number is zero. The symbol ν represents a strange particle called a neutrino. The **neutrino** has no charge and very little, if any, mass. However, the neutrino does possess energy and momentum.

The total atomic number and total mass number on both sides of the arrow are seen to be equal in our example of beta decay. However, the neutron number in this beta decay example has gone from 8 to 7. What happens in beta decay is that a neutron is changed into a proton, an electron, and a neutrino. The proton remains in the nucleus and the electron and neutrino are emitted.

Gamma decay is slightly differently from either alpha or beta decay. In **gamma decay** the atomic number, mass number, and neutron number remain the same, but the nucleus decreases its energy by emitting a ''particle'' of electromagnetic energy called a **gamma ray.** The

Figure 10.11 **Marie (Marja Sklodowska) Curie (1867–1934),** an outstanding scientist, was born in Poland and received her early scientific training from her father. While in school she became involved in a student's revolutionary organization and found it advisable to leave Poland. She completed her science degree in Paris. In 1903, she and her husband Pierre Curie shared the Nobel Prize in physics with Henri Becquerel for the discovery of radioactivity. Mme Curie was also awarded the Nobel Prize in chemistry in 1911 for the discovery of radium and the study of its properties. The Curies' daughter Irene Joliot-Curie and her husband Frederic Joliot were awarded the 1935 Nobel Prize in chemistry. (AIP Neils Bohr Library. W. F. Meggers Collection.)

gamma ray has energy but no mass or charge. A typical example of gamma decay is

$$^{204}_{82}\mathrm{Pb}^{*}_{122} \longrightarrow {}^{204}_{82}\mathrm{Pb}_{122} + \gamma$$

In this type of decay the symbol γ (Greek letter gamma) represents the particle of electromagnetic radiation that is emitted by the nucleus. The asterisk implies that the ^{204}Pb* nucleus is at a higher energy than normal ^{204}Pb. Because energy must be conserved, we get

$$\text{Energy of } {}^{204}\mathrm{Pb}^{*} = \text{Energy of } {}^{204}\mathrm{Pb} + \text{Energy of gamma ray}$$

Or, rearranged,

$$\text{Energy of gamma ray} = \text{Energy of } {}^{204}\mathrm{Pb}^{*} - \text{Energy of } {}^{204}\mathrm{Pb}$$

The energy of the gamma ray is thus seen to be equal to the difference in energy between ^{204}Pb* and ^{204}Pb. Gamma rays are different from alpha and beta particles in that gamma rays have no charge or mass associated with them.

A fourth important decay mode which a nucleus can undergo is called fission. **Fission** is the splitting of the nucleus into two nuclei of approximately equal size. It is accompanied by the emission of neutrons and the release of large amounts of energy. As an example, consider a typical fission decay of $^{236}_{92}\mathrm{U}_{144}$.

$$^{236}_{92}\mathrm{U}_{144} \longrightarrow {}^{140}_{54}\mathrm{Xe} + {}^{94}_{38}\mathrm{Sr} + 2({}^{1}_{0}\mathrm{n}_1)$$

(Note that the symbol $^{1}_{0}\mathrm{n}_1$ designates a neutron.) This is just one of the many possible fission decay reactions of ^{236}U. Many others may occur, such as

$$^{236}_{92}\mathrm{U}_{144} \longrightarrow {}^{132}_{50}\mathrm{Sn}_{82} + {}^{101}_{42}\mathrm{Mo}_{59} + 3({}^{1}_{0}\mathrm{n}_1)$$

$$^{236}_{92}\mathrm{U}_{144} \longrightarrow {}^{150}_{60}\mathrm{Nd}_{90} + {}^{81}_{32}\mathrm{Ge}_{49} + 5({}^{1}_{0}\mathrm{n}_1)$$

EXAMPLE 4 Complete the following reaction:

$$^{236}_{92}\mathrm{U} \longrightarrow {}^{88}_{36}\mathrm{Kr} + {}^{144}_{56}\mathrm{Ba} + \underline{\qquad}$$

The atomic numbers are balanced since $92 = 36 + 56$. The mass number on the left is 236, and the sum of the two mass numbers given on the right is 232. To balance the mass number there must be 4 neutrons. The reaction is then

$$^{236}_{92}\mathrm{U} \longrightarrow {}^{88}_{36}\mathrm{Kr} + {}^{144}_{56}\mathrm{Ba} + 4({}^{1}_{0}\mathrm{n}_1)$$

In all radioactive processes energy is released. This energy can be destructive if it impinges on living tissue. In alpha and beta decay the energy is in the form of kinetic energy of either alpha or beta particles. In gamma radiation it is in the form of high-energy electromagnetic radiation. In fission its most common form is in the kinetic energy of neutrons.

Alpha and beta particles ($^{4}_{2}\mathrm{He}_2$, $_{-1}^{0}\mathrm{e}$) are charged particles and can be easily stopped by materials such as paper and wood. The alpha and beta particles interact with charged particles within the materials, following Coulomb's law of electrical attraction and repulsion. Gamma rays and neutrons are not charged particles and are difficult to stop. Thick shielding is needed to protect living things from these kinds of radiation. This shielding is usually in the form of either lead or concrete for γ rays and either water or concrete for neutrons.

Some unstable nuclei take a long time to decay; others decay very rapidly. The relevant quantity that describes the rate of decay is called the half-life. The **half-life** of a given radioactive substance is the time it takes for half of the material to decay. The half-lives of typical radioactive isotopes range from millionths of a second to billions of years. The concept of a half-life is often misunderstood. An example best illustrates this concept. Consider the reaction:

$$^{90}_{38}\text{Sr} \longrightarrow {}^{90}_{39}\text{Y} + {}^{0}_{-1}\text{e} + \nu$$

This is a typical beta decay, and the half-life is 28 years. This means that if in 1979 you had 128 micrograms (0.000128 gram) of ^{90}Sr, then 28 years later (2007), you would have half as much, that is, 64 micrograms of ^{90}Sr. In 2007, you would have 64 micrograms so 28 years later, in 2035, you would have half again as much, or 32 micrograms of ^{90}Sr. This process continues as follows:

Year	*Amount of ^{90}Sr*
1979	128 micrograms
2007	64 micrograms
2035	32 micrograms
2063	16 micrograms
2091	8 micrograms
⋮	⋮

If we plot the amount of the sample of ^{90}Sr remaining versus the time, we get Fig. 10.12.

Some isotopes have half-lives of billions of years or more. Consequently, they can still be found occurring naturally. For example, the half-life of ^{238}U is 4.5×10^9 years. Since the age of the Earth is believed to be approximately 4.5×10^9 years also, we see that about half of the uranium present at the time the Earth formed would still be here today. That is, if a rock contained 2 kg of uranium when the Earth formed 4.5×10^9 years ago, 1 kg would still be present in the rock. The other kilogram of ^{238}U would have decayed into other elements. The uranium still present could be mined and processed.

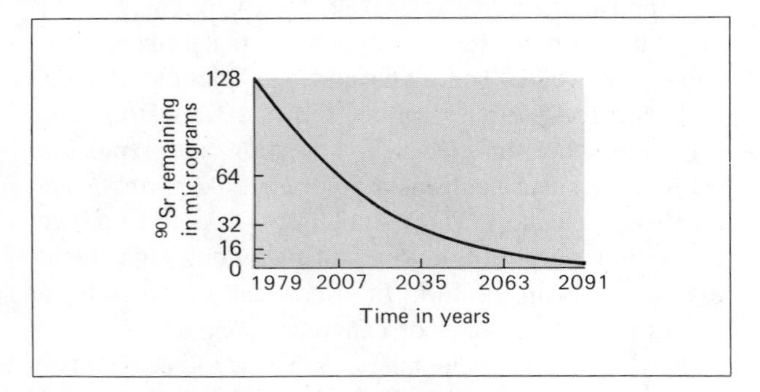

Figure 10.12 Graph of the radioactive decay of strontium-90. This graph assumes an initial sample of 128 micrograms.

When the naturally occurring isotope ^{238}U decays, it emits an alpha particle, and a ^{234}Th nucleus remains. The ^{234}Th then decays, and the series of decays continues until stable ^{206}Pb is formed. The ^{238}U radioactive series produces many additional unstable isotopes. Because of the long half-life of ^{238}U, these radioactive isotopes occur naturally.

Radioactive isotopes can also occur naturally due to the atmospheric bombardment of high-energy nuclei from outer space. These high-energy nuclei hit stable nuclei, and the collisions produce radioactive isotopes. For example, ^{14}C is formed in the atmosphere in this manner.

Some of the naturally-occurring radioactive isotopes are given in Table 10.3. Many of these, such as ^{14}C, exist in our bodies at all times. Little is known about the effects of this low-radiation dosage to which humans have been exposed throughout history.

Radioactive isotopes have many uses. One important use is found in the dating of very old artifacts. All organic matter (all matter which was once alive, such as wood, coal, bones, and parchment) contains carbon. Natural carbon occurs in three isotopes, ^{12}C, ^{13}C, and ^{14}C. The first two of these are stable, while ^{14}C decays by beta emission into ^{14}N with a half-life of 5730 years.

Table 10.3 Half-Life of Some Naturally Occurring Radioactive Nuclei

Isotope	*Half-Life*
^{8}Be	2×10^{-16} second
^{214}Po	1.6×10^{-4} second
^{26}Al	7 seconds
^{34}Cl	33 minutes
^{231}Th	26 hours
^{222}Rn	4 days
^{22}Na	2.6 years
^{3}H	12.5 years
^{14}C	5.7×10^3 years
^{235}U	0.7×10^9 years
^{238}U	4.5×10^9 years
^{40}K	1.4×10^9 years
^{87}Rb	6×10^{10} years
^{115}In	6×10^{14} years
^{124}Sn	1.7×10^{17} years

In the atmosphere, ^{13}C is converted into ^{14}C due to bombardment by cosmic rays from outer space. The ^{14}C decays at about the same rate at which it was created; so the percentage of carbon that is ^{14}C remains relatively constant. Organic materials acquire the ^{14}C in the form of carbon dioxide, and in living things there is enough ^{14}C to cause about 16 beta emissions per minute per gram of carbon. Once the organic matter dies, it ceases to take in the ^{14}C from the atmosphere, and the amount of ^{14}C that it contains slowly decays. By measuring the beta decay rate in old materials, the age of the material can be determined, since the ^{14}C half-life is known.

EXAMPLE 5 What is the age of an old bone that emits 4 beta rays per minute per gram of carbon?

Living material would give a rate of 16 beta rays per minute per gram. After one half-life, or 5730 years, this would be decreased to 8 and after two half-lives, or 11,460 years, this would be cut to 4 beta rays per minute per gram of carbon. Thus, the bone is about 11,460 years old. (See Table 10.4.)

Using similar methods, archeologists are able to date things as old as 40,000 years (see Fig. 10.13). Using isotopes with much longer half-lives enables geologists to measure the ages of meteorites from outer space or very old rock. In this manner, the approximate age of the Earth has been determined to be approximately 4.5 billion years.

Figure 10.13 Bones, 14,000 years old, that were uncovered at an ancient burial site in the Aswan area in Sudan. To determine the age of the site, the age of the bones was determined using radioactive dating techniques with ^{14}C. (Courtesy Anthropology Department, Southern Methodist University.)

Table 10.4 Decay of ^{14}C

Time since Death	*Beta rays/min/g of C*
0	16
5,730 years	8
11,460 years	4
17,190 years	2
22,920 years	1

The activity of a radioactive source is measured in a unit called the curie. One **curie** is arbitrarily defined as 3.7×10^{10} disintegrations per second.* This is approximately the number of disintegrations given off by one gram of ^{226}Ra. The curie, as defined, is a fairly large unit. Therefore, the millicurie and the microcurie are commonly used in the laboratory.

Large radioactive doses can be harmful to living cells. They can cause mutation in genes, and they are particularly harmful to cells of the fetus. For this reason the federal government maintains rigid safety standards for radioactive processes.

Radioactive isotopes have many uses in medicine, agriculture, and biology. In medicine they are often used as tracers. The movement of isotopes can easily be traced in plant and animal systems, using radiation detecting instruments called geiger counters. Tracers therefore can be used to monitor the flow of blood in biological systems, to study the metabolism of plants and animals, and to determine the structure of complex molecules. An example of tracers used in cancer research is shown in Fig. 10.14.

Radioisotopes can also be used to treat diseases, since the diseased cells are often more easily damaged by radioactive substances than normal cells, and therefore can be destroyed. As an example, the radioactive iodine isotope ^{131}I can be used in measurement connected with the thyroid gland. The patient ingests a prescribed amount of the isotope and it is absorbed by the thyroid gland. The radioactive ^{131}I then can be traced as it is released into the bloodstream in the form of protein-bound iodine.

*One disintegration per second is called a becquerel (Bq).

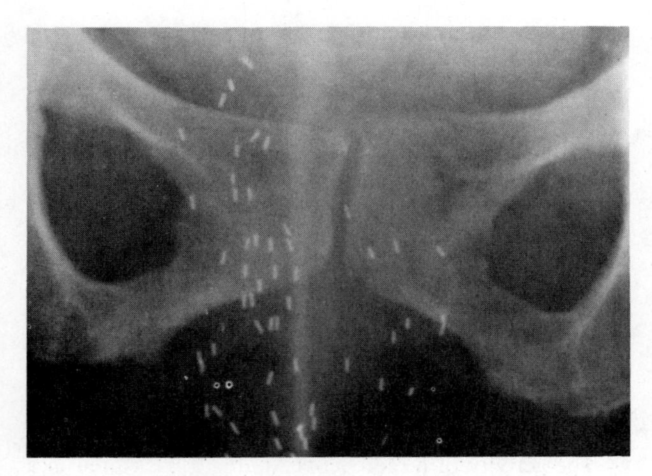

Figure 10.14 Radioactive cancer therapy. A technique for cancer therapy, using small pieces of radioactive chromium-51 wire, has been developed at Argonne Cancer Research Hospital, Argonne, Illinois. The photo shows an X-ray of the pelvis of a patient with implanted "seeds" of radioactive chromium wire that can be left in the body permanently. (Photo courtesy Argonne Cancer Research Hospital.)

10.4 Fission

In the process of fission, mass is converted into kinetic energy. If ^{235}U is bombarded with low-energy neutrons, ^{236}U is formed. This ^{236}U then fissions; that is, it splits into smaller nuclei. These two reactions are

$$^{235}_{92}U + ^{1}_{0}n \rightarrow ^{236}_{92}U$$

$$^{236}_{92}U \xrightarrow{\text{Fission}} \text{Two or more lightweight nuclei} + \text{Neutrons} + \text{Energy}$$

The three important features of fission are given by the three quantities on the right side of the fission reaction. First, the lightweight nuclei emitted are almost always radioactive, and their disposal is a troublesome problem. Second, large amounts of energy, about 200 MeV, are produced in a typical fission reaction. The third and most important feature of fission reactions is that neutrons are released. This is of importance because it was neutron bombardment of ^{235}U that started the process. There is, then, the possibility of a chain reaction.

In a **chain reaction** one initial reaction triggers a series of subsequent reactions. In the case of fission, one neutron hitting a nucleus of ^{235}U forms ^{236}U, which fissions to give off more neutrons. These neutrons then hit other ^{235}U nuclei and fission occurs again and again, each time with the release of great quantities of energy and more neutrons. The reaction continues until all of the available ^{235}U is used up. A schematic diagram of the process is shown in Fig. 10.15.

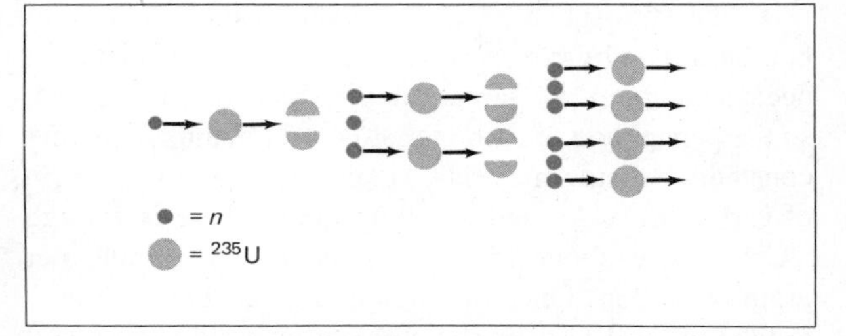

Figure 10.15 Diagram illustrating a chain reaction. A neutron hits a ^{235}U nucleus; fission then occurs and neutrons are given off. Some of these neutrons hit other ^{235}U nuclei, and the process continues until all the ^{235}U is used up.

Of course the process just described is not as simple as it appears here. In order for a chain reaction to occur, at least one neutron from each fission event must strike another ^{235}U nucleus and cause another fission event. If more than one neutron from each fission process hits more ^{235}U, then the chain reaction is uncontrolled and will proceed very quickly. The result is that an enormous amount of energy is released in a very short time. This concept led to the development of the atomic bomb.

Normally, not all the neutrons hit ^{235}U nuclei. Some may pass through the empty space between nuclei and strike nothing; others may hit other nuclei such as ^{238}U or impurities and be captured. For this reason, it is important to have enriched ^{235}U. This is necessary since natural uranium is 99.3% ^{238}U and only 0.7% ^{235}U. If the enriched uranium is put into fuel rods, a nuclear reactor can be built.

There is a logical explanation of why naturally-occurring uranium contains only 0.7% ^{235}U. The half-life of ^{235}U is only 0.7×10^9 years whereas the half-life of ^{238}U is 4.5×10^9 years. Because ^{235}U has a much shorter half-life than ^{238}U, most of the ^{235}U originally present in the Earth's crust has decayed into other elements.

A **nuclear reactor** operates on the principle of a controlled chain reaction. A simplified diagram of a nuclear reactor is shown in Fig.10.16. After the reaction has been initiated, neutron-absorbing rods are inserted so that the neutrons often hit them instead of the fissionable material. The rods are adjusted so that for every fission reaction, exactly one neutron hits more fissionable material. In this way, the chain reaction proceeds at a certain fixed rate. The energy released is maintained at a

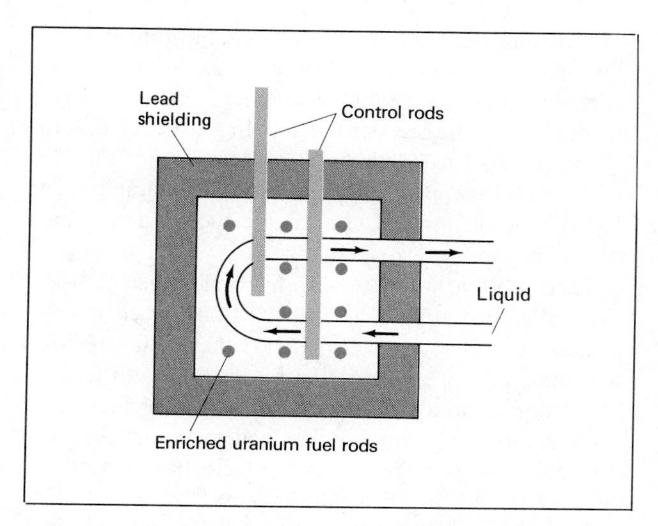

Figure 10.16 The principal components of a nuclear reactor. The enriched uranium fuel rods supply the energy. The neutron-absorbing control rods are adjusted so that the chain reaction proceeds at a constant rate. The lead shielding is necessary for protection from harmful radioactive materials produced in the fission process. A liquid flows through the reactor and becomes heated. This heat energy is then converted to electricity.

constant level. In this manner, nuclear power plants generate electricity for use by industry and in homes. Figure 10.7 (p. 162) is a photograph of one of the many electric generating plants in the United States powered by nuclear reactors.

Because small amounts of fissionable material can be transformed into vast quantities of energy, we have another important supply of energy at our disposal. Unfortunately, the disadvantages of nuclear reactors that have appeared with their increased use are thought by many people to outweigh their advantages.

Are nuclear power plants safe? This is one of the typical questions asked about nuclear power. To answer it directly is to miss the point. Nothing is absolutely safe. At this moment you could be hit by a falling airplane or meteorite, or you could be injured or killed by a violent earthquake. None of these things is very likely, so you don't fret about them. The risk of their occurring is very low.

Proponents of nuclear power argue that compared with most risks taken by our society, nuclear power reactors are very safe. To date, the number of deaths at commercially operating power plants in the United States is practically zero. In fact, the nuclear industry is probably safer in its operation than any other in the country.

One danger from nuclear power is the possibility of radiation leakage. However, fears of this possibility have so far proved to be unfounded. In fact, coal-fired power plants give off more radiation than do nuclear power plants. In both cases, however, the amount is so small as to be negligible compared with natural radiation.

Another possible problem is the disposal of radioactive wastes. However, it must be remembered that waste disposal is also a major problem in the coal industry. If a family gets its electricity from coal, its yearly waste per person is over 300 lb. Some of this is in the form of acid rain. If that same family got its electricity from a nuclear reactor, its yearly waste per person would be the size of an aspirin tablet. One solution to the radioactive waste problem is to put the waste in glassy material and bury it (see Fig. 10.17). The details of this process are still being worked out.

Another hazard is the possibility of a major accident. Such an accident occurred in 1979 at the Three Mile Island reactor in Pennsylvania. However, in spite of the severity of the problems there, no one was killed, no one was injured, and no drastic harm was done to the environment in terms of radiation contamination. The

Figure 10.17 High-level wastes. Processes to combine high-level radioactive waste in glass are under investigation by the Department of Energy. Radioactive waste from the nuclear fuel cycle can be immobilized for handling and disposal in this form. The synthetic waste glass shown is comprised of 25% high-level waste and 75% nonradioactive glass formers. The button on the left represents the annual quantity of high-level waste glass for one person if all electricity in the United States were produced by nuclear power. The cylinder at right, about a cupful, represents an individual's lifetime share of the material. (DOE photo courtesy Dick Peabody.)

1986 accident at Chernobyl (Fig. 10.18) in the Soviet Union was far more serious. Scores of people were killed and some agricultural products were contaminated. Problems of contamination of water sources will remain a hazard for years to come.

Do not forget, however, that major contamination of land and water has also occurred because of coal mining activities in the United States and elsewhere. Society must choose its risks—both coal and nuclear power pose difficult problems.

In addition to ^{235}U, the other fissionable material of importance is $^{239}_{94}Pu$. This isotope is produced by the decay of ^{239}U. The ^{239}U is produced when neutrons are captured by ^{238}U. The important reactions involved in the production of ^{239}Pu are

$$^{238}_{92}U + ^{1}_{0}n \longrightarrow ^{239}_{92}U$$

$$^{239}_{92}U \longrightarrow ^{239}_{93}Np + ^{0}_{-1}e + \nu \qquad \text{Half-life: 24 minutes}$$

$$^{239}_{93}Np \longrightarrow ^{239}_{94}Pu + ^{0}_{-1}e + \nu \qquad \text{Half-life: 2.35 days}$$

The isotope ^{239}Pu has a half-life of 24,400 years. Because natural uranium is 99.3% ^{238}U, the conversion of ^{238}U into ^{239}Pu can greatly increase the amount of fissionable material available as an energy source. This is the purpose of the breeder reactor. In a **breeder reactor** ^{238}U is bombarded with neutrons to create ^{239}Pu. The ^{239}Pu can be chemically separated from the uranium and fission by-products and used as the fuel in an ordinary nuclear reactor.

There are two ways to make fissionable material—by physically separating ^{235}U from ^{238}U or by chemically separating ^{239}Pu from the ^{238}U and other radioactive fission products in a reactor. It is very, very difficult to physically separate ^{235}U from ^{238}U. Both of these uranium isotopes have similar chemical properties. To separate them takes huge amounts of electric energy and time. Thus, it is a very costly process.

Separating ^{239}Pu from ^{238}U and other materials in a used fuel rod is not difficult chemically, but it takes a lot of careful handling because of the radioactive material present. Hence, very special and expensive procedures must be used for the safety of all concerned.

Figure 10.18 Destruction at the Chernobyl nuclear reactor in the Ukraine region of the U.S.S.R. In the spring of 1986 a major nuclear accident occurred at Chernobyl. The damage to the reactor and the surrounding buildings can be seen at the center of the photo made May 9, 1986. Many people were killed and thousands were evacuated as a result of the released radiation. (AP/Wide World Photos.)

Nuclear reactors use uranium enriched to 3% ^{235}U while nuclear weapons use uranium enriched to 30–95% ^{235}U. Thus, it is not possible to use reactor fuel for nuclear weapons. The plutonium produced in reactors, if separated from the ^{238}U, and so on, could be used in weapons. For this reason, international control of uranium and plutonium is utilized to prevent the spread of nuclear weapons material.

As technology in small nations improves, more and more countries will have the ability to produce weapons-grade nuclear material. We can only hope that no country will ever use these awesome forces.

Radioactive substances from the fission process are dangerous and must be handled with extreme caution. Radioactive nuclei with short half-lives are not too troublesome, because they decay away rather quickly. However, nuclei such as ^{14}C and ^{90}Sr, with half-lives of 5730 and 28 years, respectively, are quite hazardous and must be stored or disposed of with the utmost care.

The isotope ^{90}Sr is particularly harmful because its chemical properties are similar to calcium. Thus, it combines chemically in the bones and, through subsequent radioactivity, greatly increases the likelihood of bone cancer.

In the 1950s a number of nuclear tests were carried out in the atmosphere by the United States and Russia (see Fig. 10.19). This led to alarming increases in radioactivity in such things as milk. For this reason, an atmospheric nuclear test ban treaty between the United States and Russia was signed in the early 1960s. Since that time, all weapons testing of nuclear materials by these two countries has been carried out below the ground. However, other countries that did not sign the treaty, most notably France, China, and India, have exploded nuclear devices in the atmosphere.

The radioactive material that falls to the ground after a nuclear explosion is called fallout. This fallout is hazardous primarily because it consists of small dustlike particles that can be breathed into the lungs, where they can cause lung cancer. Once the fallout has fallen to the ground, it is not nearly so dangerous. During the days following a nuclear explosion, it is best to be inside with the doors and windows closed, so that the radioactive dust from the fallout is not inhaled.

If many nuclear explosions were to be detonated, there is a high probability that a lot of smoke would be created from fires raging out of control. Studies have shown that the smoke and dust created would circulate in the atmosphere and significantly decrease the amount of sunlight striking the Earth. This would result in a lowering of the temperature in the Northern Hemisphere by 5°C to 15°C. The resulting nuclear winter would be very difficult for the entire Northern Hemisphere. Certainly, it would drastically alter civilization as we know it.

Figure 10.19 A nuclear bomb test that took place in the 1950s. Because of the radioactive materials produced in such tests, they are no longer carried out by the United States and the Soviet Union. (AP/Wide World Photos.)

10.5 Fusion

The goal of the alchemists was to convert base metals, such as mercury and lead, into gold. We now know that this cannot be accomplished by chemical means, but it can be done by nuclear reactions. To initiate a nuclear reaction a nucleus is bombarded by another nucleus. The initial target nucleus and the projectile nucleus are called the **reactants.** The nuclei that are formed are called the **products.** The products of the reaction are determined by the energy of the bombarding nuclei and by the conservation laws of atomic number and mass number.

In every nuclear reaction total energy (mass energy + energy in other forms) must be conserved. When energy is released, mass is converted into kinetic energy. When energy is absorbed in a reaction, kinetic energy gets converted into mass.

A reaction in which energy is absorbed is called **endoergic** (or **endothermic**). A reaction is called **exoergic** (or **exothermic**) if energy is released. The energy of a reaction comes from converting mass into energy of other forms. It is easy to tell if energy is absorbed or released by examining the masses of the isotopes involved. Let us examine the nuclear reaction

$$\begin{array}{ccccccc} {}^{14}_{7}\text{N} & + & {}^{4}_{2}\text{He} & \longrightarrow & {}^{17}_{8}\text{O} & + & {}^{1}_{1}\text{H} \\ 14.0033 & & 4.0026 & & 16.9991 & & 1.0078 \end{array}$$

with the masses (in amu) of the isotopes given under each one. The masses on the left total 18.0059 amu while those on the right total 18.0069 amu. Thus, the reaction absorbs energy. The amount of energy absorbed is

$$\begin{aligned} \text{Energy absorbed} &= (18.0069 - 18.0059)\ \text{amu} \\ &\qquad \times 931\ \text{MeV/amu} \\ &= 0.931\ \text{MeV} \end{aligned}$$

This is an endoergic reaction.

EXAMPLE 6 Consider the following fusion reaction:

$$\begin{array}{ccccccccc} {}^{2}_{1}\text{H} & + & {}^{2}_{1}\text{H} & \rightarrow & {}^{3}_{2}\text{He} & + & {}^{1}_{0}\text{n} & + & \text{Energy} \\ 2.01410 & & 2.01410 & & 3.01603 & & 1.00867 & & \end{array}$$

Is this reaction exoergic or endoergic? What is the energy absorbed or released in this reaction?

In this reaction the total mass on the left is greater than the mass on the right. Thus, the reaction is exoergic. The energy released is

$$\begin{aligned} \text{Energy released} &= [(2.01410 + 2.01410) \\ &\qquad -(3.01603 + 1.00867)]\ \text{amu} \\ &\qquad \times 931\ \text{MeV/amu} \\ &= 3.26\ \text{MeV} \end{aligned}$$

Stars get their energy from fusion reactions such as the fusion reaction in Example 6. However the isotopes on the left side of Example 6 do not readily fuse. In fact, as can be seen from Fig. 10.20, they repel each other because both isotopes are positively charged. In order to fuse they must be moving at very high speeds.

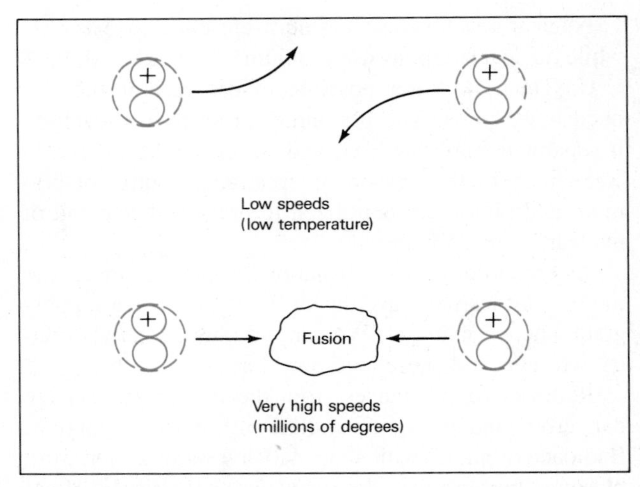

Figure 10.20 Two ${}^{2}_{1}$H nuclei moving at low speeds will repel each other. In order for them to collide and fuse, they must be moving toward each other at very high speeds. This necessitates a temperature of millions of degrees.

Thus, temperatures of millions of degrees are necessary to achieve fusion. In the fusion reactions inside the Sun, four protons fuse together in a series of reactions to form an alpha particle—a helium nucleus. The net reaction involved is due to a sequence of reactions. This net reaction is

$${}^{1}_{1}\text{H} + {}^{1}_{1}\text{H} + {}^{1}_{1}\text{H} + {}^{1}_{1}\text{H} \longrightarrow {}^{4}_{2}\text{He} + \text{other particles with negligible mass}$$

If we write this in terms of masses of the protons and helium nucleus we get (in amu)

$$\begin{aligned} 1.00783 + 1.00783 + 1.00783 + 1.00783 &= 4.00260 + \text{Mass energy} \\ \text{or} \quad \text{Mass energy} &= 4 \times 1.00783 - 4.00260 \\ \text{Mass energy} &= 0.02878 \end{aligned}$$

The percentage of mass energy converted into other forms of energy is

$$\begin{array}{l}\text{\% of mass energy} \\ \text{converted into other} \\ \text{forms}\end{array} = \frac{.02878}{4 \times 1.00783} \times 100 = 0.7\%$$

Thus, we see that in the Sun 0.7% of the hydrogen mass energy is converted into other forms of energy (such as heat and light) when hydrogen fuses together to form helium.

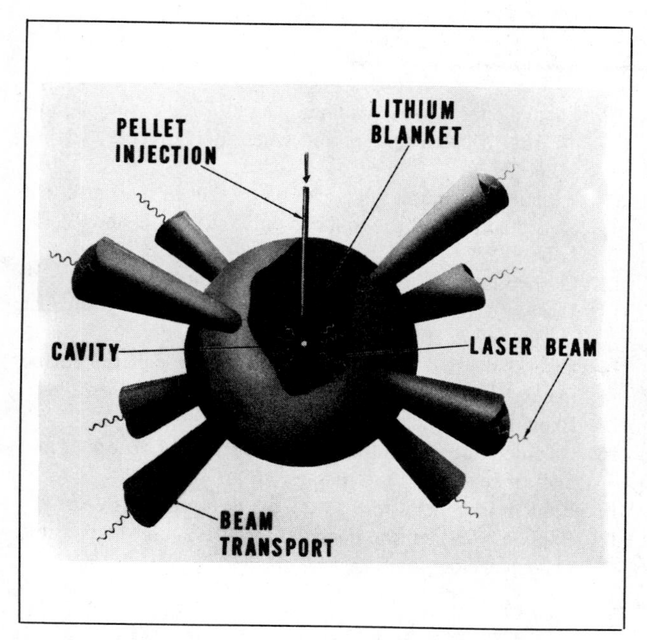

Figure 10.21 Reactor cavity. This partial cutaway view of a fusion reactor cavity shows a lithium blanket and the manner in which laser beams would interact with the target. This is a part of the work being carried on at Los Alamos (New Mexico) Scientific Laboratory in an effort to control a thermonuclear reaction—the reaction that takes place in the Sun and stars—to help meet the world's future energy demands. The laboratory is operated for the U.S. Department of Energy by the University of California. (Courtesy Los Alamos Scientific Laboratory.)

Achieving controlled fusion reactions on Earth is not easy. For one thing, extremely high temperatures are needed. A working nuclear fusion reactor has not yet been developed. One possible design (Fig. 10.21) involves the use of powerful beams of light produced by lasers. The laser beams would pulsate, striking dropped pellets of fusible material, and cause fusion. The energy of the resulting miniexplosions could then be used in the production of electricity. As yet, however, lasers are not powerful enough to provide the necessary energy to generate the high temperatures needed.

A major drawback to a commercial nuclear fusion reactor is the radiation damage done by the neutrons given off in reactions similar to that in Example 6. Neutrons easily penetrate materials because they have no charge. When they are absorbed by a nucleus such as iron, they can alter the element.

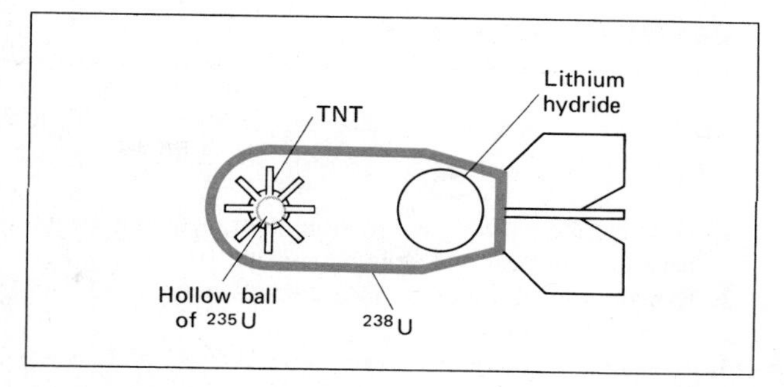

Figure 10.22 Diagram of a hydrogen bomb. The TNT explodes, pushing the ^{235}U together. This starts an uncontrolled chain reaction that heats the lithium hydride to over 180,000,000°F and creates $^{3}_{1}H$ and $^{2}_{1}H$ atoms that fuse together. The resulting explosion is followed by a fission reaction in the ^{238}U outer lining that tops off the awesome blast.

As an example, consider the following sequence of reactions:

$$^{58}_{26}\text{Fe} + \text{n} \rightarrow {}^{59}_{26}\text{Fe} \xrightarrow[\text{44.5 days}]{} {}^{59}_{27}\text{Co} + {}^{0}_{-1}\text{e} + \nu$$

Here a neutron has been absorbed by an iron-58 nucleus which in turn decays to cobalt-59. Thus, the iron atom has changed into a cobalt atom. As this happens over many months the iron in a material such as steel will become brittle and possibly crack, causing structural damage. All of the materials in a fusion reactor would be subjected to neutron damage. It is not clear that a fusion reactor could be designed to overcome this neutron-damage problem.

Twenty years ago, it was thought that fusion reactors would solve the energy problem for thousands of years to come. Since that time the development of solar cells and uranium reserves has changed the energy picture considerably. Solar power and conventional nuclear reactors now promise to be much more economical for the foreseeable future than will fusion reactors. Thus, the development of a commercial fusion reactor will most likely be far in the future.

Thermonuclear warheads or hydrogen bombs are based on the fusion process. The necessary high temperatures are first produced by exploding a fission-type bomb. The fusion process then takes place, supplying many times more energy than the fission reaction. Figure 10.22 shows a possible arrangement of the components of a hydrogen bomb.

Learning Objectives

After reading and studying this chapter, you should be able to do the following:

1. Describe the general nature of an atom including a comparison of nuclear and atomic mass and size.
2. Explain why the protons in a nucleus don't repel each other and fly apart.
3. Explain why all elements with more than 83 protons are unstable.
4. Describe the nomenclature used for an isotope.
5. Be able to convert from heat energy produced to mass energy consumed using $E = mc^2$.
6. Explain the concept of an electron volt of energy.
7. Explain the concept of binding energy and be able to work out the example of a $^4_2$He nucleus.
8. Explain how the Sun converts mass into energy of other forms.
9. Explain radioactivity and state the physical properties of alpha, beta, gamma, and neutron radiation.
10. Calculate the mass remaining in a radioactive isotope sample after a given time period when the isotope's half-life is known.
11. Explain why uranium (a) can be present in the Earth if it is radioactive and (b) contains only 0.7% ^{235}U and 99.3% ^{238}U.
12. Explain how old bones and artifacts can be dated.
13. Draw a diagram and describe the operation of a nuclear reactor in general terms.
14. Discuss the two ways in which fissionable material can be made.
15. Explain why nuclear fallout is dangerous.
16. Calculate the amount of energy released or absorbed in a nuclear reaction if the masses are known.
17. Explain in general terms how the Sun produces energy.
18. Explain what neutron damage is with regard to a possible fusion reactor.

Important Words and Terms

atom
electrons
nucleus
protons
neutrons
nucleons
element
strong nuclear force
stable
unstable
atomic number
neutron number
mass number
isotopes
deuteron
pairing effect
amu
MeV
electron volt
total binding energy
binding energy per nucleon
neutron separation energy
proton separation energy
fusion
atomic weight
radioactive
alpha decay
beta decay
neutrino
gamma decay
gamma ray
fission
half-life
curie
chain reaction
nuclear reactor
breeder reactor
reactants
products
endoergic
endothermic
exoergic
exothermic

Questions

1. How much of the mass of an atom is contained in the nucleus?
2. How does the diameter of an atom compare with the diameter of a nucleus?
3. What property of atoms specifies an element?
4. Why don't the protons in a nucleus repel each other and split apart?
5. Why are all the elements with more than 83 protons unstable?
6. What is the pairing effect?
7. What is an electron volt?
8. Does a stable nucleus weigh more or less than the free nucleons that compose it?
9. Discuss two ways in which mass energy can be converted into other forms of energy.
10. Compare the physical properties of alpha, beta, gamma, and neutron radiation in terms of charge, mass, and penetrating ability.

11. If uranium is radioactive, why does it occur naturally on Earth?
12. How can old bones and artifacts be dated?
13. Are radioactive isotopes useful?
14. State the importance of the three products occurring when ^{236}U fissions.
15. Why does naturally occurring uranium contain only 0.7% ^{235}U?
16. What are the principal features of a nuclear reactor?
17. How much waste is caused by one individual who gets electric energy from a (a) coal-fired or a (b) nuclear generating station.
18. What are the two ways to make fissionable material?
19. What would cause a nuclear winter?
20. What temperature is necessary before fusion can occur?
21. When will a commercial fusion reactor be developed?

Problems

1. Write the names of each of the following isotopes. Then give the atomic number, proton number, neutron number, electron number, and mass number for each.
(a) $^{6}_{3}X$ (b) $^{10}_{5}X$ (c) $^{12}_{6}X$ (d) $^{34}_{16}X$ (e) $^{90}_{38}X$ (f) $^{235}_{92}X$ (g) $^{239}_{93}X$ (h) $^{240}_{94}X$
Answer: (e) strontium 38, 38, 52, 38, 90
2. Give three criteria for determining when an isotope will be unstable. Apply these and other reasons to determine if the following isotopes are unstable or stable:
(a) ^{12}C (b) ^{14}N (c) ^{16}O
(d) ^{18}F (e) ^{90}Y (f) ^{96}Sn
(g) ^{22}N (h) ^{249}Cf (i) ^{226}Ra
Answer: (e) odd-odd, so unstable
(i) more than 83 protons, so unstable
3. Only two isotopes of $_{19}K$ (potassium) are stable. Pick which two they are from the following list:
^{38}K, ^{39}K, ^{40}K, ^{41}K, ^{42}K
4. How much mass energy is converted into other forms of energy in one day in a submarine's nuclear reactor which generates 1 MW of power? *Answer:* one milligram
5. Show that one electron volt (eV) is equal to 1.6×10^{-19} J.
6. Natural chlorine is 75.8% ^{35}Cl and 24.2% ^{37}Cl. The mass of ^{35}Cl is 34.97 amu and the mass of ^{37}Cl is 36.97 amu. Compute the atomic weight of chlorine. *Answer:* 35.45
7. Complete the following nuclear decay schemes and state whether the process is alpha decay, beta decay, gamma decay, or fission:
(a) $^{47}_{21}Sc \longrightarrow ^{47}_{22}Ti +$ ______
(b) $^{210}_{84}Po^* \longrightarrow ^{210}_{84}Po +$ ______
(c) $^{210}_{84}Po \longrightarrow ^{206}_{82}Pb +$ ______
(d) $^{240}_{94}Pu \longrightarrow ^{97}_{38}Sr + ^{139}_{56}Ba +$ ______
(e) $^{229}_{90}Th \longrightarrow ^{4}_{2}He +$ ______
(f) $^{47}_{21}Sc^* \longrightarrow ^{47}_{21}Sc +$ ______
(g) $^{237}_{93}Np \longrightarrow ^{0}_{-1}e +$ ______
Answer: (c) $^{4}_{2}He$, alpha decay
8. Bismuth-214 may decay by emitting an alpha or a beta particle. Write the symbol notation for the daughter nucleus in each case.
9. The half-life of ^{131}I is 8 days. If a thyroid cancer patient is given 16 micrograms of ^{131}I, how many micrograms will still be in his thyroid after 24 days?
10. A skeleton of a prehistoric woman is found to yield 2 beta rays per minute per gram of carbon. About how old is the skeleton? *Answer:* 17,190 years
11. The activity of a given sample of radioactive material at 9 A.M. was 8 microcuries. At 12 noon the activity is measured as 2 microcuries.
(a) Determine the half-life of the nuclide.
(b) How many disintegrations were given from the sample at 12 noon?
12. If the Earth had formed 0.7×10^{9} years ago with equal amounts of ^{235}U and ^{238}U, approximately what percentage of natural uranium would be ^{235}U?
13. Complete the following nuclear reaction equations by balancing atomic numbers and mass numbers:
(a) $^{27}_{13}Al + ^{4}_{2}He \longrightarrow ^{30}_{15}P +$ ______
(b) $^{48}_{20}Ca + ^{1}_{1}H \longrightarrow ^{2}_{1}H +$ ______
(c) $^{23}_{11}Na + \gamma \longrightarrow$ ______
(d) $^{236}_{92}U \longrightarrow ^{131}_{53}I + 3^{1}_{0}n +$ ______
(e) $^{20}_{10}Ne + ^{16}_{8}O \longrightarrow ^{12}_{6}C +$ ______
(f) $^{6}_{3}Li + ^{6}_{3}Li \longrightarrow ^{7}_{3}Li +$ ______
Answer: (d) $^{102}_{39}Y$
14. Given the following masses, calculate the amount of energy released in the fusion reactions below:
$^{1}_{0}n$ = 1.00867 amu $^{3}_{1}H$ = 3.01605 amu
$^{1}_{1}H$ = 1.00783 amu $^{3}_{2}He$ = 3.01603 amu
$^{2}_{1}H$ = 2.01410 amu $^{4}_{2}He$ = 4.00260 amu
(a) $^{2}_{1}H + ^{2}_{1}H \rightarrow ^{3}_{1}H + ^{1}_{1}H +$ energy
(b) $^{3}_{1}H + ^{2}_{1}H \rightarrow ^{4}_{2}He + ^{1}_{0}n +$ energy
Answer: (a) 4.02 MeV
15. Use the masses given in problem 14 for ^{3}H and ^{3}He.
(a) Which of these two is stable and which is radioactive?
(b) How much energy is given off when one of these decays into the other?
(c) Write the decay reaction.

11
The Atom

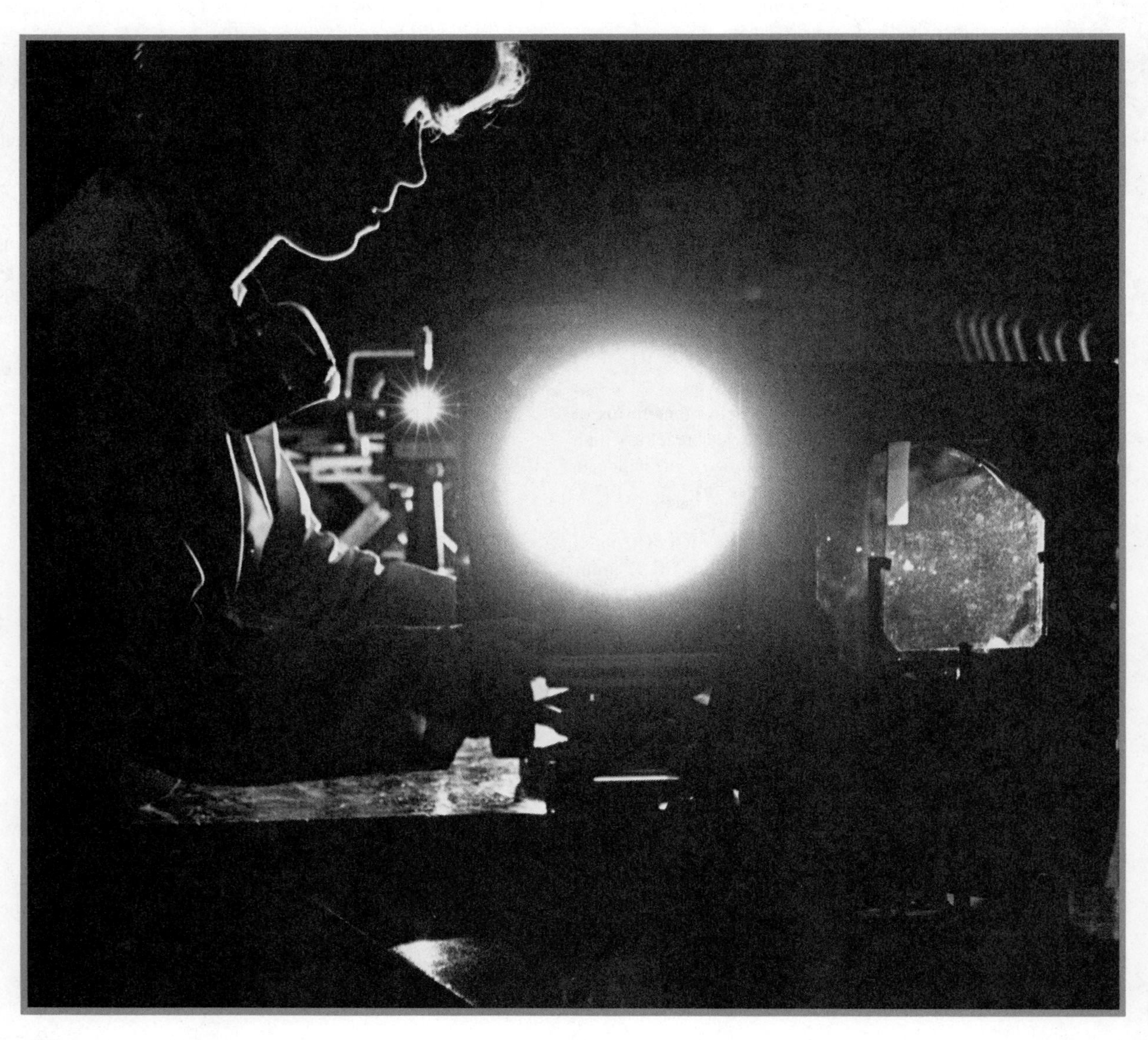

And who loves Nature more
Than he, whose painful art
Has taught and skilled his heart
To read her skill and lore?
–Robert Bridges

IN THE PREVIOUS chapter, the atom was described by using a model. According to this model, the nucleus, containing the protons and neutrons, is a small core, and the negatively charged electrons travel in almost circular orbits about the nucleus, much as the planets travel around the Sun.

In the early 1900s, this was but one of several models of the atom. Before the theory of Niels Bohr, this model was not accepted as correct for the following very good reason: In the late 1800s, much elegant theoretical work had been done in the field of electromagnetism. One well-known result of this work was the discovery that any charged particle that is accelerating radiates (gives off) energy.

Since the electron is a charged particle and is accelerating (anything traveling in a circle is accelerating), it should radiate energy. If the electron is giving off energy, then it must itself lose kinetic energy, since the law of conservation of energy will be obeyed. If it loses kinetic energy, it will also lose velocity and soon spiral into the nucleus. The time it would take for this to happen would be less than one second.

Thus, according to classical electromagnetic theory, all the atoms of the universe should have collapsed in less than one second. Obviously, this had not happened. Was something wrong with electromagnetic theory which was known to work so well, or was something wrong with the physicists' basic understanding of nature? It will be seen in this chapter that it was the physicists' concept of nature that had to be changed.

◀ A physicist checks the purity of material using a laser.

11.1 The Dual Nature of Light

In 1900 Max Planck (see Fig. 11.1) took the first step toward a new theory of physics called quantum physics. A **quantum** is a discrete amount. Planck shook the foundations of classical physics with his quantum hypothesis which states that, in most situations, energy is radiated or absorbed by atoms only in discrete amounts. This was in contrast to the classical idea that any amount of energy can be radiated or absorbed.

Figure 11.1 Max Planck (1858–1947). Planck was a professor of physics at the University of Berlin. On December 14, 1900, he founded what is now known as modern physics (as opposed to classical physics) by proposing that the energy of electromagnetic radiation exists only in discrete amounts called quanta. The important small constant that appears over and over again in quantum physics is known as Planck's constant. In 1918 Planck was awarded the Nobel Prize in physics for his contributions to quantum theory. (Photo courtesy American Institute of Physics, Meggers Gallery of Nobel Laureates.)

Planck introduced a small number now known as Planck's constant. **Planck's constant, h,** is equal to 6.625 $\times 10^{-34}$ J·s. In 1905 Albert Einstein used Planck's hypothesis to describe light in terms of particles rather than waves.

In Einstein's description of light or any electromagnetic radiation, a **photon** is a "particle" of electromagnetic radiation. Einstein wrote in 1905 that each photon carries an amount of energy proportional to its frequency with Planck's constant being the constant of proportionality. That is,

$$\text{Energy of a photon} = \text{Planck's constant} \times \text{frequency of light}$$

or

$$E = hf \qquad (11.1)$$

Since the frequency of light is related to its speed c and wavelength λ by $f = c/\lambda$, we get

$$E = hf = \frac{hc}{\lambda}$$

According to this equation, the shorter the wavelength (higher the frequency), the greater the energy. As an example, a photon of red light with a wavelength of 6500 angstroms will have less energy than a photon of blue light with a wavelength of 4400 angstroms. (Remember that an angstrom is a unit of length and 1 Å = 10^{-10} m.)

This concept of light waves carrying energy in the form of discrete packets called photons is confusing to many people. How can a wavelike entity be composed of particles? Large particles with which we have had experience (such as billiard balls) do not show wavelike effects such as diffraction, interference, and polarization. Light must be a wave because it shows these effects. But we have just commented on Einstein's description of light as particles, and this theory has been shown experimentally to be correct.

We have a confused situation. Is light a wave or a particle? The solution to this dilemma is couched in the phrase "the **dual nature of light.**" This phrase simply means that light acts sometimes like a wave and sometimes like a particle.

Our classical view of nature does not seem to work on a small scale. Light seems to be capable of having both a wave and particle behavior. This can be confusing to people who are used to large objects, but it seems to be how nature behaves.

In the **photoelectric effect,** a photon of light strikes a metal surface and ejects an electron that then flows to a positively charged plate. In this way, a light beam can produce an electric current.

Solar cells are based on the photoelectric effect. The object of **solar cells** is to convert energy from sunlight into electric energy. Photons from the Sun strike the solar cells and eject electrons that produce a usable electric current. One of the main concerns is the efficiency of the solar cells. That is, how much of the Sun's energy can be converted into electric energy.

To achieve an efficiency of approximately 10% to 30%, special materials must be used in solar cells. One fairly expensive way of making solar cells is to grow a nearly perfect crystal of silicon and then slice it into thin wafers. Currently, this method is used widely, but it is a fairly expensive technique. An example of solar cells is shown in Fig. 11.2.

Another possible material is amorphous or glass-like silicon. Amorphous silicon is a much less expensive material to produce, but it has a lower efficiency.

Figure 11.2 Solar cells convert part of the Sun's energy into usable electricity at the Mt. Laguna Air Force Station east of San Diego, California. For large-scale applications, the cost of these cells is currently too high compared with the cost of traditional electric energy sources. (Tom McHugh/Photo Researchers, Inc.)

Solar cells currently have applications in remote locations where they can provide electric energy for such tasks as radio transmission of data. As their cost decreases, more applications, such as providing power for hand-held calculators, are becoming cost-effective.

One obvious disadvantage of a home electric system based on solar cells is that they can provide power only during the daylight hours. Storage batteries are used in conjunction with solar cells in some systems, but these systems are currently more expensive than simply taking power from the nation's power grid.

Research is continuing in solar-cell production to try to achieve cheaper materials with ever better efficiencies to compete favorably with traditional forms of electric energy. As this is achieved, a gradual and partial conversion from coal and nuclear energy to solar energy can occur.

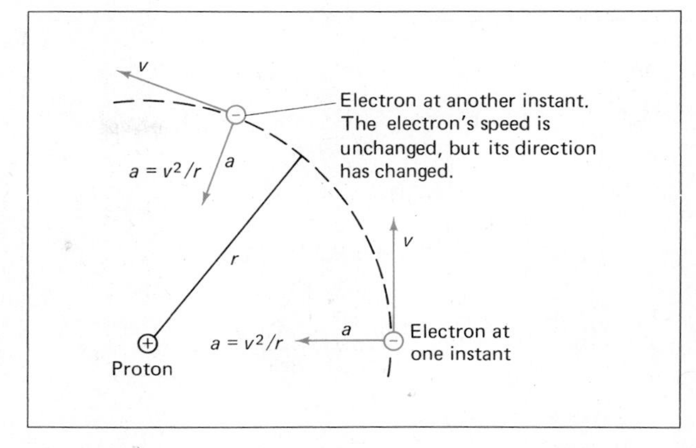

Figure 11.3 Relations between velocity and acceleration of an electron traveling in a circular path about a proton. The direction of the velocity vector is constantly changing; hence, the electron is being accelerated. The acceleration vector (a) always points toward the proton at the center of the circle.

11.2 Bohr's Equations for the Hydrogen Atom

In 1913, Niels Bohr incorporated Planck's quantum hypothesis into a theory regarding the simplest of atoms—hydrogen. His consequent theoretical results were in startlingly good agreement with experiment. His theory is summarized in the following paragraphs.

The hydrogen atom is the simplest atom. It normally consists of one proton, no neutrons, and one electron. The force between the proton and electron is due to the electric attraction of their opposite charges. The gravitational and nuclear forces present are negligible and play no role in the physics of the atom. The electric attractive force which exists between the proton and the electron is given as

$$F = k\frac{q^2}{r^2}$$

where k = a constant (9×10^9 N $\times$ m^2/C^2),
q = charge on the proton and electron, and
r = distance between proton and electron.
(See also Section 8.1.)

Any object moving in a circle at constant speed is being accelerated. This fact follows from Newton's first law, which states that an object that has no unbalanced force acting on it, and hence no acceleration, moves in a straight line. The acceleration comes about because the direction of the velocity is changing, even though the speed is not.

The acceleration for an object moving in a circle at constant speed v is called **centripetal acceleration** and is given by $a = v^2/r$. (See Section 2.4.) The direction of the acceleration is toward the center of the circle. All of these relations for an electron traveling around a proton are shown in Fig. 11.3.

Since we know both the force on the electron and the acceleration of the electron, we write

$$F = ma$$

or

$$k\frac{q^2}{r^2} = m\frac{v^2}{r}$$

This was Bohr's first equation for the hydrogen atom.

Bohr's second equation for the hydrogen atom was based on the concept of energy. We can derive the equation by considering the total energy E of the electron, which is the sum of the potential energy, E_p, and the kinetic energy, E_k. That is,

$$E = E_p + E_k$$

Before proceeding, however, it is necessary to discuss the fact that the potential energy of an object can be considered either positive or negative. Consider, for example, a ball of mass m on a stand at a height h above the ground. Its potential energy with respect to ground level is

$$E_p = mgh$$

Consider also this same ball lying in a well at a distance

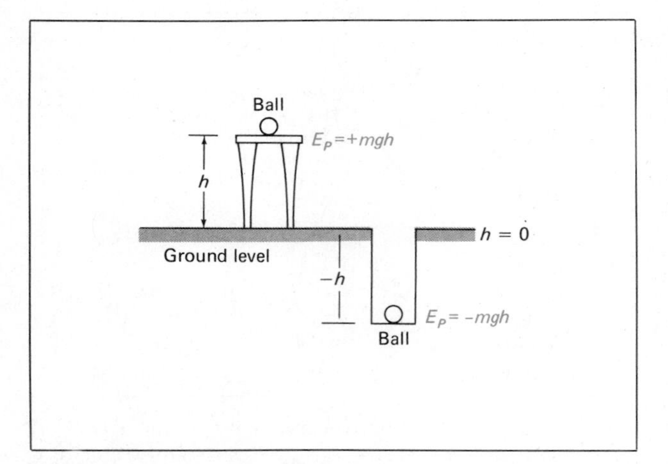

Figure 11.4 Diagram illustrating that potential energy can be either positive or negative. A ball above ground has a positive potential energy. A ball below ground has a negative potential energy relative to ground level. (See also Fig. 4.13.)

h below the ground (Fig. 11.4). Now it has a negative potential energy. When the ball is above ground, it does work by falling to the ground. When the ball is below the ground, work must be done on the ball—it must be raised—to bring it to ground level. We express this latter situation by saying the ball does negative work. Thus, it has a negative potential energy. It is common physics terminology to say that a particle with a negative potential energy is in a **potential well.** The reason for this terminology is illustrated in Fig. 11.4.

For an electron traveling around a proton, the potential energy is negative. The electron is bound to the proton in a potential well. For the ball in the well to be able to move about freely at the ground level, it must overcome the gravitational force that holds it in the well. Similarly, for the electron to be able to move about freely, it must overcome the attractive electric force that holds it close to the proton.

An object's gravitational potential energy (relative to the surface) in a well is negative and is given by $E_p = -mgh$ (Fig. 11.4). We think of the electron as having been brought from a point very far from the proton, experiencing an ever-increasing force of attraction by the proton as it moves closer and goes deeper into the well caused by the electric force. It can be shown (with calculus) that the electron's electric potential energy E_p at a distance r from the proton is given by

$$E_p = \frac{-kq^2}{r}$$

Note that the electric potential energy is zero when the electron is an infinite distance r from the proton.

The electron's kinetic energy is $E_k = mv^2/2$. From Newton's second law, we found that Bohr's first equation was

$$\frac{kq^2}{r^2} = \frac{mv^2}{r}$$

or

$$\frac{kq^2}{2r} = \frac{mv^2}{2} = E_k$$

The total energy for the electron is then

$$E = E_p + E_k$$

$$E = -k\frac{q^2}{r} + \frac{kq^2}{2r} = -\frac{kq^2}{2r}$$

This is an equation for the total energy of the electron. Notice that the total energy is negative. The larger the radius r becomes, the more positive (less negative) the total energy becomes. For an infinite radius, $E = 0$. This means, of course, that the electron is no longer bound to the proton.

So far, we have derived two equations from classical physics, which had become completely established by 1913. We now proceed to the crucial step which Bohr took.

The quantity called *angular momentum* is defined for a particle going in a circle as the linear momentum (mass × velocity) times the radius of the circle. The all-important step that Bohr took was to **quantize the angular momentum.** This means that he allowed only certain discrete values for the angular momentum of the electron. Specifically, he wrote

$$mvr = nh/2\pi$$

where m = mass of the electron moving in a circle,
v = velocity of the electron moving in a circle,
r = radius of the circle,
n = any positive integer such as 1, 2, 3, . . . , and
h = Planck's constant (6.625×10^{-34} J · s).*

This was quite a bold step. It says that not just any value of angular momentum can exist. Only integer numbers times $h/2\pi$ are allowed! The values $h/2\pi$, $2h/2\pi$, $3h/2\pi$,

*The symbol h in this equation has the units of joules × seconds and is not related to the symbol h = height used previously in explaining a potential well.

$4h/2\pi$, and so on, are possible, but no values in between these such as $1.3h/2\pi$ or $2.731h/2\pi$ are allowed.

Let us now see where this audacious assumption led Bohr. We now have three equations. They are

$$k\frac{q^2}{r^2} = \frac{mv^2}{r}$$

$$E = -\frac{kq^2}{2r}$$

$$mvr = \frac{nh}{2\pi}$$

From these three equations it is possible to show (see Appendix V) that the radius of the electron's orbit and the energy of the electron can be expressed in terms of constants and a positive integer, n. The important results are

$$r = 0.529n^2 \text{ angstroms} \tag{11.2}$$

$$E = \frac{-13.6}{n^2} \text{ eV} \tag{11.3}$$

Remember that an angstrom, Å, is 10^{-10} m and eV stands for electron volt with 1 eV = 1.6×10^{-19} J.

From Eq. 11.2 we see that the radius of a hydrogen atom with $n = 1$ is 0.529 Å, so the diameter would be 1.06 Å. Other atoms have similar sizes. Thus, one angstrom is approximately the diameter of an atom. The angstrom unit of 10^{-10} m was introduced because a typical atomic dimension is 1 Å. The volume of an atom is approximately $(1\ \text{Å})^3$ or about 10^{-30} m^3. Thus, we see that atoms are extremely small.

Even so, the radius of a typical atom is 10,000 times bigger than that of an atomic nucleus. Thus, nuclei are approximately 10^{12} times smaller in volume than an atom. Compared to nuclei, atoms are huge.

Recall that an electron volt is the kinetic energy a single electron (or proton) would gain after being accelerated through one volt of electric potential. (Also recall that one volt equals one joule per coulomb.) The unit of eV, or electron volt, is used because it is so easy to translate from electron volts to the voltage necessary for each atom to gain or lose that energy.

Thus, if a hydrogen atom has an energy of −13.60 eV, it means that a voltage of 13.60 V has the capacity to increase the energy of the atom from −13.60 eV to zero. Low, easily attainable voltages are used to work with atoms and molecules.

Table 11.1 Allowed Values of the Electron's Radius and Energy for Low Values of n

n	r_n	E_n
1	0.529 Å	−13.60 eV
2	2.116 Å	−3.40 eV
3	4.761 Å	−1.51 eV
⋮	⋮	⋮
∞	∞	0

11.3 Energy Levels for the Hydrogen Atom

Let us now examine the results of Bohr's work. As a consequence of quantizing the angular momentum, we found that both the radius of the electron's path and the energy of the electron assume only certain discrete values. Thus, the allowed values of the radius are given by

$$r_n = 0.529\ n^2\ \text{Å}$$

where n = 1, 2, 3, etc.

For $n = 1$, then $r_1 = 0.529$ Å; for $n = 2$, then $r_2 = 0.529 \times 4 = 2.116$ Å, etc.

Table 11.1 shows the allowed values of r_n for the electron. From this we can deduce a picture of the hydrogen atom as an electron going in various possible orbits around a nucleus. The larger the value of n, the greater the distance between the nucleus and the electron. This picture of the atom is sketched in Fig. 11.5.

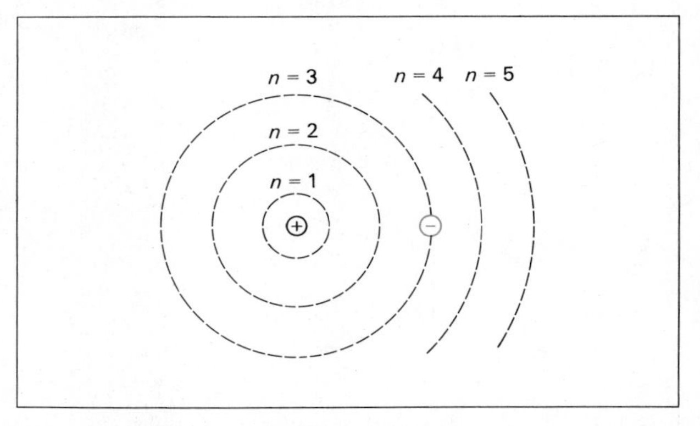

Figure 11.5 A schematic diagram of an atom according to the Bohr theory. The possible orbits are shown as dashed lines. In this diagram, the value of n is three.

The allowed values of the energy of the electron in the hydrogen atom are given by

$$E_n = \frac{-13.60}{n^2}\ \text{eV}$$

where E_n = total energy of the electron for a given value of n,
n = 1, 2, 3, etc.

This equation states that the energy of the electron in the hydrogen atom has only certain discrete values. As examples,

$$E_1 = \frac{-13.60}{1^2}\ \text{eV} = -13.60\ \text{eV}$$

$$E_3 = \frac{-13.60}{3^2}\ \text{eV} = \frac{-13.60}{9}\ \text{eV} = -1.51\ \text{eV}$$

Table 11.1 shows the allowed values of the energy of the electron for low values of n.

A plot of allowed energies, drawn to scale, is called an **energy level diagram.** The energy level diagram for the hydrogen atom is shown in Fig. 11.6. As n increases, the energy increases.

The removal of the electron from an atom is called *ionization,* and the amount of energy required to accomplish the removal is called the **ionization potential.** Thus, the ionization potential is simply the binding energy of a single electron in an atom. It is measured in electron volts. The ionization potential of the hydrogen atom is simply the energy necessary to completely separate the electron and the proton. This value varies and depends on the level (location) of the electron.

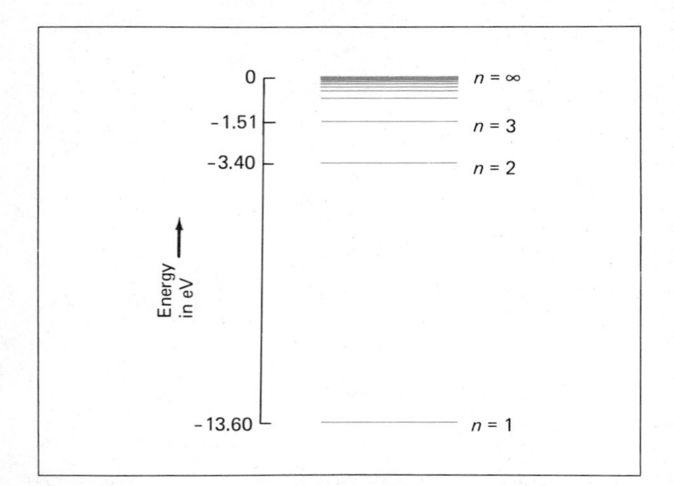

Figure 11.6 Energy level diagram for the hydrogen atom. Only certain energies are possible in the quantum theory.

From Table 11.1 it is evident that the electron will be completely separated from the proton when $n = \infty$. The energy necessary to achieve complete separation is the energy necessary to raise the electron's energy from E_n to zero. So we obtain

$$\text{Ionization energy} = -E_n$$

Thus, the ionization energy for $n = 1$ is $+13.60$ eV. To get the ionization energy for any value of n, simply change the sign of the value for E_n.

From Fig. 11.5 it is easy to see why the ionization energy decreases as n increases. As n increases, the electron gets progressively farther away from the proton, resulting in less and less electric attraction. This results in a decrease in the ionization energy—the energy necessary to completely separate the proton and the electron.

When we speak of the energy of an atom, we are speaking of the system as a whole. For the hydrogen atom the energy can be thought of as the energy of its single electron. With more complicated atoms, we need to consider all of the electrons. The energy levels are then the energy levels for the whole atom.

In the Bohr model of the hydrogen atom, the electron circles the proton with energy values corresponding to certain values of the radius. The quantity n is called the **principal quantum number.** A **quantum number** is a number that specifies a property of the system. For the hydrogen atom, the principal quantum number, n, specifies the energy of the system. In Chapter 12, we will discuss several other quantum numbers.

The description of an atomic system includes its energy, angular momentum, and other properties. The lowest energy level of an atom ($n = 1$) is called the **ground state.** The **excited states** are the states of the atom with energies above the ground state. In the hydrogen atom the state with $n = 1$ is the ground state; $n = 2$ is the first excited state; $n = 3$ is called the second excited state, and so forth.

Let us now consider in more detail the dilemma previously described in the introduction to this chapter. According to the classical theory of electromagnetism, the electron should gradually lose energy as it circles the proton and spiral in to the proton. Bohr's theory postulated that this did not occur in an atom.

By quantizing the angular momentum, Bohr's result was that the electron can occupy only certain discrete energy levels. It does not radiate energy as it circles the proton. The only way in which it can lose energy is to change energy levels. When this happens, the energy lost by the electron is emitted as a photon.

A photon is emitted every time an electron loses energy by changing energy levels. An electron must be in an excited state to lose energy. Once it is in its ground state, it cannot lose any energy. This, then, is Bohr's explanation of why the classical electromagnetic theory does not apply and why atoms do not collapse.

11.4 Photon Emission and Absorption

When an electron changes energy levels, the total energy of the system must be conserved. If the electron is initially in an excited state, it may lose energy by changing to a less excited state. When this happens the electron's energy is carried away by a photon—a light quantum. The total energy *before* must equal the total energy *after* the process, so we write

$$E_{n_i} = E_{n_f} + E_{\text{photon}}$$

where E_{n_i} = initial energy of electron,
E_{n_f} = final energy of electron,
E_{photon} = energy of photon, which is given off.

Thus,
$$E_{\text{photon}} = E_{n_i} - E_{n_f}$$

A schematic diagram of the process of photon emission is shown in Fig. 11.7.

Photon energies can be visualized in terms of the energy level diagram for the hydrogen atom. This is shown in Fig. 11.8. The electron's loss in energy—which is equivalent to the photon energy—is designated by an arrow between energy levels. The photon energy is indicated by the length of the arrow. Some of the possible ways in which an electron can lose energy are shown in Fig. 11.8.

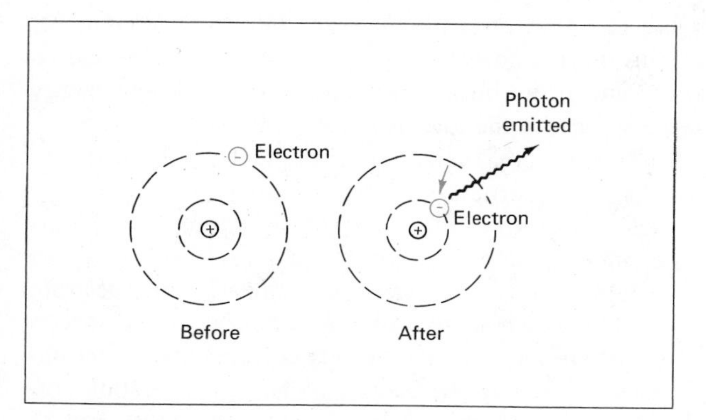

Figure 11.7 A schematic drawing of the de-excitation of an electron in an excited hydrogen atom. A photon is emitted when the electron loses energy.

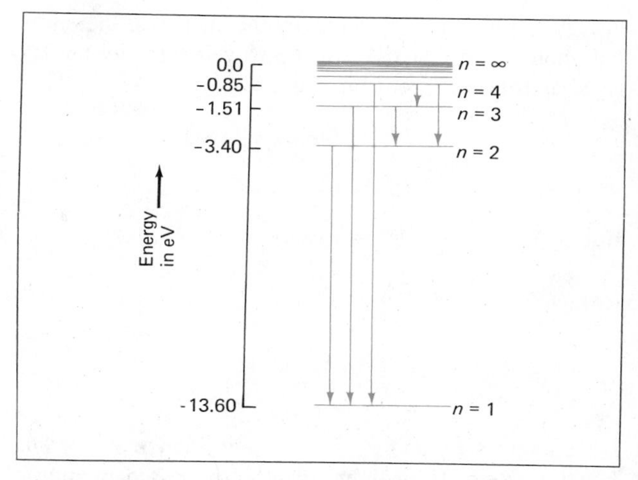

Figure 11.8 Energy level diagram for the hydrogen atom used to show some of the de-excitation energies of the electron. The lengths of the arrows indicate some of the possible energies of the photon.

The inverse reaction can also occur. A photon can be absorbed by an atom. The result is a change in an electron's energy level. A schematic diagram of this process, known as photon absorption, is shown in Fig. 11.9. The changes of energy levels of the electron are just the opposite of those shown in Fig. 11.8. For example, in photon emission an electron might go from $n = 3$ to $n = 2$, thus losing a fixed amount of energy in the form of a photon. In photon absorption a photon with the same fixed amount of energy could be absorbed, causing the electron to go from $n = 2$ to $n = 3$.

The frequency and wavelength of the electromagnetic radiation, which is emitted when an electron changes

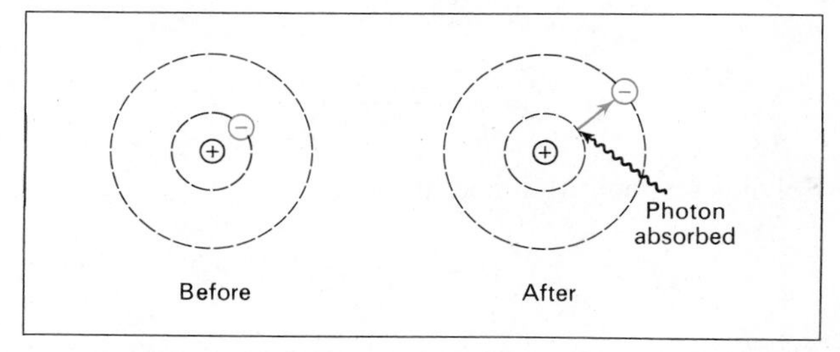

Figure 11.9 A schematic drawing of the absorption of a photon by an atom. The electron gains the photon's energy and goes to a higher energy level.

energy levels, are determined by the energy of the emitted photon. An equation for the wavelength can be derived as follows: From Eq. 11.1

$$hf = E_{\text{photon}}$$

or

$$hf = E_{n_i} - E_{n_f}$$

But since

$$f = c/\lambda$$

we have

$$\frac{hc}{\lambda} = E_{n_i} - E_{n_f}$$

or

$$\lambda = \frac{hc}{E_{n_i} - E_{n_f}}$$

If we express E_{n_i} and E_{n_f} in eV, then Planck's constant h and the velocity of light c can be expressed in appropriate units to give the simple formula

$$\lambda = \frac{12{,}400}{E_{n_i} - E_{n_f}} \text{ Å} \qquad (11.4)$$

To find the wavelength of light in angstroms, simply divide 12,400 by the difference in energy in eV between the initial and final energies of the electron.

EXAMPLE 1 What is the wavelength emitted when an excited electron goes from the $n = 3$ to the $n = 2$ level?

$$E_3 - E_2 = -1.51 \text{ eV} - (-3.40 \text{ eV})$$
$$= 1.89 \text{ eV}$$
$$\lambda = \frac{12{,}400}{1.89} \text{ Å} = 6560 \text{ Å (red)}$$

EXAMPLE 2 A blue-green color (about 4860 Å) is seen to be emitted by excited hydrogen atoms. What are the quantum numbers of the initial and final levels of the electron?

$$\lambda = 4860 \text{ Å}$$
$$= \frac{12{,}400}{E_{n_i} - E_{n_f}(\text{eV})}$$

Solving for the difference in energy

$$E_{n_i} - E_{n_f} = \frac{12{,}400}{4860} = 2.55 \text{ eV}$$

The problem now is to find a transition of 2.55 eV in the hydrogen atom, that is, we need to find two energies which differ by 2.55 eV. Referring to Fig. 11.8, the final quantum number (n_f) cannot be $n_f = 1$ because all transitions to this level have an energy greater than 10.2 eV. The final quantum number cannot be 3 or higher because all transitions to it will have an energy lower than 1.89 eV. So, the final quantum number must be $n_f = 2$. The initial quantum number is $n_i = 4$. Then, $E_{n_i} = -0.85$ eV, $E_{n_f} = -3.40$ eV, and $E_{n_i} - E_{n_f} = -0.85 \text{ eV} - (-3.40 \text{ eV}) = +2.55$ eV.

Visible light is electromagnetic radiation with wavelengths from approximately 4000 Å to 7000 Å, although the exact limits vary from person to person. The wavelength limits correspond to photon energies between 1.77 eV and 3.10 eV. The color we see depends on the wavelength.

Wavelengths of about 10 Å to 4000 Å or photon energies from 3.10 eV to 1240 eV correspond to ultraviolet rays. Wavelengths of about 7000 Å to 5×10^6 Å or photon energies of 0.00248 eV to 1.77 eV correspond to infrared rays. Of course, both ultraviolet and infrared rays are invisible.

A **spectrum** (pl. spectra) is an ordered display of the various wavelengths (or frequencies) produced by a source such as an atom, nucleus, or molecule. A partial spectrum of the hydrogen atom is shown in Fig. 11.10, and the electron transitions are illustrated in Fig. 11.11.

There are three easily visible lines in the hydrogen atom spectrum at 6560, 4860, and 4340 Å. A fourth line at 4100 Å can be seen by some people but not by others. The three readily seen lines are due to transitions from $n_i = 3$, 4, and 5 to $n_f = 2$. The 4100 Å line is from an $n_i = 6$ to $n_f = 2$ transition. The initial and final values of n are given for many of the transitions giving rise to the hydrogen spectrum in Fig. 11.10. These transitions are also shown in Fig. 11.11.

For the hydrogen atom, electron transitions from a higher n to $n_f = 1$ are all in the ultraviolet region. All transitions from $n =$ anything to $n = 3$, 4, 5, and so on, are in the infrared region or, for very low-energy transitions, in the microwave or radio region.

Each system, such as an atom, nucleus, or molecule, has its own particular energy level diagram. Therefore, each system gives off photons of different wavelengths when changing energy levels, giving rise to a spectrum characteristic of that system. By comparing the spectrum of an unknown source, such as a poison, to the spectra of known sources, the various constituents of the unknown source can be identified. This identification procedure is an extremely important and useful tool to chemists, astronomers, and physicists. This specialized branch of physical science is called **spectroscopy.**

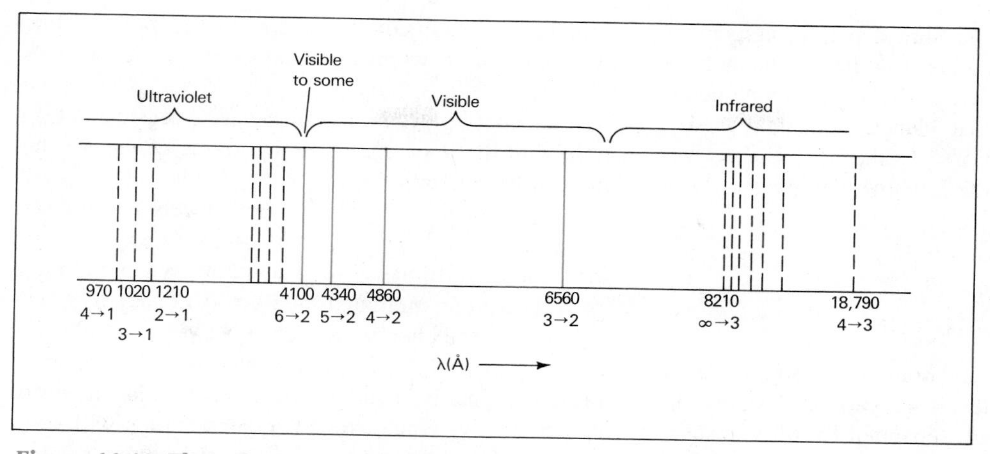

Figure 11.10 The spectrum of the hydrogen atom (not to scale). The initial and final values of n are given under the wavelength designations. Most people can see only three visible lines. All the others are in the infrared or ultraviolet (or other) regions of the spectrum. The line at 4100 Å can be seen by some people, but not by others.

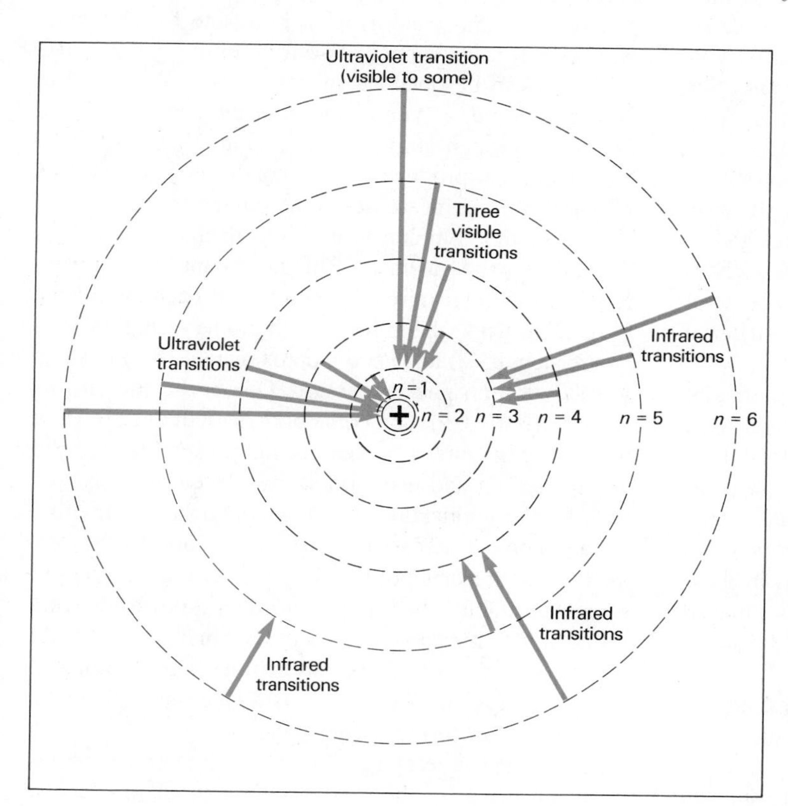

Figure 11.11 The Bohr model of the hydrogen atom illustrating various quantum jumps that give rise to various spectral lines. Transitions that end on $n = 1$ are all in the ultraviolet spectral region. The three easily visible transitions are $n = 3$, 4, and 5 to $n = 2$. All other transitions ending on $n = 2$ are in the ultraviolet region, although some people can detect $n = 6$ to $n = 2$. Transitions ending on $n = 3$, 4, and 5 are in the infrared region of the spectrum.

Spectra of several atoms, molecules, and other systems are shown in Color Plate VIII. The Sun's spectrum, containing Fraunhofer lines, is the third from the bottom. These dark Fraunhofer lines are caused by the various elements and molecules present on the surface of the Sun. By studying its spectrum, the composition of the Sun's surface can be determined.

The crucial test of the Bohr theory was whether it could explain the hydrogen atom spectrum, which was well known in 1913. Bohr's theoretical values for the photon wavelengths were in almost perfect agreement with experimental results. This was the justification of his theory, and it went a long way toward convincing the scientists of his day that Planck's quantum theory had far-reaching ramifications. In the next chapter we shall see that it gave birth to a whole new physics.

11.5 Energy Levels for Other Systems

Much of modern physics and some chemistry is based on the study of energy levels of various systems. In nuclei the energy levels are separated by MeV instead of eV. A gamma ray is a high-energy photon given off when an excited state of a nucleus spontaneously falls to a lower energy state.

Some chemists do their research in the field of molecular spectroscopy—the study of the energy levels of molecules and their associated spectra. Molecules can have quantized energies due to energies of molecular vibrations, rotations, or excited electrons. Of course, different molecules have quite different spectra.

Absorption spectra occur when a light source shines through a gas or liquid. Many of the wavelengths pass through, but photons that have the correct energy to excite the molecules to higher energy levels are absorbed. The absorption of a particular wavelength can be total or partial depending on how much absorptive material is present, and on other factors.

The various gases in the atmosphere absorb particular wavelengths. The main absorbing gases are oxygen (O_2), ozone (O_3), carbon dioxide (CO_2), and water (H_2O). In Chapter 22 the absorption properties of these gases as well as the atmosphere as a whole are discussed.

The water molecule has some rotational energy levels spaced very closely together. Water molecules in these energy levels can easily absorb microwaves—photons with low energy and wavelengths of about 0.05 to 30 cm.

The principle of a microwave oven is based on microwave absorption of energy by water molecules. The water molecules in food absorb microwave radiation, thereby heating and cooking the food, while the interior metal sides of the oven reflect radiation and remain cool.

Because it is the water content that is crucial in heating, materials such as paper plates and ceramic dishes do not get hot immediately in a microwave oven. However, they often warm up because of contact with the hot food.

As a safety precaution, microwave ovens cannot operate with the door open. Human tissue contains water and can be cooked as easily as the food.

Another device based on energy levels is the laser. The word **laser** is an acronym for Light Amplification by Stimulated Emission of Radiation. When a photon interacts with an atom and the photon's energy, hf, exactly equals the energy difference between two allowed energy states of the atom, the photon will produce a transition. That is, some photons will stimulate atoms to absorb energy. Once excited, other photons will stimulate atoms to emit radiation. Such a transition is called stimulated emission. This is in contrast to a normal transition in an excited atom which is called spontaneous emission.

Normally, more atoms exist in a state for absorbing energy than in an excited state for emitting energy. If a condition can be obtained in which more atoms are in an excited state, an amplification of the photons may be produced through stimulated emission. The laser is an instrument that produces such a condition and generates a light beam that possesses some unique properties.

One of the common lasers used in the laboratory is the helium-neon gas laser. This laser consists of a glass tube with parallel mirrors positioned at each end of the glass tube. The tube is filled with atoms of helium (approximately 85%) and neon (approximately 15%) at about 1/1000 of atmospheric pressure. One of the end mirrors is partially silvered, and some photons will emerge from this end of the tube as a beam of light. (See Fig. 11.12.)

The helium and neon atoms are placed in an excited state by the application of a high-frequency alternating voltage applied across the tube. In addition to the normally excited states produced by the addition of energy, some metastable states are produced. A **metastable state** is one in which the excited electrons remain at the higher energy level for a short period of time before falling to a lower energy state. (Most excited electrons fall almost immediately to lower energy levels.)

The electric energy supplied to the helium gas atoms raises electrons in the atoms to higher energy levels. Some of the energy levels are shown in Fig. 11.13. The 20.61 electron volt (eV) level of the helium atom is a metastable state, so there is time for stimulating a neon

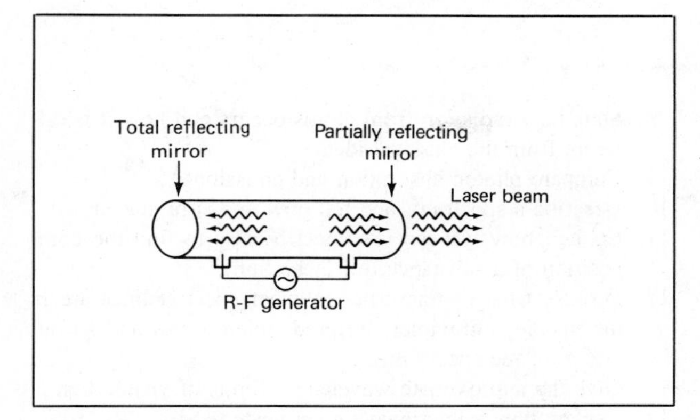

Figure 11.12 Helium-neon gas laser. The mirrors positioned at the ends of the glass tube are perfectly parallel to one another. All of the photons in the laser beam have the same wavelength, phase, and direction of travel.

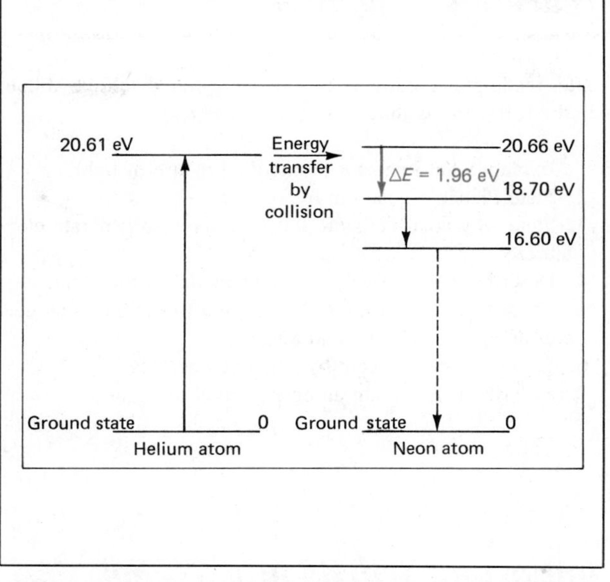

Figure 11.13 Partial energy level diagram for the helium and neon atoms, illustrating the formation of the 6328 Å red light beam. Energy is transferred from the helium 20.61 eV metastable state to the neon 20.66 eV metastable state. The 6328 Å wavelength is produced when the neon electron at the 20.66-eV level falls to the 18.70 eV level.

atom before normal spontaneous emission occurs. The 20.61 eV level of the helium atom is close enough to the 20.66 eV metastable state of the neon atom for stimulation and energy transfer to take place by atomic collisions.

This stimulation of the neon atoms produces many more electrons at the 20.66 eV level than at the 18.70 eV level. When the electrons at the 20.66 eV energy level fall to the 18.70 eV level, electromagnetic energy is radiated. The wavelength of the radiation is 6328 Å, which is in the red portion of the electromagnetic spectrum.

As the 6328 Å photons bounce back and forth between the end mirrors, they stimulate the emission of more 6328 Å photons, all in exactly the same phase and traveling in the same direction. Thus, a set of waves with a wavelength of 6328 Å is generated.

Some of the radiation passes out the end of the glass tube through the partially silvered mirror and emerges as a beam of light that retains its intensity over long distances because it diverges very little as it propagates through space.

A laser is not an energy source. Energy must be supplied to the laser. The laser converts the supplied energy into a beam of electromagnetic radiation that has three important properties, as illustrated in Fig. 11.12. All of the photons or waves have the same wavelength, phase, and direction. Waves with the same phase are called **coherent.** Thus, a beam of laser light is monochromatic, coherent, and unidirectional. Because a laser beam is so directional, it spreads very little. This has permitted a laser beam to be reflected back to Earth from a mirror placed on the moon by astronauts. Such experiments are used to measure accurately the distance between the Earth and the moon so that small fluctuations in the moon's orbit can be studied.

The unique properties of the laser beam allow a wide range of other applications. Currently the most common use of the laser is in telecommunications. In the medical profession, lasers are used as a diagnostic tool, in therapy, in surgery, in research, and for the detection and treatment of cancer, especially in the lungs.

The intense heat produced by the laser light beam over a small area can drill very small holes in hard materials and can weld and machine metals. Lasers are used in photography and holography and in optical data-storage systems. Lasers are also used in surveying and weapons systems, for purification in chemical processing, and annealing in the fabrication of large electronic circuit arrays, and in many research investigations.

Learning Objectives

After reading and studying this chapter, you should be able to do the following without referring to the text:

1. Explain what is meant by the dual nature of light.
2. State Planck's quantum hypothesis.
3. State why solar cells are not used more to generate electricity.
4. Describe quantitatively the Bohr model of the atom.
5. Explain why electron volts are used to measure energies of atoms, molecules, and nuclei.
6. Explain what is meant by negative energies.
7. Describe and explain an energy level diagram.
8. State how radiation from atoms occurs and how it is different from the classical idea.
9. Compare photon absorption and emission.
10. Describe a spectrum, and tell how one is produced.
11. Explain how we can use spectroscopy to find the composition of a substance (e.g., the Sun).
12. Discuss what distinguishes whether spectral lines are in the visible, ultraviolet, infrared, microwave, and so on, region of the spectrum.
13. Give the approximate wavelength limits of visible light.
14. Explain how a microwave oven heats food.
15. Describe the three main features of a laser.

Important Words and Terms

quantum
Planck's constant
photon
dual nature of light
photoelectric effect
solar cells
centripetal acceleration
potential well
quantize the angular momentum
energy level diagram
ionization potential
principal quantum number
quantum number
ground state
excited states
spectrum
spectroscopy
laser
metastable state
coherent

Questions

1. Is light a wave or a particle? Explain.
2. Einstein said in 1905 that light consists of "particles" called photons. What proof can you give that light consists of waves?
3. On what principle are solar cells based?
4. Why aren't solar cells used widely to generate home electricity?
5. Some hydrogen atoms have one proton and one neutron in the nucleus. How will this neutron affect Bohr's theory?
6. How big is an atom?
7. Why are the energies of the hydrogen atom all negative?
8. What is the SI unit equivalent of (a) 1 Å, (b) 1 eV?
9. Why does the ionization potential decrease as n increases?
10. Why are nuclear explosives so much more powerful than chemical explosives? (Hint: consider the binding energies of atoms and nuclei.)
11. Are the photon wavelengths absorbed by hydrogen atoms the same as those emitted?
12. What was Bohr's explanation of why the electron orbiting the proton doesn't lose energy and spiral into the proton?
13. What are the wavelength limits of visible light?
14. How are the visible lines in the hydrogen spectrum produced?
15. How do we know what the Sun is made of?
16. Why does a microwave oven heat a potato but not a ceramic plate?
17. What are the main features of a laser?

Problems

1. The human eye is most sensitive to light of approximately 5500 Å.
   (a) Determine the frequency of this electromagnetic radiation.
   (b) What is the energy in joules of such a photon?
2. Which has more energy, a photon of red light, or a photon of blue light? Give proof for your answer.
3. What is the radius of a hydrogen atom for each of the following?
   (a) $n = 2$, (b) $n = 6$ (c) $n = 10$.
   Answer: (c) 52.9 Å
4. What is the energy in electron volts of a hydrogen atom for each of the following?
   (a) $n = 2$, (b) $n = 6$, (c) $n = 10$.
   Answer: (c) -0.136 eV
5. What is the ionization energy in electron volts of a hydrogen atom for each of the following?
   (a) $n = 2$, (b) $n = 6$, (c) $n = 10$.
   Answer: (c) $+ 0.136$ eV
6. What is the energy in eV emitted in the following transitions of an electron in the hydrogen atom?
   (a) $n = 4$ to $n = 2$, (b) $n = 6$ to $n = 2$, (c) $n = 3$ to $n = 1$, (d) $n = 4$ to $n = 3$.
   Answer: (b) 3.02 eV
7. (a) What are the wavelengths corresponding to the transitions in problem 6?
   (b) Which are visible? (c) Which are in the ultraviolet region and which are in the infrared?
   Answer: (a) $n = 6$ to $n = 2$, $\lambda = 4100$ Å
8. A hydrogen atom absorbs a photon of wavelength $\lambda = 4340$ Å. (a) How much energy did the atom absorb? (b) What were the initial and final states of the electron in the hydrogen atom?
   Answer: (b) $n = 2$ to $n = 5$
9. What is the shortest wavelength present in the hydrogen atom spectrum?
10. Calculate the wavelength in Å of the radiation emitted when an excited electron in the neon atom falls from the 20.66 eV level to the 18.70 eV level.
11. There is a spontaneous emission from the 18.70 eV level to the ground state of the neon atom. Determine the wavelength of the radiation produced.
12. Determine the wavelength of a photon that has kinetic energy of 1 eV.
13. When an electron in the hydrogen atom falls from the first excited state to the ground state, radiation of frequency 2.5×10^{15} Hz is emitted. (a) Calculate the wavelength in Å; (b) Calculate the energy in electron volts.
    Answer: (a) 1200 Å
14. Determine the wavelength of a gamma ray with an energy of 4 MeV.
15. Determine the energy of a microwave photon with a wavelength of 1 cm.
    Answer: 1.24×10^{-4} eV

12

Quantum Mechanics

Matter waves is a concept conceived in the human mind and interpreted by the minds of others.

THE BEAUTIFUL AGREEMENT between Bohr's theory and the experimental hydrogen spectrum indicated that Planck's quantum hypothesis was correct. However, Bohr's theory could not adequately explain any of the elements with two or more electrons.

The great Russian chemist, Dmitri Mendeleev, had established an orderly array of the various known elements in 1868, but the theoretical basis of this arrangement was not known as late as 1925. In 1925 a whole new kind of physics, based on the quantum idea, was born. This new physics, called **quantum mechanics,** revolutionized our understanding of the submicroscopic aspects of nature. Using this new physics, the theoretical basis of Mendeleev's periodic table of the elements was discovered, and the stage was set for our modern understanding of all the physical sciences.

Quantum mechanics has applications in astronomy, nuclear physics, chemistry, meteorology, geology, and other fields of science. But the quantum idea has a much wider scope than that of science. It presented us with a precise formulation of the uncertainty involved in making any measurement. The impossibility of ever knowing anything exactly was the dawn of a new era in philosophy and theology. The old idea of a mechanistic universe, in which all things moved according to nature's laws, was replaced by a new concept of probability, with every measurement being, at least to some degree, uncertain. In this chapter some of the basic ideas of quantum mechanics, such as the uncertainty principle, will be discussed, and the modern understanding of atoms will be described.

◀ The patterns of exploding fireworks are reminiscent of the patterns made by electrons in atoms and molecules. By studying quantum mechanics, scientists seek to understand the patterns of nature.

12.1 Matter Waves

The concept of light acting both as a wave and as a particle was disturbing to physicists in the 1920s. This problem spurred the thinking of Louis de Broglie (Fig. 12.1), a French physicist, who in 1925 postulated that matter, as well as light, has properties of both waves and particles. According to the hypothesis of de Broglie, any moving particle has a wavelength, which is given by the formula

$$\lambda = \frac{h}{mv} \tag{12.1}$$

where λ = wavelength of moving particle,
m = mass of moving particle,
v = velocity of moving particle,
h = Planck's constant.

This is truly a revolutionary equation! De Broglie's hypothesis states that everything that moves has a wavelength associated with it. The waves produced by moving particles are called **matter waves.**

EXAMPLE 1 What is the wavelength of an electron traveling at 3×10^6 m/s?

$$\lambda = \frac{h}{mv}$$

$$\lambda = \frac{6.625 \times 10^{-34}\ \text{J}\cdot\text{s}}{9.1 \times 10^{-31}\ \text{kg} \times 3 \times 10^{6}\ \text{m/s}}$$

$$\lambda = 2.4 \times 10^{-10}\ \text{m} = 2.4\ \text{Å}$$

De Broglie's equation was met with skepticism at first, but it was experimentally verified in 1927 by G. Davisson and L. H. Germer in the United States. They

Figure 12.1 Louis de Broglie (1892–). Prince Louis Victor de Broglie, a French nobleman, studied medieval history at the Sorbonne. He enlisted in the French army and was a radioman in World War I. This experience created an interest in physics, and he presented his concept of matter waves in 1925 in a doctoral thesis. It was met with much skepticism at first; one cynic dubbed his theory "la comédie française." When it was shown that electrons can produce a diffraction pattern, however, de Broglie had the last laugh. He was awarded the Nobel Prize in physics in 1929. (Photo courtesy American Institute of Physics, Niels Bohr Library.)

showed that a beam of electrons exhibits a diffraction pattern. Since diffraction is a wave phenomenon, a beam of electrons must have wavelike properties, such as frequency and wavelength. To show that something diffracts, it is necessary to have the wave pass through a slit whose width is approximately the size of the wavelength. Visible light has wavelengths of about 4000–7000 Å and slits with widths of this size can be made quite easily. Electrons, as was seen in the example, have wavelengths of about 2 Å. Slits of this width cannot be made. However, nature has provided us with suitable slits in the form of crystal lattices. The atoms in these crystals are arranged in rows (or other orderly arrangements), and the rows of atoms make natural "slits."

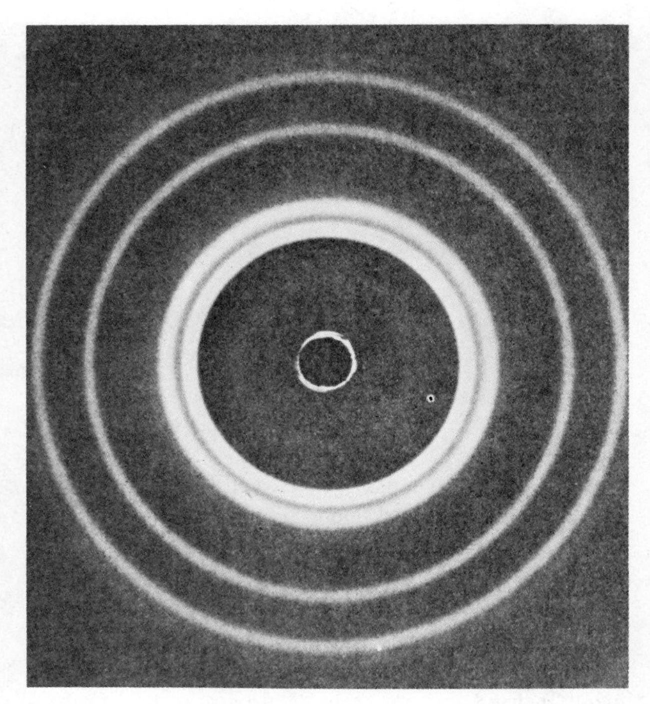

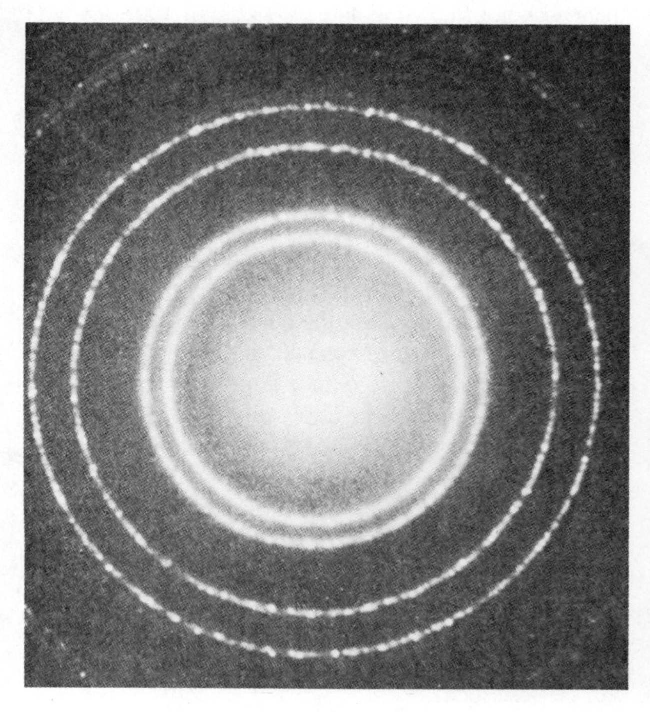

Figure 12.2 The two photographs above show diffraction patterns produced by X rays and electrons. The pattern at the top was made by X rays. The bottom pattern was made by electrons.

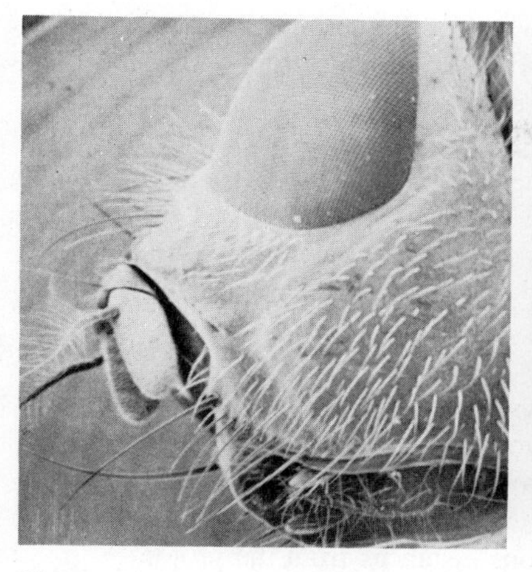

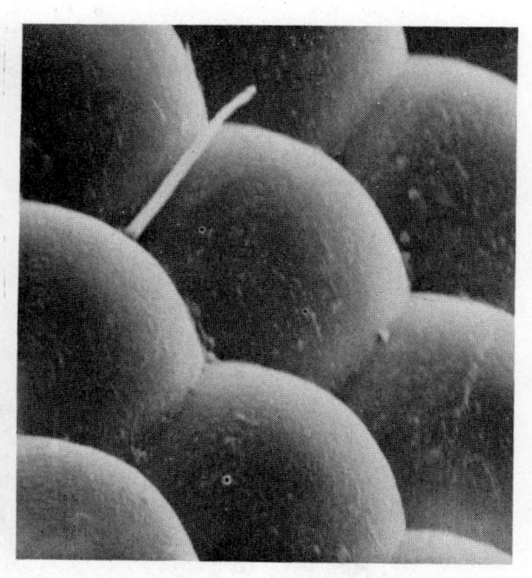

Figure 12.3 A series of electron microscope photographs showing a fly's eye. (Courtesy Richard Garten, Ohio University.)

Davisson and Germer bombarded nickel crystals with electrons and obtained a diffraction pattern on a photographic plate. Diffraction patterns made by shooting X rays (electromagnetic radiation of very short wavelength) and an electron beam at thin aluminum foil are shown in Fig. 12.2. (The top pattern was made by X rays.) The similarity in the diffraction patterns from the X rays and the electrons is evident. The fact that electrons diffract demonstrates that there is a dual nature of matter. Electrons and other particles such as protons and neutrons have wave properties similar to those of light.

An electron microscope uses the concept of matter waves by directing a beam of electrons rather than a beam of light on an object. In one technique the beam of electrons bounces off a surface. The beam spot is scanned across the specimen by means of deflecting coils, much as is done in a television tube. Surface irregularities cause directional variation in the intensity of the reflected electron beam that gives contrast to the image.

The amount of fuzziness of an image due to diffraction effects is proportional to the wavelength that is used. A typical wavelength in an electron microscope is 10 Å. This is a very short wavelength compared to visible light at $\lambda \approx 5000$ Å, so very high resolution photographs can be taken. Some examples are shown in Fig. 12.3.

12.2 The Schrödinger Equation

The next big step toward understanding the nature of atoms and nuclei was taken in 1926 when Erwin Schrödinger (Fig. 12.4, p. 194), an Austrian physicist, presented a widely applicable mathematical equation that described the matter waves of de Broglie. This equation is known as the **Schrödinger equation.** Understanding this formula takes a considerable background in the mathematics of calculus. However, the basic structure of the equation is interesting, and for curiosity's sake, let us see what it looks like. The Schrödinger equation is

$$\left(-\frac{h^2}{8\pi^2 m}\nabla^2 + E_p\right)\psi = E\psi$$

where m = mass of particle,
h = Planck's constant,
E_p = potential energy,
E = total energy,
∇^2 (pronounced "del squared") is a calculus quantity,
ψ (pronounced "sigh") is the wavefunction (to be discussed later).

The first quantity in Schrödinger's equation is related to the kinetic energy.

$$E_k = -\frac{h^2}{8\pi^2 m}\nabla^2$$

where E_k = kinetic energy. Thus, Schrödinger's equation is simply

$$(E_k + E_p)\psi = E\psi$$

So, we see that the equation can be thought of as a statement of the well-known principle that kinetic energy plus potential energy equals the total energy.

Figure 12.4 Erwin Schrödinger (1887–1961), along with Heisenberg, was one of the founders of quantum mechanics. Born in Austria, he served in World War I, afterward becoming a physics professor in Germany. In 1925 he wrote his paper which mathematically treated moving particles as waves. In 1933 he left Germany when the Nazis came to power. Later, he was a professor of physics at Dublin. (Photo courtesy Pfaundler.)

Schrödinger's equation is the basis of practically all our present knowledge of atomic and nuclear physics. It is so basic that it would be difficult to overestimate its importance. While it may look simple, using it, except in a very few cases, either is impossible or requires a high-speed computer.

One simple case that was solved almost immediately using Schrödinger's equation was that of the hydrogen atom. The possible energies were found to be exactly the same as Bohr had obtained in 1913. But additional results also came out when Schrödinger's equation was used. These additional results concerned the quantity ψ, called the **wavefunction.**

At first, it was not clear how ψ should be interpreted. After much thought and debate it was decided that ψ^2 (the wavefunction squared) represented the *probability* that the electron would be a certain distance from the nucleus. In Bohr's theory, the electron is a particle and can exist only at certain discrete distances from the proton. These distances are known as the Bohr radii and were given in Chapter 11 as $r = 0.529n^2$ Å. The Schrödinger wave equation altered the idea that the electron was a particle existing at a discrete distance from the proton. This idea was replaced by one that pictures an electron at any distance from the proton, but it is most likely to be at the radius given by Bohr.

Fig. 12.5 shows the probability, given by ψ^2, that an electron in the ground state will be at a given radius. The maximum probability occurs at the Bohr radius of 0.529 Å, but we see now that it is possible for the electron to be at various distances from the proton.

The idea that an electron has a certain probability of being at any given distance from the proton was quite new in 1926. To illustrate it, let us again refer to Fig. 12.5. If you look near zero radius (near the proton), the curve shows that there is not much chance that you will find the electron. As the radius increases, the greater is the probability, until you reach 0.529 Å. Here, at the Bohr radius, is the maximum probability. The probability then decreases as you continue to go out from the proton, until past about 2 Å, where there is very little probability that you will find it. However, at any distance, no matter how near or far from the proton, there is a chance that the electron will be there.

With the advent of quantum physics our view of nature takes a decided turn. No longer are things definite. Instead, nature acquires a look of probability. Some things are more probable than others; but, at least on an atomic or nuclear scale, almost anything is possible.

Consider, for example, the probability that you and your chair suddenly lift off the ground by 0.5 cm for 0.1 second. This could happen because for a brief mo-

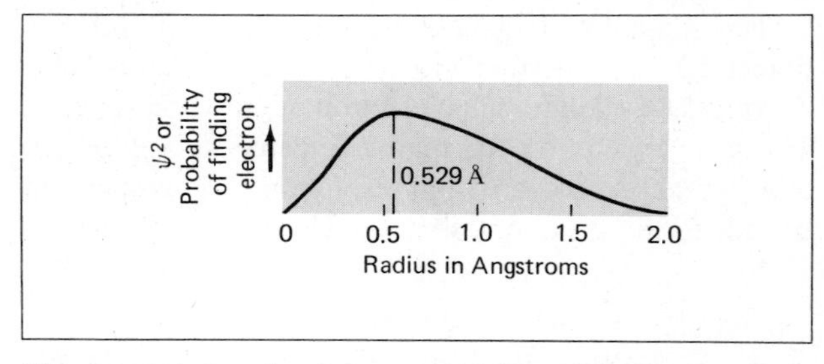

Figure 12.5 Graph of the probability of finding an electron at a given radius in the hydrogen atom in the ground state. There is a probability that the electron will be at any radius, but the greatest probability occurs at the Bohr radius of 0.529 Å.

ment fast-moving molecules hit the bottom of your chair while slow-moving molecules hit the top. Your claim of levitation would certainly be possible, but it would be so improbable that no one would believe you, and you could not repeat the event. In such rare cases, it is always more likely that the observers are lying about what happened than that the event actually occurred.

Many such occurrences in physics have very small probabilities—e.g., one chance in 10^{2000} of occurring in one year. When probabilities get this low, many would say that they are impossible. Strictly speaking, most things are not impossible, just very improbable.

12.3 The Heisenberg Uncertainty Principle

Shortly after Schrödinger's wave equation was published, Werner Heisenberg (Fig. 12.6), a German physicist, proposed what is now known as the Heisenberg uncertainty principle. The **Heisenberg uncertainty principle** states that it is impossible to know *simultaneously* the exact velocity and position of even a single particle.

The Heisenberg uncertainty principle is based on the process of making a measurement. Let us now examine this process. In order for us to see an object, photons of light reflected from the object must enter our eye. Let us consider the example of recording, visually, the position of an electron, illustrated in Fig. 12.7 (p. 196).

In order to see the electron, a photon must come from the light source and bounce off the electron into an eye or other recording device. But photons carry energy and momentum. This latter quantity must be conserved when the photon collides with the electron. Since the direction and magnitude of the photon's momentum have changed as a result of hitting the electron, the electron must have gained some momentum, thereby altering its velocity. Since the electron's velocity has changed slightly, we don't know where it will be in the future, because we don't know its new velocity. We can record only where the electron was when the photon hit it.

Heisenberg formulated his uncertainty principle mathematically. In order to understand the mathematical formulation, let us recall how we record measurements in the laboratory. If we record the length of an object measured with a meter stick, we must tell what the possible limits on our measurement are. As an example, let us say we obtain a length of 10.32 ± 0.03 m. That is, the length is somewhere between 10.29 and 10.35 m,

Figure 12.6 Werner Heisenberg (1901–1976). At the age of 23, Heisenberg wrote a paper explaining quantum mechanics. His theory was entirely different from Schrödinger's, yet it produced the same results. Schrödinger later wrote a paper showing that his and Heisenberg's theories were mathematically identical. Heisenberg's name is attached to the uncertainty principle which he conceived as a result of his quantum theory. During World War II, Heisenberg was in charge of the German nuclear energy program. (Photo courtesy American Institute of Physics, Niels Bohr Library.)

or we have an uncertainty of 0.06 m in our measurement. Heisenberg's uncertainty principle is

$$m \,\Delta v \,\Delta x = h/2\pi \qquad (12.2)$$

where m = mass of object being measured,
Δv = minimum uncertainty in velocity,
Δx = minimum uncertainty in position,
h = Planck's constant (6.625×10^{-34} J · s).

We see, then, that the Heisenberg uncertainty principle incorporates the minimum uncertainties of velocity and position as limitations. This is the best that we can ever hope to do. Most measurements are much more

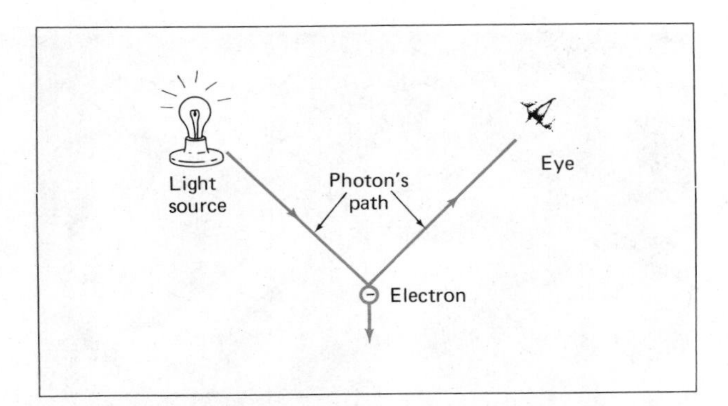

Figure 12.7 Recording the position of an electron as an illustration of the process of measurement. The photon of light bounces off the electron before it hits the eye. In so doing, it changes the velocity of the electron.

uncertain than these limits. This principle says that even if we have the most precise instruments possible, these minimum uncertainties would always exist because of the process of measurement.

From the equation it is evident that the minimum uncertainties are smaller for large masses than for small masses. Minimum uncertainties become significant only when we are dealing with very small masses, such as electrons or protons. This is because h is such a small quantity. The following example illustrates the importance of the Heisenberg uncertainty principle when dealing with electrons.

EXAMPLE 2 If the uncertainty in the velocity of an electron is 3×10^6 m/s, what is the minimum uncertainty in its position?

$$m \, \Delta v \, \Delta x = h/2\pi$$

$$(9.1 \times 10^{-31} \text{ kg})(3 \times 10^6 \text{ m/s}) \, \Delta x = \frac{6.625 \times 10^{-34}}{2(3.14)} \text{ J} \cdot \text{s}$$

Since all the units are in the mks system, we know that the answer will be in meters. Solving for Δx, the minimum uncertainty in position, we get

$$\Delta x = \frac{6.625 \times 10^{-34}}{6.28 \times 3 \times 10^6 \times 9.1 \times 10^{-31}}$$

$$= 0.386 \times 10^{-10} \text{ m} = 0.386 \text{ Å}$$

This uncertainty is rather large, when we consider that the Bohr radius is 0.529 Å.

As stated above, the uncertainty principle is a precisely formulated physical law. It can also be thought of as a statement of the fact that, on a submicroscopic scale, the very process of measurement alters what is being measured. In the social sciences this same phenomenon is met on a macroscopic scale. It is very difficult, if not impossible, to measure human behavior without affecting the behavior. For instance, the presence of news reporters at the scene of a disturbance frequently changes the very nature of the disturbance.

There are some important philosophical implications of Heisenberg's uncertainty principle. After Newton's equations of motion were published in the seventeenth century, there was considerable speculation as to whether or not the future was automatically determined by the present. Using Newton's equations of motion, the resulting motions of large objects (e.g., planets) could be determined, if it were known how fast and in what direction the objects were traveling and if the forces acting on them were known. Therefore, it was reasoned, if the velocities and positions of all the electrons, protons, and neutrons in the universe were known at some present instant of time, their future velocities and positions could be determined completely. This led some philosophers to envision a deterministic universe, that is, one whose future could never be changed.

The Heisenberg uncertainty principle refutes this view of determinism. It puts a definite limit on our ability to predict the future of a small particle, such as an electron. On a more cosmic scale, the Heisenberg uncertainty principle says that the future can never be determined by our knowledge of the present. This has revolutionized the philosopher's view of the universe.

12.4 Atomic Quantum Numbers

When the Schrödinger equation is solved for the hydrogen atom, several quantum numbers result. One of these is $\boldsymbol{n}$, the principal quantum number, which occurred in the Bohr theory. However, several other such numbers also arise, each with its limitations. The quantum number ℓ is called the orbital quantum number. It is associated with the value of the angular momentum of the electron. (Recall that the angular momentum of a particle going in a circle is given by its mass times its velocity times the radius of the circle.) The quantum number ℓ can be 0, 1, 2, 3, . . . , but it must always be less than the value for n, the principal quantum number.

EXAMPLE 3 If the principal quantum number, n, is 1, what are the possible values of ℓ?

If $n = 1$, ℓ must $= 0$, since ℓ must be less than n.

EXAMPLE 4 If $n = 3$, what are the possible values of ℓ?

If $n = 3$, ℓ can be 0, 1, or 2.

The value of the quantum number ℓ is usually designated by a letter for notational convenience. Historically, these letters are *s, p, d, f, g*, etc., for $\ell = 0, 1, 2, 3, 4$, etc., respectively. (The first four letters originally stood for the words sharp, principal, diffuse, and fundamental. These words were used by early spectroscopists.) That is,

$\ell = 0$ is designated by *s*
$\ell = 1$ is designated by *p*
$\ell = 2$ is designated by *d*
$\ell = 3$ is designated by *f*
$\ell = 4$ is designated by *g*

After *f* and *g*, the letters continue in alphabetical order for higher values of ℓ. However, in this book we shall have no occasion to use values of ℓ above 3.

A third quantum number is the magnetic quantum number designated by the symbol $\boldsymbol{m}$. The magnetic quantum number governs the orientation of the angular momentum and structures the shape of the region that the electron occupies. This number can be any positive or negative integer from $-\ell$ to $+\ell$, including zero. Therefore, there can be $2\ell + 1$ values of m. As an example, if $\ell = 2$, there are five possible values of m.

EXAMPLE 5 If $\ell = 0$, what are the possible values of m?

If $\ell = 0$, $m = 0$.

EXAMPLE 6 If $\ell = 2$, what are the possible values of m?

If $\ell = 2$, m can $= -2, -1, 0, +1$, or $+2$.

The fourth and last quantum number is the spin magnetic quantum number designated by the symbol $\boldsymbol{m_s}$. As the electron orbits around the proton, it also spins on its axis. This is analogous to the Earth's spinning as it revolves around the Sun. The electron can spin in one of two directions, just as a top can spin clockwise or counterclockwise. The quantum number m_s is arbitrarily designated either $+\frac{1}{2}$ or $-\frac{1}{2}$, depending on how the electron is spinning. No matter what the other three quantum numbers (n, ℓ, and m) are, the quantum number m_s is always either $+\frac{1}{2}$ or $-\frac{1}{2}$, for each value of m. The four quantum numbers are summarized in Table 12.1.

Relativity affects quantum mechanics in a surprising way. The quantum numbers n, ℓ, and m all come out of the solution of the Schrödinger equation for the hydrogen atom. The Schrödinger equation is a nonrelativistic wave equation with $E_k = \frac{1}{2}mv^2$. When a relativistic wave equation is used with a relativistic equation for E_k, the solution yields the additional quantum number m_s.

The quantum number m stands for the word "magnetic" and the s in m_s stands for "spin." Remember that a moving or spinning charge creates a magnetic field. The $+\frac{1}{2}$ spin of the electron is a way of saying the electron acts like a small bar magnet with the N-pole up; the $-\frac{1}{2}$ spin signifies the N-pole is down. This property of the electron, that it is a very tiny magnet, comes, amazingly, directly from the solution of the relativistic wave equation.

For the hydrogen atom, only the principal quantum

Table 12.1 Quantum Numbers for an Atomic Electron

Designation	*Meaning*	*Possible Values*
n	Principal quantum number designates in general the effective volume and determines energy	Any positive integer: 1, 2, 3, . . .
ℓ	Orbital quantum number governs magnitude of angular momentum and also helps to determine energy	Any nonnegative integer less than n: 0, 1, 2, . . . , $n - 1$
m	Magnetic quantum number governs the orientation of angular momentum	$-\ell$ to $+\ell$ in integer steps (including zero)
m_s	Spin magnetic quantum number either counterclockwise or clockwise	$+\frac{1}{2}$ or $-\frac{1}{2}$

number, n, determines the energy of the electron. However, for all other elements, those with two or more electrons, the quantum number ℓ also is necessary to designate the value of the energy of each electron. The quantum numbers m and m_s have nothing to do with the energy of the orbiting electron.

Since only n and ℓ affect the value of the energy of the electron, each energy level is labeled by these two quantities. The notation used is to write a number that stands for the value of n followed by a letter that stands for the value of ℓ. As examples, 1s means the energy level when $n = 1$ and $\ell = 0$; the 3d level means $n = 3$ and $\ell = 2$; 4p means $n = 4$ and $\ell = 1$. For each energy level there are various sets of all four quantum numbers possible, because m and m_s can have different values.

EXAMPLE 7 What are the possible sets of quantum numbers that an electron can have in the 3p energy level?

The 3 indicates the principal quantum number ($n = 3$); the p means $\ell = 1$, so m can be $+1$, 0, -1; and m_s can be $+\frac{1}{2}$ or $-\frac{1}{2}$ for each value of m. So there are six sets of quantum numbers. See Fig. 12.8.

Table 12.2 shows various sets of quantum numbers that are possible for some energy levels of interest. From this table we see also that the number of different sets of quantum numbers depends only on the value of ℓ.

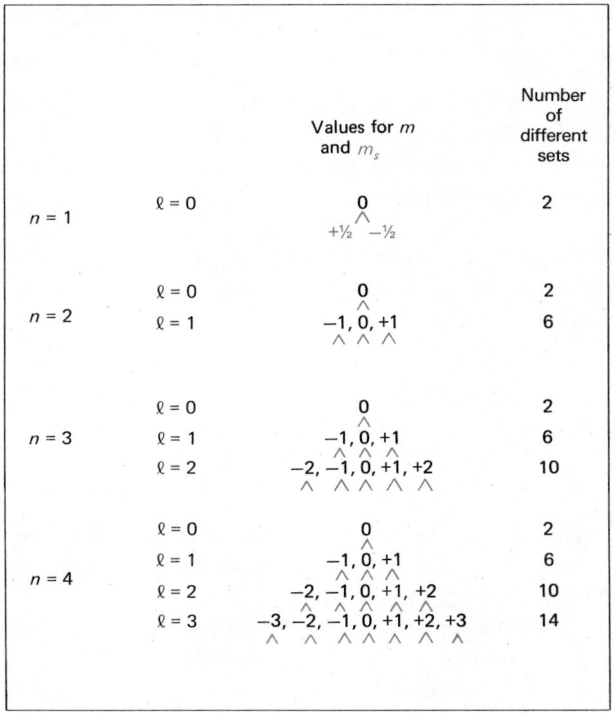

		Values for m and m_s	Number of different sets
$n = 1$	$\ell = 0$	0 (+½ −½)	2
$n = 2$	$\ell = 0$	0	2
	$\ell = 1$	−1, 0, +1	6
$n = 3$	$\ell = 0$	0	2
	$\ell = 1$	−1, 0, +1	6
	$\ell = 2$	−2, −1, 0, +1, +2	10
$n = 4$	$\ell = 0$	0	2
	$\ell = 1$	−1, 0, +1	6
	$\ell = 2$	−2, −1, 0, +1, +2	10
	$\ell = 3$	−3, −2, −1, 0, +1, +2, +3	14

Figure 12.8 Diagram illustrating the quantum numbers possible for $n = 1$ through $n = 4$. Note that the number of different sets increases by four for each additional value of ℓ. The m_s values of $+\frac{1}{2}$ or $-\frac{1}{2}$ are shown for $n = 1$ only. All others are represented by the symbol $\wedge$.

Table 12.2 *Possible Quantum Numbers for Some Energy Levels in Atoms*
(Note that after the 4p level the different sets are only indicated.)

Energy Level	n	ℓ	m	m_s	*Number of Different Sets*
1s	1	0	0	$+\frac{1}{2}$	
	1	0	0	$-\frac{1}{2}$	2
2s	2	0	0	$+\frac{1}{2}$	
	2	0	0	$-\frac{1}{2}$	2
2p	2	1	$+1$	$+\frac{1}{2}$	
	2	1	$+1$	$-\frac{1}{2}$	
	2	1	0	$+\frac{1}{2}$	
	2	1	0	$-\frac{1}{2}$	
	2	1	-1	$+\frac{1}{2}$	
	2	1	-1	$-\frac{1}{2}$	6
3s	3	0	0	$+\frac{1}{2}$	
	3	0	0	$-\frac{1}{2}$	2

Continued on following page

Table 12.2 Continued

Energy Level	n	ℓ	m	m_s	*Number of Different Sets*
3*p*	3	1	+1	$+\frac{1}{2}$	
	3	1	+1	$-\frac{1}{2}$	
	3	1	0	$+\frac{1}{2}$	
	3	1	0	$-\frac{1}{2}$	
	3	1	−1	$+\frac{1}{2}$	
	3	1	−1	$-\frac{1}{2}$	6
4*s*	4	0	0	$+\frac{1}{2}$	
	4	0	0	$-\frac{1}{2}$	2
3*d*	3	2	+2	$+\frac{1}{2}$	
	3	2	+2	$-\frac{1}{2}$	
	3	2	+1	$+\frac{1}{2}$	
	3	2	+1	$-\frac{1}{2}$	
	3	2	0	$+\frac{1}{2}$	
	3	2	0	$-\frac{1}{2}$	
	3	2	−1	$+\frac{1}{2}$	
	3	2	−1	$-\frac{1}{2}$	
	3	2	−2	$+\frac{1}{2}$	
	3	2	−2	$-\frac{1}{2}$	10
4*p*	4	1	+1	$+\frac{1}{2}$	
	4	1	+1	$-\frac{1}{2}$	
	4	1	0	$+\frac{1}{2}$	
	4	1	0	$-\frac{1}{2}$	
	4	1	−1	$+\frac{1}{2}$	
	4	1	−1	$-\frac{1}{2}$	6
5*s*	5	0	0	$+\frac{1}{2}, -\frac{1}{2}$	2
4*d*	4	2	+2, +1, 0, −1, −2	$+\frac{1}{2}, -\frac{1}{2}$	10
5*p*	5	1	+1, 0, −1	$+\frac{1}{2}, -\frac{1}{2}$	6
6*s*	6	0	0	$+\frac{1}{2}, -\frac{1}{2}$	2
4*f*	4	3	+3, +2, +1, 0, −1, −2, −3	$+\frac{1}{2}, -\frac{1}{2}$	14
5*d*	5	2	+2, +1, 0, −1, −2	$+\frac{1}{2}, -\frac{1}{2}$	10
6*p*	6	1	+1, 0, −1	$+\frac{1}{2}, -\frac{1}{2}$	6
7*s*	7	0	0	$+\frac{1}{2}, -\frac{1}{2}$	2
5*f*	5	3	+3, +2, +1, 0, −1, −2, −3	$+\frac{1}{2}, -\frac{1}{2}$	14
6*d*	6	2	+2, +1, 0, −1, −2	$+\frac{1}{2}, -\frac{1}{2}$	10

When $\ell = 0$ (s levels), there are 2 different sets of quantum numbers possible. When $\ell = 1, 2,$ or 3 (p, d, f), there are 6, 10, or 14 sets possible. The significance of the number of possible sets will be discussed in Section 12.5.

Let us now consider the energy level diagram for the hydrogen atom, shown in Fig. 12.9. This diagram is similar to Fig. 11.6 of Chapter 11 except that both n, and ℓ quantum numbers, rather than only n, are indicated. Also, in Fig. 12.9 the number of different sets of quantum numbers is shown in parentheses on the line designating the level. Figure 12.9 is not drawn to scale; this is done so that the higher levels can be identified.

The energy situation for atoms other than hydrogen is not so simple. The energies of these atoms depend on both n and ℓ. The energy levels in order of increasing energy are as follows: 1*s*, 2*s*, 2*p*, 3*s*, 3*p*, 4*s*, 3*d*, 4*p*, 5*s*, 4*d*, 5*p*, 6*s*, 4*f*, 5*d*, 6*p*, 7*s*, 5*f*, 6*d*, and so on.

The reason for this ordering is not simple. It is difficult for the following reasons: In the hydrogen atom, the only force is the electrical attraction between the nucleus and the single electron. However, when more than one electron orbits the nucleus, there are additional electrical forces between the electrons, which push the electrons away from each other. Solving this problem is extremely complicated, and a simple explanation of why the levels are ordered as they are cannot be given.

However, there is a way to remember the order: Write down the levels in a triangle as shown in Fig. 12.10.

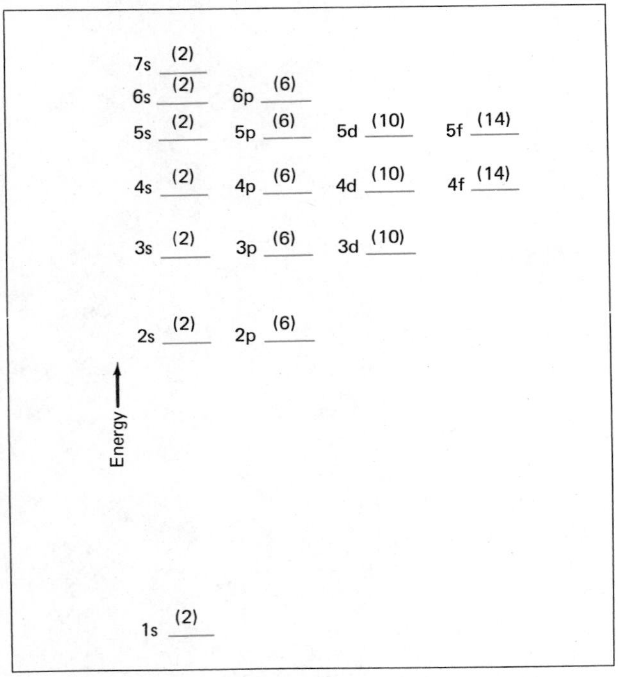

Figure 12.9 Energy level diagram for the hydrogen atom. The number of different sets of quantum numbers is shown in parentheses above each level. Note that for the hydrogen atom, the value of ℓ does not affect the energy level. The figure is not drawn to scale and only some of the levels are shown.

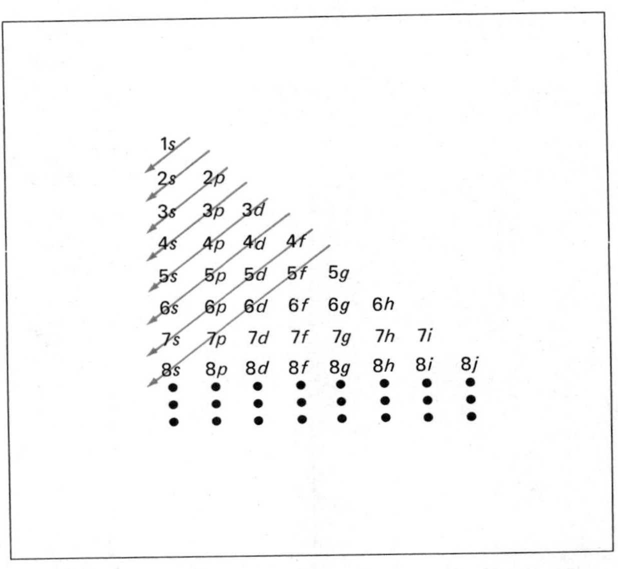

Figure 12.10 Mnemonic diagram for the ordering of energy levels in many-electron atoms. Order is from upper right to lower left.

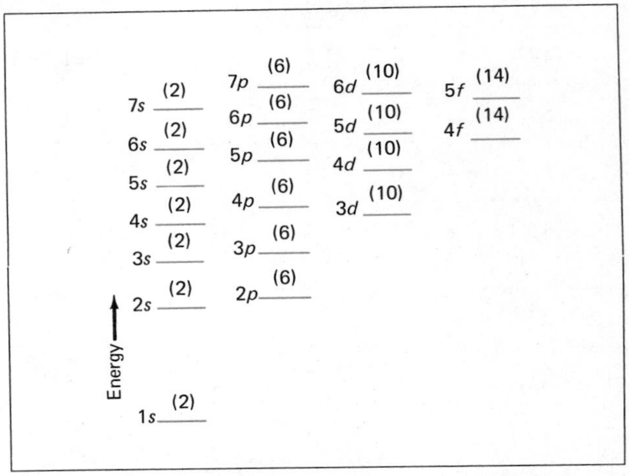

Figure 12.11 A typical energy level diagram for a many-electron atom. The number of different sets of quantum numbers is shown in parentheses above each level. (The diagram is not drawn to scale.)

Then make diagonal lines going from the upper right-hand to the lower left-hand portion as shown. The correct ordering is achieved by starting with the top diagonal line, going from the upper right to the lower left, and then continuing to the next diagonal line down and repeating the procedure.

We can now draw an energy level diagram for a typical many-electron atom. This is shown in Fig. 12.11. The various levels are now shifted to get the desired energy ordering.

12.5 Many-Electron Atoms and the Periodic Table

Solving the Schrödinger equation for the hydrogen atom is not particularly difficult. There is only one force involved—the electrostatic force between the proton and the electron. Quantum numbers arise naturally from the possible solutions of the equation, and the result is that the electron can occupy any of the various energy levels shown in Fig. 12.11.

The ground state of an atom is the combination of energy levels with the lowest total energy. For the hydrogen atom, the ground state is the 1*s* level. Almost all naturally occurring atoms are in the ground state. If an atom is not in the ground state, its electrons will give off energy in the form of photons and fall into the ground state. To understand more complicated atoms, then, it is necessary to know how the electrons distribute themselves in the ground state.

The **Pauli exclusion principle** set forth in 1928 by Wolfgang Pauli, a German physicist, brought order to the seeming chaos of many-electron atoms. The Pauli exclusion principle states that no two electrons can have the same set of quantum numbers. The underlying physical reason why this should be so is not easily understood, but application of the Pauli principle makes it fairly simple to distribute the electrons among the energy levels.

To illustrate its application, let us consider the case of aluminum, which has 13 electrons. How are we to distribute these 13 electrons to produce the ground state, that is, the level of lowest energy? We do so by putting the electrons into the lowest energy levels possible. There are only two different sets of quantum numbers in the 1*s* energy level, so only two electrons can go there. No more than two can go into this level, because if there were three or more, at least two would then have the same set of quantum numbers, in violation of the Pauli exclusion principle. Next, we put two into the 2*s* level. Then six go into the 2*p* level and two more go into the 3*s* level. The last electron then goes into the 3*p* level.

We can represent the electrons as small circles and make an energy level diagram for the ground state of aluminum. This is shown in Fig. 12.12. Because of the Pauli principle, the numbers that we put in parentheses above the various energy levels in Figs. 12.9 and 12.11 now take on a new significance. They are the maximum numbers of electrons that can go into the various energy levels.

The electrons in many-electron atoms are usually in the ground state which is made up of the lowest energy levels possible. However, one or more electrons are sometimes in a higher level. As an example, the energy level diagram for an excited state of the sodium atom is shown in Fig. 12.13. When the last electron is in the 3*p*

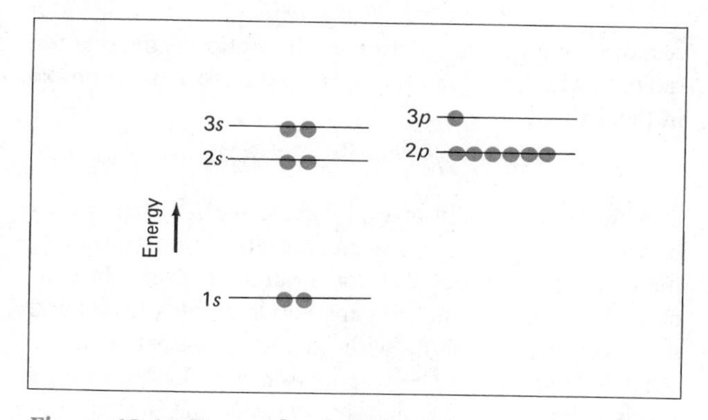

Figure 12.12 Energy level diagram for the ground state of the aluminum atom. Electrons are represented by colored circles. For the ground state, the lowest levels are filled.

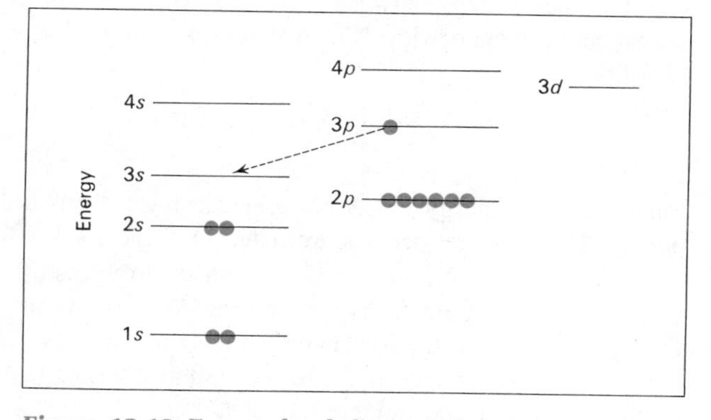

Figure 12.13 Energy level diagram for an excited state of the sodium atom. When the electron loses energy and changes to a lower level as indicated by the arrow, a photon of yellow light is emitted.

state, it can lose energy and fall to the $3s$ energy level, giving off a photon of light that looks yellow to us. Because each atom has a unique set of energy levels, each atom has a unique spectrum.

Rather than drawing energy level diagrams like Figs. 12.12 and 12.13 each time we want to represent the electron arrangement in an atom, we use a shorthand notation called the electron configuration. When writing an **electron configuration,** the levels are written in order of increasing energy and the number of electrons in each level is designated by a superscript. For example, the electron configuration of the aluminum atom in the ground state (Fig. 12.12) is

$$1s^2 2s^2 2p^6 3s^2 3p^1$$

The electron configuration for the excited state of sodium in Fig. 12.13 is

$$1s^2 2s^2 2p^6 3p^1$$

A more complicated electron configuration is the one for radium, which has 88 electrons. Radium's configuration in the ground state is

$$1s^2 2s^2 2p^6 3s^2 3p^6 4s^2 3d^{10} 4p^6 5s^2 4d^{10} 5p^6 6s^2 4f^{14} 5d^{10} 6p^6 7s^2$$

If we look closely at Fig. 12.11, we see that spaces between the energy levels are not equal. For instance, there is a big gap between the $1s$ and $2s$ levels. In general, there are big energy gaps between the s levels and the levels below them, with smaller gaps between the other levels, such as between $4s$-$3d$ and $3d$-$4p$. Energy levels (such as $4s$, $3d$, and $4p$) between the big gaps all have approximately the same energy.

An **electron period** consists of a set of energy levels, all of which have about the same energy. We can indicate the energy gaps between periods by vertical lines drawn between energy levels in an electron configuration as follows:

$$1s^2 | 2s^2 2p^6 | 3s^2 3p^6 | 4s^2 3d^{10} 4p^6 | 5s^2 4d^{10} 5p^6 |$$
$$6s^2 4f^{14} 5d^{10} 6p^6 | \text{etc.}$$

Each group of energy levels between the lines forms a period. The different periods are listed in Table 12.3.

A many-electron atom can be thought of in terms of the Bohr model of electrons orbiting the nucleus. As in the hydrogen atom, the lower the value of n, the closer the orbit to the nucleus. An **electron shell** consists of all the orbits with the same value of n.

In the Bohr model all the electrons in the same shell are represented as being in the same orbit. The first shell, or orbit, can hold two electrons; the next shells have

Table 12.3 Electron Periods

	Period Number	*Energy Levels in Period*	*Number of Electrons in Period*
Energy $\downarrow$	1	$1s$	2
	2	$2s$, $2p$	8
	3	$3s$, $3p$	8
	4	$4s$, $3d$, $4p$	18
	5	$5s$, $4d$, $5p$	18
	6	$6s$, $4f$, $5d$, $6p$	32
	7	$7s$, $5f$, $6d$, $7p$	32
	$\vdots$	$\vdots$	$\vdots$

maxima of 8, 18, 32, 50, 72, . . . , but except for the first four shells, the maxima are never reached.

Examples of the Bohr model, as applied to carbon and sodium, are shown in Fig. 12.14. Since carbon has only six electrons, two go into the first shell and four into the second. Sodium has eleven electrons: two go into the first orbit, or ring, around the nucleus, eight go into the second, and one goes into the third. The electron configurations for carbon and sodium in the ground state are

Carbon:	$1s^2 2s^2 2p^2$
Sodium:	$1s^2 2s^2 2p^6 3s^1$

These configurations should be compared with Fig. 12.14.

The Bohr model of electrons orbiting the nucleus is a simple model and easy to visualize. However, it cannot be exact, because the electrons do not travel in definite orbits. Instead, as was illustrated in Fig. 12.5 for a $1s$ electron, each electron has a probability of being at any particular distance from the nucleus. The dashed circles in Fig. 12.14 simply give a rough idea of where the electron is most likely to be found, as it goes around the nucleus.

In an atom with a high atomic number (e.g., copper with 29 protons in the nucleus), the inner shell electrons are strongly attracted to the nucleus, and the binding energies are in the range of thousands of electron volts (keV). When these inner shell electrons make a transition, e.g., from $n = 2$ to $n = 1$, a high-energy photon is emitted. Photons in the keV region can also be emitted when a free electron passes close to a nucleus with a high atomic number. In this case the electron is accelerated by the strong electric force and gives off radiation.

These high-energy photons, which are produced by atomic (as opposed to nuclear) processes, are known as

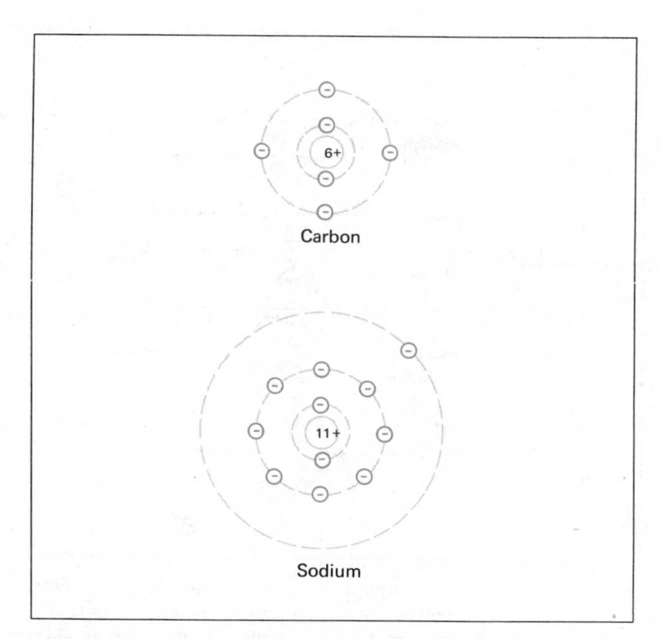

Figure 12.14 Schematic diagrams of carbon and sodium atoms. The solid ring in the center represents the nucleus. Each shell is drawn as a broken ring around the nucleus, with the appropriate number of electrons shown for each shell of the particular atom.

X rays. X rays are used routinely in medicine because these high-energy photons pass readily through water-filled tissue but not through bone. Thus, X rays of bones or teeth can be taken and used to diagnose and correct broken bones, cavities, etc. An example is shown in Fig. 12.15.

X rays in high dosages are harmful to the human body. However, with the advent of fast X-ray film, the exposure time necessary to get a good X-ray photograph has been drastically reduced from what it used to be. Now, except in extreme cases, the risk of not having an X ray is much greater than the almost negligible risk of having an X ray taken.

X rays are produced when high-energy electrons strike a metal, such as copper. Continuous X rays can have any energy below a maximum set by the energy of the bombarding electrons. They are produced when the high-energy electrons are accelerated by the strong electric forces close to a nucleus of a copper atom.

Discrete energy X rays, or characteristic X rays, are produced when a high-energy electron strikes an $n = 1$ electron and knocks it out of the atom. Once there is a vacancy in the $n = 1$ shell, an $n = 2$ (or 3 or 4) electron can spontaneously fall down to the $n = 1$ level and emit a high-energy photon (an X ray with an energy of, say, 20 keV). This two-step atomic process is shown in Fig. 12.16. Of course, the difficult and expensive part is the production of the high-energy electrons that initiate the process. To get these, you need voltages of 50 to 150 kilovolts. The high energies generated by these voltages are necessary to eject the $n = 1$ electrons.

A tube that produces X rays is shown schematically in Fig. 12.17. An electron is accelerated to high energy using a high voltage (e.g., 100,000 volts). The electrons strike a metal, such as copper, and X rays are produced in all directions. Shielding is used to limit the beam to the patient's body area that is to be x-rayed. The X rays

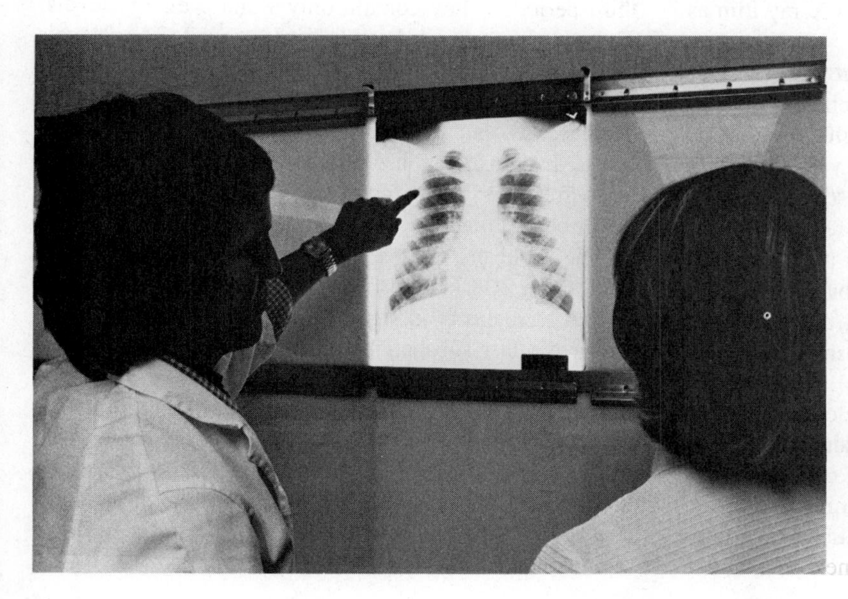

Figure 12.15 Technologists examine an X ray film of the chest. The *L* in the upper-right corner of the film indicates the patient's left side. (Courtesy Edward Golubski, Self Memorial Hospital, Greenwood, S.C.)

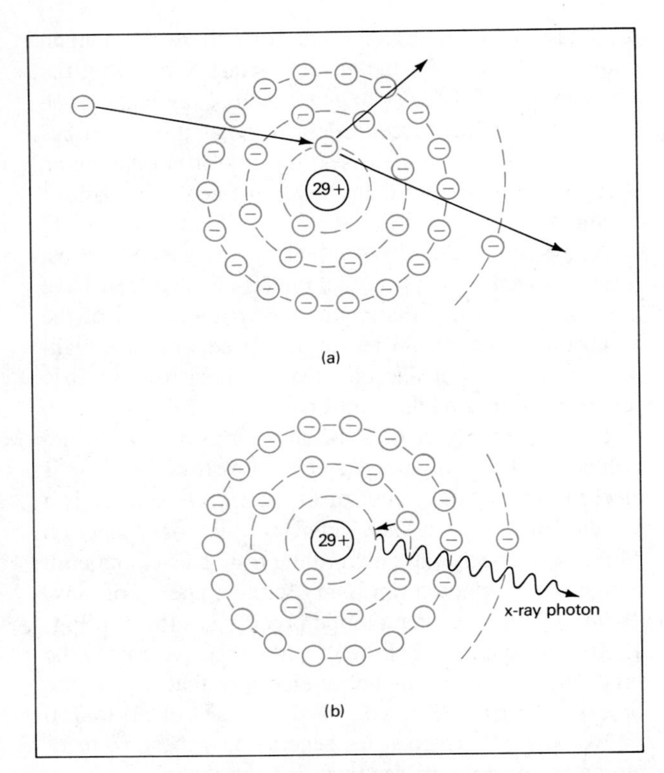

Figure 12.16 The production of characteristic X rays proceeds in two steps. First (a), a high-energy electron knocks an $n = 1$ electron out of an atom, such as copper. Then (b), an $n = 2$ electron spontaneously falls down to the $n = 1$ level, emitting an X ray (high-energy photon).

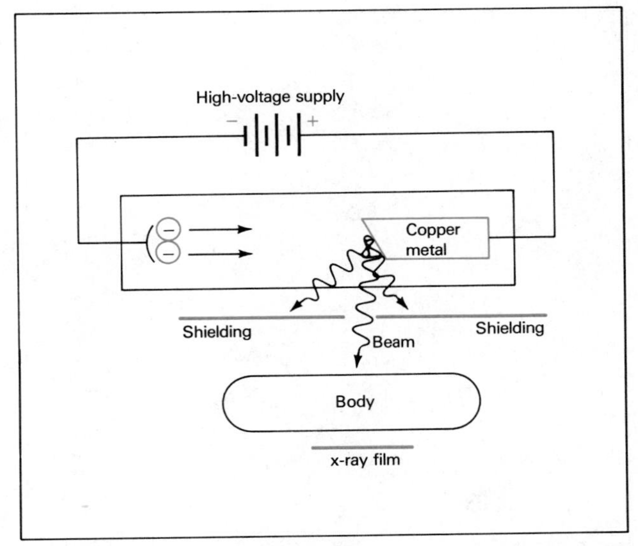

Figure 12.17 X rays are produced when high-energy (about 100-keV) electrons strike a metal. Shielding is used to limit the X-ray beam to the desired area. Film is put behind the spot to detect the X rays passing through.

that pass through the body are detected on X-ray film as shown.

If we examine Table 12.3, we see that after the first three periods, electrons from different shells can go into the same period. This is because, although the radius of an electron's path depends mostly on the value of n, the electron's energy depends on both n and ℓ.

Electrons with the same value of n are in the same shell. Electrons with the same value of n and ℓ are said to be in the same **electron subshell.** A subshell is similar to an energy level, since all electrons in the same subshell have the same energy. Several subshells make up a period, as can be seen from Table 12.3.

A table of the elements in terms of electron periods is shown in Fig. 12.18. Each row is a separate period. The table shows the order of entry of electrons with increasing atomic number. Only those subshells are given in which electrons have just entered. Four separate groups are shown in the table. The letters s, p, d, and f above each group indicate that these orbitals are being filled in the group.

There are only two electrons in the first period, which is $1s$. However, we have put the number 2 on the far right of Fig. 12.18 because it is the last number in the period. All numbers on the far right represent the last number in the period.

The second and third rows represent the second and third periods. They contain only s and p energy levels. The next four periods have d levels as well. The f levels of the sixth and seventh periods are listed below the table. Note that after numbers 56 and 88, the next number has been put into the d level instead of the f level. These discrepancies in the table are too complicated for us to fret over. After numbers 57 and 89, the f levels begin being filled.

The table ends after seven periods, since there are only about 108 known elements. We can represent the various elements by designating how many electrons there are in the last subshell being filled in the ground state of the element. For instance, in the electron configuration for carbon, the sixth and last electron is the second electron in the $2p$ subshell. Note that all elements listed in the vertical column with carbon have the p^2 orbital being filled but at a different energy level. The table in Fig. 12.18 is a representation of the **periodic table** of elements that is studied at length in chemistry.

s		*d*										*p*					
1 H $1s^1$																	2 He $1s^2$
3 Li $2s^1$	4 Be $2s^2$											5 B $2p^1$	6 C $2p^2$	7 N $2p^3$	8 O $2p^4$	9 F $2p^5$	10 Ne $2p^6$
11 Na $3s^1$	12 Mg $3s^2$											13 Al $3p^1$	14 Si $3p^2$	15 P $3p^3$	16 S $3p^4$	17 Cl $3p^5$	18 Ar $3p^6$
19 K $4s^1$	20 Ca $4s^2$	21 Sc $3d^14s^2$	22 Ti $3d^24s^2$	23 V $3d^34s^2$	24 Cr $3d^54s^1$	25 Mn $3d^54s^2$	26 Fe $3d^64s^2$	27 Co $3d^74s^2$	28 Ni $3d^84s^2$	29 Cu $3d^{10}4s^1$	30 Zn $3d^{10}4s^2$	31 Ga $4p^1$	32 Ge $4p^2$	33 As $4p^3$	34 Se $4p^4$	35 Br $4p^5$	36 Kr $4p^6$
37 Rb $5s^1$	38 Sr $5s^2$	39 Y $4d^15s^2$	40 Zr $4d^25s^2$	41 Nb $4d^45s^1$	42 Mo $4d^55s^1$	43 Tc $4d^65s^1$	44 Ru $4d^75s^1$	45 Rh $4d^85s^1$	46 Pd $4d^{10}$	47 Ag $4d^{10}5s^1$	48 Cd $4d^{10}5s^2$	49 In $5p^1$	50 Sn $5p^2$	51 Sb $5p^3$	52 Te $5p^4$	53 I $5p^5$	54 Xe $5p^6$
55 Cs $6s^1$	56 Ba $6s^2$	57 La* $5d^16s^2$	72 Hf $5d^26s^2$	73 Ta $5d^36s^2$	74 W $5d^46s^2$	75 Re $5d^56s^2$	76 Os $5d^66s^2$	77 Ir $5d^76s^2$	78 Pt $5d^96s^1$	79 Au $5d^{10}6s^1$	80 Hg $5d^{10}6s^2$	81 Tl $6p^1$	82 Pb $6p^2$	83 Bi $6p^3$	84 Po $6p^4$	85 At $6p^5$	86 Rn $6p^6$
87 Fr $7s^1$	88 Ra $7s^2$	89 Ac† $6d^17s^2$	104	105	106	107	108	109									

f

*Lanthanides	58 Ce $4f^15d^16s^2$	59 Pr $4f^36s^2$	60 Nd $4f^46s^2$	61 Pm $4f^56s^2$	62 Sm $4f^66s^2$	63 Eu $4f^76s^2$	64 Gd $4f^75d^16s^2$	65 Tb $4f^96s^2$	66 Dy $4f^{10}6s^2$	67 Ho $4f^{11}6s^2$	68 Er $4f^{12}6s^2$	69 Tm $4f^{13}6s^2$	70 Yb $4f^{14}6s^2$	71 Lu $4f^{14}5d^16s^2$
†Actinides	90 Th $6d^27s^2$	91 Pa $5f^26d^17s^2$	92 U $5f^36d^17s^2$	93 Np $5f^46d^17s^2$	94 Pu $5f^67s^2$	95 Am $5f^77s^2$	96 Cm $5f^76d^17s^2$	97 Bk $5f^97s^2$	98 Cf $5f^{10}7s^2$	99 Es $5f^{11}7s^2$	100 Fm $5f^{12}7s^2$	101 Md $5f^{13}7s^2$	102 No $5f^{14}7s^2$	103 Lw $5f^{14}6d^17s^2$

Figure 12.18 **A table of the elements in terms of electron periods. Each row is a separate period. The table shows the order of entry of electrons with increasing atomic number. Only those subshells are given that electrons have just entered.**

Learning Objectives

After reading and studying this chapter, you should be able to do the following without referring to the text:

1. State de Broglie's hypothesis concerning matter waves and give an example calculation.
2. Explain how it is possible to see electron diffraction.
3. Explain why short wavelengths are needed in high-resolution microscopes.
4. Explain the meaning of the wavefunction in Schrödinger's equation.
5. State and explain Heisenberg's uncertainty principle.
6. List the four quantum numbers generated by the wave equation for the hydrogen atom, state their meaning, and give their possible values.
7. State Pauli's exclusion principle.
8. Describe how relativity affects quantum theory for an electron in an atom.
9. Draw ground-state energy-level diagrams for a given element.
10. Draw a schematic diagram illustrating the electron structure of a given atom.
11. Write the electron configuration for any atom listed in the periodic table (ignoring exceptions).
12. Draw diagrams (before and after) to show how a discrete energy X-ray photon is produced.

Important Words and Terms

quantum mechanics
matter waves
Schrödinger equation
wavefunction
Heisenberg uncertainty principle
n, ℓ, m, m_s quantum numbers
Pauli exclusion principle
electron configuration
electron period
electron shell
X rays
electron subshell
periodic table

Questions

1. How were electrons shown to be waves?
2. Why are electron beams used in high-resolution microscopes?
3. Why is Schrödinger's equation so important?
4. Is it possible for the electron to be (a) inside the nucleus of an atom? (b) 100 Å away from the nucleus?
5. Explain, using a diagram, why the act of measuring the position of an electron will alter its position.
6. How does the Heisenberg uncertainty principle alter our philosophy of life?
7. State the four quantum numbers for an electron in a hydrogen atom and explain the meaning of each.
8. What do the letters *s*, *p*, *d*, and *f* mean?
9. Where does the quantum number m_s come from?
10. Why is the Pauli exclusion principle important?
11. What is the difference between an electron period and an electron shell?
12. How are X rays produced?
13. What is the basis of the periodic table of the elements?

Problems

1. The Bohr radius for $n = 1$ is 0.529 Å.
   (a) What is the circumference of an atom with the Bohr radius?
   (b) Set this circumference equal to a wavelength of the electron. Now use de Broglie's equation to find how fast the electron is moving. Compare your answer to the speed of an electron according to the Bohr theory, that is, $v = 2.19 \times 10^6$ m/s. ($m_e = 9.1 \times 10^{-31}$ kg).
   Answer: (b) The same as in the Bohr theory.
2. Calculate the de Broglie wavelength of a 0.5-kg baseball moving with a velocity of 26 m/s (approximately 60 mi/h).
3. Calculate the de Broglie wavelength of the Earth. ($m_E = 6 \times 10^{24}$ kg; $v_E = 3 \times 10^4$ m/s.)
4. Calculate the minimum uncertainty in the velocity (m/s) of an electron ($m_e = 9.1 \times 10^{-31}$ kg), when the uncertainty in position is 10^{-4} m.
5. Calculate the minimum uncertainty in the velocity (m/s) of a baseball ($m = 0.5$ kg), when the uncertainty in position is 10^{-4} m.
   Answer: 2.1×10^{-30} m/s
6. What are the possible values of ℓ when (a) $n = 3$, (b) $n = 5$?
7. What are the possible values of the quantum number m when quantum number (a) ℓ is 2, (b) ℓ is 3?
8. (a) List the possible sets of quantum numbers for an electron in the 4d subshell. (b) How many electrons can be put in this subshell?
9. Why is the 3f energy level impossible?
10. Consider the following electron configuration: $1s^2 2s^2 2p^6 3s^2 3p^6 4s^2 3d^9 4p^1$
    (a) What atom does this correspond to?
    (b) Is this the ground state or an excited state?
    Answer: (a) $_{30}$Zn (b) excited
11. How many electrons are in the outermost shell of the atom in problem 10?
12. What energy levels are in the 5th period?
13. Write the electron configurations for each of the following elements:
    (a) $_{17}$Cl, (b) $_{19}$K, (c) $_{53}$I.
14. Draw the ground-state energy-level diagram similar to Fig. 12.12 for each of the following elements:
    (a) $_{17}$Cl, (b) $_{19}$K.
15. Draw schematic diagrams of electrons orbiting the nucleus similar to Fig. 12.14 for each of the following elements:
    (a) $_{17}$Cl, (b) $_{19}$K.

13

The Periodic Table

Order is a lovely thing:
On disarray it lays its wing,
Teaching simplicity to sing.
It has a meek and lowly grace,
Quiet as a nun's face.
–Anna Hempstead Branch

THE OBJECTS IN our environment are composed of matter. **Matter** may be defined as anything that exists in time, occupies space, and has mass. The elements are the different kinds of atoms that make up the matter in our everyday environment. The physical, chemical, and combining properties of the various elements affect us constantly. To cite two examples, the element oxygen is necessary to sustain human life, while the diamond in an engagement ring is pure carbon. With only a few exceptions, the 108 known elements, either singularly or in combination, are the components of all matter. The science of **chemistry** deals with the transformations that this matter undergoes.

The modern understanding of matter, based on the electron structure of atoms, is relatively recent, but chemistry had its beginnings early in recorded history. Egyptian hieroglyphs that date back to 3400 B.C. show wine presses; wine making, of course, required a chemical fermentation process. By 2000 B.C. the Egyptians and Mesopotamians possessed the knowledge required to produce and work the metals gold, silver, lead, copper, iron, and bronze—a mixture of copper and tin. Dyes were discovered during this period, and methods of fixing them were known.

In the fourth century B.C., the Greek philosopher Aristotle developed the false idea that all matter was composed of four elements: earth, air, fire, and water. As the Greek culture faded, the pseudoscience of alchemy was born. Its main goals were to change base metals into gold and to find the "elixir of life," which could restore an aging human body to its youthful vigor. Alchemy was a mixture of magic and experimentation, and it was practiced in China as well as the West. Although it was practiced by many charlatans, its finest practitioners were dedicated experimentalists, who contributed to the real science of chemistry. The goals of the alchemists were never achieved, but by the time the scientific method was born in the sixteenth century, much had been discovered about the properties of matter.

In 1869, Dmitri Mendeleev (1834–1907) in Russia and Julius Meyer in Germany, working independently, published a classification of the known elements showing a relationship between the magnitude of the atomic weights and the properties of elements. This relationship led to the conclusion that the properties of elements are periodic functions of their atomic weights. In 1914, Henry Moseley, an English physicist, showed that atomic numbers rather than atomic weights are fundamental in determining the properties of elements, thus laying the foundation for the modern periodic table. In this chapter we shall learn the properties of elements and how to use the periodic table to determine many of these properties.

◀ A drop of mercury (the liquid metal) on a glass slide.

13.1 Elements

In the very earliest of civilizations, nine elements were isolated: gold, silver, lead, copper, tin, iron, carbon, sulfur, and mercury. The alchemists added five more: antimony, arsenic, bismuth, phosphorus, and zinc. As the list grew, a precise definition of an element was needed. The concept of an element as we know it today was formulated by Robert Boyle (Fig. 13.1), an English chemist, in 1661. In his book *The Skeptical Chemist,* Boyle proposed that the name "element" be applied to those simple bodies that could not be separated into components by chemical methods. In 1789, the French

Figure 13.1 Robert Boyle (1627–1691). Born in Ireland, the fourteenth child of the wealthy Earl of Cork, young Robert proved to be a child prodigy. After graduation from Oxford, Boyle used his wealth to finance his research in many facets of science. He is probably best known for Boyle's law (see Chapter 5). Considered as one of the founders of chemistry, he was the first person to define an element and the first to suggest that heat was due to molecular motion. He also proved that air transmitted sound. His work did much to advance the cause of experimentation in chemistry. He was sick much of his life, but managed to write more than 40 books on language, religion, and science. (Photo courtesy American Institute of Physics, Neils Bohr Library.)

scientist, Antoine Lavoisier (1743–1794) expanded Boyle's definition and applied the term element to the matter that remained after complete chemical decomposition had occurred.

Lavoisier, who is sometimes called the father of chemistry, introduced quantitative methods into chemistry and determined the role of oxygen in combustion. In 1789, his list of elements included 23 that we recognize today as elements. The modern definition of the word **element** is "a substance in which all the atoms have the same atomic number." Our modern listing of the elements has grown to 108, with more expected to be added.

In order to designate the different elements, a symbol notation is used. The notation used today was first conceived by the Swedish chemist Jons Jacob Berzelius. He used the first one or two letters of the Latin name for each element. Thus, sodium is designated Na for "natrium," silver is Ag for "argentum," tin is Sn for "stannum," and so forth. Since Berzelius's time, most elements have been symbolized by the first one or two letters of the English name. Examples are C for carbon, O for oxygen, H for hydrogen, and Ca for calcium. Note that the first letter is always capitalized and the second is lower case. This symbol notation of the elements has proved to be quite useful.

When the atoms of two or more elements react chemically, they often join together to form what chemists call **molecules** of a compound. The **formula** for the compound is written by putting the element symbols adjacent to each other and using subscripts to designate the number of atoms of each element in the compound. For instance, water is written H_2O. This means that in a single molecule of the compound water there are two atoms of hydrogen and one atom of oxygen. If more than one molecule is to be designated, a number, called a coefficient, is placed in front of the molecule. For instance, 3 H_2O means there are three molecules of water, each having two atoms of hydrogen and one atom of oxygen.

In Table 13.1, each element is listed in alphabetical order, along with its symbol, atomic number, and atomic weight. There are 105 elements listed in this table and numbers 106, 107, and 108 have recently been created. Efforts are constantly being made to create others.

Table 13.1 The atomic weights are based on ^{12}C = 12.0000. If the element does not occur naturally, the mass number of the most stable isotope is given in parentheses.

	Symbol	*Atomic Number*	*Atomic Weight*		*Symbol*	*Atomic Number*	*Atomic Weight*
Actinium	Ac	89	(227)	Astatine	At	85	(210)
Aluminum	Al	13	26.9815	Barium	Ba	56	137.34
Americium	Am	95	(243)	Berkelium	Bk	97	(247)
Antimony	Sb	51	121.75	Beryllium	Be	4	9.01218
Argon	Ar	18	39.948	Bismuth	Bi	83	208.9806
Arsenic	As	33	74.9216	Boron	B	5	10.81

Table 13.1 (Continued)

	Symbol	Atomic Number	Atomic Weight		Symbol	Atomic Number	Atomic Weight
Bromine	Br	35	79.90	Nitrogen	N	7	14.0067
Cadmium	Cd	48	112.40	Nobelium	No	102	(253)
Calcium	Ca	20	40.08	Osmium	Os	76	190.2
Californium	Cf	98	(251)	Oxygen	O	8	15.9994
Carbon	C	6	12.011	Palladium	Pd	46	106.4
Cerium	Ce	58	140.12	Phosphorus	P	15	30.9738
Cesium	Cs	55	132.9055	Platinum	Pt	78	195.09
Chlorine	Cl	17	35.453	Plutonium	Pu	94	(244)
Chromium	Cr	24	51.996	Polonium	Po	84	(209)
Cobalt	Co	27	58.9332	Potassium	K	19	39.102
Copper	Cu	29	63.545	Praseodymium	Pr	59	140.9077
Curium	Cm	96	(247)	Promethium	Pm	61	(145)
Dysprosium	Dy	66	162.50	Protactinium	Pa	91	(231)
Einsteinium	Es	99	(254)	Radium	Ra	88	(226)
Erbium	Er	68	167.26	Radon	Rn	86	(222)
Europium	Eu	63	151.96	Rhenium	Re	75	186.2
Fermium	Fm	100	(253)	Rhodium	Rh	45	102.9055
Fluorine	F	9	18.9984	Rubidium	Rb	37	85.4678
Francium	Fr	87	(223)	Ruthenium	Ru	44	101.07
Gadolinium	Gd	64	157.25	Rutherfordium	Rf	104	(257)
Gallium	Ga	31	69.72	Samarium	Sm	62	150.4
Germanium	Ge	32	72.59	Scandium	Sc	21	44.9559
Gold	Au	79	196.967	Selenium	Se	34	78.96
Hafnium	Hf	72	178.49	Silicon	Si	14	28.086
Hahnium	Ha	105	(260)	Silver	Ag	47	107.868
Helium	He	2	4.00260	Sodium	Na	11	22.9898
Holmium	Ho	67	164.9303	Strontium	Sr	38	87.62
Hydrogen	H	1	1.0080	Sulfur	S	16	32.06
Indium	In	49	114.82	Tantalum	Ta	73	180.9479
Iodine	I	53	126.9045	Technetium	Tc	43	(99)
Iridium	Ir	77	192.22	Tellurium	Te	52	127.60
Iron	Fe	26	55.847	Terbium	Tb	65	158.9254
Krypton	Kr	36	83.80	Thallium	Tl	81	204.37
Lanthanum	La	57	138.9055	Thorium	Th	90	232.0381
Lawrencium	Lr	103	(257)	Thulium	Tm	69	168.9342
Lead	Pb	82	207.12	Tin	Sn	50	118.69
Lithium	Li	3	6.941	Titanium	Ti	22	47.90
Lutetium	Lu	71	174.97	Tungsten	W	74	183.85
Magnesium	Mg	12	24.305	Uranium	U	92	238.029
Manganese	Mn	25	54.9380	Vanadium	V	23	50.9414
Mendelevium	Md	101	(256)	Xenon	Xe	54	131.30
Mercury	Hg	80	200.59	Ytterbium	Yb	70	173.04
Molybdenum	Mo	42	95.94	Yttrium	Y	39	88.9059
Neodymium	Nd	60	144.24	Zinc	Zn	30	65.37
Neon	Ne	10	20.179	Zirconium	Zr	40	91.22
Neptunium	Np	93	(237)				
Nickel	Ni	28	58.71				
Niobium	Nb	41	92.9064				

You will recall (Section 10.1) that the atomic number of an element refers to the number of protons in the nucleus of the atom. The atomic number also is equal to the number of electrons in the neutral atom. We will see shortly, however, that sometimes electrons are gained or lost, so that the atom is no longer neutral. It is the number of protons, not electrons, that determines what element you have. Each element can have several isotopes. **Isotopes** are forms of the same element whose nuclei have the same number of protons but different numbers of neutrons.

The element hydrogen is found in nature as three isotopes. See Fig. 13.2. The three isotopes have similar chemical properties since they all have the same electron structure. They differ in physical properties and can be separated in a mass spectrograph because they have different masses. Ordinary hydrogen ($^{1}_{1}H$) is called protium, hydrogen two ($^{2}_{1}H$) is called deuterium, and hydrogen three ($^{3}_{1}H$) is called tritium. In a given sample of hydrogen, about one atom in 5000 is deuterium and about one atom in 10,000,000 is tritium. Heavy water (D_2O) is composed of two atoms of deuterium and one atom of oxygen. Protium and deuterium are stable atoms, whereas tritium is radioactive with a half-life of 12.4 years.

The atomic weight of each element is the average weight of the isotopes contained in the naturally occurring element. Atomic weights today are based on the ^{12}C isotope, which is given a *relative* atomic weight of exactly 12.0000. The reason naturally occurring carbon has an atomic weight slightly greater than 12.0000 is that there is some ^{13}C and ^{14}C in naturally occurring carbon as well as ^{12}C.

Only a few of the elements exist in large amounts in the Earth's crust. About 75% of the weight of the Earth's crust is composed of only two elements—oxygen and silicon. Aluminum and iron, two important metals, are quite abundant also, but many elements exist naturally in only minute quantities. Several, of course, do not

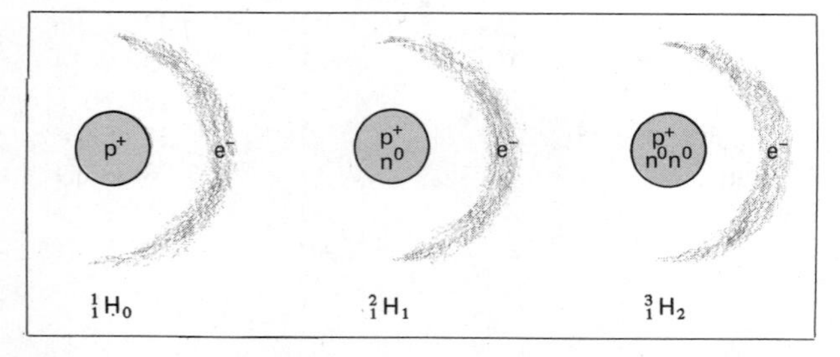

Figure 13.2 The three isotopes of hydrogen: protium, deuterium, and tritium.

Table 13.2 Relative Abundances of Elements in the Earth's Crust

Element	*Approx. Percentage (weight)*
Oxygen (O)	46.5
Silicon (Si)	27.5
Aluminum (Al)	8.1
Iron (Fe)	5.3
Calcium (Ca)	4.0
Magnesium (Mg)	2.7
Sodium (Na)	2.4
Potassium (K)	1.9
All Others	1.6
Total	100.0

occur at all in nature and can only be made artificially. Table 13.2 gives the approximate percentage by weight of the more abundant elements in the Earth's crust.

13.2 The Periodic Table

By 1870, a total of 65 elements had been discovered. Using the methods of many dedicated chemists, the atomic weight and other properties of many elements were determined. However, except for a few sketchy attempts, a system of classifying the elements had not been found. It remained for the great Russian chemist Dmitri Mendeleev to formulate a satisfactory classification scheme. Mendeleev surveyed all data on the known elements and was able to arrange them in a table. A modern version of his table, called the periodic table of the elements, is shown in Table 13.3 and on the front inside cover of this book. The **periodic table** arranges the elements in rows according to their electron periods with all the elements in the same column having similar properties. Mendeleev's original table was based on atomic weights, but H. G. J. Moseley, an English physicist, showed in 1913 that properties of the elements are determined by their atomic numbers. The modern statement of the **periodic law** is as follows:

The properties of elements are periodic functions of the atomic number.

The periodic table is most easily understood in terms of the electron configurations of the elements. The atomic number of each element corresponds to the number of electrons in a neutral atom of that element. The element's position in the table depends on the energy level

Table 13.3 Modern Periodic Table

The number of electrons in filled shells is shown in the column at the extreme left; the remaining electrons for each element are shown below the symbol and atomic number for each element. The atomic weights shown above the symbols are based on Carbon-12.

s — d — p

PERIODS	REPRESENTATIVE ELEMENTS I A (1)	(2) II A	TRANSITION ELEMENTS (3) III B	(4) IV B	(5) V B	(6) VI B	(7) VII B	VIII (8)	VIII (9)	VIII (10)	(11) I B	(12) II B	REPRESENTATIVE ELEMENTS (13) III A	(14) IV A	(15) V A	(16) VI A	(17) VII A	NOBLE GASES (18) 0
1 0	1.0080 **H**[1] 1																1.0080 **H**[1] 1	4.00260 **He**[2] 2
2 2	6.941 **Li**[3] 1	9.01218 **Be**[4] 2											10.81 **B**[5] 3	12.011 **C**[6] 4	14.0067 **N**[7] 5	15.9994 **O**[8] 6	18.9984 **F**[9] 7	20.179 **Ne**[10] 8
3 2,8	22.9898 **Na**[11] 1	24.305 **Mg**[12] 2											26.9815 **Al**[13] 3	28.086 **Si**[14] 4	30.9738 **P**[15] 5	32.06 **S**[16] 6	35.453 **Cl**[17] 7	39.948 **Ar**[18] 8
4 2,8	39.102 **K**[19] 8,1	40.08 **Ca**[20] 8,2	44.9559 **Sc**[21] 9,2	47.90 **Ti**[22] 10,2	50.9414 **V**[23] 11,2	51.996 **Cr**[24] 13,1	54.9380 **Mn**[25] 13,2	55.847 **Fe**[26] 14,2	58.9332 **Co**[27] 15,2	58.71 **Ni**[28] 16,2	63.545 **Cu**[29] 18,1	65.37 **Zn**[30] 18,2	69.72 **Ga**[31] 18,3	72.59 **Ge**[32] 18,4	74.9216 **As**[33] 18,5	78.96 **Se**[34] 18,6	79.90 **Br**[35] 18,7	83.80 **Kr**[36] 18,8
5 2,8,18	85.4678 **Rb**[37] 8,1	87.62 **Sr**[38] 8,2	88.9059 **Y**[39] 9,2	91.22 **Zr**[40] 10,2	92.9064 **Nb**[41] 12,1	95.94 **Mo**[42] 13,1	(99) **Tc**[43] 14,1	101.07 **Ru**[44] 15,1	102.906 **Rh**[45] 16,1	106.4 **Pd**[46] 18	107.868 **Ag**[47] 18,1	112.40 **Cd**[48] 18,2	114.82 **In**[49] 18,3	118.69 **Sn**[50] 18,4	121.75 **Sb**[51] 18,5	127.60 **Te**[52] 18,6	126.905 **I**[53] 18,7	131.30 **Xe**[54] 18,8
6 2,8,18	132.906 **Cs**[55] 18,8,1	137.34 **Ba**[56] 18,8,2	[57-71] *	178.49 **Hf**[72] 32,10,2	180.948 **Ta**[73] 32,11,2	183.85 **W**[74] 32,12,2	186.2 **Re**[75] 32,13,2	190.2 **Os**[76] 32,14,2	192.22 **Ir**[77] 32,15,2	195.09 **Pt**[78] 32,17,1	196.967 **Au**[79] 32,18,1	200.59 **Hg**[80] 32,18,2	204.37 **Tl**[81] 32,18,3	207.12 **Pb**[82] 32,18,4	208.981 **Bi**[83] 32,18,5	(209) **Po**[84] 32,18,6	(210) **At**[85] 32,18,7	(222) **Rn**[86] 32,18,8
7 2,8,18,32	(223) **Fr**[87] 18,8,1	(226) **Ra**[88] 18,8,2	[89-103] †	(257) **Rf**[104] 32,10,2	(262) **Ha**[105] 32,11,2	(263) [106]	(261) [107]	[108]										

f

INNER TRANSITION ELEMENTS

*LANTHANIDE SERIES 6	138.906 **La**[57] 18,9,2	140.12 **Ce**[58] 20,8,2	140.908 **Pr**[59] 21,8,2	144.24 **Nd**[60] 22,8,2	(145) **Pm**[61] 23,8,2	150.4 **Sm**[62] 24,8,2	151.96 **Eu**[63] 25,8,2	157.25 **Gd**[64] 25,9,2	158.925 **Tb**[65] 27,8,2	162.50 **Dy**[66] 28,8,2	164.930 **Ho**[67] 29,8,2	167.26 **Er**[68] 30,8,2	168.934 **Tm**[69] 31,8,2	173.04 **Yb**[70] 32,8,2	174.97 **Lu**[71] 32,9,2
†ACTINIDE SERIES 7	(227) **Ac**[89] 18,9,2	232.038 **Th**[90] 18,10,2	(231) **Pa**[91] 20,9,2	238.029 **U**[92] 21,9,2	(237) **Np**[93] 23,8,2	(244) **Pu**[94] 24,8,2	(243) **Am**[95] 25,8,2	(247) **Cm**[96] 25,9,2	(247) **Bk**[97] 26,9,2	(251) **Cf**[98] 28,8,2	(254) **Es**[99] 29,8,2	(253) **Fm**[100] 30,8,2	(256) **Md**[101] 31,8,2	(253) **No**[102] 32,8,2	(257) **Lw**[103] 32,9,2

A value (of a relative weight) given in parentheses denotes the mass number of the isotope of longest known half-life. Elements required for health are shown in color.

of its last several electrons. For example, the electron configuration of sulfur, atomic number 16, is

$$1s^2 2s^2 2p^6 3s^2 3p^4$$

The last 4 electrons are in the $3p$ level in the third period. Thus, sulfur occupies the fourth position in the $3p$ level in the third row of the periodic table. In each box of the table the element's atomic number, symbol, atomic weight (or most stable isotope, if the element is not found in nature), and number of electrons in the outermost shells are given. Thus, for sulfur, there are six electrons in the outer, unfilled shell; two are in the $3s$ subshell, and four are in the $3p$ subshell. As another example, the electron configuration of $_{76}Os$ is

$$1s^2 2s^2 2p^6 3s^2 3p^6 4s^2 3d^{10} 4p^6 5s^2 4d^{10} 5p^6 6s^2 4f^{14} 5d^6$$

It has 2 electrons in its outer, or sixth, shell, 14 electrons in its fifth shell, and 32 electrons in its fourth shell. So it has the numbers 32, 14, 2 under its symbol in the periodic table.

The rows of the periodic table are called **periods,** and they correspond to the electron periods given in Table 12.3. The obvious grouping of columns into the first 2 on the left side, the middle 10, and the last 6 on the right side, correspond to the s, d, and p energy levels, which can contain a maximum of 2, 10, and 6 electrons, respectively. Two groups of 14 elements each corresponding to the f levels, which hold a maximum of 14 electrons, are placed separately at the bottom of the table.

13.3 Classification of Elements

By examining the periodic table, we can classify the elements into four broad categories:

1. *Elements in which the s and p subshells are completely filled with electrons.* Each of these elements has electrons completely filling up all the subshells in a period. They are helium, with 2 electrons in the $1s$ level, and neon, argon, krypton, xenon, and radon, each of which has 8 electrons in the outer shell—2 in the s subshell and 6 in the p subshell. These elements are in the column at the far right labeled 0 and normally occur as gases. Only rarely do they react with other elements. They are called the **noble gases.**
2. *Elements in which an s or p subshell is incomplete.* The s and p subshells together contain a maximum of 8 electrons. Groups of elements that have their last electrons in one of these subshells are called **representative elements.** They are labeled according to the number of electrons in their s and p subshells. Elements in columns labeled IA, IB, IIA, IIB, IIIA, IVA, VA, VIA, and VIIA are included in this category. The Roman numeral indicates the number of s and p electrons in the outer shell. Columns IB and IIB are distinguished from IA and IIA as follows: Elements in columns IB and IIB have filled d subshells in the last period while elements in columns IA and IIA have empty d subshells in the last period.
3. *Elements in which the last electrons fill the d subshell in the period.* These elements are placed in the middle columns of the periodic table. The columns are labeled IIIB, IVB, VB, VIB, VIIB, and VIII. The Roman numeral designates the number of s and d electrons in the period. (The exceptions to this are the three columns labeled VIII. These have from 8 to 10 electrons in the s and d levels.) The elements in this category are called **transition elements.** Sometimes column IB or both columns IB and IIB are included in this category.
4. *Elements in which the last electrons are in the f subshell.* These elements are placed in two rows below the main table. They are in the **lanthanide,** or **rare earth, series:** cerium (Ce) through lutetium (Lu), and the **actinide series:** thorium (Th) to lawrencium (Lw). These two groups of elements make up the **inner transition elements.**

Many of the properties of elements are determined by the electrons in the outermost shell. These electrons are called **valence electrons.** In any period only the s and p electrons are in the outermost shell. The d and f electrons always have a lower value for the principal quantum number, so they are in inner shells compared to the s and p electrons in the same period. As an example, the sixth period contains the levels $6s$, $4f$, $5d$, $6p$. Only the s and p electrons are in the outer, or sixth, shell. The d and f electrons are in the inner shells 5 and 4. Since only s and p electrons occupy the outermost shell, there can only be between 0 and 8 valence electrons.

As mentioned previously, it is the valence electrons that determine many of the properties of the elements. Elements tend to gain, lose, or share valence electrons so that they will have a filled outer shell of electrons. Except for helium, elements that have one or two electrons in the s level tend to lose these electrons easily, so that there will be no electrons in the outer shell. Elements that have six or seven s and p level electrons tend to acquire electrons to fill up the s and p subshells. Elements with 3, 4, or 5 valence electrons either gain or lose electrons to get 0 or 8 in the outer shell.

One of the earliest classifications of elements was into

metals and nonmetals. This classification was originally due to certain distinctive properties. Our modern definition is that a **metal** is an element that tends to lose its valence electrons, while a **nonmetal** is an element that tends to gain electrons to complete an outer shell. The large majority of elements are metals. The reason there are so many metals can be readily seen in terms of electron structure. When the last electrons in the electron configuration occupy d and f levels, the electrons go into inner shells. The elements in which this happens (those in the transition series and inner transition series) thus have only one or two s electrons in their outermost shells. Therefore, all of these elements tend to lose their valence electrons, so they are metals. Only the elements on the far right of the periodic table can be nonmetals. The actual dividing line between metals and nonmetals cuts through the periodic table like a staircase. It is shown as a heavy line on the right side of Table 13.3.

The physical properties of metals and nonmetals can be summarized as follows:

Metals

1. Good conductors of heat and electricity
2. Malleable—capable of being beaten into thin sheets
3. Ductile—capable of being stretched into wire
4. Of metallic luster
5. Opaque as a thin sheet (translucent if extremely thin)
6. Solids at room temperature. The one exception is mercury, Hg.

Nonmetals

1. Poor conductors of heat and electricity
2. Brittle—if a solid
3. Nonductile
4. Not of a metallic luster
5. Transparent as a thin sheet
6. Solids, liquids, or gases at room temperature.

The chemical properties of metals and nonmetals can be summarized as follows:

Metals

1. Have one to four electrons in their outer energy shell
2. Lose their outer electrons easily
3. Form hydroxides that are basic
4. Are good reducing agents
5. Are electropositive—they tend to become positively charged in compounds.

Nonmetals

1. Have four to eight electrons in their outer energy shell
2. Gain electrons readily
3. Form hydroxides that are acidic
4. Are good oxidizing agents
5. Are electronegative—they tend to become negatively charged in compounds.

One of the first ways in which metals and nonmetals were distinguished was by their chemical properties. When an element combines with oxygen, the resulting compound is called an **oxide.** Many oxides of metals dissolve in water to form aqueous solutions of bases. (Bases are defined and discussed in Chapter 16.) In general, the metallic character of the elements increases as you go down and to the left of the periodic chart. The nonmetallic properties increase upward and to the right. Thus, francium is the most metallic element and fluorine the most nonmetallic. Some elements display properties of both metals and nonmetals. Elements that exhibit this hybrid behavior are called **metalloids.** Five of these elements are found on the border between the metals and nonmetals. They are boron, silicon, germanium, arsenic, and tellurium. The periodic arrangement of the metals, nonmetals, and metalloids is shown by contrasting shades of color in Fig. 13.3 (p. 216).

The elements can also be classified according to whether they are solids, liquids, or gases at room temperature and pressure. Only two, bromine and mercury, occur as liquids. Eleven occur as gases. They are hydrogen, nitrogen, oxygen, fluorine, chlorine, and the six noble gases. All the rest are solids.

13.4 Families of Elements

A row of the periodic table is called a period. The elements in a column of the table are said to be in the same **group,** or **family,** of elements. All the elements in a family have similar electron configurations; therefore, they have similar properties. In particular, if one element in a family reacts with a given substance, then all the elements in the family will react in an analogous manner with that substance. The molecular formulas of the compounds produced will also be similar. In this section we will discuss several of these families of elements.

The Noble Gases

The noble gases and their electron configurations are

$_{2}\text{He}: 1s^2$

$_{10}\text{Ne}: 1s^2 2s^2 2p^6$

$_{18}\text{Ar}: 1s^2 2s^2 2p^6 3s^2 3p^6$

$_{36}\text{Kr}: 1s^2 2s^2 2p^6 3s^2 3p^6 4s^2 3d^{10} 4p^6$

$_{54}\text{Xe}: 1s^2 2s^2 2p^6 3s^2 3p^6 4s^2 3d^{10} 4p^6 5s^2 4d^{10} 5p^6$

$_{86}\text{Rn}: 1s^2 2s^2 2p^6 3s^2 3p^6 4s^2 3d^{10} 4p^6 5s^2 4d^{10} 5p^6$
$6s^2 4f^{14} 5d^{10} 6p^6$

IA	IIA	IIIB	IVB	VB	VIB	VIIB	VIII	VIII	VIII	IB	IIB	IIIA	IVA	VA	VIA	VIIA	0
H 1																	He 2
Li 3	Be 4											B 5	C 6	N 7	O 8	F 9	Ne 10
Na 11	Mg 12											Al 13	Si 14	P 15	S 16	Cl 17	Ar 18
K 19	Ca 20	Sc 21	Ti 22	V 23	Cr 24	Mn 25	Fe 26	Co 27	Ni 28	Cu 29	Zn 30	Ga 31	Ge 32	As 33	Se 34	Br 35	Kr 36
Rb 37	Sr 38	Y 39	Zr 40	Nb 41	Mo 42	Tc 43	Ru 44	Rh 45	Pd 46	Ag 47	Cd 48	In 49	Sn 50	Sb 51	Te 52	I 53	Xe 54
Cs 55	Ba 56	* 57–	Hf 72	Ta 73	W 74	Re 75	Os 76	Ir 77	Pt 78	Au 79	Hg 80	Tl 81	Pb 82	Bi 83	Po 84	At 85	Rn 86
Fr 87	Ra 88	† 89–	Rf 104	Ha 105													

Metals Metalloids Nonmetals

*	La 57	Ce 58	Pr 59	Nd 60	Pm 61	Sm 62	Eu 63	Gd 64	Tb 65	Dy 66	Ho 67	Er 68	Tm 69	Yb 70	Lu 71
†	Ac 89	Th 90	Pa 91	U 92	Np 93	Pu 94	Am 95	Cm 96	Bk 97	Cf 98	Es 99	Fm 100	Md 101	No 102	Lr 103

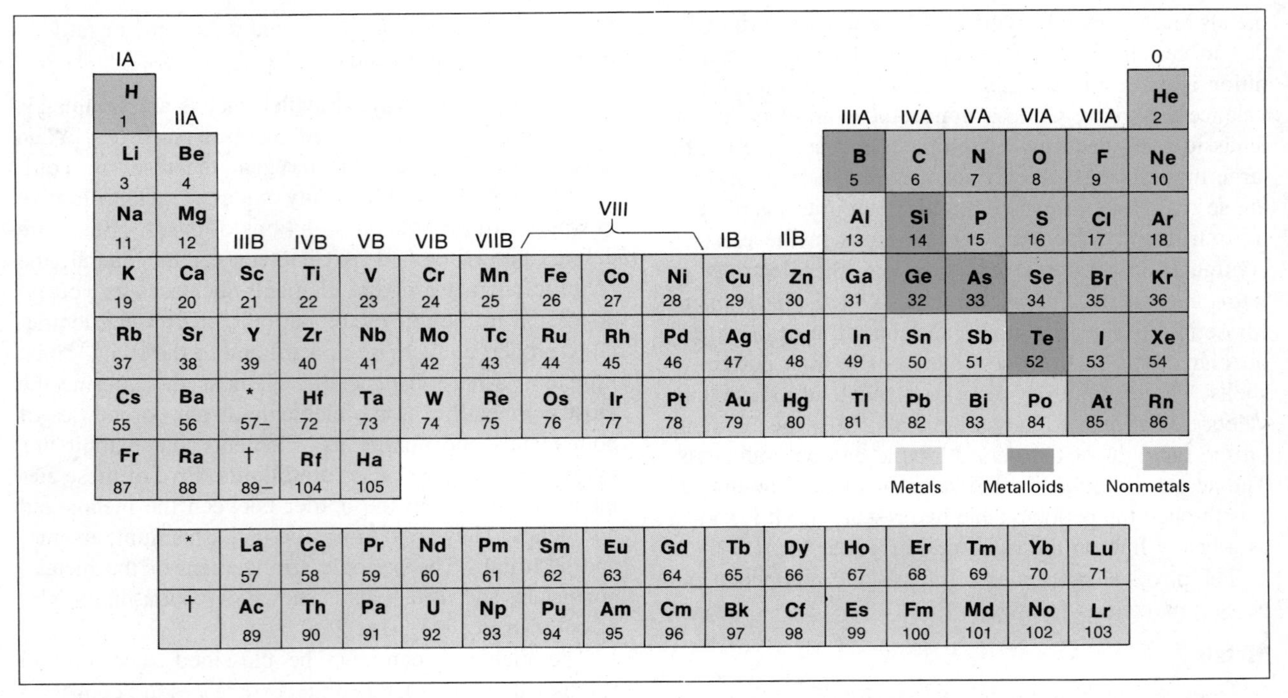

Figure 13.3 Periodic arrangement of the metals, metalloids, and nonmetals.

Each of the noble gases contains complete periods of electrons; and except for helium, they all contain eight electrons in their outermost shells. They are **monatomic;** that is, they exist as single atoms in nature. This group of elements almost never reacts with other elements to form compounds. We can conclude, then, that electron configurations with completed periods, or eight electrons in the outer shell, are quite stable. This conclusion is of great importance when considering the stability of other atoms. All atoms, except those of hydrogen and helium, tend to accept, donate, or share electrons with other atoms so as to have an octet of electrons (eight electrons) in their outer shell.

The noble gases were once called inert gases because they did not react. However, in recent years chemists have caused them to react in certain circumstances. Hence, the terminology inert gases no longer applies. The noble gases are sometimes called the **rare gases** because of their scarcity. Helium can be obtained from natural gas, and radon is a radioactive by-product of radium's decay; but the other noble gases are found only in the air. Their chief use is in ''neon signs.'' These are made by putting minute amounts of various rare gases into a sealed glass vacuum tube. When an electric current is passed through the tube, the gases glow. The particular color of the glow depends on the mixture of rare gases used.

Because the noble gases have similar electron configurations, many of their physical and chemical properties are similar. In Table 13.4 some of their physical properties are summarized. From this table we see that the melting points, boiling points, and atomic radii increase with increasing atomic number. However, the important fact here is that they have similar melting points and boiling points, particularly when compared to other atoms in the periodic table. The atomic radii will be discussed in Section 13.5.

The Alkali Metals

The elements in the far left column of the periodic table, except for hydrogen, are called the **alkali metals.** Hydrogen is omitted because of its unique spot in the table. It is the only element that can achieve a closed outer shell by either gaining or losing an electron. The electron configurations of the alkali metals all end with one *s* electron in the outer shell. Thus, they have one valence electron. The alkali metals tend to lose this outer electron quite easily so that a stable, closed-shell electron configuration will be reached. Thus, they react readily with other elements. The elements sodium and potassium are abundant in the Earth's crust, but lithium, rubidium, and cesium are rare. Francium, which is radioactive, is very

Table 13.4 Physical Properties of the Noble Gases

Element	Symbol	Atomic Number	Atomic Weight	Melting Point (°C)	Boiling Point (°C)	Atomic Radius (Å)
Helium	He	2	4.00260	−272	−268.9	0.93
Neon	Ne	10	20.179	−248.6	−246.1	1.12
Argon	Ar	18	39.948	−189.3	−186	1.54
Krypton	Kr	36	83.80	−157.2	−153.4	1.69
Xenon	Xe	54	131.30	−111.0	−108.1	1.90
Radon	Rn	86	(222)	−71	−62	2.2

rare. The most common compound containing an alkali metal is table salt, NaCl. Other alkali metal compounds known to the ancients were potash (potassium carbonate, K_2CO_3) and washing soda (sodium carbonate, Na_2CO_3).

The formulas and names of some common stable compounds of sodium are

NaCl	Sodium chloride (common table salt)
NaOH	Sodium hydroxide (caustic soda)
Na_2CO_3	Sodium carbonate (washing soda)
$NaHCO_3$	Sodium hydrogen carbonate (baking soda; bicarbonate of soda)
Na_2O	Sodium oxide

By knowing these compounds of sodium and the fact that all the elements in a family produce similar compounds, we automatically know a great deal about the compounds of the other alkali metals. Thus, we expect potassium oxide to have the molecular formula K_2O. We do not expect it to be KO, KO_2, KO_3, K_2O_5, and so on. Similarly, lithium carbonate should be, and is, Li_2CO_3.

Some of the physical properties of the alkali metals are given in Table 13.5 The concept of families can be used to estimate the melting and boiling points (at present unknown) of the radioactive element francium. We expect all the alkali metals to have similar properties, so we would guess from Table 13.5 that francium's melting and boiling points should be about 20°C and 680°C, respectively. These guesses, of course, are not exact but should be close.

The Halogens

The elements of the periodic table in the column next to the far right are called the **halogens.** They have seven electrons in the outer *s* and *p* subshells and need only one more electron to achieve a closed-shell electron configuration. Thus, they tend to react quite strongly with other elements and so are not found free in nature. The reactivity of the halogens decreases as you go down the periodic table. Generally, the halogens are all classified as nonmetals. Fluorine and chlorine occur at room temperature as gases, bromine as a liquid, and iodine and astatine as solids. The type of iodine found in a medicine cabinet is not the pure solid element. It is a tincture—solid iodine dissolved in a denatured alcohol. Some basic properties of the halogens are given in Table 13.6.

Table 13.5 Physical Properties of the Alkali Metals

Element	Symbol	Atomic Number	Atomic Weight	Melting Point (°C)	Boiling Point (°C)	Atomic Radius (Å)
Lithium	Li	3	6.941	180	1326	1.23
Sodium	Na	11	22.9898	97.5	889	1.57
Potassium	K	19	39.102	63.4	757	2.03
Rubidium	Rb	37	85.4678	38.8	679	2.16
Cesium	Cs	55	132.9055	28.7	690	2.35
Francium	Fr	87	(223)	—	—	2.7

Table 13.6 Physical Properties of the Halogens

Element	Symbol	Atomic Number	Atomic Weight	Melting Point (°C)	Boiling Point (°C)	Atomic Radius (Å)	Physical State
Fluorine	F	9	18.9984	−218	−188	0.72	Gas
Chlorine	Cl	17	35.453	−101	−34	0.99	Gas
Bromine	Br	35	79.90	−7.3	58.8	1.14	Liquid
Iodine	I	53	126.9045	114	184	1.33	Solid
Astatine	At	85	(210)	—	—	1.40	Solid

Fluorine is the most reactive of all the elements. It corrodes even platinum, a metal that withstands most other chemicals. In a stream of fluorine gas, wood, rubber, and even water burst into flame. Fluorine is a pale-yellow, highly poisonous gas. It was responsible for the deaths of several very able chemists before it was finally isolated. Chlorine is also a poisonous gas. It is used as a disinfectant in swimming pools and as a purifying agent in public water supplies. About one part chlorine is used per one million parts water. Bromine is a foul-smelling poisonous liquid. It is also used as a disinfectant. Iodine is a blue-black solid. Most table salt is now "iodized"; i.e., NaI is added, to supplement the human diet; an iodine deficiency causes thyroid trouble. Some formulas and compound names for the halogens are

NaI	Sodium iodide
$AlCl_3$	Aluminum chloride
SnF_4	Stannic fluoride
PCl_5	Phosphorus pentachloride
HBr	Hydrogen bromide
$CaCl_2$	Calcium chloride

By analogy with these formulas, we can write the correct formulas of many other stable compounds, for example, $NaBr$, AlF_3, and $SnCl_4$.

The Alkaline Earths

The **alkaline earths** are found in column IIA of the periodic table. They contain two valence electrons; hence, they are metals. They are quite active chemically, but they are not as reactive as the alkali metals. They also have higher melting points and are harder and stronger than their neighbors in column IA. Some of the properties of the alkaline earths are given in Table 13.7. Calcium is used by the bodies of mammals in the formation of bones and teeth. When radioactive strontium-90 is ingested, it goes into bone marrow because of its similarity to calcium. If enough strontium-90 is ingested, it can destroy the bone marrow or perhaps cause cancer. Magnesium is used in making lightweight metal alloys. Beryllium has a similar use. Some typical alkaline earth compounds are

$CaCO_3$	Calcium carbonate
SrO	Strontium oxide
$Mg(OH)_2$	Magnesium hydroxide
$BaSO_4$	Barium sulfate
$BeCl_2$	Beryllium chloride

There are other groups of elements which, because of similar group characteristics, can be classified as fam-

Table 13.7 Physical Properties of the Alkaline Earths

Element	Symbol	Atomic Number	Atomic Weight	Melting Point (°C)	Boiling Point (°C)	Atomic Radius (Å)
Beryllium	Be	4	9.01218	1,283	1,500	0.89
Magnesium	Mg	12	24.305	650	1,120	1.36
Calcium	Ca	20	40.08	850	1,490	1.74
Strontium	Sr	38	87.62	790	1,348	1.91
Barium	Ba	56	137.34	704	1,638	1.98
Radium	Ra	88	(226)	700	1,500	—

ilies. We have considered only four as examples. As mentioned previously, one element, hydrogen, is unique. It has one valence electron and lacks one valence electron. It acts sometimes like an alkali metal, forming HCl, H_2S, etc. (similar to NaCl, Na_2S) and sometimes like a halogen, forming NaH, CaH_2, etc. (compare with NaCl, $CaCl_2$). It is the lightest of the elements. At room temperature, it is a colorless, odorless, highly flammable gas. It has a melting point of −259°C and a boiling point of −252.7°C.

13.5 Periodic Characteristics

In this section the periodic characteristics of several properties of the elements will be discussed. One of these is the size of the atoms that compose the elements. Figure 13.4 is a plot of the atomic radius *versus* atomic number of elements in the periodic table. Generally speaking the atomic radii decrease from left to right in a period. Notice the large radius of each alkali metal, with respect to the other elements of that period. This is because the outer electron of an atom of an alkali metal is loosely bound to the nucleus. In a family of elements, the atomic radii increase from top to bottom. Each successive element of the family or group has an additional energy level containing electrons. As the charge on the nucleus increases (more protons), the outer electrons are bound more firmly, thus decreasing the atomic radius. This continues until the outer shell of an element is filled. The radius then jumps to a much higher value, since the next element has an electron in a new outer shell.

When an atom gains or loses electrons it acquires a net electric charge. The net electric charge is the number of protons minus the number of electrons. For example, if there are 8 protons and 10 electrons, the net electric charge is minus 2, written 2−. An atom or group of atoms with a net electric charge is called an **ion.** Magnesium, atomic number 12, normally has 12 protons and 12 electrons for a net electric charge of zero. When it loses its two outer (valence) electrons, it has 12 protons and 10 electrons for a net charge of 2+. The net charge on the atom is sometimes called the **valence.** The valence also can be thought of as the number of electrons an atom can give up (or acquire) to achieve a filled outer shell. Magnesium has a valence of 2+. Recall that magnesium has two valence electrons also. The valence of positive ions is frequently equal to the number of valence electrons. The negative charge of an ion can also be an indication of valence. Consider oxygen with 8 protons and 10 electrons, for a net charge of 2−. Thus, oxygen has a valence of 2−. Note that when an element gains or loses electrons it does not lose its chemical identity. This is because it is the atomic number, not the number of electrons, that determines the identity of the element.

The amount of energy necessary to remove one electron from an atom is called its **ionization potential.** The first ionization potential is the energy required to remove the most loosely held electron. The second ionization potential is the energy to remove a second electron, and so on. The periodic nature of the first ionization potential is shown for the first 90 elements in Fig. 13.5. Notice that the alkali metals have the lowest first ionization

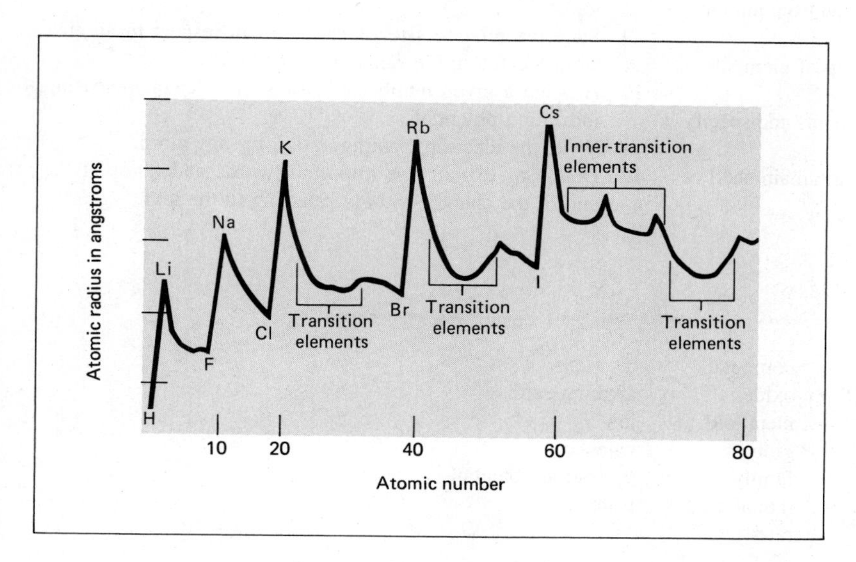

Figure 13.4 Plot of the relationship between the atomic radius and the atomic number of elements in the periodic table. At the beginning of each period the radius is greatly increased. This accounts for the relatively large radii of the alkali metals.

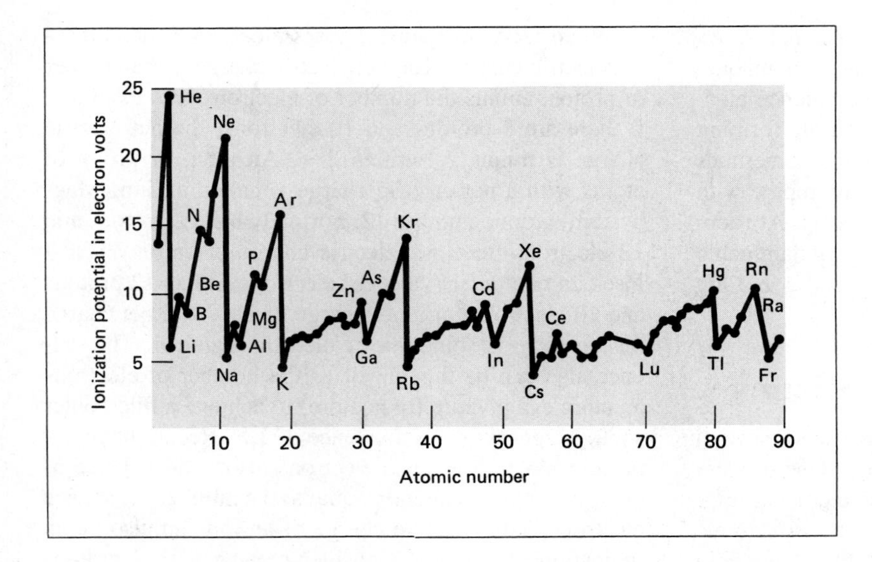

Figure 13.5 Plot showing the relationship between the first ionization potential and the atomic number for the first 90 elements in the periodic table.

potentials. This is because their last electron is in an outer shell, and it does not take much energy to remove the electron. Elements located to the right of the alkali metals in the table have more electrons added in the same shell when protons are added to the nucleus. The added protons bind the electrons more and more strongly until the shell is completely filled. In general, the ionization potential increases from left to right in the periodic table. There are a few exceptions. The noble gases have the highest ionization potentials.

Learning Objectives

After reading and studying this chapter, you should be able to do the following with the use of the periodic table:

1. Name the symbol for each element given in the periodic table and specify the element's atomic number, atomic weight, atomic mass, proton number, neutron number, electron number, and valence.
2. Name the elements in any family or group of elements.
3. Name the elements in any period.
4. Name the four main classifications of elements and specify the elements in each classification.
5. State the number of electrons located in the main shell of each atom.
6. Write the electron configuration for each element listed in the table.
7. State the phase of each element when the element is at normal atmospheric pressure and temperature.
8. Name the elements that are metals, nonmetals, or metalloids.
9. State the relative size of one atom in respect to another atom located in the same group or family.
10. Arrange a given number of atoms in order of increasing ionization potential.
11. Write the electronic configuration for any atom.
12. Define or explain the important words and terms at the end of the chapter without referring to the text.

Important Words and Terms

matter
element
molecules
formula
periodic table
periodic law
periods
noble gases
representative elements
transition elements
lanthanide series
rare earth series
actinide series
inner transition elements
valence electrons
metal
nonmetal
oxide
metalloid
group
family
monatomic
rare gases
alkali metals
halogens
alkaline earths
ion
valence
ionization potential
isotopes

Questions

1. Distinguish between an atom and an element.
2. Explain why there are many more different kinds of atoms when we know of only 108 different kinds of elements.
3. What was the view held by the Greek philosopher Aristotle concerning the composition of matter?
4. State the differences between deuterium and ordinary hydrogen.
5. Distinguish between atomic number and atomic weight.
6. What is the modern basis for the periodic law?
7. Looking at the periodic table, we notice that there are four major groupings of columns. There are 2 columns on the left, 10 in the middle, 6 on the right, and 14 below. What causes these groupings of 2, 10, 6, and 14 columns?
8. Define the terms metal and nonmetal.
9. What are metalloids? Give three examples.
10. State some physical properties that distinguish metals from nonmetals.
11. State some chemical properties that distinguish metals from nonmetals.
12. What are the two most abundant metals?
13. On which isotope are all atomic weights based?
14. What determines the chemical properties of elements?
15. List the elements that are gases and liquids at room temperature and atmospheric pressure.
16. List several elements that do not react readily. List several elements that are highly reactive.
17. A search is on for the element with atomic number 114. Which element will it most resemble?
18. Explain why the alkali metals have larger radii than the elements with one or two, more or less, protons.
19. How would you expect the ionization potential and atomic radius of francium to compare with the other alkali metals?
20. State the group number of each of the following elements: (a) scandium, (b) lead, (c) niobium, (d) hahnium, (e) uranium.
21. State the period of each of the following elements: (a) helium, (b) zinc, (c) radium, (d) europium, (e) plutonium.

Problems

1. Using the periodic table, determine the atomic number, proton number, atomic weight, mass number, neutron number, electron number, and number of valence electrons of each of the following elements and state whether each is a metal, or nonmetal, or an inert gas: (a) lithium, (b) neon, (c) calcium, (d) chlorine, (e) niobium, (f) gold, (g) holmium, (h) uranium.

   Answer: (c) 20, 20, 40.08, 40, 20, 2, metal
2. Arrange the elements cesium, lithium, potassium, rubidium, and sodium according to increasing atomic radii.
3. Arrange the elements argon, helium, krypton, and neon according to decreasing ionization potential.

   Answer: He, Ne, Ar, Kr
4. Which atom has the largest radius in each of the following pairs of elements: sodium or potassium, lead or nickel, polonium or selenium, argon or xenon?
5. Classify the following elements as noble gas, representative element, transition element, or inner transition element:

   (a) krypton, (b) iron, (c) iodine, (d) strontium, (e) hydrogen, (f) helium, (g) nitrogen, (h) mercury, (i) uranium.
6. Determine the most common valence of each of the elements in problem 5.
7. Arrange the atoms cobalt, helium, cesium, tin, and chlorine in order of increasing atomic radius.

   Answer: He, Cl, Co, Ti, Cs
8. Arrange the atoms helium, magnesium, rubidium, zinc, and xenon in order of increasing ionization potential.
9. Write the electron configuration for
   (a) The first transition element scandium (Sc).
   (b) The second transition element yttrium (Y).

   Answer: (a) $_{21}\mathrm{Sc}1s^22s^22p^63s^23p^64s^23d^1$
10. Write electron configurations for the coinage metals—copper, silver, and gold. Consult the periodic table for help. (Note: These configurations are exceptions to the rules of Chapter 12 for writing configurations. The three coinage metals only have one valence electron.)

14

Compounds, Molecules, and Ions

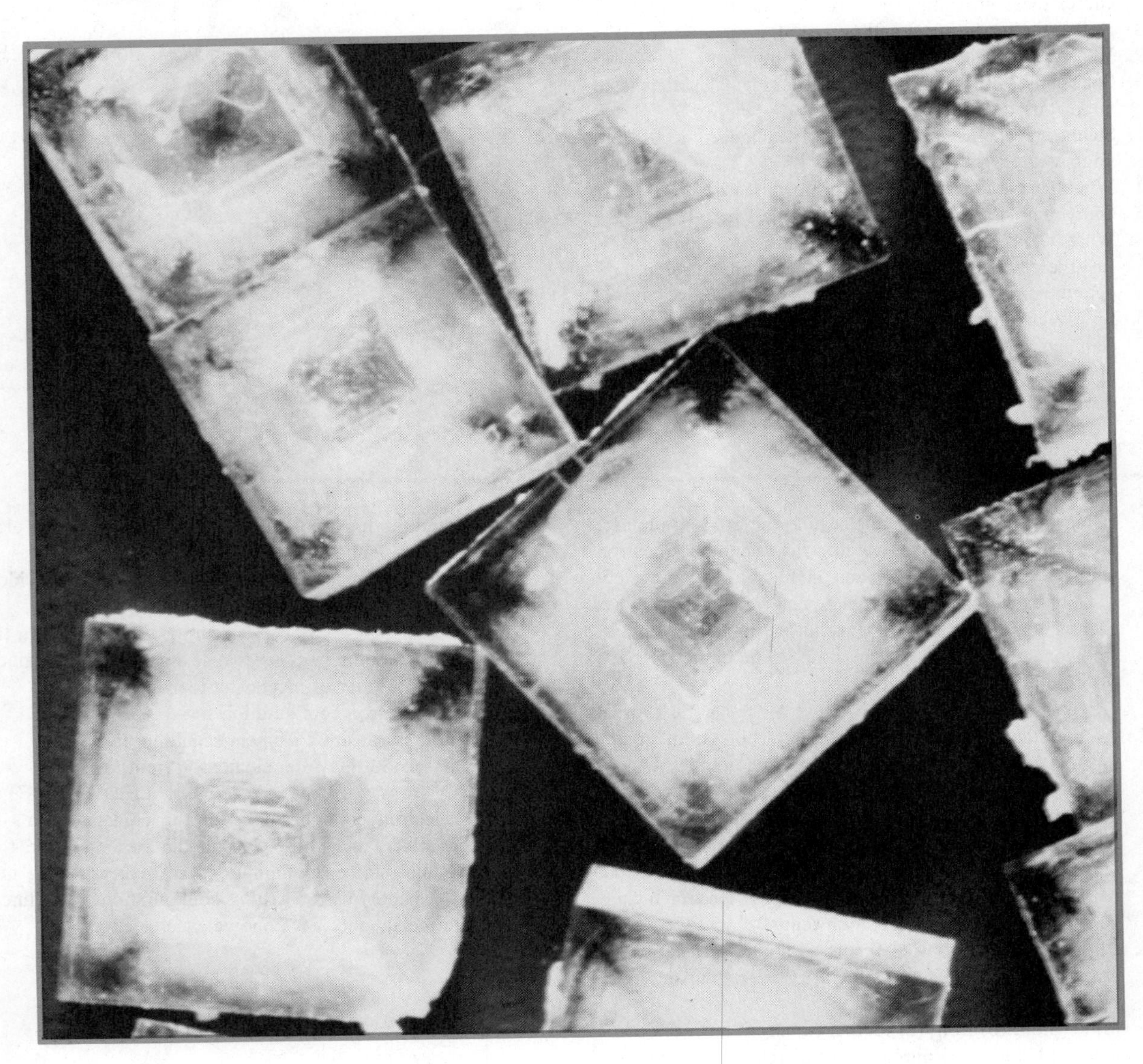

Happy the man who, studying nature's laws,
Through known effects can trace the secret cause.
–Virgil

THERE ARE 108 known elements. In Chapter 13, they were arranged in order of increasing atomic number to form the periodic table. With the use of the table, we learned many of the properties of these known elements.

Matter found in our environment is usually in the form of mixtures. A **mixture** is a nonchemical combination of two or more substances of varying proportions. When we look around, we see examples of mixtures, such as ocean water, soil, concrete, plants, and animals. Our environment is composed of millions of compounds and mixtures. In this chapter we lay the foundations for compound formation and study the properties of some of the compounds that make up our world. A **compound** is a substance composed of two or more elements chemically combined in a definite proportion. As you now know, an element is made up of only one kind of atom. Some compounds are made up of ions, others of molecules. We stated in Chapter 13 that an ion is an atom or a chemical combination of atoms with a net electric charge. A **molecule** is an uncharged particle of an element or a compound. The properties of many ions and molecules play an important role in our environment. Some molecules, such as the nitrogen oxides, sulfur dioxide, carbon dioxide, and carbon monoxide pollute the air we breathe. Others play a vital role in our being alive. For instance, we will see in Chapter 22 that the ozone, carbon dioxide, and water molecules in the air absorb much of the Sun's energy, thus keeping the Earth at a liveable temperature. Similarly, the various minerals and rocks that we find in our Earth obtain their properties from those of the various molecules and ions that constitute them.

The formation of ions and molecules is due to the electromagnetic forces between all the various electrons and the nuclei of the atoms involved. It is a very complicated problem, because there are so many forces to consider. Modern chemists use large computers and the techniques of quantum mechanics to try to understand how molecules and ions form. Although there is still a lot to be understood, chemists are now able to produce new compounds having almost any desired property. We begin our study with some of the basic principles of compound formation and the properties of some of the different types of compounds.

◀ Sodium chloride crystals.

14.1 Principles of Compound Formation

The noble gases, with their outer shells filled with electrons, are said to have closed shell electron configurations. The noble gases are monatomic, and it is extremely difficult to make them combine with any of the other elements. From these facts we can conclude that closed shell configurations result in extremely stable substances. Another important fact to remember in compound formation (see Section 13.4) is that elements in the same family form similar compounds. Since all the elements in the same family have an equal number of valence electrons, we conclude that it is the outer electrons that are important in compound formation. We now have a justification for the two basic assumptions that are made to explain compound formation.

1. The only electrons that take part in compound formation are those in the outer shell.
2. Compounds are formed when each atom in the molecule or ion achieves a closed shell electron configuration.

Both of these assumptions, however, are violated at various times in compound formation. In the transition and inner transition elements, some of the inner *d* or *f* shell electrons sometimes take part in compound formation. This violates the first assumption. There are also exceptions to the second assumption. Occasionally, a compound is formed in which one or more of the atoms does not have a closed shell configuration. However, since most compounds do obey our basic assumptions, we will use them and neglect the few exceptions that occur.

Except for elements in the first period (hydrogen and helium), there can be a maximum of eight electrons in the outer shell of elements in every period. In order for an atom to achieve a noble gas (closed shell) configuration, it must have either eight or zero valence electrons. Atoms achieve eight or zero valence electrons by gaining, losing, or sharing electrons with other atoms. The concept that atoms tend to have eight electrons in their outer shell when forming molecules or ions is called the **octet rule.** Hydrogen, of course, is the most important exception to this rule. It must have two electrons in its outer shell when it combines with other atoms. Other exceptions to the octet rule occur with elements in the two transition groups. However, in the remaining chapters we shall be dealing almost exclusively with the representative elements, for which the octet rule holds, generally.

There are two ways in which individual atoms can achieve a closed shell electron configuration. They can transfer electrons or they can share electrons. In the transfer of electrons one or more elements lose some or all of their outer electrons, and another one or more elements gain these same electrons to achieve closed shell configurations. Compounds formed by this electron transfer process are called **ionic compounds.** Because these compounds in the liquid phase conduct electricity, they are sometimes referred to as **electrovalent compounds.** The majority of compounds are formed when electrons are shared between different atoms. Compounds formed by the electron-sharing process are called **covalent compounds.** Now let us examine these two kinds of compounds.

14.2 Ionic Compounds

Compounds that are formed when electrons are lost or gained are referred to as ionic compounds. This is because when the electrons are lost or gained by the various atoms, ions are formed. The energy released when an atom gains an electron, forming a negative ion, is called **electron affinity.** The nonmetals have high electron affinities and the metals have low electron affinities. Thus only the nonmetals tend to form negative ions by gaining electrons. An atom that needs only one or two electrons to fill its outer shell can easily acquire the electrons from atoms that have low ionization energy, i.e., atoms with only one or two electrons in the outer shell such as those in groups IA and IIA of the periodic table. To illustrate this, let us consider how common table salt, NaCl, is formed. The sodium atom can be drawn schematically with two electrons in the first shell, eight in the second, and one in the third. The chlorine atom has two in the first, eight in the second, and seven in the third. Diagrams for these two atoms are shown in Fig. 14.1. The sodium atom tends to lose its outer electron so that it will have zero electrons in its third shell. The chlorine atom tends to gain an electron so that it will have eight electrons in its outer shell. When the sodium atom loses its electron, it becomes a sodium ion with 11 positive

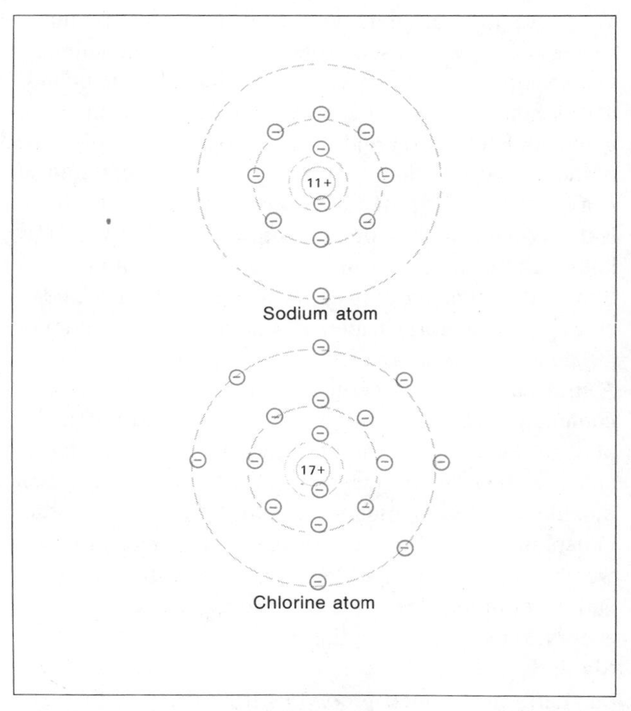

Figure 14.1 Schematic diagrams of the two atoms that comprise common table salt. The sodium atom has 11 positive charges in its nucleus, the chlorine atom 17. Each small circle represents a negatively charged electron orbiting in the shells. The charge on each atom is zero.

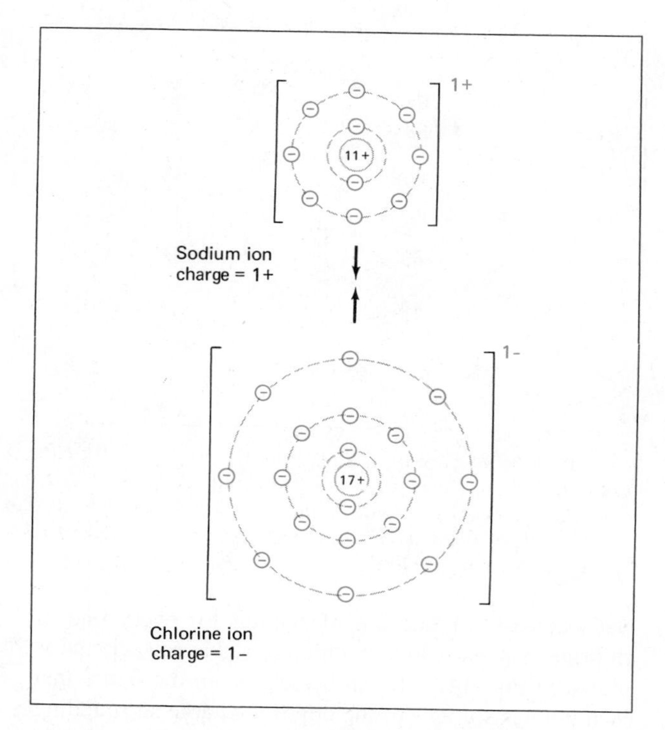

Figure 14.2 Diagram illustrating the formation of table salt, NaCl. The oppositely charged ions are attracted to each other.

charges and 10 negative charges, for a net electric charge of 1+. The chlorine atom gains the electron lost by the sodium atom and becomes a chlorine ion with 17 positive charges and 18 negative charges. It has a net charge of 1^-. The positively charged sodium ion is then attracted to the negatively charged chlorine ion. This situation is illustrated in Fig. 14.2.

Figures 14.1 and 14.2 can be represented very easily, using what is known as the electron dot notation. In this notation the nucleus and inner electrons of an element are represented by the element symbol and the valence electrons by dots. The sodium and chlorine atoms of Fig. 14.1 would be represented in the electron dot notation as

$$\mathrm{Na}\cdot \qquad :\ddot{\underset{\cdot}{\mathrm{Cl}}}:$$

The sodium and chlorine ions of Fig. 14.2 would be represented as

$$\mathrm{Na}^+ \qquad :\ddot{\underset{\cdot\cdot}{\mathrm{Cl}}}:^-$$

The charge of the ion is represented by the number of plus or minus signs. An ion with two negative charges is sometimes designated by a double minus sign (=) but most often by 2^-. Consider the following examples:

$$\mathrm{Mg}^{++} \quad \text{or} \quad \mathrm{Mg}^{2+}$$

$$:\ddot{\underset{\cdot\cdot}{\mathrm{O}}}:= \quad \text{or} \quad :\ddot{\underset{\cdot\cdot}{\mathrm{O}}}:^{2-}$$

The charge on most ions of the representative elements can easily be determined as follows: The positive charge will be equal to the number of valence electrons. The negative charge will be eight minus the number of valence electrons. For example, aluminum, in column IIIA of the periodic table, has three valence electrons. The aluminum ion thus will have a charge of 3+. Sulfur, with six valence electrons, tends to fill its outer shell. Its ion will have a negative charge of 8 minus 6, or a charge of 2^-. Some of the main group elements plus their ions are given in Table 14.1 along with an illustration of their electron dot notation. We have not included elements in group IVA of the periodic table because these elements share electrons and thus form covalent rather than ionic compounds.

In forming simple ionic compounds one element loses its electrons and the other element gains them, resulting

Table 14.1 Electron Dot Notation and Common Ions for Some Main Group Elements

I A		*II A*		*III A*		*V A*		*VI A*		*VII A*	
atom	ion	atom	ion	atom	ion	atom	ion	atom	ion	atom	ion
$\mathrm{H}\cdot$	H^+										
$\mathrm{Li}\cdot$	Li^+	$\mathrm{Be}:$	Be^{2+}							$:\ddot{\underset{\cdot\cdot}{\mathrm{F}}}:$	$:\ddot{\underset{\cdot\cdot}{\mathrm{F}}}:^-$
$\mathrm{Na}\cdot$	Na^+	$\mathrm{Mg}:$	Mg^{2+}	$\underset{\cdot}{\mathrm{Al}}:$	Al^{3+}	$\cdot\underset{\cdot\cdot}{\mathrm{N}}:$	$:\ddot{\underset{\cdot\cdot}{\mathrm{N}}}:^{3-}$	$:\underset{\cdot\cdot}{\mathrm{O}}:$	$:\ddot{\underset{\cdot\cdot}{\mathrm{O}}}:^{2-}$	$:\ddot{\underset{\cdot\cdot}{\mathrm{Cl}}}:$	$:\ddot{\underset{\cdot\cdot}{\mathrm{Cl}}}:^-$
$\mathrm{K}\cdot$	K^+	$\mathrm{Ca}:$	Ca^{2+}	$\underset{\cdot}{\mathrm{Ga}}:$	Ga^{3+}	$\cdot\underset{\cdot\cdot}{\mathrm{P}}:$	$:\ddot{\underset{\cdot\cdot}{\mathrm{P}}}:^{3-}$	$:\underset{\cdot\cdot}{\mathrm{S}}:$	$:\ddot{\underset{\cdot\cdot}{\mathrm{S}}}:^{2-}$	$:\ddot{\underset{\cdot\cdot}{\mathrm{Br}}}:$	$:\ddot{\underset{\cdot\cdot}{\mathrm{Br}}}:^-$

in ions. The ions are then held together by the electrical attraction between them. The formation of NaCl can be represented as

$$\mathrm{Na}\cdot \; + \; :\ddot{\underset{\cdot}{\mathrm{Cl}}}: \longrightarrow \mathrm{Na}^{+} \; :\ddot{\underset{\cdot\cdot}{\mathrm{Cl}}}:^{-}$$

In this example, the sodium atom lost one electron and the chlorine atom gained one. Thus, we see that for every ion of sodium there is one ion of chlorine. This one-to-one correspondence is necessary to balance the charges on the ions. Another example of an ionic compound is calcium oxide, CaO. Its formation is represented by

$$\mathrm{Ca}: \; + \; :\ddot{\mathrm{O}}: \longrightarrow \mathrm{Ca}^{2+} \; :\ddot{\underset{\cdot\cdot}{\mathrm{O}}}:^{2-}$$

In forming this compound the calcium atom has lost two electrons and the oxygen atom has gained two. Thus, to get a neutral compound, we need one ion of each element. Now, consider what happens with calcium and chlorine. Calcium has two electrons to lose, but each atom of chlorine can gain only one. There must be two atoms of chlorine to accept both of the electrons and still maintain a neutral charge on the compound. We have

$$\mathrm{Ca}: \; + \; :\ddot{\underset{\cdot}{\mathrm{Cl}}}: \; + \; :\ddot{\underset{\cdot}{\mathrm{Cl}}}: \longrightarrow :\ddot{\underset{\cdot\cdot}{\mathrm{Cl}}}:^{-}\mathrm{Ca}^{2+} \; :\ddot{\underset{\cdot\cdot}{\mathrm{Cl}}}:^{-}$$

All of the atoms in the compound have closed shells and the total charge is zero. The molecular formula of calcium chloride is $CaCl_2$. We are now beginning to see how molecular formulas arise. The numbers of atoms of the various elements involved in the compound are determined by the requirements that the total charge must be zero and that all the atoms are to have closed shell electron configurations. As another example, consider sodium and sulfur. Each sodium atom loses one electron and each sulfur atom gains two, so there must be two sodium atoms for every sulfur atom.

$$\mathrm{Na}\cdot \; + \; \mathrm{Na}\cdot \; + \; :\ddot{\mathrm{S}}: \longrightarrow \cdot\mathrm{Na}^{+} \; :\ddot{\underset{\cdot\cdot}{\mathrm{S}}}:^{2-} \; \mathrm{Na}^{+}$$

The formula for the compound formed, sodium sulfide, is thus seen to be Na_2S.

14.3 Properties of Ionic Compounds

Ionic compounds are formed by the electrical attraction of the ions. These compounds generally have higher melting and boiling points than covalent compounds. Ionic compounds occur in the solid phase as crystals. A **crystal** is an orderly arrangement of atoms. A crystal of

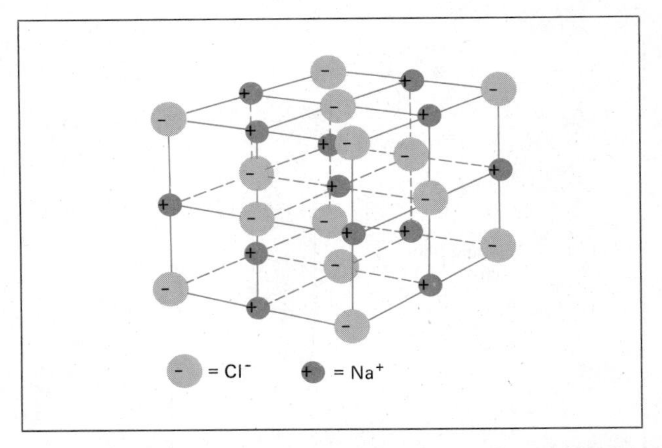

Figure 14.3 Schematic diagram of the sodium chloride crystal. The sodium ions are the smaller spheres with the plus charge indicated. The chloride ions have one more shell so they are drawn as bigger spheres with the minus charge indicated.

NaCl consists of one ion of sodium for every one of chlorine. Part of a sodium chloride (table salt) crystal is shown in Fig. 14.3. It can be seen from the figure that each ion has six ions of the opposite charge surrounding it in a three-dimensional array. Of course, Fig. 14.3 is simply a sketch of part of the crystal. An individual crystal has a large number of Na^+ and Cl^- ions. The NaCl crystal shown in Fig. 14.3 is an example of a cubic crystal; there are many other possible crystal arrangements. The cubic nature of the NaCl crystal can be seen by looking closely at a grain of ordinary table salt.

When we write the formula NaCl, it means that in the compound sodium chloride, there is one ion of Na^+ to one ion of Cl^-. On a microscopic scale it is impossible to associate any one ion of Na^+ with a particular ion of Cl^-. However, there will always be a one-to-one correspondence between the Na^+ and Cl^- ions. Similarly, the formula $CaCl_2$ means that, in the compound calcium chloride, there are two chlorine ions for every calcium ion.

One of the more important properties of ionic compounds is their behavior when an electric current is passed through the compound in its molten phase. This behavior is illustrated as follows: If a light bulb is connected in series with a battery, it lights up. This is due to electrons flowing from the negative terminal of the battery, through the light bulb, and back to the positive terminal. If the wires are cut, the electrons cannot flow (no electric current), and the light bulb will not light. Now, if the cut wires are connected to graphite (pure carbon) rods, which are inserted into a melted ionic compound, such as NaCl, the bulb lights. That is, ionic compounds in the liquid

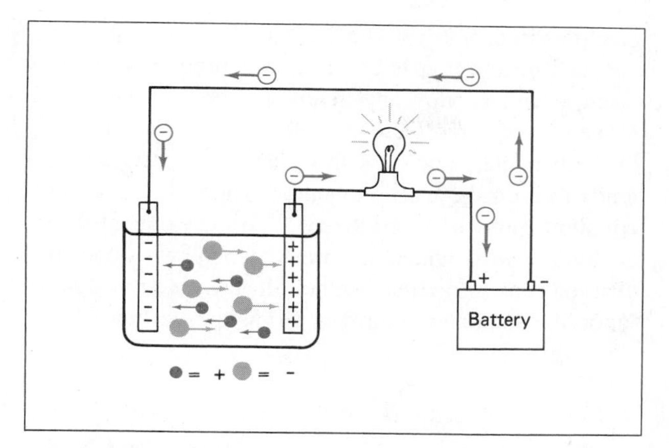

Figure 14.4 Diagram illustrating how melted NaCl conducts electricity. The lighted bulb indicates the flow of an electric current. The electric circuit is achieved because of the reactions of the ions at the electrodes.

form conduct electricity. Figure 14.4 is a diagram of this situation.

The explanation of how melted ionic compounds conduct electricity is not simple. *The electrons do not flow through the liquid*. The process is more complicated. The graphite rods that were inserted into the melt are charged positively and negatively, because they are connected to the positive and negative battery terminals. These charged rods are called *electrodes*. The positively charged rod is called the *anode;* the negatively charged rod is called the *cathode*. The positively charged sodium ions are attracted to the cathode while the negatively charged chlorine ions are drawn toward the anode. For this reason a positively charged ion is called a **cation,** and a negatively charged ion is called an **anion.** When a sodium ion reaches the cathode, an electron is taken from the cathode, and the sodium ion becomes a sodium atom. The reaction that takes place at an electrode is called a **half-reaction.** The half-reaction at the cathode is represented by

$$\mathrm{Na^+ + e^- \longrightarrow Na}$$

The net effect of this process is to take electrons off the cathode. At the anode, the negatively charged chlorine ions give up an electron to the positively charged anode. Then, two chlorine atoms form a covalent chlorine gas molecule. These appear as bubbles on the anode. The half-reaction is

$$\mathrm{2\,Cl^- \longrightarrow Cl_2 + 2\,e^-}$$

This half-reaction donates electrons to the anode. The electric current is thus achieved because of the reaction of the ions at the electrodes, and the light bulb lights up. To summarize, at the cathode electrons are removed; at the anode they are added. The net result is an electron flow, or current. Eventually, all the sodium and chlorine ions react. At this point the current ceases to flow, and the light goes out. An electric current will also be conducted if a solution of table salt in water is used instead of molten NaCl. Most ionic compounds dissolve in water, and the solutions will conduct an electric current in much the same way that molten NaCl does.

The half-reaction that occurs at the anode is an example of oxidation. **Oxidation** is the process in which electrons are lost. At the cathode, reduction takes place. **Reduction** is the process in which electrons are gained. The word oxidation comes from the word oxygen, since it is the material that frequently causes oxidation. For example, when pure iron (Fe) is exposed to the air, the oxygen atoms combine with the iron to form rust, which for our purpose can be designated by Fe_2O_3. The oxygen atoms gain electrons, so they will have a filled shell, and are said to be reduced. The iron atoms lose electrons. They are oxidized.

The total reaction in which electrons are gained and lost is called an **oxidation-reduction reaction.** If we add the two half-reactions in our example, we get

$$\begin{array}{rcl} \mathrm{2\,Na^+ + 2\,e^-} & \longrightarrow & \mathrm{2\,Na} \\ \mathrm{2\,Cl^-} & \longrightarrow & \mathrm{Cl_2 + 2\,e^-} \\ \hline \mathrm{2\,Na^+ + 2\,Cl^- + 2\,e^-} & \longrightarrow & \mathrm{2\,Na + Cl_2 + 2\,e^-} \end{array}$$

We have multiplied the half-reaction at the cathode by two before comparing it to the anode half-reaction. This is done to balance the number of electrons gained and lost. The two electrons on both sides of our oxidation-reduction reaction cancel, and we get

$$\mathrm{2\,NaCl \longrightarrow 2\,Na + Cl_2}$$

as our final reaction. By sending an electric current into melted NaCl, sodium and chlorine ions can be converted into pure sodium metal and chlorine gas. In general, the production of an oxidation-reduction reaction by means of an electric current is called **electrolysis.** The electrolysis of sodium and many other metals is an important commercial process that produces the metals in a pure form. Electrolysis is also the means by which electroplating is accomplished. The material to be plated is used as the cathode in electrolysis and a melted ionic compound or a solution of the ionic compound in water is the source of the metal to be plated.

So far, we have discussed several properties of ionic compounds. Their high binding energies, produced by the strong electrostatic attractions between ions, causes them to have high melting points. They do not exist as single molecules, but the individual ions occur in a ratio given by the molecular formula. They do not conduct electricity in the solid form but do in the liquid form. Finally, most ionic compounds dissolve in water, and in solution they conduct electricity. The mechanism by which this dissolving process occurs will be discussed after we have learned something about the structure of the water molecule.

14.4 Covalent Compounds

In the formation of ionic compounds electrons are transferred; in covalent compounds they are shared. Hydrogen gas, H_2, is the simplest example of a covalent molecule. It contains two atoms in each molecule. Molecules containing two atoms are called *diatomic*. When two hydrogen atoms are brought together, there is an attraction between each electron and proton and repulsion between the two electrons and between the two protons. This is shown schematically in Fig. 14.5. When this phenomenon is investigated using quantum mechanics and the Schrödinger equation, it is found that there is more attraction than repulsion up to a certain distance between atoms. When the positively charged nuclei get about an angstrom apart, they begin to repel each other. The hydrogen molecule consists of a total system of two nuclei and two electrons. Once the atoms are close together, the two electrons no longer orbit around individual nuclei. They both orbit around both of the protons, tracing out a circular path about one or the other. Of course, as was previously mentioned, the concept of electrons in orbits is somewhat naive. The quantum mechanical description is based on the probability that an electron will occupy any given spot around the nuclei. In the hydrogen molecule both electrons are shared equally by both nuclei, and it is this sharing of electrons that tends to hold the atoms together. There is said to be a **covalent bond** between the two atoms. Contrast this to an **ionic bond,** which is formed when two oppositely charged ions attract each other. In the electron dot notation the separated hydrogen atoms are written

H·   H·

and the H_2 molecule is written

H:H

The two dots between the hydrogen atoms indicate that these electrons are being shared.

Not all atoms will share electrons. For instance the helium atom, with a closed 1*s* shell, will not form a stable molecule with another helium atom. Two helium atoms repel each other, but two hydrogen atoms attract. The molecule H_2 is stable, but He_2 is not. It takes complicated quantum mechanical calculations to determine which molecules will be stable and which will not. However, our basic assumption serves as a good guide. Stable covalent molecules are formed when the atoms share electrons in such a way as to give all atoms a closed outer shell configuration.

Let us consider more examples of covalent bonding. Two chlorine atoms in the electron dot notation are given by

:C̤̈l·   ·C̤̈l:

Each chlorine atom needs one electron to have a closed shell. If each shares its unpaired electron, we get the following Cl_2 molecule.

:C̤̈l:C̤̈l:

In the Cl_2 molecule each chlorine atom has six electrons plus two shared electrons giving it a closed outer shell.

Covalent bonds can also be formed between unlike atoms. Consider the molecule HCl. The hydrogen and chlorine atoms have the dot structures

Ḣ   ·C̤̈l:

If each shares one electron, they both have closed outer shells.

H:C̤̈l:

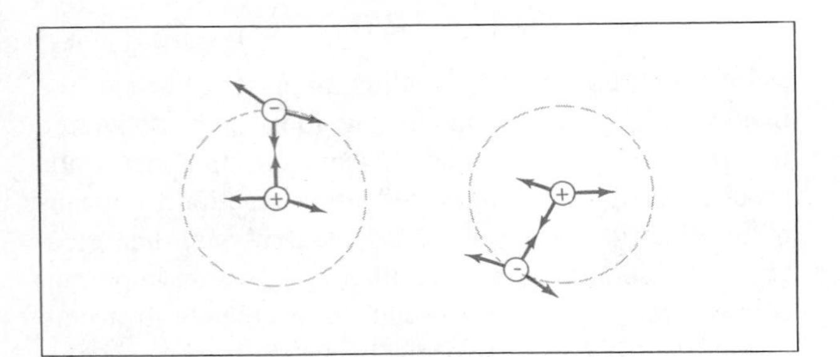

Figure 14.5 Illustrating the forces between two hydrogen atoms. Particles with unlike charges attract each other; those with like charges repel each other. Each particle has three forces on it due to charges on the other three particles. Arrows indicate the forces and the directions in which they are acting.

Sometimes more than one electron is shared by each atom. Consider how carbon dioxide, CO_2, is formed. In the dot notation two oxygen atoms and one carbon atom look like

```
:Ö:     ·Ċ·     :Ö:
          ·
```

In order for us to get a stable molecule, we need to have eight electrons around each atom. If only two electrons are shared between each oxygen atom and the carbon atom, we get the following structure:

```
 ·· ·· ··
:O:C:O:      (Unstable)
    ··
```

This structure is unstable because, although the carbon atom has eight electrons around it, each oxygen atom has only six electrons around it. In order to get a stable molecular structure, there must be four electrons shared between the carbon atom and each oxygen atom. The correct dot structure of the CO_2 molecule is

```
:Ö::C::Ö:
```

An ordinary single covalent bond consists of two shared electrons. A sharing of four electrons produces a double bond, and a sharing of six electrons produces a triple bond. Nitrogen gas, N_2, is an example of a triple bond. Two nitrogen atoms may be represented as

```
·N̈·     :N̈·
 ·
```

To satisfy the octet rule each nitrogen atom must share three electrons. The molecule N_2 looks like

```
:N:::N:
```

Another example of the triple bond is carbon monoxide, CO. Carbon and oxygen in dot notation look like

```
·Ċ·     :Ö:
 ·
```

The carbon molecule needs four more electrons and the oxygen molecule needs two to satisfy the octet rule. The only way this can be accomplished is for the oxygen atom to share four of its electrons with the carbon atom and for the carbon atom to share two of its electrons with the oxygen atom. Altogether, six electrons are shared.

```
:C:::O:
```

One of the most common molecules containing three atoms is water, H_2O. The individual atoms are, of course,

```
H·     :Ö:     ·H
```

When they combine, each hydrogen atom shares its electron with the oxygen atom, which shares two of its electrons.

```
 ··
H:O:H
 ··
```

The electron dot notation is used merely to indicate how the electrons are shared among atoms, not to indicate the spatial arrangement of the atoms. We could also have written H_2O as

```
                    H
 ··                 ··                ·O·
H:O:      or       :O:H      or      ·   ·
 ··                 ··              H     H
 H
```

The actual molecule looks more like these last three. The two O:H bonds form an angle of about 105° with each other.

The compound ammonia, NH_3, is a covalent combination of one nitrogen atom and three hydrogen atoms. Each hydrogen atom shares its electron with the nitrogen atom as follows:

```
 ··
H:N:H
 ··
 H
```

Methane gas, CH_4, is an example of a simple molecule with five atoms. Its structure can be represented by

```
  H
  ··
H:C:H
  ··
  H
```

More complicated electron dot structures arise when elements other than hydrogen are involved. The halogens combine analogously to hydrogen. Consider CH_3Cl. Its structure is just like CH_4 with a chlorine atom replacing a hydrogen atom.

```
  H
  ·· ··
H:C:Cl:
  ·· ··
  H
```

Other compounds with halogens are NH_2Cl, NI_3, $CHBrCl_2$. Their dot structures are

```
                                      ··
                                     :Br:
  ··              ··  ··  ··      ··  ··
H:N: H           :I :N: I:       :Cl:C:H
  ··              ··  ··  ··      ··  ··
 :Cl:                :I:             :Cl:
  ··                  ··              ··
```

Note that in each of the preceding structures the total number of electrons for the molecule is equal to the total number of valence electrons in the atoms making up the molecule. Also note that each atom has a closed shell of electrons around it.

It is sometimes necessary for a molecule to have single, double, and even triple bonds simultaneously in order to satisfy the octet rule. The molecules C_2H_3Br and C_2HI are examples.

```
H        :Br:
  :C::C:          H:C:::C:I:
H        H
```

In these structures the carbon atoms are bonded with double and triple bonds, while the other elements are held to the carbon atoms by single bonds.

Frequently, atoms combine together with a net charge. Such combinations are called *complex ions,* or polyatomic ions. They are also known as radicals. The atoms within the complex ion are covalently bonded. The whole aggregation then behaves like an ion in forming compounds. There is usually a strong bonding between atoms in complex ions, so that it is difficult to break them up. In chemical reactions they frequently act as a single unit. Some important examples of complex ions are given in Table 14.2.

The structure of the polyatomic ions of known charge can be determined by writing down the free atoms in their dot notation, adding or subtracting electrons depending on the charge on the ion, and then following the rules of compound formation. The structures of NO_3^- and PO_4^{3-} are examples.

```
                         :O:    3-
[:O::N:O:]-         [ :O:P:O: ]
     :O:                 :O:
```

Some compounds contain both ionic and covalent bonds. Sodium hydroxide, NaOH, is an example. Its structure is

$$Na^+[:\ddot{O}:H]^-$$

The hydroxide ion, $[OH]^-$, is covalently bonded together as a unit. In sodium hydroxide the hydroxide unit and sodium are bound by an ionic bond. Like most ionic compounds, sodium hydroxide dissolves in water. Other examples of ionic compounds formed by complex ions are silver nitrate, $Ag^+[NO_3]^-$, and barium sulfate, $Ba^{2+}[SO_4]^{2-}$.

So far, we have classified compounds as being either ionic or covalent. In ionic bonding the electrons are transferred from one element or group of elements to another. In covalent bonding the electrons are shared. Actually, most bonds are intermediate between these two extremes. For most bonds, the electrons are unequally shared between two atoms, because the atoms are unlike. Whenever two unlike atoms share electrons, one will have a greater attraction for the shared electrons than the other. The tendency to attract shared electrons is called **electronegativity,** the measure of the ability of an atom to attract electrons in the presence of another atom. Consider the polar molecule HCl. The chlorine atom needs an electron to fill its outer shell. The hydrogen atom also needs an electron to fill its outer shell, but if it loses its electron it will have a closed shell. The chlorine atom has a higher electronegativity than the hydrogen. Electronegativity increases from left to right within a period. The result is that although the electrons are shared between the two atoms, they tend to spend more time at the chlorine end than at the hydrogen end of the molecule. We get a molecule that looks like

$$\overset{+}{H}\ :\ddot{\underset{..}{Cl}}:^-$$

The + and − indicate that the chlorine end of the molecule tends to be negatively charged and the hydrogen end tends to be positively charged. A bond such as this, in which the electrons are unequally shared, is called a **polar bond.** All bonds between unlike atoms are polar bonds. Bonds between like atoms have a perfect sharing and are called **nonpolar bonds.**

Sometimes a molecule has several polar bonds arranged symmetrically. An example of this is methane, CH_4. Its structure is

```
      +H
       ..    +
  H: =C= :H
  +    ..
       H+
```

Methane has four polar bonds, but they are arranged so that no one side or end of the molecule has a more positive or negative charge than any other side or end. Because of this symmetry it is a **nonpolar molecule.** Most molecules have one end or side positively charged and one end or side negatively charged. Such molecules are called **polar molecules.** The HCl molecule is an example of a polar molecule.

Table 14.2 Some Important Complex Ions

Ammonium	NH_4^+	Carbonate	CO_3^{2-}
Hydroxide	OH^-	Sulfate	SO_4^{2-}
Nitrate	NO_3^-	Sulfite	SO_3^{2-}
Cyanide	CN^-	Phosphate	PO_4^{3-}
Nitrite	NO_2^-		

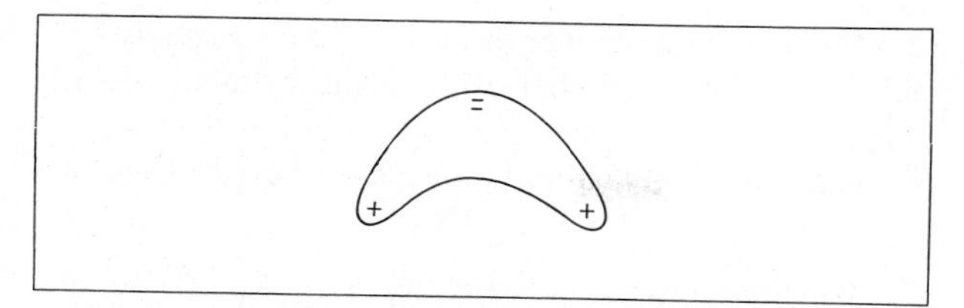

Figure 14.6 Schematic diagram of a water molecule showing its polar character.

Another example of a polar molecule is water. Its structure is

$$\overset{=}{\ddot{\mathrm{O}}}\quad {}^{+}\mathrm{H}\quad \mathrm{H}^{+}$$

In the water molecule we see that the oxygen end is negatively charged, and the hydrogen ends are positively charged. To emphasize the polar property of the water molecule it is sometimes represented as in Fig. 14.6. In ionic solutions the negative side of the water molecule attracts positive ions, while the positive ends attract negative ions.

14.5 Properties of Covalent Compounds

Because of the nature of the bonding involved, covalent compounds have quite different properties from those of ionic compounds. Although the covalent bond is strong within a molecule, the various molecules in the compound only weakly attract each other. Their binding energies in the solid and liquid forms are therefore not very high, and their melting points and boiling points are low compared to ionic compounds. Many covalent compounds occur as liquids or gases at room temperature. For instance, carbon tetrachloride, CCl_4, melts at $-23°$ C compared to 800°C for the ionic compound NaCl.

Many molecules are bonded by polar bonds. The more polar the bond, the more unequal the electron sharing. An ionic bond is actually a very polar bond. Polar molecules are partially covalent and partially ionic. In general, with the exception of the hydrogen atom, combinations of atoms from the far left and far right (ignoring the noble gases) of the periodic table form ionic compounds, for example, NaCl. Also, compounds formed from complex ions tend to be ionic. Compounds tend to be covalent if they are formed with elements from the same or adjacent columns of the periodic table, or if they are formed from any two nonmetallic atoms, such as hydrogen and nitrogen. In addition, all compounds formed with carbon are covalent.

Covalent compounds occur in the gas phase as individual molecules with a specific molecular formula. A molecule of gaseous CCl_4 consists of one carbon atom and four chlorine atoms. Ionic compounds have electric charges on the ions, and an ionic gas molecule might have any number of atoms in the ratio given by the molecular formula.

In liquid form, covalent compounds do not conduct electricity well. Benzene, a nonpolar covalent compound, does not conduct electricity at all. Pure water, a polar covalent compound, conducts electricity only very slightly. Ordinary tap water is far from pure. It usually has many ions dissolved in it, and therefore, tap water is an excellent conductor of electricity. This is why if a lightning bolt were to hit a lake or swimming pool, the occupants would be killed by the electric shock.

Water is a polar compound. When a solid ionic compound, such as table salt, is put into water, the ionic compound usually dissolves. What occurs when NaCl dissolves in water can be explained with the aid of Figs. 14.7 and 14.8. In Fig. 14.7, four ions of a salt crystal are shown surrounded by water molecules. In Fig. 14.8, the positively charged ends of the water molecules are attracted to the Cl^- ions, and the negatively charged ends of the water molecules are attracted to the Na^+ ions. The water molecules surrounding the ions tend to decrease the attraction of the Na^+ and Cl^- ions for one another. This dissolves the salt crystal. Ions surrounded

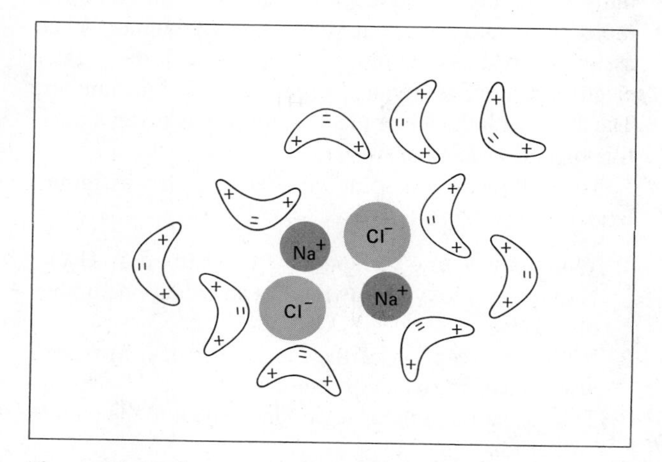

Figure 14.7 Schematic diagram of a NaCl crystal in water before the crystal is dissolved. A portion of the salt crystal is shown along with a few water molecules.

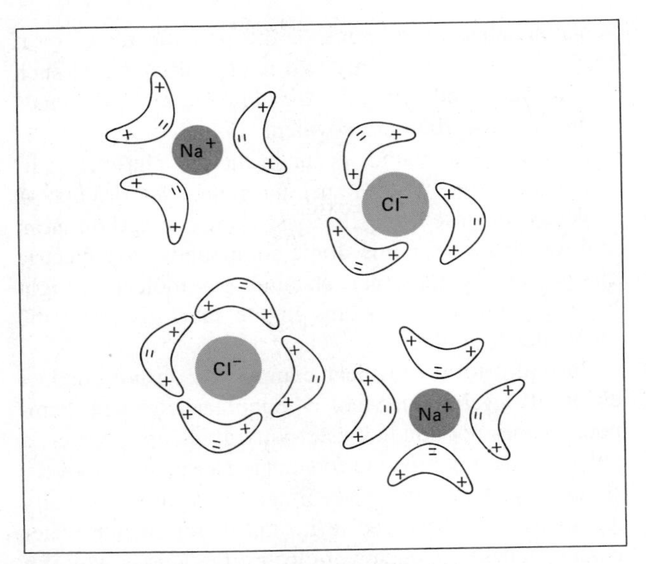

Figure 14.8 Schematic diagram of NaCl dissolved in water. The positive ends of the water molecule attract the negative chloride ions and the negative end the sodium ions. These forces act to break up the sodium chloride crystal.

by water molecules are called *solvated ions*. The dissolving process, illustrated by Figs. 14.7 and 14.8, is called *solvation*.

14.6 Oxidation Number

A useful concept in writing chemical formulas and equations is that of the oxidation number, or oxidation state. The **oxidation number,** or **oxidation state,** is essentially a measure of the electric charge that an atom in a molecule would have if it were ionically bound. When assigning oxidation numbers, the atom with the greater electron affinity is given a negative oxidation number. The atom with the lesser electron affinity is given a positive oxidation number.

The following are some general rules for assigning oxidation numbers:

1. With only a few exceptions (the peroxides: H_2O_2, Na_2O_2, etc.), oxygen always has an oxidation number of -2.
2. With the exception of the metal hydrides, hydrogen always has an oxidation number of $+1$.
3. The oxidation number of an ion is equal to the charge on the ion.
4. The oxidation number of a free element or an element combined only with itself is zero.
5. The oxidation number of the halogens is usually -1.
6. The oxidation number of the alkali metals is usually $+1$.
7. The oxidation number of the alkaline earths is usually $+2$.

Using these rules, the oxidation numbers of other elements in a neutral molecule or charged ion can be determined. The sum of the oxidation numbers of all the atoms involved in a molecule or ion must be equal to the charge on the molecule or ion. In the most basic sense, the concept of oxidation number is just a means of keeping track of the various charges involved in molecules and ions.

Some examples follow:

EXAMPLE 1 What is the oxidation number of Cl in HCl?

Since H has an oxidation number of $+1$, Cl has an oxidation number of -1.

EXAMPLE 2 What is the oxidation number of iron in Fe_2O_3?

Oxygen has an oxidation number of -2. The total negative charge on the molecule is $3 \times (-2) = -6$. Since the molecule is neutral, there must be six positive charges. There are two iron atoms so each must have $+3$. For the total charge on the molecule we get $2 \times (+3) + 3 \times (-2) = 0$.

EXAMPLE 3 What is the oxidation number of phosphorus in the $(PO_4)^{3-}$ ion?

The oxidation number of oxygen is -2. There are four oxygen atoms so there are $4 \times (-2) = -8$ negative charges. The total charge is -3, so phosphorus must have an oxidation number of $+5$. We get $+5 + [4 \times (-2)] = -3$.

EXAMPLE 4 What is the correct formula for calcium phosphate?

The phosphate ion, $(PO_4)^{3-}$, has an oxidation number of -3; the calcium ion has an oxidation number of $+2$. The total charge on the molecule must be zero. To have an equal number of positive and negative charges we need three calcium ions and two phosphate ions. Then we have $3 \times (+2) + 2 \times (-3) = 0$. The formula is then $Ca_3(PO_4)_2$.

Many elements have several possible oxidation states. In $FeCl_3$ iron has an oxidation state of +3; in $FeCl_2$, + 2. Tin has an oxidation number of +2 and +4 in $SnCl_2$ and $SnCl_4$, respectively. Chlorine usually has an oxidation number of −1, as in the preceding examples. However, it has an oxidation number of +3 in $NaClO_2$, +5 in $NaClO_3$, and +7 in $NaClO_4$.

14.7 Naming Compounds

The naming of simple compounds is straightforward. Compounds with only two elements in them are called binary compounds.

Binary compounds with one oxidation state are named by giving the name of the more metallic element, followed by the name of the more nonmetallic element but with the suffix "ide." Examples are as follows:

$NaCl$	Sodium chloride
NaH	Sodium hydride
CaI_2	Calcium iodide
HCl	Hydrogen chloride
KBr	Potassium bromide
Ba_3P_2	Barium phosphide

If there is a complex ion involved, it is given its usual name without a new ending. Some examples follow:

NH_4Cl	Ammonium chloride
$(NH_4)_2SO_4$	Ammonium sulfate
H_3PO_4	Hydrogen phosphate
$CaCO_3$	Calcium carbonate

When nonmetallic elements exist in more than one oxidation state in a binary compound with another element, the compounds are distinguished by using the prefixes "mono" (one), "di" (two), "tri" (three), "tetra" (four), "penta" (five), "hexa" (six), "hepta" (seven), and "octa" (eight). The prefixes designate the number of atoms of the element that occur in the molecule. The prefix "mono" is sometimes omitted.

CO	Carbon monoxide
CO_2	Carbon dioxide
NO_2	Nitrogen dioxide
N_2O_4	Dinitrogen tetraoxide
N_2O_5	Dinitrogen pentaoxide
SO_2	Sulfur dioxide
SO_3	Sulfur trioxide

If the metallic element has two possible oxidation numbers, these are sometimes distinguished by Roman numerals indicating the oxidation number.

$FeCl_2$	Iron(II) chloride
$FeCl_3$	Iron(III) chloride
Hg_2O	Mercury(I) oxide
HgO	Mercury(II) oxide

Under an older system of nomenclature the lower oxidation state has the ending "ous," and the higher oxidation state has the ending "ic." Under this older system, the four compounds above have the names ferrous chloride, ferric chloride, mercurous oxide, and mercuric oxide.

14.8 Some Simple Organic Compounds

One of the most important classes of compounds are the **organic compounds.** These are covalent compounds that contain the element carbon. It was originally believed that organic compounds (compounds derived from living matter) were exclusively of plant or animal origin, hence the name; but this has proved to be false, since many of the so-called organic compounds can be made in the laboratory from minerals (inorganic compounds). The study of carbon compounds is known as organic chemistry. The carbon atom has four electrons in its outer shell. This gives it the ability to form long-chain and ring-shaped molecules. There are over six million carbon compounds, and others are being added to the list constantly. Many of these carbon compounds are found in things we use daily, such as food, fuels, drugs, detergents, perfumes, and synthetic fibers. Carbon is the essential element of life, because it is the basic constituent of the complex molecules of proteins, fats, and carbohydrates, plus the nucleic acids. In this section we will examine the element carbon and some of its simpler compounds.

The element carbon exists in two crystalline structures, graphite and diamond. A third form, called amorphous carbon, has the same crystalline structure as graphite. Coke and charcoal are examples of amorphous carbon. Graphite is soft, black, slippery, and is a good conductor of electricity. Diamond is very hard (the hardest substance known, with perhaps the exception of a substance called boron carbide), colorless, transparent, and it will not conduct an electric current. Since both diamond and graphite are composed of carbon atoms that are identical in properties, the difference between graphite and diamond must be in the bonding or crystalline structure.

The carbon atoms in graphite are arranged in hexagonal rings. Each carbon atom is bonded to three other carbon atoms that lie in the same plane, and thus each carbon atom forms a part of three hexagons (Fig. 14.9). The hexagons extend in all directions within the formed plane. The remaining electron of the carbon atom is relatively free and shifts from one bond to another in an endless fashion. Thus, there is a free valence electron that gives graphite its ability to conduct an electric current. The attractive forces between planes are very weak and allow the planes to slide easily over one another. This accounts for the slippery feeling of graphite. Graphite is used in many lubricants because of this property, but it is probably best known as the black substance used in making ''lead'' pencils.

The crystalline structure of diamond is quite different from that of graphite. The four valence electrons of carbon atoms in diamond are arranged with those of four neighboring carbon atoms to form a tetrahedral structure as shown in Fig. 14.10. The carbon atoms at the vertices are in turn bonded to four other carbon atoms. Thus, the entire structure is one big crystal, which is extremely strong in all directions. Besides being very hard, diamond has a high melting point. When heated to 1000°C in the absence of air, diamond will change to graphite. (See Fig. 14.11.)

Compounds containing only carbon and hydrogen are known as **hydrocarbons.** The **alkanes** are hydrocarbons that have a composition that satisfies the general formula

$$C_nH_{2n+2}$$

where n = the number of carbon atoms,
$2n + 2$ = the number of hydrogen atoms,
n = 1, 2, 3, . . .

The formula says that the number of hydrogen atoms present in the compound is twice the number of carbon atoms plus two.

Methane (CH_4) is the first member of the alkane series. Ethane (C_2H_6) is the second member, propane (C_3H_8) the third, and butane (C_4H_{10}) the fourth. Table 14.3 lists

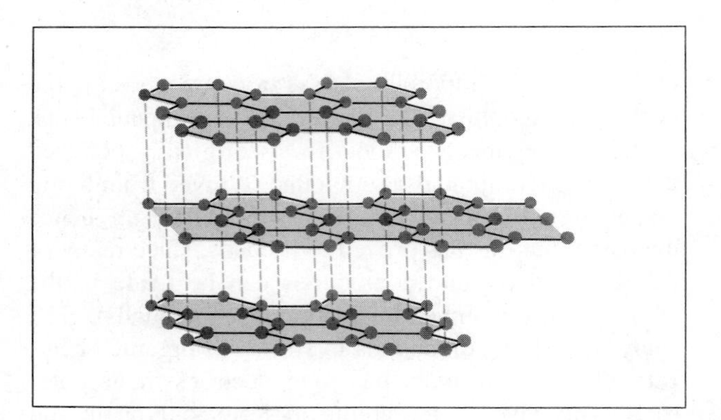

Figure 14.9 Crystalline structure of graphite. Each carbon atom is bonded to three other carbon atoms in the same plane. Atoms in one plane are weakly bonded to atoms in adjoining planes.

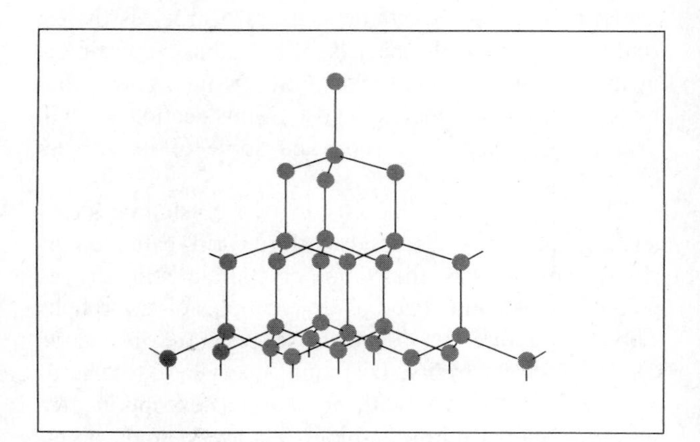

Figure 14.10 The crystalline structure of diamond. Note that each carbon atom is bonded to four others, spatially arranged at the corners of a regular tetrahedron.

Figure 14.11 Two forms of carbon: graphite and diamond. (Photo courtesy James Crouse.)

Table 14.3 Some Compounds of the Alkane Series of Hydrocarbons

CH_4	Methane
C_2H_6	Ethane
C_3H_8	Propane
C_4H_{10}	*n*-Butane
C_5H_{12}	*n*-Pentane
C_6H_{14}	*n*-Hexane
C_7H_{16}	*n*-Heptane
C_8H_{18}	*n*-Octane
C_9H_{20}	*n*-Nonane
$C_{10}H_{22}$	*n*-Decane
$C_{11}H_{24}$	*n*-Undecane
$C_{12}H_{26}$	*n*-Dodecane
$C_{15}H_{32}$	*n*-Pentadecane
$C_{20}H_{42}$	*n*-Eicosane

some other members of the series. The prefix "*n*" in front of the name, in Table 14.3 stands for the word *normal*. Normal hydrocarbons are those whose molecules are all straight chains, with no carbon-hydrogen groups branching off the chain.

The hydrocarbon's structure is easy to visualize when we write the **structural formula** (a picture representation of the way the atoms are connected to one another) instead of its molecular formula. For example, the structural formula for methane (CH_4) is written

$$\begin{array}{ccccc} & & H & & \\ & & | & & \\ H & — & C & — & H \\ & & | & & \\ & & H & & \end{array}$$

The structural formula is similar to the electron dot notation. Each dash corresponds to two shared electrons and so represents a covalent bond. Ethane and propane structural formulas are

$$\begin{array}{ccccccc} & & H & & H & & \\ & & | & & | & & \\ H & — & C & — & C & — & H \\ & & | & & | & & \\ & & H & & H & & \end{array} \qquad \begin{array}{ccccccccc} & & H & & H & & H & & \\ & & | & & | & & | & & \\ H & — & C & — & C & — & C & — & H \\ & & | & & | & & | & & \\ & & H & & H & & H & & \end{array}$$

Ethane Propane

The alkanes make up many well known petroleum products. Methane is the principal constituent in natural gas used for cooking stoves. Gasoline contains many of the alkanes from pentane to decane ($n = 5$ to $n = 10$). Kerosene contains the alkanes with $n = 10$ to 16. The alkanes with higher values of n make up other products such as diesel fuel ($n = 12$ to 20), fuel oil ($n = 14$ to 22), lubricating oil ($n = 20$ to 30), petroleum jelly ($n = 22$ to 40), and paraffin wax ($n = 25$ to 50). The alkanes are also used as starting materials for many other products such as paints, plastics, drugs, detergents, insecticides, and cosmetics.

The alkanes are highly combustible. They react with the oxygen in air to form heat, carbon dioxide, and steam. If the combustion is not complete, carbon in the form of black soot and carbon monoxide is formed. The alkanes are colorless. Recall that paraffin wax, kerosene, and "white" gasoline used in camp stoves all lack color. Most gasolines used in automobiles, however, are colored, because a little dye has been added to indicate that it contains tetraethyl lead, a deadly poison. The alkanes are nonpolar compounds. The alkane molecules thus are not attracted to the polar molecules of water and so solvation does not occur to any appreciable extent.

When two hydrogen atoms are removed from a molecule of an alkane, two electrons of the carbon atoms become unpaired. In order to satisfy the octet rule, a double bond must be formed. Hydrocarbons with a double bond are called **alkenes.** For example,

$$\begin{array}{ccccccc} & & H & & H & & \\ & & | & & | & & \\ H & — & C & — & C & — & H \\ & & | & & | & & \\ & & H & & H & & \end{array} \qquad \begin{array}{ccccccc} & & H & & H & & \\ & & | & & | & & \\ H & — & C & = & C & — & H \end{array}$$

Ethane Ethene (or Ethylene)

The general formula for the alkene series is C_nH_{2n}. The series begins with C_2H_4 (ethene, or ethylene), shown by the structural formula above. Some of the normal alkenes are listed in Table 14.4. A number, instead of

Table 14.4 Some Compounds of the Alkene Series of Hydrocarbons

CH_2CH_2	Ethene (ethylene)
CH_3CHCH_2	Propene (propylene)
$C_2H_5CHCH_2$	1-Butene
$C_3H_7CHCH_2$	1-Pentene
$C_3H_7CHCHCH_3$	2-Hexene
$C_5H_{11}CHCH_2$	1-Heptene
$C_6H_{13}CHCH_2$	1-Octene
$C_7H_{15}CHCH_2$	1-Nonene
$C_{10}H_{20}$	1-Decene
$C_{15}H_{30}$	1-Pentadecene
$C_{20}H_{40}$	1-Eicosene

n (normal), preceding the name indicates the presence of a double bond and its position. The carbons in the chain are numbered starting at the end—which end doesn't matter. For example, 1-pentene and 2-pentene have the following structural formulas:

```
      H   H   H   H   H
      |   |   |   |   |
  H—C—C—C—C═C—H
      |   |   |
      H   H   H
```

1-pentene

```
      H   H   H   H   H
      |   |   |   |   |
  H—C—C—C═C—C—H
      |   |           |
      H   H           H
```

2-pentene

The alkanes are all saturated hydrocarbons. A **saturated hydrocarbon** contains only single bonds, and its hydrogen content is at a maximum. The alkenes are known as **unsaturated hydrocarbons,** because they can take on hydrogen atoms to form saturated hydrocarbons or atoms of substances other than hydrogen to form derivatives.

When a second pair of hydrogen atoms is removed from the carbon atoms that have a double bond, a triple bond is formed. The general formula for this series, **alkyne,** is C_nH_{2n-2}, and the simplest member is ethyne (or acetylene), HC≡CH. The members of this series are more reactive than those of the alkene series. Some members of the alkyne series are listed in Table 14.5.

The general formula C_nH_{2n}, which gives the structure of the alkenes, also gives the structure for a saturated hydrocarbon series called the cycloalkanes. The prefix "cyclo-" indicates the ring structure, and the ending "-ane" the saturated condition of the chain. The carbon atoms of this series are arranged in a ring with two hydrogen atoms connected to each carbon atom in the ring. These are called the **cyclic hydrocarbons.** The first member of the series is cyclopropane, which has the structural formula

Table 14.5 Some Compounds of the Alkyne Series of Hydrocarbons

Formula	Name
HC≡CH	Ethyne (acetylene)
$CH_3C\equiv CH$	Propyne (methylacetylene)
$C_2H_5C\equiv CH$	1-Butyne
$C_3H_7C\equiv CH$	1-Pentyne
$C_3H_7C\equiv CCH_3$	2-Hexyne
$C_5H_{11}C\equiv CH$	1-Heptyne
$C_6H_{13}C\equiv CH$	1-Octyne
$C_7H_{15}C\equiv CH$	1-Nonyne
$C_{10}H_{18}$	1-Decyne
$C_{15}H_{28}$	1-Pentadecyne
$C_{20}H_{38}$	1-Eicosyne

```
        H   H
         \ /
          C
         / \
  H     /   \     H
   \   /     \   /
    C ——————— C
   /           \
  H             H
```

Cyclopropane

Various numbers of double bonds can also be introduced into cyclic structures. One of the most important cyclic hydrocarbons is benzene (C_6H_6). It has the structural formula

```
         H                       H
         |                       |
         C                       C
       ⁄⁄  \                   /  \\
  H—C       C—H      H—C       C—H
     |       ||   or      ||       |
  H—C       C—H      H—C       C—H
       \\  /                   \  ⁄⁄
         C                       C
         |                       |
         H                       H
```

Benzene ring

Since the molecule takes the structure of a ring, it is known as the **benzene ring.** The molecule has the ability to switch back and forth between the two structures shown. This property is called resonance and makes the benzene molecule more stable than the regular hydrocarbons. Sometimes the benzene structure is written

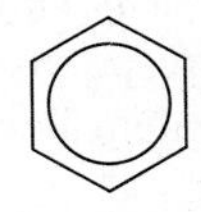

Benzene ring

Benzene is obtained from coal tar, a by-product of soft coal. When other atoms or groups of atoms are substituted for the hydrogen atoms in the benzene ring, a vast number of different compounds can be produced. These compounds include such things as perfumes, explosives, drugs, solvents, insecticides, lacquers, and a host of others. (See Fig. 14.12.)

So far we have been naively drawing structural formulas as if molecules had only two dimensions. The seemingly obvious principle that molecules exist in three dimensions was not realized until 1874, some 10 to 20

Figure 14.12 The aromatic character of the compounds of benzene gives rise to many exotic fragrances. (Photo courtesy James Crouse.)

years after the concepts of ring and chain molecules were discovered. In 1874 Jacobus van't Hoff in Holland and Joseph Le Bel in France independently reached the conclusion that many unexplained chemical mysteries could be explained if molecules were viewed in three dimensions.

The molecule CH_4 is actually a tetrahedron with a carbon atom in the center. Its three-dimensional structure looks like

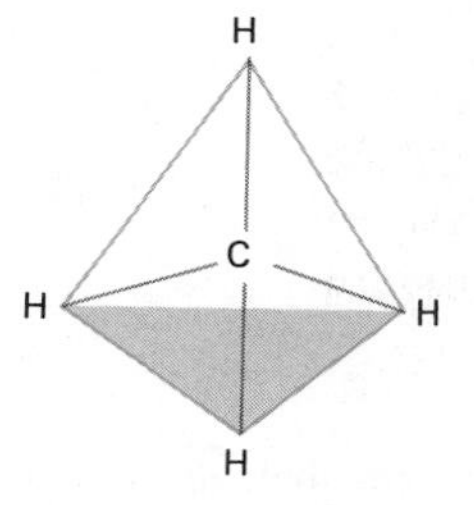

This tetrahedral arrangement is the basic geometric shape of organic molecules. The way chains are built up in three dimensions is illustrated by *n*-butane.

```
     H   H  H   H
      \ /    \ /
  H    C      C
   \  / \    / \
    C     C     H
   / \   / \
  H   H H   H
```

n-butane

From this model of *n*-butane we can see that the following structural formulas are all the same.

```
   H  H  H  H
   |  |  |  |
H--C--C--C--C--H
   |  |  |  |
   H  H  H  H
```

n-butane

```
      H
      |
   H--C--H
      |     H
      |     |
   H--C-----C--H
      |     |
      H     |
         H--C--H
            |
            H
```

n-butane

```
   H  H  H
   |  |  |
H--C--C--C--H
   |  |  |
   |  H  H
   |
H--C--H
   |
   H
```

n-butane

All of these represent the *n*-butane molecule. They all have the carbon and hydrogen atoms attached in the same way: There are three hydrogen atoms on the two end carbon atoms and two hydrogen atoms attached to the middle two carbon atoms. However, another arrangement of the atoms of the butane molecule is possible. It is an isomer of *n*-butane. **Isomers** are molecules with the same molecular formula but a different structure or arrangement of atoms. They can exist wherever the structural buildup does not violate the octet rule. Isobutane also has four carbon atoms and 10 hydrogen atoms, but its structural formula is

```
   H  H  H
   |  |  |
H--C--C--C--H
   |  |  |
   H  |  H
      |
   H--C--H
      |
      H
```

Isobutane

Isobutane's structural formula is different from the three structural formulas given for *n*-butane because it has three hydrogen atoms attached to three of the carbon atoms, while the fourth carbon atom has only one hydrogen atom bonded to it.

Learning Objectives

After reading and studying this chapter you should be able to do the following without referring to the text:

1. Define, explain, and give an example of the important terms in the following list: (a) compound, (b) mixture, (c) molecule, (d) ion.
2. State the two basic assumptions made to explain compound formation.
3. Describe ionic bonding, state the properties of ionic compounds, and give a few examples of ionic compounds.
4. Describe covalent bonding, state the properties of covalent compounds, and give a few examples of covalent compounds.
5. Distinguish between an oxidation and a reduction process and give examples of each.
6. Distinguish between polar and nonpolar molecules and give some examples of each.
7. Define *oxidation number* and state the general rules for assigning oxidation numbers.
8. State the general rules for naming compounds.
9. Describe organic compounds and state the general formula for some compounds known as hydrocarbons.
10. Draw the structural formula for benzene, state where benzene is obtained, and name some of the compounds produced by substitution.
11. Define or explain the important words and terms listed at the end of the chapter.
12. State the octet rule.

Important Words and Terms

mixture
compound
molecule
octet rule
ionic compounds
electrovalent compounds
covalent compounds
electron affinity
crystal
cation
anion
half-reaction
oxidation
reduction
oxidation-reduction reaction
electrolysis
covalent bond
ionic bond
electronegativity
polar bond
nonpolar bond
nonpolar molecule
polar molecule
oxidation number
oxidation state
organic compounds
hydrocarbons
alkanes
structural formula
alkenes
saturated hydrocarbon
unsaturated hydrocarbon
alkynes
cyclic hydrocarbons
benzene ring
isomers

Questions

1. Distinguish between a compound and a mixture. Give an example of each.
2. Distinguish between a compound and a molecule; between a compound and an ion. Give an example of each.
3. Which electrons of an atom take part in the formation of compounds?
4. What is the electron configuration for the most stable state of an atom or ion that takes part in the formation of a compound?
5. What is a chemical bond?
6. What force is responsible for molecular formation?
7. During compound formation, how is the number of valence electrons in an atom related to its tendency to lose or gain electrons?
8. Distinguish between ionic and covalent bonding.
9. Give an example of a compound not composed of molecules.
10. Why is hydrogen a diatomic gas while helium is monatomic?
11. Compare the physical properties of ionic compounds with those of covalent compounds.
12. Compare oxidation number to the number of electrons in the outermost shell of a neutral atom.
13. Where on the periodic chart do you find elements with (a) high electronegativity, (b) low electronegativity?
14. State the octet rule.
15. Why are single bonds sometimes formed while at other times double or triple bonds are formed?

16. A compound has a boiling point of $-10°C$. Is it ionic or covalent? Why?
17. A compound melts at a very high temperature of 1000°C. Would you expect it to be ionic or covalent? Why?
18. How does water dissolve ionic compounds?
19. List some of the important alkanes and their uses.
20. What is the structure of benzene? List some of the products in which benzene occurs.
21. Explain how an ionic liquid conducts electricity.
22. Write the general formula for (a) the alkanes, (b) the alkynes.
23. What are isomers?

Problems

1. Draw the electron dot notation for the following atoms and their ions:
   (a) Mg, (b) S, (c) Br, (d) K, (e) Hg, (f) Ni, (g) Al.
   Answer: (a) Mg: and Mg^{2+}
2. (a) Specify the following as either ionic or covalent substances:
   CO_2, NaCl, CaF_2, NH_4^+, $(NO_2)^-$, N_2H_3Cl
   (b) Draw the electron dot formulas for the above compounds and ions.
3. Specify the following molecules as either polar or nonpolar:
   H_2O, CO, HCl, CH_4, O_2, NH_3 *Answer:* NH_3, polar
4. Find the oxidation numbers of all elements in the following formulas:
   (a) H_2SO_4 (e) SO_3^{2-}
   (b) N_2 (f) SO_4^{2-}
   (c) CO (g) CrO_4^{2-}
   (d) CO_2 (h) $Cr_2O_7^{2-}$
   Answer: (g) O, 2−; Cr, 6+
5. Write formulas for the following compounds:
   (a) Barium nitrate (e) Calcium phosphate
   (b) Aluminum sulfate (f) Ammonium nitrite
   (c) Vanadium(II) oxide (g) Calcium iodide
   (d) Vanadium(III) oxide (h) Sodium sulfite
   Answer: (e) $Ca_3(PO_4)_2$
6. What are the names of the following compounds:
   (a) CuO (c) Na_2SO_4 (e) C_8H_{18}
   (b) Cu_2O (d) C_6H_6
7. Draw structural formulas for the following organic compounds:
   (a) benzene, (b) propane, (c) propene, (d) propyne.
8. Draw structural formulas for three isomers of pentane.
9. Given the following sets of atoms, state which will form ionic bonds and which will form covalent bonds. State the reason for your answer.
   (a) two atoms of oxygen
   (b) silicon and chlorine
   (c) magnesium and oxygen
   (d) sodium and bromine
   (e) hydrogen and phosphorus
   Answer: (b) covalent Silicon has four electrons in outer shell which it prefers to share.
10. Write the formula for each compound formed in Problem 9.
    Answer: (b) $SiCl_4$

15

Some Chemical Principles

The chemists are a strange class of mortals, impelled by an almost insane impulse to seek their pleasure among smoke and vapor, soot and flame, poisons and poverty; yet among all these evils I seem to live so sweetly that I may die if I would change places with the Persian King.

–Johann Joachim Becher

IN THIS CHAPTER some of the basic principles of chemistry will be discussed. Most of these were discovered before 1850 by chemists such as Becher (1635–1682) who earnestly pursued their research in small labs which they themselves financed. Historically, these laws and principles were known before Mendeleev constructed his periodic table in 1869. It was long after 1869, of course, that the structure of atoms and molecules was understood. In fact, the principles we shall study in this chapter helped to establish our modern understanding of molecular structure. When these principles were first discovered, the reasons behind them were unknown. The explanation for them lies in our present understanding of atoms and molecules discussed in the previous three chapters.

Many of these very old laws concerning proportions and mixtures are important to us today. For instance, if the exhaust of a car is emitting black smoke we know that an incorrect mixture of air and gasoline vapor is being burned in the engine. Such exhaust gases are one of the chief pollutants of our atmosphere. Instances of where mixtures are important are also well known to all cooks. If a recipe is not followed closely the wrong proportions of ingredients will yield an unpalatable dish. The chemists of Becher's day have left us a legacy that we would be wise to guard carefully.

15.1 Types of Matter

Matter is classified in several different ways. One of the ways, which was mentioned previously, is the classification of matter into four phases: solid, liquid, gas, and plasma. In the solid phase, matter is rigid and has a definite shape. It expands or contracts comparatively little with pressure and temperature changes. In the liquid phase, matter assumes the shape of the container and maintains a horizontal surface because of the effect of gravity. Similar to solids, liquids do not expand or contract much when pressure and temperature are varied. In the gas phase, matter fills the container completely and uniformly. Gravitational effects are small since the individual gas molecules have such a small mass. Gases expand and contract quite easily when pressure and temperature are varied. In fact, the gas laws tell us that the volume of a gas is inversely proportional to its pressure and directly proportional to its temperature.

Matter which is of uniform composition throughout is described as **homogeneous;** that is, any sample will be like any other sample. Matter which is of nonuniform composition is described as **heterogeneous.** As examples, a well-stirred cup of coffee with cream and sugar is a homogeneous mixture, while a pizza is a heterogeneous mixture. A **substance** is a homogeneous sample of matter all specimens of which have identical properties and composition. A **compound** is a substance composed of two or more elements chemically combined in a definite proportion. That is, a compound is composed of only one kind of molecule. All compounds are substances, but not all substances are compounds. Pure elements are substances, but they are not compounds. For example, pure hydrogen gas is a substance, but it is not a compound, while steam is both a substance and a compound. A sample of matter composed of two or more substances in varying amounts that are not chemically combined is called a **mixture.** Mixtures can be homogeneous or heterogeneous. A **solution** is a homogeneous mixture. From this definition a solution need not be a

◀ Radiocarbon dating.

liquid; it may also be a solid or a gas. Salt water is a liquid solution. Metal alloys are examples of solid solutions. Air is an example of a gaseous solution. In Fig. 15.1, these different types of matter are summarized.

The characteristics of a substance are known as its properties. There are two basic types of properties, physical and chemical. **Physical properties** are those that do not involve a change in the chemical composition of the substance. Among these properties are density, hardness, color, melting point, boiling point, electrical conductivity, thermal conductivity, and specific heat. These properties may change with a change in pressure or temperature. Those changes, which do not alter the chemical composition of the substance, are called **physical changes.** The processes of freezing or boiling water are good examples of physical changes brought about by changing the temperature. Other examples of physical changes are dissolving table salt in water, heating a piece of metal, and evaporating water.

The properties involved in the transformation of one substance into another are known as **chemical properties.** When wood burns, oxygen in the air unites with the different substances in the wood to form new substances. When iron corrodes, it combines with oxygen and water to form a new substance commonly known as rust. These are examples of chemical properties of wood and iron, respectively. Changes that result in the formation of new substances are known as **chemical changes.** The fermenting of wine, burning of gasoline, souring of milk, discharging of a battery, and the exploding of gun powder are all examples of chemical changes. All chemical changes involve a production or absorption of energy.

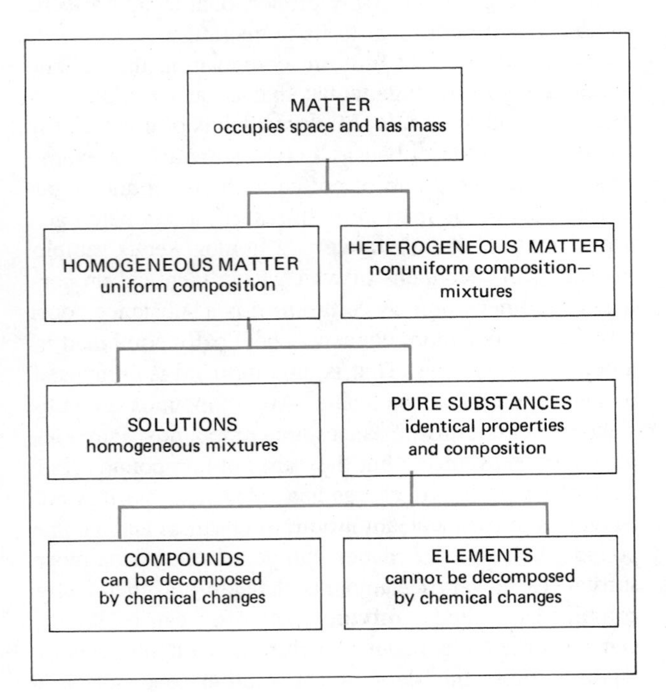

Figure 15.1 The types of matter.

15.2 Early Chemical Laws

We have seen in previous chapters that mass is a form of energy. Mass and energy are related by Einstein's formula, $E = mc^2$. Total energy, including mass energy, is always conserved in any process. In chemical reactions energy is produced or absorbed in the form of heat, kinetic, or potential energy. There is, then, a corresponding gain or loss in mass energy, but the total energy does not change. However, the change in mass in a chemical reaction is so infinitesimal that it cannot be detected. If the total mass involved in chemical reaction is precisely weighed before and after the reaction takes place, no change can be detected. This law is known as the **law of conservation of mass** and is stated as follows:

> **There is no detectable change in the total mass during a chemical process.**

This law was formulated in 1774 by Antoine Lavoisier. Lavoisier had developed balances which could measure $\frac{1}{100}$ the weight of a drop of water. He put tin in a vessel, sealed it, and carefully weighed it. The vessel was then heated so that the tin reacted with the air in the vessel to form a new compound. Lavoisier next reweighed the sealed container and found the weight to be identical with the original weight. Because the container had been sealed, no air could go in or out. When Lavoisier opened the vessel, air rushed in. This showed that the tin had reacted with the air inside, or something in the air, to form the new compound. This experiment of Lavoisier's was one of the first great quantitative chemical experiments. Now, two centuries later, Lavoisier's law of conservation of mass is still almost exactly true.

The French chemist J. L. Proust (1755–1826) was one of the first to propose, in the late 1790s, that elements combine with one another in a definite ratio by weight. Proust's proposal is now known as the law of definite composition or the **law of definite proportion.** The law states:

> **Different samples of a pure compound always contain the same elements in the same proportion by weight.**

For instance,

$$9\ g\ H_2O = 8\ g\ \text{oxygen and}\ 1\ g\ \text{hydrogen}$$

$$36\ g\ H_2O = 32\ g\ \text{oxygen and}\ 4\ g\ \text{hydrogen}$$

In both cases the ratio by weight of oxygen to hydrogen is 8 to 1. This is true, because each molecule of H_2O is composed of one atom of oxygen (atomic weight 16) and two atoms of hydrogen (atomic weight 1, total weight of hydrogen = 2 × 1 or 2). Thus, each molecule of H_2O is composed of 16 parts oxygen by weight and 2 parts hydrogen by weight or a ratio of 8 to 1. Since a pure covalent compound is simply a very large collection of identical molecules, the proportion by weight of any element in a covalent compound will be the proportion by weight that it occupies in an individual molecule of that covalent compound.

EXAMPLE 1 If 11 g of carbon dioxide, CO_2, are composed of 3 g of carbon and 8 g of oxygen, how many grams of carbon and oxygen will be in 55 g of carbon dioxide?

Since $\frac{3}{11}$ of every sample of CO_2 must be carbon and $\frac{8}{11}$ of every sample of CO_2 must be oxygen, the amount of carbon is $\frac{3}{11} \times 55\ g = 15\ g$ and the amount of oxygen is $\frac{8}{11} \times 55\ g = 40\ g$. In both the 11-g and 55-g sample of CO_2, the ratio is 3 parts carbon to 8 parts oxygen. That is, 3:8 or 15:40.

For ionic compounds, a similar submicroscopic description is easily given. Since the ions occur in a definite proportion, the weight of the elements in the compound are in the same proportion by weight as they are in the smallest combination of ions which gives the formula of the compound.

EXAMPLE 2 If 25 g of calcium carbonate ($CaCO_3$) consist of 10 g calcium, 3 g carbon, and 12 g oxygen, how many grams of each element are contained in 100 g of $CaCO_3$?

Calcium carbonate is an ionic compound with one ion of calcium for every carbonate ion (CO_3^{2-}). Using the law of definite proportions, we know that $\frac{10}{25}$ of every sample must be calcium, $\frac{3}{25}$ carbon, and $\frac{12}{25}$ oxygen. Thus in 100 g of $CaCO_3$:

$$\text{Amount of calcium} = \left(\frac{10}{25}\right) \times 100 = 40\ g$$

$$\text{Amount of carbon} = \left(\frac{3}{25}\right) \times 100 = 12\ g$$

$$\text{Amount of oxygen} = \left(\frac{12}{25}\right) \times 100 = 48\ g$$

Sometimes the elements that are to be combined to form a compound are not in the correct proportions. In this situation one of the elements will be completely used and the one in excess will be partially used in the correct proportion with some of it left over.

EXAMPLE 3 How many grams of CO_2 can be made from 36 g of carbon and 40 g of oxygen?

The ratio of carbon to oxygen in CO_2 is 3:8. So 8 g of oxygen are needed for every 3 g of carbon. If we divide 8 g into the 40 g of oxygen available, we get 5. If we divide 3 g of carbon into the 36 g available, we get 12. Since 12 is bigger than 5, the carbon is in excess. We have 5 × 8 g (= 40 g) of oxygen available. This will combine with 5 × 3 g (= 15 g) of the carbon to make 55 g of CO_2. Then 21 g of carbon are left over. In summary,

Amount of oxygen available	= 40 g
Amount of carbon available	= 36 g
Amount of carbon needed (to combine with 40 g of oxygen to produce CO_2)	= 15 g
Amount of CO_2 produced	= 40 g + 15 g = 55 g
Amount of carbon left over	= 21 g

In the preceding example we also see that the law of conservation of mass is obeyed. We began with a total of 76 g (40 g + 36 g) and we ended with a total of 76 g (55 g + 21 g).

The law of conservation of mass and the law of definite proportion, although fairly unsatisfactory at the beginning of the nineteenth century, aided John Dalton (1766–1844) (Fig. 15.2), a self-taught English school teacher, in his conception of the atomic theory of matter. Dalton's atomic theory was accepted because it explained these laws.

Our modern theory of atomic structure is based upon Dalton's concepts, which have withstood the test of time because they have explained many observed phenomena of nature. The theory has provided us with data concerning the structure and properties of atoms, the makeup of compounds and their properties, plus the mass, energy, and volume relationships of reactions between atoms.

Figure 15.2 John Dalton (1766–1844). Dalton was born in the village of Eaglesfield, England, the son of a poor weaver. He was a child prodigy, opening his own school at the age of 12. In his mid-twenties he became a teacher at the University of Manchester. His greatest contribution to science was his atomic theory of matter. He took the word "atom" from the Greek word atomos, meaning indivisible. Dalton was a bachelor all his life—because he "never had time to marry." One thing that occupied much of his time was making daily weather observations. He made thousands of these during his lifetime. (Photo courtesy Culver Pictures.)

A third law, discovered in the early 1800s, is called the law of multiple proportions. This law describes a relationship that exists for different compounds that are formed from the same elements. Carbon monoxide, CO, and carbon dioxide, CO_2, are examples. The **law of multiple proportions** is stated as follows:

> **Whenever two elements, call them E_1 and E_2, combine to form different compounds, the various amounts of E_2 combined with the same amount of E_1 are in a ratio of small whole numbers.**

Some examples will clarify the statement of this law. First, consider CO and CO_2. These two compounds are both composed of the same two elements, carbon and oxygen. The two amounts by weight of oxygen which combine with the same weight of carbon are in a ratio of small whole numbers and so conform to the law. In CO, 16 g of oxygen will combine with 12 g of carbon; in CO_2, 32 g of oxygen will combine with 12 g of carbon. The numbers 16 and 32 are in the ratio 1:2, which is a ratio of small whole numbers as predicted by the law of multiple proportions. It is easy to see from our understanding of molecules why this law must be so. In a single molecule of CO there is one oxygen atom; in a single molecule of CO_2 there are two oxygen atoms. This ratio of 1:2 is then reflected in the amounts of oxygen that will combine with equal amounts of carbon to form the two different compounds. A good example of the multiple proportion law is found in the compounds of nitrogen and oxygen. There are six of these compounds. The amounts of oxygen that unite with 100 g of nitrogen are shown in Table 15.1.

All the ratios of the amounts of oxygen that combine with 100 g of nitrogen are in small whole numbers. That is, ratios such as 1.738:1 or 1:1.23 do not occur. Again, the molecular understanding of compounds readily explains the ratios. The compounds have the formulas NO, N_2O, N_2O_3, NO_2, N_2O_5, and NO_3, and the ratios are

Table 15.1 List of Compounds Illustrating the Law of Multiple Proportions

Compound	*Amount of Nitrogen (g)*	*Amount of Oxygen (g)*	*Ratio*
NO	100	114.2	
N_2O	100	57.1	57.1 : 114.2 or 1 : 2
N_2O_3	100	171.3	171.3 : 114.2 or 3 : 2
NO_2	100	228.4	228.4 : 114.2 or 2 : 1
N_2O_5	100	285.5	285.5 : 114.2 or 5 : 2
NO_3	100	342.6	342.6 : 114.2 or 3 : 1

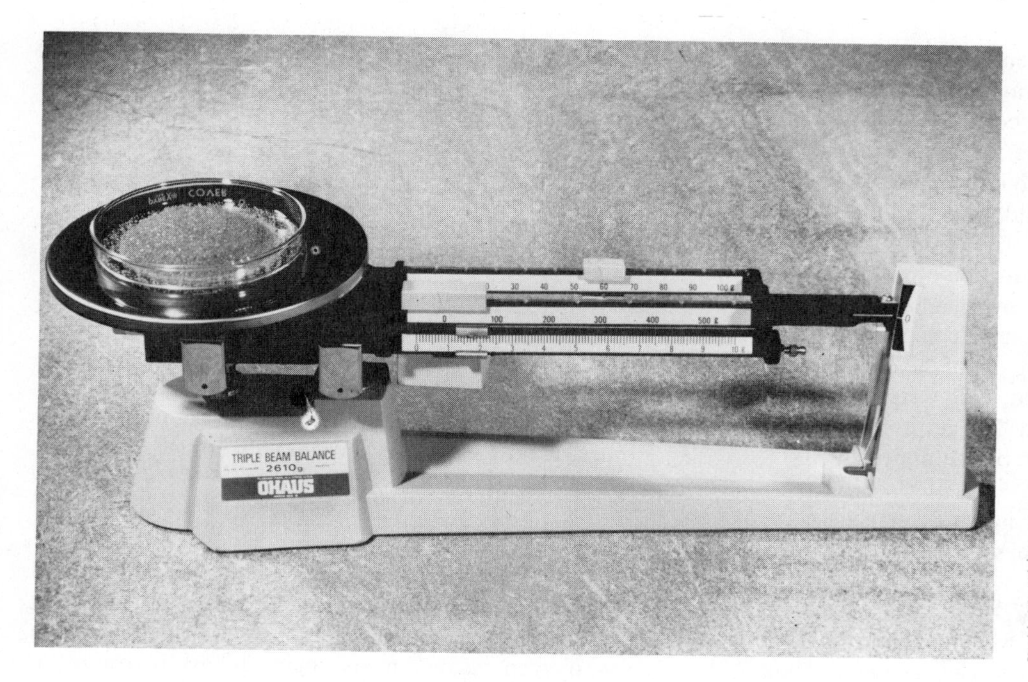

Figure 15.3 A triple beam balance, as shown, is used in the laboratory for the measurement of a substance's mass.

easily related to these molecular formulas. In the early 1800s, discovery of the law of multiple proportions was an important step toward the molecular understanding of chemical compounds. A chemical laboratory balance is shown in Figure 15.3.

15.3 Molecular Weights and Volumes

John Dalton was the first to study systematically the weight relationships among the various elements. These weight relationships are given in the periodic table as atomic weights. In the present-day table, the weight relationships are based on an arbitrary scale that assigns the ^{12}C isotope the value 12.0000. All other naturally occurring elements are given a weight value relative to that of the ^{12}C isotope (see Table 13.1). The atomic weights for most elements have been determined to several decimal places. For purposes of simplicity, however, we shall round these values off to the nearest integer or half integer. For example, we shall consider the atomic weights of hydrogen, carbon, oxygen, and chlorine to be approximately 1, 12, 16, and 35.5, respectively.

Since these are only relative weights, they do not have units. In order to give units to atomic weights, the concept of gram atomic weight is necessary. A **gram atomic weight** is defined as the weight of an atom of an element relative to ^{12}C, expressed in grams. The gram atomic weights of hydrogen and carbon are then 1 g and 12 g, respectively. The number of gram atomic weights of a given sample of an element is found by dividing the mass of the sample by one gram atomic weight. As an example, in 20 g of carbon there are $\frac{20}{12} = 1.67$ gram atomic weights.

The **formula weight** of a compound is the sum of the atomic weights given in the formula of the compound. The formula weight of $CaCO_3$ is thus $40 + 12 + (3 \times 16) = 100$. The formula weight is a relative quantity like the atomic weight. The **gram formula weight** is the formula weight of the compound expressed in grams. The gram formula weights of $CaCO_3$ and H_2O are 100 g and 18 g, respectively. A gram formula weight is also referred to as a **mole.** In 200 g of $CaCO_3$ there are 2 moles.

EXAMPLE 4 How many moles are there in (a) 9 g of H_2O, (b) 65 g of NaCl?

(a) The gram formula weight of H_2O is 18 g, so in 9 g there are $9 \text{ g}/18 \text{ g} = 0.5$ mole.
(b) The gram formula weight of NaCl is $23 \text{ g} + 35.5 \text{ g} = 58.5 \text{ g}$. Thus, in 65 g there are $65 \text{ g}/58.5 \text{ g} = 1.11$ moles.

In covalent compounds, the formula given represents a discrete molecule of the compound. Therefore, the terms **molecular weight** and **gram molecular weight** are often used for covalent compounds instead of formula weight and gram formula weight. However, in ionic compounds, distinct molecules do not exist in the crystal or in solution. Thus, the terms molecular weight and gram molecular weight do not apply to ionic compounds and so formula weight and gram formula weight are the terms used.

The **percentage composition** of an element in a compound is the percentage by weight of each element in the compound. The percentage composition can be found by computing the total weight of the element in the compound, then dividing the total weight by the formula weight of the compound. To get a percentage, the resulting answer must then be multiplied by 100.

EXAMPLE 5 What is the percentage composition of the elements in table salt, NaCl? In sulfuric acid, H_2SO_4?

First find the formula weight. The formula weight of NaCl is $23 + 35.5 = 58.5$. The percentage composition is then

$$\text{Na} = \frac{23}{58.5} \times 100 = 39.3\%$$

$$\text{Cl} = \frac{35.5}{58.5} \times 100 = 60.7\%$$

The formula weight of sulfuric acid is $2 \times 1 + 1 \times 32 + 4 \times 16 = 2 + 32 + 64 = 98$.

The percentage composition is then

$$\text{H} = \frac{2}{98} \times 100 = 2.04\%$$

$$\text{S} = \frac{32}{98} \times 100 = 32.65\%$$

$$\text{O} = \frac{64}{98} \times 100 = 65.31\%$$

In 1808, the French chemist Joseph Louis Gay-Lussac performed some experiments on gases that had a profound effect on the modern understanding of atoms and molecules. Gay-Lussac mixed different gases together to form new gases. As he did this, he took pains to measure the volumes of the initial and final gases. He found that when he mixed 2 volumes of hydrogen gas with 1 volume of oxygen gas, he obtained 2 volumes of steam. Similarly, when he mixed 1 volume of nitrogen gas with 3 volumes of hydrogen gas, he got 2 volumes of ammonia. In all the experiments, he measured the gas volumes at the same temperature and pressure. From these and many other experiments, he developed what is called **Gay-Lussac's law of combining volumes:**

When gases combine to form new gaseous compounds, all at the same temperature and pressure, the volumes of the initial and final gases are in the ratio of small whole numbers.

The explanation of Gay-Lussac's law of combining volumes was first proposed by the Italian chemist Amadeo Avogadro in 1811. Avogadro explained Gay-Lussac's experiments by making two assumptions. His first assumption, which has since been confirmed many times, is now known as **Avogadro's law:**

Equal volumes of gases at the same temperature and pressure contain equal numbers of molecules.

Avogadro's second assumption (which is also correct) was that sometimes the molecules of a gaseous element consist of two atoms. That is, hydrogen gas, nitrogen gas, oxygen gas, and many others all have two atoms in a molecule. They are diatomic gases. Of course, the noble gases are all monatomic. We have seen previously that molecules are formed in this way, because each atom needs to have eight electrons to fill its outer shell.

Avogadro's two assumptions form the basis for the modern understanding of Gay-Lussac's experiments. Figure 15.4 shows schematically what takes place when 2 volumes of hydrogen and 1 volume of oxygen unite to form 2 volumes of steam. According to Avogadro's law, equal numbers of molecules are shown in each volume of gas. Two of the hydrogen atoms combine with one oxygen atom to form one steam molecule.

Figure 15.5 is an explanation of a similar situation in which nitrogen and hydrogen combine to form ammonia gas. Since experimentally we find that 3 volumes of hydrogen combine with 1 volume of nitrogen to form 2 volumes of ammonia gas, the molecular formula of ammonia can be deduced. To do this, Avogadro's law is used, along with his assumption that H_2 and N_2 are the correct molecular formulas for hydrogen and nitrogen gas. If we then put 2 molecules of each gas into each volume, we find that the molecular formula of ammonia can only be NH_3.

Equal numbers of gas molecules occupy equal volumes at the same temperature and pressure. However, the weights of different molecules are different, and equal volumes do not, of course, weigh the same. For instance, an oxygen molecule is about 16 times as heavy as a hydrogen molecule. Thus, one volume of oxygen

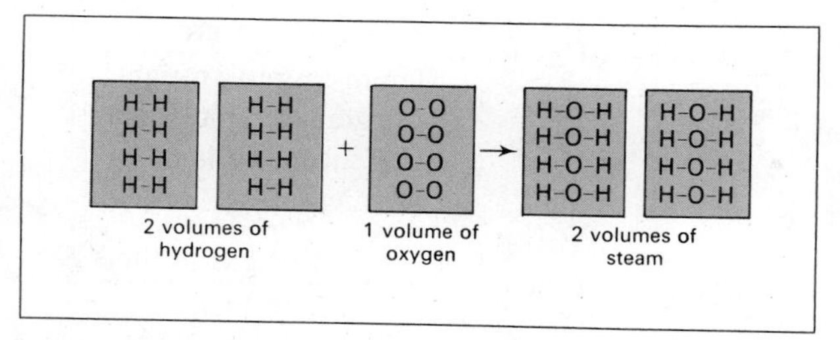

Figure 15.4 A graphic representation of Avogadro's explanation of Gay-Lussac's experiment. There are four molecules in each volume. This satisfies Avogadro's law.

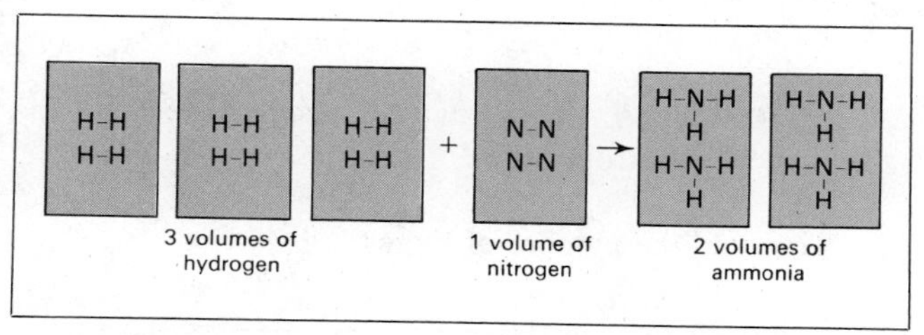

Figure 15.5 A graphic representation of Avogadro's explanation of the formation of ammonia. There are two molecules in each volume. This satisfies Avogadro's law.

gas will weigh about 16 times the weight of an equal volume of hydrogen gas. The volume unit used in chemistry is a liter. A **liter** is equal to 1000 cm^3, or roughly equivalent in volume to a quart.

One liter of oxygen gas weighs 1.43 g at standard conditions. **Standard conditions** are 0°C and sea level atmospheric pressure. Throughout the rest of the book, when a volume of gas is used the temperature and pressure will be assumed to be at the standard conditions unless otherwise specified. One liter of hydrogen gas weighs about $\frac{1}{16}$ of the weight of a liter of oxygen, or about 0.09 g. The weights of other gases can be found using the ratio of their molecular weights to that of oxygen.

EXAMPLE 6 What is the weight of one liter of nitrogen gas?

The molecular weight of N_2 is 28. The molecular weight of O_2 is 32. The weight of a volume of any gas is proportional to its molecular weight.

$$\frac{\text{Wt. of 1 liter } N_2}{\text{Mol. weight } N_2} = \frac{\text{Wt. of 1 liter } O_2}{\text{Mol. weight } O_2}$$

$$\frac{\text{Wt. of 1 liter } N_2}{28} = \frac{1.43 \text{ g}}{32}$$

or

$$\text{Wt. of 1 liter } N_2 = \frac{28}{32} \times 1.43 \text{ g} = 1.25 \text{ g}$$

One mole of oxygen gas weighs 32 g. Since one liter of oxygen gas weighs 1.43 g, the volume occupied by 32 g is given by a simple ratio

$$\frac{\text{Vol. of 32 g}}{32 \text{ g}} = \frac{\text{Vol. of 1.43 g}}{1.43 \text{ g}}$$

or

$$\frac{\text{Vol. of 1 mole } O_2}{32 \text{ g}} = \frac{1 \text{ liter}}{1.43 \text{ g}}$$

or

$$\text{Vol. of 1 mole } O_2 = \frac{32 \text{ g}}{1.43 \text{ g}} \times 1 \text{ liter}$$

$$= 22.4 \text{ liters}$$

Similarly, we find that one mole of H_2 equals about $\frac{2}{0.09} \times 1$ liter = 22.2 liters. If we had used a more exact figure for the molecular weight of hydrogen (2.016 instead of 2), we would have obtained 22.4 liters again. In fact, one mole of any gas occupies 22.4 liters. The quantity 22.4 liters is known as the **gram molecular volume.** It is the volume of one mole of any gas at standard conditions.

EXAMPLE 7 What volume will 27 g of steam occupy?

The molecular formula of steam is H_2O. The gram molecular weight is $1 \times 2 + 16 \times 1 = 18$ g. There are, therefore, $\frac{27}{18} = 1.5$ moles of steam in 27 g of steam. Now 1.5 moles of any gas occupies 1.5×22.4 liters = 33.6 liters.

EXAMPLE 8 What is the weight of 1 liter of nitrogen gas?

(This problem has been worked in a previous example but now we will do it in a simpler fashion.) First we find how many moles in 1 liter of nitrogen gas (or any gas).

$$\text{Number of moles} = \frac{1 \text{ liter}}{22.4 \text{ liters/mole}} = 0.0446 \text{ mole}$$

The weight of 0.0446 mole of nitrogen is 0.0446 × the gram molecular weight of nitrogen gas or 0.0446 mole × 28 g/mole = 1.25 g.

We see from the examples that it is important to find the number of moles of the gas. The volume can then be found by multiplying the number of moles by 22.4 liters.

$$\text{Volume of gas} = \text{No. of moles} \times 22.4 \text{ liters}$$

If the weight of the gas is desired, it can be found by multiplying the number of moles by the gram molecular weight of the substance. This works for liquids and solids as well as gases.

$$\text{Weight} = \text{Number of moles} \times \text{Gram formula weight of substance}$$

15.4 Avogadro's Number

Avogadro's law states that equal volumes of gases at the same temperature and pressure contain equal numbers of molecules. The number of molecules in one mole of a gas or liquid or solid is called **Avogadro's number.** Avogadro's number can be found by many different methods to be 6.02×10^{23} molecules per mole. Thus, in 22.4 liters of any gas at standard conditions, there will be 6.02×10^{23} molecules of that gas. The number of molecules in any covalent compound can be easily found, once the number of moles of the covalent compound is known. The formula is

$$\text{No. of molecules} = \text{No. of moles} \times \text{Avogadro's number}$$

EXAMPLE 9 How many molecules are there in 6 g of water?

In 6 g of water there are $\frac{6}{18} = \frac{1}{3}$ mole. The number of molecules is then

$$\frac{1}{3} \times 6.02 \times 10^{23} = 2.0 \times 10^{23} \text{ molecules}$$

in 6 g of water.

Again, we see that if we can find the number of moles, we can find the quantity we are looking for, in this case the number of molecules.

In order to work problems concerning volumes of gases, weights of substances, or the number of molecules in a covalent compound, we first need to find the number of moles from information given in the problem. The number of moles can be found from any of three different equations. They are

$$\text{Number of moles} = \frac{\text{Weight in grams}}{\text{Gram formula weight (or gram atomic weight if a substance is monatomic)}}$$

$$\text{Number of moles} = \frac{\text{Volume of gas in liters}}{22.4 \text{ liters}}$$

$$\text{Number of moles} = \frac{\text{Number of molecules}}{6.02 \times 10^{23}}$$

The correct equation to use depends on the information given in the problem.

EXAMPLE 10 How many molecules are there in 11.2 liters of nitrogen gas?

First find the number of moles.

$$\text{Number of moles} = \frac{11.2 \text{ liters}}{22.4 \text{ liters/mole}} = 0.5 \text{ mole}$$

$$\text{Number of molecules} = 0.5 \text{ mole} \times 6.02 \times 10^{23} \text{ molecules/mole}$$

$$= 3.01 \times 10^{23} \text{ molecules}$$

Avogadro's number also enables us to get an approximate idea of the size of molecules. For instance, water has a density of 1 g/cm^3. Using this information and Avogadro's number, the approximate size of a water molecule can be found. To do this, the number of molecules in 1 cm^3 of water must be found. There will be $\frac{1}{18}$ mole in 1 g of water and $(\frac{1}{18}) \times 6.02 \times 10^{23}$ molecules $= 0.33 \times 10^{23}$ molecules in 1 g, or 1 cm^3, of water. Next, we assume that the molecules are packed closely together in the liquid. The volume occupied by one molecule is then the total volume (1 cm^3) divided by the number of molecules in that volume.

$$\text{Vol. of one water molecule} = \frac{1 \text{ cm}^3}{0.33 \times 10^{23}}$$

$$= 3 \times 10^{-23} \text{ cm}^3$$

If we then assume that each H_2O molecule is a sphere (which we know it is not, it looks more like a triangle), we can find the radius of that sphere.

$$\frac{4}{3} \pi r^3 = 3 \times 10^{-23} \text{ cm}^3$$

We then estimate the radius of the sphere

$$r^3 = \frac{3}{4\pi} \times 3 \times 10^{-23}\ \text{cm}^3$$
$$= .72 \times 10^{-23}\ \text{cm}^3$$
$$= 7.2 \times 10^{-24}\ \text{cm}^3$$

Take the cube root to find r.

$$r = \sqrt[3]{7.2} \times 10^{-8}\ \text{cm}$$

The cube root of 8 is 2 so the cube root of 7.2 is about 2 and the radius of a water molecule is about 2×10^{-8} cm or about 2 angstroms. This seems a reasonable value since we already know that the Bohr radius of the hydrogen atom is 0.529 angstrom. We would expect the water molecule to be a few times larger than a single hydrogen atom.

15.5 Solutions

Solutions are homogeneous mixtures. They can be solid, liquid, or gaseous solutions. Metal alloys and air have previously been cited as examples of solid and gaseous solutions. The most common solutions, however, are liquids. The substance in excess in a solution is called the **solvent.** The substance dissolved is the **solute.** Solutions in which water is the solvent are called **aqueous solutions.** A solution in which the solute is present in only a small amount is called a **dilute solution.** If the solute is present in large amounts, the solution is **concentrated solution.** When the maximum amount of solute possible is dissolved in the solvent, the solution is called a **saturated solution.**

The concentration, or amount of solute dissolved, is frequently expressed in terms of the molar concentration. The **molar concentration,** or **molarity,** is the number of moles of solute per liter of solution. Thus, a one molar solution, written $1.0M$, has one gram formula weight of solute dissolved in one liter of solution. In general

$$\text{Molarity} = \frac{\text{Number of moles of solute}}{\text{Number of liters of solution}}$$

Note that it is the *number of liters of solution,* not the number of liters of solvent, that is used. For example, a one-molar solution of sodium chloride (NaCl), or table salt, in water is made by dissolving one mole (58.5 g) of NaCl in water so that the final solution is one liter. One liter of water is not added to the mole of solute; only enough water is added to make the volume of the solution one liter.

The concentration of a solution can be any value up to the saturation limit. The molarity desired can be obtained in any number of ways, depending on the amount of solution desired. For example, a $2.0M$ solution can be made by dissolving 2 moles of solute in water to make a liter of solution, or, if we wish to make a smaller quantity, by dissolving 1 mole of solute in 0.5 liter of solution. The concentration is the same either way.

EXAMPLE 11 What is the molarity of 40 g of salt dissolved in water to give 0.80 liter of solution?

First find the number of moles of salt.

$$\text{Number of moles} = \frac{40\ \text{g}}{58.5\ \text{g/mole}} = 0.68\ \text{mole}$$

$$\text{Molarity} = \frac{0.68\ \text{mole}}{0.80\ \text{liter}} = 0.85M$$

EXAMPLE 12 How many grams of H_2SO_4 are present in 0.60 liter of a $3.0M$ solution of sulfuric acid?

First find the number of moles, then convert to number of grams.

$$\text{Molarity} = \frac{\text{number of moles}}{\text{liter of solution}}$$

or

$$\text{number of moles} = \text{molarity} \times \text{liter of solution}$$
$$= 3.0\ \frac{\text{moles}}{\text{liter}} \times 0.60\ \text{liter}$$
$$= 1.8\ \text{mole}$$

Now find the gram formula weight of H_2SO_4.

$$\text{Grams } H_2SO_4 \text{ in 1 mole} = (2 \times 1) + (1 \times 32) + (4 \times 16)$$
$$= 98\ \text{g}$$
$$\text{Grams in 1.8 moles} = 1.8\ \text{mole} \times 98\ \text{g/mole}$$
$$= 176\ \text{g}$$

Remembering that Avogadro's number (6.02×10^{23}) is the number of molecules in a mole of any substance, we can find the number of molecules or ions in any concentration of solution. For example, a single liter of a $2.0M$ concentration of table sugar, a covalent compound, contains $2 \times 6.02 \times 10^{23} = 12.04 \times 10^{23}$

sugar molecules. If we have an ionic compound such as NaCl, we can compute the number of ions in the solution. A 0.5-liter quantity of a 3.0*M* NaCl solution would contain $2 \times 3.0 \times 0.5 \times 6.02 \times 10^{23} = 18.06 \times 10^{23}$ ions. The factor of two comes about because NaCl contains two ions, Na^+ and Cl^-.

When a solute is added to a solvent, it dissolves and, if thoroughly stirred, distributes itself uniformly throughout the solvent. The distribution of molecules or ions is the same at the top and bottom of the solution. As more solute is added, the solution becomes more and more concentrated; more and more solute at the top and at the bottom. Finally, after a given quantity has been added, no more solute dissolves and we say the solution is saturated. The **solubility** of a given solute is the amount of solute that will dissolve in a specified volume of solvent (at a given temperature) to produce a saturated solution. The solubility depends on the temperature of the solution. If the temperature is raised, the solubility of solids almost always increases. For this reason, hot water dissolves more of the solute than cold water.

If a gas such as CO_2 is being dissolved in water, the solubility increase is directly proportional to the pressure. This principle is used in the manufacture of soft drinks. The carbon dioxide is first forced into the beverage at high pressure. Then the beverage is bottled and tightly capped to maintain the pressure on the CO_2. Once the bottle is opened, the pressure inside the bottle is reduced to normal atmospheric pressure, and the CO_2 starts escaping from the liquid as shown in Fig. 15.6. If the bottle is allowed to remain open for some time, most of the CO_2 escapes, and the drink loses its zingy taste. The solubility of gases in most liquids decreases with increasing temperature. This effect is also demonstrated with soft drinks. If the beverage is allowed to warm before opening the bottle, the solubility of the CO_2 decreases, and when the bottle is opened, the CO_2 may escape so fast that the beverage shoots out of the bottle.

When saturated solutions of solids are prepared at high temperatures, and then cooled, the solubility drops and the excess solid crystallizes (forms crystals) and separates from the solution. However, if there are no crystals of the solid already present in the saturated solution, crystallization may not take place if the saturated solution is carefully cooled. The solution will then contain a larger amount of solute than the solubility of the solute

Figure 15.6 Dissolved carbon dioxide escapes from a carbonated beverage when pressure on the liquid surface is lowered. (Photo courtesy James Crouse.)

Figure 15.7 In sufficient amounts, an antifreeze added to water in an automobile radiator produces a solution with a freezing point low enough to prevent freezing during cold weather. (Photo courtesy James Crouse.)

dictates. Solutions that contain more than the normal maximum amount of dissolved solute are said to be **supersaturated solutions.** Such solutions are unstable. The introduction of a "seed" crystal will cause the excess solute to crystallize. As discussed later in Chapter 23, this is the basic principle behind cloud seeding to cause rain. If the air is supersaturated with water vapor, the introduction of certain types of crystals into the clouds greatly increases the probability that the water vapor will form raindrops. Under certain conditions the carbon dioxide solution comprising a soft drink also may be supersaturated. Vigorously shaking the solution causes the excess gas to be liberated.

Whenever a solid is dissolved in water, the freezing point of the water solution is lowered and the boiling point is raised. These changes are related to the number of molecules or ions of solute present in the solution. Ordinary tap water boils at a slightly higher temperature and freezes at a slightly lower temperature than pure water, because of the presence of impurities. Similarly, the salt in ocean water causes the freezing point of salt water to be lower than fresh water. A well-known application of this principle of lowering the freezing point is the use of "antifreeze" in car radiators (see Fig. 15.7). The antifreeze in use today is either an alcohol or a glycol. The more antifreeze that is present, the more the freezing point is lowered.

As stated previously, water is a polar molecule. Water dissolves ionic compounds or polar covalent compounds because of its electrical attraction for such substances. However, when a nonpolar covalent compound such as carbon tetrachloride or gasoline is combined with water, it is not attracted by the water molecules. The water molecules attract only each other, and the nonpolar molecules are forced away from the water. For this reason, nonpolar compounds will not dissolve in water. Nonpolar compounds will dissolve each other, however, but they will not dissolve ionic or polar covalent compounds. As a general rule, "like dissolves like." If you wish to dissolve stains made by a nonpolar compound such as grease, use nonpolar solvents such as gasoline and benzene. For a polar compound, such as sugar, water should be used.

Learning Objectives

After reading and studying this chapter, you should be able to do the following without referring to the text:

1. Describe the various types of matter and give examples of each.
2. Distinguish between physical and chemical properties of matter.
3. State and explain the following laws:
   (a) Conservation of mass
   (b) Definite proportions
   (c) Multiple proportions
   (d) Gay-Lussac's law of combining volumes
   (e) Avogadro's law
4. Distinguish between gram atomic weight and gram formula weight.
5. Distinguish between gram molecular weight and molecular weight.
6. State the physical characteristics required for standard conditions.
7. State the numerical value of Avogadro's number.
8. Distinguish between solvent and solute.
9. Describe the principal characteristics of solutions.
10. Distinguish between molar and molar solution.
11. Calculate the formula weight of a given compound.
12. Calculate the percentage composition of a compound.
13. Calculate the number of grams of a compound that can be obtained from a given amount of known elements.
14. Calculate the number of grams of an element that can be obtained from a given amount of a known compound.
15. Calculate the number of moles present in a given amount of a compound.
16. Calculate the number of molecules in a given amount of a compound.
17. Calculate the volume at standard conditions that will be occupied by a given amount of a known gas.
18. Calculate the molarity of a solution when the volume and the amount of the substance in the solution are given.
19. Define or explain the important words and terms listed at the end of the chapter.

Important Words and Terms

homogeneous
heterogeneous
substance
solution
physical properties
physical changes
chemical properties
chemical changes
law of conservation of mass
law of definite proportions
law of multiple proportions
gram atomic weight
formula weight
gram formula weight
mole
molecular weight
gram molecular weight
percentage composition
Gay-Lussac's law of combining volumes
Avogadro's law
liter
standard conditions
gram molecular volume
Avogadro's number
solvent
solute
aqueous solution
dilute solution
concentrated solution
saturated solution
molar concentration
molarity
solubility
supersaturated solution

Questions

1. Describe the physical properties that distinguish solids, liquids, and gases.
2. Identify each of the following as a physical or chemical change:
   (a) Melting of ice
   (b) Burning of a match
   (c) Fermenting of wine
   (d) Dissolving sugar in water
   (e) Rusting of steel
   (f) Magnetizing a sewing needle
3. Distinguish between homogeneous and heterogeneous matter.
4. Identify three pure substances and list some of the properties used to identify them as pure substances.
5. Give some examples of solutions that are not in the liquid phase.
6. State the law of conservation of mass.
7. Is the law of conservation of mass exactly true or only approximately true?
8. What is the molecular explanation of the law of definite proportions?
9. What is the molecular explanation of the law of multiple proportions?
10. State Gay-Lussac's law of combining volumes.
11. What were Avogadro's assumptions that explained Gay-Lussac's law of combining volumes?
12. Distinguish between gram atomic weight and gram molecular weight. Give an example of each.
13. How is it possible to mix one volume of a gas with three volumes of another gas and end up with only two volumes?
14. Why is the volume of a gas occupied by one mole always the same, no matter what the gas (assuming standard conditions)?
15. What is the meaning of the term ''standard conditions''?
16. What is Avogadro's number? State its numerical value.
17. Describe the following terms as applied to solutions: solute, solvent, aqueous, dilute, concentrated, saturated, supersaturated.
18. Define a 1*M* solution. Give an example.
19. Why do soft drinks contain bubbles? What are the bubbles?
20. Why does an opened soft drink go flat after a period of time?
21. Why is antifreeze added to the water in a car's radiator during the winter?
22. On what type of compounds is carbon tetrachloride a useful cleaning agent? Explain your answer.

Problems

1. Calculate the formula weight of the following compounds:
   (a) Water (H_2O)
   (b) Carbon dioxide (CO_2)
   (c) Methane (CH_4)
   (d) Benzene (C_6H_6)
2. Calculate the formula weight of the following compounds:
   (a) Sodium chloride (NaCl)
   (b) Calcium carbonate ($CaCO_3$)
   (c) Hydrochloric acid (HCl)
   (d) Sodium bicarbonate ($NaHCO_3$)
3. Determine the percentage composition of the following compounds:
   (a) Water (H_2O)
   (b) Sugar ($C_{12}H_{22}O_{11}$)
   (c) Table salt (NaCl)
   (d) Aspirin ($CH_3COOC_6H_4COOH$)
4. Determine the percentage composition of the following compounds:
   (a) Carbon monoxide (CO)
   (b) Potassium chlorate ($KClO_3$)
   (c) Manganese dioxide (MnO_2)
   (d) Iron rust (Fe_2O_3)
5. Calculate the number of grams of sodium chloride (NaCl) that can be formed from 100 g of sodium and 105 g of chlorine. What element is not completely used? How much is not used?
6. Calculate the number of grams of carbon dioxide (CO_2) that can be formed from 36 g of carbon and 100 g of oxygen. What element is not completely used? How much is not used?
7. Calculate the number of grams of sulfur that can be produced from 128 g of sulfur dioxide (SO_2).
8. Calculate the number of grams of lime (CaO) and carbon dioxide (CO_2) that can be produced from 200 g of calcium carbonate ($CaCO_3$).
9. Find the number of moles present in each of the following:
   (a) 56 g of carbon monoxide
   (b) 30 liters of carbon monoxide gas at standard conditions
   (c) 3.01×10^{23} molecules of carbon monoxide
10. Determine the number of molecules contained in the following quantities:
    (a) 1 g of water
    (b) 44.8 liters of carbon monoxide gas
    (c) 4.0 moles of sugar *Answer:* (a) 0.33×10^{23}
11. Compute the volume at standard conditions that will be occupied by the following gases:
    (a) 40 g of argon
    (b) 150 g of xenon
    (c) 48 g of oxygen
    (d) 14 g of carbon dioxide *Answer:* (c) 33.6 liters
12. Find the molarity of the following solutions:
    (a) 2.5 moles of sugar in 1 liter of solution
    (b) 49 g of H_2SO_4 in 2 liters of solution
    (c) 12 g of HCl in 0.7 liter of solution

    Answer: (c) 0.47*M*

16

Chemical Reactions

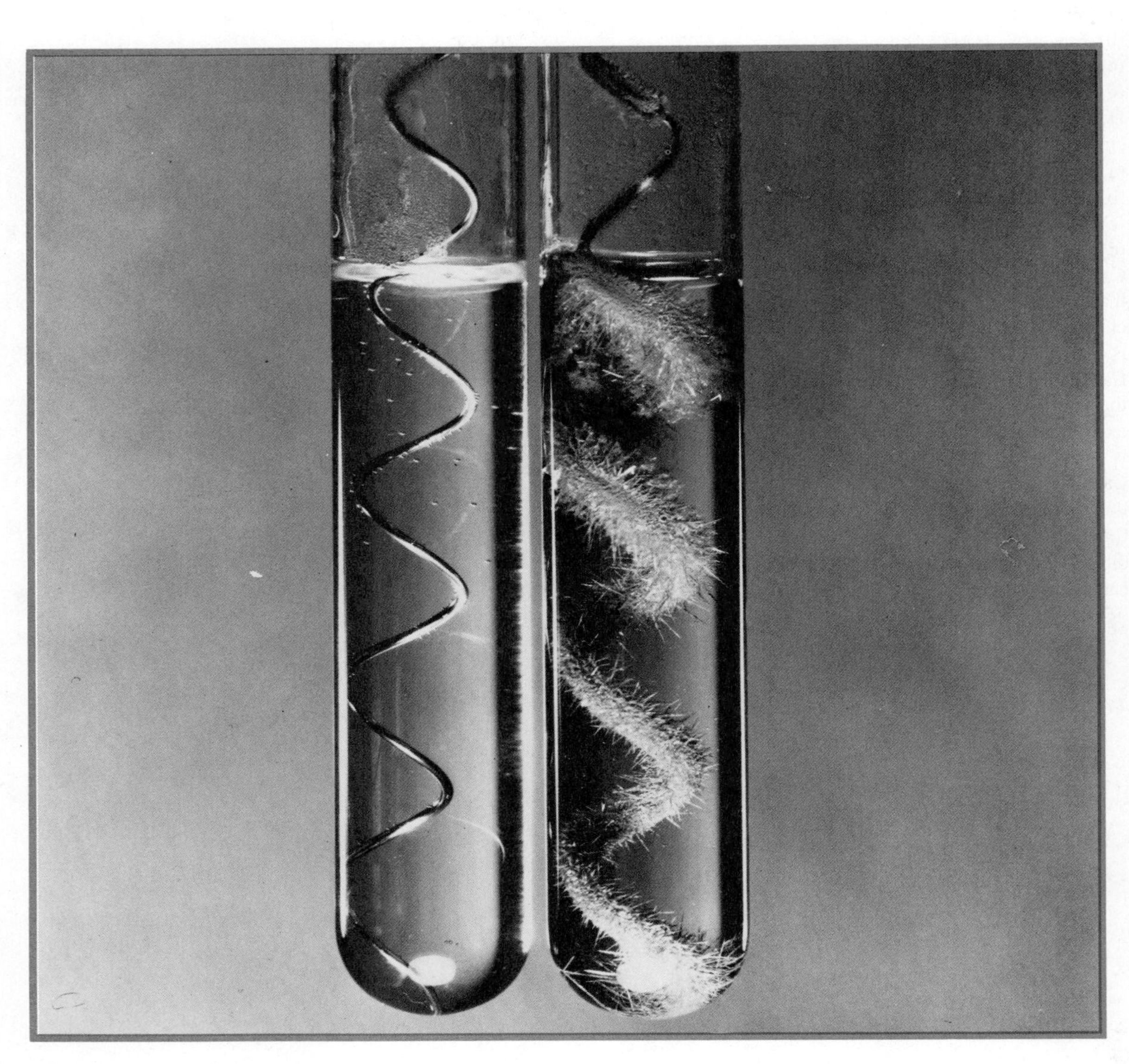

There's nothing constant in the universe,
All ebb and flow, and every shape that's born
Bears in its womb the seed of change.
—Ovid, *Metamorphoses*

CHEMICAL CHANGE IN nature takes place continuously. Green plants absorb carbon dioxide from the atmosphere and with energy from the Sun and chlorophyll as a catalyst react with water from the soil to form sugar and free oxygen. The complex chemical reaction is called photosynthesis. The animal life on our planet ingest the sugar and inhale the oxygen to obtain energy. Chemical change in the body then returns to the environment water and carbon dioxide. This interaction between plants and animals and their environment constitutes an ecosystem.

Our environment is made up of atoms that combine with one another to form molecules, which in turn react with other atoms or molecules to produce the many substances we need and use. In the production of new products, energy is released or absorbed, depending on the nature of the chemical reaction. There are countless chemical reactions, but only a few will be studied here.

Something of the nature of chemical reactions can be learned by studying examples of two specific types of reactions: oxidation-reduction and acid-base reactions. Everyone is acquainted with a match burning in air and the rusting of iron. These are examples of a chemical reaction known as an oxidation-reduction reaction. Most of us are also acquainted with a sour or acid stomach and the substances used to neutralize it. Neutralization is an example of an acid-base reaction. We will see in Chapters 17, 22, and 31 that chemical reactions are important in the chemistry of large molecules, in atmospheric processes, and in geology. The present chapter discusses the basic chemistry involved in chemical reactions, in particular, oxidation-reduction and acid-base reactions.

◀ Crystals of metallic silver forming on a helix of copper wire in a silver nitrate solution.

16.1 Basic Concepts

When two or more substances are brought together, they may react to form new substances. Consider an example in which substances A and B react to form substances C and D. This is written

$$A + B \longrightarrow C + D$$

The expression is called a chemical equation. It is read: Substances A and B react to yield the substances C and D. The initial substances A and B are called the **reactants,** and the final substances C and D are called the **products.**

In any chemical reaction three things take place.

1. The reactants disappear or are diminished.
2. Different substances appear as products. These products have chemical and physical properties that are significantly different from the original reactants.
3. Energy is either released or absorbed. This energy may take the form of heat, light, electricity, etc.

The energy characteristics of chemical reactions will be discussed more fully in Section 16.2.

Now let us suppose that we turn our original reaction around, so that when C and D are combined, A and B are formed as the products. We get

$$C + D \longrightarrow A + B$$

If we start with A and B, we get C and D; if we start with C and D, we get A and B. A reaction of this type is called a **reversible reaction.** The reversible reaction can be written

$$A + B \rightleftharpoons C + D$$

The double arrows mean that the reaction proceeds in

both directions. An example of a reversible reaction involves iron, steam, and hydrogen gas. The equation is

$$3\ Fe + 4\ H_2O \rightleftharpoons Fe_3O_4 + 4\ H_2$$

If steam is passed over hot iron, the reaction proceeds from left to right. If hydrogen gas is passed over heated Fe_3O_4, the reaction goes from right to left.

In a reversible reaction such as

$$A + B \rightleftharpoons C + D$$

all four substances (A, B, C, and D) are present in the vessel containing the substances. On a molecular scale, molecules of A and B are combining to form molecules of C and D; and at the same time, molecules of C and D are combining to form molecules of A and B. When the two competing reactions are occurring at the same rate, we say that the reactions have reached a state of equilibrium. **Equilibrium** is a dynamic process in which there is a fixed number of molecules of all reactants and products present. However, individual molecules are continuously changing back and forth, due to interactions with each other.

The amounts of A, B, C, and D present at equilibrium depend on many things. Among these are temperature, pressure, and the excess amounts of reactants and products present. As an example, consider again the reaction

$$3\ Fe + 4\ H_2O \longrightarrow Fe_3O_4 + 4\ H_2$$

and the reverse reaction

$$Fe_3O_4 + 4\ H_2 \longrightarrow 3\ Fe + 4\ H_2O$$

In both of these reactions a stream of gas is passed over a hot solid. If steam is passed over hot iron, the hydrogen is swept out by the steam, and the first reaction dominates. If hydrogen gas is passed over Fe_3O_4, the steam is swept out by the excess hydrogen, and the second reaction occurs. This example demonstrates just one of many methods by which a reversible reaction may be shifted from right to left or left to right.

Every reaction is to some extent reversible. That is, there will always be some of every kind of atom or molecule present that can possibly occur in a reaction. For example, consider the reaction

$$2\ HI \longrightarrow H_2 + I_2$$

This reaction says that hydrogen iodide decomposes into hydrogen and iodine. However, in any sample of hydrogen and iodine there will be some hydrogen iodide, even though it will be a very small amount. To show explicitly that this is the case, we would write the reaction with a small arrow going from right to left.

$$2\ HI \rightleftharpoons H_2 + I_2$$

Writing the reaction this way emphasizes the reversibility, however slight, of the reaction. However, in the discussions that follow, we will omit the small arrow from right to left. It will be understood that at equilibrium there are always some molecules of all the various reactants and products in the reaction.

Chemical equations are written for a chemical reaction and until the equation is balanced it cannot express a chemical equality. The yield sign ($\longrightarrow$) indicates the direction of the reaction and has the meaning of an equality sign. The equation is balanced by using the law of conservation of mass. That is, in a chemical reaction atoms cannot be created or destroyed. Thus, there must be the same number of atoms on each side of the arrow in any equation for a chemical reaction. As an example, consider the "roasting" of zinc ore (ZnS). In this process sulfide ore is heated in air in order to convert the zinc sulfide to an oxide from which pure zinc can be extracted. The chemical equation for the roasting of the ore zinc sulfide is

$$2\ ZnS + 3\ O_2 \longrightarrow 2\ ZnO + 2\ SO_2$$

The numbers in front of the chemical formulas in this equation are called coefficients. They indicate the relative amounts of each substance that participate in the reaction. For instance, in the above reaction, for every 3 moles of oxygen gas reacting with every 2 moles of zinc sulfide, the result will be 2 moles of zinc oxide and 2 moles of sulfur dioxide. An inspection of the equation will show that it is indeed balanced. A tabulation of the number of atoms on both sides of the equation shows

$$\left.\begin{array}{l}\text{2 atoms of zinc}\\\text{2 atoms of sulfur}\\\text{6 atoms of oxygen}\end{array}\right\} \longrightarrow \left\{\begin{array}{l}\text{2 atoms of zinc}\\\text{2 atoms of sulfur}\\\text{6 atoms of oxygen}\end{array}\right.$$

This equality of atoms is because atoms cannot be created or destroyed in chemical reactions.

In balancing chemical equations the subscripts on the various elements represent the number of atoms in each molecule, and these numbers cannot be changed. The coefficients are to be found such that there is an equal number of atoms on both sides of the arrow. As an example, consider balancing the equation for the decomposition of potassium chlorate ($KClO_3$) when heated. (This is a common method of producing oxygen in the laboratory.)

EXAMPLE 1 Balance the following:

$$KClO_3 \longrightarrow O_2 + KCl$$

To balance the equation, inspect it to see which elements have a different number of atoms on the two sides of the equation. We see that the potassium and chlorine atoms are balanced already. Only the oxygen is not. Of course, once we adjust the equation to balance the oxygen atoms, we may unbalance the potassium and chlorine atoms.

To get the oxygen atoms equal on both sides we must put a 2 in front of the $KClO_3$ (potassium chlorate) and a 3 in front of the O_2. This gives us

$$2\ KClO_3 \longrightarrow 3\ O_2 + KCl$$

The oxygen is now balanced, but the potassium and chlorine are unbalanced. However, this is easily rectified by putting a 2 in front of the KCl. Finally we have the balanced equation

$$2\ KClO_3 \longrightarrow 3\ O_2 + 2\ KCl$$

EXAMPLE 2 Balance the following:

$$\text{water} \longrightarrow \text{hydrogen} + \text{oxygen}$$

or $$H_2O \longrightarrow H_2 + O_2$$

Both hydrogen and oxygen exist as diatomic molecules in the free state.

The equation is balanced as written except for the oxygen atoms. To balance the oxygen, place the coefficient 2 in front of the water molecule.

$$2\ H_2O \longrightarrow H_2 + O_2$$

This step balances the oxygen, that is, the oxygen atoms are equal on both sides of the yield sign. But, now the hydrogen is not balanced. To balance the hydrogen place the coefficient 2 in front of the hydrogen molecule. The equation, now balanced, reads

$$2\ H_2O \longrightarrow 2\ H_2 + O_2$$

In addition to balancing the number of atoms on both sides of an equation, the electric charge must also be balanced. This is because electric charge cannot be created or destroyed. Consider the equation for the ionization of sulfuric acid.

$$H_2SO_4 \longrightarrow 2\ H^+ + SO_4^{2-}$$

In this equation the charges on each side of the arrow add up to zero. The number of atoms on both sides of this equation is also the same, as it must be.

Now let us consider balancing a more difficult equation.

EXAMPLE 3 Balance the following:

$$Pb + PbO_2 + H^+ + SO_4^{2-} \longrightarrow PbSO_4 + H_2O$$

This is a reaction that occurs in a lead storage battery when it is being used. To balance the equation we first inspect it to see what is not balanced. The lead is not balanced so we put a 2 in front of the $PbSO_4$ on the right. To balance the sulfur, we put a 2 in front of the SO_4^{2-} on the left. We then have

$$Pb + PbO_2 + H^+ + 2\ SO_4^{2-} \longrightarrow 2\ PbSO_4 + H_2O$$

Now we need a 4 in front of the H^+ to get a total charge of zero on the left (4 + charges and 4 − charges) and a 2 in front of the H_2O on the right to balance the equation. The equation is correctly written as

$$Pb + PbO_2 + 4\ H^+ + 2\ SO_4^{2-} \longrightarrow 2\ PbSO_4 + 2\ H_2O$$

16.2 Energy and Rate of Reaction

All chemical reactions involve a change in energy. The energy is either released or absorbed in the form of heat, light, electrical energy, sound, etc. If energy is released in a chemical reaction, the reaction is called an exoergic (or exothermic) reaction; if energy is absorbed, the reaction is called an endoergic (or endothermic) reaction. An example of a common exoergic reaction is the burning of natural gas, which is composed primarily of methane.

$$CH_4 + 2\ O_2 \longrightarrow CO_2 + 2\ H_2O + \text{Energy}$$

This reaction occurs when a gas stove is lighted. (See Fig. 16.1.) An example of an endoergic reaction is the production of ozone, the triatomic molecule of oxygen.

$$3\ O_2 + \text{Energy} \longrightarrow 2\ O_3$$

This reaction occurs in the upper atmosphere, where the energy is provided by the ultraviolet radiation from the Sun. It also occurs near electric discharges. Ozone has a pungent odor, which is readily detected even in very small amounts. It is the unusual odor one notices when near a toy electric train that is operating and its motor is producing sparks.

Figure 16.1 A gas flame. Visible evidence of an exoergic chemical reaction. (Photo courtesy James Crouse.)

In the two preceding equations we have explicitly included the word "energy" to show the exoergic or endoergic nature of the reactions. This is usually not done, however.

The energy associated with a chemical reaction is related to the bonding energies between the atoms that form the molecules. The process is complex, because during a chemical reaction some chemical bonds are broken while others are formed, and it is difficult to determine the energies associated with each. In general, energy is absorbed when bonds are broken and given off when they are formed. In the ozone reaction the energy balance can be summarized as follows:

O_2 bonds broken $\longrightarrow$ O_3 bonds formed
(more energy absorbed) (less energy released)
Net result: Energy is absorbed

When methane burns in air the energy balance is

CH bonds broken, O_2 bonds broken (less energy absorbed) | CO bonds formed, OH bonds formed (more energy released)
Net result: energy is released

A chemical reaction may occur when the reactants are brought together; that is, when two or more molecules or ions collide with one another, they may react chemically, but *only if the kinetic energy of the molecules is above a minimum value*. In other words, a certain amount of energy is needed to start a chemical reaction. As an example, we are all aware of the necessity of striking a match before it ignites. In this example, we must contribute some energy—through friction—to initiate the chemical reaction of a match burning. Another example is the burning of methane gas. A flame is needed to ignite the methane, because the CH and O_2 bonds must be broken initially. Once the stove is lighted the net energy released suffices to break the bonds of still more CH_4 and O_2 molecules, and the reaction proceeds continuously, giving off energy in the form of heat and light.

The energy necessary to get a chemical reaction started is called the **activation energy.** This is the kinetic energy colliding molecules must possess in order to react chemically. However, once the activation energy is supplied, an exoergic reaction can release more kinetic energy than was supplied. This is illustrated on a different scale in Fig. 16.2. In the figure a boulder is on top of a cliff. If enough energy is supplied to raise the boulder to the edge of the cliff, the boulder will roll down the cliff, releasing great amounts of energy. A similar situation exists for an exoergic reaction. Once the activation energy is supplied, the formation of new molecular bonds can release more kinetic energy than was absorbed in breaking the original bonds. In endoergic reactions, more kinetic energy is absorbed to break molecular bonds than is generated when new bonds are formed. In this case there is a net loss of kinetic energy. This is usually perceived on a large scale as a "loss of heat."

When an exoergic chemical reaction takes place so very rapidly that there is a large change in the volume of the gases and an almost instant liberation of the energy of the reaction, an explosion occurs. Some of the energy in an explosion is given off in the form of sound energy.

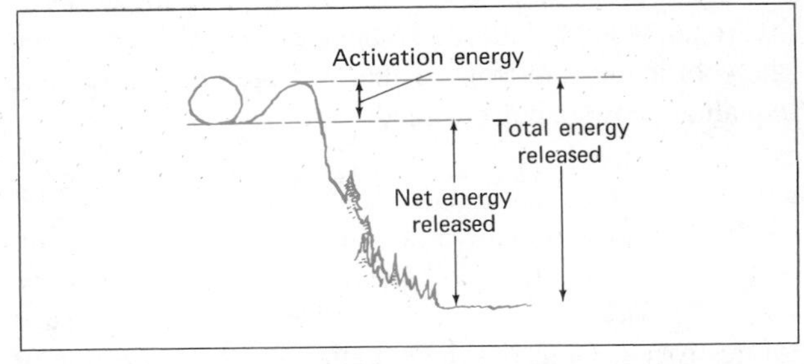

Figure 16.2 A graphic illustration of the energetics of an exoergic chemical reaction. If enough energy is supplied to activate the reaction (lifting the boulder), the total energy released (the falling boulder) will equal the activation energy plus the net energy released.

A reaction that proceeds more slowly than an explosion and yet is still quite rapid is termed combustion. In combustion, heat and light are produced in the form of fire. Common examples of combustion are the burning of natural gas, wood, paper, or coal in air. When gasoline in the form of heptane is burned, the combustion reaction is

$$C_7H_{16} + 11\ O_2 \longrightarrow 7\ CO_2 + 8\ H_2O$$

In the automobile engine there is insufficient oxygen to completely burn all of the carbon, and the reaction is more commonly

$$C_7H_{16} + 9\ O_2 \longrightarrow 4\ CO_2 + 2\ CO + C + 8\ H_2O$$

In this reaction the poisonous gas carbon monoxide, CO, is formed, along with the element carbon, which is black. The gases coming out of an automobile tailpipe consist of significant amounts of poisonous carbon monoxide. For this reason, running an automobile engine in a closed garage is very dangerous. The carbon monoxide accumulating from the exhaust can be fatal. The black color of some exhaust gases indicates the presence of large amounts of carbon, and this indicates that incomplete oxidation is occurring.

The rate of a reaction is dependent on several factors. One of the most important factors affecting the rate of reaction is the temperature. This is due to the fact that the average kinetic energy of the molecules is directly proportional to the temperature. In many cases the rate of reaction is significantly altered by even a slight change in temperature. In fact, as a general rule of thumb, a 10°C increase in temperature leads to a doubling of the reaction rate. This is sometimes referred to as the **ten-degree rule.**

The boundary of an atom or molecule is defined by its outermost orbiting electrons. These outer electrons are repelled by the electrons of neighboring atoms or molecules and thus, the particles tend not to react unless they possess enough energy to overcome this tendency. The kinetic energy possessed by any one atom or molecule may or may not be great enough for reaction. (See Fig. 16.3.) Molecules and atoms that have low activation energies readily react because a greater number of the collisions are effective. If the temperature of the reactants is raised, the kinetic energy of the molecules will be increased, and the reaction rate will increase.

In some reactions the energy of activation of the molecules or atoms can be partially altered by adding a substance called a catalyst. A **catalyst** is a substance that speeds or slows the rate of reaction but is not itself consumed in the reaction. Catalysts have no effect on the equilibrium mixture of a reaction. They only lessen the amount of time required to reach equilibrium conditions. Some catalysts act by providing a surface on which the reactants are adsorbed and concentrated. Others act as "carriers"; that is, they unite with a reactant to form an intermediate substance that takes part in the chemical process and then decomposes to release the catalyst. The use of nitrogen oxide (NO) to produce SO_3 used in the manufacture of sulfuric acid is an example of a catalyst which acts as a carrier.

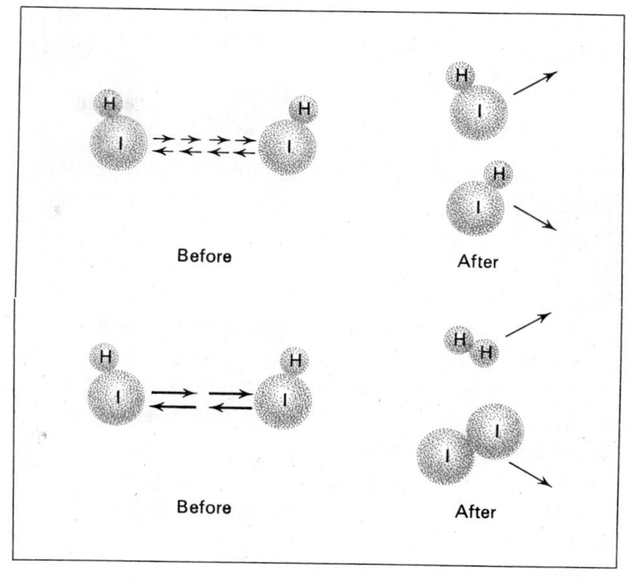

Figure 16.3 As molecules increase their speed, the probability of a collision effectively leading to a reaction increases. The top portion of the drawing shows a hydrogen iodide molecule that has insufficient energy to cause a reaction. The HI molecules merely bounce off each other. In the lower collision the molecules are moving fast enough to break their internal bonds and form H_2 and I_2 molecules, according to the equation $2\ HI \rightarrow H_2 + I_2$.

$$2\ SO_2 + O_2 \longrightarrow 2\ SO_3 \text{ (slow reaction)}$$

$$\left.\begin{array}{l} 2\ NO + O_2 \longrightarrow 2\ NO_2 \\ SO_2 + NO_2 \longrightarrow SO_3 + NO \end{array}\right\} \text{ (fast reactions)}$$

The NO combines with oxygen to form NO_2. The NO_2 then reacts with SO_2 to form SO_3, releasing NO, which can be used again.

Another example of using a catalyst occurs in the decomposition of potassium chlorate ($KClO_3$). A common method for preparing oxygen in the laboratory is the decomposition of potassium chlorate by heating. When heated to 400°C the white crystalline solid breaks down, producing potassium chloride and releasing oxygen. The reaction can be written as

$$2\ KClO_3 + 400°C \longrightarrow 2\ KCl + 3\ O_2$$

The process, however, is slow and heat requirements are high. If a small amount of manganese dioxide (MnO_2) is mixed with the $KClO_3$, the reaction takes place rapidly at 250°C. The manganese dioxide does not enter into the reaction but acts only as a catalyst.

The way in which many catalysts work is not known. A substance that acts as a catalyst for one reaction may or may not act as a catalyst for another. Catalysts are used extensively in manufacturing processes. They also play an important part in biochemical processes. The human body has many thousands of catalysts called **enzymes,** which act to control various physiological reactions. It is also believed that vitamins and hormones are catalysts.

16.3 Oxidation-Reduction Reactions

The element oxygen was discovered in 1774 by Joseph Priestley (1733–1804), an English clergyman, who referred to the element as "air." Priestley made the discovery by heating red rust of mercury (mercury(II) oxide) in a closed glass flask with a burning glass (lens). The reaction which Priestley observed was

$$2\ HgO \longrightarrow 2\ Hg + O_2$$

In this reaction, the oxide has decomposed, but it is much more common for oxygen to combine with another substance to form the oxide. We have seen that combustion is a reaction that frequently involves oxygen. When oxygen combines with another substance we call the reaction oxidation. For instance, the equation for the oxidation of carbon to form carbon dioxide is

$$C + O_2 \longrightarrow CO_2$$

Another example of oxidation is the souring of wines, that is, the changing of alcohol to acetic acid. The reaction can be written as

$$\underset{\text{Ethyl alcohol}}{CH_3CH_2OH} + O_2 \longrightarrow \underset{\text{Acetic acid}}{CH_3COOH} + H_2O$$

Acetic acid is the principal component of vinegar. Obviously, you should keep your wine sealed airtight!

When oxygen is removed from a compound, the reaction is called reduction. For example, when the ore hematite, iron(III) oxide, is heated with coke (carbon) in a blast furnace reduction occurs, the iron oxide being reduced to iron. This reaction is extremely important in the steel-making industry. It can be written as

$$2\ Fe_2O_3 + 3\ C \longrightarrow 4\ Fe + 3\ CO_2$$

We observe from the equation that oxygen is removed from the iron oxide. We say that the iron has been reduced. At the same time, we see that oxygen has reacted with the carbon to form carbon dioxide, that is, the carbon has been oxidized. Thus, the two reactions, oxidation and reduction, occur always together. A reaction in which both oxidation and reduction occur is called an **oxidation-reduction reaction.**

Oxidation-reduction reactions do not always involve oxygen. In order to apply to a larger number of reactions, oxidation-reduction reactions are also stated in terms of electrons gained or lost by atoms. By definition the loss of electrons by an atom or ion is called **oxidation.** The gain of electrons by an atom or ion is called **reduction.** When an atom or ion gives up electrons, it is **oxidized,** and when an atom or ion gains electrons, it is **reduced.** For example, when copper (Cu) combines with oxygen to form copper(II) oxide, the copper atoms lose electrons and are thus said to be oxidized. The oxygen atoms gain electrons and are reduced.

$$2\ Cu + O_2 \longrightarrow 2\ (Cu^{2+}O^{2-})$$

Since all the electrons lost by atoms or ions must be gained by other atoms or ions, it is evident that oxidation and reduction occur at the same time and at the same rate. The atom that is oxidized is the **reducing agent;** the atom that is reduced is called the **oxidizing agent.** In the example above, the metal (copper) reduced the oxygen; therefore, it was the reducing agent. Likewise, the nonmetal (oxygen) caused the oxidation to occur, so it was the oxidizing agent. Generally speaking, metals are reducing agents and nonmetals are oxidizing agents.

As just mentioned, according to the electron concept the element oxygen does not have to take part in an oxidation-reduction reaction. If we place a shiny iron nail in a copper sulfate solution, the iron nail becomes coated with a film of copper, and we find iron ions in the solution. This is an oxidation-reduction reaction. The copper has replaced the iron on the nail. The iron atoms lose electrons and are oxidized into ferrous (2+) ions. The copper ions in solution gain electrons and are reduced to copper atoms. The reaction is shown in the following equation:

$$Fe + Cu^{2+}SO_4^{2-} \longrightarrow Fe^{2+}SO_4^{2-} + Cu$$

or

$$Fe + Cu^{2+} \longrightarrow Fe^{2+} + Cu$$

In the reaction of copper sulfate with iron,

$$CuSO_4 + Fe \longrightarrow Cu + Fe^{2+}SO_4^{2-}$$

the iron displaces the copper. Such reactions are called **displacement reactions.** Thus, we see that some metals give up their valence electrons while others gain electrons. The activity of a metal, that is, its propensity to lose electrons, is due to its electron configuration. When metals compete for electrons, the more active metal yields to the less active. Thus, iron atoms, being more active than copper atoms, lose electrons to the copper ions and become iron ions. The copper ions, on the other hand, gain electrons and are converted to neutral copper atoms.

The relative activity of any metal can be discovered by merely placing the metal in an ionic solution containing another metal and seeing if the metal replaces the one in solution. If it does, it is more active, because it has given electrons to the metal in solution. If this process is carried out for all metals, an order-of-activity series can be obtained. Such a series is called the **electromotive series.** It is shown in Fig. 16.4.

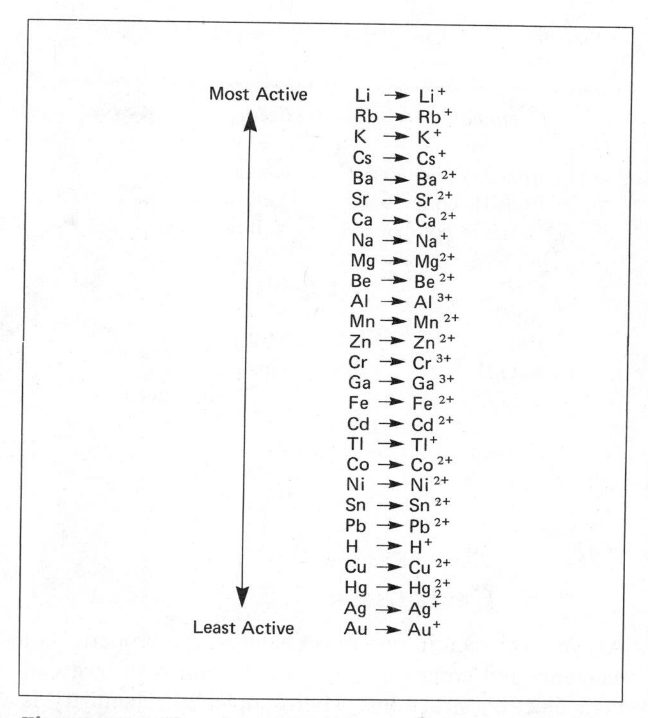

Figure 16.4 The electromotive, or activity, series. The higher the element is in the series, the more active it is. For example, Fe is more active than Cu, and Zn is more active than Fe.

There are thousands of important oxidation-reduction reactions. We will mention only a few more examples. In the commercial production of metals the metals must be separated from other elements present in the metals' ores. Frequently, the ores are sulfides or carbonates. The first step is to "roast" the ores. This consists of heating them to very high temperatures in the presence of air. Some typical reactions are

$$2\ ZnS + 3\ O_2 \longrightarrow 2\ ZnO + 2\ SO_2$$

$$PbCO_3 \longrightarrow PbO + CO_2$$

The roasting process converts the metal to its oxide. The next step is to reduce the oxide to the pure metal. Carbon and carbon monoxide are frequently used in this step. For example,

$$ZnO + C \longrightarrow Zn + CO$$

$$Fe_2O_3 + 3\ CO \longrightarrow 2\ Fe + 3\ CO_2$$

These reactions are extremely important in the commercial manufacturing of metals.

Glass consists of molten material that has cooled very rapidly. Although glass seems to be a solid, it does not have the crystalline structure typical of solids, and for this reason it is usually thought of as a liquid that flows very, very slowly. The common varieties of glass (such as that used in windows and clear bottles) are made by heating mixtures of Na_2CO_3, $CaCO_3$, and SiO_2 (sand). The reactions that occur are

$$Na_2CO_3 + SiO_2 \longrightarrow Na_2SiO_3 + CO_2$$

and

$$CaCO_3 + SiO_2 \longrightarrow CaSiO_3 + CO_2$$

A mixture of Na_2SiO_3 and $CaSiO_3$ is the basic substance in ordinary glass. Pyrex glass is ordinary glass with boron oxide, B_2O_3, added to it. The addition of B_2O_3 gives the glass a low coefficient of expansion, which enables it to withstand changes in temperatures without breaking. If a colored glass is desired, an appropriate substance is added during the manufacture. A few of the compounds used in coloring glass are listed in Table 16.1 on the next page.

Table 16.1 Compounds Used in Glassmaking to Color Glass

Compound	Color
Iron(II) compounds	Green
Iron(III) compounds	Yellow, brown
Uranium compounds	Yellow, green, fluorescence
CoO	Blue
MnO_2	Violet
CaF_2	Milky
SnO_2	Opaque
Cu_2O	Red, blue, green

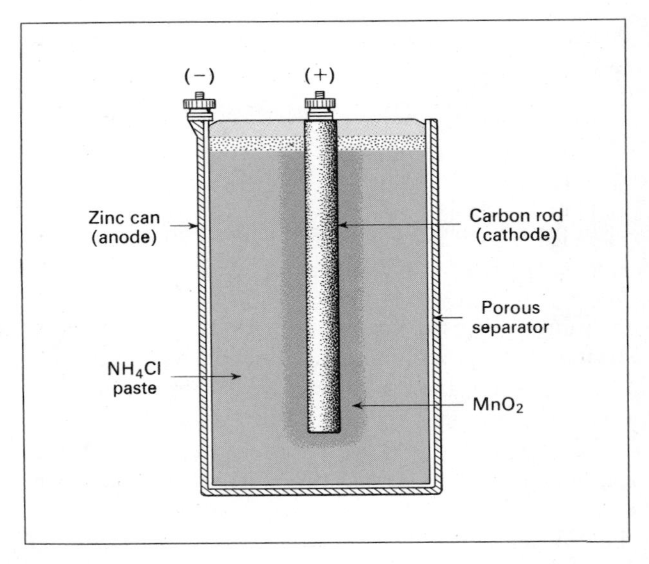

Figure 16.5 Cross section of a dry cell, showing the important parts.

16.4 Electrochemical Reactions

As we have seen in previous chapters, the properties of elements and compounds can be explained in terms of their electron structures. Microscopically, chemistry is intimately intertwined with electric charges. In this section we concern ourselves with several important chemical reactions that generate or consume electric charges.

Substances that dissolve in water to give solutions that will conduct an electric current are called **electrolytes.** In Section 14.3 we discussed how molten table salt, NaCl, conducts an electric current. All ionic compounds are electrolytes. Acids and bases are also electrolytes. They also ionize in water and conduct an electric current. Also in Section 14.3, we discussed electrolysis and identified the cathode and the anode. In electrolysis the **cathode** is the electrode at which electrons enter the electrolytic solution. The **anode** is the electrode where electrons are withdrawn from the electrolytic solution. In electrochemistry the important reactions occur at the anode and the cathode.

Electrochemical cells are extremely important. The dry cell is used in flashlights, transistor radios, etc. to provide an electrical potential. The common dry cell is shown schematically in Fig. 16.5, and some common examples are shown in Fig. 16.6. The dry cell is basically a zinc can with a carbon rod imbedded in a paste of ammonium chloride, NH_4Cl. A porous separator is inserted between the zinc can and the NH_4Cl. Some MnO_2 surrounds the carbon rod also. When the cell is connected, electrons flow externally from the zinc can to the carbon rod. The chemical reactions occurring at the cathode and the anode depend on the size of the current

Figure 16.6 Some typical dry cell batteries. (Photo courtesy James Crouse.)

required from the cell. When small currents are drawn, the following reactions probably occur:

$$Zn + 4\ NH_3 \longrightarrow Zn(NH_3)_4^{2+} + 2\ e^- \quad \text{(Anode)}$$

$$2\ MnO_2 + 2\ NH_4^+ + 2\ e^- \longrightarrow Mn_2O_3 + 2\ NH_3 + H_2O \quad \text{(Cathode)}$$

The dry cell is called an irreversible cell, because the application of an external potential will not recharge it. An example of a reversible cell is the lead storage battery of the type used in automobiles.

The lead storage battery consists of a set of lead grids in a dilute sulfuric acid solution (Fig. 16.7). One grid, or plate, contains spongy lead, the other contains lead dioxide, PbO_2. These are mounted in the solution of sulfuric acid, H_2SO_4, which acts as an electrolyte by ionizing in water to yield H^+ and SO_4^{2-} ions.

The lead grid acts as the anode, and the reaction during discharge is

$$Pb + SO_4^{2-} \longrightarrow PbSO_4 + 2\ e^- \quad \text{(Anode)}$$

The PbO_2 acts as the cathode. The reaction during discharge is

$$PbO_2 + 4H^+ + SO_4^{2-} + 2\ e^- \longrightarrow PbSO_4 + 2\ H_2O \quad \text{(Cathode)}$$

If an external electrical potential is applied to a discharged storage battery it can be recharged. The reactions that occur during recharge are just the reverse of those that occur during discharge. During discharge the sulfuric acid is consumed. Conversely, it is regenerated by the charging process. Since the presence of H_2SO_4 indicates a charged state, the condition of a lead storage battery may be determined by measuring the density of the sulfuric acid solution. The higher the density, the more the battery is charged. When the auto engine is running, the battery is being charged (provided the alternator is working properly). The battery is discharging when the car is being started. A single set of lead plates, such as shown in Fig. 16.7, generates a potential of 2.05 volts. Thus, a 12-volt battery contains six sets of plates, while a 6-volt battery contains three sets of plates.

The maintenance-free battery is similar to the lead storage battery in operation but requires no maintenance over its full life. There are three major factors that qualify a battery as maintenance-free: (a) terminal corrosion has been reduced to a point where it is no longer a factor in limiting battery life, (b) no additional water is required during the life of the battery, and (c) the battery should have a long shelf life. These characteristics have been achieved with the Sears DieHard Incredicell battery shown in Fig. 18.8. New connectors are located remote from the battery away from any source of corrosion, and harmful battery gasses are vented on the side of the battery opposite the cable terminals. These features plus low water loss of the dual alloy lead-antimony and lead-calcium construction and factory locked top results in a battery system rarely needing attention.

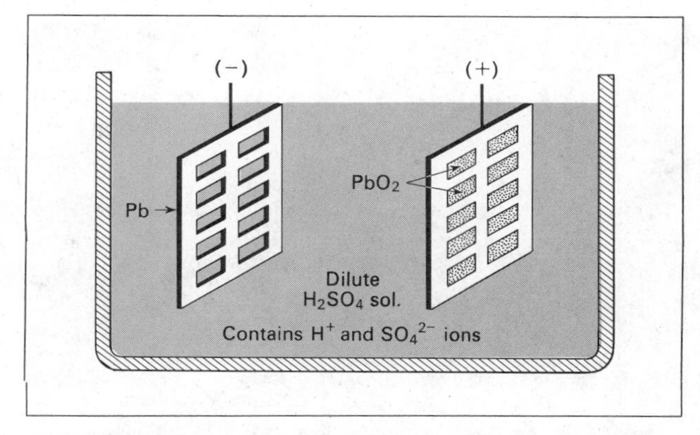

Figure 16.7 Schematic diagram of a lead storage battery.

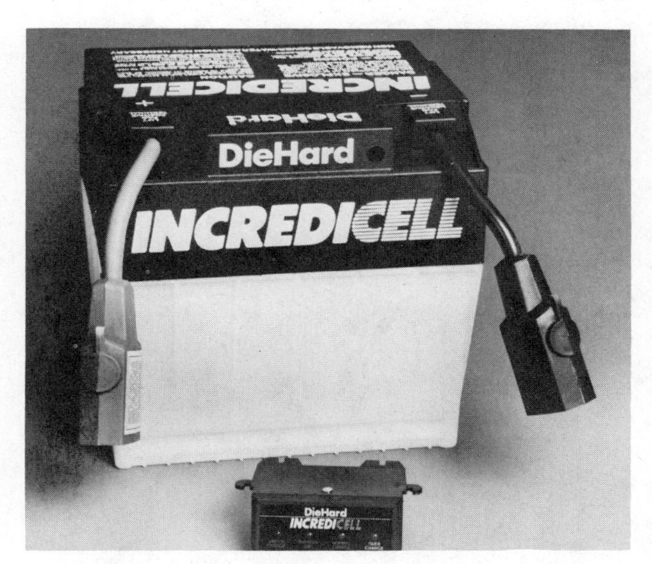

Figure 16.8 A maintenance-free battery. No filler caps are required because no additional water is ever needed; there is also minimal terminal corrosion and low "stand loss." (Photo courtesy Sears Roebuck & Co.)

The single most destructive chemical reaction that occurs in the United States is the rusting, or corrosion, of iron (see Fig. 16.9). In fact, about one out of every four workers in the steel industry is working to replace iron products that have been ruined by rusting. The corrosion of iron is brought about by the presence of both oxygen and water. Iron will not rust in dry air. The explanation of corrosion involves electrochemistry, as we shall see.

When iron is in contact with water containing an electrolyte, the iron, being very active, tends to give up its electrons according to the reaction

$$Fe \longrightarrow Fe^{2+} + 2\ e^- \quad \text{(Anode)}$$

Figure 16.9 The process of corrosion has seriously damaged this automobile exterior. (Photo courtesy James Crouse.)

This reaction can be considered a reaction at an anode. Usually, one part of the iron will be more active than other parts. At the less active part of the iron, H^+ ions react to form hydrogen gas, which collects on the iron surface.

$$2\ H^+ + 2\ e^- \longrightarrow H_2 \quad \text{(Cathode)}$$

The accumulation of hydrogen gas on the iron surface tends to inhibit the reaction. However, dissolved oxygen in the water reacts with the H_2 to form water.

$$2\ H_2 + O_2 \longrightarrow 2\ H_2O$$

The iron ions react with the OH^- ions present in the water and then with oxygen to form $Fe_2O_3 \cdot 2\ H_2O$, which is iron rust.

$$Fe^{2+} + 2\ OH^- \longrightarrow Fe(OH)_2$$

$$4\ Fe(OH)_2 + O_2 \longrightarrow \underset{\text{Iron rust}}{2(Fe_2O_3 \cdot 2\ H_2O)}$$

Figure 16.10 The burning of coal produces many pollutants. One of these is sulfur dioxide, SO_2, which forms a weakly acidic solution with water. (Photo courtesy James Crouse.)

The pollutant sulfur dioxide, SO_2, which is emitted during the burning of coal (Fig. 16.10), will combine with water to form dilute H_2SO_3. This, then, supplies H^+ ions for the cathode reaction in the rusting process. Thus, SO_2 pollution is partially responsible for the destruction of billions of dollars worth of structural materials each year.

Rusting can be prevented by coating the iron surface with a material such as paint or a ceramic enamel, like that used in sinks, stoves, or refrigerators. Some alloys of iron, such as stainless steel [a mixture of iron, chromium (15%) and nickel (8%)], are corrosion resistant.

16.5 Acids and Bases

The acid-base concept originated with the beginning of chemistry and is an important classification of chemical compounds. The word acid is derived from the Latin *acidus,* which means sour. The sour taste is one of the properties of acids, but the student should never actually taste an acid in the laboratory. An **acid** is a substance that, when dissolved in water, has the following properties:

1. Conducts electricity.
2. Changes the color of litmus dye from blue to red.
3. Tastes sour. (The student should *never* taste any acid.)
4. Reacts with a base to neutralize its properties.
5. Reacts with metals (e.g., Zn, Mg) to liberate hydrogen gas.

A **base** is a substance that, when dissolved in water, has the following properties:

1. Conducts electricity.
2. Changes the color of litmus dye from red to blue.
3. Tastes bitter and feels slippery (as with acids, the student should *never* taste a base).
4. Reacts with an acid to neutralize the acid's properties.

One of the first theories formulated to explain the acid-base concept was put forth by Svante Arrhenius (1859–1927), a Swedish chemist who proposed in 1887 that the characteristic properties of aqueous solutions of acids were due to the hydrogen ion (H^+), and the characteristic properties of aqueous solutions of bases were due to the hydroxide ion (OH^-).

According to the **Arrhenius acid-base concept,** when a substance like hydrogen chloride (HCl) is added to water, some HCl molecules dissociate into H^+ ions and Cl^- ions which exist in the solution in equilibrium with the HCl molecules that are not dissociated. The acidic properties of HCl are due to the H^+ ions. Similarly, when a substance such as sodium hydroxide (NaOH) is added to water, partial dissociation also occurs and the Na^+ ions and OH^- ions exist in equilibrium with NaOH molecules that are not dissociated. The basic properties of NaOH are due to the OH^- ions.

However, it is not necessary for a substance to have H^+ or OH^- ions in order to have acid and base properties. For example, ammonia, NH_3, and sodium carbonate, Na_2CO_3, contain no OH^- ions but in water solutions these compounds are bases. In water the reactions are

$$NH_3 + H_2O \longrightarrow NH_4^+ + OH^-$$

$$Na_2CO_3 + H_2O \longrightarrow HCO_3^- + OH^- + 2\ Na^+$$

The OH^- ions come from the water, so aqueous solutions of NH_3 and Na_2CO_3 have the same properties as other bases.

A base common to many households is sodium bicarbonate, $NaHCO_3$, known as baking soda. Its solution is weakly basic, the reaction being

$$NaHCO_3 + H_2O \longrightarrow H_2CO_3 + OH^- + Na^+$$

Acids act upon the carbonate ion of baking soda to give off carbon dioxide, a gas.

$$NaHCO_3 + H^+(\text{from an acid}) \longrightarrow Na^+ + H_2O + CO_2$$

This reaction is involved in the leavening process in baking. Baking powders contain baking soda plus an

Figure 16.11 Some common bases found around the home are ammonia, NH_3, and milk of magnesia, $Mg(OH)_2$. Baking soda, $NaHCO_3$, in water is slightly basic, and baking powder, since it contains baking soda, has basic characteristics in water solution. (Photo courtesy James Crouse.)

acidic substance such as $KHC_4H_4O_6$ (cream of tartar). When this combination is dry, no reaction occurs, but when water is added, CO_2 is given off. Some common bases are shown in Figure 16.11.

The Arrhenius theory, as we noted, is restricted to aqueous solutions containing H^+ or OH^- ions. Another theory of acids and bases was conceived independently in 1923 by J. N. Brönsted (1879–1947), a Danish physical chemist, and T. M. Lowry (1874–1936), an English chemist, and developed by others since then. The **Brönsted-Lowry theory** defines an acid as a substance that acts as a proton donor and a base as a substance that acts as a proton acceptor. According to this theory reactions involving the transfer of a proton are known as **acid-base reactions.**

Since a hydrogen ion (H^+) is in reality a proton, a proton donor must be a substance which has hydrogen atoms. For example, when hydrogen chloride (HCl) is placed in water, the hydrogen ion (H^+) is transferred from the HCl to the water (H_2O) molecule, as shown by the following equation:

$$HCl + H_2O \longrightarrow H_3O^+ + Cl^-$$

This reaction proceeds almost entirely from left to right.

The H_3O^+ ion, called the **hydronium ion,** and the Cl^- ion are formed in this reaction. In solution the H^+ ions of acids are joined to water molecules to give hydronium ions, H_3O^+.

Sulfuric acid (H_2SO_4) has two hydrogen atoms per molecule. When this acid molecule is placed in a water solution, dissociation occurs as shown.

$$H_2SO_4 + H_2O \longrightarrow \underset{\text{Hydronium ion}}{H_3O^+} + \underset{\text{Hydrosulfate ion}}{HSO_4^-}$$

The hydrosulfate ion, in turn, can act like an acid, as its proton is transferable to a water molecule.

$$HSO_4^- + H_2O \longrightarrow H_3O^+ + \underset{\text{Sulfate ion}}{SO_4^{2-}}$$

Acids are classified as strong or weak. A strong acid is one that ionizes almost completely in solution. The high propensity of an acid to ionize, and thus be classified as strong, is due to a weak binding energy of the H^+ ion. The acids HCl, HNO_3, and H_2SO_4 have weakly bound H^+ ions and therefore are strong acids. A weak acid is one that does not ionize to any great extent; that is, only a small number of the acid's protons unite with H_2O to form the H_3O^+ ion. Weak acids are formed from compounds that strongly bind the H^+ ion.

The Brönsted-Lowry concept holds that there must be a base that corresponds to every acid; that is, every acid when it donates or gives up a proton (H^+ ion) becomes a base and every base that accepts a proton becomes an acid. This can be expressed as

$$\text{Acid} + \text{Base} \longrightarrow \text{Base} + \text{Acid}$$

or, using HCl as an example, as

$$HCl + H_2O \longrightarrow Cl^- + H_3O^+$$

and

$$H_3O^+ + Cl^- \longrightarrow H_2O + HCl$$

According to the Brönsted-Lowry concept, acids and bases are classified as strong or weak according to their ability to donate or accept protons. Hydrochloric acid (HCl) is a strong acid, that is, the HCl molecule gives up its proton very readily. Accordingly, the chloride ion is a weak base, that is, it does not accept a proton readily to form a HCl molecule. Table 16.2 lists a few common acids and the base produced when the acids donate their protons. The base produced from an acid is called the **conjugate base,** and an acid produced from a base by the addition of a proton is called the **conjugate acid.** It follows from what has been said that the stronger the acid the weaker its conjugate base, and the stronger the base the weaker its conjugate acid.

Acids are very important industrially. Sulfuric acid is one of the most important chemical products in the United States. It is used in the refining of petroleum, the "pickling" of steel, and in the manufacturing of fertilizers and numerous other products. A dilute solution of hydrochloric acid is present in the human stomach and is used in the digestion of our foods. There are persons who must take a daily supplement of this acid because it is lacking in their bodies. Many acids can be found in our foods. A few examples are given in Table 16.3 and Figure 16.12.

An important property of an acid is the disappearance of its characteristic properties when brought in contact with a base and vice versa. This mixing of acid and base has to be in the proper proportions, however, for the

Table 16.2 List of Common Acids and Their Conjugate Bases

	Acid Name	*Reaction*	*Conjugate Base*	
Strong ↑	Hydrochloric	$HCl \longrightarrow H^+ + Cl^-$	Chloride ion	Weak ↑
	Nitric	$HNO_3 \longrightarrow H^+ + NO_3^-$	Nitrate ion	
	Sulfuric	$H_2SO_4 \longrightarrow H^+ + HSO_4^-$	Bisulfate ion	
	Phosphoric	$H_3PO_4 \longrightarrow H^+ + H_2PO_4^-$	Dihydrogen phosphate ion	Bases
Acids	Lactic	$HC_3H_5O_3 \longrightarrow H^+ + C_3H_5O_3^-$	Lactate ion	
	Acetic	$HC_2H_3O_2 \longrightarrow H^+ + C_2H_3O_2^-$	Acetate ion	
	Carbonic	$H_2CO_3 \longrightarrow H^+ + HCO_3^-$	Bicarbonate ion	
↓	Ammonium ion	$NH_4^+ \longrightarrow H^+ + NH_3$	Ammonia	↓
Weak	Water	$H_2O \longrightarrow H^+ + OH^-$	Hydroxide ion	Strong

Table 16.3 Some Common Acids Found in Foods

Acid	*Formula*	*Food*
Acetic	$HC_2H_3O_2$	Vinegar
Citric	$H_3C_6H_5O_7$	Citrus fruits
Carbonic	H_2CO_3	Soft drinks
Lactic	$HC_3H_5O_3$	Sour milk
Oxalic	$H_2C_2O_4$	Rhubarb
Tartaric	$H_2C_4H_4O_6$	Grapes

properties of each to disappear. When the proper proportions are mixed and the characteristics of both the acid and the base disappear, we say that neutralization has taken place. **Neutralization** is the mutual disappearance of the H^+ ions that are responsible for the characteristic properties of the acid and the OH^- ions that are responsible for the characteristic properties of the base. The other ions associated with the acid and the base undergo no chemical change and remain as ions in the aqueous solution or appear as a crystalline solid if the water is evaporated. For example, if hydrochloric acid (HCl) is added to potassium hydroxide (KOH) in the proper proportions, neutralization takes place as shown by the following equation:

$$H^+ + Cl^- + K^+ + OH^- \longrightarrow K^+ + Cl^- + HOH$$

This equation represents the typical reaction of an acid with a hydroxide base. Note that the metal ion (K^+) and the nonmetal ion (Cl^-) appear as ions on both sides of the equation, indicating that they do not take part in any chemical reaction, whereas the H^+ ion combines with the OH^- ion to form a molecule of water. The water molecule is usually written H_2O, but it was written HOH in this equation to show that the water molecule is formed by the proton donated by the acid and the hydroxide ion donated by the base. If equal numbers of H^+ ions and OH^- ions are mixed together, they will combine to form water molecules and the acid and base properties of the original substances will disappear.

Figure 16.12 Some common acids found around the home are carbonic acid, H_2CO_3, in soft drinks; acetic acid, $HC_2H_3O_2$, in vinegar, citric acid, $H_3C_6H_5O_7$, in citrus fruits; and tartaric acid, $H_2C_4H_4O_6$, in grape juice. (Photo courtesy James Crouse.)

The following examples illustrate the principles involved for neutralization to occur:

EXAMPLE 4 How many cubic centimeters of 1.0*M* HCl are needed to neutralize 10 cm^3 of 1.0*M* KOH? (Refer to Section 15.5 for molarity.)

The number of H^+ ions must equal the number of OH^- ions for neutralization to take place. Since the acid and the base are both 1.0*M* and the ion concentration is the same, that is, there is one H^+ ion for each HCl molecule and one OH^- ion for each KOH molecule, then the volumes must be equal. Thus 10 cm^3 of HCl are needed to neutralize 10 cm^3 of NaOH. They combine equally to form water (HOH) and the acid and base characteristics disappear.

EXAMPLE 5 How many cubic centimeters of 1.0*M* H_2SO_4 are needed to neutralize 10 cm^3 of 2.0*M* NaOH?

Again the number of H^+ ions must equal the number of OH^- ions for neutralization to take place. Since the acid is 1.0*M* and the base 2.0*M*, the acid is twice as strong as the base. But the H^+ ion concentration for the acid is two and the OH^- ion concentration for the base is one. Thus, 10 cm^3 of acid are needed to neutralize 10 cm^3 of the base.

More involved neutralization problems with odd values of molarity can be solved using the following:

$$M_a V_a C_a = M_b V_b C_b$$

where M_a is the molarity of the acid,
M_b is the molarity of the base,
V_a is the volume of the acid,
V_b is the volume of the base,
C_a is the concentration of H^+ ions per molecule of acid,
C_b is the concentration of OH^- ions per molecule of base.

EXAMPLE 6 How many cubic centimeters of 4.8*M* phosphoric acid, H_3PO_4, are needed to neutralize 10 cm^3 of 1.6*M* calcium hydroxide, $Ca[OH]_2$?

$$4.8M \times V_a \times 3 = 1.6M \times 10 \text{ cm}^3 \times 2$$

$$V_a = \frac{1.6 \times 10 \times 2}{4.8 \times 3} = 2.22 \text{ cm}^3$$

The positive and negative ions that remain after the H^+ ions have combined with the OH^- ions constitute a **salt.** If the water of the solution is removed by evaporation, the remaining solid is the salt in crystalline form. All salts are ionic compounds consisting, usually, of a positive metallic ion and a negative nonmetallic ion. To summarize, when an acid and a base are mixed together they produce water plus a salt. This is shown by the following equations:

$$\underset{\text{Acid}}{H^+Cl^-} + \underset{\text{Base}}{Na^+OH^-} \longrightarrow \underset{\text{Water}}{HOH} + \underset{\text{Salt}}{Na^+Cl^-}$$

$$\underset{\text{Acid}}{H^+NO_3^-} + \underset{\text{Base}}{K^+OH^-} \longrightarrow \underset{\text{Water}}{HOH} + \underset{\text{Salt}}{K^+NO_3^-}$$

The most common salt is table salt (NaCl). It is frequently called simply salt, but there are many other salts according to the strict chemical definition.

All salts are ionic compounds; most salts are crystalline solids at ordinary temperatures. Some salts have a salty taste, while others taste sweet, bitter, sour, or have no taste. Aqueous salt solutions may have either acidic, basic, or neutral characteristics. In addition to being formed when neutralization occurs, salts are produced by the direct union of a metal and a nonmetal by electron transfer, and also by reactions between metallic and nonmetallic oxides.

Some salts occur in our environment in the **hydrated** form, that is, they contain one or more molecules of water bonded with the salt. The salt and water are combined according to the law of definite proportions. As examples, sodium thiosulfate, $Na_2S_2O_3 \cdot H_2O$, (hypo), magnesium sulfate, $MgSO_4 \cdot 7\ H_2O$, (epsom salt), and sodium carbonate, $Na_2CO_3 \cdot 10\ H_2O$ (washing soda), always have 1, 7, and 10 molecules, respectively, of water bonded to the molecule of salt. **Anhydrous salts** are those that do not have water bonded to the molecule of salt. Sodium chloride, NaCl (table salt), is an example.

The partial neutralization of an acid or a base produces what are known as acid salts or basic salts. A common acid salt is potassium acid tartrate, $KHC_4H_4O_6$ (cream of tartar); a common basic salt is lead carbonate, $Pb_3CO_3(OH)_4$ (white lead).

16.6 Acids and Bases in Solution

Pure water has the unusual property of acting as either an acid or a base because it dissociates to a small extent, producing both hydrogen and hydroxide ions. That it does slightly dissociate is shown by the fact that pure water conducts a very feeble electric current. The ionization of pure water is shown in the following equation:

$$H_2O + H_2O \longrightarrow H_3O^+ + OH^-$$

Thus, we see that water can act as an acid or a base, according to the definitions given previously. We have the formation of H^+ (or H_3O^+) and OH^- ions, and one molecule acts as a proton donor and the other as a proton acceptor, as stated by the Brönsted-Lowry theory.

As was mentioned, the dissociation of water to produce H^+ and OH^- ions is very slight. The concentration is only one ten-millionth of a mole of H^+ and OH^- ions in one liter of water at 25°C, or expressed in powers of 10, the concentration is 1×10^{-7} mole/liter. This concentration of H^+ and OH^- ions is used to define a neutral solution; that is, any solution containing 1×10^{-7} mole/liter of both H^+ ions and OH^- ions at 25°C is defined as a **neutral solution.** An **acid solution** is defined as one in which the H^+ ion concentration is *greater* than 1×10^{-7} mole/liter, and an **alkaline solution** is one in which the OH^- ion concentration is greater than 1×10^{-7} mole/liter. We can also define an alkaline solution as one in which the H^+ ion concentration is *less* than 1×10^{-7} mole/liter. In all cases, the temperature is assumed to be 25°C.

The concentration of the H^+ ions is a measure of the acidity or basicity of a solution. The concentration can be expressed in powers of 10, as stated above, but this has proved to be rather awkward, and a method known as the *p*H method has been devised for expressing the H^+ ion concentration. Stated in simple terms, the ***p*H** is the exponent of the negative power to which 10 is raised when used to express the concentration. For example, a solution containing H^+ ions with a concentration of 0.0000001 (1×10^{-7}) mole/liter has a *p*H of 7. A concentration of 1×10^{-11} has a *p*H of 11. A neutral solution has a *p*H of 7. A *p*H less than 7 indicates an acid solution and a *p*H greater than 7 indicates a basic solution. A solution with a *p*H of 5 is said to be slightly acidic and one with a *p*H of 9 is slightly basic. Soda water (carbonic acid) has a *p*H of 4.5, blood 7.4, sea water 8.5 and milk of magnesia 10.5. Solutions in which the *p*H is 7 or more points from the neutral position of 7 are strongly acidic or basic. The concept of *p*H is illustrated in Table 16.4.

The *p*H of substances found in our environment varies in value. Most body fluids have a normal *p*H range, and a continued deviation from the normal range usually indicates some disorder in a body function. Thus, the *p*H value can be used as a means of diagnosis. Table 16.5 gives an approximate *p*H of some common substances. Actual values range on each side of the value given in the table.

Table 16.4 The pH Concept

pH	*Concentration of H^+ Ions*	*Acidic or Basic Property*
−1	1.0×10^{1} mole/liter	Very acidic
1	1.0×10^{-1} mole/liter	↑
4	1.0×10^{-4} mole/liter	
6	1.0×10^{-6} mole/liter	Acidic
7	1.0×10^{-7} mole/liter	Neutral
8	1.0×10^{-8} mole/liter	Basic
10	1.0×10^{-10} mole/liter	↓
15	1.0×10^{-15} mole/liter	Very basic

Table 16.5 Approximate *p*H of Some Common Substances

Substance	*pH*
Battery acid	0.0
Stomach acid	1.2
Lemons	2.3
Vinegar	2.8
Soft drinks	3.0
Apples	3.1
Grapefruit	3.1
Wines	3.2
Oranges	3.5
Tomatoes	4.2
Beer	4.5
Bananas	4.6
Carrots	5.0
Bread	5.5
Potatoes	5.8
Coffee	6.0
Rainwater	6.2
Corn	6.3
Milk (cow)	6.5
Pure water	7.0
Blood (human)	7.4
Eggs	7.8
Sea water	8.5
Clorox	9.0
Milk of magnesia	10.5
Oven cleaner	13.0

Learning Objectives

After reading and studying this chapter, you should be able to do the following without referring to the text:

1. Describe a chemical reaction and distinguish between the reactants and the products.
2. Explain chemical equilibrium and give an example.
3. Distinguish between exoergic and endoergic reactions. Give an example of each.
4. Explain activation energy.
5. Explain how temperature affects the rate of a chemical reaction.
6. Describe chemical reaction rates in the presence of a catalyst.
7. Distinguish between oxidation and reduction.
8. Explain oxidation-reduction reactions and give an example.
9. Explain the electromotive series.
10. Define and give an example of (a) an acid, (b) a base, (c) a salt.
11. Explain neutralization.
12. Describe the Arrhenius acid-base concept.
13. Describe the Brönsted-Lowry theory of acids and bases.
14. Explain the meaning of the term *p*H.
15. Define or explain the important words and terms at the end of the chapter.

Important Words and Terms

reactants
products
reversible reaction
equilibrium
exoergic reaction
endoergic reaction
activation energy
ten-degree rule
catalyst
enzymes
oxidation-reduction reaction
oxidation
reduction
oxidized
reduced
reducing agent
oxidizing agent
displacement reaction
electromotive series
electrolytes
cathode
anode
acid
base
Arrhenius acid-base concept
Brönsted-Lowry theory
acid-base reaction
hydronium ion
conjugate base
conjugate acid
neutralization
salt
hydrated
anhydrous salts
neutral solution
acid solution
alkaline solution
*p*H

Questions

1. What is a chemical reaction?
2. What three things occur in every chemical reaction?
3. List the factors that influence the rate of a chemical reaction.
4. Give an example of (a) a fast chemical reaction, (b) a slow reaction.
5. Explain the concept of dynamic equilibrium.
6. Explain why chemical reactions proceed faster as the temperature is increased.
7. Why do chemical reactions proceed at a faster rate when the concentrations of the reactants are increased?
8. What is the origin of the energy released in the chemical reaction?
9. Distinguish between exoergic and endoergic reactions.
10. How does a catalyst affect a reaction's equilibrium mixture?
11. How is the activation energy related to the total energy released by a chemical reaction?
12. Describe the role of a catalyst in a chemical reaction.
13. Why is it necessary to keep wines sealed or covered?
14. The human body converts sugar into carbon dioxide and water at body temperature, 98.6°F or 37°C. Why are much higher temperatures required for the same conversion in the laboratory?
15. Which would be easier to oxidize, sodium or chlorine? Explain.
16. Which would be easier to reduce, sodium or chlorine? Explain.
17. Which is the better oxidizing agent, nitrogen or fluorine? Why?
18. Why are the precious metals gold and silver found free in nature while the metals iron and magnesium are found in ores?
19. Identify some uses of the electromotive series.
20. Describe two theories that define an acid and a base.
21. In respect to the Arrhenius concept of acids, bases, and salts, what is meant by neutralization?
22. Distinguish between a hydrogen ion and a hydronium ion.
23. How is glass made? What causes its colors, if any?
24. Explain the difference between a reversible and an irreversible electrochemical cell.
25. Why is an automobile battery so heavy?
26. What is the most destructive chemical reaction in terms of dollars?
27. What two substances are necessary for iron to rust?
28. Explain how water can be both an acid and a base.
29. How is the strength of an acid or base expressed?

Problems

1. Balance the following chemical equations:
   (a) When an acid reacts with a metal, H_2 gas is given off.

   $$HCl + Fe \longrightarrow FeCl_2 + H_2$$

   (b) Nitric acid may be prepared by the following reaction:

   $$H_2O + NO_2 \longrightarrow HNO_3 + NO$$

   Answer:
   (b) $H_2O + 3\ NO_2 \longrightarrow 2\ HNO_3 + NO$

2. Balance the following chemical equations:
   (a) When octane gasoline is burned completely, the reaction is

   $$C_8H_{18} + O_2 \longrightarrow CO_2 + H_2O$$

   (b) Borax is a good water-softening agent. One of the important reactions of the borate ion in water is

   $$B_4O_7^{2-} + H_2O \longrightarrow H_3BO_3 + OH^-$$

3. State which element or ion is oxidized and which reduced in the following reactions:
   (a) The Haber process for manufacturing ammonia is

   $$N_2 + 3\ H_2 \longrightarrow 2\ NH_3$$

   (b) An important reaction in steel production is

   $$Fe_2O_3 + 3\ CO \longrightarrow 2\ Fe + 3\ CO_2$$

   Answer: (b) Iron is reduced, carbon is oxidized.

4. State which element or ion is oxidized and which reduced in the following reactions:
   (a) Sulfur in coal burns to form the pollutant sulfur dioxide.

   $$S + O_2 \longrightarrow SO_2$$

   (b) Hydrofluoric acid is so strong it can etch glass. It is formed as follows:

   $$H_2 + F_2 \longrightarrow 2\ HF$$

5. Using the electromotive series, use an arrow to indicate the direction of the following displacement reactions:
   (a) $Na + K^+ \qquad Na^+ + K$
   (b) $Ni + Cu^{2+} \qquad Ni^{2+} + Cu$

   Answer:
   (a) $Na + K^+ \longleftarrow Na^+ + K$

6. Using the electromotive series, use an arrow to indicate the direction of the following displacement reactions:
   (a) $Sn + Fe^{2+} \qquad Sn^{2+} + Fe$
   (b) $Pb + Fe^{2+} \qquad Pb^{2+} + Fe$

7. State whether the following solutions are acidic or basic:
   (a) 10 g of H_2SO_4 dissolved in 1 liter of water
   (b) Blood
   (c) A solution with a *p*H of 10
   (d) A solution with 1×10^{-9} mole/liter of OH^- ion concentration

8. State whether the following solutions are acidic or basic:
   (a) 10 g of $Mg(OH)_2$ dissolved in 1 liter of water
   (b) Pure water
   (c) A solution with a *p*H of 7
   (d) A solution with 1×10^{-5} mole/liter of H^+ ion concentration

9. One molar sulfuric acid (H_2SO_4) is added to 2.0*M* potassium hydroxide (KOH) to form a salt plus water, as shown by the following equation:

   $$H_2SO_4 + KOH \longrightarrow \underset{\text{salt}}{\underline{\qquad\qquad}} + H_2O$$

   (a) Write the formula for the salt.
   (b) Balance the equation.
   (c) Calculate the volume of H_2SO_4, in cubic centimeters, needed to neutralize 10 cm^3 of the KOH.

10. One molar hydrochloric acid (HCl) is added to 1.0*M* calcium hydroxide [$Ca(OH)_2$] to form a salt plus water, as shown by the following equation:

   $$HCl + Ca(OH)_2 \longrightarrow \underset{\text{salt}}{\underline{\qquad\qquad}} + HOH$$

   (a) Write the formula for the salt.
   (b) Balance the equation.
   (c) Calculate the volume of HCl, in cubic centimeters, needed to neutralize 10 cm^3 of the $Ca(OH)_2$.

11. Determine the hydrogen ion concentration in a solution with a *p*H of 8.

12. Determine the *p*H of a solution that has a hydrogen ion concentration of 1×10^{-4} mole/liter.

17

Complex Molecules

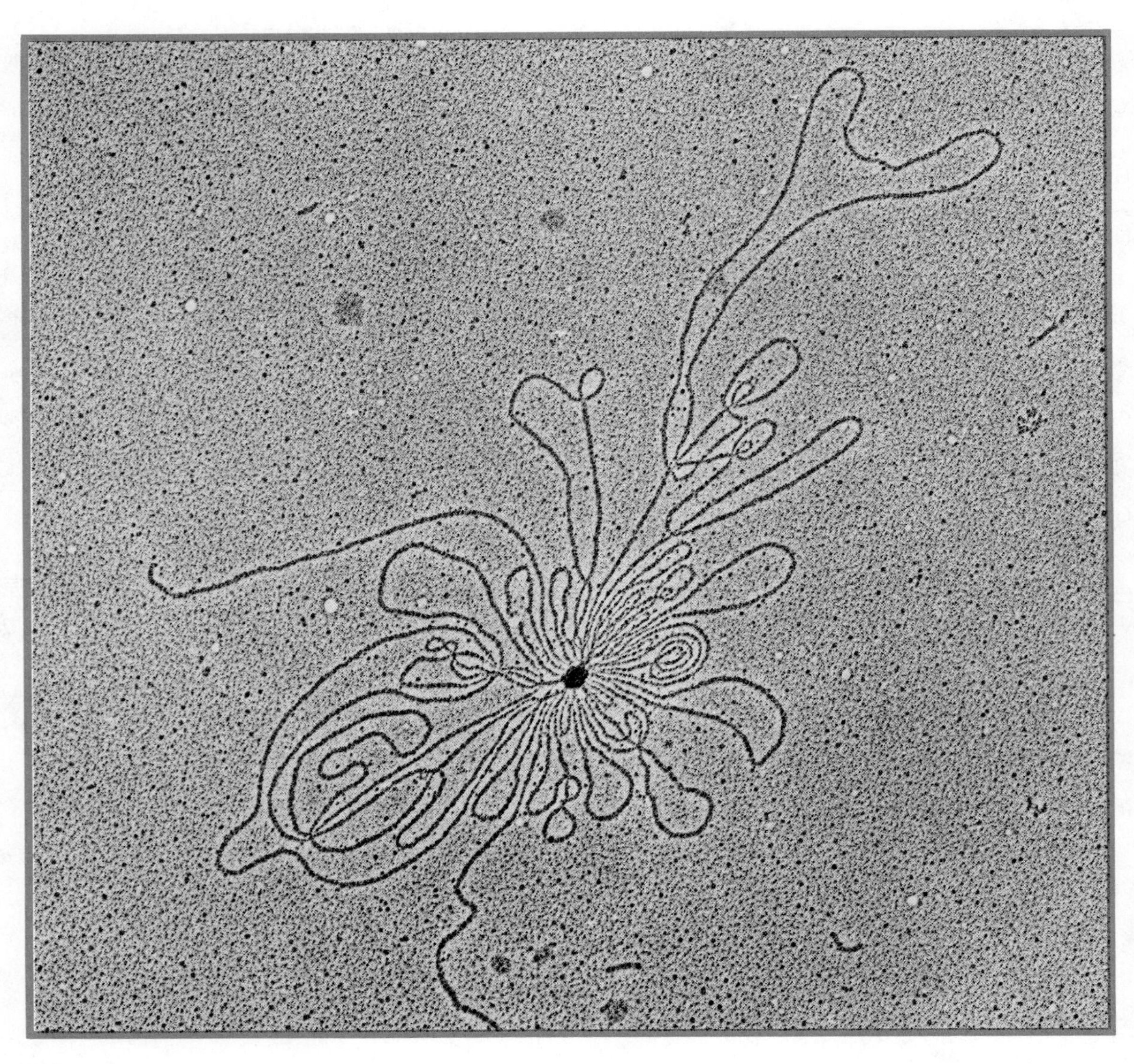

I am part of all you see
In Nature: part of all you feel:
I am the impact of the bee
Upon the blossom; in the tree
I am the sap—that shall reveal
The leaf the bloom—that flows and flutes
Up from the darkness through its roots.

–Madison Cawein

PREVIOUSLY, WE LEARNED that atoms of the chemical elements are nature's building blocks. In Chapter 14 we discovered that compounds are formed by combining the atoms of various elements. Many naturally occurring compounds are made up of very complex molecules. Since the fundamental elements are known, it would be hoped that these complex molecules could be analyzed and understood. It is true that a great degree of success has been achieved in analyzing those compounds whose molecules contain only a few atoms. However, these small molecules comprise only the inanimate portion of nature such as gases, water, and minerals. The chemical structure of the living portion of nature is grossly complex. **Complex molecules** may be either an aggregate of atoms, or repeating units of smaller molecules chemically bonded together, thus giving rise to great molecular weights and chainlike structures.

A simple example of building molecules from repeating units is the hydrocarbon series discussed in Chapter 14. Specifically, the fundamental repeating unit, or **monomer,** of the alkane series could be structurally written

$$\left[\begin{array}{c} \mathrm{H} \\ | \\ -\mathrm{C}- \\ | \\ \mathrm{H} \end{array}\right]_n$$

where it is understood that other fundamental units are attached to the open bonds, and the series is terminated by hydrogen atoms. The subscript n, which indicates the fundamental unit, may be repeated n times and the resulting chainlike structure is called a **polymer.** For example, if $n = 3$, the resulting compound is propane (see Section 14.8). This simple example is meant to introduce the concept of repeating units, and it should be kept in mind that the polymers occurring in nature, or prepared synthetically, may have repetition numbers (n) which may be in the thousands or hundreds of thousands. In addition, the fundamental unit may itself be quite complex and the chain configurations and variations may differ greatly.

Nature has been in the business of making large complex molecules for a long time. All living and growing matter is composed of these molecules, some being so complex that they are still not completely understood. The complexity of nature's molecules should come as no surprise since they compose the structure of life itself. The newest frontier of science is concerned with determining the structures and interactions of the complex molecules of which living things are made. Society someday may be forced to make some fundamental decisions regarding how far it will allow science to go in solving the mysteries of life.

◀ Linear duplex DNA molecule.

17.1 Common Organic Compounds

The number and complexity of the organic compounds in nature can, in large part, be attributed to the ability of the carbon atom to form single, double, and even triple bonds with itself and other elements. There are over six million organic compounds, and their identification, classification, and structural determination has been—and still is—the task of workers in the field of organic chemistry. Only a few of the more common compounds will be presented here, most of which should be familiar to the student because of their usage and occurrence in everyday life.

First, consider the following list of classes of compounds. A specific example from each class is given to show how the complexity of molecules can develop, starting with just a simple hydrocarbon.

$$\begin{array}{c} \text{H}\quad\text{H} \\ \text{H—C—C—H} \\ \text{H}\quad\text{H} \end{array}$$

Hydrocarbon (Ethane)

1. *Alkyl halides*

$$\begin{array}{c} \text{H}\quad\text{H} \\ \text{H—C—C—Cl} \\ \text{H}\quad\text{H} \end{array}$$

(Ethyl chloride)

2. *Alcohols*

$$\begin{array}{c} \text{H}\quad\text{H} \\ \text{H—C—C—H} \\ \text{H}\quad\text{OH} \end{array}$$

(Ethyl alcohol)

3. *Aldehydes*

$$\text{H—}\overset{\text{H}}{\underset{\text{H}}{\text{C}}}\text{—C}\begin{array}{l} {/\!\!/}\text{O} \\ \backslash\text{H} \end{array}$$

(Acetaldehyde)

4. *Acids (organic)*

$$\text{H—}\overset{\text{H}}{\underset{\text{H}}{\text{C}}}\text{—C}\begin{array}{l} {/\!\!/}\text{O} \\ \backslash\text{OH} \end{array}$$

(Acetic acid)

5. *Esters*

$$\text{H—}\overset{\text{H}}{\underset{\text{H}}{\text{C}}}\text{—C}\begin{array}{l} {/\!\!/}\text{O} \\ \backslash\text{O—C}_2\text{H}_5 \end{array}$$

(Ethyl acetate)

The colored portions of the formulas indicate the organic functional group that characterizes the general physical and chemical properties of these compounds.

Alkyl Halides

An **alkyl halide** is formed by replacing the hydrogen atom(s) in hydrocarbons with halogens—chlorine, bromine, fluorine, and iodine. Ethyl chloride, the example given in the list, is used as a local anesthetic and is the product of the reaction of ethane with chlorine.

$$\underset{\text{Ethane}}{C_2H_6} + \underset{\text{Chlorine}}{Cl_2} \longrightarrow \underset{\text{Ethyl chloride}}{\left[\begin{array}{c} \text{H}\quad\text{H} \\ \text{H—C—C—Cl} \\ \text{H}\quad\text{H} \end{array}\right]} + \text{HCl}$$

The halogen may replace more than one of the hydrogen atoms; in fact, they may all be replaced as the following series shows:

$$\underset{\text{Methyl chloride}}{\begin{array}{c} \text{H} \\ \text{H—C—Cl} \\ \text{H} \end{array}} \qquad \underset{\text{Dichloromethane}}{\begin{array}{c} \text{Cl} \\ \text{H—C—Cl} \\ \text{H} \end{array}}$$

$$\underset{\text{Chloroform}}{\begin{array}{c} \text{Cl} \\ \text{Cl—C—Cl} \\ \text{H} \end{array}} \qquad \underset{\text{Carbon tetrachloride}}{\begin{array}{c} \text{Cl} \\ \text{Cl—C—Cl} \\ \text{Cl} \end{array}}$$

The latter two compounds are common as an anesthetic and a cleaning fluid, respectively. However, the toxicity of carbon tetrachloride is a serious hazard, and caution should be exercised when using it.

Alcohols

Alcohols are organic compounds that contain one or more OH groups that have been substituted for one or more hydrogen atoms. Examples of some common alcohols are shown in Fig. 17.1. Ethyl alcohol (C_2H_5OH) is probably the most important alcohol known. It is made from sugars by the action of yeast in the process of fermentation

$$C_6H_{12}O_6 \xrightarrow{\text{Yeast}} 2\underset{\text{Ethyl alcohol}}{\left[\begin{array}{c} \text{H}\quad\text{H} \\ \text{H—C—C—OH} \\ \text{H}\quad\text{H} \end{array}\right]} + 2\,CO_2$$

or synthetically from ethylene and water

$$\underset{\text{Ethylene}}{\begin{array}{c} \text{H}\quad\text{H} \\ \text{H—C=C—H} \end{array}} + H_2O \xrightarrow{H_2SO_4} \underset{\text{Ethyl alcohol}}{\begin{array}{c} \text{H}\quad\text{H} \\ \text{H—C—C—H} \\ \text{H}\quad\text{OH} \end{array}}$$

Figure 17.1 Commercial products that contain a common alcohol. (Photo courtesy James Crouse.)

Ethyl alcohol is a colorless liquid that mixes with water in all proportions. It is the least toxic of all alcohols and is used in alcoholic beverages. Ethyl alcohol is also used as a solvent and in the production of many substances including perfumes, dyes, varnishes, antifreeze, and ethyl ether.

Alcohols are characterized by the —OH, or hydroxyl group; hence they are the organic equivalent to the inorganic bases. Many alcohols exist, some with one (—OH) group, others with two or more (—OH) groups. Ethylene glycol is an example of an alcohol with two hydroxyl groups.

$$\begin{array}{c} H \\ | \\ H—C—OH \\ | \\ H—C—OH \\ | \\ H \end{array}$$

Ethylene Glycol

Ethylene glycol is widely used as an antifreeze in automobiles.

Aldehydes

Aldehydes are characterized by the $—C\begin{array}{l} {}^{\diagup\!\!\diagup}O \\ {}_{\diagdown}H \end{array}$ group and are formed when alcohols react with oxygen (are oxidized). When ethyl alcohol is oxidized, acetaldehyde is formed.

$$2\left[\begin{array}{c} H\;\;H \\ |\;\;\;| \\ H—C—C—H \\ |\;\;\;| \\ H\;\;OH \end{array}\right] + O_2 \longrightarrow 2\left[\begin{array}{c} H\;\;\;\;\;\;O \\ |\;\;\;\;\;\diagup\!\!\diagup \\ H—C—C \\ |\;\;\;\;\;\diagdown \\ H\;\;\;\;\;\;H \end{array}\right] + 2\,H_2O$$

Ethyl alcohol — Acetaldehyde

A more common aldehyde, formaldehyde is prepared similarly from methyl alcohol (CH_3OH). Formaldehyde is used as a disinfectant and tissue preservative.

Organic Acids

The further action of oxygen on aldehydes produces a group of compounds known as **organic acids,** that are characterized by the molecular arrangement $—C\begin{array}{l} {}^{\diagup\!\!\diagup}O \\ {}_{\diagdown}OH \end{array}$ called the **carboxyl** group. There are many of these carboxylic acids. Acetic acid, whose dilute natural form is vinegar, is formed by the following reaction:

$$2\left[\begin{array}{c} H\;\;\;\;\;\;O \\ |\;\;\;\;\;\diagup\!\!\diagup \\ H—C—C \\ |\;\;\;\;\;\diagdown \\ H\;\;\;\;\;\;H \end{array}\right] + O_2 \longrightarrow 2\left[\begin{array}{c} H\;\;\;\;\;\;O \\ |\;\;\;\;\;\diagup\!\!\diagup \\ H—C—C \\ |\;\;\;\;\;\diagdown \\ H\;\;\;\;\;\;OH \end{array}\right]$$

Acetaldehyde — Acetic acid

The simplest carboxylic acid, formic acid, is common in insects and is the cause of painful discomfort from insect bites. It is prepared by the oxidation of formaldehyde.

Esters

When a carboxylic acid reacts with an alcohol, an ester is formed. For example,

$$\begin{array}{c} H\;\;\;\;\;\;O \\ |\;\;\;\;\;\diagup\!\!\diagup \\ H—C—C \\ |\;\;\;\;\;\diagdown \\ H\;\;\;\;\;\;OH \end{array} + \begin{array}{c} H\;\;H \\ |\;\;\;| \\ H—C—C—H \\ |\;\;\;| \\ HOH \end{array} \longrightarrow$$

Acetic acid — Ethyl alcohol

$$\begin{array}{c} H\;\;\;\;\;\;O \\ |\;\;\;\;\;\diagup\!\!\diagup \\ H—C—C \\ |\;\;\;\;\;\diagdown \\ H\;\;\;\;\;\;\;\; \end{array}\begin{array}{c} H\;\;H \\ |\;\;\;| \\ O—C—C—H \\ |\;\;\;| \\ H\;\;H \end{array} + H_2O$$

Ethyl acetate (an ester)

An **ester** is a compound that conforms to the general formula

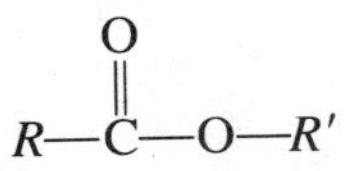

$$R\text{—}\overset{\overset{\displaystyle O}{\|}}{C}\text{—}O\text{—}R'$$

where R and R' are any alkyl groups. **Alkyl** groups have the general formula C_nH_{2n+1}; that is, they are alkanes less one hydrogen atom. R and R' may be the same group, but they are usually different. Esters are analogous to salts. Inorganic acids and bases react to form salts. Organic acids and alcohols react to form esters.

Esters possess very distinct and pleasant odors. The fragrance of many flowers and the pleasant taste of ripe fruits are due to esters. For example, bananas contain the ester amyl acetate,

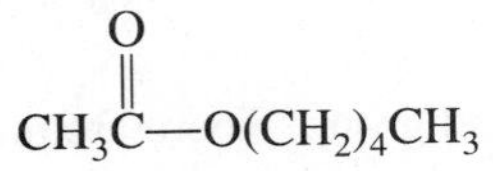

$$CH_3\overset{\overset{\displaystyle O}{\|}}{C}\text{—}O(CH_2)_4CH_3$$

Amyl acetate

and oranges contain octyl acetate,

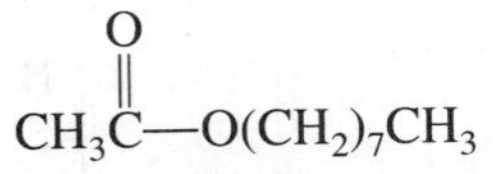

$$CH_3\overset{\overset{\displaystyle O}{\|}}{C}\text{—}O(CH_2)_7CH_3$$

Octyl acetate

The above groups of compounds may not seem to have very complex molecules; however, the examples given were selected for their simplicity. For the opposite extreme, we need look only as far as ourselves or other living matter. The molecules of living matter are extremely large and complex. These molecules contain tens of thousands of atoms (molecular weights of 1 billion) connected in three-dimensional patterns that are very complex in structure and very difficult to analyze. Common organic compounds in living matter are carbohydrates, proteins, fats (lipids), nucleic acids, and vitamins.

Carbohydrates

Carbohydrates are compounds composed of carbon, hydrogen, and oxygen with the hydrogen-to-oxygen ratio usually 2 to 1, the same as in the water molecules. The general formula for the carbohydrates is $C_n(H_2O)_x$. Thus, the name carbohydrates was given to those substances known as hydrates of carbon, although the nature of these compounds is not what the name implies. The most important carbohydrates are sugars, starches, and cellulose. The simplest sugars, glucose ($C_6H_{12}O_6$) and fructose ($C_6H_{12}O_6$), are isomers. Their structural formulas are

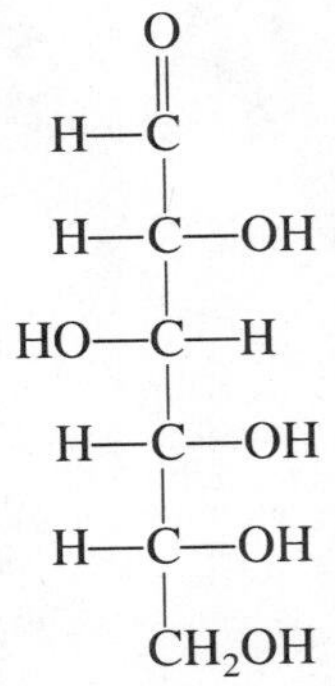

$$\begin{array}{c}
O \\
\| \\
H\text{—}C \\
| \\
H\text{—}C\text{—}OH \\
| \\
HO\text{—}C\text{—}H \\
| \\
H\text{—}C\text{—}OH \\
| \\
H\text{—}C\text{—}OH \\
| \\
CH_2OH
\end{array}
\qquad\qquad
\begin{array}{c}
H \\
| \\
H\text{—}C\text{—}OH \\
| \\
C\text{=}O \\
| \\
HO\text{—}C\text{—}H \\
| \\
H\text{—}C\text{—}OH \\
| \\
H\text{—}C\text{—}OH \\
| \\
CH_2OH
\end{array}$$

Glucose 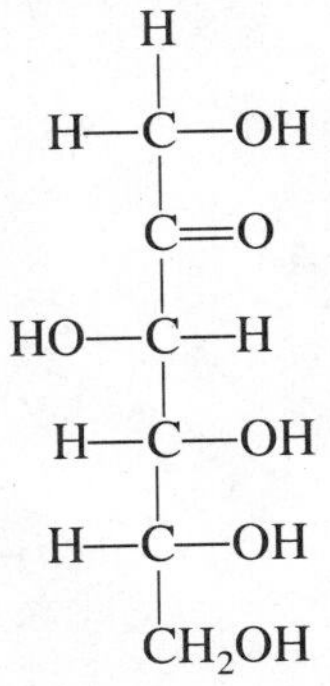 Fructose

Glucose, also known as dextrose and grape sugar, is found in sweet fruits, such as grapes and figs, and in flowers and honey. It is present in the blood to the extent of 0.1%, but is present in much greater amounts in persons suffering from diabetes. Glucose is formed in plants by the action of sunlight and chlorophyll on carbon dioxide (CO_2) from the air and water (H_2O) from the soil. The chemical reaction in simple terms is

$$6\ CO_2 + 6\ H_2O + \text{energy} \longrightarrow \underset{\text{Glucose}}{C_6H_{12}O_6} + 6\ O_2$$

with energy being absorbed in the process; that is, the energy coming from the sunlight is stored as chemical energy in the plant.

Fructose, also called fruit sugar, the sweetest of all sugars, is found in fruits and honey. Combined with glucose, it forms sucrose, also known as common sugar or cane sugar. The reaction is written as follows:

$$\underset{\text{Fructose}}{C_6H_{12}O_6} + \underset{\text{Glucose}}{C_6H_{12}O_6} \longrightarrow \underset{\text{Sucrose}}{C_{12}H_{22}O_{11}} + H_2O$$

Sucrose has two important isomers, maltose or malt sugar and lactose or milk sugar, which are sweet, crystalline solids soluble in water. Lactose is present in milk (about 4%) and is an essential ingredient in the food for infants.

Starch has the general formula $(C_6H_{10}O_5)_n$ where n may take on values up to 3000. Thus, starch is a polymer (many monomers associated together) that consists of long chains of glucose units. Starch, a noncrystalline substance, is formed by plants in seeds, tubers, and fruits.

Starch as a food is present in potatoes, grains (corn and rice), and vegetables. In the digestion process, the starches are converted to glucose. Glycogen (animal starch), a smaller and more highly branched polymer of glucose, is stored in the liver and muscles of animals as a reserve food supply that is easily converted to energy.

The energy in foods is given in terms of the energy they can deliver when oxidized to their final products. These final products are carbon dioxide, water, and nitrogenous compounds that are given off as waste products. For example, when glucose is oxidized, 674 kilocalories of energy per mole are provided. Since there are 180 grams in one mole of glucose, 1 gram of glucose provides 3.74 kilocalories of energy. This energy is used to do work and provide body heat in living organisms.

Cellulose has the same general formula $(C_6H_{10}O_5)_n$ as starch, but its structure is different; thus, it has different properties. Cellulose is the main component of the cell walls of plants and receives its name for this reason. It is a carbohydrate that cannot be digested by humans, but is used for making paper, guncotton (which is used in smokeless gun powder), rayon, and cellophane.

Proteins

Proteins are one of the essential components of living cells. Proteins are composed of carbon, oxygen, nitrogen (approximately 16% in all proteins), and hydrogen. Some also contain small amounts of sulfur, phosphorus, iron, and copper. The molecular structure of proteins is complex, but the basic chemical units are amino acids. An amino acid is an organic compound that contains an amine group and an acid group. The **amino** functional group has the structural formula

```
H
 \
  N—
 /
H
```

When an amino group is combined with a hydrocarbon group, an amine is formed. Similarly, when the carboxyl functional group

```
(      O  )
(     //  )
(  —C     )
(     \   )
(      OH )
```

is combined with a hydrocarbon group, an acid is formed. Thus, an amino acid contains both of these groups. There are over twenty known amino acids; eight of these are essential in the human diet. The simplest amino acid is glycine, whose structural formula is

```
H     H     O
 \    |    //
  N — C — C
 /    |    \
H     H     OH
```

When two molecules of glycine combine, a molecule of water is eliminated. Two molecules of glycine can be shown as

```
H     H     OH      H     H     O
 \    |    /         \    |    //
  N — C — C     +     N — C — C
 /    |    \\        /    |    \
H     H     O       H     H     OH
```

Since a water molecule is given off, a bond (called a peptide bond) is formed and the resulting dipeptide is

```
H     H          H     O
 \    |          |    //
  N — C — C — N — C — C
 /    |   ||  |  |    \
H     H   O   H  H     OH
            ↑
        Peptide bond
```

This process can be repeated by linking more glycine molecules to the protein whose structural formula appears above. Repeated again and again, long-chain proteins of very high molecular weight are formed. The molecular weight of proteins ranges from 1×10^4 up to 1×10^7 for the most complex. The human body contains more than 1×10^5 different proteins.

Lipids

Lipid is a general term that includes such substances as fats, oils, and waxes.

Fats are used in the diets of humans. In the digestive process, the fats are broken down into glycerol and acids, which are absorbed into the blood-stream and oxidized to produce energy that may be used immediately or stored for future use. Fats are also used by the body as an insulator to prevent loss of heat.

Fats are esters composed of glycerol ($C_3H_5(OH)_3$, an alcohol) and organic acids known as fatty acids. Fatty acids are acids that are present in any organic acid. A typical fatty acid is stearic acid, $C_{17}H_{35}COOH$ (a component of beef fat). When stearic acid is combined with glycerol, glyceryl stearate (a fat) is obtained. This is shown in the following reaction:

$$3\left[C_{17}H_{35}-\overset{\overset{\large O}{\|}}{C}-OH\right] + \begin{array}{c} H \\ | \\ H-C-OH \\ | \\ H-C-OH \\ | \\ H-C-OH \\ | \\ H \end{array} \longrightarrow$$

Stearic acid Glycerol

$$\begin{array}{l} \qquad\qquad\quad H \\ \qquad\qquad\quad | \\ C_{17}H_{35}COO-C-H \\ \qquad\qquad\quad | \\ C_{17}H_{35}COO-C-H \\ \qquad\qquad\quad | \\ C_{17}H_{35}COO-C-H \\ \qquad\qquad\quad | \\ \qquad\qquad\quad H \end{array} + 3\ H_2O$$

Glyceryl stearate (a fat)

The stearate structure is a large-chain hydrocarbon containing 18 carbon atoms, but written in short form here to save space.

Liquid fats, such as vegetable oils, are composed of hydrocarbon chains with double bonds between some of the carbon atoms; that is, they are unsaturated. These oils can be changed to solid fats by a process called **hydrogenation.** This is done simply by adding hydrogen to the carbon atoms that have the double bond; thus, the hydrocarbon chains become saturated. For example, when cottonseed oil (a liquid) is hydrogenated, margarine (a solid) is obtained. This is shown by the following equation:

$$\left[\begin{array}{l} CH_3(CH_2)_7CH{=}CH(CH_2)_7COOCH_2 \\ \qquad\qquad\qquad\qquad\qquad\quad | \\ CH_3(CH_2)_7CH{=}CH(CH_2)_7COOCH \\ \qquad\qquad\qquad\qquad\qquad\quad | \\ CH_3(CH_2)_7CH{=}CH(CH_2)_7COOCH_2 \end{array}\right] + 3\ H_2 \longrightarrow$$

Cottonseed oil (a liquid)

$$\left[\begin{array}{l} CH_3(CH_2)_{16}COOCH_2 \\ \qquad\qquad\qquad | \\ CH_3(CH_2)_{16}COOCH \\ \qquad\qquad\qquad | \\ CH_3(CH_2)_{16}COOCH_2 \end{array}\right]$$

Margarine (a solid)

Thus, liquid fats (oils) are esters of unsaturated acids, and solid fats are esters of saturated acids.

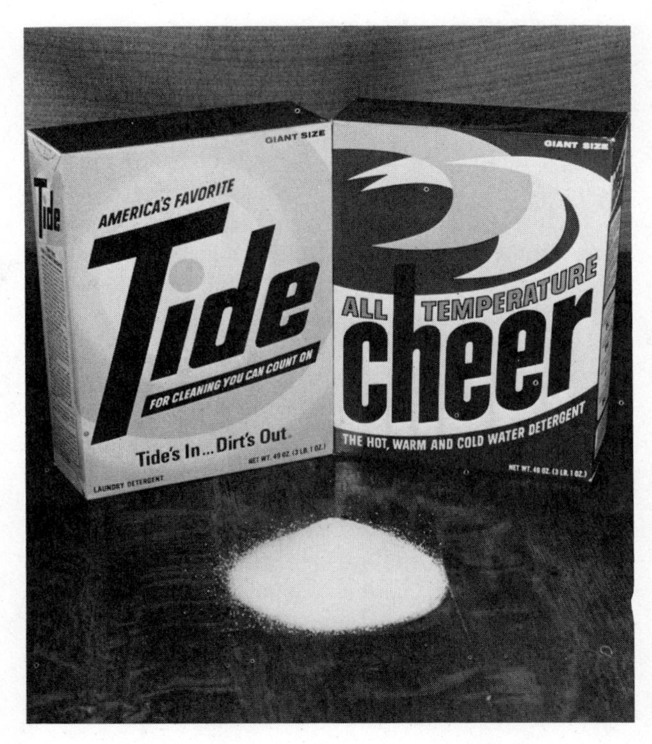

Figure 17.2 Two laundry detergents and ordinary soap powder. The soap is also a detergent. (Photo courtesy James Crouse.)

Waxes are esters derived from long-chain monohydric alcohols (an alcohol with only one OH group per molecule) and fatty acids. Beeswax from the honeycomb of the bee contains myricyl palmitate ($C_{15}H_{31}COOC_{31}H_{63}$) and is used in pharmaceutical preparations. Carnauba wax, which comes from the Brazilian wax palm, is used in commercial waxes for wood and metal surfaces. Chinese wax, which is secreted by an insect, is used in medicine.

The term **detergent** means cleansing agent and commonly refers to a soap substitute; but ordinary soap, which is a cleansing agent, is also a detergent (see Fig. 17.2). Ordinary soap is made by combining fats or oils with alkalis such as sodium or potassium hydroxide. When fats are treated with sodium hydroxide, glycerol and sodium salts of the fatty acid, sodium palmitate, sodium stearate, and sodium oleate are formed. A typical soap is sodium stearate ($C_{17}H_{35}COO^-Na^+$) whose structural formula is

$$\begin{array}{c} H \qquad\qquad\qquad\quad H \qquad O \\ | \qquad\qquad\qquad\quad | \quad // \\ H-C\cdots(C_{15}H_{30})\cdots C-C \qquad Na^+ \\ | \qquad\qquad\qquad\quad | \quad \backslash \\ H \qquad\qquad\qquad\quad H \qquad O^- \end{array}$$

Soaps have a cleansing action because the carboxyl end of the soap ion is water-soluble and the hydrocarbon end is oil-soluble. Thus, the hydrocarbon end can dissolve fats and the carboxyl or ion end surrounds the dirt particles and forms a film around them that allows them to be washed away. Thus, through a combination of emulsifying action and particle suspension, soaps serve as detergents.

Although soaps have a cleansing action, they have the disadvantage of forming precipitates when used in acidic solutions or with hard water, which contains calcium, magnesium, and iron ions.

The modern detergents, which are soap substitutes, contain a long hydrocarbon chain which is nonpolar (for example, $C_{12}H_{25}$—), and a polar group such as sodium sulfate (—OSO_3Na). This detergent has the structural formula

$$\begin{array}{ccccccccc} & & H & & H & & O & & \\ & & | & & | & & | & & \\ H- & C & \cdots(CH_2)_{10}\cdots & C & -O- & S & -O^- & Na^+ \\ & | & & | & & | & & \\ & H & & H & & O & & \end{array}$$

Sodium lauryl sulfate

Synthetic detergents such as sodium lauryl sulfate do not form precipitates with the calcium, magnesium, and iron ions. Therefore, they are effective cleansing agents in hard water.

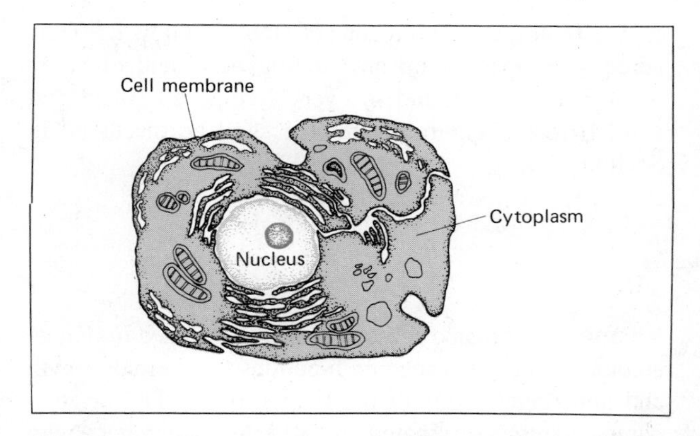

Figure 17.3 Diagram of a typical living cell. Every living cell contains a nucleus with cytoplasm surrounding it. The cell membrane encloses everything. The whole mass of living matter is called protoplasm.

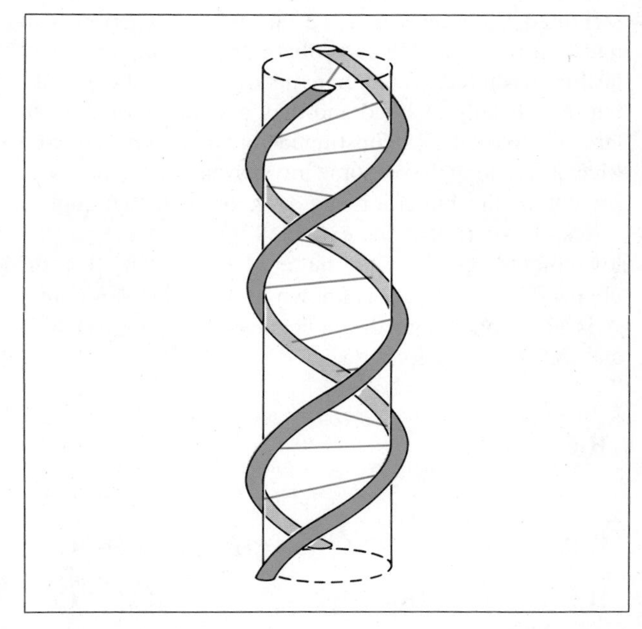

Figure 17.4 Schematic diagram of the DNA molecule. The DNA molecule has the structure of a double-stranded helix. The spiral frames are formed from sugars and phosphates. The connecting links are composed of nitrogen compounds.

Nucleic Acids

Nucleic acids are very high molecular weight polymers present in living cells. It is generally believed that they control the processes of heredity and take an active role in building proteins from amino acids. There are two types of nucleic acids; *deoxyribonucleic acid,* **DNA,** which is located primarily in the nucleus of the cell, and *ribonucleic acid,* **RNA,** which is located in the cytoplasm or outer structure of the cell. Figure 17.3 is a diagram of a living cell with some of its components. DNA has the molecular structure of a double-stranded helix (see Fig. 17.4) composed of simple sugars (deoxyribose), phosphates, and nitrogen compounds of which there are four: adenine, thymine, cytosine, and guanine, commonly labeled A, T, C, and G. The sugars and phosphates form the spiral frames of the double-stranded helix and the nitrogen bases form the connecting links, with adenine always joined with thymine and cytosine always joined with guanine to form links between the spiral frames. The molecules vary in length, but they may be several inches long and contain billions of cross linkages. The DNA molecular weight is greater than 1 million. RNA has a similar helical structure except the linkage is different, and the molecular weight is smaller, ranging from 20,000 to one million.

The biological significance of DNA and how it affects heredity is opening up an exciting new field of study because DNA is found in every living cell (plant and animal) in our environment. This will be discussed in Section 17.2.

Vitamins

Vitamins are organic substances that are needed in minute amounts to perform specific functions for normal growth and nutritional needs of the human body. The original name vitamine originated in 1912 by researchers who thought the substances were amines which were vital for good health. Since then the ''e'' has been dropped because all vitamins are not amines.

Vitamin A was the first vitamin to be discovered and identified. Elmer V. McCollum, an American scientist, identified Vitamin A in 1912 through an experiment in nutrition with rats. He noted that rats grew and remained healthy when fed a fat-source diet of butter and egg yolk, but their health declined and they eventually died when lard or olive oil was substituted for the butter. This led McCollum to believe some mysterious substance was present in the butter or egg yolk or both. Through a series of experiments he extracted this substance, a yellow-colored oil, from the butter. He called it ''fat-soluble A,'' which today is known as Vitamin A. Vitamin A is an alcohol with the molecular formula $C_{20}H_{29}OH$ and the structural formula

```
 H3C     CH3
    \   /
      C                 CH3                  CH3
    /   \                |                    |
 H2C     C—CH=CH—C=CH—CH=CH—C
  |      ||                                   ||
 H2C     C—CH3                                CH
    \   /                                     |
     CH2                                    HOCH2
```

It is found in eggs, milk, butter, and green and yellow vegetables. It increases the resistance to infections and promotes good night vision.

There are many other important vitamins, some whose functions are well known, and others whose importance is not yet understood. Table 17.1 lists some of the important vitamins with their molecular formulas.

Table 17.1 Chemical Formulas of Common Vitamins

Vitamin	*Formula*
A_1	$C_{20}H_{29}OH$
A_2	$C_{22}H_{31}OH$
B_1	$C_{12}H_{17}ON_4SClHCl$
B_2	$C_{17}H_{20}N_4O_6$
B_6	$C_8H_{11}NO_3HCl$
B_{12}	$C_{63}H_{90}N_{14}O_{14}PCo$
C	$C_6H_8O_6$
D	$C_{28}H_{44}O$
E	$C_{10}H_{25}O$
K_1	$C_{24}H_{35}O_2$

17.2 The Ingredients of Life and the Genetic Code

As we have noted, the molecular structures of living matter are quite complex. However, it is generally agreed by scientists that when the Earth was formed only simple substances such as water vapor (H_2O), hydrogen (H_2), ammonia (NH_3), and methane (CH_4) were present. How, then, did the complex molecules of life form from these simple substances? The first experimental insight to this question was gained in 1952 when Stanley Miller, a graduate student in chemistry, performed the now famous **Miller-Urey experiment.** In studying the process of forming complex molecules, Miller circulated through a glass container a mixture of water vapor, hydrogen, ammonia, and methane that he thought might be similar to our primordial atmosphere. Some form of energy must be present to initiate a chemical reaction in the mixture, so a sparking device was placed in the container. This could conceivably be likened to lightning flashes in the early atmosphere. As the gases continuously passed through the spark, drops of liquid began to collect on the bottom of the container. When this liquid was analyzed, it was found to contain several organic compounds that are found in living matter, the most significant of which were four amino acids, the building blocks of proteins. Thus the Miller-Urey experiment demonstrated a possible way in which the compounds of living matter might have been formed originally. A schematic diagram of the apparatus used in the Miller-Urey experiment is shown in Fig. 17.5.

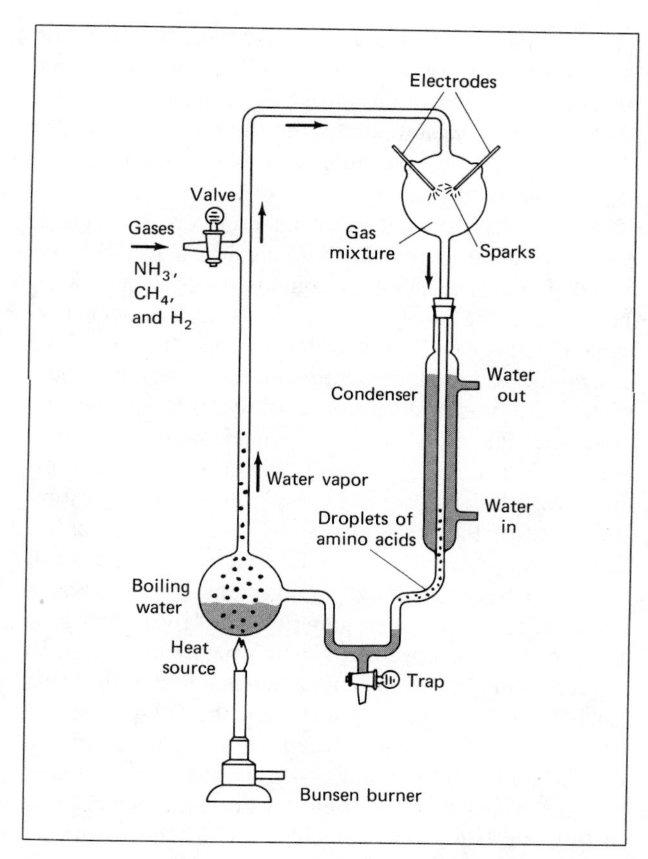

Figure 17.5 Apparatus used in the Miller-Urey experiment. Steam, ammonia, methane, and hydrogen were introduced into the reaction chamber, and the products cooled and collected in the trap. It was found that 10 amino acids could be made in this way.

Subsequent experiments yielded as many as 10 amino acids from similar reactions. This synthesis of the amino acids is extremely significant because of their relation to proteins. As discussed previously, proteins are one of the essential components of living matter. They are produced by living cells and make up approximately 75% of the dry weight of our bodies. Proteins are also the main component of plant cell interiors. Proteins are quite complex in molecular structure, being composed of chemical combinations of about 20 amino acids. The Miller-Urey experiment demonstrated how nature might have prepared the amino acids. The next logical question, ignoring that of the beginning of life itself, might be: How do living cells form the correct protein from the various combinations of amino acids? That is, how does an animal cell generate proteins characteristic of the animal, and a plant cell, proteins characteristic of the plant? In short, how are the characteristics of life reproduced? The answer to this, of course, involves heredity.

Why offspring resemble their parents has always fascinated us. It was thought for a long time that the characteristic traits were carried in the blood and were passed to a child by mixing of the parental blood. This misconception gave rise to the idea of blood lines such as royal blood, having bad blood, or being full-blooded for some race. Science however, has traced the trait-carrying agent to structures, called **chromosomes,** within the nucleus of the living cell. Chromosomes are threadlike structures that split in such a way that when a body cell divides to produce new cells, the two resulting daughter cells have the same number of identical chromosomes as the parent cell. This process of cell division is called **mitosis.** Because of mitosis, cells are reproduced with the same characteristics. All mature body cells in the same animal or plant have the same number of chromosomes.

In another type of cell division, called **meiosis,** cells result with just half the number of chromosomes as the original cell. Meiosis occurs in the reproductive organs of sexually reproducing plants and animals and involves the sex cells—sperm in the male and egg in the female. Since the sperm and the egg cells unite in the fertilization process, it is evident that for the cell of the offspring to have the proper number of chromosomes and combination of traits, the new egg and sperm cells must each have half as many chromosomes as the original cell. Thus, as a result of meiosis, each child gets a set of chromosomes from each parent and passes half of these to its own offspring. The difference between mitosis and meiosis is illustrated in Fig. 17.6 (p. 282).

Human egg cells and sperm cells have 23 chromosomes each, but humans have thousands of inherited traits and characteristics. Therefore, the chromosomes must have a multiple structure to carry all these traits. On a closer look at the threadlike structure of the chromosome, it was found that what appeared to be a thread was actually made up of small beads hooked together in a necklacelike fashion. Each human chromosome might contain as many as a thousand of the beads, or **genes,** which are the carriers of inherited traits. Thus, a cell may inherit two genes for the same characteristic in meiosis, only one of which is **dominant** or influences the trait. The other is **recessive** and does not influence the characteristics of the offspring. The recessive gene is, however, a part of the cell and may be transferred

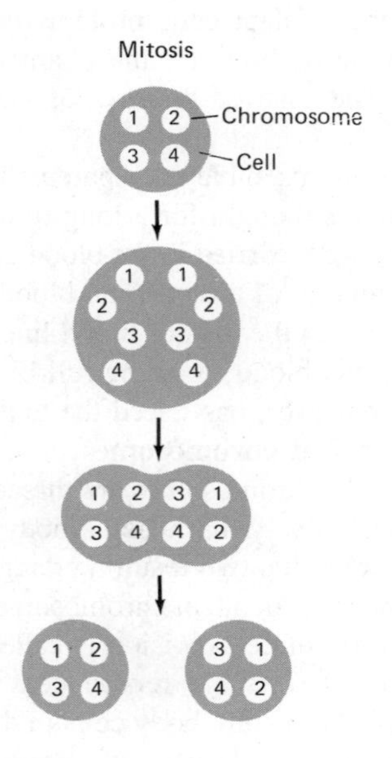

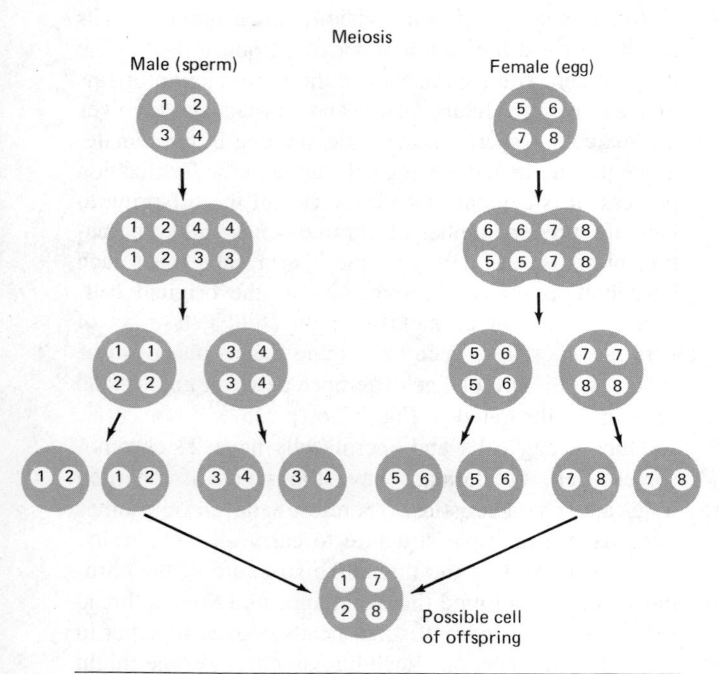

Figure 17.6 Schematic illustration of the processes of mitosis and meiosis. In mitosis, the chromosomes divide in such a way before the cell divides that each daughter cell has the original number of chromosomes. In meiosis, the immature reproductive cells of both male and female divide so that one-half of the original number of chromosomes is present in the new cell. Thus, the correct number of chromosomes is present in the body cells of the offspring when a male and female cell unite.

again in meiosis and become dominant in some later generation. The appearance of a child with red hair in a family whose generations have had all black hair is such an example. Some ancestor of the child must have added a gene for red hair that remained recessive for several generations of the family.

Science has focused much attention on the gene. It has been found to consist of the nucleic acid DNA. As noted previously, DNA is composed of sugars, phosphates, and the nitrogen compounds adenine (A), thymine (T), cytosine (C), and guanine (G). The molecule is a spiral ladderlike structure with the sugars and phosphates forming the sides and the nitrogen compounds the rungs (see Fig. 17.4). The nitrogen compounds can form in only certain combinations, A with T and C with G. However, the combined nitrogen compounds may form the rungs of the molecular DNA ladder in any order, such as AT, CG, TA, AT, GC, CG, AT, and so on. It is now believed that the arrangement of the nitrogen compound sequences, or **genetic code,** gives each gene its special characteristic. A simple gene might be a small portion of the DNA molecule a few hundred or thousand rungs long. In a single human cell, the DNA molecule is thought to be about 3 feet long when uncoiled and contain some 6 billion rungs—a truly complex molecule. J. Watson and F. Crick were awarded the Nobel Prize in 1962 for their work in elucidating the helical structure of the DNA molecule.

According to current theory, it is believed that when a chromosome of the living cell divides in mitosis, the DNA ladder literally unzips and the two individual parts pick up more raw materials from the nucleus around them. Since an A must combine with a T and a C with a G, the old substances pick up new partners such that two duplicate DNA molecules are formed with the same code, and hence the same genes. This unzipping and duplicating action is illustrated in Fig. 17.7. Should an "accident" occur in the reconstruction of the molecule that affects the order of the DNA nitrogen compound pairs, the characteristic of the gene is altered and a **mutation,** or genetic change may occur. Mutations may arise in a number of ways not fully understood. They may be caused by photons (gamma rays, X rays, light, and so on) that hit the molecule, or mutations may be caused by chance. According to quantum mechanics, events have only a probability of occurring. That is, there is a high probability the gene will exactly reproduce. Sometimes, by chance, it does not.

To complete the story and return to the original question of how the proteins are reproduced, one must con-

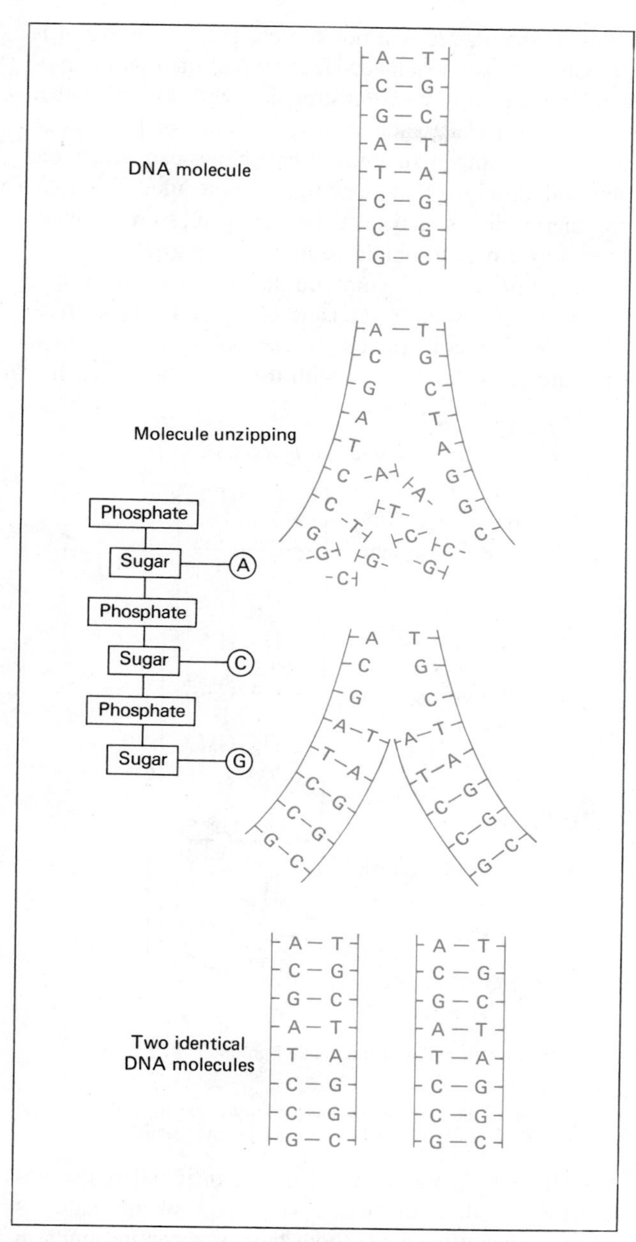

Figure 17.7 The duplication of a DNA molecule. The ladder structure "unzips" in the middle. Then the appropriate nitrogen compounds join each half of the ladder, and two new identical ladders are formed. In reality the ladders are spirals as shown in Fig. 17.4.

sider another nucleic acid, ribonucleic acid (RNA). This nucleic acid is a single-stranded molecule that is similar in structure to the DNA molecule. The biosynthesis of protein occurs on subcellular structures called ribosomes, which are found in the cytoplasm of the cell. Energy is required to form the peptide bond and specific enzymes are required to catalyze the chemical reactions that bring about the condensation of the amino acids into the peptide chain. The amino acid must be condensed into a chemical called transfer ribonucleic acid (t-RNA). The t-RNA orients an amino acid so that it will be brought to the precise adjacent amino acid to form the protein. Information providing the exact sequence of amino acids in the proteins to be synthesized is furnished by the nucleic acid called messenger ribonucleic acid (m-RNA). Thus, the amino acids are assembled in a predesignated way and the proteins and cells grow as dictated by the individual's inherited traits contained in the genetic DNA code.

The science of genetics entered a new era in the late 1970s with techniques called recombinant DNA technology or genetic engineering. This new technology came about primarily due to (1) new instrumentation and techniques for understanding molecular genetics, (2) the use of radioactive tracers that can be used to detect tiny amounts of specific macromolecules, and (3) the use of enzymes known as nucleases that hydrolyze nucleic acids.

Through genetic engineering it is now possible to transfer genes from mammals into bacteria and cause the microbes to produce large amounts of protein that can be used for the treatment of diseases. Human insulin can now be synthesized by recombining the human insulin gene into a plasmid, a small fragment of DNA found in bacterial cells. Interferon is a protein that is sensitive to certain viruses and interferes with viral replication. Only small amounts of interferon are available from natural sources. Through genetic engineering interferon can be produced in large amounts for the treatment of certain types of cancer and for medical research.

A major goal of genetic engineering is the cure of molecular diseases such as cystic fibrosis, goiter, hemophilia, and sickle-cell anemia plus many others.

17.3 Artificial Molecules—Plastics

Chemists have for a long time tried to duplicate the compounds of nature. Alchemy, or the attempt to transform chemically base metals into gold, was an early example of one such effort. As the science of chemistry progressed and the formulas and basic components of some of the natural compounds became known, chemists were able to synthesize some of these natural compounds by causing the appropriate combinations of elements or compounds to react. During this early trial and error development, there were, no doubt, as many accidental as deliberate discoveries.

From the attempts to synthesize nature's compounds, synthetics were discovered. A **synthetic** is a material whose molecule has no duplicate in nature. The first synthetic was prepared by Leo Baekeland in 1907. By mixing phenol (C_6H_5OH) and formaldehyde ($HCHO$), a polymeric combination of these two compounds, $HOCH_2C_6H_4OH$, was formed. The polymer is commercially known as Bakelite, a common electrical insulator. This discovery set off a series of investigations for synthetic materials. Necessity also helped catalyze these endeavors. During World War II many natural supplies, such as rubber, were cut off, thus providing impetus to the search for synthetic products that would take the place of the natural materials in short supply. Chemists became aware that substituting different elements or compounds in a complex molecule would change its physical properties. For example, the substitution of a chlorine atom for a hydrogen atom in ethane produces ethyl chloride, which has vastly different properties from ethane. By knowing the general properties of the substituted groups, a chemist can tailor a chemical molecule to satisfy a given requirement. As a result of this scientific approach, a host of synthetic molecules have been constructed. Probably the best known of these is the group of synthetic polymers commonly called **plastics.** Plastics have become an integral part of our modern life, being used in clothing, shoes, buildings, autos, sports, art, electrical appliances, toothbrushes, toys, and a myriad of other things.

Not only are the properties of plastics affected by their chemical composition, but the alignment of their long molecular chains can also cause differences in their qualities. For example, if a thin sheet of polymer is heated to its softening point then stretched and cooled while held fixed, the molecular chains are aligned along the direction of stretch and the material is said to be oriented, being less flexible than before. Should the material be reheated without being held, the sheet will shrink back to its original unstretched unaligned state. This property is of great use in the packaging industry where meats or other items are wrapped in stretched material which is then heated. The result is a skin-tight polymer wrapping. The next time you have occasion to have a piece of plastic wrapping, stretch it to align the molecules. If the plastic will not stretch, then the material is probably already oriented. Heating will unalign the molecular chains if it has been stretched while heated. Other chemical "tricks" such as putting short side-chains or branches on the main molecular chains can affect the physical properties of a complex molecule. The long polymeric chains may also be chemically crosslinked, or tied together, to obtain desirable properties.

A few of the more common plastics along with their monomer unit will be considered to show how these molecules are constructed. Three polymers of similar structure are listed below, with the monomer of each.

Monomers of Three Common Polymers

Polyethylene $\left[-\mathrm{CH_2}-\mathrm{CH_2}- \right]_n$ (each C bonded to two H)

Polypropylene $\left[-\mathrm{CH_2}-\mathrm{CH}(\mathrm{CH_3}^*)- \right]_n$

Polystyrene $\left[-\mathrm{CH_2}-\mathrm{CH}(\mathrm{C_6H_5}^{**})- \right]_n$

*—CH_3 is referred to as the methyl group.

** (benzene ring) is the symbol for benzene with one hydrogen missing (—C_6H_5), which is called a phenyl group.

Polyethylene is the simplest of the artificial molecules and may be thought of as a very long alkane; nature's alkanes do not have the thousands of repeating units in synthetic molecules. By replacing a hydrogen atom in the polyethylene monomer with a methyl group or a phenyl group, one gets polypropylene and polystyrene, respectively. By thus changing the molecular structure, the physical properties of the polymers are changed. For example, the melting points of the three polymers are 135°, 175°, and 220°C, respectively.

These polymers have many applications. Because of their chemical inertness, they are used for chemical storage containers as well as many other packaging applications.

Another common polymer of simple structure that has found many uses, especially in coating cooking utensils, is Teflon, a fluorocarbon resin. Its structure is

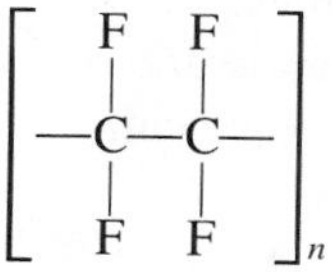

$$\left[-\overset{F}{\underset{F}{C}}-\overset{F}{\underset{F}{C}}- \right]_n$$

Teflon

In replacing the hydrogen atoms in the ethylene monomer with fluorine atoms, a compound called tetrafluoroethylene is produced. It polymerizes to form the hard, strong, high-melting-point, low-surface-friction, chemically resistant material known by the trade name *Teflon*. When a frying pan is coated with Teflon, foods will not stick. It should also be mentioned that the majority of the polymers are good electrical insulators and are used to great advantage in this application.

Some polymers may be made in fiber or thread form, hence clothing, rugs, and even artificial grass for sporting events are made from plastics. One such polymer is nylon which has found wide usage in hosiery. Nylon has a somewhat complicated monomer

$$\text{Nylon} \quad \left[-\underset{\underset{O}{\|}}{C}(CH_2)_4\underset{\underset{O}{\|}}{C}-\overset{H}{\overset{|}{N}}(CH_2)_6\overset{H}{\overset{|}{N}}- \right]_n$$

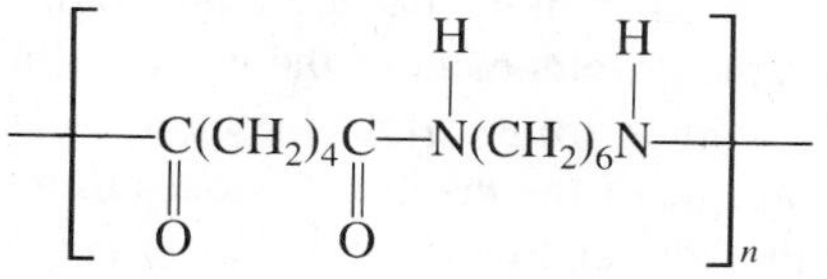

Other polymers such as rayon and dacron are well-known substitutes for natural fibers.

Technology and the development of polymers have gone hand in hand, each aiding the other. As mentioned, early wartime necessity spurred the development of synthetics. When the supply of natural rubber was cut off, a synthetic rubber called neoprene was developed. The different structures of natural rubber and neoprene are

$$\text{Rubber (polymer of isoprene)} \quad \left[-\overset{H}{\underset{H}{C}}-\overset{CH_3}{C}=\overset{H}{C}-\overset{H}{\underset{H}{C}}- \right]_n$$

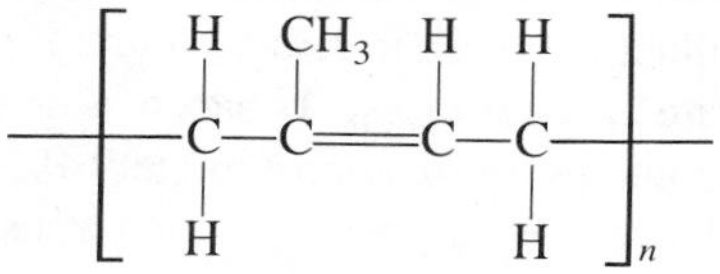

$$\text{Neoprene} \quad \left[-\overset{H}{\underset{H}{C}}-\overset{Cl}{C}=\overset{H}{C}-\overset{H}{\underset{H}{C}}- \right]_n$$

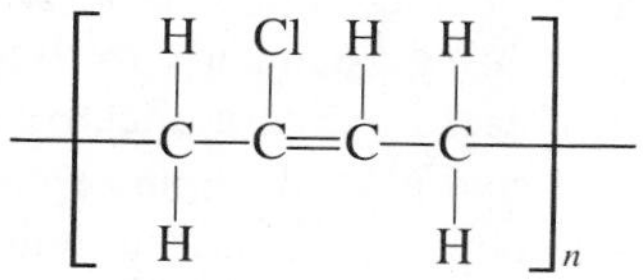

However, need did not precede discovery in all cases. Our entry into the space age has created a big demand for different and unusual materials. However, some materials that could stand the rigorous space environment were already available. Two examples are polyethylene terephthalate and polycarbonate, known commercially as Mylar polyester film and Lexan polycarbonate resin, respectively. They are extremely tough materials—the Echo satellites were constructed of thin sheets of Mylar, while the durable Lexan serves as astronauts' helmet visors. They also have everyday applications. Another group of polymers with applications from bathtub sealer to space ships is the silicones. These compounds are rubbery and flexible. One such polymer, RTV 615, is elasticlike rubber, transparent, and has a melting point four times that of steel. It has been used as heat shields for space vehicle reentry where the temperatures are in excess of 8000°C. This material transmits virtually no heat, due to a slow layer-by-layer decomposition of its molecules that uses excess heat as the latent heat of vaporization.

The list of polymers and their uses could go on and on. However, it should be evident from the few mentioned that artificial molecules are a tribute to our ingenuity.

17.4 Drugs

A **drug** is a compound that may produce a physiological change in human beings or other animals and may become a poison when used in excessive amounts. The molecular composition of drugs is usually quite complex, as the majority of drugs are extracted from natural compounds. Because of the human involvement and the complexity of the human body, the development of new or synthetic drugs is relatively slow compared to other materials such as plastics. A great degree of caution must be exercised in administering new drugs, and many experiments must be run on laboratory animals in an effort to find any side effects that may be harmful to humans. Occasionally these effects are latent and emerge only after widespread usage, as in the case of thalidomide, the sleeping compound believed to be a cause of deformities in the human fetus. Often there is no conclusive evidence of a compound being harmful to humans, but preventive action is sometimes taken, as in the case of the banning of cyclamate sweeteners when it was observed that large doses caused cancer in rats.

Figure 17.8 Five common substances containing a drug: marijuana, nicotine, alcohol, caffeine, and aspirin. (Photo courtesy James Crouse.)

From our definition of a drug, it is evident that drugs are useful for purposes other than medicinal, although they are in large part used to correct some body malfunction. One of the oldest drugs known is alcohol. As a by-product of fermented natural beverages, ethyl alcohol has been consumed by humans probably since the beginning of civilization. Some alcohols have severe physiological effects; methyl alcohol, for example, may cause blindness if taken internally. In small amounts, ethyl alcohol acts as a depressant although it is commonly, and falsely, considered to be a stimulant. Acting on the brain's control mechanisms, it can cause loss of muscular control, such as slurred speech, mental disorganization, sleep, and nausea when taken in moderate amounts. In addition, alcohol acts on the parts of the brain that control blood circulation and the kidneys. When alcohol is taken into the body the capillaries dilate, thus bringing more blood nearer the surface of the body. This causes a warm feeling but also a more rapid loss of body heat. Its effect on the kidneys is to depress the control center of the brain that ordinarily slows the kidney's excretion. Addiction to alcohol is also possible after prolonged usage. Figure 17.8 shows some common substances containing drugs.

Another very common drug is aspirin. **Aspirin** is the compound acetylsalicylic acid.

COOH

$OCOCH_3$

Aspirin
(acetylsalicylic acid)

Aspirin is used primarily as an analgesic. An **analgesic** is a drug that relieves pain without dulling consciousness. Aspirin also has the ability to reduce the body temperature, and is thus effective for reducing fevers. It is estimated that enough acetylsalicylic acid is produced in the United States to supply every person in the United States with 400 aspirin tablets each year.

Relatively recently developed drugs are the antibiotics, or ''wonder'' drugs. An **antibiotic** is an antibacterial substance that is produced by a living organism and is administered to fight bacterial infections in the body. Before the development of the wonder drugs, death was quite common from serious infections, and their discovery was one of the major milestones in medicine. Probably the best known of these drugs are the penicillins, which are derived from molds of a certain type. There are now many other wonder drugs such as the mycins, which have proved to be more effective than penicillin in many cases.

Another group of drugs called **alkaloids** are complex, nitrogen-containing compounds found in plants. These include cocaine ($C_{17}H_{24}NO_4$), morphine ($C_{17}H_{19}NO_3$), quinine ($C_{20}H_{24}N_2O_2$), nicotine ($C_{10}H_{14}N_2$), and strychnine ($C_{21}H_{22}N_2O_2$). Strychnine is best known as a heart stimulant although in large amount it is a poison, as all drugs are in overdoses. Tobacco is a source of nicotine and contains from approximately 0.5 to 5% nicotine in its leaves. Because it is a dangerous poison, nicotine finds little use as a drug, but it is an effective insecticide. Quinine was used as an early treatment for malaria, but has been replaced by more effective drugs. Morphine is a narcotic found in crude opium which is extracted from the oriental poppy. It is used to relieve pain and induce sleep. Addiction to this drug may occur with prolonged usage.

There has been increasing concern over the misuse of drugs, especially the narcotics. **Heroin,** an artificial derivative of morphine, is a habit-forming drug. Once the habit is formed the craving for drugs is insatiable, and the user will go to any extreme, including violence, to obtain a certain drug or the money to buy it, thus creating a serious social problem. **Marijuana** is the leaves and flowering tops of the Indian hemp plant. Smoking or ingestion of marijuana causes dizziness, hilarity, delusions of grandeur, heightened mental awareness, and mental confusion, depending on the user. It is generally agreed that marijuana is not habit-forming and its effects are similar to those of alcohol. However, the user becomes apathetic and may, after prolonged and heavy use, suffer permanent loss of mental acuity.

Another group of drugs are the psychedelics, or hallucinogenic drugs, one of which is **LSD** (lysergic acid diethylamide). This dangerous drug causes prolonged stimulation, which may be followed by insanity. Sensations are amplified, and exaggerations and hallucinations are experienced. The LSD molecule is quite complex. (Figure 17.9) The lower uncolored portion of the formula shows an *indole* ring; such a ring is found in a similar structure called serotonin. **Serotonin** is a compound found in the brain and is believed to transfer signals to the nerve cells. It is theorized that because of a structure similar to that of serotonin, LSD interferes with this transfer, and the signals are altered. Thus the drug could change the sensory signals, producing hallucinations and possibly almost any other reaction. Because of the risk of lasting brain disorders (including ''flashbacks''), LSD is a drug taken at risk to the user.

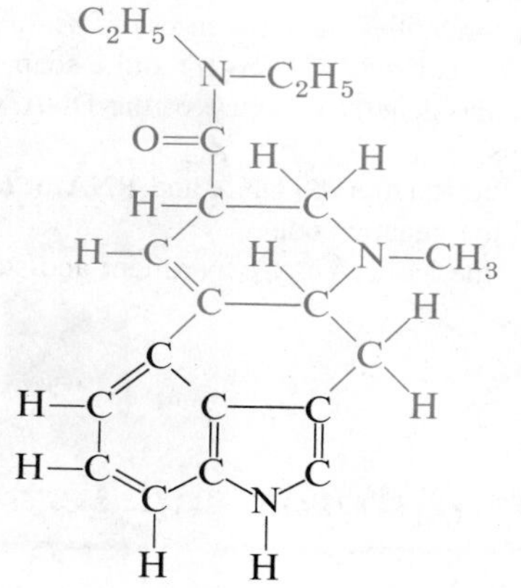

Figure 17.9 LSD molecule.

Of a somewhat different social impact are the drugs contained in birth control pills, another success with complex molecules. The drugs used in these pills are synthesized and have proved to be more effective than similar natural compounds. The first one of these drugs to be synthesized was **norethynodrel.** It was known for some time that the hormone progesterone would prevent ovulation in the human female about 85% of the time; this, however, is hardly an effective contraceptive. By synthesizing compounds similar in structure to progesterone a compound was found, norethynodrel, which turned out to be virtually 100% effective at one-third the dosage of progesterone when given with a small amount of another hormone, estrogen.

Learning Objectives

After reading and studying this chapter you should be able to do the following without referring to the text:

1. Write the organic functional group for a few types of organic compounds.
2. List one or two uses for each of the organic compounds formed using the functional groups referred to in Learning Objective 1.
3. Distinguish between a monomer and a polymer.
4. State the product formed when an organic acid reacts with an alcohol.
5. State and explain the general formula for carbohydrates.

6. Describe the chemical structure of proteins.
7. State the similarities and differences of fats, oils, and waxes.
8. Distinguish between a detergent and a soap.
9. Describe the general structure of the DNA and the RNA molecules.
10. Explain the function of DNA and RNA in cell reproduction and the genetic code.
11. Describe the Miller-Urey experiment and state its significance.
12. Distinguish between mitosis and meiosis.
13. Describe the group of synthetic polymers known as plastics.
14. Define a drug, list some examples, and give the beneficial and/or harmful effect the drug has on the human body.
15. Define or explain the important words and terms at the end of the chapter.

Important Words and Terms

complex molecules
monomer
polymer
alkyl halides
alcohols
aldehydes
organic acids
carboxyl
ester
alkyl
carbohydrates
starch
cellulose
proteins
amino
lipids
hydrogenation
waxes
detergents
nucleic acids
DNA
RNA
vitamins
Miller-Urey experiment
chromosomes
mitosis
meiosis
genes
dominant
recessive
genetic code
mutation
cytoplasm
synthetic
plastics
drug
aspirin
analgesic
antibiotic
alkaloid
heroin
marijuana
LSD
serotonin
norethynodrel

Questions

1. Describe the structure of a polymer.
2. What functional group characterizes an alkyl halide? an alcohol? an aldehyde? an acid? an ester?
3. How are aldehydes formed chemically?
4. By what general chemical reaction are alcohols formed?
5. Name the alcohol used in alcoholic beverages. Draw its structural formula.
6. What is the result of reacting oxygen with an aldehyde?
7. Write in words the general equation for the reaction of an organic acid with an alcohol.
8. What is the meaning of the word carbohydrate? Is the nature of carbohydrate compounds what the name implies? Explain.
9. What is the chemical composition of proteins?
10. What is hydrogenation?
11. Distinguish between mitosis and meiosis.
12. How is the DNA molecule constructed?
13. How does the DNA molecule reproduce itself?
14. Explain the function of DNA and RNA in cell reproduction and the genetic code.
15. What is the acetylsalicylic acid content of regular aspirins?
16. Describe the effects of LSD use by humans and give the probable cause.

Problems

1. Considering the monomer unit CH_2, what is the repetition number of each of the following:
   (a) *n*-octane (b) *n*-eicosane (c) polyethylene?
2. Complete and balance the following equation. Name the compound formed.

$$\underset{\text{Acetic acid}}{CH_3COOH} + \underset{\text{Ethyl alcohol}}{C_2H_5OH} \longrightarrow \underline{\qquad\qquad} + H_2O$$

3. Complete and balance the following equation. Name the compound formed.

$$\underset{\text{Methyl alcohol}}{CH_3OH} + O_2 \longrightarrow \underline{\qquad\qquad} + H_2O$$

4. Complete and balance the following equation:

$$CH_2O + O_2 \longrightarrow \underset{\text{Formic acid}}{\underline{\qquad\qquad}}\ .$$

5. Complete and balance the following equation:

$$CO_2 + H_2O + \text{Energy} \longrightarrow \underset{\text{Glucose}}{\underline{\qquad\qquad}} + O_2$$

6. From Table 17.1 prepare a list of the food sources of the important vitamins. Explain the effects of a deficiency of these important vitamins in a person's diet. (Outside source work will be necessary.)
7. Refer to an outside source and give the reason why termites can digest cellulose, the most abundant carbohydrate, but humans cannot.
8. Given the general formulas below, write molecular and structural formulas for the first member of each group. Then name each compound.

$$R{-}OH \quad R{-}\overset{\overset{\displaystyle O}{\|}}{C}{-}H \quad R{-}\overset{\overset{\displaystyle O}{\|}}{C}{-}OH \quad R{-}\overset{\overset{\displaystyle O}{\|}}{C}{-}O{-}R$$

18

The Solar System

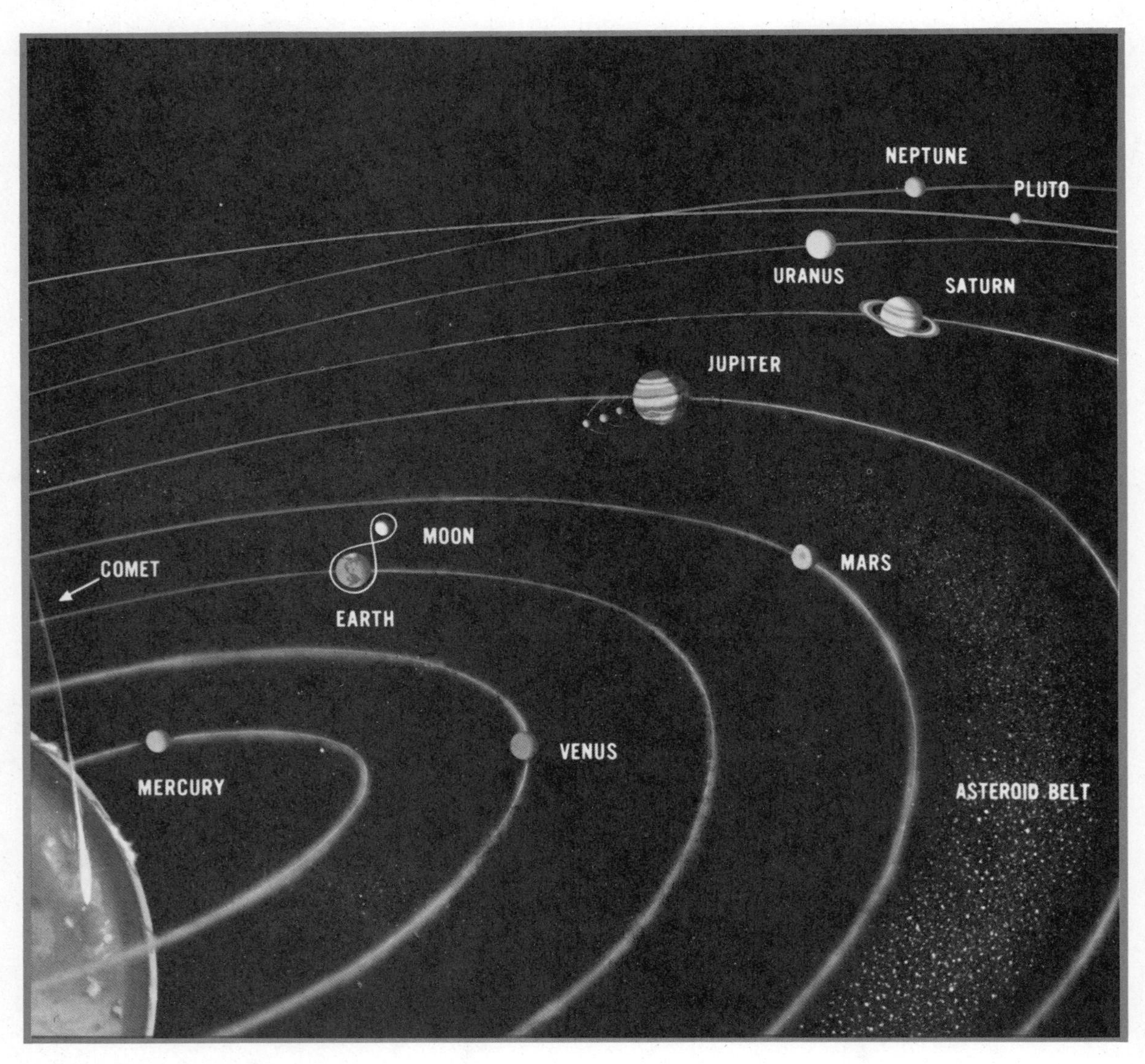

The acquisition of knowledge begins with simple fundamental ideas that expand to broader and more complex concepts.

AS WE GO about our daily lives, we often lose our sense of curiosity and wonder. But on a clear dark night when we gaze toward the heavens, we are struck by the awesome grandeur of the universe. At times like these we marvel at our seeming insignificance, and we begin to ask ourselves questions. What is the universe? How did it begin, and how will it end? When and where did we originate? Is it possible for us to know all things and control the forces of the universe?

One science concerned with such questions is astronomy, probably the oldest science. The ancient Babylonians and Greeks were two early civilizations that watched the skies and originated many of the words used today in describing our universe. They were limited in what they could observe by the ability of the unaided eye. With the invention of the telescope and the discovery and development of photography, more accurate measurements of celestial objects were made possible.

In 1957 the first artificial satellite was launched by the Soviet Union. Shortly thereafter, the United States began its own space program. The National Aeronautics and Space Administration (NASA) was formed, and astronauts and many artificial satellites were propelled into space. Most of these satellites orbited the Earth, but some were sent to the moon, Mars, Venus, Mercury, Jupiter, Saturn, and Uranus. As these satellites transmitted pictures and data back to the Earth, our knowledge of the solar system exploded. In fact, the well-read college student of today knows more about our solar system than the most distinguished scientist knew in 1960.

As we come to learn more about the moons and planets of our solar system, we also learn more about our own planet Earth. But as we find out more, our sense of curiosity increases, for there are always more and more questions needing answers.

◄ A montage of the solar system.

18.1 The Planet Earth

The Earth is one of nine planets that revolve around the Sun. These nine planets and their four dozen or so known satellites, plus thousands of asteroids and countless comets and meteors, make up the local **solar system.** Other similar planetary systems may exist in the universe. See Section 18.6.

Although we are unable to sense directly the motion of our home planet, it is undergoing several motions simultaneously. Two that have major influences on our daily lives will be explained in this section: (1) the daily rotation of the Earth on its axis, and (2) the annual revolution of the Earth around the Sun. A third motion, precession, will be discussed in Section 19.5.

As the Earth goes around the Sun, it orbits in a plane called the **ecliptic plane.** The Earth rotates once each day on an axis that is tilted 23.5° from the ecliptic plane.

It is important to know the difference between rotating and revolving. A mass is said to be in **rotation** when it spins on an axis. An example is a spinning toy top, or a ferris wheel at an amusement park. Revolving, or **revolution,** is the movement of one mass around another. The Earth revolves around the Sun, the moon revolves around the Earth, and electrons revolve around the nucleus of an atom.

The fact that the Earth rotates on its axis was not generally accepted until the nineteenth century. A few scientists had considered the possibility of a rotating Earth, but no definite proof was given to support their beliefs; therefore, their ideas were not acceptable.

In 1851, an experiment demonstrating the rotation of the Earth was performed by J. B. Leon Foucault (1819–1868), a French engineer. He used a 200-foot pendulum that today is called a **Foucault pendulum.** More noticeable results can be seen if the experiment is performed at the north or south pole.

Picture a large single-room building with a ceiling over 200 feet high located at the north pole. See Figs. 18.1 and 18.2. Fastened to the ceiling, precisely above the north pole, is a support hook having very little friction, from which a 200-ft, fine steel wire is attached. Connected to the lower end of the wire is a massive iron ball with a short sharp steel needle attached permanently to its underside. On the floor, under the pendulum, is placed a layer of fine sand which can be slightly furrowed by the sharp needle as the pendulum swings back and forth.

The pendulum is started swinging by displacing it to one side with a strong fine thread, with one end attached to the side of the ball, and the other end attached to one wall of the building where a 24-hour wall clock is mounted. To avoid any sideways motion, the iron ball is allowed to become motionless before the thread is parted by burning. Extreme care is taken to prevent any lateral external forces from being applied to the upper support point of the 200-foot wire. As the pendulum swings freely back and forth, the sharp needle point traces its path in the layer of fine sand. After a few minutes, the plane of the swinging pendulum appears to be rotating clockwise as shown by the markings in the sand. At the end of an hour, the plane of the swinging pendulum has rotated 15° clockwise from its original position. When six hours have elapsed, the plane of the swinging pendulum appears to have rotated 90° clockwise, and is parallel to the wall that holds the 24-hour clock. With the passing of each hour, the plane of the swinging pendulum appears to rotate another 15° clockwise. At the end of 24 hours, the plane has made an apparent rotation of 360°.

A person who believes in a motionless Earth would argue that the pendulum actually rotated 360°, because one rotation of the swinging pendulum has been observed by anyone stationed in the large room. A different view of the experiment can be obtained if we make the walls of the building of a transparent material such as glass, and perform the experiment sometime during the winter months for the northern hemisphere. The north pole is having 24 hours of darkness during these months, and the stars are always visible. When starting the pendulum this time, we take care to place the iron ball in direct line with the stars Alpha and Beta, the pointers in the cup of the Big Dipper. As the minutes pass, we observe as before the apparent rotation of the plane of the swinging pendulum in a clockwise direction in reference to the large room and the clock on the wall. We also observe through the transparent walls of the room that the pendulum still swings in the same direct line with the stars Alpha and Beta; that is, the pendulum, Alpha, and Beta are in the same plane. There has been no rotation of the pendulum in reference to the fixed stars. No forces have been acting on the pendulum to change its plane of swing. Only the force of gravity has been acting vertically downward to keep it swinging.

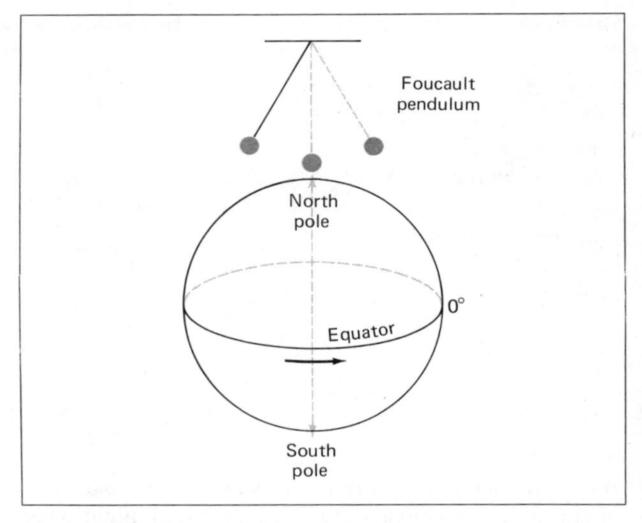

Figure 18.1 Diagram of the Earth with a Foucault pendulum positioned above the north pole. The pendulum swings in the same plane while the Earth rotates about its axis.

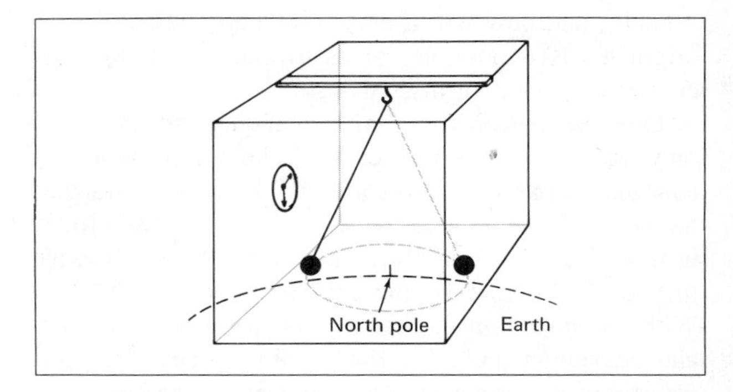

Figure 18.2 Diagram of a Foucault pendulum positioned in a room located at the north pole of the Earth. To an observer in the room, the pendulum woud appear to change its plane of swing by 360 degrees every 24 hours.

Therefore the pendulum does not rotate, but the building and the Earth rotate eastward, or turn counterclockwise, once during the 24-hour period. See Fig. 18.3 for a photograph of a Foucault pendulum. A rotating Earth has a major influence on weather, deflecting winds from their paths and causing cyclones and other cyclic storms.

The Foucault pendulum is an experimental proof of rotation. What kind of an effect can we expect due to the Earth's motion around the Sun?

As the Earth orbits the Sun once each year, the relative positions of nearby stars change with respect to faraway stars. This effect is called parallax. In general, **parallax** is the apparent motion, or shift, that occurs between two fixed objects when the observer changes position. To see parallax for yourself, hold your finger at a fixed position in front of you. Close one eye, move your head from side to side, and notice the apparent motion between your finger and some object out the window. Note also that the apparent motion becomes less as the finger is moved farther away. Figure 18.4 is an illustration of the parallax of a nearby star as measured from the Earth in relation to stars that are more distant.

As the Earth revolves around the Sun, there is an apparent shift in the positions of the nearby stars in respect to the stars that are more distant because of the motion of the Earth. Since the stars are at very great distances from the Earth, the parallax angle is very small.

Because the stars are so far away, the parallax of the stars illustrated in Fig. 18.4 cannot be seen with the naked eye. It was first observed with a telescope in 1838 by Friedrich W. Bessel (1784–1846), a German astronomer and mathematician. The observation of parallax was undisputed proof that the Earth really does go around the Sun. Today, the measurement of the parallax angle is the best method we have of determining the distances to nearby stars.

The Earth is an oblate spheroid—flattened at the poles and bulging at the equator. This is due primarily to the rotation of the Earth on its axis. Although the difference in the diameter at the poles and at the equator is about 27 miles, the difference is very small when compared with the total diameter of the Earth, which is about 8000 miles. The ratio of 27 to 8000 is approximately 1/300, which is a rather small fraction. If the Earth were represented by a basketball, which is approximately 10 inches in diameter, the eye would not detect a difference of 1/30 inch in the diameter. The Earth is a more nearly perfect sphere than the average basketball.

The full Earth appears four times as large in diameter and more than 60 times as bright as the full moon when

Figure 18.3 A Foucault pendulum. Note the small wooden sticks that have been knocked over by the swinging pendulum. Is the building turning clockwise or counterclockwise? (Courtesy COSI, Center of Science & Industry, Columbus, Ohio.)

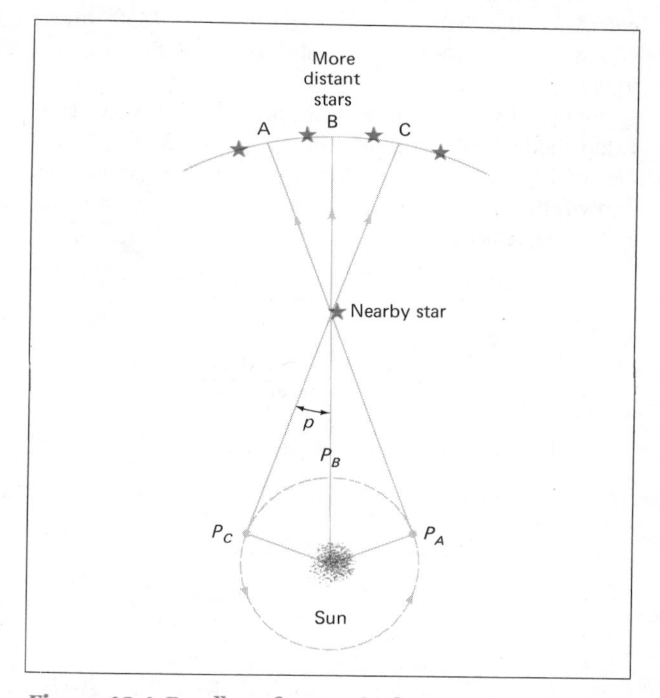

Figure 18.4 Parallax of a star is the apparent displacement of a star that is located fairly close to the Earth, in respect to more distant stars. When the observer is at P_A, the star appears in the direction A. As the Earth revolves counterclockwise, the star appears to be displaced and appears in the direction indicated for different positions of the Earth. Positions P_A' and P_C are a few months apart. The angle of parallax p is also shown.

viewed from a distance of 240,000 miles. The Earth appears much brighter because the clouds and water areas are much better reflecting surfaces than the dull, dark surface of the moon.

The Earth has the highest density of any planet. The Earth's density is 5.5 g/cm^3 or 5.5 times as dense as water.

The Greek scientist Eratosthenes calculated the circumference of the Earth in 250 B.C. He knew that the noon Sun on the first day of summer was directly overhead (on the zenith) at Syene (now Aswan), Egypt. Eratosthenes lived in Alexandria, which was located 500 miles due north of Syene. He had received reports that deep wells at Syene were lighted all the way to the bottom on the first day of summer, which meant that the Sun was directly overhead at Syene. At Alexandria, Eratosthenes discovered that a vertical stick cast a shadow on this same date, which positioned the Sun about $7\frac{1}{2}°$ south of his zenith. Assuming the Earth to be a sphere and the Sun to be such a great distance away that the rays of light would come in parallel to one another, he calculated the diameter of the Earth to be 7850 miles, which is very close to the value of 7918 that we use today.

Figure 18.5 illustrates the principles involved in Eratosthenes' calculations. Since there are 360° in a circle, and the distance between Syene and Alexandria is known, the *circumference* can be calculated by using the following relationship:

$$\frac{\text{Circumference}}{360°} = \frac{\text{Dist. Syene to Alexandria}}{7.5°}$$

$$\text{Circumference} = \frac{360° \times 500 \text{ mi}}{7.5°}$$

$$\text{Circumference} = 48 \times 500 \text{ mi}$$

$$\text{Circumference} = 24{,}000 \text{ mi}$$

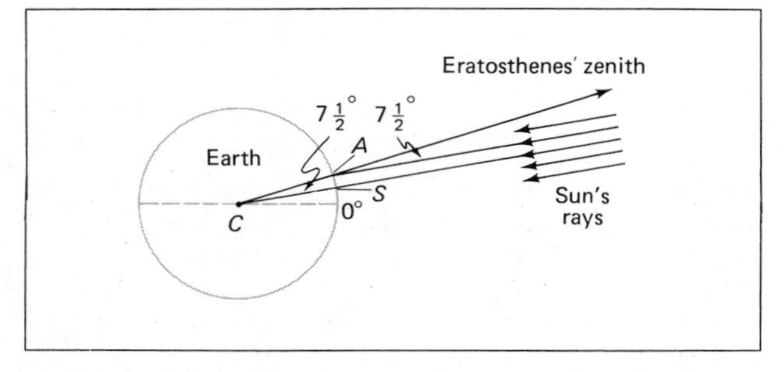

Figure 18.5 Diagram showing the method Eratosthenes used to calculate the size of the Earth. Point *A* represents Alexandria and point *S* represents Syene.

This is very close to our present value of approximately 24,900 miles.

18.2 The Solar System

The rotating motion of the planet Earth was not an easy concept to prove. In early times, most people were convinced that the Earth was motionless and that the Sun, moon, planets, and stars revolved around the Earth, which was considered the center of the universe.

This incorrect theory of the solar system was called the Earth-centered or **geocentric theory.** The geocentric theory was supported and extended by Aristotle (384–322 B.C.), who was considered a great philosopher (Fig. 18.6).

There were several important problems with the geocentric theory of Aristotle. Two outstanding faults were (1) a failure to explain the variation of brightness of the planets at different times of the year, and (2) a failure to explain the motion of some planets.

In order to solve the difficulties of varying brightness and planetary motion, Claudius Ptolemy of Alexandria (second century A.D.) modified and extended the geocentric theory. In the Ptolemaic model, the planets moved in circles within circles, called epicycles and deferents. The Ptolemaic theory was extremely complicated, but it was accepted with few questions for over 1200 years.

The first philosopher to give serious thought to the fact that the Earth rotates about an axis was Ponticus Heracleides (375–310 B.C.), a member of the Greek Academy. The Greek astronomer Aristarchus of Samos (about 310–230 B.C.) is credited with extending the idea of daily rotation of the Earth and originating the Sun-centered or **heliocentric theory,** which placed the Sun at the center of the universe with the Earth, moon, and planets revolving around the Sun in different orbits and at different speeds.

These ideas were not popular and were forgotten, probably because it was difficult for people to believe the ground upon which they lived could be in rapid motion. The theory also disagreed with the philosophic doctrine that the Earth was the center of the universe. Aristarchus also failed to support his theory with mathematical details for use in predicting future planetary positions. Lastly, he failed to satisfy his critics concerning the apparent motion stars would have if the Earth's position would change through large distances, as he proclaimed, in revolving around the Sun.

We now know the answers to these earlier objections to the heliocentric theory. The Earth's atmosphere rotates and revolves with it, and the acceleration caused

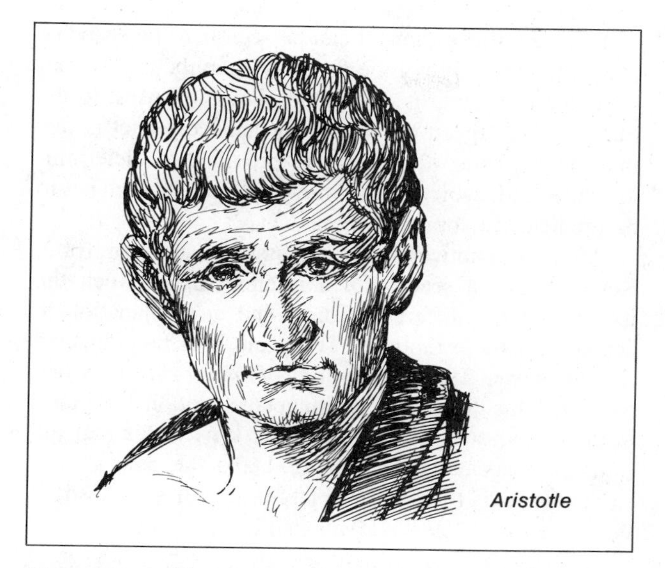

Figure 18.6 Aristotle was one of the great philosophers of the Greek era. However, his Earth-centered view of the universe was wrong.

Figure 18.7 Nicholas Copernicus (1473–1543). Polish astronomer who developed mathematical proof of the heliocentric theory.

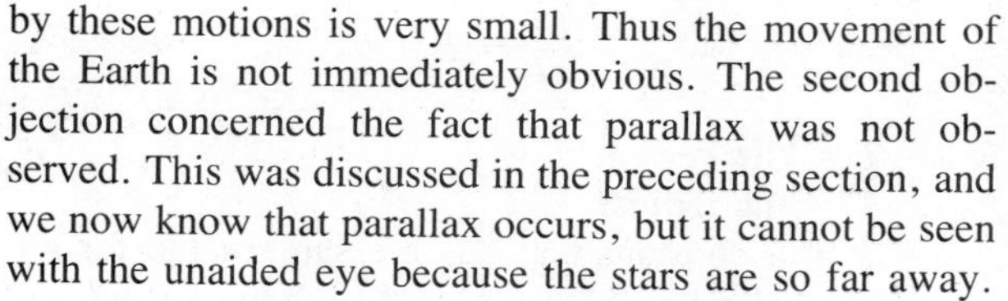

by these motions is very small. Thus the movement of the Earth is not immediately obvious. The second objection concerned the fact that parallax was not observed. This was discussed in the preceding section, and we now know that parallax occurs, but it cannot be seen with the unaided eye because the stars are so far away.

The heliocentric theory remained buried for 18 centuries. The rebirth of the theory and the development of the solar system as we think of it today began with Nicholas Copernicus (1473–1543), a Polish astronomer (Fig. 18.7), who viewed the heliocentric theory of Aristarchus as absurd when he first studied it. But much dissatisfaction was building up against the Ptolemaic, or geocentric, system. This theory, which had been useful for more than a dozen centuries for predicting positions of heavenly bodies, was then becoming unreliable. There was doubt about the exact length of the year, and the calendar needed to be revised. Copernicus was invited by the church to express his opinion on the revision, but he refused, saying that the position of the Sun and moon were not known with sufficient accuracy to make a reliable calendar.

Copernicus observed and studied the problem for many years and finally became convinced of the validity of the heliocentric theory. He published his work under the title of *Revolutions* in 1543, the year of his death. Copernicus had succeeded where Aristarchus had failed, because of mathematical proofs that could be used to predict future positions of the planets. Although the results were no more accurate than those attained with the Ptolemaic method, they were much simpler to use. Since Copernicus held to the idea of uniformly moving concentric spheres, he had to retain the epicycle motions, but they were greatly reduced in number.

There was general public opposition to Copernicus's Sun-centered theory. The same arguments that were used against Aristarchus were again applied against Copernicus, but the scientists who were calculating the positions of the planets in most cases used the heliocentric theory because of its simplicity. The full acceptance of the theory that the Earth rotates about an axis and is one of the planets that revolve around a fixed Sun had to wait another century, until more accurate predictions could be made of planetary motion and reasons given as to why the planets move as they do.

After the death of Copernicus, the study of astronomy was continued and developed by several men, three of whom made their appearance in the last half of the sixteenth century. Notable among these men was Tycho Brahe (1546–1601), a Danish astronomer who built an observatory on the island of Hven near Copenhagen and spent most of his life observing and studying the stars and planets. Brahe is considered the greatest practical astronomer since the Greeks. His measurements of the planets and stars, all made with the unaided eye (the telescope had not been invented), proved to be more accurate than any previously made. Brahe's data, published in 1603, were edited by his colleague Johannes

Kepler (1571–1630), a German mathematician and astronomer who had joined Brahe during the last year of his life. After Brahe's death, his lifetime of observations were at Kepler's disposal, and they proved very useful in providing the data necessary for the formulation of the laws we know today as Kepler's laws of planetary motion.

Kepler was very interested in the irregular motion of the planet Mars. He spent considerable time and energy before he came to the conclusion that the uniform circular orbit proposed by Copernicus was not a true representation of the observed facts. Perhaps because he was a mathematician, he saw a simple type of geometric figure that would give the correct solution to his calculations while removing all the epicycles used by Copernicus. Kepler's first law, known as the **law of elliptical paths,** states that all planets move in elliptical paths around the Sun with the Sun at one focus of the ellipse. Note that there is nothing at the other focus of the ellipse.

An ellipse is a figure that is symmetrical about two unequal diameters. See Fig. 18.8. An ellipse can be drawn by using two thumb tacks, a closed piece of string, paper, and pencil. The points where the two tacks are positioned are called the foci of the ellipse.

Kepler's first law gives the shape of the orbit, but fails to predict when the planet will be at any position in the orbit. Kepler, aware of this, set about to find a solution from the mountain of data he had at his disposal. After a tremendous amount of work, which was carried out with no indication that a solution was possible, he discovered what is now known as Kepler's second law of planetary motion, or **law of equal areas,** which states that an imaginary line (radius vector) joining a planet to the Sun sweeps out equal areas in equal periods of time (Fig. 18.9).

From the illustration, it can be seen that the speed of the revolving body will be greatest when the radius vector is the least (that is, when the planet is closest to the Sun), and its speed the slowest when the planet is farthest away. Thus, a method was provided for determining the speed, which allows the position of the planet to be predicted at some future time.

After the publication of his first two laws in 1609, Kepler began a search for a relationship between the motions of the different planets and an explanation to account for these motions. Ten years later he published *De Harmonica Mundi (Harmony of the World),* in which he stated his third law of planetary motion. Kepler's third law, known as the **harmonic law,** states that the ratio of the square of the period and the cube of the semimajor axis (one-half the larger axis of an ellipse) is the same for all planets. This can be written as

$$\frac{\text{Period}^2}{\text{Radius}^3} = \text{Constant}$$

or algebraically as

$$\frac{T^2}{R^3} = k$$

where T = the period,
R = the radius,
k = a constant that has the same value for all planets.

Planets go around the Sun in elliptical paths, but these ellipses are very close to being circles (except for Pluto). So R in Kepler's third law is approximately the distance between the planet and the Sun. For the Earth, T = 365.25 days (one year) and R = 93 million miles or 1.5×10^{11} meters.

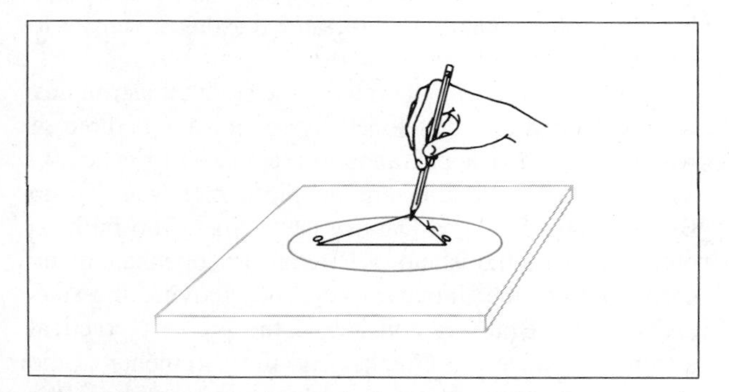

Figure 18.8 An ellipse can be drawn by using two thumb tacks, a closed piece of string, a pencil, and a sheet of paper.

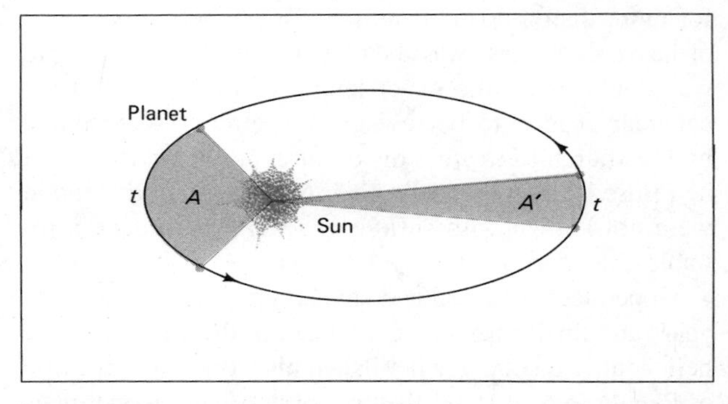

Figure 18.9 Kepler's law of equal areas. As the Earth revolves about the Sun, equal areas are swept out in equal periods of time. Area A = area A'.

Galileo Galilei (1564–1642), Italian astronomer, mathematician, and physicist who is usually called Galileo, was one of the greatest scientists of all time (Fig. 8.10). The most important of his many contributions to science were in the field of mechanics. He originated the basic ideas for the formulation of Newton's first two laws of motion, and he founded the modern experimental approach to scientific knowledge. The motion of bodies, especially the planets, was of prime interest to Galileo. His concepts of motion and the forces that produce motion opened up an entirely new approach to astronomy. In this field he is noted for his contribution to the heliocentric theory of the solar system.

In 1609 Galileo was the first person to observe the moon and planets through a telescope. With the telescope he discovered four of Jupiter's 15 or more moons, thus proving that the Earth was not the only center of motion in the universe. Equally important was his discovery that the planet Venus went through a change of phase similar to that of the moon as called for by the heliocentric theory, but contrary to the Ptolemaic theory that called for a new or crescent phase of Venus at all times.

Galileo also observed the craters of the moon, dark spots on the Sun, star clusters, and individual stars in the Milky Way. He published *Sidereus Nuntius (Starry Messenger)* in 1610, in which he reported his observations with the telescope. He published his major work, *Dialogue of the Two Chief World Systems, Ptolemaic and Copernican,* in 1632; it was received enthusiastically, but the authority of the church banned the book. Galileo, a Catholic, was forced to renouce his belief in a moving Earth. He was placed under house arrest and forbidden for the rest of his life from any work on astronomy. Thus, he had failed to convince his opponents of the correctness of the Sun-centered theory, even though the telescope had presented scientific facts to support it. Kepler did not have such conflicts with the church because he was a German Protestant.

Only the passing of time was to remove most of the prejudice against scientific knowledge and usually allow scientists the freedom to express the facts of nature as they observed them. The works of Copernicus, Kepler, and Galileo were integrated by Newton in 1687 with the publication of the *Principia*.

Sir Isaac Newton (1642–1727), English physicist regarded by many as the greatest scientist the world has known, formulated the principles of gravitational attraction between bodies and established physical laws determining the magnitude and direction of the forces that

Figure 18.10 Galileo was the first person to observe the moon and the planets with a telescope.

cause the planets to move in elliptical orbits in accordance with Kepler's laws (cf. Introduction and Chapter 3). Newton invented calculus and used it to help explain Kepler's first law. Newton also used the law of conservation of angular momentum to explain Kepler's second law.

To explain Kepler's third law, Newton showed that the constant in Kepler's equation was

$$\frac{T^2}{R^3} = \frac{4\pi^2}{Gm_{\text{Sun}}}$$

where T = period of a planet,
R = distance between the planet and the Sun,
G = gravitational constant,
m_{Sun} = mass of the Sun.

Newton's explanations of Kepler's laws unified the heliocentric theory of the solar system and brought an end to the confusion of the past. He gave us an ordered system of the Sun and planets satisfactory for the present time.

Today our solar system is known to consist of one star (the Sun, which contains 99.87% of the material of the system), nine planets (including the Earth), about four dozen satellites (our moon is an example), thousands of asteroids (Ceres is the largest with a diameter of 600 miles), billions of comets, and countless meteors. The distribution of the remaining 0.13% of the solar system's mass is shown in Table 18.1. Note that more than half the remaining mass is concentrated in Jupiter.

Table 18.1 The Solar System

Name	Mean Distance from Sun			Diameter (miles)	Mass with Respect to the Earth	Density g/cm^3 (water = 1)	Period of Revolution	Period of Rotation	Inclination of Axis with the Vertical	Inclination of Orbit with the Ecliptic	Magnetic Field	Satel-lites
	Million Miles	Astron. Units	Titius-Bode Law									
Sun				864,000	332,000.	1.4		25 days			Yes	9
Mercury	36	0.39	0.4	3,026	0.055	5.4	88 days	59 days	less than 28°	7°	Weak	None
Venus	67	0.72	0.7	7,575	0.82	5.2	225 days	243 days (retrograde)	3°	3.4°	No	None
Earth	93	1.00	1.0	7,918	1.00	5.5	365.25 days	24 h	23.5°	0°	Yes	1
Mars	142	1.52	1.6	4,216	0.11	3.9	687 days	24.6 h	24°	2°	Very weak	2
Asteroids	257	2.77	2.8				5 yr typical			10° average	No	
Jupiter	483	5.20	5.2	86,000	318.	1.3	12 yr	10 h	3.1°	1.3°	Yes	16 or more
Saturn	886	9.54	10.0	72,000	94.3	0.7	29.5 yr	10.7 h	26.7°	2.5°	Yes	17 or more
Uranus	1783	19.19	19.6	33,000	14.54	1.2	84 yr	12–24 h (retrograde)	82°	0.8°	?	5 or more
Neptune	2793	30.07	38.8	31,000	17.2	1.7	165 yr	18–22 h	29°	1.8°	?	2 or more
Pluto	3666	39.46	77.2	2,000	0.002	0.5	248 yr	6.4 days	?	17°	?	1

Planets that have orbits smaller than the Earth's are classified as inferior planets and those with orbits greater than the Earth's, as superior planets. Another method is to classify Mercury, Venus, the Earth, and Mars as the inner or **terrestrial planets** because they resemble the Earth (see color plate 1); and Jupiter, Saturn, Uranus, and Neptune as the outer or **Jovian planets** because they resemble Jupiter. The Roman god Jupiter was also called Jove. Pluto does not resemble the Earth or Jupiter and many astronomers have suggested that Pluto be classified as an asteroid.

The relative distances of the planets from the Sun are shown in Fig. 18.11. The orbits are all elliptical, but nearly circular except for Pluto. Pluto's orbit actually goes inside Neptune's orbit. The actual position of Pluto is shown for 1990. It can be seen that for several years Neptune will be farther from the Sun than Pluto! Study Fig. 18.11 for a minute and notice how far from the Sun Jupiter is compared to Mars. Note that the distance from Saturn to Neptune is greater than that from the Sun to Saturn. The distinction between the four inner planets and the five outer planets can also be seen in this illustration.

When viewed from above the solar system (i.e., looking down on the north pole of the Earth), the planets all move counterclockwise around the Sun. The planets also spin with a counterclockwise rotation when viewed from above with the exceptions of Venus and Uranus which have retrograde rotation.

The relative sizes of the planets are shown in Fig. 18.12. Note the huge size of the outer planets compared to the inner planets. Also note how small Pluto is.

The inclinations of the orbits of the planets relative to the Earth's orbit are shown in Fig. 18.13. Note that

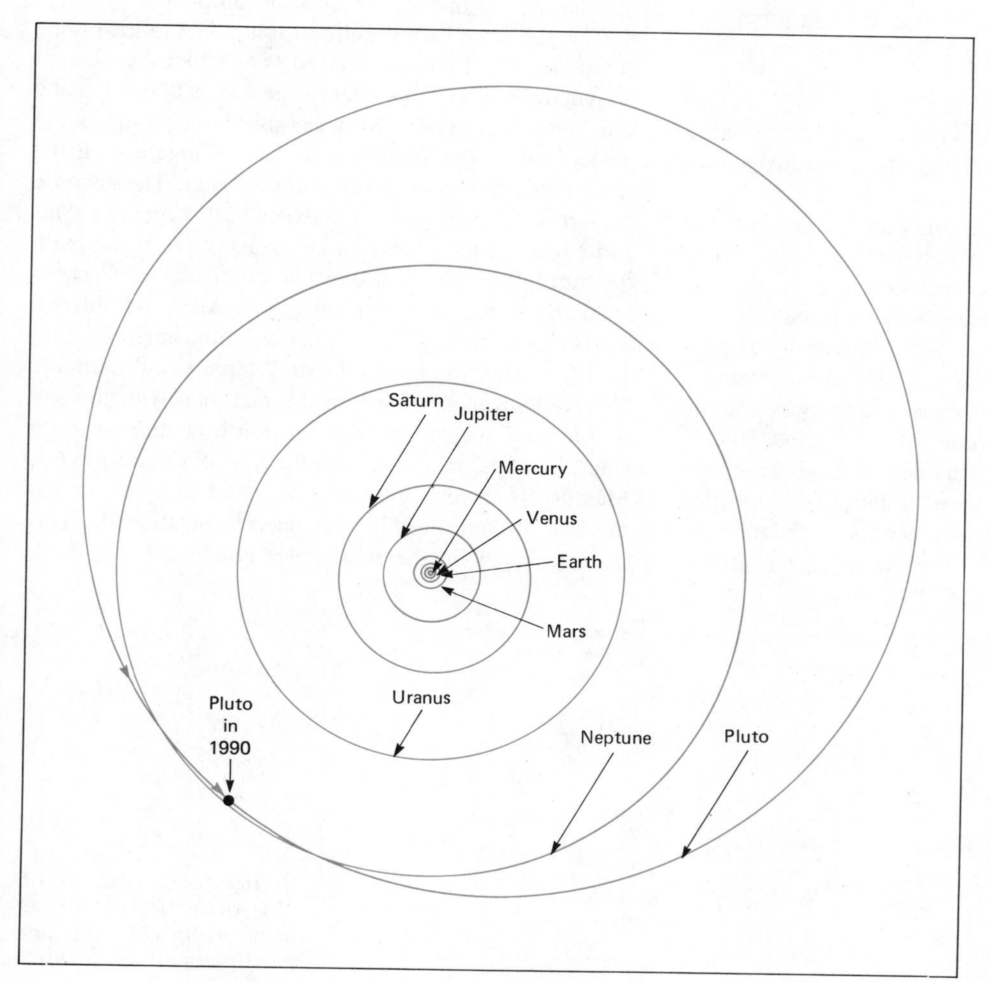

Figure 18.11 The orbits of the planets to scale are shown. The actual position of Pluto in 1990 is also shown. The orbits of the planets are all counterclockwise when viewed from above the North Pole of the Earth.

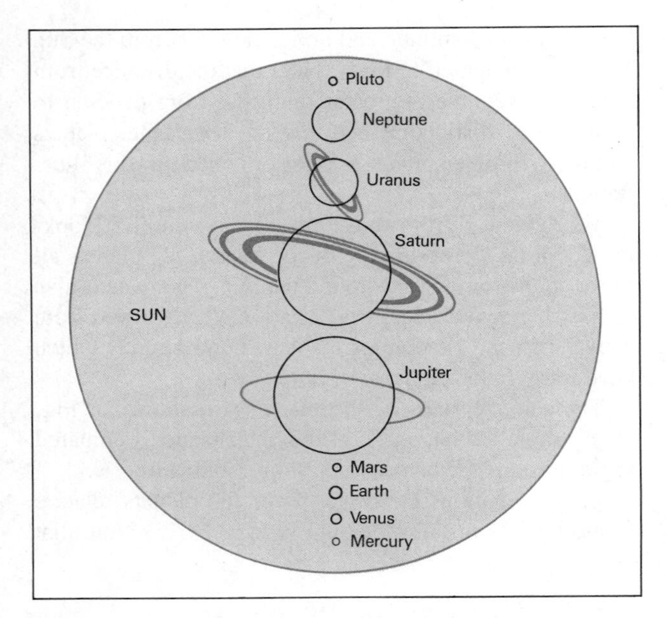

Figure 18.12 Diagram showing the relative size of the nine planets and the Sun.

the solar system is contained within a disk shape rather than a spherical shape. Also, note the large inclination of Pluto's orbit.

A simple way to remember the order of some of the planets is to realize that the first letters of the words *S*aturn, *U*ranus, and *N*eptune spell SUN.

A simple method to remember the approximate distance of the planets from the Sun is given by Bode's law. Johann Bode (1747–1826), a German astronomer and editor of a German astronomical journal, was a strong supporter, but not the discoverer, of the method. Bode's law, in reality, is not a law. That is, it does not represent a physical property of the solar system. The so-called law is a simple method to be followed in calculating the approximate radial distance of some planets from the Sun. See Table 18.1 for a comparison of the results of Bode's law to actual distances. The method was first published by Johann Daniel Titius (1729–1796), a German physicist and mathematician, in 1766. Many astronomers refer to the method more correctly as the Titius-Bode law. The method was very useful in the discovery of the asteroids (starlike objects) located between Mars and Jupiter. Values for the distances in astronomical units from the Sun to the different planets, as obtained by the Titius-Bode law, are given in Table 18.1. One **astronomical unit,** the abbreviation of which is a.u., is equal to the mean distance from the Earth to the Sun, which is 9.3×10^7 miles. The law gives the distance from the Sun to the planets when the figures 0, 3, 6, 12, 24, and so on (doubling the number each time, except for the zero) are each added to four, and the sum divided by ten. Although the law has no physical interpretation, it does provide an easy method for remembering the distance to most of the planets.

The period of time required for a planet to travel one complete orbital path is referred to as either the sidereal or synodic period. The **sidereal period** is defined as the time interval between two successive conjunctions of the planet with a star (planet and star are together on the same meridian) as observed from the Sun. The **synodic period** is the time interval between two successive **conjunctions** (either inferior or superior) of the planet with the Sun (planet and Sun on same meridian) as observed from the Earth. The relationship between the sidereal and synodic periods of a planet is illustrated in Fig. 18.14. At position P_1 the planets Mercury and Earth are observed, from the Sun, on the meridian with the star S. Mercury, observed from the Earth at this same instant, is on the meridian with the Sun. From position P_1 Mercury revolves eastward (counterclockwise) around the Sun through 360 degrees back to position P_1. This is represented by the solid color circle in Fig. 18.14.

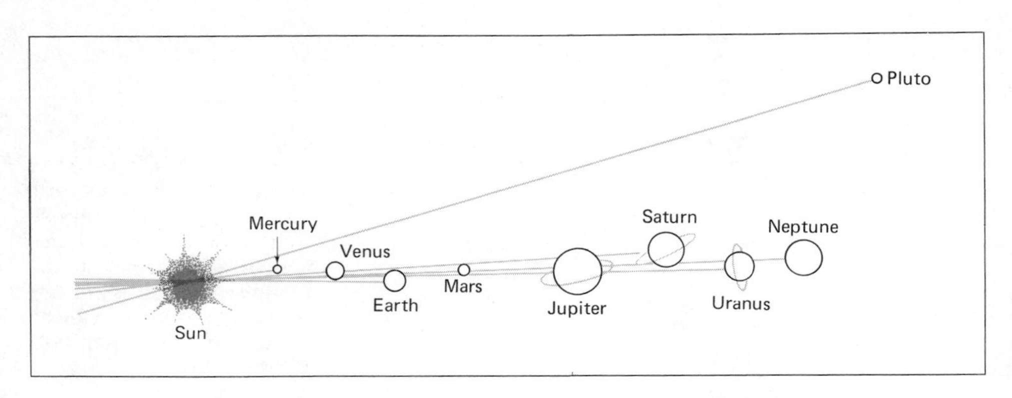

Figure 18.13 Diagram of the inclination of planetary orbits with the orbital plane of the Earth.

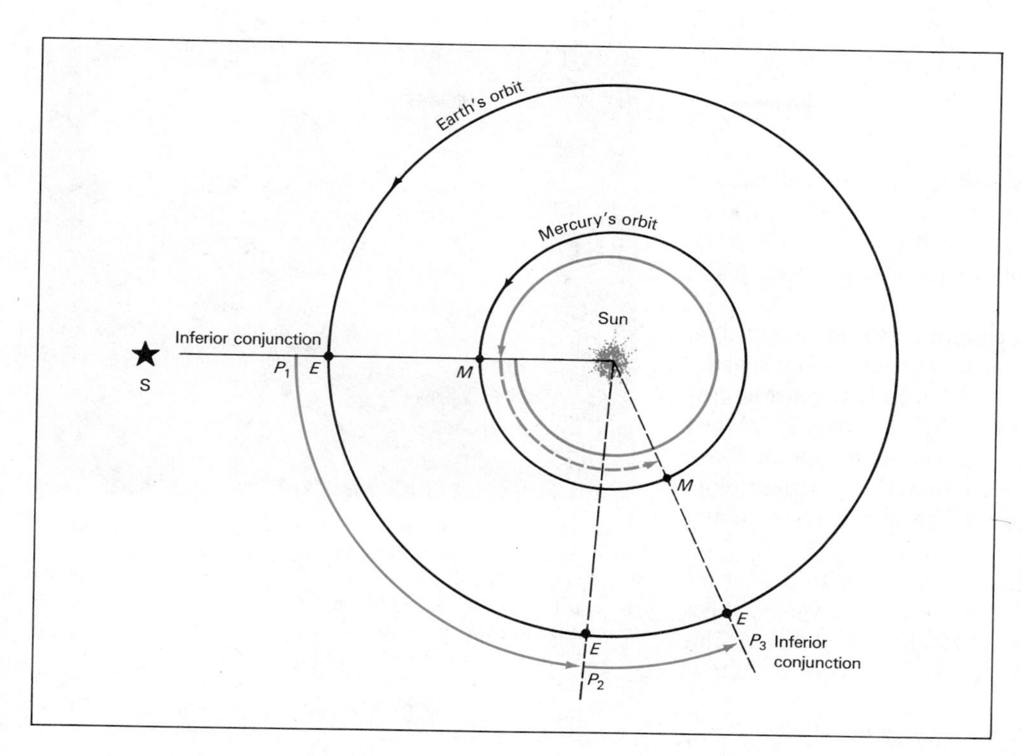

Figure 18.14 Diagram illustrating the sidereal and synodic periods of the inferior planet Mercury. Mercury and the Earth travel the distances shown in color in equal time periods. See text for explanation of diagram.

This is the sidereal period for Mercury—the time (88 Earth days) it takes the planet Mercury to make one orbit around the Sun. During this time (88 days) the Earth revolves approximately 87° eastward to position P_2. Mercury again revolves eastward from position P_1 to position P_3. During this same period the Earth revolves from position P_2 to position P_3. These motions are represented by the broken color lines in Fig. 18.14. At the position P_3 an observer on the Earth now sees Mercury on the meridian with the Sun. The total time for Mercury to revolve from P_1 back to P_1 then to P_3 is the synodic period equal to 116 Earth days. This is the time it takes the planet Mercury to make one orbit around the Sun as observed from the Earth or the time from the conjunction at position P_1 to the next conjunction at position P_3. The true period of revolution is the sidereal period.

Opposition is the term used to describe the position of a planet when the planet is 180° from the Sun; that is, the planet is on the opposite side of the Earth from the Sun. See Fig. 18.15 for an illustration of conjunction and opposition.

In the next two sections we will discuss the characteristics of the various bodies found in the solar system. Asteroids will be considered in the discussion of the inner planets because most of their orbits are inside Jupiter's. Comets will be considered with the outer planets since their highly elliptical orbits take them very far from the Sun.

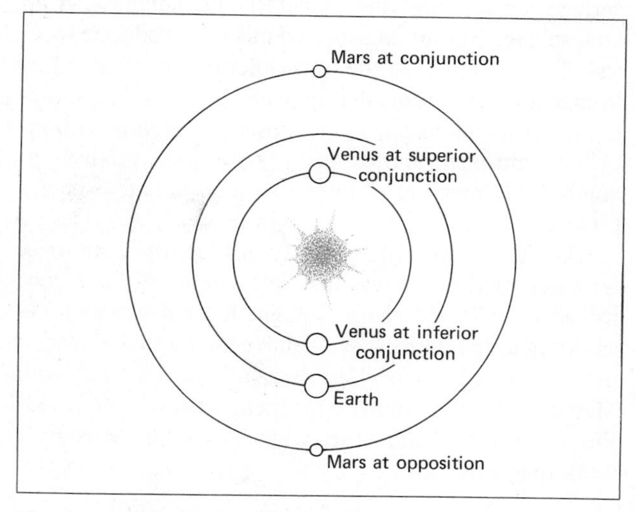

Figure 18.15 The positions of Venus and Mars at conjunction and at opposition. Venus, an inferior planet, is shown at inferior and superior conjunction with Earth. Mars, a superior planet, is shown at conjunction and opposition with Earth.

18.3 The Inner Planets

Mercury

The planet Mercury is closest to the Sun and has the shortest period of revolution. The early Greeks named it after the speedy messenger of the gods, and it is the fastest moving of the planets due to its position closest to the Sun.

Mercury, at its greatest eastern or western elongation, can be seen only just after sunset or just before sunrise. The elongation (the angular distance between Mercury and the Sun as viewed from the Earth) is only 28°. When Mercury is near eastern elongation, it will appear above the western horizon just after sunset. At western elongation Mercury will be on the eastern horizon shortly before sunrise.

The appearance of Mercury is similar to that of the moon, as can be seen from Fig. 18.16. However, it has a very high density, almost as high as the Earth's. This indicates that it probably has an inner core of iron as does the Earth.

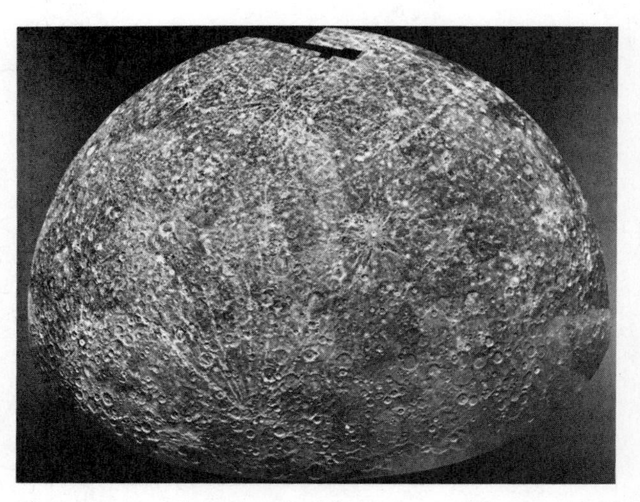

Figure 18.16 Photomosaic of Mercury. This mosaic of over 200 high-resolution Mariner 10 photographs was taken in September 1974. The mosaic shows Mercury as it would appear from 50,000 km (31,000 miles) above the surface looking at its southern hemisphere. Mercury's south pole is mid-center at the bottom of the photograph. Many craters and bright rays from later impacts dominate the surface, giving it a moonlike appearance. (Courtesy NASA.)

Mercury's rotation period is exactly two-thirds as great as its period of revolution. Thus, it rotates exactly three times while circling the Sun twice. This probably results from tidal gravitational effects from the Sun. As it rotates, the side facing the Sun has temperatures of approximately 700 K while the dark side is at about 100 K.

Because of Mercury's small size and high daylight surface temperature, the planet should not possess an atmosphere. But the Mariner 10 mission produced spectral data, operating in the ultraviolet region of the electromagnetic spectrum, that indicate Mercury has an extremely thin atmosphere composed of argon, carbon, helium, nitrogen, and oxygen. These gases produce an atmospheric pressure of only 1/10,000 the atmospheric pressure at the surface of the Earth.

The most perplexing property of Mercury is its weak magnetic field which was first detected in 1974 and confirmed in 1975. Mercury's magnetic field is about 1% as strong as the Earth's. We believe the Earth's magnetic field is caused by the Earth's rapid rotation, but with Mercury's slow rotation no magnetic field was expected. We will have to wait and see what can explain Mercury's weak magnetic field.

Venus

Venus (Fig. 18.17) is our closest planetary neighbor, approaching the Earth at a distance of 26 million miles at inferior conjunction. It is the third brightest object in

Figure 18.17 This view of Venus's clouds was taken from 760,000 km (450,000 miles) by Mariner 10 in 1974. The predominant swirl is at Venus's south pole. (Courtesy NASA.)

the sky, exceeded only by the Sun and our moon. Because of its brightness, it was named in honor of the Roman goddess of beauty.

Information about Venus has been obtained mainly from the spacecraft Venus Pioneer 1, which orbits the planet, and Venus Pioneer 2, which was a multiprobe spacecraft that penetrated Venus's atmosphere. The Venus Pioneer 2 spacecraft is shown in Fig. 18.18.

Figure 18.18 This special effect photograph shows the Pioneer Venus Multipleprobe, the launch vehicle, and the planet Venus. The multipleprobe craft is shown at the lower left with the four blunt-nosed probes attached to the carrying bus. The large probe in the center has a diameter of 4.7 ft and weighs 693 lb. The smaller probes have diameters of 2.5 ft and weigh 200 lb each. The large probe was launched from the carrying bus 24 days before arriving at Venus. Four days later the smaller probes were launched from the bus. All probes operated successfully making many measurements of Venus's atmosphere. (Courtesy NASA.)

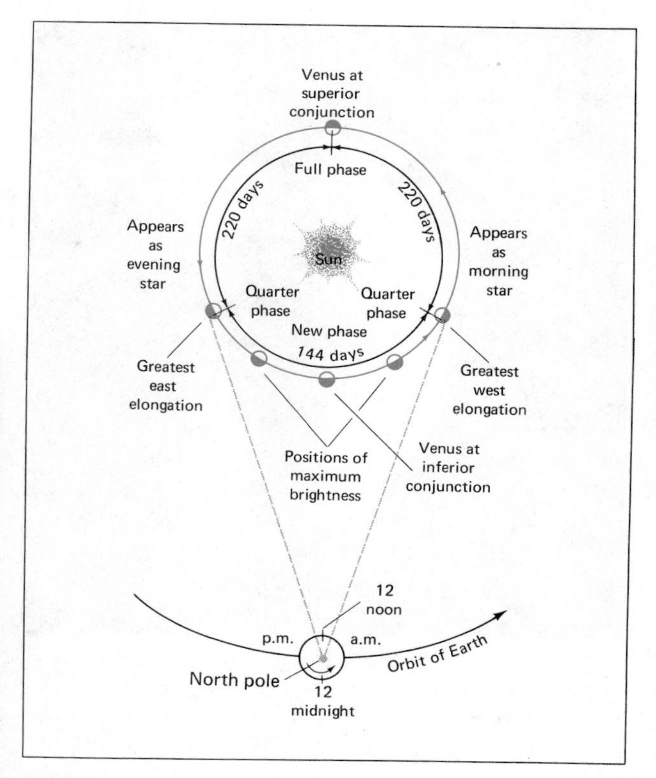

Figure 18.19 Drawing showing the positions Venus must have in order to appear as either our evening or morning star.

The relative position of Venus in respect to the Sun and the Earth and the important positions in respect to the Sun as observed from the Earth are shown in Fig. 18.19. When Venus is at superior conjunction, it is in full phase (full illumination of the side facing the Earth) for an observer located on the Earth, but is not visible to the observer at this time because of the brightness of the Sun. As Venus moves eastward, it appears to the Earth observer as the evening star until it reaches inferior conjunction. The greatest eastern elongation (greatest angular distance from the Sun) occurs 220 days after superior conjunction, but maximum brightness does not occur until Venus has about 39° elongation from the Sun. This occurs about 36 days before and after inferior conjunction.

The synodic period of Venus is 584 Earth days. Viewed from the Earth with a telescope, Venus shows phases, an example of which is shown in Fig. 18.20. An examination of Fig. 18.19 explains the phases and why the planet appears as the familiar morning and evening star. The period of rotation of Venus is 243 Earth days retrograde, that is, the motion is clockwise, or westward, the opposite of its revolving motion about the Sun.

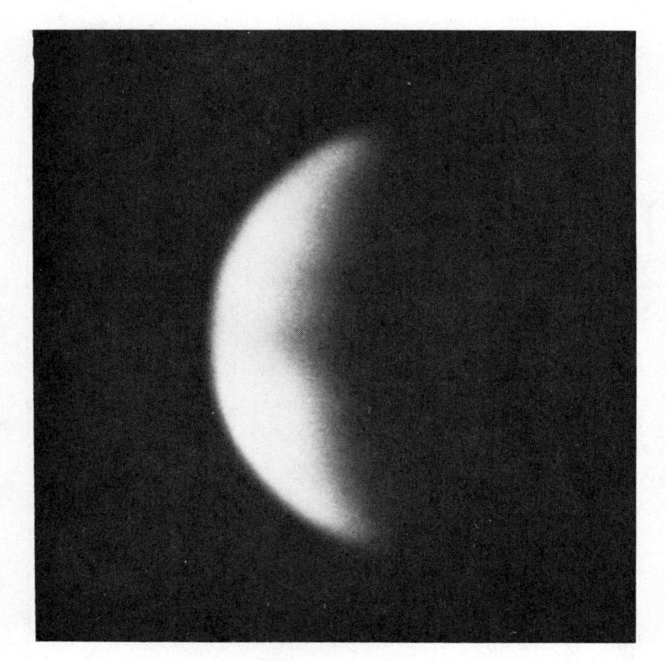

Figure 18.20 When viewed with a telescope, Venus as the morning star shows a phase like that of the last quarter moon. (Courtesy Mount Wilson and Las Campanas Observatories, Carnegie Institution of Washington.)

Venus and Earth resemble one another in many ways. They have similar properties such as average density, mass, size, and surface gravity. But the similarities end there. Venus is covered with a dense atmosphere whose composition is 96% carbon dioxide, some nitrogen (less than 4%), and traces of argon, oxygen, and water vapor. At the surface of Venus the atmospheric pressure is a tremendous 90 atmospheres and the temperature 750 K or about 900°F. The high temperature is due mainly to the large amount of carbon dioxide in the atmosphere that produces the ''greenhouse effect'' (see Section 22.4), so life as we know it cannot exist. Both temperature and pressure decrease with increase in altitude. The temperature at the top of the atmosphere is 220 K.

The surface of Venus can never be seen by an observer on the Earth because of dense thick clouds that cover the planet. The clouds are composed mainly of sulfuric acid (H_2SO_4) droplets along with some water droplets. The droplets do not fall out as rain because of the extremely high atmospheric pressure. The clouds occur in four layers beginning at about 31 kilometers (19 miles) above the surface and extending upward another 37 kilometers (23 miles). The top layer of clouds contains large amounts of yellowish sulfur dust, giving Venus its yellowish or yellow-orange color as viewed from the Earth. See color plate I.

Observation of the top layer of clouds by Venus Pioneer 1 (the Orbiter spacecraft) revealed that Venus's atmosphere makes one rotation every four Earth days in a retrograde direction. This is extremely fast compared to the 243 days for rotation of the solid planet.

Color plate III (lower photo) shows a contour map (Mercator projection) of about 83% of Venus's surface. The upper photo is an artist's conception of Venus's largest highland region. Data for the map and the artist's drawing of the highland region were obtained by the radar altimeter on the Pioneer-Venus Orbiter spacecraft.

The topography of Venus consists of highlands, lowlands, and a huge rolling plain which cover approximately 60% of the planet's surface. The highland regions appear continent-like and rest atop Venus's rolling plain. Ishtar Terra (named after the Babylonian goddess of love), located in the northern hemisphere at about 330° east longitude, is approximately the same size as Australia. The highland regions are shown in yellow and brown contours and the lower elevations in blue and violet.

East of Ishtar Terra is Maxwell Montes (named after James Clerk Maxwell, British theoretical physicist), the highest point on the planet. Maxwell Montes is 35,400 ft above ''sea level.'' Venus has no oceans; the term ''sea level'' refers to the average radius of the planet—3752 miles.

Southwest of Ishtar the Pioneer-Venus orbiter discovered two very large volcanoes in a region called Beta Regio. The two volcanoes, called Rhea Mons and Theia Mons, have altitudes of almost 20,000 ft (3.8 miles) and bases with diameters of 620 miles.

The upper photo in color plate III is an artist's conception of Venus's largest highland region, tentatively named Aphrodite Terra after the Greek equivalent of the Roman Venus. The region, about half the size of Africa, is centered at 5° south latitude and 100° east longitude. The huge land mass is about 6000 miles long and 2000 miles wide. Aphrodite Terra consists of eastern and western mountain ranges, separated by a somewhat lower region. The western mountain range rises 23,000 ft above the surrounding terrain. The eastern mountains rise 10,000 ft above the surrounding terrain. Radar imaging shows these two mountain ranges to be among the roughest and most broken-up areas on the planet.

Mars

From the Earth, Mars has a reddish color and was named after the bloody Roman god of war. Mars is about one and one-half times as far from the Sun as the Earth. It is tilted on its axis at an angle of 24° which is very close to the Earth's 23.5° angle of tilt. It rotates once every 24 hours and 37.4 minutes, which is very close to a single Earth day. It takes 687 days or about 23 Earth months to go around the Sun.

Mars has two small satellites, or moons, named Phobos and Deimos, which are about 20 and 10 miles across, respectively. Both are irregularly shaped and extensively cratered. Like our moon, they keep one side always facing the planet.

The mass of Mars is about one-tenth as large as that of the Earth. Its density is also much less (3.9 g/cm^3) than that of Mercury, Venus, and the Earth (5.5 g/cm^3). This indicates that Mars probably does not have a large iron core in the center like the Earth.

Very little is known about the internal structure of the planet. Mars does have a very weak magnetic field, less than 0.004 times as strong as the Earth's.

Many of the surface features of Mars can be seen in the topographic map of Fig. 18.21. From this map, one can observe that Mars has many craters similar to those of the moon. The largest of the so-called basins is located in the southern hemisphere from 50 to 90° east longitude. It is about 1000 miles across. Its smoothness is attributed to dust that has settled out from dust storms. The large volcanic mountain in the upper left of the figure is called *Olympus Mons,* "Mount Olympus." See color plate II. The mountain is very large, towering 15 miles above the plain and having a base with a diameter of 375 miles. The volcano is crowned with a 40-mile-wide crater. Three smaller volcanic mountains can be seen to the southeast in Figure 18.21. The large canyon farther southeast is called *Valles Marineris*. The canyon is approximately 3000 miles long and 20,000 feet deep. Its length is equivalent to the width of the United States. The depth of the canyon is almost four times as deep as the Grand Canyon. This tremendous gash in Mars's surface is thought to be a gigantic fracture formed by stress within the planet.

The topographic map shown in Fig. 18.21 does not cover the polar region. Figure 18.22 shows the northern polar cap of Mars. In the winter the polar caps are composed of dry ice (frozen CO_2) and water ice. In the summer, the dry ice changes to vapor, leaving behind a residual polar cap of water ice.

The atmosphere of Mars is much more tenuous than that of the Earth. The atmospheric pressure is slightly less than 1% of that on the Earth. The atmosphere is composed of 95% carbon dioxide, 2.7% nitrogen, 1.6% argon and traces of oxygen, carbon monoxide, krypton, and xenon. Because of the low atmospheric pressure, water cannot exist as a liquid on the Martian surface. Traces of water vapor have been measured in the atmosphere.

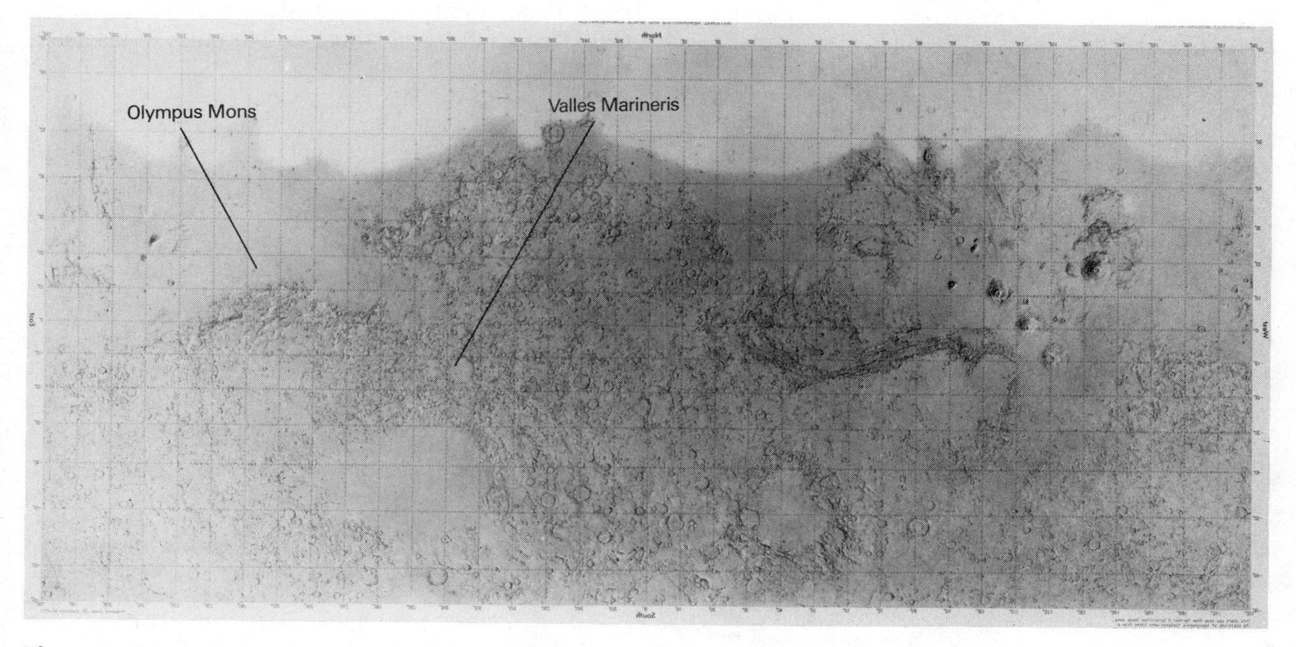

Figure 18.21 A topographic chart of Mars. Note the many craters, basins, mountains, and canyons. (Courtesy NASA.)

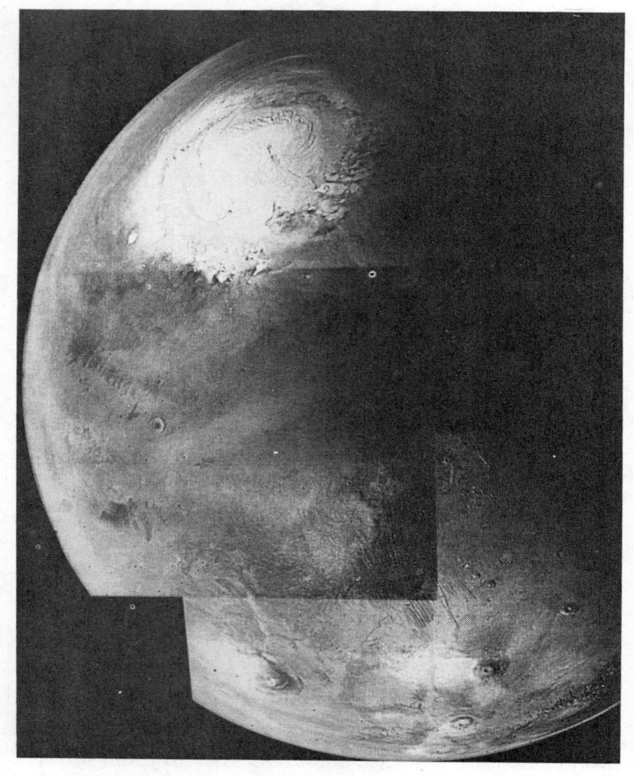

Figure 18.22 The northern hemisphere of Mars from the polar cap to a few degrees south of the equator is seen in this mosaic of three photos. The north polar ice cap (upper left) is shrinking during the late Martian spring. At the bottom, the huge Martian volcanoes are visible. The west end of the great equatorial canyon can be seen at the right. (Courtesy NASA.)

Although Mars is exceedingly dry now, there is evidence of water in the past. This is shown in Figs. 18.23 and 18.24. Figure 18.23 shows a channel and Fig. 18.24 shows teardrop features. Both of these effects are believed to indicate that water once flowed on the surface of Mars.

Our best photographs of the surface of Mars have come from the Viking landers (see color plate II). Figure 18.25 shows the first close-up photograph ever taken of the surface of Mars. These Viking missions have carried out numerous experiments searching for evidence of life on Mars. So far, evidence of life has not been found, but the possibility of life has not been ruled out.

Asteroids

The Titius-Bode law calls for the next planet beyond Mars to be 2.8 astronomical units from the Sun, but no large planetary body is found at this distance. On January 1, 1801, the first of the many planetary bodies was discovered by Giuseppi Piazzi, an Italian astronomer. This small body is named Ceres after the protecting goddess of Sicily. Ceres is slightly less than 600 miles in diameter, has an orbital period of 4.6 years, and is the largest of more than 2000 objects, which have been named and numbered, orbiting the Sun between Mars and Jupiter. These objects are called asteroids or minor planets. Only Ceres and Vesta, which is 240 miles in diameter, can be seen with the naked eye.

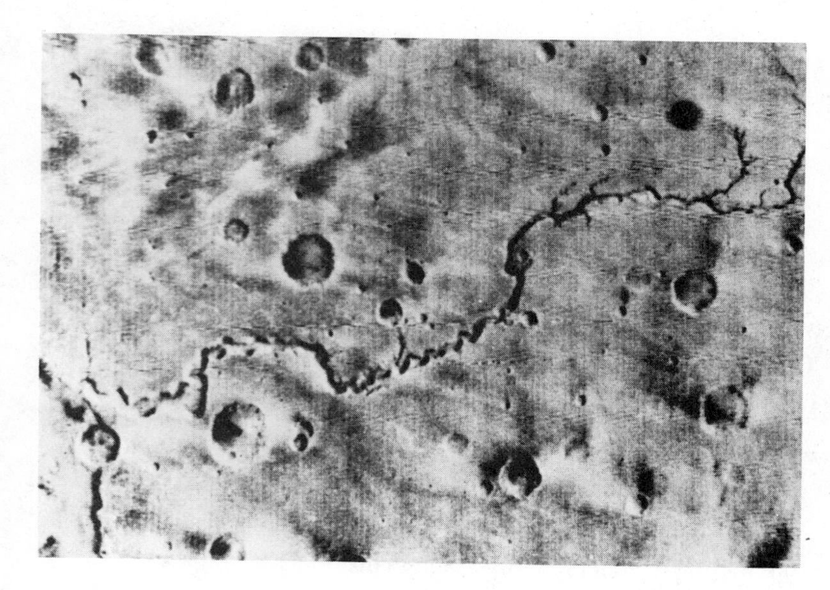

Figure 18.23 Mars channel. Though not unique on the Martian surface, this meandering "river" is the most convincing piece of evidence that a fluid once flowed over the surface of Mars, draining an extended area, and eroding a deep channel. The feature is some 575 kilometers (355 miles) long and 5–6 kilometers (3–$3\frac{1}{2}$ miles) wide. (Courtesy NASA.)

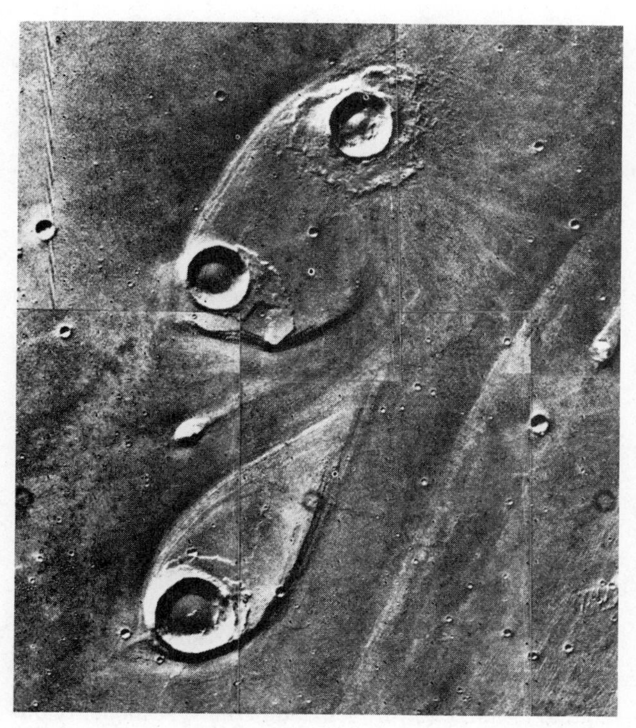

Figure 18.24 Tear-drop shaped features indicate that water flowed on the surface of Mars in the past. (Courtesy NASA.)

Asteroids are classified, according to how much light of various colors they reflect, into six major classes—C, E, M, R, S, and U. Class C refers to carbonaceous, the most common class. They are very dark with albedos (see Section 22.4) of 3 to 4%. E, S, and R refer to different "stony" compositions, M refers to "metallic," and U stands for unknown.

The diameters of the known asteroids range from that of Ceres down to only a mile. There are a dozen or more in the 100-to-200-mile range, perhaps 100 in the 50-to-100-mile range, less than 500 in the 25-to-50-mile range, and perhaps billions the size of boulders, marbles, and grains of sand. Not all asteroids are spherical in shape. Eros, which can approach the Earth to within 14 million miles, is roughly rod-shaped, being 14 miles long and 5 miles thick. Many others are odd and irregular in shape. They all revolve counterclockwise around the Sun, as viewed from above the orbits, or eastward, with an average inclination (degree of incline) to the ecliptic plane of 10°. It is estimated that 100,000 asteroids exist that can be detected with Earth-based telescopes. The total mass of all the asteroids orbiting between Mars and Jupiter is much less than the mass of the Earth's moon. Although most asteroids move in an orbit between Mars and Jupiter, some have orbits that range beyond Saturn or inside the orbit of Mercury.

The asteroids are presently believed to be early solar system material that never collected into a single planet. One piece of evidence supporting this view is that there seem to be several different kinds of asteroids. Asteroids at the inner edge of the belt seem to be stony in nature, while the ones farther out are darker, indicating more carbon content. A third group may be mostly composed of iron and nickel.

In 1977, the first asteroid was found that has an elliptical orbit lying mostly between Saturn and Uranus.

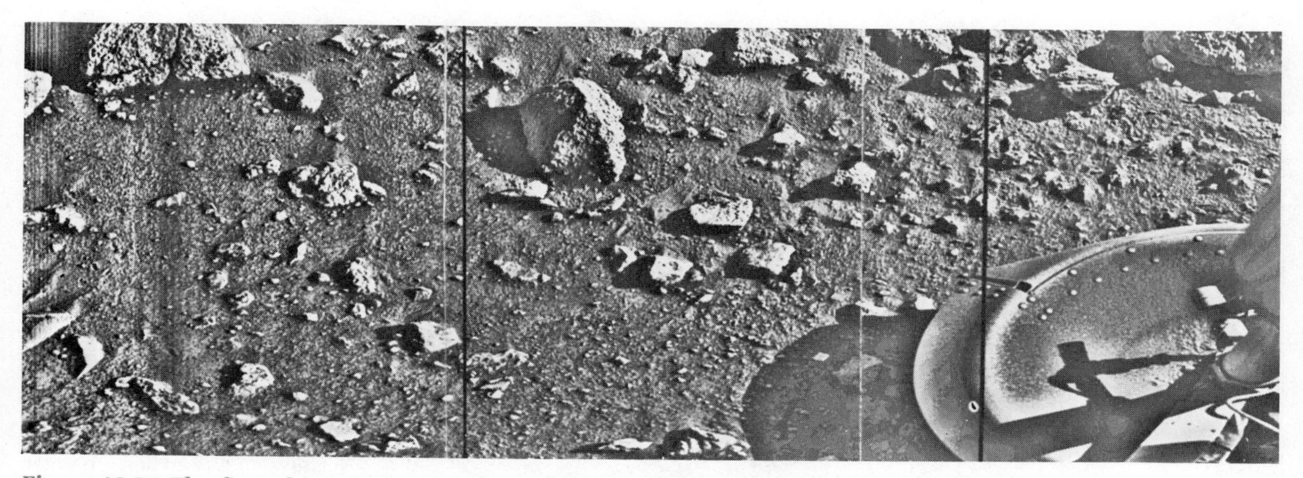

Figure 18.25 The first photograph ever taken on the surface of Mars was taken by the Viking I lander on July 20, 1976. The large rock in the center is about 4 in across. One footpad of the lander is visible on the right. (Courtesy NASA.)

This is the most remote known asteroid. It has been numbered and named Chiron (kí rŏń). Chiron is small, possessing a diameter less than 200 miles. The asteroid is in a very elliptical orbit with a period of 50 years. Because of its very elliptical orbit, Chiron may eventually collide with a planet or be ejected from the solar system.

18.4 The Outer Planets

The four large outer planets differ in two main respects from the inner planets. Their sizes are much bigger and their densities are much lower (see Table 18.1). Both of these characteristics are now generally understood as being due to their greater distances from the Sun and the corresponding lower temperatures in their atmospheres.

When the planets first began to coalesce around five billion years ago, the most predominant elements were the two least massive—hydrogen and helium. The heat from the Sun allowed these two elements to escape from the inner planets. That is, the velocities of the molecules of these elements were sufficient to allow them to escape the planets' gravitational pulls. Thus the inner planets were left with mostly rocky cores, giving them a high density. The four large outer planets were much colder, and they retained their hydrogen and helium, which now surround their rocky cores. Thus, the four large outer planets consist primarily of hydrogen and helium in various forms, and this composition gives them much lower densities. See color plate IV.

Because of their larger mass and greater gravitational pull, the large outer planets also have many more moons than the smaller inner planets do. In fact, we now know that Jupiter, Saturn, and Uranus have large satellite systems with at least 16, 17, and 5 moons, respectively. The four small inner planets have a total of only three moons—two for Mars and one for the Earth.

Jupiter

Jupiter, named after the god of the skies because of its brightness and giant size, is the largest planet of the solar system, both in volume and mass. The motion about its axis is faster than that of any other planet, since it takes only ten hours to make one rotation. Jupiter possesses more than half of the total angular momentum of the solar system.

Jupiter's diameter is 11 times as large as that of the Earth, and it has 318 times as much mass. However, its density is only 1.3 g/cm^3. Jupiter consists of a rocky core, a layer of hydrogen in metallic form (since it is at high pressure and temperature), and an outer layer of a liquid mixture of hydrogen and helium. A model of Jupiter's interior is shown in Fig. 18.26. Jupiter's atmosphere is composed of about 82% hydrogen and about 17% helium with a remaining 1% or so of methane, ammonia, water, and several other molecules.

Jupiter's temperature ranges from a very cold −200°F in the upper atmosphere to above the boiling point farther down in the atmosphere. Jupiter's atmosphere gets more and more dense until finally a liquid state is reached as shown in Fig. 18.26.

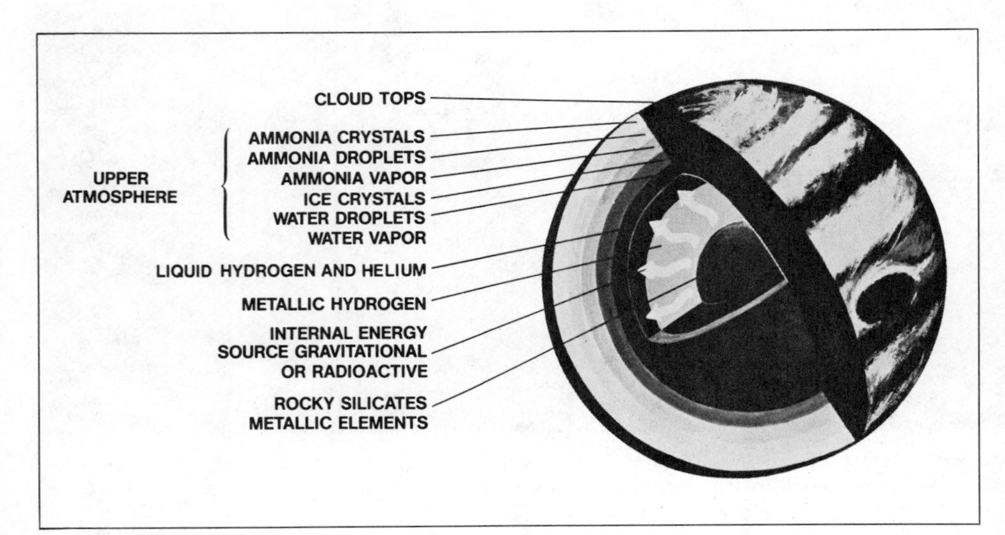

Figure 18.26 A model of Jupiter's composition and atmosphere. Jupiter has its own internal energy source, and it is so massive that it is almost a star. Its atmosphere is mostly hydrogen and helium that get more and more dense until becoming liquid and finally solid. (Courtesy NASA.)

Figure 18.27 A montage of Jupiter and its four largest moons, made by Voyager I in 1979. Size of the moons is not to scale. Io is seen in the left background, bright Europa is in the center, Ganymede is in the lower left, and Callisto is in the lower right. (Courtesy NASA.)

Jupiter has the interesting property of actually giving off twice as much heat as it receives from the Sun. This makes it, in a sense, a star as well as a planet. If Jupiter had about 75 times more mass, nuclear reactions similar to those in the center of the Sun could have started in its interior. Jupiter might be thought of as the smallest star ever observed, but it is usually considered our solar system's largest planet.

In Fig. 18.27, the cloud features are easily seen. The clouds of Jupiter show a band structure and the great red spot stands out. The red spot has an erratic movement and changes color and shape. It sometimes completely disappears. The most recent theory of the great red spot states that it is a huge cyclonic storm similar to a hurricane on Earth but much longer lasting.

Jupiter possesses a tremendous magnetic field. At the top of the atmosphere it is 10 times as strong as the Earth's field. Jupiter's magnetic poles are reversed from those on the Earth, and Jupiter's magnetic poles are 10° from its geographic poles which are defined by its rotation axis. Its rotation axis is inclined by only a few degrees, so Jupiter does not experience seasonal effects like the Earth and Mars.

Jupiter has many moons, 16 or more depending on where the line is drawn between a large rock and a small moon. In addition, Jupiter has a very faint ring around it. See Fig. 18.28.

The four largest moons of Jupiter were first discovered by Galileo in 1610. They are sometimes called the Galilean moons of Jupiter. In order of their distances from Jupiter, they are Io, Europa, Ganymede, and Callisto. See color plate I. They were photographed close up by the Voyager 1 and Voyager 2 spacecrafts in 1979. They all have diameters between about 2000 and 3300 miles; Mercury's diameter is roughly 3000 miles.

One of the most spectacular findings of the Voyager missions was that Io has many active volcanoes on it. Two of these can be seen in color plate II. The volcanoes are caused by the fact that the gravitational attraction of other nearby moons, notably Europa, causes Io's orbit to vary so that it is closer, then farther, from Jupiter. The resulting changes in Jupiter's gravitational force cause

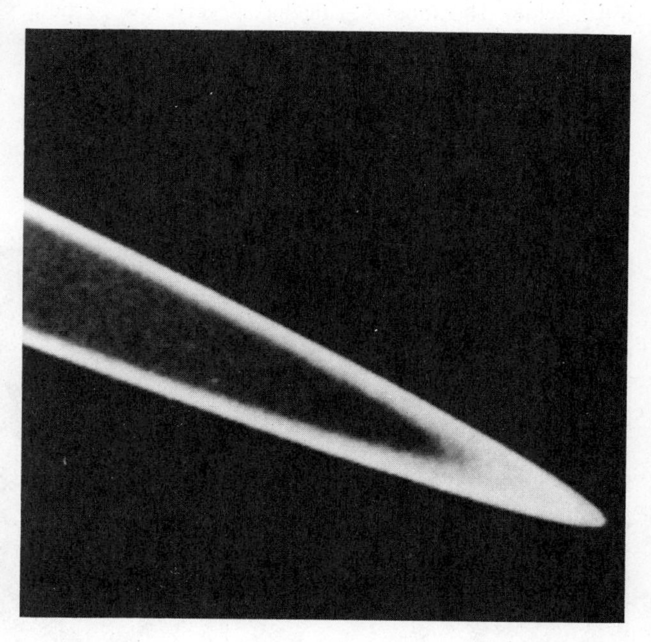

Figure 18.28 Jupiter's ring. Part of Jupiter's faint ring is shown in this photograph which was taken from Voyager 2. The ring has a distinct outer edge, but the inner edge appears unclear. Seen within the inner edge of the brighter ring is a fainter ring that may extend all the way down to Jupiter's cloud tops. (Courtesy NASA.)

stresses in the interior rock of Io, and a great deal of frictional heat is generated, resulting in volcanoes.

Europa is very bright, and this is due to the fact that its surface is covered with a mantle of ice approximately 150 miles thick. Ganymede and Callisto, the two largest satellites, have cratered surfaces similar to that of the Earth's moon.

Saturn

The most distinctive feature of Saturn is its system of three prominent rings. The three ''classic'' rings can be seen in Fig. 18.29. These rings have been viewed by Earth-based observers for many years and are the most spectacular celestial sight that can be seen with a small telescope. The rings, inclined by 27° to Saturn's orbital plane, are identified by the letters A, B, and C. The outer ring, shown in Fig. 18.29, is the A-ring and it is separated from the bright B-ring by the broad dark region known as the Cassini division. The B-ring, which contains the largest number of particles per unit volume, is the brightest ring. The Cassini division or gap is named in honor of G. D. Cassini, Italian astronomer, who discovered the dark region in 1675. The gap is about 3000 miles wide. The inner C-ring has a very low particle density and appears with less intensity than do rings A and B, because only a small amount of light is reflected from the smaller number of particles. The small dark region near the outer edge of the A-ring is known as the Encke division or as the Keeler gap by some astronomers. Although the dark regions appear to be without particles, the divisions or gaps do contain a few tiny particles which were discovered by the Voyager 1 spacecraft. The rings, which are less than 50 m (164 ft) in thickness, are believed to be composed of icy particles ranging in size from a few microns to approximately 10 m (33 ft) in diameter. When the distances of the rings are measured in respect to Saturn's center, the outer edge of the C-ring is at a radial distance of 57,000 miles. Ring B's outer edge is 73,000 miles from the center of Saturn and the radial distance of A-ring's outer edge is 85,000 miles. Four additional rings have been discovered by Pioneer 11 and the Voyager fly-bys. The D-ring is located inside the C-ring and extends from the inner edge of the C-ring down to the top of Saturn's clouds. The F-, G-, and E-rings are located beyond the A-ring. The F-ring, which is very thin and about 60 miles wide, is located next to the A-ring. The G-ring, which is extremely faint, is beyond the F-ring and has a maximum radius of 93,000 miles. The E-ring is beyond the G-ring. It is also extremely faint and extends outward to more than 150,000 miles.

The Voyager 1 and Voyager 2 flights showed the structure of the rings to be very complicated systems of many individual rings (see color plate II). This highly enhanced color view was assembled from clear, orange, and ultraviolet frames. The photograph, taken from Voyager 2 at a distance of 5.5 million miles, shows the possible variations in the chemical composition from one part of Saturn's ring system to another. Special computer processing techniques exaggerate subtle color variations in the photograph.

Figure 18.29 Saturn and its rings. This photograph of Saturn was taken by Voyager 2 when the spacecraft was 21 million miles from the planet. The broad dark region between the outer A-ring and the bright B-ring is called the Cassini division and the narrow dark region near the outer edge of the A-ring is called the Encke division. Several dark spokelike features can be seen in the broad B-ring left of the planet. The satellites Rhea and Diona appear as white dots to the south and southeast of Saturn respectively. (Courtesy NASA.)

The structure of Saturn itself is somewhat similar to that of Jupiter. That is, it has a small solid core surrounded by a layer of metallic hydrogen and an outer layer of liquid hydrogen and helium. Saturn's density is only 0.7 g/cm^3, so it would float in water. The temperature of Saturn is very cold, approximately 90 K, or −300°F in the upper atmosphere. This is about 100°F colder than Jupiter. Like Jupiter, Saturn radiates more heat than it gets from the Sun.

Saturn's mass is 95 times that of the Earth, and its diameter is 9 times larger than Earth's. It rotates about once every 11 hours. It has a magnetic field that is 1000 times stronger than that of Earth, but only 0.05 times as strong as Jupiter's.

Outside the main visible rings lie the 17 or more moons of Saturn. Many of these are quite small, with diameters between about 30 and 100 kilometers (roughly 20 to 60 miles). Most of these eight small moons were discovered in 1980 by Voyager 1. The moons Mimas, Enceladus, Tethys, Dione, Rhea, Hyperion, and Iapetus are similar in two respects. They all have diameters between 385 and 1530 kilometers (240 to 950 miles), and they are all believed to be composed of rock and ice since their densities are between 1 and 2 g/cm^3.

Except for the most distant moon, Phoebe, and possibly Hyperion, all the moons of Saturn rotate once each revolution, similar to our own moon. Thus, they always keep the same face toward Saturn. Except for the two most distant moons, Phoebe and Iapetus, all the moons of Saturn go in nearly circular orbits, revolve in the same direction, and lie along the planet's equatorial plane. Phoebe revolves in the opposite direction to that of the other 16 moons, and its orbit is inclined by 150°. Thus, Phoebe is probably a moon that was somehow captured by the Saturnian system in the distant past.

The most interesting moon of Saturn is Titan. Titan is the largest moon of Saturn, with a diameter of 5120 kilometers (3180 miles) and a density of 1.9 g/cm^3. It is the only satellite known to have a dense, hazy atmosphere. The main constituent of the atmosphere is nitrogen. Titan's surface cannot be seen, but the surface temperature is about 100 K, or −280°F. See color plate I.

Uranus

The planet Uranus was discovered in 1781 by William Herschel (1738–1832), an English amateur astronomer. When viewed through a telescope, the planet has a blue-green appearance due to the presence of methane gas in the planet's hydrogen-helium atmosphere. The methane gas absorbs red light, therefore the light reflected is largely blue-green. The planet's name, chosen in keeping with the tradition of naming planets for the gods of mythology, was first suggested by Johann Bode. Uranus was the father of the Titans and the grandfather of Jupiter.

The internal structure of Uranus and Neptune are similar but differ from those of Jupiter and Saturn. Uranus and Neptune are much smaller and less massive than Jupiter and Saturn. See Table 18.1 for a comparison of physical properties. Also their rocky cores are relatively larger compared to their total size. From a study of the physical properties, the internal structure of Uranus is calculated to be in three layers. The inner rocky core, which is about 13,000 km (8100 miles) in diameter, contains about 25% of the planet's mass and is probably composed of iron and silicon. The rocky core is surrounded by a liquid mantle approximately 8,000 km (5000 miles) deep composed of water, ammonia, and methane ice. The mantle makes up 65% of the planet's mass. The outer layer, the atmosphere, of the planet is about 11,000 km (6800 miles) thick, and is composed of mostly hydrogen plus 12% helium by volume.

Uranus has a ring system that is very thin. Before the Voyager 2 mission, Uranus was known to have nine rings, which were discovered in 1977. All nine rings are shown in false-color in color plate V. The outermost and also the brightest ring, named Epsilon, is about 43 km wide. This ring is shown at the top in color plate V and appears in a neutral color. Down from Epsilon toward the planet, the color plate shows the delta, gamma, and eta rings in shades of blue and green; the beta and alpha rings are shown in lighter tones. A final set of three rings known as 4, 5, and 6 are shown in faint off-white tones. The pastel lines seen between the rings are contributed by the process of computer enhancement. A ring newly discovered by Voyager 2 is called 1985UR1. This ring, which is very thin and hardly visible to Voyager 2 cameras, is about half-way between Epsilon and Delta. 1985UR1 is not visible in color plate V.

The rings of Uranus are composed mainly of boulder-size particles one meter in diameter and larger with very few dust size particles present. Due to the lack of dust particles the rings do not have good reflective properties like the rings of Saturn, which are filled with tiny particles one centimeter and smaller in size. Voyager 2 also recorded some very narrow sections of rings. It is believed that perhaps there are hundreds of partial rings located around Uranus.

Uranus has several other interesting features besides its rings. Its rotation axis is inclined 82° with respect to its orbital motion, which positions the Sun nearly overhead at the north and south poles as Uranus revolves

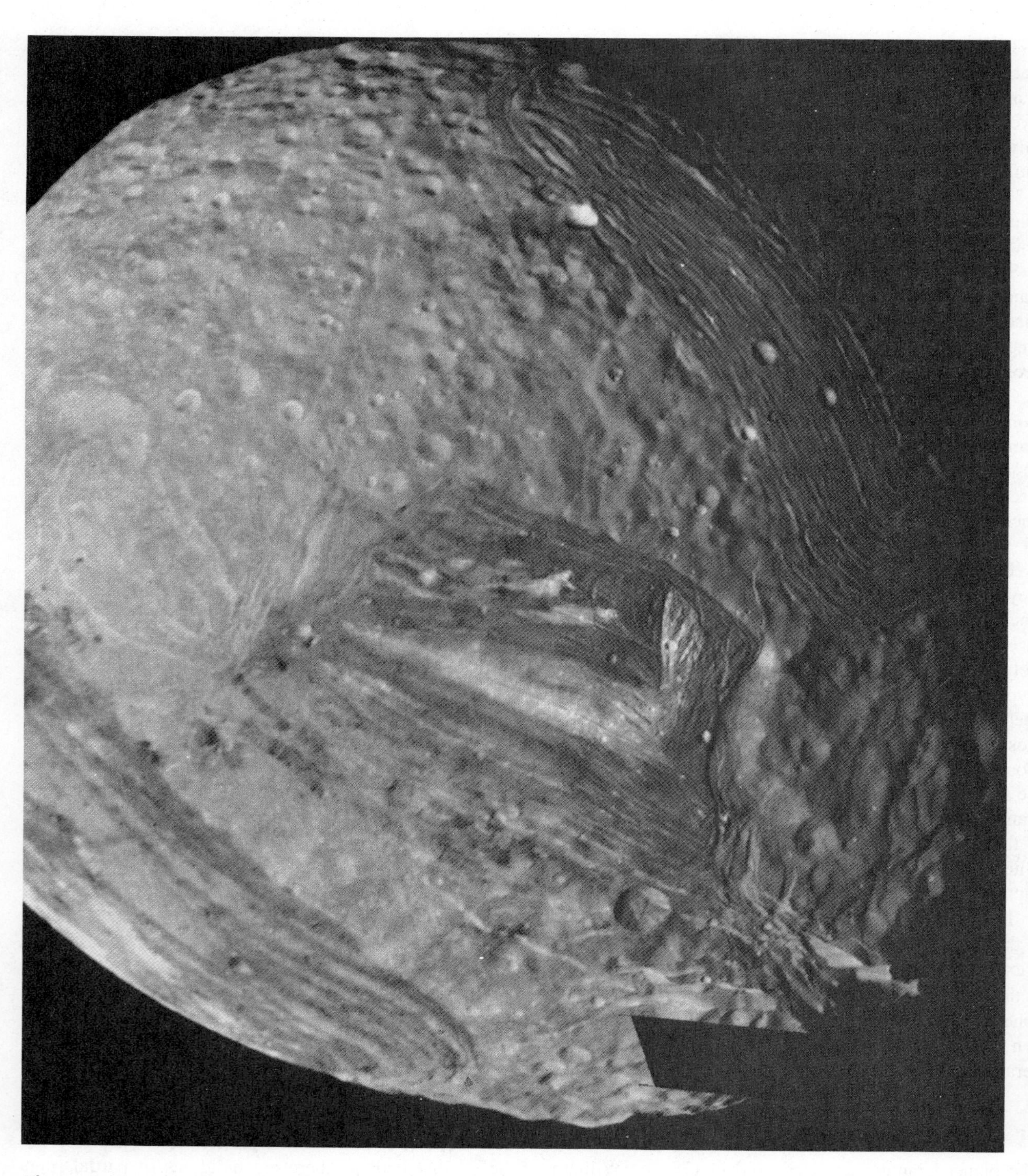

Figure 18.30 Uranus' innermost large moon, Miranda, is roughly 300 miles in diameter and exhibits a variety of geologic forms—the most bizarre forms in the solar system. Chevron shaped regions and folded ridges in circular racetrack patterns are visible on the satellite's surface. There are large scarps, or cliffs, ranging up to three miles in height. There is computer data missing for the black rectangular area in the lower right edge of the picture below the high cliffs. (Courtesy NASA.)

around the Sun. The planet rotates retrograde with a period of approximately 17 hours. Uranus has a magnetic field that is about 50 times as strong as the Earth's magnetic field. An unusual feature of the planet's magnetic field is its orientation with the rotational axis of the planet. The magnetic field is tipped 55° to the rotational axis. The clouds of Uranus are deep within its atmosphere and nearly invisible. Thus, the planet appears as a bland, almost featureless blue-green disk. The planet's atmospheric temperature, which is fairly uniform over the whole planet, is about −210°C. For some unknown reason, the planet radiates more energy than it receives from the sun.

The cameras of Voyager 2 have recorded ten new satellites circling Uranus inside the orbit of Miranda. The first of these was discovered in December 1985 by Voyager 2 and named 1985U1. The satellite has a diameter of about 165 km. The other nine satellites, discovered in 1986, are named 1986U1 through 1986U9. These nine satellites are very small, with diameters less than 100 km.

Uranus has five major satellites. They are, in order of distance from the planet: Miranda, the smallest and closest having a diameter of 298 miles and orbiting at a distance of 80,000 miles; Ariel; Umbriel; Titania, the largest with a diameter of 986 miles; and Oberon, the most distant, orbiting at 362,000 miles. The satellites have densities of about 1.6 grams per cubic centimeter. This gives them a composition of an approximate mix of ice (water) and rock. The surface features of the satellites show that, with the exception of Umbriel, the moons have been tectonically active in the past. Fig. 18.30 shows Miranda's surface with large curvilinear regions of grooves and ridges plus regions that appear chevron shaped. The satellite's surface is pock-marked with craters. The large crater shown in the lower right region of Fig. 18.30 is about 15 miles in diameter. Photos obtained from Voyager 2 show the surface with very deep valleys and high cliffs ranging in height from 0.3 to 3 miles. The geologic forms on Miranda are the most bizarre forms in the solar system. Fig. 18.31 shows the surface features of Ariel (left photo) and Titania (right).

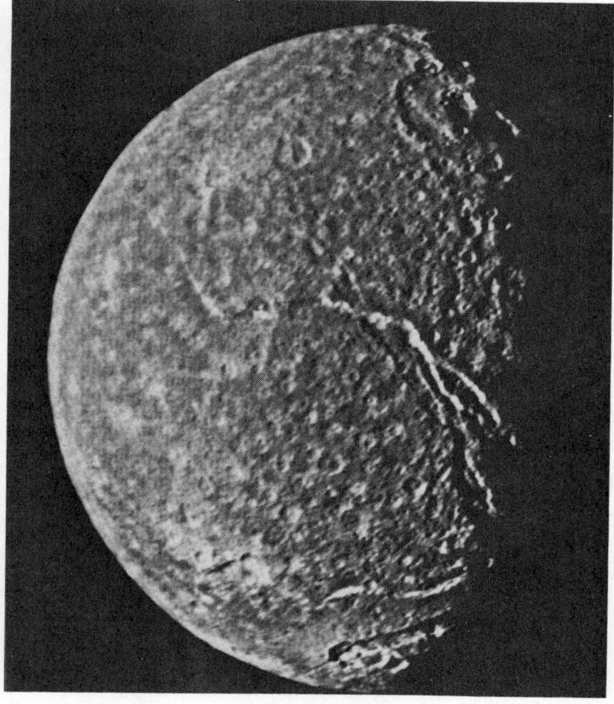

Figure 18.31 The image on the left shows Ariel, one of Uranus' large moons, densely pitted with craters three to six miles across. Numerous valleys and fault scarps crisscross the highly pitted surface. Titania (right) is Uranus' largest satellite, with a diameter of a little more than 1000 miles. Abundant impact craters of many sizes pockmark the ancient surface. (Courtesy NASA.)

The surface of both moons are pitted with craters having numerous valleys and fault scarps cutting across the highly pitted terrain. Uranus' most distant moon, Oberon, is shown in color plate V. The picture shows several large impact craters in the planet's icy surface. The large central crater has a high central peak. On the lower left edge of the photo a huge mountain about four miles high can be seen.

Neptune

Neptune was discovered in 1846 by John G. Galle (1812–1910), a German astronomer at the Berlin Observatory. Partial credit is also shared by Englishman John Couch Adams and Frenchman U. J. J. Leverrier, two mathematicians. Using Newton's law of gravitation, Adams and Leverrier made calculations that produced information on where to look for a supposed planet that was disturbing the motion of the planet Uranus. The name Neptune was proposed by D. F. Arago, a French physicist who had suggested that Leverrier begin the critical investigation of the planet. He first suggested Leverrier as the name, but later withdrew this suggestion because it received little acceptance outside France.

The planet cannot be observed with the naked eye and appears to have a greenish hue when viewed through a telescope. Neptune is shown with one of its satellites in Fig. 18.32. The physical makeup of the planet is similar to that of Uranus. Methane and hydrogen have been detected spectrographically, so the planet probably has a gaseous atmosphere.

Figure 18.32 Neptune. One of the planet's two known satellites can be seen also. (Lick Observatory Photographs.)

Figure 18.33 The Voyager mission. The Voyager 1 and 2 spacecraft have given us the remarkable views of Jupiter and Saturn seen in this section. (Courtesy NASA.)

Neptune has two known satellites. Triton, which is larger than our moon, revolves retrograde once every six days. The revolving speed is decreasing and within a few million years Triton will be torn apart by Neptune's tidal forces, and the planet will develop a ring system similar to the other Jovian planets. A faint single ring fragment around Neptune was detected by a group of American and European astronomers during an occultation of the planet on July 22, 1984. The background star dimmed briefly as the faint ring passed between the star and the receiving telescope recording infrared wavelengths. The ring fragment is calculated to be 10 to 15 kilometers wide, about 100 kilometers long, and located some 70,000 to 80,000 kilometers from the planet.

Voyager 2 (see Fig. 18.33 and color plate IV) will encounter Neptune in August 1989 and more precise data will be obtained concerning Neptune, its faint ring and its satellites, assuming the spacecraft functions properly during the encounter.

Neptune's other satellite, Nereid, is very small. Its diameter is estimated to be less than 400 miles. Nereid has the most eccentric orbit (0.749) of any satellite in the solar system. The highly elliptical orbit takes the tiny moon from 870,000 miles to over six million miles from Neptune.

Pluto

The planet Pluto, named for the god of outer darkness, is the most distant planet from the Sun. It was discovered by C. W. Tombaugh in 1930 at the Lowell Observatory in Arizona, after a thorough search near the position predicted by theoretical calculations. The planet had been predicted because discrepancies appeared in the orbital motions of Uranus and Neptune. General information concerning Pluto is given in Table 18.1. Because of Pluto's small size and great distance from the Sun very little is known about the surface features of the planet (see Fig. 18.34). Spectroscopic investigations of Pluto indicate the planet is covered with methane ice.

In June 1978 a satellite of Pluto was discovered by James W. Christy. Named Charon, Pluto's moon is about 500 to 600 miles in diameter, making it the largest satellite in relation to its parent planet. See Fig. 18.35. Simultaneously, Pluto was found to be much smaller than previously believed. Its diameter is now thought to be about 2000 miles, making it the smallest planet.

The planet is so far away that its temperature is difficult to measure. Also the diameter, density, and period of rotation given in Table 18.1 are estimated values and may change when more precise data are obtained.

Some astronomers believe that Pluto was once a moon of Neptune. There are three reasons for this:

1. Pluto is much smaller than the other four outer planets (see Fig. 18.12).
2. Pluto does not lie along the planetary disk (see Fig. 18.13).

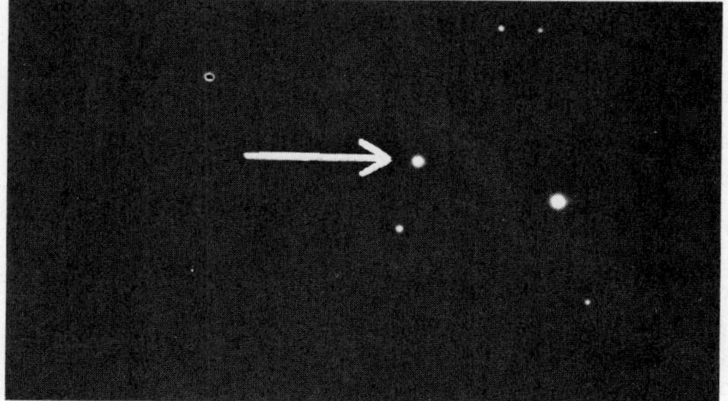

Figure 18.34 Pluto. These two photographs show the planet's change of position in 24 hours. (Palomar Observatory Photograph.)

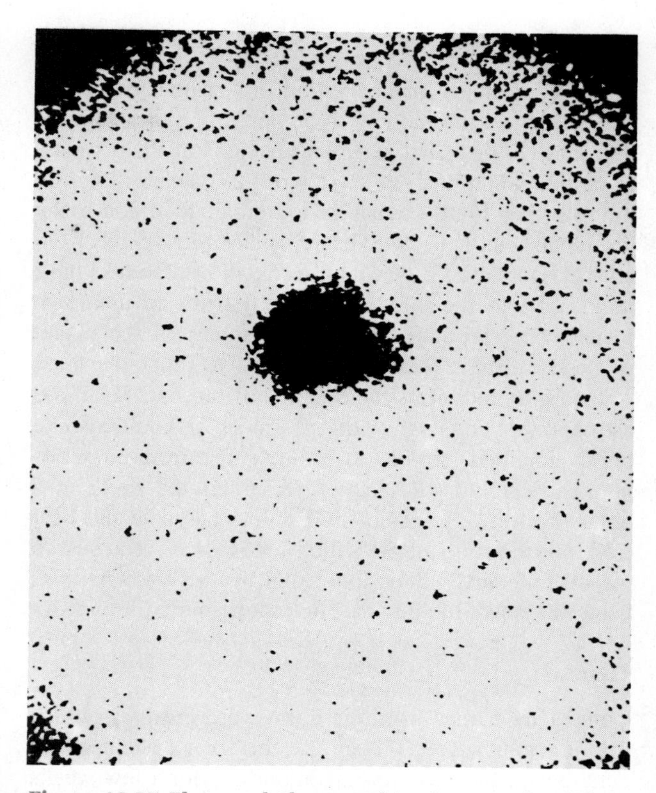

Figure 18.35 Pluto and Charon. This photograph of Pluto and Charon shows a slight elongation of the image at one edge. This slight elongation is the image of Charon. The images of the two appear together because Pluto and Charon are separated by a mere 12,000 miles and they are approximately the same size. Pluto has an estimated diameter of 2000 miles and Charon an estimated diameter of 600 miles. The name Charon comes from Greek mythology. Charon was the ferryman who carried the dead to Pluto's world. (U.S. Naval Observatory)

3. Pluto's orbit is highly elliptical. It actually goes inside the orbit of Neptune (see Fig. 18.11).

Pluto went inside Neptune's orbit in late 1978 and will exit in the year 1999. Thus, for more than 20 years Neptune will be the most distant planet from the Sun.

There is one reason to doubt that Pluto was once a moon of Neptune. For Pluto to have been a moon of Neptune it once must have been very close, and the present orbits suggest that this did not occur in the past. Detailed studies show that in their present orbits, Pluto and Neptune can never come closer than 18 astronomical units. Whether or not Pluto was once a moon of Neptune is thus not at all certain. Some astronomers believe that Pluto and Charon resemble a double asteroid, and Pluto should no longer be classified as a planet.

Planet X

The discrepancies in the motion of the planets Uranus and Neptune were thought to be due to the gravitational influence of Pluto, but recent calculations of Pluto's mass from the motion of its newly discovered satellite, Charon, indicate that Pluto's mass is too small to produce the discrepancies. Thus, the discrepancies must come from another source. U.S. Naval Observatory astronomers have indicated that the gravitational pull from an unknown planet could produce these discrepancies, if the planet possesses a mass between two and five times the mass of the Earth and orbits the Sun at a distance of 50 to 100 astronomical units. The orbit of Planet X should also be highly inclined, similar to Pluto's. The infrared astronomical satellite (IRAS), which circles the Earth in a polar orbit at a height of 560 miles, has detected heat from an object about 50 billion miles away. The search for such a planet will continue and, if it exists, scientists using the new Hubble space telescope may discover it.

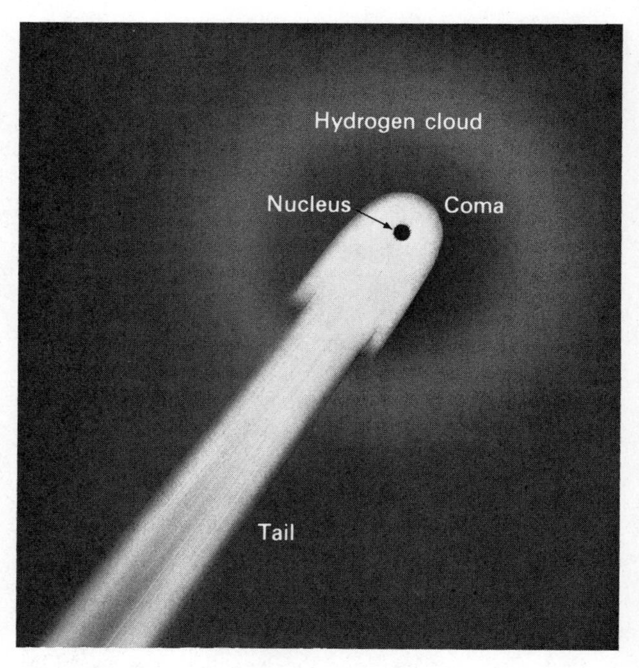

Figure 18.36 The principal parts of a comet.

Comets

Comets are named from the Latin word *cometes,* which means "longhaired." They are the solar system members that periodically appear in our sky for a few weeks or months, then disappear. A comet occupies a very large volume but has a very small mass; therefore, it must be composed of very fine particles. A **comet** consists of four parts: (1) the **nucleus,** typically a few miles in diameter and composed of rocky or metallic material, solid ices of water, ammonia, methane, and carbon dioxide; (2) the head, or **coma,** which surrounds the nucleus and which can be as much as several hundred miles in diameter, and is formed from the nucleus as it approaches within about 5 astronomical units of the Sun; (3) a long, voluminous, and magnificent **tail,** also formed from the coma by solar winds and radiation, and which can be millions of miles in length; and (4) a spherical cloud of hydrogen, believed to be formed from the dissociation of water molecules in the nucleus, surrounding the coma. The sphere of hydrogen in some comets may have a radius exceeding that of the Sun (Fig. 18.36).

Comets are seen by reflected light from the Sun and the fluorescence of some of the molecules comprising the comet. As comets approach and move around the Sun, the material in the coma and tail gets larger (Fig. 18.35). This increase in size is evidently caused by the Sun, perhaps by (1) the solar wind that consists of streams of particles (electrons, protons, and the nuclei of light elements) given off by the Sun and driven outward at speeds of hundreds of miles per second; and (2) radiation pressure generated by the radiant energy given off by the Sun. It is believed that only a thin outer shell of the comet's nucleus is heated. As it moves toward its closest approach (perihelion) to the Sun, the increasing solar radiation causes the surface materials to melt and evaporate. The evaporating surface particles form the coma and the long tail, which is driven away from the Sun by the solar wind and radiation pressure. Each time the comet passes near the Sun on its long journey through space, it loses part of its mass. Eventually the comet loses all its mass and disappears as a comet, but the rocky or metallic material continues to move around the Sun.

Halley's comet, named after Edmond Halley (1656–1742), a British astronomer, is the brightest and best known comet. Halley was the first to suggest and predict the periodic appearance of comets. He observed the comet that bears his name in 1682 and, using Newton's laws of motion, correctly predicted it would return in 76 years. Halley's comet appeared in 1910 and is shown in a composite of photographs in Fig. 18.37.

The color photograph of Halley's comet, shown in color plate V, was taken March 14, 1986 from a distance of about 18,000 km (11,000 miles) by the Giotto spacecraft. See Fig. 18.38 for a line drawing indicating the major areas of the color photograph.

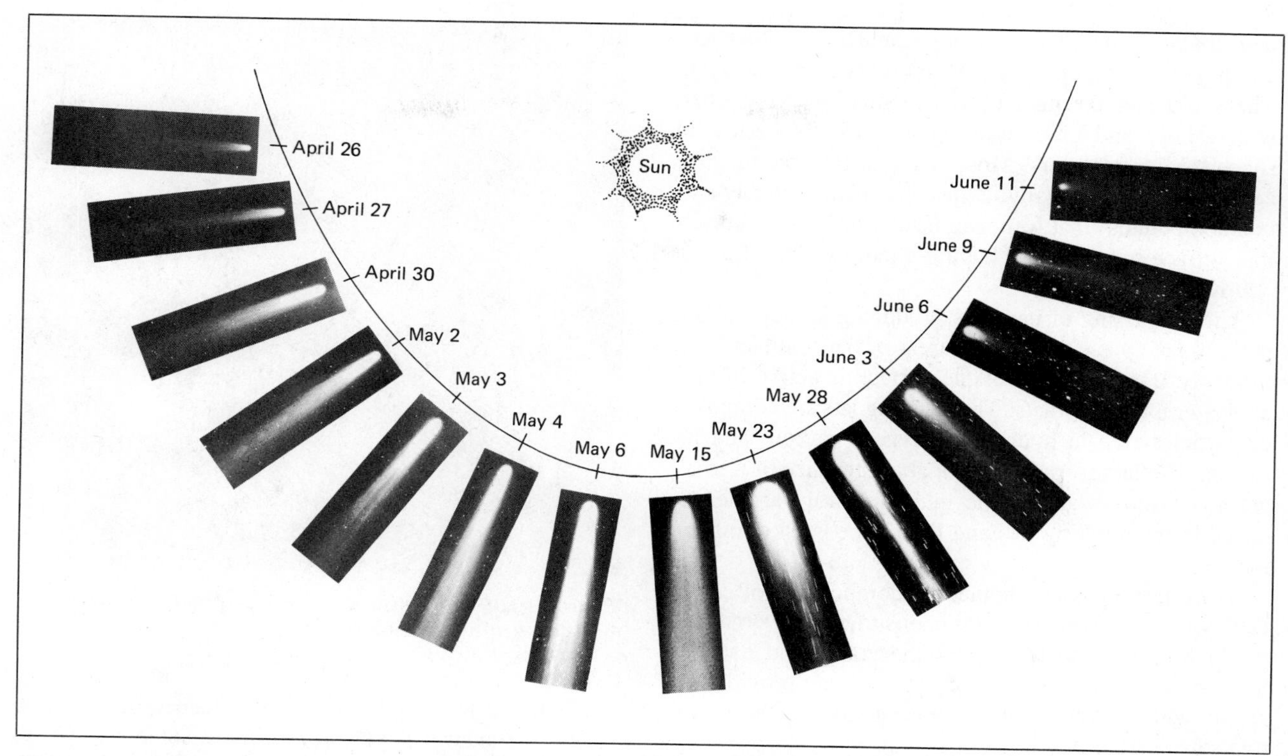

Figure 18.37 Halley's comet. Fourteen views of Halley's comet taken in 1910 between April 26 and June 11. Note the change in size of the coma and tail. (Courtesy Mount Wilson and Las Campanas Observatories, Carnegie Institution of Washington.)

The Giotto spacecraft was named after the Italian painter Giotto di Bondone, who in 1304 portrayed Halley's comet as the "star of Bethlehem" in a painting executed on plaster in the Scroegni chapel in Padua, an industrial city in northeastern Italy.

The Giotto spacecraft was launched on July 2, 1985 by an Ariana-1 rocket from Kourou, French Guiana, to encounter the comet Halley in March 1986. The 960-kg spacecraft, sponsored by the European Space Agency, was one of five space probes sent from Earth to study and obtain scientific data on Halley's comet. The closest approach to comet Halley was made by the Giotto spacecraft coming within approximately 600 km (372 miles) of the nucleus.

The data obtained by the spacecraft and sent back to the Earth confirmed the thoughts that the comet's nucleus is a "dirty snowball." This is true except the comet does not have the shape of a ball. The data received from the spacecraft indicated that the nucleus is shaped more like a peanut or potato with a major axis of about 15 km and a minor axis ranging somewhere between 7 km and 10 km. The nucleus is composed of dust particles

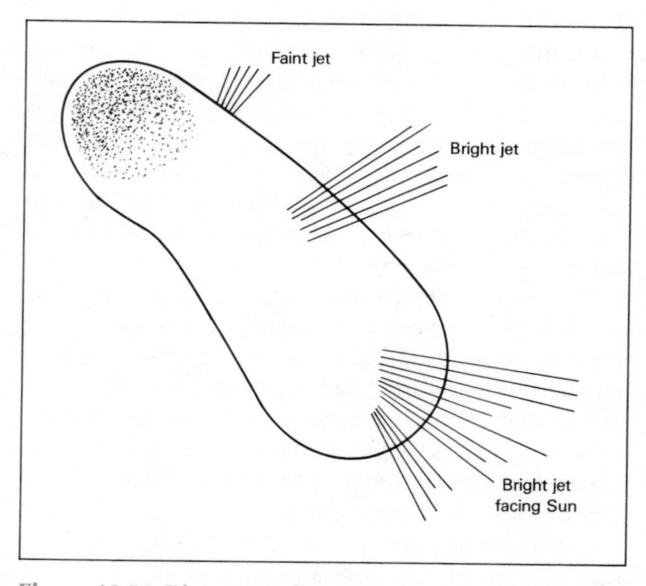

Figure 18.38 Diagram indicating the major areas of the sunlit side of comet Halley's nucleus, which is shown on color plate V. The dark area is not being illuminated by the Sun. The nucleus is thought to be more oval-shaped than indicated in the diagram.

held together mainly by water-ice and some other ices composed of carbon, oxygen, hydrogen, and nitrogen. The surface of the nucleus is irregular, black as coal, with valleys and hills, and displays relief spherical in shape similar to craters. Most of the surface seems to be covered with a dark nonvolatile substance, probably carbon compounds, that forms an insulating crust. The visible surface is fairly cool having a temperature of about 300 K.

The outgassing of particles producing jet activity appears to take place from small areas of the sunlit surface. Probably less than 10% of the surface is active in producing erupting dust jets. The erupting jets are composed of particles rich in hydrogen, oxygen, carbon, and nitrogen. The largest particles are about the size of a small grain of sand. Water is the major neutral particle detected in the erupting jets and H_3O^+ the most dominant ion.

The question as to whether the composition of comet Halley is typical for comets remains to be answered, but the lack of internal heating of the comet and external heating only in the vicinity of the Sun may provide scientists with material that is representative of the primitive solar system.

Comets originate and evolve from dirty icy objects that were part of the primordial debris thrown outward into interstellar space when the solar system was formed. These dirty icy objects reside in a spherical volume of space that begins beyond the planets and extends outward to an estimated 40,000 to 50,000 astronomical units from the Sun. This far-flung volume of space, which is the reservoir of cometary material, is called the **Oort comet cloud** in honor of Jan Hendrik Oort, the Dutch astronomer who proposed its existence in 1950. The total mass of the Oort cloud is estimated to be 1 to 10 times the mass of the Earth.

The dirty icy objects that reside within the Oort cloud are perturbed by weak gravitational force fields of passing stars. These small forces start the icy objects inward on their fall toward the Sun or outward beyond the influence of the Sun's gravitational field. Very few of these cold hard objects become captured comets that come close enough to the Sun to receive sufficient heat to vaporize the icy material. Most pass undetected beyond the orbit of Pluto. Fewer than 1000 comets have been seen and recorded by astronomers, and only one-tenth of these make repeated trips near the Sun.

Where is the outer boundary of the solar system? As indicated above, the Oort comet cloud has an estimated outer limit of 50,000 astronomical units. Does the Sun's gravitational force field have any influence beyond this limit? The answer is yes, according to some astrophysicists who have made some computer simulations. They have indicated the outer limit for the influence of the Sun's gravitational field is between 80,000 and 100,000 astronomical units. This limit or boundary is where the Sun's gravitational force field is balanced out by that of the Milky Way. Beyond this boundary the Sun's gravitation force field cannot hold any object.

Figure 18.39 Ahnighto–Cape York meteorite. Iron meteorite weighing about 79,000 pounds. (Courtesy American Museum of Natural History.)

A **meteor,** incorrectly called a shooting star, is a metallic or stony object that becomes luminous as it enters and burns in the Earth's atmosphere because of the tremendous heat generated by friction with the air. Millions of objects enter the Earth's atmosphere every day. Most of these are completely burned in the atmosphere, but many strike the Earth and are known as **meteorites.** The meteors are members of the solar system that probably come from the remains of comets and fragments of shattered asteroids. The largest known meterorite has a mass of more than 55,000 kilograms (120,000 pounds) and fell in southwest Africa. The largest known meteorite in the United States, with a mass of about 36,000 kilograms (79,000 pounds) was found near Cape York, Greenland, in 1895, and is on display at the Hayden Planetarium, New York (Fig. 18.39).

Meteorites can be of any size or shape, and are classified as metallic or stony. The metallic ones have about 91% iron, 8% nickel, small quantities of cobalt, and

Figure 18.40 Meteor crater. This crater, located near Winslow, Arizona, is 600 feet deep and 4000 feet in diameter. (Courtesy U.S. Air Force.)

phosphorus, and traces of many other elements. The stony ones have about 36% oxygen, 26% iron, 18% silicon, and 14% magnesium, with smaller quantities of several other elements. These are average compositions; individual meteorites can vary greatly. See Fig. 18.40.

18.5 The Origin of the Solar System

Any theory proposed to explain the origin and development of the solar system must account for the system as it exists now in respect to size, shape, form, substance, and change. The preceding section has been a general description of the system in its present state, which, according to our best measurements, has lasted for at least five billion years.

What are some of the major items which must be accounted for if an acceptable theory of origin is to be obtained? First, we must account for the origin of the material used to create the system. Second, we must account for the forces that were needed to form the system. Third, we must account for the size, shape, substance, positions, and motion of the Sun, planets, satellites, asteroids, comets, and meteors.

Presently, there are two types of theories that attempt to explain the origin of the solar system. One group of theories is called **catastrophe theories.** In these theories our solar system arises from a catastrophe such as the near collision of a star with our Sun. This group of theories has several drawbacks. Among these drawbacks are (1) hot gases drawn from the Sun would not collect to form planets, but would be dispersed, and (2) the angular momentum of the present planetary system requires an approaching star to be at a minimum distance to account for the angular momentum; but in order for the material to be withdrawn from the Sun, the star would have to approach closer than this minimum distance.

A second group of theories is called **evolutionary theories.** In these theories the solar system forms out of a single rotating cloud of gases due to the mutual gravitational attraction of all the particles. The main drawback of these theories is that the Sun is now spinning much more slowly than seems reasonable using these theories. Recent advances indicate that there may be a way to explain this apparent discrepancy.

The theory currently in fashion is called the **protoplanet* hypothesis.** This hypothesis proposes that initially there was a large swirling volume of cold or rarefied gases and dust positioned in space among the stars of the Milky Way. This is supported by the fact that today we observe many such nebulae throughout the universe. For some reason, such as a supernova shock wave passing through, the gases began to condense and a larger mass began to form, which produced greater force, causing the entire cloud of gases and dust to shrink (Fig. 18.41).

The collection of particles was slow at first but became increasingly faster as the central mass became larger. As the particles moved inward, the rotation of the mass had to increase to conserve angular momentum. Because of the rapid turning, the cloud began to flatten and spread out in the equatorial plane. Kepler's third law is a statement that the central part must move faster than the outer parts. This set up shearing forces which, coupled with variations in density, produced the formation of other masses that moved around the large central portion, the protosun, sweeping up more material and forming the protoplanets.

During the early stages of development, the space between the protosun and the protoplanets was filled with large amounts of gas and dust, shielding the protoplanets from the protosun, which was beginning to burn and

*"Proto" is Greek for *primitive*.

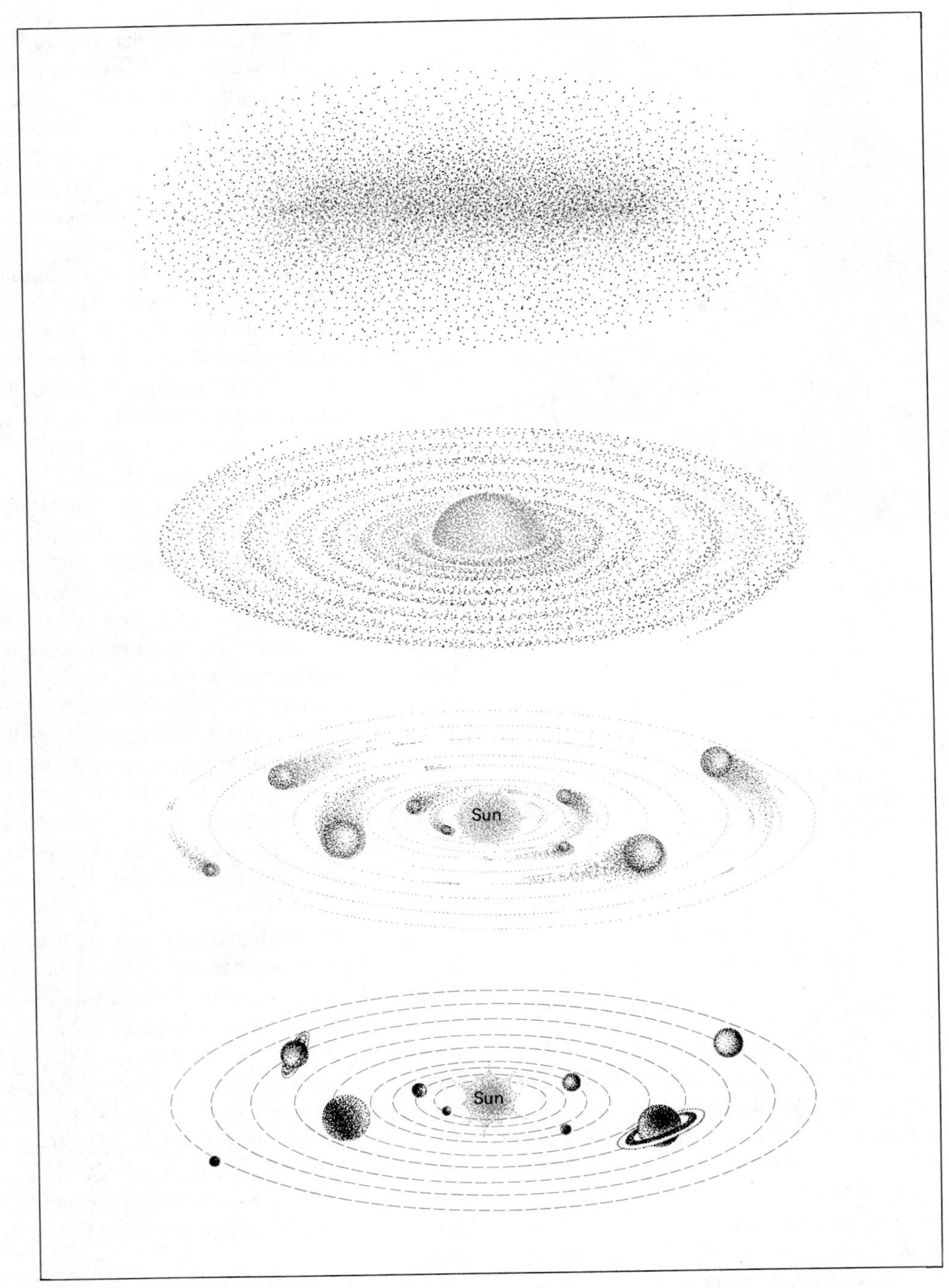

Figure 18.41 Graphic representation of the formation of the solar system according to the protoplanet hypothesis. The four sketches are not drawn to scale. The top drawing should represent a very large volume of space. Interstellar gases and dust particles, with some net angular momentum, collected because of gravitational forces. Most of this collapsed mass formed the Sun. The remainder of the mass formed the planets, which revolve around the Sun in accord with Kepler's laws.

radiate as a star. With the passing of time, the space between the protosun and protoplanets became transparent and the protoplanets' atmospheres were heated and driven off by the pressure of solar radiation. The protoearth was perhaps 1000 times more massive than the planet Earth. The inner protoplanets, receiving more heat than the outer protoplanets, were greatly affected by this heating and lost most of their atmospheres. The outer protoplanets, being at great distances and receiving little solar radiation, were less affected and appear today in a similar protoplanet stage. Planetary satellites are believed to have formed from similar condensations, surrounding their rotating protoplanets, although our moon may have been a smaller protoplanet. The formation of the solar system took place over a 100-million-year period beginning 4.3 billion years ago.

The asteroids, which have irregular size and shape, are believed to be the remnants of a planet that never formed. Meteorites are thought to be small asteroids, and by studying the composition of meteorites we have important clues to the origin of our solar system.

The comets were formed in the outer parts of the nebula, beyond the orbit of Neptune. Since very little material was available at this distance, the comets were very small in size, about one mile in diameter, composed of icy ammonia, methane, water, and some silicates. Some of the formed comets have been scattered from their birthplace by disturbances from the planets. Countless others remain in the vast space of their birthplace.

If this improved theory or hypothesis is the true picture concerning the origin of our solar system, then it is very reasonable to believe there are many, many other systems located throughout the universe, perhaps some with life similar to our own.

18.6 Other Solar Systems

The search for planets of other stars has been in progress for many years. Recent reports of discoveries of protoplanetary objects around the stars Beta Pictoris and Vega and the detection of a wobble in the motion of the star Van Biesbroeck 8 (VB8) indicating the presence of a substellar object, has aroused new interest in the search. This interest is due to recent advances in technology that increase the possibility of finding substellar objects. The star VB8 is located 21 light-years away in the constellation Ophiuchus. The discovery of a substellar object orbiting VB8 provides evidence that other solar systems may exist in our galaxy.

Detecting extra-solar planets is no easy task. Basically a star without planets will appear to move in a straight line. A star with planets will have a small wobble superimposed on its straight-line motion due to gravitational effects. Using infrared radiation and a technique called infrared speckle interferometery, scientists at the National Optical Astronomy Observatories reported in 1983 a small wobble in VB8. The detected object, called a ''brown dwarf'' star by some astronomers and a planet by others, is known as VB8B. The substellar object has a mass about 10 times that of Jupiter, a surface temperature of approximately 1400 K, and a diameter nine-tenths that of Jupiter.

Confirmation of VB8B could come in 1988 when the Hubble Space Optical Telescope is placed in a 300-mile-high orbit by the space shuttle. The optical telescope has a 2.4-m (94-in) mirror and will be able to detect objects 50 times fainter and resolve objects seven times smaller than any presently made optical telescope.

Learning Objectives

After reading and studying this chapter, you should be able to do the following without referring to the text:

1. Define *parallax* and explain why the parallax of a star cannot be seen with the naked eye.
2. State Kepler's three laws of planetary motion, and draw figures illustrating the first two laws.
3. List the planets in order of distance from the Sun.
4. Briefly describe the solar system as a whole.
5. State the distinguishing features of each planet and of Io and Titan.
6. List the terrestrial planets and the Jovian planets and give several ways in which they are different.
7. State three reasons why Pluto is thought to be an escaped moon of Neptune, and one reason why it is thought not to be.
8. Describe and differentiate between comets, meteors, and asteroids.
9. State briefly the protoplanet hypothesis for the formation of the solar system.

Important Words and Terms

solar system
ecliptic plane
rotation
revolution
Foucault pendulum
parallax
geocentric theory
heliocentric theory
Kepler's law of elliptical paths
Kepler's law of equal areas
Kepler's harmonic law
terrestrial planets
Jovian planets
Titius-Bode law
astronomical unit
sidereal period
synodic period
conjunction
opposition
asteroids
comet
nucleus
coma
tail
meteor
meteorites
catastrophe theories
evolutionary theories
protoplanet hypothesis
Oort comet cloud

Questions

1. Before 1900, what proof was there that the Earth rotated and revolved?
2. State why the parallax of a star cannot be seen with the naked eye.
3. List the major objects that comprise the solar system.
4. Briefly describe the structure of the solar system.
5. What is the explanation of Kepler's second law? (Hint: refer to a conservation law)
6. What distinguishes the terrestrial planets from the Jovian planets?
7. Which planet is presently the farthest known planet from the Sun? Explain.
8. Which planet has the most elliptical-shaped orbit?
9. Why do all the planets revolve in the same direction around the Sun?
10. What is unusual about the satellite Io?
11. What is unique about the satellite Titan?
12. Which planet is the largest? Which satellite of a planet is the largest?
13. Which planet has the greatest orbital velocity? The least?
14. Name the planets that have the highest and lowest surface temperatures.
15. How do the atmospheric pressures on Venus, Earth, and Mars differ?
16. Name the planets that are known to have rings.
17. Compare the physical characteristics of the Earth with those of the other planets of the solar system.
18. How do asteroids, comets, and meteors differ?
19. Which planet has the greatest density? The least density?
20. State the reasons why Pluto is thought to have once been a satellite of Neptune. State the opposing reasons.
21. State the characteristics of Pluto that classify it as (a) a planet, (b) an asteroid.
22. Pluto is considered one of the outer or Jovian planets. Compare its physical characteristics with those of the Jovian planets.
23. How many years were required for the formation of the solar system?
24. What is the estimated outer limit or boundary of the solar system?
25. What evidence do astronomers have concerning the possibility of another solar system?

Problems

1. Determine the distances from the Sun to the terrestrial planets and the asteroids using the Titius-Bode law. Compare your answers to the actual distances.
2. Determine the distances from the Sun to the Jovian planets using the Titius-Bode law. Compare your answers to the actual distances.

3. Using the Titius-Bode law, determine the distance from the Sun to Pluto and compare it with the actual distance.
4. Using Newton's formula $T^2/R^3 = 4\pi^2/Gm_{Sun}$, show how the mass of the Sun could be calculated in kilograms. R for the Earth is 1.5×10^{11} m.
5. Show that Kepler's third law holds (approximately) for the Earth and Mars. That is, show that (see Table 18.1):

$$\frac{(365.25)^2}{(93 \times 10^6)^3} = \frac{(687)^2}{(142 \times 10^6)^3}$$

6. Referring to problem 5 and Table 18.1, show that Kepler's third law holds (approximately) for the Earth and Venus.
7. Use Kepler's third law to show that the closer a planet is to the Sun, the shorter is its period.
8. Use Kepler's third law to show that the closer a planet is to the Sun, the faster is its speed around the Sun.
9. Use Kepler's second law to show that comets spend most of their time far from the Sun.
10. Refer to Table 18.1 and draw a diagram showing how far each inner planet goes around the Sun in 60 days.
11. Refer to Table 18.1 and draw a diagram showing how far each outer planet goes around the Sun in 10 years.
12. List the following distances in order of increasing length: (a) Sun to Earth, (b) Mars to Jupiter, (c) Jupiter to Saturn, (d) Saturn to Uranus.
13. List the planets in order of increasing distance from the Sun.
14. List the planets in order of decreasing size (largest first).
15. List the planets in order of decreasing density (densest first).
16. List the years of the next three appearances of Halley's comet.

19

Place and Time

What place have you?
What place have I?

What time have you?
What time have I?

Only here and now
Have you and I.

IN PHYSICAL SCIENCE we observe and examine events that take place in our environment. These events occur at different places and at different times. Some occur immediately and are observed, while others occur at great distances and are not observed until a later time. For example, you see a flash of lightning, but do not hear the sound of thunder until later. A star explodes at some great distance from the planet Earth and years later the event is observed as the light radiation reaches the Earth. Thus the events or happenings taking place in our environment are separated in space and time. Albert Einstein was the first to point out that space and time are related and exist as a single entity. Einstein associated gravitational fields with space and developed the concept of four-dimensional space, which has given us an entirely different concept concerning the reality of our environment. In this chapter we shall study the basic concepts in physical science that refer to our location and the objects of our environment in space and time. The questions of where and when are experiences we have in our daily lives: "The meeting of the next physical science class will be in Heath Hall at 10:00 A.M. on Tuesday." "Sally will spend the Christmas holidays in Florida." "The space shuttle was first launched from Cape Canaveral on April 12, 1981." All of these familiar phrases include place and time. Thus on each day of our lives we experience the march of time and the change of position.

◀ The periodic motion of a pendulum is used to measure time.

19.1 Cartesian Coordinates

The location of an object in our environment requires a reference system that has one or more dimensions. A one-dimensional system is pictured by the number line shown in Fig. 19.1, which illustrates two fundamental features of every coordinate system. A straight line is drawn, which may extend to plus infinity in one direction, and minus infinity in the opposite direction. For the line to represent a coordinate system, an origin must be indicated and unit length along the line must be expressed. Temperature scales, left-right, above ground–below ground, time past–time future, profit-loss are all examples of one-dimensional coordinate systems.

A two-dimensional system is shown in Fig. 19.2, in which two number lines are drawn perpendicular to each other and the origin assigned at the point of intersection. The two-dimensional system is called a **cartesian coordinate system,** in honor of the French philosopher and mathematician Rene Descartes (1596–1650), the inventor of coordinate geometry. It is also referred to as a rectangular coordinate system. The horizontal line is normally designated the x-axis, and the vertical line the y-axis. Every position or point in the plane is assigned a pair of coordinates (x and y), which gives the distance from the two lines, or axes. The x number gives the

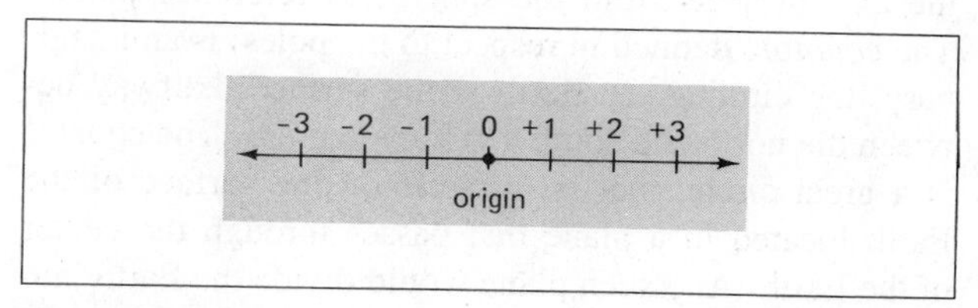

Figure 19.1 A one-dimensional reference system.

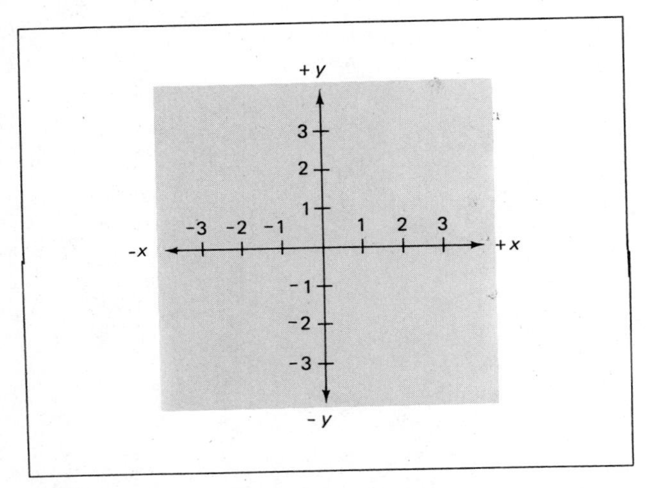

Figure 19.2 A two-dimensional reference system.

distance from the y-axis, and the y number gives the distance from the x-axis. Many of the cities in the United States are laid out in the cartesian coordinate system. Usually one street runs east and west, corresponding to the x-axis, as mentioned above, while another street runs north and south, which corresponds to the y-axis.

Our interest in the cartesian coordinate system is for determining the location of any position on the surface of the spherical Earth, and the location of any objects on the celestial sphere.

A spherical surface is a curved surface on which all points are equidistant from a point called the center. The location of any position on a spherical surface can be found using two reference circles similar to the coordinate system mentioned above.

19.2 Latitude and Longitude

The location of an object on the surface of the Earth is accomplished by means of a coordinate system known as latitude and longitude. Since the Earth is turning about an axis, we can use the *geographic poles,* which are defined as those points on the surface of the Earth where the axis projects from the sphere, as reference points. The *equator,* defined in respect to the poles, is an imaginary line circling the Earth at the surface, halfway between the north and south geographic poles. The equator is a great circle; that is, a circle on the surface of the Earth located in a plane that passes through the center of the Earth. Any such plane would divide the Earth into two equal halves.

The **latitude** of a surface position can be defined as the angular measurement, in degrees, north and south of the equator. The latitude lines are circles drawn around the surface of the sphere parallel to the equator. Any number of such circles can be drawn. The circles become smaller as the distance from the equator becomes greater (Fig. 19.3). These circles are called **parallels,** and when we travel due east or west, we follow a parallel. Latitude has a minimum value of zero degrees at the equator and a maximum value of 90° north or 90° south at the poles (Fig. 19.3).

Imaginary lines drawn along the surface of the Earth running from the north geographic pole, perpendicular to the equator, to the south geographic pole are known as **meridians.** Meridians are half circles, which are portions of a great circle, since the circle is located in the same plane as the center of the Earth. An infinite number of lines can be drawn as meridians. **Longitude** can be defined as the angular measurement, in degrees, east or west of the reference meridian, which is called the prime, or Greenwich, meridian. Longitude has a minimum value of zero at the prime meridian and a maximum value of 180° east and west (Fig. 19.4).

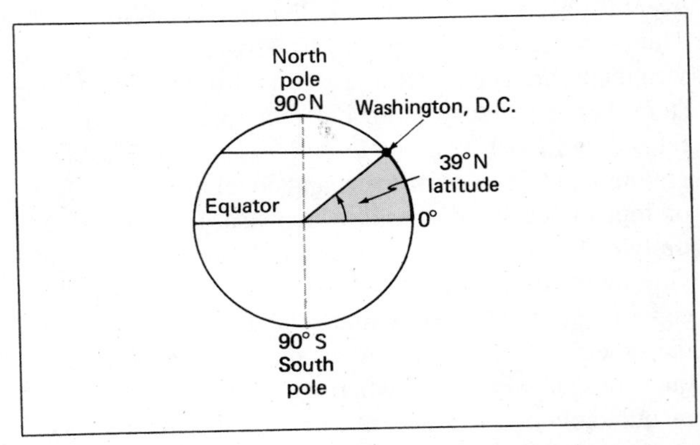

Figure 19.3 Diagram showing the latitude of Washington, D.C.

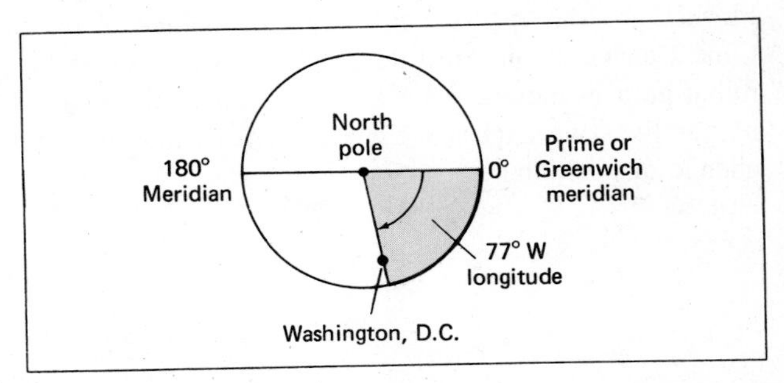

Figure 19.4 Diagram showing the longitude of Washington, D.C.

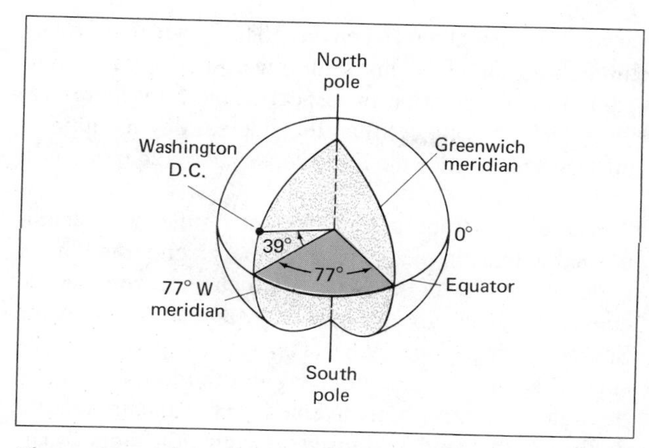

Figure 19.5 Diagram showing both the latitude and longitude of Washington, D.C.

The latitude and longitude of one point (Washington, D.C., 39° N, 77° W) are shown in Figs. 19.3 and 19.4.

The latitude and longitude shown in Figs. 19.3 and 19.4 are combined in Fig. 19.5 and shown in a cutaway of the Earth. The **Greenwich meridian** was chosen for the prime, or zero, meridian because a large optical telescope was located at Greenwich, England, and because England ruled the seas at the time the coordinate system of latitude and longitude was originated. The primary purpose of the system at the time was to determine the location of ships at sea.

19.3 Time

The continuous measurement of time requires the periodic movement of some object as a reference. On October 8, 1964, the 12th General Conference of Weights and Measures, meeting in Paris, adopted an atomic definition of the *second* as the international unit of time. The definition is based on a change in energy levels of the cesium-133 atom. The exact wording of the definition for the atomic second is as follows: "The standard to be employed is the transition between the two hyperfine levels $F = 4$, $M_F = 0$, and $F = 3$, and $M_F = 0$ of the fundamental state $2s_{1/2}$ of the atom of cesium-133 undisturbed by external fields and the value 9,192,631,770 Hz is assigned." When a transition from one energy state to another occurs, the atom emits or absorbs radiation, the frequency of which is proportional to the energy difference in the two states. The cesium-133 atom provides a highly accurate and stable reference frequency of 9,192,631,770 cycles per second, which can be referred to by well-known electronic techniques. The National Bureau of Standards Frequency Standard, NBS-III, a cesium beam with a 3.66-m interaction region is shown in Fig. 19.6.

For everyday purposes, we are interested in the Earth as a time reference, since our daily lives are influenced by the day and its subdivisions of hours, minutes, and seconds. The day has been defined in two basic ways: (1) The solar day has been defined as the elapsed time between two successive crossings of the same meridian by the Sun. This is also known as the **apparent solar**

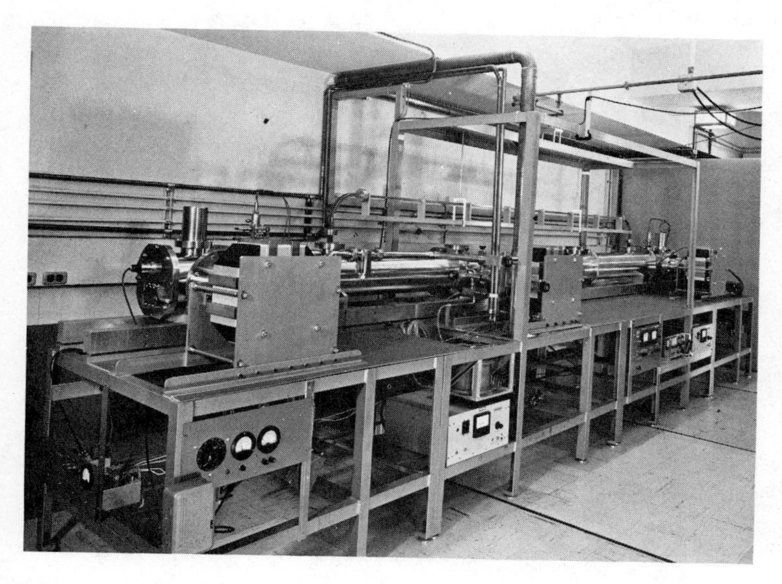

Figure 19.6 The National Bureau of Standard's NBS-111 cesium beam frequency generator for establishing the time standard. (Courtesy the National Bureau of Standards.)

day, since this is what appears to happen. Since the Earth travels in an ellipse, the orbital velocity of the Earth is not constant; therefore, the apparent days are not the same in duration. To remedy this situation, the mean solar day is computed from all the apparent days during a one-year period. The variation of the apparent solar day from the **mean solar day** is as much as 16 min. (2) The **sidereal day** has been defined as the elapsed time between two successive crossings of the same meridian by a star other than the Sun (cf. Section 1.3). Figure 19.7 illustrates the difference between the solar and sidereal days.

Since the Earth rotates 365.25 times during one revolution, the magnitude of the angle through which the Earth revolves in one day is

$$360°/365.25 \text{ days} = 0.985°$$

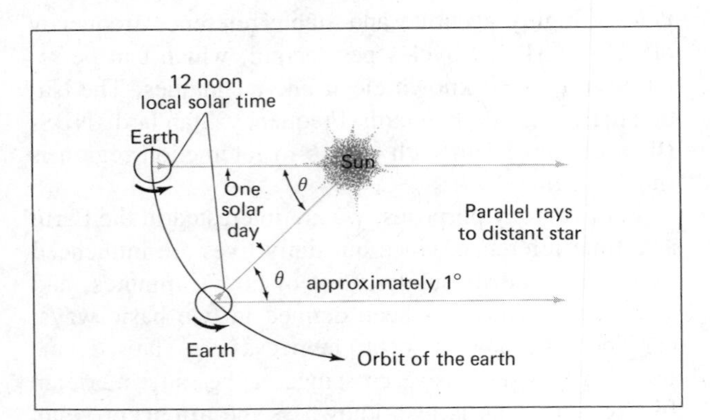

Figure 19.7 The difference between the solar day and the sidereal day. One rotation of the Earth on its axis in respect to the Sun is known as one solar day. One rotation of the Earth on its axis in respect to any other star is known as one sidereal day. Note that the Earth turns through an angle of 360° for one sidereal day and approximately 361° for one solar day.

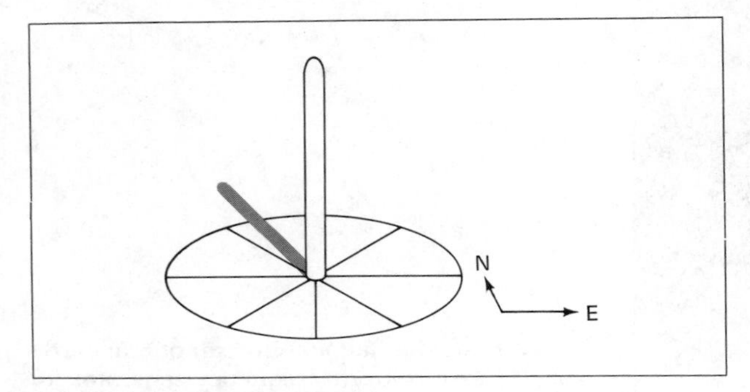

Figure 19.8 A gnomon. A vertical rod positioned so as to cast a shadow to indicate the time of day.

or slightly less than 1° per day. The Earth must rotate through an angle of this same magnitude for the completion of one rotation in respect to the Sun; therefore, the solar day is longer than the sidereal day by approximately 4 min, since the Earth rotates 360°/24 h or 15°/h or 1°/4 min.

The earliest measurement of solar time was accomplished with a simple device known as a gnomon (Greek *gnomon,* a way of knowing), which is a vertical rod erected on level ground that casts a shadow when the Sun is shining (Fig. 19.8). The vertical pointer of the sundial is called a gnomon. By the third century B.C., the water clock had been invented. In hot and dry regions of the Earth, sand was used to count the hours. The burning of candles was also used, since they burn at a fairly constant rate. It was not until the fourteenth century that the first mechanical clock was invented, and it was not accurate. In the sixteenth century, Galileo originated the plans for the construction of a pendulum clock. The story is written that while attending church, Galileo noticed a large hanging lamp swinging back and forth at regular intervals of time as measured by his pulse rate. This provided him with the incentive to experiment with the pendulum and thus pave the way for the building of the first pendulum clock by Christian Huygens after Galileo's death.

Today, we have developed the electric clock, which is controlled by the frequency of the alternating current delivered by the power company, and the quartz-crystal clocks, which are precise to one part in 10^9. Still greater precision is obtained by the atomic clocks mentioned earlier in this chapter.

The 24-h day, as we know it, begins at midnight and ends 24 h later at midnight. By definition, when the Sun is on the meridian, it is 12 noon *local solar time* at this meridian. The hours before noon are designated A.M. (**ante meridiem,** before noon) and P.M. (**post meridiem,** after noon). The time of 12 o'clock should be stated as 12 o'clock noon or 12 o'clock midnight, with the dates. For example, we write 12 o'clock midnight, December 3–4.

Our modern civilization runs efficiently because of our ability to keep accurate time. Since the late nineteenth century, most of the countries of the world have adopted the system of **standard time zones.** This scheme theoretically divides the Earth into 24 time zones, each containing 15° of longitude or 1 h. The first zone begins at the prime meridian which runs through Greenwich, England, and extends $7\frac{1}{2}°$ each side of the prime meridian. The zones continue east and west from the Greenwich meridian with the centers of the zones being multiples of 15°. The actual widths of the zones vary because of local conditions, but all places within a zone have the same time, which is the time of the central meridian of that zone. For example, Washington, D.C., is located at 77° W longitude and is within $7\frac{1}{2}$° of the 75° meridian. See Fig. 19.9.

Figure 19.10 shows the time and date on the Earth for any Tuesday at 7 A.M. (PST), 8 A.M. (MST), 9 A.M. (CST), and 10 A.M. (EST). As the Earth turns eastward the Sun appears to move westward taking 12 noon with it. Twelve midnight is 180° or 12 h eastward of the Sun, and as 12 noon moves westward, 12 midnight follows bringing the new day.

When you travel westward into a different time zone, the time kept by your watch will be one hour fast or ahead of the standard time of the westward zone; therefore, the hour hand must be moved back one hour if the watch is to have the correct time. This process would be necessary as you continue westward through additional time zones. A trip all the way around the Earth in a westward direction would mean the loss of 24 h or one complete day. When traveling eastward the opposite is true, that is, your watch will be one hour slow for each zone, and the hour hand is set ahead one hour.

A better understanding of why a day is lost in traveling around the Earth in a westward direction can be obtained if we take a make-believe trip. Suppose we leave the Dulles International Airport in Washington, D.C., by jet plane at exactly 12 noon local mean solar time on Tuesday, December 5, and fly westward at a speed equal to the apparent westward speed of the Sun. Since Washington is located at 39° N latitude, the plane must travel the 39° N parallel westward at about 800 mi/h. As we leave the airport, we observe the Sun out the left window of the plane about 32° above the southern horizon. One hour after leaving the airport, we notice the Sun can still be seen out the left window at the same altitude. Six hours later, our watches indicating 7 P.M.,

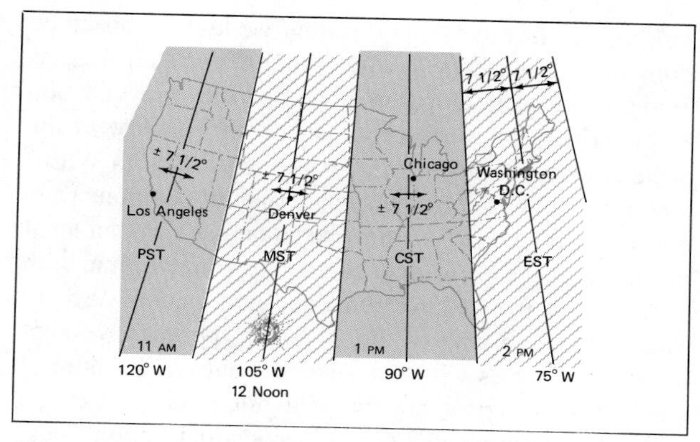

Figure 19.9 Time zones of the continental United States. Theoretical time zones are shown. Actual boundaries are irregular.

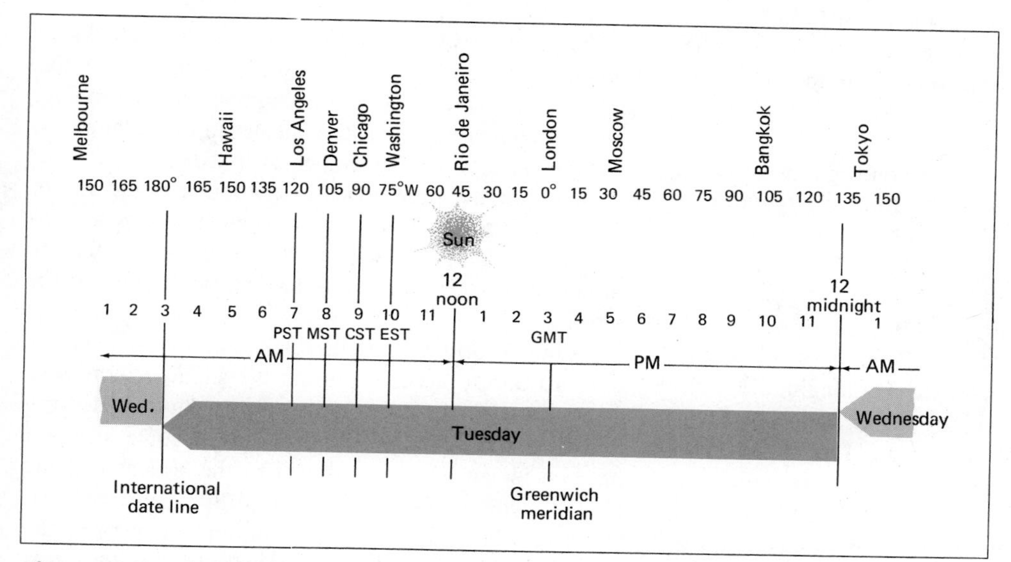

Figure 19.10 Diagram showing the times and dates on the Earth for any Tuesday at 10 A.M. EST. As time passes, the Sun moves westward; thus, 12 noon moves westward and midnight follows (180° or 12 hours) behind, bringing the new day, Wednesday, with it.

the Sun still has the same apparent position as observed from the left window of the plane. Since the plane is flying at the same apparent speed as the Sun, the Sun will continue to be observed out the left window of the plane. Twenty-four hours later we arrive back in Washington with the Sun in the same apparent position. During the 24-hour trip, our time remained at 12 noon local mean solar time. If we had been observing the time with a sundial, it would have remained at 12 noon. We are aware of the passing of the 24 hours, because we kept track of the times with our watches, but we did not see the Sun set or rise, and we did not pass through 12 midnight; therefore, the time to us is still 12 noon Tuesday. To friends meeting us at the airport, it is 12 noon Wednesday.

To remedy situations such as this, the **International Date Line** (IDL) was established at the 180° meridian. When crossing the IDL traveling westward, the date is advanced into the next day and when crossing the IDL traveling eastward, one day is subtracted from the present date.

If the local solar time is known at one longitude, the local time at another longitude can be determined by remembering there are 15° for each hour of time, or four minutes for each degree. Should the calculation extend through midnight or if the International Date Line is crossed, the date would change.

One type of problem of practical importance is to find the time and date in a distant city when you know the time and date in a given city. This problem is encountered when you are trying to make a long distance phone call and don't want to awaken someone in the middle of the night.

EXAMPLE What are the corresponding time and date in Tokyo (36° N, 140° E) when the time and date in Los Angeles (34° N, 118° W) are 6 A.M. PST, November 24?

To work problems of this sort it is simplest to draw a diagram similar to Fig. 19.11. This diagram has the north pole at the center, and the lines of longitude radiate outward from it as shown. Lines of longitude are drawn every 15° to correspond to the centers of the time zones.

Once the time zone centers have been drawn, the time zones of the cities in question should be determined. In the example, the longitude of Los Angeles (118° W) is closest to the time zone centered on 120° W, and that of Tokyo (140° E) is closest to the one centered on 135° E. The given time of 6 A.M. can be placed adjacent to the time zone center for Los Angeles (120° W), as in Fig. 19.11.

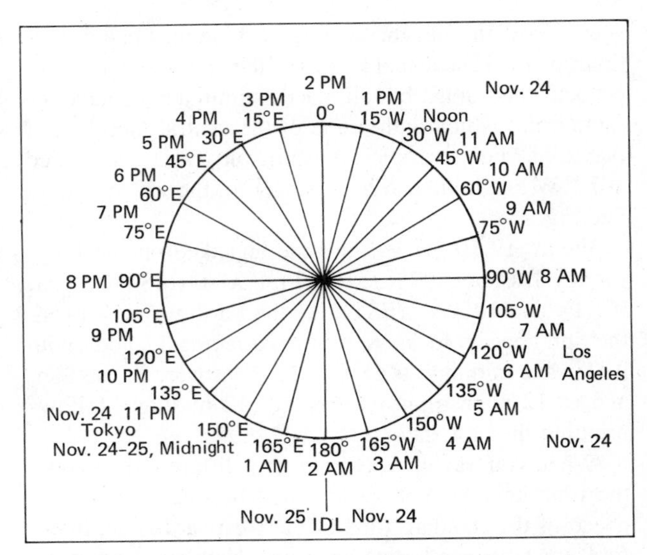

Figure 19.11 An example of finding the time and date in Tokyo, knowing that the time and date in Los Angeles is 6 A.M., Nov. 24. The North Pole is at the center of the circle.

Label each time zone center with the appropriate time and date. When going east (counterclockwise), add an hour for each time zone; when going west, subtract an hour. When crossing the IDL going west, add one day. Also be careful to change the date when going through midnight.

The problem can be solved by going either clockwise or counterclockwise. It is shortest to go clockwise in this example, but it is also more complicated. From Fig. 19.11 it can be seen that the answer in this example is that it is 11 P.M. on November 24 in Tokyo.

Note that the time zones of particular cities do not always correspond to that of the nearest time zone center. Thus, this method may yield a result that differs by an hour from the actual time in a particular city of the world.

During World War I, the clocks of many countries were set ahead one hour during the summer months to give more daylight hours in the evening, thus conserving fuel used for lighting. This has now become standard practice for all but three or four of the 50 states of the United States. The reason now is still to conserve fuel, as during the war. During the summer months in the United States, time known as **Daylight Saving Time** begins on the last Sunday of April and ends on the last Sunday of October.

19.4 The Seasons

The spinning Earth is revolving around the Sun in an orbit that is elliptical, yet nearly circular. When the Earth makes one complete orbit around the Sun, the elapsed time is known as one *year*. We are concerned with two different definitions of the year in this text. The **tropical year,** or the year of the seasons, is the time interval from one vernal equinox to the next vernal equinox; that is, the elapsed time between one northward crossing of the Sun above the equator and the next northward crossing of the Sun above the equator. In respect to the rotation period of the Earth, the tropical year is 365.2422 mean solar days.

The sidereal year is the time interval for the Earth to make one complete revolution around the Sun in respect to any particular star other than the Sun. The sidereal year has a period equal to 365.2536 mean solar days. This is approximately 20 minutes longer than the tropical year. The reason for the difference will be explained later in Section 19.5.

The axis of the spinning Earth is not perpendicular to the plane swept out by the Earth as it revolves around the Sun, but is tilted $23\frac{1}{2}°$ from the vertical as illustrated in Fig. 19.12. This position of the axis in respect to the orbital plane produces a change in the Sun's overhead position throughout the year, and causes our changing seasons. Figures 19.13 and 19.14 illustrate the apparent positions of the Sun over a period of one year.

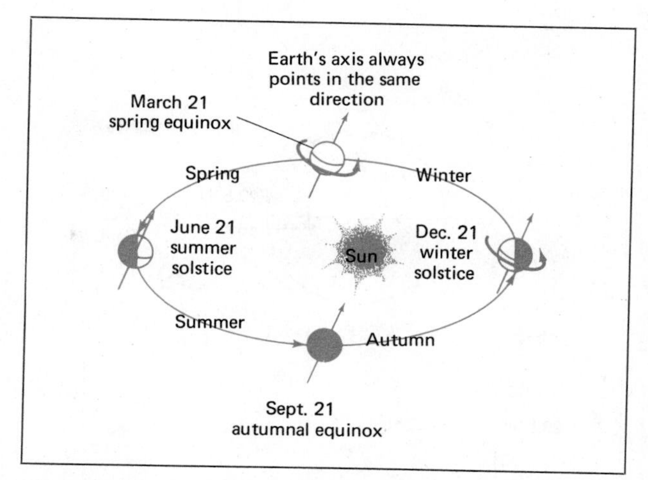

Figure 19.13 Earth's positions, relative to the Sun, and the four seasons. As the Earth revolves around the Sun, its north-south axis remains pointing in the same direction. On March 21 and September 21, the Sun is directly above the equator and all places on the Earth are having 12 hours of daylight and 12 hours of darkness. On June 21 the Sun's declination is $23\frac{1}{2}°$ N, and the northern hemisphere is having more daylight hours than dark hours. It is summer in the northern hemisphere and winter in the southern hemisphere. On December 21 the Sun's declination is $23\frac{1}{2}°$ S, and the southern hemisphere is having more daylight hours than dark hours. It is winter in the northern and summer in the southern hemisphere. The Sun is drawn slightly off center to indicate the Earth is slightly closer to the Sun during winter in the northern hemisphere than during summer.

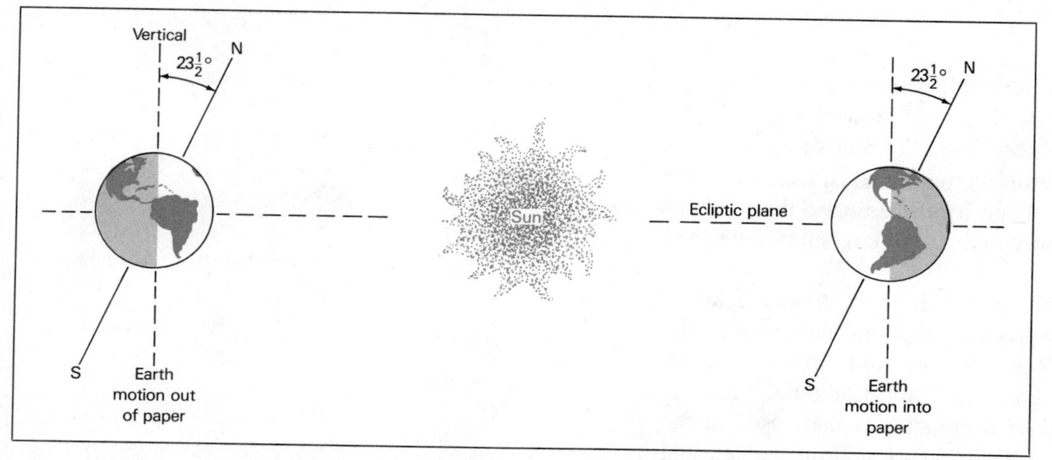

Figure 19.12 As the Earth revolves around the Sun, the Earth's axis remains tilted $23\frac{1}{2}°$ with the vertical. This inclination of the axis in conjunction with the Earth's motion causes a change in seasons on the Earth. When the Earth is at the position shown with its motion out of the paper, the northern hemisphere is having summer and the southern hemisphere is having winter. When the Earth is at the position shown with its motion into the paper, the northern hemisphere is having winter and the southern hemisphere is having summer.

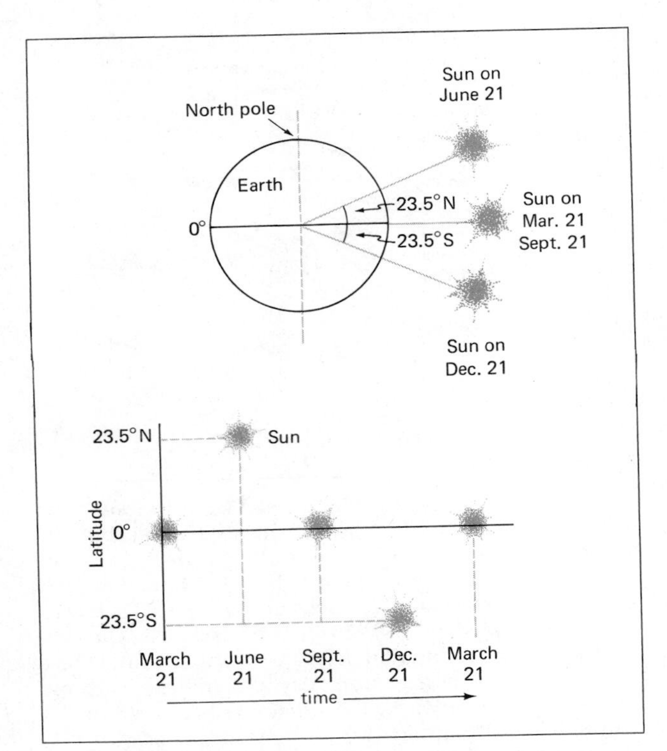

Figure 19.14 Diagrams of the declination of the Sun at different times of the year. The upper drawing is an illustration of a greatly magnified Earth in respect to the Sun showing the Sun's declination on the dates indicated. The lower graph plots the Sun's declination (degrees latitude) versus time (months).

In the summer in the northern hemisphere the Sun's rays are most direct on the northern hemisphere. Thus, it is hotter in the summer when the Sun's rays are most direct, and it is colder in the winter when the Sun's rays are the least direct. When it is summer in the northern hemisphere, it is winter in the southern hemisphere, and vice versa.

The overhead position of the Sun or any celestial object is given in degrees latitude and is known as the **declination** of the object. For example, the declination of the Sun on June 21 is $23\frac{1}{2}°$ north latitude. Later, in Section 21.2, the word declination will be used in defining the position of an object located on the celestial sphere.

The Sun's overhead position is never greater than $23\frac{1}{2}°$ latitude, and the Sun always appears due south at 12 noon local solar time for an observer located in the continental United States. When the Sun is at $23\frac{1}{2}°$ north or south, it is at its farthest point from the equator. This farthest point of the Sun from the equator is known as the solstice (Sun to stand still). The most northern point is called the **summer solstice** and the most southern position is known as the **winter solstice.** This applies to the northern hemisphere. In the southern hemisphere, the dates for the summer and winter solstices are reversed from what is shown in Fig. 19.13.

As the Earth circles the Sun, the Sun's position overhead varies from $23\frac{1}{2}°$ north to $23\frac{1}{2}°$ south of the equator. When it is directly over the equator, the days and nights have 12 h each around the world. These dates are called the equinoxes. The **vernal equinox** occurs on or about March 21 and the **autumnal equinox** occurs on or about September 21 each year. These dates are labeled in Fig. 19.13.

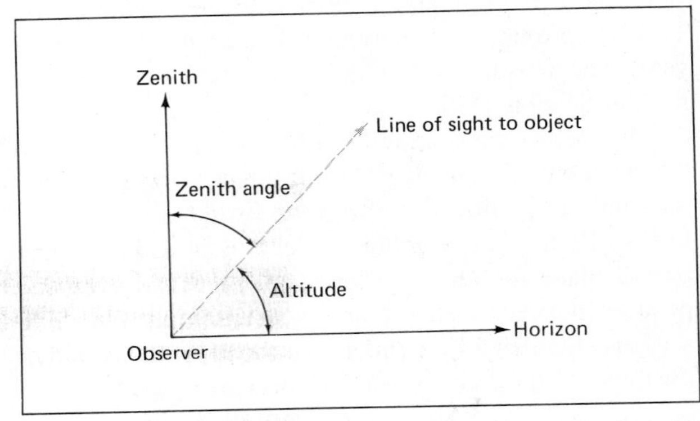

Figure 19.15 Zenith angle and altitude. Since the zenith is perpendicular to the horizon, the zenith angle plus the altitude equals 90°.

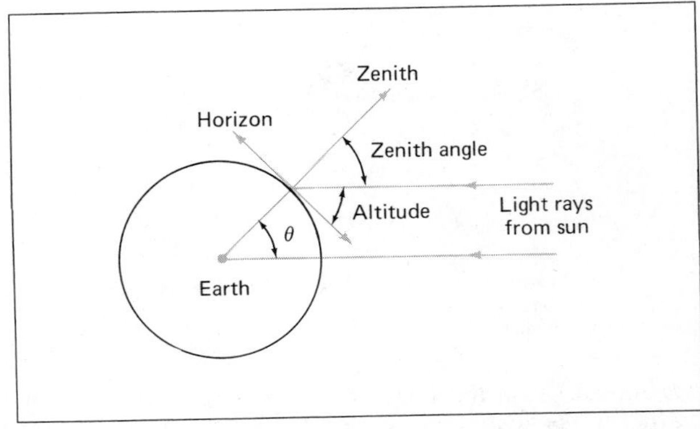

Figure 19.16 The relationship between zenith angle, altitude, horizon, and the angle θ. Since the incoming rays of light from the Sun are parallel, the angle θ and the zenith angle are equal in magnitude.

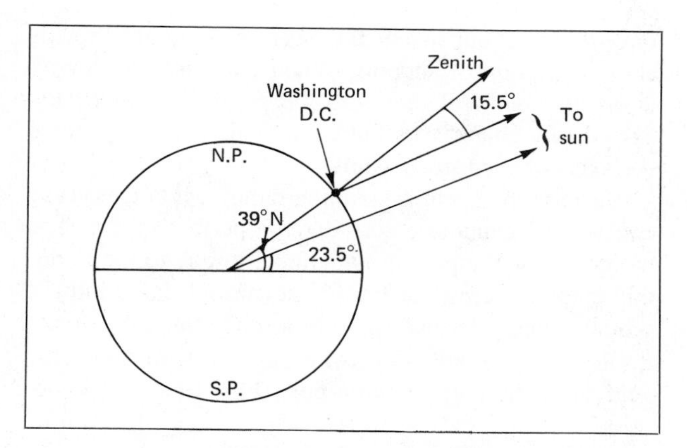

Figure 19.17 Finding the approximate altitude of the Sun as observed from Washington, D.C., on June 21. Since the angle between the Sun and the observer is 39° − $23\frac{1}{2}°$ = $15\frac{1}{2}°$, the altitude of the Sun is 90° − $15\frac{1}{2}°$ = $74\frac{1}{2}°$.

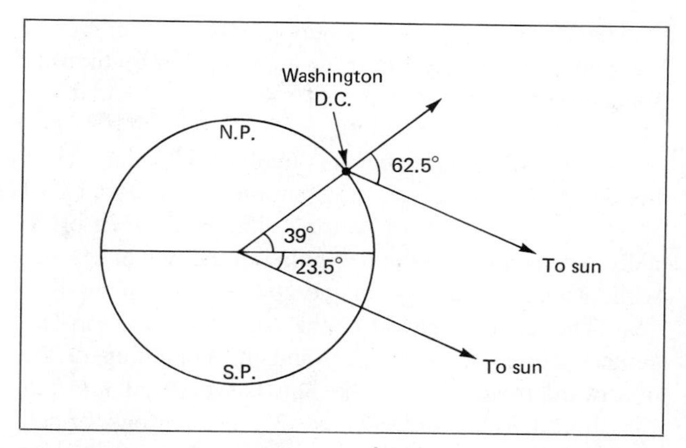

Figure 19.18 Finding the approximate altitude of the Sun as observed from Washington, D.C., on December 21. Since the angle between Sun and observer is 39° + $23\frac{1}{2}°$ = $62\frac{1}{2}°$, the altitude of the Sun is 90° − $62\frac{1}{2}°$ = $27\frac{1}{2}°$.

When the Sun is observed at 12 noon local solar time, it is on the observer's meridian and appears at its maximum above the southern horizon on that day for all observers north of the Sun. The angle measured from the horizon to the line of sight to the Sun at noon is called its **altitude,** and the angle from the **zenith** (position directly overhead) to the line of sight to the Sun at noon is called its **zenith angle.** See Fig. 19.15. The zenith is 90° from the **horizon** (the dividing line where the Earth and sky appear to meet), therefore the sum of the zenith angle and the altitude is 90°. The altitude of the Sun can easily be determined by measurement with a *sextant*. If the declination of the Sun is known, the observer's latitude can be determined. The relation between the above terms is illustrated in Fig. 19.16.

The maximum and minimum altitudes of the Sun for an observer in Washington, D.C. (39° N), can be determined by using data from Fig. 19.16. The solutions are as shown in Figs. 19.17 and 19.18. The relationship between the two solutions is illustrated in Fig. 19.19. The altitude of the Sun for all other days of the year, as observed from Washington, would be between these two values. A similar solution would give the Sun's altitude from any latitude.

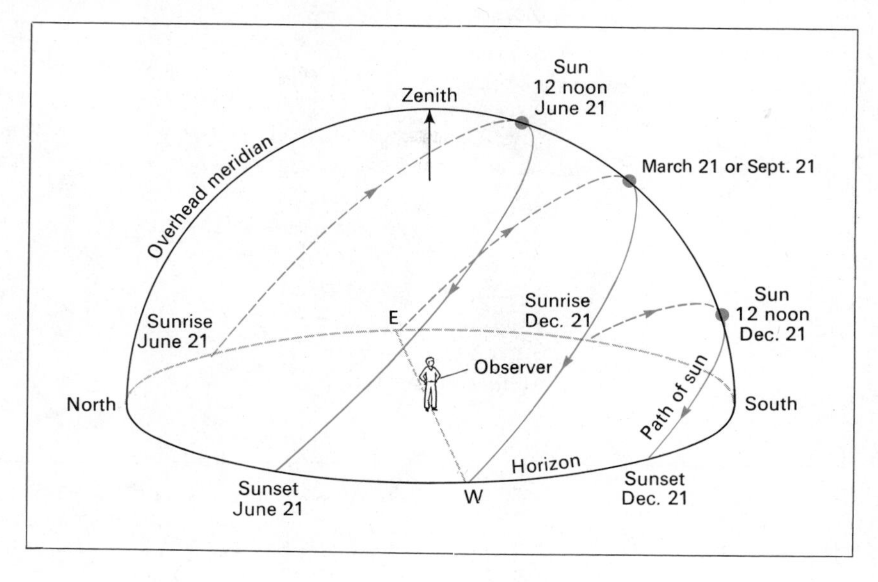

Figure 19.19 Diagram showing the apparent path of the Sun across the sky on June 21 and December 21, as observed from Washington, D.C. (39° N latitude).

The seasons have a tremendous effect on the lives of everyone. Our yearly life cycles are ordered by the season's progressions. Many of our holidays were originally celebrated as commemorating a certain season of the year. The celebration that has evolved into our Easter holiday was originally a celebration of the coming of spring and a renewal of nature's life. Halloween originally commemorated the beginning of the winter season, while Thanksgiving commemorated the end of the harvest. The ancient festival of the winter solstice (on December 21 or December 22) and the beginning of the northward movement of the Sun has evolved into our Christmas holiday on December 25. The reasons for celebrating our various holidays have changed over the course of time, but the original dates were set by nature's annual timepiece—the movement of the Earth around the Sun.

19.5 Precession of the Earth's Axis

Most of us are acquainted with the action of a toy top that has been placed in rapid motion and allowed to spin about its axis. After spinning a few seconds, the top begins to wobble or do what physicists call "precess." See Fig. 19.20. The top, a symmetrical object, will continue to spin about an axis if the center of gravity remains above the point of support. When the center of gravity is not in a vertical line with the point of support, the axis slowly changes its direction. This slow rotation of the axis is called **precession.**

Because the Earth is spinning rapidly, it bulges at the equator and cannot be considered a perfect sphere. The moon and Sun apply a gravitational torque to the Earth; this tends to bring the Earth's equatorial plane into its orbital plane. Because of this torque, the axis of the Earth slowly rotates clockwise or westward about the vertical or the north ecliptic pole (Fig. 19.21). The pe-

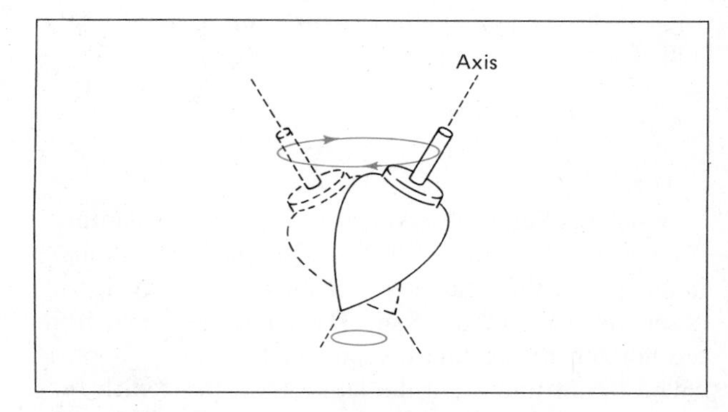

Figure 19.20 Precession of a top.

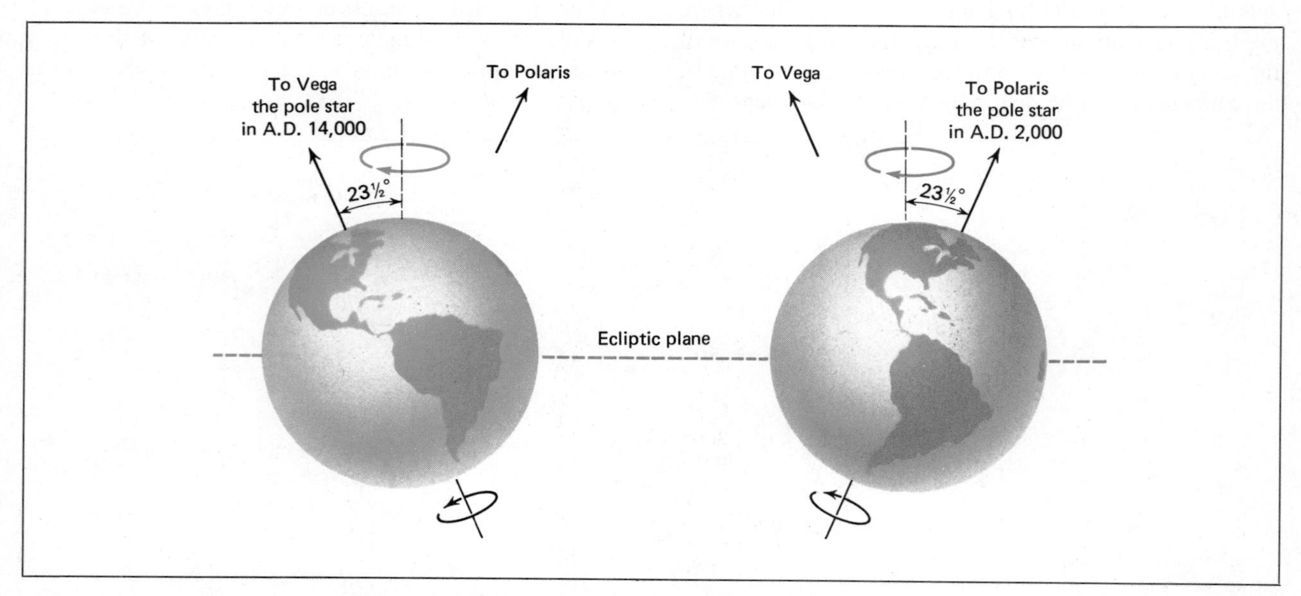

Figure 19.21 Precession of the Earth's axis clockwise about a line perpendicular to the Earth's orbit. The Earth's axis is presently pointing toward the star Polaris, which we call the pole, or north, star. In about 12,000 years the axis will be pointing toward the star Vega in the constellation Lyra.

riod of the precession is 25,800 years, that is, it takes 25,800 years for the axis to precess through 360°.

Since the precession is clockwise and the Earth is revolving counterclockwise around the Sun, the tropical year is approximately 20 min shorter than the sidereal year. As the axis precesses, Polaris will no longer be the north star. The star Vega in the constellation Lyra will be the north star some 12,000 years from now. The Southern Cross, a constellation of stars located within 27° of the present south celestial pole, will then be visible from Washington, D.C.

Our calendar is based on the seasons of the year. We want our summers to be warm and our winters to be cold. As the Earth precesses, we define June 21 to be when the northern hemisphere tilts toward the Sun at a maximum angle. This means that as the Earth precesses, the stars seen at various seasons change. This is shown in Fig. 19.22. The stars we see on summer nights will slowly change within a period of 25,800 years. In the year 14,900 A.D. stars seen on summer nights will be our present winter night stars. Due to precession, the 12 constellations in the zodiac will slowly cycle through different months with a one-month change occurring every 2150 years (25,800/12). The stars seen overhead on June 21 some 2150 years ago are seen overhead on July 21 today, and will be seen overhead on August 21 in another 2150 years.

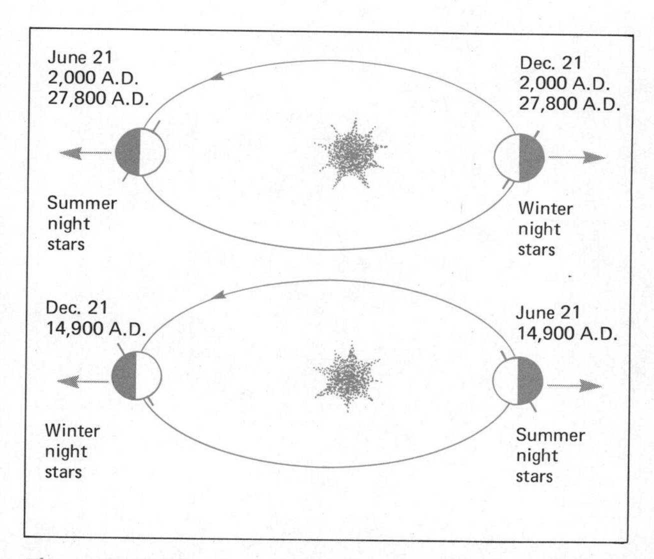

Figure 19.22 The precession of the Earth's axis will cause future generations to see different stars in the summer than are seen today. After 12,900 years our winter night stars will be their summer night stars and vice versa. After another 12,900 years, the constellations will be similar to what we see now.

In the early 1970s, a popular song called "The Age of Aquarius" took its title from the fact that the constellation Aquarius was moving from its January 21 to February 21 time slot of thousands of years ago toward the March 21 to April 21 time slot. Since older cultures started the year on March 21, the "age" or 2150-year period took on aspects of the March 21 to April 21 constellation. The song spoke of "the dawning of the age of Aquarius" which we see would slowly occur as the Earth precessed.

19.6 The Calendar

The continuous measurement of time requires the periodic movement of some object as a reference. Various lengths of time had a direct influence on the life of ancient men and women. Because of these influences, they had more than one reference for measuring the events of their lives. It is reasonable to believe that the first unit for the measurement of time was the day; because of the daily need for food people spent the daylight hours hunting. During the dark hours they slept.

A longer period of reference was based on the periodic movement of the moon. Some societies probably used the moon as a basic division of time, since they were unable to count the days over a long period. The first appearance of the crescent moon and the time of the full moon were times of worship for many primitive tribes.

Our month of today originated from the periodic movement of the moon, which requires $29\frac{1}{2}$ solar days to orbit the Earth. The plans for the first calendar seem to have originated before 3000 B.C. with the Sumerians, who ruled Mesopotamia. Their calendar was based upon the motion of the moon, which divided the year into 12 lunar months of 30 days each. Since $30 \times 12 = 360$ days, and the year actually contains $365\frac{1}{4}$ days, corrections had to be made to keep the calendar adjusted to the seasons. How the Sumerians made the corrections remains a mystery, but their successors, the Babylonians, adjusted the length of the months and added an extra month when needed. This Babylonian calendar set the pattern for many of the calendars adopted by ancient civilizations.

The calendar we use today originated with the Romans. The early Roman calendar contained only 10 months, and the year began with the coming of spring. The months were named March, April, May, June, Quintilis, Sextilis, September, October, November, and

December. The winter months of January and February did not exist. This was the period of waiting for spring to arrive. It was about 700 B.C. that Numa, who reigned in Rome, added the month of January at the beginning of the year, before March; and February at the end of the year, after December. It was about 275 years later, in 425 B.C., that the two months were changed to the present order.

The so-called Julian calendar was adopted in 45 B.C., during the reign of Julius Caesar. Augustus Caesar, who ruled the Roman Empire after Julius, renamed the month Quintilis as July in honor of Julius, and the month Sextilis as August in honor of himself. He also removed one day from February, which he added to August to make it as long as the other months.

The Julian calendar had 365 days in a year. In every year divisible by four, an extra day was added to make up for the fact that it takes approximately $365\frac{1}{4}$ days for the Earth to orbit the Sun. Thus, in 1985, 1986, and 1987 there are 365 days. In 1984 and 1988 there are 366 days. The Julian calendar was fairly accurate and was used for over 1600 years.

In 1582, Pope Gregory XIII (Fig. 19.23) realized that the Julian calendar was slightly inaccurate. The vernal equinox was not falling on March 21, and religious holidays were coming at the wrong time. A discrepancy was found, and it was decreed that 10 days would be skipped to correct it. The discrepancy was due to the fact that there are 365.2422 days in a year and not 365.25 as the Julian calendar used.

Pope Gregory set the calendar straight and devised a method to keep it correct. He decreed that every 400 years, three leap years would be skipped. The leap years to be skipped were the century years not divisible by 400. Table 19.1 shows examples of the number of days in some century years. Remember, however, that all other years divisible by four are leap years and have 366 days. Our present day calendar with these leap year designations is called the **Gregorian calendar.**

Pope Gregory's reform of 1582 was not universally accepted. Most Protestant countries did not immediately go along. England and the American colonies finally changed their calendars in 1752 when there was an 11-day discrepancy. To correct it, September 2 was followed by September 14 and some people thought they were losing 11 days of their life. Other problems also arose when landlords asked for a full month's rent and banks sought to collect a full month's interest. In fact, riots were touched off in London over the change. When Russia and China accepted the Gregorian calendar in the

Figure 19.23 Pope Gregory XIII. In 1582 he issued a proclamation to drop 10 days from the calendar and to have leap years 97 times (instead of 100) every 400 years. The Gregorian calendar is now used worldwide. (Photo courtesy Gordon Moyer.)

Table 19.1 Number of Days in Century Years

Year	Number of Days
1600	366
1700	365
1800	365
1900	365
2000	366
2100	365
2200	365
2300	365
2400	366
2500	365

Table 19.2 The Days of the Week

Heavenly Object	English Name	French Name	Saxon Name
Sun	Sunday	dimanche	Sun's day
Moon	Monday	lundi	Moon's day
Mars	Tuesday	mardi	Tiw's day
Mercury	Wednesday	mercredi	Woden's day
Jupiter	Thursday	jeudi	Thor's day
Venus	Friday	vendredi	Fria's day
Saturn	Saturday	samedi	Saturn's day

early 1900s, it marked the first time that the whole world used the same calendar. There are other calendars still in use for religious purposes, but for civil events the Gregorian calendar is used around the world.

In our calendar we have seven days in each week. The origin of the seven-day week is not definitely known, and not all cultures have had a seven-day week. One possible origin of the seven-day week is that it takes approximately seven days for the moon to go from one phase to the next (e.g., new to first-quarter phase).

A more likely explanation concerns the fact that as the ancients watched the sky night after night, there were exactly seven celestial bodies that moved relative to the fixed stars. These seven objects visible to the naked eye are the Sun, moon, and five visible planets: Mars, Mercury, Jupiter, Venus, and Saturn.

Our present days of the week can still be connected by their names to the Sun, moon, and five visible planets. Table 19.2 shows the heavenly object, English name, French name, and the Saxon name for the seven days of the week. Note that either the English or French word is similar to the name of the heavenly object.

Our English days, Tuesday, Wednesday, Thursday, and Friday, come from the words Tiw, Woden, Thor, and Fria, who were Nordic gods. Woden was the principal Nordic god; Tiw and Thor were the gods of law and war; and Fria was the goddess of love. Sunday, Monday, and Saturday retain their connection to the Sun, moon, and Saturn.

Learning Objectives

After reading and studying this chapter, you should be able to do the following without referring to the text:

1. Define latitude and longitude and give their ranges.
2. State the latitude of the equator and north and south poles.
3. State how the prime meridian and International Date Line are determined.
4. Calculate the date and time at any latitude and longitude given the date and time at another latitude and longitude.
5. Explain, with the aid of a diagram, why we have different seasons of the year.
6. Tell when we have the vernal equinox, autumnal equinox, summer solstice, and winter solstice each year.
7. Compute the altitude of the Sun at noon for 40° N latitude on March 21, June 21, September 21, and December 21.
8. Explain the origin of the dates of several holidays (e.g., Christmas, Easter).
9. Draw a diagram and use it to explain the precession of the Earth's axis.
10. Explain why we have 12 months in a year and seven days in a week in our present calendar.
11. Explain when we have leap years in the Gregorian calendar.

Important Words and Terms

cartesian coordinate system
latitude
parallels
meridians
longitude
Greenwich meridian
apparent solar day
mean solar day
sidereal day
ante meridiem
post meridiem
standard time zones
International Date Line
Daylight Saving Time
tropical year
declination
summer solstice
winter solstice
vernal equinox
autumnal equinox
altitude
zenith
zenith angle
horizon
precession
Gregorian calendar

Questions

1. On the Earth's surface, can you go north indefinitely?
2. On the Earth's surface, can you go west indefinitely?
3. How is zero degrees defined for latitude? for longitude?
4. What is the possible range of latitude? of longitude?
5. What do A.M. and P.M. mean?
6. How are standard time zones approximately determined?
7. How many time zones are there in the 48 contiguous states? In the world?
8. What is the purpose of Daylight Saving Time?
9. Determine when the following occur in the northern hemisphere: autumnal equinox, spring equinox, winter solstice, summer solstice.
10. Exactly how many days are in the tropical year?
11. What causes the seasons?
12. How would the seasons be modified if the Earth's axis tilted at 10° instead of 23.5°?
13. What are the origins of the dates for Halloween (October 31) and Christmas (December 25)?
14. How long does it take the Earth to precess one time?
15. What is the origin of the month?
16. What is the origin of the seven-day week?
17. How often was there a leap year in the Julian calendar?
18. How often is there a leap year in the Gregorian calendar?
19. What is the altitude of the north star (Polaris) for an observer at the equator (0° latitude)? At the north pole (90° N)?
20. What is the altitude of the north star for an observer at Washington, D.C. (39° N)?

Problems

1. How far away is the point at 90° N, 130° E from 90° N, 150° E?
2. Draw a diagram and explain why the points 60° N, 130° E and 60° N, 150° E are closer together than the points 30° N, 130° E and 30° N, 150° E.
3. What are the latitude and longitude of the point on the Earth that is opposite Washington, D.C. (39° N, 77° W)?
   Answer: 39° S. 103° E
4. What are the latitude and longitude of the point on the Earth that is opposite Tokyo (36° N, 140° E)?
5. A professional basketball game is to be played in Portland, Oregon. It is televised live in New York beginning at 9 P.M. EST. What time must the game begin in Portland?
   Answer: 6 P.M. PST
6. If the polls close during a presidential election at 7 P.M. EST in New York, what is the time in California?

7. If an Olympic event begins at 10 A.M. on July 28 in Los Angeles (34° N, 118° W), what time and date will it be in Moscow (56° N, 38° E)? *Answer:* 9 P.M., July 28
8. When it is 9 P.M. on November 26 in Moscow (56° N, 38° E), what time and date is it in Tokyo (36° N, 140° E)?
9. When it is 10 A.M. on February 22 in Los Angeles (34° N, 118° W), what time and date is it in Tokyo (36° N, 140° E)?
10. What is the altitude angle of the Sun on March 21 for someone at the North Pole?
11. What is the altitude angle for someone at 42° N latitude on (a) March 21, (b) June 21? *Answer:* (a) 48°
12. What is the altitude angle for someone at 34° N latitude on (a) September 21, (b) December 21?
13. What is the latitude of someone in the United States who sees the Sun at an altitude angle of 71.5° on June 21? *Answer:* 42° N
14. What is the latitude of someone in the United States who sees the Sun at an altitude angle of 31.5° on December 21?
15. How many days are in each of the following years: 1986, 1987, 1988, 1989, 1990, 1991, 1992, 2000, 2001, 2004, 2100, 2200, 2300, 2400?
16. Determine the month and day when the Sun is at maximum altitude for an observer at Washington, D.C. (39° N)? What is the altitude of the Sun at this time? *Answer:* on or about June 21
17. Determine the month and day when the Sun is at minimum altitude for an observer at Washington, D.C. (39° N)? What is the altitude of the Sun at this time? *Answer:* $27\frac{1}{2}$ degrees
18. Is the difference between the maximum and minimum altitude of the Sun as determined in problems 16 and 17 equal to twice the angle the Earth's axis is tilted from the vertical?

20

The Moon

"That's one small step for a man,
one giant leap for mankind."
–Neil Armstrong

THE EXACT ORIGIN of the word "moon" seems to be unknown, but many writers believe the original meaning related to the measurement of time. We do know that the length of our present month is based upon the motion and the phases of the moon, that primitive people worshipped the moon, and that many societies today base their religious ceremonies on the new and full phases of the moon. We also know the human reproduction cycle is synchronized to a lunar cycle with the ovaries producing ova about every 28 days.

On July 20, 1969, human beings first landed on the moon, and Apollo 11 astronauts placed a retroreflector (an optical reflector designed to return the reflected ray of a laser beam exactly parallel to the incident ray) array on the moon's surface. The retroreflector is part of a lunar-ranging experiment that measures the distance to the moon with an accuracy of 15 centimeters. Measurements with this high accuracy can be taken over long periods of time and will show the variation in the orbital distance of the moon in great detail. Such measurements can be used to (1) determine the rate of continental drift on the Earth (latitude and longitude of a place can be determined with great accuracy); (2) detect any change in the location of the north pole; (3) determine the orbit of the moon with greater accuracy; and (4) determine if the gravitational constant (G) is decreasing very slowly with time, because the universe is believed to be expanding.

The moon is a rather insignificant body if viewed from outside our solar system, but it is our largest natural satellite and appears as the second brightest object in the sky to the Earth observer because it is very close to us. Its average distance from the Earth is about 240,000 miles. Because of the moon's nearness and its influence on our lives, this chapter is devoted to its study. Fig. 20.1 on the next page shows scientist-astronaut Harrison H. Schmitt collecting samples of the lunar surface.

◀ The first landing on the moon, July 20, 1969.

20.1 General Features

The Earth's moon at its brightest is a wondrous sight as it reflects the Sun's light back to Earth. The moon appears quite large to Earth observers. In fact, our moon is the largest moon of any inner planet. Mercury and Venus have no moons and the moons of Mars are quite small. Our moon is the fifth largest in the solar system. A unique feature of the Earth and the moon is that they are nearer in size than any other planet and its satellite.

The moon revolves around the Earth once every $29\frac{1}{2}$ days, and it rotates at the same rate as it revolves. For this reason, we only see one side of the moon. An observer on the side of the moon facing the Earth would always be able to see the Earth, but the Sun would appear to rise and set and rise again once every $29\frac{1}{2}$ days. Thus, all sides of the moon are heated by the Sun's rays.

The moon is spherical in shape, with a diameter of 2160 mi, a value slightly greater than one-fourth the Earth's diameter. The slow rotation of the moon, coupled with the tidal bulge caused by the Earth's gravitational pull on the solid material, produces an oblateness that is very small. The best measurements indicate a difference of less than a mile between the polar and equatorial diameters.

The moon has mass $\frac{1}{81}$ that of the Earth, and an average density of 3.3 g/cm^3. The Earth's average density is 5.5 g/cm^3. The surface gravity of the moon is only $\frac{1}{6}$ that of the Earth. This means your weight on the surface

Figure 20.1 Scientist-Astronaut Harrison H. Schmitt collecting lunar rake samples at Station One during the first Apollo 17 extravehicular activity at the Taurus-Littrow landing site. (Courtesy NASA.)

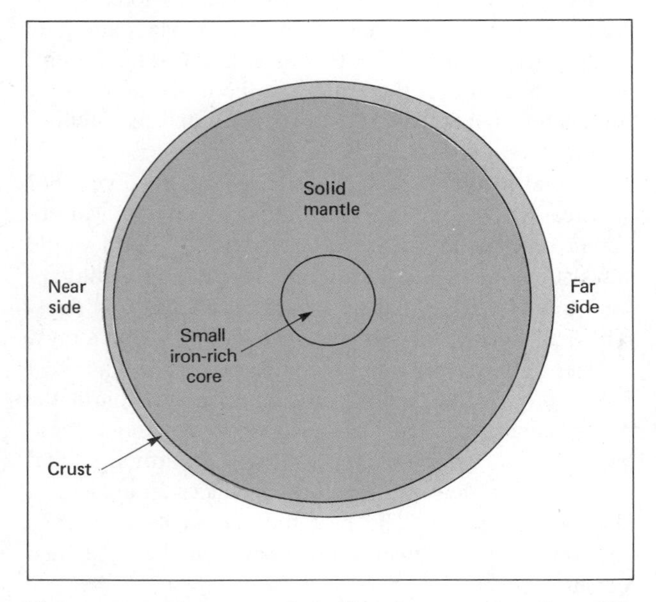

Figure 20.2 The moon is believed to consist of a solid mantle rich in silicate materials, a crust that varies in depth from about 40 miles on the near side to 80 miles on the far side from the Earth, and a small, perhaps solid, iron-rich core.

of the moon would be about $\frac{1}{6}$ of your weight at the Earth's surface. The moon's interior is thought to be made up of a small, perhaps solid, iron-rich core, a solid mantle, and a crust that is about 40 miles thick on the near side and 80 miles thick on the far side (Fig. 20.2). The moon does not possess a magnetic field. At least, no magnetic field was detected by instruments carried by the Apollo astronauts. Surface rocks brought back from the moon show some magnetism, indicating the moon had a slight magnetic field when the rocks solidified. The origin of this previous magnetic field is not known.

Except for the change in the phases of the moon, its most predominant feature is the appearance of the surface, which is marked with craters, plains, rays, rills, mountain ranges, and faults (Fig. 20.3). These features vary in size, shape, and structure. The most outstanding are the craters that are clearly visible to an Earth observer with low-power binoculars or a telescope.

Craters

The word **crater** (Greek *kratēr*) means bowl-shaped, and the lunar craters are great pits believed to be caused by the impact and explosion of small and large objects that have come from space. The craters are rather shallow (their depths are small in comparison to their diameters) and their floors are located below the lunar surface. See Fig. 20.4. The slope of a rim with a width of about 1/5 the width of the crater from crest to crest is greater on the inside than on the outside. When the measured volume of the material in the rim is compared to the volume of the crater hole, we find they are approximately the same. This tends to support the impact hypothesis.

Plains

There are thousands of craters on the lunar surface, ranging in diameter from a few feet to the 150-mi Clavius. The large flat areas called *maria* (Italian, seas), named by Galileo, are believed to be craters formed by the impact of huge objects from space and which later were filled with lava. These areas, which are now called **plains,** are very dark in appearance because the moon's surface is a poor reflector. The surface reflects only 7% of the light received from the Sun. The plains, which are similar to black asphalt, reflect very little light. There are 14 major plains on the front side which cover over 50% of the visible lunar surface. Most of the plains are located in the northern hemisphere and can be plainly seen with the naked eye during the full phase of the moon.

Figure 20.3 Lunar earthside hemisphere. USAF Lunar Reference Mosaic showing the features of the moon. Numbers 1, 2, 6, and 7 are Mare Imbrium, Mare Serenitatis, Mare Tranquillitatis, and Mare Crisium, respectively; number 3 the Apennines Mountains; number 8 the Great Straight Wall; and numbers 4, 5, 9, and 10 are the craters Kepler, Copernicus, Tycho, and Clavius, respectively. The numbers R7, R8, and R9 represent the landing positions of the U.S. lunar probes of the Ranger (hard-landing spacecraft) series. The first six flights were unsuccessful. Numbers S1 and S3 represent the landing positions of the U.S. lunar probes of the Surveyor (soft-landing spacecraft) series. Numbers A11, A12, A14, A15, A16, and A17 represent the landing positions of Apollo 11, 12, 14, 15, 16, and 17, respectively. The numbers L9 and L13 represent the landing positions of Russia's soft-landing probes Luna 9 and Luna 13. (Courtesy U.S. Air Force.)

Figure 20.4 Photograph of the lunar surface taken from Apollo 11 spacecraft in lunar orbit. The large central crater is International Astronomical Union No. 308, which has a diameter of about 50 miles. (Courtesy NASA.)

The surface of the moon is a terrain of rolling rounded knolls composed of a layer of loose debris or soil called regolith which has a depth less than 10 meters on the flat lunar plains (maria). The lunar highlands, since they are older, have a thicker layer of regolith. The rock samples brought back by Apollo astronauts are similar to the volcanic rock found on the Earth.

Rays

Some craters are surrounded by streaks, or **rays,** that extend outward over the surface. These are believed to be pulverized rock that was thrown out when the crater was formed. The rays appear much brighter than the crater, and it is known that powdered rock reflects light better than regular-sized rock. The rays also become darker with age. Photographs show that in cases where rays from one crater overlap the rays of another, the rays on top appear to be brighter. The ray system of a crater has an average diameter of about 12 times the diameter of the crater. The lunar photographs also show that the ray systems are marked with small craters called secondary craters, which are believed to have been formed by debris thrown out from the primary crater during the explosion caused by an impinging object from space.

Rills

Another feature of the lunar surface is the existence of long narrow trenches, or valleys, called **rills.** They vary from a few feet to about three miles in width and extend hundreds of miles in length with little or no variation of width. Some rills are rather straight, while others follow a circular path. The rills have very steep walls and fairly flat bottoms that are as much as $\frac{1}{2}$ mi below the lunar surface. Moonquakes are thought to cause the formation of rills. A similar separation of the Earth's surface would be produced by an earthquake.

Mountain Ranges

The **mountain ranges** on the lunar surface have peaks as high as 20,000 ft, and all formations seem to be components of circular patterns bordering the great plains, or maria. This indicates they were not formed and shaped by the same processes as mountain ranges on the Earth, which were formed by internal forces.

The formation of craters and other surface features of the moon took place a long time ago when the solar system was filled with large amounts of matter, and these features are much the same now as they were when formed, because of the absence of water and a lunar atmosphere. There has been some change due to impact of projectiles from space.

Figure 20.5 Moon at last quarter phase. Notice the Straight Wall of Mare Nubium in the lowest dark area. (Lick Observatory Photographs.)

A **fault** is a break, or fracture, in the surface of the moon in which a vertical displacement has taken place. Several of these are observed on the lunar surface. Figure 20.5 shows the last-quarter moon with a very large fault in Mare Nubium. This fault (called the Straight Wall) is about 60 mi long and 1200 ft high, and its side is inclined about 40° to the horizontal.

20.2 History of the Moon

Before the Apollo program to land a human being on the moon was begun in the early 1960s, very little was known about the origin and history of the moon. But our exploration of the moon has changed all that.

The first landing on the moon was July 20, 1969,

when the landing craft of Apollo 11 descended to the moon's surface in Mare Tranquillitatis at 0.67° North, 23.49° East. Since then, five other Apollo lunar landing missions (Apollo 12, 14, 15, 16, and 17) have been completed as shown in Fig. 20.3. Apollo 12 landed in Oceanus Procellarum on November 19, 1969; Apollo 14 landed in Fra Mauro highlands on January 31, 1971; Apollo 15 landed in Hadley-Apennines region on July 30, 1971; Apollo 16 landed in Descartes region on April 21, 1972; and Apollo 17 landed in a valley at Taurus-Littrow on December 11, 1972. The Apollo astronauts collected and brought back to Earth 379 kg of lunar material and erected on the lunar surface 2104 kg of scientific instruments that will collect data for many years. The American Apollo program has now ceased due to budgeting considerations.

The rock samples brought back from the moon, such as shown in Fig. 20.6, have enabled us to have a much better understanding of the moon's origin and history. Samples from the plains or lowlands have yielded ages considerably younger than samples from the highlands. Rocks from the highlands were formed between 3.9 and 4.4 billion years ago, whereas the rocks from plains or lowlands have ages between 3.8 and 3.1 billion years. No rocks older than 4.4 billion years, or younger than 3.1 billion years have been found.

Almost all of the craters on the moon are now known to be due to the bombardment of meteorites of various sizes. Because the moon has no atmosphere or water on its surface, very little erosion takes place on the moon. Once formed, a crater remains for billions of years or until a meteorite hits to form a crater on top of it. Compare this to the Earth, which has only a few remaining meteorite craters. The craters left by meteorites striking the Earth have been eroded away for the most part.

The plains and a few (about 1%) of the craters were produced by volcanic eruptions on the moon. The plains are composed of black volcanic lava that covered many craters. Most of the plains are on the near side of the moon. This can be seen by studying Fig. 20.7. The lower right one-third of the moon shown in this photograph is the far side of the moon. The fact that there were fewer volcanic eruptions on the moon's far side is probably correlated with the fact that the moon's crust is thicker there (see Fig. 20.2).

From the ages of moon rocks and other data, we now have a fairly solid understanding of the moon's origin and history. However, as more samples are brought back from the moon and more studies are done, changes may occur in these theories.

Figure 20.6 Lunar sample. A close-up view of a 2-centimeter chip of breccia collected by the Apollo 11 astronauts during the landing in July 1969. (Courtesy NASA.)

Figure 20.7 The full moon showing the far side of the moon in the lower right one-third of the photo. Note the scarcity of plains (black regions) on the far side compared to the near side. (Courtesy NASA.)

The Earth and moon are both believed to have originated in the same part of the solar system at about the same time 4.6 billion years ago. At this time, the Earth and moon formed from the agglomeration of many smaller pieces of rock and other matter. The Earth was a much larger body, and it was formed from higher density materials than the moon. The moon's surface was hot and molten until about 4.4 billion years ago.

The oldest rocks on the moon were formed about 4.4 billion years ago when the moon's crust was cool enough to solidify. From 4.4 to 3.9 billion years ago the moon was intensely bombarded by many meteorites that were still present near the Earth-moon system. This was the period when most of the moon's craters were formed.

During the period 3.9 to 3.1 billion years ago the moon's interior had heated up enough from radioactive effects to cause volcanic eruptions to occur that formed the many plains. The lava flows from these eruptions covered much of the moon's lowlands. During this period, meteorite bombardment was getting less intense as fewer and fewer rock fragments were left near the Earth-moon system.

After 3.1 billion years ago, the moon's mantle had become so thick that it could no longer be penetrated by molten rock, and the moon has been geologically quiet since that time. Meteorites have continued to bombard the surface and have formed a thin veneer of dust several feet thick on the moon's surface.

Much of our knowledge of the moon has been gained by the analysis of rock samples. Another important avenue of investigation has been the study of moonquakes. The seismological data gathered have enabled us to come up with Fig. 20.2. Other studies have shown that there is practically no water at all in any form on the moon. In addition, no biological life has been found.

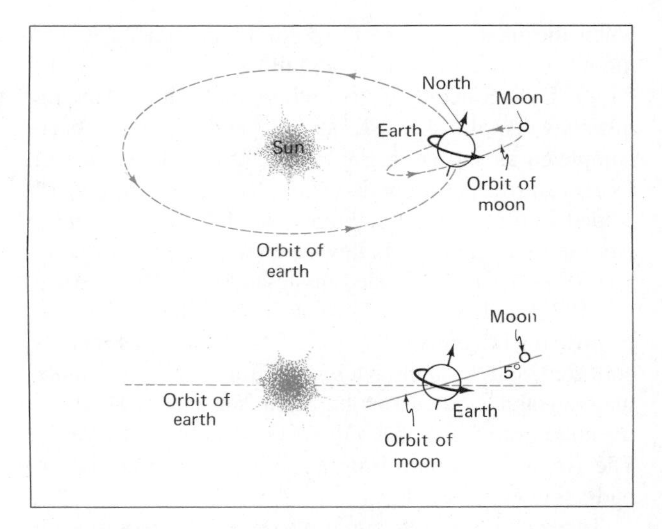

Figure 20.8 The relative motions of the moon and Earth. The top diagram is that of a view at an angle above the ecliptic plane; the lower diagram is a view parallel to the ecliptic plane.

20.3 Lunar Motion

The Earth's moon revolves eastward around the Earth in an elliptical orbit once every $29\frac{1}{2}$ solar days or $27\frac{1}{3}$ sidereal days. The orbital plane of the moon does not coincide with the orbital plane of the Earth, but is tilted at an angle of approximately 5° with respect to the Earth's orbital plane. The moon rotates eastward as it revolves, making one rotation during one revolution. Figure 20.8 is an illustration of the eastward motion of the moon and the inclination of the orbital plane to the ecliptic.

Since the moon revolves in an elliptical orbit, the distance from the Earth to the moon varies as the moon revolves. At the closest point (called **perigee**), the moon is 221,463 mi from the Earth, and at the farthest point (called **apogee**), the moon is 252,710 miles from the Earth. This change in distance of over 31,000 mi produces only a slight change in the apparent size of the moon as viewed from the Earth. The mean apparent diameter of the moon is slightly over one-half degree.

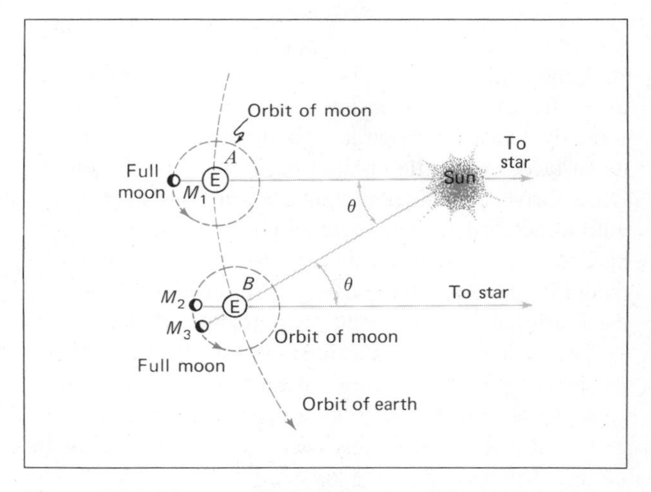

Figure 20.9 Diagram illustrating the difference between the sidereal and synodic months. With the Earth at position A and the moon at position M_1, the moon, Earth, and Sun are in the same plane. As the Earth revolves through the angle θ, arriving at position B, the moon revolves through 360° to position M_2, and one sidereal month—one revolution in respect to a star—has elapsed. The moon must move through 360° plus an angle equal to θ before arriving at position M_3, at which time the moon, Earth, and Sun will again be in the same plane, and one synodic month will have taken place.

The mean distance to the moon is 238,856 mi, but 240,000 mi will be used as the mean for solving problems in this text.

As stated above, there are two different lunar months. The period of the moon in respect to a star other than the Sun is $27\frac{1}{3}$ days; this is called the sidereal period, or **sidereal month,** and is the actual time taken for the moon to revolve 360°. The period of the moon in respect to the Sun is $29\frac{1}{2}$ days; this is called the **synodic month,** or the month of the phases. The moon revolves more than 360° during the synodic period. See Fig. 20.9.

To an observer on Earth, the moon appears to rise in the east and set in the west each day. This apparent motion of the moon is due to the spinning of the Earth on its axis once each day. The times at which it rises and sets are discussed in the following section.

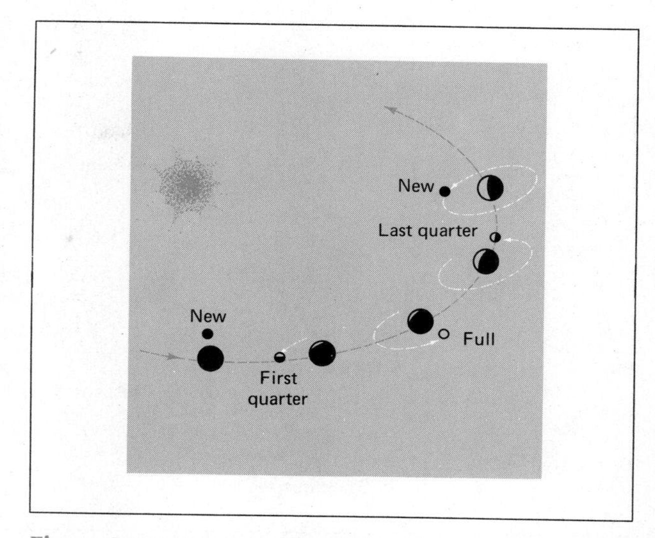

Figure 20.10 Diagram illustrating the eastward (counterclockwise, looking from above) motion of the Earth and moon around the Sun.

20.4 Phases

The most outstanding feature presented by the moon to an Earth observer is the periodic change in its appearance. One-half of the moon's surface is always reflecting light from the Sun, but only once during the lunar month does the observer on the Earth see all of the lighted half of the moon. Throughout most of the moon's period of revolution, only a portion of its lighted side is presented to us.

The starting point for the periodic, or cyclic, motion of the moon is arbitrarily taken at the new phase position. New phase of the moon occurs when the Earth, Sun, and moon are in the same plane, with the moon positioned between the Sun and Earth. They are not necessarily in a straight line. This corresponds to the position at far left in Fig. 20.10. At this position, the dark side of the moon is toward the Earth and the moon cannot be seen from this planet. Since the Sun is on the observer's meridian with the moon, the new moon occurs at 12 noon local solar time.

The **new moon** occurs just for an instant—the instant it is on the same meridian as the Sun. We often speak, however, of a phase of the moon lasting for a full day of 24 h.

The moon revolves eastward from the new phase position and for the next $7\frac{3}{8}$ solar days (one-fourth of $29\frac{1}{2}$ days) it is seen as a waxing crescent moon. The term **waxing** means that the bright portion of the moon is getting larger while **waning** means that the bright portion is getting smaller. A **crescent** moon is a moon in which less than one-quarter of the moon's surface is bright, as we observe it from the Earth. A **gibbous** moon occurs when more than one-quarter of the moon's surface appears bright, when seen by an Earth-based observer. Figures 20.11 and 20.12 illustrate how the phases occur and appear to look to an observer on the Earth.

The moon is in the *waxing crescent phase* and appears as a crescent moon to an Earth-bound observer when the moon is less than 90° east of the Sun. The moon is in **first-quarter phase** when the moon is exactly 90° east of the Sun and appears as a quarter moon on the observer's meridian at 6 P.M. local solar time. First-quarter phase occurs only for an instant, since the moon can only be 90° east of the Sun for an instant. See Figs. 20.10 and 20.11.

From the first-quarter position, the moon enters the *waxing gibbous phase* for $7\frac{3}{8}$ solar days. During this phase the moon appears larger than a quarter moon but less than a full moon. When the moon is exactly 180° east of the Sun, the moon will be in full phase and will appear as a **full moon** to the Earth-bound observer. The full moon appears on the observer's meridian at 12 midnight local solar time.

From the full phase position, the moon enters the *waning gibbous phase* and remains in the waning gibbous phase for $7\frac{3}{8}$ solar days. The appearance of the moon during this phase is the same as the waxing gibbous phase, except that the bright side of the moon is toward the east and the moon is seen in the sky at a

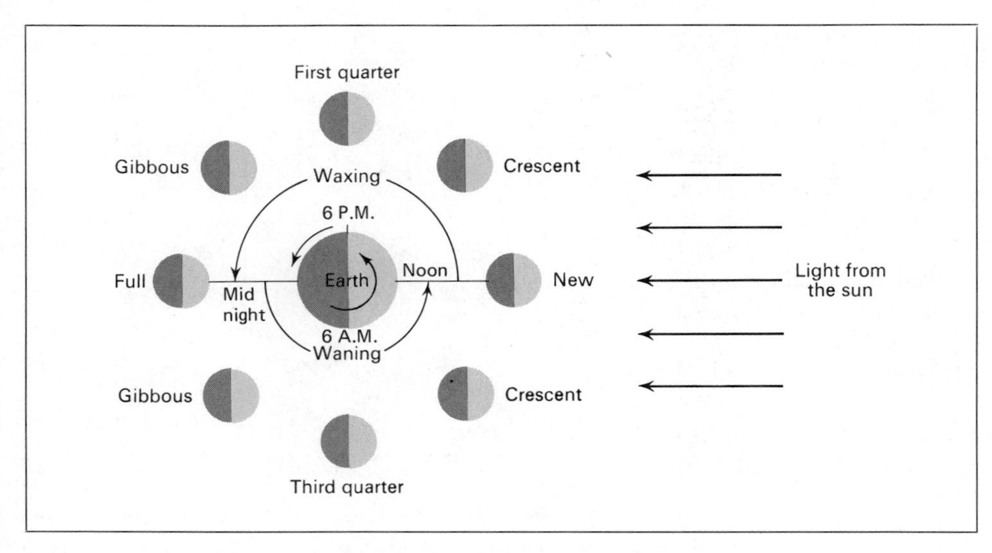

Figure 20.11 Phases of the moon as observed from a position in space above the north pole of the Earth.

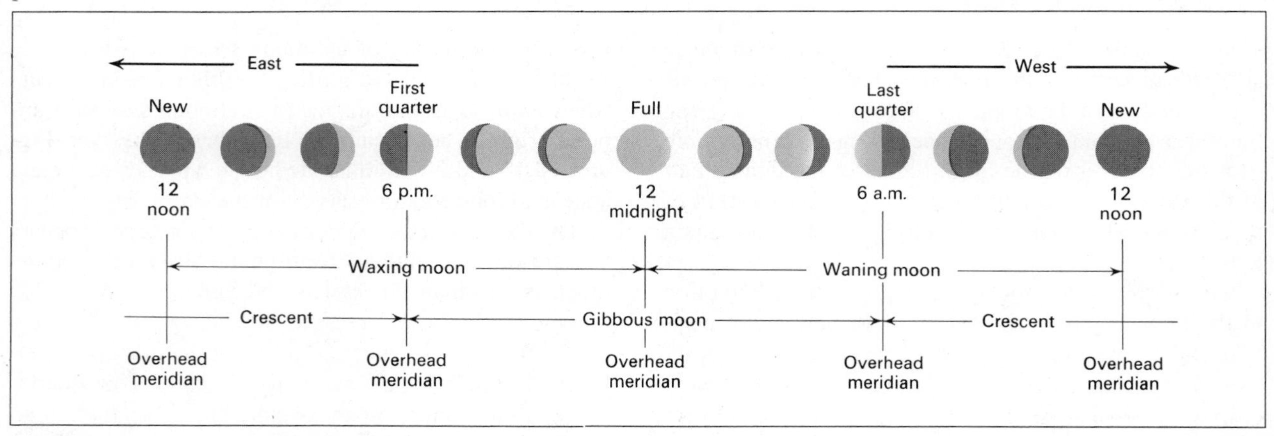

Figure 20.12 Drawing illustrating the phases of the moon as observed from the United States. The observer is looking south; therefore, east is on the left. The Sun's position can be determined by noting the time (local solar) the moon is on the overhead meridian. The time period represented in the drawing is $29\frac{1}{2}$ days, or one synodic month.

different time. When the moon is exactly 270° east of the Sun, the moon will be in the third, or last, quarter phase. The **last-quarter phase** appears on the observer's meridian at 6 A.M. local solar time, with the bright side of the moon toward the east.

From the last-quarter position, the moon enters the *waning crescent phase* and remains in this phase for $7\frac{3}{8}$ solar days. During this phase the moon appears smaller than a quarter moon. Its appearance is the same as the waxing crescent moon except the bright side of the moon is toward the east, and the moon appears in the sky at a different time.

Figure 20.12 illustrates the moon's appearance and position above the southern horizon as observed from a northern latitude greater than $28\frac{1}{2}°$ north. The moon is shown in the first drawing on the left in the new phase position. The moon is shown on the observer's meridian at 12 noon local solar time and shaded black, illustrating that it cannot be seen at this time since it is on the meridian with the Sun. The next three positions illustrate the waxing crescent phase. Note the face of the moon is appearing larger in size as the moon approaches the first-quarter phase, and that the bright side is toward the west where the Sun is located.

When the moon is in first-quarter phase, the moon is on the observer's meridian at 6 P.M. local solar time. The Sun will be at or near the western horizon at this time. The moon revolves eastward entering the waxing gibbous phase as shown by the next three positions. Note the moon is larger than a quarter moon and still increasing in size of face. The bright side is still toward the west.

The next position shows the moon at full phase and on the observer's meridian at 12 midnight. If the date is March 21 or September 21 the moon will rise on the eastern horizon at 6 P.M., when the Sun is setting in the west, and the moon will set at 6 A.M., as the Sun is seen rising on the eastern horizon.

The moon revolves on eastward entering the waning gibbous phase. Note that the size of the face is decreasing, and that the bright side of the moon is toward the east—just the opposite from the waxing gibbous moon. When the moon is 90° west of the Sun (same as 270° east of Sun), it will be on the observer's meridian at 6 A.M. local solar time as shown in the next position. The moon appears as a quarter moon, but note that the bright side is toward the east. The Sun will be rising at or near this time.

After the third-quarter phase, the moon enters the waning crescent phase and continues to decrease in size of face. Note the bright side of the waning crescent moon is toward the east. The last position shows the moon back to the new phase position.

Table 20.1 summarizes the times for the various phases of the moon to rise, be overhead, and set. An example of what an observer in the United States sees when looking at the first-quarter phase is shown in Fig. 20.13.

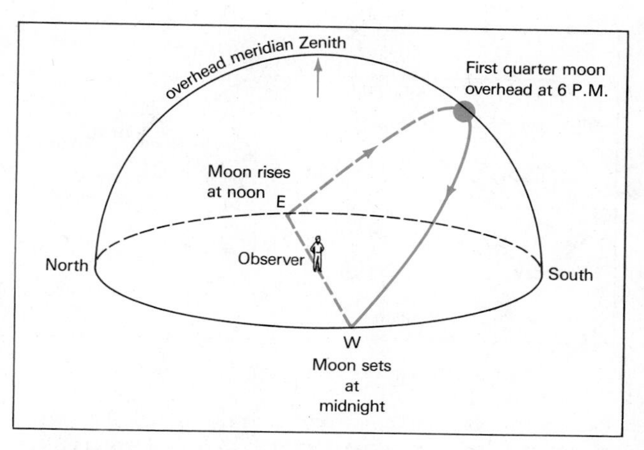

Figure 20.13 Diagram illustrating the first-quarter moon rising, on the overhead meridian, and setting for March 21 or September 21 for an observer in the United States. The side of the moon facing west is shining from reflected sunlight.

Figures similar to 20.13 for the other phases can be drawn using Table 20.1.

Since the moon revolves around the Earth every $29\frac{1}{2}$ solar days, it gains 360° in $29\frac{1}{2}$ days, or 12.2° per day on the Sun. This means the moon is on the observer's meridian about 50 min later each day, because the Earth must rotate through 360° plus 12.2° before the moon appears on the overhead meridian. (See Fig. 20.14.) The average time of moonrise is thus delayed about 50 min each day. The actual time depends on the latitude of the observer, with greater variation noted by an observer in the higher latitudes. The variation depends upon the angle between the moon's path and the horizon.

The approximate altitude of the full moon can be found by recognizing that the full moon will be on the opposite side of the Earth from the Sun (Fig. 20.15).

Table 20.1 Times for the Various Phases of the Moon to Rise, Be Overhead, and Set

Phase	*Approximate Rising Time*	*Approximate Time Overhead*	*Approximate Setting Time*
New moon	6 A.M.	Noon	6 P.M.
First-quarter moon	Noon	6 P.M.	Midnight
Full moon	6 P.M.	Midnight	6 A.M.
Last-quarter moon	Midnight	6 A.M.	Noon

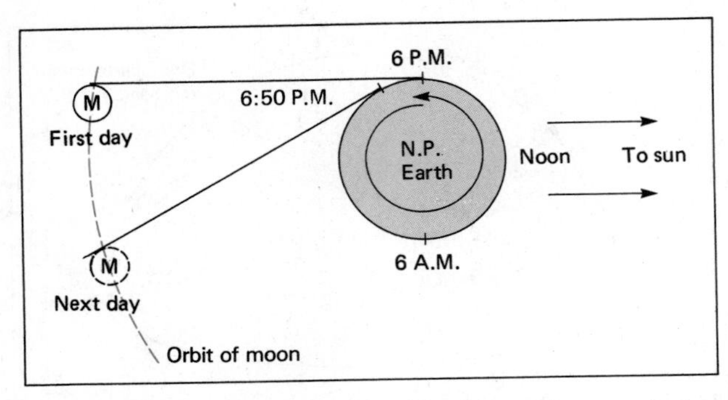

Figure 20.14 The moon rises 50 minutes later each day because as the Earth rotates, the moon is revolving around the Earth. For instance, the full moon rises in the east at about 6 P.M., but the next day it would be rising at about 6:50 P.M.

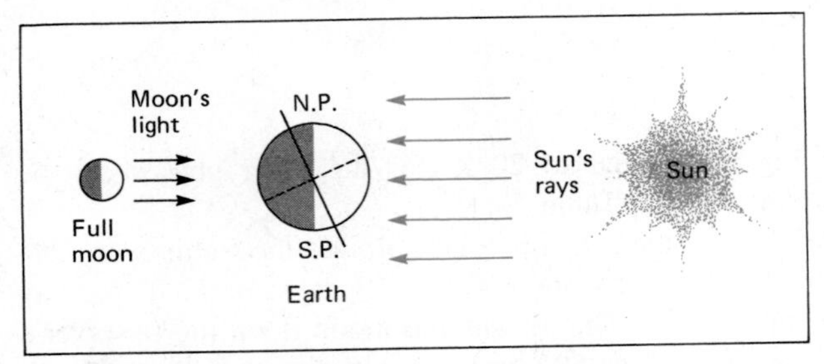

Figure 20.15 The full moon is on the opposite side of the Earth from the Sun. A side view of the Earth, full moon, and Sun in winter is shown here. The Sun's rays hit the southern hemisphere most directly. The moon's reflected light hits most directly on the northern hemisphere.

Thus, when the Sun is low in the sky in the winter, the full moon will be high in the sky. In the summer, the Sun is high in the sky and the full moon is low in the sky.

20.5 Eclipses

The word **eclipse** means the darkening of the light of one celestial body by another. The Sun provides the light by which we see objects in our solar system; that is, nonluminous objects in our solar system are observed by reflected light from the Sun. Since the light from the Sun is falling upon objects in the solar system, objects cast shadows that extend away from the Sun. The size and shape of the shadow depend on the size and shape of the object and its distance from the Sun. The Earth and moon, being spherical bodies, cast conical shadows, as viewed from space.

If we examine the shadow cast by the Earth or moon, we discover two volumes of different degrees of darkness. The darkest and smallest region is known as the **umbra.** An observer located within this region is completely blocked from the Sun during a solar eclipse. The semidark region is called the **penumbra.** An observer positioned in this region can see only a portion of the Sun during a solar eclipse. See Fig. 20.16.

A **solar eclipse** occurs when the moon is at or near new phase and also in the ecliptic plane. If these two events occur together, the Sun, moon, and Earth are all nearly in a straight line. The moon's shadow will then fall upon the Earth and the Sun's rays will be hidden from those observers in the shadow zone. A total eclipse occurs in the umbra region and a partial eclipse in the penumbra region.

The length of the moon's shadow varies as the moon's distance to the Sun varies. The average length of the moon's umbra is 233,000 mi, which is slightly less than the mean distance between the Earth and moon. Since the umbra is shorter in length than the mean distance from the Earth to the moon, an eclipse of the Sun can occur in which the umbra fails to touch the Earth. An observer positioned on the Earth's surface directly in line with the moon and Sun would see the moon's disc projected against the Sun, and a bright ring, or annulus, would appear outside the dark moon. This is called an **annular eclipse.** Around the zone of the total (very dark) annular eclipse would appear the larger semidark region of the penumbra. The penumbra region may be as large as 6000 mi in diameter at the surface of the Earth. The maximum diameter of the umbra at the Earth's surface is about 170 mi. This maximum value can exist only when the Sun is farthest from the Earth, which is early July, and the moon is at perigee, or at its closest distance to the Earth.

The motion of the moon and Earth are such that the shadow of the moon moves generally eastward during the time of the eclipse with a speed of about 1000 mi/h. Thus, the region of total eclipse does not remain long at any one place. The greatest possible value is about $7\frac{1}{2}$ min, and the average is about three or four minutes.

A **lunar eclipse** occurs when the moon is at or near full phase and also in the ecliptic plane. The Sun, Earth, and moon must be positioned in a nearly straight line,

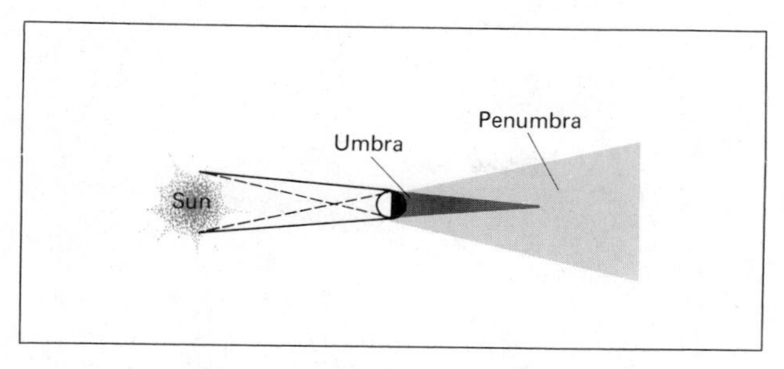

Figure 20.16 The umbra and penumbra are, respectively, the dark and semidark shadows cast by an object exposed to light rays. Here the light rays are from the Sun.

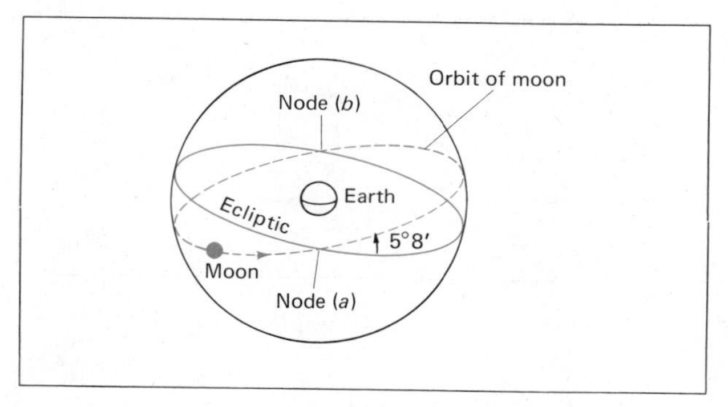

Figure 20.17 The orbital plane of the moon in respect to the ecliptic plane. The points shown in the diagram where the two planes intersect are called the nodal points. Node (a) is called the ascending node (b) the descending node.

with the Earth between the Sun and moon. Thus, the shadow formed by the Earth blots out the face of the moon. The average length of the Earth's shadow is about 860,000 mi and the diameter of the shadow at the moon's position is great enough to place the moon in total eclipse for a time slightly greater than $1\frac{1}{2}$ h. A partial eclipse of the moon can last as long as 3 h and 40 min.

The orbital plane of the moon is inclined to the ecliptic (the annual path of the Sun) at an angle slightly greater than 5°; therefore, the path of the moon crosses the ecliptic at two points as it makes its monthly journey around the Earth. The points where the moon's path crosses the ecliptic are known as **nodes.** The point of crossing going northward is called the **ascending node,** and the point of crossing going southward is called the **descending node.** See Fig. 20.17. A solar or lunar eclipse can occur only at or near the nodal points, since the Earth, moon, and Sun must be in a nearly straight line, and this occurs only at or near the points where the moon crosses the ecliptic.

The orbital plane of the moon is precessing westward or clockwise if viewed from above the orbital plane. The precession of the moon's orbit causes the nodal points to move westward along the ecliptic, making one complete cycle in 18.6 years.

20.6 Tides

Anyone who has been to the seashore for a day's visit is aware of the rising and falling of the surface of the ocean, which may be as great as 40 ft in some regions. The alternate rise and fall of the ocean surface is called the tides. People related tides to the passage of the moon in the first century A.D., but all efforts to explain the phenomenon failed until the seventeenth century, when Newton applied his law of universal gravitation to the problem. He related, and explained, the alternate rise and fall of the ocean's surface with the motions of the Earth, moon, and Sun. A few of the many factors contributing to the height that the ocean rises and falls at a particular location are (1) the rotation of the Earth on its axis; (2) the position of the Earth, moon, and Sun in respect to one another; (3) the varying distance between the Earth and moon; (4) the inclination of the moon's orbit; and (5) the varying distance between the Earth and Sun.

There are two high and two low tides, daily, because of the moon's gravitational attraction and the motion of the moon and Earth.

An understanding of the two daily tides can be clarified by visualizing what shape the Earth and its surface of water would take if there were no external gravitational forces and the Earth had no motion (Fig. 20.18).

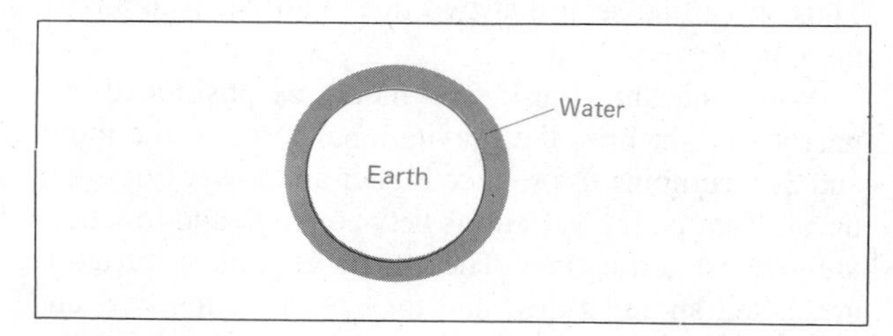

Figure 20.18 The shape of a stationary Earth with all water surfaces unaffected by any external gravitational fields.

If the Earth did not rotate, there would be no centripetal force. On a nonrotating Earth, with no external gravitational forces, there would be no forces being exerted on the Earth or the surface water and hence no tides. How then do the gravitational force of the moon and the motion of the Earth produce the tides?

The answer to this question can be found and the reason for two daily tides at a given location on the Earth's surface can be explained, if the moon's mass is considered to be concentrated at a point (Fig. 20.19). The Sun is also a factor in causing tides, but it is left out of this explanation to simplify the results. The magnitude and direction of the moon's gravitational forces acting on the Earth at points *A, B, C, D,* and *O* are shown. The magnitude is indicated by the length of the arrow drawn to represent the gravitational force. Note that the forces at *A, O,* and *B* are all in the same direction toward the moon, but the magnitudes are not the same. The force at *A* is the greatest because it is closest to the moon. Remember, Newton's law of gravitational attraction says the force between two masses is inversely proportional to the square of the distance between the two masses. Thus, the force at *O* is less than the force at *A*, and the force at *B* is less than the force at *O*. Also, forces at *C* and *D* have approximately the same magnitude but are slightly less than the force at *O*. The direction of the forces at *C* and *D* are toward the moon as shown. If all the forces shown are now added in reference to the force at *O*, at the center of the Earth, we obtain the results as shown in Fig. 20.20. The drawing shows the Earth's crust and oceans stretched out along the line *AB* and brought together along line *CD* producing an egg-shaped body, thus giving rise to the two bulges as shown with high tides at points *A* and *B*, and low tides at points *C* and *D*.

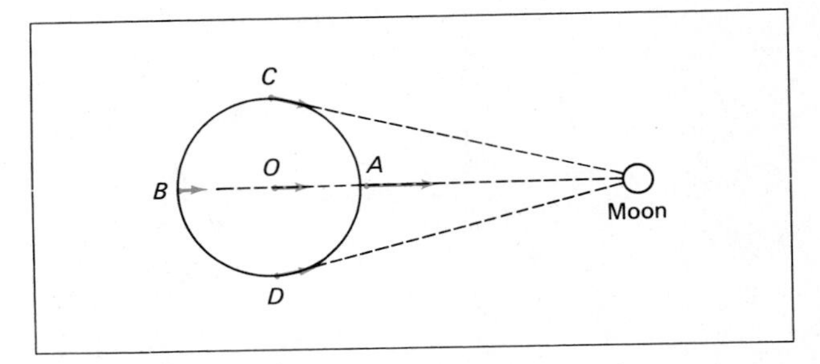

Figure 20.19 Gravitational forces (arrows) acting on the Earth by the moon, which is considered as having a mass concentrated at a point.

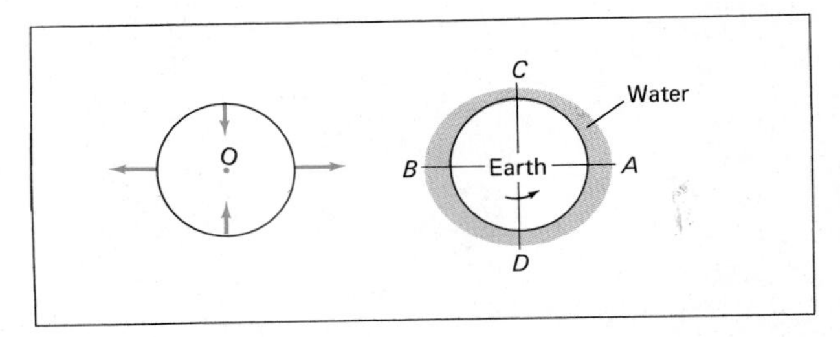

Figure 20.20 The resultant tidal forces (arrows) acting on the Earth to produce the egg-shaped contour shown at the right. View of the Earth is from above the north pole.

Since the Earth is rotating counterclockwise, completing one turn in 24 h, point *D* will move toward point *A*, point *A* toward *C*, and so on for points *C* and *B*. Thus, it can be seen that two tides will occur at any of the points every 24 h.

When the Sun, Earth, and moon are positioned in a nearly straight line, the gravitational force of the moon and Sun combine to produce higher and lower tides than usual. That is, the variations between high and low tides are greatest at this time. These tides of greatest variation are called **spring tides,** and they occur at the new and full phases of the moon. When the moon is at first- or third-quarter phase, the Sun and moon are 90° with respect to the Earth. At these times, the tidal forces of the moon and Sun tend to cancel one another, and there is a minimum difference in the height of the surface of the ocean. In this case the tides are known as **neap tides.** It should be observed that two spring tides and two neap tides take place each lunar month, since the moon passes through each of its phases once each month. There is a spring tide at new moon, a neap tide at first quarter, and another spring tide at full moon, and a second neap tide at last quarter. This is shown in Fig. 20.21.

The height of the tide also varies with latitude. See Fig. 20.22. The tide is highest at the moon's declination and at the position opposite the moon's declination. The time of high tide does not correspond to the time of the meridian crossing of the moon. Actually, the bulge is always a little ahead (eastward) of the moon because of the Earth's rotation. Since the Earth rotates faster than the moon revolves, the Earth carries the tidal bulge forward in the direction it is rotating, which is eastward.

The action of the tides produces a retarding motion on the Earth's rotation, slowing it and lengthening the

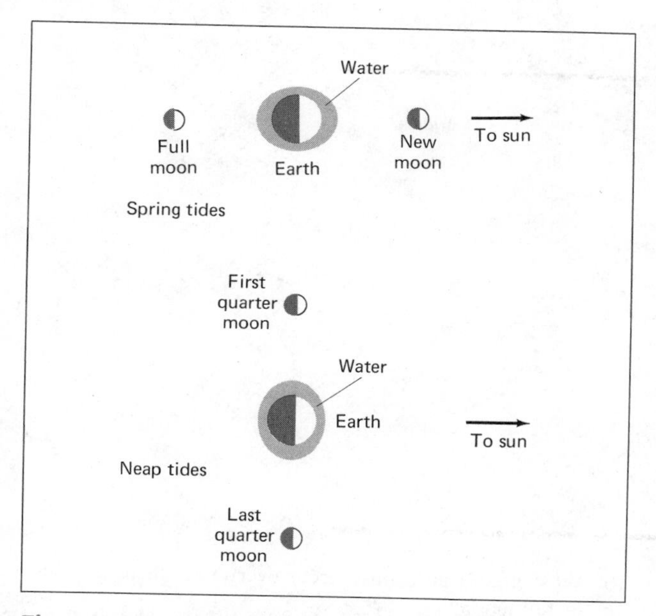

Figure 20.21 Diagram of the relative positions of the Earth, moon, and Sun at the times of spring and neap tides. During spring tides the Sun and moon are lined up and the tidal effects are more extreme. During neap tides, the Sun and moon act against each other and the tides are more moderate.

solar day about $\frac{1}{1000}$ second per century. Since the law of conservation of angular momentum must be observed, the decrease of the Earth's angular momentum must appear as an increase in the moon's angular momentum. A measurement of the moon's orbit shows that the semimajor axis is increasing about $\frac{1}{2}$ inch per year. Thus, one billion years ago the solar day was 2.8 h shorter, and the moon was 8000 mi closer to the Earth.

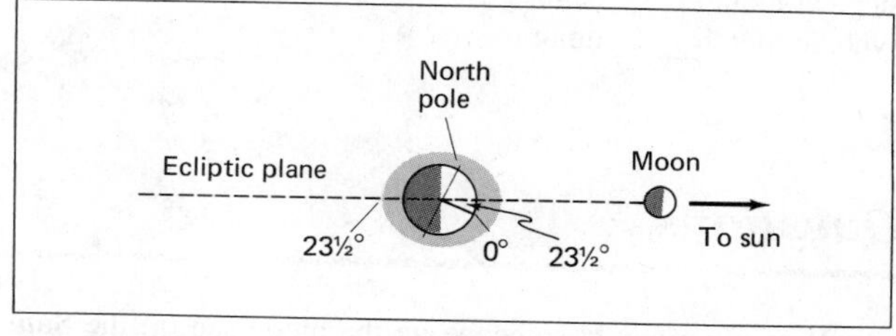

Figure 20.22 Spring tide occurring at time of summer solstice (Sun at greatest northern declination) when the moon is in new phase. Maximum height of bulge is at $23\frac{1}{2}°$ N and $23\frac{1}{2}°$ S, since the Sun's declination is $23\frac{1}{2}°$ N. If a spring tide were to occur three months later, when the Sun is on the equator, the maximum height of the bulge would be in the equatorial region.

Learning Objectives

After reading and studying this chapter, you should be able to do the following without referring to the text:

1. State the approximate distance between the Earth and moon, the approximate diameter of the moon, the time for the moon to go around the Earth, and the time for the moon's rotation.
2. Explain why the moon has so many craters compared to the Earth.
3. State what the plains are and what caused them.
4. Summarize the important periods in the moon's history.
5. Draw a diagram of the Earth, moon, and Sun when the moon is in the first-quarter phase and tell when the moon will rise, be overhead, and set.
6. Draw a diagram to explain why the moon rises 50 minutes later each day.
7. Explain what causes a solar eclipse and why we do not usually have one when the moon is in the new phase.
8. Explain what causes a lunar eclipse and why they last longer than do solar eclipses.
9. Explain why we have tides; particularly, why we have two high tides and two low tides each day.
10. Explain what causes spring tides and neap tides.

Important Words and Terms

craters
plains
rays
rills
mountain ranges
faults
perigee
apogee
sidereal month
synodic month
new moon
waxing
waning
crescent
gibbous
first-quarter moon
full moon
last-quarter moon
eclipse
umbra
penumbra
solar eclipse
annular eclipse
lunar eclipse
nodes
ascending node
descending node
spring tides
neap tides

Questions

1. How often would someone on the moon see (a) the Sun rise (b) the Earth rise?
2. What caused each of the following on the moon: (a) craters, (b) plains, (c) rays, (d) rills?
3. How long ago did the Earth and moon form?
4. How often do we have a full moon?
5. Determine when the following occur: (a) the new moon sets, (b) the full moon rises, (c) the last-quarter moon sets.
6. Why does the moon rise 50 minutes later each day?
7. Which phase of the moon (a) is overhead at 6 P.M., (b) sets at 6 A.M., (c) rises at midnight?
8. What is the difference between a waxing and a waning moon?
9. Why is the full moon higher in the sky in winter than in summer?
10. Why doesn't an eclipse occur every time there is a new or full moon?
11. Why do lunar eclipses last so much longer than solar eclipses?
12. During which phases of the moon do spring and neap tides occur?
13. Why are there two high tides and two low tides each day?
14. Why can scientists learn more about the early history of our planet by studying rocks from the moon than by studying rocks from the Earth?
15. An examination of the planets and their moons reveals something exceptional about the Earth-moon system. What is this unique feature?

Problems

1. If a person weighs 600 N on the Earth, what would the person's weight be on the moon? *Answer:* 100 N
2. If a person weighs 120 lb on the Earth, what would the person's weight be on the moon?
3. How many days are in 12 lunar months (synodic months)?
4. At what longitude would a person be who sees a full moon overhead when it is noon in Washington, D.C. (39° N, 77° W)? *Answer:* 103° E
5. Referring to problem 4, at what longitude would a person be who sees the full moon setting at the same time?
6. At what longitude would a person be who has a new moon overhead when it is noon in Washington, D.C. (39° N, 77° W)? *Answer:* 77° W
7. Referring to problem 6, at what longitude would a person be who sees the new moon rising at the same time?
8. Consider a person in the United States who sees the first-quarter phase.
   (a) Which side of the moon is bright? (East or West)
   (b) What phase would an observer in Australia see at the same time, and which side would be bright?
   Answer: (b) first quarter phase, west (left) side bright
9. Consider a person in the United States who sees the last-quarter phase.

(a) Which side of the moon is bright?
(b) What phase would an observer in Australia see at the same time, and which side would be bright?

10. A book states that the moon rises about 50 min later each day. What is a more accurate figure?

Answer: 48.8 min

11. Draw a diagram similar to Fig. 20.13 showing the times the last-quarter moon rises and sets.
12. Determine the month and day when the full moon is at maximum altitude for an observer at Washington, D.C. (39° N). What is the altitude of the full moon at this time?
13. Determine the month and day when the full moon is at minimum altitude for an observer at Washington, D.C. (39° N). What is the altitude of the full moon at this time?

Answer: June 21 28.5° ± 5°

14. Draw a diagram illustrating a total solar eclipse. Include the orbital paths of the Earth and moon, and indicate the approximate time of day the eclipse is taking place.
15. Draw a diagram illustrating a total lunar eclipse. Include the orbital paths of the Earth and moon, and indicate the approximate time of day the eclipse is taking place.
16. A high tide is occurring at Washington, D.C. (39° N, 77° W).
    (a) What other longitude is also experiencing a high tide?
    (b) What two longitudes are experiencing low tide?

Answer: (a) 103° E
(b) 13° E and 167° W

17. A low tide is occurring at Los Angeles (34° N, 118° W).
    (a) What other longitude is also experiencing a low tide?
    (b) What two longitudes are experiencing high tide?

21

The Universe

Do conservation laws dictate
A universe transformed
From spaceless, timeless, and nothing
To space, time, and something?

THE STUDY OF the stars is the oldest science. Thousands of years ago men and women under clear desert skies of the Near East watched the stars in awesome wonder. The earliest scientists plotted the positions and brightnesses of the stars as the Earth went through its calendar of seasons. The Sun and moon were worshipped as gods and the days of the week were named after the Sun, moon, and visible planets as discussed in Chapter 19. Yet, it is only recently that we have understood the most basic nature of stars.

Sixty years ago, not even Einstein himself knew what made the stars shine. Today we know that all stars go through a cycle of stages. Stars are born, they shine, and then they expand, contract, possibly explode and eventually die out. Our knowledge of how all this happens has been made possible through our study of the atomic nucleus and by applying the laws of science in many diverse fields.

Our knowledge of the universe is also growing. We are learning more and more of its secrets, and we are beginning to understand how our present civilization fits into an overall scheme in the harmony of the universe. We now have an imperfect theory of evolution—a theory concerned not so much with the origin of human beings, but with the origin of the elements of our solar system. We can now begin to understand how carbon, nitrogen, and oxygen atoms were produced in the stars. It is these atoms that were necessary to produce life itself. As these theories evolve, new mysteries appear, but the continued search for truth is invigorating.

The stars and galaxies of our universe give off many different kinds of electromagnetic radiation. These were discussed in Chapter 6 and include radio waves, microwaves, infrared, visible, ultraviolet, X rays, and gamma rays. Up until about 50 years ago, we looked only at the visible light given off by the stars. With the advent of radio telescopes, quasars and pulsars were discovered in 1960 and 1968, respectively. Most of the other regions of the electromagnetic spectrum are absorbed by our atmosphere. Satellites and balloons going above our atmosphere have enabled us to study other forms of radiation emitted by stars, and new developments are occurring frequently. There is now the strong possibility that we have detected black holes using X-ray astronomy techniques.

Recent experiments in the microwave region of the spectrum confirm the finding that we live in an expanding universe that had its beginning 13 to 20 billion years ago. But will our universe continue to expand? What is the ultimate fate of the universe? We hope the answers to these questions will be forthcoming, but for now there is much to understand. This chapter attempts to summarize many of the basic ideas about the stars and galaxies that make up our universe.

◀ **A sample of the universe; cluster of galaxies in the constellation Hercules.**

21.1 The Sun

The Sun is a star, spherical in shape, with a diameter of 865,000 mi (over 100 times greater than the Earth's diameter). Viewed from the Earth, at its mean distance, the Sun's angular diameter is approximately $\frac{1}{2}$ degree. This ordinary star of the Milky Way galaxy* is the most important object in the solar system to us because it

*For a definition of galaxy, see Section 21.4.

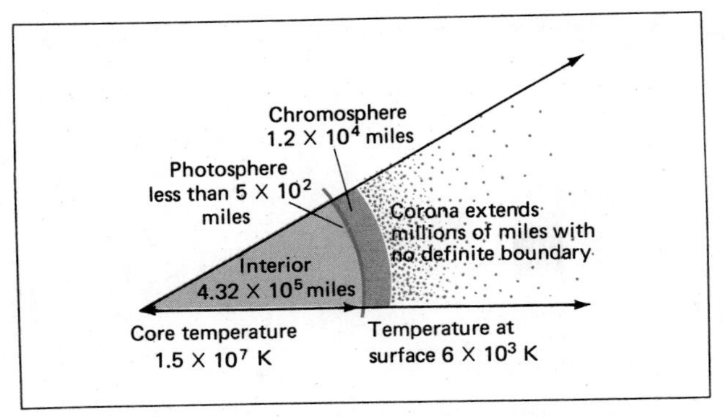

Figure 21.1 Diagram of a radial cross section of the Sun.

supplies heat, light, and other radiation for the processes of life on the Earth. The Sun rotates about an axis every 25 Earth days and moves through space with its family of planets at a speed of approximately 250 km/s in the direction of the constellation Cygnus. The Sun's equator is inclined about 7° with the orbital plane of the Earth. The rotational period of the Sun given above is the period at its equator; the period of rotation is longer at higher latitudes.

A cross-sectional view of the Sun is shown in Fig. 21.1. The Sun's temperature is believed to be about 15,000,000 K at the center, and the temperature decreases farther out to the visible surface of the Sun, called the **photosphere.** The temperature of the photosphere has been measured at about 6000 K.

The interior of the Sun is so hot that individual atoms do not exist because high speed collisions continually knock the electrons loose from the atomic nuclei. The interior is composed of high-speed nuclei and electrons moving about more or less independently, similar to a gas. A gas, you will recall, is composed of rapidly moving atoms or molecules. In the Sun, we have very rapidly moving positively charged nuclei and negatively charged electrons. These high-speed charged nuclei and electrons form a fourth phase of matter called a **plasma.** The Sun is a plasma with an average density of 1.4 g/cm^3. Note that this density is 1.4 times as great as water! We see that the Sun and the stars are definitely unlike anything we normally see on Earth.

The Sun's photosphere is composed of about 94% hydrogen, 5.9% helium, and 0.1% of heavier elements, the most abundant being carbon, oxygen, nitrogen, and neon. We believe that the interior of the Sun has a similar composition, although we have no good experimental evidence for this. Note that when we discuss elements in stars, we are really speaking about the nuclei of the elements since the electrons are stripped off the nuclei and are speeding around independently of the nuclei.

The photosphere, viewed through a telescope with appropriate filters, has a granular appearance. The granules are hot spots (about 100 K higher than the surrounding surface), a few hundred miles in diameter, that last only a few minutes. Extending more than 12,000 mi above the photosphere lies the **chromosphere** (color + sphere) which is composed mainly of hydrogen. The chromosphere can be seen as a thin red crescent for only the few seconds a total eclipse shuts out the light from the Sun. At the time of a total solar eclipse the chromosphere and photosphere are hidden by the moon, and the **corona** (outer solar atmosphere) can be seen as a white halo. See Fig. 21.2.

A very distinct feature of the Sun's surface is the appearance of sunspots. **Sunspots** are patches (some are thousands of miles in diameter) of cooler material on the surface of the Sun. Each has a central darker part, called the *umbra,* and a lighter border called the *penumbra.* Figure 21.3 is a photograph of the whole disc and an enlargement of a very large sunspot. These large sunspots last for several weeks before disappearing from view.

The number of sunspots appearing on the Sun varies over a 22-year period. A period begins with the appearance of a few spots or groups near 30° latitude in both hemispheres of the Sun. The number of spots increases, with a maximum generally between 100 and 200 occurring about four years later near an average latitude of 15°. As time passes, the number of spots decreases until, in about seven more years, only a few spots are observed near 8° latitude.

About this same time a few spots begin to appear at 30°, and the number of sunspots begins to increase again, indicating an 11-year cycle. But there is a notable difference. The sunspots have an associated magnetic field that is different in appearance from the previous ones. Studies indicate that if a sunspot has a north magnetic pole during the initial increase and decrease, the next 11-year cycle will show a south magnetic pole associated with the sunspot.

The 22-year or 11-year sunspot cycle has been observed since about 1715. Galileo saw sunspots through a telescope in 1610 and they were possibly observed even before that without using telescopes. There were reports of their observation after Galileo, but during the period 1645 to 1715 hardly any sunspots were reported. During this 70-year period very few northern lights were

Figure 21.2 Solar corona photographed at total eclipse. (Mount Wilson and Las Campanas Observatories, Carnegie Institution of Washington.)

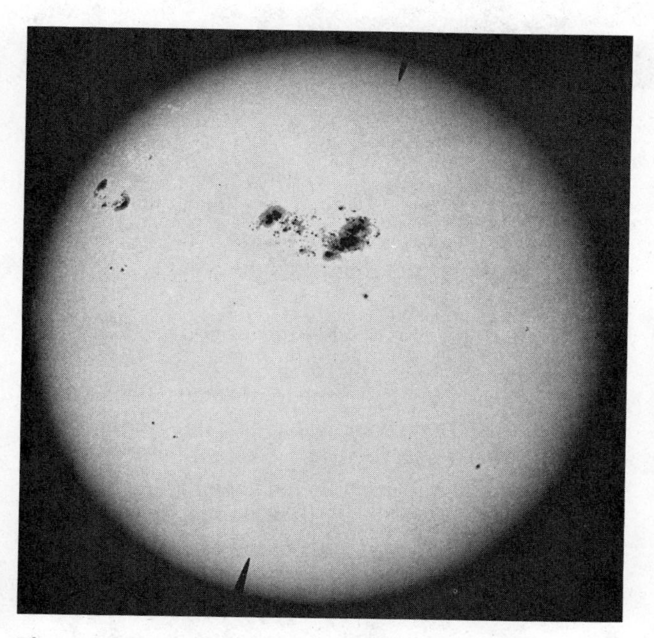

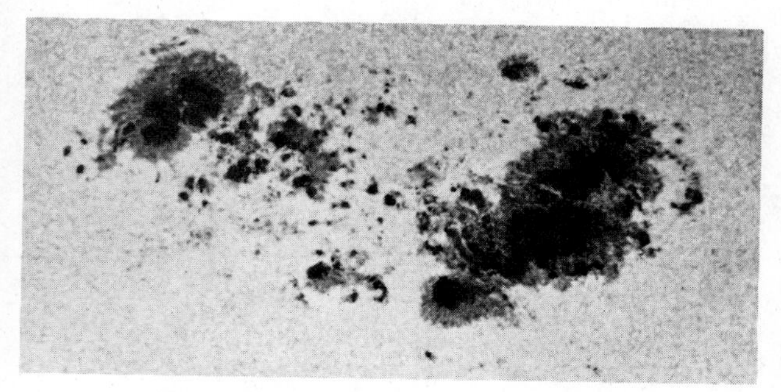

Figure 21.3 Whole solar disc (*Left*). The great sunspot group near the center of the disc is enlarged in the photograph on the right. (From a photograph taken April 7, 1947, Mount Wilson and Las Campanas Observatories, Carnegie Institution of Washington.)

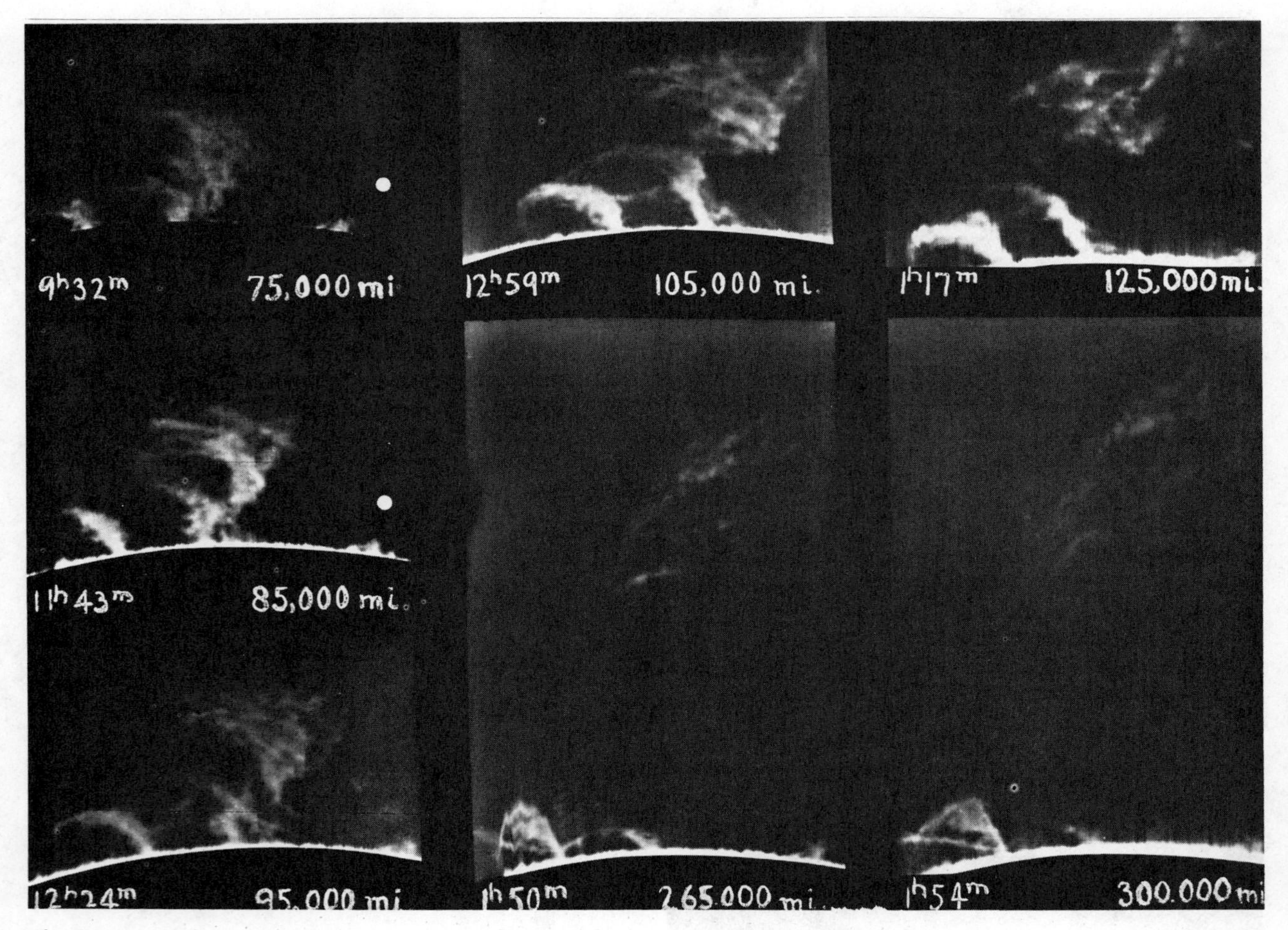

Figure 21.4 Solar prominences. (From a photograph taken October 8, 1920, courtesy Yerkes Observatory.)

seen in Northern Europe, and a "Little Ice Age" occurred during this time in Europe. The evidence seems to indicate that the Sun's activity is not always as regular as it often appears to be.

Another distinct feature of the Sun's surface is the appearance of **prominences** that seem to be connected with violent storms in the chromosphere. They are very evident to the astronomer during solar eclipses, at which time they appear as great eruptions at the edge of the Sun. They are red in color, have an associated magnetic field, and take many different shapes and forms. They may appear as streamers, loops, spiral or twisted columns, fountains, curtains, or haystacks. They extend outward for thousands of miles from the surface, occasionally reaching a height of one million miles. An extraordinarily large prominence is shown in Fig. 21.4. The elapsed time between the first and last picture in the photograph is 4 h and 22 min. Since the height and times are given, the velocity of expansion can be easily calculated. The white dot represents the relative size of the Earth.

The chief property of the Sun, of course, is the fact that it shines, or gives off energy. However, it was not until 1938 that scientists came to understand what made the Sun shine. We now know that the Sun shines because of nuclear fusion reactions inside its core.

The Sun's core is made up mostly of hydrogen nuclei or protons, or in nuclear notation ${}^1_1H$. These protons are moving at very high speeds and they occasionally fuse together as shown in Fig. 21.5. The products of this nuclear reaction are a deuteron (proton and neutron together or ${}^2_1H$) and a positive electron or positron designated ${}_{+1}e$ and a neutrino designated ν. A **neutrino** is an elementary particle that has no charge, no mass, or very

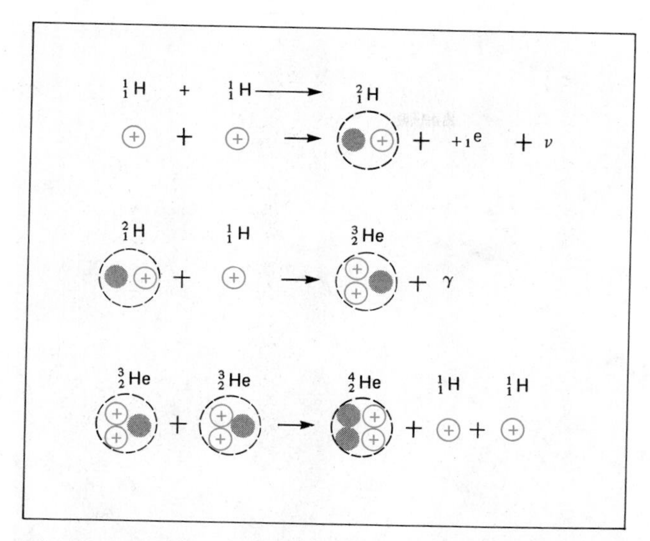

Figure 21.5 The reactions which make up the proton-proton chain.

little mass, travels at or near the speed of light, and hardly ever interacts with other particles such as electrons or protons. This first reaction is fairly rare and for this reason the Sun's hydrogen burns relatively slowly. In fact, we believe that the Sun has been shining for about 5 billion years.

Once the deuteron is formed, it quickly reacts with a proton to form a helium-3 nucleus (^3_2He) and a photon or particle of light designated γ. Next, two helium-3 nuclei fuse together to form the more common helium-4 nucleus and two protons.

In each of these three fusion reactions, mass is converted into energy and energy is liberated. These three reactions are called the **proton-proton chain** and can be written as

$$^1_1H + ^1_1H \longrightarrow ^2_1H + _{+1}e + \nu + \text{Energy} \qquad \text{(Slow)}$$

$$^2_1H + ^1_1H \longrightarrow ^3_2He + \gamma + \text{Energy} \qquad \text{(Fast)}$$

$$^3_2He + ^3_2He \longrightarrow ^4_2He + ^1_1H + ^1_1H + \text{energy} \qquad \text{(Fast)}$$

If we multiply the first two reactions by two and add up both sides, we get the net reaction which is

$$^1_1H + ^1_1H + ^1_1H + ^1_1H \longrightarrow ^4_2He + 2_{+1}e + 2\gamma + 2\nu + \text{Energy}$$

In the net reaction, four protons or hydrogen nuclei react to form a helium nucleus, two positive electrons, two high-energy photons, two neutrinos and a great deal of energy. The energy factor simply means that the particles on the right possess more kinetic and radiant energy than the particles on the left. In our reaction, we have converted mass into energy via Einstein's formula $E = mc^2$.

Every second in the Sun's interior, 630 million tons of hydrogen are being converted into 625.4 million tons of helium with the release of the equivalent of 4.6 million tons of mass energy. Yet, at this rate, we expect the Sun to shine for another 5 billion years or so.

We now have a good understanding of the basic concept of how a star shines. A star shines because of nuclear fusion reactions inside its core. But do we have any direct evidence that this is true? In an effort to get direct evidence of what goes on inside the Sun's core the solar neutrino experiment is being done. The **solar neutrino experiment** is an experiment to detect neutrinos from the Sun's core.

The experiment is being conducted by Raymond Davis, Jr., of the Brookhaven National Laboratory. The data collected thus far by Davis do not agree with theoretical calculations. He is finding fewer neutrinos than calculated by a factor of three to one.

Presently, Davis uses 100,000 gal of fluid C_2Cl_4 in a tank located deep within a South Dakota gold mine to detect the neutrinos. When a neutrino interacts with a chlorine atom (only high-energy neutrinos are able to do so) the atom is transformed into an atom of argon that can be removed from the tank by bubbling helium gas through the fluid.

Davis plans to increase the sensitivity of his equipment to detect neutrinos by using the element gallium in place of chlorine. With gallium low-energy as well as high-energy neutrinos will be detected. Perhaps the new experiment will agree more closely with calculated values.

The major difference between photons and neutrinos is that photons react much more readily with matter than neutrinos. In fact, a photon produced at the center of the Sun will interact countless times and take possibly millions of years before it finally reaches the Sun's surface. A neutrino, on the other hand, will zip right through the Sun and Earth and most neutrino detection devices without interacting at all. Of course, occasionally a neutrino will interact with a detector, and we can say that a neutrino has been detected. In the solar neutrino experiment only a few neutrinos are detected in a month's time.

So far, the results of the solar neutrino experiment have been perplexing. They indicate that fewer neutrinos are being detected than should be according to current theory. Either our understanding of neutrinos needs to be improved or there are some unsolved problems with regard to our understanding of the structure and dynamics of the interior of a star.

21.2 The Celestial Sphere

A view of the stars on a clear night makes a deep impression on an observer. As we look into the night sky, the stars appear as bright points of light on a huge dome overhead. As the time of night passes, the dome seems to turn westward as part of a great sphere, with the observer at the center. The apparent motion of the stars is due to the eastward rotation of the Earth. The unaided eye is unable to detect any relative motion among the stars on the apparent sphere or to perceive their relative distances from the Earth. The stars all appear to be mounted on a very large sphere having a very large radius.

This huge moving (apparent motion) sphere has been named the **celestial sphere,** and the way it appears to the observer depends upon the observer's position on the Earth. An observer positioned at 90° N (the north pole) would see Polaris, the north star, directly overhead. From this latitude all stars on the celestial sphere appear to move in concentric circles about the north star, never going below the horizon; that is, they never set. If the observer is located at 40° N latitude, he or she would observe the north star 40° above the northern horizon, and all stars within 40° of the north star would appear to move in concentric circles, never going below the horizon (Fig. 21.6). Detailed observations reveal the celestial sphere to rotate (apparently) about an axis that is an extension of the Earth's polar axis, with the celestial equator lying in the same plane as the Earth's equator. See Fig. 21.7.

Figure 21.6 A long exposure shows the trails of stars around the north celestial pole. The stars can be observed to go in circles, with the bright star Polaris very close to the center.

The position of a star or other object beyond our solar system is determined with the assignment of three space coordinates. The first and second are *declination* and *right ascension,* which are angular coordinates representing the direction of the star in respect to the Sun. The third coordinate is *distance,* which determines the star's linear distance from the sun (Fig. 21.8).

Declination is the angular measure north or south of the celestial equator measured in degrees. It has a minimum value of zero at the celestial equator and increases to a maximum of 90° north, and to a maximum of 90° south. The direction in respect to the equator is indicated by a plus (+) or minus (−) sign. All angles measured north of the equator have (+) values and all angles measured south of the equator have (−) values. For example, the star in Fig. 21.8 has a declination of +37 degrees.

Right ascension is the angular measure in degrees or hours, with the hours divided into minutes and seconds. Right ascension begins with zero hours at the celestial prime meridian and continues eastward to a maximum value of 24 h or 360°, either of which coincides with the starting point. The **celestial prime meridian** is an imaginary half-circle running from the north celestial pole to the south celestial pole and crossing perpendicular to the celestial equator at the intersection of the ecliptic at the point where the ecliptic passes northward.

The distance coordinate is usually measured in astronomical units, in parsecs, or in light-years. An **astronomical unit** we have defined as the mean distance of the Earth from the Sun, which is 92,955,700 miles as measured by radar. The **light-year** is the distance traveled by electromagnetic radiation in one year. One light-year equals approximately 6×10^{12} miles or 9.5×10^{12} kilometers, calculated by multiplying the speed of light by the number of seconds in one year. One **parsec** is defined as the distance to a star when the star exhibits a

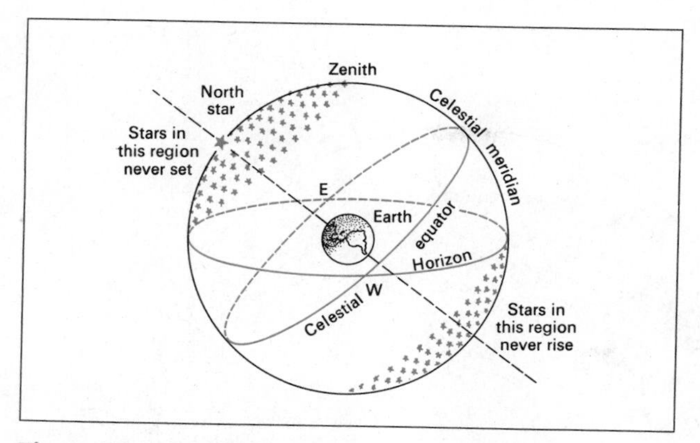

Figure 21.7 Celestial sphere, as seen by an observer on the Earth at a latitude of 40°N.

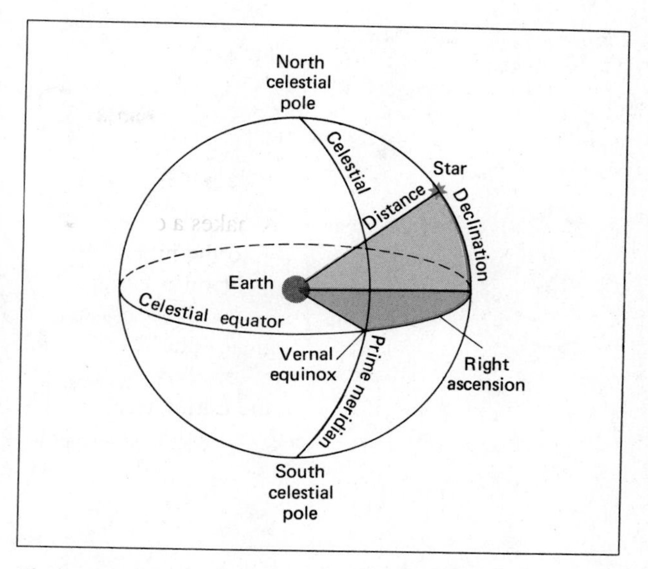

Figure 21.8 The three space coordinates: distance, declination, and right ascension.

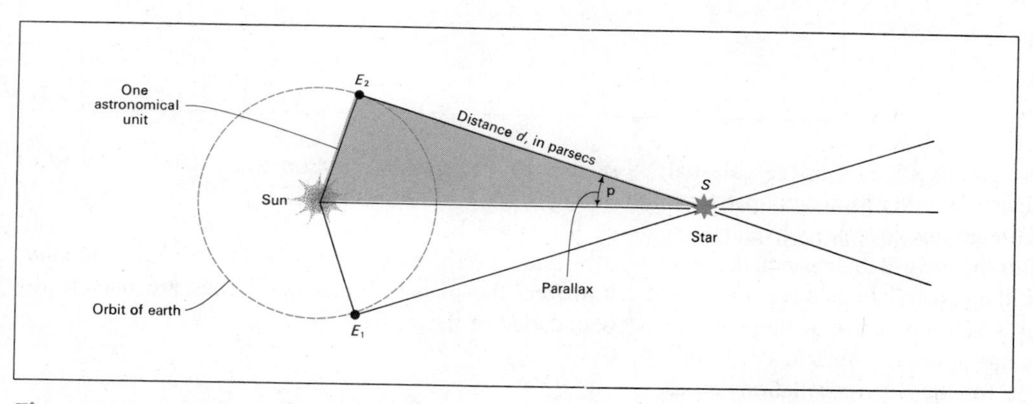

Figure 21.9 Annual parallax of a star. The annual parallax of the star S is the angle p. By definition, when angle p is 1 second of arc, the distance d is 1 parsec. The basic relation can be written: Distance in parsecs = 1/Parallax in seconds.

parallax of one second of arc (Fig. 21.9). One parsec equals 3.26 light-years, or 206,265 astronomical units. This can be written

$$1 \text{ parsec} = 3.26 \text{ light-years} = 206{,}265 \text{ astronomical units}$$

The star S in Fig. 21.9, observed from position E_1 and later from E_2, appears to move against the background of more distant stars. This apparent motion is, as you know, called parallax. The angle p measures the parallax in seconds of arc. Hence, the term parsec. The definition provides an easy method for determining the distance to a celestial object, since merely taking the reciprocal of the angle p, measured in seconds, gives the distance in parsecs.

EXAMPLE 1 What is the distance to the star Proxima Centauri if the annual parallax is 0.762 second?

$$d = \frac{1}{0.762} = 1.31 \text{ parsecs}$$

EXAMPLE 2 What is the distance to the star Sirius A if the annual parallax is 0.375 second?

$$d = \frac{1}{0.375} = 2.67 \text{ parsecs}$$

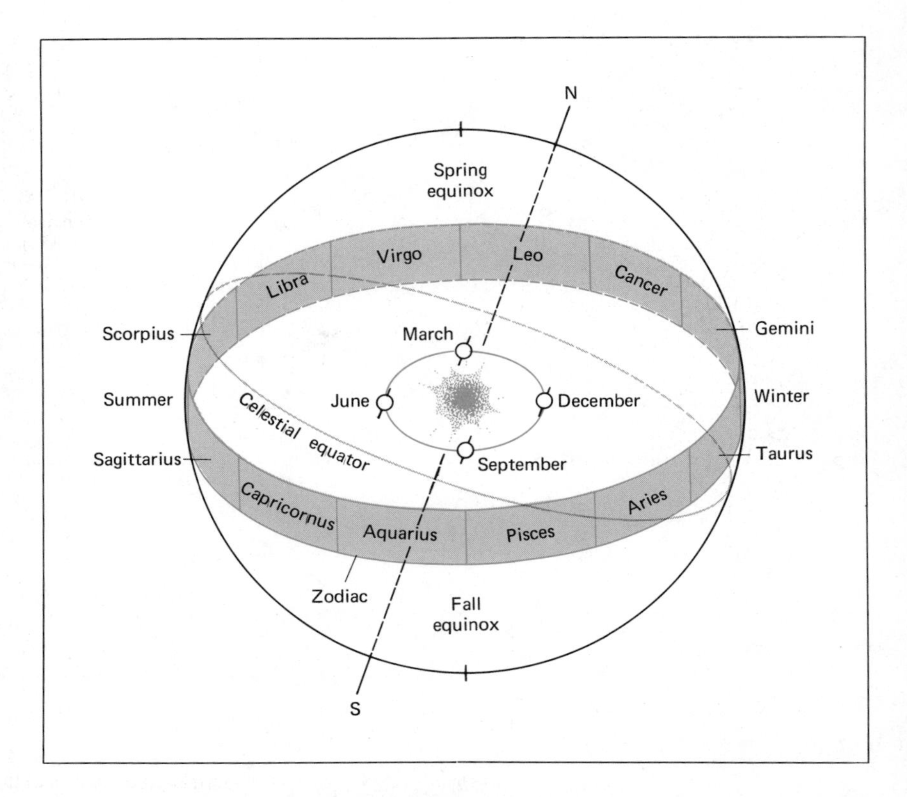

Figure 21.10 Signs of the zodiac. The drawing illustrates the boundaries of the zodiacal constellations. Each of the 12 sections of the zodiac is 30° wide and 16° high, or 8° each side of the ecliptic.

There are prominent groups of stars in the celestial sky that appear to the Earth observer as distinct patterns. These groups called **constellations** have been given names that can be traced back to the early Babylonian and Greek civilizations. Although the constellations have no physical significance, today's astronomers find them useful in referring to certain areas of the sky and, in 1927, they set specified boundaries for the 88 constellations so as to encompass the complete celestial sphere.

We are all aware of the apparent daily motion of the Sun across the sky, and when the moon is visible we are aware of its motion. The constellations also appear to move across the sky from east to west, if one observes the stars for an hour or two. Their daily motion is due to the eastward rotation of the Earth on its axis. They also have an annual motion due to the Earth's motion around the Sun. We observe the constellations Pisces, Aquarius, and Capricornus in the autumn night sky. In the winter months Orion, the Hunter, is seen along with Gemini, Taurus, and Aries. Sagittarius, the Archer, is a summer constellation. For other summer and the spring constellations, see Fig. 21.10.

The **zodiac** is a section (actually a volume) of the sky extending around the ecliptic 8° above and 8° below the ecliptic plane (Fig. 21.10). The zodiac is divided into 12 equal sections, each 30° wide and 16° high. Each section has its apex at the Sun and extends outward to infinity. The boundaries of the zodiac were specified such that the Sun, moon, visible planets, and most of the asteroids traveled within its limits. Occasionally, however, the planets Pluto and Venus are outside the boundaries of the zodiac.

21.3 Stars

Hipparchus of Nicaea, Greek astronomer and mathematician, was antiquity's greatest known observer of the stars. He measured the celestial latitude and longitude of more than 800 stars, and compiled the first star catalog, which was completed in 129 B.C. He assigned the stars, in respect to their brightness, to six **magnitudes.** The brightest ones were listed as stars of the first magnitude, those not quite so bright as second magnitude, the next less bright as third magnitude, and so on down to the sixth magnitude. A modified version of Hipparchus's scale is used today. When a comparison was made between a first magnitude star and a sixth magnitude star, it was discovered the brighter star gave off about 100 times as much radiant energy. From this, a definition of the magnitude scale was made, in which each magnitude difference is equal to the fifth root of 100. This can be written

$$\text{Magnitude difference} = \sqrt[5]{100} = 2.512$$

For example, a first magnitude star is 2.512 times as bright as a second magnitude star, and a second magnitude star is 2.512 times as bright as a third, and so on. On this scale, the brightest object in the sky, our Sun, has a magnitude of −26.7; the full moon a magnitude of −12.7; the planet Venus a magnitude of −4.2; and the brightest star, Sirius, a magnitude of −1.43.

The energy output of a star is measured by its absolute magnitude. The **absolute magnitude** of a star is defined as the apparent magnitude a star would have if it were placed 10 parsecs (32.6 light-years) from the Earth. The absolute magnitude of the Sun is about 4×10^{33} ergs/s, which is measured by determining the amount of energy falling on the Earth's surface and knowing the distance from the Earth to the Sun. If the annual parallax (from which distance is calculated) and the apparent brightness of a star can be measured, the absolute magnitude can be calculated.

The surface temperature of a star can be inferred from its color. Cool stars are red, hotter stars are yellow, then white, and the hottest stars are blue-white. Often a star's temperature is inferred from its color index.

The color index is a number indicating the relative brightness of the blue and yellow regions of the star's spectrum. The color sensitivity of the eye is greatest in the green-yellow region of the spectrum, and by using appropriate light filters, the starlight falling upon a photoelectric cell can be adjusted to that received by the eye, thus establishing a visual magnitude system. Photographic emulsions are most sensitive to the blue-violet regions of the spectrum and other filters can be used in front of the photographic plates for establishing photographic magnitudes. The number representing the color index of a star can be found by obtaining the photographic magnitude and subtracting the visual magnitude. This can be written as

$$\text{Color index} = m_p - m_v$$

where m_p = photographic magnitude
and m_v = visual magnitude.

A yellow star, for instance, might have $m_v = 4$ and $m_p = 5$ indicating that it is dimmer when seen through a blue filter. Its color index would be +1. A blue-white star would appear brighter when seen through a blue filter than through a yellow filter so m_p would be less than m_v. A blue-white star would have a negative color index. In general, stars have color indices ranging from +2.0 for cool red stars to −0.4 for hot blue-white stars.

When the absolute magnitudes or brightnesses of stars are plotted against their surface temperatures or colors, we get an **H-R diagram** named after Ejnar Hertzsprung, a Danish astronomer, and Henry Russell, an American astronomer. A typical H-R diagram for stars is shown in Fig. 21.11. Note that the temperature axis is reversed. That is, the temperature increases to the left instead of to the right.

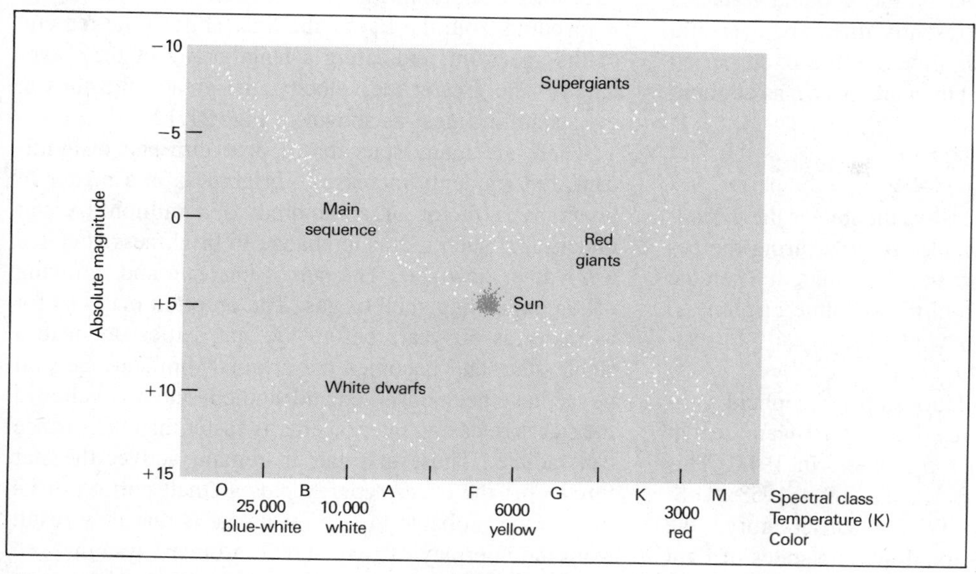

Figure 21.11 Hertzsprung-Russell diagram. The diagram is a plot of the absolute magnitude of stars versus their spectral class, temperature, or color.

Most stars on an H-R diagram get brighter as they get hotter. These stars form the **main sequence,** a narrow band going from upper left to lower right. Stars above the main sequence that are cool and yet very bright must be unusually large to be so bright. So they are called **red giants.** Stars below the main sequence that are hot yet very dim must be very small so they are called **white dwarfs.**

Stars are plasmas having a composition by number (not mass) of about 94% hydrogen, 5.9% helium, and 0.1% other elements. Their surface temperatures range from about 2000 K to 50,000 K. They vary in mass and size. Most have a mass of between 0.1 and 5 solar masses. Others range from 0.04 solar mass to 75 solar masses. Normal stars range in size from white dwarfs, about 8000 miles in diameter, to supergiants, which are over 400 million miles in diameter. Most stars appear as multiple systems. A **binary star** system consists of two stars orbiting around each other. Systems with three or more stars also occur, but they are not nearly as common as binary stars.

There are many stars in the sky that are observed to vary in brightness over a period of time. The first such star noted was Delta Cephei in the constellation Cepheus. Delta Cephei varies in brightness with a period of approximately 5.4 days. From its least bright magnitude of 5.2, it gradually becomes brighter for about two days and attains a magnitude of 4.1, after which it decreases in brightness until it reaches its minimum in another 5.4 days. The cycle begins again and goes through the same sequence. The period of other cepheid variables varies from 1 to 50 days. Presently there are over 500 known stars that vary in magnitude with a fixed period at between 1 and 50 days and they are known as **cepheid variables** after Delta Cephei.

The importance of the cepheid variables is the fact that a definite relationship exists between the period and the average brightness. Generally, the longer the period, the brighter the cepheid variable. By measuring the period, the absolute brightness can be calculated. Then the distance can be computed from the absolute brightness. Thus, the distance can be calculated to any galactic system in which cepheid variables can be detected.

The period-luminosity relationship for cepheid variables stars was discovered by Henrietta Swan Leavitt (1868–1921), an American astronomer, in 1912. This relationship provided Edwin P. Hubble (1889–1953), U.S. astronomer at the Mount Wilson Observatory, the knowledge to show that observed white patches of light (nebulae) were actually galaxies beyond the Milky Way.

The distance to more remote galaxies is calculated by means of the red-shift in the galaxy's spectrum. When a photograph is taken of an excited element in the gaseous phase by a spectrograph located in an Earth-based laboratory, a normal spectrum is obtained. When a similar photograph is taken of a galaxy containing this same element, the lines of the spectrum may show a displacement of the normal lines toward the red end of the spectrum (longer wavelengths) or toward the blue end of the spectrum (shorter wavelengths). The displacement is due to the Doppler effect and the direction of the displacement depends upon whether the galaxy is moving toward or away from the observer. Refer to Section 6.5 for a discussion of the Doppler effect.

The displacement for two lines on the calcium spectrum labeled H and K for five different galaxies is shown in Fig. 21.12. A small shift of the lines means a low velocity of recession, a greater shift means a greater velocity, and so on. The interesting point concerning the red-shift is a correlation of velocity and the apparent brightness of the galaxy. The fainter the galaxy, the greater the velocity. Generally speaking, the farther away a galaxy is located, the less its apparent brightness. Thus, a velocity and magnitude correlation yields a velocity and distance correlation. This provides us with a means for measuring great distances that are far beyond the limits of cepheid variable observation.

When a galaxy is approaching the Earth, the lines of the spectrum shift toward the blue end of the spectrum, indicating a shortening of wavelength. When the galaxy is receding from the Earth, the lines shift to the red end of the spectrum, indicating a lengthening of the wavelength. The greater the velocity, the greater the shift in the spectrum lines, as shown in Fig. 21.12.

There are many stars that appear dim and insignificant, but suddenly increase in brightness in a matter of hours by a factor of a hundred to a million. A star undergoing such a drastic change in brightness is called a **nova,** or new star. The star is unstable and throwing off an expanding shell of gas. The eruption may last for as many as 50 years before the star settles down to a stable state and becomes the original dim star. Several novae have been observed and studied, and it is believed the star is releasing internal energy faster than the surface can radiate. Thus, in order to remain active, the star throws off the excess energy plus a small portion of its mass. One probable theory for novae is that they result from the interplay of two stars in a binary system.

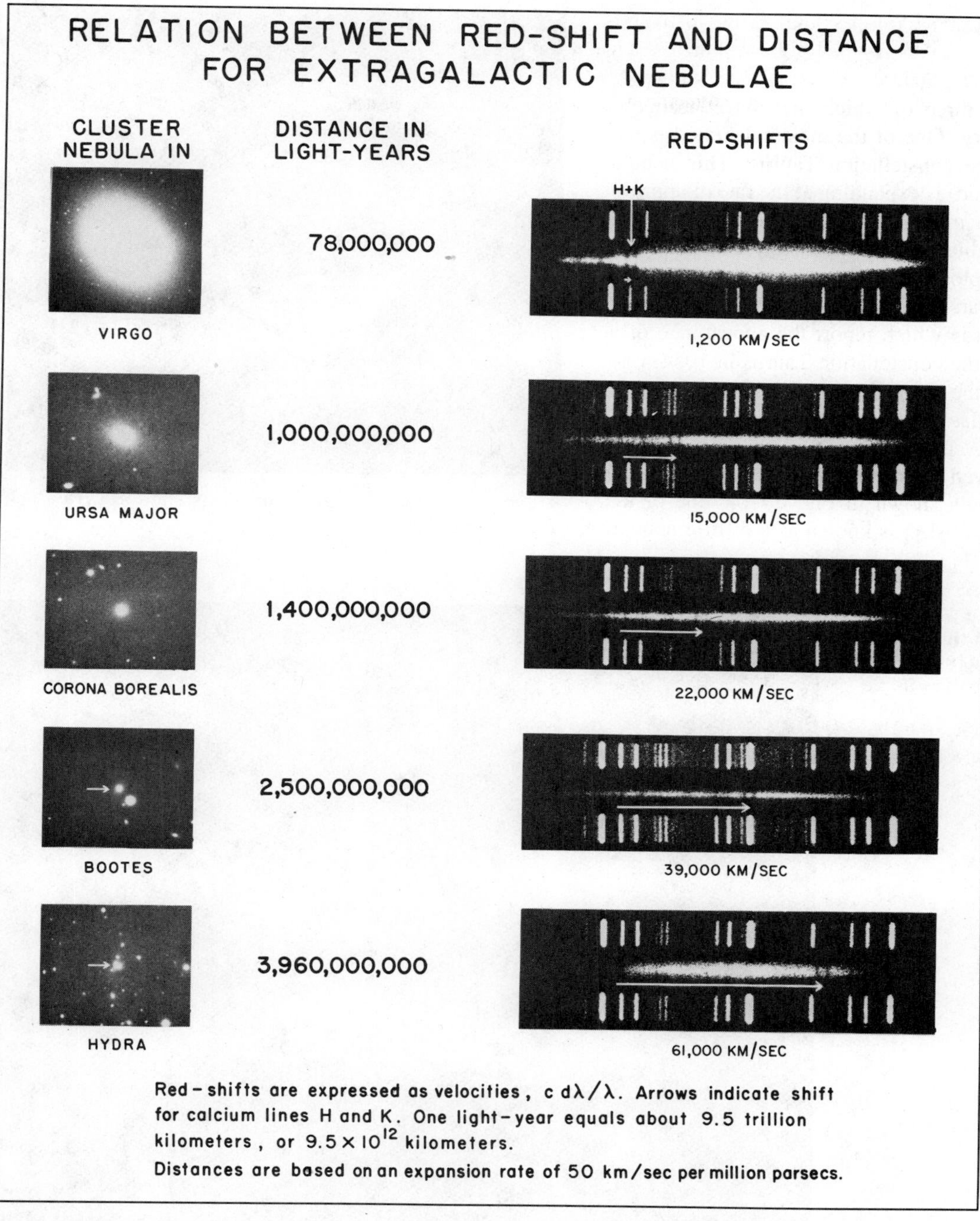

Figure 21.12 On the left are photographs of five individual, elliptical galaxies. From top to bottom, the photographs show the galaxies at increasing distance from the observer. On the right, the spectrum (the broad white band) of each galaxy is shown between an upper and lower comparison spectrum. The H and K lines of ionized calcium are the two dark vertical lines in the galaxy's spectrum. The arrows indicate the shift in the calcium lines. The red-shifts are expressed as velocities. (Palomar Observatory: California Institute of Technology)

Occasionally, a star explodes and throws off large amounts of material that may be so great that the star is destroyed. These gigantic explosions are known as **supernovae,** only three of which may have been observed in our galaxy. One of the most celebrated is the Crab Nebula in the constellation Taurus. This nebula (Latin word for cloud) is expanding at the rate of approximately 70 million miles per day. Since we know the average angular radius and the expansion rate, the original time of the explosion can be calculated. The result yields about 900 years, which agrees closely with Chinese and Japanese records which report the appearance of a bright new star in the constellation Taurus in 1054 A.D.

In addition to the novae and supernovae, there are the planetary nebulae which possess a large, slowly expanding ringlike envelope, as shown in Fig. 21.13. These nebulae, as observed with a telescope, appear greenish in color. The nebula shown in Fig. 21.14, one of the brightest in the night sky, is known as a diffuse nebula.

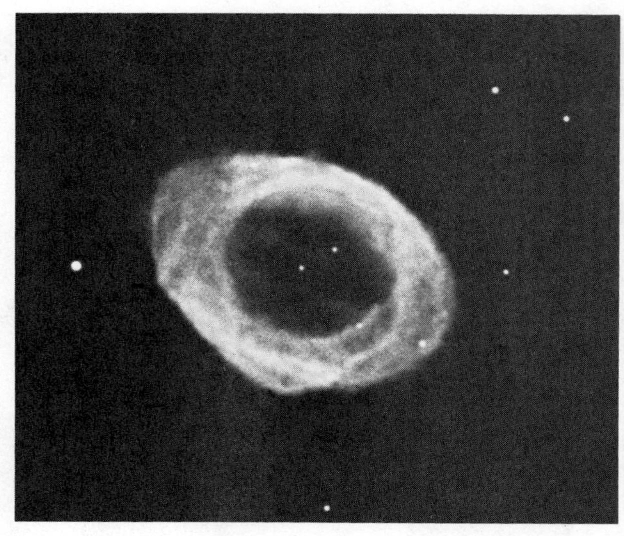

Figure 21.13 Ring nebula in the constellation Lyra. (Lick Observatory Photographs.)

Figure 21.14 Great Nebula in Orion. Considered the brightest of the diffuse nebulae, this one is located near the middle star in Orion's sword and is a gaseous and dusty nebula. (Lick Observatory Photographs.)

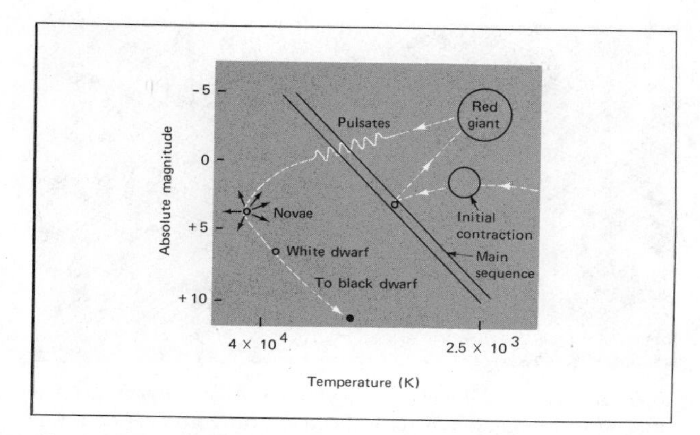

Figure 21.15 Hertzsprung-Russell diagram used to illustrate the evolution of a star with the same mass as our Sun.

This type of nebula is more irregular and turbulent than the planetary nebulae. The Great Nebula in Orion is over 25 light-years in diameter and contains hundreds of stars that illuminate the dark material. Some cosmologists (scientists who study the universe, that is, its history, present structure, and future evolution) believe that the dust and gaseous material found in clouds such as these provide the material needed for the birth of new stars.

We now know that stars are born, shine, expand, possibly explode, and then die. This is the general life cycle of a star. However, the exact details depend on a star's initial composition (the percentage amounts of hydrogen, helium, and heavier elements) and on its mass. The greater the mass of a star, the faster it moves through its life cycle.

The general evolution of a star with a mass typical of our Sun is shown in Fig. 21.15. The birth of a star, according to the best accepted theory, begins with the condensation of interstellar material, mostly hydrogen, because of radiation pressure from nearby stars, supernovae shock waves, and the gravitational attraction between the interstellar material. The size of the star formed depends upon the total mass available, which in turn determines the rate of contraction. As the interstellar mass condenses and loses gravitational potential energy, the temperature rises and the material gains thermal energy. As the star continues to decrease in size, the temperature continues to increase until a thermonuclear reaction begins and hydrogen is converted to helium, as discussed in Section 21.1. The star's position in the main sequence of the H-R diagram, its temperature, and the radiation given off (luminosity) are determined by the mass of the star.

On the main sequence of the H-R diagram, the star continues to convert hydrogen into helium. This process is called **hydrogen burning,** but it is a nuclear burning and not a chemical fire. The hydrogen-burning stage lasts for billions of years—possibly 10 to 15 billion years for a star like our Sun.

As the hydrogen in the core is converted into helium, the core begins to contract and heat up. This heats up the surrounding shell of hydrogen and causes the hydrogen burning in the shell to proceed at a more rapid rate. This rapid release of energy causes the star to expand and become a red giant.

Eventually the core gets so hot that helium can fuse into carbon, and soon other nuclear reactions occur in which all elements get created. The creation of the nuclei of elements inside stars is called **nucleosynthesis.** The physics of the problem becomes extremely complicated, but we believe that the cepheid variable stage occurs next if conditions (mass and composition) are just right. We also believe that supernovae occur sometime after the red-giant stage. The supernova stage occurs when reactions in the core proceed so rapidly that the star is blown almost completely apart, with only a remnant core remaining. The supernova stage probably does not occur for all stars.

Eventually, no more nuclear fusion reactions are possible and the star's nuclear energy has been spent. What is left of the star begins to gravitationally collapse (see Section 9.6) and go to its end stage. The end stage of a star depends upon the star's mass at the end of its active life. Average and small mass stars become white dwarfs. When a star is a white dwarf, it is very small. It has gravitationally collapsed as far down as it can go and still obey the laws of physics. It is about the size of the Earth and is so dense that a single teaspoonful of matter weighs five tons. Because it is so small it is a fairly dim star.

Large-mass stars (between 1.5 and 3.0 times the Sun's mass) have more gravitational attraction, and they collapse down to a size of approximately 20 miles in diameter. The electrons and protons in this superdense star combine to form neutrons and this **neutron star** is composed of about 99% neutrons. A teaspoonful of a neutron star would weigh one billion tons. Because the angular momentum of the star must be conserved, the small size of the neutron star dictates that it must be spinning rapidly. Rapidly rotating neutron stars give out radio radiation in pulses and are called **pulsars.**

Pulsars were first discovered in 1968 using radio telescopes similar to the one shown in Fig. 21.16. They

Figure 21.16 The 140-foot radio telescope of Green Bank, West Virginia. This telescope, completed in 1965, is the largest equatorially mounted radio telescope in the world. The immense concrete base not only serves to support the telescope but also houses the controls, hydraulic system, electric equipment, transformer vault, and electronic workshop. The instrument cost about $14 million and took seven years to design and build. (The National Radio Astronomy Observatory, operated by Associated Universities, Inc. under contract with the National Science Foundation.)

pulse with a constant period that may be between $\frac{1}{30}$ and 4 seconds. The fastest spinning pulsar is located at the center of the Crab Nebula. This pulsar is identified with the remains of a supernova that took place in 1054 A.D. This supernova was recorded by the Chinese and must have been quite bright. The period of this pulsar is slowly getting larger, indicating that the rotating neutron star is slowing down very slowly.

If the remaining core is greater than about three times the mass of the Sun, the star is believed to end up as a gravitationally collapsed object (see Chapter 9) even smaller and more dense than a neutron star. Such a star is so dense that light cannot escape from its surface due to its intense gravitational field. Thus it would appear black and is called a black hole. A **black hole** is an unbelievably dense star so massive and small that light cannot escape from its surface. Black holes can be seen by the effect they have on another star in a binary system. There is some experimental evidence for such black holes, but the facts are not yet conclusive. Nevertheless, we think that black holes do exist. Review Section 9.6.

Figure 21.17 Crab Nebula in Taurus. Also known as NGC 1952 and Messier 1, this nebula is the remains of a supernova of 1054 A.D. At the center is a pulsar. Photographed with a 200-inch telescope. (Palomar Observatory Photograph.)

After a star explodes in a supernova and the core goes into its end stage, what becomes of the ejected material? We believe that this material is thrown into space (see Fig. 21.17) and eventually becomes seed material for future stars. In fact, because of its surface composition and age, we believe our own Sun is a second-generation star. That is, one or more stars went through their life cycles and exploded as supernovae. The ejected material later coalesced again into our Sun and solar system. Since the universe is believed to be 15 to 20 billion years old and our Sun is only about 5 billion years old, this idea of the Sun being a second-generation star is generally accepted.

21.4 Galaxies

The Sun and its satellites occupy a very small volume of space in a very large system of stars known as a **galaxy** (Greek *galaxias*, Milky Way). A galaxy is an extremely large collection of stars occupying an extremely large volume of space. A galactic system is classified as irregular, spiral, or elliptical, depending upon how it appears when photographed.

The astronomer Edwin P. Hubble established a system of classifying galaxies according to their appearance in photographs. The system starts with spherical ellipticals, spreads out with increasing flatness, then branches off into a normal spiral sequence, and a barred spiral sequence, with a scattering of irregulars outside the two branches (see Fig. 21.18).

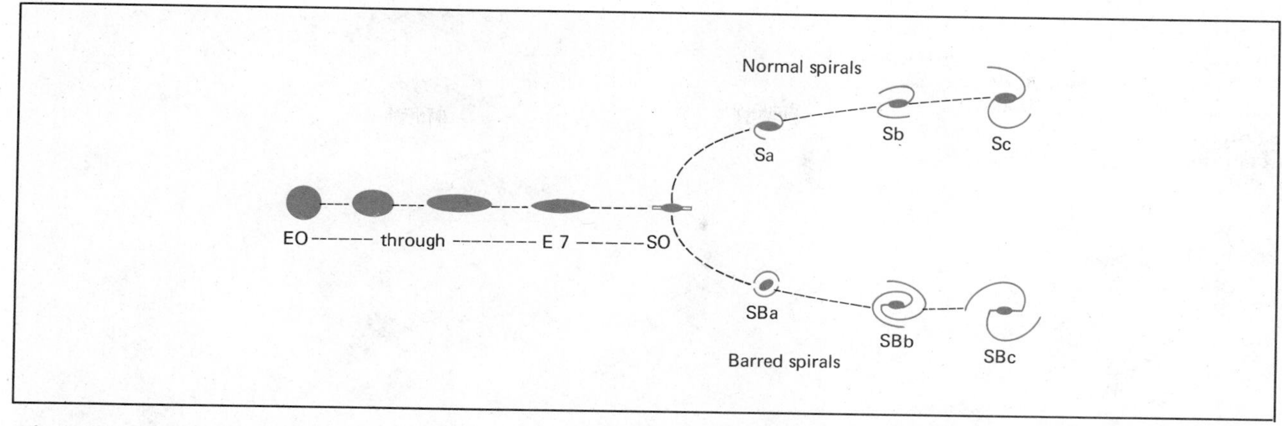

Figure 21.18 Galactic systems. Sequence of elliptical, normal spiral, and barred spiral galaxies.

The type Sa spirals have their spiral arms closely wound to the central region, whereas the Sb spirals have arms that spread out more from the center, and the Sc spirals have very loose spiral arms. The S0 type links the normal spirals to the smooth ellipticals.

The barred spirals, which are distinguished by a broad bar that extends outward from opposite sides of the central region, follow similar unwinding of the spiral arms.

The irregular type galaxies have no definite form but take on the appearance of shapeless clouds. Examples of this type are the two Magellanic Clouds that can be easily seen with the naked eye from the southern hemisphere. They are named in honor of Magellan, who reported seeing them on his famous voyage around the world.

In a given volume of space there are more elliptical galaxies than spiral galaxies, while the irregular types account for only about 3% of all galaxies. The elliptical galaxies are made up of older stars and are usually dimmer than the spirals. The spiral galaxies account for over 75% of the brighter galaxies observed.

Carl Seyfert, American astronomer of the Mount Wilson Observatory, reported in 1944 that a few spiral galaxies exhibit very bright centers and their spectra show broad emission lines that indicate that hot gas is present and expanding at very rapid rates. This gives evidence that violent activity is taking place in the central core of the galaxies. Large amounts of energy are being released from these central cores, and the best theory of the source of this energy is the gravitational collapse of an enormous amount of matter. Over 100 of these galaxies have now been reported and are known as **Seyfert galaxies.** See Fig. 21.19.

Figure 21.19 The Seyfert galaxy NGC 4151. Notice the very bright nucleus. If the galaxy were at a very extreme distance, only the bright nucleus would be visible. Perhaps quasars are the tremendous energetic sources in the nuclei of Seyfert galaxies. (Palomar Observatory Photograph)

Figure 21.20 Great Nebula in Andromeda. This galaxy, also known as Messier 31, is a spiral type galaxy similar to that of our Milky Way. Two elliptical galaxies are shown also. (Palomar Observatory Photograph.)

Our local galaxy is called the **Milky Way.** It is of the spiral Sb type. The Milky Way contains some 100 billion (10^{11}) stars and has an appearance similar to that of the Great Nebula in Andromeda (Fig. 21.20). The Milky Way (Fig. 21.21, pp. 374–375) is about 10^5 light-years in diameter and has a maximum thickness in the central region of approximately 10^4 light-years. It is rotating eastward or counterclockwise as viewed from above. The period of rotation, which is not the same for all regions of the galaxy, is more than 2×10^8 years for the region that contains our solar system. Our solar system is located near the plane of rotation about 3.0×10^4 light-years from the galactic center, and moving at a rate of approximately 250 kilometers per second in the direction of the constellation Cygnus.

The galactic equator, a great circle positioned halfway between the galactic poles, is inclined about 62° to the celestial equator (Fig. 21.22). The north galactic pole has a right ascension of 12 h, 40 min, and a declination of +28°. The south galactic pole has a right ascension of zero h, 40 min and a declination of −28°.

A galaxy is composed of such items as individual atoms and molecules, dust, planets, stars, multiple stars, and star clusters. The closest star to our solar system is Alpha Centauri, a triple star, located about 4.3 light-years away, with right ascension 14 h, 36 min and −60° 38′ declination. The next closest star is Barnard's star, named after Edward M. Barnard (1857–1923), an American astronomer who first observed the fast motion of the star in 1916. This star has a faster apparent motion than any other known star, moving at a rate of approximately 10 seconds of arc per year. The average rate for visual stars is about $\frac{1}{10}$ second per year. Barnard's star is approximately 5.9 light-years from our Sun and is located near 18 h right ascension and +4° declination.

In the immediate neighborhood of the Milky Way and

confined to an ellipsoidal volume of space—some 9×10^5 parsecs for the major axis and about 8×10^5 parsecs for the minor axis—is located a small group of galaxies known as the **Local Group.** There are at least 21 known members of the group, and others are believed to exist that have not been detected because of their low magnitude or brightness. Our Milky Way is a member located near one end of the major axis. Messier 31 (Great Nebula in Andromeda, Fig. 21.20) is also a member and is positioned near the opposite end of the major axis.

The galaxies of the Local Group seem to be moving with random motions. The two Magellanic Clouds (see Fig. 21.23) are moving away from our galaxy and several others, including M31, are moving toward our galaxy.

The galaxies astronomers photograph throughout the vast volume of the universe are lumped together in **clusters** that vary in size and number. See Figs. 21.24 and 21.25, (p. 376). The clusters, classified as regular or irregular, range in size from three to 15 million light-years in diameter. Some clusters contain a few galaxies such as our Local Group, others contain thousands. Astronomers have discovered the distribution of clusters to be isotropic and homogeneous, the same as Hubble discovered in the distribution of galaxies.

The galactic clusters lump together into what are called superclusters. That is, the **superclusters** are clusters of clusters. These superclusters have diameters as large as 300 million light-years and masses equal to or greater than 10^{15} solar masses. Our Local Group and adjoining groups and clusters such as the Virgo cluster form what is called the Local Supercluster.

When Edwin Hubble began looking at Doppler shifts of galaxies, he found nothing surprising at first. In the Local Group the Doppler shifts were small: some were blue and some were red. But as he looked at galaxies farther and farther away he found only red-shifts. In fact, the farther away the galaxy, the larger the red-shift. From these measurements, the distances to remote galaxies can be determined by converting the observed red-shift of the galaxy to radial velocity and then plotting the logarithm of the velocity against the apparent magnitude. See

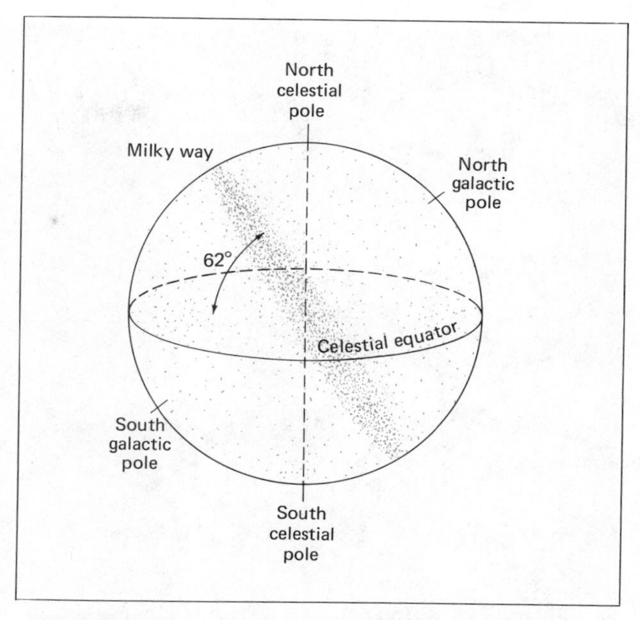

Figure 21.22 Diagram illustrating the inclination of the Milky Way with the celestial equator.

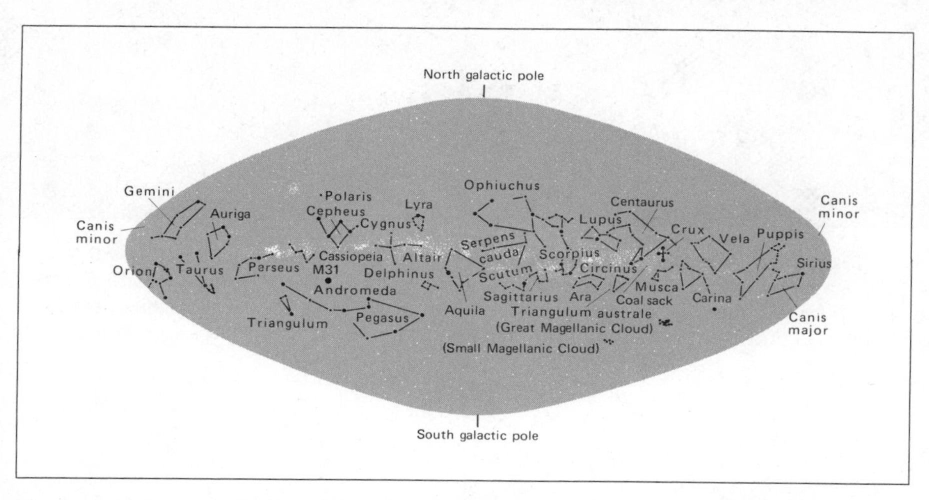

Figure 21.23 Constellations in and near the Milky Way. Compare this drawing with Fig. 21.21.

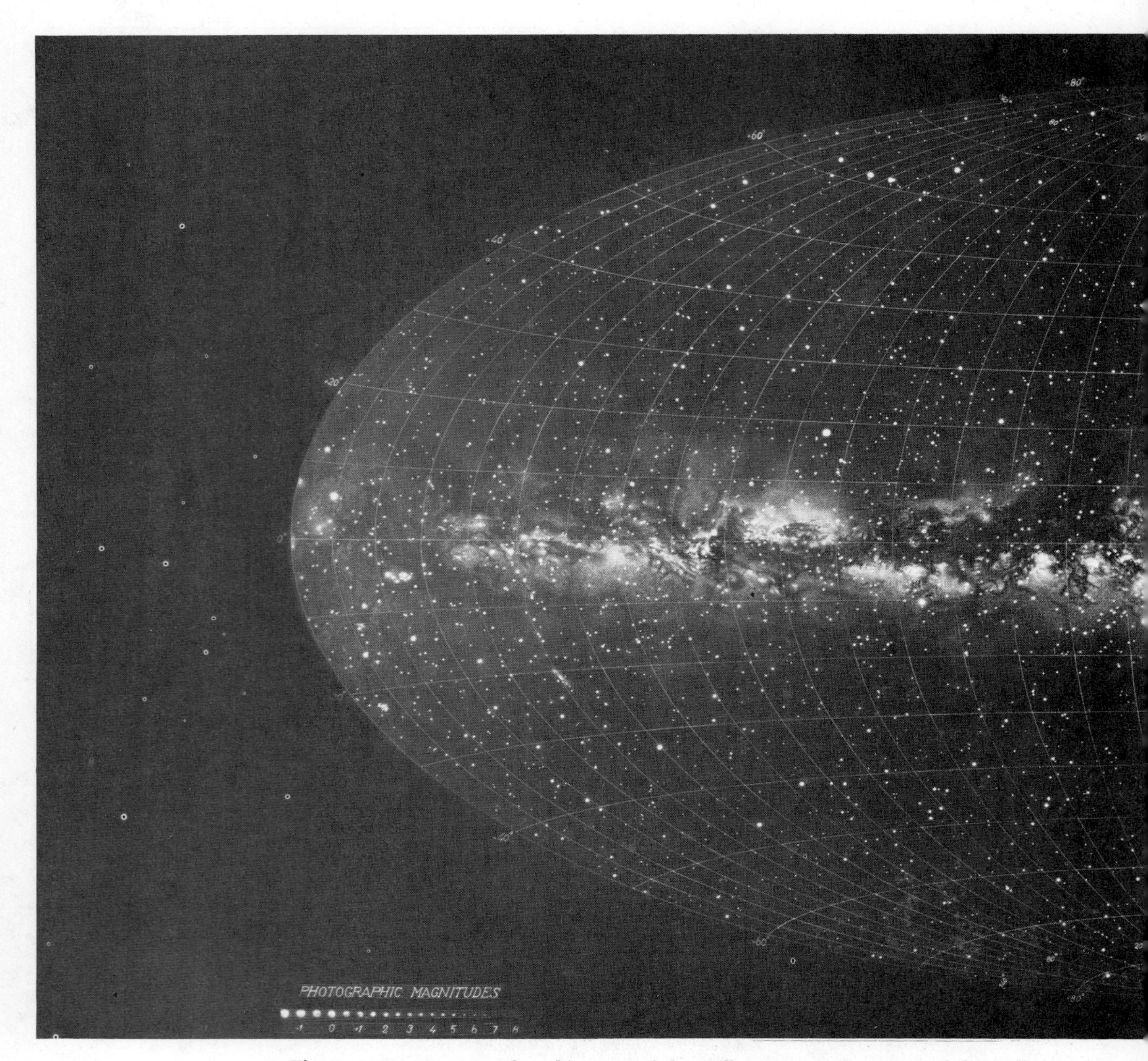

Figure 21.21 Knut Lundmark's map of the Milky Way. A panorama of the Milky Way with 7000 stars accurately drawn in with known coordinates. The composite is from many photographs and illustrates the Milky Way galaxy on an Aitoff projection. The constellation Sagittarius is near the center. The location of Sagittarius and other con-

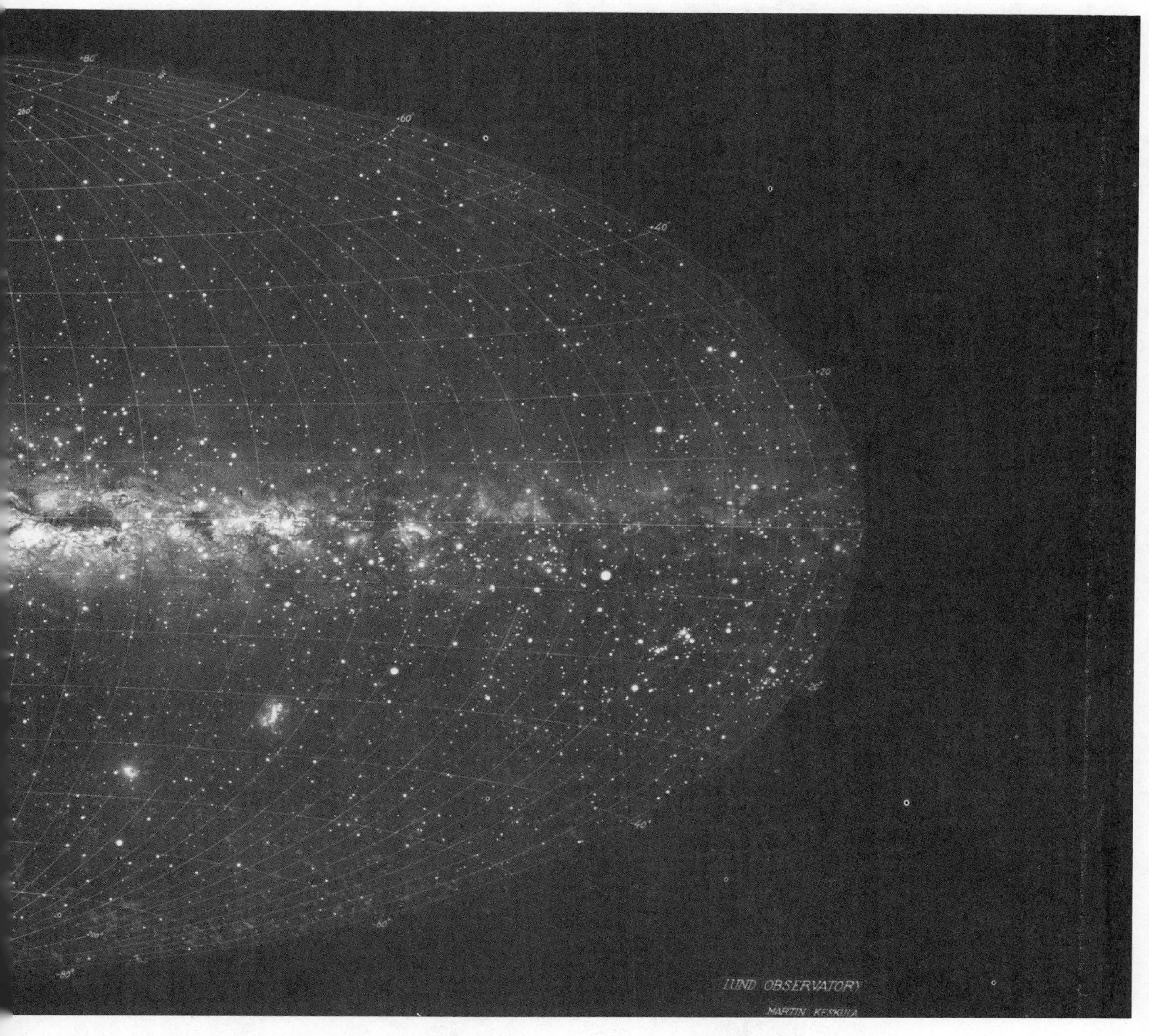

stellations can be seen by referring to Fig. 21.23. At the lower right are the Small and Large Magellanic Clouds. These are two irregular galaxies located about 200,000 light years away. At the far left, at coordinates $-20°$ and $90°$, is Messier 31, the spiral galaxy similar to the Milky Way. (Courtesy Lund Observatory, Sweden.)

Fig. 21.26. Hubble's discovery, now known as **Hubble's law,** can be written as

$$v = Hd$$

where v = recessional velocity of the galaxy,
d = distance away from the galaxy,
H = Hubble's constant

$$= \frac{50 \text{ km/s}}{10^6 \text{ parsec}}$$

The Hubble constant is believed to have a value of 50 to 100 kilometers per second per million parsecs. For

Figure 21.24 (*Left*) Central part of cluster of galaxies in Coma Berenices. The cluster is about 400 million light-years away. (Courtesy Hale Observatories.)

Figure 21.25 (*Below*) Cluster of galaxies in Virgo. The Virgo cluster, classified as irregular, contains thousands of galaxies of all types, but most are ellipticals. Two giant ellipticals, M84 and M86, are shown in the photograph. The Virgo cluster has a diameter of about 2 million parsecs and is located some 20 million parsecs from our solar system. Only the central part of the cluster is shown in the photograph. (Kitt Peak National Observatory/Cerro Tololo Inter-American Observatory)

example: if H is 50 km/s per million parsecs, the observed galaxy is moving away from our Sun 50 k/s for every one million parsecs the galaxy is from our Sun. Hubble's law is extremely important because it gives astronomers vital information about the structure of the universe.

Astronomers and other scientists determine the structure of the universe by detecting and analyzing electromagnetic waves that come from the stars and galaxies that occupy the universe. From the collected data the scientists contemplate the way in which these objects are distributed throughout the vast volume of the universe. It is estimated that there are at least 10^9 (one billion) galaxies within range of astronomers' optical telescopes that can be photographed. Even if they were inclined to do so, astronomers would not have the time to photograph the tremendous volume of the universe to obtain photographs of all galaxies. Instead, the astronomers' model of the structure of the universe is based upon a sampling of different regions of the universe. Using the 60-in and 100-in reflecting telescopes on Mount Wilson, Hubble obtained photographs of more than 1200 sample regions of space, counted some 44,000 galaxies, and estimated that 100 million were possible to photograph. After correcting his data for such things as the obstruction presented by the Milky Way and interstellar dust, he concluded that when observation of the universe is made over a large volume, the distribution of galaxies is isotropic and homogeneous. That is, when we observe a large volume of space, we observe as many galaxies in one direction as in any other and the observations are the same at all distances. This concept of the uniformity of the universe is known as the **cosmological principle** and is the basic assumption for most theories of cosmology.

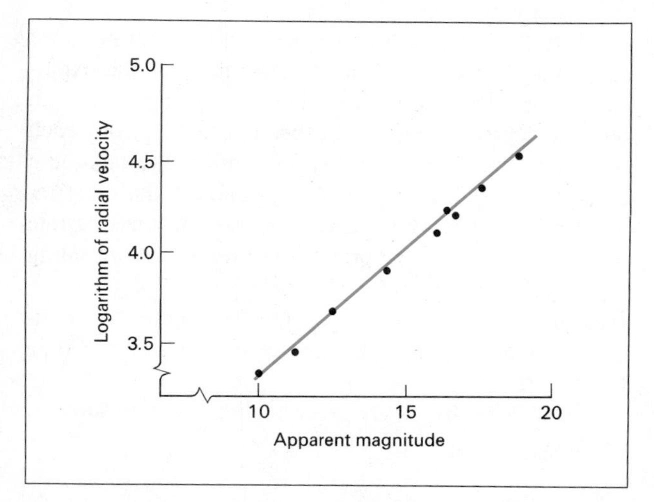

Figure 21.26 Graph of data showing the relationship (Hubble's law) between the logarithm of the radial velocity of a galaxy and its distance from the Milky Way. Recall that apparent magnitude and distance are related.

21.5 Quasars

About 1960, radio astronomers using their newly developed high-resolution (ability to distinguish the separation of two points) radio telescopes began detecting extremely strong radio signals from sources having small angular dimensions. The sources were named **quasars**—a shortened term for "quasistellar radio sources."

Hundreds of quasars have now been detected (see Fig. 21.27) and they have two important characteristics. First, it was found that quasars have extremely large red-shifts. This indicates from Hubble's law that they must be very far away. When their brightness was considered, however, it was shown that if they were indeed extremely distant, quasars must be extremely powerful sources of energy. The mystery of quasars then becomes a question of energy. Quasars have an energy output equivalent to about 10,000 times that of a typical spiral galaxy.

Quasars are considered to be the most distant objects in the universe. The quasar OH 471 has an enormous red-shift that according to Hubble's law places the quasar 18 billion light-years from our solar system. In addition to having enormous red-shifts, quasars emit tremendous amounts of electromagnetic radiation (energy) at all wavelengths, and most of the radiation varies in intensity (some regular, some random) over a range of years to a few days. Quasars are very small in size, appear blue in color, and have absolute magnitudes down to -25. Although not conclusive, the best theoretical evidence points to the theory that the "powerhouse" of quasar energy is a supermassive black hole at their center.

Recent evidence indicates that quasars are somehow related to the centers of galaxies. As mentioned in Section 21.4, Seyfert galaxies have very bright centers, and the broad emission lines present in their spectra indicate that they contain gas that is extremely hot. Since quasars are very distant objects, perhaps they are the bright nuclei of spiral galaxies. The spiral arms of a galaxy would not be visible at a great distance from us.

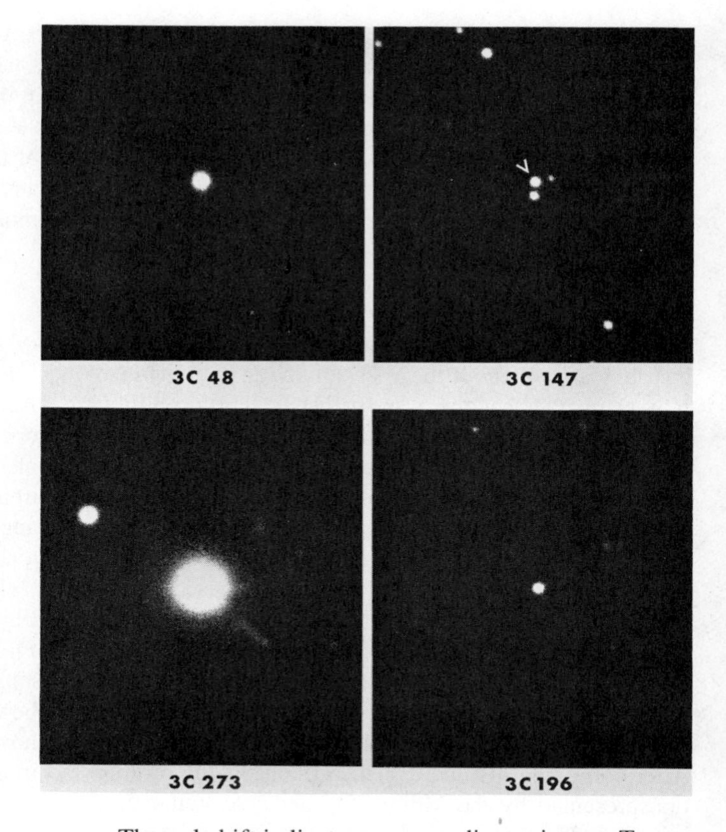

Figure 21.27 Quasistellar radio sources. Note the jetlike prong of 3C273. This quasar has an apparent visual magnitude of 12.8 and can be seen with a good 8-in telescope. (Palomar Observatory Photograph.)

21.6 Cosmology

Cosmology is a study of the entire universe. The universe as conceived by the human mind begins with concepts that generate models, based on observations, that deal with the origin, evolution, and structure of the universe. When constructing such models, it must be remembered that any model of the universe conceived by the human mind is the way the universe is conceived, not the way it actually is.

The model presently accepted by most astronomers is called the Big Bang. This model has received broad acceptance because experimental evidence has supported theoretical concepts in three major areas. First, astronomers observe galaxies that show a shift in their spectrum lines toward the low-frequency (red) end of the electromagnetic spectrum. This red-shift is known as the **cosmological red-shift.** Second, they detect a cosmic background radiation coming from space in all directions, commonly referred to as the 3-K cosmic microwave background. Third, astronomers observe a mass ratio of hydrogen to helium of three to one in stars and interstellar matter.

The red-shift indicates an expanding universe. To explain this observation, astronomers conceived the idea that the universe began with an explosion called the **Big Bang** and we are now seeing the universe expanding from that explosion.

Since the explosion came before the expansion, we can calculate the time that has elapsed since the explosion. Astronomers observe galaxies receding from each other and assume they have been receding from each other since the explosion. To determine the maximum time (t_m), let (d) represent the distance to the most remote galaxy that we observe, and assume that the rate of receding is what we presently observe. The receding velocity (v_r) is given by Hubble's law. That is, $v_r = Hd$, where H is Hubble's constant. We know, by definition, that velocity is distance divided by time ($v = d/t$).

Thus, $v_r = Hd = d/t_m$. Canceling and rearranging, we have

$$t_m = 1/H$$

Substituting the minimum value for Hubble's constant, $H = \dfrac{50 \text{ km/sec}}{10^6 \text{ parsecs}}$, we have

$$t_m = \frac{1}{\dfrac{50 \text{ km/sec}}{10^6 \text{ parsecs}}}$$

Since 3.086×10^{19} km $= 10^6$ parsecs,

$$t_m = \frac{1}{\dfrac{50 \text{ km/sec}}{3.086 \times 10^{19} \text{ km}}}$$

or $t_m = 6.173 \times 10^{17}$ seconds $= 1.957 \times 10^{10}$ years $=$ 19.57 billion years. Since we used the minimum value for Hubble's constant, this represents the maximum age of the universe.

What is the origin of all matter and radiation in the universe? One possible answer is that it came from nothing. When astronomers take a relativistic view of the universe, space and time change. At the beginning of the Big Bang, which is considered the origin of the universe, space and time did not exist. Thus, no space, no time, no anything (maybe energy) expanded violently creating space, time, and something. It seems meaningless to conceive of nothing, but out of nothing (a vacuum) comes something when we allow space to expand and time to flow. Thus, the possibility exists for the transformation from timeless, spaceless, and nothing to space, time, and something.

Presently what the something might have been at the beginning of the Big Bang is only speculation. General relativity fails to provide answers for a universe of infinite density and curvature. It is thought that all kinds of particles existed in equilibrium with radiation during the first fractional part of a second that the Big Bang took place. Particles and antiparticles were being produced in pairs from photons, and annihilating and reconverting to photons. But, by the time the explosion was one second old and the temperature about 10^{10} K matter as we know it today (electrons, protons, neutrons, and neutrinos) existed. About one million years after the Big Bang the temperature had dropped below 3000 K, and the particles could combine to form hydrogen, helium, and a trace of deuterium. The ratio of hydrogen to helium was then three to one by mass, which is the value presently observed in stars and interstellar matter.

It is often asked, "Where did the Big Bang take place?" Where the Big Bang took place has no meaning because at the moment of the explosion the makeup of the Big Bang was the entire universe. The explosion did not throw matter off into space. Space expands with time and forms the universe. Thus, time and space do not exist outside the universe.

To comprehend an expanding three-dimensional universe of galaxies, observe the expanding two-dimensional surface of a toy rubber balloon speckled with small dots (representing galaxies) as the balloon is being inflated. Think of yourself as standing where a dot is located on the surface of the balloon. As the balloon expands all dots are observed to recede from you. It makes no difference which dot you choose to observe from, the result is the same. There is no central point on the surface for the dots. Likewise, there is no central point for the galaxies in our three-dimensional universe. Note also that the surface of the balloon has no edge: You could examine every point on the surface of the balloon and find no edge. Similarly, there is no edge to our three-dimensional universe.

As early as 1948 it was predicted that residual radiation from the early universe should be present here and now, and the radiation should be at radio wavelengths. The radiation should also resemble the radiation from a black body at a temperature of a few degrees above absolute zero. Since the radiation from the early universe was everywhere at once, it should fill the entire universe and be isotropic. That is, the radiation is the same in any direction that we observe.

This radiation was first detected in 1965 by Arno Penzias and Robert Wilson at Bell Telephone Laboratories in New Jersey. They were testing a new microwave horn antenna used to make measurements of the absolute intensity of microwave radiation coming from certain regions of the Milky Way. After debugging and accounting for known static sources in their equipment, there remained a static source coming from every direction of space that they were unable to locate. The static was received equally from space at all times from any direction. After consulting with scientists at Princeton University who were designing and building equipment to detect the dying glow of the Big Bang, they concluded that the static (radiation) they were annoyed by was the residual radiation from the early universe.

Today, the dying glow is called the **3-K cosmic microwave background radiation,** and its presence is considered evidence of an ancient, extremely hot universe. This residual radiation is the greatly red-shifted radiation of the extremely hot universe that existed about one million years after the Big Bang.

Will the universe continue to expand forever? Presently, astronomers do not know the answer to this question, but the key to our understanding lies in determining the average density of matter in the universe. If the average density of matter is great enough, then space has a positive curvature and we live in a closed universe. That is, gravity in the universe is great enough to stop the expansion and eventually all matter will collapse in what is called the **Big Crunch.** If the average density of matter is too small, then space has a negative curvature and we live in an open universe that will continue to expand forever.

Present data for the value of the average density of matter are not precise enough to determine whether space has a positive or a negative curvature. Thus, present knowledge of the average density seems to indicate that we are living in a nearly flat universe.

Today, astronomers are applying the laws of physics and using the techniques of modern computer technology to probe the near and distant volumes of space seeking knowledge of its matter, radiation, composition, motion, and structure. They are using data collected by sophisticated land-based optical telescopes and others that operate from space (see Figure 21.28) plus an array of new devices operating in the X-ray, gamma ray, microwave, radio, ultraviolet, and infrared regions of the electromagnetic spectrum. Space probes have given astronomers tremendous knowledge of the solar system by sending 12 astronauts to the moon, robots to Mars to chemically

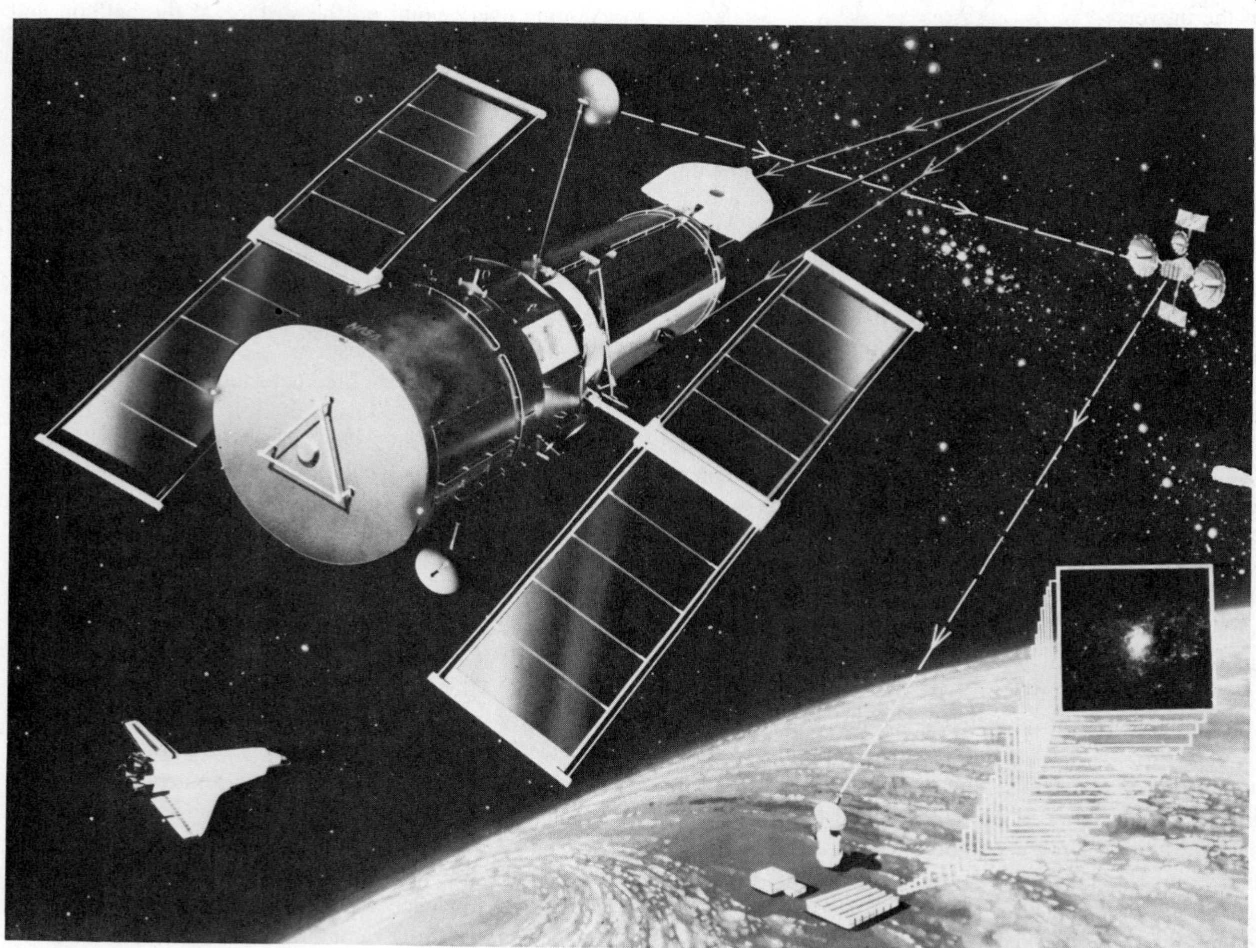

Figure 21.28 Artist's conception of NASA's space telescope in orbit with TDRSS, the Shuttle Orbiter, and the Ground Tracking Station. The space telescope, known as the Hubble Space Telescope, will allow scientists to look seven times farther into space than ever before. Astronomers will look at celestial objects such as quasars, galaxies, gaseous nebulae, and cepheid variable stars which are 50 times fainter than those seen by the most powerful telescopes on the Earth's surface. The telescope will be operated by remote control using a TV-type recording system of much higher sensitivity than photographic films. (Courtesy NASA.)

analyze the Martian surface, fly-by missions to examine the clouds and surface of Venus, spacecraft on fly-by missions to Jupiter, Saturn, and Uranus, and projecting space probes on to Neptune and interstellar space.

The search for knowledge concerning the universe will continue. By the year 2000 we may know the course our universe will follow and the distant future of galaxies.

Learning Objectives

After reading and studying this chapter, you should be able to do the following without referring to the text:

1. Define the term *plasma* and describe the composition of the Sun's interior.
2. Tell what sunspots are and describe their overall behavior.
3. Describe the basic process by which a star, such as the Sun, shines.
4. Define the term *neutrino* and tell why the solar neutrino experiment is so important.
5. Describe the motion of the stars and planets if you look at the stars for the following amounts of time: (a) One night (b) Every Monday night at midnight for one year.
6. Draw an H-R diagram and label the axes and the major features.
7. Tell why cepheid variables are important.
8. State the major events in the birth, life, and death of a star.
9. List the three possible end stages of a star and the mass a star would have to fall into each category.
10. State the Sun's age and the age of the universe.
11. State Hubble's law in words.
12. List the significant properties of quasars.
13. State and give experimental evidence to support the Big Bang theory.

Important Words and Terms

photosphere
plasma
chromosphere
corona
sunspots
prominences
neutrino
proton-proton chain
solar neutrino experiment
celestial sphere
declination
right ascension
celestial prime meridian
astronomical unit
light-year
parsec
constellations
zodiac
magnitude
absolute magnitude
H-R diagram
main sequence
red giants
white dwarfs
binary star
cepheid variables
nova
supernova
hydrogen burning
nucleosynthesis
neutron star
pulsar
black hole
galaxy
Seyfert galaxy
Milky Way
Local Group
cluster
supercluster
Hubble's law
cosmological principle
quasars
cosmology
Big Bang
Big Crunch
cosmological red-shift
3-K cosmic microwave background radiation

Questions

1. What is a plasma?
2. What is the temperature of the Sun (a) at the center, (b) at the surface?
3. Are sunspots cyclic?
4. What makes the Sun shine?
5. What are neutrinos and photons and how are they different?
6. Why are the results of the solar neutrino experiment perplexing?
7. What is the zodiac?
8. Describe each of the following: (a) light-year (b) parsec (c) astronomical unit
9. On an H-R diagram, give the position of each of the following: (a) the Sun, (b) the main sequence, (c) red giants, (d) cepheid variables, (e) white dwarfs.
10. List the following possible stages of a star in the correct order: hydrogen burning, cepheid variable, white dwarf, gravitational accretion, red giant.
11. What is a pulsar?
12. What is a black hole?
13. What is a quasar?
14. How old is (a) the Sun, (b) the universe?
15. Why is the Sun considered a second-generation star?
16. Describe the Milky Way.
17. Why is the universe thought to be expanding?
18. Explain this statement: The distribution of galaxies, over a large volume of space, is isotropic and homogeneous.
19. State the cosmological principle.
20. Explain the Big Bang model of the universe.
21. State three experimental facts that support the Big Bang model.
22. What property of the universe can be determined by taking the reciprocal of Hubble's constant?

Problems

1. The mass of the Sun is 2×10^{30} kg, or in terms of Earth weight, 2.2×10^{27} tons. Compute how many years it would take to consume one-tenth of the Sun's mass if 630 million tons are consumed every second.

   Answer: 11 billion years
2. Perform the calculations mentioned in the book to derive the net reaction

$$^1_1\text{H} + {}^1_1\text{H} + {}^1_1\text{H} + {}^1_1\text{H} \longrightarrow {}^4_2\text{He} + 2\ {}_{+1}\text{e} + 2\gamma + 2\nu + \text{Energy}$$

   from the reactions for the proton-proton chain.
3. Calculate the number of miles in a light-year using 186,000 mi/s as the speed of light. *Answer:* 5.87×10^{12} mi
4. Calculate the number of meters in a light-year using 3×10^8 m/s as the speed of light, or by referring to the answer to problem 3.
5. How many miles away is the closest star, Alpha Centauri, which is 4.3 light-years away?

   Answer: 2.47×10^{13} mi
6. How many parsecs away is Alpha Centauri, which is 4.3 light-years away?
7. How long does it take light to reach us from Alpha Centauri, which is 4.3 light-years away?
8. Find the distance in parsecs to a star with a parallax of 0.2 seconds.
9. Find the distance in light-years to a star with a parallax of 0.2 seconds.

   Answer: 16.3 light-years
10. How much brighter is a star of absolute magnitude +1 than a star of absolute magnitude (a) +2, (b) +6?
11. How much brighter is a cepheid variable of absolute magnitude −3 than a white dwarf of absolute magnitude +7?

   Answer: 10,000 times
12. (a) What is the color index of a star with a photographic magnitude of 3 and a visual magnitude of 3.3?
    (b) What is the color of this star?

   Answer: (b) blue-white
13. The Crab Nebula is expanding at a rate of 70 million miles per day. If it is the remnants of a supernova of 1054 A.D., how many miles in diameter is it today?

   Answer: 47×10^{12} mi, or 8 light-years

14. If our universe contained 100 billion galaxies with 100 billion stars in each:
    (a) How many stars would there be?
    (b) Is this number greater or less than Avogadro's number (6×10^{23})?

    Answer: (b) less
15. It takes the Sun 2×10^8 years to go around the center of the galaxy. How many times has it (and our solar system) gone around the galaxy during its life of 5 billion years?

    Answer: 25 times
16. Determine the age of the universe using a value for Hubble's constant of 100 km/sec per 10^6 parsecs.
17. Hubble's constant is estimated to be between 50 and 100 km/sec per 10^6 parsecs. Use the average of these two values and determine the age of the universe.

22
The Atmosphere

"... *this most excellent canopy, the air, look you, this brave o'erhanging firmament, this majestical roof fretted with golden fire* ..."

–Shakespeare, *Hamlet*

OUR **ATMOSPHERE** (Greek *atmos*, vapor, and *sphaira*, sphere) is the gaseous shell, or envelope, of air that surrounds the Earth. Just as certain sea creatures live at the bottom of the ocean, humans live at the bottom of this vast atmospheric sea of gases. In recent years the study of the atmosphere has expanded due to advances in technology. Every aspect of the atmosphere from the ground to outer space is now investigated in what is called **atmospheric science.** The older term of **meteorology** (Greek *meteora*, the air) is now more commonly applied to the study of the lower atmosphere. The continuously changing conditions of the lower atmosphere are what we call weather. Lower atmospheric variations are monitored daily, and changing patterns are studied to help predict future conditions.

The phenomena of the atmosphere are important to many branches of science. They produce changes on the surface of the Earth that concern the geologist, as will be discussed in later chapters. The astronomer longs for a view of the stars that is not obscured by the atmosphere and must consider the effect of atmospheric phenomena on observations. Yet, while we attempt to learn the workings of the atmosphere and to even effect local changes for our own benefit, we indiscriminately pollute the atmospheric blanket that protects and sustains life.

The condition of our atmosphere has become a topic of major interest with respect to ecology, the study of living organisms in relationship to their environment. Since the air we breathe is such an integral part of our environment, a knowledge of the atmosphere's constituents, properties, and workings is needed to understand and appreciate current environmental problems. In other words, what are some of the normal conditions and other phenomena of the atmosphere?

◀ Crepuscular rays of the Sun.

22.1 Composition

The air of the atmosphere is a mixture of many gases. In addition, the air holds many suspended liquid droplets and solid particles. However, it is somewhat amazing that only two gases comprise about 99% of the volume of air near the Earth. From Fig. 22.1 and Table 22.1, it can be seen that air is primarily composed of nitrogen (78%) and oxygen (21%) with nitrogen being about four times as abundant as oxygen. The other main constituents are argon and carbon dioxide.

Table 22.1 Composition of Air

Nitrogen	N_2	78%	(by volume)
Oxygen	O_2	21%	
Argon	Ar	0.9%	
Carbon dioxide	CO_2	0.03%	
Others (traces)		*Others (variable)*	
Neon	Ne	Water vapor (H_2O) 0–4%	
Helium	He	Carbon monoxide (CO)	
Methane	CH_4	Ammonia (NH_3)	
Nitrous oxide	N_2O	Solid particles—dust, pollen, etc.	
Hydrogen	H_2		

Many other gases of minute quantities are found in the atmosphere along with dust, pollen, salt particles, and so on. Some of these, especially water vapor and

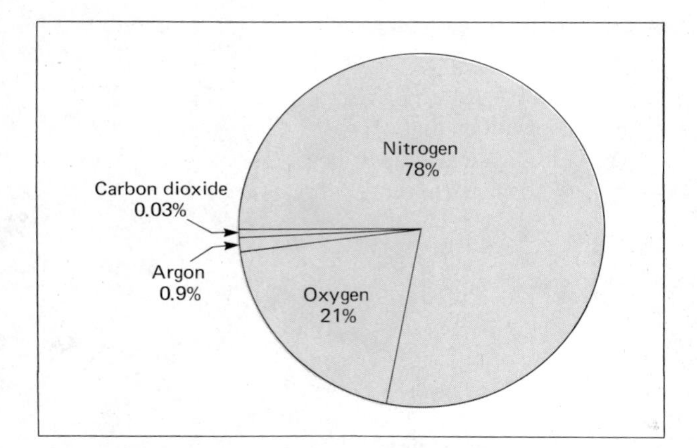

Figure 22.1 Graphic representation of the volume composition of the major constituents of air.

carbon monoxide, vary in concentration, depending on conditions and locality. The amount of water vapor in the air is dependent to a great extent on the temperature as will be discussed later. Carbon monoxide is a by-product of incomplete combustion. A sample of air taken near a freeway, for example, would contain a concentration of carbon monoxide considerably higher than that normally found in the atmosphere. It is estimated that over 10,000 tons of carbon monoxide are daily put into the atmosphere in and around Los Angeles, most of which results from automobile combustion. Carbon dioxide, another by-product of combustion, is also present in abnormally high concentrations in such populated areas.

In general, however, the relative amounts of the major constituents of the atmosphere remain fairly constant. Nitrogen, oxygen, and carbon dioxide are involved in the life cycles of plants and animals and are continually taken from the atmosphere and replenished as by-products of the various cycles. Animals breathe in oxygen and expel carbon dioxide, while plants use carbon dioxide to expel oxygen in their life cycles. Nitrogen is also assimilated by some plants and released in decay. Thus, nature operates in such a way as to maintain the balance of the various gases in the atmosphere.

On the other hand, human beings may contribute to changes in the atmosphere and perhaps eventually create imbalance. Recent concern has arisen in this regard because of increased amounts of CO_2 being poured into the atmosphere. Consider the amount of burning required to heat the houses in the United States alone in the winter. Add to this the industrial burning for heat and energy. The resultant amount of gases released into the atmosphere is considerable and gives cause for concern, as will be discussed in Chapter 26.

22.2 Origin

The origin of the atmosphere is generally associated with the origin of the Earth, which is some four to five billion years old. There are several theories of cosmogony that attempt to explain the origin of the universe. Without speculating on the validity of any one theory, it is generally agreed that at the time of the creation of our solar system, the planets were extremely hot masses, since their creation by any means would require enormous amounts of energy.

In the beginning, then, the Earth was a molten mass surrounded by hot gases, probably methane, ammonia, and hydrogen compounds. Because of their high temperatures and inherent kinetic energy, large amounts of these gases escaped into space. As the Earth cooled, gases that were dissolved in the molten mass were released, giving rise to H_2O, N_2, and CO_2 in our atmosphere. The water vapor condensed and reevaporated, causing cooling and solidification of the Earth's crust with subsequent erosion and formation of bodies of water. The gases cooled along with the planet and, being then less energetic, were retained by gravitational attraction. At this time, life began to appear.

How life occurred is still a mystery. However, scientists have shown that the fundamental building blocks of life, the amino acids, can be constructed from combinations of methane, hydrogen, ammonia, and water, all of which were available on the cooling Earth. (See Chapter 17, the Miller-Urey experiment.) In any event, the appearance of plant life allowed the formation of oxygen and the evolution of animal life. Plants produce oxygen by **photosynthesis,** the process by which CO_2 and H_2O are converted into sugars (carbohydrates) and O_2, using the energy from the Sun (Fig. 22.2).

The key to photosynthesis is the ability of chlorophyll, the green pigment in plants, to convert sunlight into chemical energy. In chemical notation, the photosynthesis process may be expressed

$$(n)\ CO_2 + (n)\ H_2O \xrightarrow[\text{chlorophyll}]{\text{sunlight}} (CH_2O)_n + (n)\ O_2$$

where the n's indicate the numbers that balance the chemical equation. Photosynthesis accounts for the liberation of approximately 130 billion tons of oxygen into the air annually, with some 2000 billion tons of CO_2 being involved in the process. Over one-half of the photosynthesis takes place in the oceans, which contain many forms of green plants. Originally, oxygen may also have come from the dissociation of water.

Figure 22.2 An appeal for photosynthesis. Oxygen is a by-product of the photosynthesis process by which plants convert carbon dioxide and water to carbohydrates (sugars).

Over millions of years, the Earth's atmosphere evolved to its present composition. The atmosphere of other planets evolved in different fashions. For example, the atmospheres of Jupiter and Saturn contain mainly ammonia and hydrogen, while Venus has a considerable amount of CO_2. Mercury, however, lost its atmosphere, as did our moon. The retention of an atmosphere depends on a balance between gravity and the thermal kinetic energy of the atmospheric gases.

The gravitational force that attracts the gas molecules of an atmosphere to a planet is dependent on the mass of the planet. (See Chapter 3, Section 3.3.) A planetary atmosphere receives energy from the Sun, and the temperature of the local atmosphere depends on the velocities of the gas molecules. If sufficient energy is acquired, some of the molecules will have enough velocity to overcome the gravitational attraction and will escape into space. As a result, a planet can gradually lose its atmosphere. Similarly, spacecraft must be given sufficient velocity (kinetic energy) to escape the Earth's gravitational attraction.

The Earth and other massive planets near the Sun have sufficient gravitational attraction to retain their atmospheres. However, because of the relatively small masses of Mercury and the moon, their gravitational attractions were not sufficient to retain their energetic atmospheres.

22.3 Vertical Structure

To distinguish different regions of the atmosphere, we look for variations that occur with height or altitude. The changes in physical properties can be used to define vertical divisions of the atmosphere. Some of the atmospheric properties that show vertical variations that we will examine are (1) density and pressure, (2) temperature, (3) homogeneity and heterogeneity of gases, (4) ozone and ion concentrations.

Density and Pressure

The atmosphere extends upward with continuously decreasing density (mass per volume). The greater density near the Earth's surface is due to gravitational attraction and compression of the air. As a result, over one-half of the mass of the atmosphere lies below an altitude of 11 km (7 mi) and almost 99% lies below an altitude of 30 km (19 mi). The air becomes quite "thin" at higher altitudes. At a height of 320 km (200 mi) above the Earth, the density of the atmosphere is such that a gas molecule may travel a distance of 1.6 km (1 mi) before encountering another gas molecule.

There is no clearly defined upper limit of the Earth's atmosphere. It simply becomes more and more tenuous and merges into the interplanetary gases, which may be thought of as part of the extensive "atmosphere" of the Sun. The continuous decrease in density makes for an outmost atmospheric limit that can be placed anywhere from 480 to 960 km (300 to 600 mi) from the surface of the Earth.

Closely related to density is atmospheric pressure (force per area). The pressure at a particular altitude is effectively a measure of the weight (or amount) of gas above that location. Just as a sea diver experiences increased pressure when descending, due to the weight of the water above him, the atmospheric pressure is greater near the Earth due to the weight of the air above. Like density, the pressure varies vertically, decreasing with increasing altitude. This occurs continuously (Fig. 22.3), and does not make for distinct boundaries. Consequently, vertical divisions of the atmosphere are not made on the basis of pressure.

Temperature

In measuring the temperature of the atmosphere versus altitude, we find that distinct changes do occur. This leads to major divisions of the atmosphere based on

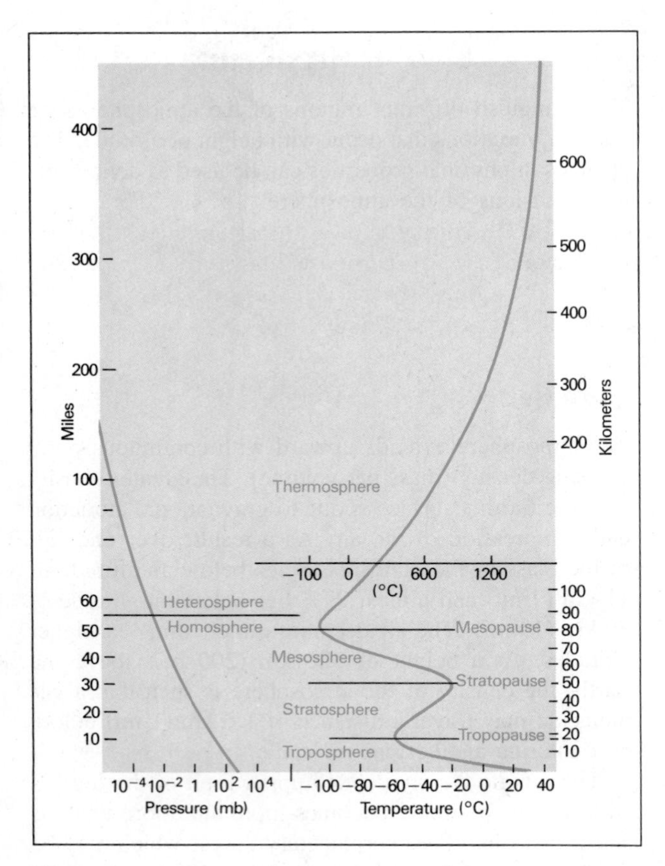

Figure 22.3 The vertical structure of the atmosphere. Variations in the physical properties of the atmosphere, such as temperature and homogeneity of the atmospheric gases, give rise to natural divisions. The gradual decrease of density and pressure with altitude, however, offers no distinct variations upon which divisions can be based.

temperature variations. Near the Earth's surface, the temperature of the atmosphere decreases with increasing altitude at an average rate of about $6\frac{1}{2}$°C/km ($3\frac{1}{2}$°F/1000 ft) up to about 16 km (10 mi), as illustrated in Fig. 22.3. This region is called the **troposphere** (Greek *tropism,* to change). The atmospheric conditions of the lower troposphere are referred to collectively as **weather.** Changes in the weather reflect the local variations of the atmosphere near the Earth's surface.

The actual depth of the troposphere varies from about 16 km (10 mi) at the equator to about 8 km (5 mi) at the poles, with an average thickness of about 11 km (7 mi). The lower troposphere was the only region of the atmosphere investigated until the twentieth century. Balloonists ascended to the upper regions of the troposphere only in the 1930s. Many unexpected problems arose. The balloonists encountered difficulty in breathing, the freezing of valves necessary for descent, and loss of the use of limbs due to frostbite. At the top of the troposphere, the temperature falls to −45° to −50°C (about −50° to −60°F). Beyond these altitudes, unmanned balloons and rockets were used for exploration. Instruments on board recorded the atmospheric properties. More recently, satellites have been used to make atmospheric observations. These are discussed in Chapter 25.

Above the troposphere, the temperature of the atmosphere increases nonuniformly up to an altitude of about 50 km (30 mi). See Fig. 22.3. This region of the atmosphere, from approximately 16 to 50 km (10 to 30 mi) in altitude, is called the **stratosphere** (Greek *stratum,* covering layer).

The temperature of the atmosphere then decreases rather uniformly to a temperature of about −95°C (−140°F) at an altitude of 80 km (50 mi). This region, between 50 to 80 km (30 to 50 mi) in altitude, is called the **mesosphere** (Greek *meso,* middle).

Above the mesosphere, the thin atmosphere is heated intensely by the Sun's rays and the temperature climbs to over 1000°C (about 1800°F). This region extending to the outer reaches of the atmosphere is called the **thermosphere** (Greek, *therme,* heat). The temperature of the thermosphere varies considerably with solar activity.

It should be realized that the boundaries of these regions are not sharply defined and extend over appreciable distances. The intermediate regions comprising the boundaries are given the suffix "pause." The top of the troposphere is referred to as the **tropopause,** and the respective upper boundaries of the stratosphere and mesosphere as the **stratopause** and **mesopause.**

Homogeneity and Heterogeneity of Gases

The mixture of atmospheric gases provides another means of atmospheric division. In the lower portion of the atmosphere, turbulence continually mixes the gases so that the air is a homogeneous mixture. Thus, this region is called the **homosphere** (*homo,* meaning "same").

Above the homosphere, beginning at about 55 to 60 mi (88 to 96 km) from the Earth's surface, is the **heterosphere** (*hetero,* meaning "different"). Because no turbulence exists there to keep the gases mixed, they settle out in a layered structure, as illustrated in Fig. 22.4. In the layer above 60 mi (96 km), the oxygen is in atomic form (O) instead of the normal diatomic form (O_2) found in the lower atmosphere.

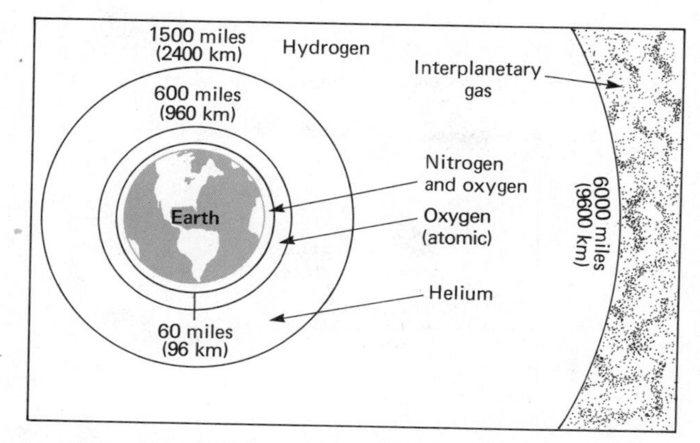

Figure 22.4 An illustration of the heterogeneous layering of the gases in the upper atmosphere based on satellite data. (Courtesy NASA.)

This settling is analogous to sand suspended in water. As long as the mixture is stirred, the sand remains in a uniform suspension. However, if the mixing is ceased, the sand settles out according to particle size and weight much the same as the gases of the heterosphere.

Ozone and Ion Concentrations

The atmosphere is also divided into the **ozonosphere** and **ionosphere.** Their boundary is approximately the same as the homosphere and heterosphere, respectively. As the names imply, there are concentrations of ozone and ions in these regions.

Ozone (O_3) is formed by the dissociation of molecular oxygen and the combining of the atomic oxygen with the molecular oxygen.

$$O_2 + \text{Energy} \longrightarrow O + O$$

$$O + O_2 \longrightarrow O_3$$

At high altitudes, energetic ultraviolet radiation from the Sun provides the energy necessary to dissociate the molecular oxygen. Oxygen, however, becomes less abundant at higher altitudes, so the production and concentration of ozone depends on the appropriate balance of ultraviolet radiation and oxygen molecules. The optimum condition occurs at an altitude of about 30 km (20 mi), and in this region the central concentration of an ozone layer is found as illustrated in Fig. 22.5. The ozone layer is a broad band of gas that extends through nearly all of the stratosphere. Ozone is very unstable in the presence of sunlight, and when it absorbs ultraviolet radiation it is again dissociated into molecular and atomic oxygen. A free oxygen atom may recombine with another oxygen molecule to again form ozone. However,

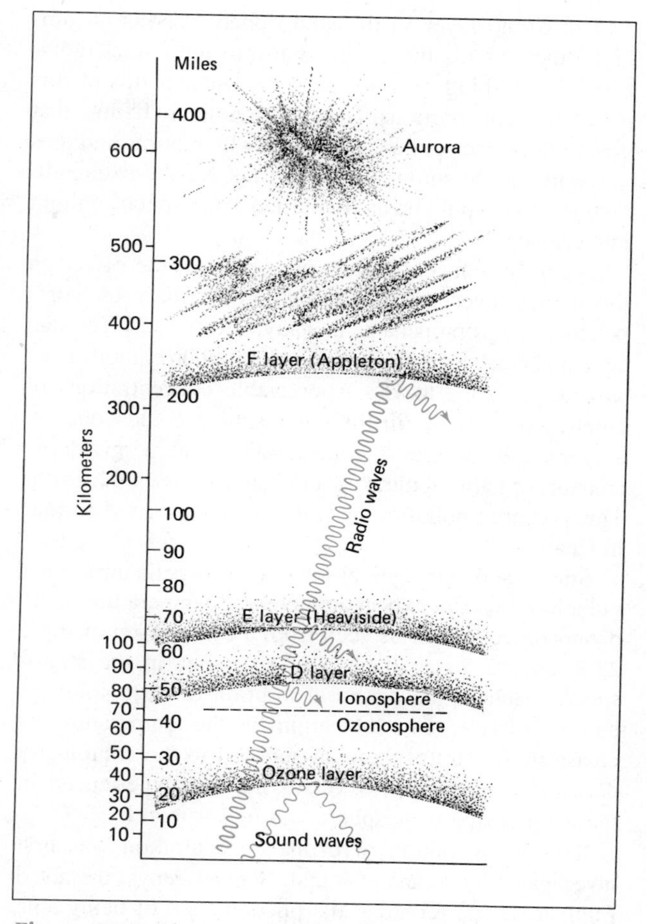

Figure 22.5 Diagram showing the regions of the ozonosphere and ionosphere. Atmospheric divisions are based on ozone and ion concentrations. The air temperature variation, due to the ozone layer, gives rise to the reflection of sound waves, while the ion layers reflect radio waves.

an oxygen atom may meet an ozone molecule and form two ordinary oxygen molecules ($O + O_3 \rightarrow O_2 + O_2$), thus destroying the ozone. These processes go on simultaneously in the ozone layer so that a balance between ozone production and destruction is maintained.

There is little ozone present near the Earth's surface. It is sometimes formed by electrical discharges and is easily detected by a distinct pungent smell by which it derives its name (Greek *ozein,* to smell). In some areas, e.g., Los Angeles, ozone is classified as an air pollutant. It is found in relatively high concentrations resulting from photochemical reactions of air pollutants. Such reactions give rise to photochemical smog, as will be discussed in Chapter 26.

The ozone layer in the stratosphere acts as an umbrella that shields life against harmful ultraviolet radiation by absorbing most of the short wavelengths of this radiation. The portion of the ultraviolet radiation that gets through the ozone layer is that which burns and tans our skins in the summer. Were it not for the ozone absorption, we would be badly burned and find the sunlight intolerable.

Recently, concern was expressed over the effect on the ozone layer by gaseous chlorofluorocarbon compounds used as propellants in spray aerosol cans. Through certain chemical reactions, these chlorofluorocarbons react to change ozone to O_2. Appreciable concentrations of chlorofluorocarbons in the ozonosphere could conceivably reduce the ozone concentration and permit large amounts of harmful ultraviolet radiation to reach the Earth. This potential pollution hazard will be discussed further in Chapter 26.

Since the ozone layer absorbs the energetic ultraviolet radiation, one can expect an increased temperature in the ozonosphere. It can be seen from a comparison of Figs. 22.3 and 22.5 that the ozone layer lies in the stratosphere. Hence, the ozone absorption of ultraviolet radiation provides an explanation for the temperature increase in the stratosphere as opposed to the continually decreasing temperatures versus altitude experienced in the neighboring troposphere and mesosphere.

This warm region of ozone concentration was first investigated by means of sound. Samuel Pepys, the noted English writer, recorded the phenomenon of being able to hear, in London, the cannons of the fleet in the English Channel during a battle of the Dutch Trade War of 1666, while people on the coast heard nothing. Similar events were observed during the Civil War and World War I. It was reasoned that the sound of the cannon must have been reflected back to Earth by something in the atmosphere. Just as light is reflected from water, perhaps the sound was reflected from the boundary of warm and cold air when incident at the proper angle. Subsequent investigations revealed the existence of the warm air layer and established the presence of ozone as the cause. In cases where sound is heard at a distance but not nearby, sound waves moving horizontally are blocked, for example by a hill, whereas those moving vertically are reflected by the layered air and heard at a distance. This explanation is illustrated in Fig. 22.6.

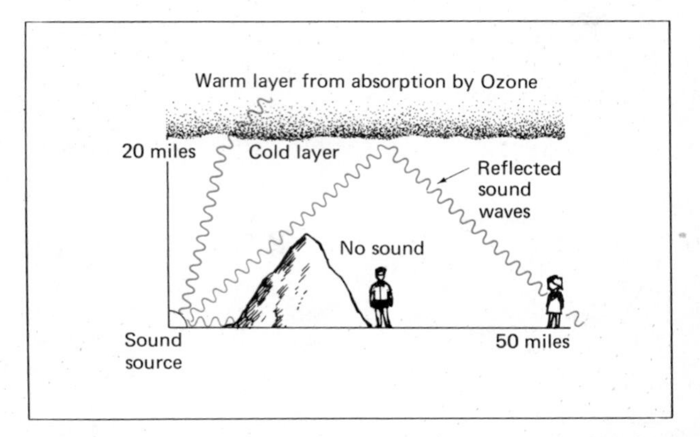

Figure 22.6 An illustration of the reflection of sound in the atmosphere. Sound waves may be reflected at the boundary of warm air associated with the ozone layer and be heard many miles away from the sound source.

In the upper atmosphere, the energetic particles from the Sun cause the ionization of the gas molecules. For example,

$$N_2 + \text{Energy} \longrightarrow N_2^+ + e^-$$

The electrically charged ions and electrons are trapped in the Earth's magnetic field and form ionic layers in the region of the atmosphere referred to as the ionosphere. Kennelley and Heaviside in 1902 independently discovered a layer at a height of 100 km or 60 mi. Appleton found another ionic layer at an altitude of 320 km or 200 mi (Fig. 22.5).

An important feature of these layers is that they reflect radio waves and hence may be used in global radio communication. Radio waves, like all electromagnetic waves, travel in straight lines, making it impossible to send them directly around the curved Earth. However, by reflecting, or "bouncing," the waves off an ionic layer, they may be transmitted around the Earth as illustrated in Fig. 22.7.

The Heaviside layer reflects longer wavelengths (10^2 to 10^3 m) than does the Appleton layer (10 to 10^2 m). These layers have been designated E and F, respectively, and are subdivided into E_1, E_2, F_1, and F_2 layers. Another layer, called D, has also been found below the E layer. The lower layers (D and E) are prominent only during the day and the division of the F layer becomes indistinguishable at night. Radio waves beamed vertically upward are reflected back to the Earth, thereby allowing calculation of the height of the reflecting layer from the knowledge of the velocity of the waves and the time elapsed.

The reflection of radio waves from the ionic layers depends on uniformity in the density of the layer. Should a solar disturbance produce a shower of energetic particles that upsets this uniformity, a communications "blackout" may occur. Radio transmission is then reduced to line-of-sight transmission as reflections from

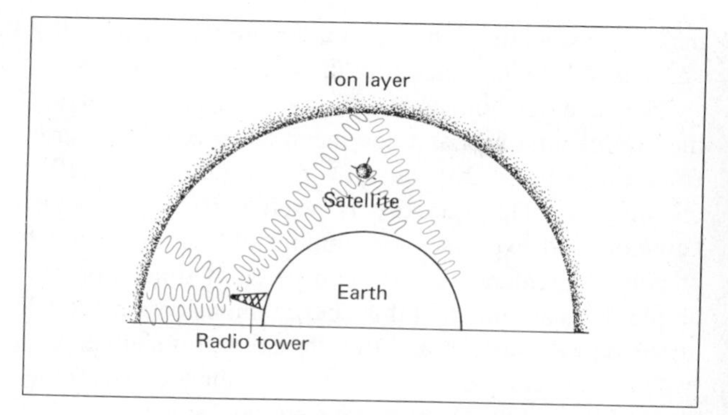

Figure 22.7 Global radio transmission. Radio waves, which normally travel in a straight line, are reflected around the curvature of the Earth by ionic layers. Ionic disturbances, however, sometimes interfere with reflections from an ionic layer and disrupt transmission. Transmission via satellite is not, in general, disrupted by ionic disturbances.

ionic layers are impaired. These disturbances caused considerable inconvenience in early worldwide communications. Cables were laid across the oceans to keep communications open, and now satellites relay radio and television communications.

Solar disturbances are also associated with beautiful displays of light in the upper atmosphere of the polar regions called northern lights, or **aurora borealis.** See Fig. 22.8. The southern hemisphere tends to be forgotten by people living north of the equator. However, light displays of equal beauty occur in the southern polar atmosphere and are called **aurora australis.**

The location and height of the aurora depend on the interaction of the incoming charged particles and the Earth's magnetic field. In general, the particles are deflected toward the Earth's magnetic poles (our polar regions) over which the majority of the auroras occur. The Norsemen thought the auroras were the reflections from shields in Valhalla. Today, however, they are associated with the excitation of oxygen and nitrogen molecules in the atmosphere by charged particles (electrons and protons) and the recombination of electrons and ions. The occurrence of auroras has been correlated with solar activity, which gives rise to an influx of ionizing radiation to the Earth's upper atmosphere.

22.4 Energy Content

The Sun is by far the most important source of energy for the Earth and its atmosphere. At an average distance of 93 million miles from the Sun, Earth intercepts only

Figure 22.8 The aurora borealis. The ions of the ionosphere give rise to brilliant displays of light called auroras. (University of Alaska photograph. Courtesy U.S. Air Force.)

a small portion of the vast amount of the solar energy emitted. This energy traverses space in the form of radiation and the portion incident on the Earth's atmosphere is called **insolation**—*in*coming *so*lar radi*ation*.

Because the Earth is tilted on its axis $23\frac{1}{2}°$ with respect to the plane of its orbit about the Sun, the insolation is not evenly distributed over the Earth's surface (Fig. 22.9). This tilt, coupled with the Earth's revolution around the Sun, gives rise to the seasons. Although the Earth is closest to the Sun in January and farthest away in July, the northern hemisphere experiences colder weather in January because this part of the Earth is tilted away from the Sun. As a result, the days are shorter and the radiation received north of the equator is distributed over a

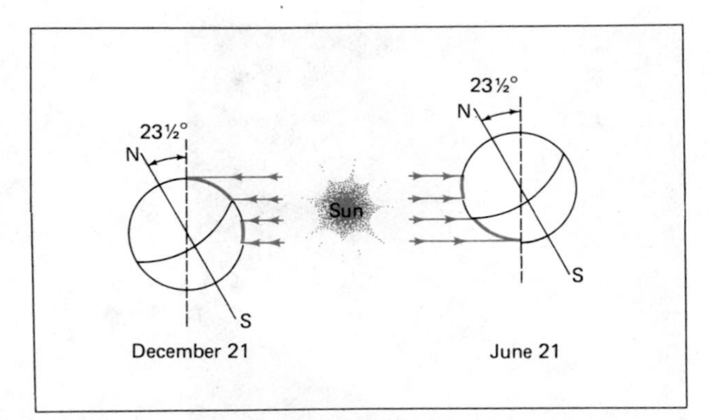

Figure 22.9 The seasonal variation of insolation. The tilt of the Earth's axis relative to the plane of its orbit causes the incoming solar radiation to be distributed differently on the Earth's surface at different times of the year. In December, the southern hemisphere receives more direct rays of the Sun than does the northern hemisphere. In June, the situation is reversed.

larger area, accounting for less warming. The southern hemisphere at this time is enjoying warm weather. Six months later, in July, the situation is reversed for the hemispheres. Thus, the tilting and resultant distribution of insolation gives rise to seasonal variations and their reversal in the northern and southern hemispheres.

The average summer temperature of the southern hemisphere is slightly higher than that of the northern hemisphere. This is, in part, due to the fact that the Earth is 5 million kilometers (about 3 million miles) closer to the Sun in January when it is summer in the southern hemisphere than in July because of its elliptical orbit. However, more water and less land mass in the southern hemisphere makes for less summer-winter temperature variation.

The maximum insolation for the Earth is received daily around noon in both hemispheres, yearly near June 21 in the northern hemisphere, and yearly near December 21 in the southern hemisphere. The maximum daily and yearly temperatures, however, lag behind these times. That is, the maximum daily temperature occurs around 2 to 3 P.M. and August is usually the hottest month in the northern hemisphere. This is readily understood since more radiation is received after 12 noon and on those days following June 21. The additional radiation contributes to the daily and average yearly temperatures, causing their maxima to occur later than the times of maximum insolation. This is much the same as slowly turning up the gas flame under a pan of water. When the maximum flame is reached and then slowly decreased, the temperature continues to rise and reaches a maximum sometime after the maximum flame, since the water is still being heated by the decreasing flame.

The average intensity of solar radiation outside the Earth's atmosphere at the mean distance of the Earth from the Sun is 2.0 cal/cm^2 · min or 3.1×10^{12} cal/mi^2 · h. This quantity is referred to as the **solar constant** and expresses the amount of energy received at normal incidence on an area per unit time. For example, let's assume that this energy intensity is received at the Earth's surface and that the area of a sunbather's back is 3000 cm^2. Every minute, he or she would receive on the average 6000 calories of energy, since

$$\frac{2.0 \text{ cal}}{\text{cm}^2 \cdot \text{min}} \times 3000 \text{ cm}^2 = 6000 \text{ cal/min}$$

If the bather remained in the Sun for 30 min, he or she would receive a total amount of energy of

$$6000 \frac{\text{cal}}{\text{min}} \times 30 \text{ min} = 180{,}000 \text{ cal}$$

This amount of energy could raise the temperature of over 9 cups of water from room temperature (20°C) to its boiling point. (See Section 5.3.) However, only about 50% or less of the insolation reaches the Earth's surface, depending on atmospheric conditions.

In considering the energy content of the atmosphere, one might think that this comes directly from insolation. However, surprisingly enough, most of the direct heating of the atmosphere comes not from the Sun, but from the Earth. To understand this, we need to examine the distribution and disposal of the incoming solar radiation (Fig. 22.10).

About 33% of the insolation received is returned to space with no appreciable effect on the atmosphere as a result of reflection by clouds, scattering by particles in the atmosphere, and reflection from terrestrial surfaces, such as water, ice, and variable ground surfaces. The reflectivity, or the average fraction of light a body reflects, is known as its **albedo** (Latin *albus,* white).

Thus, the Earth has an albedo of 0.33, or reflects about one-third of the incident sunlight. The brightness of the Earth as viewed from space depends on the amount of sunlight that is reflected, and clouds play an important role (Fig. 22.11, p. 394). In comparison, the moon with its dark surface and no atmosphere has an albedo of only 0.07, i.e., reflects only 7% of the insolation. Even if the Earth were as small as the moon, it would appear much brighter (about five times) than the moon to an astronaut in space.

Scattering of insolation occurs in the atmosphere from gas molecules of the air, dust particles, water droplets,

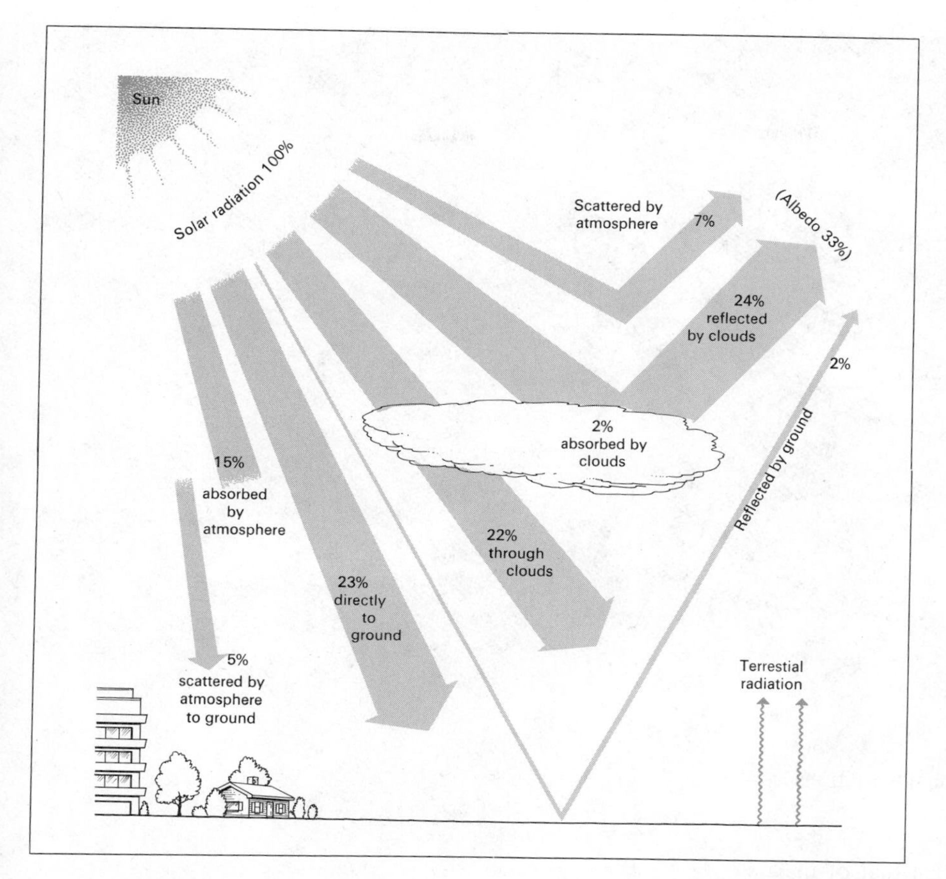

Figure 22.10 An illustration of insolation distribution. The percentages vary somewhat, depending on atmospheric conditions.

etc. By scattering, we mean the absorption of incident radiation and its reradiation in all directions. As can be seen from Fig. 22.10, some of the scattered radiation is scattered back into space, and some is scattered toward the Earth.

The important type of scattering for our discussion is the so-called **Rayleigh scattering,** named after Lord Rayleigh (1842–1919), the British physicist who developed the theory. Lord Rayleigh showed that the scattering for particles of molecular size was proportional to $1/\lambda^4$, where λ is the wavelength of the incident light. That is, the longer the wavelength of the radiation, the less the scattering.

Why the Sky is Blue and Sunsets are Red

The gas molecules of the air account for most of the scattering in the visible region of the spectrum. In the visible spectrum, the wavelength increases from violet to red. The blue end of the spectrum is therefore scattered more than the red end. (The colors of the visible spectrum—and the rainbow—may be remembered with the help of the name of ROY G. BIV—red, orange, yellow, green, blue, indigo, and violet.)

As the sunlight passes through the atmosphere, the blue end of the spectrum is preferentially scattered. Some of this scattered light reaches the Earth, where we see it as blue skylight. It should be kept in mind that all colors are present in skylight, but the dominant wavelength or color lies in the blue. You may have noticed that the skylight is more blue directly overhead or high in the sky, and less blue toward the horizon, becoming white just above the horizon. This is because there are fewer scatterers along a path through the atmosphere overhead than toward the horizon, and multiple scattering along the horizon path mixes the colors to give the white appearance. If the Earth had no atmosphere, the sky would appear black, other than in the vicinity of the Sun.

Since Rayleigh scattering is greater the shorter the wavelength, you might be wondering why the sky isn't violet since this color has the shortest wavelength in the visible spectrum. Violet light is scattered, but the eye is more sensitive to blue light than to violet light, and also,

Figure 22.11 Earth from 22,300 miles in space. The cloud cover of Earth is important in reflecting sunlight and contributes significantly to the albedo. (Courtesy NASA.)

sunlight contains more blue light than violet light. The greatest color component is yellow-green and the distribution generally decreases toward the ends of the spectrum.

The scattering of sunlight by the atmospheric gases *and* small particles give rise to red sunsets. Generally it might be thought that since the sunlight travels a greater distance through the atmosphere to an observer at sunset, most of the shorter wavelengths would be scattered from the sunlight and only light in the red end of the spectrum would reach the observer. However, it has been shown that the dominant color of this light were it due solely to molecular scattering would be orange. Hence, there must be additional scattering by small particles in the atmosphere that shifts the light from the setting (or rising) Sun toward the red. Foreign particles (natural or pollutants) in the atmosphere are not necessary to give a blue sky, and even detract from it. Yet, such particles are necessary for deep red sunsets and sunrises.

The beauty of red sunrises and sunsets is often made more spectacular by layers of pink-colored clouds. The cloud color is due to the reflection of red light.

Rayleigh scattering plays a role in the selection of amber fog lights and red tail lights for automobiles. In the case of fog lights, if the fog droplets are very small, then the red end of the visible spectrum is scattered less than the blue end, and the light from the amber fog lights penetrates the fog. Fog lights can be a definite advantage in a city where a not-too-dense fog can contain many fine smoke particles that contribute to the Rayleigh scattering. Fog lights are not too advantageous in thick fogs with large droplets. On the basis of Rayleigh scattering, you should be able to explain why automobile tail lights are red.

Larger particles of dust, smoke, haze, and those from air pollution in the atmosphere may preferentially scatter long wavelengths. These scattered wavelengths, along with the scattered blue light due to Rayleigh scattering, can cause the sky to have a milky blue appearance—white being the presence of all colors. Hence, the blueness of the sky gives an indication of atmospheric purity. Cloud droplets and raindrops scatter even longer wavelengths. This fact is used in the principle of weather radar, which is an important means of weather monitoring as we shall learn in Chapter 25.

About 15% of the insolation is absorbed directly by the atmosphere. Most of this absorption is accomplished by ozone, which removes the ultraviolet radiation, and by water vapor, which absorbs strongly in the infrared region of the spectrum. The major portion of the solar spectrum lies in the narrow visible region. Nitrogen and oxygen, which compose the bulk of the atmosphere, are practically transparent to visible radiation, so the major portion of the solar radiation passes through the atmosphere without appreciable absorption.

After reflection, scattering, and direct absorption of insolation by the atmosphere, about 50% of the total incoming solar radiation reaches the Earth. This goes into terrestrial surface heating, primarily through the absorption of visible radiation. As was noted previously, the atmosphere, in particular the troposphere, derives most of its energy content from the Earth. This is accomplished in three main ways. In order of decreasing contribution, these are (1) absorption of terrestrial radiation, (2) latent heat of condensation, and (3) conduction from the Earth's surface.

Absorption of Terrestrial Radiation

The Earth, like any warm body, radiates energy that may be absorbed by the atmosphere. The wavelength of the radiated energy is dependent on the Earth's temperature. This may be seen from the equation relating the energy (E) and wavelengths of the radiation

$$E = hf = \frac{hc}{\lambda}$$

where h is Planck's constant and the relationship for the frequency (f), wavelength (λ), and speed of light (c) is used ($f = c/\lambda$). This energy is proportional to the temperature of the source producing it. Thus

$$E = \frac{hc}{\lambda} \propto T \qquad (22.1)$$

Hence

$$\lambda \propto \frac{hc}{T}$$

Thus, the wavelength of the radiation emitted by a body is inversely proportional to its temperature.

The Earth's temperature is such that it radiates energy primarily from the infrared region. That is, the Earth absorbs energy in the shorter wavelength visible region of the solar spectrum, becomes warmed, and reradiates energy in the longer wavelength infrared portion of the spectrum.

Water vapor and carbon dioxide (CO_2) are the main absorbers of long-wavelength radiation in the atmosphere, with water vapor being the more important because it absorbs most of the infrared spectrum (see Fig.

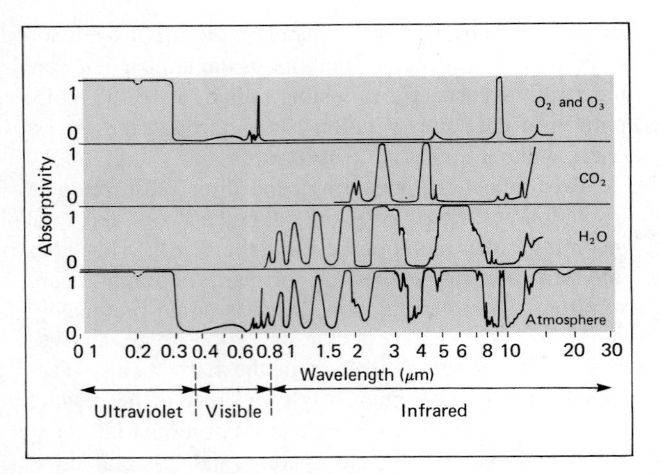

Figure 22.12 Absorption spectra for O_2 and O_3, CO_2, H_2O, and the atmosphere. CO_2 and H_2O are the chief atmospheric absorbers in the infrared region, while O_2 and O_3 absorb mainly in the ultraviolet region. Notice that the atmospheric gases are almost "transparent" or absorb very little visible radiation.

22.12). We refer to the gases as being selective absorbers, since they absorb certain wavelengths and transmit others.

The absorption of long-wavelength terrestrial radiation by water vapor and CO_2 adds to the energy content of the atmosphere. This heat-retaining process of the atmospheric gases is referred to as the **greenhouse effect,** because of a similar effect occurring in greenhouses. The absorption and transmission properties of glass are similar to those of the atmospheric gases—in general, visible radiation is transmitted, but infrared radiation is absorbed (Fig. 22.13).

We have all observed the warming effect of sunlight passing through glass, for example, in a closed car on a sunny, cold day. In a greenhouse, the objects inside become warm and reradiate long-wavelength infrared radiation, which is absorbed and reradiated by the glass. Thus, the air inside of a greenhouse heats up and is quite warm on a sunny day, even in the winter. Actually, in this case the maintained warmth is primarily due to the prevention of the escape of warm air by the glass enclosure. The temperature of a greenhouse in the summer is controlled by painting the glass panels white, which reflects the sunlight, and opening panels to allow the hot air to escape.

The greenhouse effect of the atmosphere is quite noticeable in the cooling of the Earth at night, particularly on cloudy nights. With a cloud and water-vapor cover to absorb the Earth's radiation, the night air is comparatively warm. Without this insulating effect, the night is usually "cold and clear," as the energy from the daytime insolation is quickly lost. This effect also explains the rapid nocturnal temperature drop over desert areas because of the dryness of the air over these regions.

There are several interesting sidelights associated with the greenhouse effect and terrestrial radiation. With regard to a glass greenhouse that transmits the visible portion of insolation, the short-wavelength ultraviolet transmission depends on the iron content of the glass. Ordinary glass does not appreciably transmit wavelengths below 3100 Å (0.31 μm)* or in the ultraviolet region. Hence,

*The wavelengths of light in the visible and near visible regions are also commonly expressed in micrometers (μm). $1 \mu m = 10^{-6}$ m $= 10^{-4}$ cm.

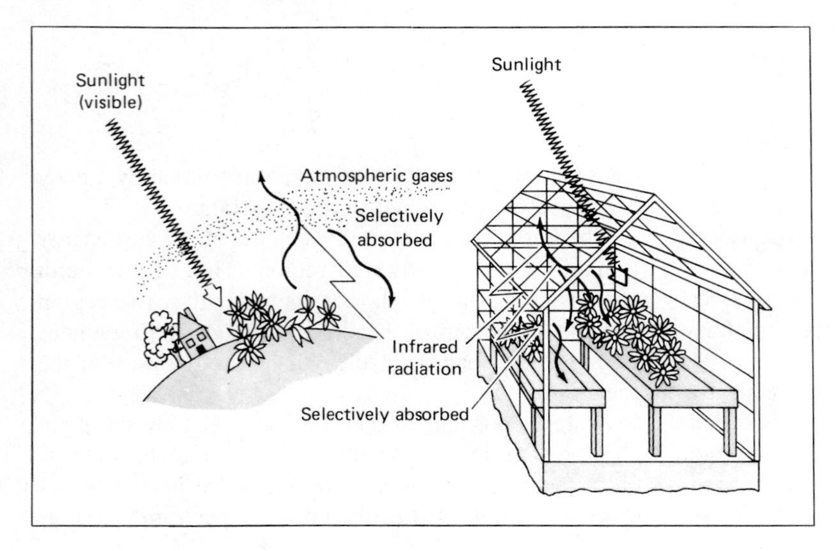

Figure 22.13 An illustration of the greenhouse effect. The gases of the lower atmosphere transmit most of the visible portion of incident sunlight as does the glass of a greenhouse. The warmed Earth emits radiation in the infrared region, which is selectively absorbed by the atmospheric gases whose absorption spectrum is similar to that of glass. This absorbed energy heats the atmosphere and helps maintain the Earth's temperature.

Montage of photographs, taken by various NASA spacecraft, displays smaller planets and larger moons of the solar system at the same scale. (Courtesy NASA)

PLATE II

Possible variations in chemical composition from one part of Saturn's ring system to another are visible in this Voyager 2 photograph as subtle color variations that can be recorded with special computer-processing techniques. This highly enhanced color view is assembled from clear, orange, and ultraviolet frames obtained at a distance of 5.5 million miles. The C Ring and the Cassini Division appear blue in the photo. (Courtesty NASA)

Volcanic explosion on Io. This photograph was taken by Voyager 1 from a distance of approximately 300,000 miles. The solid material has been thrown upward to an altitude of about 100 miles. Brightness of the eruption has been increased by computer but the color (greenish white) has been preserved. (Courtesy NASA)

Olympus Mons, "Mount Olympus," a huge Mars volcano about 15 miles high and 375 miles wide at its base. (Courtesy NASA)

The surface of the planet Mars as photographed by Viking 1 at approximately noon Mars time. (Courtesy NASA)

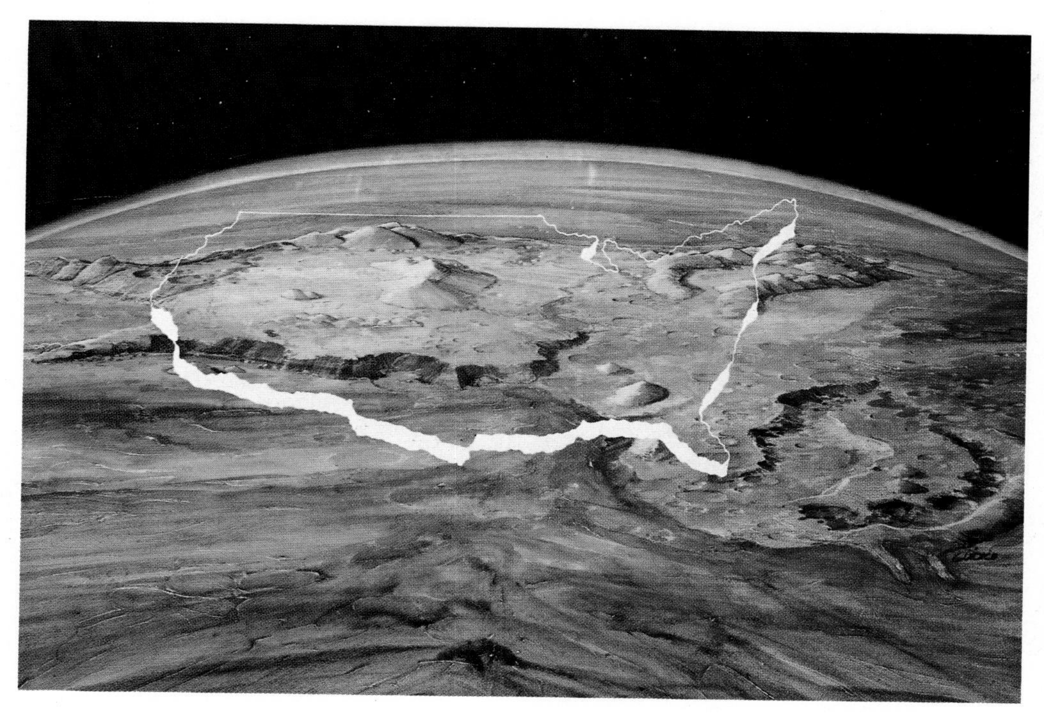

This is an artist's conception of Venus's surface. The outline of the USA is drawn in for scale. (Courtesy NASA)

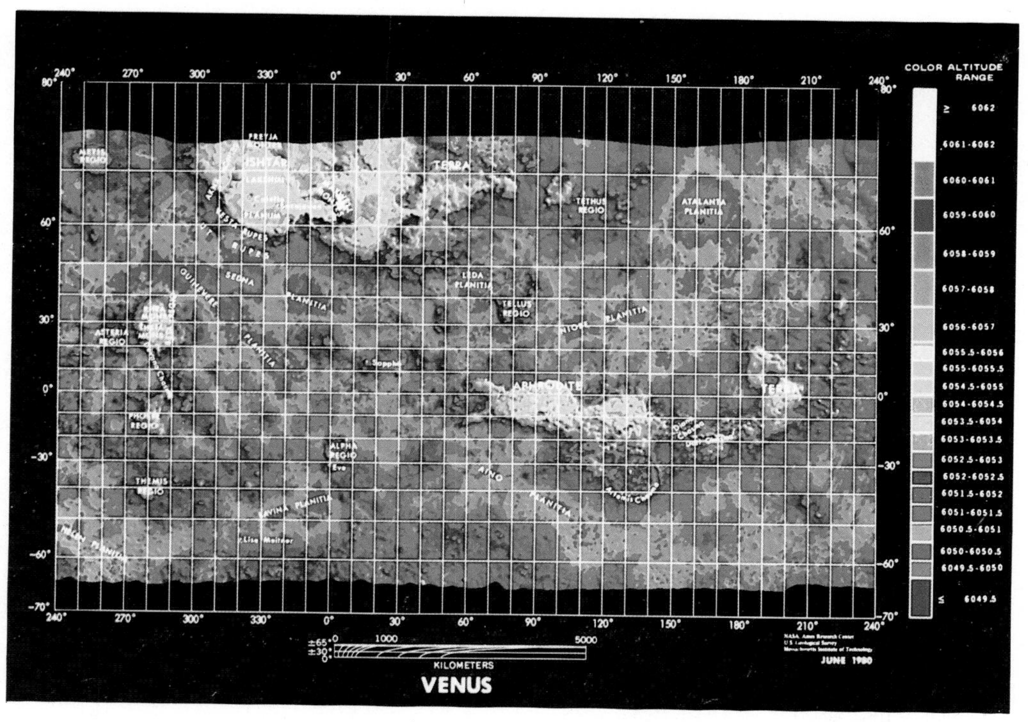

A topographic map of Venus's surface based on radar data taken by Pioneer Venus Orbiter. Most of Venus's surface is covered by rolling plain shown in blue and green. The highlands are shown in brown and yellow. (Courtesy NASA)

PLATE IV

A montage of Jupiter, Saturn, and Uranus composed from photographs taken by Voyager 1 and 2. (Courtesy NASA)

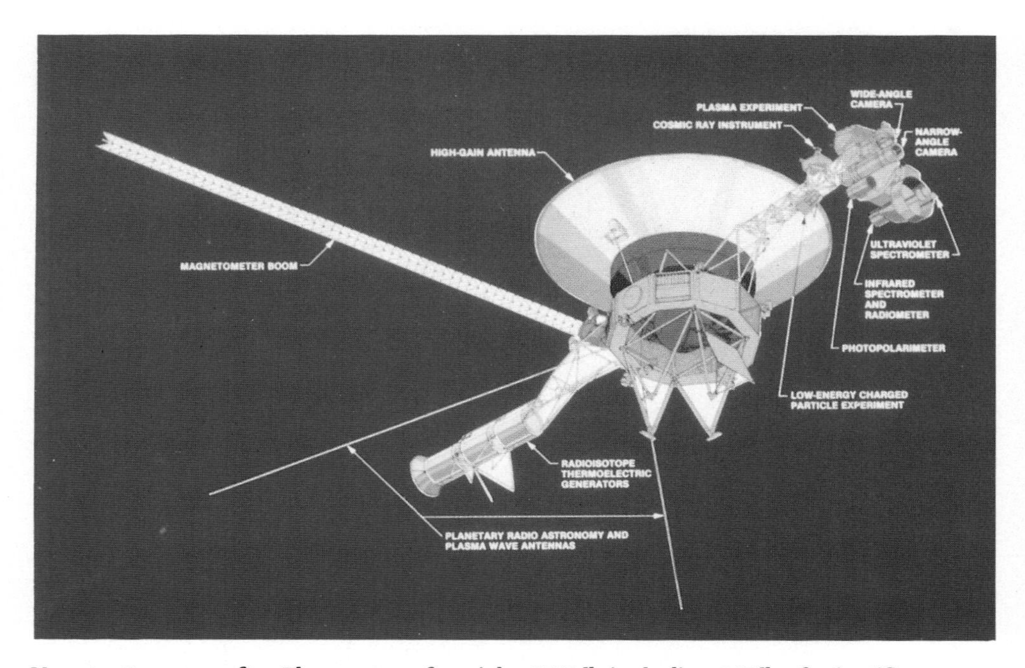

Voyager 2 spacecraft. The spacecraft weighs 1753 lb including 249 lb of scientific equipment. The magnetometer boom is 43 ft long, and the high-gain antenna is 12 ft in diameter. The spacecraft is scheduled to rendezvous with the planet Neptune on August 24, 1988. (Courtesy NASA)

The Rings of Uranus. This false-color view of the rings of Uranus was made from images by Voyager 2 at a distance of 2.59 million miles. The epsilon ring at the top is neutral in color with the eight other fainter rings showing color differences among them. (Courtesy NASA)

This computer-generated view of Uranus was made by Voyager 2 at its closest approach (66,000 miles) to the planet. The spacecraft is shown about 45 minutes after it passed through the plane of Uranus' rings. (Courtesy NASA)

Oberon, Uranus' most distant moon. Note the large mountain, about four miles high, on the lower left edge of the satellite and the very large crater near the center of the picture. (Courtesy NASA)

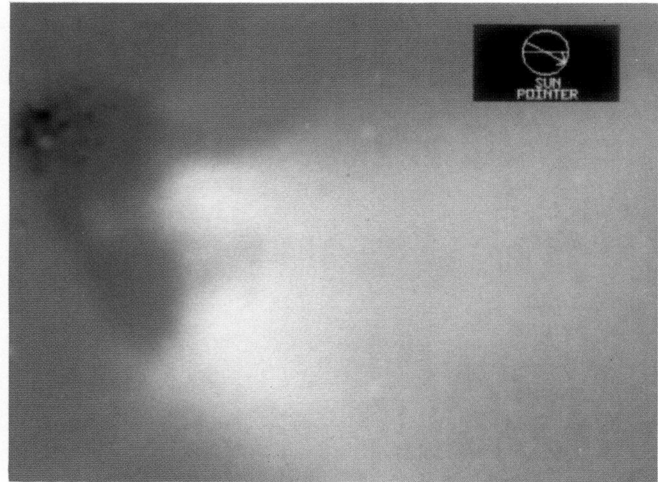

Picture of Halley's comet taken by the Halley Multicolour camera on board the Giotto spacecraft on March 14, 1986, from a distance of about 11,000 miles. See Fig. 18.38 for an explanation of the photo. (Max-Planck-Institut fuer Aeronomie, Lindau/Harz, FRG)

PLATE VI

Some Common Minerals

1. Quartz

Quartz is one of the most common minerals in igneous, sedimentary, and metamorphic rocks. Under favorable conditions, it occurs in hexagonal crystals. It is harder than glass.

2. Plagioclase

Plagioclase, a very abundant mineral, occurs in igneous, sedimentary, and metamorphic rocks. Varieties range in color from white to dark gray. It has two cleavage faces, which meet at nearly right angles. Twinning lines, seen as straight, parallel striations, are helpful in identification.

3. Pyroxene

(Var. Diopside)

Pyroxene is very abundant in igneous and metamorphic rocks. Varieties range in color from light green to dark green and black. Crystals are blocky. Two directions of cleavage meet at nearly right angles.

4. Amphibole

(Var. Hornblende)

Amphibole is very abundant in igneous and metamorphic rocks. Hornblende, the most common variety, is usually dark green to black. Two cleavage planes intersect at 55° and 125° to form cleavage fragments, which are commonly wedge-shaped.

5. Olivine

Olivine is abundant in dark igneous rocks low in SiO_2. It is usually olive green. Large crystals are rare. Only the green mineral in the photograph is olivine. Like quartz, olivine is harder than glass.

6. Mica

Mica is found in igneous, sedimentary, and especially metamorphic rocks. Biotite mica is dark brown to black. Muscovite mica is colorless to pale yellow or brown. Mica's most distinctive characteristic is perfect cleavage in one direction, which causes it to split into very thin sheets.

7. Calcite

Calcite is abundant in sedimentary and metamorphic rocks and is the principal constituent of limestone and marble. It is usually colorless or white but may be light shades of green, blue, yellow, or other colors. Calcite commonly occurs in crystals. It has three well-developed directions of cleavage inclined to one another to form rhombohedral cleavage fragments.

8. Dolomite

Dolomite is abundant in sedimentary and igneous rocks. Its color is commonly white to pink. Crystals as well-developed as those on the photograph are rare. Dolomite effervesces slowly in dilute hydrochloric acid.

9. Halite

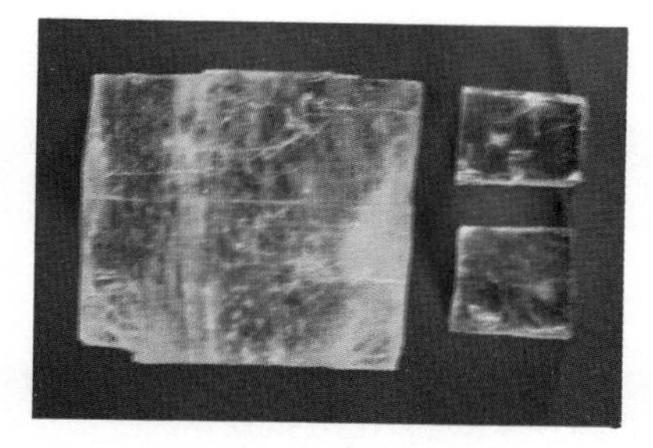

Halite occurs in sedimentary rocks. It is deposited by extreme evaporation of ocean water. Halite is usually colorless, but may be shades of blue, green, yellow, brown, or other colors. It has three very prominent directions of cleavage that meet at right angles to form cubic cleavage fragments. Its crystal form is also cubic.

10. Pyrite

Pyrite is scattered through igneous, sedimentary, and metamorphic rocks. Because of its brassy yellow color, it has been confused with gold and for this reason is called "fool's gold." It commonly occurs in crystals.

11. Hematite

Hematite occurs in igneous and metamorphic rocks but is most common in sedimentary rocks. Its crystalline varieties are black with a metallic luster. The sedimentary varieties are usually an earthy red brown. Hematite is the most important ore of iron.

12. Fluorite

Fluorite occurs in igneous and sedimentary rocks. It has an almost endless variety of colors. Fluorite commonly occurs in crystals and has several well-developed directions of cleavage.

Mineral photographs by Katherine Jensen, Pittsford, N.Y.

PLATE VIII

A continuous spectrum plus line spectra for the elements indicated. (Courtesy of Bausch and Lomb Incorporated)

Tungsten Lamp

Fluorescent Lamp

Atomic Hydrogen

Lithium

Neon

Sodium Lamp

Calcium

Iron Arc

Mercury Lamp

Fraunhofer Lines

although there is a considerable warming effect through glass from visible radiation, one cannot receive a suntan or sunburn through ordinary window glass. (Ultraviolet radiation near 3000 Å and below causes the skin to burn or tan.)

Despite the daily and seasonal gain and loss of heat, the average temperature of the Earth has remained fairly constant over a long period of time. This indicates that the Earth must lose or reradiate as much energy as it receives. If this were not the case, the continual gain of energy would cause the Earth's average temperature to rise. The selective absorption of the atmospheric gases provides a thermostatic or heat-regulating process for the Earth. For example, suppose the Earth's temperature were such that it emitted radiation with wavelengths between 4 and 6 μm. Referring to the absorption spectrum of the atmosphere in Fig. 22.12 it can be seen that the absorptivity of these wavelengths is almost 1.0 (100%); this radiation is almost completely absorbed by CO_2 and water vapor.

As a result of this absorption, the lower atmosphere becomes warmer and effectively holds in the Earth's heat, thus insulating the Earth. With additional insolation, the Earth becomes warmer and its temperature rises. However, according to Eq. 22.1, the greater the temperature, the shorter the wavelength of the emitted radiation, and so the wavelength of the terrestrial radiation is shifted to shorter wavelengths. At about 3.5 μm in the atmospheric absorption spectrum, there is a transmission "window," and the terrestrial radiation can pass through the atmosphere. Thus losing energy, the temperature of the Earth decreases and the terrestrial radiation is shifted to a longer wavelength, which is absorbed by the atmosphere. Averaged over the total spectrum, the selective absorption of the atmospheric gases plays an important role in maintaining the Earth's average temperature.

Humans radiate energy in the infrared region, as does the Earth. This radiation is not visible to the human eye, and requires special equipment to be detected. Warm, infrared emitting objects may thus be detected in the dark; hence, infrared detection has come to be associated with "seeing in the dark." Since the wavelength of the radiation is dependent on the temperature of the source, infrared photographs taken from satellites are used to study temperature variations in the Earth's crust arising from such things as volcanic activity and geothermal regions (Fig. 22.14).

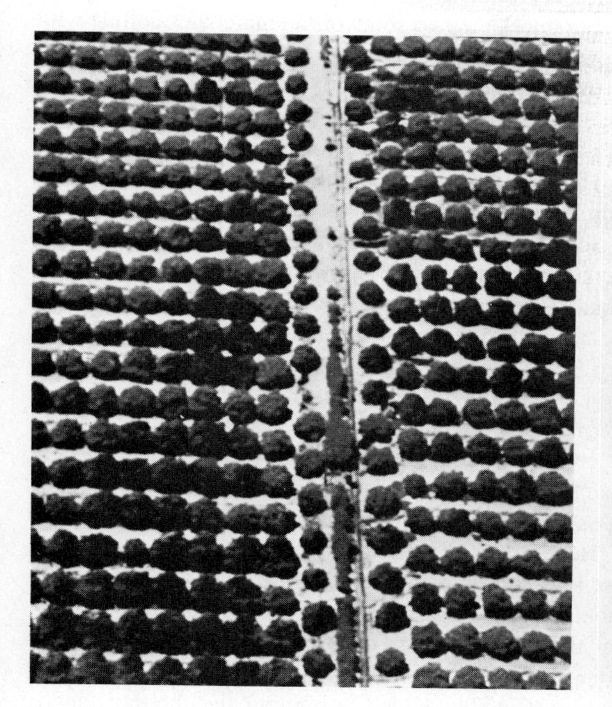

Figure 22.14 Infrared photography. The photo on the left shows how crop damage may be detected by use of infrared imagery. The photo on the right shows the infrared image of Yellowstone National Park with its geothermal areas. (Courtesy NASA.)

Latent Heat of Condensation

Approximately 70% of the Earth's surface is covered by water. Consequently, a great deal of evaporation occurs due to the insolation reaching the Earth's surface. You will recall from Chapter 5 that the latent heat of vaporization for water is 540 calories per gram. That is, 540 calories of heat energy are required to change one gram of water from a liquid to a gaseous state. Thus, a large amount of energy is transferred to the atmosphere in the form of latent heat. This energy is released in the atmosphere with the condensation of water vapor in the formation of clouds, fog, dew, etc.

Conduction from the Earth's Surface

A comparatively smaller but significant amount of heat energy is transferred to the atmosphere by conduction from the Earth's surface. Because the air is a relatively poor heat conductor, this process is restricted to the layer of air in direct contact with the Earth's surface. The heated air is then transferred aloft by convection. Thus, the temperature of the air tends to be greater near the surface of the Earth and decreases gradually with altitude.

The transfer of heat by conduction depends on the temperature difference between the air and the Earth's surface. (Heat energy may be transferred from the air to the ground if the latter is colder.) The heat transfer to the atmosphere varies for land and water surfaces because they in general have different temperatures. The specific heat, or the amount of heat required to raise a gram of mass one degree in temperature (Section 5.3), is on the order of four times greater for water than for land. Therefore, water heats more slowly than does land given equal amounts of energy. This is augmented by water being transparent, so the Sun's rays can penetrate, and by currents that affect mixing. Conversely, water cools more slowly than does land. Hence, neglecting seasonal influences, the land surface is warmer during the day, and the water is warmer at night. On a large scale, this effect also occurs seasonally, with summer and winter corresponding to day and night.

Because of the large surface area of the Earth, significant amounts of heat are transferred from land and sea to the atmosphere by conduction. The redistribution of this heat energy by atmospheric convection plays an important role in weather and climate conditions.

22.5 Atmospheric Measurements

Measurements of the atmosphere's properties and characteristics are very important in its study. Its properties are measured daily, and records have been compiled over many years. Meteorologists study these records with the hope of observing cycles and trends in the atmospheric behavior so as to better understand and predict its changes. We listen to the daily atmospheric readings to obtain a qualitative picture of the conditions we may experience that day. Fundamental measurements made in the atmosphere include those of: (1) temperature, (2) pressure, (3) humidity, (4) wind speed and direction, and (5) precipitation.

Temperature

Temperature is a measure of heat energy. In matter it is a measure of the molecular activity or the kinetic energy of the molecules. The kinetic theory of gases shows the energy of gases to be directly proportional to the temperature (Section 5.7). The higher the temperature, the more energetic the gas molecules. When measuring the temperature of a gas, some of the molecules collide with the temperature measuring device and impart to it some of their energy. The more the molecular activity, the more collisions and the greater the energy transfer, which results in a higher temperature reading.

The liquid-in-glass **thermometer** is the most common device used to measure temperature. As discussed in Section 5.1, this thermometer consists of alcohol or mercury enclosed in a glass bulb that is attached to a capillary called the bore. The liquid expands linearly up the bore when heated, and the temperature is read from a calibrated scale on the glass of the capillary. The common temperature scales are Fahrenheit and Celsius. A thermometer in contact with the gases of the atmosphere records the air temperature due to the energy transfer resulting from collisions of the air molecules with the glass bulb of the thermometer.

Heat transfer by radiation to a thermometer may give rise to a higher temperature reading. As a result, a thermometer in the Sun may read a comfortable 70°F (21°C), but the air feels quite cool. A truer air temperature is often expressed in the summer by saying it is so many degrees ''in the shade,'' implying the thermometer is not exposed to the direct rays of the Sun.

Pressure

Pressure is defined as the force per unit area ($P = F/A$). At the bottom of the atmosphere we experience the resultant weight of the gases above us. Since we experience this weight before and after birth as a natural part of our environment, little thought is given to the fact that every square inch of our bodies sustains an average weight of 14.7 lb at sea level or a pressure of 14.7 lb/in^2. We refer to 14.7 lb/in^2 as one atmosphere of pressure.

One of the first investigations of atmospheric pressure was initiated by Galileo. In attempting to pipe water to elevated heights by evacuating the air from a tube, he found that it was impossible to sustain a column of water over 34 ft (10.4 m) in height. Evangelista Torricelli (1608–1647), who was Galileo's successor as professor of mathematics in Florence, pointed out the difficulty through the invention of a device that exhibited the height of a liquid column to be dependent on atmospheric pressure. This device, a **mercury barometer** (Greek *baros*, weight), is still used to measure the atmospheric pressure. A filled tube of mercury was inverted into a pool of mercury. Although some ran out, a column of mercury 30 in (76 cm) high was left in the tube as illustrated in Fig. 22.15. A modern version of a mercury barometer is shown in Fig. 22.16. Since the column of mercury has weight, some force must hold the column up. The only available force is that of the atmospheric pressure on the surface of the mercury pool.

It can be shown that the pressure is related to the density ρ of the mercury and the height h of the column by the equation

$$P = \rho g h \qquad (22.2)$$

where g is the acceleration due to gravity. Hence, knowing the height of the column, the density, and the acceleration due to gravity, the pressure may be calculated. However, on radio and TV weather reports the barometric reading is expressed in length units, so many "inches" (of mercury), since the height of the barometer column is directly proportional to the pressure.

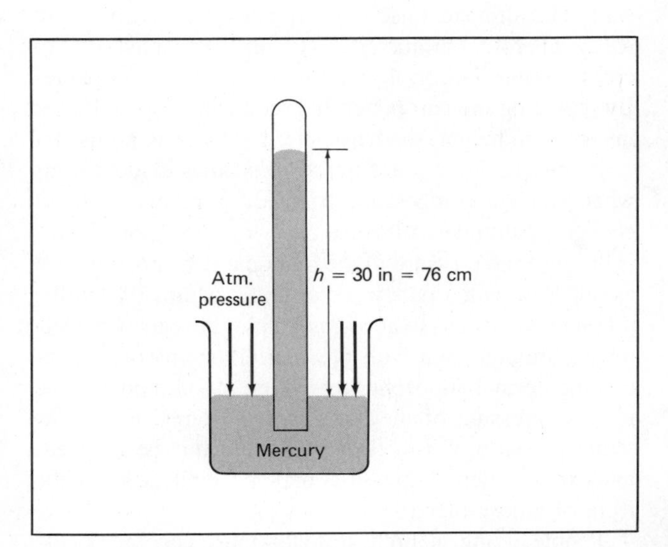

Figure 22.15 A diagram of Torricelli's barometer. The external air pressure on the surface of the pool of mercury supports the mercury column in the inverted tube. The height of the column therefore depends on atmospheric pressure and provides a way of measuring it.

Figure 22.16 A modern mercury barometer. Modern barometers operate on the same principle as Torricelli's barometer (Fig. 22.15) but are highly refined to give accurate measurements. (Courtesy NOAA.)

The density of mercury is 13.6 g/cm^3 and g = 980 cm/s^2, hence

$$P = \rho g h = (13.6)(980) \times h\ (cm)$$

One atmosphere of pressure supports a column of mercury approximately 30 in in height (actually 29.92 in). Thirty inches is approximately 76 cm; thus,

$$\begin{aligned} P &= 13.6\ \text{g/cm}^3 \times 980\ \text{cm/s}^2 \times 76\ \text{cm} \\ &= 1.0 \times 10^6\ \text{dynes/cm}^2 \end{aligned}$$

This is an approximate value because of the approximate values used in the equation.

In the SI system, pressure has the units of newton/meter squared (N/m^2) which is given the special name of pascal (Pa) in honor of the French scientist Blaise Pascal (1623–1662). For example, we might inflate automobile tires to a pressure of 30 lb/in^2 (30 "pounds" pressure), or about 200 kPa (kilopascals). The pressure of one atmosphere is about 10^5 Pa. However, this unit is not commonly used in reporting atmospheric pressure. Also, a special name has been given to a mercury length unit. One millimeter of mercury (Hg) is called a torr, in honor of Torricelli, so 760 mm Hg = 760 torr.

Summarizing these various pressure and length units, the pressure of one atmosphere is usually given as

$$\begin{aligned} 1\ \text{atmosphere (atm)} &= 1{,}013{,}250\ \text{dyn/cm}^2 \\ &= 1.013 \times 10^5\ \text{Pa} \\ &= 14.7\ \text{lb/in}^2 \\ &= 30\ \text{in Hg} \\ &= 760\ \text{mm Hg} = 760\ \text{torr} \end{aligned}$$

We are now in a position to understand Galileo's difficulty in not being able to sustain a column of water over 34 ft in height. Any liquid may be used in a barometer; however, the column heights will vary due to the different densities of the liquids. Water has a density of 1 g/cm^3, which is 13.6 times lighter than mercury. The atmospheric pressure then should support a column of water 13.6 times higher than the 30 in of mercury. Thus,

$$13.6 \times 30\ \text{in} = 408\ \text{in} \times \frac{1\ \text{ft}}{12\ \text{in}} = 34\ \text{ft}$$

Hence, the atmosphere does not have sufficient pressure to support a column of water over 34 ft, just as Galileo and Torricelli found.

Meteorologists use another unit, the **millibar** (mb), to record pressure. A bar is defined as 10^6 dyn/cm^2 or 10^5 Pa, and 1 bar = 1000 mb. Since one atmosphere is approximately 10^6 dyn/cm^2, we have

$$1\ \text{atm} \simeq 1\ \text{bar or}\ 1000\ \text{mb}$$

and normal atmospheric pressures are of the order of 1000 mb. Actually,

$$1\ \text{atm} = 1013.25\ \text{mb}$$

As would be expected, atmospheric pressure varies with altitude. This was suggested by Blaise Pascal and was proved by his brother-in-law who measured a reduced atmospheric pressure on top of a mountain. Daily atmospheric disturbances also cause the barometric pressure to vary. These variations are continually monitored by weather stations and are important in predicting weather changes as we shall learn.

Since mercury vapor is toxic and a column of 30 in of mercury awkward to handle, another type of barometer in common use is the **aneroid** (without fluid) **barometer.** This is a mechanical device having a metal diaphragm that is sensitive to pressure (Fig. 22.17). A pointer is used to indicate the pressure changes on the diaphragm. Aneroid barometers with dial faces are common around the home and are usually inlaid in wood with a dial thermometer for decorative display on the wall. The altimeters used by airplanes and skydivers are really aneroid barometers. The pressure reading decreases rather uniformly with height in the troposphere. By replacing the barometer dial face with one calibrated inversely in height, the barometer becomes an altimeter.

Atmospheric pressure quickly becomes evident to us when sudden changes are effected. A relatively small change in altitude will cause one's ears to "pop." This is because the pressure in the inner ear does not equalize as quickly, which puts force on the eardrum. When the pressure equalizes (swallowing helps), the ears pop. Airplanes are equipped with pressurized cabins that maintain the normal atmospheric pressure on our bodies. The internal pressure of the body is accustomed to the external pressure of 14.7 lb/in^2. Should this be reduced, the excess internal pressure may be evidenced in the form of a nose bleed.

People living at high altitudes must become accustomed to the reduced pressure and "thin" air. The body's metabolism changes at higher altitudes as a result of the reduced oxygen content of the air. Criticisms of the 1968 Olympic games held in Mexico City at an elevation of 7500 feet were leveled at the effect of reduced pressure and oxygen on the athletes.

Figure 22.17 The aneroid barometer. An aneroid barometer uses a metal diaphragm sensitive to pressure changes rather than a liquid, as in the Torricellian barometer. (*Left*) A recording aneroid barometer, which provides a continuous record of air pressure. (*Right*) A household aneroid barometer. Low pressure is generally associated with rainy weather and high pressure with fair weather. See Chapter 24. (Courtesy of Weather Measure Corporation and Taylor Instrument/Sybron Corporation.)

Another effect of high altitude or lower atmospheric pressure is the reduction of the boiling points of liquids. For example, at the top of Pike's Peak (elevation 14,110 ft) the atmospheric pressure is about 600 torr (mm Hg), and water boils at about 94°C rather than at 100°C. This has an effect on cooking times. A pressure cooker would help. Why? Also, cake mixes for use at high altitudes must be especially prepared with a smaller portion of ''rising'' ingredients. The sponginess and rising of a cake is due to the release of gas during baking. A cake that rises a normal amount at sea level may virtually ''blow up'' at high altitudes due to the decrease in atmospheric pressure.

A frequent use of atmospheric pressure is in drinking through a straw. It is generally believed that sucking on the straw draws the liquid up the straw. Actually, the sucking action reduces the air pressure in the straw by removing the air molecules. The atmospheric pressure on the liquid's surface, external to the straw, then pushes the liquid up the straw. It should be evident from Galileo's problem that a soda straw over 34 ft high would not work.

Humidity

Humidity is a measure of the water vapor in the air. It affects our comfort and indirectly our ambition and state of mind. In the summer many homes have the hum of a dehumidifier that removes moisture from the air, while in the winter, exposed pans of water may be strategically placed to allow the water to evaporate into the air. There are several ways to express humidity.

Absolute humidity is simply the amount of water vapor in a given volume of air and is normally measured in grains per cubic foot in the British system of units (1 lb = 7000 grains). An average value would be on the order of 4.5 gr/ft^3. Another way of expressing humidity is by **partial pressure,** or how much pressure the water vapor in the air contributes to the total atmospheric pressure.

The most common method of expressing the water vapor content of the air is in terms of relative humidity. **Relative humidity** is the ratio of the actual moisture content of a volume of air to its maximum moisture capacity at a given temperature. Expressed in terms of a percentage,

$$(\%)\ RH = \frac{AC}{MC} \quad (\times\ 100\%) \qquad (22.3)$$

where RH is the relative humidity, AC the actual moisture content of the air, and MC the maximum moisture capacity of the air.

The actual moisture content is just the absolute humidity, or the amount of water vapor in a given volume

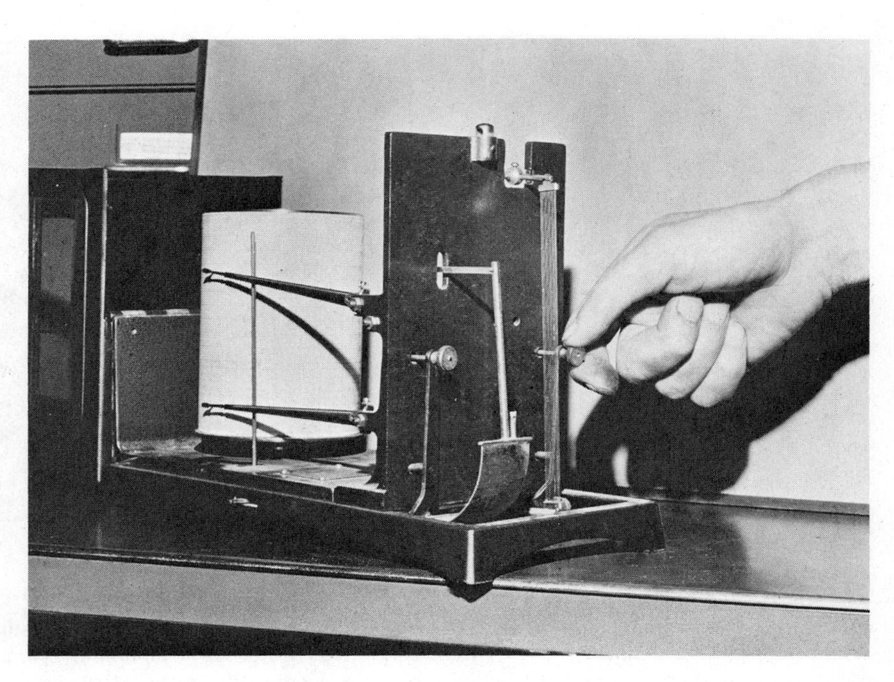

Figure 22.18 A hair hygrometer, recording type. The variation in the length of human hair due to absorbed moisture is used to measure humidity. The hair filaments of hygrometer are at the tip of the thumb. The chart on the rotating drum at the left permits continuous monitoring of the humidity. (Courtesy NOAA.)

of air. The maximum moisture capacity is the maximum amount of water vapor that the volume of air can hold *at a given temperature*.

Relative humidity is essentially a measure of how "full" of moisture a volume of air is at a given temperature. For example, if the relative humidity is 0.50 or 50%, then a volume of air is "half full" or contains one-half as much water vapor as it is capable of holding at that temperature.

To better understand how the water vapor content of air varies with temperature, consider an analogy with a salt water solution. Just as a given amount of water at a certain temperature can dissolve so much table salt, a volume of air at a given temperature can hold only so much water vapor. When the maximum amount of salt is dissolved in solution, we say the solution is saturated. This is analogous to a volume of air having its maximum moisture capacity.

The addition of more salt to a saturated solution results in salt on the bottom of the container. However, more salt may be put into solution if the water is heated. Similarly, when air is heated, it can hold more water vapor. That is, warm air has a greater capacity for holding water vapor than does colder air.

Conversely, if the temperature of a nearly saturated salt solution is lowered, at a certain temperature the solution will become saturated. Additional lowering of the temperature will cause the salt to come out of solution, as it will be oversaturated. In an analogous fashion, if the temperature of a sample of air is lowered, it will become saturated at some temperature.

The temperature to which a sample of air must be cooled to become saturated is called the **dew point.** Hence, at the dew point (temperature) the relative humidity is 100%. Why? Cooling below this causes oversaturation and may result in condensation and loss of moisture in the form of precipitation.

Measurement of Humidity. Humidity may be measured by several means. **Absorption hygrometers** employ chemicals such as $CaCl_2$ or $Ca(NO_3)_2$ that absorb water vapor from the air and may be analyzed quantitatively. Other chemicals give a qualitative indication of the humidity by a change in some physical characteristic such as color. **Hair hygrometers** make use of the fact that human hair expands about 2.5% over the range of 0–100% humidity. This expansion may be measured mechanically and calibrated in terms of humidity. See Fig. 22.18.

The most common method of measuring humidity uses the **psychrometer.** This instrument consists of two thermometers, one of which measures the air temperature while the other has its bulb surrounded by a wick that is kept wet. These thermometers are referred to as the dry bulb and wet bulb, respectively. They may be simply mounted as in Fig. 22.19 or incorporated in a **sling psychrometer** (Fig. 22.20).

For accurate measurements, it is necessary to have

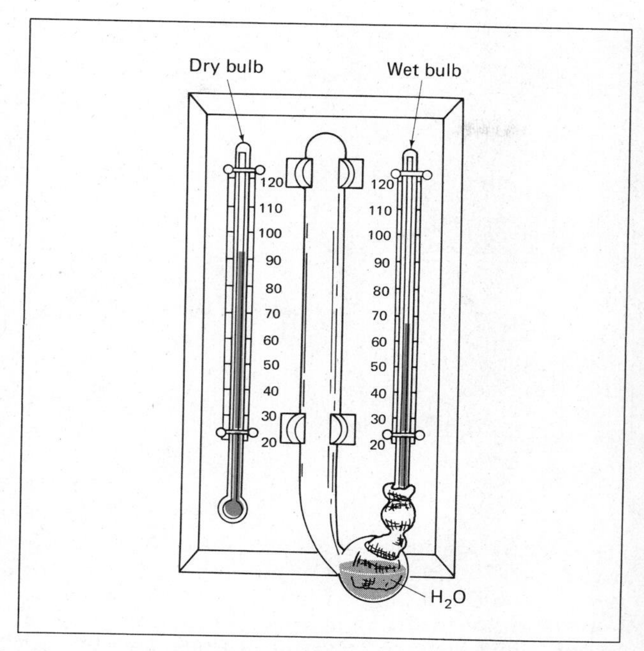

Figure 22.19 A psychrometer for relative humidity measurements. The dry bulb records the air temperature, while the wet bulb, kept moist by a cloth wick from the water reservoir, registers the lower temperature of the wick which is due to moisture evaporation. The evaporation depends on the humidity. Knowing the air temperature and the difference in the thermometer readings, one can obtain the relative humidity from reference tables.

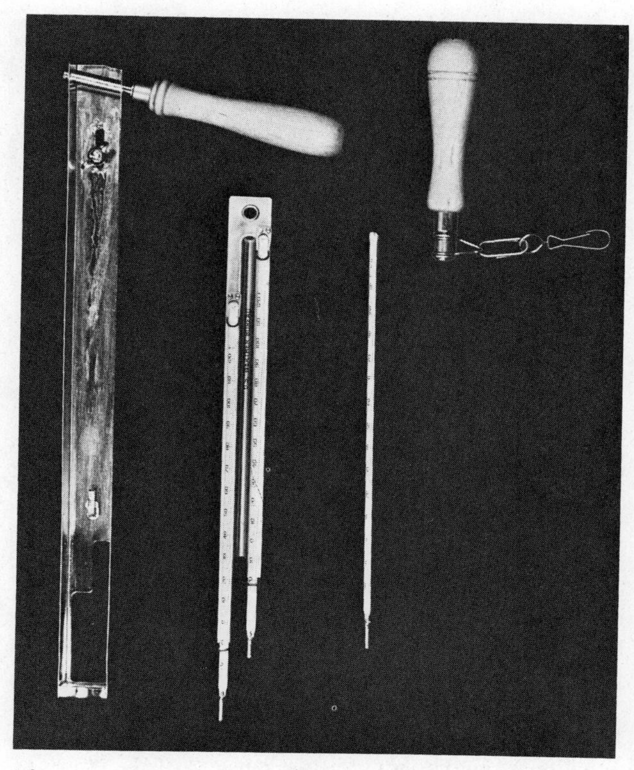

Figure 22.20 A sling psychrometer. The thermometers are whirled in a sling device. The wet bulb is covered with a tubular piece of cloth that is moistened. (Courtesy NOAA.)

air moving across the thermometer bulbs to give an average condition for both bulbs. This is easily accomplished for a sling psychrometer, which is rotated before taking a reading. Fanning may be used for the mounted stationary type. In either case, the air circulation should be continued until the wet bulb reading reaches its lowest point. The dry bulb measures the air temperature while the wet bulb has a lower reading that is a function of the amount of water vapor in the air.

Consider the human body's cooling mechanism momentarily. The body perspires and the evaporation of the perspiration causes cooling due to the removal of heat from the body to change the liquid perspiration into vapor, that is, the latent heat of vaporization (Section 5.5). If the humidity is high and the air contains a lot of water vapor, it is difficult for evaporation from the body to take place. As a result, we say it is a hot and muggy day. However, if the humidity is low, evaporation can occur easily. We then feel cool and comfortable.

The wet bulb of the psychrometer works in a similar fashion. If the humidity is high, little water is evaporated and the wet bulb is only slightly cooled. Consequently, the temperature of the wet bulb is only slightly lower than the temperature of the dry bulb, or its reading is slightly depressed. If, however, the humidity is low, a great deal of evaporation will occur accompanied by considerable cooling, and the wet bulb will be considerably depressed. Hence the temperature difference of the thermometers, or the depression of the wet bulb, is a measure of the humidity.

The relative humidity, the maximum moisture capacity, and also the dew point that depends on the moisture content and hence the relative humidity, may be read directly from Tables 1 and 2, Appendix VI, when the psychrometer readings are known. Consider an example in which the dry bulb has a reading of 80°F and the wet bulb a reading of 73°F. The depression of the wet bulb is then

$$80^\circ - 73^\circ = 7^\circ$$

Finding the dry bulb temperature in the first column of Table 1, the maximum moisture capacity for that temperature is given in the adjacent column, in this case 10.9 gr/ft^3.

The relative humidity is obtained from Table 1 by finding the depression of the wet bulb in the top row of the table and coming down that column to the row which corresponds to the dry bulb reading. Doing this, the relative humidity is found to be 72% for the above depression. The dew point temperature is obtained by the same procedure from Table 2, Appendix VI, and is 70°F.

With this information, the actual moisture content may be calculated, using Eq. 22.3

$$RH = \frac{AC}{MC}$$

where the relative humidity is expressed in decimal form; or

$$AC = RH \times MC = 0.72 \times 10.9 \text{ gr/ft}^3 = 7.8 \text{ gr/ft}^3$$

This is the actual moisture content at 80°F and should correspond to the maximum moisture capacity of the air at 70°F (dew point temperature) as can be seen from the first column in Table 1.

At the dew point the air is saturated and the relative humidity is 100%. The actual moisture content is then equal to the maximum moisture capacity, ($AC = MC$).

To visualize these concepts, consider a container of water that is partially full. The sides of this particular container are quite marvelous, inasmuch as they shrink when the temperature decreases. At 80°F the container has 7.8 gr of water but is capable of holding 10.9 gr. Hence, the container is $\frac{7.8}{10.9} \times 100 = 72\%$ full, as shown in Fig. 22.21.

But if the temperature is lowered to 70°F, the sides of the container shrink so that it can hold only 7.8 gr. At this special temperature (dew point), the container is full (or saturated) and it is 100% full. Should the temperature be lowered still further and the sides shrink further, the water would overflow. In the analogous situation, the air generally loses its moisture in the form of precipitation when cooled below the dew point.

Temperature and humidity are important factors in influencing body comfort. As mentioned earlier, the efficiency of the body's evaporation cooling mechanism depends on humidity. Hence, the apparent temperature, or how hot or cool one feels, depends on both temperature and humidity. Table 22.2 shows this relationship in terms of the apparent temperatures most people feel. For example, if the air temperature is 90°F and the relative humidity is 80%, the apparent temperature is 113°F. Note that the apparent temperature in the table is lower than the air temperature for low humidities. Why?

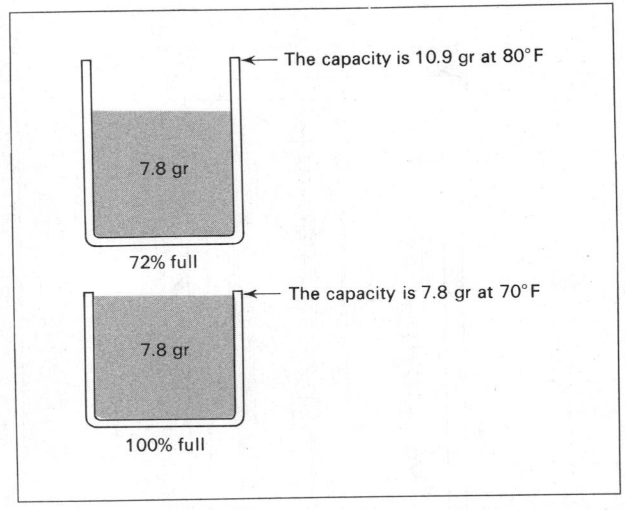

Figure 22.21 A diagram illustrating the container analogy of relative humidity. At 80°F the beaker can hold 10.9 grains of water. However, if the sides of the container were to shrink with temperature so that at 70°F the beaker's capacity were 7.8 grains, it would be 100% full (saturated). Analogously, a cubic foot of air at 80°F is capable of "holding" 10.9 grains of water vapor; but if it holds only 7.8 grains, its relative humidity (fullness) is 72%. Should the temperature of the air be decreased to 70°F (dew point), the relative humidity is then 100%.

It should be kept in mind that the relative humidity reported on the weather report is the *outdoor* relative humidity. This may bear little relationship to the indoor relative humidity, particularly during cold months when the house is heated.

Wind Speed and Direction

Wind speed is measured by an **anemometer.** This instrument consists of three or four cups each attached to a rod that is free to rotate. The cups catch the wind, and the greater the wind speed, the faster the anemometer rotates. A **wind vane** indicates the direction from which the wind is blowing. This instrument is simply a freely rotating indicator that, because of its shape, lines up with the wind and points the wind direction (the direction the wind is coming *from*). An array of anemometers and wind vanes is shown in Fig. 22.22. These instruments are common at airports where wind speed and direction are important.

Both the direction and a rough estimate of the wind's speed are given by the "wind sock," commonly seen at small airports. A rotating canvas "sock" that catches

Table 22.2 Determining the Apparent Temperature*

	Air Temperature (°F)										
	70	75	80	85	90	95	100	105	110	115	120
Relative Humidity	*Apparent Temperature (°F)*										
0%	64	69	73	78	83	87	91	95	99	103	107
10%	65	70	75	80	85	90	95	100	105	111	116
20%	66	72	77	82	87	93	99	105	112	120	130
30%	67	73	78	84	90	96	104	113	123	135	148
40%	68	74	79	86	93	101	110	123	137	151	
50%	69	75	81	88	96	107	120	135	150		
60%	70	76	82	90	100	114	132	149			
70%	70	77	85	93	106	124	144				
80%	71	78	86	97	113	136					
90%	71	79	88	102	122						
100%	72	80	91	108							

*Find the temperature in the top row and the relative humidity in the left column. Read the apparent temperature where the corresponding column and row intersect. For example, if the air temperature is 90°F and the relative humidity is 80%, the apparent temperature that one feels is 113°F.

the wind can usually be seen from small aircraft or from the ground; the way it points indicates the direction in which the wind is blowing. The sock's fullness provides a wind speed estimate. Wind vanes designed around a metal rooster were once quite popular in rural areas. Found usually atop the barn, they are commonly called weather vanes.

Figure 22.22 An array of anemometers and wind vanes. The anemometers record the wind speed while the wind vanes line up with the wind to indicate the direction from which it comes. The instruments in the photo are electronically monitored. (Courtesy NOAA.)

Precipitation

The major forms of precipitation are rain and snow. Rainfall is measured by a **rain gauge.** This may be simply a container with vertical sides and graduated in inches placed in an open area. After a rainfall, the rain gauge is read, and the amount of precipitation is reported as so many inches of rain. It is assumed that this much rainfall is distributed evenly over the surrounding area.

If the precipitation is in the form of snow, the depth of snow (where not drifted) is reported in inches. The actual amount of water received depends on the density of the snow. To record the water amount, the snow is melted and the actual amount of water is reported in inches. More elaborate rain-measuring instruments automatically measure and record the rain or snowfall.

Learning Objectives

After reading and studying this chapter, you should be able to do the following without referring to the text:

1. State the nature and general composition of air.
2. Explain why some planets have atmospheres and others do not.
3. Explain how the oxygen-carbon dioxide balance of the atmosphere is maintained.
4. Distinguish the properties that give rise to vertical divisions of the atmosphere.
5. State the divisions of the atmosphere based on temperature.
6. Explain the distribution of insolation.
7. State and explain the processes by which the lower atmosphere obtains the major portion of its energy content.
8. Explain the greenhouse effect.
9. Name the atmospheric properties that are commonly measured and describe how they are measured.

Important Words and Terms

atmosphere
atmospheric science
meteorology
photosynthesis
troposphere
weather
stratosphere
mesosphere
thermosphere
tropopause
stratopause
mesopause
homosphere
heterosphere
ozonosphere
ionosphere
ozone
aurora borealis
aurora australis
insolation
solar constant
albedo
Rayleigh scattering
greenhouse effect
thermometer
mercury barometer
millibar
aneroid barometer
absolute humidity
partial pressure
relative humidity
dew point
absorption hygrometer
hair hygrometer
psychrometer
sling psychrometer
anemometer
wind vanes
rain gauge

Questions

1. What is the difference between atmospheric science and meteorology?
2. What is the composition of the air you breathe?
3. Humans inhale oxygen and expel carbon dioxide. How is the relatively constant oxygen content of the atmosphere maintained?
4. Why do the Earth and the other planets have atmospheres, while Mercury and the moon do not?
5. Describe how the temperature of the atmosphere varies in each of the following regions: (a) the troposphere, (b) the stratosphere, (c) the mesosphere, and (d) the thermosphere.
6. What is the basic distinction between the homosphere and the heterosphere?
7. How were the ozone and ion layers detected and investigated?
8. Of what importance is the atmospheric ozone layer?
9. What is believed to be the cause of the displays of light called auroras?
10. What is meant by insolation and what is the solar constant?
11. From what source does the atmosphere receive most of its *direct* heating, and how is this accomplished (three methods)?
12. Why is the sky blue?
13. Explain what is meant by the greenhouse effect.
14. How does the selective absorption of atmospheric gases provide a thermostatic action for the Earth?
15. Explain why the temperature of the troposphere decreases with altitude and why the temperature of the stratosphere increases.
16. What is the principle of the liquid barometer, and what is the height of the barometer column for one atmosphere of pressure?
17. At high altitudes, would boiled food have to be cooked a greater or shorter amount of time than at sea level? Why? Explain the principle of a pressure cooker.
18. Why does water condense on the outside of a glass containing an iced drink?
19. How is it possible that it may be raining, yet the relative humidity is reported to be less than 100%? Explain the principle of the psychrometer.
20. Why is the apparent body temperature on a hot, humid day greater than the air temperature even when a person is in the shade?
21. Why is the apparent body temperature below that of the air temperature when the humidity is low? (See Table 22.2.)
22. Which way, relative to the wind direction, does a finned wind vane point? How about a wind sock?

Problems

1. On a vertical scale of altitude in kilometers (km) above the Earth's surface at sea level locate the heights of the following (the heights not listed may be found in the chapter).
   (a) The top of Pike's Peak
   (b) The top of Mt. Everest—29,000 ft
   (c) Commercial airline flight—35,000 ft
   (d) Supersonic transport (SST)—65,000 ft
   (e) Communications satellite—40 mi
   (f) The E and F ion layers
   (g) Auroral displays
   (h) Syncom satellite—22,000 mi
   (satellite with synchronous period to the Earth's rotation so it stays over one location)

   Note: conversion factors are found on the inside back cover.
2. If the air temperature is 70°F at sea level, what would be the temperature at the top of Pike's Peak (14,000-ft elevation)? (Hint: the temperature decreases uniformly in the troposphere.) *Answer:* 21°F
3. How much solar energy does the Earth's atmosphere intercept in one hour? Assume the thickness of the atmosphere to be 600 mi. (Hint: use the solar constant and that the cross-sectional or exposed area of a sphere is that of a circle, $A = \pi r^2$. The radius of the solid Earth is about 4000 mi.) *Answer:* 2.0×10^{20} cal
4. A satellite in orbit around the Earth near the top of the atmosphere has a solar panel array with an area of 10^4 cm^2 directed toward the Sun. Assuming that all of the incident insolation is absorbed, how much energy would the solar panels receive in one hour? Express your answer in joules. *Answer:* 5×10^6 J
5. A radio wave is directed vertically upward. The time lapse for its return is 7.0×10^{-4} s. From what ionic layer was the wave reflected? (Hint: electromagnetic waves travel at the speed of light, 3.0×10^5 km/s or 186,000 mi/s).
6. A mercury barometer has a column height of 75 cm. What is the pressure in (a) dyne/cm^2, (b) millibars, and (c) torrs? *Answer:* (a) 999,600 dyn/cm^2 (b) 999.6 mb (c) 750 torr
7. What would be the height of the barometer in the preceding problem if a liquid with a density of 6.8 g/cm^3 were used instead of mercury? *Answer:* 150 cm
8. The palm of a hand is about 3.0 in by 4.0 in. How many pounds of force is exerted on the palm by the atmosphere at sea level? *Answer:* 176 lb
9. The density of air at sea level is 1.2×10^{-3} g/cm^3. If the atmosphere had this as a constant density with altitude, what would be the approximate total height of the atmosphere in kilometers? (Hint: use the pressure-height relationship as for a barometer) *Answer:* 8.5 km
10. On a day when the air temperature is 75°F, the wet bulb reading of a psychrometer is 68°. Find each of the following:
    (a) Relative humidity
    (b) Dew point
    (c) Maximum moisture capacity of the air
    (d) Actual moisture content of the air.

    Answer: (a) 70% (b) 64°F (c) 9.4 gr/ft^3 (d) 6.6 gr/ft^3
11. A psychrometer has a dry bulb reading of 95°F and a wet bulb reading of 90°F. Find each of the quantities asked for in problem 10.
12. On a winter day, a psychrometer has a dry bulb reading of 35°F and a wet bulb reading of 29°F. (a) What is the actual moisture content of the air? (b) Would the water in the wick of the wet bulb freeze? Explain. *Answer:* (a) 1.1 gr/ft^3
13. On a very hot day with an air temperature of 105°F, the wet bulb thermometer of a psychrometer records 102°F. (a) What is the actual moisture content of the air? (b) How many degrees would the air temperature have to be lowered for the actual moisture content to be equal to the maximum moisture capacity (i.e., 100% relative humidity)? *Answer:* (a) 21.1 gr/ft^3 (b) 4°F
14. On a day when the air temperature is 85°F and the relative humidity is 80%, what are (a) the actual moisture content of the air, and (b) the apparent temperature most people would feel? *Answer:* (a) 10.2 gr/ft^3 (b) 97°F
15. What would be the apparent temperatures one would feel on days when (a) a psychrometer has a dry bulb reading of 100°F and a wet bulb reading of 91°F, and (b) a dry bulb reading of 75°F and a wet bulb reading of 60°F? *Answer:* (a) 144°F (b) 74°F
16. The dry bulb and wet bulb thermometers of a psychrometer both read 75°F. What are (a) the relative humidity, (b) the actual moisture content of the air, and (c) the apparent temperature? *Answer:* (b) 9.4 gr/ft^3 (c) 80°F

23
Winds and Clouds

Who has seen the wind?
Neither you nor I:
But when the trees bow
down their heads,
The wind is passing by.
—C. Rossetti

WINDS AND CLOUDS are common atmospheric observances. As we all know, winds involve air motion and clouds involve water in the atmosphere. However, there is a great deal more to these atmospheric phenomena that play important roles in our weather and environment.

Wind is the horizontal motion of air or air motion along the Earth's surface. Vertical air motions are referred to as updrafts and downdrafts, or collectively as **air currents.** Air movement is the agent of transportation that redistributes the energy of the lower atmosphere and brings the warming influences of spring and summer and the cold chill of winter. The effects of winds and air currents on our environment are often overlooked. Pollen carried by winds is fundamental to nature. Animals sniff the wind for scents of danger, while the transportation of radioactive particles and other contaminants may result in environmental problems. Before the invention of the combustion engine, people commonly harnessed the wind for use in transportation and developed devices that put the wind to work, for example, windmills. Occasionally, the wind causes death and destruction. Also, winds cause erosion of the Earth's surface and influence ocean currents (Chapter 30).

Winds and air currents are an integral part of cloud formation. **Clouds** are buoyant masses of visible water droplets or ice crystals. Have you ever wondered what keeps clouds afloat or how they are formed? These processes are important weather phenomena that indicate atmospheric conditions. The size, shape, and behavior of clouds are a useful key to the weather. With a little practice, one may ''read'' the sky and predict with surprising accuracy the forthcoming weather. Cloud signs were well known to observers in ancient times and cloud observations were probably the basis for the first attempts at weather forecasting.

◀ Trees bending before a strong wind.

As pointed out in Chapter 22, clouds play an important role in maintaining the Earth's energy balance by reflecting solar radiation and insulating against heat loss. Of equal importance is the precipitation process associated with clouds. Rain and other forms of precipitation originate in clouds, and this process is fundamental to life itself as all plants and animals need water to survive. Clouds and air movement form key parts of the hydrologic cycle (Chapter 30) by which moisture is distributed over the Earth.

So, as you can see, an understanding of winds and clouds is essential in the study of meteorology.

23.1 Causes of Air Motion

Winds and air currents require the air to be in motion. But what causes the air to move? As in all dynamic situations, forces are necessary to produce motion and changes in motion. The gases of the atmosphere are subject to two primary forces: (1) gravity and (2) pressure differences due to temperature variations. The force of gravity is vertically downward and acts on each gas molecule. Although this force is often overruled by forces in other directions, the downward gravity component is ever-present, and accounts for the greater density of air near the Earth.

Since the air is a mixture of gases, its behavior is governed by the gas laws of Chapter 5 and other physical principles. The pressure of a gas is directly proportional to its temperature. As defined in Section 22.5, pressure is force per unit area ($P = F/A$). Thus, a temperature

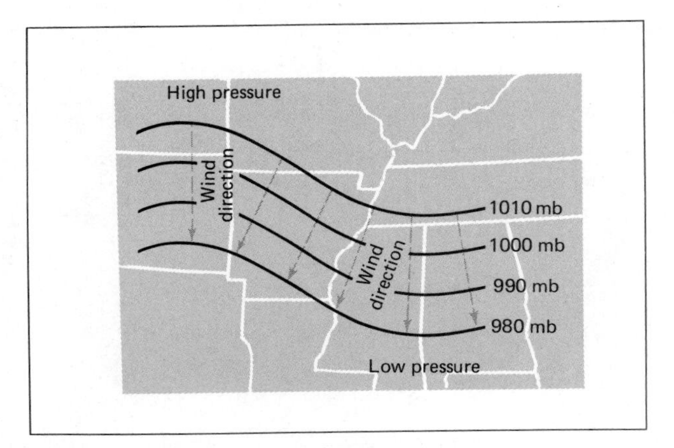

Figure 23.1 Representative isobars. Isobars are lines drawn through locations having equal atmospheric pressures. The air motion, or wind direction, is at right angles to the isobars and from a region of high pressure (greater isobars) to a region of low pressure (lower isobars).

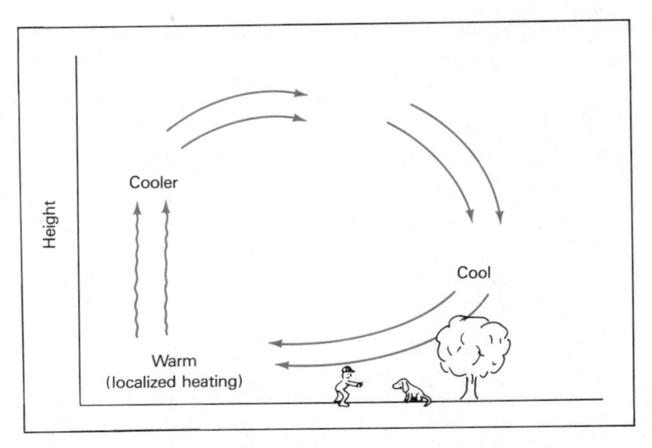

Figure 23.2 A diagram illustrating thermal circulation of air. Localized heating, which causes the air in the region to rise, initiates the circulation. As the warm air rises and cools, cool air near the ground moves horizontally into the region vacated by the rising air. The upper, still cooler air then descends to occupy the region vacated by the cool air. These motions set up a convection cycle.

variation in air generally gives rise to a difference in pressure or force. It is the differences in pressures resulting from temperature differences in the atmosphere that give rise to primary air movement, both locally and on a large scale. Hence, a pressure difference corresponds to an unbalanced force, and when there is a pressure difference, the air moves from a high- to a low-pressure region.

The pressures over a region may be mapped by taking barometric readings at different locations. A line drawn through the locations (points) of equal pressure is called an **isobar.** Since all points on an isobar are of equal pressure, there will be no air movement along an isobar. The wind direction will be at right angles to the isobar in the direction of the lower pressure, as illustrated in Fig. 23.1. Topographically, an isobar is analogous to a line of constant elevation. The isobars of greater value correspond to higher elevations. Just as a ball would move down a hill to a lower elevation, the air moves down a pressure hill, or pressure gradient, toward a lower isobar. An isobar of minimum pressure represents the bottom of a pressure "valley." This region is referred to as a low-pressure trough. It should be remembered that the isobars in Fig. 23.1 represent the pressure readings at different locations. The vertical analogy is meant only as an aid in understanding the air motion along a surface of reasonably constant elevation.

Recall from Chapter 5 that the pressure *and* volume of a gas are related to its temperature ($PV \propto T$). A change in temperature then causes a change in the pressure and/or volume of a gas. With a change in volume, there is also a change in density since $\rho = m/V$. Hence, regions of the atmosphere with different temperatures may have different air pressures and densities. As a result, localized heating sets up air motion in a **convection cycle** or gives rise to **thermal circulation.**

To understand this, consider the lower-left part of the cycle in Fig. 23.2. A warm land area may heat the immediate air by conduction. The warmed air expands, becomes less dense and buoyant, and rises, creating an updraft. For example, warm air can be "seen" rising from a blacktop road in the summer. The rising "heat waves" are seen as a result of the refraction of light. Regions with different air temperatures and densities have different indices of refraction, and the refraction or "bending" of light allows the rising air to be "seen."

The buoyant warm air rises and cools as it expands. Since the air rises from the heated region, the pressure is lowered here and cooler air moves horizontally toward this region of lower pressure. The region vacated by the horizontally moving air is then filled by descending cool air. These motions set up a convection cycle and the thermal circulation of air. Thermal circulation and convection cycles are important atmospheric mechanisms, as we shall see.

Once the air has been set into motion, velocity-dependent forces act. These secondary forces are (1) the Coriolis force and (2) friction. The **Coriolis force,** named

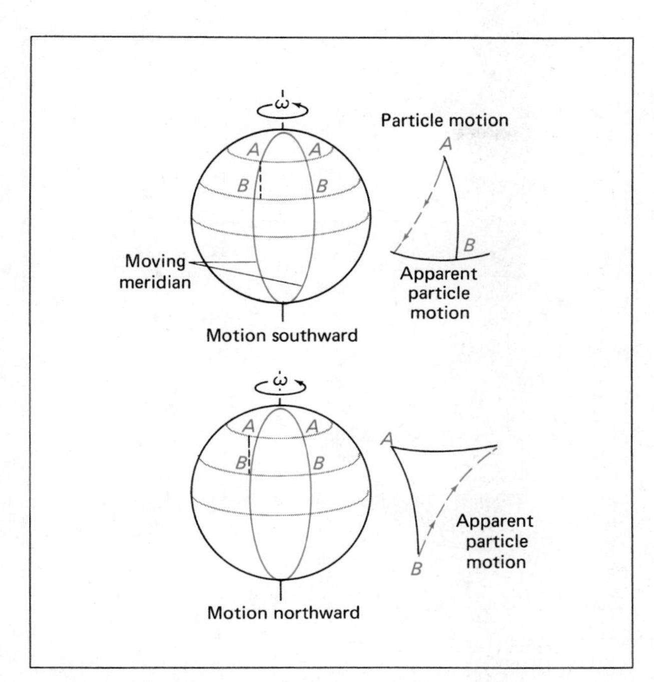

Figure 23.3 The Coriolis deflection for particle motion in the nothern hemisphere. The upper drawing shows that because B has a greater surface velocity than A, a particle traveling southward with the same surface velocity as A along the meridian of A and B will fall behind B. To an observer at A, the particle is apparently deflected to the right, viewed along the direction of particle motion. Similarly for particle motion northward from B to A, the apparent deflection is to the right.

after the French engineer and mathematician who first described it, arises from the fact that on the Earth an observer is in a rotating frame of reference. This force is sometimes referred to as a pseudoforce or false-force, since it is introduced to account for the effect of the Earth's rotation. Humans tend to be egocentric. We commonly consider ourselves to be motionless although we are on a rotating Earth that has a surface velocity of about 1000 mi/h near the equator. Newton's laws of motion apply to nonaccelerating or inertial reference frames and may be used for ordinary motions on the Earth without correction. However, for high velocities or huge masses such as the atmospheric gases, the correction for the Earth's rotation becomes important.

To understand this effect, consider Fig. 23.3. Keep in mind that objects on the surface of the rotating Earth at lower latitudes have greater surface velocities than those at greater latitudes. In Fig. 23.3, an object at A is at a greater latitude than an object at B. Since both objects follow their respective paths of rotation once every 24 hours, an object at B must have a greater surface velocity than an object at A since it covers a greater distance in the same amount of time.

Now consider particle motion southward in the northern hemisphere as shown in the upper drawing of Fig. 23.3. A particle initially at A is projected along a meridian connecting A and B. Since the particle has the same rotational surface velocity as point A, it will fall behind the moving meridian as it (the particle) approaches B since the Earth's surface has a greater velocity in this region. To an observer at A who rotates with the meridian and always looks southward along it, the particle will appear to be deflected to the right. A similar analysis shows that a particle moving northward from B toward A is also deflected to the right as illustrated in the lower drawing of Fig. 23.3.

By the same reasoning, it can be shown that moving objects in the southern hemisphere are deflected to the left in similar situations. East-west motions are more difficult to analyze, however, the results are the same as above. It is said that this deflection is the result of a Coriolis force, since a force is required to deflect an object according to Newton's laws. The Coriolis force is a pseudoforce, as it was "invented" so the situation would be consistent with the laws of motion.

Hence, due to the Coriolis force, moving objects are deflected to the right in the northern hemisphere and to the left in the southern hemisphere as observed in the direction of motion. The Coriolis force is at a right angle to the motion of the object and its magnitude varies with latitude; it is zero at the equator and becomes greater toward the poles.

Consider this effect on wind motion. Initially, air moves toward a low-pressure area (a "low") and away from a high-pressure area (a "high"). Because of the Coriolis force the wind is deflected, and in the northern hemisphere the wind rotates counterclockwise around a low and clockwise around a high as viewed from above (Fig. 23.4). These disturbances are referred to as cyclones and anticyclones, respectively. The rotations around cyclone lows and anticyclone highs are reversed in the southern hemisphere. Why? It should be noted that water motion or currents in the oceans are also affected by the Coriolis force.

Friction, or drag, can also cause the retardation or deflection of air movements. Just as a liquid has a flow resistance caused by internal friction of its molecules, moving air molecules experience frictional interactions among themselves or with terrestrial surfaces. The opposing frictional force along a surface is in the opposite

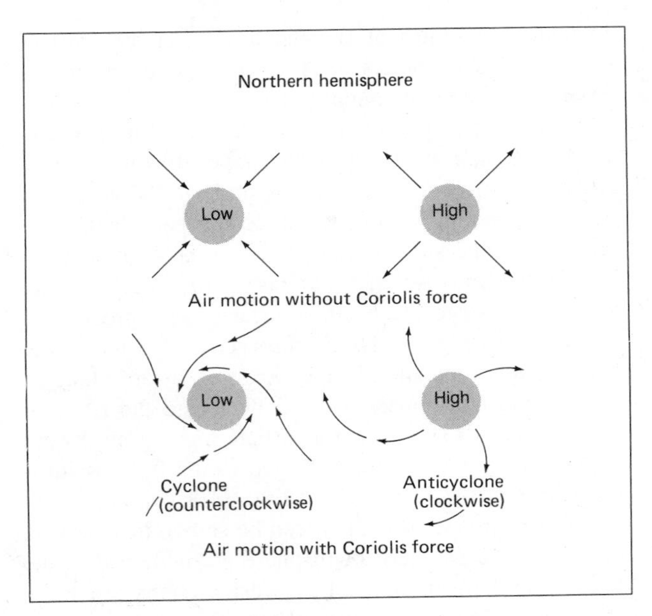

Figure 23.4 The effect of the Coriolis force on air motion. In the northern hemisphere, the Coriolis deflection to the right produces counterclockwise motion of air around a low and clockwise rotation around a high. The situation is reversed in the southern hemisphere.

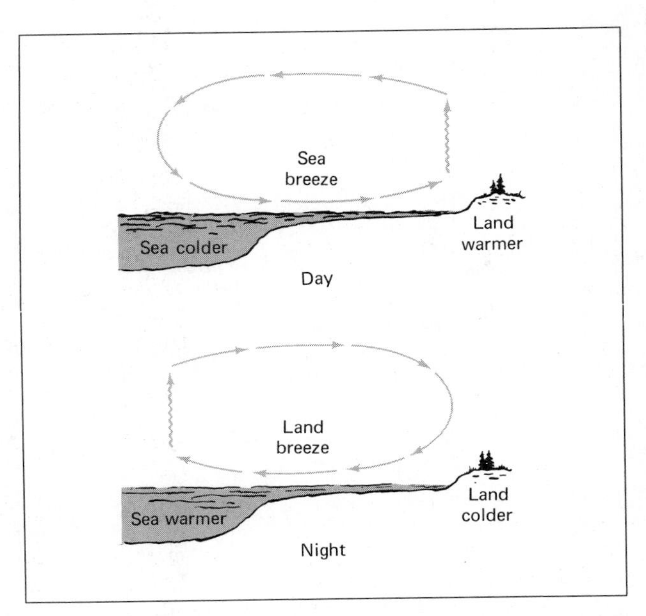

Figure 23.5 Daily convection cycles of air over land and water. During the day the land heats more quickly than water and a convection cycle is set up as illustrated in the upper drawing. At night the land cools more quickly than the water and the cycle is reversed (lower drawing).

direction of the air motion, and its magnitude depends on the air speed. As a result, winds moving into a cyclonic disturbance may be deflected in the opposite direction to that of the cyclonic rotation.

23.2 Local Winds and World Circulation

Air moves in all directions. Vertical air motion is important in cloud formation and precipitation, as will be discussed shortly. As you know, horizontal air movements near the surface of the Earth produce the winds that "blow," causing their presence to be felt and noticed. A wind is named after the direction or region *from* which it comes. For example, a wind blowing from north to south is referred to as a north wind. A wind blowing out to sea from the land is called a land breeze. The strength of the wind is determined by its speed. As previously mentioned, this is measured by an anemometer; however, an estimate of the wind speed may be obtained from Table 23.1.

Wind is an important factor in how cold we feel. At moderately low temperatures in a brisk wind we may feel extremely cold. The wind promotes the loss of body heat, which adds to the chilling effect. For example, more body heat may be lost when the temperature is 20°F and the wind speed is 18 mph than when the temperature is 0°F and there is no wind. Thus, the temperature alone does not tell us how we should dress when going out of doors.

The effect of wind on how cold we feel may be expressed by a wind-chill index, commonly called the **chill factor.** This is not to say that the chill factor indicates how "chilly" one would feel, because our feeling depends on other things besides wind and temperature, such as state of nourishment, individual metabolism, and protective clothing. However, the wind-chill index in Table 23.2, which describes the cooling power of wind on exposed flesh at various temperatures, is a good guide as to what clothing would be needed for protection on a cold, windy day.

Atmospheric pressure differences involved in thermal circulations due to geographical features give rise to local winds. As discussed in Chapter 22, land areas heat up more quickly than do water areas. The heating of the land area gives rise to a convection cycle. As a result, during the day when the land is warmer than the water, a lake or **sea breeze** is experienced as shown in Fig. 23.5. You may have noticed these daytime sea breezes at an ocean beach.

Table 23.1 Wind Speeds and Descriptions

Average Speed (mph)	*Signs*	*Terminology*
0	Smoke rises vertically	Calm
1–3	Smoke deflected	Light air
4–7	Leaves rustle	Slight breeze
8–12	Leaves and twigs move	Gentle breeze
13–18	Small branches move, dust rises	Moderate breeze
19–24	Small trees sway	Fresh breeze
25–31	Large branches move, telephone lines hum	Strong breeze
32–38	Trees in motion	Near gale
39–46	Twigs broken, walking impaired	Gale
47–54	Slight damage; e.g., to roofs and TV antennas	Strong gale
55–63	Trees uprooted, much damage	Storm
64–74	Widespread damage	Violent storm
>74	Tropical storm or blizzard	Hurricane

Table 23.2 Determining the Wind-Chill Index[a]

Windspeed (miles per hour)	*Temperature °F*																
Calm	35	30	25	20	15	10	5	0	−5	−10	−15	−20	−25	−30	−35	−40	−45
5	33	27	21	16	12	7	1	−6	−11	−15	−20	−26	−31	−35	−41	−47	−54
10	21	16	9	2	−2	−9	−15	−22	−27	−31	−38	−45	−52	−58	−64	−70	−77
15	16	11	1	−6	−11	−18	−25	−33	−40	−45	−51	−60	−65	−70	−78	−85	−90
20	12	3	−4	−9	−17	−24	−32	−40	−46	−52	−60	−68	−76	−81	−88	−96	−103
25	7	0	−7	−15	−22	−29	−37	−45	−52	−58	−67	−75	−83	−89	−96	−104	−112
30	5	−2	−11	−18	−26	−33	−41	−49	−56	−63	−70	−78	−87	−94	−101	−109	−117
35	3	−4	−13	−20	−27	−35	−43	−52	−60	−67	−72	−83	−90	−98	−105	−113	−123
40	1	−4	−15	−22	−29	−36	−45	−54	−62	−69	−76	−87	−94	−101	−107	−116	−128
45	1	−6	−17	−24	−31	−38	−46	−54	−63	−70	−78	−87	−94	−101	−108	−118	−128
50	0	−7	−17	−24	−31	−38	−47	−56	−63	−70	−79	−88	−96	−103	−110	−120	−128

VERY COLD

BITTER COLD

EXTREME COLD

Equivalent Temperatures (°F)

[a]Find the temperature in the top row and the wind speed in the left column. Read the chill factor where the corresponding column and row intersect. For example, a calm air temperature of 10°F and a wind speed of 20 mi/h together are equivalent in cooling effect to −24°F.

At night the land loses its heat more quickly than the water and the air over the water is warmer. The convection cycle is then reversed and at night a **land breeze** blows. Sea and land breezes are sometimes referred to respectively as onshore and offshore winds.

This local heating effect is also experienced in mountain valleys. During the day the air in contact with the mountain slopes heats up and rises, producing a valley breeze. At night the slopes cool faster than the air at the same elevation over the valley, and the cool air descends the slopes, giving rise to a mountain breeze.

The heating effect that produces thermal circulation also applies on a large scale. Seasonal heating and cooling of large continental land masses initiate convection cycles. In the summer the air over the heated continents rises, creating low-pressure areas into which air flows. In winter, high-pressure areas exist over the continents and air flows outward toward the warmer low-pressure areas over the oceans.

The winds of these cycles are most pronounced on the Asian continent and are called **monsoons** (Arabic *mausin,* season). During the summer, sea air moves toward the heated continent and is called the summer monsoon. As the moist sea air travels inland toward the mountains, precipitation occurs. The sea wind or summer monsoon is associated with the wet season in southern and eastern Asia. In the winter the cycle is reversed, and then a cold dry wind blows from the mountains toward the sea. This prevailing land wind is called the winter monsoon.

Other local winds depend on the local geography and acquire names particular to that area. An example of such a wind is the **chinook,** or ''snow-eater,'' experienced on the eastward slopes of the Rocky Mountains. Similar winds are called föhns in the Alps. The winds from the Pacific Ocean blow toward and up the westward slopes of the Rockies (Fig. 23.6). Since the temperature decreases with altitude, much of the moisture is lost as rain or snow on the westward slopes and near the top of the mountains. As a result, dry air descends the eastern slopes becoming warm with descent. This chinook wind of warm, dry air may cause a rapid rise in temperature and a quick melting of the winter snow, hence the name ''snow eater.''

Air motion changes locally with altitude, the geographical features, and the seasons. However, the Earth does possess a general circulation pattern. If the Earth were completely covered by water or land and did not rotate, convection cycles would circulate the surface air from the cold polar regions toward the equator. If this were the case, the prevailing winds would always assume a general north-south direction.

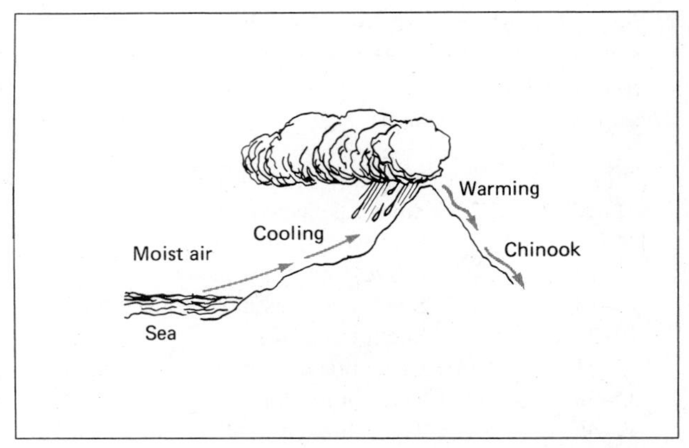

Figure 23.6 Topography for chinook formation. After the ascending air loses moisture on the windward side of the mountain, the descending dry air is warmed on the leeward side of the mountain.

However, because of the Coriolis force, land and sea variations over the Earth's surface, and other complicated reasons, the hemispheric circulation is broken up into six general convectional cycles, or pressure cells. The Earth's general circulation structure is shown in Fig. 23.7. Many local variations occur within the cells, which shift seasonally in latitude due to variations in isolation. However, the prevailing winds of this semipermanent circulation structure are important in influencing general weather movement around the world.

A low-pressure belt exists in the equatorial region due to rising warm air. Direct insolation heats this region and the air motion for the most part is rising air currents of hot, humid air. As a result, there is a vast equatorial low-pressure zone. The surface winds in this zone are light and variable, and often calm. These latitudes are referred to as the **doldrums.** Early sailing ships were often becalmed in the doldrums and floated listlessly due to lack of wind. We use the expression of ''being in the doldrums'' to describe a feeling or mood of bored listlessness.

The air rising from the equatorial low divides and travels horizontally north and south at high altitudes. At about 30° N and 30° S latitudes, the air cools, and descends toward the surface. Hence, the subtropical regions near 30° N and 30° S are characterized by descending air, which makes them high-pressure zones. There is little horizontal surface air movement. These regions became known as the **horse latitudes,** due to the fact that early sailing ships were becalmed in these latitudes and their cargoes of horses had to be eaten for food or

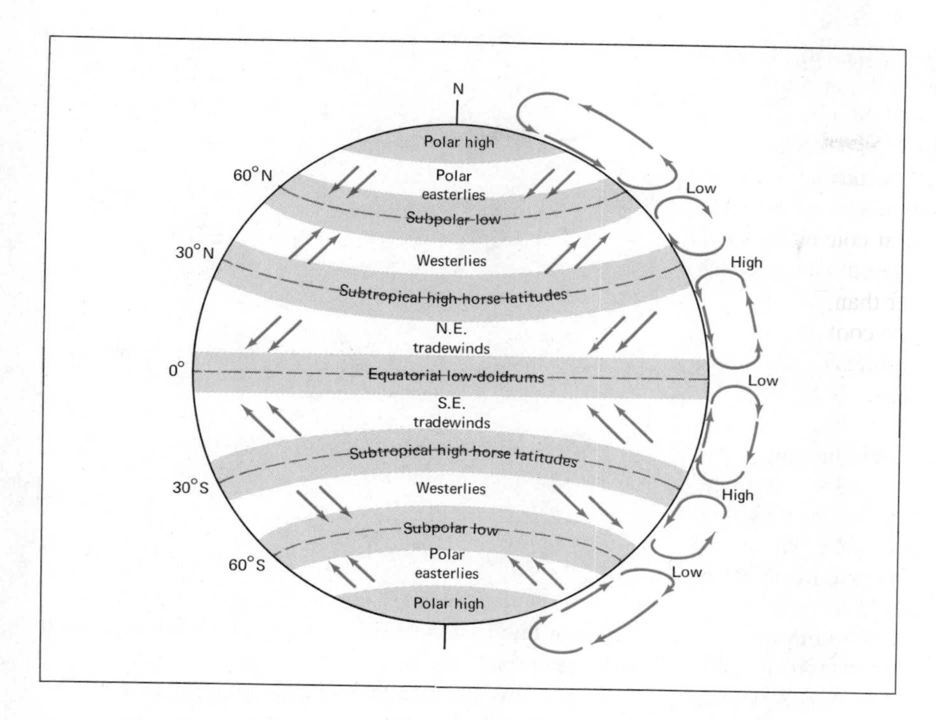

Figure 23.7 The Earth's general air circulation structure. For rather complicated reasons, the Earth's general circulation structure has six convection cycles.

thrown overboard to conserve fresh water. These high-pressure regions of descending, warming air are the locations of some of the world's great deserts, e.g., the Sahara in north Africa and the Kalahari in southwest Africa.

The air descending in the horse latitudes moves toward the equator, completing the thermal-convection cycle (see Fig. 23.7). The surface winds between the horse latitudes and the doldrums are known as the **trade winds.** These winds are generally regular and steady, and they were of considerable importance in establishing the early trade routes of sailing ships. Due to the Coriolis force, the trade winds blow in general from the northeast in the northern hemisphere and from the southeast in the southern hemisphere.

We might expect another single thermal circulation cell between the horse latitudes and the poles. However, for rather complicated reasons, there are two pressure cells. One of these is between the regions of subtropical highs of the horse latitudes (30° N and 30° S) and subpolar lows near 60° N and 60° S. In the latitudes between 30° and 60°, the **westerlies** prevail. The general directionality of these surface winds is again due to the Coriolis force. Since the 48 conterminous states of the United States lie in a westerly wind zone, the general movement of weather conditions across the country is from west to east.

These west winds blew the early sailing ships back across the Atlantic to Europe. The homeward wind was christened the ''brave West wind.'' Voyages to the New World were made with the help of the gentle northeast trade winds. Care was taken to avoid the horse latitude region near 30° N, which is the latitude of the southeastern United States, by keeping within the northeast trade wind zone. It can now be understood why Columbus in sailing from Spain arrived in the West Indies and why Spanish and Portuguese explorations were done in Central and South America.

The other sailing route to America was the northern route with the aid of the **polar easterlies.** These are the surface winds of the pressure cell between the subpolar lows at 60° latitude and the polar highs. By use of this route, the Norse, French, and English explorations were made along the northeastern coast of Canada and the United States.

The general wind circulation has a major influence on climate. Prevailing winds affect the ocean currents, which redistribute the ocean's heat energy in much the same way as do winds in the atmosphere. There is still much more to be learned about the general circulation of the air, since its movement is sparsely monitored over many regions of the Earth.

The extent of the circulation patterns was widely noticed during early nuclear test periods when the prevailing winds carried radioactive fallout around the world. Food supplies began to show unnaturally high concentrations of radioactive materials. Of chief concern was milk and its strontium-90 concentration, which could potentially affect the consumer, especially infants. Indirectly, therefore, the Earth's general air circulation was responsible for the 1963 Nuclear Test-Ban Treaty that limited nuclear testing to underground explosions (for those countries that signed it).

23.3 Jet Streams

In the upper troposphere there exist fast-moving "rivers" of air called **jet streams.** The boundaries of the jet streams appear to be reasonably well defined as they move in the atmosphere in a complex fashion. They were first noted in the 1930s, but did not receive much attention until a decade later, during World War II. Pilots reported that their aircraft had been held motionless by an "invisible hand" while at full throttle. The invisible hand was really a jet-stream headwind. Speed records were set with the help of jet-stream tailwinds.

The Japanese also knew of the jet streams. Near the end of the war they launched balloons carrying incendiary bombs in the jet stream blowing toward the United States across the Pacific Ocean. Special devices were automatically adjusted so that the balloons maintained the proper altitude until the mainland was reached. However, the balloons were widely scattered, and only a small percentage reached the west coast, doing little damage.

There are several jet streams that meander like rivers around each hemisphere. The behavior of jet streams is variable and not well understood. The so-called polar jet stream moves from west to east across the United States (Fig. 23.8). It varies in altitude and latitude with the seasons. In the summer the jet stream is found in the region of 50° latitude and at an altitude of about 10 km (6 mi). In the winter, the altitude decreases to below 10 km and the general location is in the region of 30° latitude. It is this jet stream that breeds winter storms and blizzards over the United States.

Mappings of the jet stream give its general dimensions and show that the wind speed increases toward the center of the stream. The air stream is 40–160 km (25–100 mi) wide and 3 km (about 2 mi) deep. As the center of the stream is approached, the winds may reach speeds of 400–500 km/h (250–300 mi/h). The general direction is mainly from west to east, but shows considerable variation due to deflection by the Coriolis force, as well as by frictional forces.

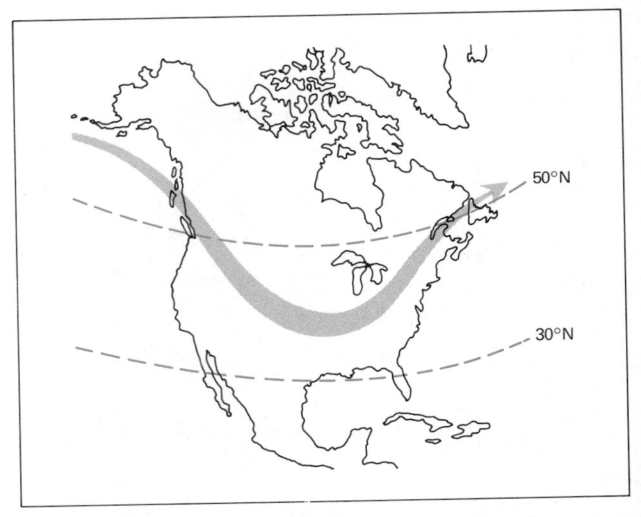

Figure 23.8 A typical jet-stream pattern across the United States.

Jet streams are thought to result from the general circulation structure in the regions where great high- and low-pressure areas meet. Under the proper conditions, the additive circulation of a high and low could produce winds of exceedingly high speeds. The jet streams are relatively recent to meteorological observation, and understanding their effects on the weather is currently somewhat vague. Variations in their seasonal migrations have been associated with the severity of winters. It has also been speculated that intersecting jet streams may contribute to the formation of destructive vortices of winds we call tornadoes (Section 24.3). As more is learned of their behavior, it is possible that these rivers of air in our atmosphere may become routes of air transportation much as the rivers of the Earth—especially downstream.

23.4 Cloud Classification

Clouds are classified according to their *shape* and appearance, and *altitude*. There are four basic root names: **cirrus** (Ci) meaning *curl* and referring to wispy, fibrous forms; **cumulus** (Cu) meaning *heap* and referring to billowy, round forms; **stratus** (St) meaning *layer* and referring to stratified or layered forms; and **nimbus** (Nb) referring to a cloud from which precipitation is occurring or threatens to occur. These root forms are then combined to describe the types of clouds and precipitation potential.

Table 23.3 Cloud Families and Types

Family	*Types*	*Illustration*	*Meteorological Symbols**
High clouds (above 20,000 ft)	Cirrus (Ci)	Fig. 23.9	
	Cirrocumulus (Cc)	Fig. 23.9	
	Cirrostratus (Cs)	Fig. 23.10	
Middle clouds (6000–20,000 ft)	Altostratus (As)	Fig. 23.11	
	Altocumulus (Ac)	Fig. 23.12	
Low clouds (ground level–6000 ft)	Stratus (St)	Fig. 23.13	
	Stratocumulus (Sc)	Fig. 23.16	
	Nimbostratus (Ns)	Fig. 23.17	
Clouds with vertical development (1500–60,000 ft; see Section 23.5)	Cumulus (Cu)	Fig. 23.18	
	Cumulonimbus (Cb)	Fig. 23.19	

*Different symbols describe variations in cloud types. Consult meteorology texts for complete listings.

When classified according to height, clouds are separated into four families: (1) **high clouds,** (2) **middle clouds,** (3) **low clouds,** and (4) **clouds with vertical development.** These are listed in Table 23.3 with their approximate heights and the cloud types belonging to each family. An illustration is provided for each cloud type in subsequent photographs.

High Clouds

The high clouds are made up of ice crystals due to the temperature at their altitudes. The **cirrus** member of this family is the well-known wispy mare's tail or artist's brush. They appear as though an artist had made short curling strokes with white paint on a blue background. The **cirrocumulus** clouds occur in layered patches and are referred to as mackerel scales due to their fish-scale–like structure. The cirrocumulus cloud pattern gives rise to the term mackerel sky. High winds at these altitudes often give the cirrocumulus clouds a wavy or ripple appearance. Cirrus and cirrocumulus clouds are shown in Fig. 23.9.

The **cirrostratus** clouds are in the form of a thin veil of ice crystals that may partially or completely cover the sky. The ice veil often gives rise to the effect of solar and lunar halos caused by the scattering of light by the ice crystals (Fig. 23.10). This effect is analogous to a light bulb placed behind a frosted pane of glass. Occasionally, there are enough brightly reflecting faces of ice crystals to produce a white circle that passes through the Sun and all the way around the sky. In some instances two bright-colored patches appear on each side of the Sun just outside the regular halo. These phenomena are known as parhelia (side suns), or more commonly are referred to as sun dogs.

Middle Clouds

The middle clouds are distinguished by names with the prefix "alto." They vary considerably in shape and thickness. The **altostratus** member of this cloud family consists of layered forms of varying thickness (Fig. 23.11). A thick layer may appear gray and dark. They often hide the Sun or moon and cast a shadow. When their thickness is such that they are translucent to the Sun or moon, an indistinctly defined halo, referred to as a corona, may be observed around the Sun or moon.

Altocumulus clouds have several varieties. They appear commonly as woolly patches or as rolled, flattened layers as shown in Fig. 23.12. The wavy air motion resulting from air passing up and over mountains may form altocumulus clouds in a lenticular, or lens, shape on the leeward side of the mountain (see Figs. 23.22 and 23.24). These clouds have a distinct "flying-saucer" appearance.

Low Clouds

The stratus clouds of the low-cloud family are thin layers of water vapor. They may appear dark and are common in the winter, giving rise to the sky's "hazy shade of

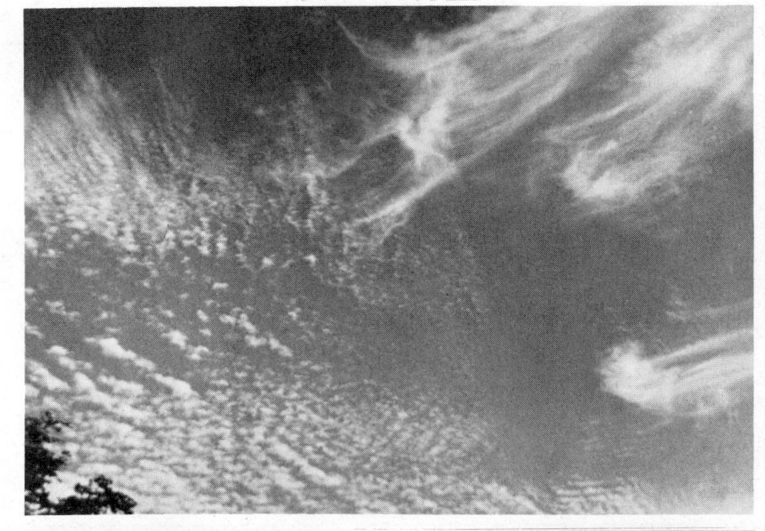

Figure 23.9 Cirrus and cirrocumulus clouds. Artist's-brush cirrus are to the upper right of the photograph and the mackerel-scale cirrocumulus are to the left. (Courtesy Paul H. Taft, Cloudcroft, N.M.)

Figure 23.10 Cirrostratus clouds. In this photograph they completely cover the sky, as evidenced by the solar halo. (Courtesy NOAA.)

Figure 23.11 Altostratus clouds. Thick, gray altostratus clouds hide the Sun. (Courtesy NOAA.)

Figure 23.12 Altocumulus clouds, rolled, and arranged in flattened layers by the moving air. (Courtesy NOAA.)

Figure 23.13 Stratus clouds. Low-lying stratus clouds are sometimes called high fogs. (Courtesy NOAA.)

winter.'' Stratus clouds are sometimes referred to as high fogs (see Fig. 23.13).

Since fog consists of visible water droplets, it falls under the definition of a cloud. In some respects ground fog may be thought of as a low-lying stratus cloud. Most fogs are either advection or radiation fogs. **Advection fog** forms when moist air moving over a colder surface is cooled below the dew point and condensation occurs. Advection fogs ''roll in,'' as shown in Fig. 23.14. **Radiation fog** results from stationary air overlying a surface that cools. This occurs typically in valleys as in Fig. 23.15.

The **stratocumulus** clouds are long layers of cotton-like masses (Fig. 23.16). The rounded masses sometimes blend together so as to give the stratocumulus clouds a wavy appearance. When the low clouds become dark and given to precipitation, they take on the generic name of **nimbostratus.** See Fig. 23.17.

Vertical Development

The clouds of massive grandeur are those with vertical development. Formed by rising air currents, the billowy cumulus clouds are a common sight on a typical summer day as shown in Fig. 23.18. When the cumulus clouds turn dark and forecast an impending storm, they are referred to as **cumulonimbus** clouds. These clouds are often called thunderheads (Fig. 23.19).

Winds and air currents cause many variations in cloud shapes. The following words are used to describe these shapes: *fracto,* meaning broken and applied to stratus and cumulus clouds; *congestus,* meaning crowded or

Figure 23.14 An advection fog. Formed when moist air moves over a cool surface. Advection fogs are sometimes considered low-lying stratus clouds. (Courtesy NOAA.)

Figure 23.15 A radiation fog that formed overnight in an Adirondack mountain valley when radiative heat loss cooled the air sufficiently to cause condensation. (Courtesy C. McNicol, Schroon Lake, N.Y.)

Figure 23.16 Stratocumulus clouds. Long layers of cottonlike masses. (Courtesy NOAA.)

Figure 23.17 Nimbostratus clouds. These dark stratocumulus clouds are given to precipitation. (Courtesy NOAA.)

Figure 23.18 Fair weather cumulus clouds. The billowy white clouds are commonly seen on a clear day. (Courtesy NOAA.)

heaped and applied to cumulus clouds; *humilis,* meaning lowly, or poorly formed, also applied to cumulus clouds; *mammatus,* referring to mammary-shaped clouds as shown in Fig. 23.20.

Cloud height, or the cloud ''ceiling,'' is important in meteorology and particularly in aviation. The height of the cloud coverage is determined by a ceilometer that operates on a reflection principle. Clouds are good reflectors of radiation. By observing the time elapsed between the emission and detection of reflected light, along with the knowledge of the speed of light, the height of a cloud may be calculated.

Figure 23.19 A cumulonimbus cloud. A massive thunderhead with rain in progress. Note the "veil" at the top of the cloud. (Courtesy Paul H. Taft, Cloudcroft, N.M.)

Figure 23.20 Cumulonimbus mammatus type clouds formed on the lower surface of the anvil of a cumulonimbus cloud. (Courtesy NOAA.)

23.5 Cloud Formation

To be visible as droplets, the water vapor in the air must condense. This requires a certain temperature, namely, the dew point. Hence, if moist air is cooled below the dew point, the water vapor contained therein will generally condense into fine droplets and form a cloud. The air is continually in motion, and when an air mass moves into a cooler region, cloud formation may take place. Since the temperature of the troposphere decreases with height, cloud formation is associated with the vertical movement of air. In general, clouds are formed by vertical air motion and are shaped and moved about by horizontal air motion, or winds.

Vertical air motion may occur by different means, such as the vertical displacement of light warm air by heavier cold air, or by the localized heating of air as in the case of a convection cycle. When a mass of air is heated locally, it rises. This occurs because the heated air is less dense than the surrounding cooler air, and it therefore is buoyant. As the warm air mass ascends, it becomes cooler as a result of further expansion. This is because internal heat energy is used to do work in expanding the rising air mass against the surrounding stationary air. A region of rising air is one of low pressure.

The rate at which the air temperature decreases with height is called the **lapse rate.** The normal lapse rate in stationary air is on the order of $3\frac{1}{2}$°F/1000 ft ($6\frac{1}{2}$°C/km). This value may vary with latitude and changing atmospheric conditions, e.g., seasonal changes.

The air temperature may be calculated for any height by the general formula

$$T = T_o - Rh$$

where T is the temperature of the air at a height h, T_o is the temperature of the air at the level from which the height is measured, e.g., $h = 0$, and R is the lapse rate. For example, if the air temperature of stationary air ($R = 3\frac{1}{2}$°F/1000 ft) at the Earth's surface is 75°F, then at 4000 ft, the stationary air temperature would be

$$\begin{aligned} T &= T_o - Rh \\ &= 75°\text{F} - (3\tfrac{1}{2}°\text{F}/1000 \text{ ft})(4000 \text{ ft}) \\ &= 75°\text{F} - 14°\text{F} = 61°\text{F} \end{aligned}$$

The formula simply says that for every 1000 ft of altitude (height), $3\frac{1}{2}$° is subtracted from the initial air temperature, in this case.

Because energy is used in the expansion of a warm air mass, a rising air column has a greater lapse rate than does the surrounding stationary air. Thus, rising air cools more quickly. When a rising air mass cools to the same temperature as the surrounding stationary air their densities become equal. The rising air mass then loses its buoyancy and is said to be in a *stable* condition. A heated air mass rises until stability is reached, and this portion of the atmosphere is referred to as a **stable layer,** that is, a layer of air of uniform temperature and density.

Clouds are formed when water vapor in rising air condenses into droplets that can be seen. If the rising air reaches its dew point before becoming stable, condensation occurs at that height, and the rising air carries the condensed droplets upward, forming a cloud. Hence, when localized heating of an air mass occurs, one of the following happens:

1. The rising air reaches stability without condensation; that is, the rising air cools to the same temperature as the surrounding stationary air and becomes stable at some height without reaching the dew point. No condensation takes place and no cloud forms.
2. The temperature of the rising air (T_r) cools to the dew point (T_d), condensation occurs ($T_r = T_d$), and a cloud begins to form. See Fig. 23.21 (p. 424). The air, still buoyant, continues to rise, but cools with a slower lapse rate due to heating from the latent heat of condensation. It rises until its temperature (T_r) becomes equal to the temperature of the surrounding stationary air (T_s), i.e., stability is reached ($T_r = T_s$).

The height at which condensation occurs is the height of the base of the cloud. The vertical distance between the level where condensation began and the level where stability is reached is the height, or thickness, of the cloud mass. It is assumed for purposes of this discussion that no precipitation occurs.

By inverse reasoning of the above cloud-forming process, it can be seen that if a cloud were carried downward by air currents, it would evaporate into the unsaturated air below the height where the cloud's temperature is equal to the dew point. Hence, cloud formation indicates updrafts and dissipating clouds indicate downdrafts. Once formed, the wind shapes the clouds. Some examples of this action are shown in Figs. 23.22, 23.23, and 23.24.

Stable layers are the limits to which air may rise. Should stability occur at a low altitude, the air is held near the surface of the Earth. Under special atmospheric conditions, such as rapid radiative cooling near the Earth,

Figure 23.21 Vertical cloud development. The cloud begins to form at the height at which the dew point is reached and condensation occurs ($T_r = T_d$). The vertical development continues until stability is reached ($T_r' = T_s$). (Courtesy NOAA.)

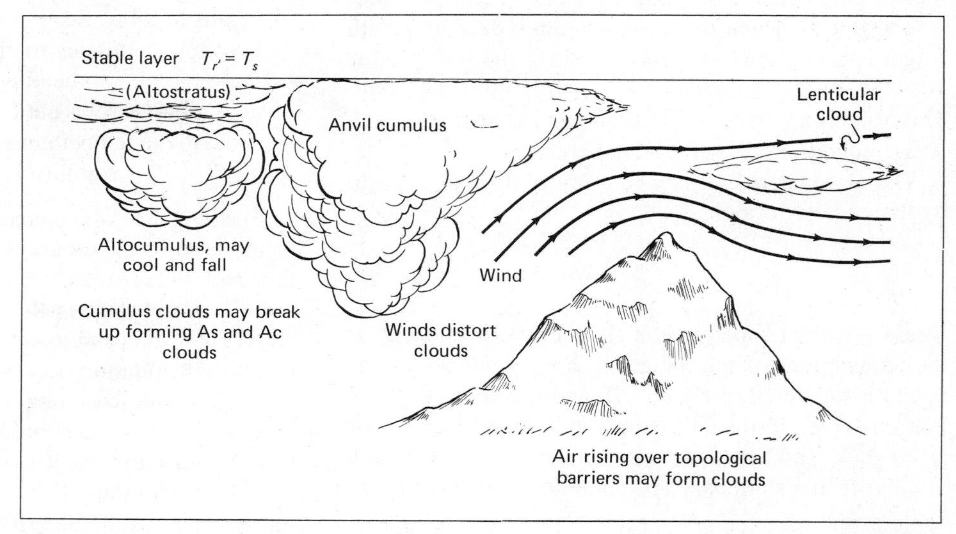

Figure 23.22 The formation of different cloud shapes. Clouds are formed and shaped by vertical ascent and air motion.

Figure 23.23 Cumulonimbus with anvil. The vertically formed cumulonimbus cloud spreads out along a stable layer, taking the shape of an anvil. At the left of the photograph the wind has distorted the top of an anvil into cirrus clouds. (Courtesy Paul H. Taft, Cloudcroft, N.M.)

Figure 23.24 Lenticular altocumulus clouds. The wavy air motion resulting from air rising over a topographical barrier sometimes forms lens-shaped clouds. Note the flying-saucer appearance of these clouds. (Courtesy NOAA.)

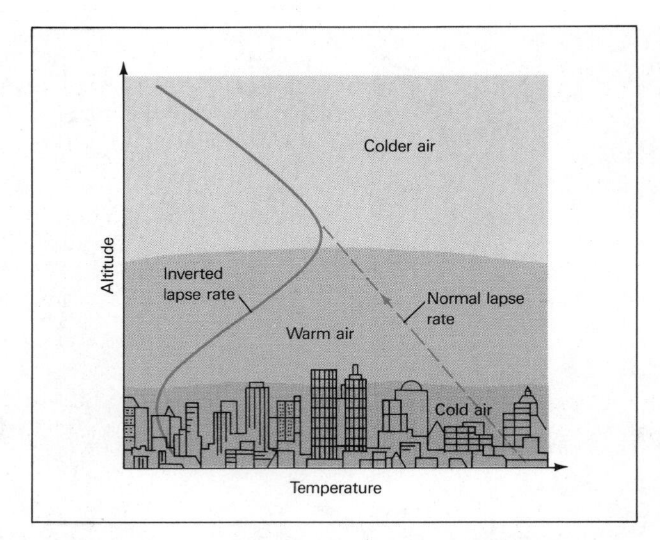

Figure 23.25 A graph illustrating a temperature inversion. Normally, near the Earth the lapse rate decreases uniformly with increasing altitude. However, radiative cooling of the Earth may cause the lapse rate to become inverted, and the temperature then increases with altitude (usually below one mile). A similar condition may occur from the subsidence of a high-pressure air mass.

the temperature may locally increase with altitude. The lapse rate is then said to be *inverted,* giving rise to a **temperature inversion** (Fig. 23.25). Local temperature inversions occur in several ways. Most common are radiation and subsidence inversions.

Radiation temperature inversions occur daily. They are associated with the Earth's radiative heat loss. The ground is heated by insolation during the day, and at night it cools by radiating heat back into the atmosphere (cf. Section 22.4). If it is a clear night, the land surface and the air near it cool quickly. The air some distance above the surface, however, remains relatively warm, thus giving rise to a temperature inversion. Radiation fogs are common evidence of this cooling effect in valleys.

Subsidence temperature inversions occur when a high-pressure air mass moves over a region and becomes stationary. As the dense air settles, it becomes compressed and heated. If the temperature of the compressed air layer exceeds the temperature of the air below it, then the lapse rate is inverted similar to that shown in Fig. 23.25.

Inverted lapse rates prohibit locally heated air from rising to high altitudes. This condition becomes a serious problem for cities and industrial areas as will be discussed in Chapter 26.

23.6 Condensation and Precipitation

Hath the rain a father? or who hath begotten the drops of dew? Out of whose womb came the ice? and the hoary frost of heaven, who hath gendered it? . . . Cans't thou lift up thy voice to the clouds, that abundance of water may cover thee?

Job 38

To answer these questions put to Job about precipitation, one must be familiar with the condensation process. In the previous section on cloud formation, it was simply stated that condensation occurs when the dew point is reached. It was assumed that all the essentials for condensation were present. However, it is quite possible for an air mass containing water vapor to be cooled below the dew point without condensation occurring. In this state the air mass is said to be **supersaturated,** or **supercooled.**

How, then, are the visible droplets of water formed? It might seem that the collision and coalescing of the water molecules would form a droplet, but this would require the collision of millions of molecules. Moreover, only after a small droplet has reached a critical size will it have sufficient binding force to retain additional molecules. The probability of a droplet forming by this process is quite remote.

Instead, the water droplets form around microscopic foreign particles already present in the air. These particles on which the droplets form are called **hygroscopic nuclei.** They are present in the air in the form of dust, combustion residue, salt from sea water evaporation, and so forth. Since foreign particles initiate the formation of droplets that eventually fall as precipitation, the previously mentioned cleansing mechanism of the atmosphere is readily understood.

Liquid water may be cooled below the freezing point without the formation of ice if it does not contain the proper type of foreign particles to act as ice nuclei. For many years it was believed that ice nuclei could be just about anything, such as dust. However, research has shown that ''clean'' dust without biological materials from plants or bacteria would not act as ice nuclei. This is an important discovery since precipitation involves ice crystals as will be discussed shortly.

The droplets formed by the nucleation process are very minute and have diameters on the order of 5–200 micrometers. (One micrometer, μm, is one-millionth of a meter.) Droplets of this size fall very slowly with an approximate speed of 15 ft/min or about $\frac{1}{6}$ mi/h and form a fine drizzle. Since the condensation is formed in updrafts, the droplets are readily suspended in the air as a cloud. To have precipitation, larger droplets or drops must form. This may be brought about by two processes: (1) **coalescence** or (2) the **Bergeron process.**

Coalescence

Coalescence is the formation of drops by the collision of the droplets, with the result that the larger droplets grow at the expense of the smaller ones. The efficiency of this process depends on the variation in the size of the droplets. Raindrops vary in size, reaching a diameter of approximately 7 mm. The volume of a raindrop is calculated by using the formula for the volume of a sphere

$$V = \tfrac{4}{3}\pi(d/2)^3$$

where V is the volume and d the diameter.

To produce a drop with a diameter of 1 mm from droplets 10 μm (10^{-2} mm) in diameter, a number of droplets n with a volume V_1 would have to coalesce to form the volume V_2 of the drop. That is,

$$V_2 = nV_1$$

Then, solving for n and expressing the volumes in terms of the diameters,

$$n = \frac{V_2}{V_1} = \left(\frac{d_2}{d_1}\right)^3 = \left(\frac{1 \text{ mm}}{10^{-2} \text{ mm}}\right)^3 = 10^6$$

Hence, it would take a million of the 10-μm diameter droplets coming together to form a drop 1 mm in diameter. However, if the droplets were 100 μm (10^{-1} mm) in diameter, a similar calculation shows that only 10^3 or 1000 droplets would be needed. From this, it is easily seen that having larger droplets greatly enhances the coalescence process.

Bergeron Process

The Bergeron process, named after the Swedish meteorologist who suggested it, is probably the more important process for the initiation of precipitation. This process involves clouds that contain ice crystals in their upper portions and have become supercooled in their lower portions. Mixing or agitation of the clouds allows the ice crystals to come into contact with the supercooled vapor. Acting as nuclei, the ice crystals grow larger from the condensing vapor. The ice crystals melt into large droplets in the lower portion of the clouds and coalesce to fall as precipitation. Air currents are the normal mixing agents. The Bergeron process is illustrated in Fig. 23.26.

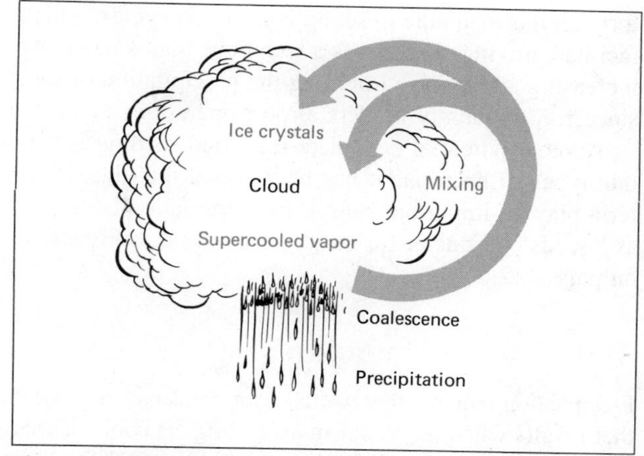

Figure 23.26 A diagram illustrating the essentials of the Bergeron process. The mixing of ice crystals and supercooled water vapor produces larger droplets and initiates precipitation.

It should be noted that there are three essentials in the Bergeron process: (1) ice crystals, (2) supercooled vapor, and (3) mixing. **Rainmaking** is based on the essentials of the Bergeron process.

The early rainmakers were somewhat charlatanic in nature. With much ceremony they would beat on drums, or more sophisticatedly, fire cannons and rockets into the atmosphere. Explosives may supply the agitation or mixing for the Bergeron process, providing the other two essentials are present.

However, modern rainmakers approach the problem differently. There are usually enough air currents present for mixing, but the ice crystal nuclei may be lacking. To correct this, the cloud is ''seeded'' with silver iodide crystals or dry ice pellets (solid CO_2). The silver iodide crystals have a crystal structure similar to that of ice and provide a substitute for the ice crystals. The silver iodide crystals are produced by a burning process. This may be done on the ground, with the silver iodide crystals being carried aloft by the rising warm air, or the burner may be attached to an airplane and the process carried out in the cloud to be seeded.

Dry ice pellets are dropped into the cloud from an airplane. The temperature of solid dry ice is -79°C, and it quickly sublimes; that is, it goes directly from the solid to the gaseous phase. Rapid cooling associated with the sublimation triggers the conversion of supercooled cloud droplets into ice crystals. Precipitation then may occur if this part of the Bergeron process has been absent. Also, the released latent heat from the ice crystal formation is available to set up convection cycles, which facilitate mixing. Seeding is a technique that is receiving increasing attention in initiating the precipitation of fog, since fog frequently hinders airport operations.

Another potential candidate for cloud seeding is certain types of bacteria. It has been found that these bacteria play an important role in frost formation by acting as "seeds" or nuclei for ice crystals (see the discussion on page 429).

Types of Precipitation

Precipitation requires that water vapor condense, a process that results when the vapor in ascending air cools. Three mechanisms by which air rises, thus allowing for precipitation, are convectional, orographic, and frontal. These are illustrated in Fig. 23.27.

Convectional precipitation is a result of convection cycles. This type of precipitation predominantly occurs in the summer, since localized heating is required to initiate the convection cycle. Condensation may occur quickly due to the convectional updrafts, and the precipitation is usually confined to the local area. The sudden summer shower is an example.

Orographic precipitation arises when air is forced to rise due to land forms such as mountain ranges. The wind blows along the surface of the Earth and ascends along geographical variations. The ascending wind may give rise to orographic precipitation on the windward side of the mountain. (See Fig. 23.6.)

Frontal precipitation results from the meeting of air masses of different temperatures. Moving warm air flows up and over the cooler air since the warm air is lighter. This ascent is associated with horizontal motion, and the warm air travels upward at an angle. As a result, the cooling is usually less rapid than in the vertical convection motion.

The boundary between two air masses is called a front and is characterized by varying degrees of precipitation and storms. Fronts are discussed in Chapter 24.

The type of precipitation depends on atmospheric conditions, and it can be in the form of rain, snow, sleet, hail, dew, or frost. **Rain** is the most common form of precipitation in the lower and middle latitudes. The formation of large water droplets which fall as rain has been described previously.

If the dew point is below 32°F, the water vapor freezes on condensing, and ice crystals result that fall as **snow.** In cold regions, these ice crystals may fall individually, but in warmer regions, the crystals become stuck together, forming a snowflake which may be as large as one inch across. Since ice crystallizes in a hexagonal pattern, most snowflakes are six-sided, or hexagonal.

Frozen rain, or pellets of ice in the form of **sleet,** occurs when rain falls through a cold surface layer of air and freezes or, more likely, when the ice pellets fall directly from the clouds without melting before striking the Earth. Large pellets of ice, or **hail,** result from successive vertical descents and ascents in vigorous convection cycles associated with thunderstorms. Additional condensation on successive cycles into supercooled regions that are below freezing may produce layered, structured hailstones the size of baseballs. See Fig. 23.28.

Dew is formed by atmospheric water vapor condensing on the Earth's surface. The land surfaces cool quickly at night and the temperature near the surface may fall below the dew point. Water vapor condenses on the available surfaces such as blades of grass, giving rise to the "early morning dew."

If the dew point is below freezing, the water vapor condenses on objects in the form of frost. The white frost, sometimes referred to as hoar frost, is composed of ice crystals. **Frost** is not frozen dew, but results from the direct change of water vapor into ice.

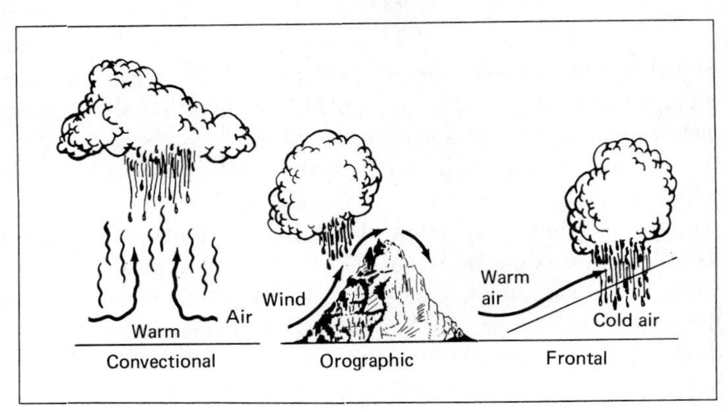

Figure 23.27 Types of air ascension and resulting precipitation. Air ascension may arise from convection heating; motion over some barrier (orographic); or the meeting of air masses having different temperatures, with the warmer, less dense air being displaced upward and over the colder air (frontal).

Figure 23.28 A hailstorm (*Left*) and hailstones (*Below*) that fell in Oklahoma. The successive vertical ascents of ice pellets into supercooled air and regions of condensation produce large, layered "stones" of ice. (Courtesy NOAA.)

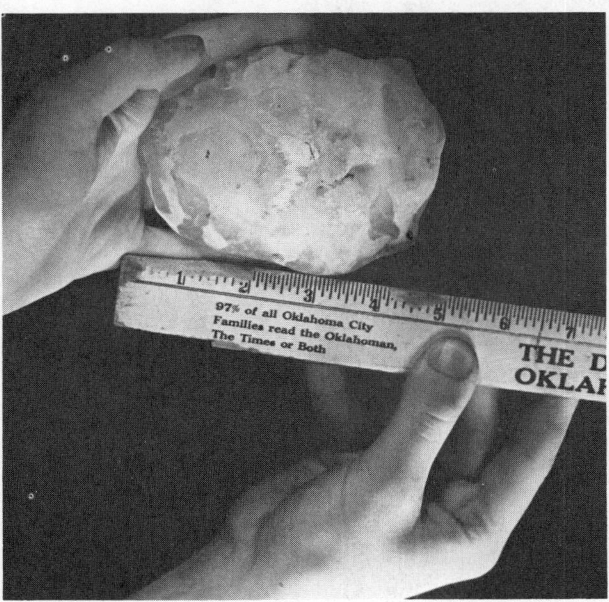

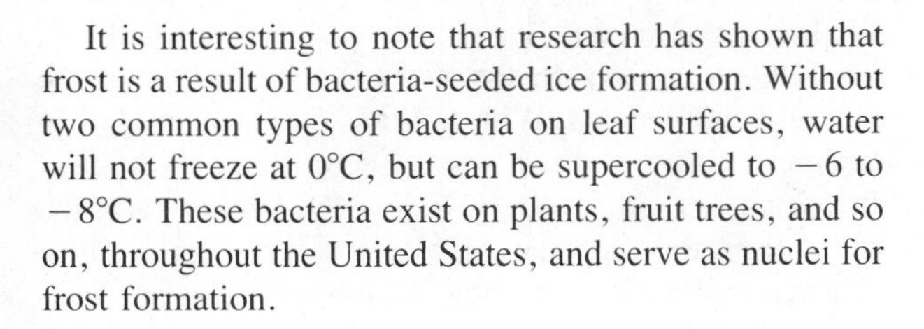

It is interesting to note that research has shown that frost is a result of bacteria-seeded ice formation. Without two common types of bacteria on leaf surfaces, water will not freeze at 0°C, but can be supercooled to −6 to −8°C. These bacteria exist on plants, fruit trees, and so on, throughout the United States, and serve as nuclei for frost formation.

With frost damage to crops and fruits exceeding $1 billion a year, techniques are being explored to scientifically prevent bacteria-seeded frost formation. One method involves the development of genetically engineered bacteria that are altered so they no longer trigger ice formation. What is responsible for the ice-seeding ability is unknown, but scientists have identified the genetic

Figure 23.29 A double rainbow. The fainter, secondary rainbow lies above the primary rainbow. (Courtesy J. Phinney.)

DNA (see Chapter 17) required for ice nucleation and have developed non-ice-forming strains of bacteria.

Field trials of the new bacteria have not been made at the time of this writing, having been blocked by legal action. Some believe that the regular ice-seeding bacteria blown into the atmosphere may be important in precipitating rain and snow. Should the genetically engineered bacteria get into the atmosphere, it is feared that they might alter the climate.

One of the most frequently observed effects accompanying precipitation in the form of rain is the rainbow. The colorful arc of a **rainbow** across the sky is the result of refraction and internal reflection of sunlight by water droplets in the air. As a result of dispersive refraction, the sunlight is separated into its component wavelengths, or colors, by the water droplets in much the same way as a prism.

Often, two rainbows are observed, a primary and secondary, as shown in Fig. 23.29. The primary rainbow results from a single internal reflection in the raindrops and the secondary rainbow from a double internal reflection as illustrated in Fig. 23.30. The colors of the primary rainbow run vertically from violet to red. The secondary rainbow's colors are reversed, running vertically from red to violet.

The observation and length of arc of a rainbow depend on the position of the Sun and the elevation of the observer. The bow is always observed in the portion of the sky opposite the Sun. As illustrated in Fig. 23.31 with the Sun's rays parallel to the Earth's surface at an observer's position O, the primary rainbow is seen above the horizon at an angular altitude between 40° and 42°. The angles depend on the dispersive refraction of the sunlight components. With millions of water droplets in the air, a colorful arc or bow is seen, running vertically in color from violet to red. Below the rainbow arc, the lights from the droplets combine to form a bright, illuminated region. The secondary rainbow occurs at an an-

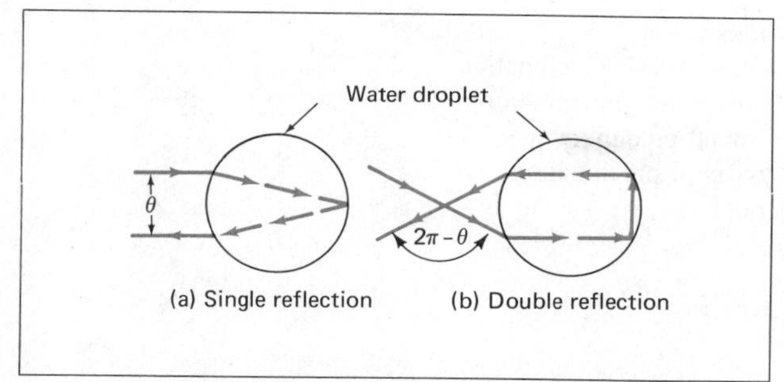

Figure 23.30 A diagram showing the internal reflections in water droplets. A single reflection gives rise to the primary rainbow and a double reflection produces the secondary rainbow, whose colors are inverted.

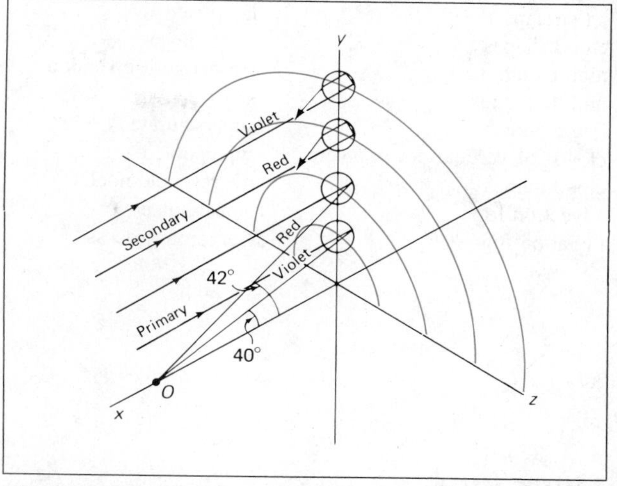

Figure 23.31 Rainbow geometry. An observer at O sees the red light in the primary rainbow from water droplets on an arc at 42° altitude and the violet light on an arc at 40° altitude. The intermediate colors come from intermediate arcs. The secondary rainbow is formed at a higher altitude.

gular altitude between 50.5° and 54°. Tertiary bows are also observed.

With the Sun higher in the sky, the arcs are lower in altitude. In fact, an observer on the ground cannot see a primary rainbow when the Sun's altitude is greater than 42°. A moment's thought should tell you why. This explains why we often do not see a rainbow after a rain when the Sun is behind us. Conversely, if the observer is elevated, more of the arc is seen. It is common to see a completely circular rainbow from an airplane, thus eliminating the possibility of finding a pot of gold at the rainbow's end. Circular rainbows are also commonly seen in the spray of a garden hose.

Learning Objectives

After reading and studying this chapter, you should be able to do the following without referring to the text:

1. Distinguish between winds and air currents.
2. State the forces that produce air motion.
3. Explain the Coriolis force and its effect on air motion around highs and lows.
4. Describe several local winds.
5. Describe the Earth's general air circulation structure.
6. Discuss jet streams.
7. Tell how clouds are classified.
8. Name and describe the various types of common clouds.
9. Explain the mechanisms by which clouds are formed.
10. State the three classifications of precipitation based on air ascension.
11. Describe the different common forms of precipitation.
12. Explain how rainbows are formed.

Important Words and Terms

wind
air current
cloud
gravity
pressure difference
isobar
convection cycle
thermal circulation
Coriolis force
friction
wind-chill factor
sea breeze
land breeze
monsoon
chinook
doldrums
horse latitudes
trade winds
westerlies
polar easterlies

jet streams
cloud shapes
high clouds
middle clouds
low clouds
clouds of vertical development
cloud names
advection fog
radiation fog
lapse rate
stable layer
temperature inversion
smog
supersaturated
supercooled
hygroscopic nuclei
coalescence
Bergeron process
rainmaking
convectional precipitation
orographic precipitation
frontal precipitation
types of precipitation
rainbow

Questions

1. What are the primary and secondary forces of air motion, and what is the basic distinction between the primary and secondary forces?
2. Explain how convection cycles are set up near the Earth.
3. (a) Explain why an object moving along a meridian in the southern hemisphere is deflected to the left. (b) Why is the Coriolis force zero at the equator?
4. Distinguish any differences between cyclones and anticyclones in the northern and southern hemispheres and explain.
5. In what direction does (a) a north wind and (b) a west wind blow?
6. Compare land and sea breezes to valley and mountain breezes. Explain seasonal monsoon effects.
7. What are the doldrums and the horse latitudes?
8. (a) What is the general wind direction for the conterminous United States and why?
   (b) Generally speaking, on which side of town would it be best to build a house so as to avoid smoke and other air pollutants generated in the town?
   (c) Should the prevailing wind direction be of any consideration in the heating plan and insulation of a house?
9. Explain how the wind zones of the Earth's general circulation pattern influenced the exploration and settlement of the Americas.
10. What are jet streams and how do they vary seasonally? Explain this variation relative to the pressure cells of the Earth's general circulation structure.
11. Name the cloud family for each of the following:
    (a) nimbostratus (b) cirrostratus (c) altocumulus (d) cumulonimbus.
12. Name the cloud type associated with each of the following:
    (a) mackerel sky (b) parhelia (c) solar or lunar corona (d) the hazy shade of winter (e) thunderhead.
13. What condition(s) is (are) necessary for cloud formation?
14. What determines the thickness of a cloud of vertical development?
15. Which is heavier, a cubic foot of dry air or a cubic foot of moist air at standard temperature and pressure? (Hint: consider the fact that some of the molecules of the gases in dry air are replaced with molecules of water)
16. How is an inverted lapse rate associated with air pollution conditions?
17. What are the principles and methods of modern rainmaking?
18. What are three types of precipitation based on mechanisms by which air rises?
19. (a) Is frost frozen dew? (b) How are large hailstones formed?
20. (a) What are the colors of the rainbow and how are rainbows formed? (b) Why aren't rainbows seen after every rain?

Problems

1. Determine the wind-chill factor for each of the following: (a) A temperature of 20°F and a wind speed of 20 mi/h. (b) A temperature of 30°F and a wind speed of 25 mi/h.
Answer: (a) −9°F
2. Sketch typical clouds from each of the four cloud families as a function of altitude.
3. If the stationary air temperature near the ground is 65°F, what is the temperature of the air at 4000 ft? Assume a normal lapse rate. *Answer:* 51°F
4. If the stationary ground air temperature is 20°C (68°F), what is the temperature of the air at an altitude of (a) 1 km, and (b) 1 mi? *Answer:* (a) 13.5°C
(b) 49.5°F
5. The stationary ground air temperature is 70°F. What is the air temperature at the top of (a) Pike's Peak (h ≃ 14,000 ft), and (b) Mt. Everest (h ≃ 29,000 ft). Assume a normal lapse rate. *Answer:* (a) 21°F
(b) −31.5°F
6. If the stationary air temperature near the ground is 65°F, what would be the air temperature outside (a) a commercial aircraft at 30,000 ft, and (b) an SST (supersonic transport) at 50,000 ft? Assume a normal lapse rate.
Answer: (a) −40°F
(b) −110°F
7. What is the temperature at the top of the troposphere (h = 16 km), assuming a normal lapse rate and a stationary ground air temperature of 25°C? *Answer:* −79°C
8. What is the normal lapse rate in °F per mile?
Answer: 18.5°F/mi
9. If the stationary ground temperature is 67°F, above what altitude will the clouds be in the form of ice crystals?
Answer: 10,000 ft
10. Clouds are observed at an altitude of 2500 m. If the stationary ground air temperature is 20°C, would you expect the clouds to be water droplets or ice crystals? Justify your answer.
11. At what altitude would the air temperature be −32°C if on a particular day the stationary ground air temperature is 20°C. Assume a normal lapse rate. *Answer:* 8 km
12. Recall that the Celsius and Fahrenheit temperatures are equal at −40°. Assuming a normal lapse rate, at what altitude would this occur if the stationary ground air temperature were 20°C (68°F)? Give answer in both kilometers and miles. *Answer:* 9.2 km
5.8 mi
13. How many 10-μm droplets would be needed to form a raindrop 2 mm in diameter? *Answer:* 8×10^6

24

Air Masses and Storms

And pleas'd the Almighty's orders to perform,
Rides in the whirlwind and directs the storm.
–Addison, *The Campaign*

IF THE AIR of the troposphere were static, there would be little change in the local atmospheric conditions that constitute the weather. The dynamics of air movement in the form of winds have been discussed in Chapter 23. In this chapter, we shall consider the movements of large masses of air that have distinguishing physical characteristics.

The air we now breathe may have been far out over the Pacific Ocean a week ago. As air moves into a region, it brings with it the temperature and humidity that are mementos of its origin and travel. Cold, dry arctic air may cause sudden drops in the temperatures of the regions in its path. Warm, moist air from the Gulf of Mexico may bring heat and humidity to some regions and make the summer seem unbearable. Thus, moving air transports the physical characteristics that produce changes and influence the weather. A large mass of air can influence the weather of a region for a considerable period of time or have only a brief effect. The movement of air masses depends a great deal on the Earth's air circulation structure and its seasonal variations.

When air masses meet, the variations of their properties may trigger storms along their boundary. The types of storms depend on the properties of the air masses. Also, local variations within an air mass can give rise to storms. Storms can be violent and sometimes destructive. They remind us of the vast amount of energy contained in the atmosphere and also of its capability. The variations of our weather are therefore closely associated with air masses and their movements and interactions.

◀ Flooding from a tropical storm.

24.1 Air Masses

As we know, the weather varies and changes a great deal. However, we often experience several days of relatively uniform weather conditions. Our general weather conditions depend in large part on large air masses that move across the country.

When a body of air takes on physical characteristics that distinguish it from the surrounding air, it is referred to as an **air mass.** The main distinguishing characteristics are temperature and moisture content. A mass of air remaining for some time over a particular region takes on the physical characteristics of the surface of the region. So an air mass forms, usually near large bodies of land or water, with rather uniform properties that are mainly dependent on the uniform surface over which it is found.

The region in which an air mass derives its characteristics is called its **source region.** The time required for an air mass to take on its source region's characteristics depends on the surface of the region. If the surface area is a warm body of water, convection currents mix the air and general uniformity is attained relatively quickly. In this case, the resulting air mass is warm and moist. However, if the surface area is a cold land mass, the air is more stable. In this situation, a longer time is required to take on the characteristics of a cold, dry air mass.

An air mass eventually moves from its source region, bringing its characteristics and a change in weather to the regions in its path. As an air mass travels, its properties may become modified due to local variations. For example, if Canadian polar air masses did not become warmer as they traveled southward, Florida might experience extremely cold winters.

Table 24.1 Air Masses that Affect the Weather of the United States

Classification	*Symbol*	*Source Region*
Maritime Arctic	mA	Arctic regions
Continental Arctic	cA	Greenland
Maritime Polar	mP	Northern Atlantic and Pacific Oceans
Continental Polar	cP	Alaska and Canada
Maritime Tropical	mT	Caribbean Sea, Gulf of Mexico, and Pacific Ocean
Continental Tropical	cT	Northern Mexico, Southwestern United States

Whether an air mass is termed cold or warm is relative to the surface over which it moves. Quite logically, if an air mass is warmer than the surface, it is referred to as a **warm air mass.** If colder, it is called a **cold air mass.** It should be remembered that this is a relative designation. The cold and warm prefixes do not always imply cold and warm weather. For example, a cold air mass in winter usually brings cold, frigid weather. However, the weather associated with a cold air mass of 70°F (21°C) traveling over an 80°F (27°C) surface may be quite pleasant. Such a cold air mass may bring welcome relief in the hot summer months.

Air masses are classified according to the surface and general latitude of their source regions:

Surface	*Latitude*
Maritime (m)	Arctic (A)
Continental (c)	Polar (P)
	Tropical (T)
	Equatorial (E)

The surface of the source region is abbreviated by a small letter and gives an indication of the moisture content of an air mass. An air mass forming over a body of water (maritime) would naturally be expected to have a greater moisture content than one forming over land (continental).

The general latitude of the source region is abbreviated by a capital letter and gives an indication of the temperature of an air mass. For example, mT designates a maritime tropical air mass. A list of the air masses that affect the weather of the United States is given in Table 24.1, along with their source regions, which are illustrated in Fig. 24.1.

The movement of air masses is influenced to a great extent by the Earth's general circulation patterns discussed in Chapter 23. Since the United States lies predominantly in the Westerlies zone, the general movement of air masses, and hence the weather, is from west to east across the continent. The circulation zones vary to some degree in latitude with the seasons, and the polar

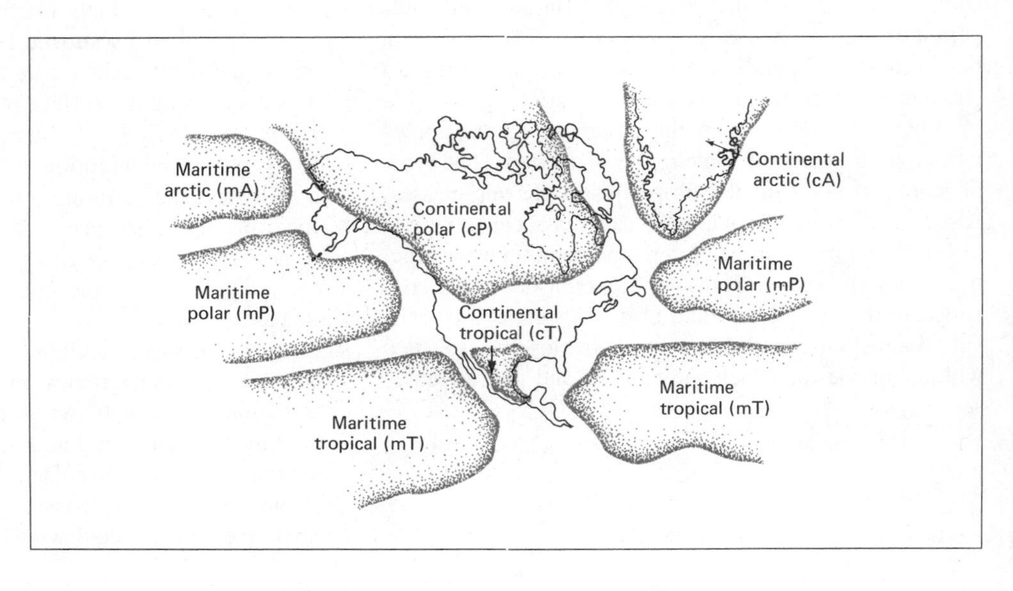

Figure 24.1 Source regions of air masses for North America.

easterlies may also move air masses into the United States during the winter.

Air masses are huge in magnitude and cover thousands of square miles. Their movement is the prime factor in the redistribution of moisture and heat energy. A maritime tropical air mass formed in the Gulf of Mexico, for example, may bring warm, wet weather along the eastern coast of the United States into New England.

24.2 Fronts and Cyclonic Disturbances

As defined previously, the boundary between two air masses is called a **front.** A **warm front** is the boundary of an advancing warm air mass over a colder surface and a **cold front** the boundary of a cold air mass moving over a warmer surface. These boundaries, called **frontal zones,** may vary in width from a few miles to a wide zone of over 100 miles.

It is along these fronts, which divide air masses of different physical characteristics, that drastic changes of weather occur. Turbulent weather and storms usually define a front. When this turbulent weather extends for some distance horizontally, it is referred to as a **squall line.** A squall line is shown in Fig. 24.2. The series of storms that form along a front are referred to as line squalls.

The degree and rate of weather change depend on the difference in temperatures of the air masses and the vertical slope of the front. A cold front moving into a warmer region causes the lighter, warm air to be displaced upward over the front. The lighter air of an advancing warm front cannot displace the heavier, colder air as readily and may move slowly over the colder air.

Cold air is associated with high pressure, and the downward divergent air flow in a high-pressure region in general gives a cold front a greater speed than a warm front. A cold front may have an average speed of 20 to 25 mi/h, while a warm front moves 10 to 15 mi/h. The

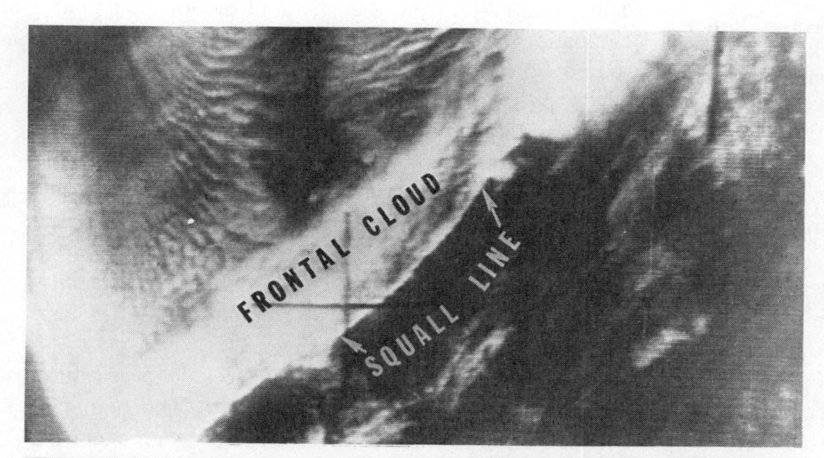

Figure 24.2 (*Top*) Squall line as seen from a satellite and (*below*) at the Earth's surface. (Courtesy NOAA.)

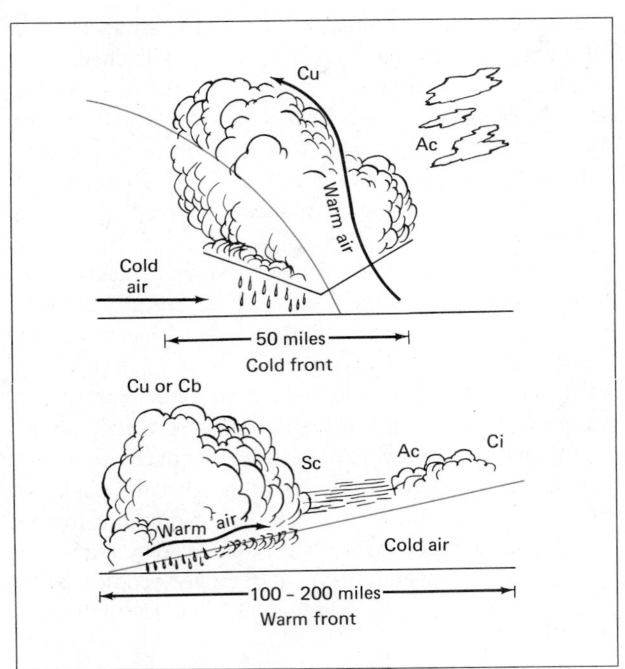

Figure 24.3 Diagrams of cold and warm fronts. Notice in the upper diagram the sharp, steep boundary (in color) characteristic of a cold front. The boundary of a warm front, as shown in the lower diagram, is less steep. As a result of this difference, different cloud types are associated with the approach of the fronts.

features of cold and warm fronts are illustrated in Fig. 24.3.

Cold fronts have sharper vertical boundaries than warm fronts, and warm air is displaced faster by an advancing cold front. As a result, cold fronts are accompanied by more violent or sudden changes in the weather. The sudden decrease in temperature is often describe as a "cold snap." Dark altocumulus clouds mark the cold front's approach. The sudden cooling and rising warm air may set off rain- or snowstorm activity along the front. The effects of an advancing cold front are described in the newspaper report below.

A warm front may also be characterized by precipitation and storms. Since its approach is more gradual it is usually heralded by a period of lowering clouds. Cirrus and mackerel scales drift ahead of the front, followed by the alto clouds. As the front approaches, cumulus or cumulonimbus clouds resulting from the rising air produce precipitation and storms. Most of this occurs before the front passes.

Storms are also associated with cyclonic, or rotational, disturbances along active fronts. **Wave cyclones** are formed by air moving in opposite directions along a front. The development of a wave cyclone is illustrated in Fig. 24.4. The graphic symbol for a cold front is ▲▲▲▲▲▲▲ and for a warm front ◠◠◠◠◠◠◠. The side of the line with the symbols indicates the direction of advance.

Friction and shear forces between the moving air masses cause a variation from their parallel motions and start a circulation between them. This action may be demonstrated by holding a pencil between the palms of the hands. When the palms are moved in opposite directions, a rotational motion of the pencil is produced. The analogous frontal motion is shown in Fig. 24.4, (a′), (b′), and (c′). Accompanying this frontal motion is the rising of warm air, and a low-pressure area develops at the crest of the "wave" as illustrated in Fig. 24.4, (d′). Air moving into this low-pressure region is deflected by the Coriolis force, producing cyclonic motion (cf. Chapter 23).

As a faster moving cold front advances, it may overtake a warm air mass and push it upward. The boundary between these two masses is referred to as an **occluded**

Cool Air Triggers Storms In Mid-U.S.

By THE ASSOCIATED PRESS

Cool air rolling across the nation's midsection dropped temperatures to record mid-June lows in the Plains today and triggered severe storms from Texas to Michigan.

Six young people were missing off lower Michigan's eastern shore after winds up to 65 miles an hour, and heavy rains, hit the Saginaw Bay area of Lake Huron. The group went out in a 13-foot boat Thursday afternoon, apparently to water ski.

Wind, hail and drenching rains were widespread along the storm belt.

Temperatures skidded 20 to 25 degrees in an hour in some areas as the cool front passed. Nearly 4 inches of rain soaked the Talpa, Tex., area, 40 miles south of Abilene. Almost 2½ inches poured into Nashville, Ill., about 50 miles southeast of St. Louis.

Small hailstones piled to a depth of 2 to 3 inches during a thunderstorm at Ardmore, Okla., late Thursday. Hailstones the size of golfballs smashed out windows in the downtown section of Winters, Tex., 45 miles northeast of Abilene.

A tornado swooped into the community of Oak Creek Lake, north of San Angelo, Tex., and overturned four house trailers. No injuries were reported.

West of the storm belt, 3 inches of wet snow fell on Helena, Mont., Thursday evening. Flurries dusted scattered areas of the northern Rockies before daybreak.

The temperature fell to 42 at North Platte, Neb., before midnight, a record low for June 12 there.

Frost or freeze warnings were in effect from Montana to northern Minnesota. Lewistown, Mont., chilled down to 29 well before dawn.

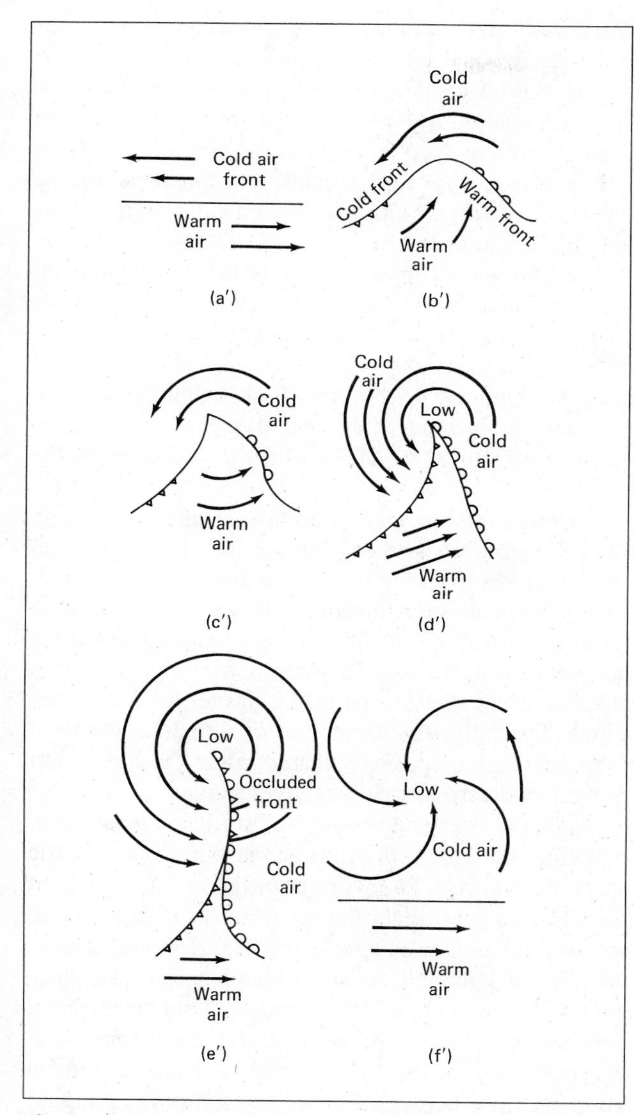

Figure 24.4 Top views of wave cyclone development. Antiparallel wind motion produces the rotational motion for the development of wave cyclones (a′). An occluded front occurs when one front overtakes another (e′). As a result of the rotational action, a low-pressure wave cyclone spins off (f′).

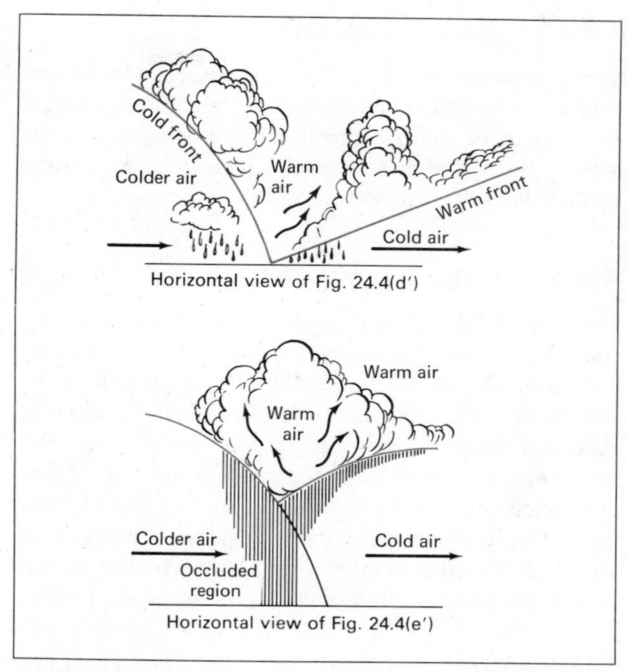

Figure 24.5 Occluded front development. The drawings show ground-level views of parts (d′) and (e′) of Fig. 24.4. The advancing colder air overtakes the retreating cold air and forces the intervening warm air upward, occluding it from the Earth's surface.

front (▲●▲●▲●). The cold front occludes, or closes off, the warm air along the occluded front. Horizontal views of Fig. 24.4 (d′) and (e′), are shown in Fig. 24.5. There are two types of occluded fronts. When the advancing cold air is colder than the air ahead of it as in Fig. 24.5, the occluded front is called a **cold front occlusion.** That is, a cold front advances under a warm front. If the situation is reversed, and the air ahead is colder than the advancing air, the occluded front is referred to as a **warm front occlusion.** In this case, a warm front advances up and over a cold front.

As the cyclonic wave fully develops, a low-pressure area or storm center spins off. See Fig. 24.4 (f′). The low moves away, carrying with it rising air currents, clouds, precipitation, and generally bad weather. Lows, or cyclones, are, therefore, generally associated with poor weather. Lows may be of a local nature or cover hundreds of square miles, depending on the extent and conditions of the frontal activity.

Highs, or anticyclones, on the other hand, are generally associated with good weather. Developing in regions of divergent air flow, highs are relatively stable due to their descending air motion. As a result, dry air and a lack of precipitation usually characterize a high. There are exceptions, however, as the characteristics of a high depend on the conditions under which it develops.

Because of their influence on the weather, the movements of highs and lows are closely observed. Opposing fronts may balance each other so that no movement occurs. This case is referred to as a **stationary front** (▲●▲●▲●).

24.3 Local Storms

Storms are atmospheric disturbances associated with vertical air motion that may develop locally within a single air mass or be due to frontal activity between two air masses. Several of these storms, which are distinguished by their intensity and violence, will be considered.

Rainstorms

The rate of rainfall is dependent upon the rate of the condensation and coalescence process discussed in Section 23.6. The duration of a rainfall is dependent on the amount of water vapor available. Should air rise rapidly, condensation and coalescence may occur quickly. The strong updraft will suspend the water droplets until they have coalesced into large drops. If there is a considerable amount of water vapor present, a heavy downpour of rain, called a **rainstorm,** results. Otherwise, the rainfall may be just a sudden shower that begins and ends quickly. Storms with rainfalls of one to two inches per hour are not uncommon.

When the convection currents cease or the cloud has an insufficient moisture content, the rainfall stops. However, if water vapor is continuously fed into the precipitating cloud, a prolonged, heavy rainfall occurs. This condition is referred to as a **cloudburst,** and the water runoff may produce a **flash flood.**

When wind accompanies a heavy rain, the storm may become violent with blowing ''sheets'' of rain. It beats against windows and quickly fills storm sewers. Wind damage in the form of broken tree limbs and downed power lines is common.

Thunderstorms

The **thunderstorm** is a rainstorm distinguished by thunder and lightning and sometimes hail. Thunderstorms may result from frontal cyclonic disturbances or strong local heating. Strong updrafts of air that produce the rain cloud are essential for thunderstorm formation.

In the upper portions of the resulting cumulonimbus cloud, or thunderhead, the water vapor becomes supercooled. With proper nucleation, raindrops form and fall; however, some may evaporate before reaching the ground. This evaporation results in a cooling of the surrounding air, and the cool air descends, producing downdrafts. These spread out along the ground and account for the cooling and the winds associated with thunderstorms. This development is illustrated in Fig. 24.6. Hail may result from successive ascents and descents of raindrops if the supercooled region is below freezing.

The **lightning** of a thunderstorm is a discharge of electrical energy. The electrical nature of lightning was demonstrated by Ben Franklin in his famous kite-flying key experiment. As he flew a kite in a thunderstorm, he drew sparks with his knuckles from a key hanging at the end of the kite cord.

Franklin was extremely lucky in the performance of this experiment in that it was a miracle that he was not electrocuted. Under no circumstances should duplication of this experiment be attempted. People have been killed trying to duplicate it—not only from lightning but from a modern hazard not extant in Franklin's day—contact of the conducting kite string with high-voltage electric lines.

It is believed that a raindrop has an outer, negatively charged, electric sheath. When a raindrop is broken up or partially evaporates, negatively charged droplets become separated from the drop, leaving the remainder positively charged. By this process or other processes not yet fully understood, a separation of charge and an electric potential develops in the turmoil of a thundercloud. Typically, the upper part of the cloud carries a preponderance of positive charge while the lower part of the cloud carries a net negative charge.

When the electric potential is of sufficient magnitude, lightning is produced in the same manner that an electric spark arcs between two wires of different electric potentials. Air is a poor conductor, but when the charge buildup or potential is great enough, the electric force ionizes the air and lightning occurs. Lightning can take place entirely within a cloud (intracloud or cloud discharges), between two clouds (cloud-to-cloud discharges), between a cloud and the Earth (cloud-to-ground or ground discharges), or between a cloud and the surrounding air (air discharges). Lightning has even been reported to occur in clear air, apparently giving rise to the expression ''a bolt from the blue.'' When it occurs below the horizon or behind clouds, lightning often illuminates the clouds with flickering flashes. This commonly occurs on a still summer night and is known as **heat lightning.**

Although the most frequently occurring form of lightning is the intracloud discharge, of major concern is the lightning between a cloud and the Earth, which is a huge reservoir of charges (Fig. 24.7). With the bottom of a thundercloud generally being negatively charged, this causes an opposite charge to be induced on the Earth's surface and an electric potential to cause the electrical breakdown of the air. Each lightning stroke begins with

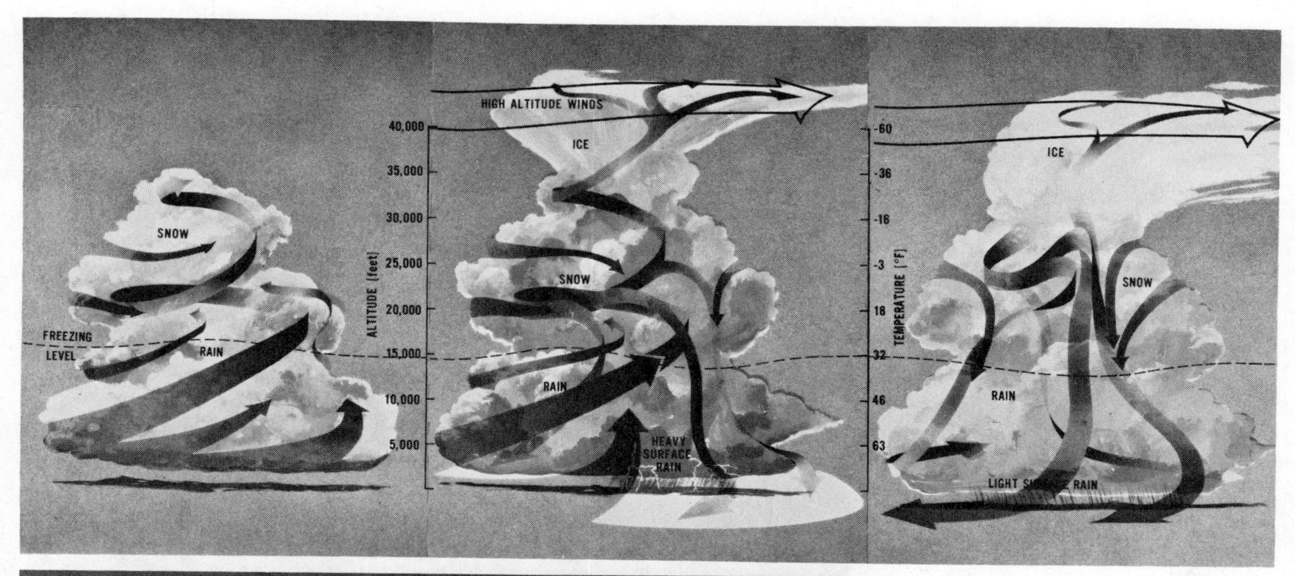

Figure 24.6 Thunderstorm development. Strong vertical air motion is essential for thundercloud development. High-altitude winds often form an anvil cumulonimbus near the stable layer. (Courtesy NOAA.)

Figure 24.7 Lightning strokes between clouds and the Earth. (Courtesy NOAA.)

a weakly luminous (nearly invisible) downward discharge called a *leader*. When this makes contact with the positive charges near the Earth, large numbers of electrons flow to the ground. This is followed immediately by the bright lightning flash or *return stroke* propagating upward as a result of electron flow from the Earth to the cloud along the original ionized path. We visually perceive the stroke to be heading downward.

The shorter the distance from the cloud to the ground, the more easily the electric discharge takes place. This is why lightning often strikes trees and high buildings. It is inadvisable, therefore, to take shelter from a thunderstorm under a tree. A person in the vicinity of a lightning strike may experience an electric shock that causes breathing to fail. In this case, mouth-to-mouth resuscitation or some other form of artificial respiration should be given immediately and the person kept warm as treatment for shock. (See boxed feature on lightning safety.)

Since barns and other tall buildings are the occasional targets of lightning, a common practice to avoid lightning damage is the installation of lightning rods. A lightning rod consists of a metal rod that extends above the structure and is connected by a conductor to the Earth. It is hoped that the higher lightning rod will make contact with the discharge leader instead of the building and conduct it harmlessly to Earth.

The old cliché that lightning never strikes twice in the same place is practically true, as there is usually nothing left of that spot for the lightning to strike again. Actually, lightning often strikes more than once in the same spot. The Empire State Building is struck by lightning on the average more than 20 times a year, but it is protected by an elaborate lightning rod system.

Lightning Safety

If you are outside during a thunderstorm and feel an electrical charge, as evidenced by hair standing on end or skin tingling, what should you do? Fall to the ground fast! Lightning may be about to strike. Statistics show that lightning kills on the average 200 people a year in the United States and injures another 550. Most deaths and injuries occur at home. Indoor casualties occur most frequently when people are talking on the telephone, working in the kitchen, doing laundry, or watching TV. During severe lightning activity, the following safety rules are recommended.

Stay indoors away from open windows and fireplaces, and electrical conductors such as sinks and stoves. Do not use electrical plug-in equipment such as radios, TVs, and lamps. Also, avoid using the telephone. Lightning may strike the telephone lines outside.

Should you be caught outside, seek shelter in a building. If no buildings are available, seek protection in a ditch or ravine.

Bolts of lightning have tremendous energies. It is estimated that the temperature in the vicinity of a lightning flash is of the order of 15,000–30,000°C. This is quite large compared to the Sun's surface temperature of 6000°C. The awesome electrical properties of lightning become clear when you compare it with ordinary household electricity, which has up to 240 volts and 100–200 amperes of current available at the main service panel. A typical lightning discharge has from 10–100 million volts and up to 300,000 amperes of current. It can leap up to a mile or more.

A lightning flash's sudden release of energy explosively heats the air, producing the compressions we hear as **thunder.** When heard at a distance of about 100 m (33 ft) or less from the discharge channel, thunder consists of one loud bang or "clap." When heard at a distance of 1 km (0.62 mi) from the discharge channel, thunder generally consists of a rumbling sound punctuated by several large claps. In general, thunder cannot be heard at distances of more than 25 km (16 mi) from the discharge channel. Presumably, the loud bang of thunder heard when one sees a nearby lightning flash is due to the strong sound wave from the channel base. For an observer at a distance of 1 km from the lightning channel, the initial loud bang is refracted overhead due to temperature variations of the air, and the discharge begins with a rumble. At distances greater than 25 km, the sound is refracted above the observer.

The association of thunder with lightning was postulated very early. In his famous treatise *Meteorologica,* Aristotle presented the general idea, but in a somewhat incorrect, reversed fashion. He thought that thunder resulted when heat condensed from clouds

> . . . moving violently and striking against the surrounding clouds delivers a blow, whose noise is called thunder. . . . Besides, the pressed out wind is usually consumed by a delicate and gentle flame, and this is what we call lightning. . . . It is produced after the blow and thus also after the thunder; but it seems to be produced before the thunder because seeing precedes hearing. We have clear proof that seeing precedes hearing when we observe the oarsmen of a trireme, for as they draw back their oars for a second stroke we hear the sound of their first stroke.

Taking the proper sequence of lightning and thunder, since lightning flashes generally occur near the storm center, the resultant thunder provides a method of easily approximating the distance to the storm. Light travels at approximately 186,000 mi/s, and the lightning flash is seen without any appreciable time lapse. Sound, however, travels at approximately $\frac{1}{5}$ mi/s, and a time lapse occurs between seeing the lightning flash and hearing the thunder. This phenomenon is also observed when watching someone at a distance fire a gun at night or hit a baseball. The report of the gun or the "crack" of the bat is always heard after the flash of the gun is observed or the baseball is well on its way. Seeing does precede hearing as Aristotle thought. By counting the seconds between seeing the lightning flash and hearing the thunder, an estimate of the distance of the storm center may be obtained. For example, if 5 s elapsed, then the storm center would be approximately 1 mi away, taking the velocity of sound as $\frac{1}{5}$ mi/s; if 8 s elapsed, the storm would be $1\frac{3}{5}$ mi away.

Ice Storms and Snowstorms

When a warm air mass overrides a cold air mass as in Fig. 24.3, rain may form. The rain falls to Earth through the underlying cold air. If the temperature of the Earth's surface is below 32°F and the raindrops do not freeze before striking the Earth, the rain will freeze on the cold objects on which it falls. A layer of ice builds up on the objects exposed to the freezing rain. The resultant glaze is referred to as an **ice storm.** The ice layer may build up to over one-half inch in thickness, depending on the magnitude of the rainfall (Fig. 24.8). Viewed in sunlight, the ice glaze produces a beautiful winter scene with the coated landscape glistening in the Sun.

Figure 24.8 Results of an ice storm. Rain freezes on objects when temperatures are below 32°F (0°C), causing damage due to the added weight of the ice. (Courtesy NOAA.)

Other aspects of an ice storm are not so beautiful. Damage can be quite severe. Tree branches and power lines snap and fall from the weight of their ice coatings; animals may become frozen statues; transportation by foot or vehicle becomes extremely hazardous; and injuries usually result.

Snow is made up of ice crystals that fall from ice clouds. A **snowstorm** is an appreciable accumulation of snow. What may be considered a severe snowstorm in some regions might be thought of as a light snow in others where snow is more prevalent. For example, 4–6 inches of snow can paralyze areas like New York City and Washington, D.C.; however, in areas like upstate New York, Michigan, and Wisconsin, this accumulation would be just an incidental snowfall.

When a snowstorm is accompanied by high winds and low temperatures, the storm is referred to as a **blizzard.** The winds whip the fallen snow into blinding swirls. Visibility may be reduced to a few inches. For this reason, a blizzard is often called a *blinding* snowstorm. The swirling snow causes a loss of one's sense of direction and people have become lost within a few feet of their homes. The wind may blow the snow across level terrain, forming huge drifts against some obstructing object as shown in Fig. 24.9. Drifting is common on the flat prairies of the western United States.

Snowstorms without appreciable winds may be equally severe due to huge quantities of snow. A single inundation of snow in western New York in 1966 measured nearly 100 in. The region to the south and east of the Great Lakes is often termed a snow belt because of the large quantities of snow it receives, especially in the early part of winter. The lakes are still comparatively warm at this time of the year and dry continental polar

Figure 24.9 A snowstorm often leaves huge quantities of snow, while accompanying high winds may blow the snow against objects forming deep drifts. (Courtesy North Dakota Highway Department.)

Weather Records

To give you an idea of weather extremes, here are a few recorded values from around the world and in the United States.

Temperature

High: 136°F (58°C), El Aziza, Libya, September 13, 1922
134°F (57°C), Death Valley, California, July 10, 1913
Low: −129°F (−89°C), Vostok, Antarctica, July 21, 1983
−80°F (−62°C), Prospect Creek, Alaska, January 23, 1971
−70°F (−57°C), Rogers Pass, Montana, January 20, 1954
Drop: 100°F in one day (44°F to −56°F), Browning, Montana, January 23–24, 1916
Rise: 49°F in 2 minutes (−4°F at 7:30 A.M. to 45°F at 7:32 A.M.), Spearfish, South Dakota, January 22, 1943

Precipitation

Rain

1 minute: 1.23 in (3.1 cm), Unionville, Maryland, July 4, 1956
24 hours: 74 in (188 cm), Cilaos La Reunion, Indian Ocean, March 15–16, 1952
43 in (109 cm), Alvin, Texas, July 26, 1979
one month: 366 in (930 cm), Cherrapunji, India, July 1981
one year: 1042 in (2647 cm), Cherrapunji, India, August 1860–July 1861
171 months (> 14 years): 0.03 in (0.08 cm), Arica, Chile, October 1903–January 1918

Snow

24 hours: 76 in (193 cm), Silver Lake, Colorado, April 14–15, 1921
one week: 189 in (480 cm), Mt. Shasta Ski Bowl, California, February 13–19, 1959
one month: 390 in (991 cm), Tamarack, California, January 1911
one season: 1122 in (2850 cm), Paradise Ranger Station, Mt. Rainier, Washington, 1971–1972
greatest depth on ground: 451 in (1146 cm), Tamarack, California, March 11, 1911

air masses moving in from the northwest pick up large quantities of water from evaporation.

Orographic ascension occurs as the air masses move eastward over the elevated land area around the lakes. Hence, the moist air masses are cooled and the conditions are favorable for snow. During the particularly severe winter of 1977, Buffalo received a total of nearly 200 inches of snow. As the winter progresses, the water of the lakes cools and evaporation is less, thus reducing the probability of snowstorms.

For some extreme weather conditions see the boxed feature on weather records above.

Tornadoes

The **tornado** is the most violent of storms. Although it may have less *total* energy than other storms (see Table 24.2), the concentration of its energy in a relatively small region gives this storm its violent distinction. Characterized by a whirling, funnel-shaped cloud that hangs

Table 24.2 Comparison of Average-Storm Characteristics

	Cyclones	*Hurricanes*	*Tornadoes*	*Thunderstorms*
Wind speed	0–50 mph	74–200 mph	100–300 mph	20–30 mph
Width	500–1000 mi	300–600 mi	$\frac{1}{8}$ mi	1–2 mi
Duration	Week or more	Week	Few minutes	An hour or less
Approx. kinetic energy, based on wind velocity (joules)[a]	10^{14}	10^{15}	10^{10}	10^{9}

[a]Energy of atomic bomb ≈ 10^{14} joules

from a dark cloud mass, the tornado is commonly referred to as a **twister.** See Fig. 24.10.

Tornadoes occur around the world, but are most prevalent in the United States and Australia. In the United States, most tornadoes occur in the Deep South and in the broad, relatively flat basin between the Rockies and the Appalachians. But no state is immune (Fig. 24.11).

Figure 24.10 Tornado funnel and damage. The high-speed winds and sudden pressure drop accompanying the awesome-looking tornado produce much damage along its path. (Courtesy NOAA.)

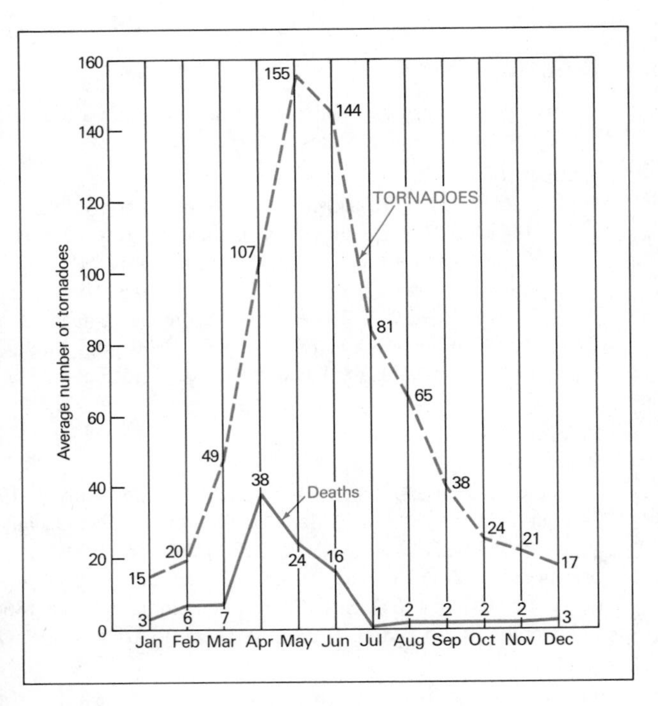

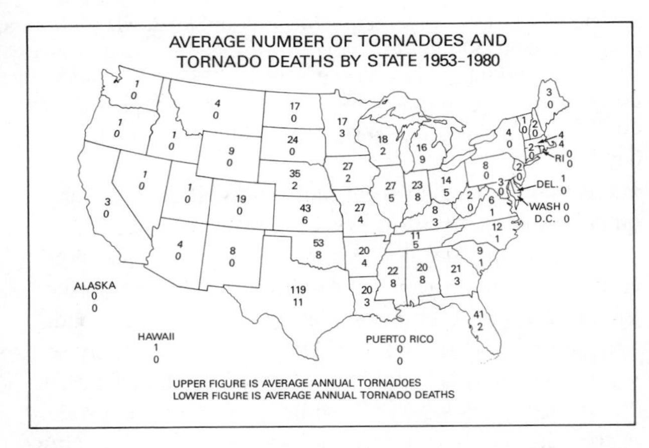

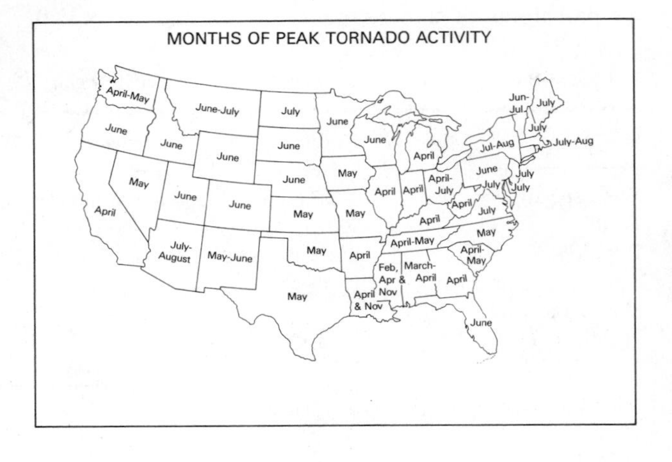

Figure 24.11 Tornado statistics in the United States, 1953–1980. (a) A graph of the average tornado incidence by month. (b) Maps showing months of peak tornado activity and annual average of number of tornadoes and deaths by state. (Courtesy NOAA.)

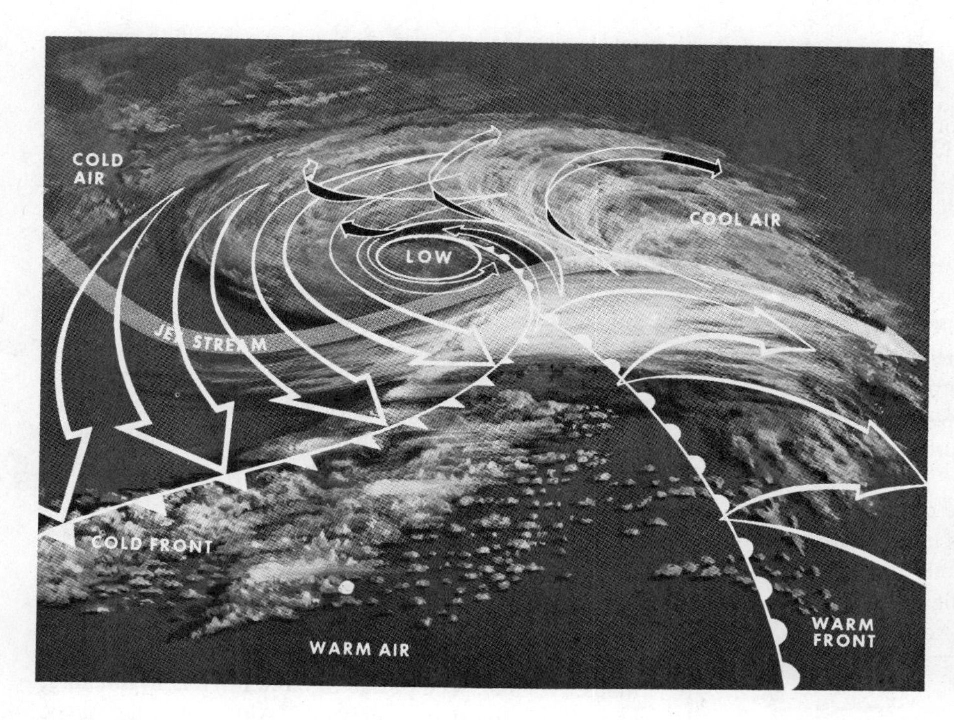

Figure 24.12 Conditions favorable for tornado formation. Cyclonic motion associated with frontal movement and jet-stream activity combine to produce tornadoes. (Courtesy NOAA.)

Peak months of activity are April, May, and June with southern states usually hit hardest in winter and spring, northern states in spring and summer. However, tornadoes have occurred in every month at all times of the day or night. A typical time of occurrence is on an unseasonably warm, sultry spring afternoon between 3 and 7 P.M.

Most tornadoes travel from southwest to northeast, but their direction of travel can be erratic and may change suddenly. They travel at an average speed of 30 mi/h, but speeds ranging from 0–70 mi/h have been reported. Tornadoes are classified into three general categories—*weak, strong,* and *violent*. Most (over 60%) of the tornadoes that occur each year fall into the weak category, with wind speeds of 100 mi/h or less. About one out of every three (30%) tornadoes is classified as strong. Wind speeds reach about 200 mi/h and strong tornadoes have an average path length of 9 mi and a width of 200 yd. Only about 2% are violent tornadoes. These extreme tornadoes can last for hours, have an average path length of 26 mi and a width of 425 yd. The largest of these may exceed a mile or more in width, with wind speeds approaching 300 mi/h.

The complete mechanism of tornado formation is not known, due to its many variables. One essential component, however, is rising air, which occurs in thunderstorm formation and in the meeting of cold and warm air masses. This condition is frequently fulfilled during the spring and summer in the midwestern United States where tornadoes are common. In the case of air masses, warm air is forced upward when it collides with a cold air mass. The rapid rising of the warm air, coupled with the frictional and Coriolis forces acting on moving cooler air, produces a circulating spiral. Intersecting jet streams are also thought to contribute to the vortex formation as shown in Fig. 24.12.

As the ascending air cools, clouds are formed that are swept to the outer portions of the cyclonic motion and outline its funnel form. Since clouds form at certain heights (Section 23.5), the funnel may appear well above the ground. Under the proper conditions, a full-fledged tornado develops. The winds increase and the air pressure near the center of the vortex is reduced as the air swirls upward. When the funnel is well developed, it may "touch down," or be seen extending up from the Earth as a result of dust and debris picked up by the whirling winds.

The alerting system for tornadoes has two phases. A **tornado watch** is issued when atmospheric conditions are ripe for tornadoes, and a **tornado warning** is issued when a tornado has actually been sighted or indicated on radar. The similarity between the terms watch and warning is sometimes confusing. Remember that you should *watch* for a tornado when the conditions are right, and when given a *warning,* the situation is dangerous and critical—no more watching.

Despite radar, satellites, and other sophisticated instruments, most tornado reports are made by volunteer spotters who relay the word of sightings to official warning centers. It is hoped that more advanced warning of these dangerous storms will be available with the installation of new Doppler radar systems (see Chapter 25).

While an individual tornado usually destroys a relatively small area, major tornado outbreaks may cause widespread damage over a more extensive area. During the afternoon and evening of April 3 and the early morning of April 4, 1974, a super outbreak of 148 tornadoes occurred across 13 states (Fig. 24.13). More than 300 people were killed and 6000 injured. The property damage exceeded $600 million.

Most of the injuries and some of the damage done by a tornado come from flying debris in the high-speed winds. For many years, it was incorrectly believed that buildings literally exploded when hit by a tornado. It was reasoned that the low-pressure tornado would cause a sudden outside pressure drop and the resulting pressure difference with the normal pressure inside a structure would push the walls outward. It was recommended that windows be opened on the approach of a tornado to equalize the pressure and prevent damage.

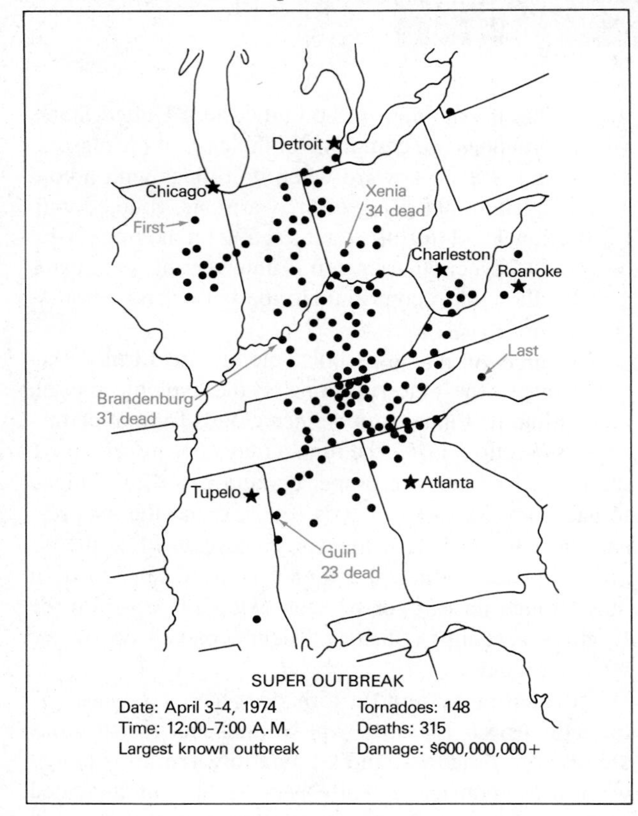

Figure 24.13 Super tornado outbreak. During the afternoon and evening of April 3 and the early morning of April 4, 1974, a series of 148 tornadoes occurred across 13 states.

However, research has shown that buildings are far from airtight even with the windows and doors closed. There are enough cracks and holes for pressure equalization. What happens in the simplest case is that the tornado's winds push the windward wall inward. The high-speed air flow over the roof lifts the roof off (much like the lift when air flows over an airplane wing), and the other walls fall outward as a result of the winds (Fig. 24.14). It is now recommended that the windows be forgotten and that you seek shelter. Flying glass from shattered windows could cause serious injury. Moreover, at night or during heavy rain, the only clue to a tornado's presence may be its roar and the time spent in opening windows may be critical in terms of seeking shelter. (See boxed feature on tornado safety.)

The wind damage accompanying a tornado is enormous. A wind speed of 250 or more mi/h is frightening, as well as almost inconceivable. Along with the havoc and destruction caused by the wind come stories of remarkable occurrences of a less serious nature. Chickens have been picked up by tornadoes, plucked clean, and redeposited otherwise unharmed. Livestock and barns have been lifted over fences into a neighbor's field. Newspapers have been driven deep into the wood of trees. Documented feats of tornadoes are endless.

A tornado occurring over water is known as a **waterspout.** Although the funnel appears to be a spout

Tornado Safety

Knowing what to do in the event of a tornado is critically important. If a tornado is sighted, if the ominous roar of one is heard at night, or if a tornado warning is issued for your particular locality, seek shelter fast! (Do not bother opening the windows. See text material.)

The basement of a home or building is one of the safest places to seek shelter. *Avoid* chimneys and windows. Get under a sturdy piece of furniture, such as an overturned couch, or in a stairwell or closet, *and cover your head.*

In a home or building without a basement, seek the lowest level in the central portion of the house and the shelter of a closet or hallway.

If you live in a mobile home, *evacuation is a must.* Seek shelter elsewhere.

Figure 24.14 The "lifting" effect of tornadoes. The roofs of buildings are forced upward, or "lifted," during a tornado and leeward walls are blown outward. (Courtesy NOAA.)

of water rising up from the water surface, the upper portions of the waterspout are chiefly composed of whirling clouds like that of a land tornado. Dust devils, or whirling columns of dust that are common to dry areas, are sometimes looked upon as mini-tornadoes, but they are not.

24.4 Tropical Storms

Tropical storms, or tropical cyclones, are more violent than the cyclones discussed previously (see again Section 24.2). A tropical storm becomes a **hurricane** when its wind speed exceeds 74 mi/h (64 knots). The hurricane is known by different names in different parts of the world. In southeast Asia, it is called a **typhoon.** In the region of the Indian Ocean **cyclone** is the term used. A particularly destructive cyclone hit Bangladesh in 1985 with the loss of thousands of lives.

Off the Australian coast, the tropical cyclone is a **willy-willy** (Fig. 24.15). Regardless of the name, the hurricane is characterized by high-speed rotating winds, the energy of which is spread over a large area. A hurricane may range from 300 to 600 mi in diameter and have wind speeds of 74–200 mi/h.

Hurricanes form over the tropical oceanic regions where the Sun heats huge masses of moist air. An ascending spiral motion results, in the same manner as described in tornado formation. When the moisture of the rising air condenses, the latent heat provides additional energy and more air rises up the column. This latent heat is a chief source of the hurricane's energy and is readily available from the condensation of the moist air of its source region. The growth and development of a hurricane is shown in the satellite photograph of Fig. 24.16.

Unlike the tornado, a hurricane gains energy from its source region. As more and more air rises, the hurricane grows, accompanied by clouds and increasing winds that blow in a large spiral around a relatively calm, low-pressure center—the eye of the hurricane (Fig. 24.17). The eye may be 20 to 30 mi wide, and ships sailing into this area have found that it is usually calm and clear with no indication of the surrounding storm. The air pressure is reduced 6–8% (to about 28 in of Hg) near the eye. Hurricanes move rather slowly at a few miles per hour.

Covering broad areas, hurricanes can be particularly destructive. There are winds of at least 74 mi/h, but these can be much greater, up to 120–130 mi/h, which are *very* dangerous. Mobile homes are particularly vulnerable to hurricane winds. The greatest threat from a

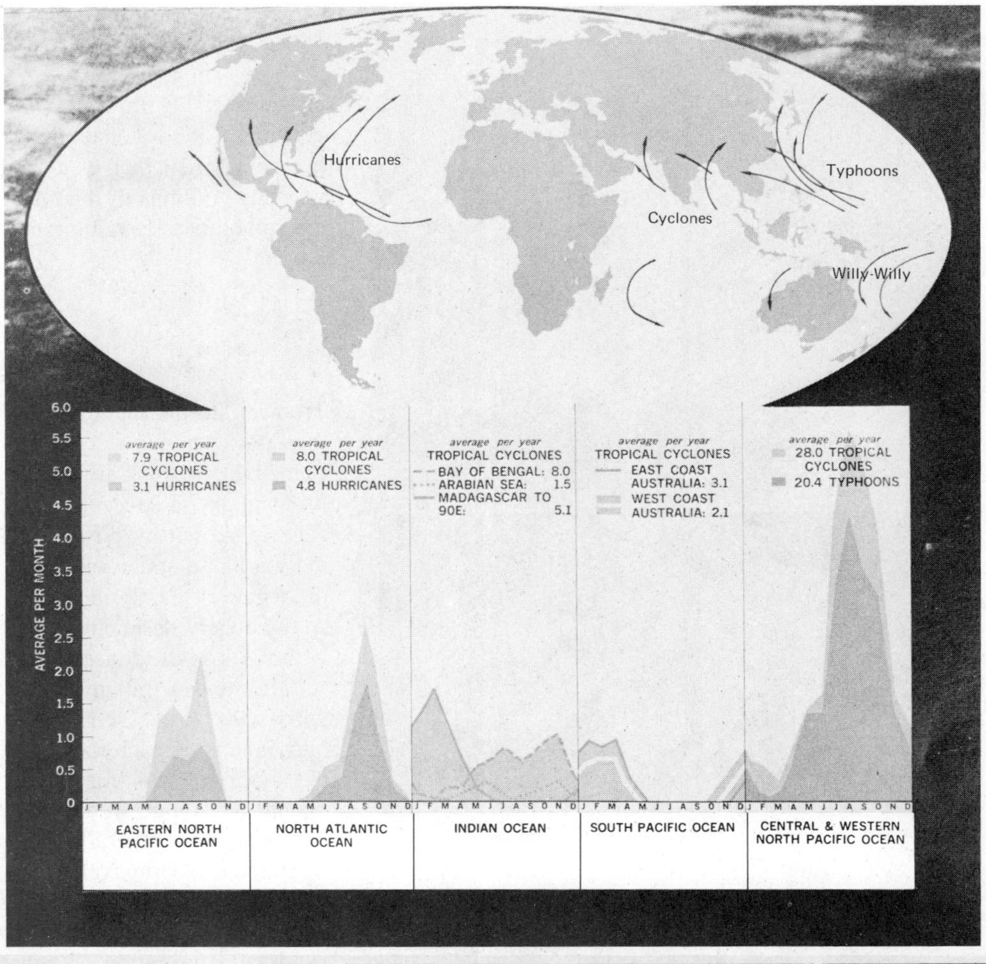

Figure 24.15 Tropical storm regions of the world. Tropical storms are known by different names in different parts of the world. (Courtesy NOAA.)

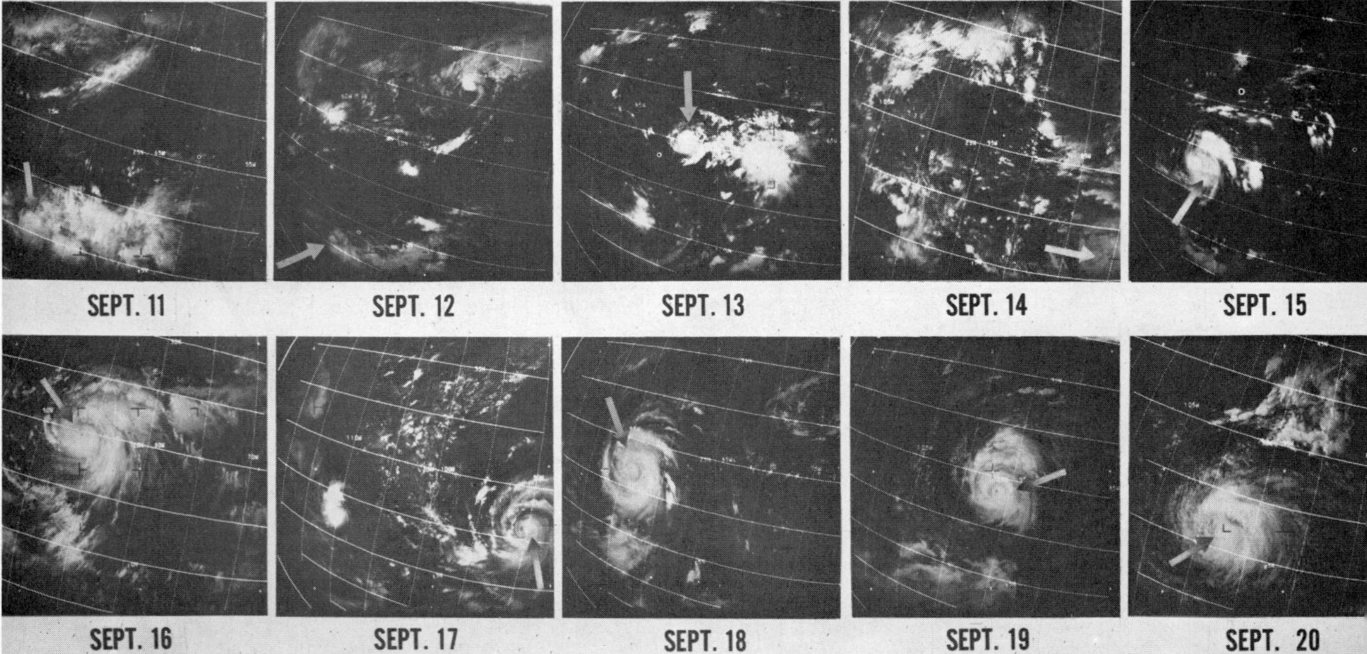

Figure 24.16 The development of a hurricane as photographed by a satellite. (Courtesy NOAA.)

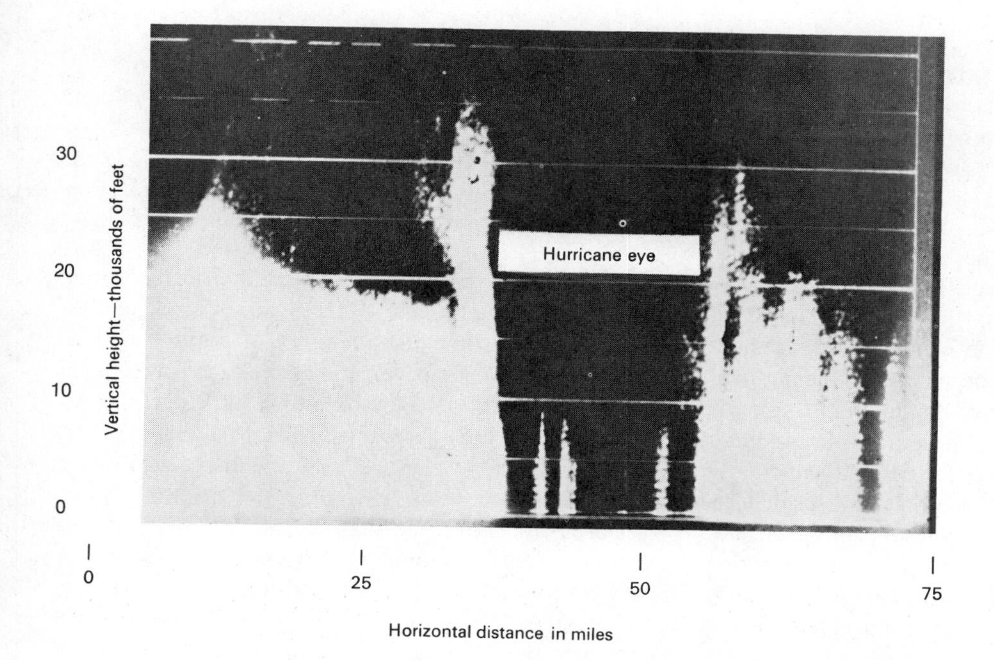

Figure 24.17 A hurricane's eye as seen by satellite and radar. The lower photo is a radar profile of a hurricane. Note the generally clear eye at about 35 to 55 mi from the radar station and the vertical buildup of clouds (over 35,000 ft) at the near side of the eye. (Courtesy NOAA and NASA.)

hurricane's winds comes from their cargo of debris—a deadly barrage of flying missiles such as lawn furniture, signs, roofing, and metal siding.

Hurricane winds do much damage, but drowning is the greatest cause of hurricane deaths. As the eye of the hurricane comes ashore or "makes landfall," a great dome of water called a storm surge, often over 50 mi wide, comes sweeping across the coast line. It brings huge waves and storm tides that may reach 25 ft or more above normal (Fig. 24.18). The rise may come rapidly, flooding coastal lowlands. Nine out of ten hurricane casualities are caused by the storm surge. The torrential rains that accompany the hurricane produce sudden flooding as the storm moves inland. As its winds diminish, rainfall floods constitute the hurricane's greatest threat.

Once cut off from the warm ocean, the storm begins to die, starved for water and heat energy, and dragged apart by friction as it moves over the land. Even though a hurricane weakens rapidly as it moves inland, the remnants of the storm can bring 6–12 in. of rain or more to the areas they cross. For example, Hurricane Diane of 1955 caused little damage as it moved into the Gulf coastal area; but long after its winds subsided, it brought floods to Pennsylvania, New York, and New England that killed 200 persons and cost an estimated $700 million in damage. In 1972, Hurricane Agnes fused with another storm system, flooding creek and river basins in the Northeast with more than a foot of rain in less than 12 hours, killing 117 people, and causing almost $3 billion damage.

Tropical storms form in the regions between 5° and 20° latitude in the late summer (Fig. 24.19). Below 20° latitude the stabilizing influence of the high-pressure horse latitudes is small and large quantities of moisture are evaporated from the warm tropical oceans. Thus, in these regions the conditions for hurricane formation are optimum.

The breeding grounds of the hurricanes that mostly affect the United States are in the Atlantic Ocean southeast of the Caribbean Sea. As the hurricane forms, it moves westward with the trade winds, traveling about 100–200 miles a day. During the summer in the northern hemisphere, the northern shift of the doldrums region is accompanied by the exchange of air from the southern to the northern hemisphere over the Mid-Atlantic.

The Coriolis force acts on this air movement, deflecting it toward the northwest. As a result, the hurricanes move away from the equator and into the Gulf of Mexico as they travel westward. Monsoon type effects may also have some influence on the hurricane movement toward land. Those hurricanes that do strike the United States usually hit along the Gulf and south Atlantic coasts.

Figure 24.18 Hurricane damage from wind and flooding. The high winds of the hurricane produce waves that do much damage along coastal areas. (Courtesy NOAA.)

These areas may be lashed by several hurricanes during the late summer hurricane season. Once over a land mass and deprived of its moist air source of energy, the hurricane dissipates. However, if the hurricane proceeds far enough northward, it may be blown back into the Atlantic by the prevailing westerlies that lie above 30° N. As a result, the coastal area may be hit by a hurricane both coming and going. (See Fig. 24.19.)

During the hurricane season, the area of their formation is constantly monitored by satellite. Because of the long time of development and migration, the hurricane requires a great deal of observation. Radar-equipped airplanes, or "hurricane hunters," seek out and track tropical storms and hurricanes. Prior to the advent of satellites, hurricanes were almost exclusively monitored by aircraft and surface ships. This is a dangerous business. However, to be able to predict a hurricane's behavior, on-the-spot weather data not furnished by satellites, such as air pressure and wind velocity, are essential.

These observations permit advanced hurricane warnings along coastal areas, which save lives and reduce property damage. The course of a hurricane is constantly observed, as it may suddenly change its path and strike elsewhere with seemingly malicious intent.

Like that for the tornado, the hurricane alerting system has two phases. A **hurricane watch** is issued for a coastal area when there is a threat of hurricane conditions within 24–36 hours. This does not mean that hurricane conditions are a certainty, but a watch does alert those people who would be affected so they can be prepared to act quickly. A **hurricane warning** indicates that hurricane conditions are expected within 24 hours and appropriate precautionary actions should be taken.

For several hundred years, many hurricanes in the West Indies were named after the particular saint's day on which the hurricane occurred. In 1953, the National Weather Service began to use female names for tropical storms and hurricanes. This practice began in World War II when military personnel named typhoons in the western Pacific after their wives and girl friends. Under this method the storms were named in alphabetical order. The first hurricane of a season received a name beginning with A, such as Alma; the second one was given a name beginning with B, such as Beulah; and so on.

The practice of naming hurricanes solely after women was changed in 1979 when men's names were included in the lists. A five-year list of names for Atlantic storms is given in Table 24.3. A similar list is available for eastern Pacific storms. Names beginning with the letters Q, U, X, Y, and Z are excluded because of their scarcity. The name lists have an international flavor because hurricanes affect other nations and are tracked by the public and the weather services of countries other than the United States. The lists are recycled. For example, the 1987 list will be used again in 1992.

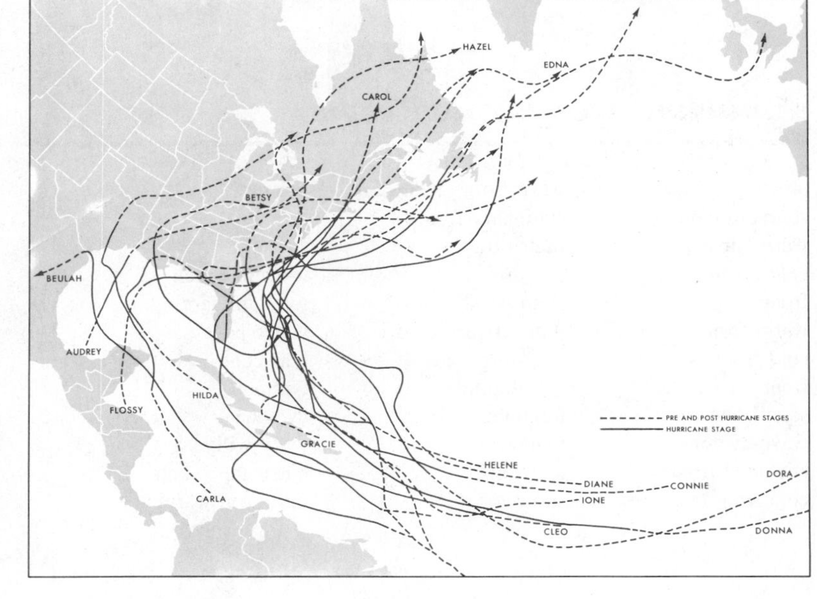

Figure 24.19 Hurricane paths. Hurricanes that form in the Atlantic travel northwest until they are blown eastward by the prevailing westerlies. As a result, these hurricanes often strike the coastal areas of the United States both coming and going. The satellite photograph above shows two hurricanes bearing down on the North American continent. One is entering the Gulf of Mexico on a straight line to the Texas coast, while the other is in the Pacific Ocean south of Baja California, Mexico. (Courtesy NOAA.)

Table 24.3 The Five-Year List of Names for Atlantic Storms*

1987		*1988*		*1989*		*1990*		*1991*	
Arlene	Lenny	Alberto	Leslie	Alicia	Luis	Ana	Larry	Allen	Lisa
Bret	Maria	Beryl	Michael	Barry	Marilyn	Bob	Mindy	Bonnie	Mitch
Cindy	Nate	Chris	Nadine	Chantal	Noel	Claudette	Nicolas	Charley	Nicole
Dennis	Ophelia	Debby	Oscar	Dean	Opal	David	Odette	Danielle	Otto
Emily	Philippe	Ernesto	Patty	Erin	Pablo	Elena	Peter	Earl	Paula
Floyd	Rita	Florence	Rafael	Felix	Roxanne	Frederic	Rose	Frances	Richard
Gert	Stan	Gilbert	Sandy	Gabrielle	Sebastien	Gloria	Sam	Georges	Shary
Harvey	Tammy	Helene	Tony	Hugo	Tanya	Henri	Teresa	Hermine	Tomas
Irene	Vince	Isaac	Valarie	Iris	Van	Isabel	Victor	Ivan	Virginia
Jose	Wilma	Joan	William	Jerry	Wendy	Juan	Wanda	Jeanne	Walter
Katrina		Keith		Karen		Kate		Karl	

*The lists are recycled. For example, the 1987 list will be used again in 1992.

Learning Objectives

After reading and studying this chapter, you should be able to do the following without referring to the text:

1. Distinguish the different air masses, particularly those that affect the weather in the United States.
2. Describe the different kinds of fronts and their associated characteristics.
3. Explain the formation of a wave cyclone.
4. Describe the formation and characteristics of a thunderstorm.
5. Explain the cause of ice storms.
6. Discuss the formation and effects of tornadoes.
7. Distinguish between tropical storms and hurricanes and describe their formation and effects.

Important Words and Terms

air mass
source region
warm air mass
cold air mass
front
warm front
cold front
frontal zone
squall line
wave cyclones
occluded front
cold front occlusion
warm front occlusion
stationary front
rainstorm
cloudburst
flash flood
thunderstorm
lightning
heat lightning
lightning rods
thunder
ice storm
snowstorm
blizzard
tornado
twister
tornado watch
tornado warning
waterspout
hurricane
typhoon
cyclone
willy-willy
hurricane watch
hurricane warning

Questions

1. How are air masses classified? Explain the relationship between air mass characteristics and source regions.
2. What air masses affect the weather in the United States?
3. What would be the abbreviated designation of an air mass that developed over Siberia? over northern Africa? over the Arabian Sea?
4. What is a front? List the meteorological symbols for four types of fronts.
5. Describe the characteristics and weather associated with cold and warm fronts. What is the significance of the sharpness of their vertical boundaries?
6. Assume a low-pressure air mass approaches and advances over your location. Give the variations in barometric readings and winds you would expect to occur.
7. Explain why the household barometer often lists descriptive adjectives, such as stormy, unsettled, and fair, on the barometer's face instead of direct pressure readings.
8. How do conditions for lightning discharges occur?
9. What type of first aid should be given to someone suffering from the shock of a lightning stroke?
10. What is thunder and why is it sometimes heard as a crash and other times as a rumble?
11. An ice storm is likely to result along what type of front? Explain.
12. What is the most violent of storms?
13. Describe the formation and characteristics of a tornado.
14. Distinguish between a tornado watch and a tornado warning.
15. For some years we have been told to open the windows of a house on the approach of a tornado. Now we are told to forget the windows and seek shelter. Why the change?
16. When a tornado approaches, what should the residents do in (a) a house with a basement, (b) a house without a basement, (c) a mobile home?
17. What is the major source of energy for a tropical storm? When does a tropical storm become a hurricane?
18. How are tropical storms and hurricanes named?

Problems

1. Locate the source regions for the following air masses that affect the United States: (a) cA (b) mP (c) cT (d) mT (e) cP
2. What would be the classifications of air masses forming over the following areas? (a) Sahara Desert (b) Antarctic Ocean (c) Greenland (d) Mid-Pacific Ocean (e) Siberia
3. About how far does the average cold front travel in one day? *Answer:* 500 mi
4. About how far does the average warm front travel in one day? *Answer:* 250 mi
5. Sketch top and horizontal views of a warm front occlusion.
6. Using the ideal gas law (Chapter 5), approximate the increase in pressure in the vicinity of a lightning flash. (Hint: assume a constant volume.) *Answer:* A factor of 50
7. If thunder is heard 4 s after the observation of a lightning flash, approximately how far away is the storm center? *Answer:* 0.8 mi
8. While picnicking on a summer day, thunder is heard 11 s after a lightning flash is seen from an approaching storm. Approximately how far away is the storm? *Answer:* 2.2 mi
9. How far does an average tornado travel in 30 min? *Answer:* 15 mi
10. Using the average speed of tornadoes and the average path length of a strong tornado, compute the touch-down time. *Answer:* 18 min
11. If a violent tornado moved with an average speed of 50 mi/h, what would be its touch-down time with an average path length? *Answer:* 31 min
12. Estimate how far a hurricane moves in one day. *Answer:* Approximately 100–120 mi

25

Weather Forecasting

When ye see a cloud rise out of the west,
straight way ye say
There cometh a shower, and so it is
And when ye see the southwind blow, ye say
There will be heat; and it cometh to pass.

–Luke 12:54–55

HUMAN BEINGS HAVE long desired to control the weather. However, little progress has been made toward achieving this desire, and as the next best thing, the preoccupation of observing and attempting to predict the weather developed. The influence of the weather on everyday activity has grown to the extent that this preoccupation has become the sole occupation of many modern meteorologists who are known as weather forecasters.

Weather forecasters are much maligned. When the weather fails to follow the forecast, they are the target of many unflattering remarks. Bad weather that spoils a planned outing is blamed on the forecaster's shortsightedness, if not on the forecaster. Although meteorologists are scapegoats, they do a commendable job and on the average are correct about 80% of the time.

The accurate prediction of atmospheric behavior is a difficult task. Changes in the weather are governed by scientific principles. However, the atmosphere is complex and contains so many variables that meteorologists must combine empirical and scientific knowledge in making reasonable forecasts. Weather data are collected and processed by the most modern means. From the current weather conditions, future conditions are projected according to established behavior patterns. This chapter will deal with a few of the many facets involved in weather forecasting.

25.1 The National Weather Service

The United States **National Weather Service** (formerly the Weather Bureau) is the federal organization that provides national weather information. It forms part of the National Oceanic and Atmospheric Administration (**NOAA**—pronounced Noah), which was created within the U.S. Department of Commerce in 1970.

The original U.S. Weather Bureau grew out of the Army Signal Corps, which maintained an early telegraph system that was used for weather reporting. It became a part of the Department of Agriculture in 1891, and then part of the Department of Commerce in 1940. The Weather Bureau's early activities were primarily directed toward weather forecasting as an aid to agriculture. However, the greatest impetus for the Bureau's growth came with the development of aviation, for which up-to-date weather reports are vital. Today, the Weather Bureau is known as the National Weather Service, and its activities are concerned with every phase of the weather.

The nerve center of the National Weather Service is the **National Meteorological Center** (NMC) located just outside Washington, D.C., in Suitland, Maryland. It is the NMC that receives and processes the raw weather data taken at numerous weather stations. Each day the NMC receives:

12,000	reports from surface stations
25,000	hourly surface aviation reports
1,500	reports from ships at sea
1,500	upper-air reports
400	weather reconnaissance reports
2,500	aircraft reports
500	radar reports

All available cloud and temperature data from weather satellites.

The NMC analyzes and makes forecasts from the received data. Other central National Weather Service organizations deal with specialized weather conditions, such

◀ Tracking hurricane Gloria.

as hurricanes, tornadoes, and severe storms. From these central organizations, data go to Forecast Offices, which have the responsibility for warnings and forecasts for states, or large portions of states, and assigned zones. State forecasts are issued twice daily for a time period up to 48 hours. An extended outlook, up to five days, is issued daily for these same areas. The forecast organizational structure is shown in Fig. 25.1.

Weather Service Offices represent the third echelon of the forecast system (Fig. 25.2). Local forecasts are adaptations of state forecasts. They are issued to meet local requirements and to provide general weather information to the public. The local forecasts are distributed by telephone services, direct radio and television broadcasts, including VHF-FM (very high frequency–frequency modulated) radio stations near major population centers that transmit weather information continuously 24 hours a day.

Let's take a closer look at how forecasts are made and distributed. The operations of the NMC are almost completely automated, and data are handled by high-speed computers, printers, and communication systems. Analyses are made twice daily at 0000 hours and 1200 hours Greenwich Mean Time (7 A.M. and 7 P.M. EST). A preliminary analysis is made after an hour and a half of data reception. An operational analysis is made after three and a half hours of data collection, when 80% of all incoming data have been received. The computer takes a half hour to analyze the information and the analysis is completed before the data are four hours old. NMC estimates the same analysis would take five people working eight hours to complete. Manual analyses are made on certain portions of the data as a check on the computer as shown in Fig. 25.3.

Once the analyses are completed, three somewhat different simulations of the weather begin. The simulations use the basic laws of physics and some statistics to calculate values of temperature, wind, humidity, and rain at equally spaced locations over the entire globe and at a number of different altitudes. The results are sent out, almost untouched by human hands, to National Weather Service field offices and other users for guidance in making forecasts. Quality control is done in the field. The forecaster must estimate possible errors and use the results which are best for his or her purpose with other knowledge to make forecasts.

The distribution phase has been done by the teletype circuits that brought in the data. Facsimile circuits transmit charts and 2500 weather maps (Fig. 25.4). However, the National Weather Service is currently installing a revolutionary new data-handling system known as AFOS (*A*utomation of *F*ield *O*perations and *S*ervices). The AFOS system uses minicomputers and TV type displays in a network of more than 200 automated weather offices.

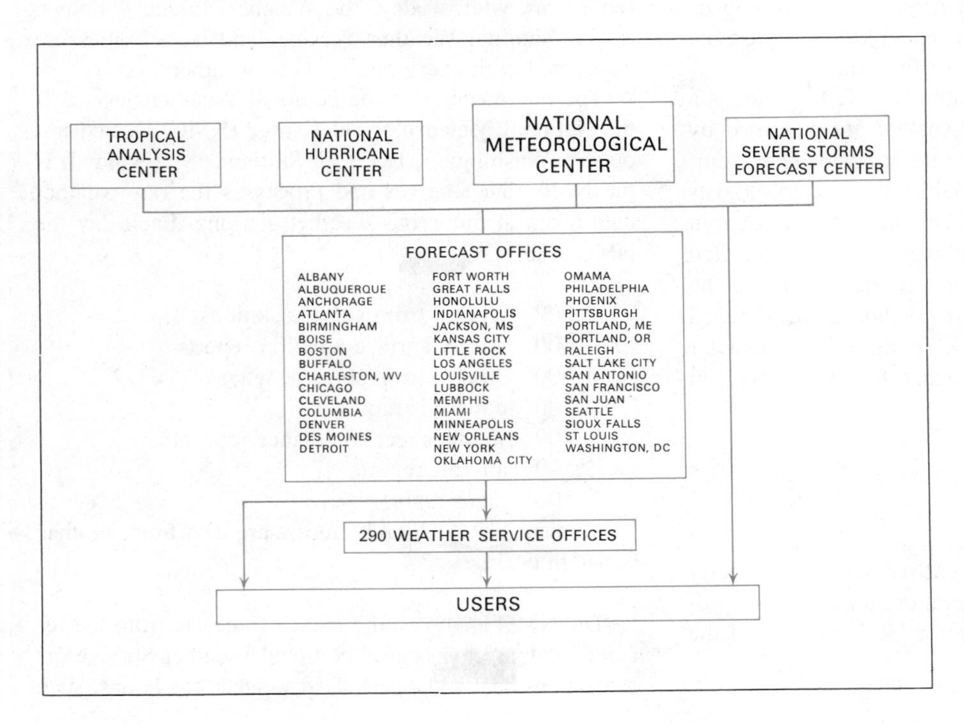

Figure 25.1 Structure of the National Weather Service's forecast organization. (Courtesy NOAA.)

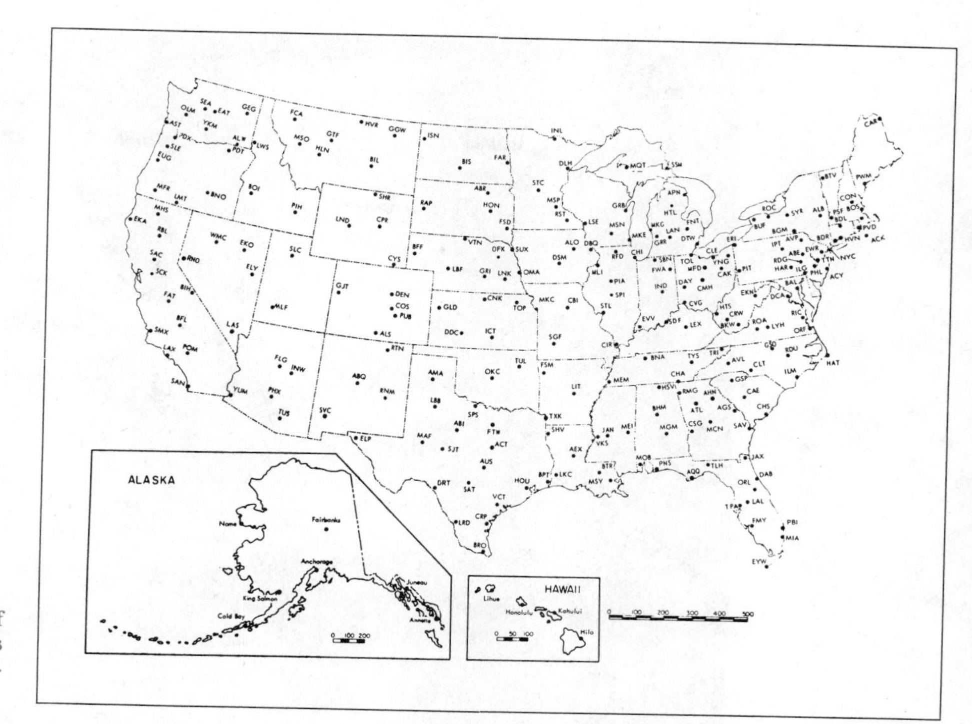

Figure 25.2 The locations of offices in the United States issuing local weather forecasts. (Courtesy NOAA.)

Figure 25.3 Manual analyses are done on portions of the weather data as a check on the computer. (Courtesy NOAA.)

Figure 25.4 (*Top*) Facsimile map transmission. Facsimile transmission of weather maps is done on teletype circuits. (*Bottom*) AFOS displays. The new AFOS system uses minicomputers and TV type displays in a network of more than 200 automated weather offices. (Courtesy NOAA.)

The network has four loop circuits in different regions of the country and centrally linked to the NMC.

AFOS will do away with the older system of teletypewriters and facsimile machines and the enormous quantities of paper they generate and will substitute an all-electronic system in which weather information from the minicomputer systems will be displayed on video screens (Fig. 25.4). No longer will it be necessary for forecasters to spend vast amounts of time tearing off, sorting, and posting paper teletype messages and paper maps. A displayed weather map will arrive on station in about one-fortieth the time it takes for a paper—15 seconds instead of 10 min. Messages will arrive 30 times as fast—3000 words per minute instead of 100.

Even as AFOS is being installed, plans are being made for a more advanced system that will integrate weather information, processing, and display. The present AFOS system was not designed to take advantage of data from satellites and the relatively new Doppler radar, which will be discussed in Section 25.2.

The National Weather Service is dedicated to getting the weather information to the public by the fastest means available. However, there is an exception to the prompt distribution of weather information. In wartime, weather data are of great military value and become classified information. During World War II weather reports were not made public except in cases of severe weather, in which warnings were necessary to prevent property damage and loss of life. The news media were prohibited from mentioning anything that pertained to the weather. Farmers were simply told it would be a good day to make hay or a rained-out ball game was called off due to conditions beyond control. As one sportscaster put it, "Well folks, I can't say anything about the weather, but that isn't perspiration on the pitcher's face."

As technology increases and facilities expand, better weather forecasts become available. It is estimated that the predictions of the National Weather Service meteorologists are about 80% accurate. This, of course, depends on the weather element and location. For example, predictions of no rain in parts of Arizona during the summer are almost 100% correct.

The method of reporting the weather has also changed over the years. In the past, forecasts of precipitation were given using such terms as probable, likely, or occasional. Precipitation forecasts are now expressed in "percent probabilities," such as a 70% chance of rain. The percent probability, given in such a forecast, results from a consideration of two quantities: (1) the probability that a precipitation-producing storm will develop in, or move into, the forecast area; and (2) the percentage of the area that the storm is expected to cover. Thus, in summer when storms (convectional) tend to be more isolated, or scattered, the probability that your immediate area will get rain tends to be less than in the winter when frontal storms are more prevalent. The percentage probabilities give the public a better indication of what the weather might be.

Although many hours of work and endeavor go into making the daily weather forecasts as accurate as possible, they sometimes fail to hold true. But after all, if the forecasts were always correct, wouldn't it take a bit of the fun out of life?

25.2 Data Collection and Weather Observation

In 1870 weather data were taken by the Army Signal Corps at 24 stations. Today, weather observations are collected from approximately 1000 land stations, six fixed ocean stations, and several hundred merchant vessels of all nationalities. Many of the land stations are associated with airport operations. Volunteer observers also supply climatology data from about 12,000 substations around the country. Weather stations are concerned with the basic meteorological measurements of temperature, pressure, humidity, precipitation, and wind direction and speed, as discussed in Chapter 22. The instruments used are also basically the same, but have official specifications and in some cases may be automated to supply continuous readings.

A typical installation for data collection is shown in Fig. 25.5. The small hut to the left is an instrument shelter. It is constructed with louvered sides, a ventilated floor, and a double roof with an air space between. This construction permits air measurements to be taken free from insolation influences. The **instrument shelter** is designed to hold a thermograph, a maximum-minimum thermometer, and a sling psychrometer. A thermograph is an automatic temperature-recording device. A hydrothermograph is sometimes available and records the temperature as well as the relative humidity by some hygrometric means, thus eliminating the need for the psychrometer. The relative humidity and temperature measurements are taken at a specified distance of six feet above ground level.

The maximum and minimum daily temperatures are important weather observations. These may be measured with a set of maximum-minimum thermometers.

Figure 25.5 A typical installation for the collection of weather data. Shown are an instrument shelter (left), containing temperature and humidity measuring devices; a wind vane and anemometer mount; and a rain gauge. (Courtesy NOAA.)

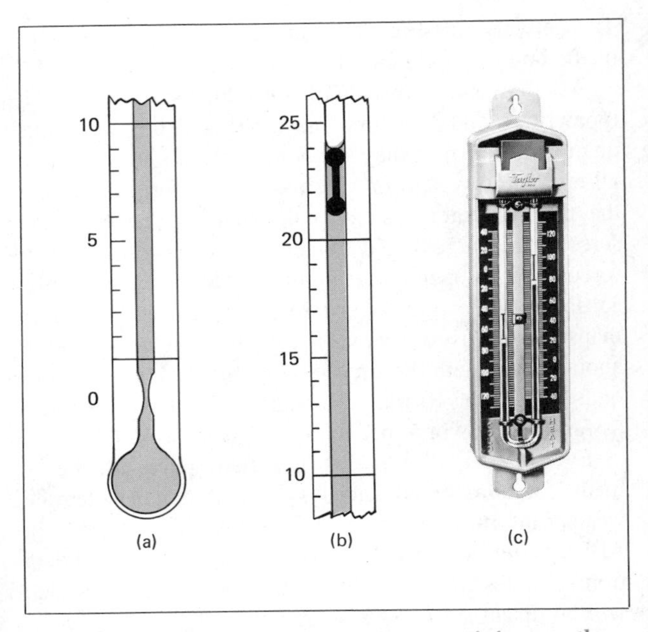

Figure 25.6 Principles of maximum-minimum thermometers. (a) The constriction in the capillary bore prevents the mercury from returning to the bulb and the thermometer retains its highest reading. (b) Surface tension draws the index rod to the lowest descent of the liquid column surface. When the temperature increases and the alcohol expands, the liquid flows around the index, leaving it at the lowest temperature reading. (c) A maximum-minimum thermometer all in one. See text for description. (Courtesy Taylor Scientific.)

The **maximum thermometer** is a mercury thermometer with a constriction in the lower part of its capillary bore (Fig. 25.6). Due to the pressure of the expanding mercury in the bulb, the mercury passes through the constriction as the temperature increases. When the temperature decreases, the mercury is unable to pass back through the narrow constriction. Thus, the column of mercury above the constriction indicates the highest temperature reached.

The maximum thermometer is reset by shaking. Shaking forces the mercury back into the bulb and readies the thermometer for a new reading. Clinical thermometers are of similar construction and are maximum thermometers. A common error in taking one's temperature at home is forgetting to reset the thermometer.

The **minimum thermometer** is an alcohol thermometer that contains a thin-colored glass rod in its bore. This dumbbell-shaped rod is called the index. The surface tension of the alcohol surface draws the index with it to the lowest point of descent of the liquid column surface. When the temperature increases and the alcohol expands, the liquid flows around the index, leaving it at the lowest position. Thus, the upper end of the index marks the lowest or minimum temperature reached. In operation, the minimum thermometer is positioned horizontally. Positioned vertically, the glass index would fall through the liquid to the bottom of the tube. The thermometer is reset by tilting it downward.

Another type of maximum-minimum thermometer that gives both readings is shown in the photograph in Fig. 25.6(c). The U-tube contains mercury in the bottom and a clear, expanding liquid above. There is a reservoir of this liquid at the top of the left side of the U-tube. As the temperature increases, the liquid inside expands and forces the mercury column *down* on the left side and *up* on the right side, carrying with it the maximum index.

As the temperature decreases, the liquid contracts, causing the mercury column on the left side of the U-tube to rise and carrying with it the minimum index. The maximum index remains at its highest graduation or maximum temperature, and the minimum index is carried to its lowest graduation or minimum temperature. (Note the inverse scale on the left side.) The indices, which are magnetic, are reset using a small ceramic magnet shown in the holder at the top.

The tall, cylindrical object on the right in Fig. 25.5 is an 8-in **rain gauge.** The gauge automatically weighs and records the accumulated precipitation. The collection bucket of the rain gauge is mounted on a weighing mechanism having a scale that converts the weight of the rain to equivalent inches. The rain gauge is also used for snow measurements. The collection bucket is sprayed with special chemicals, which melt the snow so its water equivalent can be measured.

Unlike the rain gauge, open exposure and immediate locale are not critical for barometric readings, and barometers are usually kept indoors because of their intricate construction.

Radar (*RA*dio *D*etecting *A*nd *R*anging) is used to detect and monitor precipitation, especially that of severe storms. Radar operates by sending out electromagnetic waves and monitoring the returning waves that have been reflected back from some object. In this manner, the location of the object may be determined. The objects of interest in weather observations are storms and precipitation. Continuous radar scans are now commonly seen on TV weather reports.

There are 230 conventional weather radars deployed across the United States. Radar installations are located mainly in the tornado belt of the midwestern United States and along the Atlantic and Gulf coasts where hurricanes are probable. Additional radar information is obtained from air traffic control systems at various airports. The NMC is linked to radar stations by telephone circuits, and its meteorologists can view the current radar scans across the country on television screens.

A more advanced radar system has been developed and a network of about 160 new weather radars is being planned for the 1990s. The NEXRAD (Next Generation Radar) program employs **Doppler radar.** Like conventional radar, Doppler radar measures the distribution and intensity of precipitation over a broad area. However, Doppler radar has the additional ability of measuring wind speeds. This is based on the Doppler effect (Chapter 6), the same principle used in police and highway patrol radar to measure the speeds of automobiles.

Radar waves are reflected from raindrops in storms. The direction of a storm's wind-driven rain and hence a wind ''field'' of the storm region can be mapped. This provides strong clues or signatures of developing tornadoes. Conventional radar can detect the hooked signature of a tornado only after the storm is well developed (Fig. 25.7). Doppler radar can penetrate a storm and monitor its wind speeds. With a wind field map, a developing tornado signature can be detected much earlier. In field tests, forecasters using Doppler radar were typically able to predict tornadoes 20 min before they touched down, as compared to just over 2 min for regular public warnings. The new system will no doubt save many lives with this increased advance warning time.

Upper-air observations are important in the collection of meteorological data. These observations are made chiefly with radiosondes (radio sounding). A **radiosonde** is a small package of meteorological instruments combined with a radio transmitter. These are carried aloft by balloons, and data are transmitted to ground receiving stations (Fig. 25.8). The wind direction and speed may also be obtained by tracking the flight of the radiosonde.

The carrier balloon will eventually burst, and the radiosonde descends slowly by means of a small parachute. This prevents damage to the instruments and anything in the radiosonde's path on landing. Directions on the radiosondes request the finder to return the radiosonde to the National Weather Service where they are reconditioned and flown again. Approximately 150 National Weather Service offices engage in radiosonde observations.

Rockets are also used in upper-atmosphere data collection, but probably the greatest progress in general weather observation has come with the advent of the weather satellite. Before satellites, weather observations were unavailable for more than 80% of the globe. The first weather picture was sent back from space on April 1, 1960, from the 260-lb TIROS-1 (*T*elevision *I*nfrared *O*bservation *S*atellite). The first fully operational weather satellite system was in place by 1966. These early pole-orbiting (traveling from pole to pole) satellites at altitudes of several hundred miles monitored only a limited area below the orbital path. It took almost three orbits to photograph the entire conterminous United States (Fig. 25.9).

Today, a fleet of GOES (*G*eostationary *O*rbiting *E*nvironmental *S*atellites) satellites, which orbit at fixed points above the equator, and polar-orbiting satellites—including a more recent 2,288-lb TIROS—provide an almost continuous picture of weather patterns all over the globe. At an altitude of about 23,000 mi, the GOES orbiters have the same orbital period as that of the Earth's rotation, and hence are ''stationary'' over a particular location. At this altitude, the GOES's can send back pictures of

Figure 25.7 Radar. (*Top left*) Conventional radar scan showing the characteristic hooked signature of a tornado. (*Top right*) Doppler radar installation in Oklahoma. (*Bottom*) Doppler radar display showing wind field patterns. (Courtesy NOAA.)

Figure 25.8 (*Left*) A radiosonde about to be carried aloft by a weather balloon. The man to the left holds the radiosonde instrument package. (Courtesy U.S. Air Force.)

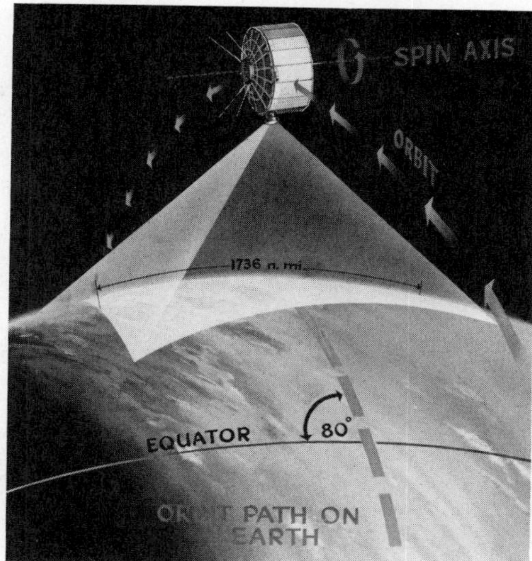

Figure 25.9 A wheel satellite. The wheel spacecraft for TOS (Trans Orbiting Satellite) rotates as its cameras scan a portion of the Earth and adjacent atmosphere. Combine scans from a TOS satellite give a wide view of the Earth's weather picture. Note the opposite cyclonic circulations in the northern and southern hemispheres.

large portions of the Earth's surface as shown in Fig. 25.10. (See also Fig. 24.19.) Geographic boundaries and longitude and latitude grids are prepared by a computer and electronically combined with the picture signal so the areas of particular weather disturbances can be easily identified.

With satellite photographs, meteorologists have a panoramic view of the weather conditions. The dominant feature of the photograph is, of course, the cloud coverage. However, with the aid of radar, which uses a wavelength that picks up only precipitation, the storm areas are easily differentiated from regular cloud coverage. Successive photographs of these storm centers give an indication of their movement, growth, and behavior, thus aiding weather forecasting.

Another important feature of the satellites is **infrared measurements.** As discussed in Chapter 22, the energy and wavelength of emitted radiation depend on the temperature of the radiator. Objects with normal temperatures emit radiation energy in the infrared region of the spectrum. Hence, using a spectrometer to determine the wavelength in this region provides a method of determining the temperature. See Fig. 25.11.

Normally, temperature measurements versus altitude are taken by conventional methods using radiosondes. However, infrared spectrometers now permit these measurements to be taken from lower altitude, pole-orbiting satellites. Fig. 25.12 shows a comparison of the temperature profiles taken by a radiosonde and a Nimbus III satellite infrared spectrometer (SIRS). The SIRS measurements offer a great advantage since a satellite covers all areas of the Earth, some of which have never

Figure 25.10 Satellite image. A GOES weather picture for the United States. A large area of thick clouds and scattered thunderstorms covers the center of the nation from the Gulf Coast to the northern Plains and western Great Lakes. Excepting the northern Rockies, the rest of the country is clear, although an intense low is anchored off the East Coast. (Courtesy NOAA.)

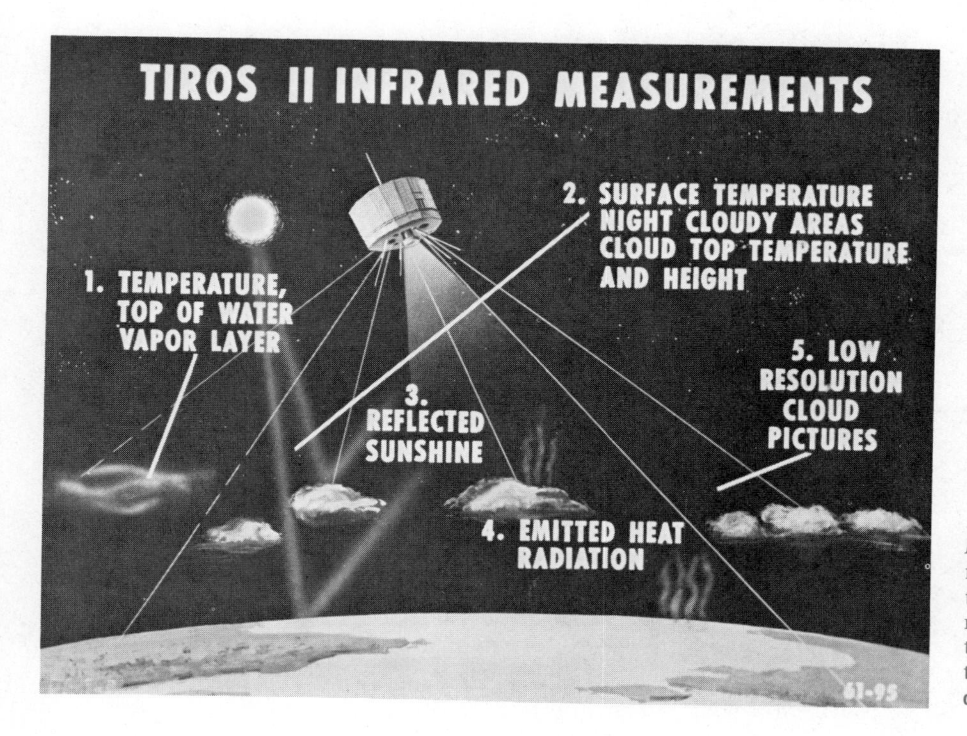

Figure 25.11 Some satellites monitor the infrared radiation of objects, thereby measuring their temperatures, which are proportional to the wavelength of the radiation. (Courtesy NOAA.)

been monitored by radiosonde due to their inaccessibility. This represents a major step in making the meteorologists' global weather picture more complete.

One of the more recent down-to-earth atmospheric observations is the **air pollution potential.** As discussed in Chapter 23, the pollution problem becomes severe when certain atmospheric conditions prohibit the dispersal of pollutants. Using upper-air observations, the NMC prepares wind and air current data that are analyzed for potential pollution conditions. Advisory forecasts are issued to areas in which stagnation conditions are expected to persist for at least 36 hours (Figs. 25.13 and 25.14). The air pollution potential provides an opportunity for air pollution control and research, particularly in studying the influence of meteorological factors on the pollution conditions observed (see Chapter 26).

Another increasingly common atmospheric observation not related to weather forecasting is the pollen count. These counts are made by health officials and are issued to give hay-fever sufferers an indication of the amount of airborne pollen that may aggravate their condition.

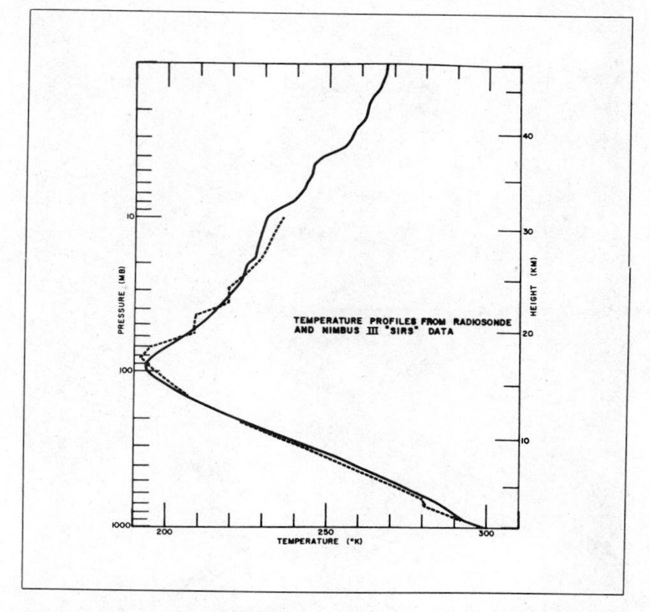

Figure 25.12 A comparison of radiosonde temperature data (dotted line) and that of Nimbus III "SIRS" infrared data (solid line). (Courtesy NOAA.)

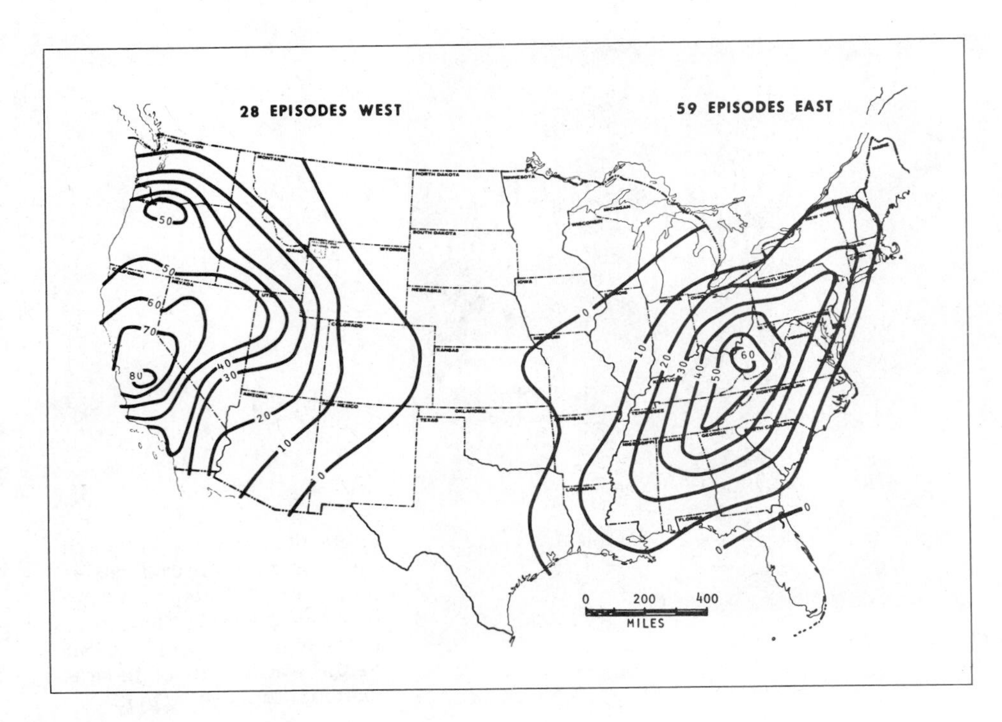

Figure 25.13 Air pollution potential. (Courtesy NOAA.)

Figure 25.14 Air pollution. A modern problem. (Courtesy NOAA.)

25.3 Weather Maps

To help analyze the data taken at the various weather stations, maps and charts are prepared that display the weather picture. Since these charts present a synopsis of the weather data, they are referred to as **synoptic weather charts.** There are a great variety of weather maps with various presentations. However, the most common are the daily weather maps issued by the U.S. National Weather Service. A set consists of a surface weather map, a 500-millibar height contour map, a map of the highest and lowest temperatures, and one showing the precipitation areas and amounts. Examine those of Fig 25.15 (pages 470 and 471) carefully.

The Surface Weather map presents station data and analysis for 7 A.M. EST, for a particular day. This map shows the major frontal systems and high- and low-pressure areas at the constant surface level. Isobars are represented by solid lines, and prevalent isotherms (the lines of constant temperature) are indicated by dashed lines (in degrees Fahrenheit).

The height contour map gives a representation of a surface of constant pressure, in this case, 500 millibars. Contours of 300- and 700-millibar pressure are also common. The temperature map gives the highest and lowest temperatures for weather stations throughout the country. The complete precipitation picture is given by the Precipitation Areas and Amounts map. Major precipitation areas are shaded.

On the Surface Weather map a station model, as shown in Fig. 25.16 (p. 472), gives the pertinent weather data at the station's location. The stations shown in Fig. 25.15 are only a fraction of those included in the National Weather Service's operational weather maps. Also, the station model information has been abbreviated.

Figure 25.17 (p. 472) shows how contour maps are prepared. A surface map shows isobars at surface level, whereas a height contour map shows height contours at a constant pressure. The contours indicate the heights of the designated pressure in feet above sea level. This is analogous to a topographical map that indicates the contour of the Earth's surface. If constant pressure maps were three-dimensional and viewed from the side, the hills and valleys represented by the contours would be the surfaces of equal barometric readings.

On the height contour map, the wind symbols show the wind directions and speeds at the 500-millibar level. Isotherms also appear on the height contour maps. Since many weather conditions and changes depend on the dynamics of the upper atmosphere, these maps are important in forecasting future conditions. Note the correspondence of the low-pressure regions on the height contour map and areas of precipitation on the precipitation map in Fig. 25.15.

From a knowledge of the general behavior of fronts, cyclonic motion, and air masses studied in the previous chapters, reasonable weather forecasts may be made from surface maps. This is not always easy, as the weather conditions may be quite complex. For example, note the complexity of the frontal systems over the northwestern United States and Canada shown in Fig. 25.18 (pages 473 and 474). The student may find it difficult to predict the weather changes for that region.

At the National Meteorological Center, a computer projects the weather conditions based on stored empirical knowledge, statistical methods, and continuous data. A **prognosis chart** (prog chart) is then printed out, based on the computer's analysis of what the weather conditions will be at some time in the future. However, some unpredictable atmospheric occurrence may take place at any time and make the projection incorrect. One should now begin to realize how difficult it is to accurately forecast the weather.

Figure 25.15 contains a classical cyclonic formation, with a stationary front extending toward the northwest into Canada. In the low-pressure occluded front region, the conditions are favorable for convective and frontal precipitation, and its occurrence is indicated by the colored region over Nebraska, Iowa, and Minnesota. Moreover, the counterclockwise circulation around the low may draw in moist air from the Great Lakes region. The clockwise rotation of the adjacent highs to the west carries the precipitation farther west into Wyoming, as shown on the precipitation areas map. The cyclonic disturbances off the east coast contribute to the precipitation in that area.

To forecast the next day's weather, one must consider the probable movement of the fronts and pressure areas. Localized forecasts are more concerned with the internal dynamics of these phenomena. Because of the general west-to-east movement of weather across the United States and the circulation around a low-pressure area, the low-pressure system over the Midwest in Fig. 25.15 should move northeast carrying its precipitation with it. The high-pressure areas would be expected to move over the middle United States, and the low-pressure region off the California coast would move inland. The speeds of the high- and low-pressure areas vary. Generally, highs

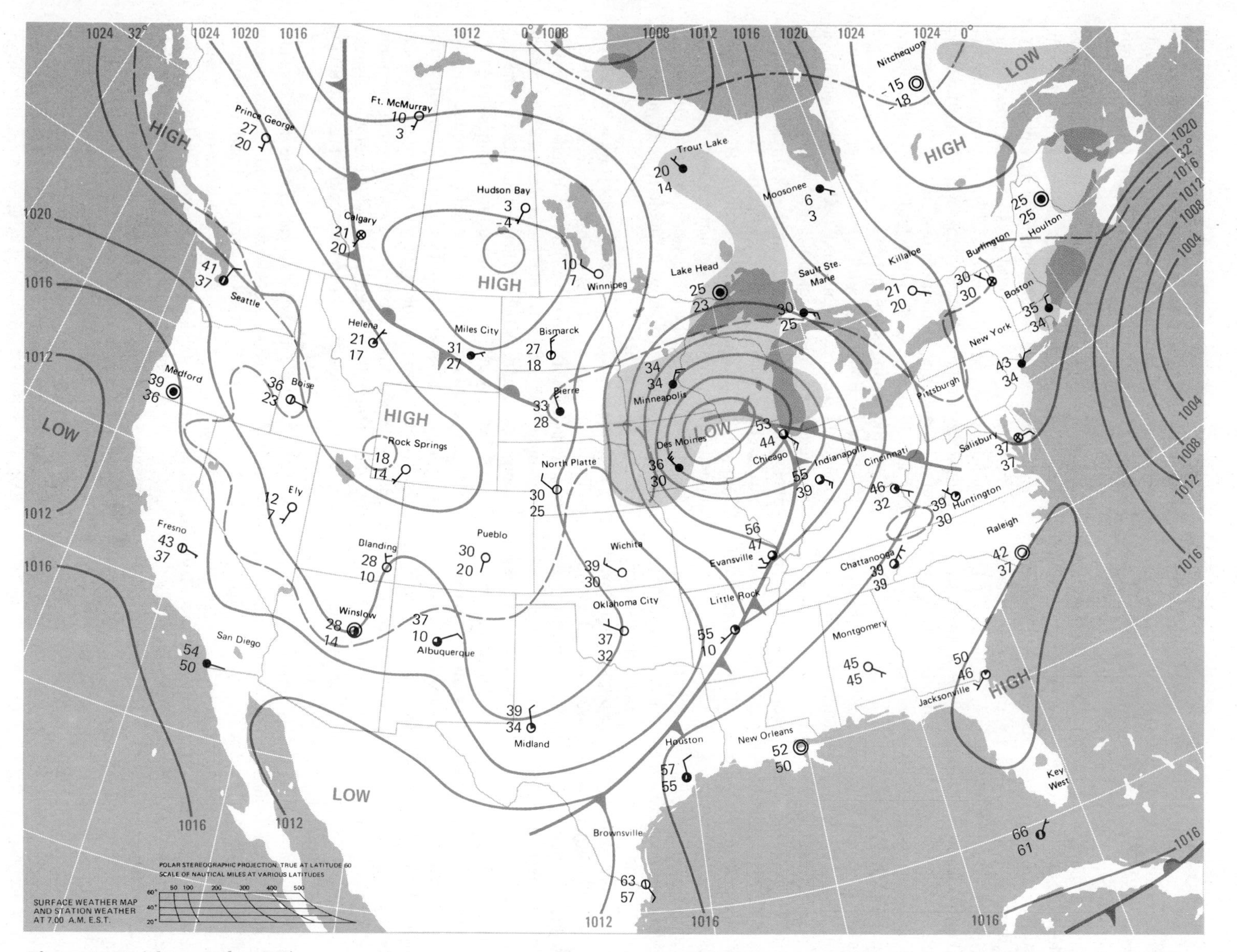

Figure 25.15 (Above and p. 471)

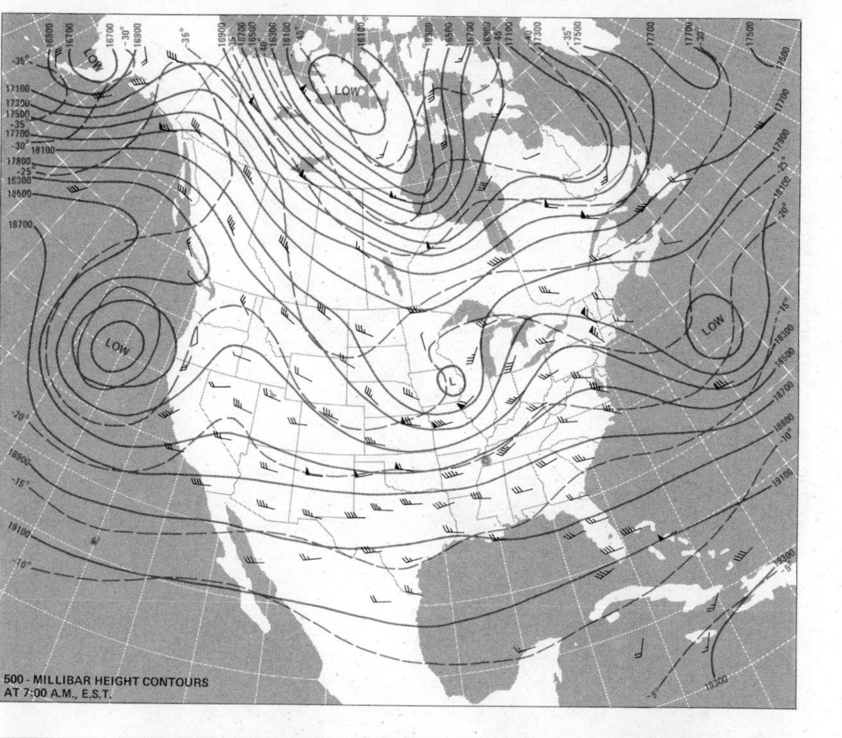

Figure 25.15 (continued) Daily weather maps. Weather maps are issued daily by the National Weather Service. The maps from which these drawings were made gave many more locations and data than are shown here. (Redrawn courtesy NOAA.)

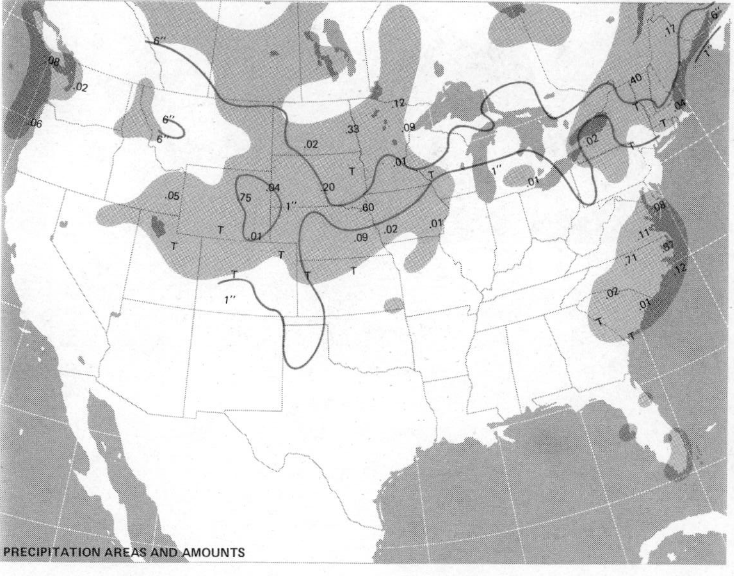

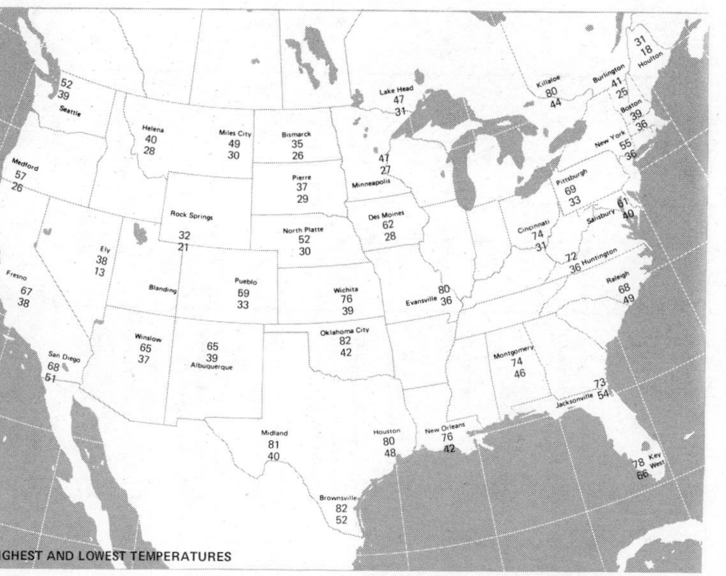

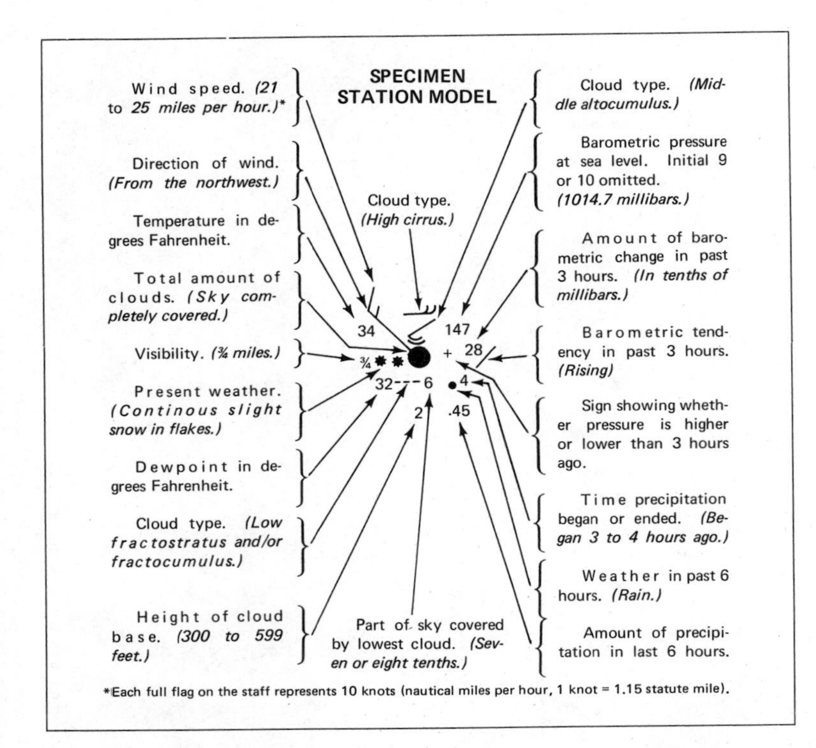

Figure 25.16 The weather data from each reporting station are arranged in this model format. (Redrawn courtesy NOAA.)

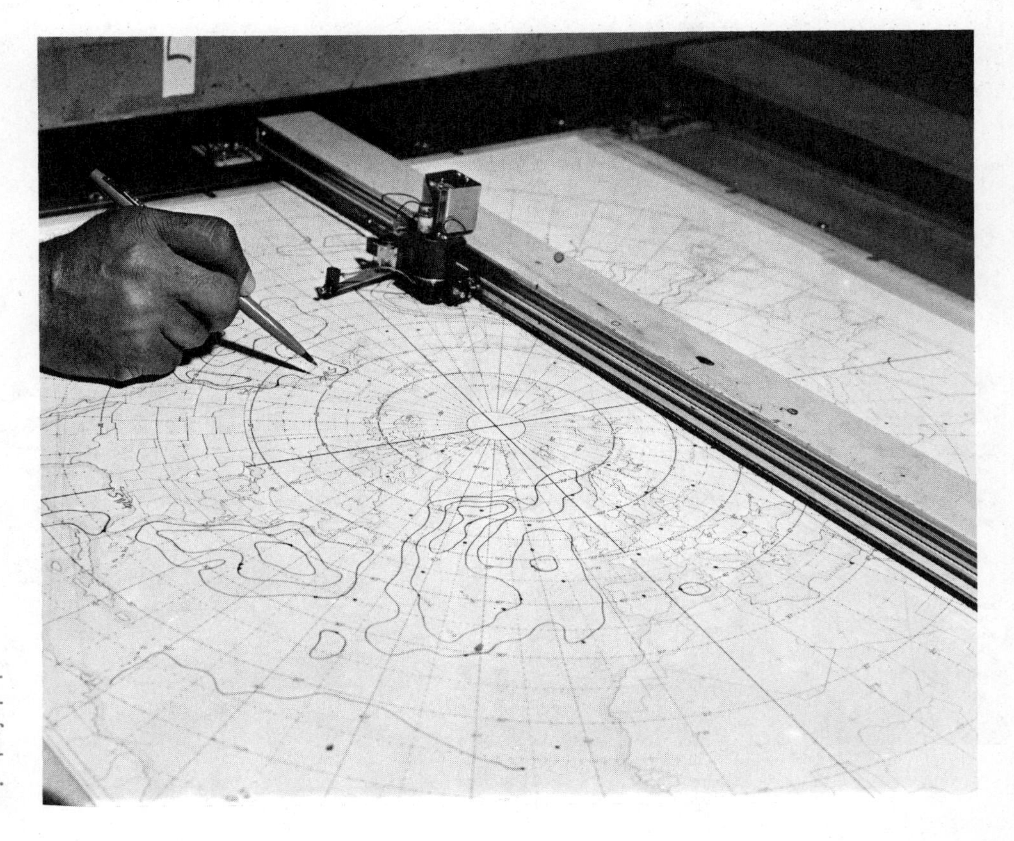

Figure 25.17 The preparation of a pressure contour map. Such contour maps are prepared by automatic curve traces. (Courtesy NOAA.)

and lows cross the United States in three to five days.

The northeastern United States would experience a warm front, which usually indicates an advancing low. The center of a low-pressure region is normally 0.5 in of mercury or more below normal atmospheric pressure and increases radially outward. Hence, aside from the small deflection of the Coriolis force, the winds are generally toward the center of the low.

As a low from the west approaches an observer, the barometer falls and the winds are generally from the east. As the low passes, the barometer begins to rise and the winds shift, coming from the west. In Fig. 25.15, these conditions should occur in the Great Lakes region over which the low should pass.

The wind conditions are reversed for the passage of a high and would be expected in the northern Plains states. The southeastern United States should feel the effect of the advancing cold front. The upward displacement of the warm air by the advancing cold front will produce a slight fall in the barometric pressure, followed by a sharp rise as the front passes, due to the incoming high. Little change would be expected for the northwest states, which will remain under the influence of high pressure. California may expect precipitation from the advancing low.

Figure 25.18 shows the next day's weather maps which prove these predictions to be fairly accurate. The entire central United States is now dominated by high pressure.

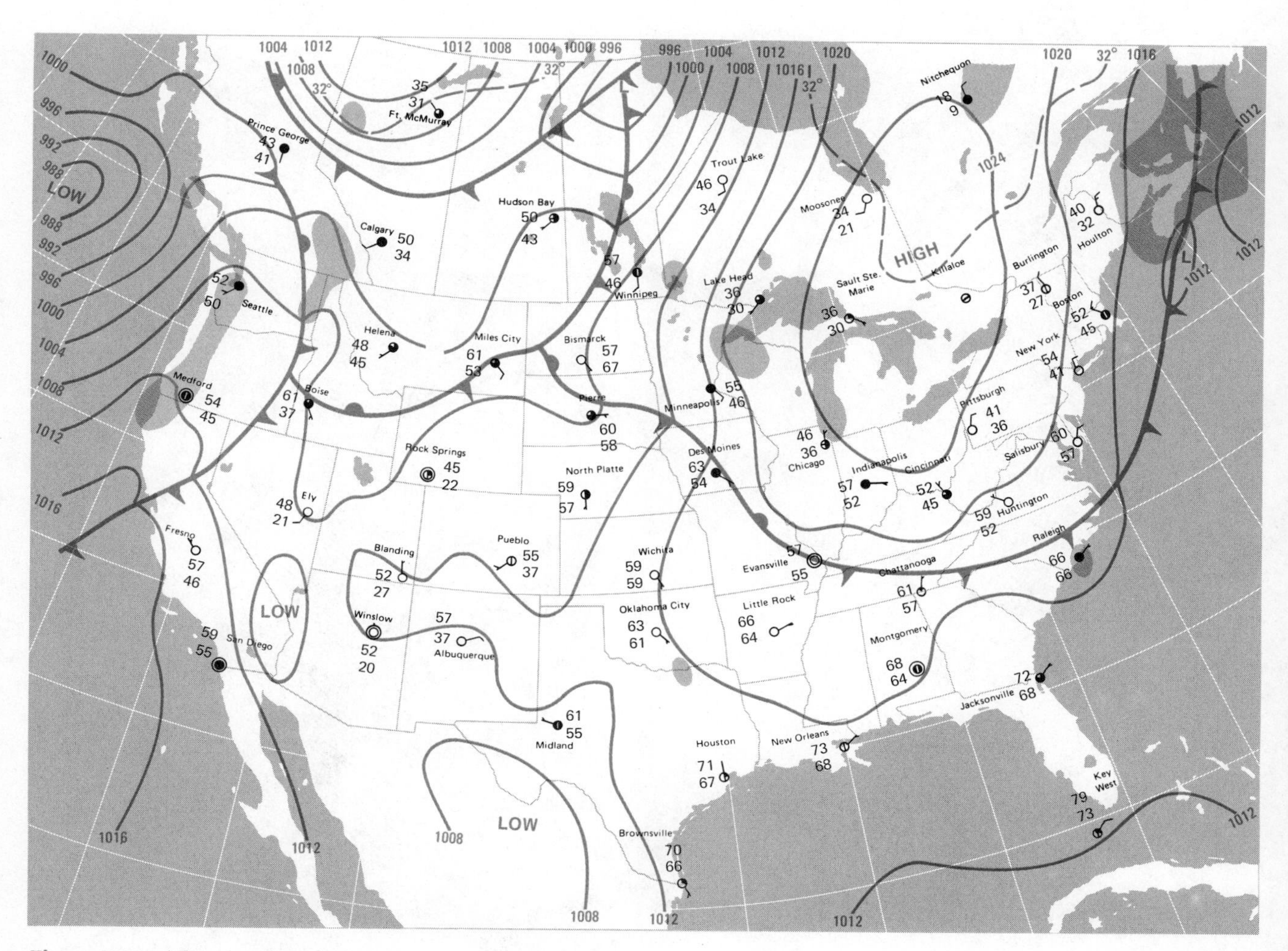

Figure 25.18 (Above and p. 474)

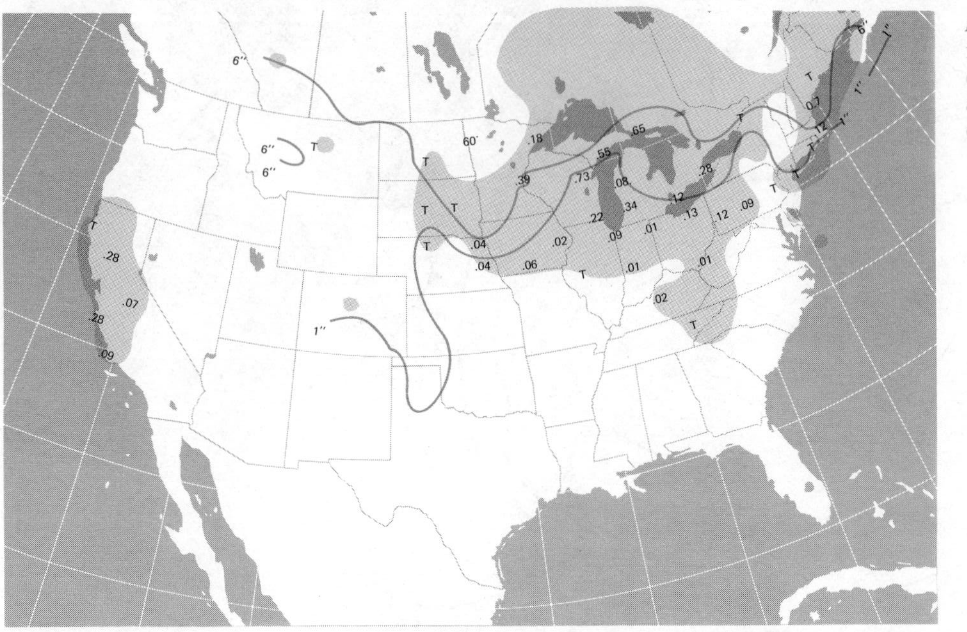

Figure 25.18 (continued) Weather maps for the day following that of Fig. 25.15. Notice the complex array of frontal systems on the surface map. Complicated atmospheric conditions make weather forecasting a difficult task. (Redrawn courtesy NOAA.)

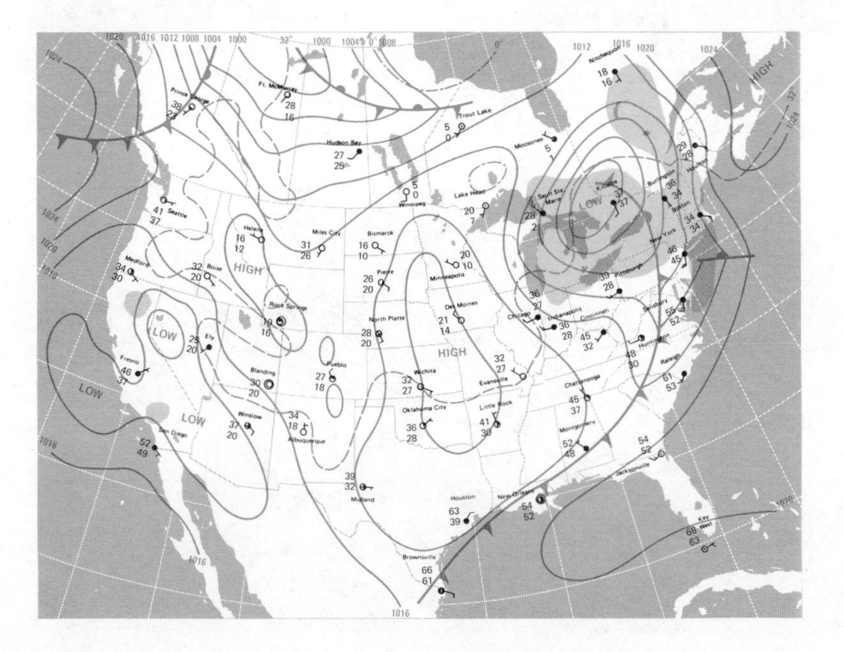

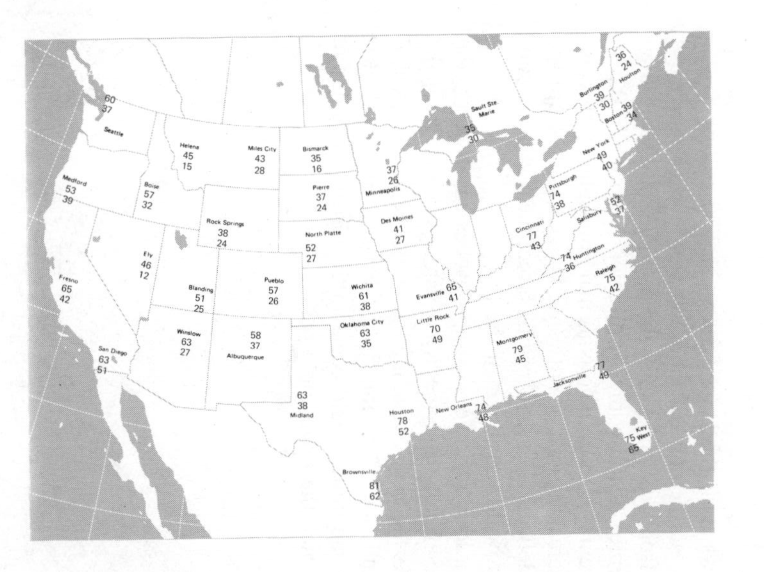

Highs are characterized by descending air. The descending air becomes warm and can hold more moisture, so that precipitation will not occur. In the summer, these conditions are responsible for the high humidity and frequent droughts in the central United States.

Note also that high-pressure areas are usually larger than low-pressure areas, with widely spaced and less streamlined isobars. In the winter, a high may enter and move across the United States in the form of a cold, cP air mass. This occurrence is evident in the maps of Fig. 25.19. Hence, it can be seen that weather forecasting is, to a large part, the projection of the movement of fronts, pressure areas, and air masses. This process is complicated by a multitude of local variations and influences.

25.4 Folklore and the Weather

Folklore and legends have been associated with the weather from the beginning of humanity. People in the early civilizations worshipped deities whose actions were manifested by weather phenomena. Myths were created to account for those occurrences that could not be otherwise explained. Thunder and lightning, for example, indicated the presence of the violent god Thor. Clouds were the cattle of the sun god Apollo, grazing in the meadows of heaven.

> Men judge by the complexion of the sky
> The state and inclination of the day.
>
> Shakespeare (*Richard II*)

As human beings progressed and gave up the mythological interpretations of the weather, they began associating its behavior with things they observed. In this manner a similar observation might allow a prediction of future weather conditions. It is these observations that have come down to us as folklore. Usually in the forms of sayings or verse, they have been handed down by word of mouth from generation to generation and often appear in literature. Some of these sayings are well founded and can be explained scientifically. Others,

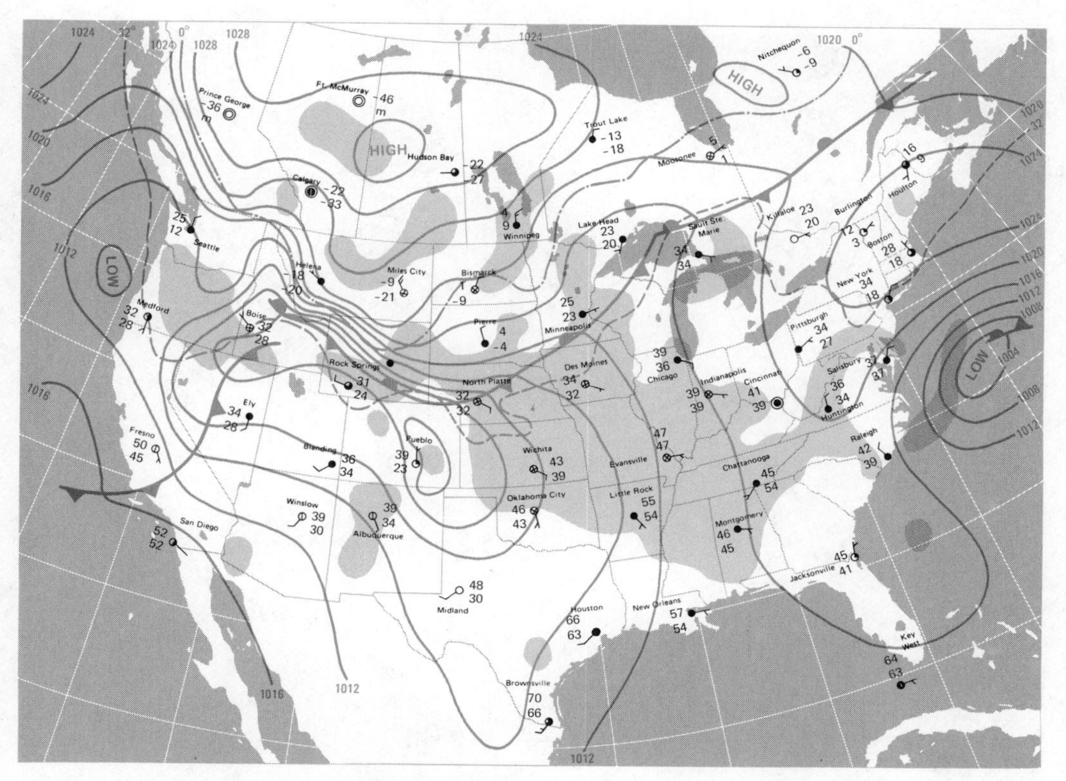

Figure 25.19 (Above and p. 476) A series of three surface maps showing the advance of a cold, high-pressure cP air mass across Canada and the United States. Notice the temperature and precipitation variations as the air mass moves across the country. (Redrawn courtesy NOAA.)

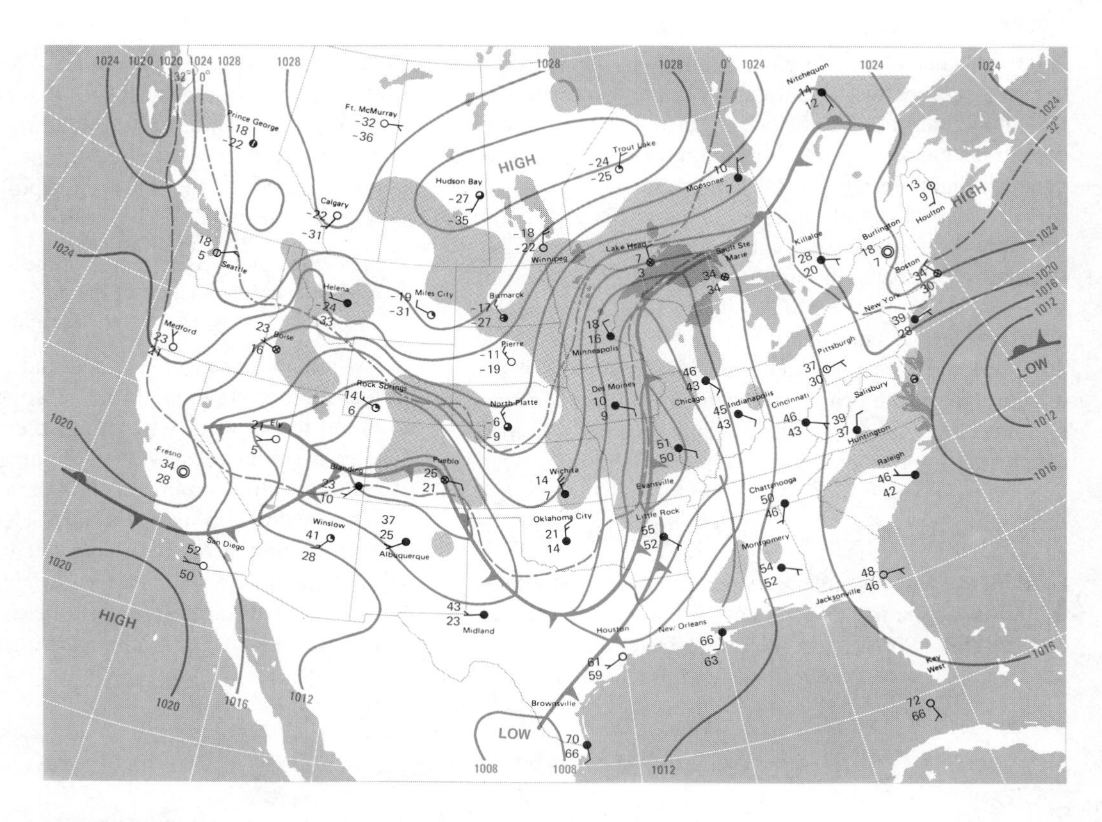

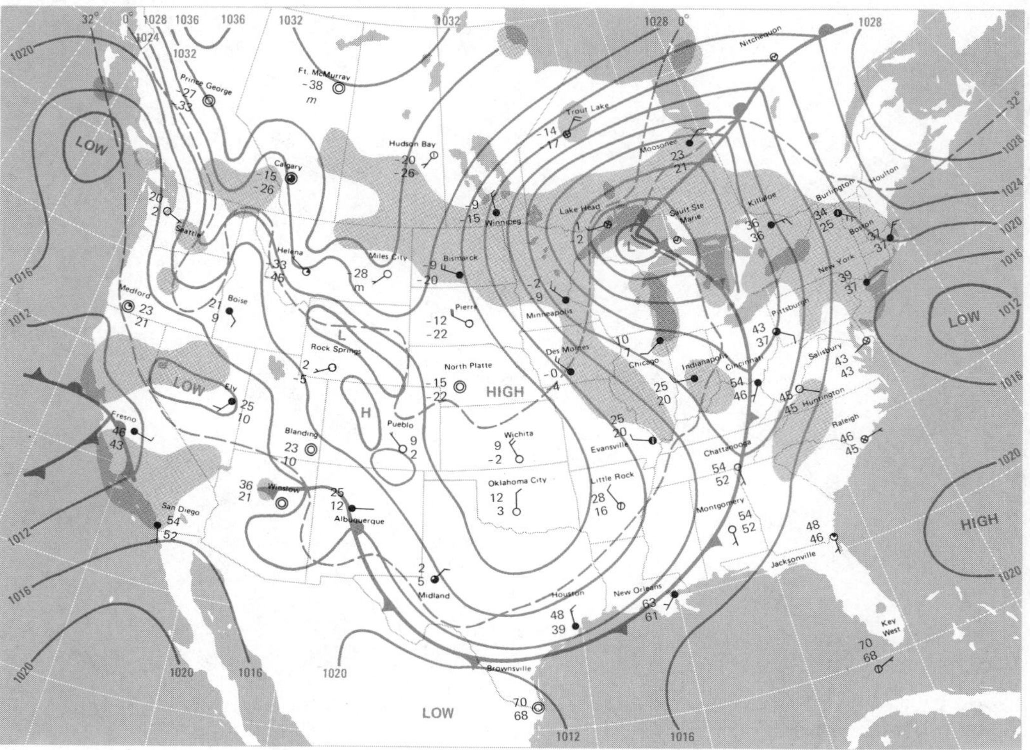

Figure 25.19 (Continued)

seemingly without any valid explanation, live on, much the same as superstitions. A few of these well-known sayings will be examined in this section. Those related to the meteorological explanations of the previous chapters will be pointed out. The others are left to the imagination of the reader.

Many amateur weather forecasters closely observe the activities of birds and animals. The antics and instincts of these creatures are often without explanation. However, they form an essential part of folklore. Ducks flapping their wings, the braying of donkeys, or the bolting of horses, for example, are thought by some to be indicative of stormy weather. Loggers in the Pacific Northwest are said to predict snow two or three days before a blizzard by watching the gathering of elks in the shelter of trees. Fiddler crabs have been observed to retreat to inland burrows days before a hurricane's arrival.

Other observations are of a seasonal nature. Marks on a caterpillar or the amount of food stored by squirrels are used to predict the severity of the coming winter. The southward flight of birds may indicate an early winter, while wild geese flying north is taken as a sure sign that warm weather is coming soon.

Probably the most celebrated weather-predicting animal is the groundhog. Legend has it that the groundhog emerges from his winter hibernation each year on February 2 to check on the approach of spring. If he sees his shadow, he returns to his burrow and six more weeks of winter are imminent. This bit of folklore is greatly publicized each year by the town of Punxatawney, Pennsylvania, where the local groundhog, "Punxatawney Phil," is said to be the superior forecaster. It is somewhat surprising that his shadow, which occurs when the Sun is shining, should warn the groundhog of continuing winter, but that's the legend.

More dependable observations result from the Sun and its effects.

> Above the rest, the sun who never lies
> Foretells the change of weather in the skies.
>
> Vergil

There are several sayings about the Sun and the weather indications it gives. Among these are the following:

> The weary sun hath made a golden set
> And by the bright tracks of his fiery car
> Gives token of a goodly tomorrow.
>
> Shakespeare (*Richard II*)

> If the red sun begins his race
> Be sure the rain will fall apace.

> The Pharisees also with the Sadducees came and tempting desired him that he would show them a sign from heaven. He answered and said unto them, When it is evening ye say, It will be fair weather for the sky is red. And in the morning, It will be foul weather today for the sky is red and lowering. Oh ye hypocrites, ye can discern the face of the sky, but can ye not discern the signs of the times.
>
> Matthew 16:1–4

The weather predictions alluded to in the Bible are more commonly stated:

> Red sky at night, sailors delight
> Red sky in the morning, sailors take warning.

"Rainbow" is often substituted for "red sky" in the above saying. All these sayings involve the red sky, which we often see at sunrise and sunset. What then causes the Sun and sky to appear red? When the Sun is on the horizon, the sunlight travels farther through the atmosphere to reach us than when the Sun is overhead. The blue portion of the sunlight is normally scattered (Section 22.4). However, if the air contains impurities such as dust, more of the longer wavelengths are scattered as the sunlight travels near the surface of the Earth, with the result that only the red portion of the spectrum may reach an observer. This, of course, makes the sky and Sun appear red. (See Chapter 22.)

The condition is enhanced if there is a stable high-pressure region between the observer and the Sun, as the high pressure holds air contaminants near the Earth and scattering is increased. As discussed in Section 24.2, highs are generally associated with good weather so if one sees a red sky in the evening, it is quite probable that there is a high-pressure area to the west, and good weather will accompany it as it moves eastward, delighting sailors in the westerly wind zone where this weather movement applies. Should the red sky occur at sunrise in the east, the high has probably passed and will usually be followed by a low-pressure system which is generally associated with poor weather.

Another saying describing an observed phenomenon of sunlight is as follows:

> When the sun draws water, rain will follow.

The Sun's rays are often blocked by dense clouds. Occasionally a cloud may be thin enough in a small area or a small break in the clouds may allow the sunlight to shine through. In the humid air associated with the clouds, the sunlight may be scattered by fine water droplets or other particles, giving rise to diffuse reflection and creating the effect of a fanlike ray of sunlight extending from the Earth to the cloud (see Chapter 22, introductory photo). This effect may also be observed when sunlight shines through dense leaf coverage in a thickly wooded area or in flashlight or searchlight beam at night. There is no drawing up of water, but clouds and particles *are* available for the production of rain.

Mackerel sky and mare's tails
Make lofty ships carry low sails.

Trace in the sky the painter's brush
The winds around you soon will rush.

The cirrus and cirrocumulus clouds referred to in the above sayings usually precede the approach of a warm front (Fig. 24.3). As the front approaches, the warm air rising over the cold air mass is likely to produce the winds predicted by these verses.

The wind in the West
Suits everyone best.

In general our good weather comes from the west as opposed to the north and east, so the saying is to some degree valid. It must be remembered, however, that some locations on the Earth receive very poor weather from the west. Hence, the saying is not universal.

When the morn is dry
The rain is nigh
When the morn is wet
No rain you get.

Or in different form and meter.

When the grass is dry at morning light
Look for rain before the night;
When the dew is on the grass
Rain will never come to pass.

The reasoning behind these rhymes is obvious. If condensation in the form of dew has occurred, the air will have a lower relative humidity and rain will be unlikely. However, a better last line to the second poem might read "Rain will *seldom* come to pass," as the air at higher elevations may have sufficient moisture to produce rain. Also, the lack of dew in the morning is not a positive indication of rain. The general validity of these sayings is, therefore, somewhat questionable.

The moon also shares considerable prominence in weather folklore. Many of its related sayings are unfounded. Among these are the indication of rain when "the new moon holds the old moon in its arms." This refers to the time when the crescent new moon is near the lower portion of the moon and appears to hold the upper darkened portion. The rain supposedly results from water being spilled from the saucer-shaped new moon. Such sayings are best answered by the countersaying,

Moon and weather may change together
But a change of the moon does not change the weather.

More reasonable predictions are given by the following:

Clear moon, frost soon.

Or:

A ring around the moon is a sure sign of rain.

Or:

When the stars begin to huddle
The Earth will soon become a puddle.

The first of the above sayings refers to the lack of cloud coverage that acts like an insulation to keep the Earth warm. In the absence of clouds, the moon is clearly seen and the land masses cool quickly, making frost likely. The latter two sayings refer to the appearance of the moon and stars as viewed through cirrostratus clouds. The moon appears with a diffuse halo, while the indistinct stars appear to be closer together. Cirrostratus clouds normally precede an approaching warm front, which is accompanied by turbulent weather, usually in the form of rain.

Although not directly related to meteorology, the moon and its phases are often referred to when planting crops. Many gardeners follow the *Farmer's Almanac,* which gives the periods of the proper phases of the moon for planting the proper crops. In general, crops that produce above ground are to be planted in the "light of the moon," or in the waxing phase. Crops that produce below the surface are to be planted in the "dark of the moon," or in the waning phase. For example, it has been reported that a certain gardener had difficulty in keeping dirt on potatoes, which were planted in the light of the moon.

Wood is also said to be affected by the phases of the moon. A board lying on the ground will supposedly curl up at the ends in the light of the moon and stick firmly to the ground in the dark of the moon. Whether or not this is true is subject to doubt, but some builders will put on wooden shingles only during the waning phase or dark of the moon.

Other weather sayings include:

Sound travelling far and wide
A stormy day will betide.

Sound waves may be reflected by air layers of different temperatures in the atmosphere. As a result, the sound is heard at a considerable distance where it is reflected back to Earth. A reflecting cold air layer may also be the source of precipitation, accounting for the stormy day.

Rain before seven, stops before eleven.

This is a reasonably safe prediction for convectional precipitation. If it were raining prior to 7 A.M., there would probably be little moisture left in the precipitating cloud by 11 A.M. Moreover, the Sun would be rising high in the sky and cause the temperature to rise above the dew point temperature. The clouds would then dissipate and the rain cease, which would bear out the prediction. However, for frontal precipitation, which is controlled by huge air masses, the prediction may not prove to be so accurate.

Leaves turn silver before a rain.

When the wind blows the leaves so that the shiny underside is exposed, they take on a silvery appearance. This turning up of the leaves may result from vertical air motion which may cause cloud formation. If a nimbus cloud develops, the leaves' prediction of rain will be fulfilled.

It smells like rain.

Before a rain there may be a musty, or earthy, smell in the air. Since convectional precipitation is associated with low pressure and rising air, air and gases in the ground may diffuse out, giving rise to the "smell of rain."

Snow on the ground for three days is waiting for another.

If snow remains on the ground for three days, it is obviously quite cold and additional precipitation is likely to be in the form of snow.

Several common sayings that are without scientific foundation are

Rain on the first Sunday,
rain every Sunday [of the month].

Rain on Easter Sunday,
rain for seven straight Sundays.

Rain on Good Friday,
the Saint is pouring water on a flat rock.

The latter demonstrates how the meanings of folklore saying may be disguised. Its interpretation is that if it rains on Good Friday, the following summer will be dry, such that the summer rains will run off the hard, dry ground as though it were a rock.

Another special day for rain is St. Swithin's Day, which is July 15. Should it rain on St. Swithin's Day, legend has it, 40 days of rain will follow. St. Swithin was an English bishop whose wish it was to be buried in the open church yard. When sainted and moved into the church, his spirit protested with 40 days of rain. A rain occurring on the anniversary of this event will supposedly arouse St. Swithin's spirit for a repeat performance.

Certain weather occurrences are given special names. A period of warm weather in October or November is sometimes referred to as **Indian Summer.** The seasonal cooling at this time of year gives rise to low-lying fogs. It is said that thin fogs lying near the tops of corn shocks reminded the early settlers of smoke coming from Indian tepees, hence the name Indian Summer. There is nothing uncommon about a brief warm "spell" this late in the year, as the season changes from autumn to winter.

The special significance of this period of warm weather is no doubt largely psychological. People are aware of the coming harsh winter and are sentimental toward this last warm reminder of the pleasant summer. The fondness for summer prompts a special name for this last trace. A wintry period in April is not so honored as Indian Winter, but is deplored and quickly forgotten. More significance is given to the break in winter called the January thaw which often occurs in February. A cold snap in the first days of spring when blackberries usually bloom is sometimes called Blackberry Winter.

Some people are able to forecast a change in the weather or rain by an ache in the knee or some other joint. This was thought to be a joke for some time; however, it is now believed that changes in the pressure and humidity affect the aches and pains of rheumatic joints. Others who are not afflicted may rely on folklore to help predict weather behavior. The date of the first snow is sometimes taken as the number of snows that will occur during the winter. Those who believe in this prediction become very discriminating between an actual snow and a noncountable flurry, as the number of snows approaches the predicted number.

Folklore, whether true or unfounded, will exist as long as people talk about the weather, and weather is one of our most talked-about subjects. When we are without anything to say or at a loss for words, it is a favorite topic that comes to our aid. Everyone comments on the weather while an old English proverb surmises:

Weather is the discourse of fools.

Learning Objectives

After reading and studying this chapter, you should be able to do the following without referring to the text:

1. Describe the organization and functions of the National Weather Service.
2. Explain how weather data are processed and how forecasts are distributed.
3. State the instruments, their operation, and the data collected in a typical weather data station.
4. Describe the operation and data collected by radar, radiosondes, and weather satellites.
5. Explain the different types of synoptic weather charts.
6. Describe how forecasts are made from weather maps.
7. Distinguish between folklore sayings with scientific merit and those without.

Important Words and Terms

National Weather Service
NOAA
National Meteorological Center
instrument shelter
maximum thermometer
minimum thermometer
rain gauge
radar
Doppler radar
radiosonde
weather satellite
infrared measurements
air pollution potential
synoptic weather charts
prognosis chart
Indian Summer

Questions

1. What is NOAA?
2. Has the United States always had a National Weather Service or a similar organization?
3. How are weather data processed and forecasts made? How often is this done?
4. What are the locations of the Forecast Office and the Office of the National Weather Service nearest your hometown? your college or university?
5. What type of data are taken at a typical weather data station?
6. Explain the operation of a maximum-minimum thermometer.
7. What does the word ''radar'' mean?
8. Distinguish between conventional and Doppler radar.
9. What is a radiosonde?
10. How are temperature profiles obtained via satellite?
11. What is a GOES and why is it ''stationary''?
12. What regions of the United States have high air pollution potentials?
13. What is a synoptic weather map and what information is given in the station models?
14. Explain a 500-millibar height contour map.
15. How can weather forecasts be made from weather maps?
16. Are folklore sayings concerning the weather reliable? Do they have any scientific merit?

Problems

1. Given the following data, plot a weather station model:
   Temperature: 68°F
   Sky: One-half covered with cirrus, cirrocumulus, and a few stratocumulus clouds
   Winds: 15 mi/h out of the northeast
   Visibility: 3 mi
   Psychrometer wet bulb: 66°F
   Barometer: 1012.7 mb, falling from 1013.4 in the last 3 hours
   Started raining 2 hours ago with current accumulation of 0.15 in
2. From Fig. 25.20, calculate the approximate average speed of the high-pressure cP air mass moving from western Canada into the central United States. *Answer:* 500 to 1000 mi/day
3. Describe the weather conditions and formations as best you can from the satellite photograph of the United States in Fig. 25.10.
4. Prepare a local weather forecast for your area from Fig. 25.15. Use Fig. 25.18 to check your predictions.
5. From the surface map in the accompanying map, what are the weather conditions across the country? Give particular attention to weather changes and variations. Check your answer with the newspaper report given in Section 24.2.
6. Prepare a general forecast for the United States from the last surface map in Fig. 25.15.
7. Prepare a 24-hour forecast for the U.S. east coast from the accompanying surface map. Give the reasons for your predictions.

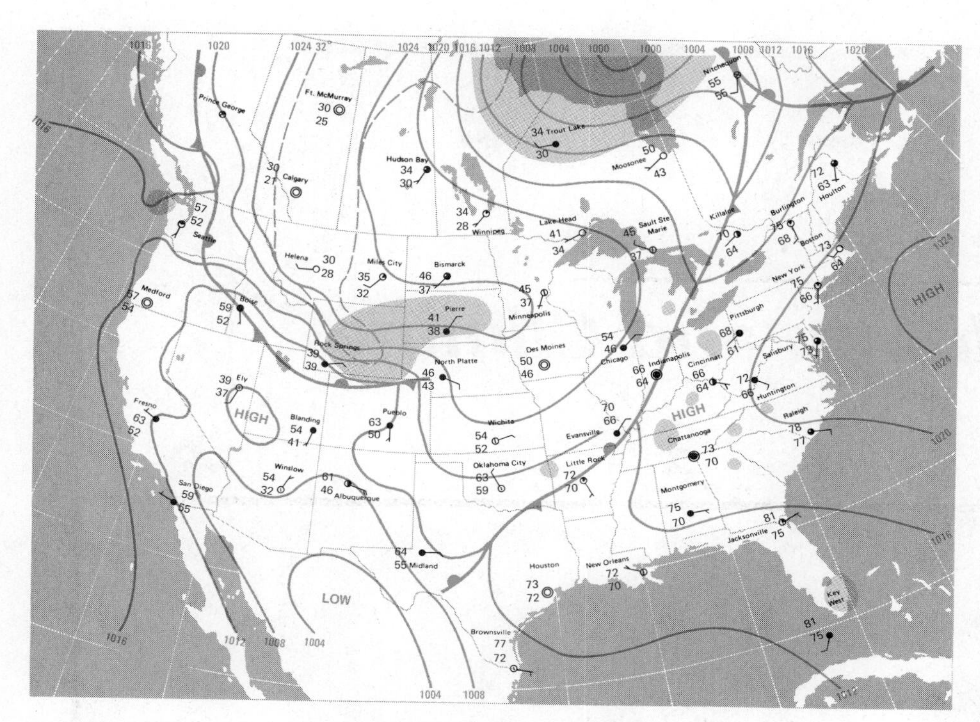

Map for Problem 5.

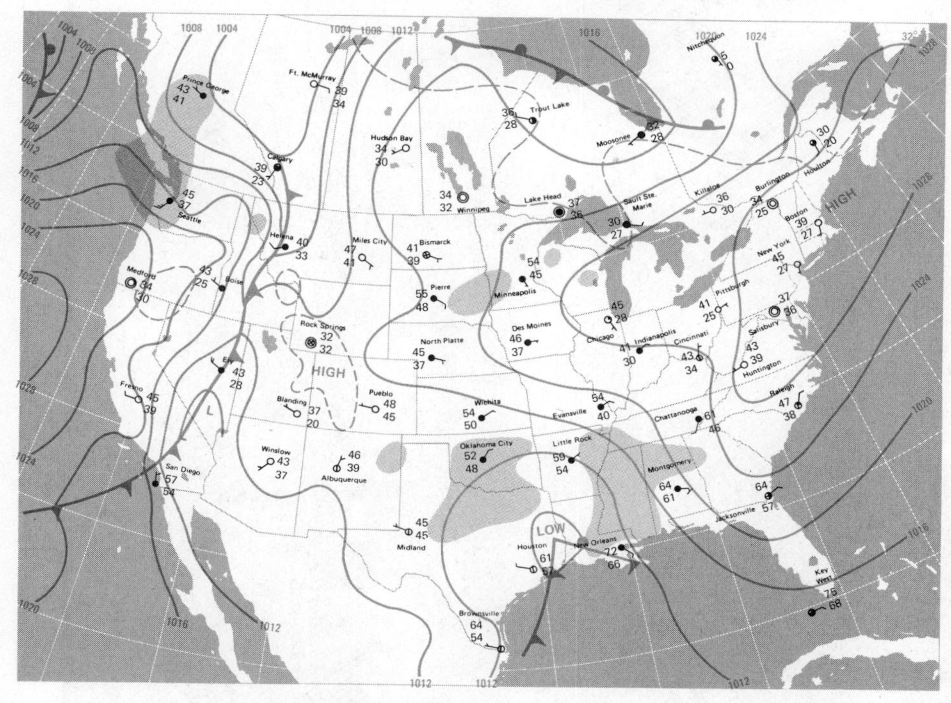

Map for Problem 7.

8. Examine the following weather sayings for meteorological merit and meaning:
   (a) A red sun has water in his eye.
   (b) Two full moons in a calendar month bring on a flood.
   (c) Mackerel clouds in the sky expect more wet than dry.
   (d) February rain is only good to fill ditches.
   (e) Candles burn dim before rain.
   (f) March comes in like a lion and goes out like a lamb.
   (g) A year of snow, a year of plenty.
   (h) Dew long on the ground, Jack Frost will be around.
   (i) It's too cold to snow.

26

Air Pollution and Climate

"... this most excellent canopy, the air, look you, this brave o'erhanging firmament, this majestical roof fretted with golden fire, why, it appears no other thing to me but a foul and pestilent congregation of vapours."

–Shakespeare—*Hamlet*

TO MAKE A complete survey of the atmosphere, one cannot neglect the current problem of air pollution. The various pollutants being expelled into the troposphere may have direct effects on our health and living conditions and may influence the local weather conditions and possibly the global climate. What, then, are the nature and sources of air pollution?

Before answering this question, we would do well to explain what we mean by air pollution. An obvious definition might be any unnatural constituent of the atmosphere. But we must then decide if the gases and particulate matter from volcanic eruptions and forest fires (started by lightning) are unnatural contributions. In order to limit and focus our discussion, let us simply define **air pollution** as any atypical contributions to the atmosphere resulting from the activities of human beings.

The air pollution resulting from these activities has received a great deal of attention in the past decade, and will probably continue to do so in the foreseeable future. The concentrations of air pollutants have become critical in many localities and preventive actions have been implemented. The belching smokestacks that were once a symbol of industrial prosperity are now frowned on as sources of pollution. It is becoming increasingly evident that the atmosphere is not limitless in the disposal of our combustive wastes, and that air pollution may have long-term effects in addition to direct and immediate effects.

In this chapter we shall consider the sources of air pollution and the effects of pollution on meteorological processes that may result in modifications of our weather and climate.

◀ Air pollution in a temperature inversion.

26.1 Air Pollutants

Air pollution results primarily from the products of combustion and industrial processes that are released into the atmosphere. It has long been a common practice to vent these wastes into the atmosphere, and the resulting problems are not new, particularly in areas of population concentrations. Smoke and soot from the burning of coal plagued England over 700 years ago. London recorded air pollution problems in the late 1200s, and particular smoky types of coal were taxed and even banned. The problem was not alleviated, and in the middle 1600s Charles II was prompted to commission one of the outstanding scholars of the day, Sir John Evelyn, to make a study of the situation. The degree of London's air pollution at that time is described in his report, *Fumifugium* (Latin, generally meaning *On Dispelling of Smoke,* see Fig. 26.1).

> . . . the inhabitants breathe nothing but impure thick mist, accompanied with a fuliginous and filthy vapor, corrupting the lungs. Coughs and consumption rage more in this one city (London) than in the whole world. When in all other places the aer is most serene and pure, it is here eclipsed with such a cloud . . . as the sun itself is hardly able to penetrate. The traveler, at miles distance, sooner smells than sees the City.

However, the Industrial Revolution was about to begin, and Sir John's report was ignored and so gathered dust (and soot). Another indication of the polluted air in parts of 17th-century England is given by the quotation from Shakespeare at the beginning of this chapter, who was drawing on his own experience.

As a result of this air pollution, London has experienced several disasters involving the loss of life. Thick fogs are quite common in this island nation, and the

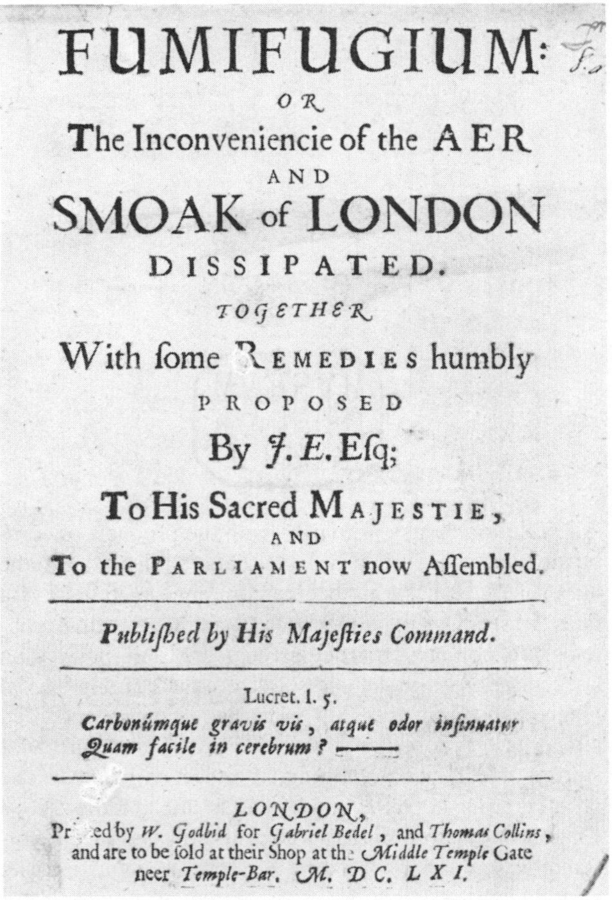

FUMIFUGIUM:
OR
The Inconveniencie of the AER
AND
SMOAK of LONDON
DISSIPATED.
TOGETHER
With ſome REMEDIES humbly
PROPOSED
By J. E. Eſq;
To His Sacred MAJESTIE,
AND
To the PARLIAMENT now Aſſembled.

Publiſhed by His Majeſties Command.

Lucret. l. 5.
Carbonúmque gravis vis, atque odor inſinuatur
Quam facile in cerebrum? ——

LONDON,
Pr[illegible]ed by *W. Godbid* for *Gabriel Bedel*, and *Thomas Collins*, and are to be ſold at their Shop at the *Middle Temple* Gate neer *Temple-Bar. M. DC. LXI.*

Figure 26.1 The title page of Sir John Evelyn's 1661 book on air pollution. The Latin quotation from the Roman poet Lucretius (97–53 B.C.) near the bottom of the page may be translated *How easily the heavy potency and odor of the carbons sneak into the brain.* (Courtesy British Museum.)

combination of smoke and fog forms a particularly noxious mixture known by the contraction of *sm*oke-f*og*, or **smog.** The presence of fog indicates the temperature of the air near the ground is below the dew point, and with the release of latent heat, there is possibly a temperature inversion (cf. Section 23.5). The gases and smoke are then held near the ground. Continued combustion causes the air to become polluted with smog (Fig. 26.2).

In London's worst smog episode in 1952, over 4000 deaths were attributed to air pollution. Such pollution conditions are particularly hazardous for persons with heart and lung ailments. Smog episodes are not confined to London. Others have occurred in Donora, Pennsylvania (Fig. 26.2), the Meuse Valley in Belgium, and to lesser degrees in New York City and Los Angeles.

It should be obvious that the major source of air pollution is the combustion of fuels, which are primarily **fossil fuels**—coal, gas, and oil. More accurately, air pollution results from the **incomplete combustion** of **impure fuels.**

Incomplete Combustion

Technically, combustion is the chemical combination (burning) of certain substances with oxygen. Fossil fuels are the remains of plant and animal life and are composed chiefly of carbon and compounds of carbon and hydrogen, or **hydrocarbons.**

If a fuel is pure and combustion is complete, the products are CO_2 and H_2O and are not usually considered pollutants, since they are part of the natural atmospheric cycles. For example, if carbon (coal) or methane (natural gas), CH_4, are burned completely, the reactions are

$$C + O_2 \longrightarrow CO_2$$

$$CH_4 + 2\,O_2 \longrightarrow CO_2 + 2\,H_2O$$

However, if fuel combustion is incomplete, the products may include carbon (soot), various hydrocarbons, and carbon monoxide (CO), all of which may contribute to air pollution. **Carbon monoxide** results from the incomplete combustion (oxidation) of carbon. That is,

$$2\,C + O_2 \longrightarrow 2\,CO$$

But, even increased concentrations of CO_2 can affect our environment. Carbon dioxide combines with water to form carbonic acid, a mild acid we all drink in the form of carbonated beverages (carbonated water).

$$CO_2 + H_2O \longrightarrow H_2CO_3$$

Carbonic acid is a natural agent of chemical weathering in geologic processes (see Chapter 30). But, as an indirect product of air pollution, increased concentrations may also aid in the corrosion of metals and react with certain materials causing decomposition (Fig. 26.3). Also, there is some concern about the increased CO_2 content of the atmosphere causing a change in global climate through the greenhouse effect. This aspect will be considered later in the chapter.

Oddly enough, some by-products of complete combustion contribute to air pollution, viz., nitrogen oxides (NO_x). **Nitrogen oxides** are formed when combustion temperatures are high enough to cause a reaction between nitrogen and the oxygen of the air. This typically occurs when combustion is nearly complete, a condition

Figure 26.2 Smog. (*Top*) Smog conditions arise from fog and confined pollution due to an inverted lapse rate. (*Bottom*) A scene in Donora, Pennsylvania, taken in 1949. During a five-day smog episode in 1948 in this Monongahela River Valley town, located 20 mi southeast of Pittsburgh, hundreds of people became ill and at least 20 died. (Courtesy EPA-DOCUMERICA.)

Figure 26.3 An effect of air pollution. The statue is that of Alexander Hamilton at the U.S. Treasury Building in Washington, D.C. (Courtesy EPA-DOCUMERICA.)

that produces the high temperature, or when combustion takes place at high pressure, e.g., in the cylinders of automobile engines. These oxides, normally NO (nitric oxide) and NO_2 (nitrogen dioxide), can combine with water vapor in the air to form nitric acid and contribute to acid rain, which will be discussed shortly. Also, the nitrogen oxides can cause lung irritation and are a key substance in the chemical reactions producing ''Los Angeles'' smog.

This is not the classical London smoke-fog variety, but a smog resulting from the chemical reactions of hydrocarbons with oxygen in the air and other pollutants in the presence of sunlight called **photochemical smog.** Since it was first identified in Los Angeles, it is often referred to as Los Angeles smog. Over 13 million people live in the Los Angeles area, which is in the form of a basin with the Pacific Ocean on one side and mountains on the other (east). This topography makes air pollution and temperature inversions a particularly hazardous combination. A temperature inversion can essentially put a ''lid'' on the city, which then becomes engulfed in its own fumes and exhaustive wastes.

Los Angeles has its share of temperature inversions, with occurrences as frequent as 320 days per year. The California coast is usually under the influence of the edge of semipermanent high-pressure center that extends over the Pacific Ocean. The descending air, coupled with the Los Angeles topography, gives rise to frequent (sometimes daily) subsidence temperature inversions. These atmospheric conditions, a generous amount of air pollution, and an abundance of sunshine, set the stage for the production of photochemical smog (see Fig. 26.4).

In contrast to the smoke fogs of London, photochemical smog produces eye irritation and contains many more dangerous contaminants. One of the products in the photooxidation of hydrocarbons is a family of organic compounds called **aldehydes.** These result from the reaction of certain hydrocarbons with oxygen (Section 17.1).

Two of the major aldehydes in photochemical-polluted air are formaldehyde (HCHO) and acrolein (CH_2CHCHO). Formaldehyde usually accounts for about 50% of the total aldehydes in polluted air and acrolein for perhaps 5%. The identities of the remaining aldehydes are unknown, but these two are problem enough. Their undesirability in our atmosphere can be inferred from the fact that formaldehyde in liquid form is used as an embalming fluid, and acrolein was used as a poison gas in World War I. The cause of eye irritation in photochemical smog may be attributed, in part, to nitrogen-hydrocarbon compounds. The main culprit is peroxyacetal nitrate, or PAN for short.

Another category of photochemically produced pollutants, nitro-olefins, also causes eye irritation. Not photochemically produced but nonetheless dangerous is the hydrocarbon from motor exhausts called benzo(a)pyrene, or more simply, benzpyrene. Benzpyrene is classified as a **carcinogen** because of its cancer-producing capabilities. There is some evidence that at least six to eight other hydrocarbons found in exhaust emissions are carcinogens.

One of the best indicators of photochemical reactions, but also a pollutant itself, is **ozone** (O_3), which is found in relatively large quantities in polluted air. In Los Angeles, air pollution warnings of various degrees are given on the basis of ozone concentrations in the air. To generate the amount of ozone in some polluted areas, a process other than dissociation of O_2 and the combination of O_2 and O, which occurs in the ozonosphere, must be involved (see Section 22.3). The energetic particles responsible for ozone production in the stratosphere do not reach the Earth's surface, so they are not responsible. Scientists think part of the answer involves nitrogen dioxide.

Nitrogen dioxide (NO_2), which is formed by the oxidation of nitric oxide (NO), does absorb energy from

Figure 26.4 Scenes of Los Angeles on a clear day and on days with inversion layers at approximately 300 ft (middle photo) and 1500 ft (bottom photo) above the ground. (Courtesy Los Angeles County Air Pollution Control District.)

the insolation and dissociates. This leads to a series of reactions in which ozone is an intermediate product.

$$NO_2 \longrightarrow NO + O$$

$$O + O_2 \longrightarrow O_3$$

$$O_3 + NO \longrightarrow NO_2 + O_2$$

However, the NO_2 cycle shown in these equations only explains the initial formation of O_3 in polluted atmospheres. It does not explain the development of concentrations as large as those measured, because NO and O_3 are formed and destroyed in equal quantities. Laboratory experiments indicate that hydrocarbons provide the necessary added reactants. Certain types of hydrocarbons from exhaust emissions enter the NO_2 cycle. Studies suggest that oxygen atoms attack the hydrocarbons and the resultant oxidized compounds react with NO to form more NO_2. Thus, the balance of O_2 consumption is upset so that O_3 and NO_2 levels build up, while NO is depleted.

Both O_3 and NO_2 are potentially dangerous to plants and animals. NO_2 has a pungent, sweet odor and is yellow-brown in color. During the peak rush hour traffic, it is often evident in a whiskey-brown haze over large cities. NO_2 reacts with water vapor in the air to form nitric acid (HNO_3), which is very corrosive.

Fuel Impurities

Fuel impurities occur in a variety of forms. Probably the most common impurity in fossil fuels and the most critical to air pollution is **sulfur.** Sulfur is present in fossil fuels in different concentrations. A low sulfur fuel has less than 1% sulfur content, and a high sulfur fuel greater than 2%. When fuels containing sulfur are burned, the sulfur combines with oxygen to form sulfur oxides (SO_x), the most common being **sulfur dioxide** (SO_2).

$$S + O_2 \longrightarrow SO_2$$

A majority of the SO_2 emissions comes from the burning of coal and an appreciable amount from the burning of fuel oils. These are the major fuels used in the generation of electricity. Almost one-half of the SO_2 pollution in the United States occurs in seven northeast industrial states.

By either one of the following double reactions, SO_2 may combine with water and oxygen to form **sulfuric acid** (H_2SO_4):

$$SO_2 + H_2O \longrightarrow H_2SO_3 \quad \text{(Sulfurous acid)}$$

$$2\,H_2SO_3 + O_2 \longrightarrow 2\,H_2SO_4$$

or

$$2\,SO_2 + O_2 \longrightarrow 2\,SO_3 \quad \text{(Sulfur trioxide)}$$

$$SO_3 + H_2O \longrightarrow H_2SO_4$$

Sulfurous acid (H_2SO_3) is mildly corrosive and is used as an industrial bleaching agent. Sulfuric acid, a very corrosive acid, is a widely used industrial chemical. However, in the atmosphere these sulfur compounds can cause considerable damage to practically all forms of life and property. Anyone familiar with sulfuric acid will be able to appreciate its undesirability as an air pollutant. Sulfuric acid is the electrolyte used in car batteries.

The sulfur pollution problem is receiving considerable attention because of the occurrence of acid rain. Rain is normally acidic as a result of carbon dioxide combining with water vapor in the air to form carbonic acid. However, sulfur oxide and nitrogen oxide pollutants cause precipitation from contaminated clouds to be even more acidic, giving rise to "acid rain" (and also acid snow, sleet, fog, and hail). The problem is most serious in New York, New England, and Canada, where pollution emissions from the industrialized areas in the midwestern United States are carried by the general weather patterns. Other areas are not immune. Acid rain is now occuring in the Southeast and acid fogs are observed on the west coast.

Rainfall with a *p*H of 1.4 has been recorded in the northeastern United States. This surpasses the *p*H of lemon juice (*p*H 2.2). In Canada, a monthly rainfall had an average *p*H of 3.5, which is more acidic than tomato juice (*p*H 3.5). The yearly average *p*H of the rain in these affected regions is of the order of 4.2 to 4.4. (See Chapter 16 for a discussion of *p*H.) In addition to acid rain, there are acid snows. Over the course of a winter, acid precipitations build up in snowpacks. During the spring thaw and runoff, the sudden release of these acids gives an "acid shock" to streams and lakes.

Acid precipitations lower the *p*H of lakes, which threatens aquatic plant and animal life. Most fish species die at a *p*H of 4.5 to 5.0. As a result, many lakes in the northeastern United States and Canada are "dead" or are in jeopardy of dying. Natural buffers in area soils tend to neutralize the acidity, so waterways and lakes in an area don't necessarily match the *p*H of the rain. However, the neutralizing capability in some regions is being taxed, and the effects of acid rain are increasing.

Thus, air pollution can be quite insidious and can consist of a great deal more than the common particulate matter in the form of smoke, soot, and fly-ash, which blackens buildings and the weekly wash if it is hung outside. Pollutants may be in the form of mists and aero-

Figure 26.5 A future advertisement?

sols. Some pollutants are metals, such as lead and arsenic. Approximately 100 atmospheric pollutants have been identified and 20 of these are metals that come primarily from industrial processes. The extent of pollution varies a great deal and depends on the conditions, sources, and types of pollutants. One hopes it will not reach the extent characterized by the ad in Fig. 26.5.

26.2 Sources of Air Pollution

Table 26.1 and Fig. 26.6 show the sources and magnitudes of atmospheric pollutants. As can be seen from the figure, transportation is the major source of total air pollutants. The United States has a mobile society, with over 100 million registered vehicles powered by internal combustion engines.

There are various types of internal combustion engines, but most common is the gasoline engine used in the automobile. The completeness of combustion depends primarily on the air-fuel mixture. The proper **air-fuel ratio** for near-complete combustion is 15:1. When accelerating, the air-fuel ratio is on the order of 12:1; while cruising 13:1; and while idling and decelerating, as in stop-and-go city traffic, 11:1. When the ratio is less than 15:1, substantial amounts of CO and hydrocarbons are formed as shown in Fig. 26.7. Note, however, that the emission of nitrogen oxides is near maximum when the air-fuel ratio is 15:1.

Considerable efforts are being made by automobile manufacturers to reduce these emissions through engine modifications and special devices. It should now be readily understood why one is urged to keep an automobile engine properly tuned. A properly tuned engine operates nearer the correct air-fuel ratio, which minimizes emissions (see Fig. 26.8). However, even with the current efforts to reduce pollution emissions from automobiles, greater pollution is predicted in the future due to the increasing number of motor vehicles as shown in Fig. 26.9.

Another source of pollution comes from the lead additive used in the gasoline as an "anti-knock" agent. The

Table 26.1 Nationwide Emission Estimates[a] (10^6 tons/year)

Source Category	SO_x	*Particle*	*CO*	*HC*	NO_x
Transportation	1.0	0.7	111.0	19.5	11.7
Fuel combustion in stationary sources	26.5	6.8	0.8	0.6	10.0
Industrial process losses	6.0	13.3	11.4	5.5	0.2
Solid waste disposal	0.1	1.4	7.2	2.0	0.4
Agricultural burning	Neg[b]	2.4	13.8	2.8	0.3
Miscellaneous	0.3	1.5	4.5	4.5	0.2
Total	33.9	26.1	149.0	34.9	22.8

[a](Courtesy EPA).
[b]Negligible (less than 0.05×10^6 tons/year).

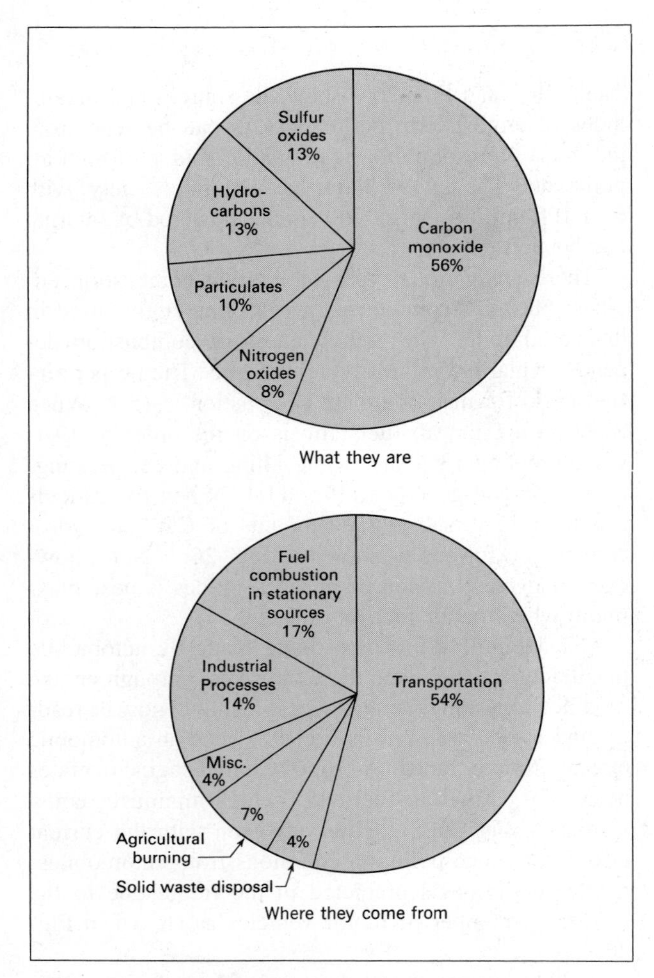

Figure 26.6 Graphic representations of data in Table 26.1 showing sources and relative magnitudes of air pollution emissions.

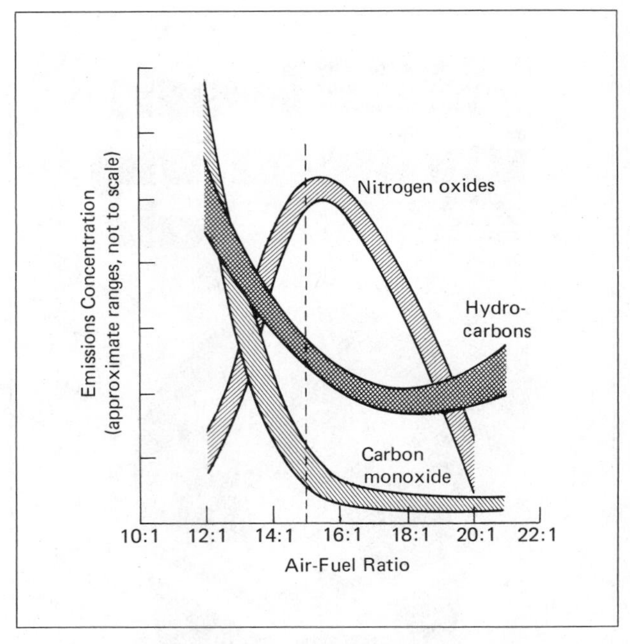

Figure 26.7 The effect of air-fuel ratios on the composition of exhaust emissions of an automobile engine. The lower the air-fuel ratio is below 15:1 (complete combustion), the greater the carbon monoxide and hydrocarbon emissions. More nitrogen oxides are formed when combustion is nearly complete, due to higher temperatures. (Courtesy U.S. DHEW.) (Source: D. A. Trayser et al., Battelle Memorial Institute, Columbus, Ohio.)

engines of new automobiles are designed to use lead-free gasoline. The levels of lead in leaded gasoline have been reduced and "regular" gasoline will eventually be phased out.

Other internal combustion engines are the diesel and jet engines. The diesel engine operates with an unrestricted air supply (no carburetor), and the fuel is injected directly into the combustion chamber; hence, the air-fuel ratio is usually higher than that for the gasoline engine. The diesel engine has no spark plug for ignition, using the heat from the high compression of the air-fuel mixture to cause ignition. Due to excess air there should be little unburned fuel, theoretically. However, improper operation and overloading, particularly in trucks, makes diesels worse polluters than gasoline engines (Fig. 26.8). With high operating temperatures, diesels also form nitrogen oxides more readily.

Jet engines operate on somewhat different principles than the diesels. Combustion in a jet engine is used to heat compressed air, which expands and escapes out of the rear of the engine. By the conservation of momentum (Section 3.4), a plane powered by jet engines receives a forward thrust from the high-speed jet of escaping gases. A jet engine is, of course, larger and more powerful than an automobile engine, and with its larger size comes increased pollution. It is estimated that a jet engine produces roughly 10,000 times the emissions of an automobile gasoline engine. In this respect, we are fortunate that there are fewer jet aircraft than automobiles.

From Fig. 26.6, it can be seen that the other major sources of air pollution are from stationary sources and industrial processes. Stationary sources refer mainly to electrical generating facilities. As can be seen from Table 26.1, these sources account for the majority of the sulfur oxide (SO_x) pollution. This results primarily from the burning of coal, which always has some sulfur content.

Figure 26.8 Air pollution from automobile, jet, and diesel engines. (Courtesy EPA. Truck photo, Los Angeles County APCD; auto photo, Ames Iowa Daily Tribune.)

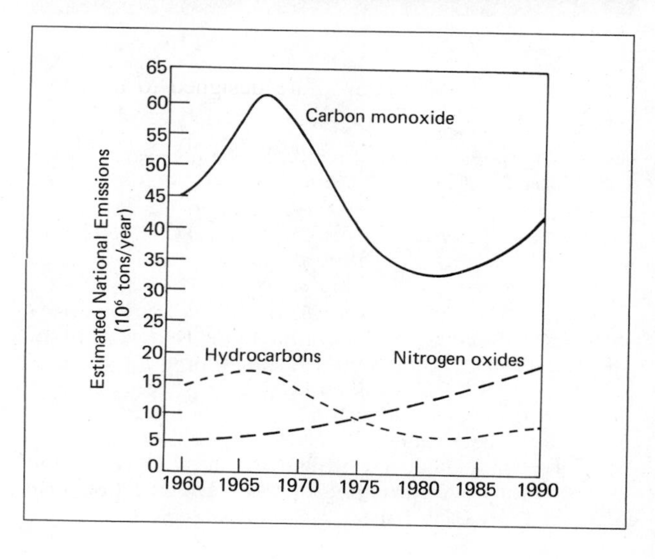

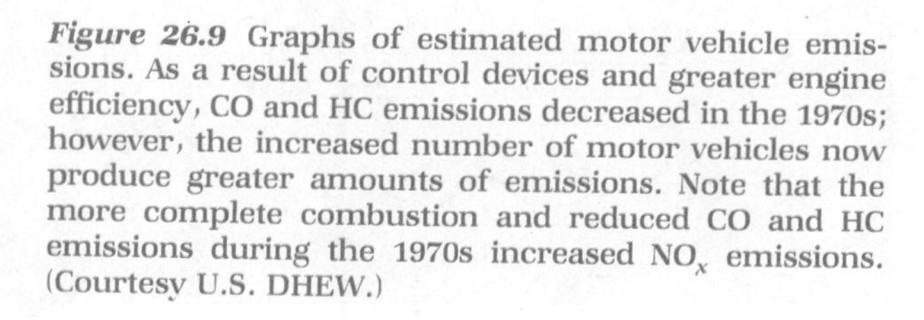

Figure 26.9 Graphs of estimated motor vehicle emissions. As a result of control devices and greater engine efficiency, CO and HC emissions decreased in the 1970s; however, the increased number of motor vehicles now produce greater amounts of emissions. Note that the more complete combustion and reduced CO and HC emissions during the 1970s increased NO_x emissions. (Courtesy U.S. DHEW.)

26.3 The Costs of Air Pollution

The costs of air pollution are difficult to assess. Visible evidence of the costs resulting from air pollution comes from reduced real estate values, repairs and replacements from corrosion, cleaning costs, animal and agricultural damages, and many more sources, not excluding the effects on human health. Some examples of air pollution damage are shown in Fig. 26.10.

The corrosion of metals and other building materials is 3 to 30 times greater in cities than in rural areas, depending on the type of pollutant. Corrosive action also depends on the amount of moisture in the air. The rate of corrosion increases sharply when the relative humidity is over 50%.

Pollution effects range from damage to seemingly permanent stone structures to damage to electrical insulators. Concentrations of ozone in photochemical smog can cause the deterioration and cracking of rubber, a common electrical insulator. This can cause electrical problems and damage. In some cities where photochemical smog is severe, car owners can purchase special anti-ozone tires for their automobiles to help alleviate the problem. Also, dust and particulate matter soil our clothes and home furnishings and can cause damage to delicate instruments. To alleviate some of this problem, we pay for the cost of filtering the air.

Crops, shrubs, trees, and other plants in general suffer from air pollution. There are many cases of industrial pollution having effects on the flora of an area. The

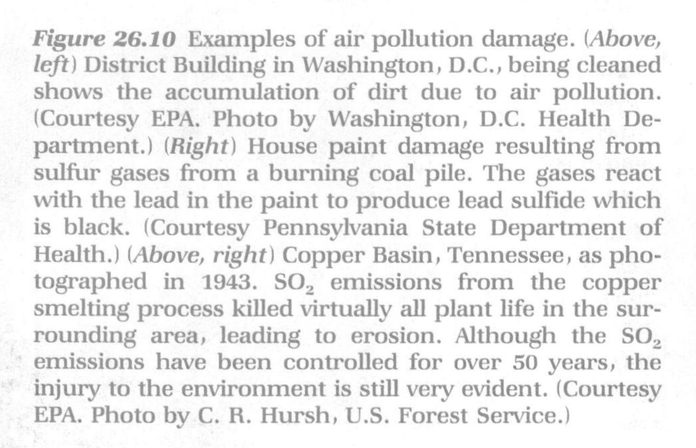

Figure 26.10 Examples of air pollution damage. (*Above, left*) District Building in Washington, D.C., being cleaned shows the accumulation of dirt due to air pollution. (Courtesy EPA. Photo by Washington, D.C. Health Department.) (*Right*) House paint damage resulting from sulfur gases from a burning coal pile. The gases react with the lead in the paint to produce lead sulfide which is black. (Courtesy Pennsylvania State Department of Health.) (*Above, right*) Copper Basin, Tennessee, as photographed in 1943. SO_2 emissions from the copper smelting process killed virtually all plant life in the surrounding area, leading to erosion. Although the SO_2 emissions have been controlled for over 50 years, the injury to the environment is still very evident. (Courtesy EPA. Photo by C. R. Hursh, U.S. Forest Service.)

classic example is Copper Basin, Tennessee, where a copper smelting plant was built shortly after the Civil War. Its production increased in the late 1800s, and consequently its SO_2 emissions. In time, these deadly fumes killed the plant life of some 30,000 acres of the surrounding timberland (Fig. 26.10). Flowers for the Rose Bowl Parade used to come from Pasadena (the location of the Rose Bowl) and the nearby Los Angeles area, but now they must come from other places too, because of the effects of smog on the floral industry of that area. Plants are particularly vulnerable to gaseous pollutants. Gases enter the microscopic openings, called stomata, on the underside of the plant leaves and attack the plant cells.

The preceding are but a few of the effects of air pollution that cause the government to estimate that the average yearly cost of air pollution in the United States is about $75 per person. With a population of over 240 million, simple arithmetic shows the total yearly cost estimate to exceed $18 billion. This includes estimates of $100 million for paint damage, $800 million for laundering, $240 million for car washing, and $500 million in damages to livestock and agriculture.

Air pollution has severe effects not only on the objects of our physical environment, but also on ourselves. The cost of air pollution on human health is difficult to estimate. How much is your health worth? Rather than conjecture about the monetary costs of air pollution on health, let's consider and become aware of the effects on the respiratory system. Humans inhale about five quarts of air per minute. Air is normally taken in through the nose and is warmed and moistened in the nasal cavity. On its route to the lungs the air passes through the pharynx, which is also connected to the mouth as illustrated in Fig. 26.11.

Below the pharynx is the funnel-shaped larynx, or voice-box, the rim of which projects outward forming the Adam's apple. Two muscular tissues across the larynx, the vocal cords, are capable of producing sound when air is exhaled. Passing through the larynx the inhaled air enters the trachea, or wind pipe, which divides into two bronchial tubes. Each of these bronchi enters

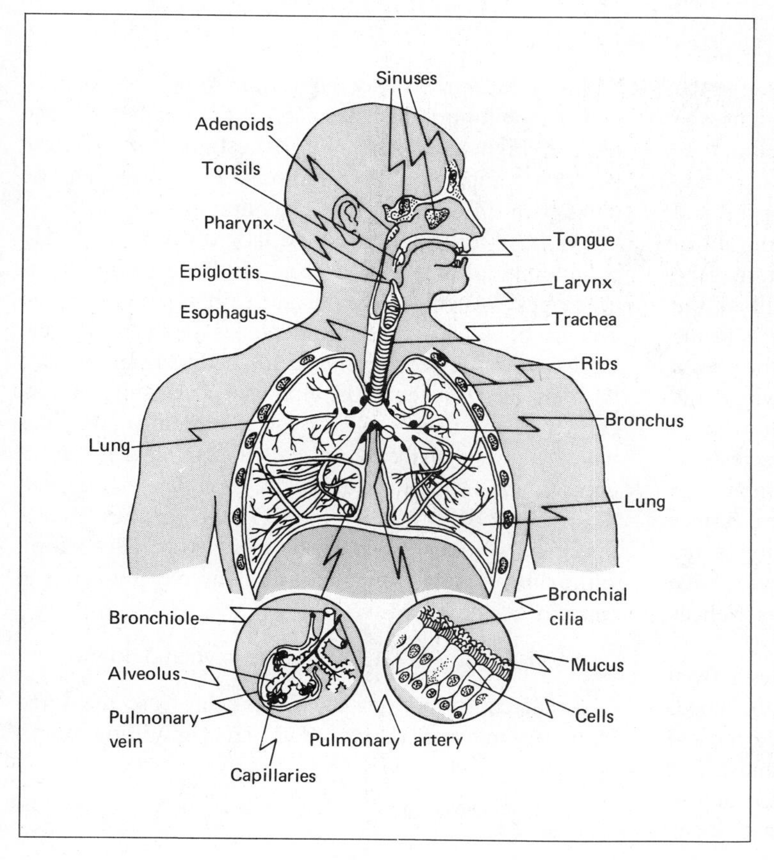

Figure 26.11 The anatomy of the human respiratory system. (From the Air Pollution Primer, The National Tuberculosis and Respiratory Disease Association.)

one of the lungs and divides again and again into fine bronchioles. At the end of the bronchioles are millions of tiny air spaces, or sacs, called **alveoli,** which are surrounded by capillaries. (See the blowup in Fig. 26.11.) The oxygen of the air is exchanged through the walls of the alveoli for the CO_2 waste from the cells which is transported to the lungs by the blood.

The respiratory system has natural protection mechanisms against foreign materials. The hair within the nostrils filters out large particles in the air. To guard the lungs against finer particles, the upper respiratory tracts—the nasal cavity, pharynx, larynx, and bronchial tubes—are lined with mucous membranes that excrete liquid mucus. The fine particles are caught in the mucus and are swept out of the respiratory system by the action of tiny hairlike projections called **cilia,** Fig. 26.11.

This natural defense is effective for particulate matter, which is nature's major contaminant in air, but the respiratory tract is vulnerable to gaseous pollutants. When attacked by SO_2 and other smog constituents, the cilia become less efficient, allowing particulate matter to penetrate deeper into the lungs. The irritation caused by pollutants causes coughing, increased mucus flow, and muscle spasms that limit the air flow to and from the lungs. This condition is symptomatic of inflamed bronchi and is referred to as bronchitis. Irritated tissue in the bronchiole may swell, and with the other conditions, the small bronchiole may be obstructed as illustrated in Fig. 26.12.

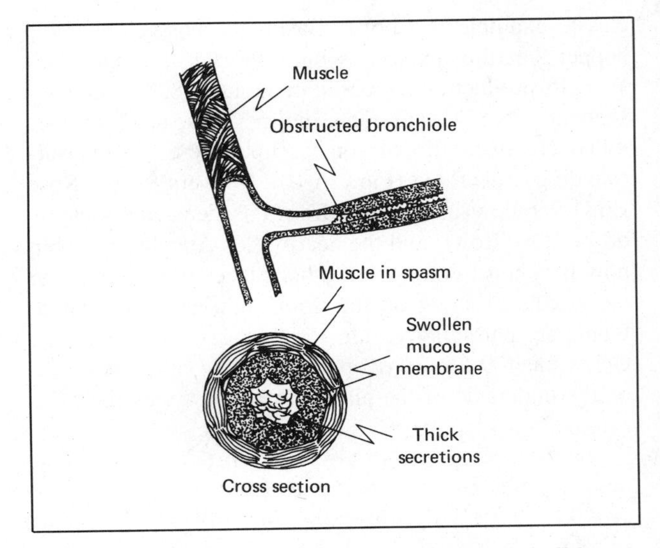

Figure 26.12 An obstructed bronchiole, or small bronchial tube. This condition, which restricts normal breathing, can arise from irritation of the respiratory tract by air pollutants. The lower drawing is a cross section of an obstructed bronchiole. (Courtesy J. D. Wilson)

The air passages to the alveoli normally expand and contract with breathing. However, if this action is impaired by pollution irritation, some of the air may be retained as new air is introduced. As a result of the increase in air volume, the air sacs get larger, and the walls of the alveoli may eventually rupture, forming a larger sac. Large pouches may form as shown in the drawing of Fig. 26.13. This condition, characterized by a shortness of breath, is called **emphysema,** a respiratory disease that is becoming increasingly common. The air pouches may enlarge as the process continues, further destroying the oxygen-CO_2 exchange capability of the lungs. Personal air pollution through smoking may give rise to respiratory conditions such as these, as well as aggravating existing ones.

One thing common to all respiratory problems from pollution is the decrease of oxygen exchange to the blood. When there is an oxygen deficiency, the body tries to rectify the situation by increased breathing and blood flow. The latter requires the heart to work faster, and so a secondary effect of air pollution is heart disease. It can be easily understood why the health problems associated with air pollution are the most critical for the sick and elderly. Figure 26.14 (p. 496) illustrates the effect of age in a comparison of emphysema cases between two cities with different degrees of pollution.

We have all experienced the lack of oxygen, if only by holding our breath. Its effects are dizziness and lightheadedness. Another air pollutant, carbon monoxide, can also account for oxygen deficiencies in the body that impose extra physical burdens. Inhaled CO can displace the combined oxygen in the blood and reduce the amount carried to the tissues. The oxygen normally combines with a protein in the blood called hemoglobin (Hb).* Oxygen and CO react with Hb in a similar manner, but the hemoglobin's affinity for CO is 200 times greater than for O_2. Carbon monoxide reacts with Hb to form carboxyhemoglobin, thus reducing the oxygen-carrying capacity of the blood.

$$CO + Hb \longrightarrow COHb \text{ (Carboxyhemoglobin)}$$

Small concentrations of contaminants are measured by the volume ratio expressed as parts per million (ppm).

*Hb is not the chemical formula for hemoglobin. Hb is used for simplicity.

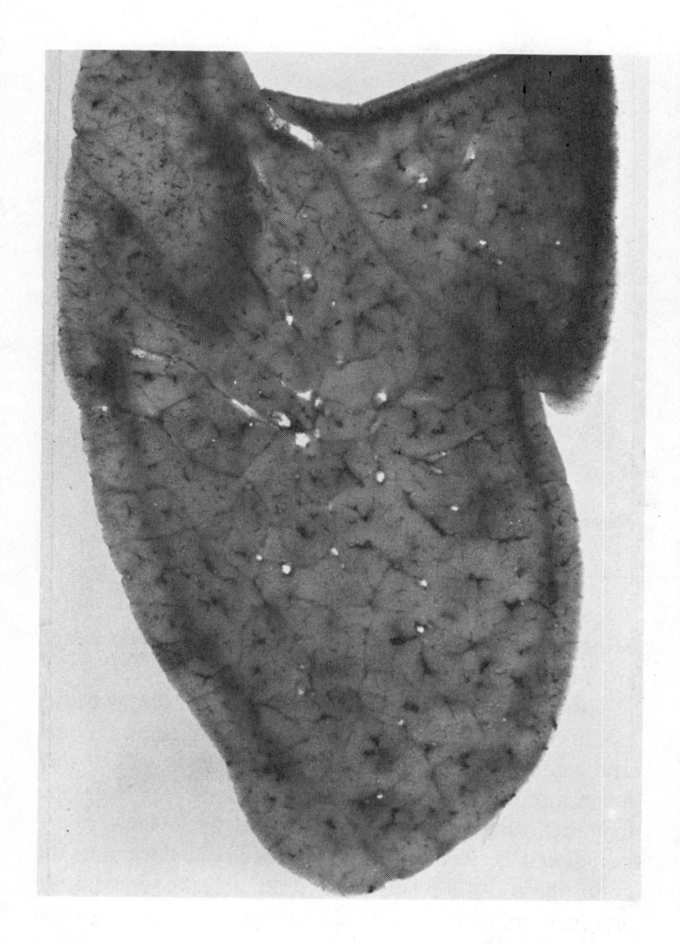

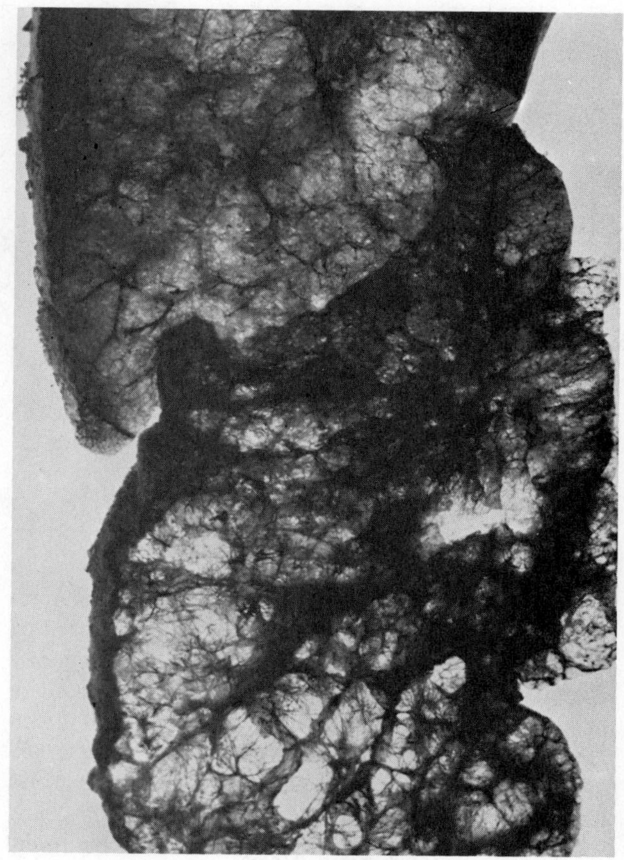

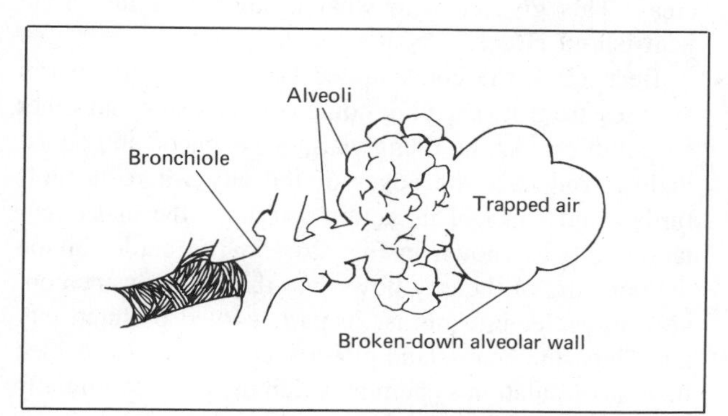

Figure 26.13 A diagram of the lung condition emphysema (*above*). Irritation of the lungs may cause air to be trapped in the alveoli. With the intake of more air, the increased volume results in the breakdown of the alveolar walls, forming large pouches of trapped air, which restricts breathing and reduces the oxygen-CO_2 exchange capacity of the lungs. The photo shows a section of normal lung (*left*) and an emphysematic lung section (*right*). (Photo courtesy EPA. Photo by Webb-Waring Institute for Medical Research.)

A concentration of 1 ppm means there is one volume of contaminant in a million such volumes. Exposure to air with 10 ppm of CO for eight hours produces mental dullness. CO concentrations of 100 ppm can produce headaches and dizziness in a short time, so there is great cause for concern when CO concentrations greater than 300 ppm have been measured in idling cars in a traffic jam. The concentration of CO in the mainstream of cigarette smoke is over 2500 times greater than the recommended maximum atmospheric concentration (15 ppm). Consequently, smokers have 5–10% less hemoglobin available for oxygen transfer due to increased CO concentrations in the blood.

The other main gaseous air contaminants, SO_2, NO_2, and O_3, are also quite dangerous in small concentrations. Sulfur dioxide (SO_2) dissolves in the mucous lining of the upper air tract, while NO_2, and O_3, being less soluble, travel farther. However, soot, which has a very chemically active surface, can carry adsorbed SO_2 deeper into the respiratory tract. Carcinogens may also use soot

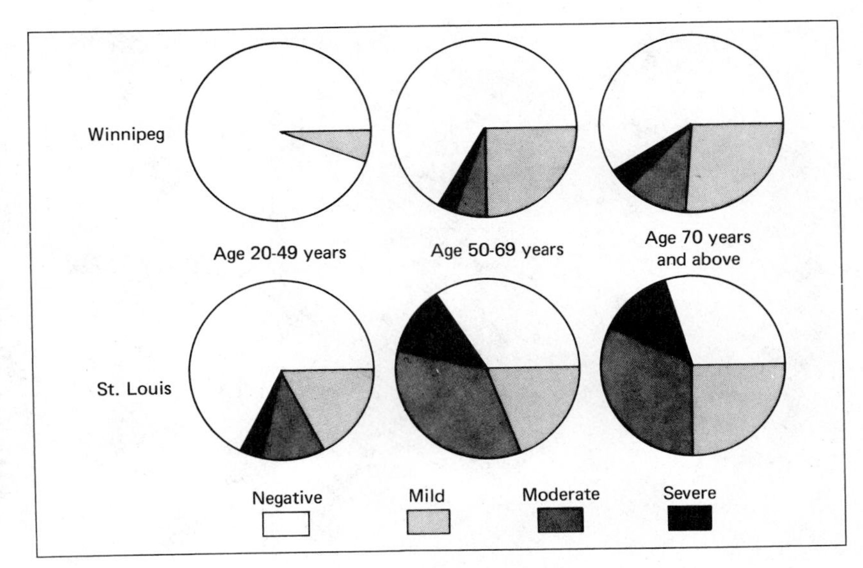

Figure 26.14 A comparison of emphysema cases in two cities with contrasting levels of air pollution. The graphs show the results of a survey of 300 residents of heavily industrialized St. Louis, Missouri, and relatively unpolluted Winnipeg, Canada. Notice the prevalence of emphysema with age. (From the Air Pollution Primer, The National Tuberculosis and Respiratory Disease Association.)

and particulate matter as a vehicle into the lungs. Aerosols may also be inhaled deeply into the lungs. You should now begin to realize the forbidding dangers of a smoky, photochemical smog. An additional hazard of many pollutants is **synergism,** or their combined effect. For example, the combinations of CO and nitrogen oxides, or CO, CO_2, and H_2O, are several times worse than CO alone.

There are many other air pollutants that create health hazards. These are usually associated with particular industrial processes and are confined to the ambient air of the factory or the atmosphere of the local region. Processes involving metallic chemicals are particularly dangerous due to metal poisoning by such metals as beryllium and arsenic. Beryllium is now being used in rocket fuels, so its pollution may be more widespread. Lead, the antiknock agent used in gasoline, may also cause metal poisoning. Like most metals, the lead accumulates in certain parts of the body, causing blood disorders and other complications.

26.4 Weather and Climatic Effects of Air Pollution

Although we sometimes purposely attempt to modify the weather, e.g., rainmaking or fog dissipation, the effects resulting from air pollution are inadvertent modifications of weather and climate. The most immediate modifications, of course, occur near the sources of pollution. Since air pollution results from the activities of human beings, the major concentrations of pollutants are in and around population centers or industrial areas.

Cities have always had somewhat different weather conditions than those of the surrounding countryside. The buildings of a city obstruct the wind and absorb much more insolation, since their surface area is greater than that of the ground they cover. Also, the heat energy is radiated between buildings rather than back into the atmosphere. As a result of such effects, the temperature and heat content of cities are higher than outlying rural areas. This gives rise to what is known as the urban **heat-island effect.**

Because of the concentrated heat of the city, warm air rises from it, carrying combustive wastes and other air pollution. As the rising warm air expands and cools, it flows out over the edges of the city where it cools further and sinks. This action sets up a thermal circulation cell, as shown in Fig. 26.15. The cooler air on the outskirts of the city flows into the center in the convection cycle, but this is, in part, cooled polluted city air. Thus, the heat-island effect sets up a self-contained thermal circulation system in which the air is continually polluted.

In the absence of winds, e.g., frontal air movement, to break up the circulation system and sweep the pollution away, serious pollution conditions can arise in cities. This is particularly true should a high-pressure air mass move over the city and become stationary. In this windless condition, along with possible topographical effects, such as a city being in a river valley, the air of a city becomes continually polluted by its own activities,

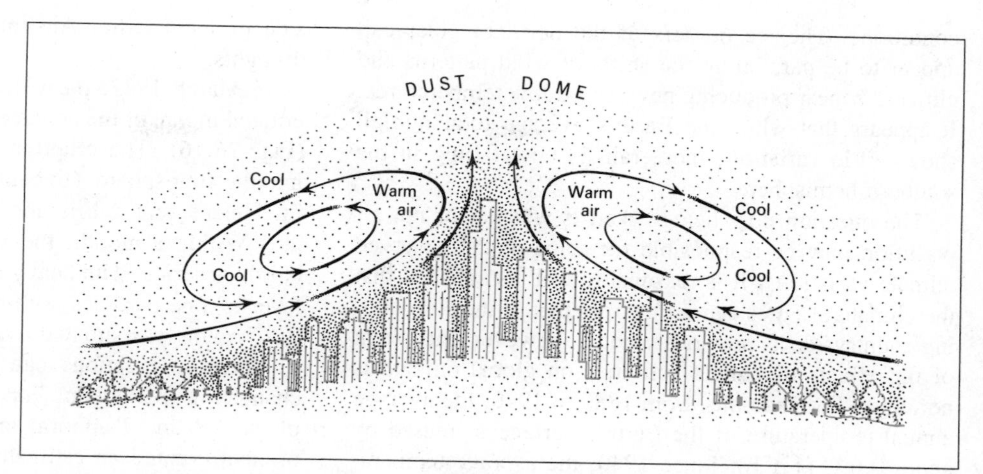

Figure 26.15 An illustration of the heat-island effect, showing the air circulation and dust dome of a city.

which may eventually have to be curtailed. This condition is not uncommon in several U.S. cities, typically in the summer as a result of a ''stagnant'' air mass.

It might be thought that additional insolation on an urban area would supply sufficient heating and convection to break up the self-contained circulatory system of the heat-island effect. This is a possibility, but it is counterbalanced by the formation of a **dust dome** over the city (Fig. 26.15). City pollutants contain large amounts of aerosols and particulate matter. As the polluted air moves up and spreads out over the city, it forms an often-discernible (dust) dome. The dust concentration may be several hundred to a thousand times that of rural air. This polluted ceiling decreases the amount of insolation that penetrates the city atmosphere and prevents the Sun from efficiently heating the metropolitan area to break up the circulation cell of the urban heat island. A city must be content to wait for outside air movements to sweep away its pollution.

Thus, air pollution causes cities to have modified weather conditions. This is manifested not only in a temperature increase and attentuated insolation, as just mentioned, but also in increased cloudiness and precipitation. The updrafts and concentration of particles give rise to cloud formation and precipitation (Chapter 23). Records show that cities receive several percent more precipitation than nearby rural areas. Also, with drains and sewer systems, the water runoff is more rapid in cities, which results in a lower relative humidity near the surface.

Climate

Although not as immediately obvious, there is a general belief that there are (and will be) changes in the global climate brought about by atmospheric pollution. **Climate** is the long-term average weather conditions of a region. Some regions are identified by their climates. For example, when Florida or California is mentioned, one usually thinks of a warm climate, and Arizona is known for its dryness and low humidity. Because of such favorable conditions, the climate of a region often attracts people to live there, and thus the distribution of population (and pollution) is affected.

Dramatic climate changes have occurred throughout the Earth's history. Probably the most familiar is that of the Ice Age, when glacial ice sheets advanced southward over the North American continent. The most recent of these, some 10,000 years ago, came as far south as the northern conterminous United States. Some evidence now exists from ocean sediment cores to support the theory that dramatic changes in global climate result from subtle, regular variations in the Earth's orbit around the Sun. For example, one cyclic variation is the advance of the Earth in its elliptical orbit. That is, the Earth's closest approach to the Sun occurs at different times of the year in a cycle of 23,000 years. The Earth and the Sun are now closest in January; in 10,000 years they will be closest in July. This would result in cooler summer temperatures, less snow melting, and a growth of the polar ice caps. Such a change could slowly lead into a new ice age.

However, climatic fluctuations have been continually occurring, but of much smaller magnitude than that of the Ice Age. For example, in the last several decades there has been a noticeable southward shift of world climate. The Sahara Desert is reportedly advancing southward in places as much as 30 mi a year, and rainfall in some regions is barely one-half of what it was 20 years ago, giving rise to great bands of land with drought conditions. The ice borders of the northern polar cap appear to be paralleling the shifts of wind patterns and climatic zones, producing new record low temperatures. It appears that while the Earth's average total rainfall shows little variation, more rain is now falling in the southern hemisphere.

The question now being asked is whether or not air pollution may be responsible for some of the observed climate variations. It is reasonably well established that the climate of cities can differ from that of the surrounding countryside as previously noted; however, the extent of the effects of human activities on global climate is not clear. For example, from 1880 to 1940, the average annual temperature of the Earth's surface increased by about 0.6°C (1.1°F). Since 1940, the average temperature has decreased by about 0.3°C (0.55°F). Associated with this lowering temperature has been the shift of the frost and ice boundaries, a weakening of zonal wind circulation, and marked variations in the world's rainfall pattern.

Global climate is sensitive to atmospheric contributions that affect the radiation balance of the atmosphere. These include the concentrations of CO_2 and other "greenhouse" gases, the particulate concentration, and the extent of cloud cover, which affect the Earth's albedo. Air pollution and other human activities do contribute to changes in these conditions. Scientists are now trying to understand climate changes using various models. These models of the workings of the Earth's atmosphere and oceans allow the comparison of theories using historical data on climate changes; however, specific data are scant.

More recent natural occurrences have received a great deal of study. In 1982–83 the "Big El's"—El Niño and El Chichón—provided data. **El Niño** is a Pacific current that sporadically rises off the coast of Peru and Ecuador (western coast of South America). El Niño is the Spanish name for "The Child," so named because the current usually occurs around Christmas. Every 8–10 years, El Niño grows more intense and its nutrient-poor waters destroy rich coastal fishing areas.

The 1982–83 El Niño was a record breaker. The ocean waters were 12–15° warmer than usual for more than a year. This El Niño was connected to a failure of the equatorial trade winds that normally push the warm waters away from the South American coastline into the Pacific. Without these winds, the warm currents pushed against the coast, bringing heavy rains to normally dry areas. The arid regions of the coastal countries received more than 100 times the normal precipitation. On the other side of the Pacific, Australia and the Philippines had droughts.

In March 1982, the volcano **El Chichón** in Mexico erupted in one of the century's most significant outbursts (Fig. 26.16). The eruption sent debris 26 mi (42 km) into the atmosphere. (In contrast, the eruptions of Mount St. Helens sent debris about 12 mi high. See Chapter 29.) Within a month, the volcanic veil had crept over North America. The heavy ash fell back to Earth, but a layer of fine particles, sulfuric acid, and salt remained. El Chichón was situated over a natural salt dome.

Volcanic eruptions can affect the climate through changes in the albedo. For example, the 1815 eruption of the volcano Tambora, located on an island just east of Java, vented an estimated 35 mi^3 of particle debris into the air. Fine volcanic dust was circulated around the Earth by global wind patterns, and the winter of 1816 was unseasonably cold due to the change in albedo. New England farmers called 1816 "the year without summer" with frosts in June and July.

A similar albedo concern has been dubbed "**nuclear winter.**" Nuclear war could plunge the Earth into severe cold. Extensive fires and massive quantities of smoke could shut off sunlight and cool the Earth.

Of course, we can't do as much about volcanic eruptions as we can about the potential of nuclear winter; however, the possible climatic effects are a concern. Such eruptions could perhaps affect the jet streams, which could lead to climatic changes. Also, with volcanic gases being projected into the stratosphere, there is concern about possible chemical reactions with and depletion of the ozone layer.

In terms of air pollution, you may recall the apprehension expressed in the 1970s over the effects of aerosol spray can propellants on the ozone layer. The common gas propellants in question were the chlorofluorocarbons $CFCl_3$ and CF_2Cl_2 (freons). It was theorized that when these gases were released, they would rise slowly into the stratosphere (a process taking over 20–30 years). In the stratosphere, the chlorofluorocarbon atoms would be broken apart by ultraviolet radiation, releasing reactive chlorine atoms. These, in turn, would react with and destroy ozone molecules in a repeating cycle:

$$\text{Chlorofluorocarbon} \xrightarrow{\text{uv}} \text{Cl}$$

$$Cl + O_3 \longrightarrow ClO + O_2$$

$$ClO + O \longrightarrow Cl + O_2$$

The depletion of the protective, ultraviolet-absorbing ozone layer in the stratosphere would have serious en-

Figure 26.16 El Chichón. The volcano, located in southeast Mexico, had been quiet for centuries before it erupted on Sunday, March 28, 1982, spewing ash miles into the atmosphere. (Courtesy AP/Wide World Photos.)

vironmental consequences. Fortunately, as a result of environmental concern, manufacturers are now using alternative, environmentally safe propellants, or have switched to mechanical pump systems. Even so, tons of chlorofluorocarbons are released into the atmosphere from refrigerant use and industrial applications. If this continues, it is estimated that there will be a 5–9% depletion of the ozone layer in the next 50 years. This would cause warming and a melting of the polar ice caps, with a rise in sea level of several feet. Such a rise would have a devastating effect on coastal cities. (The rise would be gradual, not overnight.)

Thus, **particle pollution** could contribute to changes in the Earth's thermal balance by decreasing the transparency of the atmosphere to insolation. An increase in the albedo would result in lower surface temperatures. The effect of particulate pollution depends on the number and size of the particles. Relative sizes are shown in Fig. 26.17. Small particles can also cause increased absorption of the outgoing infrared terrestrial radiation.

Contributions to an albedo change could also come from an increase in cloud coverage that might result from an abundance of particle nuclei. Recall from Section 22.4 that clouds are good reflectors of solar radiation, and thus they play a major role in the Earth's albedo. Increased aircraft activity in the upper troposphere may have resulted in an increase in cirrus clouds.

Supersonic transport (**SST**) aircraft operating in the lower stratosphere have also caused some concern because of particulate and gaseous emissions of the jet engines. It is estimated that the hourly combustion of fuel by an SST releases about 80 tons of water vapor and 70 tons of carbon dioxide into the lower stratosphere.

In the troposphere, precipitation processes act to "wash" out particle and gaseous pollutants, but there is no snow or rain washout mechanism in the stratosphere. Also, the stratosphere is a region of high chemical activity, and chemical pollutants (e.g., NO_x and hydrocarbons) could possibly give rise to climate-changing reactions.

Albedo considerations also arise from activities that affect the Earth's surface. Urbanization and agriculture affect the surface albedo.

There is also a temperature-increase pollution aspect. Vast amounts of CO_2 are being expelled into the atmosphere as a result of the combustion of fossil fuels (Table 26.1). As discussed in Section 22.4, CO_2 and water vapor play important roles in the Earth's energy balance through the greenhouse effect. An increase in the atmospheric concentration of CO_2 could alter the amount of radiation absorbed from the Earth's surface and produce an increase in the Earth's temperature. An early calculation of the effect of such an increase was made

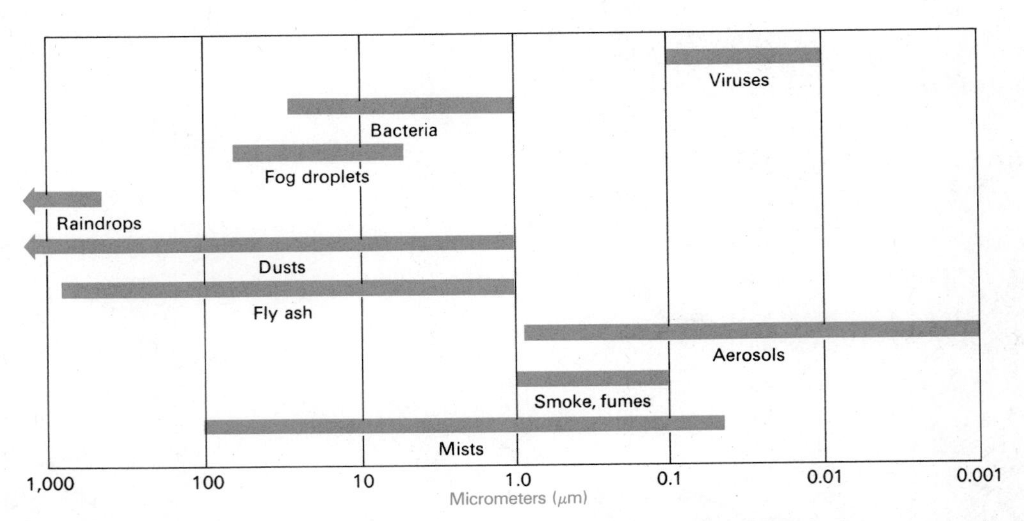

Figure 26.17 The sizes of several different particles. (One micrometer is $\frac{1}{1000}$ mm, or $\frac{1}{25000}$ in.)

in 1896 by Svante Arrhenius. Arrhenius's calculations showed that a doubling of the atmospheric **CO_2 concentration** would increase the Earth's surface temperature by 5–6°C. It might be pointed out that Arrhenius's concern arose from the CO_2 escaping from volcanoes, not from automobile exhaust emissions.

During the 19th century the atmospheric CO_2 content increased by about 10%, as determined from old records. The increase has been even greater in the 20th century with larger populations and more combustion. It appears that the Earth will be a warmer place in the 21st century, with climate changes that will affect agriculture, water resources, and sea level. Recent studies predict that the atmospheric CO_2 content will double by the year 2065, with an accompanying temperature increase of 1.5–4.5°C. It might be noted that Venus has a rich CO_2 atmosphere—200,000 times more than that of Earth—and a surface temperature of 470°C (900°F), hot enough to melt lead. Of course, Venus is closer to the Sun than is Earth.

Possible effects of increasing temperature include the melting of the polar ice caps as noted previously in the discussion on the effects of the depletion of the ozone layer. Also, there would be dryer summers in the middle latitudes. In the United States, there would be dry farmlands in the South and longer growing seasons in the North. Many other possibilities are not well understood, including the accelerated release of greenhouse gases, such as methane (CH_4), from swamps and bogs due to the warming climate.

Is it then possible that an increase in atmospheric CO_2 concentration led to the increase in global temperature (and subsequent changes in the climate) observed from 1880 to 1940? It is generally accepted that this is a possibility. However, it should be noted that a substantial portion of the change in the atmospheric CO_2 concentration has occurred in the last several decades, during which the contribution due to the activities of human beings almost doubled. At the same time (since 1940), the average global temperature has shown a decrease. Evidently, air pollution also affects other mechanisms of the atmosphere's energy balance.

Thus, air pollution may contribute to a variety of atmospheric effects—both local and global. Effects of pollution on local climatic conditions are somewhat clear, but there are too few data to understand or accurately predict global climatic effects. We are inclined to believe that increased atmospheric CO_2 concentrations and ozone layer depletions would give rise to an increase in global temperature. This would cause more water evaporation and an increase in the relative humidity. Particulate pollution could then give rise to increased cloud formation, which would increase the albedo and cause a decrease in global temperature.

Certainly little comfort can be found in this speculative, counterbalancing pollution cycle. There are too many unanswered questions on the effects of pollution, which has occurred over a relatively short time, on the natural interacting cycles of the atmosphere and biosphere, which have taken millions of years to become established.

Learning Objectives

After reading and studying this chapter, you should be able to do the following without referring to the text:

1. Discuss and describe the major air pollutants.
2. Distinguish between regular smog and photochemical smog.
3. State the major sources of the various air pollutants.
4. Discuss the effects of the air-fuel ratio of the gasoline internal combustion engine.
5. Explain some of the damaging effects of the various air pollutants to property.
6. Describe some of the effects of air pollutants on the human respiratory system.
7. Discuss the effects of air pollution on local weather conditions.
8. Discuss the effects of air pollution on global climate.

Important Words and Terms

air pollution
smog
fossil fuels
incomplete combustion
impure fuels
hydrocarbons
carbon monoxide
nitrogen oxides
photochemical smog
aldehydes
carcinogen
ozone
sulfur
sulfur dioxide (SO_2)
sulfuric acid (H_2SO_4)
acid rain
air-fuel ratio
alveoli
cilia
emphysema
synergism
heat-island effect
dust dome
climate
albedo
El Niño
El Chichón
particle pollution
CO_2 concentration
SST

Questions

1. Define *air pollution*.
2. Is air pollution a relatively new problem?
3. What gives rise to the majority of air pollution?
4. What are the products of complete combustion? incomplete combustion?
5. Are nitrogen oxides products of complete or incomplete combustion? Explain.
6. Distinguish between plain smog and photochemical smog.
7. What is one of the best indicators of photochemical smog?
8. What is the major fossil fuel impurity?
9. What is the sulfur content of a low-sulfur fuel? a high-sulfur fuel?
10. What are the causes and effects of acid rain? In which areas is acid rain a major problem and why?
11. The air-fuel mixture is important in an automobile engine. What is the proper air-fuel ratio and what are the effects of an improper air-fuel ratio?
12. Name the major source of each of the following pollutants: (a) carbon monoxide, (b) sulfur dioxide, (c) particulate matter, (d) nitrogen oxides.
13. What is the estimated yearly expense due to the effects of air pollution in the United States?
14. The human respiratory system's natural defense mechanism is primarily for what type of pollutant?
15. What are the cause and effects of emphysema?
16. Why is carbon monoxide a dangerous air pollutant?
17. What is meant by synergism?
18. What are the urban heat-island effect and dust dome, and what are their effects on urban weather conditions?
19. How has the Earth's average temperature varied over the last 100 years?
20. What is the possible effect(s) of increased atmospheric CO_2 concentrations?
21. What are possible explanations for changes in the Earth's average temperature?
22. What effect could chloroflurocarbon propellants of aerosol spray cans possibly have on the Earth's climate?
23. What is the concern about air pollution in the stratosphere?
24. How could CO_2 pollution be decreased? Consider solar and nuclear energy sources and economic effects.

27

Geology and Time

This earth, a spot, a grain, an atom.
—Milton, *Paradise Lost*

IN ADDITION TO the atmosphere, a major part of our physical environment is the Earth upon which we live. **Geology** is the study of the Earth, its processes, and its history. With the increasing emphasis on natural resources obtained from the Earth, geology has taken on a new importance. An understanding of the various geologic processes is critical in locating and developing these natural resources. Also, such an understanding may allow the prediction of catastrophic events, such as volcanic eruptions and earthquakes, saving lives and property. In addition, geological studies of meteorites and the moon have helped us understand the origin of the solar system.

The Earth is indeed a dynamic place, both externally and internally. However, the rate of most Earth processes gives geology a unique time scale known as geologic time. Rather than an actual time scale, geologic time is more of a concept, in which it is assumed that the Earth developed gradually over an incredibly long time span. The several thousand years of the recorded history of human beings is only a ''tick'' on the geologic clock.

The history of the Earth is recorded in geologic time and events. Geologists must consider processes that may have taken place over millions of years, a time interval that is very difficult to comprehend. However, if natural processes are constant and the same processes operate today on and within the Earth, then the present may be considered the key to the past. This is an important geologic concept called the **principle of uniformity** (or uniformitarianism). Changes occurring on and within the Earth today are clues to the long-term total picture.

Within the last 50 years, there has been a revolution in the science of geology that gives us a better insight into the processes occurring near the Earth's surface. This has come about with the development of the theory of plate tectonics, which considers the Earth's surface to be made up of movable ''plates.'' The assumptions of the theory of plate tectonics are surprisingly simple, but the consequences are far-reaching. With this set of concepts, a new understanding is gained of the formation of mountains and ocean basins, volcanic activity, earthquakes, and the mechanisms of operation of many geologic processes.

In an effort to give the student a contemporary view of geology, the concepts of plate tectonics will be presented in this chapter and used to explain various geologic phenomena in the following chapters. It should be kept in mind, however, that although the current enthusiasm for plate tectonics is grounded by recent experimental evidence, it must stand the test of time—perhaps geologic time. Other theories may someday replace that of plate tectonics just as it is currently replacing previous ones.

◀ Ammonite in schist.

27.1 The Earth's Structure

Is the Earth hollow? If not, is its interior solid or liquid? These are fundamental questions of geology. Some ancient beliefs considered the Earth to be hollow like a basketball. However, today we have experimental evidence to the contrary. Only the Earth's surface and a rather insignificant part of its interior in mines and drilling projects can be observed directly. To date, we have drilled about five miles into the Earth, and it is unlikely that a hole can be drilled into the Earth's deep interior. Materials extruded in volcanic eruptions obviously come from the Earth's interior, but these are from relatively shallow depths compared to the Earth's radius of approximately 4000 mi or 6400 km. How then do geologists study the Earth's deep interior structure?

Our knowledge of the Earth's interior structure comes almost exclusively from the monitoring of shock waves generated by earthquakes. The enormous releases of energy associated with movements along fractures in the solid exterior, which result in "earthquakes," generate both transverse and longitudinal waves (Chapter 6) that propagate through the Earth. By monitoring these waves at different locations on the Earth and from a knowledge of wave properties in various types of Earth materials, such as wave velocity and refraction, information about the Earth's interior structure is gained. The monitoring of these so-called seismic waves will be considered in detail in Chapter 29. For the present discussion, only the resulting theory of the Earth's internal structure will be presented.

From these indirect observations of the Earth's interior, it is believed that the Earth is made up of a series of concentric shells as illustrated in Fig. 27.1. There are three major shells: (1) the core, which is partly molten, (2) the mantle, and (3) the crust. The different shells are characterized by different composition and physical properties.

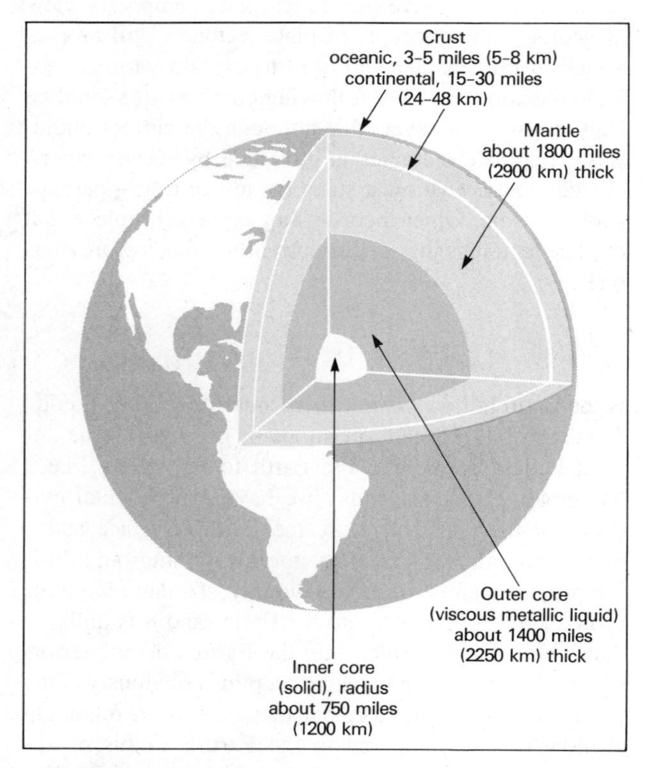

Figure 27.1 The interior structure of the Earth. The crustal thickness is greatly exaggerated.

The two innermost regions are together called the **core** and have an average density of over 10 g/cm^3. This density suggests a metallic composition, which is believed to be chiefly iron (80%) and nickel. This estimate is based on the behavior of seismic waves passing through the core, the known abundance of iron in meteorites, and the measured proportion of iron in the Sun and stars. The solid inner core has a radius of approximately 750 mi (1200 km), while the outer core, some 1400 mi (2250 km) thick, is believed to be composed of molten, highly viscous, "liquid" material. The magnetic field of the Earth is thought to be related to the liquid nature of the outer core. As pointed out in Chapter 8, the magnetic field of the Earth resembles that of a huge bar magnet within the Earth. However, the interior temperatures of the Earth are probably too high to have permanently magnetized ferromagnetic materials with the interior temperatures exceeding the Curie temperatures* of these materials. The slow change of the positions of the Earth's magnetic poles suggests that the magnetic field is due to currents, which must arise somehow from motions of the molten metals in the outer core associated with the Earth's rotation.

Around the core is the **mantle,** which is on the average about 1800 mi (2900 km) thick. The composition of the rocky mantle differs sharply from the metallic core, and their boundary is distinct. The average density of the mantle is of the order of 4.5 g/cm^3, which indicates that its composition is an iron-magnesium rock type material.

Around the mantle is a thin, rocky, outer layer upon which we live, called the **crust.** It ranges in thickness from about 3–5 miles (5–8 km) beneath the ocean basin to about 15–30 miles (24–48 km) under the continents. About 65% of the Earth's crust is oceanic crust. That is, about 65% of the Earth's surface is made up of ocean basins.

The boundary between the crust and the mantle was discovered by A. Mohorovičić, a Yugoslav seismologist, while studying a Balkan earthquake in 1909. He found that the speed of the shock waves increased suddenly several miles below the Earth's surface, indicating a change in the constituent material at this depth. This boundary is called the Mohorovičić discontinuity after its discoverer, or simply the **Moho discontinuity**.

Thus, we see that the internal structure of the Earth is somewhat analogous to that of an egg with the yolk,

*The temperature above which ferromagnetism disappears in a magnetic material.

white, and shell corresponding to the core, mantle, and crust, respectively. How did this layered structure of the Earth come about? Perhaps when the Earth was in an original molten state, the denser material settled to the bottom, in a fashion similar to the gases in the heterosphere. The liquid outer core could not be due to remanent heat from the Earth's formation, since the interior would have long since had time to cool and solidify. Although the outer core is subjected to enormous pressures that would result in high temperatures, calculations show that these temperatures are not high enough to melt the iron-nickel material of the outer core. Rather, it is believed that the heat is generated from the decay of radioactive materials.

It should be remembered that our picture of the general structure and composition of the Earth comes chiefly from indirect observations. Although materials from the upper mantle are observed in volcanic eruptions and perhaps someday exploratory drillings may be made into the mantle, scientists will probably never directly observe the Earth's interior as fantasized in Jules Verne's novel *Journey to the Center of the Earth*. However, as science and technology progress, geologists search for methods to provide a more detailed picture of the Earth's anatomy. Recently, major advances have been made in the understanding of the makeup and dynamics of the Earth's outer structure as will be described in the following sections.

27.2 Continental Drift and Sea Floor Spreading

When looking at a map of the world, we are tempted to speculate that the Atlantic coasts of Africa and North and South America could fit nicely together as though they were pieces of a jigsaw puzzle (Fig. 27.2). This observation has led scientists at various times in history to suggest that these continents, and perhaps the other continents, were once a single, giant supercontinent that broke and drifted apart. However, there was no evidence to support this theory other than the shapes of the continents.

In the early 1900s, Alfred Wegener (1880–1930), a German meteorologist and geophysicist, revived the theory of **continental drift** and brought together various geological evidences for its support. Wegener's theory gave rise to considerable controversy and only relatively recently has reasonably conclusive evidence been found that supports some aspects of the theory of continental drift.

Wegener's assumption was that the continents were once part of a single giant continent, which he called **Pangaea** (Greek, pronounced pan-jee-ah and meaning "all lands"). According to his theory, this hypothetical supercontinent somehow broke apart about 200 million years ago and its fragments drifted to their present positions and became today's continents (Fig. 27.3).

The geologic evidence supporting Wegener's theory takes on several different approaches. Some of these approaches are (a) similarities in biological species and fossils found on the various continents; (b) continuity of geologic structures such as mountain ranges and the distribution of rock types and ages; and (c) glaciation in the southern hemisphere. Let's consider each of these briefly:

(a) Similarities in biological species and fossils suggest that there was formerly an exchange of these forms

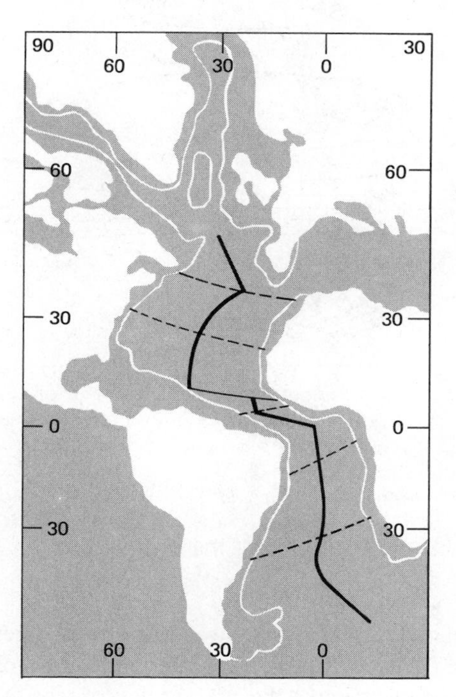

Figure 27.2 Continental jigsaw puzzle. The coasts of Africa and North and South America appear as though these continents could fit together nicely as if they were pieces of a jigsaw puzzle. The dashed lines show the paths the continents might have followed if they were once together and rifted and drifted apart. The heavy line indicating where the continents might have been once joined marks a present mid-oceanic ridge that runs the length of the Atlantic Ocean.

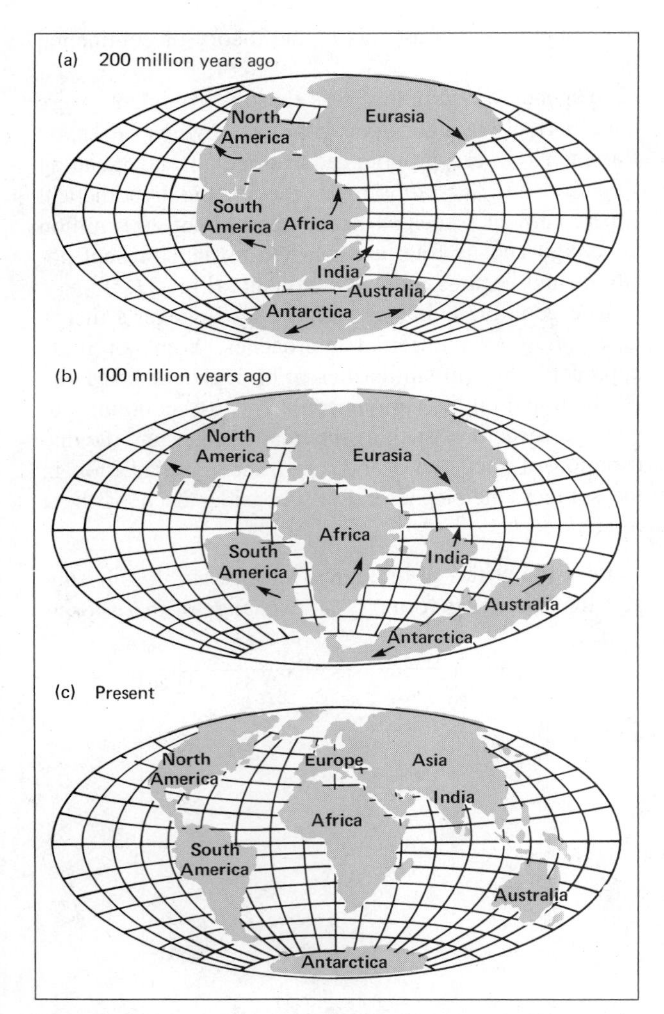

Figure 27.3 An illustration of the sequence of events in the theory of continental drift. (a) 200 million years ago the giant supercontinent Pangaea rifted and the continental fragments began to drift apart. (b) Continental positions 100 million years ago. Notice India was drifting northward on a collision course with the Asian continent. (c) Present position of the continents.

between continental regions when they were together as Pangaea. It would not be expected that these biological species could traverse the present-day oceans. For example, a certain variety of garden snail is found only in the western part of Europe and the eastern part of North America. Also, a relatively young genus of earthworm is found in the same latitudes of Japan and the Asian and European continents, while it is found in similar latitudes on the east coast, but not the west coast of North America. Similarly, fossils of identical reptiles are found in South America and Africa, and identical plant fossils have been found in South America, Africa, India, Australia, and Antarctica.

It was suggested that the similarities in biological species and fossils could be explained by "land bridges" between the continents that eventually sank or were covered by the oceans. However, there is no evidence of such land bridges in the oceans.

(b) As has been noted, it was the roughly interlocking shapes of the coastlines of the African and American continents that inspired the theory of continental drift. Imagine that the continental shapes are cut from a printed page like jigsaw-puzzle pieces and an attempt is made to put the separated pieces together again. When putting the page pieces back together, it is evident that the continuity of the printed lines would be common to the fitted pieces as shown in Fig. 27.4.

If indeed the continents had rifted and drifted apart, we might expect some similar "printed lines" common to the pieces. Such evidence does occur in the form of geologic features. If the continents were put back to-

Figure 27.4 An illustration of the continuity of continental features. Continental jigsaw pieces cut from a printed page would show the continuity of printed lines when put back together. If the continents were once together and drifted apart, we might expect some similar "printed lines" common to the separated continents in the form of continuities of geologic features such as mountains.

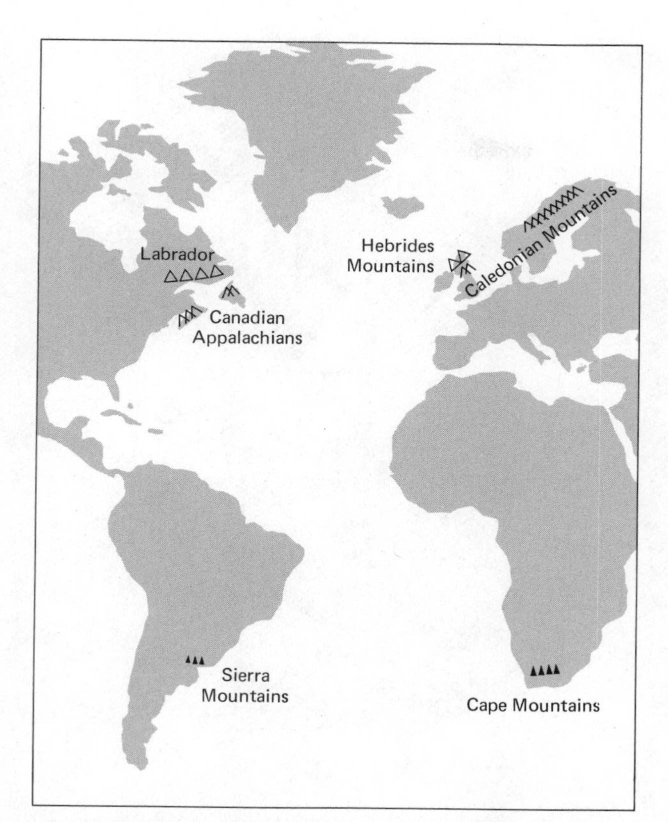

Figure 27.5 The continuity of geologic features supports the theory of continental drift. If the continents were fitted together, various mountain ranges of similar structure and rock composition on the different continents would line up analogous to the print in Fig. 27.4.

gether, the Cape mountain range in southern Africa would line up with the Sierra range near Buenos Aires, and these mountains are strikingly similar in geologic structure and rock composition (Fig. 27.5). In the northern hemisphere, the Hebrides mountains in Northern Scotland match up with similar formations in Labrador, and the Caledonian mountains in Norway and Scotland have a logical extension in the Canadian Appalachians. Various other similarities in the plateau rock formations of Africa and South America have been found. More recently, with the development of radioactive dating techniques, transatlantic areas of rocks of similar ages have been found.

(c) There is solid geological evidence that a glacial ice sheet covered the southern parts of South America, Africa, India, and Australia about 300 million years ago, similar to the one that covers Antarctica today. Hence, it is reasonable to conclude that the southern portions of these continents must have been under the influence of a polar climate at this time. There are no traces of this ice age found in Europe and North America. In fact, fossil evidence indicates that a tropical climate prevailed in these regions during that period. Yet, evidence of glaciation is found in India and in Africa near the equator. How could this be? It seems unreasonable that an ice sheet would cover the southern oceans and extend northward to the equator.

Wegener's theory suggests an answer and derives support from these observations. The direction of the glacier flow is easily determined by marks of erosion on rock floors and by moraines (deposits of rock and soil debris transported by glaciers). If the continents were once grouped together as Wegener's theory indicates, then the glaciation area was common to the various continents, as illustrated in Fig. 27.6. The glacial movements, as indicated by the arrows in the top drawing, support the idea of a single ice cap and subsequent continental drift.

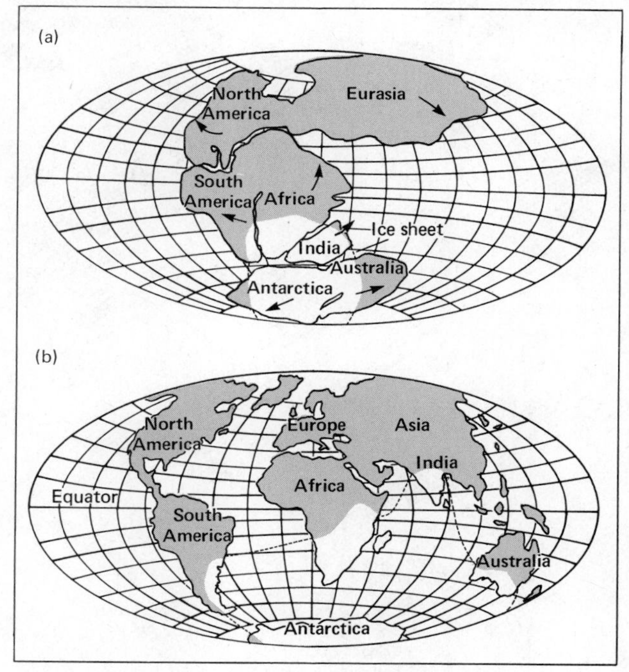

Figure 27.6 Glaciation evidence of continental drift. Geologic evidence shows that a glacial ice sheet covered parts of South America, Africa, India, and Australia some 300 million years ago, but there is no evidence of this glaciation in Europe and North America. Also, evidence of this glaciation is found near the equator in India and in Africa. This is explained if a single ice sheet covered the southern polar region of Pangaea and the continents subsequently drifted apart.

National Geographic Society
NATIONAL GEOGRAPHIC MAGAZINE
HUDSON BAY
ICELAND
NORTH SEA
BRITISH ISLES
EUROPE
ISLAND OF NEWFOUNDLAND
NORTH AMERICA
MID-OCEAN CANYON
SOHM ABYSSAL PLAIN
IBERIAN PENINSULA
ATLAS MOUNTAINS
SAHARA
AFRICA
SUDAN
CAPE VERDE ISLANDS
CENTRAL AMERICA
GUIANA HIGHLANDS
AMAZON
SOUTH AMERICA
BASIN
PERNAMBUCO ABYSSAL PLAIN
BRAZILIAN HIGHLANDS
CONGO BASIN
ANGOLA ABYSSAL PLAIN
KALAHARI DESERT
SOUTH PACIFIC OCEAN
ARGENTINE ABYSSAL PLAIN
FALKLAND PLATEAU

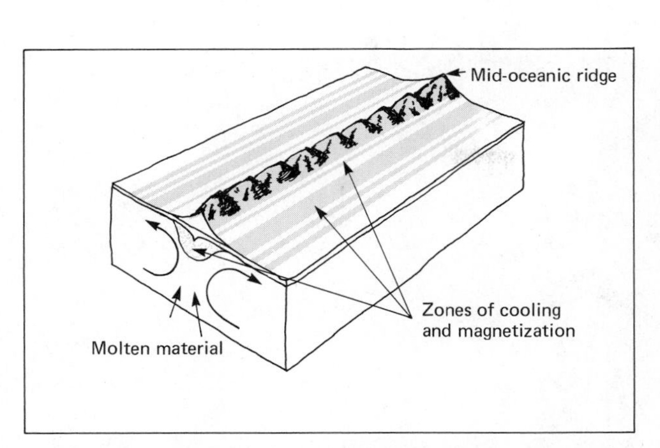

Figure 27.7 (*Left and above*) Sea floor spreading and the Mid-Atlantic Ridge. The Mid-Atlantic Ridge runs along the ocean floor between the continents. It is believed that sea floor spreading takes place along the ridge due to convection currents of molten material rising from the upper mantle and spreading outward from the ridge. This is evidenced by magnetic anomalies in the ocean floor rock (shown as strippled areas in the drawing) caused by reversals of the Earth's magnetic poles, which have occurred relatively frequently and regularly throughout geologic time. (Map courtesy National Geographic Society.)

Although there was evidence supporting Wegener's theory, it was not generally accepted, primarily because the proposed mechanism for continental drift was unsatisfactory. Wegener depicted the continents as giant rafts moving through the oceanic crust due to the Earth's rotation. This mechanism is unacceptable because the force associated with the Earth's rotation is not strong enough to overcome the measured strength of the rocks, which would be disintegrated if the continental crust slid over or moved through the oceanic crust.

A more satisfactory theory for the mechanism behind continental drift was suggested in 1960 by H. H. Hess, an American geologist. It was known that a mid-oceanic ridge system stretches through the major oceans of the world. In particular, there is the Mid-Atlantic Ridge that runs along the center of the Atlantic Ocean between the continents. This and other mid-oceanic ridges run along large fissures in the Earth's crust, as evidenced by volcanic and earthquake activity along the ridges. Hess suggested a theory of **sea floor spreading** where the sea floor spreads slowly and moves sideways away from the mid-oceanic ridges. This movement is believed to be accounted for by convection currents of subterranean molten materials that cause the formation of mid-oceanic ridges and surface motions in lateral directions from the fissure, as illustrated in Fig. 27.7.

Support for this theory has come from the studies of remanent magnetism and from the determination of the ages of the rocks on each side of the mid-oceanic ridges. **Remanent magnetism** refers to the magnetism of rocks due to a special group of minerals called ferrites, one of which is ferrous (iron) ferrite, Fe_3O_4, commonly called magnetite or lodestone. When molten material containing ferrites is extruded upward from the Earth's mantle, e.g., in a volcanic eruption, and solidifies in the Earth's magnetic field, it becomes magnetized. The direction of the magnetization indicates the direction of the Earth's magnetic field at the time.

These solidified rocks, called igneous rocks (Section 28.1), are worn down by erosion. The fragments are carried away by water, and they eventually settle in bodies of water, where they become layers in future sedimentary rocks (Section 28.2). In the settling process, the magnetized particles, which are in fact small magnets, become generally aligned with the Earth's magnetic field (Fig. 27.8). By studying the remanent magnetization of geologic rock formations and layers, scientists have learned about changes that have taken place in the Earth's magnetic field.

Measurements of the remanent magnetism of rocks on the ocean floor revealed long, narrow, symmetric bands of **magnetic anomalies** on both sides of the Mid-Atlantic Ridge (Fig. 27.7). That is, the direction of the magnetization was reversed in adjacent parallel regions. Along with data on land rocks, the magnetic anomalies indicate that the Earth's magnetic field has abruptly reversed fairly frequently and regularly throughout recent geologic time. The most recent reversal occurred about 700,000 years ago. Why this and other reversals should

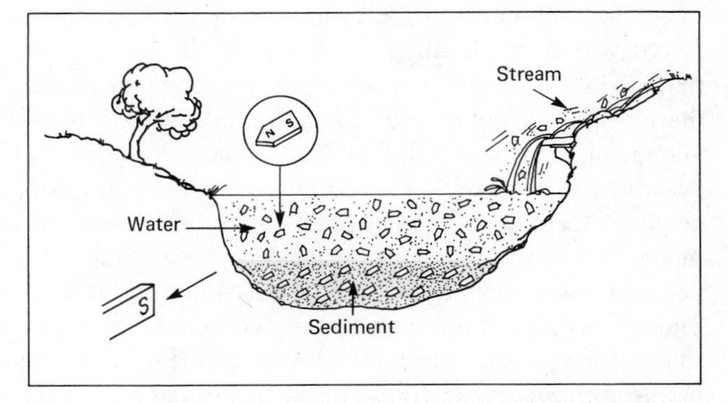

Figure 27.8 An illustration of remanent magnetism deposition in sedimentary layers. Magnetized particles from eroded rock become generally aligned with the Earth's magnetic field as they settle in bodies of water. The sediment layers eventually become sedimentary rock with remanent magnetism.

Figure 27.9 The Glomar Challenger. This 10,500-ton vessel is capable of conducting drilling operations in the open sea using dynamic positioning to maintain position over a bore hole. The drilling derrick amidship stands 194 ft above the waterline. Forward of the derrick (partially visible) is the automatic pipe racker containing 24,000 ft of 5-in drill pipe. Scientific laboratories and crew's quarters are located aft. (Courtesy Deep Sea Drilling Project/SIO, Scripps Institution of Oceanography.)

occur is not known. We are evidently living in a period between pole reversals. However, the symmetry of the anomaly bands on either side of the Ridge indicates a movement away from the Ridge at the rate of a few centimeters per year and provides evidence for sea floor spreading.

Remanent magnetism also provides other support for continental drift. The remanent magnetism of a rock remains fixed throughout the history of the rock, even if forces within the Earth move the rock. If the rock is moved, its remanent magnetism is no longer aligned with the Earth's magnetic field. This is similar to the nonalignment of printed lines on moved puzzle pieces as described in the previous analogy on the continuity of geologic features. Scientists have studied the remanent magnetism of rocks of varying ages and have found evidence of such changes of alignment. Moreover, it was found that rocks of the same age on a continent have the same misalignment, but that the misalignment is different for different continents. This suggests that the entire continent must have moved as a unit and that each continent moved or ''drifted'' in its own separate direction.

In recent years, investigative drilling into the ocean floor has been done from the oceanographic research vessel Glomar Challenger (Fig. 27.9). These drillings have shown that the ocean floor between Africa and South America is covered by relatively young sediment strata. Also, the strata thicknesses increase away from the Mid-Atlantic Ridge, which implies that the older part of the ocean floor is farther away from the Ridge. The older parts would be covered by a greater thickness of sediment because there would have been a longer time for the sediment to accumulate.

These observations support the idea of sea floor spreading as a mechanism for the theory of continental drift, which has culminated in the modern theory of plate tectonics.

27.3 Plate Tectonics

The theory of **plate tectonics*** was fully realized in the late 1960s. In this theory, the outer layer of the Earth is viewed as a series of rigid **plates** or huge slabs of rock that are in (very slow) relative motion. The basic as-

*Tectonics (Greek, *tekton,* builder) is the study of the Earth's general structural features and their changes.

sumptions of this theory are quite simple, but its explanations of basic geologic phenomena are far-reaching.

An immediate question might be: How can these plates move as a whole as though they are floating on something? Actually, they are floating in a sense. The geological term for this flotation is isostasy. Suppose the outer portions of the Earth were not rigid enough to support crustal features of different weights placed on them. These features of different weights would sink to different depths until they came to equilibrium like the blocks in Fig. 27.10. It is found that there is a similar occurrence for the Earth's crust. That is, mountains have deeper ''roots'' than plateau areas as illustrated in the figure. This is the concept of **isostasy**—that crustal material is ''floating'' in gravitational equilibrium with other crustal materials on a ''fluid'' substratum.

In terms of plate tectonics, it is convenient to divide the outer portion of the Earth as shown in Fig. 27.11. The outermost solid portion is called the **lithosphere** (Greek, *lithos,* rock). It includes the crust and some of the upper mantle, extending to a depth of about 50 mi. The material below the lithosphere, which is considered to be rocky material that is hot enough to be deformed and capable of internal ''flow,'' is called the **asthenosphere** (Greek, *astheno,* weak). The rigid plates of the lithosphere move about on the fluid substratum of the asthenosphere. The continents are like passengers on the moving plates.

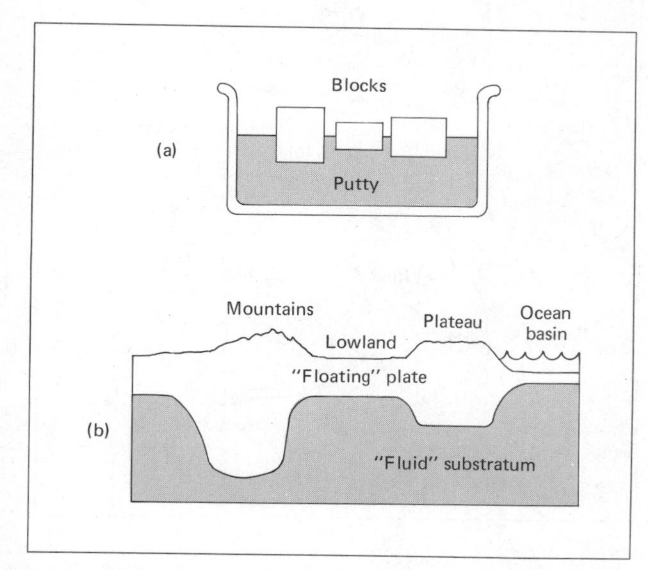

Figure 27.10 An illustration of the concept of isostasy. (a) Blocks of some heavy material will sink to different depths until they float in equilibrium in some viscous "fluid" material like putty. (b) Similarly, the solid outer portion of the Earth "floats" on a "fluid" substratum with different crustal features floating at different depths.

This relative motion of the plates is an important feature. Obviously the relative motions can be of three general types as illustrated in Fig. 27.12—divergent, convergent, and sliding motions. Dynamic geologic processes occur at the plate boundaries.

In the case of a divergent plate boundary, sea floor spreading occurs. As the plates recede from each other, hot, molten material from the mantle rises to fill the gap. The rising material cools near the surface and becomes part of the lithosphere material forming the mid-oceanic ridges and becoming part of the plates on either side of the boundary. The average rate of sea floor spreading is of the order of 4–6 centimeters per year.

If the surface area of the Earth remains constant, as it is believed to be, then as new lithospheric material is generated along divergent plate boundaries, the plates must be driven together and material ''destroyed'' at convergent plate boundaries. The types of convergent plate boundaries are illustrated in Fig. 27.13.

The first two processes in which one plate is deflected downward beneath the other into the asthenosphere is called **subduction.** When two oceanic plates collide, one is deflected downward and returns to the mantle. The subduction zone of oceanic plates is marked by a deep oceanic trench. An example of such a trench is the Marianas Trench in the western Pacific, which has the greatest known depth below sea level on the Earth's surface—

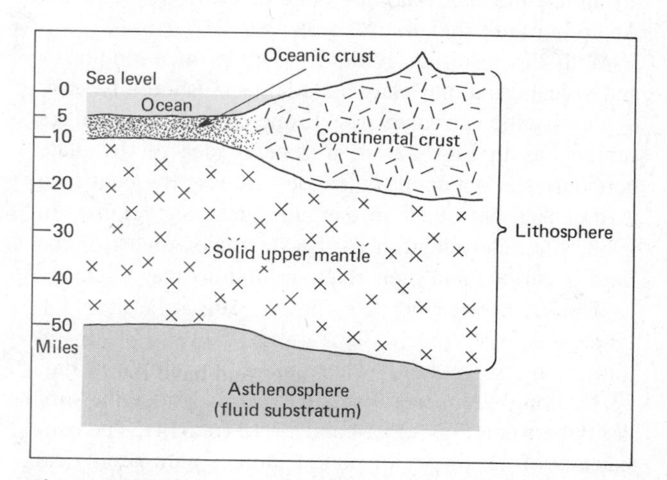

Figure 27.11 A cross section of the lithosphere. The lithosphere is made up of the oceanic and continental crusts and the solid upper portion of the mantle. The material below the lithosphere, which is considered to be hot enough to be easily deformed and capable of internal flow, is called the asthenosphere.

about 6.8 mi. The material of the oceanic crust is denser than that of the continental crust. Hence, when oceanic and continental plates collide, the oceanic plate is deflected below the continental plate.

In each of these cases, part of the descending oceanic crust in contact with the mantle begins to melt. The molten material rises into the overlying plate, and on breaking through the surface gives rise to volcanoes. In the case of the convergence of oceanic plates, the volcanoes build island arcs, of which Japan is a good example. In the case of the convergence of oceanic and continental plates, a volcanic mountain chain may be formed. The Andes mountains on the west coast of South America are a good example of such a formation. Friction between the descending and overlying plates causes stresses in the plates. The release of energy accompanying movement of the plates gives rise to earthquakes in the region of subduction zones.

When two continental plates collide, material is pushed upward to form mountain ranges. It is believed that the Himalayas were formed in this manner when the Indian plate originally broke loose from Pangaea and collided with the Eurasian plate. (See Fig. 27.3).

Finally, the third type of relative plate motion shown in Fig. 27.12 is where adjacent plates slide horizontally past each other without a gain or loss in area. This occurs along faults, which mark the plate boundaries. Movements and the release of energy along these boundaries give rise to earthquakes. Examples of such fault zones are along the San Andreas fault in California and the Anatolian fault in Turkey.

With the resulting geologic features of earthquakes and volcanoes along plate boundaries, it should be easy to identify the various plates. If one plots the earthquake centers, as in Fig. 27.14, a general idea of the plate boundaries is obtained (volcanoes are less frequent than earthquakes but occur in the same general regions). In 1968, the determination of the six major plates of the Earth's surface and their relative motions was made.

The six major plates are the African, American, Pacific, Eurasian, Indian (sometimes called the Australian), and the Antarctic. These plates are shown in Fig. 27.15 along with several smaller plates. Notice the similarity between Figs. 27.14 and 27.15 (p. 514). The consistency of the theory of plate tectonics with geological phenomena has given rise to new explanations of geological processes which will be presented in greater detail in the following chapters.

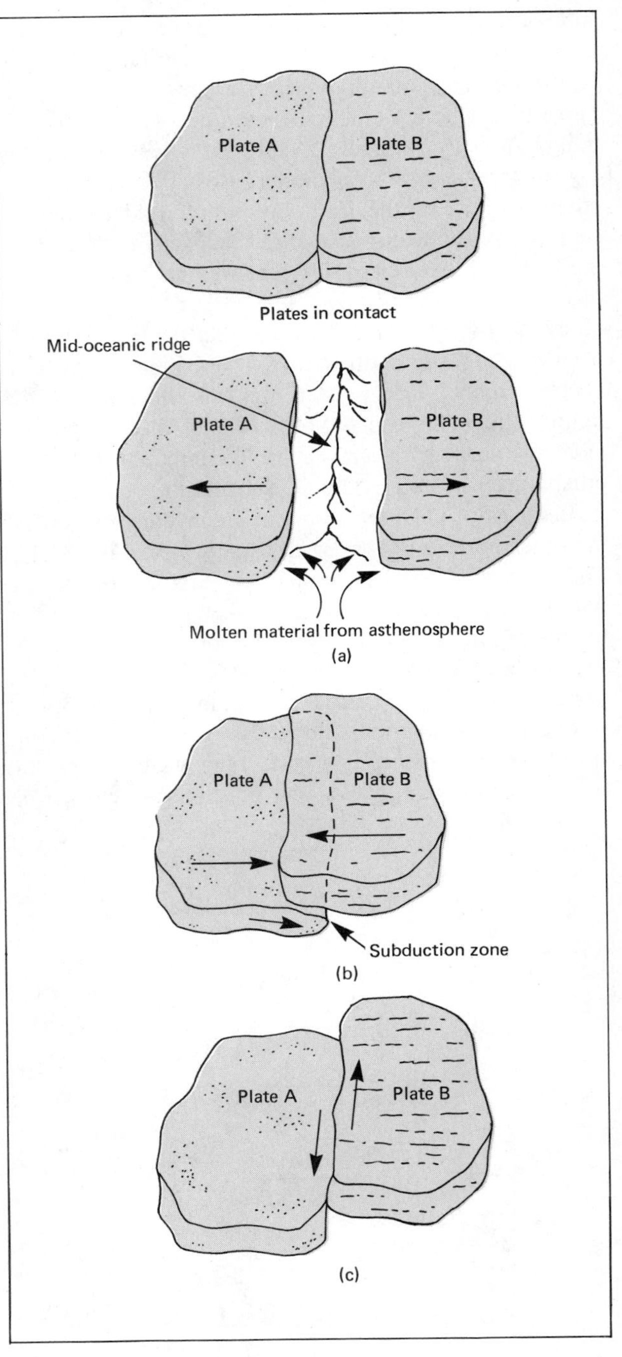

Figure 27.12 Three types of relative motions of plates. Plates in contact can (a) diverge as a result of sea floor spreading, (b) converge where plates are driven together and one plate is deflected beneath the other (subduction zone), or (c) slide past each other while in contact.

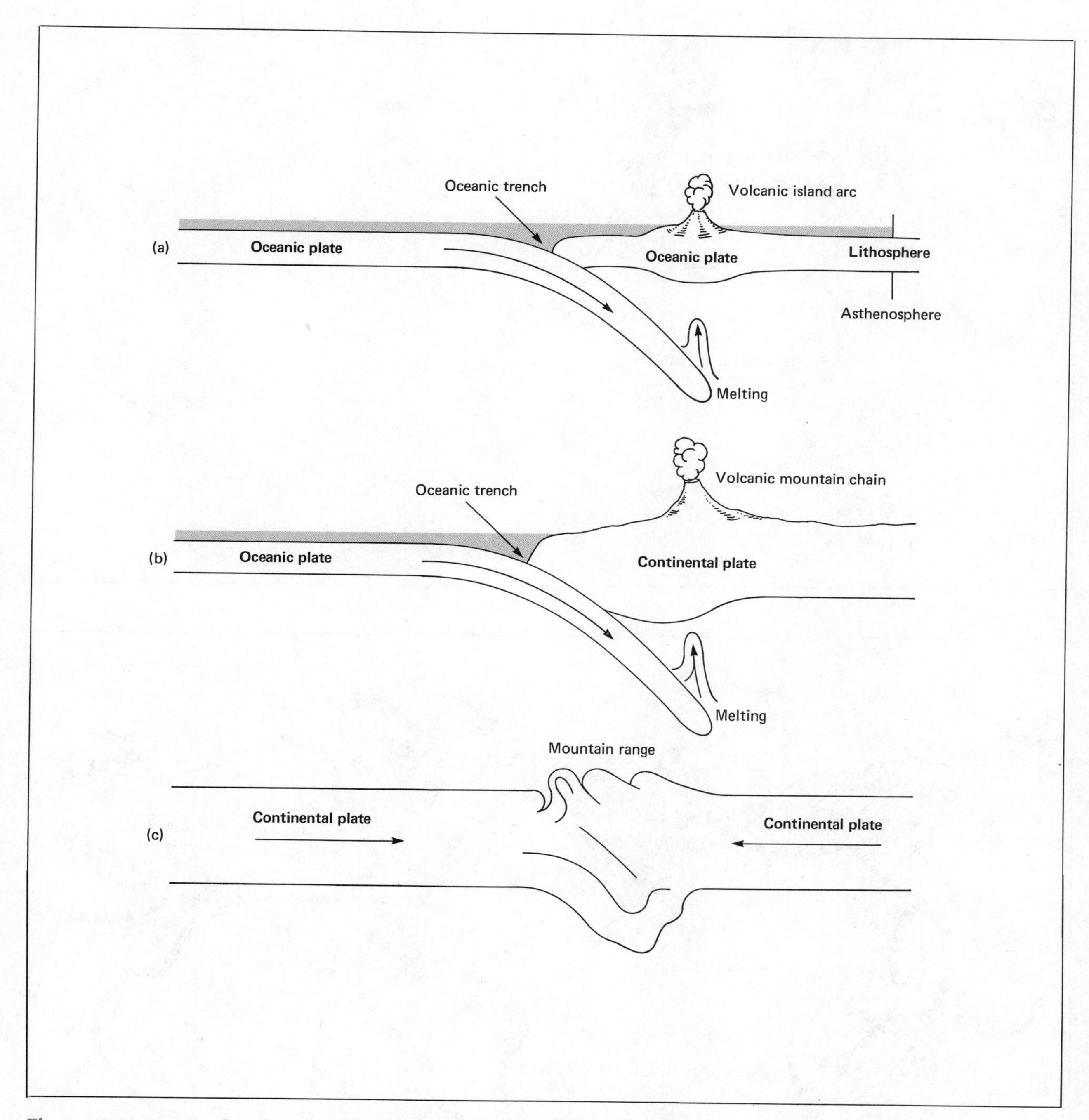

Figure 27.13 Types of convergent plate boundaries. (a) Collision of two oceanic plates. One is deflected below the other and the melted material in the subduction zone rises to form volcanic island arcs. (b) Collision of oceanic and continental plates. The less dense oceanic plate is deflected below the continental plate and the melted material in the subduction zone rises to form volcanic mountain chains. (c) Collision of two continental plates. Material is pushed upward forming mountain ranges.

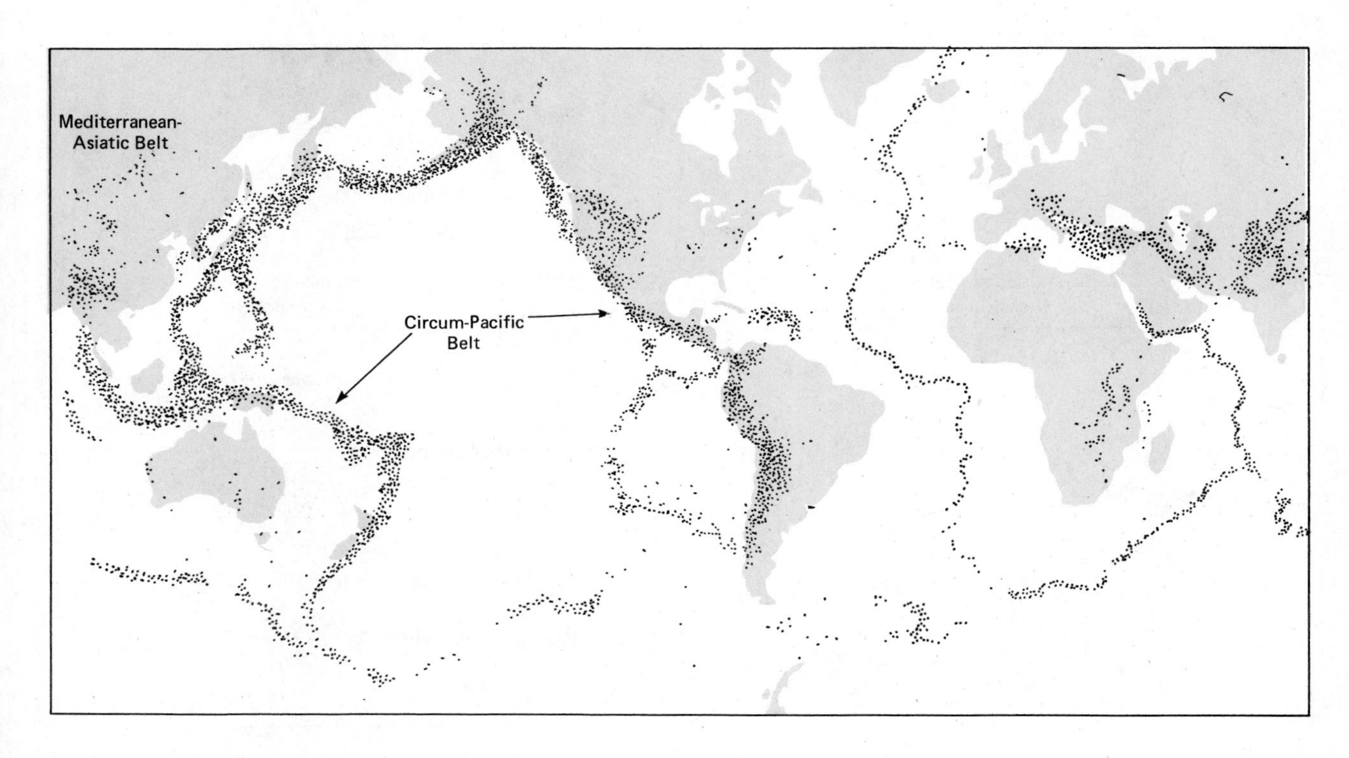
Mediterranean-
Asiatic Belt
Circum-Pacific
Belt

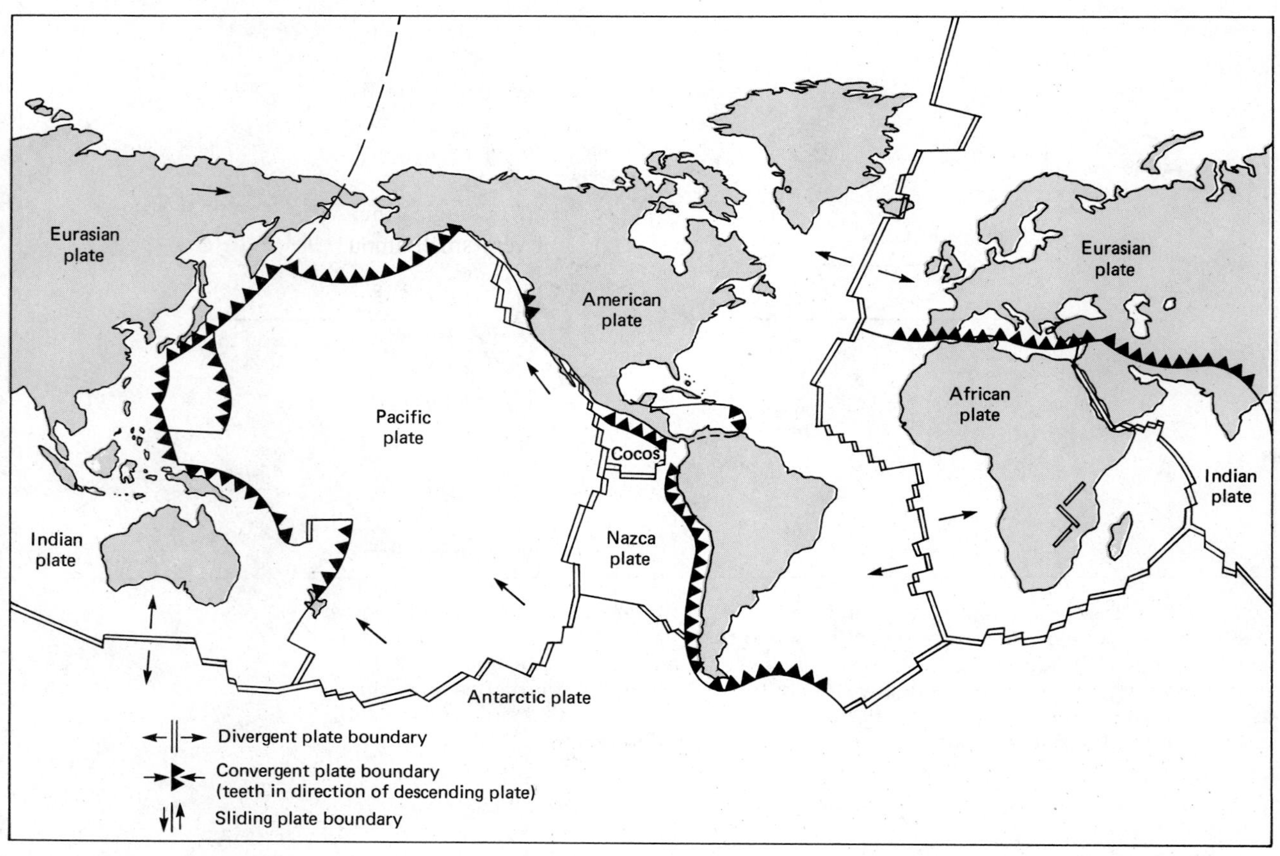
Eurasian
plate
American
plate
Eurasian
plate
Pacific
plate
African
plate
Cocos
Indian
plate
Nazca
plate
Indian
plate
Antarctic plate
Divergent plate boundary
Convergent plate boundary
(teeth in direction of descending plate)
Sliding plate boundary

Figure 27.14 (*Top left*) World map of earthquake locations from 1961–1969. As can be seen, the earthquake locations outline general boundaries. (Courtesy NOAA.)

Figure 27.15 (*Bottom left*) The six major plates of the world and several smaller ones. The relative motions along the plate boundaries are indicated. Compare this with Fig. 27.14. Notice the rifting or divergence on the interior of the African Plate in East Africa. This East African Rift marks the place of possible further plate breakup. It is uncertain whether this rifting will lead to the eventual diversion of the African Plate and the creation of a new ocean basin.

Table 27.1 Radioactive Decay Schemes in Geologic Dating

Parent Nucleus	*Daughter Nucleus*	*Half-Life*
Carbon-14	Nitrogen-14	5730 years
Potassium-40	Argon-40	1.3 billion years
Rubidium-87	Strontium-87	50 billion years
Thorium-232	Lead-208	14.1 billion years
Uranium-235	Lead-207	713 million years
Uranium-238	Lead-206	4.5 billion years

27.4 Geologic Time

In the previous sections, references were made to geologic events that happened millions of years ago. For example, Pangaea is believed to have broken up about 200 million years ago. It is fair to ask how we know, or what evidence we have, that certain geologic events took place so long ago. Early geologists had a great deal of difficulty in establishing a geologic time scale of events. As you know, the measurement of a time interval requires reference points or events, in particular a zero reference or an initial time or event.

The first "geologic clock" to be systematically used in the interpretation of Earth history was based upon the study of fossils (paleontology). Fossils are the preserved remains of former plants and animals or indications of their activities, such as tracks or burrows. Using the so-called **law of superposition,** which is the simple observation that in a succession of stratified deposits the younger layers lie on top of the older layers, geologists established a *relative* **geologic time scale** based on the characteristic fossils of each stratum. The major unit of geologic time is called an **era,** which in turn is divided into **periods** and **epochs.** The particular units in geologic time are given names as indicated in Fig. 27.16 (p. 516).

Relative time simply identifies an object or event as being younger or older than something else. From the law of superposition, we know that a set of fossils and the associated events in an upper stratum are younger or happened later in time than those in a lower stratum. However, it is difficult to assign times within a stratum. Since no accurate rate of deposition can be determined for most rock strata, the actual length of geologic time represented by a given stratum is at best an educated guess. As a result, events are assigned to a particular geologic time unit. For example, a geologist might say that dinosaurs became extinct (as evidenced by fossil remains) near the end of the Mesozoic Era, which precedes the Cenozoic Era. The boundary between these eras is *estimated* to be about 65 million years ago, so dinosaurs became extinct over 65 million years ago.

Modern scientific methods have allowed the development of an **atomic time scale** based on natural radioactivity. As presented in Chapter 8, the rate of decay of a radioactive isotope is conveniently expressed in terms of the isotope's half-life, or the time it takes for one-half of the nuclei in a sample to decay. Many isotopes have relatively short half-lives of a few days or years. However, some isotopes decay slowly and several of these are used as "atomic clocks" in measuring the ages of geologic events. The parent and daughter nuclei of the decay schemes most commonly used in geologic radioactive dating are listed in Table 27.1.

Radioactive dating is simple in theory, as described in Chapter 8 for carbon-14 dating. However, the actual laboratory procedures are complex. The principal difficulty with geologic samples lies in the accurate measurement of very small amounts of radioactive or associated elements.

Carbon-14 decay is a useful tool in dating relatively recent events. Because of its relatively short half-life, carbon-14 dating techniques are accurate for dating events that have taken place within the last 50,000 years, a very short interval in geologic time. Fortunately, other radioactive isotopes associated with geologic events have longer half-lives. Radiocarbon dating is done on once-living material, while other methods involve the dating of rocks.

The potassium-argon decay scheme is one of the most useful to the geologist because it can be used on rocks as young as a few thousand years as well as on the oldest rocks known. Potassium is a constituent of several common minerals. The half-life of potassium-40 is such that measurable quantities of argon-40 (the daughter isotope) have accumulated in potassium-bearing rocks of nearly all ages. And, even in very small quantities, the amounts of potassium and argon isotopes can be measured accurately. However, the daughter product, argon, is a gas

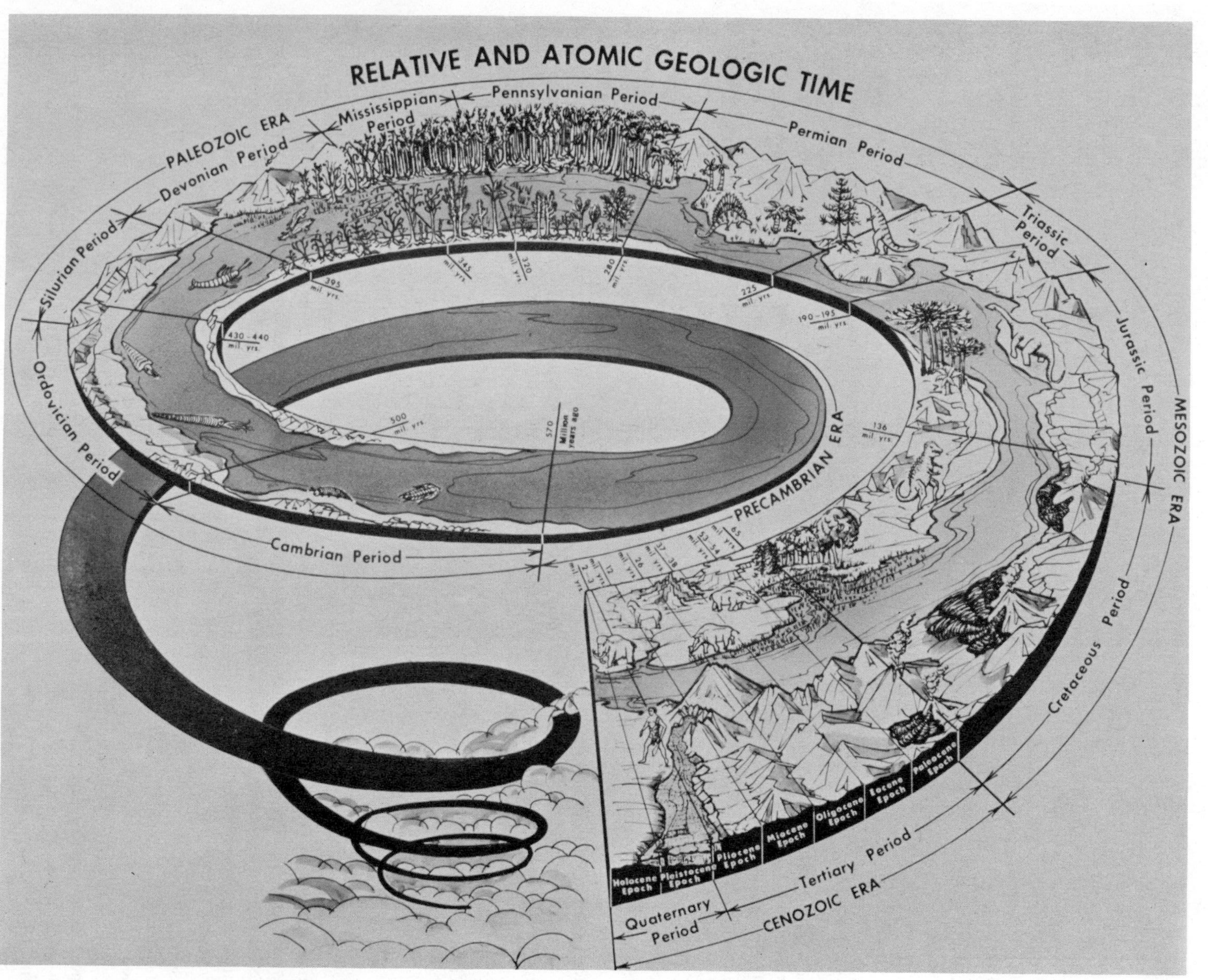

Figure 27.16 Simplified geologic time scale. (Courtesy U.S. Geological Survey.)

and can escape from the system, thereby causing the computed age of rock to be younger than it actually is. Fortunately, some rocks retain the argon quite well.

Each decay scheme dating method finds different geologic applications, but they must be applied carefully. Whenever possible, two or more methods of analysis are used on the same specimen of rock to confirm and verify the results.

The correlation of the relative geologic time scale deduced from fossil evidence with atomic time poses certain problems, chiefly because rocks solidified from molten material can be dated by atomic methods more reliably than other kinds and these rocks do not ordinarily contain fossils. Many solidified igneous rocks retain the parent and daughter isotopes within them, thereby allowing accurate determination of the relative amounts of the isotopes. Other rocks, such as sandstone and limestone (sedimentary rock formed from sediments), are related to the atomic time scale by correlating them to time intervals that are determined from the dating of appropriate igneous rocks. However, even with these problems, the "atomic clocks" of the radioactive decay processes have helped geologists relate the relative geologic time scale to one that is more absolute.

Learning Objectives

After reading and studying this chapter, you should be able to do the following without referring to the text:

1. Describe the Earth's interior structure and composition.
2. Describe the early formulation of the theory of continental drift and the geologic evidence offered for its support.
3. Explain the mechanism and evidence for sea floor spreading and its relationship to continental drift.
4. Explain the theory of plate tectonics.
5. List the general relative motions of plates and the resulting geologic implications.
6. Distinguish between the relative geologic time scale and the atomic time scale.
7. Describe several radioactive decay schemes used in geologic dating.

Important Words and Terms

geology
principle of uniformity
core
mantle
crust
Moho discontinuity
continental drift
Pangaea
sea floor spreading
remanent magnetism
magnetic anomalies
plate tectonics
plates
isostasy
lithosphere
asthenosphere
subduction
law of superposition
geologic time scale
era
period
epoch
atomic time scale

Questions

1. What is the principle of uniformity and how is it used in the study of geologic processes?
2. State the compositions and structural divisions of the Earth's interior. Are they all solid?
3. Why can there not be a permanent magnet within the Earth that gives rise to the Earth's magnetic field?
4. What is the name of the boundary between the crust and the mantle?
5. How is the Earth's internal structure analogous to an egg?
6. What is Pangaea?
7. Describe the geologic evidence that supported Wegener's theory of continental drift.
8. What is the theory of the mechanism for continental drift? Why was Wegener's explanation unacceptable?
9. What is meant by remanent magnetism?
10. What evidence supports the theory of sea floor spreading?
11. What is the average rate of sea floor spreading?
12. What is the concept of isostasy?
13. What are the lithosphere and the asthenosphere?
14. What is a "plate" in the context of plate tectonics?
15. Describe the three general types of relative plate motions and the geologic results of each.
16. What is a subduction zone?
17. How are volcanic and earthquake activity explained in the theory of plate tectonics?
18. What are the names of the six major plates?
19. What is the law of superposition?
20. On what is the relative geologic time scale based?
21. What is the atomic time scale and what "atomic clocks" are used in this scale?
22. How are the relative geologic and atomic time scales correlated?
23. What is the average length of an era, a period, an epoch?

28
Rocks and Minerals

Touch the earth, love the earth, honour the earth,
her plains, her valleys, her hills and her seas;
rest your spirit in her solitary places.

–Henry Beston

WHAT IS THE Earth made of? From our vantage point on its surface, "rocks" would be a typical answer. Even the sand and soil were once part of rocks. Since rocks are such an intimate part of the Earth, an important question in geology is, How are rocks formed and what are they made of?

The processes of rock formation provide a means of classifying rocks into three different types or families of rock: igneous, sedimentary, and metamorphic. We tend to think of rocks as permanent things; however, in the context of geologic time they are not. Geologic processes result in the creation, destruction, and alteration of rocks—the so-called rock cycle. This was alluded to in the previous chapter, where in the theory of plate tectonics parts of the lithosphere are created and other parts destroyed in a continuous cycle.

A common definition of **rock** is any naturally occurring, solid, mineral mass that makes up part of the Earth. What are rocks made of? Chemical elements and compounds, of course. However, the geologist finds it more convenient to say most rocks are made up of one or more minerals, which are defined to be inorganic, crystalline substances. With the large number of available chemical elements, it would seem that the number of such substances that could be formed would be multitudinous. But, the number of minerals is limited because only a few chemical elements make up the bulk of the Earth's crust.

◀ Queen's Garden Trail, Bryce Canyon National Park, Utah.

28.1 Igneous Rock

Igneous (Latin, *ignis,* fire) **rocks** are those formed by the cooling and solidification of hot, molten rock material called **magma.** If we accept the idea that the Earth was at one time a molten mass (magma), we would expect it to have cooled and solidified externally, so that the Earth's crust would be composed largely of igneous rock. In fact, igneous rock is the most abundant of the three rock categories of the Earth's crust, comprising as much as 80%.

The formation of magma is a continuing process. As was learned in the last chapter, plates are deflected downward toward the Earth's interior, where they melt. Investigations in deep oil wells and mines have shown the temperature of the outer portion of the Earth's crust increases about 1°F per 150 ft of depth. This variation in temperature with depth is called the **geothermal gradient.** If this rate of temperature increase were extrapolated to the center of the Earth, the central interior temperature would be greater than 200,000°F, which is unreasonable. The temperature of the surface of the Sun is only about 10,000°F.

A more reasonable estimate of 4000°F to 5000°F for the temperature of the Earth's interior comes from the study of melting points of materials on the surface that are believed to be the same as those in the central interior. Although heat is produced by the enormous pressure in the Earth's interior, the major source of the Earth's interior heat is believed to arise from the decay of radioactive materials.

Since the formation of magma takes place below the Earth's crust, it is impossible to observe the natural formation of magma. But one can observe magma and the formation of one type of igneous rock. The formation of igneous rock takes place both on and within the Earth's surface. Igneous rock is classified as being either extrusive or intrusive, depending on whether it solidified from molten magma above or below the Earth's surface, respectively.

Extrusive igneous rock is formed from the cooling and solidification of magma poured out onto the Earth's surface by volcanic eruptions. Magma that reaches the Earth's surface through a volcanic vent is called **lava.** The lava cools and solidifies, becoming igneous rock (Fig. 28.1).

When exposed to the atmosphere (and often the ocean), lava cools and solidifies relatively quickly compared to magma cooling below the Earth's surface. Ordinarily, lava cools slowly enough for crystalline grains to form. However, some lava cools and solidifies too quickly for this to happen and glassy materials—those without orderly crystalline structures—are formed. Obsidian is such a glassy volcanic rock. (See Table 28.1 and Fig. 28.5).

Figure 28.1 A solidified lava flow west of Pisgah Volcano, California. Magma reaching the Earth's surface through a volcanic vent is called lava. The lava cools and solidifies, becoming igneous rock. (Courtesy U.S. Geological Survey.)

Table 28.1 Examples of Types of Rocks*

Type	*Characteristic*
Igneous	
Intrusive	*Texture*
Granite	Coarse-grained
Extrusive	
Obsidian	Glassy
Scoria	Porous
Pumice	Glassy and porous
Sedimentary	
Clastic	*Sediment*
Conglomerate	Gravel
Sandstone	Sand
Shale (mudstone)	Mud
Breccia	Volcanic fragments
Organic	
Shell limestone (e.g., Coquina)	Compacted or cemented calcareous plant or animal remains
Coal	Compacted plant remains
Chemical	*Chemical composition*
Limestone	$CaCO_3$, calcium carbonate
Gypsum	$CaSO_4 \cdot 2\,H_2O$, hydrous calcium sulfate
Rock salt	NaCl, sodium chloride
Flint and chert	SiO_2, silicon dioxide
Metamorphic	*Parent rock*
Marble	Limestone
Quartzite	Sandstone
Slate	Shale (low grade metamorphism)
Gneiss	Shale (high grade metamorphism) or granite

*See photographs in Fig. 28.6 (pp. 524–525).

Such glassy substances tell the geologist that molten material underwent rapid cooling.

Most often, volcanic eruptions or explosions are accompanied by the escape of gases. These gases may be originally held in solution in the molten subterranean magma by high pressure, much like gas is held in a capped soft drink. When the pressure is released on a soft drink by removing the cap, gas can be seen escaping from solution. Similarly, gases are released from magma reaching the Earth's surface (lava) in volcanic eruptions. However, in the case of highly viscous lava, the gas bubbles cannot escape quickly and may be "frozen in" as the lava solidifies. Extrusive igneous rocks formed with "frozen in" gas bubbles have a frothy texture, such as scoria and pumice shown in Fig. 28.6 (pp. 524–525).

Intrusive igneous rock is formed from the cooling and solidification of magma below the Earth's surface. A distinguishing characteristic between intrusive rock and extrusive rock is their relative texture, or grain size. When an igneous mass cools and crystallizes, the size of the mineral particles depends on the rate of cooling. Extrusive rock cools relatively quickly and is characterized by fine grains. Large crystalline grains do not have time to grow. As pointed out previously, the lava rock of volcanoes is fine grained and even glassy in some cases. Intrusive rock cools slowly below the Earth's surface, and such rock is characterized by relatively larger grains and coarser texture. Granite is an example of intrusive rock (Fig. 28.6). The grain size is dependent on the rate of cooling, which involves several factors, such as depth, location, size or mass, and the specific heat of the material. Occasionally, a magma cools at varying rates, slowly at first, then more rapidly. This results in a structure of coarse mineral grains scattered through a mixture of fine mineral grains. Such a texture is said to be **porphyritic.**

Large bodies of intrusive igneous rock are called **plutons,** after Pluto, the Greek god of the underworld. These plutonic bodies are classified according to their

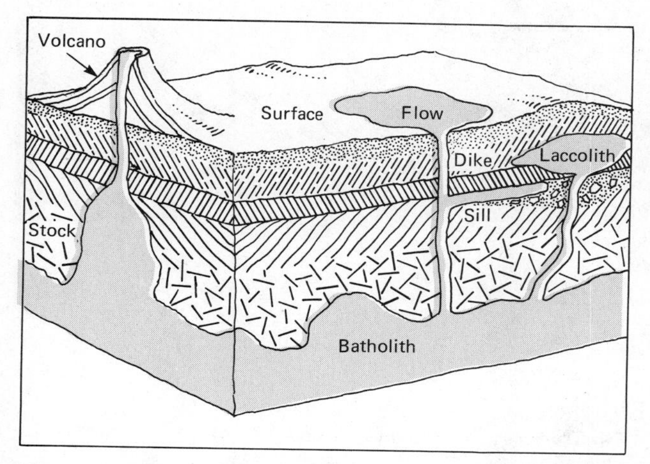

Figure 28.2 An illustration of plutonic bodies. Magma solidifying within the Earth forms intrusive igneous bodies. The batholith is the largest intrusive body. Sills and laccoliths are concordant bodies that lie parallel to existing rock formations. The discordant dike cuts across existing rock formations. A volcanic stock is shown rising from the batholith.

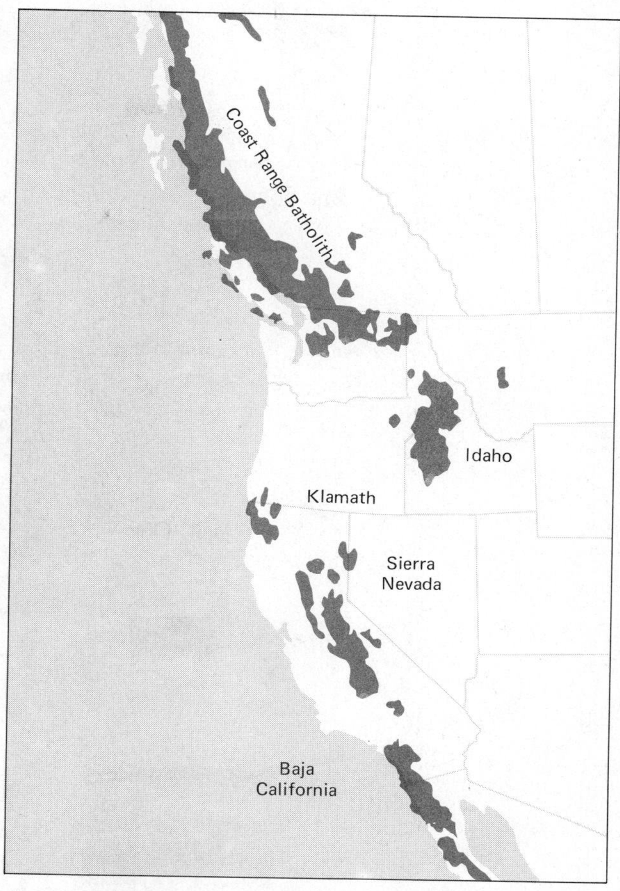

Figure 28.3 The locations of large batholiths in the western part of North America. To be classified as a batholith, the intrusive body must have an area of at least 40 mi^2.

size, shape, and orientation in surrounding rock. Intrusive rock bodies are referred to as being **concordant** when they lie more or less parallel to older formations, and as being **discordant** when they cut across older formations. The four major plutonic forms are illustrated in Fig. 28.2.

Large discordant bodies are called **batholiths.** Batholiths are enormous in size, and to be so classified, must occupy an area of at least 40 mi^2. Protruding from a batholith may be an irregular dome formation called a **stock,** which may have once supplied magma to a volcanic vent. The locations of some batholith formations are shown in Fig. 28.3. The Coastal Range batholith along the western Canadian coast is a classic example.

Another discordant plutonic body is a **dike,** which is formed when magma fills a nearly vertical fracture in rock layers. A dike is tabular with its width or thickness much smaller than its other dimensions, as shown in Fig.

Figure 28.4 (*Top*) A discordant dike formation. This exposed dike shows how the forming magma intruded across existing rock formations. (*Bottom*) A concordant sill formation. The intrusion of magma that formed this sill on Unalaska Island, Aleutian Islands, Alaska, was parallel to existing rock formations. (Courtesy U.S. Geological Survey. Sill photo by G. L. Snyder.)

28.4. Although originally formed below the Earth's surface, some intrusive bodies are now visible, due to erosion and other geologic processes.

Sills and laccoliths are intrusive concordant bodies. **Sills** are similar to dikes, but form between and parallel to existing rock beds (Fig. 28.4). A **laccolith** is a blisterlike intrusion that has pushed up the overlying rock layers.

28.2 Sedimentary Rock

In the previous section on intrusive rock formations, it was stated that magma intrudes between layers of preexisting rock. In general these rocks near the surface of the Earth are **sedimentary rocks.** These are rocks formed from sediment composed of particles of older rocks and other material. Rocks on the Earth's surface are continually being both physically and chemically worn down by environmental processes. The resulting rock particles and dissolved minerals make up the sediment that is transported by streams and rivers and eventually deposited in large bodies of water as sedimentary layers. The consolidation of this sediment forms sedimentary rock.

Sedimentary rocks are the most common rocks in our environment. Although sedimentary rocks make up only 5% of the Earth's crust, they cover about 75% of the Earth's surface. The stratified layers of sedimentary rock are commonly observed in the environment as shown in Fig. 28.5 for both natural and artificial reasons.

Because of their abundance and variety of content, sedimentary rocks are important in our everyday life. Petroleum and coal are found in sedimentary rock formations, and sedimentary rock supplies many essential building materials. The transformation of sediment into rock is referred to as **lithification** or **consolidation.** For example, when relatively fine-grained mud is subjected to high pressure such that a substantial portion of water is forced out, the mud is compacted into the sedimentary rock called shale (sometimes called mudstone, Fig. 28.6).

In some cases of compaction the grains may recrystallize to form an interlocking structure. Since the fine grains of compacted rocks are held together by forces that are not very strong, the rocks are relatively soft and break easily. If the grains of sediment are coarse, such as sand or gravel, compaction will not form a stable rock. The particles of coarse-grained sediment may be cemented together by materials dissolved in the water that permeates the sediment. Common cementing agents are silica (SiO_2), calcium carbonate ($CaCO_3$), and iron oxides. Regardless of the method of lithification, sedimentary rocks are not as hard as the fused, crystallized igneous rocks.

Sedimentary rocks are classified into two subcategories: clastic and chemical. **Clastic** (Greek *klastos,* broken) **sedimentary rocks** are formed from the particles of preexisting rocks or the skeletal remains of organisms and plants. Common examples are sandstone, shale, and shell limestone (Fig. 28.6).

Chemical sedimentary rocks are formed by the precipitation of minerals dissolved in water. Such rocks are formed in two principal ways. Indirectly, they are formed by biochemical reactions during the activities of plants and animals, e.g., certain organisms, such as coral, extract calcium carbonate ($CaCO_3$) from sea water to build skeletal material. When the coral dies, the collected skeletal deposits subsequently form biochemical limestone. Directly, they are formed when the evaporation of water with dissolved minerals leaves behind a residue of chemical sediment, e.g., calcium salt (carbonate) deposits in hot springs and sodium chloride (rock salt) from the evaporation of sea water (Fig. 28.6). The table salt we eat is mined from chemical sedimentary rock formed this way.

Another example of directly formed chemical sedimentary rock is cavern dripstone, which is formed by material precipitated from dripping water. Dripstone takes on a variety of forms, but most common are the icicle-shaped **stalactites** and **stalagmites** (Fig. 28.7, p. 526). As drops of water seep through the cavern roof and fall on its floor, minute amounts of dissolved calcium carbonate ($CaCO_3$) and other materials precipitate and cause the formations to grow. To distinguish between which formation hangs from the ceiling and which formation

Figure 28.5 Examples of exposed sedimentary rock layers in the environment due to natural (erosion) and human (road-cut) causes. (Top photo courtesy U.S. Geological Survey.)

Figure 28.6 Examples of igneous, sedimentary, and metamorphic rocks. (Courtesy Ron Docie, Athens, Ohio.)

protrudes upward from the floor, simply remember stalactites or "tites" come from the "top," and that stalagmites come from the ground. If the stalactites and stalagmites are joined, the combined formations are referred to as **columns.**

Clastic organic sediments result from living organisms, either animal or plant. The shells of marine animals may be cemented together to form organic sedimentary rocks such as coquina shown in Fig. 28.6. Chalk is another sedimentary rock formed from accumulations of shells of microscopic animals. The famed White Cliffs of Dover in England are chalk formations.

Probably the best-known organic rock of plant origin is coal. Coal is a black sedimentary rock consisting chiefly (over 60%) of partly decomposed fossil plant matter. In what was once low, swampy areas, sediment layers rich in vegetation were deposited and later became coal beds. Initially, bacterial action decomposed the plant matter with the release of gases; and the plant residue gradually became concentrated in carbon content. This carbon-rich

10. Sandstone 11. Conglomerate 12. Shale (Mudstone) 13. Breccia

14a. Limestone (chemical) 14b. Limestone (organic) 15a. Flint (black) with Chert 15b. Chert

16. Coquina (shell limestone) 17. Gypsum 18. Rock salt (Halite)

decayed plant matter is called **peat.** Existing peat beds are an important fuel source in some countries.

As peat was buried by overlying strata of sediment deposits of sand, silt, or clay, it was compressed and water was squeezed out. Volatile organic compounds such as methane (CH_4) escaped, leaving a material of increased carbon content. The peat was successively compacted into **lignite, or brown coal** (65 to 70% carbon), **sub-bituminous coal** (70 to 75% carbon), and **bituminous, or soft coal** (75 to 80% carbon). These coals are sedimentary rocks. A further change results in **anthracite, or hard coal** (approximately 90% carbon). Anthracite coal is a metamorphic rock (see Section 28.3).

The ignition of coal depends on its content of volatile materials. With a relatively high volatile content, lignite and sub-bituminous coals ignite easily, but burn with a great deal of smoke. As a result, these coals are sometimes referred to as smoky coals, and can contribute to air pollution (Chapter 26). Anthracite coal, on the other hand, is difficult to ignite and burns with little smoke.

Deposits of lignite and sub-bituminous coals are found predominantly in the western United States, and deposits of bituminous and anthracite coals are found predominantly in the eastern United States (Fig. 28.8). Coal is an important fuel resource for the highly industrialized eastern part of the country; however, the eastern coals have a relative high sulfur content compared to western coals—another major consideration in air pollution.

The decay of organic matter, both plants and animals, also produces the petroleum and natural gas resources upon which we depend so much today. The majority of the petroleum used today comes from liquid oil pools below the surface of the Earth. However, another source of petroleum, which may become increasingly important as our energy demands increase, is tar sands and oil shale. The hydrocarbons in these reserves occur in the

Figure 28.7 Chemical sedimentary cavern dripstone. Stalactites (tite-top) form from the ceiling down and stalagmites build from the floor up as a result of the precipitation of dissolved chemicals from dripping water. When the formations join, they are referred to as a column. Scientists estimate the formations in this cave have been building for more than 10 million years, and the formations are still building at the very slow rate of one cubic inch per 120 years. (Courtesy Luray Caverns, Virginia.)

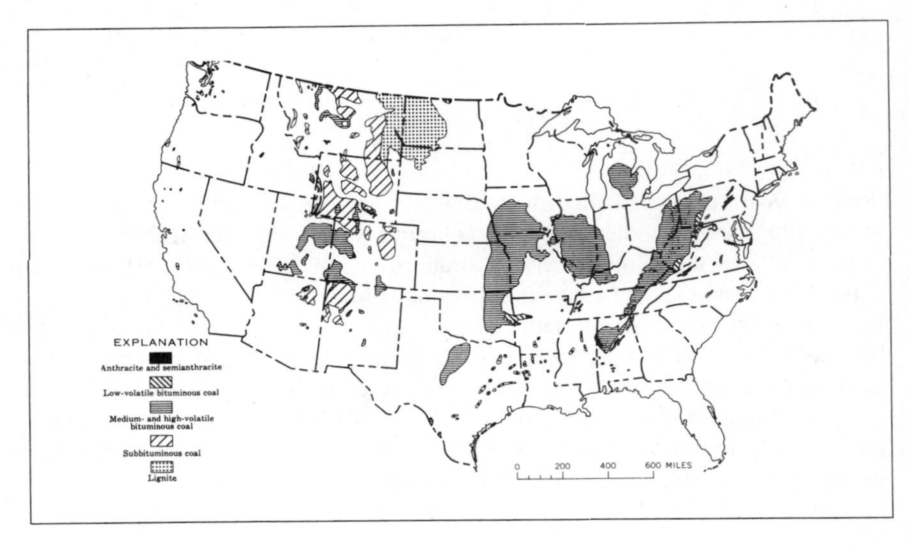

Figure 28.8 Coal deposits in the conterminous United States. (Courtesy U.S. Geological Survey.)

Figure 28.9 Bedding and folding. (*Top*) Cross-bedding in sandstone formations in Navajo County, Arizona. Sand deposited by winds that shifted directions formed these cross-bedded formations. (*Bottom*) Folded rock strata. Stresses in the Earth's crust are evidenced by folded rock strata, such as the ones shown here in a weathered shale formation in Johnson County, Tennessee. (Courtesy U.S. Geological Survey. Photos by H.E. Gregory and W.B. Hamilton.)

solid state, filling the pores of sedimentary strata. The tar sands and oil shale must be mined and heated to remove the solid hydrocarbons. Important tar sands and oil shale deposits occur in the western United States and Canada.

Perhaps the most characteristic feature of sedimentary rock formation is the stratification, or **bedding,** that results from the depositing of the layers of sediment. The sedimentary layers are not always horizontal, however. They may be either tilted, as a result of the deposition of sediment on an inclined surface, or folded as a result of pressures from within the Earth's crust (Fig. 28.9).

The layered feature of sedimentary rocks gives insight into the recent geologic history of the Earth's surface. Each layer characterizes a certain time period, much the same as the rings of a tree do on a smaller time scale. In the sedimentary layers the preserved remains of plants and animals, or fossils, are sometimes found as shown in Fig. 28.10. Fossils supply important clues to the history of life and to past features of the Earth's surface.

Figure 28.10 A clam fossil. This 570 million-year-old clam, 70 million years older than any previously known, was discovered among rocks collected near Albany, New York. (Courtesy U.S. Geological Survey.)

28.3 Metamorphic Rock

As the name implies, **metamorphic rock** results from a change or metamorphism (from the Greek word to change or transform) of preexisting rock. The metamorphism, which takes place below the Earth's surface, is a change in the composition and texture of the rock caused primarily by heat, pressure, and moisture.

When rocks of any type are formed, they are stable in the surrounding conditions of formation. However, if these conditions change, the rock may become unstable and change to conform to its new environmental conditions. For example, limestone with coral and clam shells is found in Pennsylvania and New York (indicating that this area was once covered by a shallow sea). This sedimentary limestone was formed under conditions of little heat or pressure. However, if the limestone experiences increased heat and pressure because of geologic conditions, it may undergo a change, and under proper conditions become the metamorphic rock marble, as is found in western New England, Georgia, and Alabama.

Having undergone a metamorphic process, the resulting metamorphic rock is different from the original rock. In most cases, the bulk chemical composition of the rock remains unchanged, but the new rock usually contains different chemical compounds or minerals. (Minerals or particular rock-forming chemical compounds will be discussed in Section 28.4). In some instances, the existing rock-mineral grains are recrystallized and a new texture develops. Either or both of these processes may take place in metamorphism.

Metamorphism occurs for sedimentary, igneous, and even metamorphic rocks. Due to geologic changes, rocks formed at the surface of the Earth may become buried to depths of several thousand feet. The rocks are then subject to high temperatures and pressures. Certain rock components are unstable at these temperatures and pressures and react with other rock components to form new compounds or minerals.

These metamorphic reactions occur within the solid rock. Should the temperature be great enough to cause the rock to become molten, it is then magma and its solidification would result in an igneous rock. Also, as pointed out previously, during a metamorphic process the original rock grains may recrystallize, becoming larger with perhaps slightly different forms. It is these changes in the features of rock composition, texture, and structure that distinguish metamorphic rock from sedimentary and igneous rock.

Moisture is also an important factor in metamorphism. Sedimentary rock has open spaces between its grains that are accessible to moisture. Sandstone is a good example. When metamorphism takes place, the water between the grains tend to speed up the chemical reactions. Sandstone is metamorphized to quartzite. (See Fig. 28.5.) Igneous rocks, on the other hand, are usually more dense and have relatively little open space between the rock grains and hence contain little water. To metamorphize a "dry" rock, much higher temperatures and pressures are required to initiate chemical reactions. In general, then, the metamorphism of igneous rocks requires greater temperatures and pressures than do sedimentary rocks.

As has been mentioned, metamorphic rocks may be further metamorphized into other metamorphic rocks. That is, increased temperature and pressure can cause a change in an existing metamorphic rock. An example of such a sequence begins with the metamorphism of sedimentary shale or mudstone into metamorphic slate. This platy rock was once commonly used for blackboards and as a roofing material. Under the proper conditions, slate may be changed to schist; and schist in turn may be changed to gneiss (Fig. 28.5). The examples of these types of rocks in the figure are not metamorphisms of the same rock, but are examples with different chemical contents. A type of gneiss also results from the metamorphism of granite.

Metamorphic rocks provide geologists with a means to study the forces and temperatures in the Earth's crust. The pressure for metamorphism of rock materials may arise from the weight of materials deposited on top of the rock formation or from a shift in the Earth's crust that produces a more horizontal pressure. Pressure generates heat; however, the heat associated with metamorphism may also be supplied by nearby igneous bodies. Metamorphism should not be thought of as an overnight process, rather it is a geologic process that, in general, requires a considerable time interval.

Under sufficient heat and pressure a metamorphic rock may become molten and therefore be classified as magma. This process essentially completes a cycle, which you may have noticed emerging. Magma forms igneous rocks that are worn by weathering. The particles are lithified into sedimentary rocks that are most susceptible to metamorphism. Both sedimentary and metamorphic rocks are worn away by weathering, and their particles again form sedimentary rocks. These interrelated processes by which rock is destroyed, created, and altered form what is known as the **rock cycle** and are illustrated in Fig. 28.11.

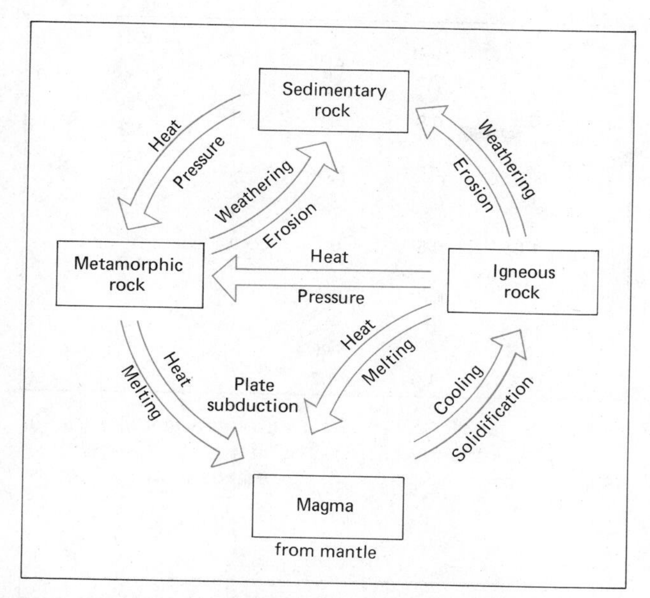

Figure 28.11 An illustration of the rock cycle. Rock material in the lithosphere follows various routes in the cycle. Plate additions along mid-oceanic ridges and plate destruction in subduction zones correspond to an entrance and exit to the rock cycle, respectively.

Within the framework of plate tectonics, the additions to the plates along the mid-oceanic ridges and the destruction of the plates in the subduction zones correspond to an entrance and exit to the rock cycle of the lithosphere.

28.4 Minerals

As has been discussed, rocks can be classified into three broad categories based on the method of their formation. However, the chemical and physical classification of rocks depends on the composition of their particles. Rock particles are made up of one or more chemical compounds; but rather than using specific compounds (or elements), geologists refer to minerals as being the building blocks of rocks. A **mineral** is any naturally occurring, inorganic, crystalline substance. Every mineral has a constant composition of elements in a definite proportion. *Most* rocks are made up of an aggregate of one or more minerals. *Some* rocks, e.g., coal, are composed of predominantly organic material. Others are a mixture of minerals and organic material, e.g., oil shale. Minerals make up a large percentage of the soil, which is composed for the most part of weathered rock material.

Everywhere around us we see minerals or the products of minerals. Some are quite valuable, while others are essentially worthless. For example, precious stones such as diamonds and rubies are valuable minerals. Also, we speak of a nation's mineral wealth when referring to natural raw materials such as ores or minerals containing iron, gold, silver, copper, etc. However, the minerals of common rock, like sandstone, have little monetary value.

The term mineral has also taken on popular meanings. For example, foods are said to contain vitamins and "minerals." In this case, the term mineral refers to compounds in food that contain elements needed in small quantities by the human body, such as Fe, Na, I, Mn, Mg, and Cu. The chemical composition of these compounds may be the same as those naturally occurring minerals that make up the Earth's crust, or they may actually be the naturally occurring minerals themselves. The names of minerals, like those of chemical elements, have historical connotations and may reflect the names of localities.

Mineral classification is advantageous because it is based on the physical and chemical properties of substances, which distinguish between different forms of minerals composed of the same element or compound. For example, graphite—a soft black, slippery substance commonly used as a lubricant—and diamond are both composed of carbon. But, because of different crystalline structures their properties are quite different. (Graphite mixed with other substances to obtain degrees of hardness is the "lead" in lead pencils.)

Also, a mineral may have a series of related compounds in which one metallic element is replaced by another. The mineral feldspar may have the compositions $NaAlSi_3O_8$, $CaAl_2Si_2O_8$, or $KAlSi_3O_8$. In this case, distinction is made between different types of feldspar.

Let's take a look at the three prerequisites of a mineral as given in the previous definition: (1) Naturally occurring. This means that a synthetic compound, such as a plastic, is not a mineral. (2) Inorganic. The implication here is in the biological sense—that a mineral was not previously part of a living thing. For example, a fragment of plant fossil is not a mineral. Coal is conveniently considered to be a rock, although it has less than 40% inorganic—mineral—content. The energy value of coal comes from its nonmineral (organic) content. (3) Crystalline. The atoms of minerals are organized in orderly arrangements, and each mineral has a unique internal structure. Substances lacking a systematic arrangement of atoms and not having a definite composition are said to be amorphous. Amorphous (noncrystalline) substances are not minerals.

There is an important class of geologic materials that falls into this nonmineral category, viz., glasses. Recall that glasses are substances that cooled so rapidly from molten material that there was not time for orderly internal structures to form before solidification. By definition, an aggregate of glass is not a rock, but it is generally treated as one. Naturally occurring glasses are relatively rare in everyday activities. The common glass materials, e.g., window glass, are not minerals on two counts—they are neither naturally occurring nor crystalline.

With the large number of known inorganic elements, the number of possible minerals that might form in the Earth's crust would seem enormous. However, the number of inorganic compounds that occur naturally in any substantial proportion is limited, because only a few elements make up the bulk of the Earth's crust. As can be seen from Table 28.2, only eight elements make up about 98% of the Earth's crust. Notice that only two elements, oxygen and silicon, make up about 75%.

There are over 2000 known minerals, but only a few minerals—less than 20—make up over 95% of the rocks in the Earth's crust. Most mineral compounds are composed of two or more elements that are held together by ionic bonding. Diamond is an example of one of the few covalent mineral compounds. In general, the ability of

Table 28.2 Relative Abundances of Elements in the Earth's Crust

Element	*Approx. Percentage (weight)*
Oxygen (O)	46.5
Silicon (Si)	27.5
Aluminum (Al)	8.1
Iron (Fe)	5.3
Calcium (Ca)	4.0
Magnesium (Mg)	2.7
Sodium (Na)	2.4
Potassium (K)	1.9
All Others	1.6
Total	100.0

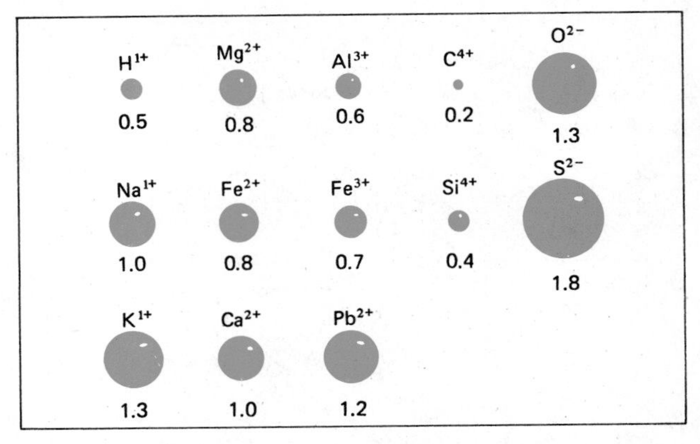

Figure 28.12 The relative sizes and charges of some of the common mineral ions. The actual radii of the ions in angstroms ($1\,\text{Å} = 10^{-10}$ m) are listed below the circles.

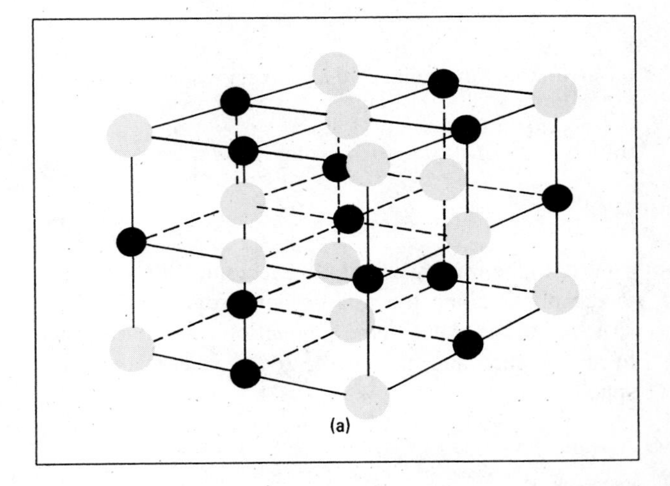

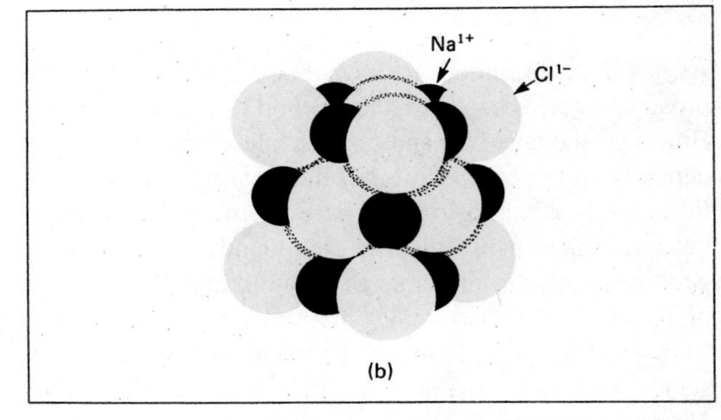

Figure 28.13 An illustration of the cubic lattice structure of NaCl. (a) A simplified expanded view. (b) A close-packed view showing the relative sizes of the sodium and chlorine ions.

an ion to combine with other ions or atoms to form the crystalline structure of minerals depends on their electric charges and their relative sizes. Figure 28.12 shows the relative sizes of several common ions.

As was learned in Chapter 14, sodium and chlorine ions form NaCl (mineral name, halite) with the internal arrangement of the ions in a crystal structure of a simple cubic lattice (Fig. 28.13). This structure depends on the electric charge and the relative sizes of the ions. As another example, let's consider the group of minerals called silicates (silicon-oxygen compounds), which are the most abundant rock-forming minerals. The positive silicon ion is Si^{4+} and the negative oxygen ion is O^{2-}. The simplest silicate compound is quartz (SiO_2, silicon dioxide).

All other important silicate minerals are combinations of silicon and oxygen and one or more other elements. The fundamental unit in these silicate crystalline structures is the **silicon-oxygen tetrahedron** (Fig. 28.14). This is a pyramid-shaped complex ion $(SiO_4)^{4-}$. How the tetrahedral units are bound together defines the various silicate crystalline types. Several of these structures are illustrated in Fig. 28.15. Olivine is a mineral with a single independent tetrahedron that is combined with either Mg^{2+} or Fe^{2+}. These metal ions can substitute for each other, depending on the conditions of mineral formation.

The silicon-oxygen tetrahedra can form single- and double-chain structures as shown in Fig. 28.15. Examples of minerals with these structures are pyroxene and hornblende, respectively. In the pyroxene chain, each tetrahedron shares two oxygens; while in the double hornblende chain, half the tetrahedra share two oxygens and the other half share three oxygens. These structures have various metallic ion components as in the case of olivine. (See Table 28.3.) The two-dimensional sheet structure of tetrahedra shown in Fig. 28.15 is the structure of mica and clay. The three-dimensional silicate structures are too complex to be shown in a simple illustration.

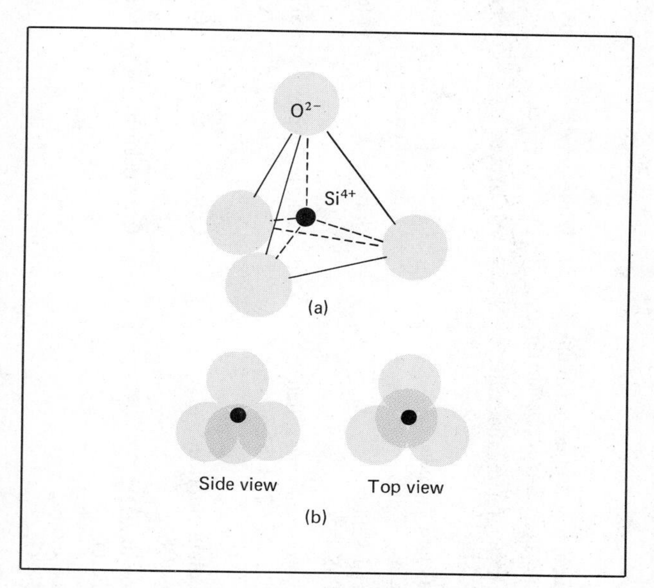

Figure 28.14 An illustration of the silicon-oxygen tetrahedron. (a) An expanded view of the complex ion $(SiO_4)^{4-}$ showing the larger oxygen ions (O^{2-}) at the four corners of the tetrahedron (three-sided pyramid) equidistant from the smaller silicon ion (Si^{4+}) at the center. (b) Close-packed side and top views.

Silicate structure	Arrangement of tetrahedra (top view)	Typical mineral
Single independent tetrahedron		Olivine
Single chain		Pyroxene
Double chain	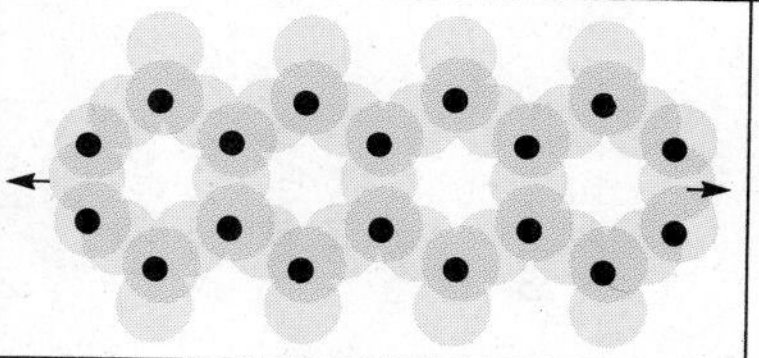	Hornblende
Continuous sheet	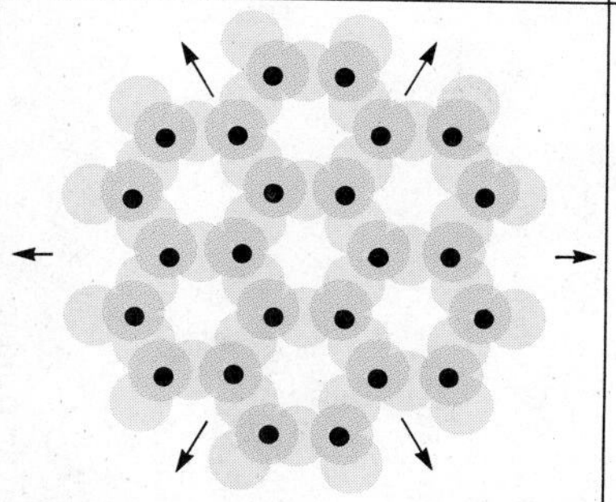	Mica
Three-dimensional network	Too complex to be shown by simple two-dimensional sketch	Quartz and feldspar

Figure 28.15 Some silicate structures

Table 28.3 Some Common Minerals and Their Properties

Minerals	*Chemical Composition*	*Color*	*Luster*	*Hardness (Mohs' Scale)*	*Specific Gravity (approx.)*	*Cleavage*	*General Use or Occurrence*
Argentite	Silver sulfide, Ag_2S	Black	Metallic	2–2.5	7.3	Yes	Silver ore
Asbestos (chrysotile)	Hydrous magnesium silicate, $H_4Mg_3Si_2O_9$	Green (different shades)	Silky	Low	2.5	Yes	Fireproofing; insulation
Calcite	Calcium carbonate, $CaCO_3$	White or colorless	Glassy to earthy	2.5–3	2.7	Yes	Cement manufacture; optical application; lime
Cinnabar	Mercury sulfide HgS	Red-brown	Dull, earthy	2–2.5	8.1	Yes	Mercury ore
Clay	Hydrous aluminum silicates containing Na, K, Fe, Mg, etc.	Light	Earthy	Very low	2.6	Fracture	Chinaware, pottery
Dolomite	Calcium (Mg, Fe, Mn, Ca) carbonate, e.g., $CaMg\ (CO_3)_2$	White-gray	Pearly	3.5–4.5	2.9	Yes	Cement manufacture; building materials; lime
Feldspar							
plagioclase	Calcium & Sodium aluminum silicate, $CaAl_2Si_2O_8NaAlSi_3O_8$	White-gray	Pearly	6–6.5	2.7	Yes	Ceramic glazes
orthoclase	Potassium aluminum silicate $KAlSi_3O_8$	White-pink	Pearly	6	2.5	Yes	Elongated crystals in igneous rock
Fluorite	Calcium fluoride, CaF_2	Clear and range of colors	Vitreous	4	3.2	Yes	Commonly found in metal ores
Galena	Lead Sulfide, PbS	Gray	Metallic	2.5	7.4	Yes	Lead ore
Gypsum	Hydrous calcium sulfate, $CaSO_4 \cdot 2\ H_2O$	Colorless-white	Silky-dull	2	2.3	Yes	Plaster of Paris; neutralizer of alkaline soils
Halite	Sodium chloride, $NaCl$	Colorless or white	Glassy	2.5	2.2	Yes	Table salt
Hematite	Iron oxide, Fe_2O_3	Gray-reddish	Metallic, dull	5.5–6.5	5.1	Fracture	Iron ore
Magnetite	Iron oxide, Fe_3O_4	Black	Metallic, dull	6	5.2	Fracture	Magnetic iron ore

Mica							
biotite	Group of hydrous silicates	Black-green;	Pearly	2.5–3	3.0	Yes	"Isinglass" heat-proof windows; electrical insulator
muscovite		Reddish-brown	Pearly	2.5–3	2.9	Yes	
Olivine	Metallic silicate, $(Mg, Fe, Mn)_2SiO_4$	Green	Glassy	6.5–7	3.2–4.4	Fracture	Igneous rock mineral
Pyrite	Iron Sulfide FeS_2	Pale yellow	Metallic	6–6.5	5.0	Fracture	Source of sulfuric acid (mine acid); fool's gold
Pyroxene	Metallic aluminum silicate (Ca, Mg, Fe, Na)	Black-green	Glassy	5–6	3.4	Yes	Igneous rock mineral
Quartz	Silicon dioxide, SiO_2	Colorless when pure	Glassy	7	2.65	Fracture	Optical applications
Sphalerite	Zinc sulfide, ZnS	Black, Yellow-brown	Resinous	3.5–4	4.0	Yes	Zinc ore
Talc	Hydrous magnesium silicate, $Mg_3(Si_4O_{10})(OH)_2$	Green-white Silvery	Pearly, greasy	1–2.5	2.7	Yes	Cosmetics

Minerals can be identified by chemical analysis, but most of these methods are detailed and costly, and are not available to the average person. More commonly, physical properties are used as the key to mineral identification. These properties are well known to all serious rock and mineral collectors. Some of the physical properties used in mineral identification are as follows.

Crystal Form. Crystals have definite geometric arrays of atoms or ions that are often recognizable from the external crystal configurations of some minerals (Fig. 28.16). Crystalline forms in minerals can be studied in detail by means of X-ray analysis.

Figure 28.16 Mineral crystal forms. Certain minerals are characterized by natural crystal forms. Quartz (SiO_2) crystals are shown below, fluorite (CaF_2) crystals above. (Courtesy Ron Docie, Athens, Ohio.)

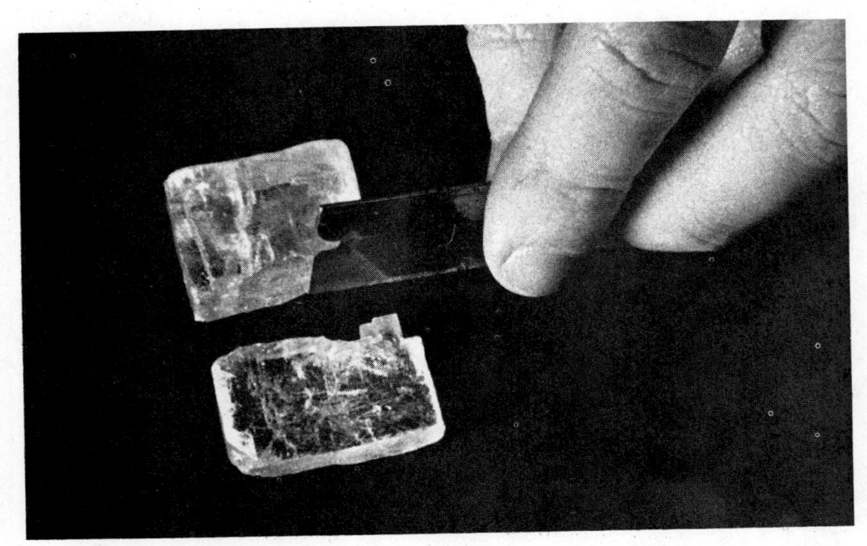

Figure 28.17 Cleavage of an NaCl crystal. Some minerals are characterized by cleavage planes along which they may be easily cleaved. (Courtesy Ron Docie, Athens, Ohio.)

Cleavage and Fracture. Cleavage refers to the splitting of a mineral along directions of weak bonding between planes of atoms (Fig. 28.17). Fracture is the breaking up of a mineral other than along a cleavage plane.

Specific Gravity. This is the ratio of the mass of body to the mass of an equal volume of water. Specific gravity indicates the compactness of the mineral structure.

Hardness. This property is comparative and so is indicated by a harder mineral being able to scratch a less-hard mineral. The varying degrees of hardness are represented on **Mohs' scale of hardness,** which runs from 1 to 10, soft to hard. This arbitrary scale is expressed by the 10 minerals listed in Table 28.4. Talc is the softest and diamond is the hardest. A particular mineral on the scale is harder than (can scratch) all those with lower numbers. Using these minerals as standards, one finds the following on the hardness scale: fingernail, 2–3; a penny, 3; window glass, 5–6.

Table 28.4 Mohs' Scale of Hardness

1. Talc	6. Feldspar (orthoclase)
2. Gypsum	7. Quartz
3. Calcite	8. Topaz
4. Fluorite	9. Corundum
5. Apatite	10. Diamond

Luster. The characteristics of light on the surface of the mineral due to absorption, reflection, and/or refraction give rise to a mineral's look or luster. Common descriptions of a mineral's luster are metallic, glassy, greasy, and dull.

Color and Streak. Color is just what it implies, the color of a mineral. Color depends a great deal on the type of metal atoms a mineral contains. Streak refers to the color of the powdered material on a streak plate (unglazed porcelain). The streaks of some metallic ores differ from the color of the unpowdered mineral. Other properties used to identify minerals include magnetism, radioactivity, and fluorescence.

Minerals are the natural source of many commercially important chemicals. Metal-rich rocks (ores) and mineral deposits are mined and processed for our use. Because of the vast extent of the Earth's crust, it was once thought that the supply of many common minerals was virtually inexhaustible. However, at the current rates of use, some mineral resources may last only a few more generations.

Learning Objectives

After reading and studying this chapter, you should be able to do the following without referring to the text:

1. State the three types of rocks and how they are formed.
2. Give several examples of the three types of rocks.
3. Describe the major intrusive rock formations.
4. State the difference between clastic and chemical rocks.
5. Explain the different types of metamorphism.
6. Explain the rock cycle.
7. Discuss minerals and mineral structures.
8. Describe the methods of mineral identification.

Important Words and Terms

rock
igneous rock
magma
geothermal gradient
extrusive igneous rock
lava
intrusive igneous rock
porphyritic
pluton
concordant
discordant
batholith
stock
dike
sill
laccolith
sedimentary rock
lithification
consolidation
clastic sedimentary rock
chemical sedimentary rock
stalactite
stalagmite
column
peat
lignite coal
sub-bituminous coal
bituminous coal
anthracite coal
bedding
metamorphic rock
rock cycle
mineral
silicon-oxygen tetrahedron
crystal form
cleavage
fracture
specific gravity
hardness
Mohs' scale of hardness
luster
color
streak

Questions

1. How are igneous rocks formed?
2. What is the geothermal gradient? Does this gradient extend to the center of the Earth?
3. What is the source of the Earth's internal heat?
4. Distinguish between intrusive and extrusive igneous rocks.
5. What is the difference between concordant and discordant igneous formations?
6. Describe each of the following: (a) batholith, (b) laccolith, (c) dike, (d) sill.
7. What is lithification and how does it take place?
8. Distinguish between clastic and chemical sedimentary rocks and give examples of each.
9. What are stalactites, stalagmites, and columns? How are they formed?
10. Give the four general types of coal and their differences. Are all coals sedimentary rocks?
11. What is meant by "bedding"?
12. Describe how metamorphic rocks are formed.
13. What types of rocks are metamorphized? Give an example of each.
14. What is the rock cycle, and what geologic processes correspond to an entrance and exit to the cycle?
15. How is mineral defined?
16. What two elements make up the majority of the Earth's crust and in what percentages?
17. On what does the formation of the crystalline structure of minerals depend?
18. What is the silicon-oxygen tetrahedron? Sketch its structure.
19. The mineral beryl has a ring arrangement of six silicon-oxygen tetrahedra. Sketch this arrangement.
20. Give six physical properties used in mineral identification.
21. What are the hardest and softest substances on Mohs' scale?

29

Internal Processes

It is useful to be assured that the heavings of the earth are not the work of angry deities. These phenomena have causes of their own.

–Seneca

WE HAVE EXPLORED the atmosphere and beyond—landing on the moon and probing the nearby planets. Yet, the direct exploration of the Earth has gone no more than a few miles below its surface. The vast majority of our knowledge about the Earth's interior composition and structure is the result of indirect observations. That is, we must be content to observe the manifestations of internal processes from our surface vantage point and then form models of the Earth below us. The theory of plate tectonics discussed in Chapter 27 is an example of such a model.

Occasionally materials of the upper mantle can be observed through the eruptions of volcanoes when mantle magma is vented to the Earth's surface. The dynamics of the Earth's internal processes is also evidenced on the surface in the form of earthquakes, which are the result of movements in the Earth's lithosphere. It is the shock waves from these enormous releases of energy traveling through the Earth's interior that give us our internal picture of the Earth's structure. Land forms, such as mountains, also provide clues as to what has happened within the Earth. By the principle of uniformity, these processes are still taking place.

In this chapter we shall examine these occurrences and features that give us our current ''view'' of the Earth's interior structure and processes.

29.1 Volcanoes

The word **volcano** comes from the volcanic island of Vulcano in the Mediterranean Sea. It was once believed that Vulcano was the chimney of the forge of Vulcan, the blacksmith of the Roman gods. Eruptions were thought to be the by-products of the forge of Vulcan as he beat out thunderbolts for Jupiter, king of the gods, and weapons for Mars, the god of war. However, we now know that a volcano is formed when magma from deep within the Earth's crust or the upper part of the mantle is vented to the surface. Above the surface, the molten **lava** with a temperature range of 1000–1200°C cools and solidifies into extrusive igneous rocks.

There have been several hundred recorded volcanic eruptions, and geologic evidence exists for many other cases of volcanic activity. Some volcanic eruptions make their mark in history. A classic example is the eruption of the dormant volcano Mount Vesuvius in A.D. 79 that buried the Roman cities of Pompeii and Herculaneum.

Other violent eruptions include the volcano Tamboro located on a small island in the East Indies. This most violent volcanic explosion in recorded history occurred in 1815 and blew as much as 35 mi^3 of particle debris into the air. As discussed in Chapter 26, the large volume of fine dust carried around by the global atmospheric circulation affected the Earth's albedo and 1816 was an uncommonly cold year. In 1883, a violent eruption on the island of Krakatoa, located between Java and Sumatra in the East Indies, literally blew apart this and a neighboring island. Fortunately these eruptions occurred in a relatively sparsely populated area.

In this century, the volcano hall of fame includes Mount Pelee on the island of Martinique in the West Indies. This volcano was known to have been previously active, but without serious incident. However, in April of 1902, more activity was observed; gases were being emitted from **fumaroles**—volcanic vents that typically exhibit such activity. The activity increased and in May the volcano exploded, emitting a huge cloud of gas and

◀ Old Faithful, Wyoming.

dust that engulfed the city of St. Pierre on the volcano's lower slopes. In a matter of minutes, an estimated 30,000 to 40,000 people lost their lives.

In 1912, Mount Katmai, a dormant volcano on the Alaskan peninsula, erupted with a similar explosion that sent great quantities of dust and glassy volcanic ash into the air. This eruption occurred in an uninhabited area, so there was no loss of human life as in the case of Mt. Pelee. The fine dust of the ejected material was carried around the world, but the larger particles dropped out within a few hundred miles. A hot blanket of material, almost 100 ft thick, collected around the volcano. Areas of Kodiak Island, some 60 mi away, were covered with several feet of dust. It took many years for this blanket to cool.

Steam and gaseous emissions from millions of fumaroles (gas vents) led to the name of "Valley of Ten Thousand Smokes" being applied to the area. Studies of this region have given much information about volcanic gaseous emissions and their contents. A great deal of the gas from Katmai was hydrogen chloride, which gave rise to raindrops of hydrochloric acid.

From more recent times are the eruptions of Paricutin, Surtsey, Mount St. Helens, and El Chichón. In 1943, Paricutin erupted in a Mexican farmer's cornfield. In the first year, the volcanic cone rose to a height of over 1000 ft (Fig. 29.1). After about 10 years of subsiding activity, Paricutin is now a dormant volcano with a cone of a height over 1300 ft.

Figure 29.1 The volcano Paricutin in Mexico, which erupted in a farmer's cornfield in 1943. (Courtesy U.S. Geological Survey, photo by F.O. Jones.)

In 1963, volcano Surtsey (after *Surtr,* the subterranean god of fire in Icelandic mythology) boiled up from the ocean floor off the coast of Iceland. Having sufficient lava to form a barrier against the sea, Surtsey is now a permanent volcanic island that may be found on the world map (Fig. 29.2). Geologists were able to study Surtsey from its "birth."

In almost everyone's recent memory is the eruption of Mount St. Helens. This volcano is located in southwestern Washington, about 40 mi from Portland, Oregon. Prior to 1980, Mount St. Helens was a placid, snow-capped, dormant volcano, one of many such volcanoes in the Cascade Mountains in Oregon and Washington. Its last period of activity had been between 1800 and 1857.

In March 1980 a series of minor earthquakes occurred near Mount St. Helens, and the first eruptions took place. On May 18, 1980, a massive eruption occurred, devastating an area of more than 150 mi^2 and leaving more than 60 persons dead or missing (Fig. 29.3, p. 540). A column of ash rose to an altitude of more than 12 mi. Nearby cities were blanketed with ash, and a light dusting of ash fell as far away as 900 mi to the east.

Huge mudflows from the ash caused flooding and silting in the rivers near the volcano, destroying and damaging many homes. Mount St. Helens became relatively quiet in 1981. However, it has shown minor activity in the years since, and it is expected that intermittent activity may continue for years, and even decades, if the behavior of the volcano follows the pattern of its eruptions in the 1800s. Volcano-triggered mudflows also had a devastating effect in the 1985 eruption of a volcano in Colombia, South America. Over 20,000 people were killed.

Another significant volcanic eruption in this decade was that of El Chichón in Mexico (Fig. 29.3). In late

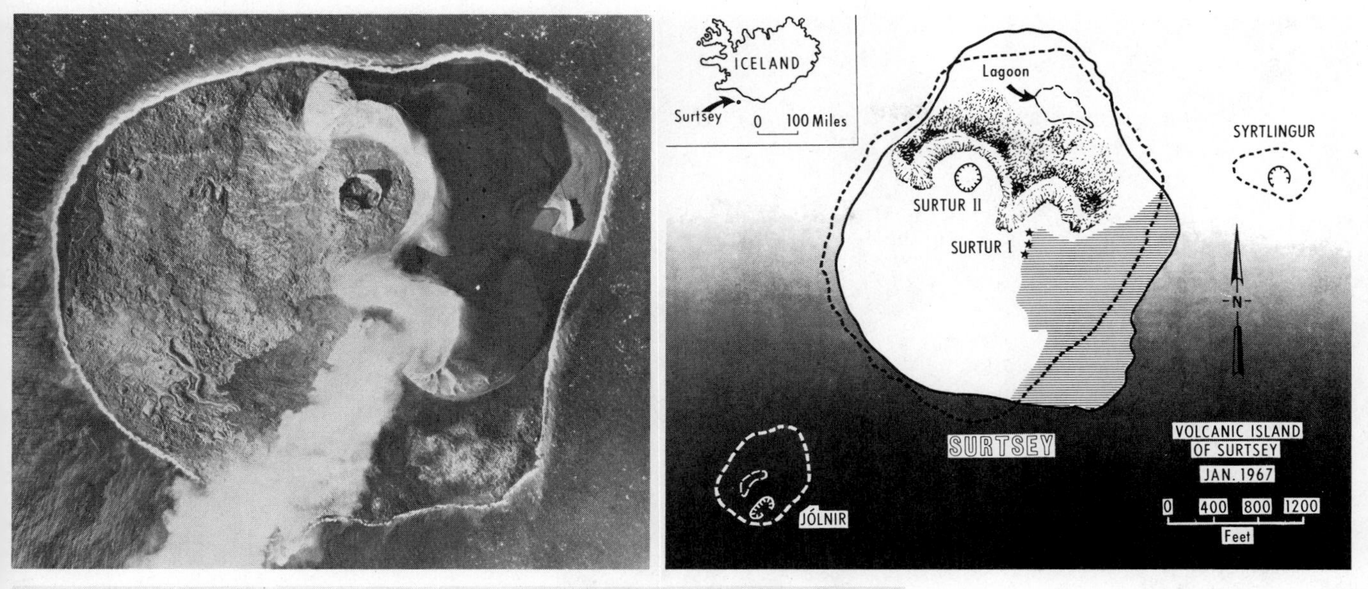

Figure 29.2 The volcano and island Surtsey off the coast of Iceland several months after eruption. Notice the lava fountains in the left part of the crater. The molten lava flows to the right into the ocean. (Courtesy U.S. Geological Survey.)

March 1982 El Chichón roared into life with a tremendous explosion that sent a column of ash and gases 10 mi high within an hour. Although the eruption of El Chichón was not very impressive compared to its earlier Mexican neighbor Paricutin, its significance arises from the projection of debris some 26 mi into the atmosphere. This debris in the stratosphere may have long-term climatic effects. See Chapter 26.

Not all volcanoes erupt so dramatically. There are various types and degrees of volcanic eruption. Some volcanoes are typically nonexplosive and erupt relatively quietly with an outflow of lava. One of the most noted examples of such eruptions is in the chain of volcanoes that forms the foundation of the Hawaiian Islands (Fig. 29.4). Eruptions of huge quantities of lava from cracks in the floor of the Pacific Ocean have caused the formation

Figure 29.3 (*Top*) Mount St. Helens in eruption on May 18, 1980, showing the violence of the eruption in contrast to the quiet countryside. Mount Adams is seen in the background. (*Bottom*) Aerial view of El Chichón showing the new crater formed in place of the original summit dome, which was blown off during the March–April 1982 eruption. (Courtesy U.S. Geological Survey.)

Figure 29.4 Lava from a fissure eruption. Fiery, molten lava partly obscured by fumes cascades over a cliff near the volcano Kilauea on the island of Hawaii. (Courtesy U.S. Geological Survey.)

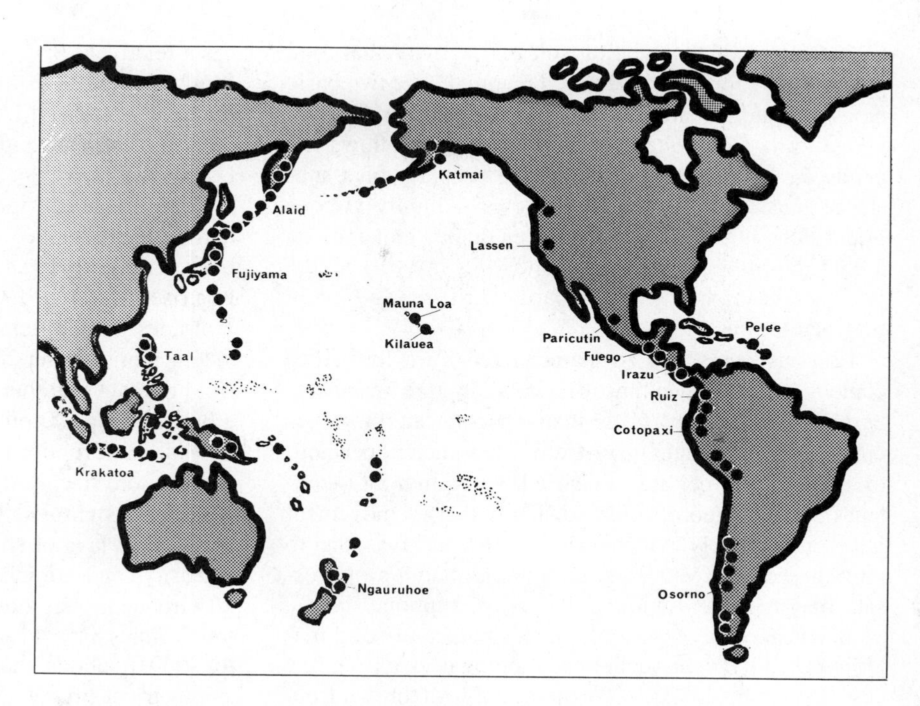

Figure 29.5 The Pacific "Ring of Fire." The outer rim of the Pacific Ocean is circumscribed by a ring of volcanoes that tend to have violent eruptions, hence the name "Ring of Fire." (Courtesy U.S. Geological Survey.)

of one of the most extensive mountain chains on Earth, most of which is beneath the ocean. The Hawaiian chain is nearly 1600 mi long and the volcanic formations on the ocean's floor, nearly three miles below sea level, rise to form island projections above sea level.

The occurrence of volcanic activity is for the most part unpredictable. New volcanoes may be formed unexpectedly while existing volcanoes lying dormant may suddenly erupt with practically no warning. However, the locations of eruptions and potential eruptions are known. From the theory of plate tectonics (Chapter 27), volcanic activity takes place predominantly at plate boundaries above subduction zones where one plate is deflected downward into the Earth's interior. (See Fig. 27.13.) For example, the area of the Pacific Ocean is one of widespread volcanic activity. The outer rim of the Pacific Ocean is marked by a ring of volcanoes known as the **"Ring of Fire"** (Fig. 29.5). Comparing Fig. 29.5 with Fig. 27.15, the correlation with plate boundaries is easily seen.

The formation of the Hawaiian Islands in the central Pacific Ocean and on the central portion of the Pacific Plate is explained by the so-called **hot spot theory.** This theory visualizes that as one of the Earth's crustal plates moves over a fixed source of heat beneath it, a "hot spot," a succession of volcanoes are generated. Each volcano slides away or moves with the plate, leaving room for the next.

According to the theory, volcanic island chains trace out the plates' movement over a hot spot. The Hawaiian Islands are cited as a classic example of this process, and studies have yielded data in support of the theory (age dating of sea floor samples). However, not every island chain appears to fit the theory, so geologists have more work to do in explaining the Earth's internal processes.

Volcanism in the United States has been confined mainly to the Hawaiian Islands, Alaska and the Aleutian Islands, and the Cascade Mountains in Oregon and Washington. Most of these volcanoes, particularly in the latter region, are now dormant. However, in addition to Mount St. Helens, an active volcano can be seen in Hawaii. The volcano Kilauea, on the island of Hawaii (the largest island in the Hawaiian chain) is one of the most active in the world (Fig. 29.4).

Because of its frequent, yet mild eruptions, the U.S. Geological Survey maintains a Volcano Observatory on the summit of Kilauea. It is of special interest to geologists because the erupted material comes from great depths and thus provides clues to the geochemical composition of the Earth's interior. The observatory is located in the Hawaii Volcanoes National Park where over 500,000 visitors come each year to view an active volcano.

When a volcano erupts, a cone is built up around its central vent. The shape and size of a volcanic cone give an indication of the type of volcanic activity that took place and the form of the emitted material. The two basic types of materials are lava and pyroclastic debris. Lava, as we know, is molten rock. How fast lava flows depends on its **viscosity**—the internal property of a substance that offers resistance to flow. A highly viscous liquid does not flow readily, for example, molasses on a cold day is highly viscous. The viscosity of lava depends on its composition, primarily its silica (SiO_2) content, and on its temperature.

Low-viscosity lava is characterized by a low silica content and is called **basaltic lava.** In rare instances, basaltic lava can flow faster than a person can run. More commonly, flow rates vary from a few meters per hour to a few meters per day. **Felsitic lava,** with a SiO_2 content of 70% or more, flows so slowly that its movement is not immediately evident. Highly viscous lavas tend to plug the volcano vent, and the internal buildup of pressure may result in a violent explosive eruption.

A volcano may also emit large volumes of solid material called **pyroclastic debris,** ranging in size from fine dust to huge boulders. Large masses of lava thrown from a volcano harden in mid-air, forming large rocks called **volcanic bombs.** These boulder-size rocks can be heard whistling through the air as they descend.

As was pointed out previously, many volcanoes do not erupt violently, but relatively calmly eject liquid magma from fissures in the Earth's surface. If the lava has a relatively low viscosity such that it flows easily, a gently sloping low profile **shield volcano** is formed by frequently repeated lava flows.

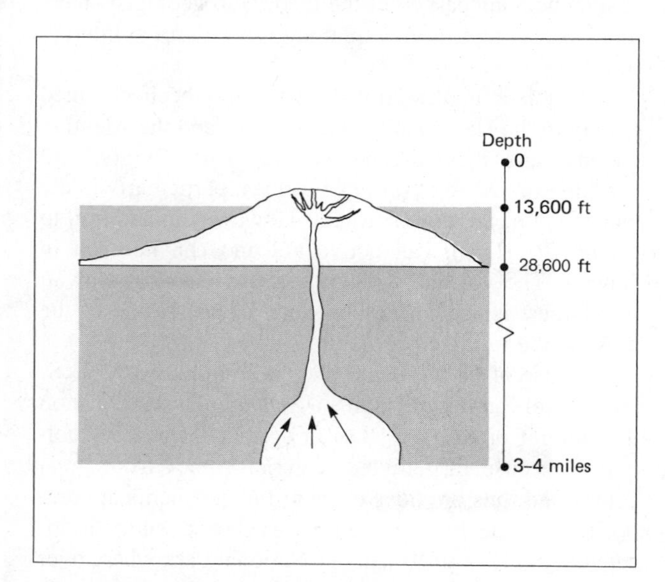

Figure 29.6 An idealized profile of Mauna Loa. The magma from depths of three to four miles rises to a shallow reservoir, from which it erupts onto the surface. Mauna Loa's gently sloping profile is characteristic of shield volcanoes.

The classic example of a shield volcano is Mauna Loa in the Hawaiian island chain. In fact, Mauna Loa is a huge volcanic mountain—the largest single mountain on Earth in sheer bulk. Although not as tall as Mt. Everest (slightly over 29,000 ft above sea level), Mauna Loa rises 15,000 ft from the ocean floor to sea level and protrudes an additional 13,600 ft above sea level for a total height of about 28,600 ft (Fig. 29.6). Its bulk comes from the fact that this partially submerged mountain has a base almost 100 mi in diameter.

Volcanic eruptions of both lava and pyroclastic debris form a more steeply sloping, layered composite cone that is called a **stratovolcano** (also called a **composite volcano**). The lava of stratovolcanoes has a relatively high viscosity, and eruptions are more violent and generally less frequent than those of shield volcanoes. Many stratovolcanoes have an accumulation of material up to 6000 to 8000 ft above their base and have a characteristic symmetrical profile. Mount St. Helens (Fig. 29.3) is a stratovolcano. Dormant stratovolcanoes include Mt. Fuji in Japan (Fig. 29.7), and Mts. Shasta, Hood, and Rainier in the Cascade Mountains in Washington and Oregon.

A volcanic eruption may also consist primarily of pyroclastic debris. In this case, steeply sloped **cinder cones** are formed, which rarely exceed 1000 ft in height. A cinder cone is shown in Fig. 29.8 rising above a previous lava flow. Sunset Crater near Flagstaff, Arizona, is another example of these small, steep, symmetric cones of pyroclastic debris.

Volcanic activity is usually not confined to the region of the central vent of a volcano. Fractures may split the cone with volcanic material being emitted along the flanks of the cone. Also, material and gases may emerge from small auxiliary vents forming small cones on the slope of the main central vent. A funnel-shaped depression called a **volcanic crater** exists near the summit of most volcanoes from which material and gases are ejected (Fig. 29.9).

Many volcanoes are marked by a much larger depression called a **caldera.** These roughly circular, steep-walled depressions may be up to several miles in diameter. Calderas result primarily from the collapse of the chamber at the volcano's summit from which lava and ash were emitted. The weight of the ejected material of the partially empty chamber causes its roof to collapse much like the collapse of the snow-ladened roof of a building. Crater Lake on top of Mount Mazama in Oregon occupies the caldera formed by the collapse of the volcanic

Figure 29.7 Mount Fuji in Japan. Mount Fuji is a dormant stratovolcano that was formed from eruptions of both lava and pyroclastic debris. Seen in the foreground of the photograph is Lake Kawaguchi, one of the five lakes around the base. (Japan Air Lines Photo.)

Figure 29.8 Cerro Negro Volcano, near Leon, Nicaragua (1968). Notice the initial lava flows to the side and in back of the steep symmetrical cinder cone. The shape or profile of a volcano gives a good indication of the type of volcanic activity. Also notice the huge volume of smoke and particulate matter being vented into the atmosphere. (Courtesy U.S. Geological Survey.)

Figure 29.9 Crater of Mauna Loa in the Hawaiian Islands. A funnel-shaped crater exists near the summit of most volcanoes from which gases and material are ejected. (Courtesy U.S. Geological Survey.)

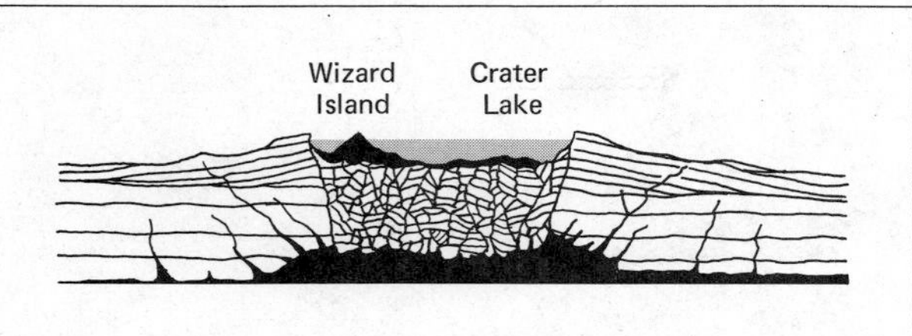

Figure 29.10 Crater Lake caldera. Large depressions or caldera are formed when an evacuated volcano chamber collapses. This occurred to a stratovolcano in Oregon (now called Mount Mazama) over 6000 years ago. The caldera is now filled with water and forms Crater Lake with a small protrusion called Wizard Island (small flat-topped mound near the center of lake foreground in photo). The lake is approximately 5 miles in diameter. (Courtesy Oregon State Highway Department.)

chamber of this once-active stratovolcano (Fig. 29.10). In some instances, calderas are formed by a violent explosion, rather than a collapse. Felsitic lava may clog the volcanic vent, and the resulting buildup of pressure causes the volcano to "blow its top," so to speak.

29.2 Earthquakes

As was learned in Chapter 6, the sudden release or transfer of energy gives rise to the propagation of waves in elastic media. Sound waves and water are commonly observed examples. The Earth's internal processes involve transfers of energy, which may occur either quickly or slowly. Although the Earth is not ordinarily thought of as a wave-propagating medium, when a sudden release or transfer of energy occurs, the Earth vibrates and waves propagate, resulting in what we call **earthquakes.**

An earthquake, as everyone knows who has experienced a substantial one, is manifested by the vibrating and sometimes violent movement of the Earth's surface. But waves are also propagated through the Earth. It is these waves that provide one of the most useful ways of studying the Earth's interior. **Seismology** is the branch of geophysics that studies these waves and uses them as probes to "see" below the Earth's surface to determine its internal structure.

Earthquakes may be caused by explosive volcanic eruptions or even explosions caused by humans, but most earthquakes appear to be associated with movements in the Earth's crust along a fault or fracture. According to the theory of plate tectonics, the optimum place for such movement is at the plate boundaries, and the major earthquake belts of the world are observed in these regions. (See Fig. 29.11, also Figs. 27.14 and 27.15.) Notice how the earthquake region around the Pacific Ocean is similar to that of the volcanic "Ring of Fire" (Fig. 29.5).

Movements of the Earth's plates would certainly put strains on rock formations in the Earth's crust near the plate boundaries. Rocks possess elastic properties, and energy is stored in the elastic deformations of the lithosphere. If the forces causing this deformation are great enough to overcome the force of friction along a plate boundary or some nearby fault, then the fault walls move suddenly and the stored energy in the rocks is released, causing an earthquake.

If the elastic limit of a stressed rock formation is not exceeded, the rock may rebound elastically and snap back to its original shape after a shift along a fault has relieved the stress—much like a spring. However, if the elastic limit of the rock is exceeded, the rock cannot fully recover when the strain is removed. If the elastic limit is greatly exceeded by the deformation force, the rock formation may even rupture.

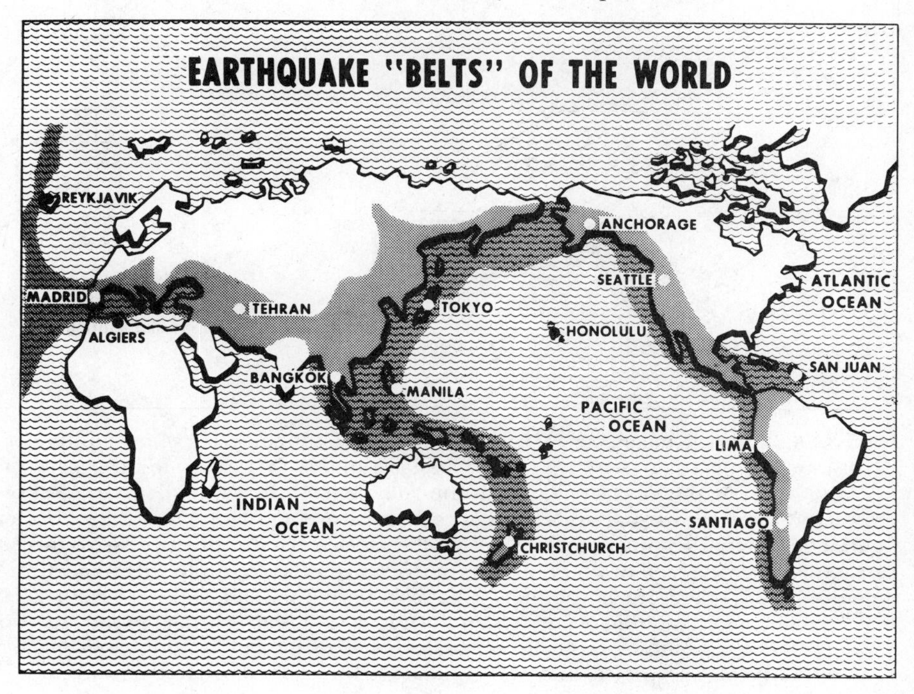

Figure 29.11 The earthquake belts of the world. Notice the belt that circumscribes the Pacific Ocean in a manner similar to that of the volcanic "Ring of Fire" in Fig. 29.5. (Courtesy U.S. Geological Survey.)

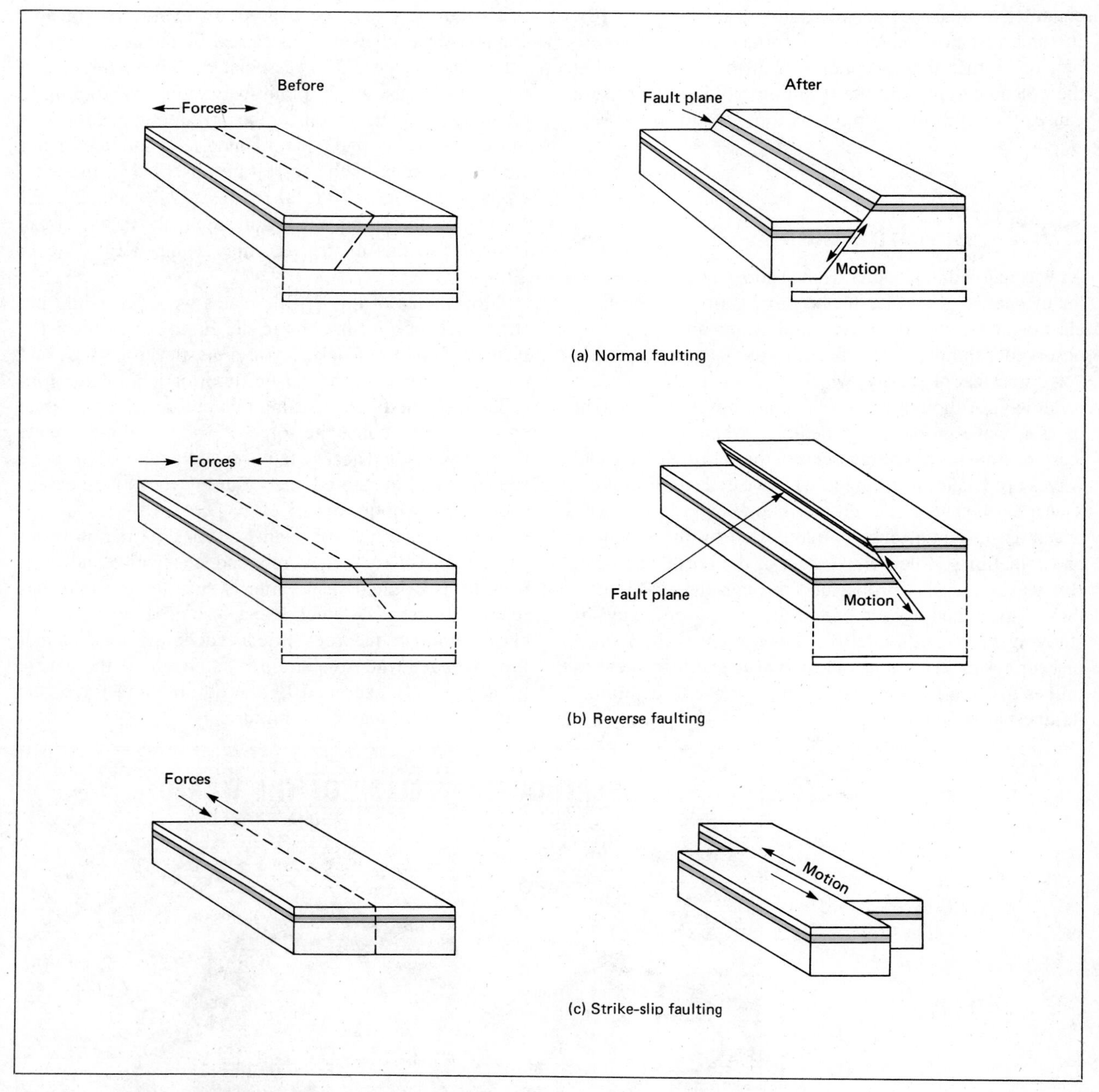

Figure 29.12 Block diagrams of general types of faulting. (a) Normal faulting occurs along a nonvertical fault as a result of expansive forces that cause the overlying side of the fault to move downward relative to the side beneath it and tends to expand or pull the material apart. Normal faulting occurs near diverging plate boundaries. (b) Reverse faulting also occurs along a nonvertical fault, but as a result of compressive forces that cause the overlying side of the fault to move upward relative to the side beneath it. This type of faulting occurs near subduction zones where one plate slides over a descending plate. (c) Strike-slip faulting occurs when the stress forces are parallel to the fault boundary such that the resulting fault slip is horizontal. This type of faulting takes place along the boundary of two plates that slip by each other—along what is commonly called transform faults.

The relative motions along fault boundaries give rise to three general types of faulting—normal, reverse, and strike-slip. These three types of faulting correspond to the three relative motions of plate boundaries (cf. Fig. 27.12).

Normal faulting occurs along a nonvertical fault as the result of expansive forces that cause the overlying side of the fault to move downward relative to the side beneath it (Fig. 29.12). In this case, the stress forces are in opposite directions and the faulting tends to expand or pull the material apart. Normal faults occur near diverging plate boundaries in the regions of sea floor spreading. (See the plate boundaries in Fig. 27.15).

Reverse faulting also occurs along a nonvertical fault and is the result of compressional stress forces that cause the overlying side of the fault to move upward relative to the side beneath it. A special case of reverse faulting, called **thrust faulting,** describes the faulting when the fault plane is at a small angle to the horizontal. As might be expected, this type of faulting occurs in subduction zones where one plate slides over a descending plate, or along convergent plate boundaries.

Finally, **strike-slip faulting** occurs when the stresses are parallel to the fault boundary, such that the fault slip is horizontal. This type of faulting takes place along the boundary of two plates that *strike* and *slip* by each other without an appreciable gain or loss of surface area. Strike-slip faulting occurs along what are commonly called **transform faults.** Transform faults occur near plate boundaries that do not move directly away from or toward each other.

An example of a transform fault is the famous San Andreas fault in California. It is the master fault of an intricate network of faults that runs along the coastal regions of California (Fig. 29.13). This huge fracture in the Earth's crust is 600 or more miles long and extends vertically to depths of at least 20 mi. Over much of its length, a linear trough of narrow ridges reveals the fault's presence, as shown in Fig. 29.13.

As can be seen in Fig. 29.13, the San Andreas fault system lies on the boundary of the Pacific Plate and the American Plate (cf. Fig. 27.15). Movement along the transform fault arises from the relative motion between these plates. The Pacific Plate is moving northward relative to the American Plate at a rate of several centimeters per year. At this current rate and direction, in about 10 million years Los Angeles will have moved northward to the same latitude as San Francisco. In another 50 million years, the segment of continental crust on the Pacific Plate will have become completely separated from the continental land mass of North America.

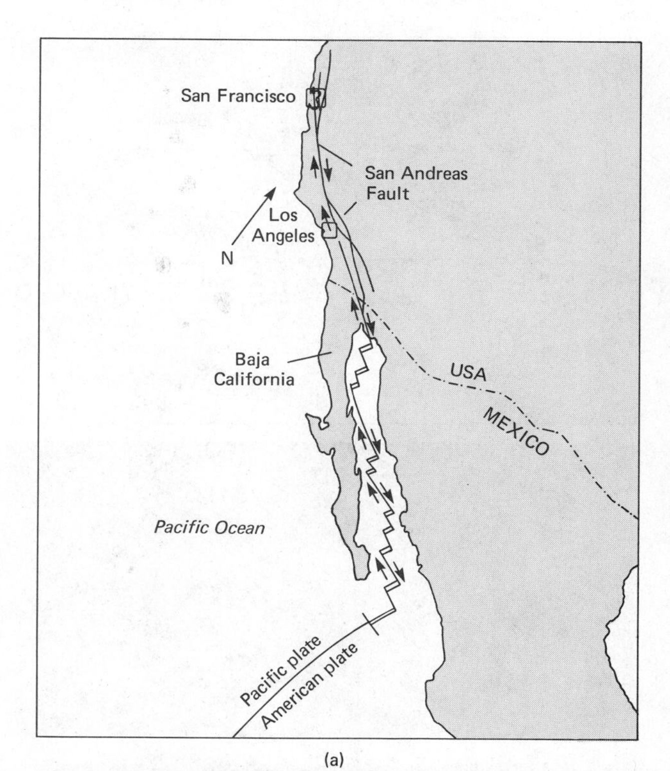

(a)

(b)

Figure 29.13 (a) A map of the San Andreas fault system in California. The San Andreas fault forms part of the boundary between the Pacific and American plates. (b) An aerial view of the San Andreas fault in the Carrizo Plain area of central California. (Courtesy U.S. Geological Survey.)

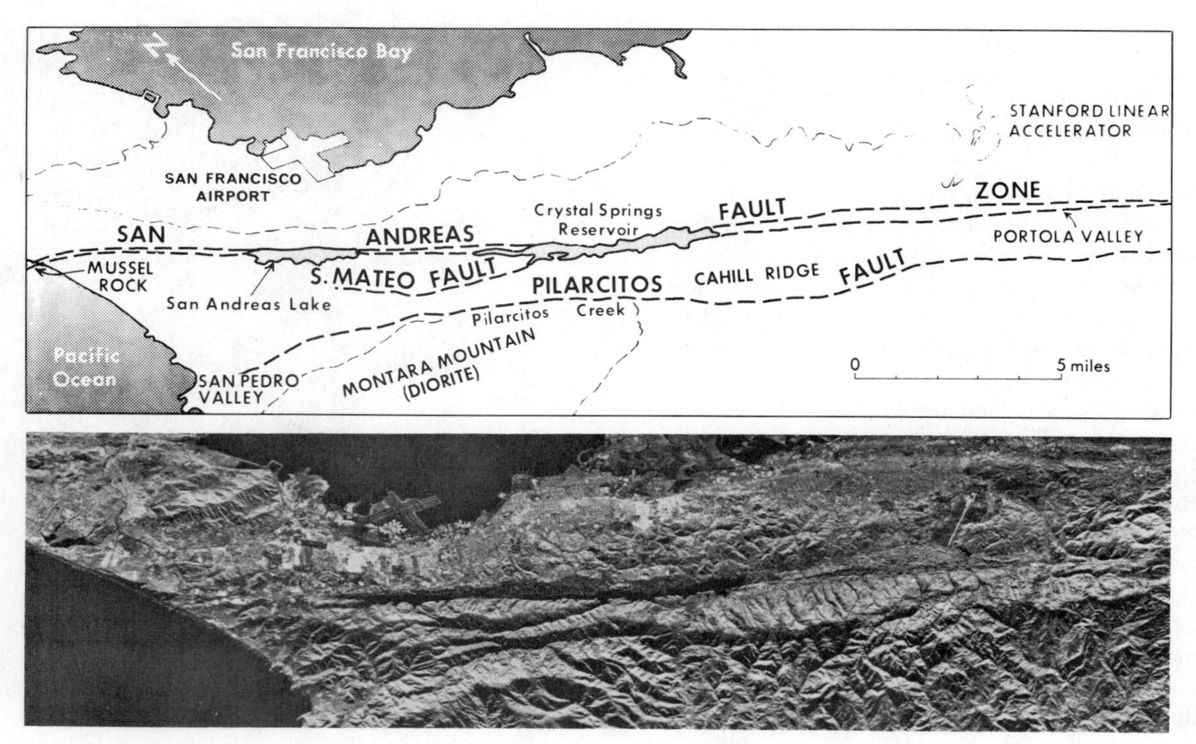

Figure 29.14 The San Andreas fault and the City by the Bay. The image shown here of the San Francisco Peninsula is not a regular photograph, but a radar image produced with equipment mounted on an airplane. In this image, the geological structural elements, especially those of the San Andreas fault system, are clearly delineated. (Courtesy NASA and USDA-SCS.)

The horizontal movement along the San Andreas fault is a cause for considerable concern for the populated San Francisco Bay area through which it runs (Fig. 29.14). The famous San Francisco earthquake of April 18, 1906, which measured 8.3 on the Richter scale (earthquake scales will be discussed shortly), resulted in the loss of approximately 700 lives and in millions of dollars of property damage.

There have been many other milder earthquakes since then along the San Andreas and other branch faults. In 1971, an earthquake in the San Fernando Valley near Los Angeles resulted in the deaths of over 60 people and considerable damage (Fig. 29.15). Although this fault region had been dormant for hundreds of years, the occurrence of this recent earthquake serves as a warning and reminder that another quake as strong as that of 1906 could happen again.

There is little that can be done about the San Andreas fault except to learn to live with it. Geologists measure the stresses along the fault in hope of predicting future earthquakes. Also, building codes in the area are strict and are concerned with construction that will lessen earthquake damage. However, increased population and an ever-increasing need for housing have caused land developers to use any available land, resulting in rows of houses that straddle the trace fault of 1906 and sit on hills where earthquake-induced landslides could occur. The people in these houses live there with the knowledge of the fracture in the Earth below and the hope that any future movement along the fault will occur elsewhere.

When an earthquake does occur, the point or region of the initial energy release or slippage is called the **focus** of the ''quake.'' The vast majority of earthquakes originate in the crust or upper mantle, so an earthquake's focus generally lies at some depth—from a few to several hundred miles. Because of this, it is convenient to designate the location on the Earth's surface directly above the focus. This point is called the **epicenter** of the earthquake.

The energy released at the focus of an earthquake propagates outwardly as **seismic waves.** There are two general types of seismic waves produced by earthquake

Figure 29.15 California earthquakes. (*Left*) 1906 San Francisco earthquake. A view of the earthquake- and fire-wrecked area of Nob Hill from the corner of Van Ness and Washington Streets. (*Right*) 1971 San Fernando earthquake. Aerial view of collapsed overpass of the Golden State Freeway, Los Angeles County. (Courtesy U.S. Geological Survey. Left photo by W.C. Mendenhall.)

vibrations: **surface waves,** which travel along the Earth's surface or a boundary within it; and **body waves,** which travel through the Earth. Surface waves cause most earthquake damage as they move along the Earth's surface. Body waves propagate through the Earth's interior and so do less damage.

There are two types of body waves, P (for primary) or compressional waves and S (for secondary) or shear waves. The **P waves** are longitudinal compressional waves that are propagated by particles in the propagating material moving longitudinally back and forth in the same direction as the wave is traveling (cf. Chapter 6). Sound waves are an example of longitudinal compressional waves. In the case of **S waves,** the particles move at right angles to the direction of the wave travel, and hence are transverse waves. An example of transverse shear waves is the vibration in a plucked guitar string.

There are two important differences between P and S waves. S waves can travel only through solids. Liquids and gases cannot support a shear stress; that is, they have no elasticity in this direction and therefore their particles will not oscillate in the direction of a shearing force. For example, little or no resistence is felt when one shears a knife through a liquid or gas. The compressional P waves, on the other hand, can travel through any kind of material—solid, liquid, or gas. The other important difference is the velocity of the waves. P waves derive their name ''primary'' because they always travel faster than the ''secondary'' S waves in any particular solid material and hence arrive earlier at a seismic station. It is these differences between P and S waves that allow seismologists to locate the focus of an earthquake and learn about the Earth's internal structure.

The speeds of the body waves depend on the density of the material, which generally increases with depth. As a result, the waves are curved or refracted. Also, the waves are refracted when they cross the boundary or discontinuity between different media—in the same manner light waves are refracted (Chapter 7). The refraction of seismic waves and the fact that S waves cannot travel through liquid media provide our present view of the Earth's structure as shown in Fig. 29.16. Because of the so-called *shadow zones* of the S and P waves, the outer core of the Earth is believed to be a highly viscous liquid.

The seismic waves of an earthquake are monitored by an instrument called a **seismograph,** the principle of which is illustrated in Fig. 29.17. The recorded seismogram gives the time delay of the arrival of different types of waves, as well as an indication of their energy. (The greater the energy, the greater the amplitude of the traces on the seismograph.)

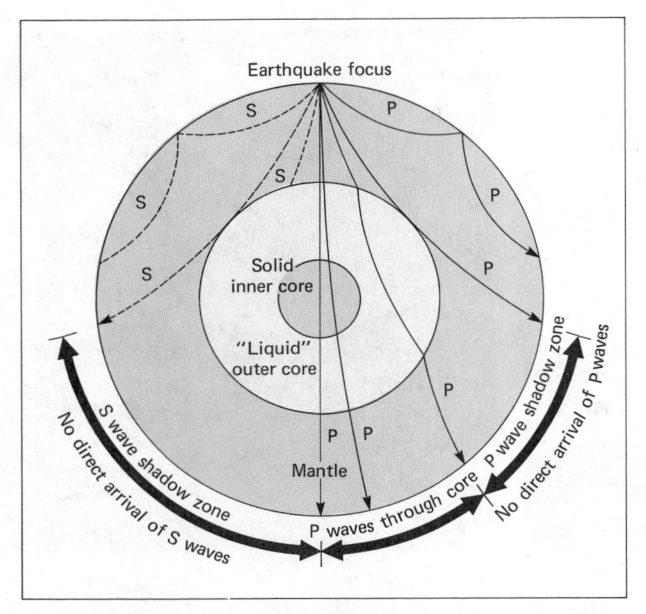

Figure 29.16 An illustration of seismic waves traveling through the Earth's interior. Because of the S and P wave shadow zones, the Earth is believed to have a liquid outer core. The S wave shadow zone arises because the liquid outer core will not transmit the S shear waves. The P wave shadow zone results from the refraction the compressional P waves have upon entering the liquid outer core. The S and P waves are separated on opposite sides of the Earth in the drawing for simplicity.

The severity of earthquakes is represented on different scales. The two most common are the **Richter scale,** which gives an absolute measure of the energy released by calculating the energy of seismic waves at a standard distance, and the modified **Mercalli scale,** which describes the severity of an earthquake by its observed effects. A comparison of these two scales is shown in Table 29.1. Notice that the 1985 Mexico earthquake was less severe on the Richter scale, but the death and destruction it caused gave it a high ranking on the Mercalli scale. The epicenter of the quake was off the Pacific coast of Mexico; however, it had devastating effects in Mexico City 250 mi away. This city, the second most populated on Earth (Shanghai, China is first), suffered thousands of deaths and injuries, and billions of dollars of damage. The first earthquake was followed by an aftershock 36 hours later that measured 7.5 on the Richter scale.

Of these two scales the Richter scale* is used more often, with magnitudes expressed in whole numbers and decimals usually between 3 and 9. However, the scale is a logarithmic function, that is, each whole number step represents about 31 times more energy than the preceding number step. For example, an earthquake that

*Developed in 1935 by Charles Richter of the California Institute of Technology.

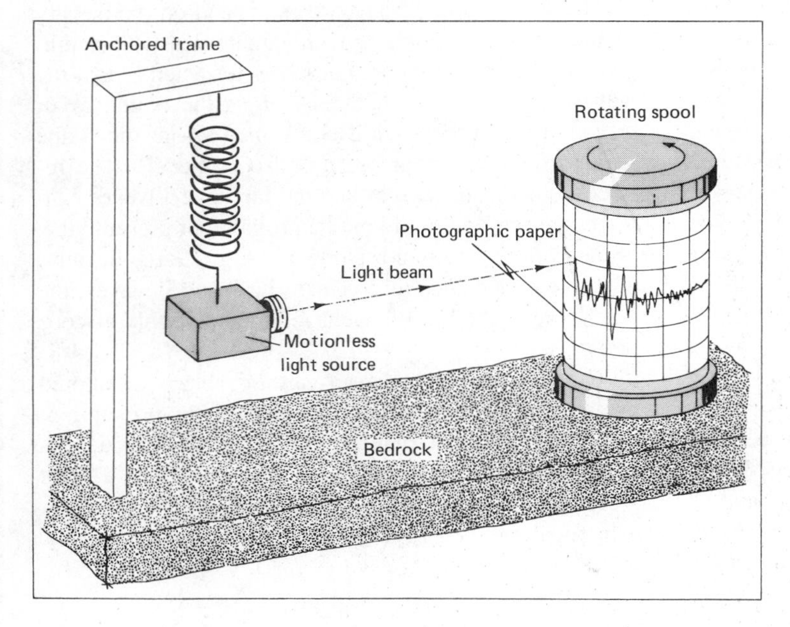

Figure 29.17 An illustration of the principle of the seismograph. The rotating spool (anchored in bedrock) vibrates during the quake. A light beam from the relatively motionless source on a spring traces out a record or seismograph of the earthquake's energy on the light-sensitive photographic paper. (Redrawn courtesy U.S. Geological Survey.)

Table 29.1 Scales of Earthquake Activity

Richter Scale	*Modified Mercalli Scale*	
Magnitude	*Maximum Expected Intensity (at epicenter)*	
1–2	I.	Usually detected only by instruments
3–4	II–III:	Can slightly be felt
4–5	IV–V:	Generally felt; slight damage
6–7	VI–VIII:	Moderately destructive
7–8	IX–X	Major earthquake
8+	XI–XII:	Great earthquake

Specific Earthquakes

	Richter	Mercalli
1906 San Francisco	8.3	XI
1964 Alaska	8.4	X
1985 Mexico	8.1	XI (estimated)

registers a magnitude of 5.5 on the Richter scale indicates the measuring seismograph received about 31 times as much energy as for an earthquake that registers a magnitude of 4.5.

Similarly, an earthquake with a magnitude of 8 does not represent twice as much energy as one with a magnitude of 4, but about one million times as much! (See Fig. 29.18.) As can be seen from Table 29.1, an earthquake with a magnitude of 2 to 3 is the smallest tremor felt by human beings. The largest recorded earthquakes are in the magnitude range of 8.7 to 8.9.

The Richter scale gives no indication of the damage caused by an earthquake, only its potential for damage. Earthquake damage depends not only on the magnitude of the quake but also on the location of its focus and epicenter and the environment of that region, specifically, the local geologic conditions, the density of population, and the construction designs of buildings. The modified Mercalli scale gives a better indication of earthquake effects, since the scale is based on actual observations.

The Mercalli scale was developed by Giuseppe Mercalli in the 1890s before the advent of seismographs. This scale has a range of values from I, which indicates that the earthquake is so slight that it is not felt by human beings (except by a few under especially favorable circumstances), to XII, which designates extensive damage with practically all works of construction damaged greatly or destroyed. Waves are seen on the ground surface and objects are thrown upward into the air for an earthquake with a modified Mercalli rating of XII. The general indications of this scale are given in Table 29.1. The modified Mercalli scale is more illustrative with its classifying descriptions. For example,

VI. Felt by all, many frightened and run outdoors. Some heavy furniture moved; a few instances of fallen plaster or damaged chimneys. Damage slight.

VII. Everybody runs outdoors. Damage negligible in buildings of proper design and construction; slight to moderate in well-built ordinary structures; considerable in poorly built or badly designed structures; some chimneys broken. Noticed by persons driving motor cars.

Earthquake damage may result directly from the vibrational tremors or indirectly from landslides and subsidence, as shown in Fig. 29.19. In populated areas, a great deal of the property damage is caused by fires because of a lack of ability to fight them, due to disrupted water mains and so on. When the energy release of a quake occurs in the vicinity of, or beneath the ocean

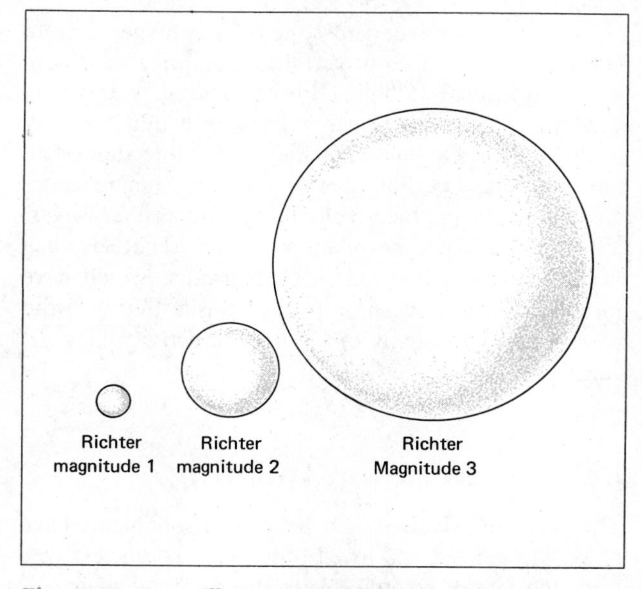

Figure 29.18 An illustration of the relationship between earthquake magnitude on the Richter scale and energy. The volumes of the spheres are roughly proportional to the amount of energy released by earthquakes of the magnitudes given, and illustrate the exponential relationship between magnitude and energy. At the same scale, the energy released by the San Francisco earthquake of 1906 (Richter magnitude 8.3) would be represented by a sphere with a diameter of 220 ft. (Redrawn courtesy U.S. Geological Survey.)

Figure 29.19 Earthquake and tsunami damage. The photo at the left shows subsidence in Anchorage, Alaska, resulting from the "Good Friday" earthquake of March 27, 1964. The quake generated a tsunami that washed many fishing boats into the heart of Kodiak, Alaska. A section of the harbor can be seen in the upper left of the photo. (Courtesy U.S. Geological Survey.)

floor, huge waves called **tsunamis** are sometimes generated.

These waves travel across the oceans at speeds up to 600 mi/h. In the open ocean, the waves may be 100 mi long and only 3 ft high. However, as they travel in shallower water, the tsunamis grow in height and may be 50 ft high or higher when they smash into shore with immense force (see Fig. 29.19). Tsunamis coming ashore are commonly and incorrectly referred to as *tidal waves*, although they have no relation to tides. The Hawaiian Islands have experienced many tsunamis, which have raced across the ocean from the Alaska-Bering Strait area or other locations of earthquake activity that circumscribe the Pacific Ocean.

29.3 Mountain Building

The sheer massiveness and heights of mountains have always fascinated and awed observers. To ancient cultures their lofty summits were the dwelling places of gods. Throughout history, and even today, superstitions surround mountains. To some, mountains present a challenge that must be met by scaling the highest peak. However, to the geologist, mountains are a key to the Earth's geologic history. In an effort to know this history, geologists study and theorize how mountains were built or formed.

Mountains differ greatly. Some are singular masses rising majestically above plains, such as the volcanic cone of Mt. Kilimanjaro in Africa. Other mountains belong to a geologic unit or series of mountains that make up a **mountain range,** such as the Sierra Nevada range in the western United States. A group of similar ranges form a **mountain system.** The Sierra Nevada range, along with several other ranges, form the Rocky Mountain system. An elongated unit of ranges is referred to as a **mountain chain.**

Mountains are generally classified into three principal kinds based on characteristic features: (a) volcanic mountains, (b) fault-block mountains, and (c) folded mountains.

Volcanic mountains are those that have been built by volcanic eruptions. As has been noted, most volcanoes, and hence volcanic mountains, are located above the subduction zones of plate boundaries. If the colliding plates are both oceanic, volcanic mountain chains are formed on the ocean foor of the overlying plate. New mountains are being formed beneath the sea today by this process. Chains of oceanic volcanic mountains are evidenced above sea level by island arcs. Japan and the islands of the West Indies are good examples of island arcs. Island arcs lie adjacent to ocean trenches that mark the boundaries of oceanic plates.

Continental volcanic mountain chains occur when the overlying plate of the subduction zone is continental crust.

Figure 29.20 Volcanic mountains. A view of the Chile-Argentina lake district of the Andes mountains. The Andes are continental volcanic mountains that were formed from continental crust overlying a subduction zone. (Courtesy Lan-Chile Airlines.)

The Andes mountains along the western coast of South America are an example of such a continental mountain range (Fig. 29.20). The Cascade mountains in Washington and Oregon are the only volcanic mountain range in the United States.

Fault-block mountains are believed to have been built by normal faulting in which giant pieces of the Earth's crust were tilted and uplifted (Fig. 29.21). These mountains evidence great stresses within the Earth's crustal plates. Fault-block mountains rise sharply above the surrounding plains. The Sierra Nevada range in California,

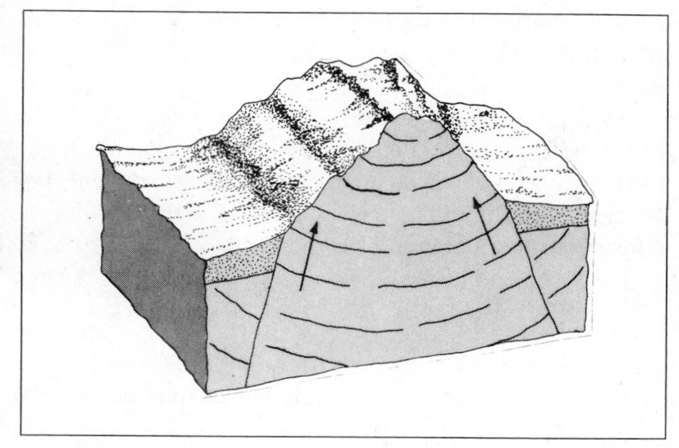

Figure 29.21 Fault-block mountains. The diagram illustrates how fault-block mountains are formed due to the uplifting of normal faulting. The Grand Teton mountains in Wyoming shown in the photograph are examples of fault-block mountains. (Courtesy U.S. Geological Survey.)

the Grand Teton mountains in Wyoming, and the Wasatch range in Utah are examples of fault-block mountains in the United States.

Folded mountains, as the name implies, are characterized by folded rock strata. Examples of folded mountains include the Alps, Himalayas, and the Appalachian mountains (Fig. 29.22). Although folding is the main feature of folded mountains, these complex structures also contain external evidence of faulting and central evidence of igneous metamorphic activity. Folded mountains are also characterized by exceptionally thick sedimentary strata, which indicates that the material of these mountains was once associated with ocean basins. Indeed, marine fossils have been found at high elevations in the Himalayas and other folded mountain systems.

Two eminent American geologists, James D. Dana (1813–1895) and his contemporary James Hall (1811–1898), advanced many of the original concepts of how such mountains were formed. These concepts considered the mountains on the Earth's surface to be a result of the cooling and contracting of the Earth during its formation—much like the wrinkling of the skin of an apple as it dries out. A good statement of this theory is presented in the conclusion of an article by Dana that appeared in the *American Journal of Science and Arts* in 1873 (Series 3, Vol. 106).

> The views on mountain building now sustained suppose the existence, through a large part of geologic time, of a thin crust, and of liquid rock beneath that crust so as to make its oscillations possible and refers the chief oscillations, whether of elevation or subsidence (sinking), to lateral pressure from the contraction of that crust; and this accords with my former view, and with that earlier presented by the clear-sighted French geologist, Prevost. I hold also, as before, that the prevailing position of mountains on the borders of continents, with the like location of volcanoes and of the greater earthquakes, is due to the fact that the oceanic areas were much the largest, and were the areas of greatest subsidence under the continued general contraction of the globe.

However, Dana and Hall disagreed on how the subsidence of ocean basins occurred. From a study of fossils once contained in **geosynclines,** which are long, narrow,

Figure 29.22 Folded mountains. A view of the Alps near Berchtesgaden in Germany. In the theory of plate tectonics, the Alps are believed to have formed when the African and Eurasian plates collided. (Courtesy Lufthansa German Airlines.)

subsiding ocean troughs containing large accumulations of sediment, Hall found evidence that over geologic time periods a basin sank at approximately the same rate as the sediment accumulated. He believed that the basin sank because of the accumulating burden of sediment.

Dana, on the other hand, thought that the sinking was the result of **diastrophism,** which is a term meaning the movement of the Earth's crust. Most of the present-day geological data support Dana's view. Certainly, the weight of the sediment contributes to subsidence, but the fundamental mechanism of mountain building appears to be diastrophism.

The theory of plate tectonics offers an explanation of the mechanism of diastrophism and the building of folded mountains. Oceanic geosynclines containing large accumulations of sediment occur along the margin of continents, which supply the eroded sediment. If these occur in regions of descending ocean plates, the downward movement of a plate pushes the stratified layers in the geosyncline against the continental crust (Fig. 29.23). The deep-water strata are dragged downward by the moving plate, where they are heated and subjected to metamorphism. Some of the deepest strata may melt and rise to form igneous intrusions in the strata above, becoming the core of a new chain of folded mountains. The shallow-water strata are pushed aside and are folded and faulted along the marginal regions of the folded chain.

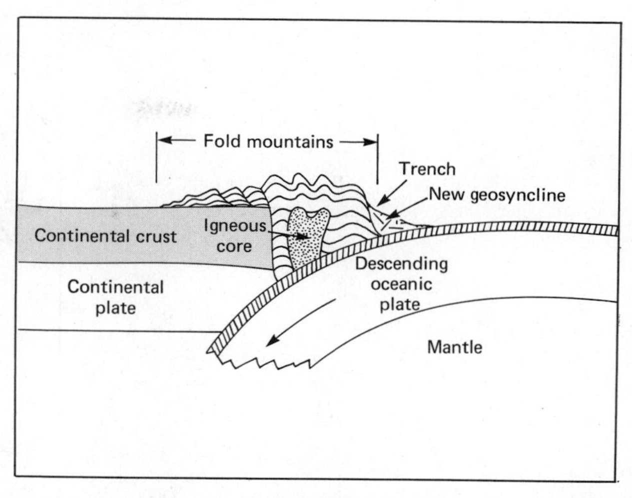

Figure 29.23 An illustration of the formation of folded mountains. The downward movement of the descending oceanic plate pushes the geosyncline strata against the continental plate. The deep-water strata are dragged downward by the moving plate, where they are heated. The deepest strata melt and rise to form the core of the folded mountain chain, while the shallow-water strata are folded and faulted along the marginal regions of the chain. Accumulations of sediment eventually form a new geosyncline in the oceanic trench.

Such mountain building occurs today along the western coasts of North and South America. Large geosynclines lie along the Atlantic coasts of North and South America, but these regions are not along plate boundaries. If perhaps the growing accumulations of sediment in the geosynclines cause a downward plunge into the lithosphere at some future geologic time, then a folded mountain chain may form along the Atlantic coast.

Let's consider the formation of some of the folded mountain ranges of the Earth within the framework of plate tectonics. The Himalayas are believed to have formed after the supercontinent Pangaea broke up some 200 million years ago. Essentially in terms of surface features, India broke away from Africa and ran into Asia with the collision resulting in the Himalayas.

In more detail, the northward movement of the Indian Plate after its separation from Africa resulted in the loss of oceanic lithosphere formerly separating India and Asia, with the eventual collision of these two continental masses. The Indian Plate descended under the Eurasian Plate and the geosynclines along the Eurasian Plate were folded into mountains, as illustrated in Fig. 29.24. When the two continental crusts met, the edge of the Eurasian Plate was lifted with the spectacular result of the highest mountain range on Earth. Present-day analyses of plate movements indicate that the Indian Plate is still moving northward relative to the Eurasian Plate causing the slow, continuing uplift of the Himalayas.

The Alps are another example of what is believed to be the result of plate collision. In this case, the colliding plates were the African Plate and the Eurasian Plate.

The Appalachian mountains are quite old compared to the Himalayas and the Alps, and were apparently once joined to mountains in northern Europe (see Fig. 27.5). How did these folded mountains form? That is, where was the geosyncline? It is speculated that before Pangaea an ancient ocean existed between North America and Europe. The ocean disappeared as the plates of these land masses converged to form Pangaea. In the collision, the Appalachians were formed. Thus, the Appalachians were once an interior mountain range of Pangaea. When Pangaea broke up, a plate boundary cut across the mountain range and part went with the new American Plate and part with the new Eurasian Plate.

A similar explanation of the collision of pre-Pangaea plates is given for the Ural mountain range, which runs

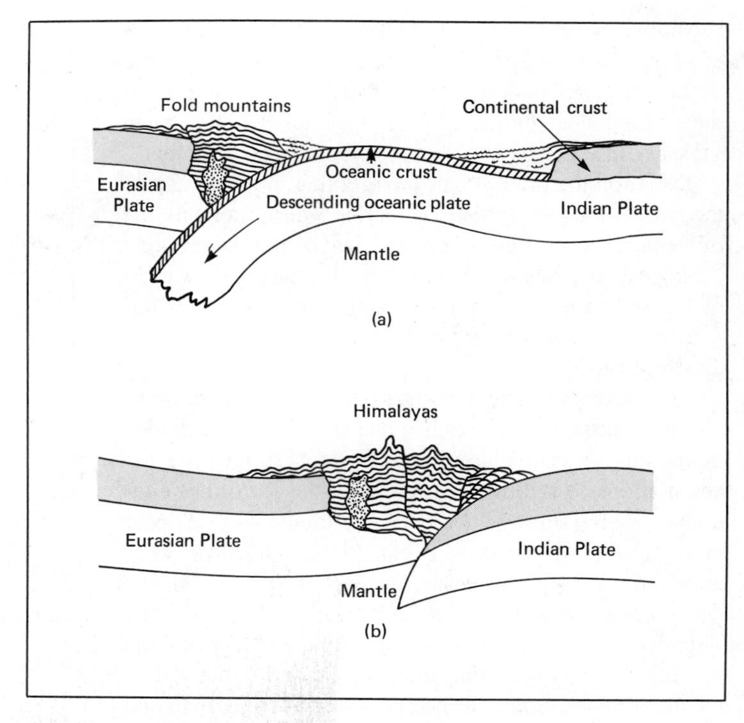

Figure 29.24 An illustration of the formation of the Himalayas. (a) After its separation from Africa, the northward-moving Indian Plate descended under the Eurasian Plate and the geosynclines along the Eurasian Plate were folded into mountains. (b) When the two continental crusts met, the edge of the Eurasian Plate was lifted, giving rise to the lofty Himalayas.

north-south between Europe and Asia. In this case, the European and Asian plates remained intact as the Eurasian Plate when Pangaea broke up.

Thus we see that the theory of plate tectonics provides a beautiful and simplified explanation of mountain building. However, it must be kept in mind that this theory is relatively new: and although there is a great deal of supporting evidence, it has not really been examined and tested fully. There are other theories of mountain building that also give acceptable explanations.

One such theory suggests that mountains can arise independently of horizontal plate movements. According to this theory, gravity causes heavier rocks to sink in the lithosphere while lighter rocks rise to form mountains such as the Alps. Any theory must stand the test of time and evidence; and becaue of the complex nature of the Earth, geologic evidence is slow in coming. There is a great deal of work to be done before it is fully established whether or not the sequence of events described by the theory of plate tectonics happened in this manner.

Learning Objectives

After reading and studying this chapter, you should be able to do the following without referring to the text:

1. Explain the cause and general locations of volcanic activity.
2. Describe the characteristics and different types of volcanic structures.
3. Explain the causes of earthquakes.
4. Distinguish the different types of faulting.
5. State the difference between S and P waves and explain how these waves provide information about the Earth's interior.
6. Define the *Richter scale* and the modified *Mercalli scale*.
7. Describe the three principal types of mountains and give an example of each.
8. Explain mountain building in the framework of plate tectonics.

Important Words and Terms

volcano
lava
fumaroles
"Ring of Fire"
hot spot theory
viscosity
basaltic lava
felsitic lava
pyroclastic debris
volcanic bombs
shield volcano
stratovolcano
composite volcano
cinder cone
volcanic crater
caldera
earthquake
seismology
normal faulting
reverse faulting
thrust faulting
strike-slip faulting
transform fault
focus
epicenter
seismic waves
surface waves
body waves
P waves
S waves
seismograph
Richter scale
Mercalli scale
tsunami
mountain range
mountain system
mountain chain
volcanic mountains
fault-block mountains
folded mountains
geosyncline
diastrophism

Questions

1. What is the difference between lava and magma?
2. Give the locations of the following volcanoes: (a) Mt. Vesuvius, (b) Mt. Pelee, (c) Katmai, (d) Krakatoa, (e) Paricutin, (f) Surtsey, (g) Mount St. Helens, (h) El Chichón.
3. What is the "Ring of Fire" and how is it explained by plate tectonics?
4. Describe the Hawaiian island chain. What is the "hot spot" theory?
5. Are there any active volcanoes in the United States? If so, where?
6. What is the difference between basaltic and felsitic lava?
7. Distinguish between (a) shield volcanoes, (b) stratovolcanoes, and (c) cinder cones.
8. What is a caldera and how are they formed?
9. What causes an earthquake?
10. Explain each of the following: (a) normal faulting, (b) reverse faulting, (c) thrust faulting, (d) strike-slip faulting.
11. What types of relative plate motions correspond to the different types of faulting?
12. What type of faulting is associated with a transform fault? Give an example of an active transform fault.
13. What are the focus and epicenter of an earthquake?
14. Give the two general types of seismic waves and the subdivision of one of these types.
15. What is the difference between S and P waves? How do these waves allow seismologists to locate the focus of an earthquake and to investigate the Earth's interior structure?
16. What is the basis of (a) the Richter scale, and (b) the modified Mercalli scale?
17. Why is *tidal wave* an incorrect and misleading term for a huge ocean wave generated by an earthquake? What is the correct term?
18. What are the three principal kinds of mountains and how are they distinguished? Give an example of each.
19. How is the formation of each kind of mountains explained by the theory of plate tectonics?
20. What is a geosyncline?
21. What is diastrophism?
22. Describe how the formation of each of the following mountain ranges is explained by the theory of plate tectonics: (a) the Himalayas, (b) the Alps, (c) the Appalachians.

30

Surface Processes

The same regions do not remain always seas or always land, but all change their condition in the course of time.

–Aristotle

NOTHING IS PERMANENT on the Earth's surface—in the context of geologic time. Since early times, people have erected edifices and monuments as memorials to persons, peoples, and as symbols of various cultures. In doing so, the most durable rock materials were used; however, even recent erections show the inevitable signs of deterioration. Buildings, statues, and tombstones will eventually crumble when exposed to the elements; and, on a larger scale, nature's mountains are continually being leveled.

An integral part of the rock cycle, as discussed in Chapter 28, is the decomposition of rocks by weathering and other natural processes. As soon as rock is exposed at the Earth's surface, its destruction begins. Many of the resulting particles of rock are washed away by surface water and are transported in streams and rivers as sediment toward the oceans. Over geologic time, this constitutes a leveling of the Earth by the wearing away of high places and transporting of sediment to lower elevations.

We say that the wearing away and leveling of the Earth's surface is due to **erosion,** a group of interrelated processes by which rock is broken down and the products are removed. While volcanism and diastrophism account for the uplifting of land masses, erosion accounts for their leveling. In this chapter we shall consider some of the processes and agents of erosion, including "unnatural" agents—human beings.

Also, the oceans and ocean currents and sea floor topography will be considered. Important geologic processes take place on the ocean floor, which accounts for about 65% of the Earth's surface.

◀ Delicate Arches, Utah.

30.1 Weathering and Mass Wasting

Weathering

The erosion process begins with weathering, which is the physical disintegration and chemical decomposition of rock. Weathering depends on a number of factors, such as the type of rock, moisture, temperature, and overall climate.

Physical weathering involves the mechanical disintegration or fracture of rock, primarily as a result of pressure. For example, a common type of physical weathering in some regions is **frost wedging.** When rocks are formed by solidification, lithification, or metamorphism, internal stresses are produced in the rocks. One common result of these stresses is cracks or crevices in the rock, called joints. Joints provide an access route for water to penetrate into the rock. If the water freezes, the less dense ice requires more space than the water and exerts a strong pressure on the surrounding rock. As a result, the rock may break apart, as illustrated in Fig. 30.1. The expansive force of freezing water is readily observed by freezing water in a glass container, which usually cracks. Frost wedging is most effective in regions where freezing and thawing occur daily.

The action of freezing water is also effective in loosening the soil, which promotes its erosion. In the winter, water in the soil freezes as the ground cools. An accumulation of freezing water from the upward migration of water by capillary action or from rain water or melting snow can give rise to an upward expansion that heaves up the material above and creates a bulge at the surface. This so-called **frost heaving** can result in fractures and bulges in street pavement. Highway joints and cracks

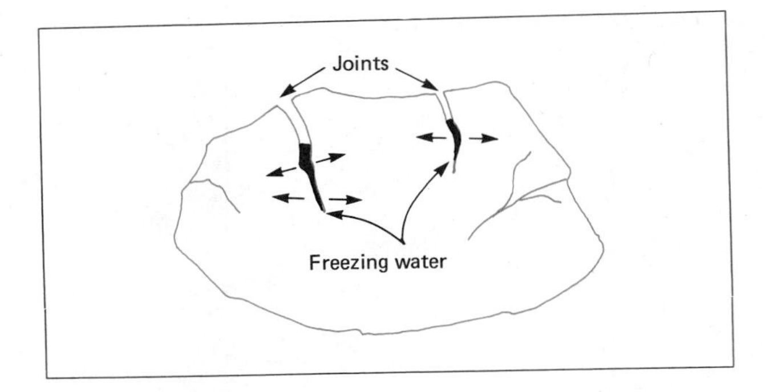

Figure 30.1 Frost wedging. The expansive forces of freezing water in rock joints cause the rock to fracture. The photo shows an example of this type of physical weathering in a glacial cirque or depression in the Big Horn Mountains of Wyoming. (Courtesy U.S. Geological Survey. Photo by N. H. Darton.)

are usually "sealed" with waterproofing materials to prevent accumulations of water and frost heaving. Also building foundations are laid below the surface frost zone to prevent damage from frost heaving.

In cold upper latitudes, the subsurface soil may remain frozen permanently, giving rise to a **permafrost** layer. During a few weeks in the summer, the top soil may thaw to a depth of a few inches to a few feet. The subsurface permafrost provides a stable base and prohibits the melted water from draining. As a result, the ground surface becomes wet and spongy (Fig. 30.2). The permafrost in Alaska caused many problems in the construction of the Alaskan oil pipeline. On the North Slope where the oil fields were discovered, the thawed surface is especially unstable because of a relatively high moisture content. The thawed surface does not recover easily if disturbed and, if removed or compressed, the subsurface permafrost is susceptible to progressive thawing and erosion.

Frost wedging and frost heaving account for little if any physical weathering in hot, arid climates, but a similar process can. This arises from the growth of salt crystals rather than water crystals. Groundwater with high salt concentrations seeps into porous rock—sandstone is a prime example. When the water evaporates and the salt crystallizes, the salt crystals exert pressure on the surrounding rock. This process of **salt wedging** can loosen and fracture the cementing material holding the sand grains together. The loosened sand grains can then be carried away by wind and water from rainstorms.

Plants and animals play a relatively small role in physical weathering. Most notable is the fracture of rock by plant root systems when they invade and grow in rock crevices. Burrowing animals, earthworms in particular, loosen and bring soil to the surface. This promotes aeration and access to moisture, which are important factors in chemical weathering. Thus, one type of weathering promotes another. The activities of human beings also give rise to weathering and erosion that are often unwanted, as we will learn later in the chapter.

Figure 30.2 A "roller-coaster" railroad near Strelna, Alaska. This condition was caused by differential subsidence stemming from the thawing of top soil over subsurface permafrost. (Courtesy U.S. Geological Survey.)

Figure 30.3 A sinkhole collapse in Barstow, Florida. Water flowing through limestone formations can carve out large underground caverns. A cavern ceiling can collapse, causing a sinkhole to appear on the land surface. In this case, the collapse caused considerable property damage. (Courtesy U.S. Geological Survey.)

In all cases of physical weathering the disintegrated rock still has the same chemical composition. **Chemical weathering,** however, involves a chemical change in the rock's composition. Since heat and moisture are two important factors in chemical reactions, this type of weathering is most prevalent in hot, moist climates. One of the most common types of chemical weathering involves limestone, which is made up of the mineral calcite ($CaCO_3$). Rain can absorb and combine with carbon dioxide (CO_2) in the atmosphere to form a weak solution of carbonic acid.

$$H_2O + CO_2 \longrightarrow \underset{\text{Carbonic acid}}{H_2CO_3}$$

Also, as water moves downward through soil, it can take up even more carbon dioxide that is released by soil bacteria involved in plant decay. Recall that carbonic acid (carbonated water) is the weak acid in carbonated drinks.

When carbonic acid comes in contact with limestone, it reacts with the limestone to produce calcium hydrogen carbonate or calcium bicarbonate.

$$\underset{\text{Carbonic acid}}{H_2CO_3} + \underset{\text{Limestone}}{CaCO_3} \longrightarrow \underset{\text{Calcium bicarbonate}}{Ca(HCO_3)_2}$$

Calcium bicarbonate dissolves readily in water and is carried away in solution. Since limestone is generally impermeable to water (and dilute carbonic acid), this type of chemical weathering acts primarily on the surfaces of limestone rocks along which water flows.

Water flowing through underground limestone formations can carve out large caverns over millions of years. The cavern ceilings may collapse, causing **sinkholes** to appear on the land surface (Fig. 30.3). In many stable caverns, as water seeps through the cavern ceiling and drips to its floor, it loses its carbon dioxide and minute amounts of calcium carbonate are precipitated. These precipitations build up to form icicle-shaped stalactite, stalagmite, and column dripstone (see Fig. 28.7). Do you recall which formation extends down from the ceiling and which protrudes up from the floor?

The rate of chemical weathering depends on many factors. The principal factors are climate and the mineral content of rock, with humidity and temperature being their chief climate controls. Chemical weathering is relatively rapid in hot, humid climates as compared to chemical decomposition in polar regions. On the other hand, chemical weathering is quite slow in hot, dry climates. Egyptian pyramids and statues have stood for centuries with relatively minimal weathering. Some rock minerals are more susceptible to chemical weathering than others. This is evidenced in the weathering of tombstones (Fig. 30.4). Marble tombstones, which consist of soluble calcite, may show a great deal of chemical weathering as compared to sandstone tombstones, which consist of relatively insoluble quartz. However, the weathering of sandstone tombstones can cause a loosening of the cementing materials holding the sand grains together, and give rise to the erosion of large blocks of rock.

As mentioned previously, burrowing animals, both large and small, bring soil or rock particles to the surface where the exposed particles are more susceptible to chemical weathering. The mounds of soil particles built up by ants and earthworms are common sights.

Mass Wasting

The weathered material that accumulates on base rock is called **overburden.** This general term refers to rock and rock fragments or soil that is free of the base rock and so may start its descent toward lower elevations. The downslope movement of overburden under the influence of gravity is called **mass wasting.** If you drop a rock from a cliff, it is, in the geological sense, mass wasted. However, mass movement usually occurs with the aid of a transport medium such as running water. It is convenient to divide mass wasting into two categories, fast and slow, which are based on the comparative times involved in mass movement process. As might be expected, the angle of the land slope is a critical factor in the distinction between fast and slow mass wasting.

Two important types of *fast mass wasting* are landslides and mudflows. **Landslides** involve the downslope movement of large blocks of weathered materials. Spectacular landslides occur in mountainous areas when large quantities of rock break off and move rapidly down the steep slopes. This type of landslide is termed a **rockslide.**

A comparative slower form of landslide is a **slump.** This is the downslope movement of an unbroken block of overburden, which leaves a curved depression on the slope (Fig. 30.5). Slumps are commonly accompanied by debris flows that consist of a mixture of rock fragments, mud, and water that flows downslope as a viscous liquid. Small slumps are commonly observed on the bare slopes of new road constructions.

Mudflows are the movements of large masses of soil that have accumulated on steep slopes and become unstable with the absorption of large quantities of water from melting snows and heavy rains. Mudflows are likely to occur on slopes lacking vegetation, which hinders mass movement. Consequently, mudflows are common in hilly and mountainous regions where land is cleared for development without regard to soil conservation.

Mudflows are also associated with volcanoes. Volcanic mudflows result from loose ash on the slopes of the volcano that absorbs water from thunderstorms commonly associated with volcanic activity. The mudflows

Fig. 30.4 Examples of chemical weathering. Marble tombstones, which consist of soluble calcite, are susceptible to chemical weathering. The weathering of sandstone, which consists of relatively insoluble quartz, is primarily due to the loosening of cementing materials holding the sand grains together. As a result, large blocks of material are eroded.

Figure 30.5 A large slump that threatens an apartment complex and debris flow. The missing building was torn down a short time before the landslide occurred and the other buildings were evacuated. Notice the damage to the road by the debris flow in the foreground. (Courtesy David Kantner, Marietta, Ohio.)

Figure 30.6 A result of creep. Tilted fence posts indicate the imperceptively slow mass wasting due to creep.

of ash can flow long distances and build up thick deposits of sediment. In some instances, volcanic mudflows have buried nearby towns.

Fast mass wasting is quite dramatic, but *slow mass wasting* is a more effective geological transport process. In contrast to fast mass wasting, the rates of slow mass wasting processes are generally imperceptible. An important type of slow mass wasting is called creep. **Creep** is the slow particle-by-particle movement of weathered debris down a slope, taking place year after year. It cannot be seen happening, but the manifestations of creep are evident, such as that shown in Fig. 30.6. Although the spectacular landslides and slumps involve large quantities of mass movement, this is but a fraction of the cumulative total of mass movement by creep over a period of time.

Another type of slow mass wasting is **solifluction** or soil flow, which is the ''flow'' of weathered material over a solid, impermeable base. This type of mass wasting is common in cold climates where permanently frozen ground (permafrost) forms a solid base for the solifluction of a surface layer that thaws during the short summers. Moisture in the thawing surface and from summer rains is trapped in the surface layer due to the frozen ground below. As a result, the surface layer becomes saturated and the water-soaked soil ''flows'' slowly downhill.

30.2 Agents of Erosion

The interacting factors in erosion processes are quite complex. To simplify matters, it is convenient to talk about the **agents of erosion,** that is, the major physical phenomena and mechanisms that supply the energy for erosion. These are running water, ice, wind, and waves.

Running Water. Running water refers primarily to the waters of streams and rivers that erode the land surface and transport and deposit eroded materials. Rainfall that is not returned to the atmosphere by evaporation either sinks into the Earth as groundwater or flows over

Figure 30.7 Sheet erosion. The overland flow of water removes a large amount of sediment. This field, which was plowed in the fall and left without any cover winter crops, shows the signs of sheet erosion. (Courtesy USDA, Soil Conservation Service.)

the surface as runoff. Runoff or overland flow occurs whenever rainfall exceeds the amount of water that can be immediately absorbed in the ground. Overland flow usually occurs only for short distances before the water ends up in a stream. The erosion performed by overland flow is referred to as **sheet erosion.** The flowing ''sheets'' of water can move a surprisingly large amount of sediment, particularly from unirrigated slopes and cultivated fields (Fig. 30.7).

Stream flow is the flow of water occurring between well-defined banks. The eroded material in a stream is referred to as **stream load.** The load of a stream varies from dissolved minerals and fine particles to large rocks, and the transportation process, as well as erosion, depends on the volume and swiftness (discharge) of the stream's current. A stream's load is divided into three components: a dissolved load, a suspended load, and a bed load.

The *dissolved load* consists of dissolved water-soluble minerals that are carried along by a stream in solution. As much as 20% of the material reaching the oceans is transported in solution. Fine particles not heavy enough to sink to the bottom are carried along in suspension. The *suspended load* is quite evident after a heavy rain when the stream appears muddy (Fig. 30.8). Coarse particles and rocks along or close to the bed of the stream constitute its *bed load*. These rocks and particles are rolled and bounced along by the current. This transport mechanism of bed load is referred to as **traction.**

As it is moved along by traction, the coarser load material near the bottom of a stream is further broken into smaller rocks and particles by **abrasion.** Stream abrasion is evidenced by smooth, rounded rocks and pebbles (Fig. 30.8). During the transport process, the bed load is worn finer and finer and eventually may be transported by suspension. The final products of a long journey are sand (rock particles) or a mixture of finer particles and organic material picked up along the way which is collectively referred to as **silt** and clay.

The action of the flowing water in a river causes the erosion of its bed. By studying the development of this process, the age of a river may be determined. Over geologic time a river goes through three general stages: youth, maturity, and old age. In its youth, it is characterized by a V-shaped valley or canyon which reflects downcutting erosion. The depth of an eroded valley is limited by the level of the body of water into which the river flows. The limiting level below which a stream cannot erode the land is called its **base level.** In general, the ultimate base level for rivers is sea level.

As a stream or river flows overland, it twists and turns, following the path of least resistance. A looplike bend in a river channel is called a **meander** (Fig. 30.9, p. 566). Meanders shift or migrate due to greater erosion

Figure 30.8 Stream erosion and transport. The upper photo shows the cutting of a road embankment that occurred when the stream flooded due to heavy rains. Fine soil particles are carried away as suspended load. The lower photo shows the result of such erosion in an aerial view of the Loosahatchie River pouring suspended sediment into the Mississippi River one mile north of Memphis, Tenn. (Courtesy EPA and USDA, Soil Conservation Service.)

on the side of the stream bed on the outside of the curved loops. The speed of the stream flow is greater in this region than near the inside bank of the meander where sediment can deposit. When the erosion along two sharp meanders causes the stream to meet itself, it may abandon the water-filled meander, which is then known as an oxbow lake.

As the river matures it becomes **graded,** which is the condition when its erosion and transport capabilities are in balance. A mature graded river no longer has much tendency to deepen its channel. Instead, it begins to widen its valley, forming a **flood plain.** This widening continues until in old age the river has a very wide flood plain.

Figure 30.9 Meandering. (*Top*) An aerial view of a classic case of river meanders in the Shenandoah River near Woodstock, Virginia. The 1 to $1\frac{1}{2}$-mile-wide meander belt cuts into the adjoining shale to depths up to 150 ft. (*Bottom*) When erosion along two sharp meanders causes a stream to meet itself, it may abandon the meander loop, leaving behind an oxbow lake as occurred here in the White River in Arkansas. (Courtesy U.S. Geological Survey. Bottom photo by J. R. Balsley.)

The amount of material eroded and transported by streams and rivers is enormous. The oceans receive billions of tons of sediment each year as a result of the action of running water. The Mississippi River alone discharges approximately 500 million tons of sediment yearly into the Gulf of Mexico. A river's suspended and bed loads may accumulate at its mouth and form a **delta,** such as the Nile River Delta (Fig. 30.10). The rich sediment makes delta areas important for agriculture.

However, some sediment deposits are not always to our benefit as shown in Fig. 30.11. Many harbors must be dredged to remove sediment deposits to keep them navigable, and the sediment buildup behind dams poses operational problems. Our activities that remove erosion-preventing vegetation from the land may give rise to sediment pollution in rivers and streams and result in environmental problems. Some of these activities and problems will be discussed shortly.

Figure 30.10 A satellite photograph of the Nile Delta. The dark areas are cultivated lands. The Mediterranean Sea is in the left foreground and the Suez Canal and the Red Sea are above the dark delta region. The rich sediment load deposited at the mouth of the river makes the delta area an important agricultural region. (Courtesy NASA.)

Figure 30.11 A sediment problem. To reach the water, the men must drag their boat over a large area of silt deposits in Lake Calhoun near Lafayette, Ill. The erosion of land fills harbors and lakes with unwanted sediment. (Courtesy EPA.)

Ice. One of the eroding actions of ice, frost wedging, has been mentioned previously. Many of us are familiar with the ice of winter, but parts of the Earth are covered with large masses of ice the year round. These large masses of ice are called **glaciers.** To most of us, the term glacier usually brings to mind the thought of the Ice Age. Indeed, large areas of Greenland and Antarctica are presently covered with glacial ice sheets or **continental glaciers** similar in size to those that covered Europe and North America during the last Ice Age—over 10,000 years ago.

But are there any glaciers in the United States today? The answer is yes. In fact, you may be surprised to learn that there are about 1100 glaciers in the western part of the conterminous (48 states) United States and that 3% of the land area of Alaska is covered by glaciers (17,000 square miles). However, these glaciers differ greatly in size and form from ice sheets of Greenland and Antarctica.

Glaciers are formed when, over a number of years, more snow falls than melts. As the snow accumulates and becomes deeper, it is compressed by its own weight into solid ice. When enough ice accumulates, the glacier "flows" downhill or out from its center if on a flat region. The icebergs commonly found in the North Atlantic and Antarctic oceans are huge chunks of ice that have broken off from the edges of the glacial ice sheets of Greenland and Antarctica, respectively.

Small glaciers, called **cirque glaciers,** form in hollow depressions along mountains that are protected from the Sun. A majority of the glaciers in the United States are of this variety. The ice movement further erodes the land and forms an amphitheatrelike depression called a *cirque* (see Fig. 30.12). When the ice melts, these glacier-eroded cirques often become lakes.

If snow accumulates in a valley, the valley floor may be covered with compressed glacial ice, and a **valley glacier** or mountain glacier is formed, which flows down the valley. At lower and warmer elevations the ice melts, and where the melting rate equals the glacier's flow rate,

Figure 30.12 Glaciers. The top photo shows the Blue Glacier in Olympic National Park, Washington. The lower photo shows the snout of Wolverine Glacier on the Kenai Peninsula near Seward, Alaska. A glacier advances until the melting rate equals its flow rate. (Courtesy U.S. Geological Survey.)

the glacier becomes stationary (Fig. 30.12). The flow rate of a glacier may be a few inches to over 100 ft per day, depending on the glacier's size and other conditions. The end of a glacier is called its *snout,* which may advance and recede with the seasons.

The erosion action of a valley glacier is not unlike that of a stream. As the ice flows, it loosens and carries away material, or bed load, that will be ground fine by abrasion. Although much slower than a stream, a glacier can pick up huge boulders and gouge deep holes in the valley. The paths of vigorous, preexisting mountain glaciers are well marked by the deep U-shaped valleys they leave (Fig. 30.13).

Glaciers, like streams, also deposit the material they carry. The general term **drift** is applied to any type of glacial sediment deposit. Material that is transported and deposited by ice, in contrast to meltwater, is called **till.** Till deposits are not layered or sorted as would be sediment carried away by the melted water of a glacier. Near the end and sides of a glacier the till may form ridges known as **moraines.** The terminal moraine marks the farthest advance of the glacier. Terminal moraines give us an indication of the extent and advance of the glacial ice sheet in North America which retreated about 10,000 years ago. These moraines lie as far south as Indiana, Ohio, and on Long Island, New York.

Not only do glaciers erode the land surface, but they supply an estimated 560 billion gal of water to stream flow during the summer months in the conterminous United States alone. These huge fresh water reservoirs are being eyed as water sources for the heavily populated western areas. The glacier cycle fits well into our needs. Glaciers accumulate and store water in the winter when most areas have sufficient water supplies and release it in the summer when many city reservoirs are low.

Wind. The action of wind as an agent of erosion is a slow process, but nevertheless wind contributes significantly to the leveling of the Earth. Dust particles that are small enough are transported great distances by the wind, while larger particles are moved short distances by rolling or bouncing along the surface. The transport action of the wind is not unlike that of traction in a stream, only on a broader and less confined scale. This action is quite evident in areas with large quantities of loose, weathered debris. Dust storms may darken the sky and be of such intensity that visibility is reduced to almost zero. During the 1930s, drought conditions created areas known as *dust bowls,* in which layers of fertile topsoil were blown away by the wind.

Because sand grains are relatively heavy, the transport action of sand by the wind is usually within a few

Figure 30.13 (*Left*) A U-shaped glacial valley with moraine in the foreground in Glacier National Park, Montana. The valleys formed by glacial action mark the paths of preexisting mountain glaciers. (Courtesy U.S. Geological Survey.)

Figure 30.14 (*Left*) A sand and dust storm. At the height of the storm the sand drifts have almost covered the railroad tracks. (*Right*) Isolated sandhills are evidence of the eroding action of the wind. (Courtesy USDA, Soil Conservation Service.)

Figure 30.15 Marine terraces. The eroding action of waves is evident in this view along the California coast near Los Angeles. (Courtesy U.S. Geological Survey. Photo by K. S. W. Kew.)

feet of the ground, similar t[illegible]arse sediment transportation near the bed of a s[illegible]. On the ground the sand forms drifts, as shown in Fig. 30.14. The resulting abrasion from such sandblast action is evident in nature.

Waves. The waves of large bodies of water also erode the land along their shorelines. This is most evident in the oceans where the surf pounds the shoreline. Some coasts are rocky and jagged, which evidences that only the hardest materials can withstand the unrelenting wave action over long periods of time. Along other coastlines, cliffs are formed on the shoreline, terraced from the eroding action of waves (Fig. 30.15).

30.3 The Earth's Water Supply

All rivers run into the sea, yet the sea is not full; into the place whence the rivers come, thither they return again.

Eccles. 1:7

Water is often referred to as the basis of life. The human body is composed of 65–70% water by weight, and water is necessary to maintain our body functions. This common chemical compound is an essential part of our physical environment, not only in life processes, but also in other areas, such as agriculture, industry, sanitation, firefighting, and even religious ceremonies. Early civilizations developed in valleys where water was in abundance, and even today the distribution of water is a critical issue. Consider how your life would be affected without an adequate water supply.

The Earth's water supply, some 300 million mi^3, may seem inexhaustible, since it is one of our most abundant natural resources. Approximately 70% of the Earth's surface is covered with water. However, about 98% of the water on the Earth is salt water, and only about 2% is fresh water (Fig. 30.16). Most of the fresh water is frozen in the glacial ice sheets of Greenland and Antarctica. Even so, there are about 10^{15} gal of fresh water available each year. The oceans contain another 10^{20} gal of water that is easily accessible, but this water has 3–4% salt content—principally sodium chloride—which makes it unfit for human consumption without a costly desalination process.

Although the daily withdrawal of fresh water in the United States is over 450 billion gal, the supply of fresh water is generally adequate in most areas. Regional problems arise due to population concentrations, but these are human problems that involve water management.

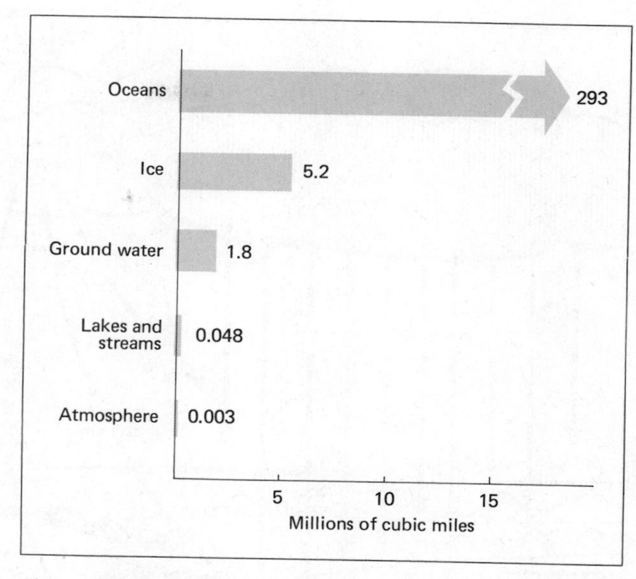

Figure 30.16 Distribution of the Earth's water supply. (Courtesy U.S. Geological Survey.)

The Hydrologic Cycle

The Earth's water supply is a reusable resource that is constantly being redistributed over the Earth. There are many factors that enter into this redistribution, but in general it is a movement of moisture from large reservoirs of water, such as oceans and seas, to the higher elevated inland regions. As discussed in the preceding section, the water flows back to the sea, eroding as it goes. This gigantic cyclic process is known as the **hydrologic cycle** (Fig. 30.17)

Moisture evaporated from the oceans moves over the continents through atmospheric processes and falls as precipitation. Some of this water evaporates and returns to the oceans via atmospheric processes, but a large part of it soaks into the Earth to become groundwater.

Some of the precipitation falling on the land may be lost in direct runoff. How much water goes into the ground depends on the permeability of the soil and rocks near the Earth's surface. **Permeability** is a measure of a material's capacity to transmit fluids. Naturally, loosely packed soil components, such as sand and gravel, permit a greater movement of water. Clay, on the other hand, has fine openings and relatively low permeability.

Porosity is closely related to permeability. This is the percentage volume of unoccupied space in the total volume of a substance. The porosity of rocks and soil near the Earth's surface determines the capacity of the ground

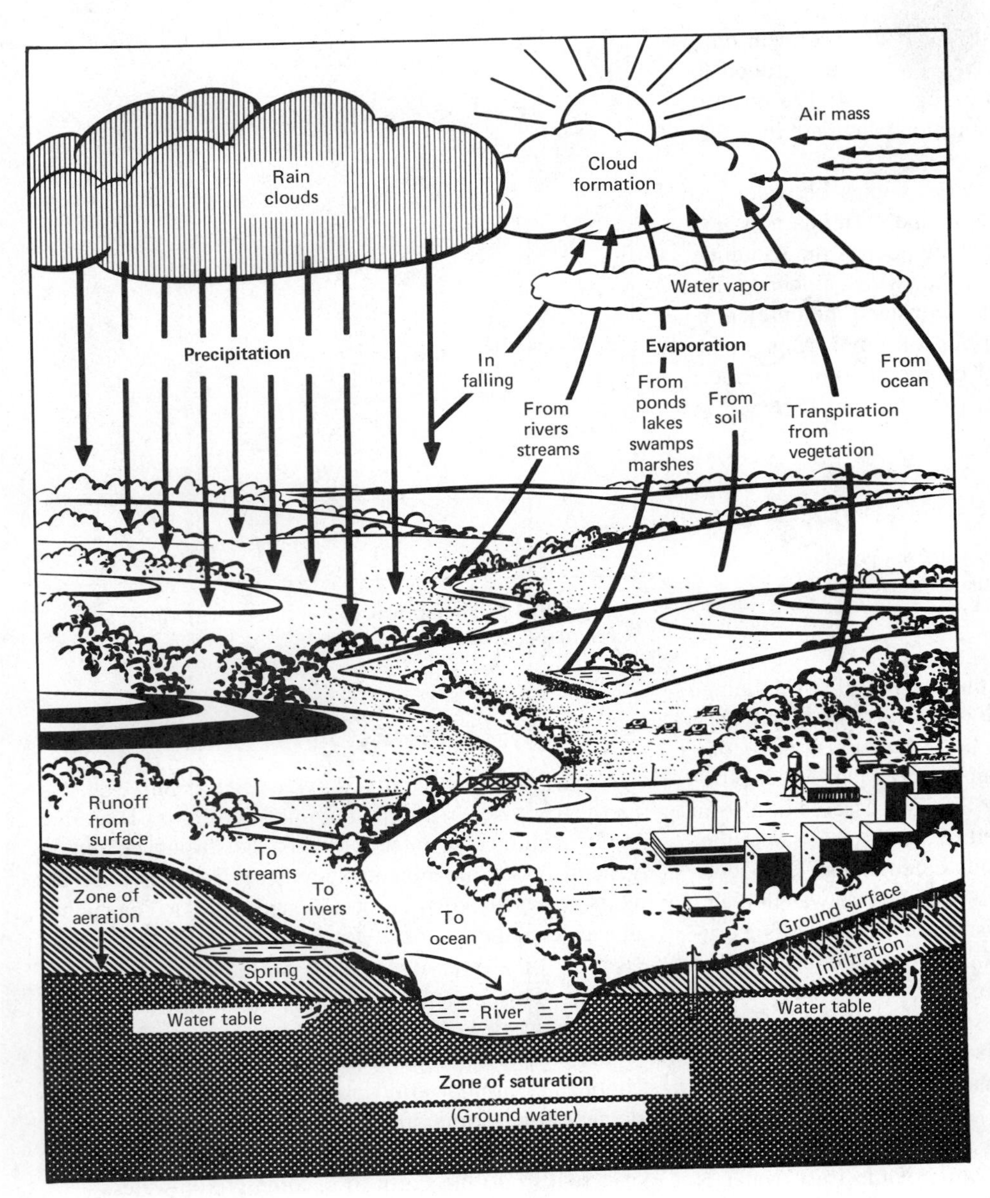

Figure 30.17 An illustration of the hydrologic cycle. Evaporated moisture from oceans, rivers, lakes, and soil is distributed by atmospheric processes. It eventually falls as precipitation and returns to the soil and bodies of water. The groundwater percolates downward through the zone of aeration until it reaches the level (water table) where the ground is saturated with water. (Courtesy USDA.)

for water storage. In total, ground porosity is so great that, except for glaciers, the ground is our largest reservoir of fresh water.

Under the influence of gravity, water percolates downward through the soil until at some level the ground becomes saturated with water. The upper boundary of this **zone of saturation** is called the **water table.** The unsaturated zone above the water table is called the **zone of aeration** (Fig. 30.17).

In the zone of aeration the pores of the soil and rocks are partially or completely filled with air. However, in the zone of saturation all the voids are saturated with groundwater, forming a reservoir from which we obtain part of our water supply by drilling wells to depths below the surface of the water table. Lakes, rivers, and springs occur where the water table is at the same level as the Earth's surface. Springs may also form above the water table as a result of impervious rock layers, but these tend to dry up in seasons of light precipitation. Also, the level of the water table shows seasonal variations, and shallow wells may go dry in late summer.

A body of permeable rock through which groundwater moves is called an **aquifer** (Latin, water carrier). Sand, gravel, and loose sedimentary rock are good aqui-

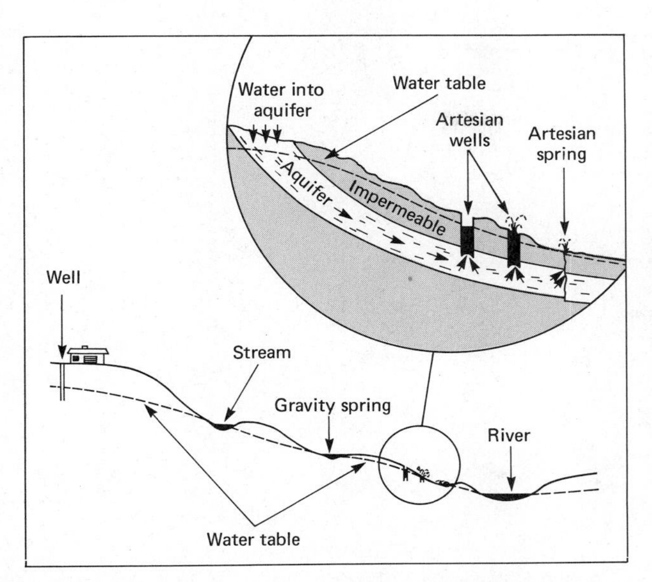

Figure 30.18 An illustration of wells, streams, and springs relative to the water table. As shown in the blowup, a special geometry of impermeable rock layers gives rise to artesian wells and springs. If the groundwater is not recharged by precipitation, the level of the water table will fall; then wells, springs, and streams can go dry. This often occurs during a hot, dry summer.

fer materials. Aquifers are found under more than half of the area of the conterminous United States, and it is in these aquifers that we find wells and springs. Ordinary wells and springs fill with water because they intersect the water table. A *spring* is simply a flow of groundwater that emerges naturally on the Earth's surface. Ordinary springs result from the gravitational flow of water and are hence sometimes called gravity springs (Fig. 30.18).

Special geometry of impermeable rock layers gives rise to what are called **artesian wells** and **artesian springs.** These are wells and springs in which water rises above the aquifer. As illustrated in Fig. 30.18, this can occur when an aquifer is sandwiched between sloping impermeable rock strata. The pressure due to gravity can cause water to spurt or bubble onto the surface above the aquifer. The name artesian comes from the French province of Artois where such wells and springs are common.

Thermal or hot springs are found in many areas. These involve heat energy from the Earth's interior. Groundwater seeping several thousand feet underground is heated as a result of the Earth's geothermal gradient (1°F per 150 ft of depth). Also at these depths, the water can come in contact with hot igneous rocks and may be heated to 400°F or higher (superheated) without boiling because of the great underground pressures.

The superheated water expands and rises to the Earth's surface through fissures, where it forms thermal springs. Heat energy is still lost during the water's journey to the surface, but some springs are boiling hot. The reduction of pressure near the surface can cause the water to boil. The resulting stream causes some hot springs to erupt in the form of a **geyser** (Icelandic, *geysir,* gush or rage).

Geysers from time to time spurt water into the air to heights of 150 ft or more. Most geysers have very irregular eruptions, but a few, such as Old Faithful in Yellowstone National Park, are regular enough to satisfy the impatient tourists (Fig. 30.19). Old Faithful's eruptions vary between 30 and 90 min apart with an average time interval of 65 min.

Geothermal areas are found around the world, principally in volcanic regions associated with plate boundaries (Fig. 30.20). Hot springs have been used as baths or spas down through history. In Iceland, hot water from geyser fields is used for domestic heating. A significant commercial application occurred in 1905 with the building of the first geothermal electric generation plant in

Figure 30.19 Old Faithful geyser. Eruptions of hot water vary between 30 and 90 min apart with an average time interval of 65 min. (Courtesy National Park Service.)

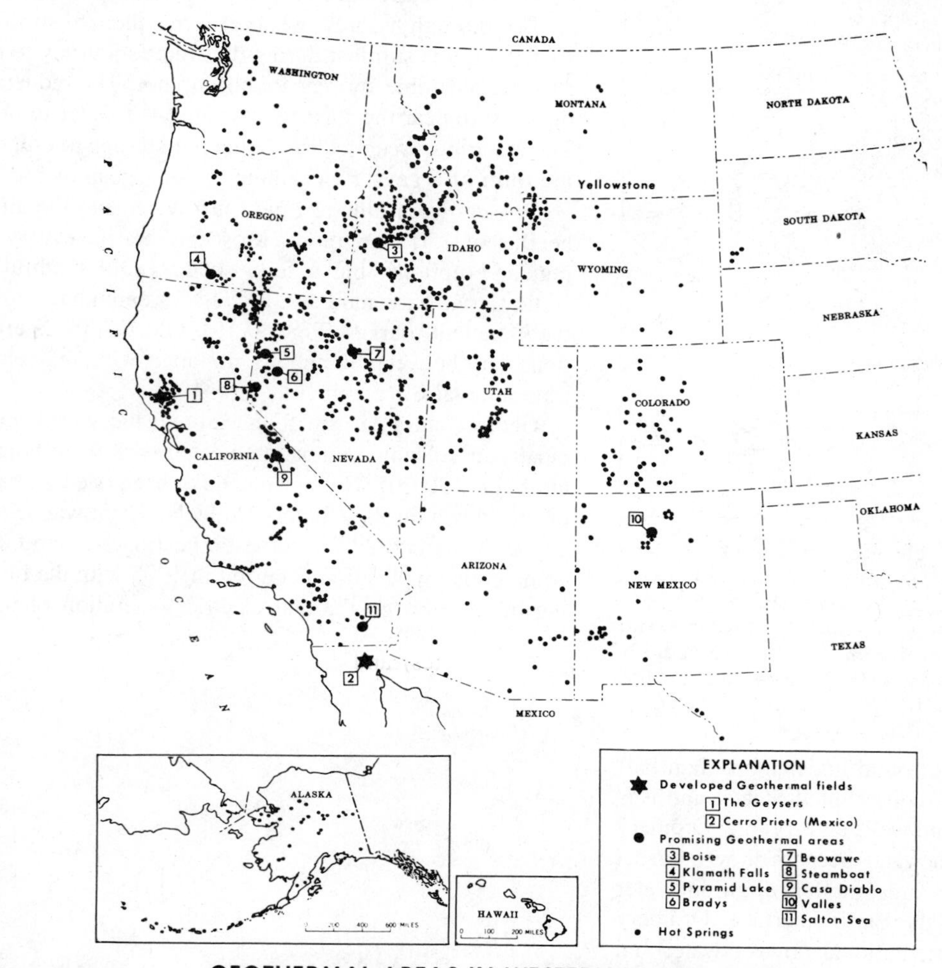

GEOTHERMAL AREAS IN WESTERN UNITED STATES

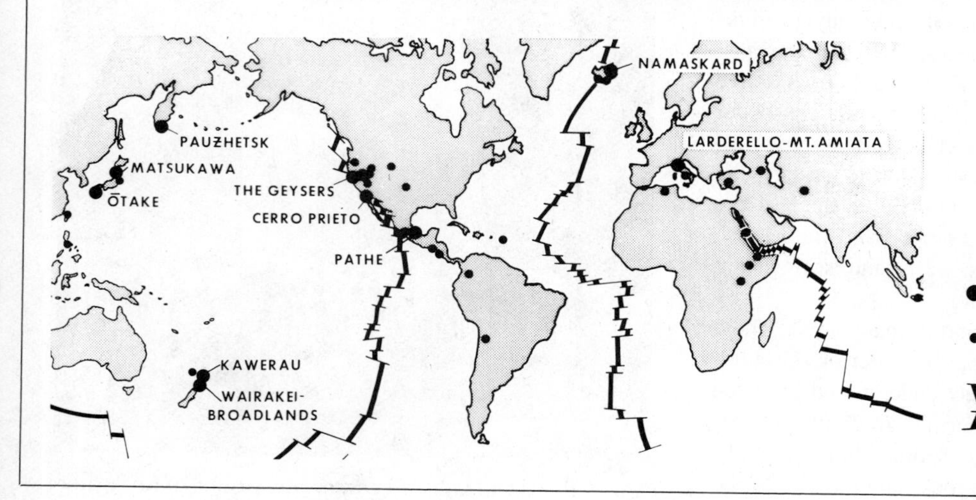

Figure 30.20 (*Left*) The geothermal areas of the world. Notice how the areas are associated with the regions of divergent plate boundaries. In some places, such as The Geysers, California (shown in Figure 30.21), steam from geysers is used to generate electricity. (Courtesy U.S. Geological Survey.)

Larderello, Italy. The first geothermal generation facility in the United States was built by the Pacific Gas and Electric Company in 1960 at The Geysers, California, some 90 mi north of San Francisco (Fig. 30.21). Other geothermal areas in the western United States are being eyed as possible energy sources.

Geothermal fields are of two general types—hot springs at the surface and deep, insulated, superheated reservoirs with little surface leakage. It is the latter type that is important in electrical generation application. When such reservoirs are tapped by drillings, the water boils and the expanding stream causes the hot water to be propelled to the surface. The drilled well acts as a continuously erupting geyser, so there is very little pumping cost. In some wells "flash" boiling occurs and only steam erupts with very little water discharge. This type of well is suitable for electrical generation, as is the case with the wells at The Geysers and Larderello.

Although these geothermal sources eliminate the air pollution associated with combustive electrical generation, they are not without problems. Gases in the steam and dissolved minerals in the hot water cause equipment to erode rapidly. The cooling and disposal of the salty, mineral-rich water from the spring may also present a problem. This water could, and possibly should, be returned to the water table of the area rather than allowing it to run off. There is evidence that the groundwater in areas where geothermal energy is used must be replaced to prevent subsidence.

Thus, we see that our water supply, within the context of the hydrologic cycle, is an intricate system that depends on many factors. The quality of the local water supply depends to a great extent on the mineral composition of the soil and rocks through which it percolates in the zone of aeration. Water-soluble minerals are dissolved and carried along with the water in solution.

In some cases, water containing bicarbonates (hydrogen carbonates) and chlorides is sold commercially as

Figure 30.21 The Geysers, California. A commercial geothermal electric generation plant is located at the Geysers in Sonoma County, 90 mi north of San Francisco. Production of electricity was started in 1960. (Courtesy U.S. Geological Survey.)

"mineral water." However, dissolved minerals in "hard" water have undesirable effects. **Hard water** is due to dissolved calcium and magnesium salts (bicarbonates, chlorides, and sulfates). Iron salts also contribute to hard water. Such dissolved minerals not only affect the taste of the water, but also cause usage problems. The salts combine with the organic acids in soaps to form insoluble compounds, thereby reducing the lather and cleansing quality of the soap. It is because of these insoluble compounds from soap that clothing does not come out "whiter white" on wash day and also that there is a ring around the bathtub.

At high temperatures, the metal ions present may form insoluble compounds (e.g., carbonates and sulfates) and precipitate. This precipitation tends to clog pipes and forms scale in teakettles and boilers. Boiler scale is a poor conductor of heat and reduces the boiler efficiency and wastes fuel. To avoid these conditions, the hard water may be "softened" by removing the mineral salts prior to use. Water softening is an active business in many parts of the country.

30.4 The Oceans and Sea Floor Topography

The Oceans

The vastness of the restless oceans imparts an awesome and humbling feeling to most observers. And rightly so, since the oceans cover about 65% of the Earth's surface. There are five major oceans. In order of decreasing size, these are the Pacific, the Atlantic, the Indian, the Antarctic, and the Arctic oceans. The average depth of the oceans is about 2.5 mi, with the greatest measured depth of about 7 mi in the western Pacific.

To the novice beachgoer, one of the first things noticed on an initial swim is the saltiness of the sea water. About 3.5% of the average sea water by weight consists of dissolved mineral salts. Through the hydrologic cycle, the dissolved salts of over 2 billion years of erosion have found their way into the ocean waters. Some ions of the dissolved salts are effectively removed from the oceans by biological processes and some by precipitation or absorption by other minerals. (Recall from Chapter 16 that dissolved chemical salts dissociate into metallic and nonmetallic ions.)

These salts become part of the ocean sediment and reenter the rock cycle from which they originated. It is believed that during the course of geologic time these processes have reached a steady state, such that the present salt composition of sea water has been maintained for millions of years. The quantities of salts in the seas may vary at some locations due to variations in the size of the mineral deposits of the drainage regions whose rivers feed the oceans. However, the same salts are found everywhere, generally in the same proportions.

The saltiness of the sea is measured in terms of its **salinity.** This is commonly expressed in parts per thousand (ppt) rather than percentage or parts per hundred. Parts per thousand is the same as the grams of dissolved salt per kilogram of sea water (g/kg). The salinity of the average sea water is therefore 35 ppt or 35 g/kg (equivalent to 3.5%). The range of sea water salinity in the open sea is from 33 ppt to 38 ppt. As can be seen from Table 30.1 the principal substances that contribute to the salinity of sea water are sodium and chloride ions.

The sea is sometimes referred to as a mineral storehouse from which we might extract minerals. This is true in part, but one must not overlook the concentrations and the difficulties of extracting these minerals from solution. It is a fact that NaCl has been taken from the ocean since ancient times, and that needs prompted the extraction of magnesium and bromine from sea water during World War II. Even today much of our output of these substances still comes from sea water. The extraction of metals like gold, however, is a different matter. There are 10^{10} tons of gold in the oceans, but the concentration is only about 6×10^{-11} oz/gal. This means that with 100% recovery about 17,000 million gal of sea

Table 30.1 **Some Chemical Constituents of Sea Water*** **(average concentrations)**

Ion or Element	*Concentration (ppt)*	*Ion or Element*	*Concentration (ppt)*
Chloride	19.3	Nitrate	3.5×10^{-4}
Sodium	10.7	Iodine	5×10^{-5}
Sulfate	2.7	Iron	1×10^{-5}
Magnesium	1.3	Copper	5×10^{-6}
Calcium	0.4	Zinc	5×10^{-6}
Potassium	0.39	Lead	4×10^{-6}
Bicarbonate	0.14	Silver	3×10^{-7}
Bromine	0.070	Nickel	1×10^{-7}
Strontium	0.013	Mercury	3×10^{-8}
Fluorine	0.001	Gold	6×10^{-9}

*Over 60 chemical elements have been found in sea water.

water would have to be processed to obtain one ounce of gold. This would hardly pay for the effort.

Another feature noted when at the beach is the restless motion of the ocean waters. There are three types of sea water movements that are quite noticeable at the beach or along the coast. These are surface waves, long shore currents, and tidal currents. **Surface waves** continually lap the shore. In general, the waves near the surface are a combination of longitudinal and transverse wave motions. It is tempting to think of water waves as being transverse since the sinusoidal profile is clearly evident. However, the water ''particles'' move in more or less circular paths as illustrated in Fig. 30.22. Circular particle motion is a combination of longitudinal and transverse motions.

The diameter of the circular paths of the water particles decreases rapidly with depth and the wave motion is hardly observed when one dives several feet below the surface. As a wave approaches shallower water near the shore, the water particles experience difficulty in completing their circular paths and are forced into more elliptical paths. The surface wave then grows higher and steeper. Finally, when the depth becomes too shallow, the water particles can no longer move through the bottom part of their paths and the wave breaks, with the crest of the wave falling forward to form a **surf.**

When at the beach, you may have noticed debris moving along the shore as it bobs up and down. This movement is an indication of a **long shore current** flowing along the shore. These currents arise from incoming ocean waves that break at an angle to the shore. The component of water motion along the shore causes a current in that direction. Waves and their resulting long shore currents are important agents of erosion along coastlines.

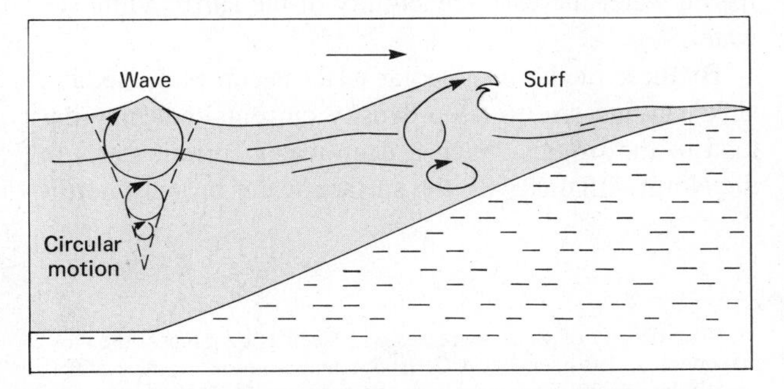

Figure 30.22 An illustration of a surface wave approaching the shore. The water particles move in more or less circular paths, the diameters of which decrease with increasing depth. In shallow water near the shore, the water particles are forced into more elliptical paths. This causes the wave to grow higher and steeper and to eventually break, forming a surf.

The periodic rise and fall of the tides are also quite evident at the beach. These **tidal currents** result from the two tidal bulges that ''move'' around the Earth daily as a result of the gravitational attractions of the moon and Sun and the rotation of the Earth (cf. Chapter 20). The water level may rise as much as 40 ft in some regions at high tide.

In the open ocean, tidal currents are of little significance, but then as they approach the shore, the rise of the water level is evident, particularly when they are confined within the boundaries of a bay or a river outlet. The water level can rise rapidly and sea water can back up into rivers that open to the sea. In France, tidal currents are used to generate electrical power on the Rance River. Not only is power generated from the incoming tidal current as the sea water flows rapidly into the Rance River estuary, but also from the outgoing tidal current as the sea water flows back into the Gulf of St. Malo.

In the open ocean the major movement of sea water results from two general types of currents: surface currents and density currents. **Surface currents** in the ocean are broad drifts of surface water that are set in motion by the prevailing surface winds. These currents rarely extend more than 100 meters in depth (~ 100 yards—one football field length). As might be expected, surface currents are influenced primarily by the prevailing winds of the Earth's atmospheric semipermanent circulation structure as discussed in Section 23.2.

A comparison of the global surface currents shown in Fig. 30.23 with the prevailing wind zones in Fig. 23.7 should make this evident. Surface currents, like air movements, are affected by the Coriolis force (cf. Section 23.1). As a result, the general circulations of surface currents are clockwise in the northern hemisphere (deflection to the right) and counterclockwise in the southern hemisphere (deflection to the left). The surface current circulations are larger in the southern hemisphere since there is more open ocean—two-thirds of the Earth's land surface is north of the equator. The lack of land forms in high southern latitudes makes possible a global circulating surface current—the West-Wind Drift. Water can move from one ocean to another in this drift.

The surface current circulation patterns are important climatic factors. As can be seen from Fig. 30.23, the circulations form sea water convection cycles and the currents carry warm water from equatorial regions to higher latitudes. For example, in the Atlantic Ocean the North Equatorial Current carries warm water westward

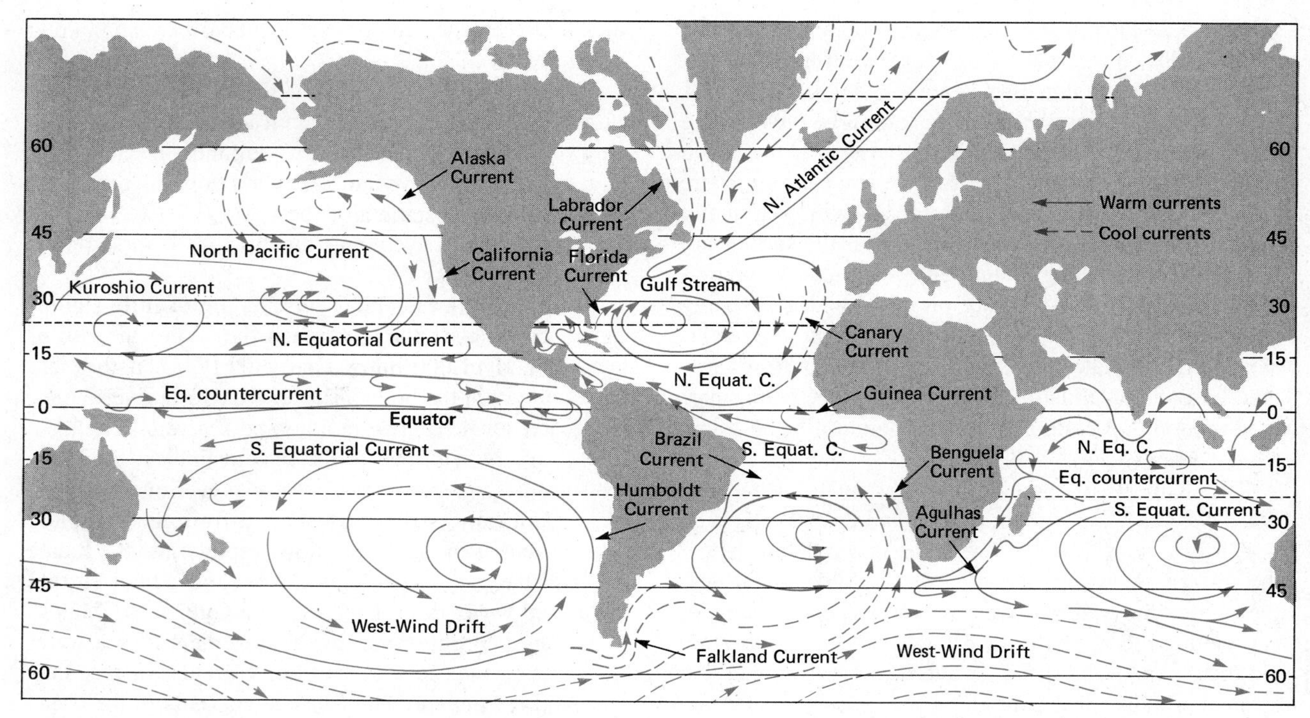

Figure 30.23 Major ocean surface currents. Notice that the general circulations of the surface currents are clockwise in the northern hemisphere and counterclockwise in the southern hemisphere. Like atmospheric circulations, this results from the Coriolis force. (Courtesy Hydrographic Center, DOD.)

into the Gulf of Mexico. It emerges as the Florida Current and flows northward as the Gulf Stream. The circulation continues as the North Atlantic Current flowing toward Europe, where it is diverted to the north and south.

The warm water that the northward-deflected current brings to the British Isles and to the northern European coast is an important factor in moderating the climate of this northern region. The current deflected to the south cools and returns to the equatorial regions via the Canary Current, where it is warmed and starts on another cycle. In a similar manner, the surface current circulation in the northern Pacific brings a warming influence to Japan via the Kuroshio Current, and the returning cool California Current moderates the climate along the coast of southern California.

Deep ocean currents are called density currents. A **density current** is the motion of sea water caused by dense water sinking through less dense water. The density of sea water is dependent on its temperature, salinity, and sediment load. The cooler the water and the greater its salinity, the greater its density.* In the polar regions, the surface water in contact with the atmosphere is cooled and its density increases. Also, when sea water freezes, the dissolved salts are not incorporated in the ice, which is fresh water. The residual salts from the frozen water increase the salinity of the surrounding sea water.

By these methods the polar water becomes denser and sinks, giving rise to deep-density currents that may extend to the tropics. Such a deep-water current starts in the North Atlantic with the surface water brought north-

*The density of water increases with decreasing temperatures to a maximum of 1.0 g/cm^3 at 4°C. Above and below 4°C, the water is slightly less dense. For this reason, lakes freeze at the top rather than at the bottom. As the water near the surface loses heat to the atmosphere, it becomes denser and sinks (causing a density current), displacing the warm water below. This continues until the air and surface water temperature is below 4°C. The cool surface water is then less dense than the water below and hence remains at the surface where freezing begins when the temperature goes below 0°C.

ward by the surface North Atlantic Current. This water already has a high salinity due to evaporation in the equatorial region where it began its journey. This North Atlantic Deep-Water Current flows southward along the ocean bottom far past the equator before mixing decreases its density and halts its flow.

Deep-water currents flow northward from Antarctica, where cooling and freezing give rise to some of the densest water in the oceans. These currents flow northward into the Atlantic, Pacific, and the Indian oceans. However, there is no counterpart of the south-bound North Atlantic Deep-Water Current in the Pacific. This is because there is no large, deep, cold-water source in the northern Pacific. A shallow barrier in the Bering Strait prevents the cold Arctic water from entering this region.

Sea Floor Topography

It was once thought that the surface features or topography of the ocean basins consisted of an occasional mountainous island arc on a relatively smooth sediment-covered floor. This incorrect view resulted from the lack of direct observation. Surprisingly enough, the surface of a major portion of the Earth—the oceanic crust—was not explored in great detail until after World War II. With advent of modern technology, sounding and drilling operations revealed that the ocean floor is about as irregular as the surfaces of the continents, if not more so (Fig. 30.24, pp. 580–581).

We now know that the sea floor has a system of mid-oceanic ridges, which mark divergent plate boundaries. These rocky submarine mountain chains are along fracture zones through which magma rises from below to form new oceanic crust. Oceanic volcanism is widespread. Large volcanic mountains also rise from the ocean floor in the midst of plates, such as those in the Hawaiian island chain. As pointed out in Chapter 29, these are believed to arise from plates moving over internal "hot spots."

Many isolated submarine volcanic mountains have also been discovered. These are known as **seamounts.** They are individual mountains that may extend to heights of over a mile above the sea floor. Some seamounts are found to have flat tops, and are given the special name of **guyots**† (Fig. 30.24). Their shapes suggest that the tops were once islands that were eroded away by wave action. However, many of the guyot tops are several thousand feet below sea level. Evidently, the eroded seamounts must have subsided and sunk below sea level. This subsidence in the oceanic crust perhaps occurs as the oceanic crust moves away from a spreading ridge or results from an isostatic adjustment due to the weight of the large volcanic seamount.

Another marked feature of sea floor topography are **sea floor trenches,** which mark the locations of plunging plate boundaries. These trenches are as much as 150 mi in width and 15,000 mi or more in length. The deepest trenches show little evidence of sediment accumulation, whereas other trenches near land areas are partially filled with sediment.

The huge volumes of sediment flowing into the oceans from continental regions do have an effect on sea floor topography. Distributed by ocean currents, sediment accumulates in some regions such that a sedimentary layer covers and masks the irregular features of the rocky ocean floor. The resulting large flat areas are called **abyssal plains.** Abyssal plains are most common near the continents, which supply the sediment.

Although 70% of the Earth's surface is covered with water, the oceanic crust basins account for only about 65% of the surface area. This means that the continental crust comprises 35% of the Earth's surface area; but since only about 30% of the Earth's surface is land area, 5% of the continental crust must be submerged. This occurs along the continental margins, and the shallowly submerged borders of the continental masses are called **continental shelves** (Fig. 30.25).

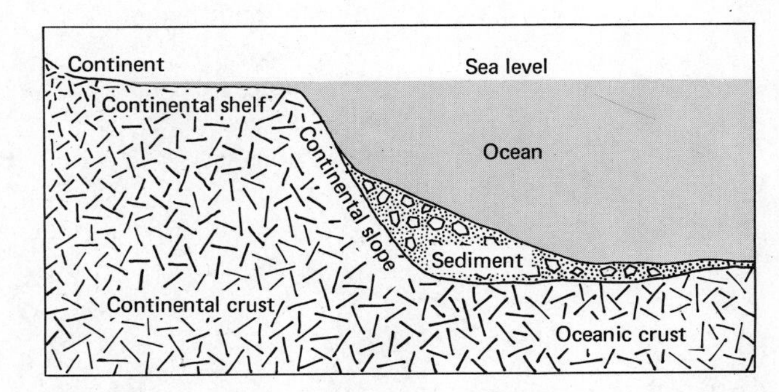

Figure 30.25 A cross-sectional illustration of a continental shelf. Continental shelves are the shallowly submerged borders of the continents. Continental slopes define the true edges of the continents. Sediment from continental erosion collects at the bases of the continental slopes.

†Named in honor of Arnold Guyot, the first geologist at Princeton University, by Professor Harry Hess, a geologist at Princeton University who discovered the first flat-topped seamounts in the 1950s.

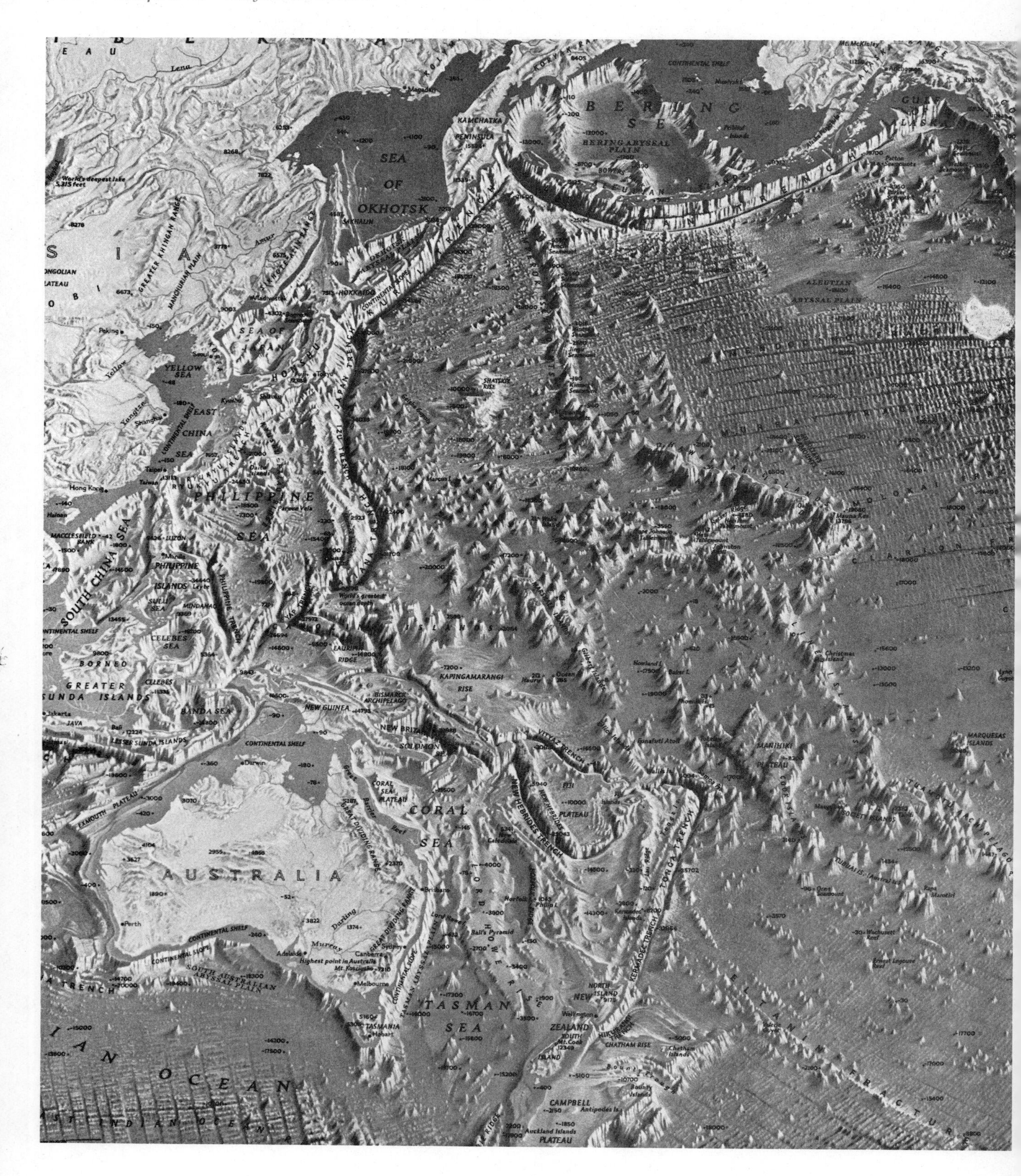
SEA OF OKHOTSK
BERING SEA
BERING ABYSSAL PLAIN
KAMCHATKA PENINSULA
SAKHALIN
HOKKAIDO
SEA OF JAPAN
HONSHU
YELLOW SEA
EAST CHINA SEA
PHILIPPINE SEA
SOUTH CHINA SEA
PHILIPPINE ISLANDS
LUZON
MINDANAO
SULU SEA
CELEBES SEA
BORNEO
GREATER SUNDA ISLANDS
BANDA SEA
NEW GUINEA
BISMARCK ARCHIPELAGO
NEW BRITAIN
SOLOMON
CORAL SEA
AUSTRALIA
TASMAN SEA
TASMANIA
NEW ZEALAND
CAMPBELL PLATEAU
CHATHAM RISE
ALEUTIAN ABYSSAL PLAIN
SHATSKY RISE
KAPINGAMARANGI RISE
MANIHIKI PLATEAU
MARQUESAS ISLANDS
JAPAN TRENCH
IZU TRENCH
KERMADEC TRENCH
TONGA TRENCH
NEW HEBRIDES TRENCH
VITYAZ TRENCH
PHILIPPINE TRENCH
CONTINENTAL SHELF
CONTINENTAL SLOPE
World's deepest lake 5,315 feet
World's greatest ocean depth
Highest point in Australia Mt. Kosciusko
Perth
Adelaide
Melbourne
Sydney
Canberra
Brisbane
Darwin
Hobart
Wellington
Manila
Taipei
Hong Kong
Shanghai
Peking
Seoul
Tokyo
Vladivostok
Magadan
Jakarta
Amur
Lena
Yangtze
Yellow

The widths of these shelves vary greatly. Along the Pacific coast of South America, there is almost no continental shelf—only a relatively sharp, abrupt continental slope. However, off the north coast of Siberia, the continental shelf extends outward into the ocean for about 800 mi. The average width of the continental shelves is of the order of 40–50 mi.

The continental shelves have recently become a point of international interest and dispute. The majority of commercial fishing is done in the waters above the continental shelves. Also, the continental shelves are the locations of oil deposits that are now being tapped by offshore drilling. As a result, many countries including the United States have extended their territorial claims to an offshore 200-mi limit. It should be mentioned that one of the reasons for the Argentine-British conflict over the Falkland Islands was potential offshore oil deposits.

Beyond a continental shelf, the surface of the continental land mass slopes downward to the floor of the ocean basin. The **continental slopes** define the true edges of the continental land masses. Erosion along these slopes gives rise to deep submarine canyons that extend downward toward the ocean basins. And near the edges of the ocean basins, the sediment collects in geosyncline troughs. Thus, we come back to one of the factors in continental mountain building.

Undersea exploration is a relatively new phase of scientific investigation. Indeed, there is a great deal more to be learned about this vast region that makes up almost 70% of the Earth's surface. As advances in technology provide more data on this previously inaccessible part of the Earth's surface, we may expect our knowledge of the Earth and its geologic processes to grow.

Figure 30.24 The Pacific Ocean Floor. The sea floor topography was once thought to be smooth due to the accumulation of sediment deposits. However, relatively recent explorations have shown the sea floor surfaces to be as irregular as the surfaces of the continents. There are many isolated submarine volcanic seamounts, some of which are flat-topped guyots. See also Fig. 27.7 for the topography of the Atlantic Ocean floor. (Courtesy National Geographic Society.)

Learning Objectives

After reading and studying this chapter, you should be able to do the following without referring to the text:

1. Define *erosion*.
2. Distinguish between physical and chemical weathering and give examples of each.
3. Define *mass wasting* and give two important types of mass wasting with examples of each.
4. Describe the agents of erosion.
5. Distinguish between the three components of a stream's load.
6. Explain how glaciers are formed and distinguish between the different types of glaciers.
7. Discuss the distribution of the Earth's water supply.
8. Describe the hydrologic cycle.
9. Explain why the sea is salty.
10. State and explain the types of ocean currents.
11. Explain how ocean currents affect climate.
12. Describe the major features of sea floor topography.

Important Words and Terms

erosion
physical weathering
frost wedging
frost heaving
permafrost
salt wedging
chemical weathering
sinkhole
overburden
mass wasting
landslides
rockslides
slump
mudflows
creep
solifluction
agents of erosion
sheet erosion
stream load
traction
abrasion
silt
base level
meander
graded river
flood plain
delta
glacier
continental glacier
cirque glacier
valley glacier
drift
till
moraine
hydrologic cycle
permeability
porosity
zone of saturation
water table
zone of aeration
aquifer
artesian wells
artesian springs
geyser
hard water
salinity
surface wave
surf
long shore current
tidal current
surface current
density current
seamount
guyot
sea floor trench
abyssal plains
continental shelves
continental slopes

Questions

1. What is erosion?
2. Distinguish between physical and chemical weathering.
3. What are frost wedging and frost heaving?
4. What type of weathering takes place in hot, arid climates that is similar to frost wedging?
5. How do plants and animals contribute to weathering?
6. On what factors does chemical weathering depend?
7. Describe in detail the chemical weathering process of limestone and the formation of caverns and dripstone.
8. What is a sinkhole?
9. What is mass wasting?
10. Explain each of the following and state whether it is a fast or slow type of mass wasting. (a) rockslide, (b) creep, (c) slump, (d) solifluction, (e) mudflow.
11. State the four major agents of erosion.
12. What is sheet erosion?
13. State and explain the three components of a stream's load.
14. Explain stream abrasion and traction.
15. What is silt?

16. State and describe the three general stages of river development.
17. Why does a river meander and what is a graded river?
18. Distinguish between continental and valley glaciers.
19. Are there any glaciers in the United States? Explain.
20. Describe each of the following: (a) drift, (b) till, and (c) a moraine.
21. How much of the Earth's water supply is fresh water and where is most of it located?
22. Explain the hydrologic cycle.
23. Explain each of the following: (a) permeability, (b) porosity, (c) the zone of aeration, (d) the zone of saturation, and (e) the water table.
24. What is an aquifer?
25. Distinguish between gravity springs and artesian wells and springs.
26. What causes hot springs and geysers?
27. What are the two general types of geothermal fields? Which of the types is more important in the generation of electricity?
28. What is the cause of ''hard'' water and what are some of its effects?
29. What is salinity and in what units is it measured? What is the average salinity of the oceans?
30. Describe the three types of sea water movements along a coast.
31. Why do waves form a surf?
32. What causes surface currents in the open ocean and what are some of the major surface currents?
33. Why is the surface current circulation larger in the southern hemisphere? Are there any global circulating surface currents?
34. Explain how the climate of the British Isles and northern Europe is moderated by surface currents.
35. What are density currents? Describe their effect on ocean circulation.
36. Why is there no counterpart of the North Atlantic Deep-Water Current in the Pacific Ocean?
37. Define and explain the formation of each of the following: (a) seamounts, (b) guyots, (c) sea floor trenches, (d) abyssal plains.
38. What are continental shelves and continental slopes? What defines the true edges of the continental land masses?
39. Why are the continental shelves the focus of current international interest?

31

Land and Water Pollution

You can't get pure water and put sewage into the lake.
I say this on behalf of your children.

–President Theodore Roosevelt
1910, at Ashtabula, Ohio (on Lake Erie)

AS IN THE case of air pollution, land and water pollution result primarily from the indiscriminate disposal of waste products. Land and water pollution are closely related through the hydrologic cycle. Through decomposition and erosion processes, many materials contributing to land pollution, or their by-products, eventually find their way into the Earth's water supply. Many of the waste products were once natural parts of the environment, but were processed by human beings into different forms for their use. Other pollution results directly from our activities.

Having studied the physical and chemical processes of the environment, you should now realize that what may not be a problem in a particular situation or location may contribute to or give rise to environmental problems elsewhere. Water is used to carry away waste, as is the air, but such pollution is better defined within waterway boundaries. That is, waste dispersion in water is in general not as great as in the atmosphere, and the path of waste in surface waterways is easily charted. Groundwater pollution is less easily observed, but its effects can be obvious and dangerous. In this chapter some of the sources and effects of land and water pollution will be examined.

31.1 Land Pollution

As was mentioned, it is difficult to separate land and water pollution because of the interaction hydrologic cycle. However, we shall first look at some of the **land pollution** or land surface problems that may give rise to water pollution problems. Later, we will consider water pollution in more detail.

One of the chief sources of pollution of our countryside is improper solid waste disposal. **Solid waste** is a general term that is applied to any normally solid material resulting from human or animal activities that is useless or unwanted. More than 12 million tons of solid waste are produced each day in the United States. The major sources of solid wastes and annual productions are listed in Table 31.1

Modern packaging technology and disposable items have contributed greatly to urban wastes. ''Waste not'' is no longer a popular adage. The domestic solid waste per capita in this country has more than doubled from its 1920 value of 2.9 lb per person per day. If you have had to empty the garbage and trash for an average family, you are aware of the magnitude of the solid waste generated.

The terms refuse, garbage, trash, and rubbish are used to describe solid waste. Their definitions vary slightly in

◀ A soil conservation service technician runs water quality tests on a trout pond in Price County, Wisconsin.

Table 31.1 Major Sources of Solid Waste

Source	*Production (millions of tons)*
Residential, municipal, and commercial	300
Industrial	125
Agriculture	2300
Mineral (mining)	1700
Federal	50
Total	4475

Source: Bureau of Solid Waste Management.

different regions. Refuse is a rather general term, while garbage usually refers to food wastes, rubbish, trash, and combustible (paper cartons, etc.) and noncombustible (junk) items. Regardless of the term used to describe solid waste, the problem is what to do with it (Fig. 31.1).

Solid waste management begins with proper disposal. However, when situations occur such as those shown in Fig. 31.2, the pollution is in the form of litter. **Littering** is the indiscriminate disposal of solid waste on public and private property. The litterbug's disrespect is quite costly. The cost of litter removal from public property alone exceeds an estimated $500 million of tax money annually. All states have litter laws, but evidence of their violation can be seen everywhere.

Tons of litter are strewn along streets and highways and in parks and along beaches. Some wastes, such as food scraps and paper items, are **biodegradable** (capable of being decomposed by bacteria) and will relatively quickly decompose and return to the soil. However, other items such as plastics are nonbiodegradable and are semipermanent. Some show little or no sign of weathering, as in the case of chemical weathering of metal wastes and plastics.

Solid waste disposal is a serious problem, particularly in metropolitan areas where the concentrations of solid waste are enormous. At one time, the **open dump** was the common method of solid waste disposal (Fig. 31.3). Dump sites were usually land of little economic value and at some distance from the center of population that used them. However, open dumps are now forbidden by law for obvious reasons. Solid waste disposal now takes place in **sanitary landfills.** In a sanitary landfill, refuse is covered with a layer of soil within a day after being deposited. This action eliminates some of the nuisances and hazards of the open dump. The landfill method is more sightly, and odor-breeding places for vermin and air pollution from burning are eliminated.

Figure 31.1 Solid waste can be a form of land pollution.

Figure 31.2 (*Above and right*) Littering, the indiscriminate disposal of solid waste on public and private property. (Courtesy USDA-SCS.)

Figure 31.3 (*Left and above*) Past and present. Open dumps were once a common method of solid waste disposal. Today, most solid wastes are disposed of in sanitary landfills. (Courtesy EPA and USDA-SCS.)

However, sanitary landfills still have potential pollution danger. As with the open dump, the seepage of water through the wastes can leach out undesirable components that pollute the groundwater. The extent of pollution from leaching is largely dependent on the geologic features of the area in which the wastes are deposited.

Some cities along the ocean use the convenience of this body of water to dispose of their solid waste. Solid wastes from New York City are hauled offshore by barge and dumped into the ocean. Incineration is also a common method of combustible solid waste disposal, but air pollution is associated with this method.

The activities of human beings often lead to erosion and water pollution. Housing and land developments are necessary to accommodate our growing population. Many prefer country settings, so housing developments are frequently built in wooded areas. Poor planning and the removal of trees and vegetation that prevent erosion can give rise to landslide situations as shown in Fig. 31.4.

Another likely result is sediment or silt concentrations in streams and rivers. Silting is a natural process, as was learned in the last chapter. However, increased silt concentrations from unnatural silting can have a profound effect on the environments of lakes and ponds into which streams drain (Fig. 31.5). The suspended sediment reduces the extent of sunlight penetration into the water. This reduces the photosynthesis of submerged vegetation, thus reducing the oxygen production and the food supply for fish that eat the vegetation. Also, silt can foul the spawning beds of fish.

Mining operations can give rise to land and water pollution. The minerals and fuels needed by society are mined from the Earth. Many such deposits lie well below the surface and are reached by deep mines. Other deposits lying near the surface are surface mined. Large excavating equipment of the type shown in Fig. 31.6 remove the overlying earth (overburden) to expose the commercially valuable materials. Surface mining is faster and cheaper than deep mining. Properly treated and managed, the land after being surface mined can be returned to its natural condition. Unclaimed, it remains an ugly scar on the Earth's surface and a source of stream-fouling sediment and leached pollutants.

The extent of surface mining in the United States is illustrated in Fig. 31.7, along with the proportionate acreage of particular mining operations. Erosion is the most obvious problem associated with surface mining. Sediment yields from surface-mined areas average nearly 300,000 tons per square mile annually. This is as much as 60 times the amount of sediment removed from agricultural lands.

Another pollution problem arising from mining is **acid mine drainage.** Acid formations occur when oxygen (from the air) and water react with sulfur-bearing minerals to form sulfuric acid and iron compounds. For example,

$$\underset{\text{Pyrite}}{2\ FeS_2} + 7\ O_2 + 2\ H_2O \longrightarrow \underset{\text{Ferrous sulfate}}{2\ FeSO_4} + \underset{\text{Sulfuric acid}}{2\ H_2SO_4}$$

Figure 31.4 Land development and a result of poor planning in the removal of erosion-preventing trees and vegetation. (Courtesy California State Dept. of Conservation, Division of Soil Conservation.)

Figure 31.5 Silting from people's activities can foul streams and lakes, causing severe damage to animal and plant life. (Courtesy California State Dept. of Conservation, Division of Soil Conservation.)

Figure 31.6 Surface mine operation. The large dragline shovel in the background removes the overburden and the smaller shovel removes the coal. This strip mine is near Oak Creek, Colorado. (Courtesy USDA-SCS. Photo by N. P. McKinstry.)

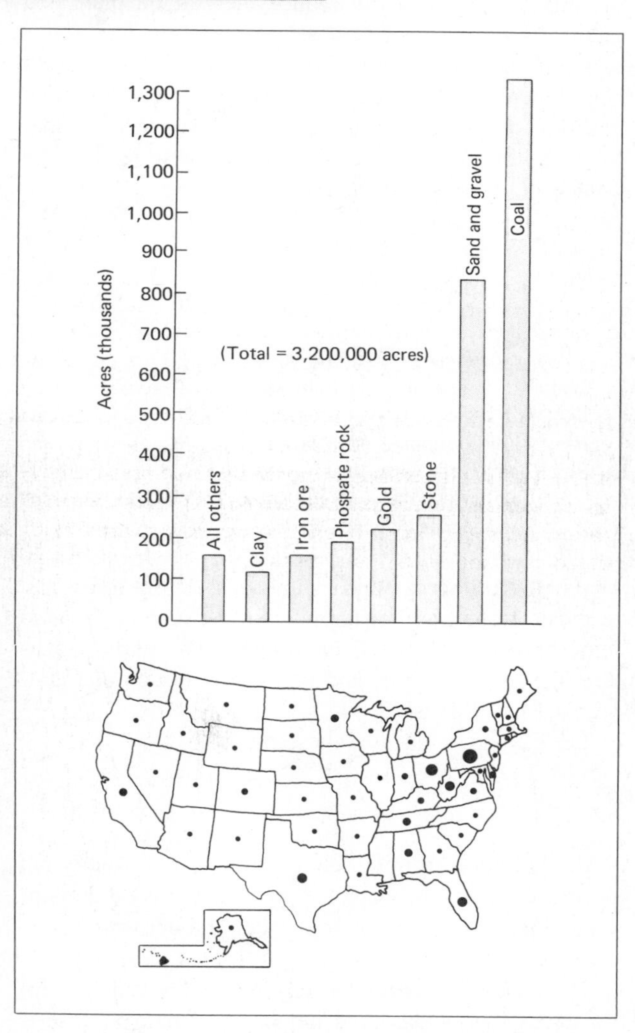

Figure 31.7 Acreage disturbed by surface mining in the United States. The dots on the map show the proportionate surface-mined acreage in each state. About only one-third of the mined area has been reclaimed. The bar graph shows the extent of surface mining by commodities. (Courtesy USDI and USDA.)

About 70% of the acid mine drainage comes from deep coal mines—chiefly from inactive mines. The major problems of acid drainage occur in the coal regions of Appalachia and in some specific western regions. Coal formations in these regions have a high sulfur content (>2%) primarily due to the sulfur-bearing mineral iron pyrite (FeS_2—a yellowish mineral sometimes called fool's gold).

Acid mine waste may directly affect the plant life that absorbs it or have an indirect effect on the flora of a region by affecting the soil minerals and microorganisms. The acid wastes that drain off into ponds and streams can dramatically reduce the aquatic plant and animal life. Changes in the *p*H of water may cause fish to grow slowly or even to die. To reduce pollution from acid mine drainage, reclamation projects include sealing abandoned mine shafts and other accesses to the mine. Sealing prevents the circulation of air and so reduces the oxidation of the sulfur minerals.

31.2 Water Pollution

I counted two-and-seventy stenches,
All well defined, and several stinks.
The river Rhine, it is well known,
Doth wash your city of Cologne;
But tell me, nymphs! what power divine
Shall henceforth wash the river Rhine?

(From the poem *Cologne* by Samuel Taylor Coleridge, 1828)

Water pollution may occur in many different forms and originate from many different sources, such as the acid mine drainage discussed previously. To simplify matters, the major sources and types of pollution are summarized in Table 31.2.

Let's consider each of these briefly.

Table 31.2 Sources and Types of Water Pollution

Agricultural	*Industrial*	*Domestic*
Pesticides	Chemical	Detergents
Fertilizers	Thermal	Organic wastes
Animal wastes	Radioactive	

Agriculture

Water pollution from agricultural sources is usually not direct, but enters the water supply via runoff and groundwater absorption in areas of agricultural activity.

Pesticides. **Pesticides** may be subdivided into insecticides, herbicides, and fungicides. Their purpose is to kill some unwanted insect and plant species, usually for the promotion of an agricultural crop. There are a variety of other uses. For example, in the home, insecticides have all but replaced the pollutionless fly-swatter.

In modern agriculture, pesticides are essential to food production and other crop products to produce the yield required by our population. Without them, tons of agricultural products would be lost each year to pests. Prior to World War II mostly natural organic pesticides were used, which were easily decomposed in the environment. In 1939, DDT (dichlorodiphenyl-trichloroethane) was discovered and used extensively during the war to combat insects in tropical areas. Afterward, this effective agent was expanded to general agricultural use, but in 1972 the government banned the use of DDT in the United States. DDT is a chlorinated hydrocarbon that, unlike natural pesticides, is not biodegradable.

As a result of its extensive usage (worldwide) prior to this, DDT had accumulated in the environment. DDT residue was found in appreciable concentrations in animals, entering through the biological food chains. Even penguins in Antarctica were found to have DDT concentrations. DDT is only one of several chlorinated hydrocarbons used as pesticides. Other pesticides include arsenic compounds of calcium, lead, and mercury.

Pesticide pollution of water in appreciable concentrations poses critical health hazards. There are some 34,500 registered pesticide products composed of one or more of 900 chemical compounds. Although pesticides are usually sold for a specific pest, they often kill nonpest species and have side effects on the growth and reproduction of birds and fish.

Fertilizers. Like pesticides, **fertilizers** are used to increase food production. Farmers in the United States apply millions of tons of fertilizers to their fields each year. Water pollution results from the **phosphates** (PO_3^{4-}) and **nitrates** (NO_3^-) present in the fertilizers. These substances enter the water supply primarily through water runoff and through the erosion of topsoil.

In sufficient concentration, nitrates are toxic to animals and humans. Nitrates can be reduced to nitrites (NO_2^-) which interfere with the transport of oxygen by

Figure 31.8 A cattle feed lot. Commercial beef cattle no longer roam the open range. Instead, they are raised in feed lots such as the one shown here near Omaha, Nebraska, which has facilities to feed about 6000 cattle. Such concentrations of livestock create a problem with animal wastes and water pollution. (Courtesy USDA.)

hemoglobin in the blood. Normal water-purifying procedures do not remove the nitrate contaminants. Nitrates and phosphates are both believed to contribute to the excessive growth of microscopic plant algae in lakes, a condition that affects the local ecosystem. This effect will be considered in the discussion of domestic detergents, the major contributor of phosphates to water pollution.

Animal wastes. Animal wastes are a source of water pollution with potential health hazards. With the trend toward raising animals in feed lots, disposal of these wastes becomes a problem (Fig. 31.8). Table 31.3 gives an indication of the magnitude of the problem. From the data in the table, we see that a feed lot handling 1000 head of cattle would have the same weight-equivalent waste disposal problem as that of a city with a population of 157,000. The city would probably have a sewage treatment plant, the feed lot would probably not.

Table 31.3 Animal Waste Production

Source	*Pounds/Day*
Human being	0.33
Cow	52.0*
Pig	6.0
Sheep	2.5
Chicken	0.4

Source: Soil Conservation Service, U.S. Dept. of Agriculture.

*Perhaps you, like the author, thought this was an inordinate amount, but a county agricultural agent said it is a good figure.

Traditionally, the disposal of animal waste has been as a fertilizer. However, studies show that compared to modern chemical fertilizers the benefits of soil fertilization from animal wastes do not justify the costs of hauling and application. Thus, the livestock owner must contend with a daily supply of a product that cannot be sold, given away, or burned, and that may cause water pollution in water runoff. The effects of this pollutant will be discussed with those of domestic waste.

Industrial

Industrial pollution, one of the more serious forms of water pollution, arises from the disposal of noxious industrial wastes into rivers and streams (Fig. 31.9). This has been a common industrial practice for many years, and at one time streamflows were adequate to dilute and carry away the wastes without environmental damage. But with the increase of industry and population, industrial pollution has become serious.

Most responsible companies are aware of the problems and have instituted programs to reduce or to eliminate their pollution activities. Just as the pollution problems did not develop overnight, neither can the solutions be developed and applied overnight. While solutions are being sought, industry continues to be the largest user of water with some 240,000 industries in the United States using this resource. The amount of water usage,

Figure 31.9 A case of water pollution before effective environmental control. The photographs show paper pulp waste being discharged at its source (not to mention the air pollution) and its effects downstream. (Courtesy Bureau of Sport Fisheries and Wildlife.)

of course, depends on the nature of the industry. A comparison of amounts used is shown in Fig. 31.10.

Chemical. The number of chemical compounds discharged by industry is legion, and it would be hopeless to attempt to list and discuss them here. Instead, let us consider only one as an example, mercury. Mercury (Hg) is a readily available element obtained from the abundant ore cinnabar (HgS). Mercury has many industrial uses (as many as 3000), and compounds of mercury are used extensively as pesticides.

Metallic mercury in liquid form is not overly dangerous unless ingested in large amounts. However, mercury vapor is highly toxic and readily absorbed through the lungs. Also, organic compounds of mercury, such as methylmercury, are quite dangerous. The action of microorganisms in water on metallic mercury can produce methylmercury, which in sufficient quantities harms the central nervous system of animals, including human beings. Methylmercury is retained in the body for long periods of time, so there can be a cumulative effect from otherwise small, harmless doses.

Mercury compounds were utilized to treat the beaver pelts used in the early hat industry, and the early symptoms of mercury poisoning (giddiness, silliness, and strange behavior) were common among the workers; hence the term ''mad as a hatter.'' It is said that the madness of the Mad Hatter in Lewis Carroll's *Alice in Wonderland* was the result of mercury poisoning.

As a form of water pollution, mercury compounds find their way into the food chains of wildlife, particularly fish. Fish are a food supply for many species, including human beings. **Mercury pollution** poses a severe health threat to the inhabitants of areas in which fish forms a substantial part of the diet. Such a case occurred in the coastal town of Minamata, Japan, where

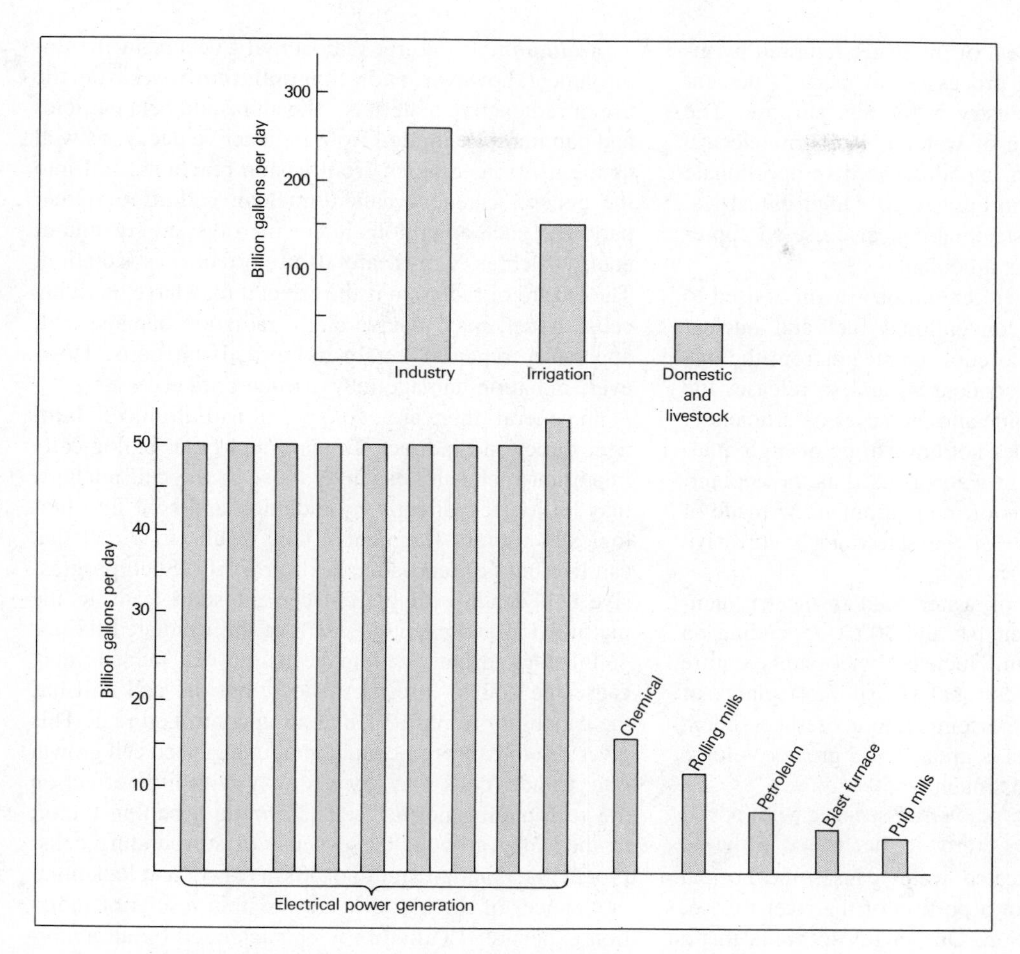

Figure 31.10 The use of water in the United States. (a) By sector. (b) By industry. (Courtesy USDI.)

a plastics manufacturing plant discharged quantities of wastes containing mercury into Minamata Bay. Between 1953 and 1960, 110 people, mostly from the families of fishermen, died or were severely disabled after eating fish caught in the mercury-polluted waters.

Another source of chemical pollution that has come to light in recent years is chemical waste burial. At one time, chemical companies buried their wastes without standards or specifications. Chemical wastes seeping from the containers have contaminated the groundwater as well as being directly toxic to nearby inhabitants. In some cases, homes have been built unknowingly on the sites of old waste dumps. One such incident at Love Canal in western New York was publicized in 1980. Two years later, Times Beach, Missouri, was in the public spotlight when the federal government had to evacuate and resettle its entire population. The extent of these hazardous waste sites is illustrated in Fig. 31.11.

Acid rain might also be classified as chemical water pollution. However, since this results from air pollution, it was discussed in Chapter 26.

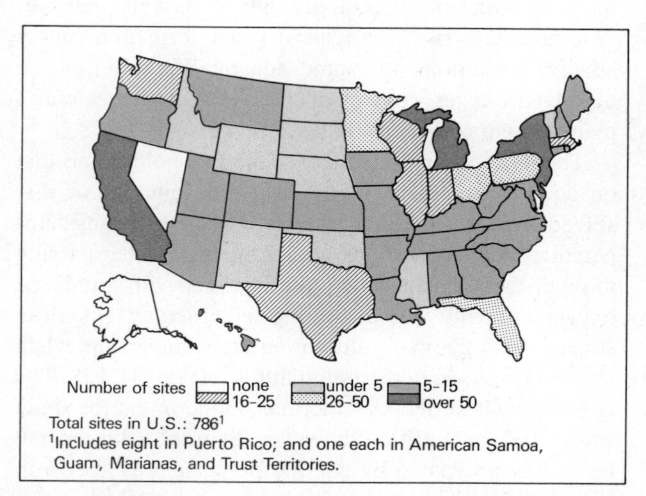

Figure 31.11 Hazardous waste sites. The number of targeted waste sites identified for cleanup under the national trust "Superfund" in 1984. (Courtesy EPA.)

Thermal. A great deal of the water required by industry is used in cooling processes. In fact, 94 percent of the water used by industry is for this purpose. The major cooling application of water is in thermoelectric power generation, which explains the disproportionate share of this user shown in Fig. 31.10. Other industries, such as oil refining and steelmaking, also use an appreciable amount of water as a coolant.

In thermoelectric power generation, steam is used to drive turbines in both conventional fuel and nuclear methods. Water is used to cool the steam from the turbine. The cooled steam condenses, and so releases the back pressure on the turbine and increases its efficiency. The water for cooling does not have to be of high quality. In some locations, salt water is used as the coolant, but this requires that parts of the equipment be made of special materials because of the salt water's corrosive properties.

The temperature rise of water used in electric generation cooling is between 10° and 30°C, depending on plant design and operation. Nuclear power plants require more cooling than conventional (fossil fuel) plants of equivalent size. Either the temperature of the effluent water from nuclear plants is greater or a greater volume of water must be used, as much as 50% more.

The **thermal pollution** problem associated with power generation is due to the return of the heated water to rivers and lakes. The heated water causes the normal temperature of the lake or a portion of the river to rise, which is a type of pollution. One might not think that a few degrees rise in the temperature of a body of water would create an environmental problem. However, the thermal aspect of an ecologic balance is very delicate. A temperature rise in a natural water formation causes adverse conditions for some aquatic life, reduces the dissolved oxygen content of the water, and accelerates plant growth and other biologic processes.

The most observable effect of thermal pollution is that on aquatic life, particularly fish. The amount of dissolved oxygen in water decreases with increasing temperature. Fish absorb oxygen from the water passing through their gills and may die in water with a reduced oxygen content. This possibility is particularly likely in streams with thermal pollution in the summertime when the water temperature rises naturally (Fig. 31.12). The environmental effects of thermal pollution can be quite severe and far-reaching; however, thermal pollution can be readily prevented by cooling processes, as shown in Fig. 31.12.

Radiation. Natural radioactivity occurs in the environment; however, **radiation pollution** arises from the use of radioactive materials. The alpha and beta particles and gamma rays emitted from radioactive decay, as well as the neutrons emitted from fission reactions, fall into the general category called **nuclear radiation.** Other particles, such as protons in cosmic rays, and quanta of energy such as X rays, are also referred to as radiation. The hazard of radiation is the effect it may have on living cells. A cell may receive slight radiation damage and, apparently repaired, retain its normal functions. However, radiation damage may also be critical.

In general, there are two types of **radiation cell damage,** direct and indirect. In the direct type, biologically important molecules, such as those in the cell nucleus, may be ionized directly by radiation and split into biologically useless fragments. This results in a cell that can live but not reproduce itself properly. Such progressive cell death without replacement soon leads to the malfunction and eventual death of the irradiated tissue.

In other instances, damage to the cell nucleus may cause the cell to lose its identity; but the cell still has the capability to reproduce at an uncontrolled rate. This gives rise to **cancer,** a condition of unregulated cell growth rate. Cancer cells may grow slowly with little effect on the surrounding normal cells, forming a benign tumor; or they may grow at the expense of surrounding cells, producing a malignant tumor. Skin cancer and leukemia, or "cancer of the blood," are possible results of radiation exposure. **Leukemia** is characterized by an abnormal increase of white blood cells (leucocytes).

In the indirect type of damage, a less critical molecule, such as a water molecule in the cytoplasm (that part of the cell exterior to the nucleus), may be ionized by radiation into reactive complex ions. The ions and the removed electrons may simply recombine, but some of the reactive fragments may also drift within the cell to react with critical molecules of the nucleus. The damage will be much the same as if the nuclear molecules had been affected directly. On the other hand, complete cell destruction by radiation may also have a useful effect. Radioisotopes are used in medicine to kill diseased cells, such as the uncontrolled cancer cells.

Three main sources of radiation pollution are coal-fired power plants, nuclear power plants, and nuclear explosions. Radiation from coal-fired power plants results from natural radioactive materials in coal. The radiation escaping from nuclear power plants is less than that from coal-fired plants and is continuously monitored

Figure 31.12 An effect and prevention of thermal pollution. (*Right*) Thousands of dead shad, victims of hot weather and river pollution, washed ashore on the Anacosta River near Washington, D.C. The oxygen content of the river water was reduced to the point where the fish could not survive. (*Upper left*) Although hot weather cannot be prevented, thermal pollution can be minimized by the spray air cooling of heated water before it is discharged into nearby waterways. (*Lower left*) Cooling towers may be employed in locations where bodies of water are not accessible. The cooled water is recycled and used again. The towers are emitting water vapor, not smoke. (Courtesy USDA, Bureau of Sport Fisheries and Wildlife, EPA, and ERDA.)

to be sure it is under the maximum permissible levels. However, permissible levels are the subject of much debate and disagreement. Some persons feel that any radiation is too much. The possibility of radiation pollution resulting from accidents in nuclear power plants is extremely low, but it is always a possibility. Probably the best-known examples of nuclear reactor accidents occurred at Three Mile Island in the United States (Pennsylvania) and, more recently, at Chernobyl in the Soviet Union.

Direct sources of radiation pollution are nuclear explosions. In the past, test explosions were usually done in the atmosphere. These are now banned by the Nuclear Test Ban Treaty, signed by the major nuclear powers in

1963. Most test explosions are presently carried out underground, which reduces the risk of radiation pollution. But, not all nations signed the Test Ban Treaty. Several exceptions are France, India, and China.

Domestic

Pollution from domestic sources increases as our population increases. No one wants to be labeled a polluter, particularly at home. But we all contribute, inadvertently and unknowingly. We are now finding that even with accepted disposal methods there are long-term effects of water pollution that are evident only after several years. Of course, there can be immediate health hazards with improper and unsanitary disposal practices. In the case of domestic pollution it may be said, quite truthfully, that pollution prevention begins at home. Two major considerations in domestic pollution are detergents and human organic wastes.

Detergents. A **detergent** in the general sense is a cleansing agent. Ordinary soap, which is a cleansing agent and detergent, is made by combining either fats or oils with alkalies such as sodium or potassium hydroxide. The complex soap molecules in water can effectively interact with grease and oil so that both are carried away with dirt in this solution. This is the detergent action, or detergency, of soap.

Few problems occur with ordinary soap in wastes, since it is normally degraded into harmless substances by bacterial action in sewage. However, the detergent action of soap is impaired by dissolved minerals in water. To accommodate the consumer, synthetic detergents were invented that react with the dissolved minerals without greatly impairing the cleansing action. But, it was quickly found that the synthetic detergents were nonbiodegradable and not removed by accepted waste treatment methods.

As a result, the synthetic detergents persisted when discharged into streams, sometimes to reappear as shown in Fig. 31.13. The manufacturing of these "hard" detergents was voluntarily stopped in the mid-1960s. Through chemical modification of the molecular structures, other synthetic detergents were made that were biodegradable. After sufficient time, these "soft" detergents are completely broken down by bacterial action. There is still some concern, however, about possible yet unknown effects of the decomposition products on the environment. Detergent sales in the United States are in excess of $1 billion.

Another more obvious pollution problem arises from **detergent builders.** These are substances added to the detergent to make its cleaning action more efficient. In some cases, detergent builders account for as much as 40% of the detergent weight. Phosphate compounds are

Figure 31.13 Suds from detergents. The use of "hard" detergents caused many such scenes as the one shown here in Montgomery County, Pennsylvania. The use of hard detergents has been discontinued and replaced by "soft" biodegradable detergents, since this is obviously not the way to "clean up" our rivers and streams. (Courtesy USDA.)

commonly used as detergent builders. The phosphates form compounds with the Ca, Mg, and Fe ions in hard water. This prevents the ions from reacting with the soap and reducing its detergency. As expected, the waste water has high phosphate content. Unfortunately, the phosphates are not removed by ordinary waste treatment processes.

Environmental problems arise from phosphate waste water in lakes, where the phosphates act as a nutrient for microscopic plant algae. A normal amount of algae is good for a lake, because oxygen is added to the water by algae through photosynthesis. Algae also serve as food for fish, which means the algae population is kept in a natural balance. However, the phosphates from detergents cause the algae to multiply rapidly, or "bloom." The excess algae bloom dies and in time decays as the result of bacterial action. Decay reduces the oxygen content of the water, and so the aquatic animal life of the lake is affected. Thus, with the natural balance upset, the lake begins to "die." It becomes covered with algae mats and is so deficient in oxygen that only low animal forms can survive.

The process by which this occurs is called **eutrophication** (Greek, *eutrophos,* well nourished). Eutrophication is also a natural process and takes place over thousands of years, gradually turning lakes into marshes. However, human beings hasten the natural aging process of lakes by introducing phosphates, nitrates (fertilizers), and other pollutants. The classic example is Lake Erie, which has aged as much in the past 50 years as it might have normally aged in 15,000 years.

Organic wastes. Large quantities of human wastes along with the other unwanted liquids of human living are generally referred to as sewage. Sewage is composed of over 99% water and only 0.02 to 0.04% organic solids. Any concentration of people creates a sewage disposal problem, and rivers and streams are usually relied upon to carry away waste products (Fig. 31.14). Ancient civilizations were not exempt from these problems. Rome had an adequate sewage disposal system because of the ample water supply provided by aqueducts. Surplus water was used to continually flush the city sewers into the Tiber River. Some 1500 to 1700 years later the practice of washing away human and animal wastes from city streets was reinstituted in Europe and was considered a major breakthrough in sewage disposal.

The flushing of raw sewage into rivers and streams, however, causes water pollution and contamination problems as would be expected—perhaps not in the location of the sewage dumping, but somewhere farther downstream. Rivers may be able to handle a certain amount of sewage effectively, but with high concentrations of population, problems arise. Even if sewage disposal is attempted by underground storage and drainoff, the water supply may easily become contaminated if there are large volumes of waste.

Figure 31.14 Sewage pollution. (*Top*) Raw sewage being discharged into the Brandywine River near Wilmington, Delaware. (*Bottom*) An isolated case of direct water pollution. (Courtesy EPA and USDA-SCS.)

Figure 31.15 Oil-soaked birds due to an oil spill. (Courtesy USDI.)

When sewage is flushed into rivers and streams the organic matter is broken down by microorganisms, which use up oxygen in the process. Streams usually replenish the oxygen supply by aeration, but a high degree of organic pollution causes the depletion of the water's oxygen supply. This creates problems for most aquatic life.

No doubt the most critical effect of water pollution on human beings is the possibility of waterborne bacterial diseases such as cholera, typhoid fever, and dysentery. A healthy human being lives in balance with bacteria, since intestinal processes depend on certain bacterial action. However, some strains of bacteria are pathogenic and affect some people but not others. As a result, healthy individuals may be carriers of pathogenic (disease-producing) bacteria.

Although there may be no known cases of a disease in a locality, its bacterial agent can be introduced into sewage along with normal waste. This could create a health hazard should a water supply become polluted by the sewage waste. In a public system, the water is treated and its quality checked prior to consumption, but this is usually not the case in private water supplies.

Other Sources

Another source of water pollution has come about through the demand for petroleum. Our nation runs on oil, which accounts for 96% of the fuel used for transportation and 85% of the organic chemicals produced in this country. Domestic production does not keep up with demand. As a result, about 30% of the oil currently used in our country is imported.

The transportation of oil poses an environmental problem. Nearly half of the world's oil production is transported by ocean-going tankers. Some 4000 tankers throughout the world form a petroleum supply chain. The potential hazard to the environment is not from the transportation of the oil, but from oil spills. Oil spills may occur accidentally—or purposefully, since it has been common practice in the past for ships to discharge unwanted oil or oil residues at sea. In 1971, the U.S. Coast Guard reported approximately 8700 known oil spills in the United States waters.

The spilled oil is not readily degradable and floats on the water, being carried by currents. Aquatic life is stifled by the black crude oil, which coats practically everything with which it comes into contact. Should the currents carry the oil spill ashore, beaches are coated with the black ooze and are useless for recreation. An oil spill's toll of sea birds can be enormous (Fig. 31.15). One of the most notorious of all oil spills happened in 1967 when the tanker Torrey Canyon went aground off the shore of southern England and discharged some 100,000 tons of oil into the sea. The oil washed ashore

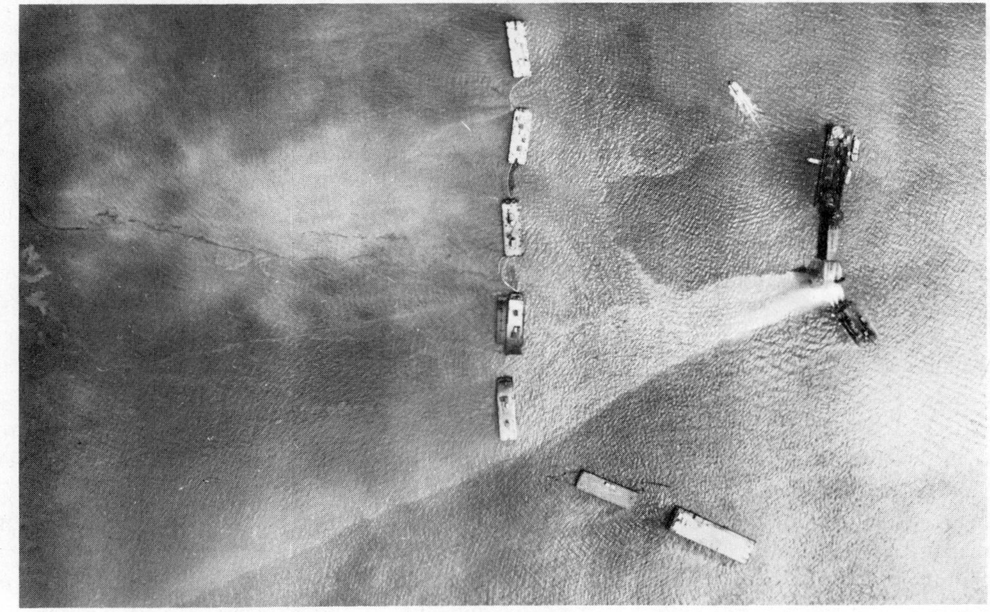

Figure 31.16 Oil spills. (*Top*) The Santa Barbara oil spill, January 1969. Before the break in the well was repaired, an estimated 200,000 gal of oil had escaped. Currents carried the oil ashore in the Santa Barbara area. (*Bottom*) Infrared photo of an oil spill in the Gulf of Mexico, March 1970, just after an accompanying fire was extinguished. A line of booms and barges are in position to attempt to contain the spill, some of which had already floated past the barges. Fortunately, the oil did not reach the Gulf coast. (Courtesy U.S. Geological Survey.)

and ruined the many recreational beaches and boating facilities along England's southern coast, as well as laying waste the sea life in the vicinity.

To satisfy the demand for oil, wells are being drilled offshore, even though it is two to three times more costly. Since 1960, over 9700 wells have been drilled on the continental shelf of the United States. Offshore oil wells increase oil production, but they also increase the possibility of oil spills. In 1969, a well off the coast of California near Santa Barbara discharged into the ocean some 200,000 gal of oil, which later drifted ashore (Fig. 31.16). The consequences were particularly severe for animal life. People tried to clean the oil-covered birds, but only 10% of 1700 birds treated survived. In 1970, fire broke out in eight offshore wells in the Gulf of Mexico, and 800,000 gal of oil were released into the water. A similar situation occurred in 1984. Fortunately, in these instances the wind and currents kept the oil from the beaches and wildlife refuges along the shore.

31.3 The Cost of Clean Water

As in the case of determining the cost of clean air (Chapter 26), determining the cost of clean water is also an insurmountable task. Millions of dollars are spent each year in the treatment of municipal water supplies to make the water safe for human consumption. However, this amount is minor compared to the monetary expenditures needed to alleviate the water pollution problems discussed in the previous section. These problems are interwoven with land management and waste disposal. The sheer magnitude of these factors alone makes the analysis of even the indirect costs of clean water enormously complex. Monetary valuations cannot be placed on the effects of water pollution on the environment and on human health.

Because of the large variety of water pollutants, many measures are required to ensure clean water. Specific treatment may be required for some pollutants, particularly toxic industrial wastes. Space limitations in this text do not permit a detailed consideration of the various methods of pollution prevention and water treatment, so we will consider only the general aspects of two major problems: land management and sewage treatment.

Land Management. With proper planning, the unnatural loads of sediment and silt in our rivers and streams can be prevented. The material from natural erosion processes is part of a balanced environment, but the activities of human beings often upset this balance, as previously pointed out. Human beings disturb the land for their agricultural and other domestic needs, such as construction, and obtaining natural resources.

Proper land management can virtually eliminate stream-fouling sediment and other by-products. For example, **contour farming** prevents the erosion of valuable top soil and increases food production without causing environmental problems (Fig. 31.17). Perhaps it would be somewhat easier to prepare a hillside for cultivation by going up and down the hill instead of around as in the case of contour farming. However, the farmer would increase erosion resulting from the natural water runoff. Planted in contours the crops hold the soil in place and prevent it from being washed away into streams and rivers. In housing developments, similar considerations of proper land management prevent erosion.

A great deal of water pollution comes from coal mining operations. Nearby streams receive an overload of sediment from surface strip mines. Some 3.2 million acres of land in the United States have been disturbed by surface mining operations. Of this figure, one-half of the acreage needs no treatment to prevent sediment and other damage to adjacent land and water. However, the remainder is subject to various degrees of erosion and does need treatment.

Strip mined land can be put back into productive usefulness through **reclamation.** For reclamation to be effective the mounds of spoil need to be leveled and graded and sometimes treated so that trees, shrubs, and grass can take root (Fig. 31.18). A government project which properly reclaimed 650 acres of stripped land cost an average of $1650 per acre.

Sewage Treatment. Sewage constitutes a large part of water pollution and is an important consideration because of potential health hazards from organic wastes. How is sewage handled and how is it treated to avoid pollution problems? In rural areas vaults and septic tanks are the normal means of waste disposal and treatment in rural homes. These consist of storage containers in the ground into which water and human wastes flow or are deposited. Bacterial action breaks down the organic matter, and the water seeps into the ground directly or through subsurface leach beds. However, the lifetimes of these systems are finite since the soil eventually becomes clogged with nonbiodegradable particulate matter. Also, the septic tank-leach bed is not a good system in areas of heavy clay soils.

As the water percolates through the ground toward the water table, it is further purified by natural processes; however, proper planning is required so that the effluent does not seep into and contaminate nearby sources of drinking water.

In cities and muncipalities, sewers are used to collect waste water (and organic contents) and deliver it to treatment plants for processing before it is discharged into oceans, lakes, or rivers. Some city dwellers may be unaware of how this is done, but all the householders are intimately aware of the sewage assessments in terms of the bills they must pay. There are two general types of sewer systems in use—combined and separated systems. The **combined sewer system** uses the same network as both a sanitary and storm sewer.

Figure 31.17 Land management. Through proper land management, such as the contour strip farming shown here, soil erosion from water runoff can be prevented. (Courtesy USDA.)

Figure 31.18 Reclamation. (*Top*) The checking of weeping lovegrass on a heavily limed test plot on a strip mine spoil in White Oak Mountain, West Virginia. The highly acid soil originally had a pH of 2.8. The bare spots were seeded, but not limed. Vegetation is needed to prevent erosion as shown in the background. (*Bottom*) Properly reclaimed land can be put back into productive use. (Courtesy USDA and Consolidated Coal Co.)

Many older cities and towns have a combined system because the original sewer installation was for storm and rain water runoff. Human waste disposal was usually an individual home problem and handled as described previously. However, when the population became sufficiently large that water supplies were threatened with contamination, the original storm sewer system was pressed into combined service. With small populations, the receiving streams sufficiently diluted the sewage, and microorganisms consumed or converted the sewage into less noxious products. Large populations, however, produced excessive sewage loads and upset and impaired this process. Today, sewage treatment plants have been installed in most cities, but some raw sewage still finds its way into our waterways.

Combined sewers usually have overflow bypasses that divert part of the load during a rainstorm when the water volume is much greater than usual. This safeguards treatment plants from overload, but the part of the water bypassed usually flows directly into the receiving streams and, of course, includes amounts of human wastes.

Separated sewer systems provide separate networks for sanitation and storm water. The storm sewer water does not require treatment and may be run directly into the receiving system. The sanitary sewer runs to the treatment plant.

Through applied technology, waste treatment plants essentially speed up the natural processes by which water purifies itself. The two basic methods of treating ordinary municipal wastes are called primary and secondary treatments. In the **primary sewage treatment** there are four main steps, as illustrated in Fig. 31.19. First, the large objects are removed by passing the effluent through screens with openings of a fraction of an inch. This debris is usually removed from the screens and discarded, but some plants have a disintegrator that grinds or breaks the coarse material into smaller pieces. Beyond the first stage screening process is a grit chamber where sand, gravel, and heavy, coarse materials settle out. The sediment of the grit chamber may be used for landfill or construction purposes. The grit chamber stage is particularly important in combined sewer systems where the grit load is heavy.

The third step of primary treatment is a sedimentation tank in which suspended solids are removed. This sediment is known as raw sludge and has a high organic content. In some plants, the raw sludge is dried and sold or given away as fertilizer. In more sophisticated plants the sludge is put into a heated digestion tank where bacteria digest the solids, producing gaseous by-products. About two-thirds of the gas is methane that may be sold or used as fuel in generating electrical power for the treatment plant. The final step is the chlorination of the remaining liquid. Chlorination kills disease-causing bacteria and also reduces odors. The product of this primary treatment is then placed into the receiving steam which dilutes and carries it away.

About 30% of the cities in the United States give only primary treatment to their sewage. However, in dealing with sewage in large amounts and with a variety of wastes, secondary treatment is recommended. **Secondary sewage treatment** removes up to 90% of the suspended organic matter. This is accomplished by adding an additional step to the primary treatment between the sedimentation tank and the chlorination. This secondary step is done by either of two processes—trickling filters and activated sludge.

In the **trickling filter process** the sewage effluent is trickled through a bed of stones several feet deep. A trickle filter bed is shown in Fig. 31.20. Bacteria gather

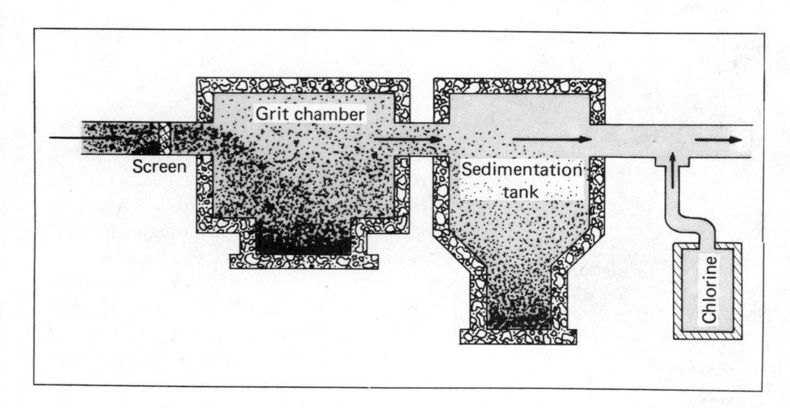

Figure 31.19 A diagram of the four steps of primary sewage treatment. (1) Large objects are removed by screens. (2) Heavier materials settle out in a grit chamber. (3) Suspended solids are removed in a sedimentation tank. (4) Bacteria and odors are destroyed by chlorination. (Courtesy EPA.)

Figure 31.20 Sewage treatment plant with trickle filter bed. Bacteria gather and multiply on the stones until they have the capacity to consume most of the organic matter in the sewage effluent. (Courtesy USDA-SCS.)

and multiply on the stones until they have the capacity to consume most of the organic matter in the sewage effluent. A relatively new type of trickle filter is built above ground and uses a plastic material instead of rocks.

The current trend favors the use of an **activated sludge process** for the secondary treatment step. In this process, the sewage effluent passes into an aeration tank and then into a large container where it mixes with sludge heavily laden with bacteria (Fig. 31.21). The aeration oxygenates the effluent and speeds up the bacterial breakdown of the organic matter. Though more expensive, the activated sludge process is more efficient and is a closed system that reduces odors.

The sewage treatment methods just described are generally adequate for normal sewage. However, other treatments are necessary for sewage with complex industrial wastes. Some treatments may be specific for a particular type of chemical wastes that cannot be broken

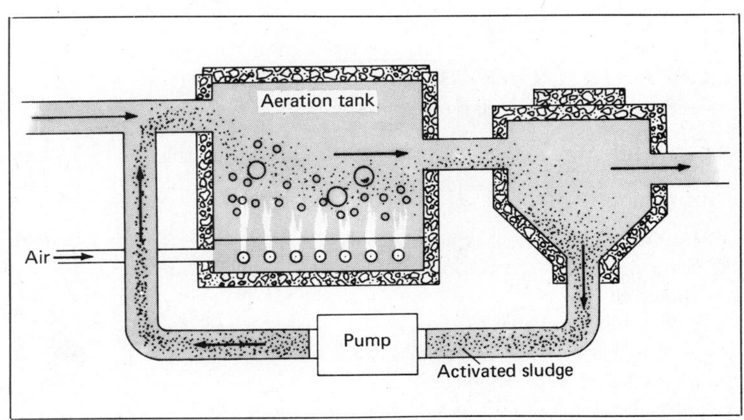

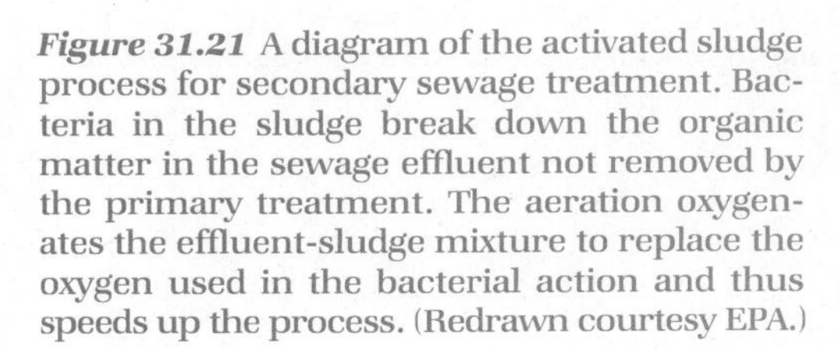

Figure 31.21 A diagram of the activated sludge process for secondary sewage treatment. Bacteria in the sludge break down the organic matter in the sewage effluent not removed by the primary treatment. The aeration oxygenates the effluent-sludge mixture to replace the oxygen used in the bacterial action and thus speeds up the process. (Redrawn courtesy EPA.)

Figure 31.22 The ultimate goal—unpolluted water. (Courtesy EPA.)

down by the bacterial action (nonbiodegradable) of conventional methods. Precipitation and coagulation methods have been used for years to treat industrial wastes. Chemicals are added to precipitate dissolved materials or to coagulate fine, suspended particles. The precipitate or coagulate is then allowed to settle out in a sedimentation tank.

Another special treatment for organic matter that resists normal bacterial action is adsorption. The waste effluent is passed through beds of activated charcoal which adsorbs as much as 98% of the organic material. It is obvious that all of these treatment methods add to the cost of clean water (Fig. 31.22).

Learning Objectives

After reading and studying this chapter, you should be able to do the following without referring to the text:

1. Describe the different types of land pollution.
2. State the major sources of solid waste production in the United States.
3. Define biodegradable.
4. Distinguish between deep and surface mining.
5. Explain acid mine drainage.
6. State the sources and types of water pollution.
7. Explain eutrophication and the effects of pollution.
8. Discuss land management.
9. Distinguish between (a) combined and separated sewer systems, and (b) primary and secondary sewage treatments.

Important Words and Terms

land pollution
solid waste
litter
biodegradable
open dump
sanitary landfill
acid mine drainage
water pollution
pesticides
fertilizers
phosphates
nitrates
mercury pollution
thermal pollution
radiation pollution
nuclear radiation
radiation cell damage
cancer
leukemia
detergent
detergent builders
eutrophication
contour farming
reclamation
combined sewer system
separated sewer system
primary sewage treatment
secondary sewage treatment
trickling filter process
activated sludge process

Questions

1. What is one of the chief sources of land pollution?
2. What is the greatest source of solid waste in the United States?
3. Approximately how much domestic solid waste is generated per capita in the United States?
4. Distinguish between open dumps and sanitary landfills.
5. What is the potential pollution danger of sanitary landfills?
6. How have human activities given rise to land and water pollution?
7. What is acid mine drainage? Describe its cause, source, and effects.
8. State the sources and types of water pollution.
9. What are the effects of pesticide pollution?
10. How do fertilizers contribute to water pollution?
11. Why are animal wastes a current potential pollution problem?
12. What are the effects of mercury poisoning?
13. What is thermal pollution, and what are its sources and effects?
14. Describe the effects of radiation damage on living cells.
15. What are the major sources of radiation pollution?
16. What are detergents and how do they contribute to water pollution?
17. What are detergent builders?
18. Describe the causes and effects of eutrophication due to pollution.
19. What is the composition of sewage?
20. Describe the sources and effects of oil spills.
21. Give some examples of land management that prevent pollution.
22. What is the difference between combined and separated sewer systems?
23. What are the four main steps in primary sewage treatment?
24. How does secondary sewage treatment differ from primary sewage treatment?
25. Describe the operation of trickling filters and the activated sludge process.
26. The average amounts of domestic water use are as follows:

	(avg. gal)
Flushing a toilet (once)	4
Washing dishes	10
A washing machine load	25
A shower (5 min)	25
A tub bath	35
Watering lawn (1 h)	300

Using these values, estimate the (a) daily, (b) monthly, and (c) yearly amounts of water used by a single person and a family of four.

Appendixes

Appendix I

The Seven Base Units of the International System of Units (SI)

Meter-m (length)	The meter is defined in reference to the standard unit of time. One meter is the length of the path traveled by light in a vacuum during a time interval of 1/299,792,458 of a second. That is, the speed of light is a universal constant of nature whose value is defined to be 299,792,458 meters per second.
Kilogram-kg (mass)	The kilogram is a cylinder of platinum-iridium alloy kept by the International Bureau of Weights and Measures, at Paris. A duplicate in the custody of the National Bureau of Standards serves as the mass standard for the United States. This is the only base unit still defined by an artifact.
Second-s (time)	The second is defined as the duration of 9,192,631,770 cycles of the radiation associated with a specified transition of the cesium-133 atom.
Ampere-A (electric current)	The ampere is defined as that current that if maintained in each of two long parallel wires separated by one meter in free space, would produce a force between the two wires (due to their magnetic fields) of 2×10^{-7} newton for each meter of length.
Kelvin-K (temperature)	The kelvin is defined as the fraction 1/273.16 of the thermodynamic temperature of the triple point of water. The temperature 0 K is called *absolute zero*.
Mole-mol (amount of substance)	The mole is the amount of substance of a system that contains as many elementary entities as there are atoms in 0.012 kilogram of carbon-12.
Candela-cd (luminous intensity)	The candela is defined as the luminous intensity of 1/600,000 of a square meter of a blackbody at the temperature of freezing platinum (2045 K).

Appendix II

Measurement and Significant Figures

A **measurement** is a comparison of an unknown quantity with a precisely specified quantity called a standard unit. Measurements should be made and recorded as accurately as possible. The term **accuracy** refers to how close the measurement comes to the exact value. Accuracy depends upon the precision of the measuring instrument and the ability of the individual taking the measurement. Errors will be made in every measurement, but the magnitude of the errors can be kept small when the observer is careful and uses an instrument with high precision. **Precision** refers to the degree of reproducibility of a measurement; that is, to the maximum possible error of the measurement. Precision may be expressed as a plus or minus correction. The accuracy of a measurement depends on the precision of the measuring instrument. All measurements are approximate. It is impossible to know the ''exact'' length, mass or amount of anything. The observer and the measuring instrument place a limit on the accuracy of all measurements.

Instruments for taking measurements are constructed with a calibrated scale for obtaining numerical values concerning the property being measured. The smallest division on the calibrated scale that can be read by the observer, without guessing, is known as the **least count** of the instrument. For example: a meter stick is divided into 100 equal divisions and marked on the stick as centimeters. Each centimeter is further divided into 10 equal divisions. Thus, the least count of the meter stick is one tenth of one centimeter or one thousandth of one meter.

When the meter stick is used to take a measurement of an unknown quantity (for example, length), the observer can always obtain a value within one thousandth (0.001) of a meter of the exact value of the unknown length, without guessing. But, an additional step taken can increase the accuracy by one doubtful digit. This is accomplished because the observer can estimate a fractional part of the smallest division on the meter stick. For example, if one end of the meter stick is placed at one end of the object to be measured, and the other end of the object falls between two of the smallest divisions, the observer estimates this additional value and adds it to the known scale reading. This estimated digit is significant and is the last digit recorded, when taking a measurement.

When the digits are recorded, the number recorded contains all known digits of the measurement plus the doubtful digit. This recorded number is a significant figure. By definition, a **significant figure** is a number that contains all known digits plus one doubtful digit.

Recorded numbers may contain the digit zero (0), which may or may not be significant. When a zero digit is used to locate the decimal point, it is not significant. For example, the numbers 0.048, 0.0032, and 0.00057 each have two significant digits. When a zero appears between two nonzero digits in a number, it is significant. For example, 2.04 has three significant digits; 8.002 has four significant digits. Zeros appearing at the end of a number may or may not be significant. For example, in the number 5480 if the digit eight is a doubtful digit, then the zero is not significant. If the eight is a known digit and the zero is a doubtful digit, than the zero is significant.

The powers of 10 notation can be used to remove the ambiguity concerning a number like 5480. When using the powers of 10 notation, we first write all the significant digits of the number. This is followed by 10 to the correct power to locate the decimal point. For example, for three significant digits we write 5.48×10^3, for four significant digits we write 5.480×10^3.

The following procedure is usually used to round off significant figures to fewer digits. If the last significant digit on the right is less than 5, drop it and insert zero instead. If the last significant digit on the right is 5, drop it and round the measurement to the nearest even number. If the last significant digit is greater than 5, drop it and increase the preceding digit by one.

EXAMPLES Round off the following numbers to two significant digits

247. Since the last digit on the right is greater than 5, drop it and increase the preceding digit by one, for the result 250.

243. Since the last digit on the right is less than 5, drop it and insert zero instead, for the result 240.

245. Since the last digit on the right is 5, drop it and round the measurement to the nearest even number, for the result 240.

275. Since the last digit on the right is 5, drop it and round the measurement to the nearest even number, for the result 280.

As a general rule, the number of significant digits of the product or the division of two or more measurements should be no greater than that of the measurement with the least number of significant digits. For example, suppose the area of a table is to be determined. This is accomplished by measuring the length and width of the table, then multiplying one value by the other. In this precedure, it is inaccurate and meaningless to calculate and give an answer indicating greater accuracy than justified by the original data. For example: the length of a table is measured with a meter stick as 1.8245 m and the width as 0.3672 m. The area $A = 1.8245 \text{ m} \times 0.3672 \text{ m} = 6.678714 \text{ m}^2$ as shown on a calculator. This six-figure number is not justified as the correct area of the table as given by the two measurements. The correct value for the area is 0.6787 m^2. This value has four significant digits corresponding to the least number of significant digits in the original data.

When adding or subtracting significant figures use the following two rules;

Rule 1. A known digit plus or minus a doubtful digit will give a doubtful digit.

Rule 2. Only one doubtful digit is allowed in a significant figure.

EXAMPLE

2.34	the 4 is doubtful
+ 16.5	the 5 is doubtful
28.84	the 8 and the 4 are doubtful

From Rule 1, the known digit 3 plus the doubtful digit 5 equals a doubtful digit 8. therefore, the correct answer is 28.8. This is a significant figure with only one doubtful number, which satisfies rule two.

Appendix III

Rules and Examples of the Four Basic Mathematical Operations Using Powers of Ten Notation

1. In addition or subtraction, only like terms can be used; that is, the exponent for the number 10 must be of the same value. Thus,

$$\begin{array}{r} 4.6 \times 10^{-8} \\ +1.2 \times 10^{-8} \\ \hline 5.8 \times 10^{-8} \end{array} \quad \text{and} \quad \begin{array}{r} 4.8 \times 10^{-8} \\ -2.5 \times 10^{-8} \\ \hline 2.3 \times 10^{-8} \end{array}$$

2. In multiplication, the exponents are added. Thus,

$$2 \times 10^4 \text{ multiplied by } 4 \times 10^3 = 8 \times 10^7$$

$$1.2 \times 10^{-2} \text{ multiplied by } 3 \times 10^6 = 3.6 \times 10^4$$

3. In division, the exponents are subtracted. Thus,

$$\frac{4.8 \times 10^8}{2.4 \times 10^2} = 2.0 \times 10^6$$

and

$$\frac{3.4 \times 10^{-8}}{1.7 \times 10^{-2}} = 2.0 \times 10^{-6}$$

4. A power of 10 may be transferred from the numerator to the denominator, or vice versa, by changing the sign of the exponent. Thus,

$$\frac{3.4 \times 10^{-8}}{1.7 \times 10^{-2}} = \frac{3.4 \times 10^{-8} \times 10^2}{1.7} = 2.0 \times 10^{-6}$$

Appendix IV

Problem Solving: Five Major Steps

1. Read the problem carefully to find out what is given and what is inferred.
2. Determine the unknown or unknowns; that is, what are you looking for?
3. Apply the mathematical relationship that connects the unknown quantity or quantities with the given data.
4. Check data for correct units.
5. Solve mathematically for the unknown value or values.

EXAMPLE I A ball is thrown upward from the ground with an original velocity of 128 ft/s. How long does the ball remain in the air? What was the maximum height attained by the ball?

Step 1. Write down the given data:

$$v_o = 128 \text{ ft/s}$$

and inferred data:

(acceleration) $a = -32$ ft/s^2 (The minus sign is used because v_o and the acceleration are in opposite directions.)
(final velcity) $v_f = 0$ at maximum height.

Step 2. Write down the unknown term or terms:

t = ? seconds (total time the ball is in the air)
s = ? feet (maximum height or displacement of ball).

Since the original velocity is in ft/s, find the unknown time in seconds and the unknown height in feet.

Step 3. The terms v_o, v_f, and a are known; therefore, use the relation

$$a = \frac{v_f - v_o}{t}$$

which contains these terms, plus the time, which is unknown.

Step 4. All the quantities used in solving a problem should be in the same system of units. The units of the result will then also be in that system. In this example, the units are in the British system, hence the height will be in feet (Step 5). Units are often mixed in a problem. For example, velocity may be given in feet per second while an acceleration may be given in meters per second^2. One of these must then be converted before being used in an equation.

Step 5. Solve the equation for t, and obtain

$$t = \frac{v_f - v_o}{a}$$

Substitute the given data and solve to obtain

$$t = \frac{0 - 128 \text{ ft/s}}{-32 \text{ ft/s}^2}$$

$$t = 4 \text{ s (time for ball to reach maximum height)}$$

The total time in the air is twice this value since the ball will require the same amount of time to return to the ground.

After you know the value of t, v_o, v_f, and a, the unknown height can be determined by

$$s = v_o t + \frac{at^2}{2}$$

Since v_o at the top of the ball's flight $= 0$

$$s = \frac{at^2}{2}$$

Since a in this case is equal to g, which is 32 ft/s^2, we can substitute the data and find

$$s = \frac{32 \text{ ft/s}^2 \times 4^2 \text{ s}^2}{2}$$

therefore

$$s = \frac{512}{2} \text{ ft}$$

or

$$s = 256 \text{ ft}$$

In solving problems in grade school, you learned not to add apples to oranges, because doing so gives an answer that has no meaning. In physical science, you must remember that only like terms can be added or subtracted. When you have a mathematical relationship (i.e., an equation), terms on the left side of the equals sign must be equal to the terms on the right side, not only in numerical value, but also in units. For example, in the equation

$$a = \frac{v_f - v_o}{t}$$

both sides of the equation must have like dimensions. A **dimensional analysis** of an equation is done by substituting the fundamental units of length, mass, and time in the equation for the given terms.

EXAMPLE 2 We can use dimensional analysis to show the above is an equality. Although it has no mass term, the expression has terms that can be traced back to the fundamental units of length and time. Therefore, by substituting

$$\frac{L}{t^2} = \frac{\frac{L}{t} - \frac{L}{t}}{t}$$

Solving just for the numerator on the right, and remembering that we are subtracting units rather than numerical or algebraic values, we obtain

$$\frac{L}{t} - \frac{L}{t} = \frac{L}{t}$$

The top expression now can be written

$$\frac{L}{t^2} = \frac{\frac{L}{t}}{t}$$

Using negative exponents to bring our denominators above the line, we find

$$Lt^{-2} = Lt^{-1} \times t^{-1}$$

or

$$Lt^{-2} = Lt^{-2}$$

The left side of the expression has the same dimensions as the right side, showing the original to be an equality. If the two sides were not the same dimensionally, it would not be an equation that would give a correct solution.

Data found in physics problems often must be converted to different units in order to give the answer desired. For instance, measurements for rates of travel may be given in miles per hour, and the answer requested in feet per second. How can we obtain the proper result?

EXAMPLE 3 What is the distance in feet traveled by an automobile in 10 s if the average speed of the car is 60 miles/hour? Since we want to know the distance traveled in feet, we must convert the mi/h to ft/s. This can be done as follows:

1. Write down the term you want to convert and set it equal to itself. At the same time, make a mental or written note of the terms to which you wish to convert.

$$\frac{60\text{ mi}}{\text{h}} = \frac{60\text{ mi}}{\text{h}}$$

2. Multiply the right side of the equation by 1 or 1/1, which is the same thing.

$$\frac{60\text{ mi}}{\text{h}} = \frac{60\text{ mi}}{\text{h}} \times \frac{1}{1}$$

3. Place a unit in the numerator or denominator of the 1/1 which will cancel out an unwanted unit in the original term. In this case, we would place the unit "mile" in the denominator. At the same time, place an equal term above our newly introduced denominator unit, in order to keep the value equal to 1/1. Since we are after feet in our answer for this problem, it would be well in this case to place 5280 ft in the numerator of the 1/1, since that is equal to 1 mi. Then cancel like units.

$$\frac{60\text{ mi}}{\text{h}} = \frac{60\ \cancel{\text{mi}}}{\text{h}} \times \frac{5280\text{ ft}}{1\ \cancel{\text{mi}}}$$

4. The next step is to eliminate the unit hour in the denominator. This can be accomplished by multiplying again by 1/1 and placing the unit hour in the numerator of the 1/1 term. At the same time, an equal unit should be placed in the denominator of the 1/1 term in order to keep the value equal to 1/1. Since 60 minutes equal 1 hour, we can use it, then cancel both "hour" terms.

$$\frac{60\text{ mi}}{\text{h}} = \frac{60\ \cancel{\text{mi}}}{\cancel{\text{h}}} \times \frac{5280\text{ ft}}{1\ \cancel{\text{mi}}} \times \frac{1\ \cancel{\text{h}}}{60\text{ min}}$$

5. At this point, we have feet per minute, but we want feet per second; therefore we must convert the minutes to seconds. Again we multiply by 1 or 1/1 and add the appropriate units for conversion. Since 60 seconds equal 1 minute, we obtain

$$\frac{60\text{ mi}}{\text{h}} = \frac{\cancel{60}\ \cancel{\text{mi}}}{\cancel{\text{h}}} \times \frac{528\cancel{0}\text{ ft}}{1\ \cancel{\text{mi}}} \times \frac{1\ \cancel{\text{h}}}{\cancel{60}\ \cancel{\text{min}}} \times \frac{1\ \cancel{\text{min}}}{\cancel{60}\text{ s}}$$

6. After all like units are canceled, the numbers are canceled insofar as is possible; we arrive at

$$\frac{60\text{ mi}}{\text{h}} = \frac{528\text{ ft}}{6\text{ s}}$$

or, after dividing, $\dfrac{60\text{ mi}}{\text{h}} = \dfrac{88\text{ ft}}{\text{s}}$

The problem can now be solved using 88 ft/s for 60 mi/h. Knowing that

$$s = \bar{v}t$$

and substituting $s = \dfrac{88\text{ ft}}{\text{s}} \times 10\text{ s}$

we obtain $s = 880\text{ ft}$

Derivation of Displacement Equation

The displacement of an object moving with constant acceleration can be determined by

$$s = v_o t + \frac{at^2}{2}$$

This relationship is derived as follows:

$$\bar{v} = \frac{s}{t}$$

therefore, $s = \bar{v}t$

We know that $\bar{v} = \dfrac{v_f + v_o}{2}$

when the acceleration is constant. Substituting, we obtain

$$s = \left(\frac{v_f + v_o}{2}\right)t$$

Knowing that $v_f = v_o + at$ we may substitute for v_f and obtain

$$s = \left(\frac{v_o + at + v_o}{2}\right)t$$

or

$$s = \left(\frac{2v_o + at}{2}\right)t$$

or

$$s = v_o t + \frac{at^2}{2} \qquad \text{(A.1)}$$

The simplest way to obtain an equation of motion which does not include time is to take two equations that include time and eliminate the time factor by cancellation. By definition

$$a = \frac{v_f - v_o}{t}$$

From the derivation of Eq. A.1 we know that

$$s = \left(\frac{v_f + v_o}{2}\right)t$$

Multiplying these equations

$$\left(a = \frac{v_f - v_o}{t}\right) \times \left[s = \left(\frac{v_f + v_o}{2}\right)t\right]$$

and canceling the time, we obtain

$$as = (v_f - v_o)\left(\frac{v_f + v_o}{2}\right)$$

or

$$2as = v_f^2 - v_o^2 \qquad \text{(A.2)}$$

Derivation of Work-Kinetic Energy Relationship

In Chapter 4 we use the fact that a change in kinetic energy is related to work done against inertia. This equation can be derived as follows: First, we write the equation for the work done by an unbalanced force in accelerating a mass through a horizontal distance.

$$W = Fs \qquad \text{(A.3)}$$

The unbalanced force, F, obtained from Newton's second law, is

$$F = ma$$

Substituting in Eq. A.3 for F, we have

$$W = mas \qquad \text{(A.4)}$$

Acceleration has been defined as

$$a = \frac{v_f - v_o}{t}$$

Substituting in Eq. A.4 for a, we have

$$W = m\left(\frac{v_f - v_o}{t}\right)s \qquad \text{(A.5)}$$

The distance traveled can be obtained from

$$s = \bar{v}t$$

or

$$s = \left(\frac{v_f + v_o}{2}\right)t$$

Substituting in Eq. A.5 for s, we obtain

$$W = m\left(\frac{v_f - v_o}{t}\right) \times \left(\frac{v_f + v_o}{2}\right)t$$

Canceling the t, we have

$$W = m\left(\frac{v_f - v_o}{1}\right) \times \left(\frac{v_f + v_o}{2}\right)$$

Multiplying $(v_f - v_o)$ by $(v_f + v_o)$ we have

$$W = \left(\frac{m}{2}\right)(v_f^2 - v_f v_o + v_f v_o - v_o^2)$$

or

$$W = m\left(\frac{v_f^2 - v_o^2}{2}\right)$$

or

$$W = \left(\frac{mv_f^2}{2}\right) - \left(\frac{mv_o^2}{2}\right)$$

which is the equation used in Chapter 4.

Derivation of Acceleration in Circular Motion

The acceleration of an object going in circular motion at a constant speed is

$$a = \frac{v^2}{r}$$

This can be derived by referring to Figs. 2.12 and 2.13 as follows: Lines CP_1 and CP_2 are drawn parallel to (and represent) velocities v_1 and v_2 in Fig. 2.12 at positions P_1 and P_2, respectively. Triangles $v_1 \Delta v v_2$ in Fig. 2.13 and P_1OP_2 in Fig. 2.12 are similar, since they are isosceles triangles and have equal angles at C and O.

Therefore, $$\frac{\Delta v}{v} = \frac{P_1P_2}{OP_1}$$

and $$\Delta v = \frac{P_1P_2}{OP_1} v$$

If angle θ is very small, the distance from P_1 to P_2 is very small

or $$\Delta v = \frac{s}{r} v$$

Dividing each side of this equation by t

$$\frac{\Delta v}{t} = \left(\frac{s}{t}\right)\frac{v}{r}$$

or $$a = v \times \frac{v}{r}$$

Then, $$a = \frac{v^2}{r}$$

Appendix V

Derivation of Bohr's Equations

Niels Bohr's quantization of angular momentum led to two useful equations (see Chapter 11, Eqs. 11.2 and 11.3). These equations are derived as follows:

Bohr's set of equations were

$$(1)\ \frac{kq^2}{r^2} = \frac{mv^2}{r}$$

$$(2)\ E = -\frac{kq^2}{2r}$$

and $$(3)\ mvr = \frac{nh}{2\pi}$$

The third equation can be restated as

$$v = \frac{nh}{2\pi mr}$$

Then, substituting this in the first yields

$$k\frac{q^2}{r^2} = \frac{m}{r}(v^2)$$

$$k\frac{q^2}{r^2} = \frac{m}{r}\left(\frac{n^2h^2}{4\pi^2m^2r^2}\right)$$

or $$r = \left[\frac{h^2}{kq^2 4\pi^2 m}\right] n^2 \quad \text{(A.6)}$$

We can now solve for E

$$E = -\frac{kq^2}{2}\left(\frac{1}{r}\right)$$

$$E = -\frac{kq^2}{2}\left(\frac{kq^2 4\pi^2 m}{h^2n^2}\right)$$

$$E = -\left[\frac{k^2q^4 4\pi^2 m}{2h^2}\right]\frac{1}{n^2} \quad \text{(A.7)}$$

The quantities in brackets in Eq. A.6 and A.7 are all constants and can be evaluated. The results are Eqs. 11.2 and 11.3:

$$r = 0.529n^2 \text{ angstroms}$$

$$E = \frac{-13.6}{n^2}\text{ eV}$$

Appendix VI

Psychrometric Tables (Pressure: 30 in Hg)

Table 1 Relative Humidity (%) and Maximum Mositure Capacity*

Air Temp. (°F) (dry bulb)	Max. Moisture Capacity (gr/ft³)	Degrees Depression of Wet-Bulb Thermometer (°F)													
		1	2	3	4	5	6	7	8	9	10	15	20	25	30
25	1.6	87	74	62	49	37	25	13	1						
30	1.9	89	78	67	56	46	36	26	16	6					
35	2.4	91	81	72	63	54	45	36	27	19	10				
40	2.8	92	83	75	68	60	52	45	37	29	22				
45	3.4	93	86	78	71	64	57	51	44	38	31				
50	4.1	93	87	80	74	67	61	55	49	43	38	10			
55	4.8	94	88	82	76	70	65	59	54	49	43	19			
60	5.7	94	89	83	78	73	68	63	58	53	48	26	5		
65	6.8	95	90	85	80	75	70	66	61	56	52	31	12		
70	7.8	95	90	86	81	77	72	68	64	59	55	36	19	3	
75	9.4	96	91	86	82	78	74	70	66	62	58	40	24	9	
80	10.9	96	91	87	83	79	75	72	68	64	61	44	29	15	3
85	12.7	96	92	88	84	80	76	73	69	66	62	46	32	20	8
90	14.8	96	92	89	85	81	78	74	71	68	65	49	36	24	13
95	17.1	96	93	89	85	82	79	75	72	69	66	51	38	27	17
100	19.8	96	93	89	86	83	80	77	73	70	68	54	41	30	21
105	23.4	97	93	90	87	83	80	77	74	71	69	55	43	33	23
110	26.0	97	93	90	87	84	81	78	75	73	70	57	46	36	26

*To use table, determine air temperature with dry-bulb thermometer and degrees depressed on wet-bulb thermometer. Read maximum capacity directly. Read relative humidity (in percent) opposite and below these values.

Table 2 Dew Point (°F)*

Air Temp. (°F) (dry bulb)	Degrees Depression of Wet-Bulb Thermometer (°F)													
	1	*2*	*3*	*4*	*5*	*6*	*7*	*8*	*9*	*10*	*15*	*20*	*25*	*30*
25	22	19	15	10	5	−3	−15	−51						
30	27	25	21	18	14	8	2	−7	−25					
35	33	30	28	25	21	17	13	7	0	−11				
40	38	35	33	30	28	25	21	18	13	7				
45	43	41	38	36	34	31	28	25	22	18				
50	48	46	44	42	40	37	34	32	29	26	0			
55	53	51	50	48	45	43	41	38	36	33	15			
60	58	57	55	53	51	49	47	45	43	40	25	−8		
65	63	62	60	59	57	55	53	51	49	47	34	14		
70	69	67	65	64	62	61	59	57	55	53	42	26	−11	
75	74	72	71	69	68	66	64	63	61	59	49	36	15	
80	79	77	76	74	73	72	70	68	67	65	56	44	28	−7
85	84	82	81	80	78	77	75	74	72	71	62	52	39	19
90	89	87	86	85	83	82	81	79	78	76	69	59	48	32
95	94	93	91	90	89	87	86	85	83	82	74	66	56	43
100	99	98	96	95	94	93	91	90	89	87	80	72	63	52
105	104	103	101	100	99	98	96	95	94	93	86	78	70	61
110	109	108	106	105	104	103	102	100	99	98	91	84	77	68

*To use table, determine air temperature with dry-bulb thermometer and degrees depressed on wet-bulb thermometer. Find dew point opposite and below these values.

Glossary

Abyssal plain a large flat area on the ocean floor where layers of sediment have covered the original sea floor topography.

Acceleration the change in velocity divided by the change in time: $a = \Delta v/\Delta t$.

Acceleration of gravity usually given as the symbol g; equal to 32 ft/s^2, 980 cm/s^2, or 9.8 m/s^2.

Accuracy refers to how close the measurement comes to the exact or true value.

Acid a substance which acts as a proton donor.

Acid mine drainage water drainage from mining areas that contain sulfuric acid due to the reaction in the air (oxygen) and water with sulfur-bearing minerals.

Acid rain rain that has a relatively low *p*H (acid) due to air pollution.

Actinides the 14 elements following actinium in the periodic table—the second inner transition series.

Activated sludge process a sewage treatment process in which aerated sewage effluent is mixed with bacteria-laden sludge for the removal of organic matter.

Activation energy the energy necessary to get a chemical reaction started.

Adiabatic a process in which no heat is added or removed from the system.

Advection fog a fog formed as a result of convectional air movement.

Air current vertical air movement.

Air-fuel ratio the ratio of air to fuel mixture that is important in the operation of the gasoline internal combustion engine.

Air mass a mass of air with physical characteristics that distinguish it from other air.

Air pollution any atypical contributions to the atmosphere resulting from human activities.

Air pollution potential weather conditions that are favorable for potential pollution conditions, e.g., stagnation conditions as a result of a temperature inversion.

Albedo the reflectivity or average fraction of light a body reflects.

Alcohol any compound that contains an —OH group and a hydrocarbon group.

Aldehydes any compound that contains a —C(H)=O group.

Alkanes hydrcarbons which have a composition that satisfies the general formula C_nH_{2n+2}.

Alkenes a hydrocarbon containing a double bond. The general formula for the alkenes is C_nH_{2n}.

Alkynes a hydrocarbon containing a triple bond. The general formula for the alkynes is C_nH_{2n-2}.

Alpha decay the disintegration of a nucleus into an alpha particle ($^4_2$He nucleus) and the nucleus of another element.

Alpha particle the nucleus of a helium-4 atom ($^4_2$He).

Altitude the angle measured from the horizon to a celestial object.

Alto the prefix associated with the middle cloud family.

Alveoli tiny air spaces, or sacs, in the lungs.

Amino acids compounds that contain both $—NH_2$ and —COOH groups.

Ampere the unit of electric current defined as that current which, if maintained in each of two long parallel wires separated by one meter in free space, would produce a magnetic force between the two wires of 2×10^{-7} newton for each meter of length.

Amplitude the maximum displacement of a wave from its equilibrium position.

amu one atomic mass unit—$\frac{1}{12}$ the mass of the ^{12}C isotope.

Anemometer an instrument used to measure wind speed.

Angstrom a unit of length equal to 10^{-8} cm.

Angular momentum *mvr* for a mass *m* going at a speed *v* in a circle of radius *r*.

Anion a negatively charged ion.

Annular eclipse an eclipse of the Sun in which the moon blocks out all of the Sun except for a ring around the outer edge of the Sun.

Anticyclone a high-pressure area characterized by clockwise air circulation (in the northern hemisphere).

Aquifer a body of permeable rock through which ground water moves.

Artesian wells and springs wells and springs in which water rises above an aquifer. The pressure due to gravity can cause the water to bubble or spurt onto the surface.

Asteroids large chunks of matter that orbit the Sun (usually between Mars and Jupiter) and which are too small to be labeled as planets.

Asthenosphere the rocky substratum below the lithosphere that is hot enough to be deformed and capable of internal flow.

Astronomical unit the mean distance between the Earth and the Sun, 93,000,000 mi.

Atmosphere the gases that surround a planet or celestial object.

Atmospheric science the investigation of every aspect of the atmosphere, from the ground to the edge of outer space.

Atom the smallest particle of an element that can enter into a chemical combination.

Atomic number the number of protons in the atom.

Atomic time scale a geologic time scale based on radioactive dating.

Atomic weight the weight of an average atom of an element with reference to the ^{12}C isotope. ^{12}C is 12 exactly.

Autumnal equinox the time (near September 21) when the Sun's declination crosses the equator moving south (for the northern hemisphere).

Avogadro's law equal volumes of gases at the same temperature and pressure contain equal number of molecules.

Avogadro's number the number of molecules in a mole of gas, liquid or solid—6.02×10^{23} molecules per mole.

Barometer a device used to measure pressure, viz., atmospheric pressure.

Basaltic lava lava with a low silica content and low viscosity.

Base a substance that acts as a proton acceptor.

Batholith a large intrusive igneous rock formation that has an area of at least 40 mi^2.

Bedding the stratification of sedimentary rock formations.

Beta decay the disintegration of a nucleus into a beta particle (electron) and the nucleus of another element.

Beta particle an electron.

Big Bang Theory theory of the universe that states that the known universe was concentrated in a massive glob of extremely dense material which exploded approximately 15 billion years ago.

Binding energy total binding energy—the amount of energy necessary to completely separate the protons and neutrons of the nucleus of an atom.

Biodegradable capable of being broken down or degraded by bacterial action.

Black hole a very dense collapsed star from which no light can escape.

Blizzard a snow storm accompanied by high winds that whip the snow into blinding swirls and drifts.

Boiling point the temperature at which a substance changes from the liquid to the gas phase.

Boyle's gas law the volume of a perfect gas varies inversely as the absolute pressure, if the temperature remains constant.

British system the system of measurement used in the United States with the foot as the length standard, the pound as the weight standard, the second as the time standard, and the coulomb as the electric charge standard.

Btu the amount of heat required to raise one pound of water one degree Fahrenheit at normal atmospheric pressure.

Caldera a roughly circular, steep-walled depression formed as a result of the collapse of a volcanic chamber.

Calorie the amount of heat necessary to raise one gram of pure liquid water one degree Celsius at normal atmospheric pressure.

Cancer a condition of unregulated cell growth rate.

Carbohydrate compounds composed of carbon, hydrogen, and oxygen with the hydrogen-to-oxygen ratio usually 2 to 1; sugars and starches are typical.

Carcinogen a cancer-producing agent.

Catalyst a substance that changes the rate of a chemical reaction without undergoing a permanent change itself.

Cation a positively charged ion.

Ceilometer an instrument used to measure the heights of clouds by light reflection.

Celestial sphere the imaginary sphere on which all the stars seem to be.

Centripetal force a ''center-seeking'' force that causes an object to travel in a circle.

Cepheid variables stars that vary in magnitude with a fixed period of between 1 and 100 days.

cgs system the metric system that has the centimeter, gram, and the second as the standard units of length, mass, and time, respectively, and the coulomb as the standard of electric charge.

Chain reaction a self-propagating process in which product neutrons from one fission are used to induce fission in other nuclei.

Charles' gas law the volume of a perfect gas varies in direct proportion to the absolute temperature, if the pressure remains constant.

Chemical properties the properties involved in the transformation of one substance into another.

Chemical sedimentary rocks rocks formed by the precipitation of minerals dissolved in water.

Chinook the name applied (literally, snow eater) to the warm leeward winds on the eastern slopes of the Rocky Mountains that give rise to the rapid melting of snow. In Europe the term *föhn* is used.

Chromosphere an outer layer of the Sun which lies just outside the photosphere.

Cilia tiny, hairlike projections in the respiratory tract.

Cinder cone a volcano with a steeply sloped cinder cone formed by eruptions of pyroclastic debris.

Cirque glacier a small glacier formed in a hollow depression along a mountain.

Cirrus a root name used to describe wispy, fibrous cloud forms.

Clastic sedimentary rocks rocks formed from sediment composed of fragments of preexisting rocks.

Cleavage the splitting of a mineral along an internal molecular plane.

Climate the long-term average weather conditions of a region or the world.

Cloud buoyant masses of visible droplets of water vapor and ice crystals in the lower troposphere.

Coalescence the combining of small droplets of water vapor to make larger drops.

Cold front the boundary of an advancing cold mass over a warmer surface.

Column the cavern dripstone formation consisting of a joined stalactite and stalagmite.

Combined sewer system a system that uses the same network as both a sanitary and storm sewer.

Comet a chunk of matter that displays a long tail as it passes near the Sun and has a highly elliptical orbit.

Composite volcano a volcano with steeply sloping symmetric cone formed by eruptions of high viscosity lava and pyroclastic debris. Also called a stratovolcano.

Compound a substance composed of two or more elements chemically combined in a definite proportion.

Concave mirror a mirror shaped like the inside of a small section of a sphere.

Concept a meaningful idea that can be used to describe and explain phenomena.

Conduction the transfer of heat energy by molecular transfer.

Conductor a material that easily conducts an electric current because some electrons in the material are free to move.

Conjunction the time at which a planet and sun occur on the same meridian.

Conservation of angular momentum, law of the angular momentum of an object remains costant unless acted upon by an external torque.

Conservation of energy, law of the total energy of an isolated system remains constant.

Conservation of linear momentum, law of the total linear momentum of a system remains constant if there are no external unbalanced forces acting on the system.

Conservation of mass, law of there is no detectable change in the total mass during a chemical process.

Consolidation the process of forming sedimentary rock from sediment. Also called lithification.

Continental drift the theory that continents move or drift apart.

Continental glacier (ice sheet) a large mass of ice that covers a large surface region and flows outwardly.

Continental shelf the shallowly submerged margin of a continental land mass.

Continental slope the seaward slope beyond the continental shelf that extends downward to the ocean basin.

Convection the transfer of heat through the movement of a substance.

Convection cycle the cyclic movement of matter, e.g., air, due to localized heating and convectional heat transfer.

Convex mirror a mirror shaped like the outside of a small section of a sphere.

Core the innermost region of the Earth, which is composed of two parts, a solid inner core and a molten, highly viscous, ''liquid'' outer core.

Coriolis force a pseudoforce arising in an accelerated reference frame on the rotating (accelerating) Earth. The apparent deflection of objects is attributed to the Coriolis force.

Coulomb the unit of electric charge equal to one ampere-second ($A \cdot s$).

Coulomb's law the force of attraction or repulsion between two charged bodies is directly proportional to the product of the two charges and inversely proportional to the square of the distance between them.

Covalent bond a chemical bond in which electron pairs are shared by atoms.

Covalent compounds compounds formed by the electron-sharing process.

Crater the funnel-shaped depression of the summit of a volcano.

Creep a type of slow mass wasting involving the slow particle-by-particle movement of weathered debris down a slope which takes place year after year.

CRT a cathode ray tube, or picture tube. It consists of an electron beam that is deflected by electric or magnetic fields and strikes a phosphorescent screen to form a picture.

Crust the thin outer layer of the Earth.

Crystal an orderly arrangement of atoms.

Cumulus a root name used to describe billowy, round cloud forms.

Curie a unit of radioactivity from a radioactive source. One curie is arbitarily defined as 3.7×10^{10} disintegrations per second.

Curie temperature a high temperature above which ferromagnetic materials cease to be magnetic.

Current rate of flow of electric charge.

Cyclone a low-pressure area characterized by counterclockwise air circulation (in the northern hemisphere).

Declination the overhead position of a celestial object such as the Sun measured in degrees latitude.

Definite proportions, law of different samples of a pure compound always contain the same elements in the same proportions by weight.

Degree a unit of angle. There are 360 degrees in a circle.

Density current a deep ocean current that is due to dense water sinking through less dense water.

Detergent a general term for a cleansing agent.

Detergent builders substances added to a detergent to make its cleansing action more efficient. Phosphates are common detergent builders.

Deuteron the nucleus of a deuterium atom (^2_1H).

Dew point the temperature at which a sample of air becomes saturated, i.e., has a relative humidity of 100 percent.

Diastrophism a geologic term meaning the movement of the Earth's crust.

Diffraction the bending of waves when an opening or obstacle has a size smaller than or equal to the wave length.

Dike a discordant pluton formation that is formed when magma fills a nearly vertical fracture in rock layers.

Diode an electronic device that allows current to flow in only one direction.

Dispersion the fact that different frequencies are refracted at slightly different angles.

DNA deoxyribonucleic acid, a nucleic acid located primarily in the nucleus of a cell. DNA has the molecular structure of a double-stranded helix.

Doldrums the low-pressure region near the equator.

Doppler effect an apparent change in frequency resulting from the relative motion of the source or the observer.

Drift glacial deposits of eroded material that are deposited either by ice or meltwater.

Dust dome a concentration of dust over a city due to the self-contained thermal circulation cell set up as a result of the heat-island effect.

Dyne a unit of force, 1 $g \cdot cm/s^2$.

Earthquake the sudden release or transfer of energy resulting from sudden movement due to stresses in the Earth's lithosphere.

Ecliptic plane the plane of the Earth's orbit around the Sun.

Electric charge a fundamental property of matter that can be either positive or negative and gives rise to electrical forces.

Electric field a set of imaginary lines that indicate the direction of a small positive charge would move if it were placed at a particular spot.

Electricity the effects produced by moving charges.

Electrolysis the production of an oxidation-reduction reaction by means of an electric current.

Electrolyte a compound in the molten or dissolved form that conducts an electric current.

Eletromagnet a current-carrying coil of insulated wire wrapped around a piece of soft iron that creates a magnetic field inside the iron only when the wire conducts a current.

Electromagnetic wave a wave caused by oscillations of electric and magnetic fields.

Electromagnetism the interaction of electric and magnetic effects.

Electromotive series a list of standard oxidation potentials at a given temperature.

Electron an elementary subatomic particle, with a very small mass of 9.01×10^{-31} kilograms and a negative charge of 1.602×10^{-19} coulombs, that goes in orbit around the atomic nucleus.

Electronegativity the measure of the ability of an atom to attract electrons in the presence of another atom.

Electron period a set of energy levels all of which have approximately the same energy.

Electron shell consists of all the orbits of electrons with the same principal quantum number (n).

Electron subshell similar to an energy level, all electrons in the same subshell have the same energy.

Electron volt the amount of kinetic energy an electron (or proton) acquires when it is accelerated through an electric potential of one volt.

Element a substance that has the same number of protons in all of its atoms.

Emphysema a lung condition characterized by large air sacs due to the breakdown of the walls of the alveoli.

Endothermic reaction a reaction where energy is absorbed as in the melting of ice.

Endoergic a reaction in which energy is absorbed.

Energy the capacity to do work.

Entropy a measure of the disorder of a system.

Enzymes organic substances of high molecular weight that catalyze reactions in living organisms.

Epicenter the point on the surface of the Earth directly above the focus of an earthquake.

Epoch an interval of geologic time that is a subdivision of a period.

Era an interval of geologic time made up of periods and epochs.

Erg a unit of energy, 1 $dyne \cdot cm$ or 1 $g \cdot cm^2/s^2$.

Erosion the group of interrelated processes by which rock is broken down and the products removed.

Ester any compound that conforms to the general formula $R{-}\overset{\displaystyle O}{\overset{\|}{C}}{-}O{-}R'$.

Eutrophication the natural aging process of lakes that is accelerated by phophate pollution.

Event horizon the position in space where the escape velocity from a black hole equals the speed of light.

Excited state a state of the atom with energies above the ground state. (*See* ground state.)

Exoergic a reaction in which energy is released.

Exothermic reaction a reaction in which energy is released, as in the burning of a candle.

Fault a fracture along which a relative displacement of the sides has occurred.

Fault-block mountains mountains that were built by normal faulting in which giant pieces of the Earth's crust were uplifted.

Faulting the relative motion along a fracture of fault that results in the displacement of rock masses on one side of the fracture relative to those of the other side.

Felsitic lava lava with a high silica content and high viscosity.

First quarter moon the phase of the moon between the new and full moon in which an observer in the United States sees the right half of the moon bright and overhead at 6 P.M. local solar time.

Fission the splitting of the nucleus of an atom into two nuclei of approximately equal size with the release of energy.

Focal length the point at which light rays from a distant source will converge after being reflected from a mirror or refracted by a lens.

Focus (earthquake) the point within the Earth where the initial energy release of an earthquake occurs.

Fog a maze of visible droplets of water vapor near the Earth's surface.

Föhn the name applied to the warm leeward winds on mountain slopes that give rise to the rapid melting of snow. In the Rocky Mountain region, the term *chinook* is used.

Folded mountains mountains characterized by folded rock strata, wiht external evidence of faulting and central evidence of igneous metamorphic activity. Folded mountains are believed to be formed at covergent plate boundaries.

Foliation the orientation characteristic of some metamorphic rocks due to formational directional pressures.

Force any quantity capable of producing motion.

Formula weight the sum of the atomic weights given in the formula of the compound. If the formula is the molecular formula, the formula weight is also its molecular weight.

Fossil fuels fuels of organic origin, viz., coal, gas, and oil.

Foucault pendulum a pendulum with a very long length that swings for a long period of time and demonstrates the rotation of the Earth.

Frequency the number of oscillations of a wave in a given period of time.

Front the boundary between two air masses.

Frost heaving the uplifting of rock and soil due to accumulations of freezing water.

Frost wedging a form of physical weathering by which rocks are pushed apart and fractured due to the pressure of freezing water within the rocks.

Full moon the phase of the moon that occurs when the moon is on the opposite side of the Earth from the Sun.

Fundamental properties the four fundamental or basic properties of nature are length, time, mass, and electric charge.

Fusion the transmutation of lightweight atomic nuclei into heavier nuclei with the release of energy.

G the universal gravitational constant—$G = 6.67 \times 10^{-11}$ $N \cdot m^2/kg^2$.

Galaxy an extremely large collection of stars occupying an extremely large volume of space.

Gamma decay the emission of electromagnetic energy from the nucleus of an atom. The atomic number, mass number, and neutron number of the atom remain the same, but the nucleus decreases in energy.

Gamma ray high-energy electromagnetic radiation with wavelength ranging from 3×10^{-14} to 3×10^{-12} m.

Gas matter that has no definite volume or shape.

Gay-Lussac's law when gases combine to form new gaseous compounds, all at the same temperature and pressure, the volumes of the initial and final gases are in the ratio of small whole numbers.

General theory of relativity a theory of relativity true for systems that are accelerating with respect to one another.

Generator a device that converts mechanical work or energy into electrical energy.

Geocentric theory the old false theory of the solar system which placed the Earth at the center of the universe.

Geologic time scale a relative time scale based on the fossil index of rock strata.

Geology the study of the Earth, its processes and history.

Geosyncline long, narrow ocean troughs containing large accumulations of sediment.

Geothermal gradient the increase of temperature with depth that occurs in the outer portion of the Earth's crust. The rate of increase is about 1°F per 150 feet of depth.

Geyser a surface opening through which steam and boiling water erupt intermittently.

Glacier a large ice mass, consisting of recrystallized snow, that flows on a land surface under the influence of gravity.

Graded river the condition of a river when its erosion and transport capabilities are in balance.

Gram a unit of mass in the metric cgs system of units. One gram is the mass of one cubic centimeter of pure water at its maximum density. (4°C)

Gram atomic weight the weight of an atom of an element relative to ^{12}C, expressed in grams.

Gravitational collapse the collapse of a very massive body because of its attraction for itself.

Greenhouse effect the heat-retaining process of atmospheric gases, viz., water vapor and CO_2, due to the selective absorption of long-wavelength terrestrial radiation.

Greenwich meridian the meridian of half-circle on the Earth's surface which has zero longitude.

Ground state the lowest energy level of an atom.

Group a number of elements appearing in any one column of the periodic table.

Guyot a seamount with a flat top.

Half-life the time required for half of any sample of radioactive element to disintegrate.

Hard water water containing dissolved calcium and magnesium salts.

Heat a form of energy; energy in transit.

Heat capacity the amount of heat energy in calories required to raise the temperature of a substance one degree Celsius.

Heat engine a device that uses heat energy to perform useful work.

Heat-island effect the condition of the temperature and heat content of a city being higher than the surrounding rural areas due to greater radiation absorption, etc.

Heat lightning lightning occurring below the horizon or behind a cloud which illuminates the cloud with flickering flashes of light.

Heat pump a device used to transfer heat from a low-temperature reservoir to a high-temperature reservoir.

Heisenberg uncertainty principle it is impossible to know simultaneously the exact velocity and position of a particle.

Heliocentric theory the current theory of the solar system which places the Sun in the center.

Hertz one cycle per second.

Heterogeneous matter that is of nonuniform composition.

Heterosphere a region of the atmosphere based on the heterogeneity of the atmospheric gases between approximately 60 and several hundred miles in altitude.

Homogeneous matter that is of uniform composition throughout.

Homosphere a region of the atmosphere based on the homogeneity of the atmospheric gases between approximately 0 to 60 miles in altitude.

Horse latitudes the high-pressure region near 30° latitude.

Horsepower a unit of power, 550 ft·lb/s.

Hot spot theory the theory that explains central plate volcanic chains as being formed as a result of plate movement over a hot spot beneath it.

H-R diagram a plot of absolute magnitude versus temperature of stars.

Hubble's law the recessional speed of a distant galaxy is proportional to its distance away.

Humidity a measure of the water vapor in the air.

Hurricane a tropical storm with winds of 75 mi/h or greater.

Hurricane warning an alert that hurricane conditions are expected within 24 hours.

Hurricane watch an advisory alert that hurricane conditions are a definite possibility.

Hydrocarbons compounds composed of carbon and hydrogen, e.g., methane, CH_4.

Hydrologic cycle the cyclic movement of the Earth's water supply from the oceans to the mountains and back again to the oceans.

Hydronium ion H_3O^+

Hygroscopic nuclei particulate matter that acts as nuclei in the condensation process.

Ice point the temperature of a mixture of ice and air-saturated water at normal atmospheric pressure.

Igneous rock rock formed by the cooling and solidification of hot, molten material.

Incomplete combustion combustion in which the carbon element of the fuel is not completely reacted to form CO_2, but rather CO.

Index of refraction the ratio of the speed of light in a vacuum and the speed of light in a medium.

Indian summer a period of warm weather in the late fall.

Inertia the property of matter to resist any change of motion.

Inner planets the four planets closest to the Sun (Mercury, Venus, Earth, and Mars).

Insolation the solar radiation received by the Earth and its atmosphere—*in*coming *so*lar radi*ation*.

Insulator a material that does not conduct an electric current because the electrons in the material are not free to move.

International date line the meridian that is 180° E or W of the prime meridian.

Inter-transition elements elements in which the third shell from the outer shell is increasing from 18 to 32 electrons.

Ion an atom or group of atoms with a net electric charge.

Ionic bond a chemical bond in which electrons have been trandferred from atoms of low ionization potential to atoms of high electron affinity.

Ionic compounds compounds that are formed by the process of electron transfer.

Ionization potential the electrical voltage required to remove an electron from an atom.

Ionosphere a region of the atmosphere characterized by ion concentrations between 50 to several hundred miles in altitude.

Isobar a line drawn through points of equal pressure.

Isomers molecules with the same molecular formula but a different structure or arrangement of atoms, hence slightly different physical properties.

Isostasy the concept that the Earth's crustal material ''floats'' in gravitational equilibrium on a ''fluid'' substratum.

Isothermal a constant-temperature process.

Isotope atoms whose nuclei have the same number of protons but different number of neutrons.

Jet streams fast-moving ''rivers'' of air in the upper troposphere.

Joule a unit of energy, 1 N·m or $kg{\cdot}m^2/s^2$.

Kepler's harmonic law the ratio of the square of the period to the cube of the semimajor axis (one-half the larger axis of an ellipse), is the same for all the planets.

Kepler's law of elliptical paths all planets (asteroids, comets, etc.) go in elliptical orbits around the Sun.

Kepler's law of equal areas as a planet (or asteroid or comet) goes around the Sun, an imaginary line joining the planet to the Sun sweeps out equal areas in equal periods of time.

Kilo prefix that means 10^3 or one thousand.

Kilogram the unit of mass in the mks system. One kilogram has an equivalent weight of 2.2 pounds.

Kinetic energy energy of motion equal to $\frac{1}{2}mv^2$.

Laccolith a concordant pluton formation formed from a blisterlike intrusion that has pushed up the overlying rock layers.

Landslide a type of fast mass wasting that involves the downslope movement of large blocks of weathered material.

Laser an acronym for "light amplification by stimulated emission of radiation." It is coherent, monochromatic light.

Lathanides the 14 elements following lanthanum in the periodic table—the first inner-transition series.

Lapse rate the rate of temperature change with altitude. In the troposphere the normal lapse rate is $-3\frac{1}{2}$°F per 1000 ft.

Last quarter moon (or third quarter moon) the phase of the moon between the full and new moon in which an observer in the United States sees the left half of the moon bright and overhead at 6 A.M. local solar time.

Latent heat of fusion the amount of heat required to change one gram of a substance from the solid to the liquid phase at the same temperature.

Latent heat of vaporization the amount of heat required to change one gram of a substance from the liquid to the gas phase at the same temperature.

Latitude for a point on the surface of the Earth, the angular measurement, in degrees, north or south of the equator.

Lava magma that reaches the Earth's surface through a volcanic vent.

Length the measurement of space in any direction.

Leukemia a cancerous condition characterized by an abnormal increase in the white cells (leucocytes) of the blood.

Lightning an electric discharge in the atmosphere.

Light-year the distance light travels in one year.

Linear momentum mass × velocity.

Linearly polarized transverse waves that only vibrate in one direction.

Line squalls a series of storms along a front.

Lipid a general term that includes such substances as fats, oils, and waxes.

Liquid matter that has a definite volume but no definite shape.

Lithification the process of forming sedimentary rock from sediment. Also called consolidation.

Lithosphere the outermost solid portion of the Earth, which includes the crust and part of the upper mantle.

Local Group the cluster of galaxies including our own Milky Way galaxy.

Longitude for a point on the surface of the Earth, the angular measurement, in degrees, east or west of the prime meridian.

Longitudinal wave a wave in which the vibrations are in the same direction as the wave velocity.

Long shore current a current along a shore due to the waves that break at an angle to the shore line.

Lorentz-Fitzgerald contraction the length of a moving object appears to be shorter to an observer who is at rest than to an observer moving with the object.

Lunar eclipse an eclipse of the moon caused by the Earth's blocking of the Sun's rays to the moon.

Magma hot, molten rock material.

Magnetic anomalies adjacent regions of rocks with remanent magnetism of opposite polarities. That is, the directions of the magnetism are reserved.

Magnetic declination the angular variation of a compass from geographic north.

Magnetic field a set of imaginary lines that indicate the direction a small compass needle would point if it were placed at a particular spot.

Magnetic monopole a single magnetic north or south pole without the other—as yet undiscovered.

Main sequence a narrow band on the H-R diagram on which most stars fall.

Mantle the interior region of the Earth between the core and the crust.

Maria the large dark areas on the moon.

Mass a quantity of matter and a measurement of the amount of inertia that a body possesses.

Mass number the number of protons plus the number of neutrons in an atom.

Mass wasting the downslope movement of overburden under the influence of gravity.

Matter anything that exists in time, occupies space, and has mass.

Matter waves the waves produced by moving particles.

Meandering the looping, ribbonlike path of a river channel due to accumulated deposits of eroded material that diverts the stream flow.

Mean solar day the average length of a solar day. One solar day is the elapsed time between two successive crossings of the same meridian by the Sun.

Measurement the comparison of an unknown quantity to a standard.

Mega prefix that means 10^6 or one million.

Mercalli scale a scale of earthquake severity based on the physical effects produced by an earthquake.

Mesosphere a region of the atmosphere based on temperature between approximately 35 to 60 miles in altitude.

Metal an element that tends to lose its valence electrons.

Metalloid an element that exhibits the properties of both metals and nonmetals.

Melting point the temperature of a substance at which the substance changes from a solid to a liquid.

Metamorphic rock rock that results from a change or metamorphism in pre-existing rock due to heat and pressure.

Meteor a small chunk of matter that burns up as it goes through the Earth's atmosphere and appears to be a shooting star.

Meteorites chunks of matter from the solar system that fall through the atmosphere and strike the Earth's surface.

Meteorology the study of atmospheric phenomena.

Meter the standard of length in the mks system. It is equal to 39.37 inches or 3.28 feet.

MeV million electron volts, a unit of energy.

Micro prefix that means 10^{-6} or one one-millionth.

Milky Way the name of our galaxy.

Milli prefix that means 10^{-3} or one one-thousandth.

Millibar a unit of pressure. 1 mb = 10^3 dyne/cm^2.

Mineral any naturally occurring, inorganic, crystalline substances.

Mixture a nonchemical combination of two or more substances of varying proportions.

mks system the metric system that has the meter, kilogram, and second as the standard units of length, mass, and time and the coulomb as the standard of electric charge.

Moho discontinuity the boundary between the Earth's crust and mantle.

Moh's scale a 10-point scale of mineral hardness based on minerals standards—diamond being the hardest and talc being the softest.

Molarity the number of moles of solute in one liter of solution. Molarity is designated by the letter *M*.

Mole a gram formula weight.

Molecule an uncharged particle of an element or compound.

Molecular weight the weight of one molecule of a substance relative to ^{12}C, expressed in grams.

Monomer the fundamental repeating unit of a long chain molecule.

Monsoons winds associated with seasonal convectional cycles set up between continents and oceans.

Motion the changing of position.

Moraine a ridge of glacial till.

Motor a device that converts electrical energy into mechanical energy.

Mountain range a geologic unit or series of mountains.

Mountain system a group of similar mountain ranges.

Mudflow the movement of large masses of soil that have accumulated on steep slopes and become unstable with the absorption of large quantities of water from melting snows and heavy rains.

Multiple proportions, law of whenever two elements, say E_1 and E_2, combine to form different compounds, the various amounts of E_2 combined with the same amount of E_1 are in a ratio of small whole numbers.

Muons mu mesons, which are particles with properties like the electron but are more massive than the electron—1 mu meson mass = 207 electron masses.

National Meteorological Center the nerve center of the National Weather Service where weather data are received and processed and forecasts are made.

National Weather Service a federal organization under NOAA that provides national weather information.

Neap tide moderate tides with the least variation between high and low.

Nephanalysis a satellite cloud-cover photograph with a map grid.

Neutralization the mutual disappearance of the H^+ and the OH^- ions. They combine to form water (HOH).

Neutrino a subatomic particle that has no rest mass or electric charge but does possess energy and momentum.

Neutron number the number of neutrons in the atom.

Neutrons elementary particles that have approximately the same mass (1.6×10^{-27} kg) as protons but have no charge. They are one constituent of the atomic nucleus.

New moon the phase of the moon that occurs when the moon is between the Earth and the Sun.

Newton 1 kg × m/s^2.

Newton's first law of motion a body will move at a constant velocity unless acted upon by an external unbalanced force.

Newton's law of universal gravitation $F = Gm_1m_2/r^2$.

Newton's second law of motion the acceleration of an object is equal to the unbalanced force on the object divided by the mass of the object ($a = F/m$).

Newton's third law of motion whenever one mass exerts a force upon a second mass, the second mass exerts an equal and opposite force upon the first mass.

NOAA the National Oceanic and Atmospheric Administration of the U.S. Department of Commerce, which is responsible for the study and monitoring of oceanic and atmospheric phenomena.

Node a point where the moon's path crosses the ecliptic.

Nonmetal an element that tends to gain electrons to complete its outer shell.

Normal faulting movement along a nonvertical fault in which the overlying side moves downward relative to the side beneath it.

Novae stars that suddenly increase dramatically in brightness for a brief period of time.

Nucleic acid very high molecular weight polymers that are present in living cells.

Nucleus the central core of the atom.

Occluded front a front that is occluded, or closed off, from the Earth's surface due to forced ascension.

Octet rule the tendency of atoms to have eight electrons in their outer shell when forming molecules or ions. There are many exceptions to the rule.
Ohm the unit of resistance equal to one volt per ampere.
Ohm's law the voltage across two points is equal to the current flowing between the points times the resistance between the points.
Opposition the time at which a planet is on the opposite side of the Earth from the Sun.
Organic acid any compound that contains the molecular

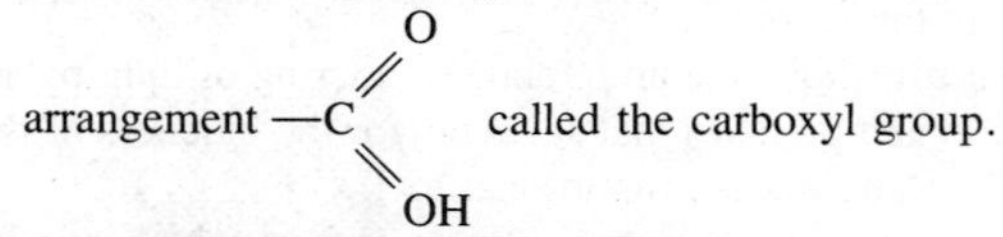
O
arrangement —C
called the carboxyl group.
OH

Organic compounds compounds of carbon.
Outer planets the five planets farthest from the Sun (Jupiter, Saturn, Uranus, Neptune, and Pluto).
Overburden the weathered material that accumulates on base rock.
Oxidation the process in which electrons are lost.
Oxidation number a positive or negative number assigned to an atom as a measure of the electric charge that an atom in a molecule would have if it were ionically bound.
Oxidation state a term used to indicate the oxidation number.
Oxide a compound containing oxygen and one or more other elements.
Ozone the compound O_3. It is found naturally in the atmosphere in the ozonosphere and is also a costituent of photochemical smog,
Ozonosphere a region of the atmosphere characterized by an ozone concetration between approximately 0 to 50 miles in altitude.

Pangaea the single giant supercontinent that is believed to have existed over 200 million years ago.
Parhelia two bright-colored patches that appear on each side of the Sun due to scattering by the ice crystals of cirrostratus clouds. (Sometimes called side suns.)
Parallax the apparent motion, or shift, that occurs between two fixed objects when the observer changes position.
Parsec the distance to a star where the star exhibits a parallax of one second. This distance is equal to 3.26 light years or 206,265 astronomical units.
Pauli exclusion principle no two electrons can have the same set of quantum numbers.
Penumbra a region of partial shadow. During an eclipse, an observer in the penumbra sees only a partial eclipse.
Perfect gas law the relationship that exists between the volume, absolute pressure, and temperature of a gas. $PV/T = K$.
Period an interval of geologic time that is a subdivision of an era and made up of epochs; the time for a complete cycle of motion; a horizontal row of the periodic table, elements that have approximately the same energy.
Periodic law the properties of elements are periodic functions of the atomic number.
Permafrost ground that is permanently frozen.
Permeability a measure of a material's capacity to transmit fluids.
Pesticides chemicals used to kill unwanted insect and plant species.
***p*H** the exponent of the negative power to which 10 is raised when used to express H^+ ion concentration.
Photochemical smog air pollution conditions resulting from the photochemical reactions of hydrocarbons with oxygen in the air and other pollutants in the presence of sunlight.
Photon a ''particle'' of electromagnetic radiation.
Photosphere the Sun's outer surface visible to the eye.
Photosynthesis the process by which plants convert CO_2 and H_2O to sugars with the release of oxygen.
Physical properties those that do not involve a change in the chemical composition of the substance.
Pi(π) the ratio of the circumference of a circle to its diameter. It is equal to 3.14159. . . .
Pitch the highness or lowness of a sound. It is a consequence of the frequency of the sound waves received by the ear.
Planck's constant a constant of proportionality relating the energy and frequency of a photon. The constant has the value of 6.225×10^{-27} erg·s.
Plasma matter that is a disordered population of ionized atoms where the ion cores and the free electrons are not in thermal equilibrium—a state in which the temperature is so high that electrons and atomic nuclei are separated and move very rapidly, constantly colliding with each other.
Plate a huge slab of rock that makes up a portion of the outer layer of the Earth and is in relative motion with respect to other plates.
Plate tectonics the theory that the outer layer of the Earth is made up of rigid plates that are in relative motion with respect to each other.
Pluton a large body of intrusive igneous rock.
Polar easterlies the prevailing winds in the latitudes from 60° to 90°.
Polymer the resulting chainlike structure produced by many monomers.
Porosity the percentage of unoccupied space in the total volume of a substance.
Porphyritic a rock texture characterized by a structure of coarse mineral grains scattered through a mixture of fine mineral grains.
Position the location of an object in respect to another object.
Potential energy the energy a body possesses due to its position in a force field.
Potential well a term used to indicate a negative potential energy.

Power work per unit time.
Precession the slow rotation of the axis of spin of the Earth around an axis perpendicular to the ecliptic plane.
Precision refers to the degree of reproducibility of a measurement, that is, to the maximum possible error of the measurement.
Pressure force per unit area.
Pressure gradient a variation of pressure with position.
Pressure trough a region with a pressure minimum.
Primary sewage treatment the basic method of sewage treatment consisting of four main steps—screening, grit removal, sedimentation, and chlorination.
Principal quantum number the numbers (1, 2, 3, . . .) used to designate the various principal energy levels that an electron may occupy in an atom.
Protein the complex combination of amino acids. Proteins are composed mainly of carbon, oxygen, nitrogen, and hydrogen.
Protons elementary particles that are 1836 times as massive (1.67×10^{-27} kg) as electrons and have a positive charge (1.6×10^{-19} coulomb). They are one constituent of the atomic nucleus.
Psychrometer an instrument used to measure relative humidity.
Pulsar a star that has a radio signal that pulses regularly and very rapidly.
P waves primary (P) waves, so-called because they reach a seismic station before the S waves. P waves are longitudinal or compressional waves, i.e., their particle oscillations are in the direction of propagation and are transmitted by solids, liquids, and gases.
Pyroclastic debris solid material emitted by volcanoes that range in size from fine dust to large boulders.

Quantum a discrete amount.
Quantum mechanics (or wave mechanics) a field of physics used to solve problems when geometrical sizes are comparable with the wavelengths of particles. Schrödinger's equations forms the basis of wave mechanics.
Quantum number a number assigned to one of the various values of a quantized quantity.
Quarks particles with fractional charges such as $+\frac{2}{3}$ or $-\frac{1}{3}$ that combine together (three at a time) to form protons and neutrons. So far, a single free quark has not been found.
Quasar a shortened term for "quasi-stellar radio source."

Radar an instrument that sends out electromagnetic (radio) waves, monitors the returning wave that is reflected by some object, and thereby locates the object. Radar stands for RAdio Detecting And Ranging. Radar is used to detect and monitor precipitation and severe storms.
Radian a unit of angle equal to 57.4°. There are 2π radians in a circle.
Radiation the transfer of energy by means of electromagnetic waves.
Radiation fog a fog formed as a result of radiative heat loss (sometimes called a valley fog).
Radioactive the spontaneous disintegration of certain atomic nuclei, generating one or more of three types of radiation: alpha, beta, and/or gamma.
Radiosonde a small package of meteorological instruments with a radio transmitter that is carried aloft by a balloon.
Rays streaks of light-colored material extending outward from craters on the moon.
Rayleigh scattering the preferential scattering of light by air molecules and particles that accounts for the blueness of the sky. The scattering is proportional to $1/\lambda^4$.
Real image an image from a mirror or lens which can be brought to focus on a screen.
Red giant a red star that has a much larger diameter than average.
Red shift a Doppler effect caused when a light source, such as a galaxy, moves away from the observer and shifts the light frequencies lower or toward the red.
Reduction the process in which electrons are gained.
Reflection the change in the direction of a wave due to a boundary.
Refraction the bending of light waves caused by a velocity change as light goes from one medium to another.
Remanent magnetism the magnetism retained in rocks containing ferrite minerals after solidifying in the Earth's magnetic field.
Representative elements elements in which the added electron enters the outermost shell in which the outermost shell is incomplete—the A groups in the periodic table.
Resistance the opposition to the flow of electric current.
Resonance a wave effect that occurs when an object has a natural frequency that corresponds to an external frequency.
Rest mass the mass of an object at zero velocity.
Retina the part of the human eye onto which light is focused by the lens of the eye.
Reverse faulting movement along a nonvertical fault in which the overlying side moves downward relative to the side beneath it.
Revolution the movement of one mass around another.
Richter scale a scale of earthquake severity based on the amplitude or intensity of seismic waves.
Rill a narrow trench or valley on the moon.
Ring of Fire the area generally circumscribing the Pacific Ocean that is characterized by volcanic activity.
RNA ribonucleic acid, a nucleic acid located primarily in the cytoplasm or outer structure of the cell. RNA has the structure of a double-stranded helix.
Rock any naturally occurring, solid, mineral mass that makes up part of the Earth's lithosphere.
Rock cycle the cyclic movement of rock, during which the

rock is created, destroyed, and metamorphized by the Earth's internal and external geologic processes.

Rockslide a type of landslide occurring in mountain areas when large quantities of rock break off and move rapidly down steep slopes.

Rotation a spinning motion.

Salinity a measure of the saltiness of water.

Salt an ionic compound containing the cation of a base and the anion of an acid.

Sanitary landfill a method of solid waste disposal in which refuse is covered with a layer of soil within a day after being deposited at the landfill site.

Saturated hydrocarbons contain only single bonds and their hydrogen content is at a maximum.

Saturated solution when the maximum amount of solute possible is dissolved in the solvent.

Scalar a quantity that has a magnitude and units but no direction is associated with it.

Sea floor spreading the theory that the sea floor slowly spreads and moves sideways away from mid-ocean ridges. The spreading is believed to be due to convection cycles of subterranean molten material that causes the formation of the ridges and a surface motion in a lateral direction from the ridges.

Seamount an isolated submarine volcanic structure.

Second the standard unit of time. It is now defined in terms of the frequency of a certain transition in the cesium atom.

Secondary sewage treatment the treatment that adds an additional step of either a trickling filter or an activated sludge process to the primary treatment for the removal of suspended organic matter.

Sedimentary rock rock formed from the consolidation of layers of sediment.

Seismic waves the waves generated by the energy release of an earthquake.

Seismograph an instrument that records the intensity of seismic waves.

Separated sewer system a system with separate networks for sanitation and storm water transportation.

Sheet erosion erosion resulting from runoff or the overland flow of water.

Shield volcano a volcano with a low, gently sloping profile formed by a fissure eruption of low-viscosity lava.

Silicon tetrahedron the pyramid-shaped structure of the complex silicate ion $(SiO_4)^{4-}$.

Sill a pluton formation that lies between and parallel to existing rock layers.

Silt a mixture of fine particles and organic material delivered by a stream.

Singularity the center of a black hole. The point where the entire mass of a star has contracted.

Slug a unit of mass, 1 $lb/ft/s^2$.

Slump a type of landslide involving the downslope movement of an unbroken block of overburden that leaves a curved depression on the slope.

Smog a contraction of smoke-fog used to describe the combination of these conditions.

Solar constant the average amount of solar energy received per area at the top of the atmosphere per time, 2.0 $cal/cm^2 \cdot min$.

Solar eclipse an eclipse of the Sun caused by the moon's blocking of the Sun's rays to an observer on Earth.

Solid matter that has a definite volume and a definite shape.

Solid waste any normally solid material resulting from human or animal activities that is useless or unwanted.

Solifluction (soil flow) a type of slow mass wasting involving the ''flow'' of weathered material over a solid, impermeable base, permafrost.

Solubility the amount of solute that will dissolve in a specified volume of solvent (at a given temperature) to produce a saturated solution.

Solute the dissolved substance in a solution.

Solution a homogeneous mixture.

Solvent the substance in excess in a solution.

Sound a wave phenomenon caused by variations in pressure in a medium such as air.

Source region the region or surface from which an air mass derives its physical characteristics.

Special theory of relativity a theory of relativity true for systems that are moving at a constant velocity with respect to one another.

Specific heat the amount of heat energy in calories necessary to raise the temperature of one gram of the substance one degree Celsius.

Spectroscope an instrument used to separate electromagnetic waves into their component wavelengths.

Spectrum an ordered arrangement of various frequencies or wavelengths of electromagnetic radiation.

Speed of light 186,000 mi/s or 3×10^{10} cm/s or 3×10^8 m/s.

Spring tide the tides of greatest variation between high and low.

Squall line a region along a front characterized by storms and turbulent weather.

SST an acronym for ''supersonic transport'' aircraft.

Stable layer a layer of air with uniform temperature and density.

Stalactite an icicle-shaped dripstone formation extending downward from a cavern roof.

Stalagmite a blunt icicle-shaped dripstone formation extending upward from a cavern floor.

Standard conditions 0°C and sea level atmospheric pressure.

Standard unit a fixed and reproducible reference value used for the purpose of taking accurate measurements.

Steam point the temperature at which pure water, at normal atmospheric pressure, boils.
Stratosphere a region of the atmosphere based on temperature between approximately 10 to 35 miles in altitude.
Stratovolcano a volcano with a steeply sloping symmetric cone formed by eruption of high viscosity lava and pyroclastic debris. A stratovolcano is also called a composite volcano.
Stratus a root name used to describe stratified of layered cloud forms.
Streak the color of a powdered mineral on a streak plate (unglazed porcelain).
Stream load the eroded material transported by a stream.
Strike-slip faulting movement along a horizontal transform fault in which the two sides of the fault strike and slip by each other.
Subduction the process in which one plate is deflected downward beneath another plate into the asthenosphere.
Substance a homogeneous sample of matter all specimens of which have identical properties and identical composition.
Summer solstice the farthest point of the Sun's declination north of the equator (for the northern hemisphere).
Sunspots patches of cooler, darker material on the surface of the Sun.
Supercooled same as supersaturated—see below.
Supernova an exploding star.
Supersaturated the condition of air when its temperature is lowered below the dew point without condensation.
Supersaturated solution a solution that contains more than the normal maximum amount of dissolved solute.
Surf the breaking of surface waves along a shore.
Surface current a broad drift of surface water in the ocean that is set in motion by prevailing winds.
Surface wave (oceanic) a wave on the surface of the ocean near a shore line.
Surface wave (seismic) a seismic wave that travels along the Earth's surface or a boundary within it.
Superposition, law of the principle that in a succession of stratified deposits the younger layers lie on the older layers.
S waves secondary (S) seismic waves, so-called because they reach a seismic station after the P waves. S waves are transverse or shear waves, i.e., their particle oscillations are at right angles to the direction of propagation, and are transmitted only by solids.
Synergism the combined effect of two or more pollutants.
Synoptic weather charts charts or maps that present a synopsis of weather data.

Tectonics the study of the Earth's general structural features and their changes.
Temperature a measure of the average kinetic energy of the molecules.
Temperature inversion a condition charcterized by an inverted lapse rate.
Ten-degree rule a 10-degree Celsius increase in temperature leads to a doubling of the chemical reaction rate.
Thermal circulation the cyclic movement of matter (e.g., air) due to localized heating and convectional heat transfer.
Thermal conductivity a measure of the ability of a substance to conduct heat energy.
Thermal pollution pollution resulting from warm water, chiefly from industrial cooling processes, that is discharged into natural waterways.
Thermodynamics, first law of the heat energy added to a system must go into increasing the internal energy of the system, or any work done by the system, or both. The law also states that heat energy removed from a system must produce a decrease in the internal energy of the system, or any work done on the system, or both. The law is based upon the law of conservation of energy.
Thermodynamics, second law of it is impossible for heat to flow spontaneously from an object having a lower temperature to an object having a higher temperature.
Thermodynamics, third law of a temperature of absolute zero can never be attained.
Thermosphere a region of the atmosphere based on temperature between approximately 60 to several hundred miles in altitude.
Thrust faulting a special case of reverse faulting in which the fault plane is at a small angle to the horizontal.
Thunder the sound associated with lightning that arises from the explosive release of electrical energy.
Tidal current a current due to the tidal movement of sea water.
Till eroded material that is transported and deposited by glacial ice rather than meltwater.
Time the continuous forward flowing of events.
Time dilation (the stretching out of time) the passage of time for a moving object appears to be longer to an observer who is at rest than to an observer moving with the object. All motions appear to be slower than to an observer at rest.
Tornado a violent storm characterized by a funnel-shaped cloud and high winds.
Tornado warning the alert issued when a tornado has actually been sighted or indicated on radar.
Tornado watch the alert issued when conditions are favorable for tornadoes.
Total internal reflection a phenomenon in which light is totally reflected because refraction is impossible.
Traction the transport of a stream's bed load as a result of rocks and particles being rolled and bounced along by the stream's current.
Trade winds the prevailing winds in the latitudes from 0° to 30°.

Transformer a device that increases or decreases the voltage of alternating current.
Transform fault a fault with a horizontal fault plane along which strike-slip faulting occurs.
Transistor an electronic device whose primary purpose is to amplify an input signal.
Transition elements elements in which the second shell from the outer shell is increasing from 8 to 18 electrons—the B groups in the periodic table.
Transverse wave a wave in which the vibrations are perpendicular to the wave velocity.
Trickling filter a bed of bacteria-laden stones through which sewage effluent trickles for the removal of organic matter.
Tropical year the time interval from one vernal equinox to the next vernal equinox.
Troposphere a region of the atmosphere based on temperature between the Earth's surface and 10 miles in altitude.

Umbra a region of total darkness in a shadow. During an eclipse, an observer in the umbra sees a total eclipse.
Uniformitarianism *see* Uniformity.
Uniformity, principle of the principle that the same processes operate today on and within the Earth as in the past. Hence, the present is considered the key to the past.
Unsaturated hydrocarbons compounds that can take on hydrogen atoms to form saturated hydrocarbons or atoms of substances other than hydrogen to form derivatives.

Valence the net electric charge of an atom or the number of electrons an atom can give up (or acquire) to achieve a filled outer shell.
Valence electrons the electrons that are involved in bond formation, usually the outermost electrons.
Valley glacier a glacier that flows downward creating a valley.
Vector a quantity that has not only a magnitude and units but also a direction associated with it.
Velocity, average the change in displacement divided by the change in time, $v = \Delta s/\Delta t$.
Velocity, instantaneous the velocity at a particular instant of time.
Vernal equinox the time (near March 21) when the Sun's declination crosses the equator moving north (for the northern hemisphere).
Virtual image an image from a lens or mirror that cannot be brought to focus on a screen.
Viscosity the internal property of a substance that offers resistance to flow.
Vitamins organic substances that are needed in minute amounts to perform specific functions for normal growth and nutritional needs of the human body.
Volcanic mountains mountains that have been built by volcanic eruptions.
Volt the unit of voltage equal to one joule per coulomb.
Voltage the amount of work it would take to move an electric charge between two points, divided by the value of the charge, i.e., work per unit charge.

Warm front the boundary of an advancing warm air mass over a colder surface.
Waterspout a tornado over water.
Water table the boundary between the zone of aeration and zone of saturation.
Watt a unit of power, 1 $kg{\cdot}m^2/s^3$ or 1 J/s.
Wave cyclone cyclonic or rotational disturbances along a front that are formed by air motion in opposite directions.
Wavelength the distance from any point on a wave to an identical point on the adjacent wave.
Wave motion the emanation of energy from the disturbance of matter.
Weather the atmospheric conditions of the lower troposphere.
Weathering the physical disintegration and chemical decomposition of rock.
Weight the force of gravity on the Earth's surface.
Westerlies the prevailing winds in the latitudes from 30° to 60°.
White dwarf a white star that has a much smaller diameter than average. It is believed to be the end stage of small and average mass stars.
Wind horizontal air motion.
Winter solstice the farthest point of the Sun's declination south of the equator (for the northern hemisphere).
Work the product of a force and the parallel distance through which it acts.

Zenith the positon directly overhead of an observer on Earth.
Zenith angle the angle between the zenith and the Sun at noon.
Zodiac a section of the sky extending around the ecliptic 8° above and below the ecliptic plane.
Zone of aeration the zone of rock and soil above the water table whose open spaces are filled mainly with air.
Zone of saturation the subsurface zone of rock and soil in which all the openings are completely filled or saturated with water.

Index

A

Abrasion, 564
Absolute magnitude of a star, 365
Absolute temperature, 68, 76
Absolute zero, 76
Abyssal plains, 579
Acceleration, 22–27
  centripetal, 26
  in circular motion, 25–27
  defined, 22
  due to gravity, 23, 38
Accuracy of measurement, 13
Acid, 264
  amino, 277
  conjugate, 266
  mine drainage, 588
  nucleic, 279
  organic, 275
  solution, 268
  strong, 266
  weak, 266
Acid-base
  Arrhenius concept, 265
  Brönsted-Lowry theory, 265
  reaction, 265
Acids and bases, 264–269
Actinides, 214
Activated sludge process, 603
Activation energy, 258
Adiabatic process, 72
Affinity, electron, 224
Air
  composition of, 385
  friction, 411
  fuel ratio, 489
  masses, 435–437
Air pollutants, 483–489
  aldehydes, 486
  carbon dioxide, 484
  carbon monoxide, 484
  carcinogens, 486
  chlorofluorocarbons, 498
  metals, 489, 496
  nitrogen oxides, 484, 486
  ozone, 486
  particles, 499
  sulfur dioxide, 488
Air pollution, 483–496
  cost of, 492
  potential, 467
  sources of, 489–491
  weather and climatic effects of, 496–500
Albedo, 392
Alcohols, 274
Aldehydes, 275
Alkali metals, 216
Alkaline earths, 218
Alkaline solution, 268
Alkaloid, 286
Alkanes, 234
Alkenes, 235
Alkyl, 276
Alkyl halides, 274
Alkynes, 236
Alpha Centauri, 372
Alpha decay, 164
Alternating current, 127
Altitude, 333
Altocumulus clouds, 417
Altostratus clouds, 417
Alveoli, 494
Amino acid, 277
Ampere (unit), 122
Amplitude, 87
amu, 162
Analgesic, 286
Anaxagoras, xiv
Anemometer, 404
Angstrom (unit), 89
Angular momentum, 44
Anhydrous salts, 268
Anion, 227
Annular eclipse, 350
Anode, 262
Ante meridiem, 328
Antibiotic, 286
Anticyclone, 411, 412

Antinodes, 93
Apogee (moon), 346
Apparent solar day, 327, 328
Appleton layer, 390
Aqueous solution, 249
Aquifier, 572, 573
Archimedes, xv
Aristarchus, xv, 294
Aristotle, xiv, 294, 295
Arrhenius, Svante, 265
Arrhenius acid-base concept, 265
Artesian
  spring, 573
  well, 573
Ascending node (moon), 351
Aspirin, 286
Asteroids, 306, 307
Asthenosphere, 511
Astronomical unit, 300, 362
Atmosphere, 385–407
  composition, 385
  energy content, 391–393
  humidity of planet, 402–404
  origin, 386
  pressure, 399
  temperature, 398
  vertical structure, 387–391
Atom, 158
Atomic
  nucleus, 158–160
  number, 159
  quantum numbers, 196–201
  time scale, 515
  weight, 164
Aurora
  australis, 391
  borealis, 391
Autumnal equinox, 322
Average
  speed, 20
  velocity, 20
Avogadro, Amadeo, xvii, 246
Avogadro's
  law, 246
  number, 248

B

Barnard, Edward M., 372
Barnard's star, 372
Barometer
  aneroid, 400
  mercury, 399
Base, 265
  conjugate, 266
Batholith, 521
Battery, lead storage, 263, 264
Becher, Johann J., 241
Bedding, 527
Bed load, 564
Benzene ring, 236
Bergeron process, 427
Berzelius, Jons Jacob, 210
Bessel, Friedrich W., 293
Beta decay, 164
Big Bang Theory, 378
Big Crunch, 380
Binary star, 366
Binding energy
  per nucleon, 162
  total, 162
Biodegradable, 586
Black hole, 153, 370
Blizzard, 444
Bohr, Niels, xvii, 179
Bohr's equation for the hydrogen atom, 179–181
Boiling point, 71
Bombs, volcanic, 542
Bond
  covalent, 228
  ionic, 228
  nonpolar, 230
  polar, 230
Boyle, Robert, xvii, 210
Boyle's law, 77
Brahe, Tycho, xv, xvi, 295
Breeder reactor, 170
Breeze
  land, 414
  mountain, 414
  sea, 412
  valley, 414
British system, 6
Brönsted, J. N., 265
Brönsted-Lowry theory, 265
Btu, 69

C

Caldera, 542
Calendar, 335, 336
  Gregorian, 336
  Julian, 336
  Roman, 335
Calorie, 69
Cancer, 594

Candela, 8
Carbohydrates, 276
Carbon dioxide
  in greenhouse effect, 396
  pollution, 484, 498
Carbonic acid, 484
Carbon monoxide, 484, 496
Carboxyl, 275
Carcinogen, 486
Cartesian coordinate system, 325
Catalyst, 259
Catastrophe theories, 319
Cathode, 262
Cathode ray tube, 137–139
Cation, 227
Caverns, 561
Celestial
  prime meridian, 362
  sphere, 362
Cellulose, 277
Celsius scale, 68
Centripetal
  acceleration, 26, 179
  force, 42
Cepheid variables, 366
Cesium, 7, 327
cgs system, 7
Chadwick, James, xviii
Chain reaction, 168
Charles' gas law, 77
Charon, 315
Chemical
  change of matter, 242
  equation balancing, 255–257
  properties of matter, 242
  reactions, 255–269
Chemistry, 209
Chill factor, 412
Chinook, 414
Chiron, 308
Chromosomes, 281
Chromosphere, 358
Cilia, 494
Circular motion, 25–27
Cirque, 568
  glacier, 568
Cirrocumulus clouds, 417
Cirrostratus clouds, 417
Cirrus clouds, 416
Cleavage, 534
Climate, 497
Cloudburst, 440
Clouds, 408–433
  classification, 416–419
  formation, 423–426
Coal, 525, 526
  anthracite (hard), 525
  bituminous (soft), 525
  lignite (brown), 525
  sub-bituminous, 525
  sulfur content of, 526, 590
Coalescence, 427
Coherent waves, 187
Color (mineral), 532–534
Columns, 524
Coma, of a comet, 316
Combustion
  complete, 484
  incomplete, 484–488
Comet, 316–318
Complex
  ions, 230
  molecules, 273
Compound pendalum, 57
Compound(s), 223
  covalent, 223
  electrovalent, 223
  ionic, 223
  naming, 233
  organic, 273
Compression, 89
Concave mirror, 110
Concentrated solution, 249
Concept, xx
Conduction, 80
  from the Earth's surface, 398
Conductor, 124
Conjugate
  acid, 266
  base, 266
Conjunction, 300
Consolidation, 523
Constellation, 364
Continental
  drift, 503–510
  glaciers, 567
  shelves, 579
  slopes, 581
Contour farming, 600
Convection, 79
  cycle, 410
Convex
  lens, 114
  mirror, 110
Copernicus, Nicholas, xv, 295
Core, Earth's, 561

Coriolis force, 410
Corona, 358
Cosmological principle, 377
Cosmological red-shift, 378
Cosmology, 378–381
Coulomb, Charles, 122
Coulomb (unit), 122
Coulomb's law, 123
Covalent
  bond, 228
  compounds, 224, 228
    properties of, 231
Craters (moon), 342
Creep, 563
Crescent moon, 347
Crust, Earth's, 504
Crystal, 226
  form (mineral), 533
Cumulonimbus clouds, 419
Cumulus clouds, 416
Curie (one) defined, 167
Curie, Marie, 165
Curie temperature, 132, 504
Current
  alternating, 127
  defined, 122
  density, 578
  electric, 122
  long shore, 577
  surface, 577
  tidal, 577
Cyclic hydrocarbons, 236
Cyclone
  low, 411, 412, 439
  tropical storm, 449–453
Cyclonic disturbances, 437–440

D

Dana, James D., 554
Davisson, G., 191
Daylight Savings Time, 330
DDT, 590
De Broglie, Louis, xvii, 192
Decay
  alpha, 164
  beta, 164
  gamma, 164
Decibel, 91
Declination, 121, 332, 362
Degree, 11
  Celsius, 68
  Fahrenheit, 68
  Kelvin, 68
Delta, 566
De Maricount, Petrus P., 121
Democritus, xiv
Density, 9
Derived units, 9
Detergent(s), 278, 596
  builders, 596
Deuteron, 160
Dew, 428
  point, 402
Dextrose, 276
Diastrophism, 555
Diffraction, 107
  grating, 108
Diffuse reflection, 101
Dike, 521, 522
Dimensional analysis, 610
Diode, 138
Dip, magnetic, 133
Direct current, 127
Dispersion, 105, 106
Displacement, 20
  reaction, 261
DNA, 279
Doldrums, 414
Domains, 131
Doppler effect, 97, 98
Doppler radar, 463
Drift, glacial, 569
Drug, 285
Dry cell, 263
Dual nature of light, 178
Dump, open, 586
Dust
  bowls, 569
  dome, 497
Dyne, 37

E

Earth,
  core, 504
  crust, 504
  internal processes of, 537–557
  mantle, 504
  the planet, 291–294
  revolution, 291
  rotation, 291
  size of, 293, 294
  structure of, 503–505
  surface processes of, 559–583
  water supply of, 571–576

Earthquake(s), 545–552
belts, 545
epicenter, 548
focus, 548
scales, 550, 551
Eclipse
annular, 350
lunar, 350
solar, 350
Ecliptic plane, 291
Einstein, Albert, xvii, xviii, 5, 142, 146, 178, 325
Einstein's postulates, 146
El Chichón, 498
Electric
charge, 6, 122
current, 122
field, 130
generator, 135
motor, 136
potential energy, 125
Electricity, 125–129
Electrochemical reactions, 262–264
Electrolysis, 227
Electrolytes, 262
Electromagnet, 133
Electromagnetic
force, 158
spectrum, 88
waves, 87–89
Electromagnetism, 133–137
Electromotive series, 261
Electron, 122, 158
affinity, 224
configuration, 202
period, 202
shell, 202
subshell, 204
volt, 162
wavelength of, 191
Electronegativity, 230
Electronics, 137–139
Electrovalent compounds, 224
Element(s), 209–212
abundance in Earth's crust, 212
alphabetical listing of, 210, 211
classification of, 214, 215
periodic characteristics of, 214, 219, 220
El Niño, 498
Empedocles, xiv
Emphysema, 494
Endoergic, 172
Endothermic, 172
reaction, 257
Energy, 53–63
activation, 258
binding per nucleon, 162
consumption, 61–63
electric potential, 125
gravitational potential, 55
kinetic, 54
level diagram, 182–187
levels for the hydrogen atom, 181–183
mass relationship, 149
potential, 54
and rate of chemical reactions, 257
requirements in the United States, 61–63
rest, 149
Entropy, 74
Enzymes, 260
Epicenter, 548
Epoch, 515
Equilibrium, 256
Equinox
autumnal, 352
vernal, 352
Era, 515
Eratosthenes, xv, 294
Erg, 50
Erosion, 559–563
agents of, 563–571
sheet, 564
Ester, 275, 276
Estrogen, 287
Eutrophication, 597
Evelyn, Sir John, 483
Event horizon, 152
Evolutionary theories, 319
Excited states, 182
Exoergic, 172
Exothermic, 172
reaction, 257
Experimental error, 12–14

F

Fahrenheit scale, 68
Family of elements, 215
Faraday, Michael, 121
Farsightedness, 117
Fault-block mountains, 553
Faulting, 546–548
normal, 547
reverse, 547
strike-slip, 547
transform, 547
Faults (moon), 344

Fermi, E., 157
Ferromagnetic, 131
First law of thermodynamics, 72
First quarter moon, 347
Fission, 165, 168–171
Flash flood, 440
Flood plain, 565
Focus, earthquake, 548
Fog, 419
  advection, 419
  radiation, 419
Föhn, 414
Folded mountains, 554
Folklore (weather), 475–479
Foot, 6, 7
  -pound, 50
Force(s), 34
  centripetal, 42
  electromagnetic, 158
  gravitational, 39, 158
  and motion, 33–45
  in nature, 158, 159
  nuclear, 159
Formula
  chemical, 210
  structural, 235
  weight, 245
Fossil fuels, 484
Foucault, J. B. Leon, 292
Foucault pendulum, 292
Fracture (mineral), 534
Franklin, Ben, 440
Frequency, 86
Friction, air, 411
Frontal zone, 437
Fronts, 437–439
Frost, 428
  heaving, 559
  wedging, 559
Fructose, 276
Fuel(s)
  fossil, 484
  impurities, 488
Full moon, 347
Fumaroles, 537
Fundamental frequency, 93
Fundamental properties, 6
Fusion, 163, 171–173

G

g, 23, 38
G, 39, 40
Galaxies, 370–377
  clusters, 373
  composition of, 370
  Local Group, 373
  naming of, 371
  recession of, 378
  Seyfert, 371
  superclusters, 373
  types of, 370, 371
Galileo, xvi, 23, 33, 297, 399, 400
Gamma
  decay, 164
  ray(s), 88, 164
Gas, 76
Gay-Lussac's law of combining volumes, 246
General theory of relativity, 150, 151
Generator, 135
Genes, 281
Genetic code, 282
Geocentric theory, 294
Geologic time, 515–517
  scale, 515
Geology, defined, 503
Geometrical optics, 107
Geosynclines, 554
Geothermal
  areas, 573–576
  gradient, 519
Germer, L. H., 191
Geyser, 573
Gibbous moon, 347
Gilbert, Dr. William, 121
Glacier
  cirque, 568
  continental (ice sheet), 567
  drift, 569
  moraine, 569
  snout, 569
  till, 569
  valley, 568
Glomar Challenger, 510
Glucose, 276
Gram, 7
  atomic weight, 245
  formula weight, 245
  molecular volume, 247
  molecular weight, 245
  unit of mass, 7
Gravitational
  collapse, 152
  force, 39, 158
  potential energy, 55
Greenhouse effect, 396

Greenwich meridian, 327
Gregorian calendar, 336
Ground state, 182
Group of elements, 215
Guyots, 579

H

Hahn, Otto, xviii
Hail, 428
Half-life, 165
Half-reaction, 227
Hall, James, 554
Halley's comet, 316, 317
Halogens, 217, 218
Hardness (mineral), 534
Hard water, 576
Heat, 69–72
  capacity, 70
  engine, 73, 74
  of fusion, 70
  -island effect, 496
  pump, 73
  specific, 70
  transfer, 79–81
  of vaporization, 70
Heaviside layer, 390
Heisenberg, Werner, 195
Heisenberg uncertainty principle, 195
Heliocentric theory, 294
Henry, Joseph, 121
Heracleides, Ponticus, 294
Heroin, 287
Hertz (unit), 86
Hertzsprung, Ejnar, xviii, 365
Hess, H. H., xix, 509
Heterogeneous, 241
Heterosphere, 388
High clouds, 417
Highs (anticyclone), 439
Hipparchus, xv
Homogeneous, 241
Homosphere, 388
Horizon, 333
Horse latitudes, 414
Horsepower, 52
Hot spot theory, 541
H-R diagram, 365, 369
Hubble, Edwin P., xviii, xix, 370, 373
Hubble's law, 376
Humidity, 401–404
  absolute, 401
  relative, 401
Hurricane, 449–453
  warning, 453
  watch, 453
Hutton, James, xix
Huygens, Christian, xvii, 53
Hydrocarbons, 234
Hydrogen
  burning, 369
  isotopes of, 212
Hydrogenation, 278
Hydrologic cycle, 571–576
Hydrometer, 9, 10
Hydronium ion, 266
Hydrothermograph, 461
Hygrometer, 402
  adsorption, 402
  hair, 402
Hygroscopic nuclei, 426

I

Ice
  point, 68
  storm, 443
Ideal gas law, 76
Igneous rocks, 519–522
  extrusive, 520
  intrusive, 521
Index of refraction, 103
Indian summer, 479
Inertia, 35
Inertial mass, 35
Infrared
  measurements (satellite), 466
  rays, 88
Inner planets, 299
Inner transition elements, 214
Instantaneous
  speed, 21
  velocity, 22
Insolation, 391
Insulator, 124
Intensity of sound wave, 91
Interference effects, 107–109
International Date Line, 330
International system of units, 8
Io, 309
Ion(s), 224–226
  complex, 230
  layers, 389–391
Ionic
  bond, 228
  compounds, 224

Ionization potential, 182, 219
Ionosphere, 389
Isobar, 410
Isomers, 237
Isostasy, 511
Isothermal, 77
Isotope, 159, 212

J

Jet stream, 416
Joule, James Prescott, 67
Joule (unit), 50
Julian calendar, 336
Jupiter (planet), 308–310

K

Kelvin, 8, 68
  scale, 68
Kennelley-Heaviside layer, 390
Kepler, Johannes, xvi, 296
Kepler's
  harmonic law, 296, 297
  law of elliptical paths, 296
  law of equal areas, 296
Kilo, 14
Kilocalorie, 69
Kilogram, 7
Kilowatt-hour, 53
Kinetic
  energy, 54
  theory, 77–79
  theory of gas, 77

L

Laccolith, 522
Land
  breeze, 414
  management, 600
  pollution, 535–590
    littering, 586
    mining, 588–590
    solid waste, 585–590
  reclamation, 600
Landfill, sanitary, 586
Landslides, 562
Lanthanides, 214
Lapse rate, 423
Laser, 186, 187
Last quarter moon, 348
Latent heat
  of condensation, 423
  of fusion, 71
  of vaporization, 71
Latitude, 326
Lattice, 75
Lava, 520, 537
  basaltic, 542
  felsitic, 542
Lavoisier, Antoine, 210
Law(s)
  Boyle's, 77
  Charles', 77
  conservation
    angular momentum, 44
    energy, 56
    linear momentum, 44
    mass, 242
  Coulomb's, 123
  definite proportions, 242
  Gay-Lussac's, 246
  Hubble's, 376, 377
  Ideal gas, 76
  Kepler's, 296
  multiple proportions, 244
  Newton's, 33–44
  Ohm's, 125
  superposition, 515
Leavitt, Henrietta Swan, 366
Le Bel, Joseph, 237
Length, 4
Length contraction, 147
Lens, convex, 114
Lenses, 114–117
Light
  dual nature of, 177–179
  speed, 88
  waves, 88
Lightning, 440
  heat, 440
Light year, 362
Lignite, 525
Linear momentum, 44
Linearly polarized wave, 109
Line squall, 437
Lipids, 277
Liquid, 75
Liter, 247
Lithification, 523
Lithosphere, 511
Littering, 586
Load
  bed, 564
  dissolved, 564

stream, 564
suspended, 564
Local Group, 373
Longitude, 326
Longitudinal wave, 86
Lorentz-Fitzgerald contraction, 147
Lorentz transformation, 146, 147
Los Angeles smog, 486, 487
Lowry, T. M., 265
ℓ-quantum number, 197
LSD, 287
Lunar eclipse, 350
Lyell, Charles, xix, xx

M

Magma, 580, 581
Magnetic
anomalies, 509
declination, 133
dip, 133
field, 129
monopole, 130
Magnetism, 129–133
remanent, 509
Magnification, 112
Magnitude, of a star, 364
absolute, 365
Main sequence (of stars), 366
Mantle, Earth's, 504
Maria, 342
Marijuana, 287
Marina trench, 511
Mars (planet), 304–306
Mass, 5, 35
inertial, 35
number, 159
rest, 150
wasting, 562, 563
mass, 562
slow mass, 563
Matter, 209
classification, 241, 242
phases, 75, 76
waves, 191
Maxwell, James Clerk, 121, 122
Meander, 564
Mean solar day, 328
Measurement, 12
Mega, 14
Meiosis, 281
Melting point, 71
Mendeleev, Dmitri, 191, 209, 241
Mercalli scale, 550, 551
Mercury (planet), 302
Meridians, 326
Mesopause, 388
Mesosphere, 388
Metalloids, 215
Metals
alkali, 216, 217
properties of, 215
Metamorphic rock, 528, 529
Metamorphism, 528
Metastable state, 186
Meteor, 318
Meteorites, 318
Meteorology, 385
Meter, 7
MeV, 162
Meyer, Julius, 209
Micro, 14
Mid-Atlantic Ridge, 509
Milky Way, 372–375
Miller-Urey experiment, 281
Milli, 14
Millibar, 400
Mineral(s), 529–534
color, 532, 534
crystal form, 533
defined, 529
fracture, 534
indentification, 529–534
Mining operations, 588
Mirage, 475
Miranda, 312
Mirror
concave, 110
convex, 110
Mitosis, 281
Mixture, 223
MKS system, 6
Moho discontinuity, 504
Mohorovicic, A., 504
Mohs' scale of hardness, 534
Molar concentration, 249
Molarity, 249
Mole, 245
Molecular weight, 245
Molecule(s), 67, 210, 223
complex, 273–289
Momentum
angular, 44
linear, 44
Monatomic, 216
Monomer, 273

Monsoon, 414
Moon, 340–355
  apogee, 346
  ascending node, 351
  craters, 342
  descending node, 351
  eclipses, 350
  fault, 344
  general features, 341
  history of, 344–346
  maria, 342
  motion of, 346
  mountain ranges, 344
  nodes, 351
  perigee, 346
  phases, 347–349
  plains, 342
  rays, 344
  rills, 344
  sidereal month, 347
  size and shape, 341
  synodic month, 347
Moraine, 569
Moseley, Henry, 209
Motion, 19–31
  circular, 25–27
  and force, 33–47
  projectile, 27–29
  straight line, 19
  wave, 85
Motor, 136
Mountain(s)
  building, 552–556
  chain, 552
  fault-block, 553
  folded, 554
  moon, 344
  range, 552
  system, 552
  volcanic, 552
m_s-quantum number, 197–200
Mudflows, 562
Muons, 147
Mutation, 282

N

National Meteorological Center (NMC), 457
National Oceanic and Atmospheric Administration (NOAA), 457
National Weather Service, 457–461
Neap tides, 352
Nearsightedness, 117
Neptune (planet), 314
Neutralization, 267
Neutral solution, 268
Neutrino, 164, 360
Neutron, 122, 158
  number, 159
  separation energy, 162
  star, 369
Newton, Sir Isaac, xvi, xvii, 297
Newton (unit), defined, 37
Newton's
  first law of motion, 33–35
    defined, 37
  law of universal gravitation, 39–42
  second law of motion, 36–39
  third law of motion, 42–44
Nimbostratus clouds, 419
Nimbus clouds, 416
Nitrogen oxides, 484, 486
Noble gases, 214
  properties of, 215, 216
Nodes (moon), 351
  in standing waves, 93
Nonmetals, properties of, 215
Nonpolar
  bond, 230
  molecule, 230
Noretynodrel, 287
Normal faulting, 547
Nova, 366
n quantum number, 196–201
Nuclear
  force, 159
  products, 171
  reactions, 168–173
  reactor, 168
  stability, 160–164
  winter, 498
Nucleic acids, 279
Nucleon, 158
Nucleosynthesis, 369
Nucleus
  of an atom, 158, 159
  of a comet, 316

O

Occlusion, 439
Oceans, 576–581
Octave, 95
Octet rule, 224
Oersted, Hans Christian, 121
Ohm (unit), 125

Ohm's law, 125
Oil shale, 526
Oort, Jan H., 318
Oort comet cloud, 318
Opposition, 301
Organic
  acids, 275
  chemistry, 233–237
  compounds, 233
Outer planets, 308–315
Overburden, 562
Overtones, 93
Oxidation, 227
  number, 232
  state, 232
Oxidation-reduction reaction, 227, 260
Oxide, 215
Oxidized, 260
Oxidizing agent, 260
Ozone, 389
  as air pollutant, 486
  layer, 389, 390
Ozonosphere, 389

P

Pacific Ocean floor, 580
Pairing effect, 160
Pangaea, 505
Parallax, 293
Parallel circuit, 128
Parallels, 326
Parhelia, 417
Parsec, 362
Particle pollution, 499
Pascal, B., 400
Pauli, Wolfgang, xviii, 201
Pauli exclusion principle, 201
Peat, 525
Pendulum
  compound, 57
  simple, 57
Penumbra, 350
Percentage
  composition, 246
  difference, 14
  error, 14
Perigee, moon, 346
Period(s), 86, 214
  of an electron, 214
  geologic time, 515
  row of periodic table, 213
  of a wave, 87
Periodic
  law, 212
  table, 204, 212–214
    alphabetical listing, 210, 211
Permeability, 560, 571
*p*H, 268
Phases of matter, 75, 76
Photochemical smog, 486
Photoelectric effect, 178
Photon, 178
  absorption, 183–186
  emission, 183–186
Photosphere, 358
Photosynthesis, 386
Physical
  changes of matter, 242
  optics, 107
  properties of matter, 242
  science [defined], xi
Pi (π), 11
Pipe
  closed, 96
  open, 96
Pitch, 90
Plains (moon), 342
Planck, Max, xvii, 177
Planck's constant, 178
Planets
  inner, 299, 302–306
  Jovian, 299
  Jupiter, 308–310
  Mars, 304–306
  Mercury, 302
  Neptune, 314
  outer, 299, 308–316
  Pluto, 314, 315
  Saturn, 310, 311
  terrestrial, 299
  Uranus, 311–314
  Venus, 302–304
  X, 316
Plasma, 358
Plastics, 283
Plate
  names, 512
  relative motion, 511–513
  tectonics, 510–515
Pluto (planet), 314, 315
Pluton, 521, 522
  concordant, 521
  discordant, 521
Polar
  bond, 230

Polar (*continued*)
- easterlies, 415
- molecule, 230

Polarized wave, 109
Pollution
- air, 467, 468
- land, 585–590
- particle, 499
- water, 590–598

Polymer, 273
Porosity, 571
Position, 19
Post meridiem, 328
Potential
- energy, 54
- well, 190

Pound, 6
Power, 52, 53
Powers of ten, 14, 609
Precession, 334
Precipitation, 428–431
- convectional, 428
- frontal, 428
- orographic, 428

Precision, 12
Pressure, 76
- atmospheric, 487
- gradient, 410
- trough, 410

Principal quantum number, 182
Principle of uniformity, 503
Products
- in chemical reactions, 255
- nuclear, 171–173

Projectile motion, 27–29
Prominences, 360
Proteins, 277
Proton, 122, 158
- separation energy, 162

Proton-proton chain, 361
Protoplanet hypothesis, 319
Psychrometer, 402–404
- sling, 402

Ptolemy, xv, 294
Pulsars, 369
P waves, 549
Pyroclastic debris, 542
Pythagoras, xiv

Q

Quantization of angular momentum, 180
Quantum, 177
- mechanics, 191
- number, 182, 196–201

Quarks, 122
Quasars, 377

R

Radar, 463
Radiation, 80
- absorption of terrestrial, 392, 393
- 3-K cosmic microwave, 379
- temperature inversion, 426

Radioactive
- dating, 515
- decay, 164–167

Radiosonde, 463
Radio wave, 88
Rain, 428
- gauge, 463

Rainbow, 430, 431
- colors, 393

Rainmaking, 427
Rainstorm, 440
Rare
- earths, 214
- gases, 216

Rayleigh scattering, 393
Rays
- gamma, 88, 164
- moon, 344
- visible, 88

Reactants
- in chemical reactions, 255
- nuclear, 168, 171

Real image, 111
Red giants, 366
Red shift, 98
Reduced (atoms), 260
Reducing agent, 260
Reduction, 227, 260
Reference system, 143
Reflection, 101–103
- diffuse, 101
- regular, 101
- total internal, 105

Refraction, 103–106
- index of, 103

Relativity, 143–153
- general theory, 150, 151
- special theory, 146

Remanent magnetism, 509
Representative elements, 214
Resistance, 125

Resonance, 91
Respiratory system, 493–496
Rest
  energy, 149
  mass, 150
Retina, 117
Reverse faulting, 547
Reversible reaction, 255
Revolution of Earth, 291
Richter scale, 550, 551
Right ascension, 362
Rills (moon), 344
"Ring of Fire," 541
River
  base level, 564
  delta, 566
  flood plain, 565
  graded, 565
  meander, 564
RNA, 279
Rock(s), 519–529
  cycle, 529
  defined, 519
  igneous, 519–522
  metamorphic, 528, 529
  sedimentary, 522–527
Rockslide, 562
Roman calendar, 335
Rotation of Earth, 291
Rumford, Count, 67
Russell, Henry Norris, xviii, 365

S

Salinity, 576
Salt, 268
Salt wedging, 560
San Andreas fault, 547, 548
Sanitary landfill, 586
Satellites, weather, 463–466
Saturated
  hydrocarbons, 236
  solution, 249
Saturn (planet), 310, 311
Scalar, 20
Schrödinger, Erwin, 194
Schrödinger equation, 193
Scientific method, xx
Sea floor
  breeze, 412
  spreading, 509
  topography, 579–581
  trenches, 579
Seamounts, 579
Seasons, 331–334
Second, 7
Second law of thermodynamics, 72
Sedimentary rocks, 522–527
  chemical, 523
  clastic, 523
Seismic waves, 548–550
  body, 549
  P (primary), 549
  S (secondary), 549
  surface, 549
Seismograph, 549, 550
Seismology, 545
Semiconductor, 125
Senses, 3, 4
Series circuit, 127
Serotonin, 287
Sewage treatment, 600–604
  primary, 602
  secondary, 602
Sewer systems
  combined, 600
  separated, 602
Seyfert galaxy, 371
Shadow zones, 549
Shapley, Harlow, xviii
SI (units), 8, 607
Sidereal
  day, 328
  month (moon), 347
  period, 300
Significant figures, 607, 608
Silicon tetrahedron, 531
Sill, 522
Silt, 564
  pollution, 588
Simple pendulum, 57, 58
Singularity, 152
Sinkholes, 561
Sleet, 428
Slug, 7
Slump, 562
Smith, William, xix, xx
Smog, 484–488
  photochemical, 486
Snowstorm, 428, 443
Solar
  constant, 392
  day, 327, 328
  eclipse, 350
  neutrino experiment, 361

Solar (*continued*)
system, 291, 295, 298, 321
origin of, 319–321
Solid, 75
waste, 585–590
Solifluction, 563
Solstice
summer, 332
winter, 332
Solubility, 250
Solute, 249
Solution(s), 241, 249
aqueous, 249
dilute, 249
Solvated ions, 232
Solvent, 249
supersaturated, 251
Sound, 89–93
intensity levels in decibels, 91
Source region, 436
Special theory of relativity, 146
Specific
gravity, 534
heat, 70–72
Spectroscope, 106
Spectroscopy, 106, 184
Spectrum, 184
electromagnetic, 88
Speed, 20
average, 20
instantaneous, 21
of light, 88
wind, 413
Spring, 573
artesian, 573
tides, 352
Squall line, 437
Stable
cloud layer, 423
nucleus, 159
Stalactites, 523
Stalagmites, 523
Standard
conditions, 247
time zones, 328
unit, 6
units, table, 7
Starch, 276
Stars, 364–370
Alpha Centauri, 372
binary, 366
black hole, 153, 370
cepheid variables, 366
magnitudes, 364
nova, 366
neutron, 369
pulsars, 369
red giants, 366
supernovae, 368
white dwarfs, 366
Station model (weather), 472
Steam point, 68
Steno, Nicolaus, xix
Sterotonin, 287
Stock, 521
Storms
average characteristics, 445
local, 440–449
tropical, 449–454
Strassman, Fritz, xviii
Stratocumulus clouds, 417
Stratopause, 388
Stratosphere, 388
Stream
base level, 564
erosion, 564–566
load, 564
meander, 564
Strike-slip faulting, 547
Structural formula, 235
Subduction, 511
Subsidence temperature inversion, 426
Substance, 241
Sulfur dioxide, 488
Summer solstice, 332
Sun, 357–362
annular eclipse, 350
chromosphere, 358
corona, 358
prominences, 360
size and shape, 357
temperature, 358
Sunspots, 358
Supercooled air, 426
Supernova, 368
Superposition, law of, 515
Supersaturated
air, 426
solution, 251
Surf, 577
S waves, 549
Synergism, 496

Synodic
  month (moon), 347
  period, 300
Synthetic, 284
System of reference, 143

T

Tail of a comet, 316
Tambora, 537
Temperature, 67–69, 76
  absolute, 68, 76
  atmospheric, 398
  Curie, 132
  defined, 76
  global variations, 497, 498
  inversion, 426
    Los Angeles, 486
    radiation, 424
    subsidence, 426
  scales, 69
Ten-degree rule, 259
Thales of Miletus, xiv, 121
Theory,
  Brönsted-Lowry, 265
  defined, xx
  geocentric, 294
  heliocentric, 294
  hot spot, 541
  of relativity, 143–155
Thermal
  circulation, 410
  conductivity, 80
  insulator, 80
Thermodynamics, 72–75
  First Law of, 72
  Second Law of, 72
  Third Law of, 75
Thermograph, 461
Thermometer, 68
  maximum-minimum, 462
Thermosphere, 388
Third law of thermodynamics, 75
Thunder, 443
Thunderstorms, 440
"Tidal wave," 552
Tides, 351–353
  neap, 352
  spring, 352
Till, glacial, 569
Time, 5, 327–330
  dilation, 147
  scales
    atomic, 515
    geologic, 515
Titan (moon), 311
Titius-Bode law, 300
Tone quality, 97
Topography
  ocean, 576–579
  sea floor, 579–581
Tornado, 445–449
  safety, 449
  warning, 447
  watch, 447
Torque, 44
Torricelli, Evangelista, 399
Total
  binding energy, 162
  internal reflection, 105
Traction, 564
Trade winds, 415
Transformer, 127
Transform faults, 547
Transistor, 138
Transition elements, 214
Transverse wave, 86
Tropical year, 331
Tropopause, 388
Troposphere, 388
Tsunami, 552
Twins paradox, 148
Typhoon, 449

U

Ultrasound, 90
Ultraviolet rays, 88
Umbra, 350
Uniformitarianism, 503
Uniformity, principle of, 503
Units, international system, 8, 607
Universe, 357–383
Unpolarized wave, 109
Unsaturated hydrocarbons, 236
Unstable nucleus, 159
Uranus (planet), 311–314

V

Valence, 219
  electrons, 214

Van't Hoff, Jacobus, 237
Vector, 20
Velocity
  average, 20
  instantaneous, 22
Venus (planet), 302–304
Vernal equinox, 352
Versorium, 121
Virtual image, 111
Viscosity, 542
Visible rays, 88
Vitamins, 280
Volcanic
  bombs, 542
  crater, 542
  mountains, 552
Volcano(s), 537–545
  cindercone, 542
  Mount St. Helens, 538
  shield, 542
  strato- (composite), 542
Volt (unit), 125
Voltage, 125

W

Water, 571–576
  hard, 576
  table, 572
Water pollution, 590–600
  acid mine drainage, 588
  agricultural, 590
  animal waste, 591
  chemical, 592
  detergent, 596
  domestic, 596
  fertilizer, 590
  industrial, 591
  mercury, 592
  oil spills, 598
  organic waste, 597
  pesticide, 590
  radiation, 594
  thermal, 594
Waterspout, 448
Watt (unit), 52
Wave(s)
  as an agent of erosion, 571
  amplitude, 87
  coherent, 187
  cyclone, 438
  electromagnetic, 87
  function, 194
  longitudinal, 86
  matter, 191
  motion, 85
  period, 87
  polarized, 86, 109
  properties, 86, 87
  radio, 88
  seismic, 548–552
  shock, 98
  sound, 89–93
    intensity of, 91
  standing, 93–97
  surface
    oceanic, 577
    seismic, 614
  transverse, 86
  unpolarized, 109
  velocity, 86
Wavelength, 87
  of an electron, 191
Waxes, 278
Weather
  charts, 469
  data collection, 461
  defined, 388
  folklore, 475–479
  forecasting, 456–481
  maps, 469–475
  observations, 461
  records, 445
  satellites, 463–467
Weathering, 559–563
  chemical, 561
  physical, 559
Wegener, Alfred, xix, 505, 507
Weight, 37
Westerlies, 415
White dwarfs, 366
Willy-willy, 449
Wind, 404, 405, 409, 412
  chill index, 404, 405
  direction, 404
  erosion, 569–571
  speed, 404, 413
  vane, 404
Winter solstice, 332
Work, 49–52
  against friction, 51
  against gravity, 51
  against inertia, 50

X

X (planet), 316
X rays, 203

Z

Zenith, 333
  angle, 333
Zero point energy, 78
Zodiac, 364
Zone
  of aeration, 572
  frontal, 437
  of saturation, 572

2 3 4 5 6 7 8 9 0

Conversion Factors

Mass

1 gram = 10^{-3} kg = 6.85×10^{-5} slug
1 kg = 10^3 g = 6.85×10^{-2} slug
1 slug = 1.46×10^4 g = 14.6 kg
1 amu = 1.66×10^{-24} g = 1.66×10^{-27} kg

Length

1 cm = 10^{-2} m = 0.394 in
1 m = 10^{-3} km = 3.28 ft = 39.4 in
1 km = 10^3 m = 0.62 mi
1 in = 2.54 cm = 2.54×10^{-2} m
1 ft = 12 in = 30.48 cm = 0.3048 m
1 mi = 5280 ft = 1609 m = 1.609 km
1 Å = 10^{-10} m = 10^{-8} cm
1 parsec = 3.26 light years = 205,265 a.u.

Time

1 h = 60 min = 3600 s
1 day = 24 h = 1440 min = 8.64×10^4 s
1 year = 365 days = 8.76×10^3 h
= 5.26×10^5 min = 3.16×10^7 s

Energy

1 joule = 10^7 erg = 0.738 ft · lb
= 0.239 cal = 9.48×10^{-4} Btu
= 6.24×10^{18} eV
1 kcal = 4186 J = 4.186×10^{10} erg = 3.968 Btu
1 Btu = 1055 J = 1.055×10^{10} erg
= 778 ft · lb = 0.252 kcal
1 cal = 4.186 J = 3.97×10^{-3} Btu
= 3.09 ft · lb
1 ft · lb = 1.36 J = 1.36×10^7 erg
= 1.29×10^{-3} Btu
1 eV = 1.60×10^{-19} J = 1.60×10^{-12} erg
1 kWh = 3.60×10^6 J = 3.413×10^3 Btu

Speed

1 m/s = 3.6 km/h = 3.28 ft/s = 2.24 mi/h
1 km/h = 0.278 m/s = 0.621 mi/h = 0.911 ft/s
1 ft/s = 0.682 mi/h = 0.305 m/s = 1.10 km/h
1 mi/h = 1.467 ft/s = 1.609 km/h = 0.447 m/s
60 mi/h = 88 ft/s

Force

1 newton = 10^5 dynes = 0.225 lb
1 dyne = 10^{-5} N = 2.25×10^{-6} lb
1 lb = 4.45×10^5 dynes = 4.45 N
Equivalent weight of 1 kg mass = 2.2 lb
= 9.8 N

Pressure

1 atm = 14.7 lb/in^2 = 1.013×10^5 N/m^2
= 1.013×10^6 dyne/cm^2
= 30 in Hg = 76 cm Hg
1 bar = 10^6 dyne/cm^2 = 10^5 Pa
1 millibar = 10^3 dyne/cm^2 = 10^2 Pa
1 Pa = 1 N/m^2 = 10^{-2} millibar

Power

1 watt = 0.738 ft · lb/s = 1.34×10^{-3} hp
= 3.41 Btu/h
1 ft · lb/s = 1.36 W = 1.82×10^{-3} hp
1 hp = 550 ft · lb/s = 745.7 watt
= 2545 Btu/h